DATE DUE

Methods in Enzymology

Volume 399
UBIQUITIN AND PROTEIN DEGRADATION
Part B

METHODS IN ENZYMOLOGY

EDITORS-IN-CHIEF

John N. Abelson Melvin I. Simon

DIVISION OF BIOLOGY
CALIFORNIA INSTITUTE OF TECHNOLOGY
PASADENA, CALIFORNIA

FOUNDING EDITORS

Sidney P. Colowick and Nathan O. Kaplan

Methods in Enzymology

Volume 399

Ubiquitin and Protein Degradation

Part B

EDITED BY

Raymond J. Deshaies

HOWARD HUGHES MEDICAL INSTITUTE
DIVISION OF BIOLOGY
CALIFORNIA INSTITUTE OF TECHNOLOGY
PASADENA, CALIFORNIA

AMSTERDAM • BOSTON • HEIDELBERG • LONDON
NEW YORK • OXFORD • PARIS • SAN DIEGO
SAN FRANCISCO • SINGAPORE • SYDNEY • TOKYO
Academic Press is an imprint of Elsevier

Elsevier Academic Press
525 B Street, Suite 1900, San Diego, California 92101-4495, USA
84 Theobald's Road, London WC1X 8RR, UK

This book is printed on acid-free paper. ∞

Copyright © 2005, Elsevier Inc. All Rights Reserved.

No part of this publication may be reproduced or transmitted in any form or by any means, electronic or mechanical, including photocopy, recording, or any information storage and retrieval system, without permission in writing from the Publisher.

The appearance of the code at the bottom of the first page of a chapter in this book indicates the Publisher's consent that copies of the chapter may be made for personal or internal use of specific clients. This consent is given on the condition, however, that the copier pay the stated per copy fee through the Copyright Clearance Center, Inc. (www.copyright.com), for copying beyond that permitted by Sections 107 or 108 of the U.S. Copyright Law. This consent does not extend to other kinds of copying, such as copying for general distribution, for advertising or promotional purposes, for creating new collective works, or for resale. Copy fees for pre-2005 chapters are as shown on the title pages. If no fee code appears on the title page, the copy fee is the same as for current chapters.
0076-6879/2005 $35.00

Permissions may be sought directly from Elsevier's Science & Technology Rights Department in Oxford, UK: phone: (+44) 1865 843830, fax: (+44) 1865 853333, E-mail: permissions@elsevier.co.uk. You may also complete your request on-line via the Elsevier homepage (http://elsevier.com), by selecting "Customer Support" and then "Obtaining Permissions."

For all information on all Elsevier Academic Press publications
visit our Web site at www.books.elsevier.com

ISBN-13: 978-0-12-182804-2
ISBN-10: 0-12-182804-2

PRINTED IN THE UNITED STATES OF AMERICA
05 06 07 08 09 9 8 7 6 5 4 3 2 1

Working together to grow libraries in developing countries

www.elsevier.com | www.bookaid.org | www.sabre.org

ELSEVIER   BOOK AID International   Sabre Foundation

Table of Contents

CONTRIBUTORS TO VOLUME 399 xi
ACKNOWLEDGMENTS xvii
VOLUMES IN SERIES xix

Section I. Ubiquitin and Ubiquitin Derivatives

1. Chemical and Genetic Strategies for Manipulating Polyubiquitin Chain Structure
   SARA VOLK, MIN WANG, AND CECILE M. PICKART   3

2. Controlled Synthesis of Polyubiquitin Chains
   CECILE M. PICKART AND SHAHRI RAASI   21

3. Derivitization of the C-Terminus of Ubiquitin and Ubiquitin-Like Proteins Using Intein Chemistry: Methods and Uses
   KEITH D. WILKINSON, TUDEVIIN GAN-ERDENE, AND NAGAMALLESWARI KOLLI   37

4. Preparation, Characterization, and Use of Tagged Ubiquitins
   JUDY CALLIS AND RICHARD LING   51

5. Knocking out Ubiquitin Proteasome System Function *In Vivo* and *In Vitro* with Genetically Encodable Tandem Ubiquitin
   Y. SAEKI, E. ISONO, M. SHIMADA, H. KAWAHARA, H. YOKOSAWA, AND A. TOH-E   64

6. Production of Antipolyubiquitin Monoclonal Antibodies and Their Use for Characterization and Isolation of Polyubiquitinated Proteins
   MASAHIRO FUJIMURO AND HIDEYOSHI YOKOSAWA   75

7. Application of Ubiquitin Immunohistochemistry to Diagnosis of Disease
   JAMES LOWE, NEIL HAND, AND JOHN MAYER   86

8. Mechanism-Based Proteomics Tools Based on Ubiquitin and Ubiquitin-Like Proteins: Crystallography, Activity Profiling, and Protease Identification
   PAUL J. GALARDY, HIDDE L. PLOEGH, AND HUIB OVAA   120

Section II. Ubiquitin Binding Domains

9. Identification and Characterization of Modular Domains That Bind Ubiquitin — MICHAEL FRENCH, KURT SWANSON, SUSAN C. SHIH, ISHWAR RADHAKRISHNAN, AND LINDA HICKE — 135

10. Analysis of Ubiquitin Chain-Binding Proteins by Two-Hybrid Methods — JENNIFER APODACA, JUNGMI AHN, IKJIN KIM, AND HAI RAO — 157

11. Quantifying Protein Protein Interactions in the Ubiquitin Pathway by Surface Plasmon Resonance — RASMUS HARTMANN-PETERSON AND COLIN GORDON — 164

12. Using NMR Spectroscopy to Monitor Ubiquitin Chain Conformation and Interactions with Ubiquitin-Binding Domains — RANJANI VARADAN, MICHAEL ASSFALG, AND DAVID FUSHMAN — 177

13. Analysis of Ubiquitin-Dependent Protein Sorting Within the Endocytic Pathway in *Saccharomyces cerevisiae* — DAVID J. KATZMANN AND BEVERLY WENDLAND — 192

Section III. Methods to Study the Proteasome

14. Preparation of Ubiquitinated Substrates by the PY Motif-Insertion Method for Monitoring 26S Proteasome Activity — Y. SAEKI, E. ISONO, AND A. TOH-E — 215

15. Large- and Small-Scale Purification of Mammalian 26S Proteasomes — YUKO HIRANO, SHIGEO MURATA, AND KEIJI TANAKA — 227

Section IV. Identification and Characterization of Substrates and Ubiquitin ligases

16. Is This Protein Ubiquitinated? — PETER KAISER AND CHRISTIAN TAGWERKER — 243

17. Experimental Tests to Definitively Determine Ubiquitylation of a Substrate — JOANNA BLOOM AND MICHELE PAGANO — 249

18. Identification of Ubiquitination Sites and Determination of Ubiquitin-Chain Architectures by Mass Spectrometry — PETER KAISER AND JAMES WOHLSCHLEGEL — 266

19. Mapping of Ubiquitination Sites on Target Proteins — SHIGETSUGU HATAKEYAMA, MASAKI MATSUMOTO, AND KEIICHI I. NAKAYAMA — 277

20. Identification of Substrates for F-Box Proteins	JIANPING JIN, XIAOLU L. ANG, TAKAHIRO SHIROGANE, AND J. WADE HARPER	287
21. Fusion-Based Strategies to Identify Genes Involved in Degradations of a Specific Substrate	RANDOLPH Y. HAMPTON	310
22. Bi-Substrate Kinetic Analysis of an E3-Ligase–Dependent Ubiquitylation Reaction	DAVID C. SWINNEY, MICHAEL J. ROSE, AMY Y. MAK, INA LEE, LILIANA SCARAFIA, AND YI-ZHENG XU	323
23. Screening of Tissue Microarrays for Ubiquitin Proteasome System Components in Tumors	NORMAN L. LEHMAN, MATT VAN DE RIJN, AND PETER K. JACKSON	334
24. Structure-Based Approaches to Create New E2–E3 Enzyme Pairs	G. SEBASTIAN WINKLER AND H. TH. MARC TIMMERS	355
25. Proteomic Analysis of Ubiquitin Conjugates in Yeast	JUNMIN PENG AND DONGMEI CHENG	367

Section V. Genome and Proteome-wide Approaches to Identify Substrates and Enzymes

26. Two-Step Affinity Purification of Multiubiquitylated Proteins from *Saccharomyces cerevisiae*	THIBAULT MAYOR AND RAYMOND J. DESHAIES	385
27. Identification of SUMO-Protein Conjugates	MEIK SACHER, BORIS PFANDER, AND STEFAN JENTSCH	392
28. Identification of Ubiquitin Ligase Substrates by *In Vitro* Expression Cloning	NAGI G. AYAD, SUSANNAH RANKIN, DANNY OOI, MICHAEL RAPE, AND MARC W. KIRSCHNER	404
29. *In Vitro* Screening for Substrates of the N-End Rule-Dependent Ubiquitylation	ILIA V. DAVYDOV, JOHN H. KENTEN, YASSAMIN J. SAFIRAN, STEFANIE NELSON, RYAN SWENERTON, PANKAJ OBEROI, AND HANS A. BIEBUYCK	415
30. Genome-Wide Surveys for Phosphorylation-Dependent Substrates of SCF Ubiquitin Ligases	XIAOJING TANG, STEPHEN ORLICKY, QINGQUAN LIU, ANDREW WILLEMS, FRANK SICHERI, AND MIKE TYERS	433

31. Yeast Genomics in the Elucidation of Endoplasmic Reticulum (ER) Quality Control and Associated Protein Degradation (ERQD) — ANTJE SCHÄFER AND DIETER H. WOLF — 459

32. Mechanism-Based Proteomics Tools Based on Ubiquitin and Ubiquitin-Like Proteins: Synthesis of Active-Site Directed Probes — HUIB OVAA, PAUL J. GALARDY, AND HIDDE L. PLOEGH — 468

Section VI. Real Time/Non-Invasive Technologies

33. Application and Analysis of the GFP^u Family of Ubiquitin-Proteasome System Reporters — NEIL F. BENCE, ERIC J. BENNETT, AND RON R. KOPITO — 481

34. Monitoring of Ubiquitin-Dependent Proteolysis with Green Fluorescent Protein Substrates — VICTORIA MENÉNDEZ-BENITO, STIJN HEESSEN, AND NICO P. DANTUMA — 490

35. Monitoring Proteasome Activity *In Cellulo* and in Living Animals by Bioluminescent Imaging: Technical Considerations for Design and Use of Genetically Encoded Reporters — SHIMON GROSS AND DAVID PIWNICA-WORMS — 512

36. Bioluminescent Imaging of Ubiquitin Ligase Activity: Measuring Cdk2 Activity *In Vivo* Through Changes in p27 Turnover — GUO-JUN ZHANG AND WILLAM G. KAELIN, JR. — 530

37. Monitoring the Distribution and Dynamics of Proteasomes in Living Cells — TOM A. M. GROOTHUIS AND ERIC A. J. REITS — 549

Section VII. Small Molecule Inhibitors

38. Identifying Small Molecules Inhibitors of the Ubiquitin-Proteasome Pathway in *Xenopus* Egg Extracts — ADRIAN SALIC AND RANDALL W. KING — 567

39. Development and Characterization of Proteasome Inhibitors — KYUNG BO KIM, FABIANA N. FONSECA, AND CRAIG M. CREWS — 585

40. Screening for Selective Small Molecule Inhibitors of the Proteasome Using Activity-Based Probes — MATTHEW BOGYO — 609

41. Development of E3-Substrate (MDM2-p53)–Binding Inhibitors: Structural Aspects — DAVID C. FRY, BRADFORD GRAVES, AND LYUBOMIR T. VASSILEV — 622

42. Druggability of SCF Ubiquitin Ligase–Protein Interfaces — TIMOTHY CARDOZO AND RUBEN ABAGYAN — 634

43. Overview on Approaches for Screening for Ubiquitin Ligase Inhibitors	YI SUN	654
44. A Homogeneous FRET Assay System for Multiubiquitin Chain Assembly and Disassembly	TARIKERE L. GURURAJ, TODD R. PRAY, RAYMOND LOWE, GUOQIANG DONG, JIANING HUANG, SARKIZ DANIEL-ISSAKANI, AND DONALD G. PAYAN	663
45. Assays for High-Throughput Screening of E2 and E3 Ubiquitin Ligases	JOHN H. KENTEN, ILIA V. DAVYDOV, YASSAMIN J. SAFIRAN, DAVID H. STEWART, PANKAJ OBEROI, AND HANS A. BIEBUYCK	682
46. Quantitative Assays for Mdm2 Ubiquitin Ligase Activity and Other Ubiquitin-Utilizing Enzymes for Inhibitor Discovery	KURT R. AUGER, ROBERT A. COPELAND, AND ZHIHONG LAI	701
47. High-Throughput Screening for Inhibitors of the Cks1–Skp2 Interaction	KUO-SEN HUANG AND LYUBOMIR T. VASSILEV	717
48. In Vitro $SCF^{\beta\text{-Trcp1}}$–Mediated $I\kappa B\alpha$ Ubiquitination Assay for High-Throughput Screen	SHUICHAN XU, PALKA PATEL, MAHAN ABBASIAN, DAVID GIEGEL, WEILIN XIE, FRANK MERCURIO, AND SARAH COX	729
49. High-Throughput Screening for Inhibitors of the E3 Ubiquitin Ligase APC	JIANING HUANG, JULIE SHEUNG, GUOQIANG DONG, CHRISTINA COQUILLA, SARKIZ DANIEL-ISSAKANI, AND DONALD G. PAYAN	740

Section VIII. Generally Applicable Technologies

50. The Split-Ubiquitin Sensor: Measuring Interactions and Conformational Alterations of Proteins In Vivo	CHRISTOPH REICHEL AND NILS JOHNSSON	757
51. Ubiquitin Fusion Technique and Related Methods	ALEXANDER VARSHAVSKY	777
52. Heat-Inducible Degron and the Making of Codintional Mutants	R. JÜRGEN DOHMEN AND ALEXANDER VARSHAVSKY	799
53. Ectopic Targeting of Substrates to the Ubiquitin Pathway	JIANXUAN ZHANG AND PENGBO ZHOU	823

54. Chimeric Molecules to Target Proteins for   KATHLEEN M. SAKAMOTO   833
    Ubiquitination and Degradation

AUTHOR INDEX   849
SUBJECT INDEX   889

Contributors to Volume 399

Article numbers are in parentheses and following the name of Contributors. Affiliation listed are current.

RUBEN ABAGYAN (42), *The Scripps Research Institute, Department of Molecular Biology, La Jolla, California*

MAHAN ABBASIAN (48), *Celgene Corporation, San Diego, California*

JUNGMI AHN (10), *Institute of Biotechnology, Department of Molecular Medicine, University of Texas Health Science Center, San Antonio, Texas*

XIAOLU L. ANG (20), *Department of Pathology, Harvard Medical School, Boston, Massachusetts*

JENNIFER APODACA (10), *Institute of Biotechnology, Department of Molecular Medicine, University of Texas Health Science Center, San Antonio, Texas*

MICHAEL ASSFALG (12), *Department of Chemistry and Biochemistry, Center of Biomolecular Structure and Organization, University of Maryland, College Park, Maryland*

KURT R. AUGER (46), *GlaxoSmith-Kline Pharmaceuticals, Collegeville, Pennsylvania*

NAGI G. AYAD (28), *Department of Systems Biology, Harvard Medical School, Boston, Massachusetts*

NEIL F. BENCE (33), *Department of Biological Sciences, Stanford University, Stanford, California*

ERIC J. BENNETT (33), *Department of Biological Sciences, Stanford University, Stanford, California*

HANS A. BIEBUYCK (29, 45), *Meso Scale Discovery, Gaithersburg, Maryland*

JOANNA BLOOM (17), *Department of Pathology, New York University School of Medicine and NYU Cancer Institute, New York*

MATT BOGYO (40), *Department of Pathology, Microbiology and Immunology, Stanford University School of Medicine, Stanford, California*

JUDY CALLIS (4), *Section of Molecular and Cellular Biology, University of California-Davis, Davis, California*

TIMOTHY CARDOZO (42), *Department of Pathology, NYU School of Medicine, New York*

DONGMEI CHENG (25), *Department of Human Genetics, Center for Neurodegenerative Diseases, Emory University, Atlanta, Georgia*

ROBERT A. COPELAND (46), *Glaxo-SmithKline Pharmaceuticals, Collegeville, Pennsylvania*

CHRISTINA COQUILLA (49), *Rigel Pharmaceuticals, Inc., South San Francisco, California*

SARAH COX (48), *Celgene Corporation, San Diego, California*

CRAIG M. CREWS (39), *Department of Molecular, Cell, and Developmental Biology, Yale University, New Haven, Connecticut*

SARKIZ DANIEL-ISSAKANI (44, 49), *Rigel Pharmaceuticals, Inc., South San Francisco, California*

NICO P. DANTUMA (34), *Department of Cell and Molecular Biology Karolinska Institute, Stockholm, Sweden*

xi

ILLIA V. DAVYDOV (29, 45), *Meso Scale Discovery, Gaithersburg, Maryland*

RAYMOND J. DESHAIES (26), *Division of Biology, California Institute of Technology, Pasadena, California*

R. JÜRGEN DOHMEN (51), *Institute for Genetics, University of Cologne, Cologne, Bermany*

GUOQIANG DONG (44, 49), *Rigel Pharmaceuticals, Inc., South San Francisco, California*

FABIANA N. FONSECA (39), *Department of Molecular, Cell, and Developmental Biology, Yale University, New Haven, Connecticut*

MICHAEL FRENCH (9), *Department of Biochemistry, Molecular Biology, and Cell Biology, Northwestern University, Evanston, Illinois*

DAVID C. FRY (41), *Structural Chemistry Group, Hoffmann-La Roche, Inc., Nutley, New Jersey*

MASAHIRO FUJIMURO (6), *Department of Biochemistry, Graduate School of Pharmaceutical Sciences, Hokkaido University, Sapporo, Japan*

DAVID FUSHMAN (12), *Department of Chemistry and Biochemistry, Center of Biomolecular Structure and Organization, University of Maryland, College Park, Maryland*

PAUL J. GALARDY (8, 32), *Department of Pathology, Harvard Medical School (NRB), Boston, Massachusetts*

TUDEVIIN GAN-ERDENE (3), *Department of Biochemistry, Emory University, Atlanta, Georgia*

DAVID GIEGEL (48), *Celgene Corporation, San Diego, California*

COLIN GORDON (11), *Medical Research Council Human Genetics Unit, Western General Hospital, Edinburgh United Kingdom*

BRADFORD GRAVES (41), *Structural Chemistry Group, Hoffmann-La Roche, Inc., Nutley, New Jersey*

TOM A. M. GROOTHUIS (37), *Department of Tumor Biology, Netherlands Cancer Institute, Amsterdam, The Netherlands*

SHIMON GROSS (35), *Molecular Imaging Center, Mallinckrodt Institute of Radiology, Washington University School of Medicine, St. Louis, Missouri*

TARIKERE L. GURURAJ (44), *Rigel Pharmaceuticals, Inc., South San Francisco, California*

RANDOLPH Y. HAMPTON (21), *UCSD Division of Biologcal Sciences, La Jolla, California*

NEIL HAND (7), *School of Molecular Medical Sciences, University of Nottingham Medical School, Queens Medical Centre, Nottingham, United Kingdom*

J. WADE HARPER (20), *Department of Pathology, Harvard Medical School, Boston, Massachusetts*

RASMUS HARTMANN-PETERSON (11), *Department of Biochemistry, Institute for Molecular Biology and Physiology, Copenhagen University, Copenhagen, Denmark*

SHIGETSUGU HATAKEYAMA (19), *Department of Molecular Biochemistry, Hokkaido University Graduate School of Medicine, Sapporo, Japan*

STIJN HEESSEN (34), *Ludwig Institute for Cancer Research, Stockholm, Sweden*

LINDA HICKE (9), *Department of Biochemistry, Molecular Biology, and Cell Biology, Northwestern University, Evanston, Illinois*

YUKO HIRANO (15), *Laboratory of Frontier Science The Tokyo Metropolitan Institute of Medical Science, Tokyo, Japan*

JIANING HUANG (44, 49), *Rigel Pharmaceuticals, Inc., South San Francisco, California*

Kuo-Sen Huang (47), *Discovery Technologies, Hoffmann-La Roche, Inc., Nutley, New Jersey,*

E. Isono (5, 14), *Department of Biological Sciences, Graduate School of Science, The University of Tokyo, Tokyo, Japan*

Peter K. Jackson (23), *Stanford University School of Medicine, Programs in Biophysics, Chemical Biology, and Cancer Biology, Palo Alto, California*

Stefan Jentsch (27), *Department of Molecular Cell Biology, Max Planck Institute of Biochemistry, Martinsried, Germany*

Jianping Jin (20), *Department of Pathology, Harvard Medical School, Boston, Massachusetts*

Nils Johnsson (50), *Forschungszentrum Karlsruhe, Institut für Toxikologie, Eggenstein-Leopoldshafen, Germany*

Willam G. Kaelin (36), *Howard Hughes Medical Institute, Dana-Farber Cancer Institute and Brigham and Women's Hospital, Harvard Medical School, Boston, Massachusetts*

Peter Kaiser (16, 18), *Department of Biological Chemistry, University of California, Irvine, California*

David J. Katzmann (13), *Department of Biochemistry and Molecular Biology, Mayo Clinic College of Medicine, Rochester, Minnesota*

H. Kawahara (5), *Department of Biological Sciences, Graduate School of Science, The University of Tokyo, Tokyo, Japan*

John H. Kenten (29, 45), *Meso Scale Discovery, Gaithersburg, Maryland*

Ikjin Kim (10), *Institute of Biotechnology, Department of Molecular Medicine, University of Texas Health Science Center, San Antonio, Texas*

Kyung Bo Kim (39), *Department of Molecular, Cell, and Developmental Biology, Yale University, New Haven, Connecticut*

Randall W. King (38), *Department of Cell Biology, Harvard Medical School, Boston, Massachusetts*

Marc W. Kirschner (28), *Department of Systems Biology, Harvard Medical School, Boston, Massachusetts*

Nagamalleswari Kolli (3), *Department of Biochemistry, Emory University, Atlanta, Georgia*

Ron R. Kopito (33), *Department of Biological Sciences, Stanford University, Stanford, California*

Zhihong Lai (46), *GlaxoSmith-Kline Pharmaceuticals, Collegeville, Pennsylvania*

Ina Lee (22), *Biochemical Pharmacology, Roche Bioscience, Palo Alto, California*

Norman L. Lehman (23), *Department of Pathology, Stanford University School of Medicine, Palo Alto, California*

Richard Ling (4), *Section of Molecular and Cellular Biology, University of California-Davis, Davis, California*

Qingquan Liu (30), *Samuel Lunenfeld Research Institute, Mount Sinai Hospital, Toronto, Ontario*

James Lowe (7), *School of Molecular Medical Sciences, University of Nottingham Medical School, Queens Medical Centre, Nottingham, United Kingdom*

Raymond Lowe (44), *Acumen Pharmaceuticals, Inc., South San Francisco, California*

Amy Y. Mak (22), *Biochemical Pharmacology, Roche Bioscience, Palo Alto, California*

Masaki Matsumoto (19), *Department of Molecular and Cellular Biology, Medical Institute of Bioregulation, Kyushu University, Fukuoka, Japan*

JOHN MAYER (7), *School of Biomedical Sciences, University of Nottingham Medical School, Queens Medical Centre, Nottingham, United Kingdom*

THIBAULT MAYOR (26), *Division of Biology, California Institute of Technology, Pasadena, California*

VICTORIA MENENDEZ-BENITO (34), *Department of Cell and Molecular Biology, Karolinska Institute, Stockholm, Sweden*

FRANK MERCURIO (48), *Celgene Corporation, San Diego, California*

SHIGEO MURATA (15), *Laboratory of Frontier Science, The Tokyo Metropolitan Institute of Medical Science, Tokyo, Japan*

KEIICHI I. NAKAYAMA (19), *Department of Molecular and Cellular Biology Medical Institute of Bioregulation, Kyushu University, Fukuoka, Japan*

STEFANIE NELSON (29), *Meso Scale Discovery, Gaithersburg, Maryland*

PANKAJ OBEROI (29, 45), *Meso Scale Discovery, Gaithersburg, Maryland*

DANNY OOI (28), *Department of Systems Biology, Harvard Medical School, Boston, Masschusetts*

STEPHEN ORLICKY (30), *Samuel Lunenfeld Research Institute, Mount Sinai Hospital, Toronto, Ontario*

HUIB OVAA (8, 32), *Netherlands Cancer Institute, Division of Cellular Biochemistry, Amsterdam, The Netherlands*

MICHELE PAGANO (17), *Department of Pathology, New York University School of Medicine and NYU Cancer Institute, New York, New York*

PALKA PATEL (48), *Celgene Corporation, San Diego, California*

DONALD G. PAYAN (44, 49), *Rigel Pharmaceuticals, Inc., South San Francisco, California*

JUNMIN PENG (25), *Department of Human Genetics, Center for Neurodegenerative Diseases, Emory University, Atlanta, Georgia*

BORIS PFANDER (27), *Department of Molecular Cell Biology, Max Planck Institute of Biochemistry, Martinsried, Germany*

CECILE M. PICKART (1, 2), *Department of Biochemistry and Molecular Biology, Johns Hopkins University, Baltimore, Maryland*

DAVID PIWNICA-WORMS (35), *Molecular Imaging Center, Mallinckrodt Institute of Radiology, Washington University School of Medicine, St. Louis, Missouri*

HIDDE L. PLOEGH (8, 32) *Department of Pathology, Harvard Medical School (NRB), Boston, Massachusetts*

TODD R. PRAY (44), *Acumen Pharmaceuticals, Inc., South San Francisco, California*

SHAHRI RAASI (2), *Department of Biochemistry and Molecular Biology, Johns Hopkins University, Baltimore, Maryland*

ISHWAR RADHAKRISHNAN (9), *Department of Biochemistry, Molecular Biology, and Cell Biology, Northwestern University, Evanston, Illinois*

SUSANNAH RANKIN (28), *Department of Systems Biology, Harvard Medical School, Boston, Massuchusetts*

HAI RAO (10), *Institute of Biotechnology, Department of Molecular Medicine, University of Texas Health Science Center, San Antonio, Texas*

MICHAEL RAPE (28), *Department of Systems Biology, Harvard Medical School, Boston, Massachusetts*

CHRISTOPH REICHEL (50), *Forschungszentrum Karlsruhe, Institut für Toxikologie und Genetik, Eggenstein-Leopoldshafen, Germany*

ERIC A. J. REITS (37), *Department of Cell Biology and Histology, Academic Medical Center, University of Amsterdam, Amsterdam, The Netherlands*

MATT VAN DE RIJN (23), *Stanford University School of Medicine, Programs in Biophysics, Chemical Biology, and Cancer Biology, Palo Alto, California*

MICHAEL J. ROSE (22), *Department of Chemistry and Biochemistry, University of California, Santa Cruz, Santa Cruz, California*

MEIK SACHER (27), *Department of Molecular Cell Biology, Max Planck Institute of Biochemistry, Martinsried, Germany*

Y. SAEKI (5, 14), *Department of Biological Sciences, Graduate School of Science, The University of Tokyo, Tokyo, Japan*

YASSAMIN J. SAFIRAN (29, 45), *Meso Scale Discovery, Gaithersburg, Maryland*

KATHLEEN M. SAKAMOTO (54), *Department of Pediatrics, Division of Hematology-Oncology Department of Pathology and Laboratory Medicine, David Geffen School of Medicine, Los Angeles, California*

ADRIAN SALIC (38), *Department of Systems Biology, Harvard Medical School, Boston, Massachusetts*

LILIANA SCARAFIA (22), *Biochemical Pharmacology, Roche Bioscience, Palo Alto, California*

ANTJE SCHAFER (31), *Institut für Biochemie, Universität Stuttgart, Stuttgart, Germany*

JULIE SHEUNG (49), *Rigel Pharmaceuticals, Inc., South San Francisco, California*

SUSAN C. SHIH (9), *Department of Biochemistry, Molecular Biology, and Cell Biology, Northwestern University, Evanston, Illinois*

M. SHIMADA (5), *Department of Biological Sciences, Graduate School of Science, The University of Tokyo, Tokyo, Japan*

TAKAHIRO SHIROGANE (20), *Department of Pathology, Harvard Medical School, Boston, Massachusetts*

FRANK SICHERI (30), *Samuel Lunenfeld Research Institute, Mount Sinai Hospital, Toronto, Ontario*

DAVID H. STEWART (45), *Director of Proteomics, Meso Scale Discovery, Gaithersburg, Maryland*

YI SUN (43), *Division of Cancer Biology, Department of Radiation Oncology, University of Michigan, Ann Arbor, Michigan*

KURT SWANSON (9), *Department of Biochemistry, Molecular Biology, and Cell Biology, Northwestern University, Evanston, Illinois*

RYAN SWENERTON (29), *Meso Scale Discovery, Gaithersburg, Maryland*

DAVID C. SWINNEY (22), *Biochemical Pharmacology, Roche Bioscience, Palo Alto, California*

CHRISTIAN TAGWERKER (16), *Department of Biological Chemistry, University of California, Irrine, California*

KEIJI TANAKA (15), *Laboratory of Frontier Science, The Tokyo Metropolitan Institute of Medical Science, Tokyo, Japan*

XIAOJING TANG (30), *Samuel Lunenfeld Research Institute, Mount Sinai Hospital, Toronto, Ontario*

H. TH. MARC TIMMERS (24), *Department of Physiological Chemistry, University Medical Center—Utrecht, Utrecht, The Netherlands*

A. TOH-E (5, 14), *Department of Biological Sciences, Graduate School of Science, The University of Tokyo, Tokyo, Japan*

MIKE TYERS (30), *Samuel Lunenfeld Research Institute, Mount Sinai Hospital, Toronto, Ontario*

RANJANI VARADAN (12), *Department of Chemistry and Biochemistry, Center of Biomolecular Structure and Organization, University of Maryland, College Park, Maryland*

ALEXANDER VARSHAVSKY (51, 52), *Division of Biology, California, Institute of Technology, Pasadena, California*

LYUBOMIR T. VASSILEV (41, 47), *Discovery Oncology, Hoffmann-La Roche, Inc., Nutley, New Jersey*

SARA VOLK (1), *Department of Biochemistry and Molecular Biology, Johns Hopkins University, Baltimore, Maryland*

MIN WANG (1), *Department of Biochemistry and Molecular Biology, Johns Hopkins University, Baltimore, Maryland*

BEVERLY WENDLAND (13), *Department of Biology, Johns Hopkins University, Baltimore, Maryland*

KEITH D. WILKINSON (3), *Department of Biochemistry, Emory University, Atlanta, Georgia*

ANDREW WILLEMS (30), *Samuel Lunenfeld Research Institute, Mount Sinai Hospital, Toronto, Ontario*

G. SEBASTIAN WLNKLER (24), *Department of Physiological Chemistry, University Medical Center—Utrecht, Utrecht, The Netherlands*

JAMES WOHLSCHLEGEL (18), *Department of Cell Biology, The Scripps Research Institute, La Jolla, California*

DIETER H. WOLF (31), *Institut für Biochemie, Universität Stuttgart, Stuttgart, Germany*

WEILIN XIE (48), *Celgene Corporation, San Diego, California*

SHUICHAN XU (48), *Celgene Corporation, San Diego, California*

YI-ZHENG XU (22), *Elan Pharmaceuticals, South San Francisco, California*

H. YOKOSAWA (5, 6), *Department of Biological Sciences, Graduate School of Science, The University of Tokyo, Tokyo, Japan*

GUO-JUN ZHANG (36), *Howard Hughes Medical Institute, Dana-Farber Cancer Institute and Brighan and Women's Hospital, Harvard Medical School, Boston, Massachusetts*

JIANXUAN ZHANG (53), *Department of Pathology and Laboratory Medicine, Weill Medical College of Cornell University, New York*

PENGBO ZHOU (53), *Department of Pathology and Laboratory Medicine, Weill Medical College of Cornell University, New York*

Acknowledgments

These volumes, which I hope will be valuable to both rookie and veteran researchers in the ubiquitin field, would not exist were it not for the the many authors who invested considerable time and energy. I wish to thank them for their excellent contributions and for their patience in dealing with the editor. I also wish to acknowledge the many people who assisted in the preparation of these volumes, including Drs. Avram Hershko, Stefan Jentsch, Cecile Pickart, and Keiji Tanaka for their help in selecting chapters, my assistant Daphne Shimoda, for her help with soliciting and processing manuscripts, Gracy Noelle and Cindy Minor at Elsevier for their overall help in pulling this project together and keeping it on track, and Jamey Stegmaier at SPI for his efforts on the page proofs. Finally, I wish to thank Linda Silveira, who inevitably bears greater burdens as a result of projects such as this.

Ray Deshaies

METHODS IN ENZYMOLOGY

VOLUME I. Preparation and Assay of Enzymes
Edited by SIDNEY P. COLOWICK AND NATHAN O. KAPLAN

VOLUME II. Preparation and Assay of Enzymes
Edited by SIDNEY P. COLOWICK AND NATHAN O. KAPLAN

VOLUME III. Preparation and Assay of Substrates
Edited by SIDNEY P. COLOWICK AND NATHAN O. KAPLAN

VOLUME IV. Special Techniques for the Enzymologist
Edited by SIDNEY P. COLOWICK AND NATHAN O. KAPLAN

VOLUME V. Preparation and Assay of Enzymes
Edited by SIDNEY P. COLOWICK AND NATHAN O. KAPLAN

VOLUME VI. Preparation and Assay of Enzymes *(Continued)*
Preparation and Assay of Substrates
Special Techniques
Edited by SIDNEY P. COLOWICK AND NATHAN O. KAPLAN

VOLUME VII. Cumulative Subject Index
Edited by SIDNEY P. COLOWICK AND NATHAN O. KAPLAN

VOLUME VIII. Complex Carbohydrates
Edited by ELIZABETH F. NEUFELD AND VICTOR GINSBURG

VOLUME IX. Carbohydrate Metabolism
Edited by WILLIS A. WOOD

VOLUME X. Oxidation and Phosphorylation
Edited by RONALD W. ESTABROOK AND MAYNARD E. PULLMAN

VOLUME XI. Enzyme Structure
Edited by C. H. W. HIRS

VOLUME XII. Nucleic Acids (Parts A and B)
Edited by LAWRENCE GROSSMAN AND KIVIE MOLDAVE

VOLUME XIII. Citric Acid Cycle
Edited by J. M. LOWENSTEIN

VOLUME XIV. Lipids
Edited by J. M. LOWENSTEIN

VOLUME XV. Steroids and Terpenoids
Edited by RAYMOND B. CLAYTON

VOLUME XVI. Fast Reactions
Edited by KENNETH KUSTIN

VOLUME XVII. Metabolism of Amino Acids and Amines (Parts A and B)
Edited by HERBERT TABOR AND CELIA WHITE TABOR

VOLUME XVIII. Vitamins and Coenzymes (Parts A, B, and C)
Edited by DONALD B. MCCORMICK AND LEMUEL D. WRIGHT

VOLUME XIX. Proteolytic Enzymes
Edited by GERTRUDE E. PERLMANN AND LASZLO LORAND

VOLUME XX. Nucleic Acids and Protein Synthesis (Part C)
Edited by KIVIE MOLDAVE AND LAWRENCE GROSSMAN

VOLUME XXI. Nucleic Acids (Part D)
Edited by LAWRENCE GROSSMAN AND KIVIE MOLDAVE

VOLUME XXII. Enzyme Purification and Related Techniques
Edited by WILLIAM B. JAKOBY

VOLUME XXIII. Photosynthesis (Part A)
Edited by ANTHONY SAN PIETRO

VOLUME XXIV. Photosynthesis and Nitrogen Fixation (Part B)
Edited by ANTHONY SAN PIETRO

VOLUME XXV. Enzyme Structure (Part B)
Edited by C. H. W. HIRS AND SERGE N. TIMASHEFF

VOLUME XXVI. Enzyme Structure (Part C)
Edited by C. H. W. HIRS AND SERGE N. TIMASHEFF

VOLUME XXVII. Enzyme Structure (Part D)
Edited by C. H. W. HIRS AND SERGE N. TIMASHEFF

VOLUME XXVIII. Complex Carbohydrates (Part B)
Edited by VICTOR GINSBURG

VOLUME XXIX. Nucleic Acids and Protein Synthesis (Part E)
Edited by LAWRENCE GROSSMAN AND KIVIE MOLDAVE

VOLUME XXX. Nucleic Acids and Protein Synthesis (Part F)
Edited by KIVIE MOLDAVE AND LAWRENCE GROSSMAN

VOLUME XXXI. Biomembranes (Part A)
Edited by SIDNEY FLEISCHER AND LESTER PACKER

VOLUME XXXII. Biomembranes (Part B)
Edited by SIDNEY FLEISCHER AND LESTER PACKER

VOLUME XXXIII. Cumulative Subject Index Volumes I-XXX
Edited by MARTHA G. DENNIS AND EDWARD A. DENNIS

VOLUME XXXIV. Affinity Techniques (Enzyme Purification: Part B)
Edited by WILLIAM B. JAKOBY AND MEIR WILCHEK

VOLUME XXXV. Lipids (Part B)
Edited by JOHN M. LOWENSTEIN

VOLUME XXXVI. Hormone Action (Part A: Steroid Hormones)
Edited by BERT W. O'MALLEY AND JOEL G. HARDMAN

VOLUME XXXVII. Hormone Action (Part B: Peptide Hormones)
Edited by BERT W. O'MALLEY AND JOEL G. HARDMAN

VOLUME XXXVIII. Hormone Action (Part C: Cyclic Nucleotides)
Edited by JOEL G. HARDMAN AND BERT W. O'MALLEY

VOLUME XXXIX. Hormone Action (Part D: Isolated Cells, Tissues, and Organ Systems)
Edited by JOEL G. HARDMAN AND BERT W. O'MALLEY

VOLUME XL. Hormone Action (Part E: Nuclear Structure and Function)
Edited by BERT W. O'MALLEY AND JOEL G. HARDMAN

VOLUME XLI. Carbohydrate Metabolism (Part B)
Edited by W. A. WOOD

VOLUME XLII. Carbohydrate Metabolism (Part C)
Edited by W. A. WOOD

VOLUME XLIII. Antibiotics
Edited by JOHN H. HASH

VOLUME XLIV. Immobilized Enzymes
Edited by KLAUS MOSBACH

VOLUME XLV. Proteolytic Enzymes (Part B)
Edited by LASZLO LORAND

VOLUME XLVI. Affinity Labeling
Edited by WILLIAM B. JAKOBY AND MEIR WILCHEK

VOLUME XLVII. Enzyme Structure (Part E)
Edited by C. H. W. HIRS AND SERGE N. TIMASHEFF

VOLUME XLVIII. Enzyme Structure (Part F)
Edited by C. H. W. HIRS AND SERGE N. TIMASHEFF

VOLUME XLIX. Enzyme Structure (Part G)
Edited by C. H. W. HIRS AND SERGE N. TIMASHEFF

VOLUME L. Complex Carbohydrates (Part C)
Edited by VICTOR GINSBURG

VOLUME LI. Purine and Pyrimidine Nucleotide Metabolism
Edited by PATRICIA A. HOFFEE AND MARY ELLEN JONES

VOLUME LII. Biomembranes (Part C: Biological Oxidations)
Edited by SIDNEY FLEISCHER AND LESTER PACKER

VOLUME LIII. Biomembranes (Part D: Biological Oxidations)
Edited by SIDNEY FLEISCHER AND LESTER PACKER

VOLUME LIV. Biomembranes (Part E: Biological Oxidations)
Edited by SIDNEY FLEISCHER AND LESTER PACKER

VOLUME LV. Biomembranes (Part F: Bioenergetics)
Edited by SIDNEY FLEISCHER AND LESTER PACKER

VOLUME LVI. Biomembranes (Part G: Bioenergetics)
Edited by SIDNEY FLEISCHER AND LESTER PACKER

VOLUME LVII. Bioluminescence and Chemiluminescence
Edited by MARLENE A. DELUCA

VOLUME LVIII. Cell Culture
Edited by WILLIAM B. JAKOBY AND IRA PASTAN

VOLUME LIX. Nucleic Acids and Protein Synthesis (Part G)
Edited by KIVIE MOLDAVE AND LAWRENCE GROSSMAN

VOLUME LX. Nucleic Acids and Protein Synthesis (Part H)
Edited by KIVIE MOLDAVE AND LAWRENCE GROSSMAN

VOLUME 61. Enzyme Structure (Part H)
Edited by C. H. W. HIRS AND SERGE N. TIMASHEFF

VOLUME 62. Vitamins and Coenzymes (Part D)
Edited by DONALD B. MCCORMICK AND LEMUEL D. WRIGHT

VOLUME 63. Enzyme Kinetics and Mechanism (Part A: Initial Rate and Inhibitor Methods)
Edited by DANIEL L. PURICH

VOLUME 64. Enzyme Kinetics and Mechanism
(Part B: Isotopic Probes and Complex Enzyme Systems)
Edited by DANIEL L. PURICH

VOLUME 65. Nucleic Acids (Part I)
Edited by LAWRENCE GROSSMAN AND KIVIE MOLDAVE

VOLUME 66. Vitamins and Coenzymes (Part E)
Edited by DONALD B. MCCORMICK AND LEMUEL D. WRIGHT

VOLUME 67. Vitamins and Coenzymes (Part F)
Edited by DONALD B. MCCORMICK AND LEMUEL D. WRIGHT

VOLUME 68. Recombinant DNA
Edited by RAY WU

VOLUME 69. Photosynthesis and Nitrogen Fixation (Part C)
Edited by ANTHONY SAN PIETRO

VOLUME 70. Immunochemical Techniques (Part A)
Edited by HELEN VAN VUNAKIS AND JOHN J. LANGONE

VOLUME 71. Lipids (Part C)
Edited by JOHN M. LOWENSTEIN

VOLUME 72. Lipids (Part D)
Edited by JOHN M. LOWENSTEIN

VOLUME 73. Immunochemical Techniques (Part B)
Edited by JOHN J. LANGONE AND HELEN VAN VUNAKIS

VOLUME 74. Immunochemical Techniques (Part C)
Edited by JOHN J. LANGONE AND HELEN VAN VUNAKIS

VOLUME 75. Cumulative Subject Index Volumes XXXI, XXXII, XXXIV–LX
Edited by EDWARD A. DENNIS AND MARTHA G. DENNIS

VOLUME 76. Hemoglobins
Edited by ERALDO ANTONINI, LUIGI ROSSI-BERNARDI, AND EMILIA CHIANCONE

VOLUME 77. Detoxication and Drug Metabolism
Edited by WILLIAM B. JAKOBY

VOLUME 78. Interferons (Part A)
Edited by SIDNEY PESTKA

VOLUME 79. Interferons (Part B)
Edited by SIDNEY PESTKA

VOLUME 80. Proteolytic Enzymes (Part C)
Edited by LASZLO LORAND

VOLUME 81. Biomembranes (Part H: Visual Pigments and Purple Membranes, I)
Edited by LESTER PACKER

VOLUME 82. Structural and Contractile Proteins (Part A: Extracellular Matrix)
Edited by LEON W. CUNNINGHAM AND DIXIE W. FREDERIKSEN

VOLUME 83. Complex Carbohydrates (Part D)
Edited by VICTOR GINSBURG

VOLUME 84. Immunochemical Techniques (Part D: Selected Immunoassays)
Edited by JOHN J. LANGONE AND HELEN VAN VUNAKIS

VOLUME 85. Structural and Contractile Proteins (Part B: The Contractile Apparatus and the Cytoskeleton)
Edited by DIXIE W. FREDERIKSEN AND LEON W. CUNNINGHAM

VOLUME 86. Prostaglandins and Arachidonate Metabolites
Edited by WILLIAM E. M. LANDS AND WILLIAM L. SMITH

VOLUME 87. Enzyme Kinetics and Mechanism (Part C: Intermediates, Stereo-chemistry, and Rate Studies)
Edited by DANIEL L. PURICH

VOLUME 88. Biomembranes (Part I: Visual Pigments and Purple Membranes, II)
Edited by LESTER PACKER

VOLUME 89. Carbohydrate Metabolism (Part D)
Edited by WILLIS A. WOOD

VOLUME 90. Carbohydrate Metabolism (Part E)
Edited by WILLIS A. WOOD

VOLUME 91. Enzyme Structure (Part I)
Edited by C. H. W. HIRS AND SERGE N. TIMASHEFF

VOLUME 92. Immunochemical Techniques (Part E: Monoclonal Antibodies and General Immunoassay Methods)
Edited by JOHN J. LANGONE AND HELEN VAN VUNAKIS

VOLUME 93. Immunochemical Techniques (Part F: Conventional Antibodies, Fc Receptors, and Cytotoxicity)
Edited by JOHN J. LANGONE AND HELEN VAN VUNAKIS

VOLUME 94. Polyamines
Edited by HERBERT TABOR AND CELIA WHITE TABOR

VOLUME 95. Cumulative Subject Index Volumes 61–74, 76–80
Edited by EDWARD A. DENNIS AND MARTHA G. DENNIS

VOLUME 96. Biomembranes [Part J: Membrane Biogenesis: Assembly and Targeting (General Methods; Eukaryotes)]
Edited by SIDNEY FLEISCHER AND BECCA FLEISCHER

VOLUME 97. Biomembranes [Part K: Membrane Biogenesis: Assembly and Targeting (Prokaryotes, Mitochondria, and Chloroplasts)]
Edited by SIDNEY FLEISCHER AND BECCA FLEISCHER

VOLUME 98. Biomembranes (Part L: Membrane Biogenesis: Processing and Recycling)
Edited by SIDNEY FLEISCHER AND BECCA FLEISCHER

VOLUME 99. Hormone Action (Part F: Protein Kinases)
Edited by JACKIE D. CORBIN AND JOEL G. HARDMAN

VOLUME 100. Recombinant DNA (Part B)
Edited by RAY WU, LAWRENCE GROSSMAN, AND KIVIE MOLDAVE

VOLUME 101. Recombinant DNA (Part C)
Edited by RAY WU, LAWRENCE GROSSMAN, AND KIVIE MOLDAVE

VOLUME 102. Hormone Action (Part G: Calmodulin and Calcium-Binding Proteins)
Edited by ANTHONY R. MEANS AND BERT W. O'MALLEY

VOLUME 103. Hormone Action (Part H: Neuroendocrine Peptides)
Edited by P. MICHAEL CONN

VOLUME 104. Enzyme Purification and Related Techniques (Part C)
Edited by WILLIAM B. JAKOBY

VOLUME 105. Oxygen Radicals in Biological Systems
Edited by LESTER PACKER

VOLUME 106. Posttranslational Modifications (Part A)
Edited by FINN WOLD AND KIVIE MOLDAVE

VOLUME 107. Posttranslational Modifications (Part B)
Edited by FINN WOLD AND KIVIE MOLDAVE

VOLUME 108. Immunochemical Techniques (Part G: Separation and Characterization of Lymphoid Cells)
Edited by GIOVANNI DI SABATO, JOHN J. LANGONE, AND HELEN VAN VUNAKIS

VOLUME 109. Hormone Action (Part I: Peptide Hormones)
Edited by LUTZ BIRNBAUMER AND BERT W. O'MALLEY

VOLUME 110. Steroids and Isoprenoids (Part A)
Edited by JOHN H. LAW AND HANS C. RILLING

VOLUME 111. Steroids and Isoprenoids (Part B)
Edited by JOHN H. LAW AND HANS C. RILLING

VOLUME 112. Drug and Enzyme Targeting (Part A)
Edited by KENNETH J. WIDDER AND RALPH GREEN

VOLUME 113. Glutamate, Glutamine, Glutathione, and Related Compounds
Edited by ALTON MEISTER

VOLUME 114. Diffraction Methods for Biological Macromolecules (Part A)
Edited by HAROLD W. WYCKOFF, C. H. W. HIRS, AND SERGE N. TIMASHEFF

VOLUME 115. Diffraction Methods for Biological Macromolecules (Part B)
Edited by HAROLD W. WYCKOFF, C. H. W. HIRS, AND SERGE N. TIMASHEFF

VOLUME 116. Immunochemical Techniques (Part H: Effectors and Mediators of Lymphoid Cell Functions)
Edited by GIOVANNI DI SABATO, JOHN J. LANGONE, AND HELEN VAN VUNAKIS

VOLUME 117. Enzyme Structure (Part J)
Edited by C. H. W. HIRS AND SERGE N. TIMASHEFF

VOLUME 118. Plant Molecular Biology
Edited by ARTHUR WEISSBACH AND HERBERT WEISSBACH

VOLUME 119. Interferons (Part C)
Edited by SIDNEY PESTKA

VOLUME 120. Cumulative Subject Index Volumes 81–94, 96–101

VOLUME 121. Immunochemical Techniques (Part I: Hybridoma Technology and Monoclonal Antibodies)
Edited by JOHN J. LANGONE AND HELEN VAN VUNAKIS

VOLUME 122. Vitamins and Coenzymes (Part G)
Edited by FRANK CHYTIL AND DONALD B. MCCORMICK

VOLUME 123. Vitamins and Coenzymes (Part H)
Edited by FRANK CHYTIL AND DONALD B. MCCORMICK

VOLUME 124. Hormone Action (Part J: Neuroendocrine Peptides)
Edited by P. MICHAEL CONN

VOLUME 125. Biomembranes (Part M: Transport in Bacteria, Mitochondria, and Chloroplasts: General Approaches and Transport Systems)
Edited by SIDNEY FLEISCHER AND BECCA FLEISCHER

VOLUME 126. Biomembranes (Part N: Transport in Bacteria, Mitochondria, and Chloroplasts: Protonmotive Force)
Edited by SIDNEY FLEISCHER AND BECCA FLEISCHER

VOLUME 127. Biomembranes (Part O: Protons and Water: Structure and Translocation)
Edited by LESTER PACKER

VOLUME 128. Plasma Lipoproteins (Part A: Preparation, Structure, and Molecular Biology)
Edited by JERE P. SEGREST AND JOHN J. ALBERS

VOLUME 129. Plasma Lipoproteins (Part B: Characterization, Cell Biology, and Metabolism)
Edited by JOHN J. ALBERS AND JERE P. SEGREST

VOLUME 130. Enzyme Structure (Part K)
Edited by C. H. W. HIRS AND SERGE N. TIMASHEFF

VOLUME 131. Enzyme Structure (Part L)
Edited by C. H. W. HIRS AND SERGE N. TIMASHEFF

VOLUME 132. Immunochemical Techniques (Part J: Phagocytosis and Cell-Mediated Cytotoxicity)
Edited by GIOVANNI DI SABATO AND JOHANNES EVERSE

VOLUME 133. Bioluminescence and Chemiluminescence (Part B)
Edited by MARLENE DELUCA AND WILLIAM D. MCELROY

VOLUME 134. Structural and Contractile Proteins (Part C: The Contractile Apparatus and the Cytoskeleton)
Edited by RICHARD B. VALLEE

VOLUME 135. Immobilized Enzymes and Cells (Part B)
Edited by KLAUS MOSBACH

VOLUME 136. Immobilized Enzymes and Cells (Part C)
Edited by KLAUS MOSBACH

VOLUME 137. Immobilized Enzymes and Cells (Part D)
Edited by KLAUS MOSBACH

VOLUME 138. Complex Carbohydrates (Part E)
Edited by VICTOR GINSBURG

VOLUME 139. Cellular Regulators (Part A: Calcium- and Calmodulin-Binding Proteins)
Edited by ANTHONY R. MEANS AND P. MICHAEL CONN

VOLUME 140. Cumulative Subject Index Volumes 102–119, 121–134

VOLUME 141. Cellular Regulators (Part B: Calcium and Lipids)
Edited by P. MICHAEL CONN AND ANTHONY R. MEANS

VOLUME 142. Metabolism of Aromatic Amino Acids and Amines
Edited by SEYMOUR KAUFMAN

VOLUME 143. Sulfur and Sulfur Amino Acids
Edited by WILLIAM B. JAKOBY AND OWEN GRIFFITH

VOLUME 144. Structural and Contractile Proteins (Part D: Extracellular Matrix)
Edited by LEON W. CUNNINGHAM

VOLUME 145. Structural and Contractile Proteins (Part E: Extracellular Matrix)
Edited by LEON W. CUNNINGHAM

VOLUME 146. Peptide Growth Factors (Part A)
Edited by DAVID BARNES AND DAVID A. SIRBASKU

VOLUME 147. Peptide Growth Factors (Part B)
Edited by DAVID BARNES AND DAVID A. SIRBASKU

VOLUME 148. Plant Cell Membranes
Edited by LESTER PACKER AND ROLAND DOUCE

VOLUME 149. Drug and Enzyme Targeting (Part B)
Edited by RALPH GREEN AND KENNETH J. WIDDER

VOLUME 150. Immunochemical Techniques (Part K: *In Vitro* Models of B and T Cell Functions and Lymphoid Cell Receptors)
Edited by GIOVANNI DI SABATO

VOLUME 151. Molecular Genetics of Mammalian Cells
Edited by MICHAEL M. GOTTESMAN

VOLUME 152. Guide to Molecular Cloning Techniques
Edited by SHELBY L. BERGER AND ALAN R. KIMMEL

VOLUME 153. Recombinant DNA (Part D)
Edited by RAY WU AND LAWRENCE GROSSMAN

VOLUME 154. Recombinant DNA (Part E)
Edited by RAY WU AND LAWRENCE GROSSMAN

VOLUME 155. Recombinant DNA (Part F)
Edited by RAY WU

VOLUME 156. Biomembranes (Part P: ATP-Driven Pumps and Related Transport: The Na, K-Pump)
Edited by SIDNEY FLEISCHER AND BECCA FLEISCHER

VOLUME 157. Biomembranes (Part Q: ATP-Driven Pumps and Related Transport: Calcium, Proton, and Potassium Pumps)
Edited by SIDNEY FLEISCHER AND BECCA FLEISCHER

VOLUME 158. Metalloproteins (Part A)
Edited by JAMES F. RIORDAN AND BERT L. VALLEE

VOLUME 159. Initiation and Termination of Cyclic Nucleotide Action
Edited by JACKIE D. CORBIN AND ROGER A. JOHNSON

VOLUME 160. Biomass (Part A: Cellulose and Hemicellulose)
Edited by WILLIS A. WOOD AND SCOTT T. KELLOGG

VOLUME 161. Biomass (Part B: Lignin, Pectin, and Chitin)
Edited by WILLIS A. WOOD AND SCOTT T. KELLOGG

VOLUME 162. Immunochemical Techniques (Part L: Chemotaxis and Inflammation)
Edited by GIOVANNI DI SABATO

VOLUME 163. Immunochemical Techniques (Part M: Chemotaxis and Inflammation)
Edited by GIOVANNI DI SABATO

VOLUME 164. Ribosomes
Edited by HARRY F. NOLLER, JR., AND KIVIE MOLDAVE

VOLUME 165. Microbial Toxins: Tools for Enzymology
Edited by SIDNEY HARSHMAN

VOLUME 166. Branched-Chain Amino Acids
Edited by ROBERT HARRIS AND JOHN R. SOKATCH

VOLUME 167. Cyanobacteria
Edited by LESTER PACKER AND ALEXANDER N. GLAZER

VOLUME 168. Hormone Action (Part K: Neuroendocrine Peptides)
Edited by P. MICHAEL CONN

VOLUME 169. Platelets: Receptors, Adhesion, Secretion (Part A)
Edited by JACEK HAWIGER

VOLUME 170. Nucleosomes
Edited by PAUL M. WASSARMAN AND ROGER D. KORNBERG

VOLUME 171. Biomembranes (Part R: Transport Theory: Cells and Model Membranes)
Edited by SIDNEY FLEISCHER AND BECCA FLEISCHER

VOLUME 172. Biomembranes (Part S: Transport: Membrane Isolation and Characterization)
Edited by SIDNEY FLEISCHER AND BECCA FLEISCHER

VOLUME 173. Biomembranes [Part T: Cellular and Subcellular Transport: Eukaryotic (Nonepithelial) Cells]
Edited by SIDNEY FLEISCHER AND BECCA FLEISCHER

VOLUME 174. Biomembranes [Part U: Cellular and Subcellular Transport: Eukaryotic (Nonepithelial) Cells]
Edited by SIDNEY FLEISCHER AND BECCA FLEISCHER

VOLUME 175. Cumulative Subject Index Volumes 135–139, 141–167

VOLUME 176. Nuclear Magnetic Resonance (Part A: Spectral Techniques and Dynamics)
Edited by NORMAN J. OPPENHEIMER AND THOMAS L. JAMES

VOLUME 177. Nuclear Magnetic Resonance (Part B: Structure and Mechanism)
Edited by NORMAN J. OPPENHEIMER AND THOMAS L. JAMES

VOLUME 178. Antibodies, Antigens, and Molecular Mimicry
Edited by JOHN J. LANGONE

VOLUME 179. Complex Carbohydrates (Part F)
Edited by VICTOR GINSBURG

VOLUME 180. RNA Processing (Part A: General Methods)
Edited by JAMES E. DAHLBERG AND JOHN N. ABELSON

VOLUME 181. RNA Processing (Part B: Specific Methods)
Edited by JAMES E. DAHLBERG AND JOHN N. ABELSON

VOLUME 182. Guide to Protein Purification
Edited by MURRAY P. DEUTSCHER

VOLUME 183. Molecular Evolution: Computer Analysis of Protein and Nucleic Acid Sequences
Edited by RUSSELL F. DOOLITTLE

VOLUME 184. Avidin-Biotin Technology
Edited by MEIR WILCHEK AND EDWARD A. BAYER

VOLUME 185. Gene Expression Technology
Edited by DAVID V. GOEDDEL

VOLUME 186. Oxygen Radicals in Biological Systems (Part B: Oxygen Radicals and Antioxidants)
Edited by LESTER PACKER AND ALEXANDER N. GLAZER

VOLUME 187. Arachidonate Related Lipid Mediators
Edited by ROBERT C. MURPHY AND FRANK A. FITZPATRICK

VOLUME 188. Hydrocarbons and Methylotrophy
Edited by MARY E. LIDSTROM

VOLUME 189. Retinoids (Part A: Molecular and Metabolic Aspects)
Edited by LESTER PACKER

VOLUME 190. Retinoids (Part B: Cell Differentiation and Clinical Applications)
Edited by LESTER PACKER

VOLUME 191. Biomembranes (Part V: Cellular and Subcellular Transport: Epithelial Cells)
Edited by SIDNEY FLEISCHER AND BECCA FLEISCHER

VOLUME 192. Biomembranes (Part W: Cellular and Subcellular Transport: Epithelial Cells)
Edited by SIDNEY FLEISCHER AND BECCA FLEISCHER

VOLUME 193. Mass Spectrometry
Edited by JAMES A. MCCLOSKEY

VOLUME 194. Guide to Yeast Genetics and Molecular Biology
Edited by CHRISTINE GUTHRIE AND GERALD R. FINK

VOLUME 195. Adenylyl Cyclase, G Proteins, and Guanylyl Cyclase
Edited by ROGER A. JOHNSON AND JACKIE D. CORBIN

VOLUME 196. Molecular Motors and the Cytoskeleton
Edited by RICHARD B. VALLEE

VOLUME 197. Phospholipases
Edited by EDWARD A. DENNIS

VOLUME 198. Peptide Growth Factors (Part C)
Edited by DAVID BARNES, J. P. MATHER, AND GORDON H. SATO

VOLUME 199. Cumulative Subject Index Volumes 168–174, 176–194

VOLUME 200. Protein Phosphorylation (Part A: Protein Kinases: Assays, Purification, Antibodies, Functional Analysis, Cloning, and Expression)
Edited by TONY HUNTER AND BARTHOLOMEW M. SEFTON

VOLUME 201. Protein Phosphorylation (Part B: Analysis of Protein Phosphorylation, Protein Kinase Inhibitors, and Protein Phosphatases)
Edited by TONY HUNTER AND BARTHOLOMEW M. SEFTON

VOLUME 202. Molecular Design and Modeling: Concepts and Applications (Part A: Proteins, Peptides, and Enzymes)
Edited by JOHN J. LANGONE

VOLUME 203. Molecular Design and Modeling:
Concepts and Applications (Part B: Antibodies and Antigens, Nucleic Acids, Polysaccharides,
and Drugs)
Edited by JOHN J. LANGONE

VOLUME 204. Bacterial Genetic Systems
Edited by JEFFREY H. MILLER

VOLUME 205. Metallobiochemistry (Part B: Metallothionein and Related Molecules)
Edited by JAMES F. RIORDAN AND BERT L. VALLEE

VOLUME 206. Cytochrome P450
Edited by MICHAEL R. WATERMAN AND ERIC F. JOHNSON

VOLUME 207. Ion Channels
Edited by BERNARDO RUDY AND LINDA E. IVERSON

VOLUME 208. Protein–DNA Interactions
Edited by ROBERT T. SAUER

VOLUME 209. Phospholipid Biosynthesis
Edited by EDWARD A. DENNIS AND DENNIS E. VANCE

VOLUME 210. Numerical Computer Methods
Edited by LUDWIG BRAND AND MICHAEL L. JOHNSON

VOLUME 211. DNA Structures (Part A: Synthesis and Physical Analysis of DNA)
Edited by DAVID M. J. LILLEY AND JAMES E. DAHLBERG

VOLUME 212. DNA Structures (Part B: Chemical and Electrophoretic Analysis of DNA)
Edited by DAVID M. J. LILLEY AND JAMES E. DAHLBERG

VOLUME 213. Carotenoids (Part A: Chemistry, Separation, Quantitation, and Antioxidation)
Edited by LESTER PACKER

VOLUME 214. Carotenoids (Part B: Metabolism, Genetics, and Biosynthesis)
Edited by LESTER PACKER

VOLUME 215. Platelets: Receptors, Adhesion, Secretion (Part B)
Edited by JACEK J. HAWIGER

VOLUME 216. Recombinant DNA (Part G)
Edited by RAY WU

VOLUME 217. Recombinant DNA (Part H)
Edited by RAY WU

VOLUME 218. Recombinant DNA (Part I)
Edited by RAY WU

VOLUME 219. Reconstitution of Intracellular Transport
Edited by JAMES E. ROTHMAN

VOLUME 220. Membrane Fusion Techniques (Part A)
Edited by NEJAT DÜZGÜNEŞ

VOLUME 221. Membrane Fusion Techniques (Part B)
Edited by NEJAT DÜZGÜNEŞ

VOLUME 222. Proteolytic Enzymes in Coagulation, Fibrinolysis, and Complement Activation (Part A: Mammalian Blood Coagulation Factors and Inhibitors)
Edited by LASZLO LORAND AND KENNETH G. MANN

VOLUME 223. Proteolytic Enzymes in Coagulation, Fibrinolysis, and Complement Activation (Part B: Complement Activation, Fibrinolysis, and Nonmammalian Blood Coagulation Factors)
Edited by LASZLO LORAND AND KENNETH G. MANN

VOLUME 224. Molecular Evolution: Producing the Biochemical Data
Edited by ELIZABETH ANNE ZIMMER, THOMAS J. WHITE, REBECCA L. CANN, AND ALLAN C. WILSON

VOLUME 225. Guide to Techniques in Mouse Development
Edited by PAUL M. WASSARMAN AND MELVIN L. DEPAMPHILIS

VOLUME 226. Metallobiochemistry (Part C: Spectroscopic and Physical Methods for Probing Metal Ion Environments in Metalloenzymes and Metalloproteins)
Edited by JAMES F. RIORDAN AND BERT L. VALLEE

VOLUME 227. Metallobiochemistry (Part D: Physical and Spectroscopic Methods for Probing Metal Ion Environments in Metalloproteins)
Edited by JAMES F. RIORDAN AND BERT L. VALLEE

VOLUME 228. Aqueous Two-Phase Systems
Edited by HARRY WALTER AND GÖTE JOHANSSON

VOLUME 229. Cumulative Subject Index Volumes 195–198, 200–227

VOLUME 230. Guide to Techniques in Glycobiology
Edited by WILLIAM J. LENNARZ AND GERALD W. HART

VOLUME 231. Hemoglobins (Part B: Biochemical and Analytical Methods)
Edited by JOHANNES EVERSE, KIM D. VANDEGRIFF, AND ROBERT M. WINSLOW

VOLUME 232. Hemoglobins (Part C: Biophysical Methods)
Edited by JOHANNES EVERSE, KIM D. VANDEGRIFF, AND ROBERT M. WINSLOW

VOLUME 233. Oxygen Radicals in Biological Systems (Part C)
Edited by LESTER PACKER

VOLUME 234. Oxygen Radicals in Biological Systems (Part D)
Edited by LESTER PACKER

VOLUME 235. Bacterial Pathogenesis (Part A: Identification and Regulation of Virulence Factors)
Edited by VIRGINIA L. CLARK AND PATRIK M. BAVOIL

VOLUME 236. Bacterial Pathogenesis (Part B: Integration of Pathogenic Bacteria with Host Cells)
Edited by VIRGINIA L. CLARK AND PATRIK M. BAVOIL

VOLUME 237. Heterotrimeric G Proteins
Edited by RAVI IYENGAR

VOLUME 238. Heterotrimeric G-Protein Effectors
Edited by RAVI IYENGAR

VOLUME 239. Nuclear Magnetic Resonance (Part C)
Edited by THOMAS L. JAMES AND NORMAN J. OPPENHEIMER

VOLUME 240. Numerical Computer Methods (Part B)
Edited by MICHAEL L. JOHNSON AND LUDWIG BRAND

VOLUME 241. Retroviral Proteases
Edited by LAWRENCE C. KUO AND JULES A. SHAFER

VOLUME 242. Neoglycoconjugates (Part A)
Edited by Y. C. LEE AND REIKO T. LEE

VOLUME 243. Inorganic Microbial Sulfur Metabolism
Edited by HARRY D. PECK, JR., AND JEAN LEGALL

VOLUME 244. Proteolytic Enzymes: Serine and Cysteine Peptidases
Edited by ALAN J. BARRETT

VOLUME 245. Extracellular Matrix Components
Edited by E. RUOSLAHTI AND E. ENGVALL

VOLUME 246. Biochemical Spectroscopy
Edited by KENNETH SAUER

VOLUME 247. Neoglycoconjugates (Part B: Biomedical Applications)
Edited by Y. C. LEE AND REIKO T. LEE

VOLUME 248. Proteolytic Enzymes: Aspartic and Metallo Peptidases
Edited by ALAN J. BARRETT

VOLUME 249. Enzyme Kinetics and Mechanism (Part D: Developments in Enzyme Dynamics)
Edited by DANIEL L. PURICH

VOLUME 250. Lipid Modifications of Proteins
Edited by PATRICK J. CASEY AND JANICE E. BUSS

VOLUME 251. Biothiols (Part A: Monothiols and Dithiols, Protein Thiols, and Thiyl Radicals)
Edited by LESTER PACKER

VOLUME 252. Biothiols (Part B: Glutathione and Thioredoxin; Thiols in Signal Transduction and Gene Regulation)
Edited by LESTER PACKER

VOLUME 253. Adhesion of Microbial Pathogens
Edited by RON J. DOYLE AND ITZHAK OFEK

VOLUME 254. Oncogene Techniques
Edited by PETER K. VOGT AND INDER M. VERMA

VOLUME 255. Small GTPases and Their Regulators (Part A: Ras Family)
Edited by W. E. BALCH, CHANNING J. DER, AND ALAN HALL

VOLUME 256. Small GTPases and Their Regulators (Part B: Rho Family)
Edited by W. E. BALCH, CHANNING J. DER, AND ALAN HALL

VOLUME 257. Small GTPases and Their Regulators (Part C: Proteins Involved in Transport)
Edited by W. E. BALCH, CHANNING J. DER, AND ALAN HALL

VOLUME 258. Redox-Active Amino Acids in Biology
Edited by JUDITH P. KLINMAN

VOLUME 259. Energetics of Biological Macromolecules
Edited by MICHAEL L. JOHNSON AND GARY K. ACKERS

VOLUME 260. Mitochondrial Biogenesis and Genetics (Part A)
Edited by GIUSEPPE M. ATTARDI AND ANNE CHOMYN

VOLUME 261. Nuclear Magnetic Resonance and Nucleic Acids
Edited by THOMAS L. JAMES

VOLUME 262. DNA Replication
Edited by JUDITH L. CAMPBELL

VOLUME 263. Plasma Lipoproteins (Part C: Quantitation)
Edited by WILLIAM A. BRADLEY, SANDRA H. GIANTURCO, AND JERE P. SEGREST

VOLUME 264. Mitochondrial Biogenesis and Genetics (Part B)
Edited by GIUSEPPE M. ATTARDI AND ANNE CHOMYN

VOLUME 265. Cumulative Subject Index Volumes 228, 230–262

VOLUME 266. Computer Methods for Macromolecular Sequence Analysis
Edited by RUSSELL F. DOOLITTLE

VOLUME 267. Combinatorial Chemistry
Edited by JOHN N. ABELSON

VOLUME 268. Nitric Oxide (Part A: Sources and Detection of NO; NO Synthase)
Edited by LESTER PACKER

VOLUME 269. Nitric Oxide (Part B: Physiological and Pathological Processes)
Edited by LESTER PACKER

VOLUME 270. High Resolution Separation and Analysis of Biological Macromolecules (Part A: Fundamentals)
Edited by BARRY L. KARGER AND WILLIAM S. HANCOCK

VOLUME 271. High Resolution Separation and Analysis of Biological Macromolecules (Part B: Applications)
Edited by BARRY L. KARGER AND WILLIAM S. HANCOCK

VOLUME 272. Cytochrome P450 (Part B)
Edited by ERIC F. JOHNSON AND MICHAEL R. WATERMAN

VOLUME 273. RNA Polymerase and Associated Factors (Part A)
Edited by SANKAR ADHYA

VOLUME 274. RNA Polymerase and Associated Factors (Part B)
Edited by SANKAR ADHYA

VOLUME 275. Viral Polymerases and Related Proteins
Edited by LAWRENCE C. KUO, DAVID B. OLSEN, AND STEVEN S. CARROLL

VOLUME 276. Macromolecular Crystallography (Part A)
Edited by CHARLES W. CARTER, JR., AND ROBERT M. SWEET

VOLUME 277. Macromolecular Crystallography (Part B)
Edited by CHARLES W. CARTER, JR., AND ROBERT M. SWEET

VOLUME 278. Fluorescence Spectroscopy
Edited by LUDWIG BRAND AND MICHAEL L. JOHNSON

VOLUME 279. Vitamins and Coenzymes (Part I)
Edited by DONALD B. MCCORMICK, JOHN W. SUTTIE, AND CONRAD WAGNER

VOLUME 280. Vitamins and Coenzymes (Part J)
Edited by DONALD B. MCCORMICK, JOHN W. SUTTIE, AND CONRAD WAGNER

VOLUME 281. Vitamins and Coenzymes (Part K)
Edited by DONALD B. MCCORMICK, JOHN W. SUTTIE, AND CONRAD WAGNER

VOLUME 282. Vitamins and Coenzymes (Part L)
Edited by DONALD B. MCCORMICK, JOHN W. SUTTIE, AND CONRAD WAGNER

VOLUME 283. Cell Cycle Control
Edited by WILLIAM G. DUNPHY

VOLUME 284. Lipases (Part A: Biotechnology)
Edited by BYRON RUBIN AND EDWARD A. DENNIS

VOLUME 285. Cumulative Subject Index Volumes 263, 264, 266–284, 286–289

VOLUME 286. Lipases (Part B: Enzyme Characterization and Utilization)
Edited by BYRON RUBIN AND EDWARD A. DENNIS

VOLUME 287. Chemokines
Edited by RICHARD HORUK

VOLUME 288. Chemokine Receptors
Edited by RICHARD HORUK

VOLUME 289. Solid Phase Peptide Synthesis
Edited by GREGG B. FIELDS

VOLUME 290. Molecular Chaperones
Edited by GEORGE H. LORIMER AND THOMAS BALDWIN

VOLUME 291. Caged Compounds
Edited by GERARD MARRIOTT

VOLUME 292. ABC Transporters: Biochemical, Cellular, and Molecular Aspects
Edited by SURESH V. AMBUDKAR AND MICHAEL M. GOTTESMAN

VOLUME 293. Ion Channels (Part B)
Edited by P. MICHAEL CONN

VOLUME 294. Ion Channels (Part C)
Edited by P. MICHAEL CONN

VOLUME 295. Energetics of Biological Macromolecules (Part B)
Edited by GARY K. ACKERS AND MICHAEL L. JOHNSON

VOLUME 296. Neurotransmitter Transporters
Edited by SUSAN G. AMARA

VOLUME 297. Photosynthesis: Molecular Biology of Energy Capture
Edited by LEE MCINTOSH

VOLUME 298. Molecular Motors and the Cytoskeleton (Part B)
Edited by RICHARD B. VALLEE

VOLUME 299. Oxidants and Antioxidants (Part A)
Edited by LESTER PACKER

VOLUME 300. Oxidants and Antioxidants (Part B)
Edited by LESTER PACKER

VOLUME 301. Nitric Oxide: Biological and Antioxidant Activities (Part C)
Edited by LESTER PACKER

VOLUME 302. Green Fluorescent Protein
Edited by P. MICHAEL CONN

VOLUME 303. cDNA Preparation and Display
Edited by SHERMAN M. WEISSMAN

VOLUME 304. Chromatin
Edited by PAUL M. WASSARMAN AND ALAN P. WOLFFE

VOLUME 305. Bioluminescence and Chemiluminescence (Part C)
Edited by THOMAS O. BALDWIN AND MIRIAM M. ZIEGLER

VOLUME 306. Expression of Recombinant Genes in
Eukaryotic Systems
Edited by JOSEPH C. GLORIOSO AND MARTIN C. SCHMIDT

VOLUME 307. Confocal Microscopy
Edited by P. MICHAEL CONN

VOLUME 308. Enzyme Kinetics and Mechanism (Part E: Energetics of
Enzyme Catalysis)
Edited by DANIEL L. PURICH AND VERN L. SCHRAMM

VOLUME 309. Amyloid, Prions, and Other Protein Aggregates
Edited by RONALD WETZEL

VOLUME 310. Biofilms
Edited by RON J. DOYLE

VOLUME 311. Sphingolipid Metabolism and Cell Signaling (Part A)
Edited by ALFRED H. MERRILL, JR., AND YUSUF A. HANNUN

VOLUME 312. Sphingolipid Metabolism and Cell Signaling (Part B)
Edited by ALFRED H. MERRILL, JR., AND YUSUF A. HANNUN

VOLUME 313. Antisense Technology (Part A: General Methods, Methods of
Delivery, and RNA Studies)
Edited by M. IAN PHILLIPS

VOLUME 314. Antisense Technology (Part B: Applications)
Edited by M. IAN PHILLIPS

VOLUME 315. Vertebrate Phototransduction and the Visual Cycle (Part A)
Edited by KRZYSZTOF PALCZEWSKI

VOLUME 316. Vertebrate Phototransduction and the Visual Cycle (Part B)
Edited by KRZYSZTOF PALCZEWSKI

VOLUME 317. RNA–Ligand Interactions (Part A: Structural Biology Methods)
Edited by DANIEL W. CELANDER AND JOHN N. ABELSON

VOLUME 318. RNA–Ligand Interactions (Part B: Molecular Biology Methods)
Edited by DANIEL W. CELANDER AND JOHN N. ABELSON

VOLUME 319. Singlet Oxygen, UV-A, and Ozone
Edited by LESTER PACKER AND HELMUT SIES

VOLUME 320. Cumulative Subject Index Volumes 290–319

VOLUME 321. Numerical Computer Methods (Part C)
Edited by MICHAEL L. JOHNSON AND LUDWIG BRAND

VOLUME 322. Apoptosis
Edited by JOHN C. REED

VOLUME 323. Energetics of Biological Macromolecules (Part C)
Edited by MICHAEL L. JOHNSON AND GARY K. ACKERS

VOLUME 324. Branched-Chain Amino Acids (Part B)
Edited by ROBERT A. HARRIS AND JOHN R. SOKATCH

VOLUME 325. Regulators and Effectors of Small GTPases (Part D: Rho Family)
Edited by W. E. BALCH, CHANNING J. DER, AND ALAN HALL

VOLUME 326. Applications of Chimeric Genes and Hybrid Proteins (Part A: Gene Expression and Protein Purification)
Edited by JEREMY THORNER, SCOTT D. EMR, AND JOHN N. ABELSON

VOLUME 327. Applications of Chimeric Genes and Hybrid Proteins (Part B: Cell Biology and Physiology)
Edited by JEREMY THORNER, SCOTT D. EMR, AND JOHN N. ABELSON

VOLUME 328. Applications of Chimeric Genes and Hybrid Proteins (Part C: Protein–Protein Interactions and Genomics)
Edited by JEREMY THORNER, SCOTT D. EMR, AND JOHN N. ABELSON

VOLUME 329. Regulators and Effectors of Small GTPases (Part E: GTPases Involved in Vesicular Traffic)
Edited by W. E. BALCH, CHANNING J. DER, AND ALAN HALL

VOLUME 330. Hyperthermophilic Enzymes (Part A)
Edited by MICHAEL W. W. ADAMS AND ROBERT M. KELLY

VOLUME 331. Hyperthermophilic Enzymes (Part B)
Edited by MICHAEL W. W. ADAMS AND ROBERT M. KELLY

VOLUME 332. Regulators and Effectors of Small GTPases (Part F: Ras Family I)
Edited by W. E. BALCH, CHANNING J. DER, AND ALAN HALL

VOLUME 333. Regulators and Effectors of Small GTPases (Part G: Ras Family II)
Edited by W. E. BALCH, CHANNING J. DER, AND ALAN HALL

VOLUME 334. Hyperthermophilic Enzymes (Part C)
Edited by MICHAEL W. W. ADAMS AND ROBERT M. KELLY

VOLUME 335. Flavonoids and Other Polyphenols
Edited by LESTER PACKER

VOLUME 336. Microbial Growth in Biofilms (Part A: Developmental and Molecular Biological Aspects)
Edited by RON J. DOYLE

VOLUME 337. Microbial Growth in Biofilms (Part B: Special Environments and Physicochemical Aspects)
Edited by RON J. DOYLE

VOLUME 338. Nuclear Magnetic Resonance of Biological Macromolecules (Part A)
Edited by THOMAS L. JAMES, VOLKER DÖTSCH, AND ULI SCHMITZ

VOLUME 339. Nuclear Magnetic Resonance of Biological Macromolecules (Part B)
Edited by THOMAS L. JAMES, VOLKER DÖTSCH, AND ULI SCHMITZ

VOLUME 340. Drug–Nucleic Acid Interactions
Edited by JONATHAN B. CHAIRES AND MICHAEL J. WARING

VOLUME 341. Ribonucleases (Part A)
Edited by ALLEN W. NICHOLSON

VOLUME 342. Ribonucleases (Part B)
Edited by ALLEN W. NICHOLSON

VOLUME 343. G Protein Pathways (Part A: Receptors)
Edited by RAVI IYENGAR AND JOHN D. HILDEBRANDT

VOLUME 344. G Protein Pathways (Part B: G Proteins and Their Regulators)
Edited by RAVI IYENGAR AND JOHN D. HILDEBRANDT

VOLUME 345. G Protein Pathways (Part C: Effector Mechanisms)
Edited by RAVI IYENGAR AND JOHN D. HILDEBRANDT

VOLUME 346. Gene Therapy Methods
Edited by M. IAN PHILLIPS

VOLUME 347. Protein Sensors and Reactive Oxygen Species (Part A: Selenoproteins and Thioredoxin)
Edited by HELMUT SIES AND LESTER PACKER

VOLUME 348. Protein Sensors and Reactive Oxygen Species (Part B: Thiol Enzymes and Proteins)
Edited by HELMUT SIES AND LESTER PACKER

VOLUME 349. Superoxide Dismutase
Edited by LESTER PACKER

VOLUME 350. Guide to Yeast Genetics and Molecular and Cell Biology (Part B)
Edited by CHRISTINE GUTHRIE AND GERALD R. FINK

VOLUME 351. Guide to Yeast Genetics and Molecular and Cell Biology (Part C)
Edited by CHRISTINE GUTHRIE AND GERALD R. FINK

VOLUME 352. Redox Cell Biology and Genetics (Part A)
Edited by CHANDAN K. SEN AND LESTER PACKER

VOLUME 353. Redox Cell Biology and Genetics (Part B)
Edited by CHANDAN K. SEN AND LESTER PACKER

VOLUME 354. Enzyme Kinetics and Mechanisms (Part F: Detection and Characterization of Enzyme Reaction Intermediates)
Edited by DANIEL L. PURICH

VOLUME 355. Cumulative Subject Index Volumes 321–354

VOLUME 356. Laser Capture Microscopy and Microdissection
Edited by P. MICHAEL CONN

VOLUME 357. Cytochrome P450, Part C
Edited by ERIC F. JOHNSON AND MICHAEL R. WATERMAN

VOLUME 358. Bacterial Pathogenesis (Part C: Identification, Regulation, and Function of Virulence Factors)
Edited by VIRGINIA L. CLARK AND PATRIK M. BAVOIL

VOLUME 359. Nitric Oxide (Part D)
Edited by ENRIQUE CADENAS AND LESTER PACKER

VOLUME 360. Biophotonics (Part A)
Edited by GERARD MARRIOTT AND IAN PARKER

VOLUME 361. Biophotonics (Part B)
Edited by GERARD MARRIOTT AND IAN PARKER

VOLUME 362. Recognition of Carbohydrates in Biological Systems (Part A)
Edited by YUAN C. LEE AND REIKO T. LEE

VOLUME 363. Recognition of Carbohydrates in Biological Systems (Part B)
Edited by YUAN C. LEE AND REIKO T. LEE

VOLUME 364. Nuclear Receptors
Edited by DAVID W. RUSSELL AND DAVID J. MANGELSDORF

VOLUME 365. Differentiation of Embryonic Stem Cells
Edited by PAUL M. WASSAUMAN AND GORDON M. KELLER

VOLUME 366. Protein Phosphatases
Edited by SUSANNE KLUMPP AND JOSEF KRIEGLSTEIN

VOLUME 367. Liposomes (Part A)
Edited by NEJAT DÜZGÜNEŞ

VOLUME 368. Macromolecular Crystallography (Part C)
Edited by CHARLES W. CARTER, JR., AND ROBERT M. SWEET

VOLUME 369. Combinational Chemistry (Part B)
Edited by GUILLERMO A. MORALES AND BARRY A. BUNIN

VOLUME 370. RNA Polymerases and Associated Factors (Part C)
Edited by SANKAR L. ADHYA AND SUSAN GARGES

VOLUME 371. RNA Polymerases and Associated Factors (Part D)
Edited by SANKAR L. ADHYA AND SUSAN GARGES

VOLUME 372. Liposomes (Part B)
Edited by NEJAT DÜZGÜNEŞ

VOLUME 373. Liposomes (Part C)
Edited by NEJAT DÜZGÜNEŞ

VOLUME 374. Macromolecular Crystallography (Part D)
Edited by CHARLES W. CARTER, JR., AND ROBERT W. SWEET

VOLUME 375. Chromatin and Chromatin Remodeling Enzymes (Part A)
Edited by C. DAVID ALLIS AND CARL WU

VOLUME 376. Chromatin and Chromatin Remodeling Enzymes (Part B)
Edited by C. DAVID ALLIS AND CARL WU

VOLUME 377. Chromatin and Chromatin Remodeling Enzymes (Part C)
Edited by C. DAVID ALLIS AND CARL WU

VOLUME 378. Quinones and Quinone Enzymes (Part A)
Edited by HELMUT SIES AND LESTER PACKER

VOLUME 379. Energetics of Biological Macromolecules (Part D)
Edited by JO M. HOLT, MICHAEL L. JOHNSON, AND GARY K. ACKERS

VOLUME 380. Energetics of Biological Macromolecules (Part E)
Edited by JO M. HOLT, MICHAEL L. JOHNSON, AND GARY K. ACKERS

VOLUME 381. Oxygen Sensing
Edited by CHANDAN K. SEN AND GREGG L. SEMENZA

VOLUME 382. Quinones and Quinone Enzymes (Part B)
Edited by HELMUT SIES AND LESTER PACKER

VOLUME 383. Numerical Computer Methods (Part D)
Edited by LUDWIG BRAND AND MICHAEL L. JOHNSON

VOLUME 384. Numerical Computer Methods (Part E)
Edited by LUDWIG BRAND AND MICHAEL L. JOHNSON

VOLUME 385. Imaging in Biological Research (Part A)
Edited by P. MICHAEL CONN

VOLUME 386. Imaging in Biological Research (Part B)
Edited by P. MICHAEL CONN

VOLUME 387. Liposomes (Part D)
Edited by NEJAT DÜZGÜNEŞ

VOLUME 388. Protein Engineering
Edited by DAN E. ROBERTSON AND JOSEPH P. NOEL

VOLUME 389. Regulators of G-Protein Signaling (Part A)
Edited by DAVID P. SIDEROVSKI

VOLUME 390. Regulators of G-protein Sgnalling (Part B)
Edited by DAVID P. SIDEROVSKI

VOLUME 391. Liposomes (Part E)
Edited by NEJAT DÜZGÜNEŞ

VOLUME 392. RNA Interference
Edited by ENGELKE ROSSI

VOLUME 393. Circadian Rhythms
Edited by MICHAEL W. YOUNG

VOLUME 394. Nuclear Magnetic Resonance of Biological Macromolecules (Part C)
Edited by THOMAS L. JAMES

VOLUME 395. Producing the Biochemical Data (Part B)
Edited by ELIZABETH A. ZIMMER AND ERIC H. ROALSON

VOLUME 396. Nitric Oxide (Part E)
Edited by LESTER PACKER AND ENRIQUE CADENAS

VOLUME 397. Environmental Microbiology
Edited by JARED R. LEADBETTER

VOLUME 398. Ubiquitin and Protein Degradation (Part A)
Edited by RAYMOND J. DESHAIES

VOLUME 399. Ubiquitin and Protein Degradation (Part B)
Edited by RAYMOND J. DESHAIES

VOLUME 400. PHASE II Conjugation Enzymes and Transport Systems (in preparation)
Edited by HELMUT SIES AND LESTER PACKER

VOLUME 401. Glutathione Transferases and Gamma Glutamyl Transpeptidases (in preparation)
Edited by HELMUT SIES AND LESTER PACKER

VOLUME 402. Biological Mass spectrometry (in preparation)
Edited by A. L. BURLINGAME

VOLUME 403. (In Preparation)

VOLUME 404. GTPases Regulating Membrane Dynamics (in preparation)
Edited by WILLIAM E. BALCH, CHANNING J. DER, AND ALAN HALL

VOLUME 405. (In Preparation)

Section I

Ubiquitin and Ubiquitin Derivatives

[1] Chemical and Genetic Strategies for Manipulating Polyubiquitin Chain Structure

By SARA VOLK, MIN WANG,* and CECILE M. PICKART*

Abstract

Ubiquitin can be conjugated to lysine residues of other ubiquitin molecules to form polymers called polyubiquitin chains. Ubiquitin has seven lysine residues, creating the potential for seven distinct types of chains, at least five of which have been observed *in vitro* or *in vivo*. A subset of these chains mediates substrate targeting to proteasomes, whereas other types of chains have been implicated in nonproteolytic signaling pathways. In this chapter, we outline chemical and genetic strategies that can be used to deduce (or control) the structures of polyubiquitin chains *in vitro* and in living cells.

Introduction

Ubiquitin modifications occur through a sequence of reactions in which ubiquitin's C-terminal carboxyl group (G76) is activated by E1, transferred to an E2, and finally transferred to a substrate lysine residue through the action of that substrate's cognate E3 enzyme (Pickart, 2001). Ubiquitin is conjugated to a host of proteins in eukaryotic cells. It is a multifunctional signal whose attachment can modulate the stability, localization, or activity of its target proteins. Among the factors that influence the interpretation of ubiquitin signals are their structures, which can be defined using two parameters.

First, a protein can be modified by one or by many ubiquitin molecules. When multiple ubiquitins are present, they are often arranged in the form of an isopeptide-linked polymer or polyubiquitin chain. Therefore, polyubiquitin modifications can be further classified on the basis of their chemical structures. This feature reflects the presence of seven lysine residues in the ubiquitin sequence (Fig. 1A). In theory, any of these lysines could be used in the synthesis of polyubiquitin chains; in practice, all seven ubiquitin–ubiquitin linkages have been observed in budding yeast (Peng *et al.*, 2003).

Functions of Polyubiquitin Chains

Monoubiquitination and polyubiquitination often lead to different functional outcomes; for example, a single ubiquitin is an efficient signal for endocytosis (Shih *et al.*, 2000) but a poor signal for targeting to

*Equal contributors.

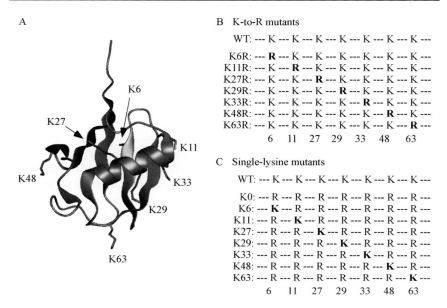

FIG. 1. Ubiquitin mutants frequently used in the analysis of polyubiquitin chain structure. (A) Structure of ubiquitin. The seven lysine residues are indicated. (B) Lysine-to-arginine point mutants. (C) Single-lysine mutants. Only ubiquitin's lysine residues are shown in (B) and (C).

proteasomes (Chau et al., 1989; Thrower et al., 2000). Moreover, there is evidence to suggest that structurally distinct polyubiquitin chains represent functionally distinct signals. At the moment, this model is best supported for K48-linked versus K63-linked chains [reviewed in Pickart and Fushman (2004) and Sun and Chen (2004)]. The former type of chain targets substrates to proteasomes (Finley et al., 1994), but the known targets of K63-linked chains are stable proteins (Deng et al., 2000; Hoege et al., 2002). Nonproteolytic signaling by K63-linked chains is best understood in a pathway of IκBα kinase (IKK) activation that operates in mammalian cells (Sun and Chen, 2004). Specific kinase adaptor proteins are responsible for transducing the K63-chain signal in this pathway (Kanayama et al., 2004).

A number of observations hint at distinctive functionalities for other types of polyubiquitin chains. K6-linked chains assigned to the activity of the E3 enzyme Brca1 have been implicated in nuclear focus formation during DNA repair (Morris and Solomon, 2004), whereas K27-linked chains are observed in a nonproteolytic signaling pathway that involves protein kinase C (Okumura et al., 2004). K29-linked and K11-linked chains, in contrast, have been implicated in specific cases of proteasomal targeting (Baboshina and Haas, 1996; Johnson *et al.*, 1995). The factors downstream of these chain signals are largely unknown, but several families of ubiquitin/polyubiquitin-binding domains have recently been identified [see Buchberger (2002), Hicke and Dunn (2003), and Pickart and Fushman (2004)], and a few of these domains are known to interact preferentially with specifically structured polyubiquitin chains (Kanayama *et al.*, 2004; Raasi *et al.*, 2004). These developments have created strong interest in the signaling potentials of "noncanonical" (*i.e.*, non-K48-linked) polyubiquitin chains, leading to a demand for methods that can be applied to assess or control chain structure.

Polyubiquitin Chain Metabolism

Polyubiquitin chains are very unstable *in vivo* and in cell extracts because of the presence of multiple deubiquitinating enzymes (DUBs) (Amerik and Hochstrasser, 2004). This feature of ubiquitin metabolism is dramatically demonstrated by the rapid disappearance of conjugated ubiquitin in living cells transiently depleted of ATP (Riley *et al.*, 1988) and it has important consequences for discerning or controlling chain structure. For example, ubiquitin's K48 residue is essential in budding yeast; expression of Ub-K48R as the sole form of ubiquitin causes cell death as a result of global inhibition of proteasome proteolysis (Finley *et al.*, 1994). Because short K48-linked chains are inefficient proteolytic signals (Thrower *et al.*, 2000), one might expect Ub-K48R to act as a dominant inhibitor of proteolysis through chain termination effects. However, there is only mild inhibition of cell growth and proteasomal proteolysis by Ub-K48R as long as wild-type ubiquitin is present (Finley *et al.*, 1994). The absence of dominant inhibition in part reflects the rapid disassembly and resynthesis of K48-linked chains, as indicated by strengthened inhibition when the Ub-G76A mutation, which inhibits DUBs (Hodgins *et al.*, 1996), is introduced into Ub-K48R (Finley *et al.*, 1994).

At present, there is no cell-permeable DUB inhibitor. In extracts, thiol-alkylating agents such as *N*-ethylmaleimide can be used against DUBs that act by a cysteine protease mechanism, whereas metal chelators inhibit a smaller family of zinc-dependent DUBs (Amerik and Hochstrasser, 2004). Blocking chain disassembly through the use of these agents can facilitate the use of mutant forms of ubiquitin as probes of chain structure in cell extracts. However, because E1, all E2s, and many E3s are irreversibly inactivated by thiol alkylating agents, this approach is impractical if one is assaying an unidentified activity in an extract. Therefore, it is usually preferable to use a high concentration ($\sim 5 \mu M$) of the commercially available transition state inhibitor ubiquitin aldehyde, which is broadly effective against thiol protease DUBs (Hershko and Rose, 1987).

In this chapter, we outline methods that can be used to assess (or control) polyubiquitin chain linkage *in vitro* and *in vivo*. Rather than presenting detailed methods, we give an overview of the approaches that can be used to address a commonly asked question: "What is the structure of the polyubiquitin chain that acts as a signal in a given system?" Although we cite selected reports to illustrate specific principles, this chapter is not intended to be a comprehensive review of the relevant literature.

Chain structure is most commonly addressed through the use of ubiquitin mutants. This general approach is applicable with purified conjugating factors, in DUB-inhibited extracts, and, to a degree, in living cells. Different types of ubiquitin mutants have specific advantages and disadvantages, but when used properly, they can provide powerful probes (or guarantors) of chain structure.

Chemical Strategies for Manipulating Polyubiquitin Chain Structure *In Vitro*

Lysine-to-Arginine Point Mutants

We define "homopolymeric" chains as structurally uniform; that is, they are linked by the same lysine residue throughout the chain. To determine the linkage site in such chains, one usually analyzes a complete panel of K-to-R mutants (Fig. 1B) for competence in substrate polyubiquitination. In the simplest case, the polyubiquitin ladder collapses to a monoubiquitin modification on alteration of the relevant ubiquitin lysine (Fig. 2A, left). Blocking chain extension should also inhibit downstream consequences, such as proteasome degradation, that depend on polyubiquitination. This approach was first used to study canonical chains; it was found that the Ub-K48R mutation blocked polyubiquitination and proteasome-mediated degradation of a model substrate (Chau *et al.*, 1989). Similarly, Ub-K63R cannot be polymerized by the conjugation factors involved in IKK activation and DNA damage tolerance, which are processes that depend on K63-linked chains (Deng *et al.*, 2000; Hofmann and Pickart, 1999). Supplementation of extracts with Ub-K63R inhibits IKK activation (Deng *et al.*, 2000), whereas the Ub-K63R mutation inhibits DNA damage tolerance in intact yeast cells (Spence *et al.*, 1995). These selected examples show that the K-to-R ubiquitin mutants can be excellent diagnostic tools for assessing chain structure *in vitro* (and in special cases, *in vivo*), but there are several factors to keep in mind when using these mutant proteins.

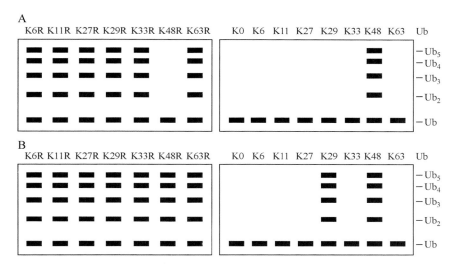

FIG. 2. Diagnosing polyubiquitin chain structure *in vitro*. The two panels depict two possible outcomes with simple interpretations. The diagrams are schematic representations of SDS-PAGE gels that have been analyzed by anti-ubiquitin Western blot. The results shown are for synthesis of unanchored polyubiquitin chains; if the chains were substrate-linked, the result would be qualitatively similar, but all molecular masses would be increased by an increment equal to the substrate's mass. (A) Conjugation factors synthesize only one type of polyubiquitin chain. For the case shown, only K48 is used in chain synthesis. Therefore, Ub-K48R cannot be used as a substrate (left panel), whereas among the single-lysine proteins, *only* K48-Ub is competent. (B) Conjugation factors synthesize more than one type of polyubiquitin chain. For the case shown, either K29 *or* K48 is used in chain synthesis (see You and Pickart, 2001). Therefore, all of the K-to-R mutants can be used in chain synthesis, whereas both K29- and K48-Ub are competent.

The Interpretation of In Vitro *Experiments Depends Critically on the Integrity of the Mutant Protein.* If the arginine mutation was introduced using an AGA codon, there is a potential for lysine to be incorporated in place of arginine during ubiquitin expression in *Escherichia coli*, leading to false-positive chain synthesis. This problem arises because of the scarcity of the AGA-specific tRNA [see You *et al.* (1999)] and can be averted by overexpressing this tRNA with a helper plasmid (You *et al.*, 1999) or through the use of tRNA-supplemented expression strains (Stratagene and Novagen). When expressing an AGA-containing gene for the first time, it is advisable to confirm by mass spectrometry that the protein consists of a single species with the expected molecular mass. Most of the K-to-R ubiquitin mutants are now available commercially; data confirming protein integrity should be available from the supplier.

Crude Extracts Contain a Significant Level of Wild-Type Ubiquitin, Which Can Dilute the Point Mutant and Diminish its Impact. Wild-type ubiquitin can often be outcompeted by using a high concentration of the mutant. Alternatively, endogenous ubiquitin can be depleted in advance of the assay. This can be achieved by ion exchange fractionation to produce "fraction II" (Hershko *et al.*, 1983), but this manipulation could deplete conjugation factors of interest. A second strategy, ubiquitin immunodepletion, is feasible in principle, but we have found it to be difficult in practice. Yet another approach uses a tagged E2 enzyme to deplete endogenous ubiquitin through an autoubiquitination reaction (Flierman *et al.*, 2003). Budding yeast offers a final, unique approach; because His_6-Ub supports viability, the tagged endogenous ubiquitin can be removed from an extract prepared from the appropriate strain (Ling *et al.*, 2000). Six of seven K-to-R point mutants can also be expressed as the sole form of ubiquitin in budding yeast (see later).

Some Substrates Can Be Modified by Monoubiquitin on Multiple Lysines (Hershko and Heller, 1985), Particularly in Long Incubations and/ or at High Concentrations of Conjugation Factors. Thus, blocking the relevant lysine of ubiquitin could fail to limit the modification to simple monoubiquitination. In a few cases, multiple monoubiquitination seems to be a favored mode of modification and can be difficult to distinguish from polyubiquitination in SDS-PAGE analysis. To test for this scenario, one can use a lysine-less version of ubiquitin (see later) that is incapable of chain formation; if robust "polyubiquitination" is still seen, it is most likely to represent the conjugation of multiple monoubiquitins [for example, Lai *et al.* (2001) and Petroski and Deshaies (2003)].

If the Conjugation Factors of Interest Synthesize More Than One Kind of Homopolymeric Chain, or if They Produce Heteropolymers (Chains Containing Several Different Lysine Linkages), True Polyubiquitination Will Be Seen with all of the K-to-R Mutants. There is one report of a heteropolymeric linkage in yeast (Peng *et al.*, 2003), but little is known about the abundance or properties of heteropolymeric chains. In contrast, certain conjugation factors can use more than one lysine residue in the synthesis of homopolymers *in vitro* (see later). The latter interpretation could be supported by findings that (1) none of the K-to-R mutations blocks polyubiquitin chain synthesis (Fig. 2B, left) and (2) a lysine-less ubiquitin supports only monoubiquitination. A different set of ubiquitin mutants can be used to confirm this interpretation and identify the lysines used in chain synthesis, as discussed further later.

Because the Arginine and Lysine Side Chains Have Similar Properties (aside from Nucleophilicity), Enzyme/Ubiquitin Binding Interactions That

Are Critical for Chain Synthesis Should Be Maintained with the K-to-R Mutant Proteins. Such interactions could be impaired, however, by more drastic mutations. In this case, the failure to produce chains would not reflect the use of the mutated lysine as a conjugation site.

Single-Lysine Ubiquitin Mutants

Single-lysine ubiquitin mutants, in which all lysine residues except one are mutated to arginine, currently provide the most rigorous tool for defining chain structure (Fig. 1C). Bacterial expression constructs specifying each of the seven single-lysine proteins, plus a lysine-less ubiquitin (called K0-Ub), were originally developed by the Ellison laboratory (Arnason and Ellison, 1994). Since then, the single-lysine genes have been adapted to incorporate a variety of N-terminal epitope tags. They are also widely used in mammalian cells.

If a substrate is modified with many molecules of a particular single-lysine ubiquitin but with only one (or a few) molecule of the other single-lysine proteins (Fig. 2A, right), this is indicative of polyubiquitination by means of the lysine residue present in the first ubiquitin mutant. Ideally, single mutation of this lysine residue to arginine will also impede substrate polyubiquitination as discussed previously (Fig. 2A, left). This specific combination of positive and negative results provides a compelling definition of chain structure, although confirmation of structure by mass spectrometry is still desirable (see below).

If not just one, but instead a subset, of the single-lysine mutants is permissive for polyubiquitination, this suggests that the factor in question can synthesize more than one type of homopolymeric chain (Fig. 2B, right). For example, the mammalian/yeast E3 enzyme KIAA10/Hul5 synthesizes unanchored K48- or K29-linked homopolymers *in vitro* (You and Pickart, 2001). Similarly, Brca1 synthesizes homopolymeric K6-linked chains (Nishikawa *et al.*, 2004; Wu-Baer *et al.*, 2003), but if K6 is blocked, K29 is used with a reduced efficiency (Nishikawa *et al.*, 2004). In a final example, the yeast E3 Rsp5 has been implicated in biological pathways that depend on monoubiquitination, K48-polyubiquitination, and K63-polyubiquitination [see Galan and Haguenauer-Tsapis (1997) and Pickart (2001)]. Because it is likely a significant number of conjugation factors can use more than one lysine in polyubiquitin chain synthesis, it is important to show that novel chain structures observed *in vitro* are actually relevant for signaling *in vivo*.

Besides serving as diagnostic tools, the single-lysine proteins can be used to produce chains of rigorously defined structure for purposes such as binding studies. Importantly, a single-lysine ubiquitin can be used to produce homogeneous chains even in conjunction with a linkage-promiscuous conjugating factor.

Although the single-lysine ubiquitins are exceptionally useful for characterizing chain structure *in vitro* and *in vivo* (see later), they present several potential pitfalls. First, the widely used single-lysine genes are rich in AGA codons (Arnason and Ellison, 1994), so precautions should be taken to block lysine misincorporation during bacterial expression (see earlier). Because the single-lysine ubiquitins show a reduced solubility in acid, they cannot be purified by a simple procedure based on perchloric acid precipitation of nonubiquitin proteins (see Chapter 2). We have also found that the single-lysine proteins can display aberrant substrate properties. For example, the monoubiquitination of a specific substrate by one E3, as well as self-ubiquitination by the same enzyme, proceeded inefficiently in assays containing K0-Ub. We were able to bypass this problem by substituting K0-Ub with Ub-K48R, because the enzyme in question synthesized only K48-linked chains (MW and CP, unpublished data). Finally, some of the single-lysine mutants (*e.g.*, K29-Ub) have a tendency to precipitate at high concentration (CP, unpublished data).

Purification of Recombinant Single-Lysine Ubiquitins

An expression protocol for the original pET3a plasmids produced by the Ellison laboratory, which specify untagged proteins (Arnason and Ellison, 1994), can be found in Chapter 2, which also describes cell lysis and extract preparation. Slowly reduce the pH of the clarified *E. coli* extract to 4.5 by the dropwise addition of 7% perchloric acid, remove the precipitated proteins by centrifugation, and then dialyze the (soluble) single-lysine mutant protein into neutral pH buffer such as 50 mM Tris-HCl (pH 7.6). After concentration, the mutant protein is further purified by gel filtration on Sephacryl-200 (Amersham-Pharmacia Biotech) in the same buffer. The resulting protein is usually 80–95% homogeneous by SDS-PAGE. Among the single-lysine mutants, we find K33-Ub to be the most difficult to obtain in good yield.

Chemical Modification Approach for Generating "Single-Lysine" Ubiquitins

There are two alternatives to expressing the single-lysine ubiquitins. If only small amounts of protein are needed, one can purchase them (BostonBiochem), or one can generate a functional equivalent starting from a

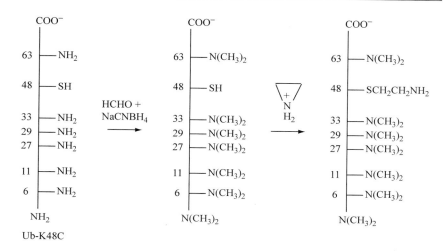

FIG. 3. Method to produce single-lysine mutants by chemical modification. The "single-lysine" mutant is produced from a lysine-to-cysteine point mutant (here, Ub-K48C). Its remaining six lysines (and the α-amino group) are first blocked by reductive methylation. This modification is followed by alkylation of the cysteine using ethyleneimine (see the text). For simplicity, the native and dimethylated lysines are shown as uncharged.

K-to-C ubiquitin point mutant (Fig. 3). The latter mutants are easily purified by standard procedures (see Chapter 2). After purification, the cysteine mutant is reductively methylated to block its six lysine residues and its α-amino group (Jentoft and Dearborn, 1979). (A structural distinction between the chemically modified and recombinant "single-lysine" ubiquitins is that the chemically modified proteins have a blocked α-amino group, whereas the recombinant proteins do not. We are not aware of a case in which this feature has proved to be functionally significant.) The cysteine side chain is then reacted with ethyleneimine to introduce an aminoethyl group (Piotrowski et al., 1997). S-aminoethylcysteine has a lower pK_a value than lysine (Gloss and Kirsch, 1995), but it mimics lysine effectively in the conjugation systems so far tested (Hofmann and Pickart, 1999; Mastrandrea et al., 1999; Piotrowski et al., 1997).

For reductive methylation, add 2–6 mg of the protein to 1 ml of buffer containing 0.1 M HEPES (pH 7.5) and 6 M urea. The buffer should lack primary amino groups, or it will undergo reductive methylation. Add formaldehyde (12 mM), followed by ∼20 mM sodium cyanoborohydride (Sigma, add ∼2 mg of solid). After 12–16 h at room temperature, again add the same amounts of formaldehyde and cyanoborohydride and let the incubation proceed for 1 h more. Dialyze the protein overnight in the cold against a buffer of 5 mM ammonium acetate (pH 5.5), 0.1 mM EDTA, and 1 mM DTT, using a device with a mass cutoff of 7 kDa or less. Note that this procedure can also be applied to wild-type ubiquitin, yielding so-called RM-Ub (Hershko and Heller, 1985), which is also commercially available (Biomol or BostonBiochem).

To alkylate the cysteine residue of the reductively methylated K-to-C mutant, use buffer containing 0.1 M Tris-HCl (pH 8) and 1 mM EDTA. Initiate the reaction by adding ethyleneimine (Chemservice) to a final concentration of 55 mM. Alkylation is complete in 1 h at 37°. For quantitative alkylation, the cysteine must be fully reduced beforehand. If there is any doubt on this point, preincubate the protein with 1 mM DTT (in reaction buffer) for 10 min at 37° before adding ethyleneimine. After the reaction, dialyze the protein to remove ethyleneimine or concentrate and dilute repeatedly in a centrifugal concentrator. Ethyleneimine is rather toxic. Consult Chapter 2 for recommendations about handling.

Mass Spectrometry for Confirmation of Chain Structure

As discussed previously, certain combinations of results obtained with K-to-R mutants and single-lysine ubiquitins lead to straightforward interpretations. But extrapolation from such results should be done with caution, because enzymes can respond in unexpected ways to apparently innocuous alterations of ubiquitin's structure. For example, when assayed with single-lysine ubiquitins, the KIAA10 E3 showed activity only with the K29- and K48-containing proteins (You and Pickart, 2001). Therefore, to synthesize K29-linked chains on a large scale, we used Ub-K48R mutants. Unexpected chain products, produced at a high level, proved to be linked through Ub-K6 as determined by tryptic digestion and LC/MS/MS analysis of the peptides to determine which carried the glycylglycine dipeptide signature that derives from the next ubiquitin in the chain [J. Peng, MW, and CP, unpublished data obtained according to Peng et al. (2003)]. Further assays with single-lysine ubiquitins revealed that K6-polyubiquitination could be observed, but at a rate that was <10% that of polymerization through K48 or K29. Evidently, a secondary reaction that normally occurred at a slow rate became more pronounced when one of the favored conjugation sites (K48) was blocked. On the basis of these experiences, we recommend that mass spectrometry be done to confirm the structural integrity of polyubiquitin chains whose structures are enforced through the introduction of additional ubiquitin mutations.

Mass spectrometry can also be used for *de novo* determination of polyubiquitin chain structure (Peng *et al.*, 2003) in samples recovered from cells by stringent immunoprecipitation or affinity purification of specific substrates. This approach was used to show that the budding yeast transcription factor Met4 is modified with a K48-linked chain (Flick *et al.*, 2004). As an added benefit, if the chain is linked to a unique substrate lysine residue, then mass spectrometry has the potential to identify this site, although the yield of substrate site peptide will be lower than the yield of ubiquitin–ubiquitin junction peptides (Flick *et al.*, 2004). Note that if more than one polyubiquitin linkage is detected in this experiment, one cannot distinguish between substrate modification with a heteropolymeric chain versus two homopolymeric chains. (It could also be difficult to exclude contamination by a low level of a different substrate modified with a different type of chain.) The interested reader should consult Chapter 25 for more information about how mass spectrometry can be used to analyze ubiquitin modifications.

Linear Polyubiquitin Chains and N-Terminal Ubiquitination

The units of linear polyubiquitin chains are linked through ubiquitin's α-amino group, that is, through peptide rather than isopeptide bonds. Such chains are produced through translation of stress-induced polyubiquitin genes (such as *UBI4* in yeast), but they are cotranslationally processed to monoubiquitin (Finley *et al.*, 1987). Although posttranslational synthesis of a linear polyubiquitin chain has been reported *in vitro* (Hodgins *et al.*, 1996), we are not aware of physiological examples of this chain structure. Note that although certain targets of the ubiquitin-proteasome pathway seem to be modified on the substrate's α-amino group, the multiple ubiquitins conjugated to this site appear to take the form of a K48-linked chain [see Chen *et al.* (2004) and Ciechanover and Ben-Saadon (2004)]. *In vitro*, the presence of a linear structure could be deduced on the basis of retention of polyubiquitination with all seven K-to-R mutants *and* all seven single-lysine ubiquitins; K0-Ub would be competent, but RM-Ub, which is blocked at its α-amino group, would not.

Genetic Strategies for Manipulating Polyubiquitin Chain Structure

The difficulties inherent in manipulating chain structure are magnified in the *in vivo* setting, where DUB activity is high and ubiquitin is abundant [estimated to be $\sim 20~\mu M$ in many cell types (Haas and Bright, 1985)]. Although some of the strategies discussed previously have been used successfully in living cells, a number of points need to be considered in experimental design and interpretation.

Expression of Lysine-to-Arginine Point Mutants as the Sole Form of Ubiquitin in S. cerevisiae

The most rigorous use of K-to-R mutants involves expressing a given mutant as the sole form of the ubiquitin protein. So far, this can be done only in budding yeast, where the Finley laboratory has deleted the complete complement of (four) ubiquitin genes. The deletions are covered by a plasmid, which is selected against to effect gene replacement (Finley *et al.*, 1994). A phenotypic survey of the complete set of (single) K-to-R ubiquitin mutants revealed that the K48R mutation is lethal, and the K63R mutation causes hypersensitivity to DNA damage (Finley *et al.*, 1994; Spence *et al.*, 1995). It should be noted that, although K48 is the only essential lysine, a ubiquitin that contains *only* K48 does not support viability (C. Tagwerker and P. Kaiser, personal communication, 2004).

The lysine point-mutant strains represent a unique resource for establishing chain structure and probing the relationship between structure and signaling. For example, proliferating cell nuclear antigen (PCNA) is polyubiquitinated in yeast expressing wild-type ubiquitin but not in cells expressing Ub-K63R (Hoege *et al.*, 2002). The same is true of a specific plasma membrane receptor (Galan and Haguenauer-Tsapis, 1997). In both cases, specific downstream consequences (DNA damage tolerance and receptor endocytosis, respectively) are inhibited by the K63R mutation, confirming the relevance of this specific chain structure for biological signaling.

Dominant Negative Approaches Involving Lysine-to-Arginine Mutants

Although complicated by the dynamic qualities of polyubiquitin chains (earlier), overexpression approaches have been used in phenotypic analyses *in vivo*. (For simplicity, we use the term "overexpression" to describe this approach, although, as discussed later, the level of the mutant protein may not exceed the level of endogenous ubiquitin.) In budding yeast, overexpression of Ub-K63R or Ub-K29R (but not Ub-K48R) in cells deleted for the (stress-induced) *UBI4* gene causes hypersensitivity to stress (Arnason and Ellison, 1994). Overexpression of Ub-K63R in mammalian cells harboring their full complement of endogenous ubiquitin genes causes cadmium sensitivity above that seen after overexpression of Ub-K48R (Tsirigotis *et al.*, 2001b). Overexpression of Ub-K48R has also been found to impede the proteasome-mediated turnover of several specific substrates in mammalian cells (Table I). On the other hand, we have received anecdotal reports of other cases in which overexpression of Ub-K48R (by means of transfection) failed to inhibit the turnover of specific proteasome substrates. The ratio of mutant to wild-type ubiquitin is likely to be a

TABLE I
Proteasome Substrates Whose Proteolysis Is Inhibited by Dominant-Negative Interventions in Mammalian Cells (Partial List)

Substrate	Ubiquitin	Cell type/method	Reference
Cystic fibrosis trans-membrane receptor	His_6-myc-Ub-K48R	HEK293/transient transfection	(Ward et al., 1995)
T-cell receptor subunit	His_6-myc-Ub-K48R	HEK293/transient transfection	(Yu and Koptio, 1999)
IκBα	Ub-K48R	Macrophages/RBC-mediated delivery	(Antonelli et al., 1999)
E2–F1	His_6-Ub-K48R	HT4 neuroblastoma/stable transfection	(Tsirigotis et al., 2001b)
GFP$^{u a}$	His_6-Ub-K48R	HT4 neuroblastoma/stable transfection	(Tsirigotis et al., 2001b)
Inducible nitric oxide synthase	Ub-K48R	HEK293/transient transfection	(Kolodziejski et al., 2002)
p21	K0-Ub	HEK293T/transient transfection	(Bloom et al., 2003)

^{a} GFP^u is a model proteasome substrate (Bence et al., 2001).

critical factor in the success of this type of experiment. Viral transduction may afford higher levels of the mutant protein than DNA transfection; in one recent study, Sindbis virus transduction produced Ub-K48R at a 10-fold higher concentration than the endogenous wild-type protein (Patrick et al., 2003).

As in the *in vitro* setting, one expects overexpression of the relevant K-to-R mutant to diminish the number of ubiquitins conjugated to the substrate. However, reports of this outcome are unexpectedly sparse. Overexpression of Ub-K48R (but not Ub-K63R) in budding yeast diminished the ubiquitination of Met4, which is modified by K48-liked chains, even though it is metabolically stable (Flick et al., 2004). Similarly, overexpression of K0-Ub in cultured mammalian cells stabilized p21 and diminished its ubiquitination (Bloom et al., 2003). However, overexpression of Ub-K48R in mammalian cells has frequently been found to *augment*, rather than reduce, the overall ubiquitination of proteasomal substrates (Kolodziejski et al., 2002; Mehle et al., 2004; Ward et al., 1995). Overexpression of Ub-K48R in murine HT4 neuroblastoma cells and human HEK293 cells dramatically increased the *total* pool of conjugated ubiquitin as well (Kolodziejski et al., 2002; Patrick et al., 2003; Tsirigotis et al., 2001b). The basis of these effects is uncertain (Tsirigotis et al., 2001b).

One possibility is that inefficient degradation of oligo-ubiquitinated species (produced through chain-termination effects) causes these intermediates to accumulate dramatically. Alternatively, a reduced flux through K48-chain synthesis pathways could promote substrate modification with novel chains that are deficient in proteasomal targeting ability. Either scenario greatly complicates the interpretation of results obtained through this type of dominant-negative approach; if the substrate of interest does not show diminished ubiquitination, it is hard to exclude that the ubiquitin point mutant is modulating substrate turnover through an indirect mechanism.

The Gray laboratory has created transgenic mouse lines that express His_6-tagged forms of wild-type, K48R, or K63R ubiquitin from a constitutive promoter (Gray *et al.*, 2004; Tsirigotis *et al.*, 2001a). Although levels of the transgene-derived proteins are ~10-fold substoichiometric relative to endogenous ubiquitin (Gilchrist *et al.*, 2005; Tsirigotis *et al.*, 2001a), mice expressing Ub-K48R display significant protection against diverse stresses, including canavanine, viral infection, and lipopolysaccharide challenge (Gray *et al.*, 2004). These properties are puzzling and, at present, defy explanation (Gray *et al.*, 2004), because the same Ub-K48R gene caused hypersensitivity to canavanine when stably transfected into mouse neuroblastoma cells (Tsirigotis *et al.*, 2001b). Similarly, overexpression of Ub-K48R in budding yeast causes canavanine sensitivity (Finley *et al.*, 1994). The contradictory effects of Ub-K48R expression in cultured mammalian cells versus transgenic mice may reflect higher expression of the mutant ubiquitins in the transfected cell setting (Carter *et al.*, 2004; Gilchrist *et al.*, 2005; Tsirigotis *et al.*, 2001a,b).

The points discussed previously suggest that results obtained through dominant-negative approaches in mammalian cells should be interpreted with caution. In particular, if overexpression of the K-to-R ubiquitin mutant does not predictably alter substrate ubiquitination, one cannot confidently conclude that effects on signaling are due to blockade of polyubiquitination. Moreover, the failure of point mutant overexpression to elicit an effect cannot be taken as evidence *against* a particular chain structure without a positive control of a validated substrate of the same polyubiquitin modification in the same cells. When endogenous (poly)ubiquitinated forms of the substrate can be directly purified, mass spectrometry may well be the most direct way to determine chain structure (see preceding).

Single-Lysine Ubiquitins as Probes of Chain Structure in Living Cells

The single-lysine ubiquitins can be useful probes of chain structure in intact cells. For example, the signal transduction protein RIP can be modified with either K48- or K63-linked chains (Ravid and Hochstrasser,

2004). In a recent study, the presence of each tagged single-lysine ubiquitin in RIP immunoprecipitates was predictably altered by modulating deubiquitination (specific for K63-linked chains) or ubiquitination (specific for K48-linked chains) (Wertz et al., 2004). Even though either single-lysine ubiquitin can be incorporated at the ends of all types of polyubiquitin chains or added to proteins as a monomer, the results of Wertz et al. (2004) suggest that this uninformative background may not prevent mutant enrichment in specific chains of interest. However, because of concerns about mutant protein level and chain dynamics, it is difficult to interpret a negative result in the absence of a positive control.

Epitope Tagging of Ubiquitin

Many of the mammalian-cell studies mentioned previously used ubiquitin mutants tagged at the N-terminus with one or more epitopes, including myc, His$_6$, hemagglutinin (HA), or FLAG (Ellison and Hochstrasser, 1991; Treier et al., 1994; Tsirigotis et al., 2001b; Ward et al., 1995). Epitope tags distinguish the mutant protein from its endogenous wild-type counterpart and facilitate the purification of (poly)ubiquitinated proteins. Essential conjugating factors use tagged ubiquitins comparably to the wild-type protein, as indicated by the ability of His$_6$- and myc-tagged ubiquitin support viability in budding yeast (Ling et al., 2000; Spence et al., 2000). However, some conjugating factors are known to discriminate against tagged ubiquitins (Hofmann and Pickart, 2001; You and Pickart, 2001). Controls with tagged wild-type ubiquitin should be done to establish whether this is a cause for concern.

Acknowledgments

We thank members of the Pickart laboratory and Ray Deshaies for helpful discussions. Our work on polyubiquitin chains is funded by grants from the NIH. S.V. was supported by a training grant from the NIEHS (T32 ES07141).

References

Amerik, A. Y., and Hochstrasser, M. (2004). Mechanism and function of deubiquitinating enzymes. *Biochem. Biophys. Acta* **1695,** 189–207.

Antonelli, A., Crinelli, R., Bianchi, M., Cerasi, A., Gentilini, L., Serafini, G., and Magnani, M. (1999). Efficient inhibition of macrophage TNF-α production upon targeted delivery of K48R ubiquitin. *Br. J. Haematol.* **104,** 475–481.

Arnason, T., and Ellison, M. J. (1994). Stress resistance in *Saccharomyces cerevisiae* is strongly correlated with assembly of a novel type of multiubiquitin chain. *Mol. Cell. Biol.* **14,** 7876–7883.

Baboshina, O. V., and Haas, A. L. (1996). Novel multiubiquitin chain linkages catalyzed by the conjugating enzymes E2$_{EPF}$ and RAD6 are recognized by 26 S proteasome subunit 5. *J. Biol. Chem.* **271,** 2823–2831.

Bence, N. F., Sampat, R. M., and Kopito, R. R. (2001). Impairment of the ubiquitin-proteasome system by protein aggregation. *Science* **292**, 1552–1555.

Bloom, J., Amador, V., Bartolini, F., DeMartino, G., and Pagano, M. (2003). Proteasome-mediated degradation of p21 via N-terminal ubiquitinylation. *Cell* **115**, 71–82.

Buchberger, A. (2002). From UBA to UBX: New words in the ubiquitin vocabulary. *Trends Cell Biol.* **12**, 216–221.

Carter, S., Urbe, S., and Clague, M. J. (2004). The Met receptor degradation pathway: Requirement for Lys^{48}-linked polyubiquitin independent of proteasome activity. *J. Biol. Chem.* **279**, 52835–52839.

Chau, V., Tobias, J. W., Bachmair, A., Marriott, D., Ecker, D. J., Gonda, D. K., and Varshavsky, A. (1989). A multiubiquitin chain is confined to specific lysine in a targeted short-lived protein. *Science* **243**, 1576–1583.

Chen, X., Chi, Y., Bloecher, A., Aebersold, R., Clurman, B. E., and Roberts, J. M. (2004). N-acetylation and ubiquitin-independent proteasomal degradation of p21^{Cip1}. *Mol. Cell.* **16**, 839–847.

Ciechanover, A., and Ben-Saadon, R. (2004). N-terminal ubiquitination: More protein substrates join in. *Trends Cell Biol.* **14**, 103–106.

Deng, L., Wang, C., Spencer, E., Yang, L., Braun, A., You, J., Slaughter, C., Pickart, C., and Chen, Z. J. (2000). Activation of the IkB kinase complex by TRAF6 requires a dimeric ubiquitin-conjugating enzyme complex and a unique polyubiquitin chain. *Cell* **103**, 351–361.

Ellison, M. J., and Hochstrasser, M. (1991). Epitope-tagged ubiquitin. A new probe for analyzing ubiquitin function. *J. Biol. Chem.* **266**, 21150–21157.

Finley, D., Ozkaynak, E., and Varshavsky, A. (1987). The yeast polyubiquitin gene is essential for resistance to high temperatures, starvation, and other stresses. *Cell* **48**, 1035–1046.

Finley, D., Sadis, S., Monia, B. P., Boucher, P., Ecker, D. J., Crooke, S. T., and Chau, V. (1994). Inhibition of proteolysis and cell cycle progression in a multiubiquitination-deficient yeast mutant. *Mol. Cell. Biol.* **14**, 5501–5509.

Flick, K., Ouni, I., Wohlschlegel, J. A., Capati, C., McDonald, W. H., Yates, J. R., and Kaiser, P. (2004). Proteolysis-independent regulation of the transcription factor Met4 by a single Lys48-linked ubiquitin chain. *Nat. Cell Biol.* **6**, 634–641.

Flierman, D., Ye, Y., Dai, M., Chau, V., and Rapoport, T. A. (2003). Polyubiquitin serves as a recognition signal, rather than a ratcheting molecule, during retrotranslocation of proteins across the endoplasmic reticulum membrane. *J. Biol. Chem.* **278**, 34774–34782.

Galan, J. M., and Haguenauer-Tsapis, R. (1997). Ubiquitin lys63 is involved in ubiquitination of a yeast plasma membrane protein. *EMBO J.* **16**, 5847–5854.

Gilchrist, C. A., Gray, D. A., Stieber, A., Gonatas, N. K., and Kopito, R. R. (2005). Effect of ubiquitin expression on neuropathogenesis in a mouse model of familial amyotrophic lateral sclerosis. *Neuropathol. Appl. Neurobiol.* **31**, 20–33.

Gloss, L. M., and Kirsch, J. F. (1995). Decreasing the basicity of the active site base, Lys-258, of *Escherichia coli* aspartate aminotransferase by replacement with gamma-thialysine. *Biochemistry* **28**, 3990–3998.

Gray, D. A., Tsirigotis, M., Brun, J., Tang, M., Zhang, M., Beyers, M., and Woulfe, J. (2004). Protective effects of mutant ubiquitin in transgenic mice. *Ann. N Y Acad. Sci.* **1019**, 215–218.

Haas, A. L., and Bright, P. M. (1985). The immunochemical detection and quantitation of intracellular ubiquitin-protein conjugates. *J. Biol. Chem.* **260**, 12464–12473.

Hershko, A., and Heller, H. (1985). Occurrence of a polyubiquitin structure in ubiquitin-protein conjugates. *Biochem. Biophys. Res. Commun.* **128**, 1079–1086.

Hershko, A., Heller, H., Elias, S., and Ciechanover, A. (1983). Components of ubiquitin-protein ligase system. *J. Biol. Chem.* **258**, 8206–8214.

Hershko, A., and Rose, I. A. (1987). Ubiquitin-aldehyde: A general inhibitor of ubiquitin-recycling processes. *Proc. Natl. Acad. Sci. USA.* **84,** 1829–1833.

Hicke, L., and Dunn, R. (2003). Regulation of membrane protein transport by ubiquitin and ubiquitin-binding proteins. *Annu. Rev. Cell Dev. Biol.* **19,** 141–172.

Hodgins, R., Gwozd, C., Arnason, T., Cummings, M., and Ellison, M. J. (1996). The tail of a ubiquitin-conjugating enzyme redirects multi-ubiquitin chain synthesis from the lysine 48-linked configuration to a novel nonlysine-linked form. *J. Biol. Chem.* **271,** 28766–28771.

Hoege, C., Pfander, B., Moldovan, G. L., Pyrowolakis, G., and Jentsch, S. (2002). RAD6-dependent DNA repair is linked to modification of PCNA by ubiquitin and SUMO. *Nature* **419,** 135–141.

Hofmann, R. M., and Pickart, C. M. (1999). Noncanonical *MMS2*-encoded ubiquitin-conjugating enzyme functions in assembly of novel polyubiquitin chains for DNA repair. *Cell* **96,** 645–653.

Hofmann, R. M., and Pickart, C. M. (2001). *In vitro* assembly and recognition of K63 poly-ubiquitin chains. *J. Biol. Chem.* **276,** 27936–27943.

Jentoft, N., and Dearborn, D. G. (1979). Labeling of proteins by reductive methylation using sodium cyanoborohydride. *J. Biol. Chem.* **254,** 4359–4365.

Johnson, E. S., Ma, P. C., Ota, I. M., and Varshavsky, A. (1995). A proteolytic pathway that recognizes ubiquitin as a degradation signal. *J. Biol. Chem.* **270,** 17442–17456.

Kanayama, A., Seth, R. B., Sun, L., Ea, C.-K., Hong, M., Shaito, A., Deng, L., and Chen, Z. J. (2004). TAB2 and TAB3 activate the NF-kB pathway through binding to polyubiquitin chains. *Mol. Cell.* **15,** 535–548.

Kolodziejski, P. J., Musial, A., Koo, J.-S., and Eissa, N. T. (2002). Ubiquitination of inducible nitric oxide synthase is required for its degradation. *Proc. Natl. Acad. Sci. USA* **99,** 12315–12320.

Lai, Z., Ferry, K. V., Diamond, M. A., Wee, K. E., Kim, Y. B., Ma, J., Yang, T., Benfield, P. A., Copeland, R. A., and Auger, K. R. (2001). Human mdm2 mediates multiple mono-ubiquitination of p53 by a mechanism requiring enzyme isomerization. *J. Biol. Chem.* **276,** 31357–31367.

Ling, R., Colon, E., Dahmus, M. E., and Callis, J. (2000). Histidine-tagged ubiquitin substitutes for wild-type ubiquitin in *Saccharomyces cerevisiae* and facilitates isolation and identification of *in vivo* substrates of the ubiquitin pathway. *Anal. Biochem.* **282,** 54–64.

Mastrandrea, L. D., You, J., Niles, E. G., and Pickart, C. M. (1999). E2/E3-mediated assembly of lysine 29-linked polyubiquitin chains. *J. Biol. Chem.* **274,** 27299–27306.

Mehle, A., Strack, B., Ancuta, P., Zhang, C., McPike, M., and Gabuzda, D. (2004). Vif overcomes the innate antiviral activity of APOBEC3G by promoting its degradation in the ubiquitin-proteasome pathway. *J. Biol. Chem.* **279,** 7792–7798.

Morris, J. R., and Solomon, E. (2004). BRCA1:BARD1 induces the formation of conjugated ubiquitin structures, dependent on K6 of ubiquitin, in cells during DNA replication and repair. *Hum. Mol. Genet.* **13,** 807–817.

Nishikawa, H., Oooka, S., Sato, K., Arima, K., Okamoto, J., Klevit, R. E., Fukuda, M., and Ohta, T. (2004). Mass spectrophotometric and mutational analyses reveal Lys-6-linked polyubiquitin chains catalyzed by BRCA1-BARD1 ubiquitin ligase. *J. Biol. Chem.* **279,** 3916–3924.

Okumura, F., Hatakeyama, S., Matsumoto, M., Kamura, T., and Nakayama, K. I. (2004). Functional regulation of FEZ1 by the U-box-type ubiquitin ligase E4B contributes to neurotogenesis. *J. Biol. Chem.* **279,** 53533–53543.

Patrick, G. N., Bingol, B., Weld, H. A., and Schuman, E. M. (2003). Ubiquitin-mediated proteasome activity is required for agonist-induced endocytosis of GluRs. *Curr. Biol.* **13,** 2073–2081.

Peng, J., Schwartz, D., Elias, J. E., Thoreen, C. C., Cheng, D., Marsischky, G., Roelofs, J., Finley, D., and Gygi, S. P. (2003). A proteomics approach to understanding protein ubiquitination. *Nat. Biotechnol.* **21,** 921–926.

Petroski, M. D., and Deshaies, R. J. (2003). Context of multiubiquitin chain attachment influences the rate of Sic1 degradation. *Mol. Cell.* **11,** 1435–1444.

Pickart, C. M. (2001). Mechanisms underlying ubiquitination. *Annu. Rev. Biochem.* **70,** 503–533.

Pickart, C. M., and Fushman, D. (2004). Polyubiquitin chains: Polymeric protein signals. *Curr. Op. Chem. Biol.* **8,** 610–616.

Piotrowski, J., Beal, R., Hoffman, L., Wilkinson, K. D., Cohen, R. E., and Pickart, C. M. (1997). Inhibition of the 26 S proteasome by polyubiquitin chains synthesized to have defined lengths. *J. Biol. Chem.* **272,** 23712–23721.

Raasi, S., Orlov, I., Fleming, K. G., and Pickart, C. M. (2004). Binding of polyubiquitin chains to ubiquitin-associated (UBA) domains of HHR23A. *J. Mol. Biol.* **341,** 1367–1379.

Ravid, T., and Hochstrasser, M. (2004). NF-kappaB signaling: flipping the switch with polyubiquitin chains. *Curr. Biol.* **14,** R898–R900.

Riley, D. A., Bain, J. L. W., Ellis, S., and Haas, A. L. (1988). Quantitation and immunocytochemical localization of ubiquitin conjugates within rat red and white skeletal muscles. *J. Histochem. Cytochem.* **36,** 621–632.

Shih, S. C., Sloper-Mould, K. E., and Hicke, L. (2000). Monoubiquitin carries a novel internalization signal that is appended to activated receptors. *EMBO J.* **19,** 187–198.

Spence, J., Gali, R. R., Dittmar, G., Sherman, F., Karin, M., and Finley, D. (2000). Cell cycle-regulated modification of the ribosome by a variant multiubiquitin chain. *Cell* **102,** 67–76.

Spence, J., Sadis, S., Haas, A. L., and Finley, D. (1995). A ubiquitin mutant with specific defects in DNA repair and multiubiquitination. *Mol. Cell. Biol.* **15,** 1265–1273.

Sun, L., and Chen, Z. J. (2004). The novel functions of ubiquitination in signaling. *Curr. Opin. Cell Biol.* **16,** 119–126.

Thrower, J. S., Hoffman, L., Rechsteiner, M., and Pickart, C. M. (2000). Recognition of the polyubiquitin proteolytic signal. *EMBO J.* **19,** 94–102.

Treier, M., Staszewski, L. M., and Bohmann, D. (1994). Ubiquitin-dependent c-Jun degradation *in vivo* is mediated by the delta domain. *Cell* **78,** 787–798.

Tsirigotis, M., Thurig, S., Dube, M., Vanderhyden, B. C., Zhang, M., and Gray, D. A. (2001a). Analysis of ubiquitination *in vivo* using a transgenic mouse model. *Biotechniques* **31,** 120–126.

Tsirigotis, M., Zhang, M., Chiu, R. K., Wouters, B. G., and Gray, D. A. (2001b). Sensitivity of mammalian cells expressing mutant ubiquitin to protein-damaging agents. *J. Biol. Chem.* **276,** 46073–46078.

Ward, C. L., Omura, S., and Koptio, R. R. (1995). Degradation of CFTR by the ubiquitin-proteasome pathway. *Cell* **83,** 121–127.

Wertz, I. E., O'Rourke, K. M., Zhou, H., Eby, M., Aravind, L., Seshagiri, S., Wu, P., Wiesmann, C., Baker, R., Boone, D. L., Ma, A., Koonin, E. V., and Dixit, V. M. (2004). De-ubiquitination and ubiquitin ligase domains of A20 downregulate NF-kappaB signalling. *Nature* **430,** 694–699.

Wu-Baer, F., Lagrazon, K., Yuan, W., and Baer, R. (2003). The BRCA1/BARD1 heterodimer assembles polyubiquitin chains through an unconventional linkage involving lysine residue K6 of ubiquitin. *J. Biol. Chem.* **278,** 34743–34746.

You, J., Cohen, R. E., and Pickart, C. M. (1999). Construct for high-level expression and low misincorporation of lysine for arginine during expression of pET-encoded eukaryotic proteins in *Escherichia coli*. *BioTechniques* **27,** 950–954.

You, J., and Pickart, C. M. (2001). A HECT domain E3 enzyme assembles novel polyubiquitin chains. *J. Biol. Chem.* **276,** 19871–19878.

Yu, H., and Koptio, R. R. (1999). The role of multiubiquitination in dislocation and degradation of the a subunit of the T cell antigen receptor. *J. Biol. Chem.* **274,** 36852–36858.

[2] Controlled Synthesis of Polyubiquitin Chains

By CECILE M. PICKART and SHAHRI RAASI

Abstract

Many intracellular signaling processes depend on the modification of proteins with polymers of the conserved 76-residue protein ubiquitin. The ubiquitin units in such polyubiquitin chains are connected by isopeptide bonds between a specific lysine residue of one ubiquitin and the carboxyl group of G76 of the next ubiquitin. Chains linked through K48-G76 and K63-G76 bonds are the best characterized, signaling proteasome degradation and nonproteolytic outcomes, respectively. The molecular determinants of polyubiquitin chain recognition are under active investigation; both the chemical structure and the length of the chain can influence signaling outcomes. In this article, we describe the protein reagents necessary to produce K48- and K63-linked polyubiquitin chains and the use of these materials to produce milligram quantities of specific-length chains for biochemical and biophysical studies. The method involves reactions catalyzed by linkage-specific conjugating factors, in which proximally and distally blocked monoubiquitins (or chains) are joined to produce a particular chain product in high yield. Individual chains are then deblocked and joined in another round of reaction. Successive rounds of deblocking and synthesis give rise to a chain of the desired length.

Introduction

Ubiquitin modifies a broad spectrum of proteins in eukaryotic cells (Peng *et al.*, 2003). All of these modified proteins share a common structural feature: at least one molecule of ubiquitin is linked through its C-terminus (the carboxyl group of G76) to an amino group of the target protein. Usually, the linkage site is a lysine side chain of the target protein; less frequently, it is the α-amino group (Chen *et al.*, 2004; Ciechanover and Ben-Saadon, 2004). The covalently linked ubiquitin(s) modulates the stability, location, or function of the target protein. Such regulation can follow directly from protein modification with a single ubiquitin, as in many instances of ubiquitin-dependent endocytosis and trafficking (Hicke and Dunn, 2003). In other cases, appropriate functional regulation requires the initially conjugated ubiquitin to be extended into a polymer (a polyubiquitin chain) through the repeated use of a lysine residue of ubiquitin as the conjugation site. Abundant evidence indicates that monoubiquitin and polyubiquitin can be functionally distinct signals (Pickart and Fushman, 2004).

Ubiquitin has seven lysine residues. Polyubiquitin chains linked through K48 and K63 have well-characterized and distinct roles; at least some other polyubiquitin chains probably serve novel signaling functions (Peng et al., 2003; Pickart and Fushman, 2004). Recent appreciation of the structural and functional diversity of polyubiquitin chain signals, in conjunction with the discovery of multiple ubiquitin-interacting domains, has generated a demand for polyubiquitin chains in quantities necessary for biochemical, structural, and biophysical studies. In this chapter we outline methods for the preparation of milligram amounts of K48- and K63-linked polyubiquitin chains of defined length.

Functions of Polyubiquitin Chains

Ubiquitin's best-characterized function is to direct cellular proteins to the 26S proteasome for degradation. The first studies of this role showed that proteasomal targeting requires target protein modification with a polyubiquitin chain linked through K48-G76 isopeptide bonds (Chau et al., 1989). Because such chains are the principal proteasomal targeting signal and proteasomal proteolysis is an essential function, the K48R mutation in ubiquitin is lethal in budding yeast (Finley et al., 1994). The extensive scope of this function is further emphasized by the recent finding that ubiquitin itself is the most abundant ubiquitinated protein in budding yeast (Peng et al., 2003). Although this mass spectrometric study used a nonquantitative method, the data suggested that K48 was the predominant linkage (Peng et al., 2003). This conclusion is consistent with the failure of K-to-R mutations in ubiquitin, except K48R, to strongly alter the pattern or abundance of ubiquitinated proteins in the same species (Spence et al., 1995).

The Ub-K63R mutation has no observable effect on proteasome function. Instead, this mutation causes hypersensitivity to DNA damage (Spence et al., 1995). This phenotype reflects the modification of proliferating cell nuclear antigen (PCNA) with a K63-linked polyubiquitin chain, which in turn promotes a specific mode of DNA lesion bypass (Hoege et al., 2002). K63-linked chains also act as signals in ribosomal translation, certain endocytosis events, and (in mammals) kinase activation [reviewed in Pickart and Fushman (2004) and Sun and Chen (2004)]. Although the receptors that transduce K63-chain signals remain to be identified in most cases, it is unlikely that the proteasome is the destination of most substrates modified by K63-linked chains. A recent study detected all seven ubiquitin–ubiquitin isopeptide linkages in the budding yeast proteome (Peng et al., 2003), and although definitive functional information is lacking for the remaining linkages, some hints are available. Chains containing K29 and K11 linkages have been implicated in the targeting of certain proteins to proteasomes (Baboshina and Haas, 1996; Johnson et al., 1995). Chains containing K6 and K27 linkages seem more likely be nonproteolytic signals (Morris and Solomon, 2004; Nishikawa *et al.*, 2004; Okumura *et al.*, 2004).

Polyubiquitin Chain Structure

At least two levels of structure are relevant when considering polyubiquitin chains. The first is chemical structure; that is, which of ubiquitin's seven lysines is/are used within the polymer? The significance of chemical structure has been rigorously analyzed for canonical K48-linked chains. Here, *in vitro* analyses of the chain ligated to a substrate by its cognate E3 enzyme showed that only K48-G76 bonds were present (Chau *et al.*, 1989), whereas studies that use preassembled chains of defined structure demonstrated the signaling competence of K48-linked homopolymers (Thrower *et al.*, 2000). Certain noncanonical chain signals are probably also homopolymers. The factors that catalyze K63-chain synthesis in DNA damage tolerance and kinase activation produce homopolymers *in vitro* (Deng *et al.*, 2000; Hofmann and Pickart, 1999), and enzymes that produce K6- and K29-linked homopolymers have been described (Nishikawa *et al.*, 2004; Wu-Baer *et al.*, 2003; You and Pickart, 2001). Does a single chain ever contain more than one linkage? Proteomic evidence shows that the answer can be yes (Peng *et al.*, 2003), but it is uncertain whether such chains arise purposefully or adventitiously. Very little is known concerning the signaling properties of heteropolymeric chains (Pickart and Fushman, 2004).

The second level of chain structure is conformational. Because ubiquitin is a globular protein, chains with different chemical structures might possess distinctive ubiquitin–ubiquitin interfaces. Solution structural studies indicate that K48- and K63-linked chains indeed have different conformations in solution. At neutral pH, there are extensive ubiquitin–ubiquitin contacts in K48-linked Ub_2 (Varadan *et al.*, 2002), whereas K63-Ub_2 adopts an extended conformation in which the covalent bond is the only significant inter-subunit contact (Varadan *et al.*, 2004).

Enzymes Used for In Vitro *Chain Synthesis*

Because polyubiquitin chains are assembled through isopeptide (versus peptide) bonds, enzymatic synthesis is necessary. Therefore, one's ability to make a given chain presupposes the availability of a conjugating enzyme(s) that is (1) linkage-specific and (2) displays robust activity toward free ubiquitin. So far, we have identified such factors for K48- and K63-linked chains. Although a number of other linkage-specific factors have been reported (see preceding), the activity of most of these factors toward free ubiquitin has not been carefully investigated. Consequently the suitability of these enzymes for the method outlined in the following remains uncertain.

The conjugating enzyme should be highly linkage-specific. This issue is often addressed using mutant forms of ubiquitin (see Chapter 1). However, mass spectrometry provides the most rigorous criterion of linkage specificity (Peng *et al.*, 2003; Pickart and Fushman, 2004). Application of a semiquantitative version of this method to K48- and K63-linked chains synthesized by the methods described in the following has shown that only the desired linkage is detectable (J. Peng and C. Pickart, unpublished data). In contrast, this criterion was not met when we used a specific E3 enzyme to synthesize K29-linked chains (M. Wang, J. Peng, and C. Pickart, unpublished data).

For the synthesis of K48-linked chains, we use the mammalian enzyme E2-25K (Chen *et al.*, 1991). The biological function of E2-25K is uncertain, but its activity is well characterized in the *in vitro* setting. Of particular importance, although E2-25K binds free ubiquitin weakly, it is very active at the high concentrations of ubiquitin used in chain synthesis (Haldeman *et al.*, 1997). The same is true of the yeast Mms2/Ubc13 complex (a UEV/ E2 heterodimer), which participates in K63-chain synthesis in DNA damage tolerance *in vivo* (Hofmann and Pickart, 2001). These conjugating factors are conveniently expressed in *Escherichia coli*.

Polyubiquitin Chain Recognition

The past several years have seen the discovery of a small group of protein domains that bind ubiquitin and/or polyubiquitin chains, including the ubiquitin interacting motif (UIM), ubiquitin-associated domain (UBA), NPL4 zinc finger (NZF), and CUE domain (coupling of ubiquitin to endoplasmic reticulum degradation) [see Buchberger (2002); Hicke and Dunn (2003); and Pickart and Fushman (2004)]. Each domain occurs in a modest number of proteins in a given species. If polyubiquitin chains with different structures represent unique signals, there should be downstream binding factors that can discriminate among different chains. Recent studies identified a UBA domain and a zinc finger domain that preferentially bind K48- and K63-linked chains, respectively (Kanayama *et al.*, 2004; Raasi *et al.*, 2004). The ability to generate large quantities of structurally defined polyubiquitin chains will aid in the discovery of new linkage-specific binding proteins and should facilitate structural biology aimed at explaining the molecular basis of such linkage specificity. Enabling such studies has been an important motivation for developing the methods described in the following.

A Method for Controlled Synthesis of Polyubiquitin Chains

The method is outlined in Fig. 1 (Piotrowski *et al.*, 1997). It involves a series of enzymatic reactions catalyzed by the linkage-specific enzymes

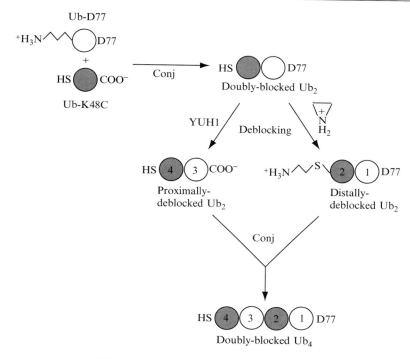

FIG. 1. Synthesis of K48-linked Ub_4. This scheme outlines the steps in the synthesis of K48-Ub_4 (see the text). The circles denote Ub molecules; the shading lets the reader keep track of the different ubiquitins in the chain. In the doubly blocked Ub_4 molecule, D77 of Ub-1 is the proximal chain terminus and C48 of Ub-4 is the distal terminus.

discussed previously. We refer to the end of the chain that would normally carry unconjugated G76 as the proximal end, whereas the end that would normally carry unconjugated K48 is the distal end (see Fig. 1). In each round of reaction, proximally and distally blocked monoubiquitins (or chains) are joined to produce a doubly blocked chain. The proximal block consists of an extra C-terminal residue (D77) that is labile to ubiquitin carboxyl–terminal hydrolases (UCHs). The distal block consists of a cysteine residue, placed at the normal conjugation site, that is later converted to a lysine mimic (S-aminoethylcysteine) through alkylation. Successive rounds of deblocking and conjugation give rise to a chain of any desired length. The method is presented in detail for K48-linked chains and in outline form for K63-linked chains. All of the plasmids mentioned in this chapter are available to academic researchers on request. When protein reagents are commercially available, we mention current suppliers.

Expression and Purification of Proximally and Distally Blocked Ubiquitin Monomers

Ub_2 linked through K48 or K63 is synthesized from a matched pair of monomeric ubiquitin reactants. One monomer, which cannot conjugate through its C-terminus, carries an extra residue (D77). The other monomer cannot conjugate through its lysine, because it carries a mutation to cysteine (or arginine) at this position. The blocked ubiquitins are produced in *E. coli*. We use pET3a plasmids that specify untagged ubiquitins, which yield 50–100 mg of purified ubiquitin per liter of culture. The procedures below are for a 2-liter scale expression. To scale up, all volumes should be increased in a directly proportional manner. Tagged version of ubiquitin can also be used to synthesize K48-linked chains.

Protocol for Ubiquitin Expression

1. Two days before expression, streak *E. coli* cells carrying the appropriate plasmid on selective medium. We express the ampicillin-marked ubiquitin plasmids (pET3a) in the BL21(DE3) strain carrying a chloramphenicol-marked helper plasmid, pJY2, that specifies the AGA-decoding arginine tRNA, as well as T7 lysozyme (LysS) (You *et al.*, 1999). Alternately, one can use commercial *E. coli* strains that overexpress rare tRNAs and carry the LysS gene. Because the synthetic genes that form the basis of most ubiquitin cDNAs are rich in AGA codons (this codon is also used for lysine-to-arginine mutations), failure to supply additional tRNA can result in misincorporation of lysine for arginine at AGA codons [see You *et al.* (1999)]. The presence of LysS blocks premature ubiquitin expression that could otherwise be toxic (You *et al.*, 1999).

2. To make the starter culture, inoculate 25 ml of 2 × YT medium (containing appropriate antibiotics) with a single colony. For pET3a plasmids, the starter culture should not reach saturation to avoid loss of cell viability. It is convenient to start this culture in the morning 1 day before large-scale expression, growing at 37° to $OD_{600} \sim 0.6$ (requires 5–8 h). This starter culture is refrigerated overnight.

3. Dilute 20 ml of starter culture into 2 liters of medium. Culture with good aeration at 37° to $OD_{600} \sim 0.6$, then induce with 0.4 mM IPTG. Culture for 4 h more, then harvest cells and freeze at −80°. Cell pellets can be stored indefinitely.

Protocol for Ubiquitin Purification

The following procedure describes a simple purification of untagged ubiquitin that takes advantage of ubiquitin's solubility in perchloric acid. Although the protein is >90% pure after the acid precipitation step

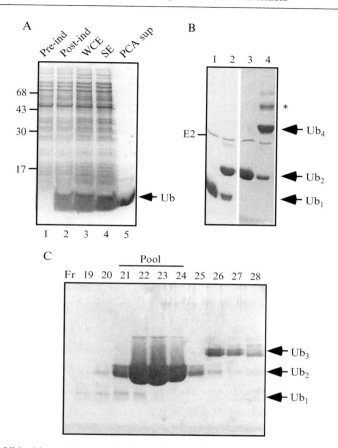

FIG. 2. Ubiquitin and chain synthesis/purification (Coomassie-stained gels). (A) Ubiquitin expression. Proportional aliquots of stages in the purification of Ub-K29R,D77 were analyzed. WCE, whole-cell extract; SE, extract after centrifugation; PCA sup, supernatant of perchloric acid precipitation. (B) Synthesis of K48-linked chains. Lanes 1 and 2, 48-mg scale synthesis of K48-Ub$_2$ at time zero and 4h, respectively. Lanes 3 and 4, 20-mg scale synthesis of K48-Ub$_4$ at time zero and 2h, respectively. *Asterisk* denotes a small amount of Ub$_6$ that was probably formed after carboxypeptidase-catalyzed removal of D77 from the distally deblocked dimer. (C) Purification of K48-Ub$_2$. The 50-mg scale synthetic reaction was applied to a 6-ml Resource S column (Amersham-Pharmacia Biotech) preequilibrated with 50mM ammonium acetate, pH 4.5. The column was eluted with a 60-ml gradient of NaCl (0–0.6M). Fractions (40 × 1.5ml) were collected; 2μl aliquots of the indicated fractions were analyzed. The bar highlights the fractions that were pooled and concentrated. A total of 44mg of Ub$_2$ was recovered (88%).

(Fig. 2A, lane 5), gradient cation exchange chromatography will remove UV-absorbing contaminants that would otherwise prevent the use of UV absorbance to determine protein concentrations.

1. Prepare 50 ml of lysis buffer consisting of 50 mM Tris-HCl (pH 7.6), 1 mM phenylmethylsulfonyl fluoride, 50 μM tosyllysylchloromethyl ketone, 2.5 μg/ml leupeptin, 5 μg/ml soybean trypsin inhibitor, 0.02% (v/v) NP-40, and 0.4 mg/ml lysozyme. If the ubiquitin contains a cysteine residue, include 1 mM DTT in the lysis buffer. (Protease inhibitors and detergent can be purchased from Sigma or another biochemical supply company.)

2. Add the buffer to the frozen cell pellet. To lyse, pipet up and down gently with a plastic transfer pipet. As the cells lyse, the suspension will become viscous from released DNA.

3. Digest the DNA by adding $MgCl_2$ to 10 mM and DNAseI (Sigma) to 20 μg/ml. Pipet up and down (or place tube on a rocker) until the DNA is lysed as judged by normal viscosity; this usually requires 10–20 min at room temperature.

4. Centrifuge the lysate at 8000 rpm for 20 min in a Sorvall SS-34 rotor. Carefully transfer the supernatant to a chilled beaker with a small stir bar (on ice).

5. Slowly add 0.35 ml of 70% perchloric acid while stirring vigorously. The solution will immediately turn milky. Continue stirring for 10 min after all the acid has been added.

6. Centrifuge the suspension as in step 4.

7. Transfer the supernatant to dialysis tubing with a molecular weight cutoff of 3.5 kDa. It is simplest to use extra-wide tubing.

8. Dialyze the perchloric acid supernatant in the cold against 2 liters of 50 mM ammonium acetate (pH 4.5) for at least 4 h. Repeat with 2 liters of fresh buffer. If the ubiquitin contains a cysteine residue, include 1 mM EDTA and 1 mM DTT in the buffer.

9. While the protein is dialyzing, pour a 15-mL; column of S-Sepharose Fast Flow (Amersham-Pharmacia Biotech) and equilibrate it with 10 column volumes of 50 mM ammonium acetate, pH 4.5 (plus 1 mM each of EDTA and DTT if the ubiquitin contains a cysteine).

10. Apply the dialysate to the column. All of the ubiquitin should bind. Wash the loaded column with 1 volume of equilibration buffer, and then elute at 2–3 ml/min with a 300-ml linear gradient of 0–0.5 M NaCl in the same buffer, collecting fractions of 4 ml. (*Note:* one can do this step with an open column and a peristaltic pump or with an automated chromatography device.) Locate the peak by SDS-PAGE/Coomassie staining of fraction aliquots or by UV absorbance. Ubiquitin elutes at ∼0.24 M NaCl. Pool the peak fractions and concentrate in a centrifugal concentrator (we use 15 ml Amicon Ultra devices with a 5-kDa mass cutoff).

11. Exchange the ubiquitin into a buffer compatible with subsequent enzymatic reactions. Tris-HCl (pH 7.6), 10 mM, is convenient. If there is a

cysteine residue, include EDTA (0.5 mM) and DTT (1 mM) in this buffer. Buffer exchange can be done by diluting and reconcentrating several times. To minimize protein losses during freezing and thawing, concentrate the ubiquitin to at least 50 mg/ml (all of the ubiquitin mutants used here are soluble to at least 110 mg/ml). The concentration of ubiquitin is determined by UV absorbance. A 1 mg/ml solution of ubiquitin has $OD_{280} = 0.16$. Concentrated stocks of ubiquitin can be stored indefinitely at $-20°$.

Expression and Purification of Conjugating Enzymes

For K48-chain synthesis catalyzed by E2-25K, we use untagged recombinant enzyme purified by conventional chromatography (Haldeman et al., 1997). However, most researchers will find it easier to produce a GST-tagged version of E2-25K using pGEX*E225K (Haldeman et al., 1997). In this case, release bound E2-25K from GSH beads (Sigma) using thrombin. Use 1 U thrombin (Amersham-Pharmacia Biotech) per 100 μg of fusion protein, incubating for several hours at room temperature. Cleavage is recommended, because GST-fused E2-25K is a less robust conjugating enzyme than free E2-25K. This E2 (also called UbcH1) is commercially available from BostonBiochem and Biomol. It is only active with mammalian E1 enzymes, which are commercially available from the same suppliers. One cautionary note: on occasion, commercial E1 preparations have been found to be contaminated with deubiuqitinating activity (M. Petroski and R. Deshaies, personal communication, 2004). The interested reader should consult the chapter by Beaudenon and Huibregtse (2005) and colleagues for information about recombinant E1 enzymes.

For K63-chain synthesis catalyzed by the Mms2/Ubc13 complex, we use the budding yeast versions of these proteins. We use mammalian E1, but yeast E1 can be substituted. Ubc13 is expressed as a GST fusion and released from GSH beads by thrombin cleavage (Hofmann and Pickart, 2001). PolyHis-tagged Mms2 is expressed using pET16b-Mms2; it is primarily insoluble. We purify the protein under mild denaturing conditions (4 M urea) using nickel beads (Novagen) and renature it by dialysis (VanDemark et al., 2001). The human Mms2/Ubc13 complex is commercially available from BostonBiochem.

Synthesis of K48-Linked Chains

Protocol for Synthesis and Purification of K48-Ub$_2$

The experienced protein chemist can also consult a more concise version of the protocol in the following (Raasi and Pickart, 2005).

1. Prepare PBDM8 buffer containing 250 mM Tris-HCl (50% base, pH 8.0), 25 mM MgCl$_2$, 50 mM creatine phosphate (Sigma P7396), 3 U/ml inorganic pyrophosphatase (Sigma I1891), and 3 U/ml creatine phosphokinase (Sigma C3735). Also prepare a neutral 0.1 M ATP solution (Sigma A2383). Both solutions are stable at −20°.

2. To synthesize K48-Ub$_2$, combine Ub-D77 and Ub-K48C at 7.5 mg/ml each in an incubation containing one-fifth volume of PBDM8 plus 2.5 mM ATP, 0.5 mM DTT, and 20 μM E2-25K. Avoid higher concentrations of DTT, which can trap ubiquitin by reacting with the E2-ubiquitin thiol ester. Remove 1 μl for SDS-PAGE analysis. (*Note:* do not boil polyubiquitin chains, because this can cause nonspecific cross-linking to yield products that resemble chains.) The conjugation reaction, initiated by adding 0.1 μM mammalian E1, requires 4 h at 37° (lane 2, Fig. 2B). To avoid having unreacted ubiquitin monomer, use precisely equal concentrations of Ub-K48C and Ub-D77. At the end of the incubation, add DTT (5 mM, freshly prepared) and EDTA (1 mM), then incubate at room temperature for 20 min to reduce disulfide-linked chains that could precipitate later. Remove 1 μl for SDS-PAGE.

3. Add one-fifth volume of 2 N acetic acid to the reduced reaction; check that the pH is ∼4 by spotting 1 μl onto pH paper. Apply the acidified mixture to a cation exchange column preequilibrated with buffer A (50 mM ammonium acetate, pH 4.5, 1 mM EDTA, 5 mM DTT). Recommended media include Mono S, Resource S, or S-Sepharose Fast Flow (all from Amersham-Pharmacia Biotech), using 1-ml beads per 20 mg of total ubiquitin. Wash the loaded column with ∼3 volumes of Buffer A (save to verify that Ub$_2$ is absent). Elute the column with a linear gradient of NaCl (0–0.6 M) in Buffer A, using 10–40 column volumes and collecting 40–60 fractions. Ub$_2$ elutes at ∼0.33 M NaCl. The fractions should be examined by SDS-PAGE to reject those that contain significant Ub$_1$ or Ub$_3$ (Fig. 2C); often, a small amount Ub$_3$ is formed during the synthetic reaction because of carboxypeptidase-catalyzed removal of D77 from Ub-D77 (a low level of carboxypeptidase activity contaminates some preparations of E2-25K). Good resolution of Ub$_2$ from Ub$_1$ and Ub$_3$ (Fig. 2C) is best obtained with an automated chromatography device. Pool and concentrate the peak fractions of Ub$_2$, then exchange the sample into storage buffer (20 mM Tris-HCl, pH 7.6, 0.5 mM EDTA, 2 mM DTT). To reduce losses caused by nonspecific absorption, concentrate to 30–80 mg/ml before storing at −80°. We use Amicon Ultra 4-ml concentrators with a 5-kDa mass cutoff. Polyubiquitin chains tend to precipitate when left at pH 4.5. Therefore, we recommend pooling, concentrating, and exchanging into storage buffer on the day that the column is run. Should precipitation occur, the chains can be dissolved in a buffer of 10 mM Tris (pH 7.6), 1 mM EDTA, 5 mM DTT, and 8 M urea. Remove the urea by dialysis.

Protocols for Deblocking Reactions

The Ub_2 resulting from the procedures described in the preceding is doubly blocked; it carries D77 at its proximal terminus and a C48 residue at its distal terminus (Fig. 1). To generate Ub_4, half of the Ub_2 is deblocked at its proximal terminus, exposing G76, whereas the remainder of the Ub_2 is deblocked at its distal terminus, by alkylating to introduce a lysine mimic. The two singly blocked dimers are later conjugated to produce Ub_4.

1. D77 is removed enzymatically by treating Ub_2 with ubiquitin C-terminal hydrolase (UCH). We use yeast ubiquitin hydrolase-1 (YUH1), which we express and purify according to published procedures (Johnston *et al.*, 1999), but commercially available UCH enzymes can be substituted (available from BostonBiochem or Biomol). To the doubly blocked Ub_2 (30–80 mg/ml), add: 50 mM Tris-HCl (pH 7.6), 1 mM EDTA, and 1 mM fresh DTT. Initiate deblocking by adding purified UCH to a final concentration of 16 μg/ml. Quantitative removal of D77 occurs in 60 min at 37°. Add 4 mM more DTT and incubate for 10 min at room temperature (to reduce any disulfide bonds). Remove the UCH by passing the reaction mixture through a 0.5-ml Q-Sepharose Fast Flow column preequilibrated with Q buffer (50 mM Tris-HCl, pH 7.6, 1 mM EDTA, 5 mM DTT). Collect the unbound fraction together with four washes (0.5 ml Q buffer each), reconcentrate to the original volume, and determine the concentration of the proximally deblocked dimer by UV absorbance.

2. To alkylate the remaining doubly blocked dimer (30–80 mg/ml) add 0.2 M Tris-HCl, pH 8.0, and 1 mM EDTA. Initiate alkylation by adding ethyleneimine to 55 mM (available from Chemservice). The reaction requires 60 min at 37°. Ethyleneimine must be removed to prevent subsequent inactivation of E1 and E2-25K. One can dialyze the incubation against 1 liter of 10 mM Tris-HCl, pH 8.0 (overnight at 5°). Or one can repeatedly concentrate and dilute with 10 mM Tris-HCl, pH 8.0, 2 mM DTT in a centrifugal concentrator until [DTT] = [ethyleneimine]. Concentrate the distally deblocked dimer to 30–80 mg/ml before freezing. Ethyleneimine is toxic and should be handled with care. Vials should be opened only in a fume hood, and manipulations involving the concentrated stock should be performed there as well. Unused ethyleneimine can be diluted into 10–50 volumes of alkaline DTT and allowed to sit for 24 h before disposal.

Protocol for Synthesis and Purification of K48-Ub_4

Conditions are the same as in the synthesis of K48-Ub_2 (above), except as follows. First, the reactants are the proximally and distally deblocked Ub_2 molecules (above). Second, each reactant is added at 10 mg/ml. Finally, the incubation can be shortened to 2 h (Fig. 2B, lanes 3 and 4). The purification method is the same as for Ub_2 (above), except that Ub_4 binds more tightly to the cation exchange column, so a gradient of 0–0.7 M NaCl is used.

Synthesis of K48-Linked Chains of Other Lengths

The principles are as described previously. Because resolution during cation exchange purification is only possible if the chains differ significantly in their lengths, long chains should be made by joining two chains of similar lengths. For example, make Ub_{12} by linking Ub_6 to Ub_6 (or Ub_4 to Ub_8) rather than by linking Ub_2 to Ub_{10}. Longer chains require higher salt concentrations to elute, so gradients should be adjusted accordingly. We have successfully made chains up to $n = 12$ (Raasi and Pickart, 2005; Raasi et al., 2004; Thrower et al., 2000). Once the chain has reached its final length, the distal C48 residue can be alkylated with ethyleneimine (above) or iodoacetamide, if desired, to reduce the potential for precipitation. We add iodoacetamide ~threefold excess over total thiol and incubate for 1–2 h at 37° (pH 7.6). Excess iodoacetamide should be removed by repeated concentration/dilution or dialysis.

Synthesis of K63-Linked Chains

The principles and procedures are similar to those outlined for K48-linked chains, with several differences. First, the synthetic reaction contains 8 μM each of yeast Mms2 and Ubc13 in place of E2-25K. Second, the buffer is PBDM7.6. (PBDM7.6 is the same as PBDM8, except that (1) Tris, pH 7.6 (24% base), is substituted for Tris, pH 8, and (2) 10 mM ATP is included in PBDM7.6. Accordingly, it is not necessary to add ATP independently to the conjugation reaction). Third, yeast or mammalian E1 can be used. Fourth, tagged ubiquitins should not be used with the yeast Mms2/Ubc13 complex, because they are inefficiently conjugated (Hofmann and Pickart, 2001). Finally, this complex discriminates against alkylated ubiquitin, making it impractical to deblock the distal terminus with ethyleneimine (Hofmann and Pickart, 2001). Therefore, we build up K63-linked chains one ubiquitin at a time. K63-Ub_2 can be synthesized from Ub-K63R and Ub-D77 at 10 mg/ml each (use of Ub-K63R instead of Ub-K63C eliminates precipitation problems), purified, and deblocked with UCH according to the same procedures used for K48-Ub_2. Then conjugate the proximally deblocked Ub_2 to Ub-D77, yielding K63-Ub_3. After purification and deblocking, K63-Ub_3 is conjugated to Ub-D77 to yield K63-Ub_4 (Hofmann and Pickart, 2001).

Factors Contributing to Yield and Recovery

In our original procedures, we removed the conjugating enzymes by subtractive anion exchange before acidifying the reaction mixture (Hofmann and Pickart, 2001; Piotrowski et al., 1997; Raasi and Pickart, 2005). This step is dispensable for chains up to $n = 4$ but may be necessary to ensure maximum purity of longer chains. In general, good normalization of the molar concentrations of the chain reactants will maximize conversion, simplify purification, and optimize the yield. However, when adding a single ubiquitin to a preexisting chain, use a 10% stoichiometric excess of the monomer reactant to force complete conversion of the chain to the $n + 1$ product; this will improve resolution during cation exchange. For the purification scheme discussed previously, we usually recover 70–90% of the input ubiquitins in the specific chain product in 30- to 60-mg scale reactions. Recovery during UCH-dependent deblocking is 80–90%, whereas recovery during ethyleneimine alkylation is 75–90%; both recoveries are better in large-scale reactions. Ubiquitin and polyubiquitin chains are rather sticky; nonspecific absorption to surfaces significantly reduces recovery. Reusing columns, avoiding glass tubes, and maximizing protein concentrations will counteract this problem.

Specialized Applications

It is straightforward to modify the preceding protocols for selected purposes. For example, one can make structurally defined heteropolymers. We used the Mms2/Ubc13 complex to conjugate a preassembled K48-chain to Ub-K63 in a ubiquitin-dihydrofolate reductase fusion protein (M. Ajua-Alemanji and C. Pickart, unpublished data). Here we synthesized the K48-linked chain from monomers carrying the K63R mutation to avert chain–chain conjugation in the final step. There are other instances in which the appropriate use of additional mutations can be used to avoid self-ligation of chains by other conjugation factors. Thus, to conjugate preassembled K63-Ub_4 to an internal lysine residue of a model substrate using an E3, we synthesized the chain from monomers carrying the K48R mutation to block self-ligation of the chain in the E3-catalyzed step (P.-Y. Wu and C. Pickart, unpublished data).

Any doubly blocked, K48-linked chain (Fig. 1) carries a unique cysteine at its distal terminus that can be used to introduce a thiol-reactive cross-linker, spin label, or fluorescent group (Fig. 3B). Alternately, such a group can be introduced at the proximal end by reacting the distal cysteine (if present) with iodoacetamide and then conjugating the chain to Ub-G76C (Fig. 3C). UCH is included in the latter reaction to remove the

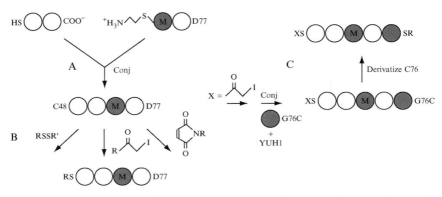

FIG. 3. Specialized chains: representative examples. (A) Chimeric mutant chains. In this example, the Ub-K48C that was used to make the distally deblocked dimer on the right carried an additional mutation (indicated by the shading and the letter "M"). (B) Distally modified chains. The C48 thiol group at the distal terminus of this doubly blocked tetramer can be modified in a site-specific manner (left to right) by disulfide exchange or reaction with iodoacetyl or maleimide derivatives. (C) Proximally modified chains. C48 of the doubly blocked tetramer from (A) is reacted with iodoacetamide, followed by another round of conjugation to introduce Ub-G76C at the proximal terminus. The C76 thiol group can then be modified as shown in (B).

chain's original proximal block; the G76C mutation creates a new proximal block. Finally, one can introduce a mutant ubiquitin at any point in the chain by incorporating additional mutations into the singly blocked ubiquitins (Fig. 3A). For example, to test the importance of Ub-L8 in K48-chain binding to proteasomes and UBA domains, we made a series of chimeric tetramers in which the L8A mutation was introduced into pairs of monomer units in all possible combinations (Raasi et al., 2004; Thrower et al., 2000). We also synthesized a K48-Ub$_4$ whose proximal ubiquitin carried the P37C mutation with or without the L8A mutation. We introduced a cross-linker at C37 and tested how the L8A mutation affected cross-linking of the chain to proteasomes (Lam et al., 2002). The available spectrum of such "designer chains" is limited principally by the effort involved in cloning and expressing the necessary ubiquitin mutants.

Acknowledgments

We thank Matt Steele for providing the data shown in Fig. 2A and members of the Pickart laboratory for comments on the manuscript. Our research on polyubiquitin chains is funded by grants from the NIH.

References

Baboshina, O. V., and Haas, A. L. (1996). Novel multiubiquitin chain linkages catalyzed by the conjugating enzymes E2$_{EPF}$ and RAD6 are recognized by 26 S proteasome subunit 5. *J. Biol. Chem.* **271,** 2823–2831.

Beaudenon, S. L., and Huibregtse, J. M. (2005). High-level expression and purification of recombinant E1 enzyme. *Methods Enzymol.* **398,** 3–8.

Buchberger, A. (2002). From UBA to UBX: New words in the ubiquitin vocabulary. *Trends Cell Biol.* **12,** 216–221.

Chau, V., Tobias, J. W., Bachmair, A., Marriott, D., Ecker, D. J., Gonda, D. K., and Varshavsky, A. (1989). A multiubiquitin chain is confined to specific lysine in a targeted short-lived protein. *Science* **243,** 1576–1583.

Chen, X., Chi, Y., Bloecher, A., Aebersold, R., Clurman, B. E., and Roberts, J. M. (2004). N-acetylation and ubiquitin-independent proteasomal degradation of p21^{Cip1}. *Mol. Cell.* **16,** 839–847.

Chen, Z., Niles, E. G., and Pickart, C. M. (1991). Isolation of a cDNA encoding a mammalian multi-ubiquitinating enzyme (E2-25K), and overexpression of the functional enzyme in *E. coli. J. Biol. Chem.* **266,** 15698–15704.

Ciechanover, A., and Ben-Saadon, R. (2004). N-terminal ubiquitination: More protein substrates join in. *Trends Cell Biol.* **14,** 103–106.

Deng, L., Wang, C., Spencer, E., Yang, L., Braun, A., You, J., Slaughter, C., Pickart, C., and Chen, Z. J. (2000). Activation of the IkB kinase complex by TRAF6 requires a dimeric ubiquitin-conjugating enzyme complex and a unique polyubiquitin chain. *Cell* **103,** 351–361.

Finley, D., Sadis, S., Monia, B. P., Boucher, P., Ecker, D. J., Crooke, S. T., and Chau, V. (1994). Inhibition of proteolysis and cell cycle progression in a multiubiquitination-deficient yeast mutant. *Mol. Cell. Biol.* **14,** 5501–5509.

Haldeman, M. T., Xia, G., Kasperek, E. M., and Pickart, C. M. (1997). Structure and function of ubiquitin conjugating enzyme E2-25K: The tail is a core-dependent activity element. *Biochemistry* **36,** 10526–10537.

Hicke, L., and Dunn, R. (2003). Regulation of membrane protein transport by ubiquitin and ubiquitin-binding proteins. *Annu. Rev. Cell Dev. Biol.* **19,** 141–172.

Hoege, C., Pfander, B., Moldovan, G.-L., Pyrowolakis, G., and Jentsch, S. (2002). *RAD6*-dependent DNA repair is linked to modification of PCNA by ubiquitin and SUMO. *Nature* **419,** 135–141.

Hofmann, R. M., and Pickart, C. M. (1999). Noncanonical *MMS2*-encoded ubiquitin-conjugating enzyme functions in assembly of novel polyubiquitin chains for DNA repair. *Cell* **96,** 645–653.

Hofmann, R. M., and Pickart, C. M. (2001). *In vitro* assembly and recognition of K63 polyubiquitin chains. *J. Biol. Chem.* **276,** 27936–27943.

Johnson, E. S., Ma, P. C., Ota, I. M., and Varshavsky, A. (1995). A proteolytic pathway that recognizes ubiquitin as a degradation signal. *J. Biol. Chem.* **270,** 17442–17456.

Johnston, S. C., Riddle, S. M., Cohen, R. E., and Hill, C. P. (1999). Structural basis for the specificity of ubiquitin C-terminal hydrolases. *EMBO J.* **18,** 3877–3887.

Kanayama, A., Seth, R. B., Sun, L., Ea, C.-K., Hong, M., Shaito, A., Deng, L., and Chen, Z. J. (2004). TAB2 and TAB3 activate the NF-kB pathway through binding to polyubiquitin chains. *Mol. Cell.* **15,** 535–548.

Lam, Y. A., Lawson, T. G., Velayutham, M., Zweier, J. L., and Pickart, C. M. (2002). A proteasomal ATPase subunit recognizes the polyubiquitin degradation signal. *Nature* **416,** 763–767.

Morris, J. R., and Solomon, E. (2004). BRCA1:BARD1 induces the formation of conjugated ubiquitin structures, dependent on K6 of ubiquitin, in cells during DNA replication and repair. *Hum. Mol. Genet.* **13,** 807–817.

Nishikawa, H., Oooka, S., Sato, K., Arima, K., Okamoto, J., Klevit, R. E., Fukuda, M., and Ohta, T. (2004). Mass spectrophotometric and mutational analyses reveal Lys-6-linked polyubiquitin chains catalyzed by BRCA1-BARD1 ubiquitin ligase. *J. Biol. Chem.* **279,** 3916–3924.

Okumura, F., Hatakeyama, S., Matsumoto, M., Kamura, T., and Nakayama, K. I. (2004). Functional regulation of FEZ1 by the U-box-type ubiquitin ligase E4B contributes to neuritogenesis. *J. Biol. Chem.* **279,** 53533–53543.

Peng, J., Schwartz, D., Elias, J. E., Thoreen, C. C., Cheng, D., Marsischky, G., Roelofs, J., Finley, D., and Gygi, S. P. (2003). A proteomics approach to understanding protein ubiquitination. *Nature Biotechnol.* **21,** 921–926.

Pickart, C. M., and Fushman, D. (2004). Polyubiquitin chains: polymeric protein signals. *Curr. Opin. Chem. Biol.* **8,** 610–616.

Piotrowski, J., Beal, R., Hoffman, L., Wilkinson, K. D., Cohen, R. E., and Pickart, C. M. (1997). Inhibition of the 26 S proteasome by polyubiquitin chains synthesized to have defined lengths. *J. Biol. Chem.* **272,** 23712–23721.

Raasi, S., Orlov, I., Fleming, K. G., and Pickart, C. M. (2004). Binding of polyubiquitin chains to ubiquitin-associated (UBA) domains of HHR23A. *J. Mol. Biol.* **341,** 1367–1379.

Raasi, S., and Pickart, C. M. (2005). Ubiquitin chain synthesis. *In* "Ubiquitin-Proteasome Protocols" (C. Patterson and D. M. Cyr, eds.), Vol. 301, pp. 47–56. Humana Press, Totowa, New Jersey.

Spence, J., Sadis, S., Haas, A. L., and Finley, D. (1995). A ubiquitin mutant with specific defects in DNA repair and multiubiquitination. *Mol. Cell. Biol.* **15,** 1265–1273.

Sun, L., and Chen, Z. J. (2004). The novel functions of ubiquitination in signaling. *Curr. Opin. Cell Biol.* **16,** 119–126.

Thrower, J. S., Hoffman, L., Rechsteiner, M., and Pickart, C. M. (2000). Recognition of the polyubiquitin proteolytic signal. *EMBO J.* **19,** 94–102.

VanDemark, A. P., Hofmann, R. M., Tsui, C., Pickart, C. M., and Wolberger, C. (2001). Molecular insights into polyubiquitin chain assembly: Crystal structure of the Mms2/Ubc13 heterodimer. *Cell* **105,** 711–720.

Varadan, R., Assfalg, M., Haririnia, A., Raasi, S., Pickart, C. M., and Fushman, D. (2004). Solution conformation of Lys^{63}-linked di-ubiquitin chain provides clues to functional diversity of polyubiquitin signaling. *J. Biol. Chem.* **279,** 7055–7063.

Varadan, R., Walker, O., Pickart, C. M., and Fushman, D. (2002). Structural properties of polyubiquitin chains in solution. *J. Mol. Biol.* **324,** 637–647.

Wu-Baer, F., Lagrazon, K., Yuan, W., and Baer, R. (2003). The BRCA1/BARD1 heterodimer assembles polyubiquitin chains through an unconventional linkage involving lysine residue K6 of ubiquitin. *J. Biol. Chem.* **278,** 34743–34746.

You, J., Cohen, R. E., and Pickart, C. M. (1999). Construct for high-level expression and low misincorporation of lysine for arginine during expression of pET-encoded eukaryotic proteins in *Escherichia coli*. *BioTechniques* **27,** 950–954.

You, J., and Pickart, C. M. (2001). A HECT domain E3 enzyme assembles novel polyubiquitin chains. *J. Biol. Chem.* **276,** 19871–19878.

[3] Derivitization of the C-Terminus of Ubiquitin and Ubiquitin-like Proteins Using Intein Chemistry: Methods and Uses

By KEITH D. WILKINSON, TUDEVIIN GAN-ERDENE, and
NAGAMALLESWARI KOLLI

Abstract

Here we describe a general method to synthesize and use a panel of reagents with selectivity for deubiquitinating enzymes exhibiting specificity for each ubiquitin-like protein. A substrate (Ubl-AMC), a reversible inhibitor (Ubl-aldehyde), and an active-site–directed irreversible inhibitor (Ubl-vinylsulfone) are described for each Ubl. Because of space constraints, we give details for only the Nedd8 derivatives, but these methods have been used in our laboratory to produce these derivatives of Ubiquitin, Nedd8, Sumo-1, Sumo-2, and Isg15. These reagents are useful in defining the specificity of DUBs, as well as in purifying and identifying these important proteins. The reagents are selective and useful in crude extracts. The reactivity of these reagents reveals differences in both the S1 and S1' sites of deubiquitinating enzymes. Only active enzymes are efficiently detected with these reagents. Published results indicate that specificity is not strictly defined by the evolutionary relationships of these DUBs.

Background

The ubiquitin family consists of several members sharing a similar fold and chemistry (Huang *et al.*, 2004; Jentsch and Pyrowolakis, 2000; Mayer *et al.*, 1998; Schwartz and Hochstrasser, 2003; Yeh *et al.*, 2000). The C-terminus of Ubl (ubiquitin or ubiquitin-like protein) is activated by adenylation and thiol ester formation. This activated Ubl is then reacted with side chain amines on the target protein, resulting in a conjugate between the Ubl and the target protein (Huang *et al.*, 2004; Kim *et al.*, 2002; Ohsumi and Mizushima, 2004; Parry and Estelle, 2004; Staub, 2004).

One of the most challenging questions regarding ubiquitin-like protein modification pathways is the one of specificity. Which enzymes work on which substrates to make what products? The problem is even more complex when one considers that many of the enzyme families of the system contain dozens to hundreds of genes and that they function in multiprotein complexes. For instance, there may be several hundred ring finger and cullin-based ubiquitin ligases, each exhibiting some specificity for the E2 ubiquitin conjugating enzyme, the substrate adapter protein (F-box, BTB, etc.), and/or the protein scaffold used (Cardozo and Pagano, 2004; Fang and Weissman, 2004; Hatakeyama and Nakayama, 2003; Murata *et al.*, 2003; Passmore and Barford, 2004). A similar level of complexity is observed with the deubiquitinating enzymes (DUBs), proteases that reverse ubiquitination. There are at least 80 of these important enzymes in mammals encoded by at least six different gene families (Borodovsky *et al.*, 2002; Fischer, 2003; Kim *et al.*, 2003; Soboleva and Baker, 2004; Wilkinson, 2000).

These combinatorial possibilities for ubiquitination and deubiquitination are, indeed, staggering, and it has become apparent that we need specific tools that can give us information about the expression, specificity, and activity of various enzymes of the pathway. This chapter presents a panel of such reagents, all made with intein fusion technology. The intein is a self-splicing protein sequence that can be used to produce proteins and peptides activated at the C-terminus by formation of a thiol ester (Chong *et al.*, 1997; Cottingham *et al.*, 2001; Evans *et al.*, 1998).

Production of Ubl Thiol Esters

Vector Constructs and Protein Expression

General Strategy. The general strategy is to produce a ubiquitin domain chemically activated at its C-terminus by formation of a thiol ester (Cottingham *et al.*, 2001). Reaction of this thiol ester with a nucleophile then produces the desired derivatives (Fig. 1). This approach works in good yield with ubiquitin and all the Ubl proteins. Primers containing a 5′ Nde1 site and a 3′ Sma1 site were used to amplify by polymerase chain reaction (PCR) the coding sequence of ubiquitin, Nedd8, Sumo-1, Sumo-2, and Isg15 lacking the C-terminal glycylglycine of the mature products. The PCR product is then digested with Nde1 and cloned into the Nde1 and Sma1 sites of the vector pTYB2 (New England Biolabs Inc.). Use of the Sma1 cloning site restores a single glycine residue at the C-terminus of the ubiquitin-like protein. The presence of a C-terminal glycine enhances chemical ligation, thus reducing the chances of side reactions.

The final construct consists of the ubiquitin-like domain (lacking the C-terminal glycine) as a fusion protein to an intein and a chitin-binding domain (CBD). The second glycine is added by reaction with the nucleophile to make the desired derivative. When necessary, the Ubl domain can also be epitope-tagged. For construction of the epitope-tagged fusion

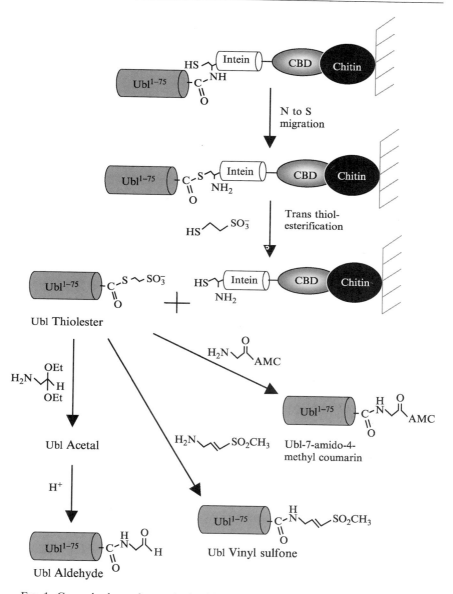

FIG. 1. General scheme for synthesis of Ubl C-terminal derivatives (Gan-Erdene et al., 2003).

proteins, the pTYB2-Sumo-1/2, -Ub, -Nedd8, or Isg15 vectors were digested with Nde1 and ligated to a cassette that introduces the epitope tag. Alternatively, the epitope can be introduced by use of PCR primers extended to code for the epitope.

The fusion proteins (Ub/Sumo-1/Sumo-2/Nedd8/Isg15) are expressed in BL21-DE3 cells (CodonPlus, Stratagene). Use of the CodonPlus cells reduces problems with poorly used codons. Cultures are generally grown to an OD_{600} of 0.8, and the temperature is reduced to 30° before adding IPTG to induce expression of the Ubl-intein-chitin binding domain fusion protein. A lower temperature and low inducer concentration often result in a more soluble fusion protein (Galloway et al., 2003) and were used for expression of FLAG-Nedd8 and HA-Isg15 fusion proteins. After induction for 3 hours at 30° (or 18 hours at 15°), cells are harvested by centrifugation.

Specific Example: FLAG-Nedd8 Fusion Protein

A cDNA encoding residues 1–75 of human Nedd8 was amplified from pRSNedd8p (Krantz and Wilkinson, unpublished) by PCR using the primers 5′-pCTATAGGGAGACCACAACGG-3′ and 5′-pTCTCA-GAGCCAACACCAGG-3′. The PCR product was digested with NdeI and inserted into pTYB2 (New England, Biolabs) digested with NdeI and SmaI before ligation. For construction of the Flag-tagged fusion protein, a cassette was inserted using the primers 5′-pTATGATCGACTA-CAAAGACGATGACGATAAACA-3′ and 5′-pTATGTTTATCGT-CATCGTCTTTGTAGTCGATCA-3′. The sequences of the PCR products were confirmed by automated sequencing.

Bl21-DE3 cells transformed with pTYB2-Nedd8 or Flag-Nedd8-pTYB2 were grown in LB media to an OD_{600nm} of 0.6–0.8. Expression of the $Nedd8^{1-75}$-intein-CBD fusion protein was induced with 0.5 mM IPTG for 3 h at 30°. Expression of the Flag-$Nedd8^{1-75}$-intein-CBD fusion protein was induced with 50 μM IPTG at 15° for 18 hrs. Cell pellets from 1 liter of cells were resuspended in 50 ml lysis buffer (20 mM HEPES, 50 mM sodium acetate, pH 6.5, 75 mM NaCl). After sonication to break the cells, debris was removed by centrifugation at 20,000g for 30 min.

Application Notes and Potential Complications

The use of BL21 CodonPlus cells (Stratagene) for expression gave moderately higher yields of fusion protein, perhaps because of suppression of problems caused by unfavorable codon use. No toxicity resulting from protein expression was observed, but if toxicity is a problem, steps should be taken to reduce leaky expression of the fusion protein.

Expression at low temperature (15–30°) and low levels of IPTG reduced the tendency of some fusion proteins to aggregate on expression. Insolubility was a particular problem with NEDD8 fusion proteins: NEDD8 fusion proteins are >80% soluble when expressed at 30°, whereas FLAG-Nedd8 fusion protein has to be expressed at 15° to obtain soluble protein. The HA-NEDD8 fusion protein is insoluble even when expressed at 15°.

The lysis buffer should not contain lysozyme since it will digest the chitin affinity column used to purify the fusion protein.

Splicing and Purification of the Ubl-Thiol Ester

The Splicing Reaction. Excision of the Ubl protein occurs by trapping the intermediate of the normal self-splicing reaction. In the first step of splicing, an *N-S* acyl rearrangement results in the formation of a small amount of thioester linkage between the target protein and intein (Fig. 1). Because the intein used in these expression vectors is mutated such that it cannot complete the splicing step, the splicing intermediate (the thiol ester between the side chain thiol of a cysteine at the N-terminus of the intein and the C-terminus of the fused protein) is stable. This intermediate can be reacted with free thiol groups to liberate the fused protein as a thiol ester (Chong *et al.*, 1997; Evans *et al.*, 1998). Because the intein-CBD portion of the fusion protein is bound to the chitin column, only the Ubl-thiol ester is released and washed from the column upon incubation with thiols.

Specific Example: Affinity Purification of Nedd8- and Flag-Nedd8 Thiol Esters. The clarified lysate (50 ml) was mixed with 2.5 ml of chitin beads (New England BioLabs) equilibrated with lysis buffer and incubated with the beads at 37° for 2 h. The beads were then poured into a column, and the resin was washed with 25 ml of 20 mM HEPES, 50 mM CH$_3$COONa, 75 mM NaCl, pH 6.5. The fusion protein was cleaved on the column by applying 2.5 ml of 100 mM 2-mercaptoethanesulfonic acid (MESNA) and incubating at 37° overnight. The Nedd8$^{1-75}$ thiol ester (or the Flag-tagged version) was eluted with 4 column volumes of 20 mM HEPES, 50 mM CH$_3$COONa, 75 mM NaCl, pH 6.5. The eluted fractions were combined and concentrated 10-fold. When frozen, the thiol esters are stable for several weeks.

Affinity binding, cleavage, and purification were monitored by SDS-PAGE and/or high-performance liquid chromatography (HPLC) on a C8 reverse phase column with buffers A (25 mM NaClO$_4$, 0.07% HClO$_4$) and B (buffer A in 75% acetonitrile). Protein was monitored by absorbance at 205 nm and eluted with a 0–80% gradient over 25 min.

Application Notes and Potential Complications. Some steps of the original intein method have been modified. Because ubiquitin and ubiquitin-like proteins (ubl) are thermostable, the temperature for binding and cleavage can be increased to 37° to more quickly bind and cleave the fusion protein. However, it is preferable to do the cleavage at 4° to minimize hydrolysis of the ester. At either temperature, the C-terminal thiol ester is efficiently generated in an overnight reaction. Another important factor is the pH of cleavage. The optimal pH during cleavage is 6.5. If pH is higher than 6.5, the Nedd8 thiol ester generated will be partially lost because of hydrolysis.

In the case of SUMO derivatives, the N-terminal methionine is removed by the action of bacterial methionine aminopeptidase. Partial processing of the ubiquitin N-terminus was also noted. We have not noticed any differences in the reactivity of full-length and truncated products. This processing should not present any problem, because it has been shown that even an N-terminal epitope tag does not prevent function.

Derivatization of the C-Terminus

In general, C-terminal derivatives are generated by the attack of a nucleophile on the C-terminus of the excised fusion protein thiol ester (Cottingham *et al.*, 2001). The terminal glycine, or an equivalent spacer, is provided by the nucleophile in most cases. However, the C-terminus of the Ubl can be further truncated or extended as dictated by the structure of the nucleophile. The compounds discussed in the following are all substrates or inhibitors of deubiquitinating enzymes, with Nedd8 as the specific example.

Solubility of Reagents and Use of Catalysts

Some of the nucleophiles one would like to use are only minimally soluble in water. We have used two approaches to circumvent this limitation, both of which result in increased concentrations of the nucleophile. Glycyl-7-amido-4-methylcoumarin (gly-AMC) is insoluble but forms 2:2 host–guest complexes with hydroxypropyl β-cyclodextrin. The crystal structure of similar complexes indicates that the amino groups are pointing out such that the glycyl moiety is accessible for reaction (Brett *et al.*, 2000). Accordingly, inclusion of 30% (w/v) of hydroxypropyl β-cyclodextrin allows the use of at least 65 mM gly-AMC in the coupling reaction. Alternately, we have found that the reaction can be conducted in 8 M urea, a chaotrope that is similarly effective in solubilizing gly-AMC. Refolding of the Ubl domain on removal of the urea by dialysis has been effective in most cases.

It is also advantageous to increase the reactivity of the ester by use of NHS (N-hydroxysuccinimide) as a catalyst. Initial attack of NHS on the thiol ester generates an "active ester" that is more reactive to nucleophiles and increases the yield of desired products.

Minimizing Hydrolysis and Other Side Reactions

To minimize hydrolysis of the thiol ester during coupling, it is necessary to keep the pH below about 8. Under the conditions we usually use for coupling, the half-life of the thiol ester in the absence of nucleophile is 60 min. Thus, it is seldom necessary, or desirable, to extend the reaction time beyond 2–4 h. This is especially true when bivalent nucleophiles such as glycyl vinyl sulfone are used. At longer times of incubation, we observe the accumulation of Ubl derivatives that have reacted with two molecules of glycyl vinyl sulfone. Thus, we stop the reaction as soon as all the original Ubl thiol ester has been consumed.

Specific Example: Nedd8 Derivatives

Nedd8-Amidomethyl Coumarin. Nedd8-amidomethyl coumarin (AMC) was synthesized by reacting 0.5 ml Nedd8$^{1-75}$ thiol ester (1–3 mg/ml) with 2 ml 50 mM glycyl 7-amido-4-methyl coumarin, 30% hydroxypropyl β-cyclodextrin in 50 mM Tris, pH 10, and 0.1 ml 2 M N-hydroxysuccinimide. The final pH of the reaction was 8, and the mixture was incubated at 37° overnight. Alternately, Nedd8-amidomethyl coumarin (AMC) was synthesized by reacting 1 ml Nedd8$^{1-75}$ thiol ester (1–3 mg/ml) with 50 mg glycyl 7-amido-4-methyl coumarin in 8 M urea, 0.04 ml 2 M N-hydroxysuccinimide, and 0.1 ml 1 M Tris base. The reaction mixture was incubated at 37° overnight. In both cases, the overall yield of Ubl-AMC is 10–15%, and this is also typical of the yields obtained with other Ubl domains.

Nedd8-Aldehyde. Nedd8-aldehyde was synthesized by reacting 1 ml Nedd8$^{1-75}$ thiol ester (1–3 mg/ml) with 0.2 ml 4 M aminoacetaldehyde diethyl acetal, pH 8.5, and 0.005 ml 2 M NHS, pH 7.2, at 37° for 2 h. After purification of the product (see later), the resultant acetal is deprotected to yield the free aldehyde by incubation with 0.15 M HCl at 37° for 60 min. Overall yield for Ubl-aldehyde is 50%.

Flag-Nedd8-Vinylsulfone. Flag-Nedd8-vinylsulfone was synthesized by reacting 1 ml Flag-Nedd8$^{1-75}$ thiol ester (1–3 mg/ml) with 0.1 ml 2 M glycine vinylmethylsulfone tosylate pH 7.0; 0.1 ml 2 M NHS, pH 7.2; and 0.04 ml 0.5 M Tris base at 37° for 3 h. The overall yield is about 30%.

Monitoring the Synthesis and Purification of Ubl Derivatives

All steps of synthesis and purification of Nedd8 derivatives were monitored by HPLC (Fig. 2). Ubl derivatives were separated on C8 column, 5 × 460 mm (Altech). Buffer A contained 25 mM sodium perchlorate

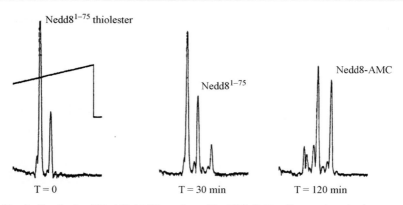

FIG. 2. Synthesis of Nedd8-AMC monitored by HPLC. Details are given in the text.

and 0.07% (w/w) perchloric acid, and buffer B contained 25 mM sodium perchlorate and 0.07% (w/w) perchloric acid in 75% (v/v) acetonitrile. The flow rate was 1 ml/min, and derivative was detected by monitoring the 205-nm absorbance. The elution gradient was either linear (0–80% over 25 min) or nonlinear (0 min 40% buffer B; 3 min, 60% buffer B; 18 min, 90% buffer B; 18.01 min, 40% buffer B, and 25 min, 40% buffer B).

Purification of Ubl Derivatives

In some circumstances, the Ubl derivatives can be used without further purification but will be most useful if they are first purified. The major contaminant to be removed is the hydrolyzed thiol ester that results from incomplete reaction with the nucleophile. Thus, methods that bind the Ubl or epitope tag are generally not useful. Two methods are suggested in the following: ion exchange chromatography to exploit the charge differences between the Ubl-derivative and the hydrolyzed thiol ester and hydrophobic interaction chromatography to make use of the different polarity of these two molecules. Other methods of purification may also prove useful. All purified derivatives were verified by MALDI mass spectroscopy, SDS-PAGE, and functional assays.

Ion Exchange Chromatography

Nedd8 and ubiquitin derivatives were purified by ion exchange chromatography on CM 52 Cellulose (Whatman, UK). The crude reaction mixture was desalted using a PD-10 desalting column (Amersham Biosciences) or dialyzed against 50 mM sodium acetate, pH 4.5, the starting buffer for CM 52 cellulose used in the next step of purification. A 2–3 ml desalted sample was applied to 1 ml CM 52 Cellulose equilibrated in 50 mM sodium acetate, pH 4.5, and the column was washed with 5 ml starting buffer. Nedd8$^{1-75}$ (generated by hydrolysis of the thiol ester) was eluted with 20 ml 50 mM sodium acetate, pH 5.5. The derivatized Nedd8$^{1-75}$ was subsequently eluted with 10 ml 50 mM sodium acetate, pH 7.0.

Sumo-1 and Sumo-2 derivatives were either desalted on PD 10 columns or dialyzed into 50 mM TEA, pH 6.5. The derivatives were further purified by adsorbing on a Mono Q HR anion exchange column (Amersham Biosciences) equilibrated with 50 mM TEA, pH 6.5, and eluted with a linear gradient of NaCl from 0–0.5 M (flow rate, 0.75 ml/min; fraction size, 1 ml). Alternately, we have used 50 mM Bis-Tris, pH 5.5, buffer with a gradient from 0–0.15 M NaCl.

Protein was monitored by absorbance at 205 nm. The pooled fractions were dialyzed against 10 mM HCl, concentrated on Centricon, YM-3, (3000 MWCO, Millipore Corp., MA.) and stored in aliquots at $-80°$.

Hydrophobic Interaction Chromatography

Because some of the derivatives contain nonpolar groups, hydrophobic interaction chromatography can separate the derivatives from the hydrolyzed thiol ester. For example, Sumo-1-AMC was purified by chromatography on High Trap Phenyl Sepharose (high substitution) from Amersham Biosciences. The crude reaction mixture was equilibrated with buffer A (20 mM Tris-HCl, pH 7.0, containing 1 M NaSO$_4$) by gel filtration, and 3.5 ml was applied to 1 ml of phenyl sepharose resin equilibrated with buffer A. The column was washed with 15 ml buffer A at a flow rate of 0.5 ml/min and then eluted with a 40-ml gradient from buffer A to buffer B (20 mM Tris-HCl, pH 7.0). Sumo-1-AMC eluted at about 0.7 M salt and was then concentrated and desalted into 10 mM HCl by chromatography on Sephadex G25. Sumo-1-AMC was stored in aliquots at $-80°$.

Uses of the Ubl Derivatives

Assays of Deubiquitinating Enzymes

Verification of the Reagents. To verify the identity and quality of the derivatives, standard enzyme preparations can be used. Ub-AMC was efficiently hydrolyzed in a reaction that was catalyzed by UCH-L3 (Boston Biochem), and this cleavage was inhibited by the addition of small amounts of ubiquitin aldehyde or vinyl sulfone (Fig. 3). Nedd8 derivatives were also verified with UCH-L3 or Den1, whereas yeast Ulp1 or mammalian Senp1

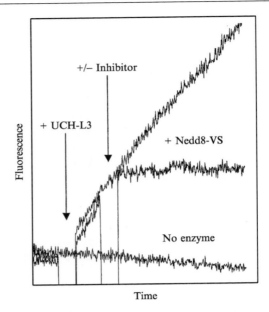

FIG. 3. Hydrolysis of Nedd8-AMC by UCH-L3 and its inhibition by Nedd8-VS. Nedd8-AMC (92 nM) was incubated with buffer or 15 pM UCH-L3 as indicated. At 190 s, Nedd8-VS (100 nM) or buffer was added. The reaction was monitored by fluorescence as described previously (Dang et al., 1998; Gan-Erdene et al., 2003).

are suitable for verifying Sumo derivatives. To verify the Isg-15 derivatives we have used a mouse thymus homogenate.

Steady-state Assays. Standard steady-state kinetic analysis was carried out with the AMC derivatives (Dang et al., 1998; Stein et al., 1995) by taking advantage of the large increase in fluorescence observed on hydrolysis of this class of substrates. Hydrolysis of Ubl-AMC substrates was determined spectrofluorometrically in a final volume of 120 μl. The assays contained Ubl-AMC at concentrations from 50 nM to 50 μM as needed, 50 mM Tris HCl, pH 7.6, 0.1 mg/ml ovalbumin, and enzyme at a concentration of 15 pM–50 nM as needed. The excitation wavelength was 340 nm, and emission was monitored at 440 nm. Absolute concentrations were determined by reference to a standard curve of fluorescence versus concentration of 7-amino-4-methyl coumarin.

Application Notes and Potential Complications. The AMC chromophore is light sensitive, and Ubl-AMC should be protected from light. Similarly, excessive excitation intensity during the fluorescence assay leads to photodegradation of the substrate and a slowly decreasing baseline fluorescence. Thus, it is advisable to use the minimum possible slit width (light intensity) for excitation.

Active Site-Directed Labeling of Deubiquitinating Enzymes

General Strategy. Most known deubiquitinating enzymes are thiol proteases that can be efficiently and specifically labeled with Ubl-vinyl sulfones or other derivatives containing an electrophile at the C-terminus (Borodovsky *et al.*, 2001, 2002; Gan-Erdene *et al.*, 2003; Hemelaar *et al.*, 2004). The labeling seems to be highly selective. For instance, an extract can be reacted with Ub-VS to block all the reactive Ub-specific DUBs with little or no blocking of Den1, a specific deneddylating enzyme. The same protocol can be used with Sumo-1 VS to preblock Sumo-2–specific DUBs before reacting the lysates with epitope-tagged-Sumo-2 VS.

Specific Example: Labeling of Den1, a Nedd8-Specific DUB. Deneddylases in L-M(TK$^-$) cell lysates were labeled by reacting with the indicated amount of FLAG-Nedd8-VS with lysates (10 mg/ml total protein) for 30 min at 37° (Gan-Erdene *et al.*, 2003). For prelabeling with Ub-VS, 36 ml of L-M(TK$^-$) cell lysate was incubated with 0.54 ml of 0.2 mg/ml Ub-VS for 30 min at 37°. After prelabeling, 0.72 ml of 0.2 mg/ml FLAG-Nedd8-VS was added for an additional 30 min at 37°.

Application Notes and Potential Complications. Although the labeling afforded by these derivatives is very specific, one often sees slow reactivity of distantly related DUBs. Isopeptidase T, a DUB that disassembles free polyubiquitin chains, is rapidly labeled by Ub-VS but can be slowly labeled with HA-Isg15-VS (Gan-Erdene and Wilkinson, unpublished). As a rule of thumb, enzymes that are truly specific for any particular Ubl will be labeled within minutes by a modest excess of the Ubl-VS, whereas labeling that requires a large excess of reagent or progresses over a period of hours may be indicative of less specificity.

Even with a "specific" DUB/Ubl pair, the specificity and efficiency of labeling varies with the identity of the nucleophile and the activation state of the enzyme. Thus, some DUBs are efficiently labeled with Ub-VS but only weakly with Ub-vinylmethyl ester (Ub-VME) or other derivatives (Borodovsky *et al.*, 2002). This is probably indicative of the specificity of the S1 site (recognizing the Ubl moiety) and the S1' site that allows access of only some electrophiles. A DUB that is not active or appropriately localized may also be difficult to label. Recombinant yeast UBP6 is unreactive with Ub-VS and Ub-AMC but becomes very reactive with these reagents when it is bound to the proteasome.

Unfortunately, few of the reagents necessary for synthesis of this series of active site–directed reagents are commercially available.

Affinity Purification of Labeled Proteins

By equipping the Ubl-electrophile with an epitope tag, we can affinity purify and identify the reactive DUBs. Both FLAG and HA epitopes have been used, and it is likely that others, including Hexa histidine, will also work.

Specific Example: Purification of Den1, a Nedd8-Specific DUB. FLAG-Nedd8-VS-labeled proteins were isolated (Gan-Erdene *et al.*, 2003) using anti-FLAG M2 affinity gel (Fig. 4). The affinity gel (0.6 ml) was equilibrated by washing three times with 0.6 ml of 50 mM Tris-HCl, 150 mM NaCl, pH 7.45, then once with 0.6 ml of 0.1 M glycine HCl, pH 3.0, then five times with 0.6 ml of 50 mM Tris-HCl, 150 mM NaCl, pH 7.45. Labeled proteins were adsorbed onto the gel at 4° for 1 h with gentle mixing, and the supernatant was removed from the gel. After washing the gel 10 times with 0.6 ml 50 mM Tris-HCl, 150 mM NaCl, pH 7.45, it was incubated with 0.6 ml of 0.225 mg/ml FLAG peptide in the same buffer at 4° for 30 min with gentle mixing. The eluted protein was removed, and the affinity gel was eluted two more times with 0.6 ml of 0.15 mg/ml FLAG peptide. After elution, the affinity gel was stripped with 0.6 ml of 0.1 M glycine HCl, pH 3.0, for 10 min and

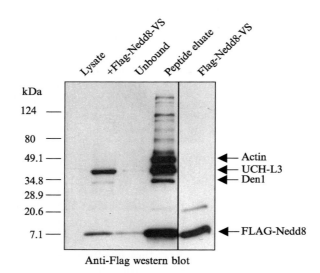

FIG. 4. Labeling and purification of Nedd8 specific proteases of L-M(TK⁻) cell lysates. Crude L-M(TK⁻) cell lysate was labeled with Flag-Nedd8-VS at 37° for 30 min. Flag-Nedd8-VS–labeled proteins were adsorbed to anti-Flag M2 affinity gel and eluted with FLAG peptide. Purification of Flag-Nedd8-VS–labeled proteins was monitored by 4–20% gradient SDS-PAGE and Western blotting analysis with Anti-Flag M5 antibody (Gan-Erdene *et al.*, 2003).

immediately washed with starting buffer. All steps of the purification were monitored by Western blotting using the enhanced chemiluminescence (ECL) detection system (Amersham Biosciences). The secondary antibody was horseradish peroxidase–conjugated anti-mouse IgG. Primary and secondary antibodies were prepared in Tris-buffered saline–Tween 20 containing 5% dry milk.

Application Notes and Potential Complications

If the labeled DUB is present in low concentrations, the background of nonspecifically adsorbed protein may be significant. Control reactions lacking the Ubl-VS or in the presence of excess competitor Ubl during labeling should always be conducted to assess the degree of nonspecific binding. It is also possible that adapters or scaffold proteins may be co-purified with the labeled DUB. Finally, it is possible that the reagent may react with a DUB that is not specific for the Ubl. Conducting the labeling for short periods of time with only a small excess of reagent would be expected to minimize this complication.

Acknowledgments

We thank Huib Ovaa and Hidde Ploegh for generously furnishing the glycyl vinylsulfone used in these studies. This work was supported by grants NIH R01-GM66355 to KDW and NIH F05 TW05461 to TG.

References

Borodovsky, A., Kessler, B. M., Casagrande, R., Overkleeft, H. S., Wilkinson, K. D., and Ploegh, H. L. (2001). A novel active site-directed probe specific for deubiquitylating enzymes reveals proteasome association of USP14. *EMBO J.* **20,** 5187–5196.

Borodovsky, A., Ovaa, H., Kolli, N., Gan-Erdene, T., Wilkinson, K. D., Ploegh, H. L., and Kessler, B. M. (2002). Chemistry-based functional proteomics reveals novel members of the deubiquitinating enzyme family. *Chem. Biol.* **9,** 1149–1159.

Brett, T. J., Alexander, J. M., and Stezowski, J. J. (2000). Chemical insight from crystallographic disorder-structural studies of supramolecular photochemical systems. Part 2. The b-cyclodextrin–4,7-dimethylcoumarin inclusion complex: A new b-cyclodextrin dimer packing type, unanticipated photoproduct formation, and an examination of guest influence on b-CD dimer packing. *Perkin Trans.* **2,** 1095–1103.

Cardozo, T., and Pagano, M. (2004). The SCF ubiquitin ligase: Insights into a molecular machine. *Nat. Rev. Mol. Cell. Biol.* **5,** 739–751.

Chong, S., Mersha, F. B., Comb, D. G., Scott, M. E., Landry, D., Vence, L. M., Perler, F. B., Benner, J., Kucera, R. B., Hirvonen, C. A., Pelletier, J. J., Paulus, H., and Xu, M. Q. (1997). Single-column purification of free recombinant proteins using a self-cleavable affinity tag derived from a protein splicing element. *Gene* **192,** 271–281.

Cottingham, I. R., Millar, A., Emslie, E., Colman, A., Schnieke, A. E., and McKee, C. (2001). A method for the amidation of recombinant peptides expressed as intein fusion proteins in *Escherichia coli*. *Nat. Biotechnol.* **19,** 974–977.

Dang, L. C., Melandri, F. D., and Stein, R. L. (1998). Kinetic and mechanistic studies on the hydrolysis of ubiquitin C-terminal 7-amido-4-methylcoumarin by deubiquitinating enzymes. *Biochemistry* **37,** 1868–1879.

Evans, T. C., Jr., Benner, J., and Xu, M. Q. (1998). Semisynthesis of cytotoxic proteins using a modified protein splicing element. *Protein Sci.* **7,** 2256–2264.

Fang, S., and Weissman, A. M. (2004). A field guide to ubiquitylation. *Cell Mol. Life Sci.* **61,** 1546–1561.

Fischer, J. A. (2003). Deubiquitinating enzymes: Their roles in development, differentiation, and disease. *Int. Rev. Cytol.* **229,** 43–72.

Galloway, C. A., Sowden, M. P., and Smith, H. C. (2003). Increasing the yield of soluble recombinant protein expressed in *E. coli* by induction during late log phase. *Biotechniques* **34,** 524–526, 528, 530.

Gan-Erdene, T., Nagamalleswari, K., Yin, L., Wu, K., Pan, Z. Q., and Wilkinson, K. D. (2003). Identification and characterization of DEN1, a deneddylase of the ULP family. *J. Biol. Chem.* **278,** 28892–28900.

Hatakeyama, S., and Nakayama, K. I. (2003). U-box proteins as a new family of ubiquitin ligases. *Biochem. Biophys. Res. Commun.* **302,** 635–645.

Hemelaar, J., Borodovsky, A., Kessler, B. M., Reverter, D., Cook, J., Kolli, N., Gan-Erdene, T., Wilkinson, K. D., Gill, G., Lima, C. D., Pliegh, H. L., and Ovaa H. (2004). Specific and covalent targeting of conjugating and deconjugating enzymes of ubiquitin-like proteins. *Mol. Cell Biol.* **24,** 84–95.

Huang, D. T., Walden, H., Duda, D., and Schulman, B. A. (2004). Ubiquitin-like protein activation. *Oncogene* **23,** 1958–1971.

Jentsch, S., and Pyrowolakis, G. (2000). Ubiquitin and its kin: How close are the family ties? *Trends Cell Biol.* **10,** 335–342.

Kim, J. H., Park, K. C., Chung, S. S., Bang, O., and Chung, C. H. (2003). Deubiquitinating enzymes as cellular regulators. *J. Biochem. (Tokyo)* **134,** 9–18.

Kim, K. I., Baek, S. H., and Chung, C. H. (2002). Versatile protein tag, SUMO: Its enzymology and biological function. *J. Cell Physiol.* **191,** 257–268.

Mayer, R. J., Landon, M., and Layfield, R. (1998). Ubiquitin superfolds: Intrinsic and attachable regulators of cellular activities? *Fold Des.* **3,** R97–99.

Murata, S., Chiba, T., and Tanaka, K. (2003). CHIP: A quality-control E3 ligase collaborating with molecular chaperones. *Int. J. Biochem. Cell Biol.* **35,** 572–578.

Ohsumi, Y., and Mizushima, N. (2004). Two ubiquitin-like conjugation systems essential for autophagy. *Semin. Cell Dev. Biol.* **15,** 231–236.

Parry, G., and Estelle, M. (2004). Regulation of cullin-based ubiquitin ligases by the Nedd8/RUB ubiquitin-like proteins. *Semin. Cell Dev. Biol.* **15,** 221–229.

Passmore, L. A., and Barford, D. (2004). Getting into position: The catalytic mechanisms of protein ubiquitylation. *Biochem. J.* **379,** 513–525.

Schwartz, D. C., and Hochstrasser, M. (2003). A superfamily of protein tags: Ubiquitin, SUMO and related modifiers. *Trends Biochem. Sci.* **28,** 321–328.

Soboleva, T. A., and Baker, R. T. (2004). Deubiquitinating enzymes: Their functions and substrate specificity. *Curr. Protein Pept. Sci.* **5,** 191–200.

Staub, O. (2004). Ubiquitylation and isgylation: Overlapping enzymatic cascades do the job. *Sci. STKE.* **245,** pe43.

Stein, R. L., Chen, Z., and Melandri, F. (1995). Kinetic studies of isopeptidase T: Modulation of peptidase activity by ubiquitin. *Biochemistry* **34,** 12616–12623.

Wilkinson, K. D. (2000). Ubiquitination and deubiquitination: Targeting of proteins for degradation by the proteasome. *Semin. Cell Dev. Biol.* **11,** 141–148.

Yeh, E. T., Gong, L., and Kamitani, T. (2000). Ubiquitin-like proteins: New wines in new bottles. *Gene* **248,** 1–14.

[4] Preparation, Characterization, and Use of Tagged Ubiquitins

By JUDY CALLIS and RICHARD LING

Abstract

The low abundance and heterogeneity of ubiquitinated proteins has led to the development and use of tagged forms of ubiquitin. Additional residues present at the ubiquitin amino terminus provide immunological and/or affinity sites to facilitate visualization, identification, and purification of ubiquitinated substrates by virtue of their covalent attachment to the tagged ubiquitin. The use of tagged ubiquitin to understand the scope, nature, and biological relevance of this conserved modification system has been demonstrated in multiple ways. Unknown substrates can be identified, or a previously identified substrate can be analyzed with tagged ubiquitin *in vitro* or *in vivo* to determine the specificity, regulation, and type of ubiquitin linkages formed. This contribution describes the generation and use of multiple types of modified ubiquitins: biotinylated ubiquitin produced *in vitro*, or GST-, myc-, HA-, and hexahistidine-tagged ubiquitins produced *in vivo*.

Introduction

The use of a peptide "tag" to expedite the purification of an introduced protein is a familiar experimental approach (Kolodziej and Young, 1991). A tag, consisting of additional amino acids placed in the open reading frame, provides either an immunological or affinity binding site that can be used to purify the tagged protein with either antibodies or affinity resins, respectively. This approach has been extended further to isolate other molecules associated with the tagged protein in a noncovalent complex (for example, coimmunoprecipitation criteria for complex formation). The covalent modification of proteins by small protein modifiers such as ubiquitin provides yet another variation on isolation of associated proteins using tags. Here, the tagged protein (ubiquitin) is covalently attached to substrates of the ubiquitin pathway, either *in vivo* or *in vitro*, and the ubiquitinated substrates are enriched for by virtue of the tag on ubiquitin.

Immunological tags, such as influenza virus hemagglutinin (HA) (Treier et al., 1994) and myc (Ellison and Hochstrasser, 1991) epitopes that are recognized by commercially available monoclonal antibodies (Kolodziej and Young, 1991) have been used to tag ubiquitin, and HA-Ub and myc-Ub are competent to conjugate to proteins *in vivo* and *in vitro*. In addition, affinity tags such as biotin (Corsi et al., 1995; Mitsui and Sharp, 1999), glutathione-*S*-transferase (GST) (Scheffner et al., 1993), and hexahistidine (Beers and Callis, 1993; Ling et al., 1998; Treier et al., 1994) have been successfully used as N-terminal tags with ubiquitin.

For *in vitro* assays, many of the tagged ubiquitin proteins mentioned here are available commercially (Boston Biochem Inc., Cambridge MA, BIOMOL International LP, Plymouth Meeting, PA, and Calbiochem, San Diego CA-GST-Ub and His$_6$Ub only), but all can be produced *in situ*.

Tagged Ubiquitins

Biotin is added *in vitro* to purified ubiquitin (Corsi et al., 1995), using commercially available modification protocols (Pierce Biotechnology, Inc, Rockford IL), precluding the general use of biotinylated ubiquitin for *in vivo* conjugation. Because biotinylation modifies lysyl -amino groups, seven of which are potential sites for ubiquitin-ubiquitin linkages (Peng et al., 2003), care should be taken to determine the extent of modification. Several ubiquitin lysyl -amino groups, notably Lys-48, -63, and -11 are the major sites of ubiquitin chain formation (Arnason and Ellison, 1994; Chau et al., 1989; Peng et al., 2003; Spence et al., 1995), and biotinylation at these sites could block polyubiquitylation of substrates.

All other tagged ubiquitins mentioned previously are produced by ribosome-catalyzed translation. Codons for these tags added to a ubiquitin open reading frame (ORF) must be placed upstream of the amino terminus of the ubiquitin open reading frame, leaving the carboxy-terminus intact for activation by ubiquitin E1, thioester formation, and finally isopeptide linkage to substrates [Fang and Weissman (2004) and articles in the same volume for reviews]. The amino acid sequence encoding the HA-, myc-, and His$_6$-epitopes is short and can be added to ubiquitin ORFs using PCR and standard cloning procedures. GST is a 26-kDa protein and is most easily added by ligating a ubiquitin ORF into a commercially available expression vector such as pGEX-2T and pGEX-2TK (Amersham Biosciences Piscataway, NJ). pGEX-2TK has been used for synthesis of a GST-Ub (Scheffner et al., 1993) ORF by introduction of an in-frame ubiquitin ORF into the *Bam*HI/*Eco*RI sites of pGEX-2TK. This vector also contains a cAMP-dependent phosphorylation site. GST-Ub is produced in *Escherichia coli*, purified on glutathione agarose, and then added to conjugation reactions. After incubation to allow conjugation, ubiquitinated proteins can be purified on glutathione agarose or visualized immunologically using anti-GST antibodies (Santa Cruz Biotechnology, Santa Cruz, CA). Alternately, GST-Ub can be labeled with ^{32}P using PKA (Protein Kinase A, Sigma Chemical Co., St. Louis, MO) and $[\gamma\text{-}^{32}P]$-ATP while bound to glutathione agarose, and after removal of unincorporated nucleotides, released from the beads. Proteins ubiquitinated with ^{32}P-GST-Ub can be visualized after SDS-PAGE and autoradiography (Scheffner et al., 1993).

His$_6$Ub, as referred to here, designates a protein containing an initiator methionine residue immediately followed by six histidine residues in front of a complete ubiquitin including its own initiator methionine. His$_6$Ub is a versatile tag. His$_6$Ub can be expressed in and purified from *E. coli* (see below) and used for *in vitro* conjugation, but also, like myc-Ub and HA-Ub, His$_6$Ub can be expressed in eukaryotic cells from introduced genes, where it conjugates to proteins *in vivo*. What distinguishes His$_6$Ub from HA-Ub and myc-Ub is that, although affinity purification under native conditions can be performed for all three tagged ubiquitins, only His$_6$Ub can be purified in the presence of strong protein denaturants, because binding to nitriloacetic agarose (Ni-NTA) affinity resins is retained under these conditions. Strong protein denaturants can remove cellular proteins noncovalently associated with His$_6$Ub, substrates and/or the affinity resin. Strong denaturants also provide for more efficient solubilization of proteins, improving yield. For large-scale enrichments, Ni-NTA resin is more economical than immunoprecipitations. Finally, an antibody to the hexahistidine moiety has been developed (Qiagen, Chatsworth, CA), allowing the visualization of His$_6$Ub and its conjugates, although in our hands, this method of detection is not nearly as sensitive as visualization with anti-myc or anti-HA antibodies. Thus, His$_6$Ub can be use for a variety of purposes and His$_6$Ub-ubiquitylated proteins visualized or purified under different conditions.

Some background binding to Ni-NTA can remain even under denaturing conditions from histidine-rich and/or Ni+ binding proteins. For this reason, the presence of a combination of tags such as myc together with hexahistidine may be advantageous (Spence et al., 1995). The advantages of myc-Ub and HA-Ub are the ability to express these *in vivo*, and the existence of very sensitive detection with monoclonal antibodies, 9E10 and 12CA5, respectively (Roche Molecular Biochemicals, Indianapolis, IN). If a combination of His$_6$ and myc or HA is preferred, we recommend the combination of His$_6$ and the myc epitope as reported (Peng et al., 2003) rather than His$_6$ and the HA epitope, because we have observed HA-immunoreactive proteins from yeast retained nonspecifically on Ni-NTA (unpublished observations).

Considerations for Construction of Tagged Ubiquitin Expression Constructs

Because ubiquitin is highly conserved and cross-kingdom proteins seem to function equivalently (Ling *et al.*, 1998; Wilkinson *et al.*, 1986), any ubiquitin ORF can serve as a template for creation of HA-, myc-, and His$_6$Ub. The important considerations for construction of expression cassettes are the inclusion/exclusion of codons for amino acids C-terminal to the codon for glycine-76 and whether to express tagged monomeric or polymeric ubiquitin. Many ubiquitin ORFs encode additional amino acids C-terminal to gly-76 at the end of the ORF and additionally are encoded as protein fusions (Callis and Vierstra, 1989; Lee *et al.*, 1988; Schlesinger and Bond, 1987; Swindle *et al.*, 1988; Wiborg *et al.*, 1985), In eukaryotic cells, ubiquitin fusions and the additional C-terminal amino acids are quickly cleaved by endogenous ubiquitin-specific proteases, releasing monomer ubiquitin with an intact C-terminus essential for ubiquitylation (Bachmair *et al.*, 1986). For *E. coli* expression, unless a ubiquitin-specific protease is coexpressed or added to the lysate (Catanzariti *et al.*, 2004), a monomeric tagged ubiquitin ORF terminating after the codon for gly-76 should be used. However, for expression in any eukaryotic organism, either monomeric or polymeric forms can be synthesized, and the additional amino acids included, because endogenous ubiquitin-specific proteases will process these precursors to the active monomer.

The expression of multimeric His$_6$Ub, encoding multiple His$_6$Ub proteins in a single ORF, allows for a higher level of protein production from a single expression unit. However, construction of such an ORF is challenging, because the endogenous ubiquitin-specific proteases do not tolerate substitutions at the ubiquitin C-terminus. Adopting a strategy used to produce multimeric ubiquitin (Bachmair *et al.*, 1990), we describe a method to construct a multimeric His$_6$Ub ORF.

To be a successful visualization and purification aid, the tagged ubiquitin must have the following properties: (1) It must bind to a purification reagent, either an antibody or affinity resin; (2) it must be competent to conjugate to proteins; (3) when conjugated to proteins, it must retain its ability described in No. 1; and (4) it must conjugate to the same subset of intracellular proteins as wild-type protein modifier. Although the last property is impossible to prove *en toto*, these criteria have been fulfilled for myc-Ub (Ellison and Hochstrasser, 1991), His$_6$Ub (Ling *et al.*, 1998), and myc-His$_6$Ub (Peng *et al.*, 2003).

This chapter describes the binding properties of His$_6$Ub on Ni-NTA resins after expression in *E. coli*. This chapter also describes the synthesis of an ORF for multimeric His$_6$Ub that can be expressed in eukaryotic cells for *in vivo* conjugation or expressed in *E. coli* for use as a substrate for deubiquitylating enzymes (Rao-Naik *et al.*, 2000). This chapter also describes expression of His$_6$Ub and purification of His$_6$Ub protein conjugates from a eukaryotic cell, *Saccharomyces cerevisiae*, to identify ubiquitinated proteins by immunological methods. His$_6$Ub, conjugated to proteins *in vivo* can also be used to identify previously unknown substrates of the ubiquitin pathway (see other issues in this volume for proteomic uses of tagged ubiquitin).

Production of a His$_6$Ub Open Reading Frame for Expression in E. coli or for Multimeric Expression in Eukaryotic Cells

Although there are commercially available vectors with hexahistidine-encoding nucleotides followed by restriction endonuclease sites to facilitate cloning ORFs in-frame, many contain additional nucleotides that add amino acids to the ORF. These extra amino acids contribute either an immunological epitope or seem to be residual from vector construction. We recommend minimizing the number of amino acids added to the N-terminus of ubiquitin. To construct such an ORF, the specific amino acids desired can be easily added using PCR-based methods by including the additional codons in the primer sequence with appropriate restriction enzyme sites at the primer termini for cloning into standard expression vectors. The template can be any ubiquitin-coding region isolated from any organism. Alternately, a ubiquitin ORF can be assembled from chemically synthesized oligonucleotides (Ecker *et al.*, 1987), with insertion of double-stranded oligonucleotide containing the codons for the hexahistidine tag.

In the following is our approach for creation of a multimeric His$_6$Ub-encoding ORF in which every ORF produces His$_6$Ub based on a previously described method (Bachmair *et al.*, 1990). The cassette is flexible, so that a different number of ORFs can be included; the optimal number for maximal expression can be determined empirically, and the resulting multimeric His$_6$Ub-encoding ORF can be easily cloned into a mammalian, fungal, or plant expression vector.

The coding region for His$_6$Ub was created by PCR amplification of a ubiquitin ORF (lacking translation start and stop codons) from *Arabidopsis thaliana* with a 5' primer of sequence
5'-CG<u>GAATTC</u>CT<u>GCGCA</u>T**CACCATCACCATCAC**ATGCAGAT CTTCGTAAAG-3', with the underlined sequences representing sites for *Eco*RI, and *Fsp*I, left to right, required for the cloning process, and the double underline representing the codons for six histidines (note that multiple codons are used to minimize repeated sequences; be sure to check codon use for host organism); and a 3'-primer of sequence 5'-GC<u>GTCGAC</u>AA

GGGTTGATATCCCACCACGGAGACGGAGG-3' with the underlined sequences representing sites for *Sal*I and *Eco*RV, respectively. For amplification of other ubiquitin coding regions, the nucleotides that are indicated in bold for both primers have to be changed to match the ubiquitin template used for PCR. The *Eco*RI and *Sal*I sites in the 5' and 3' primer sequence, respectively, were used to ligate the PCR fragment into a basic cloning plasmid containing those sites, such as pUC18 (Vieira and Messing, 1982). After confirmation of the DNA sequence, the resulting plasmid, pHub1, was digested with *Fsp*I and *Sal*I, and the His_6Ub ORF-containing fragment ligated into the same plasmid digested with *Eco*RV and *Sal*I, creating pHub2. The former restriction enzymes of each set create blunt ends that when ligated together create codons for the exact junction between two ubiquitin ORFs (Fig. 1). This process can be repeated with pHub2, ligating a His_6Ub_2 ORF-containing *Fsp*I/*Sal*I fragment from pHub2 into pHub2 digested with *Eco*RV and *Sal*I, creating pHub4. Alternately, an *Fsp*I/*Sal*I fragment from pHub1 can be ligated into pHub2 digested with *Eco*RV and *Sal*I, creating pHub3. Plasmids encoding a larger number of His_6Ub repeats can be created by the same process using these plasmids: pHub1, pHub2, pHub3, pHub4, encoding 1, 2, 3, and 4 His_6Ub repeats, respectively.

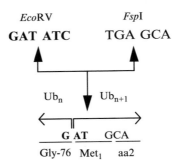

FIG. 1. The *Eco*RV/*Fsp*I junction critical to the construction of tagged Ub-Ub repeats. Nucleotides corresponding to the *Eco*RV recognition site are in bold (left) to distinguish them from those of the *Fsp*I recognition site (right), and the sites of phosphodiester bond cleavage by the restriction enzymes are indicated by the upward arrowheads. Lower, The product after ligation is shown with the predicted amino acid sequence and their relationship to the ubiquitin coding region. The second amino acid of the repeat Ub_{n+1} is denoted by aa2, which may be either histidine (in His_6Ub) or glutamine (as found in wild-type ubiquitin), depending on the next nucleotide 3' of the *Fsp*I site (not shown). Codons for both histidine and glutamine begin with CA- (followed by U or C for His, A or G for Gln). This method of creating authentic Ub–Ub junctions was first devised by Bachmair et al. (1990).

When the desired number of His_6Ub repeats have been produced, translation initiator and terminator sequences have to be added to the termini. The initiator sequences were introduced by cleaving any of the His_6Ub-containing plasmids with *Eco*RI/*Fsp*I and ligating to those sites a double-stranded oligonucleotide (the 5′ adaptor) of sequence 5′-AATTC<u>GGATCC</u>AATTACTATTTACAATTACA<u>CATAT</u>-3′ with a 5′ AATT overhang and a 3′ blunt end (created by independent synthesis of two primers followed by annealing the two together before adding them to the ligation reaction). The underlined sequences from left to right represent a *Bam*HI recognition site for subsequent cloning and a segment, which, on ligation with the *Fsp*I blunt end, becomes part of a *Nde*I site. The 3′ AT nucleotides when ligated to the 3′ half of a *Fsp*I site, GCA, generate the initiator ATG codon. This exact 5′ adapter can be used for any initial template; however, the sequences between the two restriction sites can be modified to optimize translation/mRNA stability for a specific organism. To complete the ORF at the C-terminus, a second double-stranded oligonucleotide (the 3′ adapter) of sequence 5′-TCGACTGAT-CA <u>GGTACC</u>**TCA** *GAA G*- 3′ with a 5′ TCGA overhang and a blunt 3′ end, encoding a translation terminator is ligated at the *Eco*RV/*Sal*I sites of a plasmid containing the ORF and the 5′ adapter. Underlined sequences represent *Bcl*I and *Kpn*I sites downstream of the ORF, either one of which can be used for ligating the ORF into expression plasmids. This exact oligonucleotide can also be used for any initial template, but downstream sites can be modified for subsequence cloning purposes. The anticodon for translation termination is in bold type. The last His_6Ub ORF has to include one or more additional amino acids after glycine-76 because of the method used to create the multimers. One nucleotide must be added to complete the last codon; here, codons for two amino acids were added (in italic type). But as described previously, these amino acids are quickly removed *in vivo* in eukaryotic cells by ubiquitin-specific proteases (Baker *et al.*, 1992; Chandler *et al.*, 1997; D'Andrea and Pellman, 1998; Yan *et al.*, 1997). The complete His_6Ub_n coding region cassette containing a translation start and stop can then be ligated into any expression cassette containing a promoter and poly A addition site, using *Eco*R1, *Bam*HI, or *Nde*I at the 5′ end and *Kpn*I, *Bcl*I, or *Sal*I at the 3′ end.

Behavior of His_6Ub on Ni-NTA Resins

To maximize the usefulness of His_6Ub as a purification tag, the elution properties of this protein were examined in detail under both native and denaturing conditions. Elution from a nickel-chelating column can be accomplished by reduction in pH or by imidazole competition (Hochuli *et al.*, 1987). When an *E. coli* lysate from cells expressing His$_6$Ub was applied to a Ni-NTA column (Ni-NTA Superflow, Qiagen) and challenged sequentially with PT buffers (100 mM sodium phosphate, 10 mM Tris) of pH 7.0, 6.5, 6.0, 5.5, 5.0, and 4.5, most of the His$_6$Ub remains bound to the column until PT buffer at pH 4.5 is applied, with a trace amount of eluted His$_6$Ub at PT pH 5.0 (Table I). Alternately, when bound His$_6$Ub was challenged with competition using 10 mM to 120 mM imidazole, the 10 mM fraction contained numerous endogenous bacterial proteins but was devoid of His$_6$Ub. The major His$_6$Ub peak appeared at 80 mM imidazole, although some also eluted at 20 mM and 40 mM imidazole as well. The amount of His$_6$Ub dislodged by 20 mM imidazole was not significant, but the amount freed by 40 mM imidazole was.

The same elution experiments for His$_6$Ub bound to Ni-NTA agarose were performed under denaturing conditions. *E. coli* cells expressing His$_6$Ub were lysed in a buffer of 100 mM sodium phosphate, 10 mM Tris, and 8 M urea (buffer PTU) and loaded onto a Ni-NTA column at pH 8.0. When challenged with the same stepwise gradient of decreasing pH, denatured His$_6$Ub eluted earlier than the native protein at pH 5.5 rather than pH 4.5, with substantially lower amounts of His$_6$Ub in the pH 6.0 and pH 5.0 fractions as well (Table I). In the denaturing PTU buffer, His$_6$Ub's elution behavior under imidazole competition was identical to its behavior under native conditions (Fig. 2).

From these data, it was determined that optimal elution under both native and denaturing conditions could be achieved by loading the His$_6$Ub-containing sample at pH 8.0, followed by an intermediate wash at 10–20 mM imidazole to remove background proteins, followed by elution at 80 mM imidazole. An identical elution profile could also be achieved with a pH elution method by washing at pH 6.3 and eluting at pH 4.5 (summarized in Table I). These same binding and elution properties were also observed for His$_6$Ub-modified proteins.

TABLE I
CHROMATOGRAPHIC PROPERTIES OF His$_6$Ub ON NI-NTA[a]

Ubiquitin	Buffer	Stepwise pH gradient						Stepwise mM imidazole				
		7.0	6.5	6.0	5.5	5.0	4.5	10	20	40	80	120
His$_6$Ub	PT	−	−	−	−	+/−	++	−	+/−	+	++	−
His$_6$Ub	PTU	−	−	+/−	++	+/−	−	−	+/−	+	++	−

[a] Buffer PT is 100 mM sodium phosphate, 10 mM Tris, and buffer PTU is buffer PT with 8 M urea. "−" represents no His$_6$Ub protein eluted, "+" and "++" represent a significant amount eluted, and "+/−" represents trace amount eluted, all as detected by Coomassie blue staining. Unless stated otherwise, buffers were used at pH 8.0.

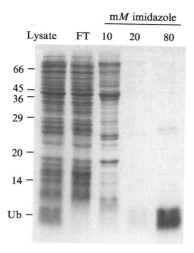

FIG. 2. Chromatographic properties of His$_6$Ub. *E. coli* cells expressing His$_6$Ub were disrupted by sonication in buffer PTU (see text), clarified by centrifugation, the clarified lysate loaded onto a gravity flow Ni-NTA column (Qiagen), and eluted sequentially with 3–5 × bed volumes of PT buffer with 10, 20, and 80 mM imidazole. Fractions were precipitated with 6 volumes of acetone, redissolved in 10 mM Tris, 1% SDS, pH 7.5, and analyzed by 15% SDS-PAGE and Coomassie blue staining.

Creation of a Yeast Strain Encoding His$_6$Ub as the Only Source of Ubiquitin

To increase the yield of His$_6$Ub-containing conjugates from eukaryotic cells, either the level of protein produced from His$_6$Ub-expressing cassettes can be increased or the level of competing nontagged ubiquitin can be decreased. His$_6$Ub may compete poorly compared with wild-type ubiquitin when both are present (M. Hochstrasser, personal communication, 2004). Thus, the most effective *in vivo* use of His$_6$Ub in yeast is as the sole source of ubiquitin. Yeast strains created by D. Finley and colleagues (Finley *et al.*, 1987, 1994) were used to create a strain with His$_6$Ub as the sole source of ubiquitin. However, in other eukaryotes, where disruption of endogenous genes is not so facile and the number of ubiquitin-encoding genes large, the use of multimeric forms of tagged ubiquitin may be very important to increase expression levels.

In the yeast *Saccharomyces cerevisiae*, ubiquitin is encoded by a family of four genes (Finley *et al.*, 1987). Whereas *UBI1*, *UBI2*, and *UBI3* encode both ubiquitin ribosomal protein fusions, *UBI4* encodes a polyubiquitin gene (Finley *et al.*, 1987, 1989). In yeast strain SUB288, the entire family of four ubiquitin genes was inactivated by gene replacement. Because ubiquitin is essential, cell viability was maintained by a plasmid-encoded *UBI4* gene containing the *URA3* marker (Finley *et al.*, 1987), and ubiquitin-independent expression of the essential ribosomal proteins was achieved by means of integration of DNA sequence encoding the ribosomal extension proteins into the genome as tandem repeats or by means of expression from a strong promoter (Finley *et al.*, 1989). Transformation of SUB288 by a His$_6$Ub-expressing plasmid, using the *LYS2* marker for positive selection, followed by negative selection with 5-fluroorotic acid (Guthrie and Fink, 1991) against the endogenous *UBI4*-bearing plasmid produces a cell in which the only functional ubiquitin is a His$_6$Ub (Ling *et al.*, 1998; Peng *et al.*, 2003). Tagged ubiquitin vectors for expression in yeast use YEp96 containing the "cassette adapted" ubiquitin under control of the *CUP1* promoter (Ecker *et al.*, 1987; Ellison and Hochstrasser, 1991) as the backbone. The wild-type ubiquitin ORF was replaced by a His$_6$Ub ORF (created as described previously) using *Eco*R1 and *Kpn*I restriction sites that border both the His$_6$Ub ORF and the wild-type ubiquitin ORF in YEp96. A 5.4-kb *Cla*I fragment encoding the *LYS2* selection marker from pUB148 (Finley *et al.*, 1994) was inserted into the *Cla*I site of Yep96 and its derivatives for plasmid selection.

Alternately, SUB280 can be used as the starting host strain (Finley *et al.*, 1994). SUB280 has the same chromosomal background as SUB288, but the plasmid encoding ubiquitin carries the *LYS2* marker. In this case, *URA3*-based plasmids expressing tagged ubiquitin are introduced, followed by negative selection with α-aminoadipic acid to select against retention of the *LYS2*-containing plasmid.

Ni-NTA Enrichment of Ubiquitinated Proteins from Yeast Using His$_6$Ub

For this example, yeast cultures were grown beyond the log phase in standard enriched yeast media with 2% dextrose, and cells were harvested by centrifugation at 5000*g* for 10 min at 4°. Cell pellets were washed once with PT buffer at pH 7.2 supplemented with 5 mM N-ethylmaleimide (NEM) to inhibit the ubiquitin-specific proteases that are capable of cleaving ubiquitin from ubiquitinated proteins *in vitro*. Washed pellets were resuspended in PTU buffer at pH 7.2, supplemented with 5 mM NEM (Sigma) and 1 mM PMSF (phenyl-methane-sulfonyl fluoride, Sigma). Lysis was achieved by agitation with glass beads, and bulk protein was precipitated from the clarified lysate with three volumes of saturated ammonium sulfate. This step removed cellular debris and other insoluble material that proved problematic during chromatography. The NH$_4$SO$_4$ pellet was then dissolved in PTU buffer, the pH adjusted to 8.0 before loading onto a Ni-NTA column, and then after loading, the column was rinsed with PTU buffer until protein concentration as indicated by A_{280} returned to baseline. Subsequent buffers contained reduced urea, 2 M urea instead of the 8 M present during lysis. This reduction did not affect binding or elution but reduced the impact on FPLC pumps. Weakly bound proteins were collected in a wash fraction containing 10 mM imidazole in PTU$_{2M}$ buffer, again rinsing with the same buffer until A_{280} returned to baseline. A fraction enriched in His$_6$Ub-conjugated proteins was obtained by elution with 80 mM imidazole in PTU$_{2M}$ buffer. Alternately, an identical elution profile could be obtained by washing with PTU$_{2M}$ buffer at pH 6.3 until A_{280} returned to baseline and eluting with the same buffer at pH 4.5.

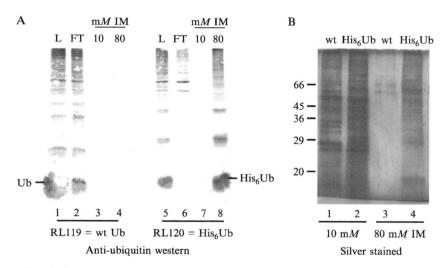

FIG. 3. Enrichment of ubiquitylated proteins from *Saccharomyces cerevisiae* expressing His$_6$Ub. (A) Yeast strains expressing only His$_6$Ub, or untagged ubiquitin in the same expression vector, RL120 and RL119 (Ling *et al.*, 1998), respectively, were lysed, loaded onto a Ni-NTA column, the column washed with 3–5× bed volume with 10 mM and 80 mM imidazole. Equivalent portions of each fraction were separated by SDS-PAGE, transferred to a membrane, and reacted with antiubiquitin antibodies. Reprinted by permission. L, lysate; FT, flow through. (B) The 10 mM and 80 mM fractions from RL119 (wt) and RL120 (H) shown in (A) were separated by SDS-PAGE and total protein stained with silver reagent (Invitrogen, Carlsbad CA). IM, imidazole. Molecular weight standards in kDa are indicated to the left in (B). (A) Reprinted from *Analytical Biochemistry*, Vol. 282, Ling, R., Colon, E., Dahmus, M. E. and Callis, J., Histidine-tagged ubiquitin substitutes for wild-type ubiquitin in *Saccharomyces cerevisiae* and facilitates isolation and identification of *in vivo* substrates of the ubiquitin pathway, page No., 54–64, copyright 1999, with permission from Elsevier.

Fractions from the Ni-NTA column were concentrated by ultrafiltration and stored at $-20°$ in the PTU$_{2M}$ buffer. Alternately, proteins in each fraction were precipitated with six volumes of acetone and redissolved in 10 mM Tris, 1% SDS, pH 7.5. Two percent of each eluted fraction was loaded onto a 15% SDS-PAGE and analyzed with antiubiquitin Western blot demonstrating enrichment of ubiquitinated proteins (Fig. 3A). Enrichment of a significant amount of protein as well was seen only in the strain expressing His$_6$Ub (Fig. 3B).

Acknowledgments

The authors would like to thank D. Finley and M. Hochstrasser for useful information. Research on tagged ubiquitin is supported by NSF (IBN-0212659) and NSF 2010 (MSB-00115870).

References

Arnason, T., and Ellison, M. J. (1994). Stress resistance in *Saccharomyces cerevisiae* is strongly correlated with assembly of a novel type of multiubiquitin chain. *Mol. Cell. Biol.* **14**, 7876–7883.

Bachmair, A., Becker, F., Masterson, R. V., and Schell, J. (1990). Perturbation of the ubiquitin system causes leaf curling, vascular tissue alteration and necrotic lesions in a higher plant. *EMBO J.* **9**, 4543–4549.

Bachmair, A., Finley, D., and Varshavsky, A. (1986). *In vivo* half-life of a protein is a function of its amino-terminal residue. *Science* **234**, 179–186.

Baker, R. T., Tobias, J. W., and Varshavsky, A. (1992). Ubiquitin-specific proteases of *Saccharomyces cerevisiae*. *J. Biol. Chem.* **267**, 23364–23375.

Beers, E., and Callis, J. (1993). Utility of polyhistidine-tagged ubiquitin in the purification of ubiquitin-protein conjugates and as an affinity ligand for the purification of ubiquitin-specific hydrolases. *J. Biol. Chem.* **268**, 21645–21649.

Callis, J., and Vierstra, R. D. (1989). Ubiquitin and ubiquitin genes in higher plants. *Oxford Surv. Plant Mol. Biol.* **6**, 1–30.

Catanzariti, A., Soboleva, T., Jans, D., Board, P., and Baker, R. (2004). An efficient system for high-level expression and easy purification of authentic recombinant proteins. *Protein Science* **13**, 1331–1339.

Chandler, J., McArdle, B., and Callis, J. (1997). *At*UBP3 and *At*UBP4 are two closely related *Arabidopsis thaliana* ubiquitin-specific proteases present in the nucleus. *Mol. Gen. Genet.* **255**, 302–310.

Chau, V., Tobias, J., Bachmair, A., Marriott, D., Ecker, D., Gonda, D., and Varshavsky, A. (1989). A multiubiquitin chain is confined to specific lysine in a targeted short-lived protein. *Science* **243**, 1576–1583.

Corsi, D., Galluzzi, L., Crinelli, R., and Magnani, M. (1995). Ubiquitin is conjugated to the cytoskeletal protein alpha-spectrin in mature erythrocytes. *J. Biol. Chem.* **270**, 8928–8935.

D'Andrea, A., and Pellman, D. (1998). Deubiquitinating enzymes: A new class of biological regulators. *Crit. Rev. Biochem. Mol. Biol.* **33**, 337–352.

Ecker, D. J., Khan, M. I., Marsh, J., Butt, T. R., and Crooke, S. T. (1987). Chemical synthesis and expression of a cassette adapted ubiquitin gene. *J. Biol. Chem.* **262**, 3524–3527.

Ellison, M. J., and Hochstrasser, M. (1991). Epitope-tagged ubiquitin. *J. Biol. Chem.* **266,** 21150–21157.
Fang, S., and Weissman, A. (2004). A field guide to ubiquitylation. *Cell. Mol. Life Sci.* **61,** 1546–1561.
Finley, D., Bartel, B., and Varshavsky, A. (1989). The tails of ubiquitin precursors are ribosomal proteins whose fusion to ubiquitin facilitates ribosome function. *Nature* **338,** 394–401.
Finley, D., Ozkaynak, E., and Varshavsky, A. (1987). The yeast polyubiquitin gene is essential for resistance to high temperatures, starvation and other stresses. *Cell* **48,** 1035–1046.
Finley, D., Sadis, S., Monia, B. P., Boucher, P., Ecker, D. J., Crooke, S. T., and Chau, V. (1994). Inhibition of proteolysis and cell cycle progression in a multiubiquitination-deficient yeast mutant. *Mol. Cell. Biol.* **14,** 5501–5509.
Guthrie, C., and Fink, G. (1991). Guide to yeast genetics and molecular biology. *In* "Methods in Enzymology" (J. Abelson and M. Simon, eds.), Vol. 194, pp. 302–318. Academic Press, Pasadena.
Hochuli, E., Dobeli, H., and Schacher, A. (1987). New metal chelate adsorbent selective for proteins and peptide containing neighboring histidine residues. *J. Chromatogr.* **411,** 177–184.
Kolodziej, P., and Young, R. A. (1991). Epitope tagging and protein surveillance. *Methods Enzymol.* **194,** 508–519.
Lee, H., Simon, J. A., and Lis, J. T. (1988). Structure and expression of ubiquitin genes of *Drosophila melanogaster*. *Mol. Cell. Biol.* **8,** 4727–4735.
Ling, R. L., Colon, E., Dahmus, M. E., and Callis, J. (1998). Histidine-tagged ubiquitin substitutes for wild-type ubiquitin, identifies RNAP II large subunit as an *in vivo* ubiquitin pathway substrate, and facilitates purification of ubiquitinated proteins from *Saccharomyces cerevisiae* and *Arabidopsis thaliana*. *Anal. Biochem.* **282,** 54–64.
Mitsui, A., and Sharp, P. A. (1999). Ubiquitination of RNA polymerase II large subunit signaled by phosphorylation of carboxyl-terminal domain. *Proc. Natl. Acad. Sci. USA* **96,** 6054–6059.
Peng, J. M., Schwartz, D., Elias, J. E., Thoreen, C. C., Cheng, D. M., Marsischky, G., Roelofs, J., Finley, D., and Gygi, S. P. (2003). A proteomics approach to understanding protein ubiquitination. *Nature Biotech.* **21,** 921–926.
Rao-Naik, C., Chandler, J. S., McArdle, B., and Callis, J. (2000). Ubiquitin-specific proteases from *Arabidopsis thaliana*: Cloning of AtUBP5 and analysis of substrate specificity of AtUBP3, AtUBP4, and AtUBP5 using *Escherichia coli in vivo* and *in vitro* assays. *Arch. Bioch. Biophys.* **379,** 198–208.
Scheffner, M., Huibregtse, J. M., Vierstra, R. D., and Howley, P. M. (1993). The HPV-16 E6 and E6-AP complex functions as a ubiquitin-protein ligase in the ubiquitination of p53. *Cell* **75,** 495–505.
Schlesinger, M. J., and Bond, U. (1987). Ubiquitin genes. *Oxford Surv. Eukaryotic Genes* **4,** 77–89.
Spence, J., Sadis, S., Haas, A. L., and Finley, D. (1995). A ubiquitin mutant with specific defects in DNA repair and multiubiquitination. *Mol. Cell. Biol.* **15,** 1265–1273.
Swindle, J., Ajioka, J., Eisen, H., Sanwai, B., Jacquemot, C., Browder, Z., and Buck, G. (1988). The genomic organization and transcription of the ubiquitin genes of *Trypanosoma cruzi*. *EMBO J.* **7,** 1121–1127.
Treier, M., Staszewski, L. M., and Bohmann, D. (1994). Ubiquitin-dependent c-Jun degradation *in vivo* is mediated by the δ domain. *Cell* **78,** 787–798.
Vieira, J., and Messing, J. (1982). The pUC plasmids an M13mp7-derived system for insertion mutagenesis and sequencing with synthetic universal primers. *Gene* **19,** 259–268.

Wiborg, O., Pedersen, M. S., Wind, A., Berglund, L. E., Marckcker, K. A., and Vuust, J. (1985). The human ubiquitin multigene family: Some genes contain multiple directly repeated ubiquitin coding sequences. *EMBO J.* **4,** 755–759.

Wilkinson, K. D., Cox, M. J., O'Conner, L. B., and Shapira, R. (1986). Structure and activities of a variant ubiquitin sequence from baker's yeasts. *Biochemistry* **25,** 4999–5004.

Yan, N., Falbel, T. G., Thoma, S., Walker, J., and Vierstra, R. D. (1997). A large family of ubiquitin-specific proteases in *Arabidopsis* potentially regulating cell growth and development. *Plant Physiol.* **114,** 318.

[5] Knocking out Ubiquitin Proteasome System Function *In Vivo* and *In Vitro* with Genetically Encodable Tandem Ubiquitin

By Y. SAEKI, E. ISONO, M. SHIMADA, H. KAWAHARA, H. YOKOSAWA, and A. TOH-E

Abstract

At present, the 26S proteasome–specific inhibitor is not available. We constructed polyubiquitin derivatives that contained a tandem repeat of ubiquitins and were insensitive to ubiquitin hydrolases. When these artificial polyubiquitins (tUbs, tandem ubiquitins) were overproduced in the wild-type yeast strain, growth was strongly inhibited, probably because of inhibition of the 26S proteasome. We also found that several substrates of the ubiquitin-proteasome pathway were stabilized by expressing tUbs *in vivo*. tUbs containing four units or more of the ubiquitin monomer were found to form a complex with the 26S proteasome. We showed that tUb bound to the 26S proteasome inhibited the *in vitro* degradation of polyubiquitinylated Sic1 by the 26S proteasome. When *tUB6* (six-mer) messenger RNA was injected into *Xenopus* embryos, cell division was inhibited, suggesting that tUb can be used as a versatile inhibitor of the 26S proteasome.

Introduction

Several types of low molecular weight inhibitors of the proteasome have been available (Lee and Goldberg, 1998; Tsubuki *et al.*, 1992). The most widely used are peptide aldehydes, such as CDZ-Leu-Leu-Leucinal (MG132). This inhibitor is incorporated into the lumen of the 20S proteasome, and aldehyde moiety interacts reversibly with the catalytic threonine residue of the proteasome. Lactacystin (Omura *et al.*, 1991) inhibits the proteasome by covalently linking with the hydroxyl groups of the active site threonine (Fenteany *et al.*, 1995). This inhibitor shows high specificity toward the proteasome. Another inhibitor functioning by a similar mechanism to that of lactacystin is vinylsufone (Bogyo *et al.*, 1997). All these inhibitors are valuable tools for revealing the proteasome participation in various phenomena in cultured cells or tissues. However, it is impossible to deliver these drugs to a specific site or tissue or organ to inhibit the proteasome activity at a targeted location.

We devised gene constructs encoding ubiquitin-hydrolase–insensitive tandem ubiquitins consisting of 2–8 units in a head-to-tail ubiquitin fusion. These polyubiquitin analogs, designated as tandem ubiquitins (tUbs), were found to be toxic in yeast when they were overproduced (Saeki *et al.*, 2004). From *in vivo* and *in vitro* experiments, we show that tUbs inhibit the 26S proteasome–mediated protein degradation and propose that tUbs can be used as versatile inhibitors of the 26S proteasome *in vivo*.

Construction of the Tandem Ubiquitin Genes

Procedure

The procedure for the construction of the *tUB* genes is summarized in Fig. 1. Polymerase chain reaction (PCR) was carried out by use of a set of primers (1-*Eco*RI: 5′-GGAAGAATTCATGCAAATTTTCGT-CAAAACTCTAAAGGG-3′, 2-*Xho*I: 5′-GGAACTCGAGCAGCCCT-CAACCTCAAGACAAGG-3′) and W303D genomic DNA as template. By this reaction, a series of ubiquitin multimer genes were amplified from the *UBI4* genes. The DNA segment containing two tandem ubiquitin units was excised from a 1% agarose gel after electrophoresis. In this diubiquitin, the 75G and 76G of the most downstream ubiquitin were changed to A's. The nucleotide sequence of this diubiquitin gene was determined to confirm the correct PCR amplification. Next, the G residue in the 76G-1M linkage, the joint of the C-terminal glycine residue of the first ubiquitin and first methionine residue of the second ubiquitin, was changed to V by PCR mutagenesis using two sets of primers (1-*Eco*RI, 4: 5′-GACAAAAATTTG-CATAACACCTCTCAGTC-3′) and 2-*Xho*I, 3: 5′-GACTGAGAGGTGT-TATGCAAATTTTTGTC-3′) and the diubiquitin gene as template. The mutant diubiquitin gene, designated *tUB2*, thus obtained was cloned into the *Eco*RI-*Xho*I gap of Bluescript KS+ (Stratagene). The *tUB2* gene was amplified by PCR using a set of primers (5-*Bam*HI: 5′-GGAAGGATCCATGCAAATTTTCGTCAAAACTCTAACAGGG-3′ and 6-*Mun*I: 5′-GGAACAATTGAGCAGCCCTCAACCTCAAGACAAGG-3′) and the *tUB2* gene as template. The resultant DNA segment

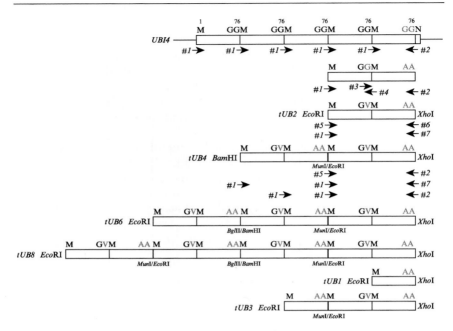

FIG. 1. Procedure for construction of the tandem ubiquitin genes.

was joined with the *tUB2* gene to generate the *tUB4* gene. The *tUB2* gene and *tUB4* genes were amplified by PCR using a set of primers (1-*Eco*RI and 7-*Bgl*II: 5'-AAGGAGATCTAGCAGCCCTCAACCTCAAGACAAGG-3'), and the product was joined with *tUB4* to generate the *tUB6* gene and *tUB8* gene, respectively. The terminal restriction sites of the *tUB4* gene just described were changed to be compatible for the subsequent cloning experiments. The monoubiquitin and *tUB3* genes were constructed by PCR using a set of primers (1-*Eco*RI, 2-*Xho*I) and the *tUB4* gene as template. DNA segments corresponding to the monoubiquitin and triubiquitin genes were excised from a 1% agarose gel. Each of the *tUB1–tUB8* genes was inserted into the *Eco*RI-*Xho*I gap of pKT10 (GAL1) (Kawamura *et al.*, 1996). Plasmid possessing one of the *tUB* genes was introduced into W303-1B, and a representative transformant was subjected to growth test. The result indicates that tUb2–tUb8 inhibit yeast growth (Fig. 2).

In Vivo Inhibition of the Ubiquitin-Proteasome Pathway by tUb

To examine the inhibition of the ubiquitin-proteasome pathway by tUbs *in vivo*, we used Gcn4 as a native substrate, whose degradation by the ubiquitin-proteasome pathway has been well documented (Chi *et al.*,

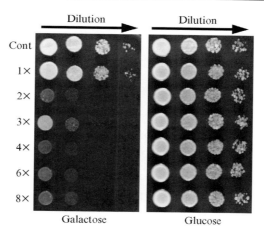

FIG. 2. tUb inhibits yeast growth. Wild-type cells carrying the indicated *tUB* gene were first grown in SC-URA medium until the stationary phase. Cell density was adjusted to $OD_{600nm} = 1.0$, $5 \mu l$ of serial 10-fold dilutions was spotted on SC-Ura and SGal-Ura, and the plates were incubated at 25° for 3 days. tUb2–tUb8 were toxic. For unknown reasons, tUb3 (trimer) displayed a weaker inhibition.

2001; Meimoun *et al.*, 2000). We also analyzed Ub-X-β-galactosidase (X = S, R, or P) as a model substrate (Bachmair *et al.*, 1986). It is known that the R-β-galactosidase is polyubiquitinylated by the N-end rule pathway, whereas the Ub-P–galactosidase is a substrate of the ubiquitin-fusion degradation pathway.

Materials and Reagents

Yeast strains used in this study were isogenic to W303D (*MATαMATa leu2/leu2 his3/his3 trp1/trp1 ura3/ura3 ssd1/ssd1 ade2/ade2 can1/can1*). W303-1B *GCN4^{HA}* (*MATα gcn4::GCN4-HA-LEU2 leu2 his3 trp1 ura3 ade2 ssd1 can1*) in which the *GCN4* gene had been replaced with the *GCN4-HA* gene, W2329-4A (*MATα nin1-1 gcn4::GCN4-HA-LEU2 leu2 his3 trp1 ura3*). The following media were used; SC-Ura (0.67% yeast nitrogen base without amino acid, 2% glucose, appropriate supplements without uracil), SGal-Ura (2% glucose in SC-Ura was replaced with 2% galactose), and SRaf-Ura-Trp (0.67% yeast nitrogen base without amino acid, 2% raffinose, appropriate supplements without uracil and tryptophan). Plasmids: P_{GAL1}-*tUB2* (*URA3*, 2μ), P_{GAL1}-*tUB6* (*URA3*, 2μ), P_{GAL1}-*FLAG-tUB1* (*TRP1*, 2μ), P_{GAL1}-*FLAG-tUB2* (*TRP1*, 2μ), P_{GAL1}-*FLAG-tUB3* (*TRP1*, 2μ), P_{GAL1}-*FLAG-tUB4* (*TRP1*, 2μ), P_{GAL1}-*FLAG-tUB6* (*TRP1*, 2μ), P_{GAL1}-*FLAG-tUB8* (*TRP1*, 2μ),

Ub-A-LacZ (P_{GAL1}-Ub-A-lacZ URA3, 2 μ), Ub-R-LacZ (P_{GAL1}-Ub-R-lacZ URA3, 2 μ), Ub-P-LacZ (P_{GAL1}-Ub-P-lacZ URA3, 2 μ).

Procedure

Inhibition of Gcn4 Degradation. Plasmid carrying the *tUB6* gene and the empty vector were separately introduced into the $GCN4^{HA}$ strain by the lithium acetate method (Ito *et al.*, 1983). Each transformant was grown in SGal-Ura or SC-Ura for 3.5 h. As a reference, a proteasome-defective strain, the *nin1-1/rpn12-1* $GCN4^{HA}$ strain, was incubated at 37° for 3.5 h in SC-Ura. Total cell extracts were prepared by the mild alkali method (Kushnirov, 2000). Then, proteins were separated by SDS-PAGE and subjected to Western blotting using mouse anti-HA antibody (12CA5, Rocshe). $Gcn4^{HA}$ was accumulated in the culture expressing the *tUB6* gene (Fig. 3A).

Inhibition of Model Substrates. W303D strain containing two plasmids, one carrying a *FLAG-tUB* gene (TRP) and the other carrying a *Ub-X-lacZ*

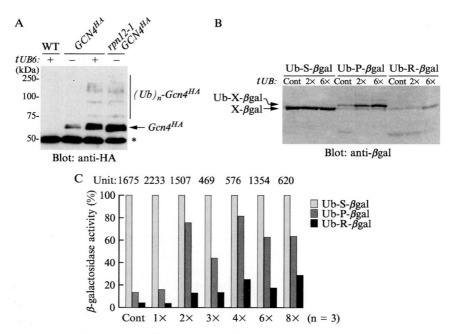

FIG. 3. Inhibition of the ubiquitin-proteasome pathway by expressing tUbs *in vivo*. (A) Accumulation of Gcn4 by tUb. *, A product of a nonspecific reaction. (B) Stabilization of an N-end rule substrate and a UFD substrate. A steady-state level of β-galactosidase analyzed by Western blotting. (C) β-Galactosidase activity.

gene (URA), was grown overnight in SRaf-Ura-Trp and refreshed in the same medium. After 3 h incubation, 2% galactose was added, and incubation was continued for another 4 h. Total cell extracts were prepared by the mild alkali method. Proteins were separated by SDS-PAGE and then subjected to Western blotting using mouse anti-β-galactosidase antibody (Invitrogen) (Fig. 3B). Also, β-galactosidase activity was assayed using o-nitrophenyl-β-D-galactoside as substrate as described previously (Isono et al., 2004) (Fig. 3C). Data were an average of three independent experiments. Enzyme activity corresponding to 100% was defined as that expressed by extract prepared from cells without tUb and is shown in Fig. 3.

Comments. In the case of Gcn4^{HA}, ubiquitinylated species were also seen, suggesting that tUb inhibits the proteolysis at the postubiquitinylation step. Also, tUb does not seem to inhibit deubiquitinating enzymes, because the processing of the ubiquitin-moiety of both Ub-R-β-galactosidase and Ub-S-β-galactosidase was detected.

In Vitro Inhibition of 26S Proteasome Activity by tUb

To examine the effect of tUb on the 26S proteasome activity, we carried out two different experiments. In one experiment, the 26S proteasome and the 26S proteasome-tUb6 complex were purified, and their activity was measured using polyubiquitinylated Sic1^{PY} (Procedure 1). In the second experiment, degradation assays were carried out with or without tUb4 expressed in and purified from *E. coli* (procedure 2).

Note: The preparation of polyubiquitinylated Sic1^{PY} and the purification of yeast 26S proteasome are described Chapter 14.

Materials and Reagents

S. cerevisiae RPN11^{3FLAG} strain (YYS40, *MATa rpn11::RPN11-3FLAG-HIS3 leu2 his3 ura3 trp1 ade2 can1 ssd1*); medium, SRaf-Ura (0.67% yeast nitrogen base without amino acid, 2% raffinose, appropriate supplements without uracil and tryptophan); plasmids, P_{GAL1}-*tUB6* (*URA3*, 2 μ), pQE-*tUB4* (for expression of hexahistidine tag (His$_6$) fused-tUb4 in *E. coli*, the *tUB4* gene was inserted into the pQE vector (QIAGEN)).

Procedure 1

The *RPN11*^{3FLAG} strain carrying P_{GAL1}-*tUB6* or control plasmid was grown to OD$_{600}$ of 0.7 in SRaf-Ura medium, then, galactose was added to 1% and further cultured for 6 h at 25°. Cells were collected, washed twice with water and once with buffer A (50 mM Tris-HCl, pH 7.5, 100 mM NaCl, 10% glycerol) and stored at −80° until use. One gram of cells was resuspended with 1 ml of buffer B, buffer A containing 4 mM ATP, 10 mM MgCl$_2$, and 2 × ATP regenerating system, and lysed by glass beads. Extract was collected and added to an equal volume of buffer B and centrifuged at 15,000 rpm for 20 min at 4°. Eighty microliters of M2-agarose (SIGMA) were added to the supernatant and rotated for 1 h at 4°. Then, the beads were washed five times with 1 ml of proteasome buffer, buffer A containing 2 mM ATP and 5 mM MgCl$_2$, two times with 1 ml of proteasome buffer containing 0.2% Triton X-100, and washed again two times with 1 ml of proteasome buffer. Then, the beads were collected into a mini-column and incubated with 100 μl of 3 × Flag peptide (200 μg/ml) in proteasome buffer for 30 min at 4°. Purified 26S proteasomes were divided to small aliquots and stored at −80°. The formation of the tUb6-26S proteasome complex was confirmed by native-PAGE (Elsasser et al., 2002) and SDS-PAGE. Then, the degradation assay was performed as follows: Equal amounts of two proteasome preparations (1 pmol) was separately incubated with polyubiquitinylated Sic1^{PY} (2 pmol) in 10 μl of proteasome buffer at 30°. The reaction was terminated by the addition of SDS-loading buffer. Then, the samples were subjected to Western blotting with anti-T7 antibody (Novagen).

Comments

The tUb6-26S proteasome complex showed much lower degrading activity toward polyubiquitinylated Sic1^{PY} than did the 26S proteasome (Fig. 4A). It should be noted that cleavage of polyubiquitin chains from polyubiquitinylated Sic1^{PY} was inhibited in the reaction mixture containing the 26S proteasome-tUb6 complex.

Procedure 2

His$_6$-tUb4 was expressed and purified from *E. coli* BL21 (DE3) cells transformed with pQE-*tUB4* by standard protocols. Purified His$_6$-tUb4 was dialyzed against buffer A overnight and stored at −80°. The purified 26S proteasome (final concentration 12 nM) was preincubated with or without tUb4 (1 μM) for 15 min at 30°, and then the degradation assay was initiated by the addition of polyubiquitinylated Sic1^{PY} (50 nM). The reaction products were analyzed by SDS-PAGE followed by Western blotting using anti-T7 antibody as described previously.

Comments

A high dose of tUb4 is required for the inhibition in this assay (about 100-fold and 20-fold excess of the 26S proteasome and substrate, respectively) (Fig. 4B). Probably, the binding efficiency of tUb4 toward the 26S

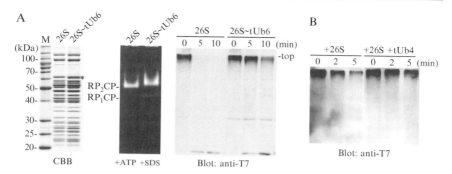

FIG. 4. tUb inhibits the 26S proteasome *in vitro*. (A) The tUb6-26S proteasome complex was not able to degrade ubiquitinylated Sic1^{PY}. The purified 26S proteasome and the tUb6-26S proteasome complex were subjected to SDS-PAGE followed by CBB staining. The protein band corresponding to tUb6 is indicated by an asterisk (left panel), native PAGE followed by in-gel assays (middle panel), and incubation with polyubiquitinylated Sic1^{PY} for 0, 5, and 10 min followed by Western blotting with anti-T7 antibody (right panel). (B) Inhibition by recombinant tUb4.

proteasome is lower than K48-linked Ub chain, known as the recognition signal of the 26S proteasome (Thrower *et al.*, 2000). We attempted a similar assay using the larger tUbs, but we failed to purify tUb6 or tUb8 from the *E. coli* expression system.

Inhibition of the Ubiquitin-Proteasome Pathway by tUb in Xenopus Embryos

To examine whether tUb can be used more generally, we examined the effect of overexpression of *tUB* on mitosis in *Xenopus* embryos.

Procedure

To produce translatable *tUB* messenger RNA, *tUB6* and *tUB6-HA* and monomer ubiquitin genes were separately inserted into an RN3 vector (Lemaire, 1995) in an appropriate direction. Full-length mRNAs for *tUB6*, *tUB6-HA*, and monomer ubiquitin genes were synthesized *in vitro* using an mMESSAGEmMACHINE T3 kit (Ambion). Synthesized mRNA was dissolved in RNase-free water, and 5 ng of mRNA was injected in a volume of 9.2 nl into a blastomere of a two-cell stage *Xenopus* embryo. Embryos were cultured in 0.2 × MMR (20 mM NaCl, 0.4 mM KCl, 0.4 mM CaCl$_2$, 0.2 mM MgCl$_2$, 1 mM HEPES, pH 7.4) solution at 20°, and photographs were taken at appropriate time points after fertilization. Forced expression of *tUB6* in a blastomere at the two-cell stage caused marked

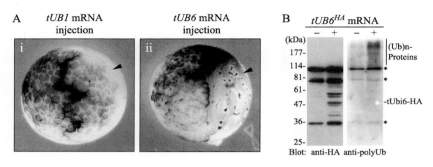

FIG. 5. Overexpression of tUB6 leads to impaired mitosis in *Xenopus* embryos. (A) Effect of overexpression of *tUB6* on the cell division of *Xenopus* blastomeres. Monomer ubiquitin mRNA (i) and *tUB6* mRNA (ii) were injected into a blastomere (indicated by an arrowhead) at the two-cell stage. Photographs were taken at 5.5 h (blastula stage) after fertilization. (B) Detection of tandem ubiquitin protein expressed in *Xenopus* embryos after mRNA injection. *tUB6-HA* mRNA was injected into a blastomere at the two-cell stage, and Western blot analyses with anti-HA antibody and antipolyubiquitin antibody were performed using the extracts prepared from the cells harvested at the indicated times. Asterisks indicate artifact bands.

abnormalities of the injected cell and its progeny (Fig. 5A). The cleavage pattern of the injected blastomere became progressively more abnormal, so that at the equivalent of the blastula stage, cleavage furrows were either absent or incomplete in the *tUB* mRNA–injected blastomere (Fig. 5A, ii), indicated by an arrowhead. Injection of 5 ng of monoubiquitin mRNA did not appreciably affect cell-cycle progression or subsequent developmental processes compared with those in noninjected control blastomeres (Fig. 5A, i). Western blot analysis showed that the tUb6-HA protein became detectable within 1 h after the injection of mRNAs. As in yeast, tUb6-HA was produced as native and monoubiquitinylated forms (Fig. 5B). In tUb6-producing embryos, polyubiquitinylated proteins were accumulated. These results suggest that tUb6 production interfered with the degradation of polyubiquitinylated proteins by the 26S proteasome in *Xenopus* embryos. The protein levels remained unchanged until mid-blastula stage.

Conclusion

We described here the inhibition of the ubiquitin-proteasome pathway by expressing tUb in yeast. tUb seemed to inhibit the activity of the proteasome, most probably by binding it by way of the ubiquitin receptors such as Rpn10 and Rad23/Dsk2. In fact, tUb4 binds directly with Rpn10 and Dsk2 *in vitro* (Saeki *et al.*, unpublished result). Because recognition of the polyubiquitinylated substrate by the 26S proteasome is critical for the ubiquitin-proteasome pathway, tUb might interfere with this step. Recently, it was reported that small-molecule inhibitors called ubistatins also interfere with this step by binding with the polyubiquitin chain (Verma *et al.*, 2004).

We found that the overexpression of *tUB* impaired mitosis in *Xenopus* embryos and was toxic in mammalian cells (Saeki *et al.*, unpublished observation). tUb seems to bind with the proteasome in mammalian cells (Iwai *et al.*, unpublished observation). These results indicate that tUb can be widely used as inhibitor of 26S proteasomes. The fact that tUb is produced by the cellular protein synthesis system indicates that tUb can be delivered and expressed in a targeted tissue or organ by using an appropriate promoter and can inhibit proteasomes there. Such inhibition is impossible with currently available inhibitors. Thus, tUb is a promising tool for the study of the ubiquitin-proteasome pathway in a wide range of organisms.

References

Bachmair, A., Finley, D., and Varshavsky, A. (1986). *In vivo* half life of a protein is a function of its amino-terminal residue. *Science* **234,** 179–186.

Bogyo, M., McMaster, J. S., Gaczynska, M., Tortorella, D., Goldberg, A. L., and Ploegh, H. (1997). Covalent modification of the active site threonine of proteasomal beta subunits and the *Escherichia coli* homolog HslV by a new class of inhibitors. *Proc. Natl. Acad. Sci. USA* **94,** 6629–6634.

Chi, Y., Huddleston, M. J., Zhang, X., Young, R. A., Annan, R. S., Carr, S. A., and Deshaies, R. J. (2001). Negative regulation of Gcn4 and Msn2 transcription factors by Srb10 cyclin-dependent kinase. *Genes Dev.* **15,** 1078–1092.

Elsasser, S., Gali, R. R., Schwickart, M., Larsen, C. N., Leggett, D. S., Muller, B., Feng, M. T., Tubing, F., Dittmar, G. A., and Finely, D. (2002). Proteasome subunit Rpn1 binds ubiquitin-like protein domains. *Nature Cell Biol.* **4,** 725–730.

Fenteany, G., Standaert, R. F., Lane, W. S., Choi, S., Corey, E. J., and Schreiber, S. L. (1995). Inhibition of proteasome activities and subunit-specific amino-terminal threonine modification by lactacystin. *Science* **268,** 726–731.

Isono, E., Saeki, Y., Yokosawa, H., and Toh-e, A. (2004). Rpn7 is required for the structural integrity of the 26S proteasome of *Saccharomyces cerevisiae*. *J. Biol. Chem.* **279,** 28807–28816.

Ito, H., Fukuda, Y., Murata, K., and Kimura, A. (1983). Transformation of intact yeast cells treated with alkali cations. *J. Bacteriol.* **153,** 163–168.

Kawamura, M., Kominami, K., Takeuchi, J., and Toh-e, A. (1996). A multicopy suppressor of *nin1-1* of the yeast *Saccharomyces cerevisiae* is a counterpart of the *Drosophila melanogaster* diphenol oxidase A2 gene, *Dox-A2*.. *Mol. Gen. Genet.* **251,** 146–152.

Kushnirov, V. V. (2000). Rapid and reliable protein extraction from yeast. *Yeast* **16,** 857–860.

Lee, D. H., and Goldberg, A. L. (1998). Proteasome inhibitors: Valuable new tools for cell biology. *Trend Cell Biol.* **8,** 397–403.

Lemaire, P., Garrett, N., and Gurdon, J. B. (1995). Expression cloning of Siamois, Xenopus homeobox gene expressed in dorsal-vegetal cells of blastulae and able to induce a complete secondary axis. *Cell* **81,** 85–94.

Meimoun, A., Holtzman, T., Weissman, Z., McBride, H. J., Stillman, D. J., Fink, G. R., and Kornitzer, D. (2000). Degradation of the transcription factor Gcn4 requires the kinase Pho85 and the SCF (CDC4) ubiquitin-ligase complex. *Mol. Biol. Cell.* **11,** 915–927.

Omura, S., Fujimoto, T., Otoguro, K., Matsuzaki, K., Moriguchi, R., Tanaka, H., and Sasaki, Y. (1991). Lactacystin, a novel microbial metabolite, induces neuritogenesis of neuroblastoma cells. *J. Antibiot. (Tokyo)* **44,** 113–116.

Saeki, Y., Isono, E., Oguchi, T., Sone, T., Shimada, M., Kawahara, H., Yokosawa, H., and Toh-e, A. (2004). Intracellularly inducible, ubiquitin hydrolase-insensitive tandem ubiquitins inhibit the 26S proteasome activity and cell division. *Genes Genet. Syst.* **79,** 77–86.

Thrower, J. S., Hoffman, L., Rechsteiner, M., and Pickart, C. M. (2000). Recognition of the polyubiquitin proteolytic signal. *EMBO J.* **19,** 94–102.

Tsubuki, S., Kawasaki, H., Saito, Y., Miyashita, N., Inomata, M., and Kawashima, S. (1992). Purification and characterization of a Z-Leu-Leu-Leu-MCA degrading protease expected to regulate neurite formation: A novel catalytic activity in proteasome. *Biochem. Biophys. Res. Commun.* **196,** 1195–1201.

Verma, R., Peters, N. R., D'Onofrio, M., Tochtrop, G. P., Sakamoto, K. M., Varadan, R., Zhang, M., Coffino, P., Fushman, D., Deshaies, R. J., and King, R. W. (2004). Ubistatins inhibit proteasome-dependent degradation by binding the ubiquitin chain. *Science* **306,** 117–120.

[6] Production of Antipolyubiquitin Monoclonal Antibodies and Their Use for Characterization and Isolation of Polyubiquitinated Proteins

By MASAHIRO FUJIMURO and HIDEYOSHI YOKOSAWA

Abstract

Formation of a Lys48-linked polyubiquitin chain is required for destruction of targeted proteins by the 26S proteasome, whereas formation of a Lys63-linked polyubiquitin chain is required for modulation of protein–protein interaction, enzyme activity, and intracellular localization. In addition, monoubiquitination plays key roles in endocytosis and protein trafficking. To gain a better understanding of the role of polyubiquitination, we attempted to produce monoclonal antibodies against the polyubiquitin chains, two of which were designated as FK1 and FK2 and were extensively characterized. Both FK1 and FK2 antibodies recognize the polyubiquitin moiety but not free ubiquitin, whereas FK2 antibody, but not FK1 antibody, can recognize monoubiquitinated proteins. The FK1/FK2 antibodies can be applied to ELISA for quantification of polyubiquitin chains, to immunocytochemistry for staining of intracellular polyubiquitin chains, and also to immunoaffinity chromatography for isolation of polyubiquitinated proteins. Thus, these two antibodies are useful for isolating polyubiquitin chain–tagged proteins and for probing proteins that are modified through polyubiquitination or monoubiquitination in various cells and tissues under physiological and pathological conditions.

Introduction

Posttranslational protein modifications play key roles in the regulation of cellular events. Ubiquitin-conjugation (i.e., modification with polyubiquitin chains or monoubiquitin) is involved in various cellular events, including cell cycle progression, signal transduction, endocytosis, transcription, DNA repair, apoptosis, and immune response (Glickman and Ciechanover, 2001; Hershko and Ciechanover, 1998; Hicke, 2001; Pickart, 2001). Polyubiquitin chains linked through Lys48 function as a signal for degradation of target proteins by the 26S proteasome; chains linked through Lys63 are involved in proteasome-independent events such as DNA repair, endocytosis, and signal transduction; and monoubiquitination is involved in endocytosis and protein trafficking. We succeeded in producing several monoclonal antibodies against the polyubiquitin chains, including FK1 and FK2 antibodies (Fujimuro et al., 1994). The FK1 and FK2 antibodies recognize the polyubiquitin moiety but not free ubiquitin, and FK2 antibody shows affinity to proteins tagged with monoubiquitin. Both antibodies have been applied to ELISA for quantification of polyubiquitin chains (Takada et al., 1995), to immunocytochemistry for staining of intracellular polyubiquitin chains (Fujimuro et al., 1997), and to immunoaffinity chromatography for isolation of ubiquitin–protein conjugates and thioester-linked ubiquitin-intermediates (Takada et al., 2001). Thus, these antibodies are useful tools for the identification, purification, and quantification of polyubiquitinated proteins and polyubiquitin chains that have been formed in various cells and tissues under physiological and pathological conditions. In this chapter, we describe in detail methods for the production of antipolyubiquitin monoclonal antibodies and their use for characterization and isolation of polyubiquitinated proteins.

Production of Antipolyubiquitin Monoclonal Antibodies

We prepared ubiquitin-conjugated proteins using a partial purified enzyme mixture of E1, E2s, and E3s, and we produced antipolyubiquitin monoclonal antibodies using the thus-prepared polyubiquitin-conjugated proteins as antigen (Fujimuro et al., 1994).

Preparation of Polyubiquitin-Conjugated Proteins

Fractions containing E1, E2s, and E3s were isolated from rabbit reticulocyte lysate according to the method of Tamura et al. (1991). The reticulocyte lysate (about 200 ml) was ultracentrifuged at 100,000g for 6 h to precipitate the 26S proteasome, and the resulting supernatant was subjected to DEAE-cellulose chromatography. After washing, fraction II was eluted from the column with 50 mM Tris-HCl (pH 7.5), 1 mM dithiothreitol (DTT), and 0.5 M KCl and was then subjected to ubiquitin-Affi-Gel 10 chromatography (Ubiquitin-Affi-Gel 10 was prepared according to the manufacturer's protocol: 10 mg of bovine ubiquitin [Sigma] immobilized/ml of Affi-Gel 10 [Bio-Rad]). After washing, a mixture of E1 and E2s was first eluted from the column with 50 mM Tris-HCl (pH 7.5) and 20 mM DTT, and a mixture of E3s was next eluted with 50 mM Tris-HCl (pH 9.0), 2 mM DTT, and 1 M KCl. Both mixtures, separately isolated, were then dialyzed against 50 mM Tris-HCl (pH 7.5), 1 mM DTT, 0.2 mM ATP, and 20% glycerol.

Lysozyme denatured at 100° for 20 min was used as a substrate for E1/E2/E3 enzymes. Polyubiquitin conjugates were prepared in a reaction mixture consisting of 100 mM Tris-HCl (pH 9.0), 5 mM MgCl$_2$, 1 mM DTT, 2 mM ATP, 20 µg/ml heat-denatured chicken egg white lysozyme, 6 mg/ml bovine ubiquitin, 0.15 mg/ml E1/E2 mixture, 0.45 mg/ml E3 mixture, 2.4 unit/ml yeast inorganic pyrophosphatase, 10 µg/ml rabbit creatine phosphokinase, and 10 mM phosphocreatine at 37° for 1 h.

Immunization and Screening of Hybridoma Cells

A reaction mixture consisting of the thus-prepared polyubiquitin-conjugated lysozyme, free ubiquitin, ubiquitin-ligating enzymes, and the ATP-regenerating system was used as antigen (50 µg of lysozyme injected/mouse), and the fusion of mouse spleen cells with myeloma cells was performed according to the standard polyethylene glycol method (Kohler and Milstein, 1976). Antibody production was screened by the sandwich ELISA method as follows.

Positive Screening (Fig. 1A). Each well of an ELISA plate (96 wells) was coated with 100 µl of purified antiubiquitin polyclonal antibody (5 µg of antibody/ml of phosphate-buffered saline [PBS]). Antiubiquitin polyclonal antibody was produced in rabbits using bovine ubiquitin cross-linked with γ-globulin as antigen and purified by protein A-immobilized Sepharose. After washing with PBS and blocking with 200 µl of 3% skim milk in PBS, the reaction mixture (100 µl) containing newly formed polyubiquitin-conjugated lysozyme and free ubiquitin, diluted with PBS (300-fold), was added to each well, and the plate was incubated at room temperature for 2 h to allow the polyubiquitin-conjugated lysozyme and ubiquitin to be adsorbed on the antiubiquitin polyclonal antibody that had been bound to the well. After the wells had been extensively washed with PBS, a culture medium of hybridoma cells was added to each well, and the plate was incubated at room temperature for 2 h to allow antibodies in the culture medium to be adsorbed on the polyubiquitin-conjugated lysozyme bound to the antiubiquitin polyclonal antibody. After extensive washing with PBS, biotin-conjugated anti-mouse IgG and then the avidin–biotinated peroxidase complex were added to each well, and the plate was incubated at room temperature for 30 min. Color was developed using 0.02% H$_2$O$_2$ and 0.2 mg/ml 2,2'-azinobis(3-ethylbenzothiazoline-6-sulfonic acid) as peroxidase substrates.

Negative Screening (Fig. 1B, C). Positive culture media of hybridoma cells, tested by positive screening as described previously, were further subjected to negative screening by ELISA using either free ubiquitin (50 µg/ml) (Fig. 1B) or the reaction mixture for ubiquitin conjugation that lacked only ubiquitin (Fig. 1C) in place of the polyubiquitin-conjugated lysozyme-containing reaction mixture.

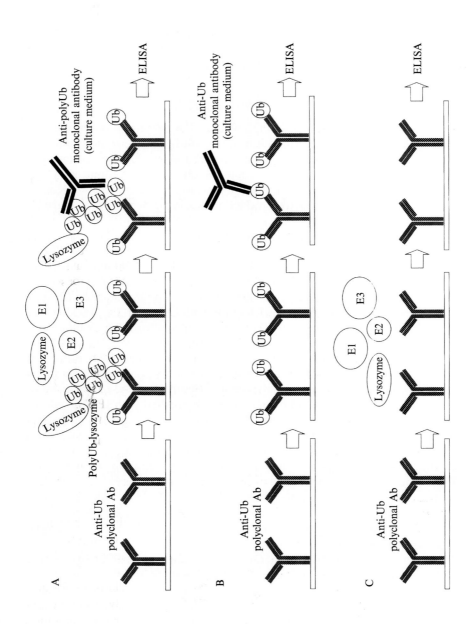

Hybridoma cells, which produced antibodies that were cross-reactive with the polyubiquitin-conjugated lysozyme but not with free ubiquitin or the preceding negative control reaction mixture, were cloned by the method of limiting dilution, and 10 antibody-producing clones were established (Fujimuro et al., 1994). Among them, two antibodies designated as FK1 (subclass, IgM) and FK2 (subclass, IgG1) were subjected to further characterization.

Cross-Reactivity of Antipolyubiquitin Monoclonal Antibody

To demonstrate antibody specificity for monoubiquitinated proteins, we prepared methyl ubiquitin-conjugated lysozyme using methyl-ubiquitin in place of ubiquitin. Methyl ubiquitin was prepared as previously reported (Fujimuro et al., 1994). Western blot analysis revealed that both FK1 and FK2 antibodies react with the polyubiquitin-conjugated lysozyme but not with free ubiquitin, whereas FK2 antibody but not FK1 antibody can recognize the methyl ubiquitin–conjugated lysozyme (Fujimuro et al., 1994). Neither antibody was cross-reactive with the original substrate lysozyme.

Characterization of Polyubiquitin Chains by the Use of Antipolyubiquitin Monoclonal Antibodies

The use of immunoblotting and immunofluorescent staining with FK1 and FK2 antibodies has been reported to enable detection of dynamic change in the amount, localization, and distribution of the ubiquitin-conjugate moiety (for example, see Lelouard et al. [2002], Sawada et al. [2002] and Schubert et al. [2000]), as well as identification of ubiquitin-conjugates (for example, see Takada et al. [2001]). Furthermore, these antibodies have been used to investigate the formation of polyubiquitin

---

Fig. 1. Procedures for screening of hybridoma cells producing antipolyubiquitin monoclonal antibodies. A 96-well ELISA plate was coated with antiubiquitin (Ub) polyclonal antibody (Ab), and then to the plate was added (A) a reaction mixture containing newly formed polyubiquitin (polyUb)-conjugated lysozyme, original lysozyme, free ubiquitin, ubiquitin conjugating enzymes, inorganic pyrophosphatase, and the ATP-regenerating enzyme system (procedure A), (B) only ubiquitin (procedure B), and (C) the same reaction mixture as that just described but lacking ubiquitin (procedure C). Polyubiquitin- or monoubiquitin-conjugated lysozyme and free ubiquitin could be adsorbed to the antiubiquitin antibody on the plate in procedure A, whereas only ubiquitin could be adsorbed to the antiubiquitin antibody on the plate in procedure B. Next, culture media of hybridoma cells were added to the wells in procedure A and subjected to ELISA. Positive culture media detected by procedure A were further tested by negative screening procedures B and C. Hybridoma cells, culture media of which were positive by procedure A but negative by procedures B and C, were cloned by the method of limiting dilution.

conjugates under pathological conditions, such as Parkinson's disease (Ardley et al., 2004) and viral infection (Burch and Weller, 2004).

Here, we describe the use of FK1 or FK2 antibody for detection of change in polyubiquitin conjugates under heat shock stress conditions (Fujimuro et al., 1997).

Immunoblotting

Heat-shocked HeLa cells were subjected to Western blotting with FK1 or FK2 antibody as follows. Cells (1×10^6 cells) were washed with PBS and solubilized in 500 μl of SDS-sample buffer containing 1 mM phenylmethylsulfonyl fluoride, 5 μg/ml aprotinin, 1 μg/ml pepstatin, and 2 mM N-ethylmaleimide. The cell lysate was sonicated for 15 s with an immersible tip-type sonicator to shear the chromosomal DNA and was then boiled for 5 min. It should be noted that in cases of Lys48 and Lys63-linked polyubiquitin chain–containing proteins, boiling of samples should be avoided to prevent them from aggregating, and incubation for 30 min at 25° instead of boiling will give a clearer result. The resulting sample (20 μl) was then subjected to Western blotting with FK1 or FK2 antibody. It was found that the heat-shock treatment induced accumulation of FK1 antibody–immunoreactive materials of high molecular masses, the level of which decreased during subsequent treatment at normal temperature (37°) (Fujimuro et al., 1997). On the basis of the specificity of FK1 antibody, it is concluded that polyubiquitinated proteins are accumulated under heat-shock conditions.

Immunofluorescent Staining

Heat-shocked HeLa cells were further subjected to immunofluorescent staining with FK1 or FK2 antibody as follows. The cells were seeded at 1×10^5 cells per well in two-well slide chambers. After heat shock treatment, cells were fixed with cold methanol for 15 min at $-20°$ and rinsed with PBS. Fixed cells were incubated with FK1 or FK2 antibody for 1 h at 25°. After washing, the cells were treated with a secondary antibody, such as fluorescein isothiocyanate (FITC)- or rhodamine-conjugated donkey IgG for 1 h and were subsequently stained with 5 mg/ml Hoechst 33258 in PBS for 15 min for DNA staining: All antibodies were used at a 1:200 dilution in 1% bovine serum albumin–PBS. Immunofluorescence-labeled cells were observed under a fluorescence microscope or confocal microscope. It was found that the heat-shock treatment induced increase in the cytoplasm and decrease in the nucleus of FK1-immunoreactive materials compared with the case of pretreated cells (Fujimuro et al., 1997). On the basis of the specificity of FK1 antibody, it is again concluded that polyubiquitinated proteins undergo changes in subcellular distribution under stress conditions.

Isolation of Polyubiquitinated Proteins by the Use of Antipolyubiquitin Monoclonal Antibodies

FK2 antibody (IgG1) is more suitable than FK1 antibody for use in immunoprecipitation together with protein A/G-Sepharose beads, because the subclass of FK1 antibody is IgM. Western blotting and ELISA revealed that FK2 antibody recognizes polyubiquitin and monoubiquitin conjugates but not ubiquitin molecules that had been immobilized on a nitrocellulose membrane in the immunoprecipitation or a microtiter well in the ELISA (Fujimuro et al., 1994; Takada et al., 1995), but it should be noted that in the immunoaffinity chromatography, this antibody shows affinity toward free ubiquitin in solution, together with polyubiquitin and monoubiquitin conjugates in solution (Takada et al., 2001).

Immunoprecipitation of Polyubiquitinated Proteins

Immunoprecipitation of polyubiquitinated proteins with FK2 antibody was performed as follows. HeLa cells (1×10^6) were first treated with 10 μM MG132 (a proteasome inhibitor) for 6 h, leading to accumulation of polyubiquitinated proteins. The cells were then solubilized in 1 ml of ice-cold lysis buffer (PBS containing 0.1% SDS, 0.5% deoxycholic acid, 1% Nonidet P-40, 0.5 mM EDTA, 2 mM N-ethylmaleimide, 50 μM MG132, 0.5 mM phenylmethylsulfonyl fluoride, 1 μg/ml pepstatin, 5 μg/ml aprotinin, 0.2 mM Na$_4$P$_2$O$_4$, and 1 mM NaF). The lysate was precleared with 100 μl of Sepharose beads and centrifuged at 13000 rpm for 5 min. The resultant supernatant was further centrifuged at 15000 rpm for 10 min. The cell extract was incubated with 5 μg of FK2 antibody for 2 h at 4° and successively with protein A/G-Sepharose beads (30 μl) for 1 h at 4°. The beads were washed with ice-cold lysis buffer five times. The resulting beads were suspended in sample buffer (30 μl) and subjected to SDS-PAGE followed by Western blotting with FK1 antibody. FK1 antibody–immunoreactive materials of high molecular masses were detected, indicating that high-molecular-mass polyubiquitinated proteins that have accumulated in HeLa cells are able to be immunoprecipitated with FK2 antibody.

Isolation of Polyubiquitinated Proteins

Polyubiquitinated proteins that have intracellularly accumulated can be isolated by use of FK1 or FK2 antibody–immobilized beads as an affinity adsorbent. FK1 antibody was purified by Sephadex G200 gel filtration and was then covalently coupled to Affi-Gel 10 (5 mg of IgM immobilized/ml of gel) according to the manufacturer's instructions, and the resulting beads were used as the affinity adsorbent. Alternately, FK2 antibody was

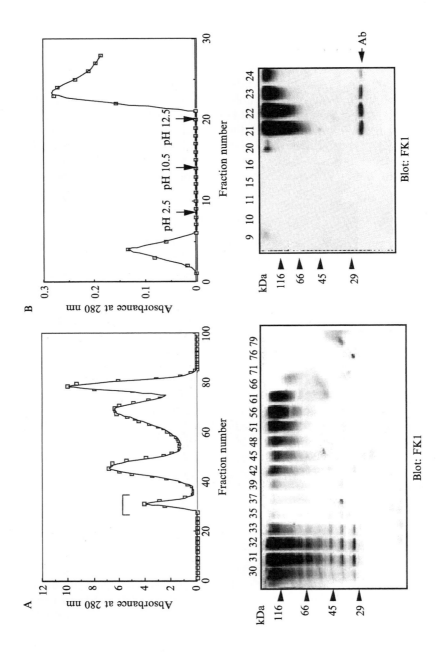

covalently coupled to Affi-Gel 10 (2.3 mg of IgG immobilized/ml of gel) or HiTrap NHS-activated Sepharose (Amersham) (2.6 mg /ml), and the resulting beads were also used as the affinity adsorbent.

Here, we describe the isolation by gel filtration (Fig. 2A) and subsequent affinity chromatography on FK1 antibody–immobilized Affi-Gel 10 (Fig. 2B) of polyubiquitinated proteins from the temperature-sensitive *nin1-1* mutant YK109 of *Saccharomyces cerevisiae*. *NIN1* encodes a regulatory subunit Rpn12 of the 26S proteasome. Early exponential YK109 cells (*nin1-1*) grown in 1 liter of YPD medium at 25° were transferred to a restrictive temperature, 37°, and further incubated for 4 h. Polyubiquitinated proteins accumulated in *nin1-1* cells at a restrictive temperature (Kominami *et al.*, 1995). The cells were harvested and suspended in 5 ml of lysis buffer (50 mM Tris-HCl (pH 7.5) containing 0.1% CHAPS, 0.5 mM EDTA, 5 mM N-ethylmaleimide, 50 μM MG132, 1 mM phenylmethylsulfonyl fluoride, 10 μg/ml pepstatin, and 10 μg/ml aprotinin) and disrupted by a French Press. The cell extract was centrifuged at 10,000 rpm for 30 min at 4° and applied to a Sepharose CL-4B column previously equilibrated with lysis buffer (Fig. 2A). Each fraction was subjected to Western blotting with FK1 antibody (Fig. 2A, lower). Fractions (number 30–33) containing ubiquitin conjugates were pooled and applied to an FK1 antibody–immobilized Affi-Gel 10 column (6 ml) previously equilibrated with 50 mM Tris-HCl (pH 7.5) containing 0.1% CHAPS, 0.1 mM EDTA, and 0.5 mM N-ethylmaleimide (Fig. 2B). After the column had been successively washed with equilibration buffer, 50 mM glycine-HCl (pH 2.5) containing 0.1% CHAPS, and 50 mM Tris-HCl (pH 10.5) containing 0.1% CHAPS, adsorbed proteins were eluted from the column with 50 mM triethylamine-NaOH (pH 12.5) containing 0.1% CHAPS. Immediately after elution, the affinity beads should be neutralized with 1 M Tris-HCl (pH 7.0). Each fraction was subjected to Western blotting with FK1 antibody (Fig. 2B, lower). On the basis of the specificity of FK1 antibody used for immunoblotting, it is concluded that high-molecular-mass

---

Fig. 2. Purification of polyubiquitinated proteins from the yeast *S. cerevisiae*. The yeast cell extract was subjected to gel filtration on a Sepharose CL-4B column (2.7 × 100 cm) (A, upper panel). Fractions of 4 ml were collected, and each fraction was subjected to SDS-PAGE (12.5%) followed by Western blotting with FK1 antibody (A, lower panel). Fractions (number 30–33) containing ubiquitin conjugates were subjected to immunoaffinity chromatography on an FK1 antibody–immobilized Affi-Gel 10 column (6ml) (B, upper panel). The column was washed with equilibration buffer (fractions 1–8), 50mM glycine-HCl (pH 2.5)-0.1% CHAPS (fractions 9–14), and 50mM Tris-HCl (pH 10.5)-0.1% CHAPS (fractions 15–19). Adsorbed proteins were then eluted from the column with 50mM triethylamine-NaOH (pH 12.5)-0.1% CHAPS (fractions 20–24). Each fraction was subjected to Western blotting with FK1 antibody (B, lower panel).

polyubiquitinated proteins, accumulated in *nin1-1* cells, can be isolated by immunoaffinity chromatography on FK1 antibody-immobilized beads. Fractions containing ubiquitin conjugates were further purified by Mono Q HPLC and were then subjected to lysylendopeptidase digestion followed by reverse-phase HPLC and peptide sequence analysis to identify target proteins ubiquitinated.

Conclusions

Antipolyubiquitin monoclonal antibodies described in this chapter are useful for isolating polyubiquitinated proteins and for probing proteins that are modified through polyubiquitination or monoubiquitination under various conditions. As described in yeast *nin1-1* cells, impairment of the 26S proteasome results in the accumulation of polyubiquitinated proteins, and, conversely, when the accumulation of polyubiquitinated proteins, which can be detected only by antipolyubiquitin monoclonal antibodies, is observed in yeast mutants at a restrictive temperature, it can be concluded that the corresponding gene products are involved in ubiquitin-dependent proteolysis. In fact, yeast mutants of proteasomal subunits, Rpn1 (Tsurumi *et al.*, 1996), Rpn2 (Yokota *et al.*, 1996), Rpn3 (Kominami *et al.*, 1997), Rpn7 (Isono *et al.*, 2004), Rpn9 (Takeuchi *et al.*, 1999), and Sem1 (Rpn15) (Sone *et al.*, 2004), accumulated polyubiquitinated proteins at a restrictive temperature, confirming that the corresponding proteins are proteasomal subunits. Recently, several ubiquitin-binding proteins have been identified. In these cases too, antipolyubiquitin monoclonal antibodies are useful for detecting polyubiquitinated proteins that can be pulled down by the respective GST-fusion proteins or immunoprecipitated together with the respective FLAG-tagged proteins by an anti-FLAG antibody (for example, Rpn10, Rad23, and Dsk2 [Saeki *et al.*, 2002], and Tom1 [Yamakami *et al.*, 2003]). In addition, results obtained by using FK1 antibody, which recognizes polyubiquitinated proteins but not monoubiquitinated proteins, led to the conclusion that receptor tyrosine kinases, such as EGF and PDGF receptors, are monoubiquitinated on multiple lysine residues, but not polyubiquitinated, after their activation (Haglund *et al.*, 2003). FK1 and FK2 antibodies recognize both Lys48- and Lys63-linked polyubiquitin chains. Production of antipolyubiquitin monoclonal antibodies that recognize specific ubiquitin chain linkage types will be useful to study the proteasome-dependent and proteasome-independent roles of polyubiquitination.

Note: FK1 and FK2 antibodies are commercially available from BIO-MOL International LP (URL:http://www.biomol.com or URL: http://www.affiniti-res.com), COSMO BIO Co., Ltd, Japan (URL: http://www.cosmobio.co.jp), Wako Pure Chemical Industries, Ltd, Japan (URL:

http://www.wako-chem.co.jp), and Medical & Biological Laboratories Co., Ltd, Japan (MBL) (URL: http://www.mbl.co.jp).

References

Ardley, H. C., Scott, G. B., Rose, S. A., Tan, N. G. S., and Robinson, P. A. (2004). UCH-L1 aggresome formation in response to proteasome impairment indicates a role in inclusion formation in Parkinson's disease. *J. Neurochem.* **90**, 379–391.

Burch, A. D., and Weller, S. K. (2004). Nuclear sequestration of cellular chaperone and proteasomal machinery during herpes simplex virus type 1 infection. *J. Virol.* **78**, 7175–7185.

Fujimuro, M., Sawada, H., and Yokosawa, H. (1994). Production and characterization of monoclonal antibodies specific to multi-ubiquitin chains of polyubiquitinated proteins. *FEBS Lett.* **349**, 173–180.

Fujimuro, M., Sawada, H., and Yokosawa, H. (1997). Dynamics of ubiquitin conjugation during heat-shock response revealed by using a monoclonal antibody specific to multi-ubiquitin chains. *Eur. J. Biochem.* **249**, 427–433.

Glickman, M. H., and Ciechanover, A. (2001). The ubiquitin-proteasome proteolytic pathway: Destruction for the sake of construction. *Physiol. Rev.* **82**, 373–428.

Haglund, K., Sigismund, S., Polo, S., Szymkiewicz, I., Di Fiore, P. P., and Dikic, I. (2003). Multiple monoubiquitination of RTKs is sufficient for their endocytosis and degradation. *Nature Cell Biol.* **5**, 461–466.

Hershko, A., and Ciechanover, A. (1998). The ubiquitin system. *Annu. Rev. Biochem.* **67**, 425–479.

Hicke, L. (2001). Protein regulation by monoubiquitin. *Nat. Rev. Mol. Cell. Biol.* **2**, 195–201.

Isono, E., Saeki, Y., Yokosawa, H., and Toh-e, A. (2004). Rpn7 is required for the structural integrity of the 26S proteasome of *Saccharomyces cerevisiae*. *J. Biol. Chem.* **279**, 27168–27176.

Kohler, G., and Milstein, C. (1976). Derivation of specific antibody-producing tissue culture and tumor lines by cell fusion. *Eur. J. Immunol.* **6**, 511–519.

Kominami, K., DeMartino, G. N., Moomaw, C. R., Slaughter, C. A., Shimbara, N., Fujimuro, M., Yokosawa, H., Hisamatsu, H., Tanahashi, N., Shimizu, Y., et al. (1995). Nin1p, a regulatory subunit of the 26S proteasome, is necessary for activation of Cdc28p kinase of *Saccharomyces cerevisiae*. *EMBO J.* **14**, 3105–3115.

Kominami, K., Okura, N., Kawamura, M., DeMartino, G. N., Slaughter, C. A., Shimbara, N., Chung, C. H., Fujimuro, M., Yokosawa, H., Tanaka, K., and Toh-e, A. (1997). Yeast counterparts of subunits S5a and p58 (S3) of the human 26S proteasome are encoded by two multicopy suppressors of *nin1-1*. *Mol. Biol. Cell.* **8**, 171–187.

Lelouard, H., Gatti, E., Cappello, F., Gresser, O., Camosseto, V., and Pierre, P. (2002). Transient aggregation of ubiquitinated proteins during dendritic cell maturation. *Nature* **417**, 177–182.

Pickart, C. M. (2001). Mechanisms underlying ubiquitination. *Annu. Rev. Biochem.* **70**, 503–533.

Saeki, Y., Saitoh, A., Toh-e, A., and Yokosawa, H. (2002). Ubiquitin-like proteins and Rpn10 play cooperative roles in ubiquitin-dependent proteolysis. *Biochem. Biophys. Res. Commun.* **293**, 986–992.

Sawada, H., Sakai, N., Abe, Y., Tanaka, E., Takahashi, Y., Fujino, J., Kodama, E., Takizawa, S., and Yokosawa, H. (2002). Extracellular ubiquitination and proteasome-mediated degradation of the ascidian sperm receptor. *Proc. Natl. Acad. Sci. USA* **99**, 1223–1228.

Schubert, U., Anton, L. C., Gibbs, J., Norbury, C. C., Yewdell, J. W., and Bennink, J. R. (2000). Rapid degradation of a large fraction of newly synthesized proteins by proteasomes. *Nature* **404,** 770–774.

Sone, T., Saeki, Y., Toh-e, A., and Yokosawa, H. (2004). Sem1p is a novel subunit of the 26S proteasome from *Saccharomyces cerevisiae*. *J. Biol. Chem.* **279,** 28807–28816.

Takada, K., Nasu, H., Hibi, N., Tsukada, Y., Ohkawa, K., Fujimuro, M., Sawada, H., and Yokosawa, H. (1995). Immunoassay for the quantification of intracellular multi-ubiquitin chains. *Eur. J. Biochem.* **233,** 42–47.

Takada, K., Hirakawa, T., Yokosawa, H., Okawa, Y., Taguchi, H., and Ohkawa, K. (2001). Isolation of ubiquitin-E2 (ubiquitin-conjugating enzyme) complexes from erythroleukaemia cells using immunoaffinity techniques. *Biochem. J.* **356,** 199–206.

Takeuchi, J., Fujimuro, M., Yokosawa, H., Tanaka, K., and Toh-e, A. (1999). Rpn9 is required for efficient assembly of the yeast 26S proteasome. *Mol. Cell. Biol.* **19,** 6575–6584.

Tamura, T., Tanaka, K., Tanahashi, N., and Ichihara, A. (1991). Improved method for preparation of ubiquitin-ligated lysozyme as substrate of ATP-dependent proteolysis. *FEBS Lett.* **292,** 154–158.

Tsurumi, C., Shimuzu, Y., Saeki, M., Kato, S., DeMartino, G. N., Slaughter, C. A., Fujimuro, M., Yokosawa, H., Yamasaki, M., Hendil, K. B., Toh-e, A., Tanahashi, N., and Tanaka, K. (1996). cDNA cloning and functional analysis of the p97 subunit of the 26S proteasome, a polypeptide identical to the type-1 tumor-necrosis-factor-receptor-associated protein-2/55.11. *Eur. J. Biochem.* **239,** 912–921.

Yamakami, M., Yoshimori, T., and Yokosawa, H. (2003). Tom1, a VHS domain-containing protein, interacts with Tollip, ubiquitin, and clathrin. *J. Biol. Chem.* **278,** 52865–52872.

Yokota, K., Kagawa, S., Shimuzu, Y., Akioka, H., Tsurumi, C., Noda, C., Fujimuro, M., Yokosawa, H., Fujiwara, T., Takahashi, E., Ohba, M., Yamasaki, M., DeMartino, G. N., Slaughter, C. A., Toh-e, A., and Tanaka, K. (1996). cDNA cloning of p112, the largest regulatory subunit of the human 26S proteasome, and functional analysis of its yeast homologue, Sen3p. *Mol. Biol. Cell.* **7,** 853–870.

[7] Application of Ubiquitin Immunohistochemistry to the Diagnosis of Disease

By JAMES LOWE, NEIL HAND, and R. JOHN MAYER

Abstract

Ubiquitin immunohistochemistry has changed understanding of the pathophysiology of many diseases, particularly chronic neurodegenerative diseases. Protein aggregates (inclusions) containing ubiquitinated proteins occur in neurones and other cell types in the central nervous system in afflicted cells. The inclusions are present in all the neurological illnesses, including Alzheimer's disease, Parkinson's disease, amyotrophic lateral sclerosis, polyglutamine diseases, and rarer forms of neurodegenerative disease. A new cause of cognitive decline in the elderly, "dementia with

Lewy bodies," accounting for some 15–30% of cases, was initially discovered and characterized by ubiquitin immunocytochemistry. The optimal methods for carrying out immunohistochemical analyses of paraffin-embedded tissues are described, and examples of all the types of intracellular inclusions detected by ubiquitin immunohistochemistry in the diseases are illustrated. The role of the ubiquitin proteasome system (UPS) in disease progression is being actively researched globally and increasingly, because it is now realized that the UPS controls most pathways in cellular homeostasis. Many of these regulatory mechanisms will be dysfunctional in diseased cells. The goal is to understand fully the role of the UPS in the disorders and then therapeutically intervene in the ubiquitin pathway to treat these incurable diseases.

Introduction

The understanding of the pathophysiology of disease is hampered by a scarcity of molecular markers, particularly markers that can be identified in paraffin-embedded tissues, which constitute most patient-derived archived tissue samples. The detection of protein aggregates by ubiquitin immunocytochemistry in a variety of diseases, particularly chronic neurodegenerative diseases, has dramatically changed understanding of these illnesses and has allowed the identification and characterization of new pathological lesions in diseases caused by the disease-associated aggregation of abnormal proteins. The wide recognition of dementia with Lewy bodies, which is the second most common cause of dementia after Alzheimer's disease, and the characterization of different forms of frontotemporal lobar degeneration were driven by the application of ubiquitin immunohistochemistry.

Ubiquitin is abundant in cells, perhaps accounting for 1% of cellular protein, and the advances in molecular neuropathology arising from ubiquitin immunocytochemistry have been based on (1) the fact that the antiubiquitin antibodies used in the immunohistochemical procedures predominately detect ubiquitin protein conjugates (ubiquitinated proteins) and (2) the fact that the disease-causing proteins in many chronic degenerative conditions (*e.g.*, neurodegenerative diseases) have a propensity to form intracellular aggregates containing the ubiquitinated proteins. Ubiquitin immunocytochemistry, therefore, easily detects these aggregates (inclusions) in cells in a wide range of diseases. Such protein aggregation disorders are not limited to the nervous system and are also seen in skeletal muscle and liver. Ubiquitin immunohistochemistry remains a highly effective method of "screening" tissues for the presence of abnormal protein aggregates. For a review of the biology of the ubiquitin system, the article by Glickman and Ciecanover details the mechanistic aspects of ubiquitination (Glickman and Ciechanover, 2002).

Antibodies to Ubiquitin

Several polyclonal and monoclonal antibodies directed to either free ubiquitin or ubiquitin-protein conjugates are available.

- The most widely used rabbit polyclonal antibodies are those raised against SDS-denatured conjugated ubiquitin attached to keyhole limpet hemocyanin (Haas and Bright, 1985). These tend to have a greater affinity for ubiquitin-protein conjugates compared with free ubiquitin.
- Monoclonal antibodies are generally raised against native ubiquitin or synthetic peptides and tend to have a high affinity for free ubiquitin compared with that for ubiquitin-protein conjugates.
- Monoclonal antibodies are available that detect ubiquitin–ubiquitin chains.
- Antibodies are available that detect UBB(+1), a mutant ubiquitin carrying a 19-amino acid C-terminal extension generated by a transcriptional dinucleotide deletion and seen in diseases such as Alzheimer's disease (Fischer *et al.*, 2003; van Leeuwen *et al.*, 2000).

These differences have practical implications for immunohistochemical detection of ubiquitin in tissue sections. The main use for immunohistochemical detection of ubiquitin is to detect pathological accumulations of proteins in different cell compartments. In such applications, polyclonal anti-ubiquitin antibodies, detecting ubiquitin-protein conjugates, are more routinely used than monoclonal antibodies. Antibodies to ubiquitin may detect ubiquitin fusion proteins and ubiquitinated proteins. Certain antibodies, particularly those raised against ubiquitin-protein conjugates, also detect ubiquitin–cross-reactive protein (UCRP), a 15-kDa protein that has sequence homology with ubiquitin (Ahrens *et al.*, 1990) Cross-reactivity with UCRP must be taken into account in studies using antiubiquitin antibodies as a detection system (Lowe *et al.*, 1995).

Technical Aspects of Immunohistochemical Staining

Fixation

Immunohistochemical detection of ubiquitin in tissue sections may be done in fresh-frozen or fixed material. Ubiquitin is an abundant protein in most cells. If fresh-frozen material is stained, background detection of ubiquitin is noted in cytoplasm and nucleus corresponding to the normal locations of this protein. After fixation of tissue in formaldehyde-containing aqueous fixatives and wax-embedding, the degree of background staining is greatly reduced, allowing pathological lesions to be better seen.

Immunohistochemical Demonstration of Ubiquitin

The histological demonstration of ubiquitin may be easily and reliably achieved using immunohistochemistry on paraffin wax sections of formalin-fixed tissue. Although several different types of immunohistochemical staining procedures are available, a generic protocol using a conventional triple-step procedure such as a labeled avidin/avidin-biotin technique is described (Miller, 2002). The most common staining techniques used are those that rely on peroxidase-conjugated systems in which the enzyme may be detected by the chromogen diaminobenzidine (DAB).

The dilution of primary antibody should be determined by experiments, because there are variations in products according to the supplier. Techniques for immunohistochemical staining using monoclonal antibodies, fluorescent labeling, and different signal amplification steps are well covered in standard textbooks (Miller, 2002).

Immunohistochemical staining should be performed in an enclosed chamber with a moist atmosphere. Sections are often counterstained, usually with hematoxylin.

The methods described are based on the use of polyclonal antibodies. For the demonstration of Mallory bodies in formalin-fixed tissue, pretreatment of the sections using heat-induced epitope retrieval before immunostaining is required. There are several ways this may be achieved, but the use of a microwave oven and a sodium citrate solution are described, because this has generally proved successful.

Methods

Fixation

Ten percent formol saline, for 24–48 h.
Paraffin Wax Sections. Sections are cut between 3 and 10 μm in thickness and mounted on slides coated with 3-aminopropylethoxy-silane to reduce section loss. The sections are dried at room temperature for at least 30 min, followed by 15 min on a hotplate at approximately 62° to melt the wax.

Labeled Avidin/Avidin-Biotin Immunohistochemical Technique

1. De-wax sections in xylene for 5 min (adequate for 5-μm thick sections).
2. Rinse in a fresh bath of xylene.
3. Rinse in three volumes of ethanol or industrial alcohol (IMS) methcol.
4. Wash well in running tap water.
4a. For the demonstration of Mallory bodies, insert the appropriate pretreatment method here (see below).
5. Rinse in ethanol or industrial alcohol (IMS) methcol.
6. Block endogenous peroxidase by incubating in peroxidase-blocking solution (hydrogen peroxide) for 15 min.
7. Rinse in ethanol or industrial alcohol (IMS) methcol.
8. Wash well in deionized water.
9. Rinse in 0.005 M Tris/HCl buffered saline (TBS) pH 7.6.
10. Incubate in 1:5 dilution of normal swine or goat serum made up in 0.005 M TBS for 20 min.
11. Drain and wipe off excess serum and incubate in optimally diluted unconjugated primary antibody for 1 hour.
12. Wash vigorously in 0.005 M Tris/HCl buffered saline (TBS) pH 7.6.
13. Wash in TBS for 5 min.
14. Incubate in secondary antibodies conjugated to biotin for 30 min.
15. Jet wash in TBS.
16. Wash in TBS for 5 min.
17. Incubate in horseradish peroxidase (HRP) conjugated to Streptavidin for 30 min.
18. Jet wash in TBS.
19. Wash in TBS for 5 min.
20. Incubate in DAB solution for 10 min.
21. Wash in running tap water.
22. Counterstain in hematoxylin. (If necessary, differentiate staining and then blue sections.)
23. Wash in running tap water.
24. Dehydrate through three baths of ethanol or industrial alcohol (IMS) methcol.
25. Clear in two baths of xylene.
26. Mount in a synthetic mountant (*e.g.*, DPX).

The overall outcome is that nuclei are stained blue and sites of peroxidase activity are brown.

Notes: Do not allow the sections to dry out at any stage.

Steps 10–20 should be performed in an incubating chamber, except when washing in TBS.

Antigen Retrieval with 10 mM Sodium Citrate Solution Using a Microwave Oven

1. Wash slides in deionized water.
2. Place the slides in a plastic rack into a suitable (deep) plastic dish containing sufficient 10 mM sodium citrate buffer at pH 6.0 to ensure that the slides are completely immersed.
3. Place the container into the microwave oven in the center of the turntable with the lid on so that there is a small gap for air/vapor to escape and shut the oven door.
4. Plug in the microwave oven and select the power and temperature switches for operation. Key in a suitable time. This is related to the volume of solution used and the power output, but for routine use, 560 W with a volume of 850 ml for 23 min is successful.
5. Press the "START" button. The temperature will now slowly rise, and the solution should be seen to be boiling after approximately 12 min. The level of fluid should not fall below the level of the slides if the preceding procedure is followed.
6. When the time set has been completed, the microwave oven will sound. Wearing heat-resistant/waterproof gloves, carefully remove the container, take it to a sink, and gently immerse the container in cold running water until slides can be safely removed from the container and left in cold running water for 20 min.
7. Remove the rack of slides and wash in deionized water.

Note: Do not allow the sections to dry out at any stage.

Solutions

Peroxidase Blocking Solution (Methanolic Hydrogen Peroxide). Methanol (50 ml), hydrogen peroxide (6%:5 ml). Pour the solution into a glass dish that will allow the sections to be fully covered. Be careful, because hydrogen peroxide is corrosive.

Primary Antibody Dilution. A working dilution is suggested of 1 in 500 for antiubiquitin for use in neurodegenerative diseases and 1 in 100 for liver Mallory bodies, both in 5% serum, with the antiubiquitin antibody from DakoCytomation (Z0458) http://www.dakocytomation.com/ The serum is made up in 0.005 M Tris-buffered saline (TBS).

Secondary Antibody. Swine anti-rabbit secondary antibodies conjugated to biotin diluted 1 in 500 made up in 5% serum is suggested. The serum is made up in 0.005 M TBS.

Avidin-Biotin Complex (ABC) Solution. Several commercial ABC kits are available (*e.g.*, http://www.vectorlabs.com http://www.dakocytomation.com.

Diaminobenzidine (DAB) Solution. Several commercial DAB solutions are available (*e.g.*, http://www.sigmaaldrich.com/.

Tris/HCl-Buffered Saline. TBS: 0.005 M, pH 7.6, sodium chloride (8.1g), Tris (hydroxymethyl) methylamine (0.6 g), hydrochloric acid (1 N:4.2 ml), deionized water (1l). Adjust to pH 7.6 with 1 M sodium hydroxide.

Sodium Citrate Solution, pH 6.0 (10 mM). Citric acid (2. 1 g), deionized water (1l). Adjust to pH 6.0 with 1 M sodium hydroxide.

Results

Generalities of Immunolocalization

The most well-documented aspect of ubiquitin function is its role in the degradation of proteins, which is performed in association with enzymes for ubiquitin ligation and conjugation of the proteasome. In many of the localizations of ubiquitin, it is assumed that proteins have been ubiquitinated as a signal targeting them for degradation, generally poly ubiquitination (Pickart, 2000). Under some circumstances, monoubiquitylation is a signal for stabilizing a protein (Dantuma and Masucci, 2002).

The detection of ubiquitin in a cell by immunohistochemistry may reflect localization because of the normal function of ubiquitin or it may indicate a pathological accumulation. There are several common themes related to the pathological accumulation of ubiquitin (Bossy-Wetzel *et al.*, 2004; Ciechanover and Brundin, 2003; Ehlers, 2004; Hatakeyama and Nakayama, 2003; Hernandez *et al.*, 2004; Korhonen and Lindholm, 2004; Kostova and Wolf, 2003; Layfield *et al.*, 2001; Lindsten and Dantuma, 2003; Mayer 2003; Petrucelli and Dawson, 2004).

Ubiquitin and the Cytoskeletal System. There is a constant association between ubiquitin immunostaining and abnormalities associated with the cytoskeletal system. Ubiquitin is associated with filamentous inclusions derived from proteins of the cytoskeleton and usually in association with chaperone proteins. A family of intermediate filament-ubiquitin inclusions has been described that also associate with αB-crystallin and enzymes of the ubiquitin system (Lowe et al., 2001; Mayer 2003; Mayer et al., 1989a,b, 1992, 1996).

Ubiquitin and the Aggresomal Response. The aggresomal response has been defined as a redistribution of intermediate filament proteins, driven by microtubule-mediated transport, to form a cage surrounding a pericentriolar core of aggregated, ubiquitinated protein. It has been proposed that the aggresome is a general response that occurs in cells when the production of misfolded proteins exceed the capacity of the proteasome. Positive ubiquitin immunostaining is a constant association with the aggresomal response. This type of response is now believed to be involved in several conditions in which there is abnormal accumulation of proteins in cells (Johnston et al., 1998; McNaught et al., 2002; Olanow et al., 2004; Riley et al., 2002; Taylor et al., 2003).

Ubiquitin and the Endosome-Lysosome System. Ubiquitin protein conjugates are taken up by the endosome-lysosome system. Monoubiquitination can act as a signal for sorting transmembrane proteins into intraluminal vesicles of multivesicular endosomes for later delivery to lysosomes (Raiborg et al., 2003). Immunostaining of tissues may reveal punctate staining of the cytoplasm that corresponds to localization to the endosome-lysosome system. This localization may also be seen in association with disease.

Ubiquitin immunoreactivity can be seen in the normal aging brain in several species as dot-like bodies in both cortex and white matter (Fig. 1)

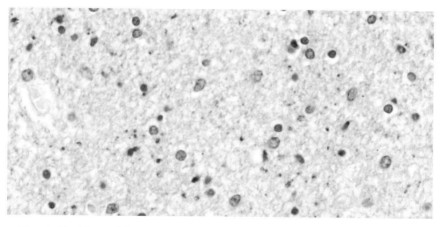

FIG. 1. Dot-like staining in the white matter of the brain. This is an age-related abnormality seen in most mammalian species. The intensity of this type of staining increases in many neurodegenerative diseases. (See color insert.)

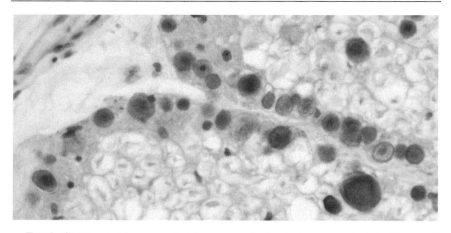

FIG. 2. Corpora amylacea are mainly composed of polyglucosan material with a small protein component. They are spherical bodies with a lamellar structure. It is common for these to be stained with antiubiquitin. It remains uncertain whether this is nonspecific antibody binding to polyglucosan or represents a biologically significant accumulation. The main importance of recognizing this pattern of staining is to avoid misinterpretation with other inclusions. (See color insert.)

(Dickson et al., 1990; Pappolla et al., 1989). Ultrastructural studies have shown some of this staining is in relation to axonal lysosomal dense bodies, whereas other staining is associated with the myelin sheath (Dickson et al., 1990; Migheli et al., 1992). The intensity of this dot-like staining appears increased in relation to a wide range of neurodegenerative conditions. Other structures stained by antiubiquitin that increase in aging and in neurodegenerative diseases are corpora amylacea (Fig. 2), which are accumulations of polyglucosan material mainly in astroglial cells.

Traumatic Damage to Axons and Axonal Spheroids

Normal axons do not exhibit ubiquitin-conjugate immunoreactivity in formalin-fixed wax-embedded material. After traumatic severance of nerves, in some neurodegenerative diseases and in aging, axons develop swelling of their proximal terminal portion to form axonal spheroids. Such spheroids contain microtubules, abundant neurofilament protein, synaptic vesicles, mitochondria, and numerous lysosome-related bodies that accumulate in the dilated end as a result of impaired axonal transport systems. Immunohistochemical studies indicate that spheroids are ubiquitin immunoreactive (Fig. 3) (Li and Farooque, 1996; Schultz et al., 2001; Schweitzer et al., 1993; Yamada et al., 1991). Axonal spheroids are a feature of rare

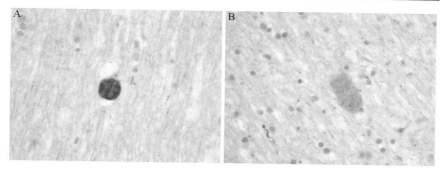

FIG. 3. (A) An axonal spheroid seen as a spherical swelling in the white matter. This example shows relatively dense immunostaining for antiubiquitin. (B) Many spheroids show less intense ubiquitin immunoreactivity and have a granular pattern of staining, possibly related to localization in lysosome-related vesicles. (See color insert.)

degenerative diseases of the nervous system, termed *neuroaxonal dystrophies*. Ubiquitin immunoreactivity is seen in spheroids in all such cases (Bacci *et al.*, 1994; Bonin *et al.*, 2004; Cochran *et al.*, 1991; Malandrini *et al.*, 1995; Moretto *et al.*, 1993; Wu *et al.*, 1993).

Neuronal Swelling

Ballooned neurons, seen in a variety of pathological conditions, inconsistently stain with antiubiquitin (Fig. 4) (Dickson *et al.*, 1996; Freiesleben *et al.*, 1997; Halliday *et al.*, 1995; Josephs *et al.*, 2003; Kato and Hirano, 1990; Kato *et al.*, 1992a,b; Lowe *et al.*, 1992; Smith *et al.*, 1992; Mori *et al.*, 1996). Such neuronal swelling can be seen as part of an axonal response in Pick's disease, corticobasal degeneration, or in some causes of frontotemporal lobar degeneration. If this pattern of staining is seen, a sensitive way of confirming neuronal swelling is by immunostaining for α B-crystallin.

Inclusion Bodies in Disease

Abnormal protein aggregation is now believed to be an important part of the pathogenesis of many diseases, especially those that are classified as neurodegenerative disorders. In these situations, ubiquitin antibodies frequently decorate protein aggregate–containing inclusions. It is also possible to immunostain inclusions for the underlying abnormal aggregated species. Table I shows the main cell types in which inclusions develop. Inclusions can be seen in a variety of different compartments and cell types. In some diseases, for example, the neurological disease multiple system

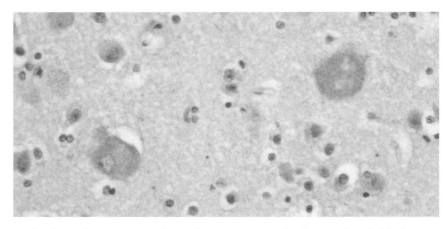

FIG. 4. Swollen neurons in the cerebral cortex generally show weak ubiquitin immunoreactivity. In some instances, swollen neurons do not show enhanced ubiquitin immunoreactivity. (See color insert.)

atrophy, inclusions can be seen in several cell types and several cell compartments at the same time.

Ubiquitin Immunoreactivity in Alzheimer's Disease (AD)

Ubiquitin immunoreactivity is seen in association with several pathological lesions of Alzheimer's disease (Figs. 5 and 6). Immunohistochemical staining of nerve cell processes (neurites) is seen around amyloid plaques (Fig. 5). These are of two types (He *et al.*, 1993; Yasuhara *et al.*, 1994):

1. Bulbous neuritic profiles (Fig. 5B), which are not usually associated with T protein immunoreactivity
2. Linear neuritic profiles (Fig. 5C), which are associated with T protein immunoreactivity

Neurofibrillary tangles are stained with antiubiquitin (Fig. 6A, B), including end-stage or "ghost" tangles (Bancher *et al.*, 1989; Brion *et al.*, 1989; Mori *et al.*, 1987; Perry *et al.*, 1987). The proportion of tangles detected by antiubiquitin is lower than that seen with other staining methods (He *et al.*, 1993). Antibodies to UBB +1 also detect neurofibrillary tangles (Fergusson *et al.*, 2000; van Leeuwen *et al.*, 2000, 2002). Neurofibrillary tangles in other disorders such as progressive supranuclear palsy and postencephalitic parkinsonism are also detected by antiubiquitin

TABLE I
HISTOPATHOLOGICAL OCCURRENCE OF UBIQUITINATED PROTEINS IN CELLULAR INCLUSIONS

Cell type	Location	Protein	Inclusion
Neurons	Endosome/lysosome system	Storage products	Dotlike bodies
	Nerve cell processes	α-synuclein	Lewy neurites in Parkinson's disease and dementia with Lewy bodies
		T protein	Neurites in Alzheimer's disease and inherited tauopathies
		Unknown protein	Neurites in FTLD
	Cytosolic	Neurofilament protein	Neurofilament inclusions
		T protein	Tangles in Alzheimer's disease
			Pick bodies
			Tangles in sporadic and inherited tauopathies
		Unknown protein	ALS inclusions
			Inclusions of FTLD
		α-synuclein	Lewy bodies
			Neuronal inclusions in MSA
	Nuclear inclusions	Polyglutamine expansion proteins	Inclusions in Huntington's disease and several hereditary ataxias
		α-synuclein	Nuclear inclusions of multiple system atrophy
		Unknown protein	Nuclear inclusions in FTLD
			Marinesco body
			Neuronal nuclear inclusion disease
Glial cell	Cytosolic	α-synuclein and T protein	Glial inclusions in MSA
Astrocytes		αB-crystallin	Rosenthal fiber
Hepatocyte	Cytosol	Cytokeratin	Mallory body
Skeletal muscle	Cytosol	Desmin and αB-crystallin	Spheroid bodies
			Desmin myopathies
		Unknown protein	Inclusion body myopathy
	Nucleus	poly(A) binding protein 2	Oculopharyngeal muscular dystrophy (OPMD)

FTLD, frontotemporal lobe dementia; MSA, multiple system atrophy.

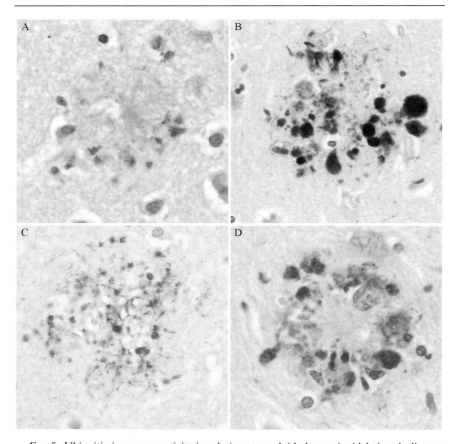

FIG. 5. Ubiquitin immunoreactivity in relation to amyloid plaques in Alzheimer's disease. The amyloid component is not specifically stained in these preparations. Ubiquitin immunoreactivity is either in lysosome-related structures or with accumulated T protein in nerve cell processes (neurites). (A) In early and loose deposits of amyloid, dot-like staining is seen together with fine linear neurites. (B) Large bulbous, dense-staining neurites are seen in relation to some amyloid plaques. (C) This pattern of thin, wispy neurite staining in relation to plaques is generally associated with T immunoreactivity. (D) A classical cored neuritic plaque showing a mixture of bulbous and linear ubiquitin-immunoreactive neurites. (See color insert.)

antibodies (Dale *et al.*, 1991; Lennox *et al.*, 1988; Love *et al.*, 1995; Lowe *et al.*, 1993; Murphy *et al.*, 1990; Wong *et al.*, 1996).

Neuropil threads are detected by antiubiquitin (Fig. 6C). These are nerve cell processes containing abnormally phosphorylated T protein, not associated with extracellular amyloid (Iwatsubo *et al.*, 1992).

Pyramidal neurons in the hippocampus in AD may show granulovacuolar degeneration (GVD), which commonly immunostains for ubiquitin (Fig. 6D) in addition to a variety of other proteins. GVD is believed to represent an autophagic phenomenon (Dickson et al., 1993; Okamoto et al., 1991)

CA1 pyramidal cells in AD, aging, and some other degenerative brain diseases are commonly surrounded by ubiquitin-positive *granules* measuring 1–4 μm in diameter, termed *perisomatic granules* (PSG) (Fig. 6E) (Probst et al., 2001) These are identical to previously described nonplaque dystrophic dendrites detected by antibodies to glutamate receptors (Aronica et al., 1998).

Ubiquitin Immunoreactivity in Parkinson's Disease and the Lewy Body Disorders

Parkinson's disease and the Lewy body disorders are united by pathological accumulation of α-synuclein in inclusion bodies termed *Lewy bodies*, as well as in nerve cell processes (Lewy body neurites) (Fig. 7A) (Dickson, 2001; Galvin et al., 2001; Goedert, 2001; Ince, 2001; McKeith and Mosimann, 2004; McKeith et al., 2004; Spillantini and Goedert, 2000).

Ubiquitin immunostaining is seen in a high proportion of Lewy bodies and in Lewy neurites. In brainstem Lewy bodies, with a characteristic core and halo structure, ubiquitin immunoreactivity is seen in the halo region. In cortical Lewy bodies, there is typically uniform ubiquitin immunoreactivity (Fig. 7B, C).

Trinucleotide Repeat Disorders (Including Spinocerebellar Ataxias, Huntington's Disease, Kennedy's Syndrome)

This family of diseases is associated with a common genetic mechanism, that of expansion of a trinucleotide repeat sequence in the genes of unrelated proteins. In each disease, there is production of an abnormal protein that may accumulate as an inclusion body either in the nucleus or in the cytoplasm (Cummings and Zoghbi, 2000a,b; Everett and Wood, 2004; Rubinsztein et al., 1999). The most common repeat accounting for disease is CAG, leading to polyglutamine tracts. These disorders have therefore been termed *polyglutamine diseases*.

The most studied condition associated with a trinucleotide repeat is Huntington's disease. Intranuclear ubiquitin-immunoreactive inclusions are seen in cases with a long expansion repeat length (Fig. 8). Similar inclusions are seen in transgenic mouse models of disease. In Huntington's disease, abnormal protein associated with ubiquitin also accumulates in nerve cell processes (Jackson et al., 1995) (Fig. 8B).

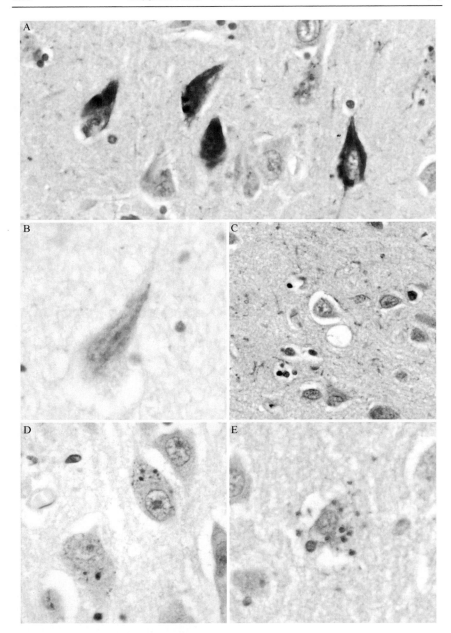

FIG. 6. Ubiquitin immunostaining in Alzheimer's disease. (A) Neurofibrillary tangles can stain strongly with antiubiquitin. Most are not stained. (B) In many cases of Alzheimer's disease, tangles that stain for ubiquitin do so weakly. (C) Neuropil threads are nerve cell

Several of the hereditary cerebellar ataxias are also examples of trinucleotide repeat diseases (Table II).

In each case, intranuclear inclusions can be stained for ubiquitin, which is believed to be conjugated with the abnormal polyglutamine-containing protein in association with other nuclear proteins.

Aging and Rarer Neurological Illnesses

Nuclear Filamentous Inclusions. Marinesco bodies are ubiquitinated filamentous inclusions seen in neuronal nuclei in association with aging (Fig. 8C) (Dickson *et al.*, 1990); similar inclusions are observed in a rare condition termed *nuclear hyaline inclusion disease* (Funata *et al.*, 1990). It has been shown that polyglutamine-containing proteins derived from genes with normal CAG repeats are incorporated into such inclusions (Fujigasaki *et al.*, 2000, 2001; Kettner *et al.*, 2002).

Ubiquitin and Prion Encephalopathies. In prion diseases, ubiquitin immunostaining has been found in lysosome-related vesicles both in human and animal disease (Ironside *et al.*, 1993; Laszlo *et al.*, 1992; Lowe *et al.*, 1990, 1992; Migheli *et al.*, 1991). Ubiquitin immunoreactivity is seen in a dot-like distribution at the periphery of prion protein amyloid plaques and in a fine punctate pattern in the neuropil around areas of spongiform change. Some neurons show ubiquitin-reactive dot-like structures, and some microglia in areas of spongiform change are ubiquitin immunoreactive (Cammarata and Tabaton, 1992; Ironside *et al.*, 1993). Ubiquitination of PrP has been demonstrated but is thought to be a late event that occurs after the formation of protease-resistant PrP(Sc) (Kang *et al.*, 2004).

Amyotrophic Lateral Sclerosis. Amyotrophic lateral sclerosis is a neurodegenerative disease characterized by loss of motor neurons in the brain and spinal cord, leading to muscle weakness and death in 2–5 y (Lowe, 1994). Motor neurons contain ubiquitin-immunoreactive inclusions in both sporadic and familial cases (Leigh *et al.*, 1988; Lowe *et al.*, 1988, 1989; Migheli *et al.*, 1990; Murayama *et al.*, 1990; Schiffer *et al.*, 1991). In a minority of familial cases, inclusions contain superoxide dismutase. However, the abnormal proteins underlying other familial and inherited cases remain uncertain (Wood *et al.*, 2003). In histological sections, inclusions can appear as granular punctuate ubiquitin staining, loose wispy skeins

---

processes containing abnormally accumulated τ protein. Ubiquitin immunostaining can detect a proportion of such neurites. (D) Granulovacuolar degeneration seen in pyramidal cells of the hippocampus can be stained for ubiquitin, but this is inconsistently present. (E) A pyramidal cell in the hippocampus surrounded by ubiquitin-immunoreactive perisomatic granules. (See color insert.)

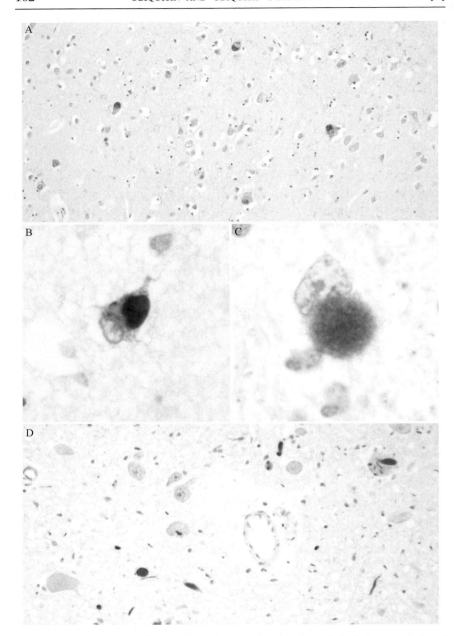

FIG. 7. Lewy body pathology. Lewy bodies are related to the pathological accumulation of α-synuclein. (A) Cortical Lewy bodies are readily detected with antiubiquitin. (B) At high magnification, some cortical Lewy bodies are densely stained for ubiquitin. (C) Many Lewy

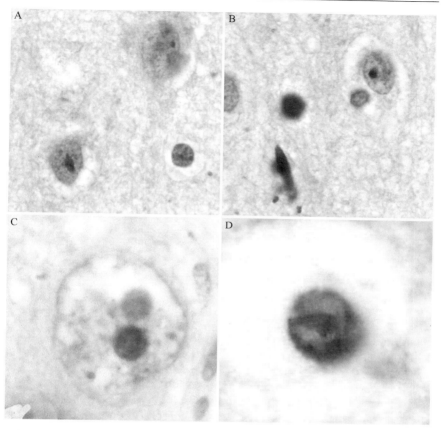

Fig. 8. (A) Huntington's disease neuronal nuclei contain ubiquitin-immunoreactive inclusions. These would also immunostain with antibodies to huntingtin or polyglutamine. (B) Huntington's disease showing a nuclear inclusion and accumulation of ubiquitinated protein in nerve cell processes as Huntington neurites. (C) A neuronal nucleus showing a Marinesco body immunoreactive for ubiquitin. The blue-stained round profile next to it is the nucleolus. (D) Neuronal inclusions immunoreactive for ubiquitin in a case of frontotemporal lobar degeneration associated with amyotrophic lateral sclerosis. (See color insert.)

bodies show less intense staining with a somewhat granular pattern. (D) Ubiquitin staining of the dorsal vagal nucleus showing Lewy neurites as linear, ovoid, and beaded structures. Such neurites are seen in regions affected by Lewy bodies and are also based on the accumulation of α-synuclein. (See color insert.)

TABLE II
TRINUCLEOTIDE REPEAT DISORDERS WITH INCLUSIONS DETECTED BY
UBIQUITIN IMMUNOHISTOCHEMISTRY

Disease	Protein	Aggregates
Huntington	Huntingtin	Nuclear
Huntington 2	Junctophilin-3	Nuclear
SCA1	Ataxin-1	Nuclear
SCA2	Ataxin-2	Cytoplasmic
SCA3	Ataxin-3	Nuclear
SCAT	Ataxin-7	Nuclear
SBMA (Kennedy's disease)	Androgen receptor	Nuclear
DRPLA	Atrophin-1	Nuclear

SCA, Spinocerebellar ataxia; SBMA, spinobulbar muscular atrophy; DRPLA, dentato-rubropallidonigroluysial atrophy.

of filaments, or rounded dense-stained inclusions often with a frayed, filamentous margin (Fig. 9A–F).

Similar inclusions are a characteristic of the most common type of frontotemporal dementia, where they are seen in neurons of the hippocampal dentate granule cells and in the small neurons in layer 2 of the frontal and temporal cortex (Fig. 10) (Chang et al., 2004; Hodges et al., 2004; Jackson and Lowe, 1996; Josephs et al., 2004a, b; Kertesz et al., 2000; Mackenzie and Feldman, 2004; Rosso et al., 2001; Rossor et al., 2000; Toyoshima et al., 2003; Woulfe et al., 2001).

In some patients with frontotemporal lobar degeneration, ubiquitin immunoreactive neurites are seen in the cerebral cortex with or without neuronal inclusions characteristic of motor neuron disease (Fig, 10) (Tolnay and Probst, 1995, 2001).

Neurofilament Inclusion Disease. Ubiquitin immunoreactivity is present but generally weak in cytoplasmic neuronal inclusions seen in a small number of patients with frontotemporal dementia. This condition has been termed *neurofilament inclusion disease* and has been cited as an example of a neurofilamentopathy (Bigio et al., 2003; Cairns and Armstrong, 2004; Cairns et al., 2003; 2004; Josephs et al., 2003; Mackenzie and Feldman, 2004).

Multiple System Atrophy. Multiple system atrophy is a neurodegenerative disease characterized by the presence of inclusions in neuronal cytoplasm, neuronal nuclei, and the cytoplasm of oligodendroglial cells. Patients develop autonomic failure, cerebellar ataxia, or parkinsonism. The inclusions have been shown to contain α-synuclein, α-B crystallin, and T protein. They are immunoreactive for ubiquitin, but this is generally faint

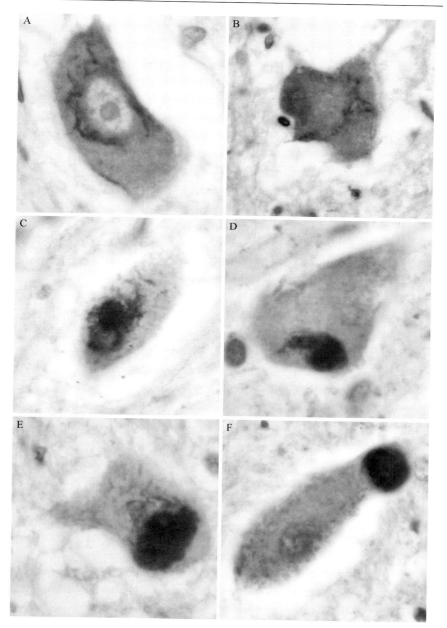

FIG. 9. Inclusions in amyotrophic lateral sclerosis. (A and B) Filamentous inclusions termed *skeins* are seen in affected motor neurons. (C–E) Many inclusions appear as solid masses of material with frayed filamentous margins. (F) A minority of inclusions have a dense-stained solid spherical appearance. (See color insert.)

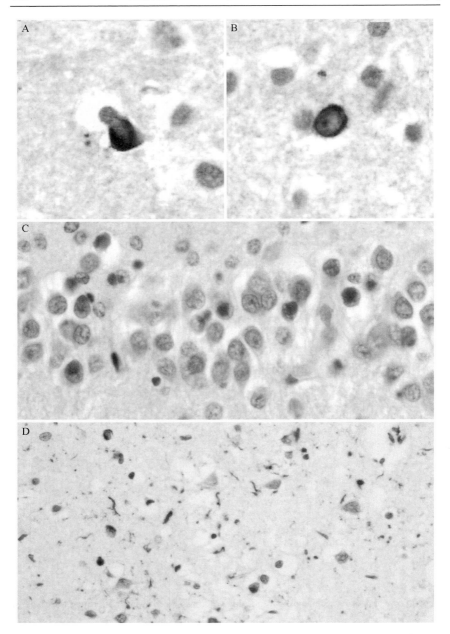

FIG. 10. In some cases of frontotemporal lobar degeneration, patients have pathological changes detected with antiubiquitin in nonmotor areas of the brain that are similar to those seen in amyotrophic lateral sclerosis. (A and B) Ubiquitin-immunoreactive inclusions seen in

(Fig. 11) (Arai *et al.*, 1994; Gai *et al.*, 1999; Papp and Lantos, 1992; Papp *et al.*, 1989; Tamaoka *et al.*, 1995).

Inclusions in Inferior Olivary Neurons. Ubiquitin-immunoreactive inclusions can be seen as an age-related finding in the neurons of the inferior olive. This is a nucleus found in the lower part of the medulla oblongata. These typically have a multilobed appearance resembling a blackberry (Fig. 12) (Kato *et al.*, 1990).

Glial Inclusions in Alexander's Disease and in Rosenthal Fibers. Rosenthal fibers are inclusions seen in astrocyte processes and contain glial fibrillary acidic protein, α-B crystallin, and ubiquitin. These inclusions are seen in certain types of astrocytic brain tumor and in a rare inherited brain disease called Alexander's disease. Ubiquitin immunostaining characteristically stains the periphery of Rosenthal fibers, leaving the hyaline central region unstained (Fig. 13) (Goldman and Corbin, 1991; Lach *et al.*, 1991; Lowe *et al.*, 1988, 1989, 1992, 1993; Manetto *et al.*, 1989; Tomokane *et al.*, 1991; Tuckwell *et al.*, 1992).

Inclusion Body Myositis. Inclusion body myositis is a degenerative and inflammatory muscle disease usually seen in the elderly. The histology of affected muscle shows autophagic vacuoles termed *rimmed vacuoles* adjacent to abnormal intracellular filamentous inclusions. Immunohistochemical studies have shown that rimmed vacuoles contain ubiquitin and a series of other proteins characteristically seen in other neuronal degenerative diseases such as prion protein and T protein (Albrecht and Bilbao, 1993; Askanas *et al.*, 1991, 1992, 1994; Leclerc *et al.*, 1993; Prayson and Cohen, 1997; Semino-Mora and Dalakas, 1998).

Desmin Myopathies and Myopathies with Spheroids. Several myopathic diseases of muscle are characterized by the accumulation of desmin either as inclusion bodies termed cytoplasmic and spheroid bodies or as less well-defined masses of granulofilamentous material. These inclusions are associated with ubiquitin immunoreactivity. A familial form of desmin-related myopathy is associated with a missense mutation (R120G) in αB-crystallin. Some cases have also been associated with mutations in the desmin gene. In addition to desmin and ubiquitin, other proteins have been found in this type of inclusion, including αB-crystallin, amyloid precursor protein, and gelsolin (Fidzianska *et al.*, 1999; Goebel, 1995; Goebel and Warlo, 2000; Vajsar *et al.*, 1993; Wanschit *et al.*, 2002).

---

layer 2 cortical neurons. Some are paranuclear, whereas others form rings around the nucleus. (C) Many small ubiquitin-immunoreactive inclusions in the neurons of the hippocampal dentate granule cells. (D) Ubiquitin-immunoreactive neurites seen in the outer cortical layers of frontal and temporal lobe from patients with frontotemporal lobar degeneration. (See color insert.)

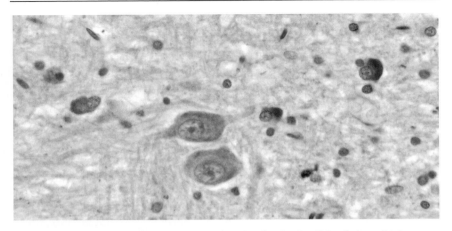

Fig. 11. Ubiquitin staining showing inclusions in oligodendroglial cells in multiple system atrophy. (See color insert.)

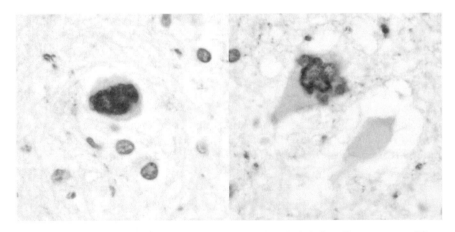

Fig. 12. Ubiquitin immunostaining showing inclusions in inferior olivary neurons. These are age-related and have no disease-specific association. The nature of these inclusions remains uncertain. (See color insert.)

Mallory's Hyaline. Mallory bodies are inclusions found in hepatocytes based on the accumulation of cytokeratin intermediate filaments. There are associated proteins including αB-crystallin and ubiquitin.

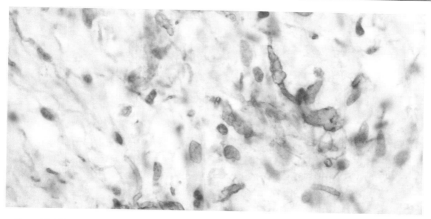

Fig. 13. Rosenthal fibers showing staining of the periphery with antiubiquitin. These inclusions are seen in astrocytic cells and are based on accumulation of αB-crystallin and glial fibrillary acidic protein. (See color insert.)

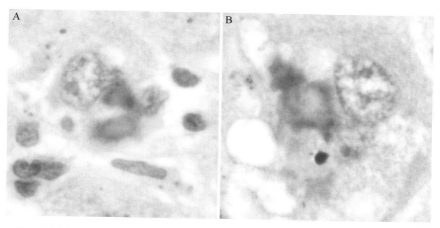

Fig. 14. Mallory's hyaline (A and B) seen in mouse liver after induction by griseofulvin treatment. Ubiquitin immunoreactivity is located at the periphery of inclusions. (See color insert.)

(Bardag-Gorce *et al.*, 2003, 2004; French *et al.*, 2001; Jensen and Gluud, 1994; Lowe *et al.*, 1988, 1992; Manetto *et al.*, 1989; Muller *et al.*, 2004; Ohta *et al.*, 1988; Riley *et al.*, 2002; Stumptner *et al.*, 2002; Vyberg and Leth,

1991). This type of inclusion is associated with alcoholic liver disease and less common forms of liver pathology. Ubiquitin immunostaining especially stains the periphery of inclusions (Fig. 14). Mallory's hyaline corresponds to one of the group of ubiquitin-intermediate filament-crystallin containing inclusions.

Discussion

Ubiquitin immunohistochemistry remains an extremely useful method for detecting pathological accumulations of protein in disease. However, interpretation of patterns of immunostaining requires care and would usually involve the further characterization of any pathology by an extended panel of antibodies to characterize underlying aggregated protein.

The award of the 2004 Nobel Prize for Chemistry to the founders of the ubiquitin proteasome system of intracellular proteolysis (UPS) is a timely highlight of the importance of the UPS for cell physiology and disease pathogenesis. There are now several diseases associated with genetic defects in the UPS that span a range of pathologies to include tumors, inflammatory disorders, and neurodegenerative diseases. Chains of ubiquitin linked to each other by isopeptide bonds by way of lysine 48 of each ubiquitin are attached to target proteins as a signal for degradation by the 26S proteasome. Ubiquitins can also be linked to each other through other ubiquitin lysine residues (*e.g.*, lysine 29- or lysine 63-linked ubiquitins). The linkage of these chains to target proteins may not be signals for recognition by the 26S proteasome. These linked chains of ubiquitins act as signals for downstream receptors and adaptors in signal transduction pathways (*e.g.*, in NFKB signaling). Appropriate antibodies to these chains will be of great value in molecular neuropathology. In addition, ubiquitin has several paralogs in mammalian cells (*e.g.*, SUMO and Urml) with incompletely characterized functions. It is predictable that with appropriate antibodies to these paralogs immunohistochemistry will identify new disease modalities.

Acknowledgments

We wish to acknowledge the MRC, Alzheimer's Research Trust, Motor Neurone Disease Association, and the Parkinson's Disease Society for support of some of the work.

References

Ahrens, P. B., Besancon, F., Memet, S., and Ankel, H. (1990). Tumour necrosis factor enhances induction by beta-interferon of a ubiquitin cross-reactive protein. *J. Gen. Virol.* **71**(Pt. 8), 1675–1682.

Albrecht, S., and Bilbao, J. M. (1993). Ubiquitin expression in inclusion body myositis. An immunohistochemical study. *Arch. Pathol. Lab. Med.* **117**, 789–793.

Arai, N., Papp, M. I., and Lantos, P. L. (1994). New observation on ubiquitinated neurons in the cerebral cortex of multiple system atrophy (MSA). *Neurosci. Lett.* **182**, 197–200.

Aronica, E., Dickson, D. W., Kress, Y., Morrison, J. H., and Zukin, R. S. (1998). Non-plaque dystrophic dendrites in Alzheimer hippocampus: A new pathological structure revealed by glutamate receptor immunocytochemistry. *Neuroscience* **82**, 979–991.

Askanas, V., Engel, W. K., Bilak, M., Alvarez, R. B., and Selkoe, D. J. (1994). Twisted tubulofilaments of inclusion body myositis muscle resemble paired helical filaments of Alzheimer brain and contain hyperphosphorylated tau. *Am. J. Pathol.* **144**, 177–187.

Askanas, V., Serdaroglu, P., Engel, W. K., and Alvarez, R. B. (1992). Immunocytochemical localization of ubiquitin in inclusion body myositis allows its light-microscopic distinction from polymyositis. *Neurology* **42**, 460–461.

Askanas, V., Serdaroglu, P., King Engel, W., and Alvarez, R. B. (1991). Immunolocalization of ubiquitin in muscle biopsies of patients with inclusion body myositis and oculopharyngeal muscular dystrophy. *Neurosci. Lett.* **130**, 73–76.

Bacci, B., Cochran, E., Nunzi, M. G., Izeki, E., Mizutani, T., Patton, A., Hite, S., Sayre, L. M., Autilio-Gambetti, L., and Gambetti, P. (1994). Amyloid beta precursor protein and ubiquitin epitopes in human and experimental dystrophic axons. Ultrastructural localization. *Am. J. Pathol.* **144**, 702–710.

Bancher, C., Brunner, C., Lassmann, H., Budka, H., Jellinger, K., Seitelberger, F., Grundke-Iqbal, I., Iqbal, K., and Wisniewski, H. M. (1989). Tau and ubiquitin immunoreactivity at different stages of formation of Alzheimer neurofibrillary tangles. *Prog. Biol. Res.* **311**, 837–840.

Bardag-Gorce, F., Riley, N., Nguyen, V., Montgomery, R. O., French, B. A., Li, J., van Leeuwen, F. W., Lungo, W., McPhaul, L. W., and French, S. W. (2003). The mechanism of cytokeratin aggresome formation: The role of mutant ubiquitin (UBB+1). *Exp. Mol. Pathol.* **74**, 160–167.

Bardag-Gorce, F., Riley, N. E., Nan, L., Montgomery, R. O., Li, J., French, B. A., Lue, Y. H., and French, S. W. (2004). The proteasome inhibitor, PS-341, causes cytokeratin aggresome formation. *Exp. Mol. Pathol.* **76**, 9–16.

Bigio, E. H., Lipton, A. M., White, C. L., 3rd, Dickson, D. W., and Hirano, A. (2003). Frontotemporal and motor neurone degeneration with neurofilament inclusion bodies: Additional evidence for overlap between FTD and ALS. *Neuropathol. Appl. Neurobiol.* **29**, 239–253.

Bonin, M., Poths, S., Osaka, H., Wang, Y. L., Wada, K., and Riess, O. (2004). Microarray expression analysis of gad mice implicates involvement of Parkinson's disease associated UCH-L1 in multiple metabolic pathways. *Brain Res. Mol. Brain Res.* **126**, 88–97.

Bossy-Wetzel, E., Schwarzenbacher, R., and Lipton, S. A. (2004). Molecular pathways to neurodegeneration. *Nat. Med.* **10**(Suppl.), S2–S9.

Brion, J. P., Power, D., Hue, D., Couck, A. M., Anderton, B. H., and Flament Durand, J. (1989). Heterogeneity of ubiquitin immunoreactivity in neurofibrillary tangles of Alzheimer's disease. *Neurochem. Int.* **14**, 121–128.

Cairns, N. J., and Armstrong, R. A. (2004). Quantification of the pathological changes in the temporal lobe of patients with a novel neurofilamentopathy: Neurofilament inclusion disease (NID). *Clin. Neuropathol.* **23,** 107–112.

Cairns, N. J., Jaros, E., Perry, R. H., and Armstrong, R. A. (2004). Temporal lobe pathology of human patients with neurofilament inclusion disease. *Neurosci. Lett.* **354,** 245–247.

Cairns, N. J., Perry, R. H., Jaros, E., Burn, D., McKeith, I. G., Lowe, J. S., Holton, J., Rossor, M. N., Skullerud, K., Duyckaerts, C., Cruz-Sanchez, F. F., and Lantos, P. L. (2003). Patients with a novel neurofilamentopathy: Dementia with neurofilament inclusions. *Neurosci. Lett.* **341,** 177–180.

Cammarata, S., and Tabaton, M. (1992). Ubiquitin-reactive axons have a widespread distribution and are unrelated to prion protein plaques in Creutzfeldt-Jakob disease. *J. Neurol. Sci.* **110,** 32–36.

Chang, H. T., Cortez, S., Vonsattel, J. P., Stopa, E. G., and Schelper, R. L. (2004). Familial frontotemporal dementia: A report of three cases of severe cerebral atrophy with rare inclusions that are negative for tau and synuclein, but positive for ubiquitin. *Acta Neuropathologica* **108,** 10–16.

Ciechanover, A., and Brundin, P. (2003). The ubiquitin proteasome system in neurodegenerative diseases: Sometimes the chicken, sometimes the egg. *Neuron* **40,** 427–446.

Cochran, E., Bacci, B., Chen, Y., Patton, A., Gambetti, P., and Autilio Gambetti, L. (1991). Amyloid precursor protein and ubiquitin immunoreactivity in dystrophic axons is not unique to Alzheimer's disease. *Am. J. Pathol.* **139,** 485–489.

Cummings, C. J., and Zoghbi, H. Y. (2000a). Fourteen and counting: Unraveling trinucleotide repeat diseases. *Hum. Mol. Genet.* **9,** 909–916.

Cummings, C. J., and Zoghbi, H. Y. (2000b). Trinucleotide repeats: Mechanisms and pathophysiology. *Annu. Rev. Genomics Hum. Genet.* **1,** 281–328.

Dale, G. E., Leigh, P. N., Luthert, P., Anderton, B. H., and Roberts, G. W. (1991). Neurofibrillary tangles in dementia pugilistica are ubiquitinated. *J. Neurol. Neurosurg. Psychiatry* **54,** 116–118.

Dantuma, N. P., and Masucci, M. G. (2002). Stabilization signals: A novel regulatory mechanism in the ubiquitin/proteasome system. *FEBS Lett.* **529,** 22–26.

Dickson, D. W. (2001). Alpha-synuclein and the Lewy body disorders. *Curr. Opin. Neurol.* **14,** 423–432.

Dickson, D. W., Feany, M. B., Yen, S. H., Mattiace, L. A., and Davies, P. (1996). Cytoskeletal pathology in non-Alzheimer degenerative dementia: New lesions in diffuse Lewy body disease, Pick's disease, and corticobasal degeneration. *J. Neural Transm.* **47**(Suppl.), 31–46.

Dickson, D. W., Liu, W. K., Kress, Y., Ku, J., De, J. O., and Yen, S. H. (1993). Phosphorylated T immunoreactivity of granulovacuolar bodies (GVB) of Alzheimer's disease: Localization of two amino terminal tau epitopes in GVB. *Acta Neuropathol. (Berl.)* **85,** 463–470.

Dickson, D. W., Wertkin, A., Kress, Y., Ksiezak Reding, H., and Yen, S. H. (1990). Ubiquitin immunoreactive structures in normal human brains. Distribution and developmental aspects. *Lab. Invest.* **63,** 87–99.

Ehlers, M. D. (2004). Deconstructing the axon: Wallerian degeneration and the ubiquitin-proteasome system. *Trends Neurosci.* **27,** 3–6.

Everett, C., and Wood, N. W. (2004). Trinucleotide repeats and neurodegenerative disease. *Brain.* **127,** 2385–2405.

Fergusson, J., Landon, M., Lowe, J., Ward, L., van Leeuwen, F. W., and Mayer, R. J. (2000). Neurofibrillary tangles in progressive supranuclear palsy brains exhibit immunoreactivity to frameshift mutant ubiquitin-B protein. *Neurosci. Lett.* **279,** 69–72.

Fidzianska, A., Drac, H., and Kaminska, A. M. (1999). Familial inclusion body myopathy with desmin storage. *Acta Neuropathologica* **91,** 509–514.

Fischer, D. F., De Vos, R. A., Van Dijk, R., De Vrij, F. M., Proper, E. A., Sonnemans, M. A., Verhage, M. C., Sluijs, J. A., Hobo, B., Zouambia, M., Steur, E. N., Kamphorst, W., Hoi, E. M., and Van Leeuwen, F. W. (2003). Disease-specific accumulation of mutant ubiquitin as a marker for proteasomal dysfunction in the brain. *FASEB* **17,** 2014–2024.

Freiesleben, W., Soylemezoglu, F., Lowe, J., Janzer, R. C., and Kleihues, P. (1997). Wernicke's encephalopathy with ballooned neurons in the mamillary bodies: An immunohistochemical study. *Neuropathol. Appl. Neurobiol.* **23,** 36–42.

French, B. A., van Leeuwen, F., Riley, N. E., Yuan, Q. X., Bardag-Gorce, F., Gaal, K., Lue, Y. H., Marceau, N., and French, S. W. (2001). Aggresome formation in liver cells in response to different toxic mechanisms: Role of the ubiquitin-proteasome pathway and the frameshift mutant of ubiquitin. *Exp. Mol. Pathol.* **71,** 241–246.

Fujigasaki, H., Uchihara, T., Koyano, S., Iwabuchi, K., Yagishita, S., Makifiichi, T., Nakamura, A., Ishida, K., Toru, S., Hirai, S., Ishikawa, K., Tanabe, T., and Mizusawa, H. (2000). Ataxin-3 is translocated into the nucleus for the formation of intranuclear inclusions in normal and Machado-Joseph disease brains. *Exp. Neurol.* **165,** 248–256.

Fujigasaki, H., Uchihara, T., Takahashi, J., Matsushita, H., Nakamura, A., Koyano, S., Iwabuchi, K., Hirai, S., and Mizusawa, H. (2001). Preferential recruitment of ataxin-3 independent of expanded polyglutamine: An immunohistochemical study on Marinesco bodies. *J. Neurol. Neurosurg. Psychiatry* **71,** 518–520.

Funata, N., Maeda, Y., Koike, M., Yano, Y., Kaseda, M., Muro, T., Okeda, R., Iwata, M., and Yokoji, M. (1990). Neuronal intranuclear hyaline inclusion disease: Report of a case and review of the literature. *Clin. Neuropathol.* **9,** 89–96.

Gai, W. P., Power, J. H., Blumbergs, P. C., Culvenor, J. G., and Jensen, P. H. (1999). Alpha-synuclein immunoisolation of glial inclusions from multiple system atrophy brain tissue reveals multiprotein components. *J. Neurochem.* **73,** 2093–2100.

Galvin, J. E., Lee, V. M., and Trojanowski, J. Q. (2001). Synucleinopathies: Clinical and pathological implications. *Arch. Neurol.* **58,** 186–190.

Glickman, M., and Ciechanover, A. (2002). The ubiquitin-proteasome proteolytic pathway: Destruction for the sake of construction. *Physiol. Rev.* **82,** 373–428.

Goebel, H. H. (1995). Desmin-related neuromuscular disorders. *Muscle Nerve.* **18,** 1306–1320.

Goebel, H. H., and Warlo, I. (2000). Gene-related protein surplus myopathies. *Mol. Genet. Metab.* **71,** 267–275.

Goedert, M. (2001). Alpha-synuclein and neurodegenerative diseases. *Nat. Rev. Neurosci.* **2,** 492–501.

Goldman, J. E., and Corbin, E. (1991). Rosenthal fibers contain ubiquitinated alpha B-crystallin. *Am. J. Pathol.* **139,** 933–938.

Haas, A. L., and Bright, P. M. (1985). The immunochemical detection and quantitation of intracellular ubiquitin-protein conjugates. *J. Biol. Chem.* **260,** 12464–12473.

Halliday, G. M., Davies, L., McRitchie, D. A., Cartwright, H., Pamphlett, R., and Morris, J. G. (1995). Ubiquitin-positive achromatic neurons in corticobasal degeneration. *Acta Neuropathologica* **90,** 68–75.

Hatakeyama, S., and Nakayama, K. I. (2003). Ubiquitylation as a quality control system for intracellular proteins. *J. Biochem.* **134,** 1–8.

He, Y., Delaere, P., Duyckaerts, C., Wasowicz, M., Piette, F., and Hauw, J. J. (1993). Two distinct ubiquitin immunoreactive senile plaques in Alzheimer's disease: Relationship with the intellectual status in 29 cases. *Acta Neuropathol. (Berl.)* **86,** 109–116.

He, Y., Duyckaerts, C., Delaere, P., Piette, F., and Hauw, J. J. (1993). Alzheimer's lesions labelled by anti-ubiquitin antibodies: Comparison with other staining techniques. A study of 15 cases with graded intellectual status in ageing and Alzheimer's disease. *Neuropathol. Appl. Neurobiol.* **19,** 364–371.

Hernandez, F., Diaz-Hernandez, M., Avila, J., and Lucas, J. J. (2004). Testing the ubiquitin-proteasome hypothesis of neurodegeneration *in vivo*. *Trends Neurosci.* **27,** 66–69.

Hodges, J. R, Davies, R. R., Xuereb, J. H., Casey, B., Broe, M., Bak, T. H., Kril, J. J., and Halliday, G. M. (2004). Clinicopathological correlates in frontotemporal dementia. *Ann. Neurol.* **56,** 399–406.

Ince, P. (2001). Dementia with Lewy bodies. *Adv. Exp. Med. Biol.* **487,** 135–145.

Ironside, J. W., McCardle, L., Hayward, P. A., and Bell, J. E. (1993). Ubiquitin immunocytochemistry in human spongiform encephalopathies. *Neuropathol. Appl. Neurobiol.* **19,** 134–140.

Iwatsubo, T., Hasegawa, M., Esaki, Y., and Ihara, Y. (1992). Lack of ubiquitin immunoreactivities at both ends of neuropil threads. Possible bidirectional growth of neuropil threads. *Am. J. Pathol.* **140,** 277–282.

Jackson, M., Gentleman, S., Lennox, G., Ward, L., Gray, T., Randall, K., Morrell, K., and Lowe, J. (1995). The cortical neuritic pathology of Huntington's disease. *Neuropathol. Appl. Neurobiol.* **21,** 18–26.

Jackson, M., and Lowe, J. (1996). The new neuropathology of degenerative frontotemporal dementias. *Acta Neuropathologica* **91,** 127–134.

Jensen, K., and Gluud, C. (1994). The Mallory body: Theories on development and pathological significance (Part 2 of a literature survey). *Hepatology* **20,** 1330–1342.

Johnston, J. A., Ward, C. L., and Kopito, R. R. (1998). Aggresomes: A cellular response to misfolded proteins. *J. Cell Biol.* **143,** 1883–1898.

Josephs, K. A., Holton, J. L., Rossor, M. N., Braendgaard, H., Ozawa, T., Fox, N. C., Petersen, R. C., Pearl, G. S., Ganguly, M., Rosa, P., Laursen, H., Parisi, J. E., Waldemar, G., Quinn, N. P., Dickson, D. W., and Revesz, T. (2003). Neurofilament inclusion body disease: A new proteinopathy? *Brain* **126**(Pt. 10), 2291–2303.

Josephs, K. A., Holton, J. L., Rossor, M. N., Godbolt, A. K., Ozawa, T., Strand, K., Khan, N., Al-Sarraj, S., and Revesz, T. (2004a). Frontotemporal lobar degeneration and ubiquitin immunohistochemistry. *Neuropathol. Appl. Neurobiol.* **30,** 369–373.

Josephs, K. A., Jones, A. G., and Dickson, D. W. (2004b). Hippocampal sclerosis and ubiquitin-positive inclusions in dementia lacking distinctive histopathology. *Dementia Geriatr. Cogn. Dis.* **17,** 342–345.

Kang, S. C., Brown, D. R., Whiteman, M., Li, R., Pan, T., Perry, G., Wisniewski, T., Sy, M. S., and Wong, B. S. (2004). Prion protein is ubiquitinated after developing protease resistance in the brains of scrapie-infected mice. *J. Pathol.* **203,** 603–608.

Kato, S., and Hirano, A. (1990). Ubiquitin and phosphorylated neurofilament epitopes in ballooned neurons of the extraocular muscle nuclei in a case of Werdnig-Hoffmann disease. *Acta Neuropathologica* **80,** 334–337.

Kato, S., Hirano, A., Suenaga, T., and Yen, S. H. (1990). Ubiquitinated eosinophilic granules in the inferior olivary nucleus. *Neuropathol. Appl. Neurobiol.* **16,** 135–139.

Kato, S., Hirano, A., Umahara, T., Kato, M., Herz, F., and Ohama, E. (1992a). Comparative immunohistochemical study on the expression of alpha B crystallin, ubiquitin and stress-response protein 27 in ballooned neurons in various disorders. *Neuropath. Appl. Neurobiol.* **18,** 335–340.

Kato, S., Hirano, A., Umahara, T., Llena, J. F., Herz, F., and Ohama, E. (1992b). Ultrastructural and immunohistochemical studies on ballooned cortical neurons in

Creutzfeldt-Jakob disease: Expression of alpha B-crystallin, ubiquitin and stress-response protein 27. *Acta Neuropathologica* **84,** 443–448.

Kertesz, A., Kawarai, T., Rogaeva, E., St. George-Hyslop, P., Poorkaj, P., Bird, T. D., and Munoz, D. G. (2000). Familial frontotemporal dementia with ubiquitin-positive, tau-negative inclusions. [see comment]. *Neurology* **54,** 818–827.

Kettner, M., Willwohl, D., Hubbard, G. B., Rub, U., Dick, E. J., Jr., Cox, A. B., Trottier, Y., Auburger, G., Braak, H., and Schultz, C. (2002). Intranuclear aggregation of nonexpanded ataxin–3 in Marinesco bodies of the nonhuman primate substantia nigra. *Exp. Neurol.* **176,** 117–121.

Korhonen, L., and Lindholm, D. (2004). The ubiquitin proteasome system in synaptic and axonal degeneration: A new twist to an old cycle. *J. Cell Biol.* **165,** 27–30.

Kostova, Z., and Wolf, D. H. (2003). For whom the bell tolls: protein quality control of the endoplasmic reticulum and the ubiquitin-proteasome connection. *EMBO J.* **22,** 2309–2317.

Lach, B., Sikorska, M., Rippstein, P., Gregor, A., Staines, W., and Davie, T. R. (1991). Immunoelectron microscopy of Rosenthal fibers. *Acta Neuropathol.* **81,** 503–509.

Laszlo, L., Lowe, J., Self, T., Kenward, N., Landon, M., McBride, T., Farquhar, C., McConnell, I., Brown, J., Hope, J., *et al.* (1992). Lysosomes as key organelles in the pathogenesis of prion encephalopathies. *J. Pathol.* **166,** 333–341.

Layfield, R., Alban, A., Mayer, R. J., and Lowe, J. (2001). The ubiquitin protein catabolic disorders. *Neuropathol. Appl. Neurobiol.* **21,** 171–179.

Leclerc, A., Tome, F. M., and Fardeau, M. (1993). Ubiquitin and beta-amyloid-protein in inclusion body myositis (IBM), familial IBM-like disorder and oculopharyngeal muscular dystrophy: An immunocytochemical study. *Neuromuscul. Disord.* **3,** 283–291.

Leigh, P. N., Anderton, B. H., Dodson, A., Gallo, J. M., Swash, M., and Power, D. M. (1988). Ubiquitin deposits in anterior horn cells in motor neurone disease. *Neurosci. Lett.* **93,** 2–3.

Lennox, G., Lowe, J., Morrell, K., Landon, M., and Mayer, R. J. (1988). Ubiquitin is a component of neurofibrillary tangles in a variety of neurodegenerative diseases. *Neurosci. Lett.* **94,** 1–2.

Li, G. L., and Farooque, M. (1996). Expression of ubiquitin-like immunoreactivity in axons after compression trauma to rat spinal cord. *Acta Neuropathologica.* **91,** 155–160.

Lindsten, K., and Dantuma, N. P. (2003). Monitoring the ubiquitin/proteasome system in conformational diseases. *Ageing Res. Rev.* **2,** 433–449.

Love, S., Bridges, L. R., and Case, C. P. (1995). Neurofibrillary tangles in Niemann-Pick disease type C. *Brain* **118,** 119–129.

Lowe, J. (1994). New pathological findings in amyotrophic lateral sclerosis. *J. Neurol. Sci.* **124,** 38–51.

Lowe, J., Aldridge, F., Lennox, G., Doherty, F., Jefferson, D., Landon, M., and Mayer, R. J. (1989). Inclusion bodies in motor cortex and brainstem of patients with motor neurone disease are detected by immunocytochemical localisation of ubiquitin. *Neurosci. Lett.* **105,** 1–2.

Lowe, J., Blanchard, A., Morrell, K., Lennox, G., Reynolds, L., Billett, M., Landon, M., and Mayer, R. J. (1988). Ubiquitin is a common factor in intermediate filament inclusion bodies of diverse type in man, including those of Parkinson's disease, Pick's disease, and Alzheimer's disease, as well as Rosenthal fibres in cerebellar astrocytomas, cytoplasmic bodies in muscle, and Mallory bodies in alcoholic liver disease. *J. Pathol.* **155,** 9–15.

Lowe, J., Errington, D. R., Lennox, G., Pike, I., Spendlove, I., Landon, M., and Mayer, R. J. (1992). Ballooned neurons in several neurodegenerative diseases and stroke contain alpha B crystallin. *Neuropathol. Appl. Neurobiol.* **18,** 341–350.

Lowe, J., Fergusson, J., Kenward, N., Laszlo, L., Landon, M., Farquhar, C., Brown, J., Hope, J., and Mayer, R. J. (1992). Immunoreactivity to ubiquitin-protein conjugates is present early in the disease process in the brains of scrapie-infected mice. *J. Pathol.* **168,** 169–177.

Lowe, J., Lennox, G., Jefferson, D., Morrell, K., McQuire, D., Gray, T., Landon, M., Doherty, F. J., and Mayer, R. J. (1988). A filamentous inclusion body within anterior horn neurones in motor neurone disease defined by immunocytochemical localisation of ubiquitin. *Neurosci. Lett.* **94,** 1–2.

Lowe, J., Mayer, J., Landon, M., and Layfield, R. (2001). Ubiquitin and the molecular pathology of neurodegenerative diseases. *Adv. Exp. Med. Biol.* **487,** 169–186.

Lowe, J., Mayer, R. J., and Landon, M. (1993). Ubiquitin in neurodegenerative diseases. *Brain Pathol.* **3,** 55–65.

Lowe, J., McDermott, H., Kenward, N., Landon, M., Mayer, R. J., Bruce, M., McBride, P., Somerville, R. A., and Hope, J. (1990). Ubiquitin conjugate immunoreactivity in the brains of scrapie infected mice. *J. Pathol.* **162,** 61–66.

Lowe, J., McDermott, H., Loeb, K., Landon, M., Haas, A. L., and Mayer, R. J. (1995). Immunohistochemical localization of ubiquitin cross-reactive protein in human tissues. *J. Pathol.* **177,** 163–169.

Lowe, J., McDermott, H., Pike, I., Spendlove, I., Landon, M., and Mayer, R. J. (1992). alpha B crystallin expression in non-lenticular tissues and selective presence in ubiquitinated inclusion bodies in human disease. *J. Pathol.* **166,** 61–68.

Lowe, L., Morrell, K., Lennox, G., Landon, M., and Mayer, R. J. (1989). Rosenthal fibres are based on the ubiquitination of glial filaments. *Neuropathol. Appl. Neurobiol.* **15,** 45–53.

Mackenzie, I. R., and Feldman, H. (2004a). Neurofilament inclusion body disease with early onset frontotemporal dementia and primary lateral sclerosis. *Clin. Neuropathol.* **23,** 183–193.

Mackenzie, I. R., and Feldman, H. (2004b). Neuronal intranuclear inclusions distinguish familial FTD-MND type from sporadic cases. *Dementia Geriatr. Cogn. Disord.* **17,** 333–336.

Malandrini, A., Cavallaro, T., Fabrizi, G. M., Berti, G., Salvestroni, R., Salvadori, C., and Guazzi, G. C. (1995). Ultrastructure and immunoreactivity of dystrophic axons indicate a different pathogenesis of Hallervorden-Spatz disease and infantile neuroaxonal dystrophy. *Virchows Arch.* **427,** 415–421.

Manetto, V., Abdul-Karim, F. W., Perry, G., Tabaton, M., Autilio-Gambetti, L., and Gambetti, P. (1989). Selective presence of ubiquitin in intracellular inclusions. *Am. J. Pathol.* **134,** 505–513.

Mayer, R. J. (2003). From neurodegeneration to neurohomeostasis: The role of ubiquitin. *Drug News Perspect.* **16,** 103–108.

Mayer, R. J., Landon, M., Laszlo, L., Lennox, G., and Lowe, J. (1992). Protein processing in lysosomes: The new therapeutic target in neurodegenerative disease. *Lancet* **340,** 156–159.

Mayer, R. J., Lowe, J., Lennox, G., Doherty, F., and Landon, M. (1989a). Intermediate filaments and ubiquitin: A new thread in the understanding of chronic neurodegenerative diseases. *Prog. Clin. Biol. Res.* **317,** 809–818.

Mayer, R. J., Lowe, J., Lennox, G., Landon, M., Mac Lennan, K., and Doherty, F. J. (1989b). Intermediate filament-ubiquitin diseases: Implications for cell sanitization. *Biochem. Soc. Symp.* **55,** 193–201.

Mayer, R. J., Tipler, C., Arnold, J., Laszlo, L., Al-Khedhairy, A., Lowe, J., and Landon, M. (1996). Endosome-lysosomes, ubiquitin and neurodegeneration. *Adv. Exp. Med. Biol.* **389,** 261–269.

McKeith, L, Mintzer, J., Aarsland, D., Burn, D., Chiu, H., Cohen-Mansfield, J., Dickson, D., Dubois, B., Duda, J. E., Feldman, H., Gauthier, S., Halliday, G., Lawlor, B., Lippa, C., Lopez, O. L., Carlos Machado, J., O'Brien, J., Playfer, J., and Reid, W.D. L. B. International Psychogeriatric Association Expert Meeting. (2004). Dementia with Lewy bodies. *Lancet Neurol.* **3,** 19–28.

McKeith, I. G., and Mosimann, U. P. (2004). Dementia with Lewy bodies and Parkinson's disease. *Parkinsonism Rel. Disord.* **10**(Suppl. 1), S15–S18.

McNaught, K. S., Shashidharan, P., Perl, D. P., Jenner, P., and Olanow, C. W. (2002). Aggresome-related biogenesis of Lewy bodies. *Eur. J. Neurosci.* **16,** 2136–2148.

Migheli, A., Attanasio, A., Pezzulo, T., Gullotta, F., Giordana, M. T., and Schiffer, D. (1992). Age-related ubiquitin deposits in dystrophic neurites: An immunoelectron microscopic study. *Neuropathol. Appl. Neurobiol.* **18,** 3–11.

Migheli, A., Attanasio, A., Vigliani, M. C., and Schiffer, D. (1991). Dystrophic neurites around amyloid plaques of human patients with Gerstmann-Straussler-Scheinker disease contain ubiquitinated inclusions. *Neurosci. Lett.* **121,** 1–2.

Migheli, A., Autilio Gambetti, L., Gambetti, P., Mocellini, C., Vigliani, M. C., and Schiffer, D. (1990). Ubiquitinated filamentous inclusions in spinal cord of patients with motor neuron disease. *Neurosci. Lett.* **114,** 5–10.

Miller, K. (2002). Immunocytochemical Techniques. In "Theory and Practice of Histological Techniques." (J. Bancroft and M. Gamble, eds.). Churchill Livingstone, London.

Moretto, G., Sparaco, M., Monaco, S., Bonetti, B., and Rizzuto, N. (1993). Cytoskeletal changes and ubiquitin expression in dystrophic axons of Seitelberger's disease. *Clin. Neuropathol.* **12,** 34–37.

Mori, H., Kondo, J., and Diara, Y. (1987). Ubiquitin is a component of paired helical filaments in Alzheimer's disease. *Science* **235,** 1641–1644.

Mori, H., Oda, M., and Mizuno, Y. (1996). Cortical ballooned neurons in progressive supranuclear palsy. *Neurosci. Lett.* **209,** 109–112.

Muller, T., Langner, C., Fuchsbichler, A., Heinz-Erian, P., Ellemunter, H., Schlenck, B., Bavdekar, A. R., Pradhan, A. M., Pandit, A., Muller-Hocker, J., Melter, M., Kobayashi, K., Nagasaka, H., Kikuta, H., Muller, W., Tanner, M. S., Sternlieb, I., Zatloukal, K., and Denk, H. (2004). Immunohistochemical analysis of Mallory bodies in Wilsonian and non-Wilsonian hepatic copper toxicosis. *Hepatology* **39,** 963–969.

Murayama, S., Mori, H., Ihara, Y. Y., Bouldin, Y., Suzuki, K., and Tomonaga, M. (1990). Immunocytochemical and ultrastructural studies of lower motor neurons in amyotrophic lateral sclerosis. *Ann. Neurol.* **27,** 137–148.

Murphy, G. M., Jr., Eng, L. F., Ellis, W. G., Perry, G., Meissner, L. C., and Tinklenberg, J. R. (1990). Antigenic profile of plaques and neurofibrillary tangles in the amygdala in Down's syndrome: A comparison with Alzheimer's disease. *Brain Res.* **537,** 1–2.

Ohta, M., Marceau, N., Perry, G., Manetto, V., Gambetti, P., Autilio Gambetti, L., Metuzals, J., Kawahara, H., Cadrin, M., and French, S. W. (1988). Ubiquitin is present on the cytokeratin intermediate filaments and Mallory bodies of hepatocytes. *Lab. Invest.* **59,** 848–856.

Okamoto, K., Hirai, S., Iizuka, T., Yanagisawa, T., and Watanabe, M. (1991). Reexamination of granulovacuolar degeneration. *Acta Neuropathol.* **82,** 340–345.

Olanow, C. W., Perl, D. P., De Martino, G. N., and McNaught, K. S. (2004). Lewy-body formation is an aggresome-related process: A hypothesis. *Lancet Neurol.* **3,** 496–503.

Papp, M. I., Kahn, J. E., and Lantos, P. L. (1989). Glial cytoplasmic inclusions in the CNS of patients with multiple system atrophy (striatonigral degeneration, olivopontocerebellar atrophy and Shy-Drager syndrome). *J. Neurol. Sci.* **94,** 79–100.

Papp, M. I., and Lantos, P. L. (1992). Accumulation of tubular structures in oligodendroglial and neuronal cells as the basic alteration in multiple system atrophy. *J. Neurol. Sci.* **107,** 172–182.

Pappolla, M. A., Omar, R., and Saran, B. (1989). The 'normal' brain. 'Abnormal' ubiquitinylated deposits highlight an age-related protein change. *Am. J. Pathol.* **135,** 585–591.

Perry, G., Friedman, R., Shaw, G., and Chau, V. (1987). Ubiquitin is detected in neurofibrillary tangles and senile plaque neurites of Alzheimer disease brains. *Proc. Natl. Acad. Sci. USA* **84,** 3033–3036.

Petrucelli, L., and Dawson, T. M. (2004). Mechanism of neurodegenerative disease: Role of the ubiquitin proteasome system. *Ann. Med.* **36,** 315–320.

Pickart, C. M. (2000). Ubiquitin in chains. *Trends Biochem. Sci.* **25,** 544–548.

Prayson, R. A., and Cohen, M. L. (1997). Ubiquitin immunostaining and inclusion body myositis: Study of 30 patients with inclusion body myositis. *Hum. Pathol.* **28,** 887–892.

Probst, A., Herzig, M. C., Mistl, C., Ipsen, S., and Tolnay, M. (2001). Perisomatic granules (non-plaque dystrophic dendrites) of hippocampal CA1 neurons in Alzheimer's disease and Pick's disease: A lesion distinct from granulovacuolar degeneration. *Acta Neuropatholo.* **102,** 636–644.

Raiborg, C., Rusten, T. E., and Stenmark, H. (2003). Protein sorting into multivesicular endosomes. *Curr. Opin. Cell Biol.* **15,** 446–455.

Riley, N. E., Li, J., Worrall, S., Rothnagel, J. A., Swagell, C., van Leeuwen, F. W., and French, S. W. (2002). The Mallory body as an aggresome: *In vitro* studies. *Exp. Mol. Pathol.* **72,** 17–23.

Rosso, S. M., Kamphorst, W., de Graaf, B., Willemsen, R., Ravid, R., Niermeijer, M. F., Spillantini, M. G., Heutink, P., and van Swieten, J. C. (2001). Familial frontotemporal dementia with ubiquitin-positive inclusions is linked to chromosome 17q21–22. *Brain* **124**(Pt. 10), 1948–1957.

Rossor, M. N., Revesz, T., Lantos, P. L., and Wanington, E. K. (2000). Semantic dementia with ubiquitin-positive tau-negative inclusion bodies. *Brain* **123**(Pt. 2), 267–276.

Rubinsztein, D. C., Wyttenbach, A., and Rankin, J. (1999). Intracellular inclusions, pathological markers in diseases caused by expanded polyglutamine tracts? *J. Med. Genet.* **36,** 265–270.

Schiffer, D., Autilio Gambetti, L., Chio, A., Gambetti, P., Giordana, M. T., Gullotta, F., Migheli, A., and Vigliani, M. C. (1991). Ubiquitin in motor neuron disease: Study at the light and electron microscope. *J. Neuropathol. Exp. Neural.* **50,** 463–473.

Schultz, C., Dick, E. J., Cox, A. B., Hubbard, G. B., Braak, E., and Braak, H. (2001). Expression of stress proteins alpha B-crystallin, ubiquitin, and hsp27 in pallido-nigral spheroids of aged rhesus monkeys. *Neurobiol. Aging.* **22,** 677–682.

Schweitzer, J. B., Park, M. R., Einhaus, S. L., and Robertson, J. T. (1993). Ubiquitin marks the reactive swellings of diffuse axonal injury. *Acta Neuropathol. (Berl).* **85,** 503–507.

Semino-Mora, C., and Dalakas, M. C. (1998). Rimmed vacuoles with beta-amyloid and ubiquitinated filamentous deposits in the muscles of patients with long-standing denervation (postpoliomyelitis muscular atrophy): Similarities with inclusion body myositis. *Hum. Pathol.* **29,** 1128–1133.

Smith, T. W., Lippa, C. F., and de Girolami, U. (1992). Immunocytochemical study of ballooned neurons in cortical degeneration with neuronal achromasia. *Clin. Neuropathol.* **11,** 28–35.

Spillantini, M. G., and Goedert, M. (2000). The alpha-synucleinopathies: Parkinson's disease, dementia with Lewy bodies, and multiple system atrophy. *Ann. N. Y. Acad. Sci.* **920,** 16–27.

Stumptner, C., Fuchsbichler, A., Heid, H., Zatloukal, K., and Denk, H. (2002). Mallory body«a disease-associated type of sequestosome. *Hepatology* **35,** 1053–1062.

Tamaoka, A., Mizusawa, H., Mori, H., and Shoji, S. (1995). Ubiquitinated alpha B-crystallin in glial cytoplasmic inclusions from the brain of a patient with multiple system atrophy. *J. Neurol. Sci.* **129**, 192–198.

Taylor, J. P., Tanaka, F., Robitschek, J., Sandoval, C. M., Taye, A., Markovic-Plese, S., and Fischbeck, K. H. (2003). Aggresomes protect cells by enhancing the degradation of toxic polyglutamine-containing protein. *Hum. Mol. Genet.* **12**, 749–757.

Tolnay, M., and Probst, A. (1995). Frontal lobe degeneration: Novel ubiquitin-immunoreactive neurites within frontotemporal cortex. *Neuropathol. Appl. Neurobiol.* **21**, 492–497.

Tolnay, M., and Probst, A. (2001). Frontotemporal lobar degeneration. An update on clinical, pathological and genetic findings. *Gerontology* **47**, 1–8.

Tomokane, N., Iwaki, T., Tateishi, J., Iwaki, A., and Goldman, J. E. (1991). Rosenthal fibers share epitopes with alpha B-crystallin, glial fibrillary acidic protein, and ubiquitin, but not with vimentin: Immunoelectron microscopy with colloidal gold. *Am. J. Pathol.* **138**, 875–885.

Toyoshima, Y., Piao, Y. S., Tan, C. F., Morita, M., Tanaka, M., Oyanagi, K., Okamoto, K., and Takahashi, H. (2003). Pathological involvement of the motor neuron system and hippocampal formation in motor neuron disease-inclusion dementia. *Acta Neuropathologica* **106**, 50–56.

Tuckwell, D. S., Laszlo, L., and Mayer, R. J. (1992). 2,5-Hexanedione-induced intermediate filament aggregates contain ubiquitin-protein conjugate immunoreactivity and resemble Rosenthal fibres. *Neuropathol. Appl. Neurobiol.* **18**, 593–609.

Vajsar, J., Becker, L. E., Freedom, R. M., and Murphy, E. G. (1993). Familial desminopathy: Myopathy with accumulation of desmin-type intermediate filaments. *J. Neurol. Neurosurg. Psychiatry* **56**, 644–648.

van Leeuwen, F. W., Fischer, D. F., Benne, R., and Hoi, E. M. (2000). Molecular misreading. A new type of transcript mutation in gerontology. *Ann. N. Y. Acad. Sci.* **908**, 267–281.

van Leeuwen, F. W., Gerez, L., Benne, R., and Hoi, E. M. (2002). +1 Proteins and aging. *Int. J. Biochem. Cell Biol.* **34**, 1502–1505.

Vyberg, M., and Leth, P. (1991). Ubiquitin: An immunohistochemical marker of Mallory bodies and alcoholic liver disease. *Apmis.* **99**(Suppl.), 46–52.

Wanschit, J., Nakano, S., Goudeau, B., Strobel, T., Rinner, W., Wimmer, G., Resch, H., Jaksch, M., Akiguchi, I., Vicart, P., and Budka, H. (2002). Myofibrillar (desmin-related) myopathy: Clinico-pathological spectrum in 3 cases and review of the literature. *Clin. Neuropathol.* **21**, 220–231.

Wong, K. T., Alien, I. V., McQuaid, S., and McConnell, R. (1996). An immunohistochemical study of neurofibrillary tangle formation in post-encephalitic parkinsonism. *Clin. Neuropathol.* **15**, 22–25.

Wood, J. D., Beaujeux, T. P., and Shaw, P. J. (2003). Protein aggregation in motor neurone disorders. *Neuropathol. Appl. Neurobiol.* **29**, 529–545.

Woulfe, J., Kertesz, A., and Munoz, D. G. (2001). Frontotemporal dementia with ubiquitinated cytoplasmic and intranuclear inclusions. *Acta Neuropathologica* **102**, 94–102.

Wu, E., Dickson, D. W., Jacobson, S., and Raine, C. S. (1993). Neuroaxonal dystrophy in HTLV-1-associated myelopathy/tropical spastic paraparesis: Neuropathologic and neuroimmunologic correlations. *Acta Neuropathologica* **86**, 224–235.

Yamada, T., Akiyama, H., and McGeer, P. L. (1991). Two types of spheroid bodies in the nigral neurons in Parkinson's disease. *Can. J. Neurol. Sci.* **18**, 287–294.

Yasuhara, O., Kawamata, T., Aimi, Y., McGeer, E. G., and McGeer, P. L. (1994). Two types of dystrophic neurites in senile plaques of Alzheimer disease and elderly non-demented cases. *Neurosci. Lett.* **171**, 73–76.

[8] Mechanism-Based Proteomics Tools Based on Ubiquitin and Ubiquitin-Like Proteins: Crystallography, Activity Profiling, and Protease Identification

By PAUL GALARDY, HIDDE L. PLOEGH, and HUIB OVAA

Abstract

Isopeptidases that specifically remove ubiquitin or ubiquitin-like molecules from polypeptide adducts are emerging as key regulatory enzymes in a multitude of biochemical pathways. We have developed a set of tools that covalently target the active site of ubiquitin or ubiquitin-like deconjugating enzymes. We have used epitope-tagged ubiquitin and ubiquitin-like derivatives in immunoprecipitation assays to identify active proteases by mass spectrometry (MS/MS). The epitope tag confers the ability to conduct an immunoblot-based profiling assay for active isopeptidases in cell extracts. We have applied a ubiquitin-based probe in the structural analysis of the ubiquitin hydrolase UCH-L3 in its ligand-bound state. We describe the use of these electrophilic derivatives of ubiquitin and ubiquitin-like molecules in the identification, activity profiling, and structural analysis of these proteases. These tools can be used to rapidly profile activity of multiple Ub/UBL-specific proteases in parallel in cell extracts. We also show that *in vitro* these probes can be conjugated onto parts of the Ub/UBL conjugating machinery.

The Use of Ub/UbL-Derived Proteomics Probes: Detection and Identification of Ub/UBL Deconjugating Activities

Many biochemical pathways are regulated in part by posttranslational modification of proteins with ubiquitin (Ub) and ubiquitin-like (UBL) molecules, including nuclear transport, protein degradation, autophagy, and embryonic development (Glickman and Ciechanover, 2002; Hicke, 2001; Jentsch and Pyrowolakis, 2000). Precise regulation is achieved through the opposing actions of Ub/UBL-specific conjugating and deconjugating enzymes. There are predicted to be more than 100 Ub/UBL-specific proteases encoded by the human genome, suggesting great specificity in substrate selection. Given the breadth of processes that involve modification by Ub or UBLs, it is not surprising that malfunction of several members is known to cause diseases ranging from neurodegeneration (Liu *et al.*, 2002) to neoplastic processes (Bignell *et al.*, 2000; Graner *et al.*, 2004;

Gray *et al.*, 1995; Jensen *et al.*, 1998; Nakamura *et al.*, 1992; Naviglio *et al.*, 1998; Oliveira *et al.*, 2004; Ovaa *et al.*, 2004; Sasaki *et al.*, 2001; Tezel *et al.*, 2000; Yamazaki *et al.*, 2002). Despite the importance of these enzymes in health and disease, relatively few tools have been generated to facilitate their biochemical analysis.

Although much emphasis is placed on the analysis of gene expression by transcriptional profiling, the presence or absence of enzymatic activity is ultimately the most relevant information when studying biochemical pathways. Although transcriptional regulation of ubiquitin-specific protease (USP) activity has been described (Zhu *et al.*, 1996, 1997), the activity of USPs is also regulated by posttranslational events, including the formation of protein–protein interactions. We (Borodovsky *et al.*, 2001; Leggett *et al.*, 2002) and others (Cohen *et al.*, 2003) have described USPs that require the formation of protein–protein interactions to become active. In these examples, USP activities change dramatically despite constant transcription and protein levels. Recent studies have also implicated calcium-based intracellular signals in the activation of USP9 (FAM), presumably through a phosphate-dependent intermediate (Chen *et al.*, 2003). These examples underscore the importance of obtaining a readout of enzyme activity in addition to merely determining enzyme presence.

In an accompanying chapter, we describe the synthesis of mechanism-based probes that target Ub/UBL deconjugating enzymes in a substrate-specific manner. These probes consist of the full-length sequences of Ub or UBL molecules, modified at their C-terminus with an electrophilic trap that can form a covalent adduct with active proteases specific for the given Ub/UBL molecule. The mechanism of adduct formation between the ubiquitin hydrolase UCH-L3 and UbVME has been studied by x-ray crystallography, directly revealing a covalent bond between UbVME and the active site cysteine of UCH-L3 (Misaghi *et al.*, 2005). This crystal structure identified specific contacts responsible for substrate specificity and showed conformational changes present in the liganded form of UCH-L3.

When equipped with an appropriate epitope tag, these probes allow the isolation and identification of reacting proteases by immunoprecipitation, followed by tandem mass spectrometry (MS/MS)–based polypeptide sequencing (Borodovsky *et al.*, 2002; Ovaa *et al.*, 2004). We have made use of the mechanism-based nature of these probes to "profile" the complement of active cellular isopeptidases from different tissue types (Borodovsky *et al.*, 2002; Ovaa *et al.*, 2004). These studies confirm a tissue-specific pattern of Ub/UBL protease activities.

Here we describe in detail how to use Ub/UBL-based functional proteomics tools to characterize active USPs or UBL-specific protease activity in cell extracts. We have successfully generated probes specific for Ub, SUMO-1, Nedd8, GATE-16, MAP1-LC3, GABARAP, Apg8L MAP1-LC3, and ISG15 (Borodovsky et al., 2001, 2002; Hemelaar et al., 2004). We describe how to generate protease activity profiles and how to immunoprecipitate and sequence labeled proteases and associated proteins. Either radiolabeled or epitope-tagged versions of the Ub/UBL derivatives may be used to visualize modified enzymes in cell lysates. Through the use of these probes, specificity of a given protease for Ub or UBL molecules may be investigated (Fig. 1). Depending on the structure of the reactive group used, different labeling profiles are obtained, suggesting subtle differences in active-site geometries within protease families. Taken to the extreme, unique reactivity patterns may be exploited to generate enzyme-specific inhibitors.

We also describe the use of the ubiquitin derivative UbVME for crystallographic purposes. Reaction of the probe UbVME with UCH-L3 is described in detail, as well as the purification and crystallization of the resulting complex. Furthermore, we characterize the reactivity of these probes toward purified Ub/UBL-activating (E1), Ub/UBL-conjugating (E2), and Ub/UBL-deconjugating enzymes. Epitope-tagged versions of Ub/UBL probes were used to immunoprecipitate and to identify proteases with specificity for Ub and the UBL modifiers Nedd8, MAP1-LC3, and ISG15. The approach described here will be valuable for the further characterization of enzymatic pathways involved in Ub/UBL modification.

Labeling of UBL-Conjugating Enzymes

Probe specificity was investigated through the analysis of probe adducts formed with HAUbVS and HAUbVME in extracts of the murine cell line EL4 (Borodovsky et al., 2002). For reasons that are not entirely clear, the EL4 cell line has been a particularly robust source of USPs and related enzymes. Making use of the epitope tag, we retrieved probe-enzyme adducts by immunoprecipitation and identified the targeted enzymes with MS/MS. The only labeled targets identified in multiple experiments with the extracts of different cell types were USPs, including two previously unknown USPs enzymes, otubain1 and CYLD. Specific labeling of deconjugating enzymes by UBL-based probes was observed as well, including SUMO-1, Nedd8, ISG15, and the autophagy-related UBLs GATE-16, MAP1-LC3, GABARAP, and Apg8L (Hemelaar et al., 2003a,b). Although we did not retrieve a single conjugating enzyme from cell lysates, their reliance on an active site cysteine residue and affinity for their cognate Ub/UBL made this prospect intriguing.

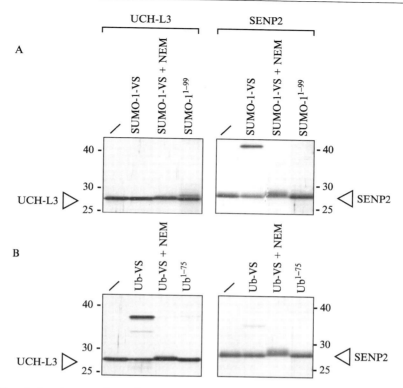

FIG. 1. *In vitro* reaction using purified enzymatic components. The SUMO specific protease SENP2 reacts with a SUMO-based but not with an Ub-based site-directed probe. All labeling is inhibited by *N*-ethylmaleimide (NEM) and, therefore, cysteine dependent. Reversely, the USP UCH-L3 reacts with an Ub probe but not with a SUMO-based one. (See color insert.)

We, therefore, sought to address the question of the reactivity of conjugating enzymes with Ub/UBL probes by use of high concentrations of purified enzyme and Ub/UBL derivatives *in vitro*. The E1-activating enzyme of SUMO1 is a heterodimer of Aos1 and Uba2, in which the Uba2 subunit harbors the active site cysteine. The SUMO-1 E2-conjugating enzyme is Ubc9. Aos1/Uba2 and Ubc9 were produced in recombinant form in *Escherichia coli*. Incubation of the SUMO1 specific protease SENP2 with SUMO1VS leads to the formation of a SENP2-SUMO-1VS adduct, which was absent from reactions that included either of the two components alone. Similarly, incubation of SUMO-1VS with Aos1/Uba2 leads to the appearance of a species with a molecular mass consistent with that of a Uba2-SUMO1VS adduct, although this occurs with reduced efficiency compared with SENP2 (Hemelaar *et al.*, 2003a). A species of identical size was observed when radioiodinated SUMO-1VS was incubated with Aos1/Uba2. Because Uba2 contains the E1 active-site cysteine residue, this species likely represents an Uba2-SUMO1VS adduct. In addition, incubation of SUMO-1 probe with the E2 enzyme Ubc9 led to the formation of a species with a molecular mass consistent with that of an Ubc9-SUMO-1-VS adduct (Hemelaar *et al.*, 2003a). A species of identical size was also observed by autoradiography. Thus, although the Ub/UBL probes characterized to date are chemoselective for isopeptidases in cell extracts, Ub/UBL probes may be complexed with conjugating enzymes (E1 and E2) using high concentrations of purified components, thus providing a set of new biochemical tools. These adducts may be useful in the structural analysis of Ub/UBL conjugating enzymes.

Materials and Methods

Slide-a-lyzer™ dialysis membranes were from Pierce. Analysis of tryptic peptides: samples were separated using a nanoflow liquid chromatography system (Waters Cap LC, Medford, MA) equipped with a picofrit column (75 μm inner diameter (ID), 10 cm, NewObjective, Woburn, MA) at a flow rate of approximately 150 nl/min using a nanotee (Waters, Medford, MA) 16/1 split (initial flow rate 5.5 μl/min). The LC system was directly coupled to a tandem mass spectrometer (Q-TOF micro, Micromass, Manchester, UK). Analysis was performed in survey scan mode, and parent ions with intensities greater than 6 were sequenced in MS/MS mode using MassLynx 3.5 Software (Micromass, UK). MS/MS data were processed and subjected to database searches using ProteinLynx Global Server 1.1 Software (Micromass, UK) against Swissprot, TREMBL/New (http://www.expasy.ch), or using Mascot (Matrixscience) against the NCBI nonredundant (nr) or mouse EST databases.

Protein Expression and Purification

Full-length recombinant human Ubc9, Aos1, Uba2, and UCH-L3 (158 amino acids, 18 kDa) were expressed and purified as previously described (Hemelaar *et al.*, 2003a; Larsen *et al.*, 1996).

Synthesis and Purification of the UCH-L3/UbVME Complex

The pH of the UbVME solution (in 50 mM sodium acetate, pH 4.5) was adjusted to 7.5 by addition of 1 M Tris, pH 7.5, UCH-L3 (in 50 mM Tris, pH 7.5, 0.5 mM EDTA, 1 mM DTT, and 3% glycerol) was added so that the final UCH-L3:UbVME molar ratio was 2:1 (2 ml of UbVME at 0.3 mg/ml, 1.9 ml of 1 M Tris, pH 7.5, and 200 μl of UCH-L3 at 16 mg/ml). The reaction was allowed to proceed for 4 h at 4°, after which no more unreacted probe remained as judged by SDS-PAGE (Fig. 2). An 8-kDa shift in MW of UCH-L3 was observed, with no unreacted probe remaining. The reaction mixture was dialyzed against buffer A (50 mM Tris, pH 7.5, 1 mM EDTA, 3 mM DTT, and 5% glycerol), and the UCH-L3/UbVME complex was separated from unreacted UCH-L3 on a Mono Q column (1 ml, Pharmacia), using linear gradient from 0–500 mM NaCl in buffer A. Fractions containing the UCH-L3/UbVME complex were pooled and dialyzed against 10 mM Na-MOPS, pH 6.8, 10 mM DTT, 1 mM EDTA and concentrated to 17 mg/ml in a Centricon dialysis concentration unit (10,000 MWCO).

Crystallization and Data Collection

UCH-L3/UbVME crystals were grown by vapor diffusion at 4°C from hanging drops containing 0.5 μl of protein solution (17 mg/ml in 10 mM NaMOPS, pH 6.8, 10 mM DTT, 1 mM EDTA) and 0.5 μl of precipitant solution (100 mM Tris, pH 8.5, 23.5% PEG 4000, and 260 mM MgCl$_2$). Rod-shaped crystals grew over a week. Crystals were soaked briefly in a cryoprotectant solution (100 mM Tris, pH 8.5, 22% PEG 4000, 20% PEG 400, and 260 mM MgCl$_2$) and flash-cooled in liquid nitrogen. X-ray data were collected to 1.45 Å with an ADSC Q315 charge-coupled device detector at beamline 8-BM at the Advanced Photon Source of Argonne National Laboratory (Argonne, IL). The crystals belong to space group P1, with unit cell dimensions a = 46.11 Å, b = 49.29 Å, c = 67.62 Å, α = 86.12°, β = 75.03°, γ = 76.78°, with two UCH-L3/UbVME complexes per asymmetric unit. Diffraction data were processed using HKL2000 (HKL Research, Charlottesville, VA).

Structure Determination and Refinement

The structure was determined by molecular replacement in CNS using the ligand-free human UCH-L3 and human ubiquitin structures as search models (Brunger *et al.*, 1998). Cross-rotation and translation functions identified two solutions each for both UCH-L3 and ubiquitin, corresponding to the two complexes in the asymmetric unit. Refinement was performed with CNS version 1.1. O was used for model building (Jones *et al.*, 1991). The final model includes residues 2–230 and 4–230 for the first and second copies of UCH-L3, respectively, residues 1–75 of ubiquitin and the glycine-vinyl methylester ligand for both copies of the complex. Residues 1 (copy one) and 1–3 (copy two) of UCH-L3 were disordered.

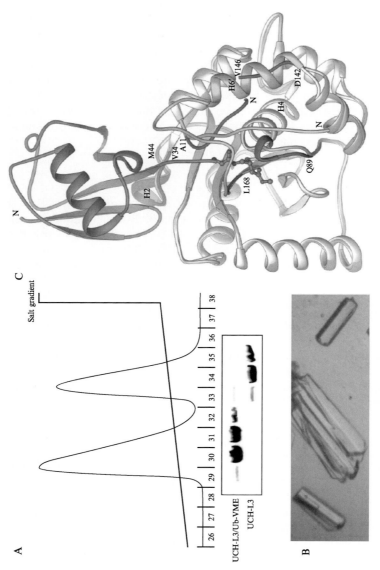

FIG. 2. Reaction and crystallization. (A) Mono-Q profile of the separation of UCH-L3/UbVME complex and unreacted UCH-L3. (B) Crystals. (C) Crystal structure.

Structural analysis used LIGPLOT, CONTACT, DYNDOM, and AREA-IMOL (CCP4, Daresbury Laboratory, Warrington, UK) and LSQMAN (Uppsala Software Factory, Uppsala, Sweden). Figures were generated with RIBBONS (Carson, 1997).

Preparation of EL4 Cell Extracts and Labeling with Active Site-Directed Probes

EL-4 cells (cultured in RPMI1640-HEPES supplemented with 10% FCS, and 1% penicillin/streptomycin) were harvested and washed 3× with PBS. Cell pellets were lysed with 1 pellet volume of glass beads (106 μm acid washed) in 2–3 pellet volumes of buffer HR (50 mM Tris, pH 7.4, 5 mM MgCl$_2$, 250 mM sucrose, 1 mM DTT, 2 mM ATP). Nuclei were removed by centrifugation, and 40 μg protein extract was used for labeling with HAUb derivatives. Approximately 2 μg of epitope-tagged probe was incubated with cell extracts for 1 h at 37°. Immunoblots with the anti-HA monoclonal antibody 12CA5 were carried out according to standard protocols; 8% SDS-PAGE gels were used to resolve high molecular weight USPs (Fig. 3).

Anti-HA Immunoprecipitation for Tandem Mass Spectrometry Analysis

EL4 cell lysates were prepared as previously, except 0.5–2 × 10^9 cells were used, and 50 μM PMSF was included in the lysis buffer. Lysates (at 5 mg/ml) were incubated with the desired HAUb-derived probe (5 mg lysate and 6.6 μg of the probe for silver stains, 14–20 mg lysate and 20 μg of probe for Coomassie stains) for 2 h at 37°. SDS was added to "denatured" samples to the final concentration of 0.4% and then diluted to less then 0.1% with NET buffer (50 mM Tris, pH 7.5, 5 mM EDTA, 150 mM NaCl, 0.5% NP40) before the addition of anti-HA agarose. Anti-HA Ig agarose (Sigma) was incubated with the samples overnight at 4°, the immunoprecipitate was washed extensively with NET buffer, and bound proteins were eluted with 50 mM glycine, pH 2.5, at 4° for 30 min. All samples were evaporated to dryness and dissolved in 50 μl of 1× reducing SDS-PAGE sample buffer. The pH was adjusted with 1 M Tris, pH 8, if necessary. Samples were resolved by 8% reducing SDS-PAGE and stained with silver or Coomassie stain using standard procedures.

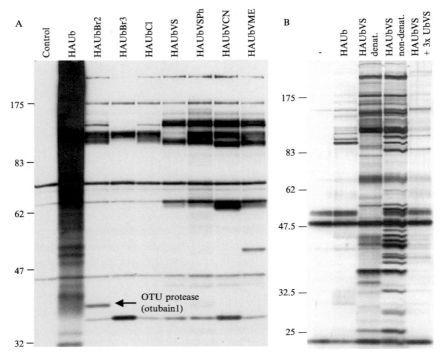

Fig. 3. Probe-based immunoblots and IPs. (A) Anti-HA immunoblot showing subtly different labeling profiles obtained in cell extracts of EL-4 (murine thymoma) cells on labeling with HAUb-based probes equipped with different reactive groups. Efficient polyubiquitination of proteins in cell extracts is observed on incubation with HAUb (second lane). (B) Silver-stained gel obtained after anti-HA immunoprecipitation under different conditions. First lane, no probe added; second lane, HAUb added; third lane, probe added, denaturing before IP; fourth lane, probe added, not denatured; fifth lane, competition with unlabeled probe.

Labeling Reactions in Cell Extracts and Detection

Radiolabeled Probes. Reactions using purified enzymes and iodinated Ub/UBL-based probes were performed at 37° for 1 to 2 h in 75 mM Tris–50 mM NaCl–5 mM MgCl$_2$–2 mM DTT–2 mM ATP. Enzymes were typically used at a 250-nM final concentration, and crude UBL-VS probes were used at approximately 3 to 5 μg per reaction. ^{125}I-Ub/UBL probe (2.5×10^5–1×10^6 cpm) was incubated with 20 to 40 μg of cell lysate for 1 h at 37°. Reactions were terminated by addition of sodium dodecyl sulfate (SDS) sample buffer with β-mercaptoethanol followed by boiling for 5 min. When indicated, enzymes or cell lysates were preincubated with 1 mM PMSF or 10 or 20 mM N-ethylmaleimide (NEM) (Sigma) for 15 min on ice before addition of the UBL probe.

Epitope-Tagged Probes

Polypeptides were resolved by 8% SDS-polyacrylamide gel electrophoresis (PAGE) and electroblotted onto polyvinylidenedifluoride (PVDF) membrane. Immunoblot detection of hemagglutinin (HA)-tagged proteins was with the mouse monoclonal anti-HA antibody 12CA5, followed by horseradish peroxidase–conjugated goat anti-mouse secondary antibody followed by visualization by chemiluminescence (Western Lightning, Perkin Elmer) according to the manufacturer's instructions.

References

Bignell, G. R., Warren, W., Seal, S., Takahashi, M., Rapley, E., Barfoot, R., Green, H., Brown, C., Biggs, P. J., Lakhani, S. R., Jones, C., Hansen, J., Blair, E., Hofmann, B., Siebert, R., Turner, G., Evans, D. G., Schrander-Stumpel, C., Beemer, F. A., van Den Ouweland, A., Halley, D., Delpech, B., Cleveland, M. G., Leigh, I., Leisti, J., and Rasmussen, S. (2000). Identification of the familial cylindromatosis tumour-suppressor gene. *Nat. Genet.* **25**, 160–165.

Borodovsky, A., Kessler, B. M., Casagrande, R., Overkleeft, H. S., Wilkinson, K. D., and Ploegh, H. L. (2001). A novel active site-directed probe specific for deubiquitylating enzymes reveals proteasome association of USP14. *EMBO J.* **20**, 5187–5196.

Borodovsky, A., Ovaa, H., Kolli, N., Gan-Erdene, T., Wilkinson, K. D., Ploegh, H. L., and Kessler, B. M. (2002). Chemistry-based functional proteomics reveals novel members of the deubiquitinating enzyme family. *Chem. Biol.* **9**, 1149–1159.

Brunger, A. T., Adams, P. D., Clore, G. M., DeLano, W. L., Gros, P., Grosse-Kunstleve, R. W., Jiang, J. S., Kuszewski, J., Nilges, M., Pannu, N. S., Read, R. J., Rice, L. M., Simonson, T., and Warren, G. L. (1998). Crystallography & NMR system: A new software suite for macromolecular structure determination. *Acta Crystallogr. Dev. Biol. Crystallogr.* **54**(Pt 5), 905–921.

Carson, M. (1997). Ribbons. *Methods Enzymol.* **277**, 493–505.

Chen, H., Polo, S., Di Fiore, P. P., and De Camilli, P. V. (2003). Rapid Ca^{2+}-dependent decrease of protein ubiquitination at synapses. *Proc. Natl. Acad. Sci. USA* **100**, 14908–14913.

Cohen, M., Stutz, F., Belgareh, N., Haguenauer-Tsapis, R., and Dargemont, C. (2003). Ubp3 requires a cofactor, Bre5, to specifically de-ubiquitinate the COPII protein, Sec23. *Nat. Cell Biol.* **5**, 661–667.

Glickman, M. H., and Ciechanover, A. (2002). The ubiquitin-proteasome proteolytic pathway: Destruction for the sake of construction. *Physiol. Rev.* **82**, 373–428.

Graner, E., Tang, D., Rossi, S., Baron, A., Migita, T., Weinstein, L. J., Lechpammer, M., Huesken, D., Zimmermann, J., Signoretti, S., and Loda, M. (2004). The isopeptidase USP2a regulates the stability of fatty acid synthase in prostate cancer. *Cancer Cell.* **5**, 253–261.

Gray, D. A., Inazawa, J., Gupta, K., Wong, A., Ueda, R., and Takahashi, T. (1995). Elevated expression of Unph, a proto-oncogene at 3p21.3, in human lung tumors. *Oncogene* **10**, 2179–2183.

Hemelaar, J., Borodovsky, A., Kessler, B. M., Cook, D. R. J., Kolli, N., Gan Erdene, T., Wilkinson, K. D., Gill, G., Lima, C. D., Ploegh, H. L., and Ovaa, H. (2003a). Specific and covalent targeting of conjugating and deconjugating enzymes of ubiquitin-like proteins. *Mol. Cell. Biol.* **24**, 84–95.

Hemelaar, J., Borodovsky, A., Kessler, B. M., Reverter, D., Cook, J., Kolli, N., Gan-Erdene, T., Wilkinson, K. D., Gill, G., Lima, C. D., Ploegh, H. L., and Ovaa, H. (2004). Specific and covalent targeting of conjugating and deconjugating enzymes of ubiquitin-like proteins. *Mol. Cell. Biol.* **24**, 84–95.

Hemelaar, J., Lelyveld, V. S., Kessler, B. M., and Ploegh, H. L. (2003b). A single protease, Apg4B, is specific for the autophagy-related ubiquitin-like proteins GATE16, MAP1 LC3, GABARAP and Apg8L. *J. Biol. Chem.* **278**, 51841–51850.

Hicke, L. (2001). Protein regulation by monoubiquitin. *Nat. Rev. Mol. Cell Biol.* **2**, 195–201.

Jensen, D. E., Proctor, M., Marquis, S. T., Gardner, H. P., Ha, S. I., Chodosh, L. A., Ishov, A. M., Tommerup, N., Vissing, H., Sekido, Y., Minna, J., Borodovsky, A., Schultz, D. C., Wilkinson, K. D., Maul, G. G., Barlev, N., Berger, S. L., Prendergast, G. C., and Rauscher, F. J., 3rd (1998). BAP1: A novel ubiquitin hydrolase which binds to the BRCA1 RING finger and enhances BRCA1-mediated cell growth suppression. *Oncogene* **16**, 1097–1112.

Jentsch, S., and Pyrowolakis, G. (2000). Ubiquitin and its kin: How close are the family ties? *Trends Cell Biol.* **10**, 335–342.

Jones, T. A., Zou, J. Y., Cowan, S. W., and Kjeldgaard (1991). Improved methods for building protein models in electron density maps and the location of errors in these models. *Acta Crystallogr. A.* **47**(Pt 2), 110–119.

Larsen, C. N., Price, J. S., and Wilkinson, K. D. (1996). Substrate binding and catalysis by ubiquitin C-terminal hydrolases: identification of two active site residues. *Biochemistry* **35**, 6735–6744.

Leggett, D. S., Hanna, J., Borodovsky, A., Crosas, B., Schmidt, M., Baker, R. T., Walz, T., Ploegh, H., and Finley, D. (2002). Multiple associated proteins regulate proteasome structure and function. *Mol. Cell* **10**, 495–507.

Liu, Y., Fallon, L., Lashuel, H. A., Liu, Z., and Lansbury, P. T., Jr. (2002). The UCH-L1 gene encodes two opposing enzymatic activities that affect alpha-synuclein degradation and Parkinson's disease susceptibility. *Cell* **111**, 209–218.

Misaghi, S., Galardy, P. J., Meester, W. J., Ovaa, H., Ploegh, H. L., and Gaudet, R. (2005). Structure of the ubiquitin hydrolase UCH-L3 complexed with a suicide substrate. *J. Biol. Chem.* **280**(2), 1512–1520.

Nakamura, T., Hillova, J., Mariage-Samson, R., Onno, M., Huebner, K., Cannizzaro, L. A., Boghosian-Sell, L., Croce, C. M., and Hill, M. (1992). A novel transcriptional unit of the tre oncogene widely expressed in human cancer cells. *Oncogene* **7**, 733–741.

Naviglio, S., Mattecucci, C., Matoskova, B., Nagase, T., Nomura, N., Di Fiore, P. P., and Draetta, G. F. (1998). UBPY: A growth-regulated human ubiquitin isopeptidase. *EMBO J.* **17**, 3241–3250.

Oliveira, A. M., Hsi, B. L., Weremowicz, S., Rosenberg, A. E., Dal Cin, P., Joseph, N., Bridge, J. A., Perez-Atayde, A. R., and Fletcher, J. A. (2004). USP6 (Tre2) fusion oncogenes in aneurysmal bone cyst. *Cancer Res.* **64**, 1920–1923.

Ovaa, H., Kessler, B. M., Rolen, U., Galardy, P. J., Ploegh, H. L., and Masucci, M. G. (2004). Activity-based ubiquitin-specific protease (USP) profiling of virus-infected and malignant human cells. *Proc. Natl. Acad. Sci. USA* **101**, 2253–2258.

Sasaki, H., Yukiue, H., Moriyama, S., Kobayashi, Y., Nakashima, Y., Kaji, M., Fukai, I., Kiriyama, M., Yamakawa, Y., and Fujii, Y. (2001). Expression of the protein gene product 9.5, PGP9.5, is correlated with T-status in non-small cell lung cancer. *Jpn. J. Clin. Oncol.* **31,** 532–535.

Tezel, E., Hibi, K., Nagasaka, T., and Nakao, A. (2000). PGP9.5 as a prognostic factor in pancreatic cancer. *Clin. Cancer Res.* **6,** 4764–4767.

Yamazaki, T., Hibi, K., Takase, T., Tezel, E., Nakayama, H., Kasai, Y., Ito, K., Akiyama, S., Nagasaka, T., and Nakao, A. (2002). PGP9.5 as a marker for invasive colorectal cancer. *Clin. Cancer Res.* **8,** 192–195.

Zhu, Y., Carroll, M., Papa, F. R., Hochstrasser, M., and D'Andrea, A. D. (1996). DUB–1, a deubiquitinating enzyme with growth-suppressing activity. *Proc. Natl. Acad. Sci. USA* **93,** 3275–3279.

Zhu, Y., Lambert, K., Corless, C., Copeland, N. G., Gilbert, D. J., Jenkins, N. A., and D'Andrea, A. D. (1997). DUB–2 is a member of a novel family of cytokine-inducible deubiquitinating enzymes. *J. Biol. Chem.* **272,** 51–57.

Section II

Ubiquitin Binding Domains

[9] Identification and Characterization of Modular Domains That Bind Ubiquitin

By Michael French, Kurt Swanson, Susan C. Shih, Ishwar Radhakrishnan, and Linda Hicke

Abstract

To receive and transmit the information carried by ubiquitin signals, cells have evolved an array of modular ubiquitin-binding domains. These domains bind directly and noncovalently to monoubiquitin and polyubiquitin chains and are found within proteins that function in diverse biological processes. Ubiquitin-binding domains characterized thus far are generally small and structurally diverse, yet they all interact with the same hydrophobic patch on the surface of ubiquitin. The rapid identification and characterization of ubiquitin-binding domains has been accomplished through the extensive use of bioinformatics, biochemistry, molecular biology, and biophysics. Here, we discuss the strategies and tools that have been most successful in the identification and characterization of ubiquitin-binding domains.

Introduction

Ubiquitin-binding domains are found in a surprising variety of proteins and in combination with many different sequence motifs. They bind to ubiquitin either to transmit and interpret ubiquitin signals attached to proteins or to facilitate ubiquitin conjugation and deconjugation. There are currently eight characterized ubiquitin-binding domains, ranging in size from 20–150 amino acids (reviewed in Di Fiore *et al.*, 2003; Schnell and Hicke, 2003; see Table I). The existence of ubiquitin surface residues that are important for function, but as yet have no known binding partners, and the multitude of biological processes that are regulated by ubiquitination, suggest that even more ubiquitin-binding domains remain to be discovered. Described in the following pages are effective strategies that have been used to identify and biochemically characterize ubiquitin-binding domains.

Strategies to Identify Monoubiquitin-Binding Proteins

Bioinformatics

The ubiquitin-binding domains first characterized were discovered using bioinformatics approaches. Iterative database searches and profile

TABLE I
Characterized Ubiquitin-Binding Domains

Domain	Length (a. a.)	Structural features	Method of identification	Selected references
UEV	~145	Similar to E2 catalytic domain	Binding to immobilized GST-Ub	Katzmann et al. (2001); Pornillos et al. (2002)
NZF (ZnF-RBZ)	~35	Zinc finger, four antiparallel β-strands	Binding to immobilized Ub-GST; binding to ubiquitinated proteins; binding to polyubiquitin chains	Kanayama et al. (2004); Meyer et al. (2002); Wang et al. (2003)
PAZ (ZnF-UBP)	~58	Not available	Two-hybrid analysis; binding to immobilized GST-Ub	Hook et al. (2002)
UBA	45–55	Three-helix bundle	Two-hybrid analysis; binding to Ub-Sepharose	Bertolaet et al. (2001); Dieckmann et al. (1998)
UIM	~20	Amphipathic helix	Bioinformatics; multiple binding assays	Fisher et al. (2003); Hofmann and Falquet (2001); Polo et al. (2002); Shih et al. (2002); Swanson et al. (2003)
CUE	42–43	Three-helix bundle	Two-hybrid analysis; multiple binding assays	Davies et al. (2003); Kang et al. (2003); Prag et al. (2003); Shih et al.. (2003)
GAT C-terminus	135	Three-helix bundle	Two-hybrid analysis; binding to immobilized GST-Ub	Scott et al. (2004); Zhu et al. (2003)
VHS	150	Right-handed superhelix of eight helices	Binding of polyubiquitin chains to immobilized domain; binding to Ub-agarose	Mao et al. (2000); Mizuno (2003)

Hidden Markov Models (HMMs) were used to identify both the ubiquitin-associated (UBA) domain and the ubiquitin-interacting motif (UIM) (Hofmann and Bucher, 1996; Hofmann and Falquet, 2001). The UBA domain was defined using iterative database searches as a sequence motif common to proteins involved in ubiquitination pathways before its characterization as a ubiquitin-binding domain (Hofmann and Bucher, 1996). By contrast, the UIM was the result of directed bioinformatics searches with the aim of identifying ubiquitin-binding domains. The UIM searches (Donaldson et al., 2003a; Hofmann and Falquet, 2001) started with a conserved 20-amino acid sequence from the Rpn10/S5a proteasomal subunit that was known to bind directly to ubiquitin (Young et al., 1998). By use of a set of ubiquitin-binding sequences derived from the S5a family, related sequences were identified with generalized profiles and profile HMMs built from previously established members of this family (Hofmann and Falquet, 2001). These probabilistic models allow one to make predictions about a set of uncharacterized sequences based on conservation information derived from a set of related sequences (e.g., ubiquitin-binding domains of the S5a family). For a more detailed description regarding profiles and profile HMMs, see Durbin et al. (1998). Similar bioinformatics approaches can be used to identify families of new domains using novel ubiquitin-binding sequences isolated by the techniques discussed in the following sections.

Yeast Two-Hybrid Screens

Yeast two-hybrid screens using ubiquitin as bait have been a valuable method for the identification of ubiquitin-binding domains. To bias screens toward the identification of proteins that bind to monoubiquitin, two different ubiquitin mutants were used. In one case, the bait was a ubiquitin mutant in which the primary site of polyubiquitin chain formation, Lys48, was mutated to prevent chain assembly at this location (Ub^{K48R}) (Shih et al., 2003). However, for future studies, a mutant ubiquitin lacking all seven lysine residues may be a more appropriate bait, because it is now known that chains can be formed though all seven ubiquitin lysine residues (Peng et al., 2003). In a second case, a mutant ubiquitin that lacked the two C-terminal glycine residues ($Ub^{\Delta GG}$) was used as bait (Donaldson et al., 2003b). The ΔGG mutation was used to prevent the covalent attachment of the ubiquitin bait to lysine residues in other proteins, including other ubiquitin molecules. Screens using these different ubiquitin mutants identified a yeast CUE domain–containing protein, Vps9 (vacuolar protein sorting) (Donaldson et al., 2003b; Shih et al., 2003), as well as other proteins (L. Hicke and S. Shih, unpublished data). Several additional ubiquitin-binding proteins (Gga2, Tom1) were identified serendipitously when ubiquitin was isolated in a two-hybrid screen with a bait protein not previously known to bind to ubiquitin (Scott *et al.*, 2004; Shiba *et al.*, 2004; Yamakami *et al.*, 2003).

Several reporter genes and strains have been developed for yeast two-hybrid screening (Bartel and Fields, 1995). The Gal4 and LexA transcriptional activation systems have been most commonly used. In these systems, the Gal4 or LexA DNA-binding domain is usually fused to a "bait" protein, whereas the Gal4 or LexA transactivation domain is fused to the "prey," proteins encoded by fragments of a cDNA library (Fashena *et al.*, 2000; Fields and Song, 1989). An interaction between a bait protein (*e.g.*, ubiquitin) and a protein expressed from a cDNA library is detected by assaying for transcriptional activation of a reporter gene. One strain, pJ69–4a, has the advantage of carrying three reporter genes, *HIS3*, *ADE2*, and *LACZ*, allowing screening for an interaction by three independent assays (James *et al.*, 1996). Generally, activation of *ADE2* transcription requires a stronger interaction between bait and prey proteins than activation of *HIS3* transcription, thus growth on medium lacking adenine is a more stringent test of binding between bait and prey than growth on medium lacking histidine. As with all two-hybrid screens, ubiquitin bait plasmids should be tested for activation of gene transcription in the presence of an empty prey vector.

When using growth on medium lacking histidine as an assay for gene activation, background growth can often be reduced by the addition of 3-aminotriazole to the growth medium. This drug acts as a competitive inhibitor of histidine binding to the *HIS3* gene product, causing the cell to require a higher level of this enzyme to grow. Thus, in the presence of 3-aminotriazole, a stronger two-hybrid interaction is required to promote high levels of transcription from *HIS3* and correspondingly higher expression levels of gene product (Bartel *et al.*, 1993). For a detailed description on using the yeast two-hybrid method to identify protein–protein interactions, see Bartel and Fields (1995).

The two-hybrid system has been successfully used to define the limits of a ubiquitin-binding domain within a protein, both by constructing truncations of an interacting gene fragment in a prey vector (Shih *et al.*, 2003) and by reverse two-hybrid screening (Shiba *et al.*, 2004). When mapping a ubiquitin-binding domain by constructing truncations, a quantitative measure of the strength of the two-hybrid interaction is useful. Generally, qualitative analysis of growth on media lacking histidine or adenine is used as an indication of transcriptional activation of *HIS3* or *ADE2*, respectively. Quantitative measurements of interaction by both the Gal4 or LexA two-hybrid system can be determined by assaying β-galactosidase activity in yeast lysates, a measure of the transcription occurring from a *LACZ* reporter gene (Bartel and Fields, 1995).

In reverse two-hybrid screening (White, 1996), mutations in a gene encoding an identified ubiquitin-binding protein are generated by error-prone polymerase chain reaction. The resulting products are subsequently ligated into a prey vector. Mutants defective in ubiquitin-binding are isolated by selecting against an interaction with the ubiquitin bait. This is accomplished by use of a reporter gene that is toxic to the cell when transcribed. For example, transcription from the *URA3* reporter is toxic in the presence of 5-fluoroorotic acid. Thus, cells carrying mutations in the gene encoding a ubiquitin-binding protein that abolish interaction with ubiquitin will survive growth in the presence of this drug. The mutations are identified by rescuing plasmids from these cells and determining the sequence of the mutant gene encoding the defective ubiquitin-binding protein.

Affinity Chromatography

Ubiquitin-binding proteins may not be detected with the two-hybrid method for a number of reasons. Expression of a fusion protein may disrupt a protein's structure, or, as mentioned previously, a fusion protein may activate the transcriptional reporter system on its own and be discarded, because activation does not require the ubiquitin bait. Furthermore, some proteins are not well represented in libraries used for two-hybrid screening.

Affinity chromatography is another technique that has proved useful for the identification of ubiquitin-binding proteins. With this method, monoubiquitin is immobilized on an affinity resin. The direct coupling of ubiquitin to Sepharose resin and the immobilization of ubiquitin on glutathione beads as a glutathione-S-transferase fusion protein (GST-Ub [Shih *et al.*, 2002] or Ub-GST [Koegl *et al.*, 1999]) are two methods that have been successfully used (Fig. 1). Immobilization of ubiquitin as a GST fusion protein requires expression in bacteria and binding to glutathione beads, whereas ubiquitin-Sepharose or ubiquitin-agarose can be purchased directly from several sources (Boston Biochem, Cambridge MA; A.G. Scientific, San Diego, CA; Sigma, St. Louis, MO). In either case, ubiquitin immobilized on a matrix is then incubated with cell lysates expressing endogenous levels of cellular proteins. Bound proteins are typically eluted, fractionated by SDS-PAGE, and visualized by a sensitive protein staining method such as silver staining. The proteins bound to immobilized monoubiquitin should be compared with those bound to immobilized GST or Sepharose resin alone. Ubiquitin-specific bands can then be excised and

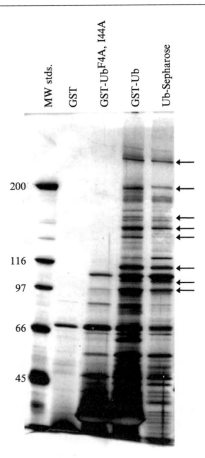

FIG. 1. Identification of monoubiquitin-binding proteins by affinity chromatography. Lysate prepared from wild-type yeast cells (6.7 mg protein) was incubated with 50 μg of immobilized GST, GST-Ub fusion proteins, or Ub-Sepharose. Bound proteins were eluted by incubation with 10 mM reduced glutathione for 30 min at 4°, separated by 5–12.5% SDS-PAGE, and visualized by silver staining. Several unique bands (arrows) present in the eluates of wild-type ubiquitin but not a ubiquitin mutant (F4A, I44A) were analyzed by mass spectrometry.

identified by peptide sequencing by means of mass spectrometry (reviewed in Steen and Mann [2004]).

A similar technique can be used to identify proteins that bind to specific surface residues of ubiquitin. This can be accomplished by immobilizing tagged ubiquitin mutants on an affinity resin and selecting protein bands specifically eluted from the wild-type, but not the mutant ubiquitin, for excision and sequencing (Fig. 1). Because different ubiquitin surface residues have been implicated in distinct cellular processes (Beal et al., 1996; Sloper-Mould et al., 2001), this approach may be useful in identifying proteins that function in specific ubiquitin-mediated processes.

Affinity chromatography has also been used to identify proteins that bind specifically to polyubiquitin chains linked through different lysine residues. The discovery of proteins that bind to polyubiquitin chains has been hindered by technical difficulties with chain synthesis and stability. Recently, stable Lys29-linked polyubiquitin chain analogs have been used to identify proteins that specifically bind to Lys29-linked chains. These chain analogs were synthesized by cross-linking ubiquitin molecules through Lys29 with 1,3-dichloroacetone to make a tetrameric chain. The chains were immobilized on a Sepharose resin and incubated with yeast lysates. Bound proteins were eluted and identified using two different types of mass spectrometry methods. USP5, a human deubiquitinating enzyme previously known to bind to ubiquitin, and two novel Lys29-linked polyubiquitin-binding proteins from yeast, Ufd3 and Ubp14, were identified by this approach (Russell and Wilkinson, 2004). It is likely that proteins that specifically bind to chains linked through other lysine residues may be identified using analogs prepared in a similar manner.

Characterization of Ubiquitin-Binding Properties

Assays for Biochemical Interactions Between Ubiquitin and Ubiquitin-Binding Domains

To confirm that a ubiquitin-binding protein or domain identified by the yeast two-hybrid system binds to ubiquitin through a biochemical assay, several different approaches have been used (Fig. 2). To demonstrate that a protein binds to ubiquitin in a specific eukaryotic cell type, lysates from cells expressing an epitope-tagged protein can be incubated with ubiquitin immobilized on beads (GST-Ub, Ub-Sepharose, etc.). Bound proteins can then be eluted and analyzed on an immunoblot with antiserum that recognizes the epitope tag or the protein itself (Katzmann et al., 2001; Parcellier et al., 2003; Shih et al., 2002; 2003). To ensure that a protein or fragment does not bind nonspecifically to GST or Sepharose, lysates should also be incubated with naked Sepharose beads or immobilized GST. In addition, a protein or fragment that does not bind to ubiquitin can be used as a negative control. It should be noted that by use of this approach one cannot demonstrate that a protein or domain binds directly to ubiquitin, because other proteins may mediate binding through the formation of a complex.

To test whether a protein or domain directly binds to ubiquitin, recombinant proteins can be expressed in and purified from *Escherichia coli*,

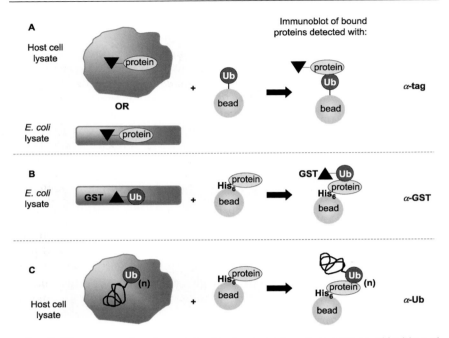

FIG. 2. Three approaches for assaying biochemical interactions between ubiquitin and ubiquitin-binding proteins. (A) Ubiquitin immobilized on a matrix can be incubated with lysates from host cells or bacteria that express an epitope-tagged version of a putative ubiquitin-binding protein. After washing and elution, binding of the protein is determined by probing immunoblots of the eluates with appropriate antiserum. (B) To demonstrate that a protein or domain directly binds to ubiquitin, *E. coli* lysates from cells expressing a GST-Ub or Ub-GST fusion protein can be incubated with a recombinant, tagged protein immobilized on beads. Bound proteins are analyzed as described for (A). (C) To test whether a protein binds to ubiquitinated protein conjugates, eukaryotic cell lysates can be incubated with a recombinant, tagged protein immobilized on beads, or the protein can be immunoprecipitated from cell lysates under native conditions. Immunoblots of bound or coprecipitated proteins are then probed with ubiquitin antiserum.

because this organism does not produce ubiquitin or other host-associated proteins. Expression of proteins or protein fragments in a His_6-tagged form from a pET vector (Novagen, Madison, WI) allows for the purification and immobilization of these proteins on a nickel or cobalt resin (BD Biosciences, San Jose, CA). Immobilized proteins can then be incubated with bacterial lysates from cells expressing a GST-Ub fusion protein and bound proteins analyzed by α-GST immunoblotting (Shih *et al.*, 2003). Conversely, it is possible to incubate GST-fusion proteins expressed in *E. coli* with ubiquitin-agarose beads and detect binding by analysis of eluates with α-GST. The use of naked beads and a protein that does not bind to ubiquitin can help to control for nonspecific binding (Shiba et al., 2004). In addition, lysates from cells expressing a GST protein alone can help to control for nonspecific binding of GST to naked or protein-coated beads. As an alternate approach, ubiquitin can be immobilized on glutathione beads as a GST fusion and then incubated with lysates from bacteria expressing a His_6-tagged protein. In this case, eluates from glutathione beads would be analyzed by α-His_6 immunoblotting. Some ubiquitin-binding proteins seem to bind to ubiquitin better when the N-terminus is exposed, and, therefore, it may be beneficial to use both Ub-GST and GST-Ub in these assays.

To demonstrate that a protein or domain binds to ubiquitinated protein conjugates, in addition to free ubiquitin, two different approaches have been used. In one method, a recombinant ubiquitin-binding protein or domain is immobilized on an affinity matrix and incubated with cell lysates (Meyer et al., 2002; Polo et al., 2002; Seibenhener et al., 2004). In another approach, an epitope-tagged protein is immunoprecipitated from cell lysates under native conditions (Yamakami et al., 2003). In either case, bound proteins are analyzed by immunoblotting with an antibody that recognizes ubiquitin to detect bound ubiquitin conjugates. Proteasomal inhibitors can be added to lysates to stabilize ubiquitin conjugates and increase the likelihood of detecting binding to the protein of interest (Yamakami et al., 2003).

Identification of Ubiquitin Residues Required for Binding

There are two hydrophobic patches on the surface of ubiquitin that are vital to its cellular function. One patch surrounds Ile44 of ubiquitin and is important for proteasomal targeting and endocytosis. The other surrounds Phe4 of ubiquitin and is important for endocytosis and essential for vegetative growth of yeast but is not required for proteasomal targeting (Beal et al., 1996; Sloper-Mould et al., 2001). For all characterized ubiquitin-binding domains where the mode of interaction with ubiquitin is known, the Ile44 patch seems to be crucial to the interaction (reviewed in Schnell and Hicke [2003]).

To characterize ubiquitin residues required for binding of a protein or domain to ubiquitin, ubiquitin mutants carrying point mutations in individual residues have been successfully used. For this experiment, an entire protein or domain is expressed in a recombinant form, immobilized on beads, and incubated with lysates from E. coli cells expressing a GST-Ub or Ub-GST fusion protein carrying a point mutation in a surface residue (Shih et al., 2002, 2003). Lysates should be adjusted by dilution so that all GST-Ub fusion proteins are at an equivalent concentration when incubated with the immobilized protein. Bound proteins are then eluted and analyzed by α-GST immunoblotting. Binding of ubiquitin mutants can then be compared with binding of wild-type ubiquitin to assess the contribution of a specific ubiquitin residue to the interaction. Alternately, different ubiquitin mutants can be purified and equivalent amounts of purified mutant or wild-type ubiquitin incubated with the protein of interest (Shiba *et al.*, 2004). Under these conditions, the affinities of different mutants can be directly compared with wild-type ubiquitin.

Assays for Binding of Proteins to Different Types of Ubiquitin Modifications

In addition to testing for an interaction with monoubiquitin, it is possible to assay binding of a purified protein or domain to polyubiquitin chains. Purified Lys48-linked tetraubiquitin chains and chain mixtures of different lengths linked through Lys48 or Lys63 are commercially available from several sources (Affiniti Research, Exeter, U.K.; Boston Biochem, Cambridge, MA). These purified chains can be incubated with a protein or domain immobilized on beads, and bound chains can be detected by immunoblotting eluted proteins with ubiquitin antiserum. Another technique for determining chain-binding specificity is a variation of the previously mentioned approach to determine whether ubiquitin-binding proteins interact with ubiquitinated protein conjugates. In this approach, cells are transfected with epitope-tagged ubiquitin carrying mutations in one or more lysine residues. A purified ubiquitin-binding protein immobilized on beads is then used to precipitate epitope-tagged chains from lysates. Eluted chains are analyzed by immunoblotting with an antibody that recognizes the epitope tag (Seibenhener *et al.*, 2004). Finally, it is also possible to assay binding of a protein or domain to polyubiquitin chains linked through different lysine residues by use of ubiquitin chain analogs. Because chains can be formed between all seven of ubiquitin's lysine residues both *in vitro* (Baboshina and Haas, 1996) and *in vivo* (Peng *et al.*, 2003), the identification of proteins that bind to different types of chain linkages is an increasingly important objective. The use of polyubiquitin chain analogs linked through different lysine residues may prove to be a useful tool in identifying proteins that bind to specific types of chains (Russell and Wilkinson, 2004).

Detection of Monoubiquitination Mediated by a Ubiquitin-Binding Domain

The finding that a subset of ubiquitin-binding domains promotes intramolecular monoubiquitination indicates that there is a close association between ubiquitin-binding and ubiquitination (Davies *et al.*, 2003; Klapisz *et al.*, 2002; Oldham *et al.*, 2002; Polo *et al.*, 2002; Shiba *et al.*, 2004; Shih *et al.*, 2003). There are several different approaches to determine whether a protein carrying a ubiquitin-binding domain is monoubiquitinated. Initially, putative ubiquitinated forms of proteins may be observed on immunoblots of cell lysates performed with an antibody that recognizes the protein. Monoubiquitinated forms of a protein typically migrate 8–12 kDa higher than unmodified forms. To definitively demonstrate that a protein is ubiquitinated, a protein can be immunoprecipitated from cell lysates and analyzed on an immunoblot with ubiquitin antiserum (Polo *et al.*, 2002). On the other hand, α-ubiquitin immunoprecipitates of a cell lysate can be probed on an immunoblot with antiserum against a protein suspected to be ubiquitinated. Epitope-tagged ubiquitin has also been used to demonstrate that a band with a shifted mobility is a ubiquitinated form of a protein. For example, a strain overexpressing myc-Ub has been used to demonstrate that Vps9 is monoubiquitinated by showing that a modified form of Vps9 exhibits an increased mobility shift consistent with the size of a c-myc tag (Shih *et al.*, 2003). The recent development of proteomics approaches to identify ubiquitinated proteins has facilitated the rapid identification of ubiquitinated proteins on a large scale (Gururaja *et al.*, 2003; Peng *et al.*, 2003). However, proteins identified as ubiquitinated by proteomics approaches should still be confirmed using one or more of the biochemical approaches discussed earlier.

After determining that a protein is monoubiquitinated, a mutant defective in its ubiquitin-binding domain can be used to test whether monoubiquitination is dependent on ubiquitin-binding. To do this, a mutation in the domain that abolishes or significantly diminishes ubiquitin binding must be identified. To determine whether ubiquitin-binding mediates a protein's monoubiquitination, these mutants can be assayed for ubiquitination using one of the techniques described previously.

Biophysical Methods for Analysis of Monoubiquitin Interactions

Both nuclear magnetic resonance (NMR) spectroscopy and x-ray crystallography have been used successfully to analyze monoubiquitin interactions at the atomic level (for examples, see Table I). Here we focus on the analyses of ubiquitin interactions by NMR, which have provided insights ranging from quantitative measures of binding affinities to the delineation of interacting surfaces and residues at varying levels of resolution. The success of NMR analyses can be attributed to the relatively small molecular size of ubiquitin complexes studied thus far (*ca.* 10–20 kDa) and to the rapid dissociation kinetics of these complexes, two conditions that ensure high sensitivity of NMR spectra. In the following sections, we provide an overview of the NMR techniques that have been used to monitor ubiquitin and ubiquitin-like protein interactions and discuss some of the practical aspects of quantifying binding affinities and determining high-resolution structures.

NMR Chemical Shift Perturbation Experiments

The chemical shift of an NMR-active atomic nucleus is exquisitely sensitive to its local environment, and changes in chemical shifts can serve as an indication of a noncovalent interaction with a ligand (Chen *et al.*, 1993; Foster *et al.*, 1998; Gronenborn and Clore, 1993; Grzesiek *et al.*, 1996; van Nuland *et al.*, 1993). To analyze protein interactions, one of the interacting components is enriched with a suitable NMR-active isotope, ^{15}N, and NMR spectra are recorded as a function of increasing amounts of the other "unenriched" component(s). The ^{15}N isotope is used for enrichment because it not only simplifies the NMR spectrum but also ensures at least one reporter signal per residue because of the nitrogen-containing amide groups in the polypeptide backbone. Further enhancement of spectral sensitivity and resolution is achieved by recording correlated spectra in a two-dimensional format. Interpretation of ligand-induced chemical shift perturbations on the NMR spectrum is relatively straightforward, although the level of rigor associated with the analysis can vary.

Broadly, three types of ligand-induced spectral changes over the course of a titration can be distinguished: (1) disappearance of resonances with concomitant appearance of resonances at new positions, (2) progressive broadening and disappearance of resonances, and (3) migration of resonances to new positions. The dissociation kinetics (and thus the equilibrium dissociation constant) of the complex determines which of these effects occurs. In NMR jargon, complexes that exhibit these changes are characterized as being in slow, intermediate, and fast exchange, respectively (Freeman, 1987). Complexes that are in slow exchange are readily amenable to structural characterization by NMR. Although less widely appreciated, this is also the case for complexes in fast exchange, provided the dissociation constant is in the millimolar range or less. However, complexes in intermediate exchange are virtually impossible to characterize structurally, because resonances likely to arise from residues at the protein-ligand interface are also those that disappear from the spectrum because of line broadening. Almost all ubiquitin/ubiquitin-like protein complexes that have been analyzed by NMR to date fall into the fast exchange category and, thus, have been amenable to rigorous structural analysis.

In addition to the type of exchange, the number of resonances affected on ubiquitin addition can provide insights into the nature of a ubiquitin-binding interaction. For example, when only a subset of resonances is affected, the formation of a specific complex can be inferred (*i.e.*, a unique ubiquitin-binding site exists on the protein). In contrast, if extreme line broadening equivalently perturbs all the resonances in the spectrum, a nonspecific complex (*i.e.*, infinite binding sites on the surface of the protein) is indicated. A lack of perturbations in the spectrum indicates no association.

Insight into the exact location of a ubiquitin-binding site can be obtained from chemical shift perturbation experiments. This requires prior knowledge of sequence-specific ^{1}H and ^{15}N resonance assignments for both the free and ubiquitin-bound forms of the protein if the complex is in slow exchange, but only that of the free protein if the complex is in fast exchange, because resonances migrate to new positions during the course of a titration. It is common to quantify the chemical shift change for each backbone amide as

$$\Delta\delta = \sqrt{\{0.5[(\delta_{H,bound} - \delta_{H,free})^2 + 0.04(\delta_{N,bound} - \delta_{N,free})^2]\}}$$

where δ_H and δ_N denote ^{1}H and ^{15}N chemical shifts (Grzesiek *et al.*, 1996). Potential ubiquitin-interacting residues can be identified from the magnitude of the chemical shift change, typically indicated by a greater than one standard deviation from average (Fig. 3A). However, knowledge of the three-dimensional structure of the protein or that of a suitable homolog can aid this process on the basis of the spatial relationships of strongly perturbed residues (Fig. 3B, C). Given the ease with which chemical shift perturbation experiments can be conducted, most studies of ubiquitin/ubiquitin-like protein interactions have taken advantage of this approach (Bilodeau *et al.*, 2003; Fisher *et al.*, 2003; Liu *et al.*, 1999; Merkley and Shaw, 2004; Miura *et al.*, 1999; Mueller *et al.*, 2004; Pornillos *et al.*, 2002; Rajesh *et al.*, 1999; Ryu *et al.*, 2003; Sakata *et al.*, 2003; Scott *et al.*, 2004; Shekhtman and Cowburn, 2002; Varadan *et al.*, 2002, 2004; Walters *et al.*, 2003; Wang *et al.*, 2003; Wilkinson *et al.*, 1999; Yuan *et al.*, 2004).

A comparison of the results obtained from chemical shift perturbation analysis with that from a more rigorous, albeit time-consuming, structure determination effort shows good agreement between the surface defined by strongly perturbed residues and the actual interaction surface (Fig. 3C; Swanson *et al.*, 2003). There are now efforts to mine chemical shift perturbation data more extensively, especially when structural and perturbation data are available for both interacting components (Dominguez *et al.*, 2003). In this approach, residues are deemed to be interacting on the basis of certain criteria such as the degree of perturbation and solvent accessibility. The interacting surfaces are then docked, and configurations that yield

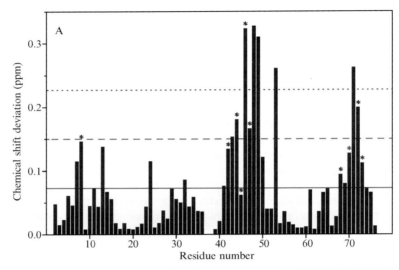
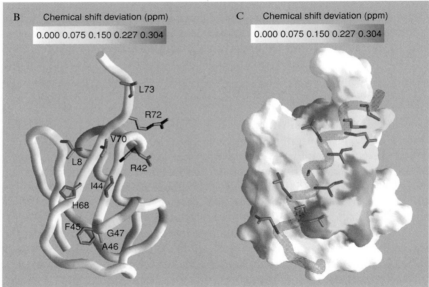

FIG. 3. Chemical shift perturbation analysis of the Vps27 UIM-1–ubiquitin complex and comparison with the solution structure. (A) The ^{1}H and ^{15}N chemical shift deviations ($\Delta\delta$) for ubiquitin residues on titration with one equivalent of Vps27 UIM-1 peptide plotted as a function of residue number. The solid, dashed, and dotted horizontal lines correspond to the average chemical shift deviation and one and two standard deviations above the average chemical shift deviation, respectively. Residues that contact the UIM are identified with

low values for a potential energy function are selected as plausible models. This approach has been used to model several ubiquitin complexes (Mueller et al., 2004; Yuan et al., 2004). Although these models serve as good working hypotheses, the results should be interpreted with caution, because there are significant limitations in the accuracy of the potential energy functions, and the basis for chemical shift perturbation is poorly understood. Indeed, it is difficult to distinguish *a priori* perturbations caused by direct ligand binding effects from those caused by indirect conformational effects. It is also worth noting that chemical shift perturbation data are usually obtained for backbone nuclei, whereas intermolecular interactions usually involve side chains to a large extent.

In a few ubiquitin complexes, structural information in the form of ^{1}H–^{1}H NOEs (nuclear Overhauser effects), the mainstay of NMR structure determination, has been difficult to observe. For these complexes residual dipolar couplings (RDCs) in conjunction with molecular docking approaches have been used to generate plausible models (Varadan et al., 2002, 2004; Walters et al., 2003). RDCs constitute a unique set of NMR restraints and provide a powerful means of defining relative domain orientations (De Alba and Tjandra, 2004). A principal limitation of these approaches is that the relative position of the interacting domains is undefined and relies on the accuracy of molecular docking approaches.

Quantifying Binding Affinities for Complexes in Fast Exchange

For an ideal, single-site binding ubiquitin/ubiquitin-binding protein complex, the equilibrium dissociation constant K_d is given by the expression:

$$K_d = [P][Ub]/[PUb]$$

where the square brackets denote concentrations of the respective species. The expression can be rearranged to yield

$$K_d = ([P]_T - [PUb])([Ub]_T - [PUb])/[PUb] \quad (1)$$

where $[P]_T = [P] + [PUb]$ and $[Ub]_T = [Ub] + [PUb]$ are the total protein and ubiquitin concentrations, respectively. By introducing three new

---

asterisks. (B) The same chemical shift deviations as in (A) but mapped on to the backbone of ubiquitin. Side chains of interacting residues are shown. Increasing color shade represents increasing chemical shift. (C) Chemical shift deviations mapped onto the molecular surface of ubiquitin. The backbone and ubiquitin interacting side chains of Vps27 UIM-1 from the solution structure of the UIM-ubiquitin complex (PDB accession: 1Q0W) are also shown. Increasing color shade represents increasing chemical shift.

dimensionless parameters, $B = [PUb]/[Ub]_T$, $M = [P]_T/[Ub]_T$, and $C = K_d/[Ub]_T$, corresponding to the molar fraction of the ligand–protein complex, molar ratio of the interacting components, and a constant, respectively, equation (1) can be rearranged to yield

$$B = 0.5[(M + 1 + C) \pm \sqrt{\{(M + 1 + C)^2 - 4M\}}] \tag{2}$$

For a protein nucleus experiencing distinct local environments in the free and ligand-bound states, the apparent chemical shift in the fast exchange regimen is a population-weighted average of the chemical shifts of the respective states:

$$\delta_{apparent} = (1 - B)\delta_{free} + B\delta_{bound} \tag{3}$$

where δ denotes the chemical shift. Equation (3) can be rearranged to yield

$$B = (\delta_{apparent} - \delta_{free})/(\delta_{bound} - \delta_{free}) = \Delta\delta/\Delta\delta_{max} \tag{4}$$

By titrating a constant amount of protein with increasing amounts of ubiquitin, the migration of protein resonances can be monitored (Fig. 4A) and $\Delta\delta$ can be measured as a function of M (compare equations [2] and [4]), yielding a binding isotherm that is amenable to nonlinear regression analysis. The fitted parameters are C (K_d in units of $[Ub]_T$) and $\Delta\delta_{max}$, whereas M and $\Delta\delta$ are the independent and dependent variables, respectively (Fig. 4B). Standard graphing software such as XMGR/Grace, KaleidaGraph (Synergy Software Inc., Essex Junction, VT) or Origin (MicroCal Inc., Northampton, MA) can be used to facilitate this process. Unix awk scripts for quantifying chemical shift deviations based on Felix (Accelrys Inc., San Diego, CA) cross peak lists and for generating input files for curve fitting programs such as XMGR/Grace are available on the Internet at http://monster.northwestern.edu/radhakrishnan/htdocs/publications.html.

Key practical considerations for a successful K_d measurement using NMR include choice of the initial ubiquitin and ^{15}N-enriched ubiquitin-binding protein concentrations, actual values of molar ratios (M values) to sample, and an estimate of the approximate K_d. The protein concentration must be just high enough, severalfold to an order of magnitude higher than the K_d, so that during the titration saturation with ubiquitin is readily approached. A coarse titration using a standard set of parameters (see later) can be performed to get an initial estimate for the K_d. The starting ubiquitin-binding protein concentration must be high enough so that M values of approximately five can be attained without significantly diluting the protein concentration (twofold at most). Because the binding isotherms are hyperbolic in nature, we favor sampling M values at closely spaced

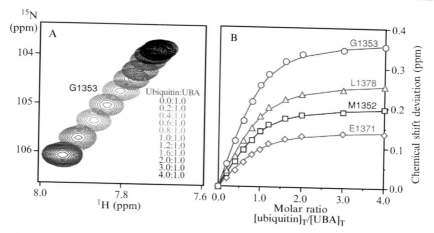

Fig. 4. Quantitative evaluation of the equilibrium dissociation constant of the Ede1 UBA–ubiquitin complex. (A) Overlaid, expanded contour plots illustrating changes in the ^{1}H and ^{15}N chemical shifts of Ede1 Gly1353, a key residue within the highly conserved MGF motif in UBA domains, as a function of added ubiquitin. The titration was performed following the protocol described in the text. (B) Plots of the chemical shift deviation ($\Delta\delta$) as a function of molar ratio (M) for Met1352, Gly1353, Glu1371, and Leu1378 of the Ede1 UBA domain. The square, circle, diamond, and triangle symbols denote actual experimental data, and the solid lines trace curves resulting from the nonlinear least-squares fitting procedure. Statistical analysis of the fits yields an average K_d of 83 ± 9 μM for the Ede1UBA-ubiquitin interaction.

intervals initially and increasing the sampling interval as saturation is approached. The ubiquitin-binding protein concentration should not be too high to introduce significant pipetting errors for the initial M values. The same considerations apply when ^{15}N-labeled ubiquitin is titrated with different concentrations of unlabeled ubiquitin-binding protein.

For an experiment in which labeled ubiquitin is titrated with increasing amounts of ubiquitin-binding protein, we typically use 500 μl of a 0.5 mM uniformly ^{15}N-enriched ubiquitin sample at a pH that is guided by the solubility characteristics of the ubiquitin-binding protein. A 2 mM stock solution of the unenriched ubiquitin-binding protein in the same solution conditions as the ubiquitin sample is used. A typical set of M values used for sampling is {0.0, 0.2, 0.4, 0.6, 0.8, 1.0, 1.2, 1.6, 2.0, 3.0, 4.0}, so that the minimum volume of added ubiquitin-binding protein is 25 μl. After each addition and before recording the NMR spectrum, the pH is checked and adjusted to the starting value, if necessary. The NMR sample is transferred to a microcentrifuge tube between additions so as to ensure proper delivery and mixing of ubiquitin with protein through gentle agitation. In doing so, it is crucial to carefully recover as much of the sample as possible through centrifugation.

All ^1H–^{15}N correlated spectra should be acquired with identical parameters during the course of a titration to simplify subsequent data processing and analysis. The spectral width parameter merits special attention, because it needs to be large enough to accommodate all the resonances in the spectrum without aliasing, besides providing room (2 ppm from edges for ^1H and 5 ppm for ^{15}N) for any large chemical shift perturbations. This is particularly important, because the most upfield and downfield backbone ^{15}N-ubiquitin resonances of Ala46 and Gly47, respectively, typically undergo large chemical shift changes because of the proximity of these residues to the Ile44 hydrophobic patch (a common binding surface on ubiquitin). For K_d determinations, data for only those resonances that undergo large chemical shift perturbations and are otherwise not broadened or obscured by another resonance during the course of the titration should be used. Nonlinear regression must be performed on multiple curves so as to enable statistical evaluation of the fitted values of the K_d.

Challenges in NMR Structure Determination of Complexes in Fast Exchange

The approach used to determine three-dimensional solution structures for transient complexes is essentially the same as that for stable complexes. Intermolecular, interproton distance restraints are especially crucial to the success of these endeavors, and there are two key issues pertaining to transient complexes that merit discussion. The first relates to the reliable detection of intermolecular NOEs, which depends on how well the complex is populated. This, in turn, is determined by the equilibrium dissociation constant and the concentration of the interacting components. A concentration equal to or greater than the dissociation constant for each interacting component is a good choice, because it ensures that at least 50% of the molecules are in the complex. Given the intrinsically low sensitivity of detection of NOEs, protein concentrations need to be in the millimolar range for accurate quantification of intensities.

The second issue relates to the interpretation of intermolecular NOEs, which requires prior knowledge of ^1H resonance assignments. Resonances of each interacting component are typically assigned independently by way of through-bond approaches using samples suitably enriched with NMR-active isotopes (an exception to this are protein-peptide complexes, which are, therefore, not subject to the concerns raised in this paragraph). Because chemical shifts are sensitive to both the concentration and the molar ratio of the interacting components for transient complexes (cf. equations

[2] and [3]), these parameters should be preserved across samples, because they can have profound implications on the interpretation of intermolecular NOEs. One way to ensure that these parameters are identical across samples is by recording and carefully comparing ^{15}N- and ^{13}C-decoupled 1D ^{1}H NMR spectra. We have been reasonably successful in determining structures of multiple ubiquitin complexes by maintaining equimolar ratios of interacting components and concentrations (typically in the 1 mM range) across samples (Kang et al., 2003; Swanson et al., 2003).

Acknowledgments

This work was supported by grants from the National Institutes of Health to I. R. and L. H. K. A. S. was supported by an NIH Molecular Biophysics training grant. I. R. is a Scholar of the Leukemia and Lymphoma Society. Support from the Robert H. Lurie Comprehensive Cancer Center for structural biology research at Northwestern University is gratefully acknowledged.

References

Baboshina, O. V., and Haas, A. L. (1996). Novel multiubiquitin chain linkages catalyzed by the conjugating enzymes E2$_{EPF}$ and RAD6 are recognized by 26S proteasome subunit 5. *J. Biol. Chem.* **271,** 2823–2831.

Bartel, P., Chien, C. T., Sternglanz, R., and Fields, S. (1993). Elimination of false positives that arise in using the two-hybrid system. *Biotechniques.* **14,** 920–924.

Bartel, P. L., and Fields, S. (1995). Analyzing protein-protein interactions using two-hybrid system. *Meth. Enzymol.* **254,** 241–263.

Beal, R., Deveraux, Q., Xia, G., Rechsteiner, M., and Pickart, C. (1996). Surface hydrophobic residues of multiubiquitin chains essential for proteolytic targeting. *Proc. Natl. Acad. Sci. USA* **93,** 861–866.

Bertolaet, B. L., Clarke, D. J., Wolff, M., Watson, M. H., Henze, M., Divita, G., and Reed, S. I. (2001). UBA domains of DNA damage-inducible proteins interact with ubiquitin. *Nat. Struct. Biol.* **8,** 417–422.

Bilodeau, P. S., Winistorfer, S. C., Kearney, W. R., Robertson, A. D., and Piper, R. C. (2003). Vps27-Hse1 and ESCRT-I complexes cooperate to increase efficiency of sorting ubiquitinated proteins at the endosome. *J. Cell Biol.* **163,** 237–243.

Chen, Y., Reizer, J., Saier, M. H., Jr., Fairbrother, W. J., and Wright, P. E. (1993). Mapping of the binding interfaces of the proteins of the bacterial phosphotransferase system, HPr and IIAglc. *Biochemistry* **32,** 32–37.

Davies, B. A., Topp, J. D., Sfeir, A. J., Katzmann, D. J., Carney, D. S., Tall, G. G., Friedberg, A. S., Deng, L., Chen, Z., and Horazdovsky, B. F. (2003). Vps9p CUE domain ubiquitin binding is required for efficient endocytic protein traffic. *J. Biol. Chem.* **278,** 19826–19833.

De Alba, E., and Tjandra, N. (2004). Residual dipolar couplings in protein structure determination. *Methods Mol. Biol.* **278,** 89–106.

Dieckmann, T., Withers-Ward, E. S., Jarosinski, M. A., Liu, C. F., Chen, I. S., and Feigon, J. (1998). Structure of a human DNA repair protein UBA domain that interacts with HIV-1 Vpr. *Nat. Struct. Biol.* **5,** 1042–1047.

Di Fiore, P. P., Polo, S., and Hofmann, K. (2003). When ubiquitin meets ubiquitin receptors: A signalling connection. *Nat. Rev. Mol. Cell. Biol.* **4,** 491–497.

Dominguez, C., Boelens, R., and Bonvin, A. M. (2003). HADDOCK: A protein-protein docking approach based on biochemical or biophysical information. *J. Am. Chem. Soc.* **125,** 1731–1737.

Donaldson, K. M., Li, W., Ching, K. A., Batalov, S., Tsai, C. C., and Joazeiro, C. A. (2003a). Ubiquitin-mediated sequestration of normal cellular proteins into polyglutamine aggregates. *Proc. Natl. Acad. Sci. USA* **100,** 8892–8897.

Donaldson, K. M., Yin, H., Gekakis, N., Supek, F., and Joazeiro, C. A. (2003b). Ubiquitin signals protein trafficking via interaction with a novel ubiquitin binding domain in the membrane fusion regulator, Vps9p. *Curr. Biol.* **13,** 258–262.

Durbin, R., Eddy, S., Krogh, A., and Mitchison, G. (1998). "Biological Sequence Analysis: Probabilistic Models of Proteins and Nucleic Acids." Cambridge University Press, Cambridge, UK.

Fashena, S. J., Serebriiskii, I. G., and Golemis, E. A. (2000). LexA-based two-hybrid systems. *Methods Enzymol.* **328,** 14–26.

Fields, S., and Song, O. (1989). A novel genetic system to detect protein-protein interactions. *Nature* **340,** 245–246.

Fisher, R. D., Wang, B., Alam, S. L., Higginson, D. S., Robinson, H., Sundquist, W. I., and Hill, C. P. (2003). Structure and ubiquitin binding of the ubiquitin interacting motif. *J. Biol. Chem.* **278,** 28976–28984.

Foster, M. P., Wuttke, D. S., Clemens, K. R., Jahnke, W., Radhakrishnan, I., Tennant, L., Reymond, M., Chung, J., and Wright, P. E. (1998). Chemical shift as a probe of molecular interfaces: NMR studies of DNA binding by the three amino-terminal zinc finger domains from transcription factor IIIA. *J. Biomol. NMR.* **12,** 51–71.

Freeman, R. (1987). "A Handbook of Nuclear Magnetic Resonance." Harlow, Essex, England.

Gronenborn, A. M., and Clore, G. M. (1993). Identification of the contact surface of a streptococcal protein G domain complexed with a human Fc fragment. *J. Mol. Biol.* **233,** 331–335.

Grzesiek, S., Bax, A., Clore, G. M., Gronenborn, A. M., Hu, J. S., Kaufman, J., Palmer, I., Stahl, S. J., and Wingfield, P. T. (1996). The solution structure of HIV-1 Nef reveals an unexpected fold and permits delineation of the binding surface for the SH3 domain of Hck tyrosine protein kinase. *Nat. Struct. Biol.* **3,** 340–345.

Gururaja, T., Li, W., Noble, W. S., Payan, D. G., and Anderson, D. C. (2003). Multiple functional categories of proteins identified in an *in vitro* cellular ubiquitin affinity extract using shotgun peptide sequencing. *J. Proteome Res.* **2,** 394–404.

Hofmann, K., and Bucher, P. (1996). The UBA domain: A sequence motif present in multiple enzyme classes of the ubiquitination pathway. *Trends Biochem. Sci.* **21,** 172–173.

Hofmann, K., and Falquet, L. (2001). A ubiquitin-interacting motif conserved in components of the proteasomal and lysosomal protein degradation systems. *Trends Biochem. Sci.* **26,** 347–350.

Hook, S. S., Orian, A., Cowley, S. M., and Eisenman, R. N. (2002). Histone deacetylase 6 binds polyubiquitin through its zinc finger (PAZ domain) and copurifies with deubiquitinating enzymes. *Proc. Natl. Acad. Sci. USA* **99,** 13425–13430.

James, P., Halladay, J., and Craig, E. A. (1996). Genomic libraries and a host strain designed for highly efficient two-hybrid selection in yeast. *Genetics* **144,** 1425–1436.

Kanayama, A., Seth, R. B., Sun, L., Ea, C. K., Hong, M., Shaito, A., Chiu, Y. H., Deng, L., and Chen, Z. J. (2004). TAB2 and TAB3 activate the NF-kappaB pathway through binding to polyubiquitin chains. *Mol. Cell* **15,** 535–548.

Kang, R. S., Daniels, C. M., Francis, S. A., Shih, S. C., Salerno, W. J., Hicke, L., and Radhakrishnan, I. (2003). Solution structure of a CUE-ubiquitin complex reveals a conserved mode of ubiquitin-binding. *Cell* **113,** 621–630.

Katzmann, D. J., Babst, M., and Emr, S. D. (2001). Ubiquitin-dependent sorting into the multivesicular body pathway requires the function of a conserved endosomal sorting complex, ESCRT-I. *Cell* **106**, 145–155.

Klapisz, E., Sorokina, I., Lemeer, S., Pijnenburg, M., Verkleij, A. J., and van Bergen en Henegouwen, P. M. (2002). A ubiquitin-interacting motif (UIM) is essential for Eps15 and Eps15R ubiquitination. *J. Biol. Chem.* **277**, 30746–30753.

Koegl, M., Hoppe, T., Schlenker, S., Ulrich, H., Mayer, T., and Jentsch, S. (1999). A novel ubiquitination factor, E4, is involved in multiubiquitin chain assembly. *Cell* **96**, 635–644.

Liu, Q., Jin, C., Liao, X., Shen, Z., Chen, D. J., and Chen, Y. (1999). The binding interface between an E2 (UBC9) and a ubiquitin homologue (UBL1). *J. Biol. Chem.* **274**, 16979–16987.

Mao, Y., Nickitenko, A., Duan, X., Lloyd, T. E., Wu, M. N., Bellen, H., and Quiocho, F. A. (2000). Crystal structure of the VHS and FYVE tandem domains of Hrs, a protein involved in membrane trafficking and signal transduction. *Cell* **100**, 447–456.

Merkley, N., and Shaw, G. S. (2004). Solution structure of the flexible class II ubiquitin-conjugating enzyme Ubc1 provides insights for polyubiquitin chain assembly. *J. Biol. Chem.* **279**, 478139–47147.

Meyer, H. H., Wang, Y., and Warren, G. (2002). Direct binding of ubiquitin conjugates by the mammalian p97 adaptor complexes, p47 and Ufd1-Npl4. *EMBO J.* **21**, 5645–5652.

Miura, T., Klaus, W., Gsell, B., Miyamoto, C., and Senn, H. (1999). Characterization of the binding interface between ubiquitin and class I human ubiquitin-conjugating enzyme 2b by multidimensional heteronuclear NMR spectroscopy in solution. *J. Mol. Biol.* **290**, 213–228.

Mizuno, E., Kawahata, K., Kato, M., Kitamura, N., and Komada, M. (2003). STAM proteins bind ubiquitinated proteins on the early endosome via the VHS domain and ubiquitin-interacting motif. *Mol. Biol. Cell* **14**, 3675–3689.

Mueller, T. D., Kamionka, M., and Feigon, J. (2004). Specificity of the interaction between ubiquitin-associated domains and ubiquitin. *J. Biol. Chem.* **279**, 11926–11936.

Oldham, C. E., Mohney, R. P., Miller, S. L., Hanes, R. N., and O'Bryan, J. P. (2002). The ubiquitin-interacting motifs target the endocytic adaptor protein epsin for ubiquitination. *Curr. Biol.* **12**, 1112–1116.

Parcellier, A., Schmitt, E., Gurbuxani, S., Seigneurin-Berny, D., Pance, A., Chantome, A., Plenchette, S., Khochbin, S., Solary, E., and Garrido, C. (2003). HSP27 is a ubiquitin-binding protein involved in I-kappaBalpha proteasomal degradation. *Mol. Cell. Biol.* **23**, 5790–5802.

Peng, J., Schwartz, D., Elias, J. E., Thoreen, C. C., Cheng, D., Marsischky, G., Roelofs, J., Finley, D., and Gygi, S. P. (2003). A proteomics approach to understanding protein ubiquitination. *Nat. Biotechnol.* **21**, 921–926.

Polo, S., Sigismund, S., Faretta, M., Guidi, M., Capua, M. R., Bossi, G., Chen, H., De Camilli, P., and Di Fiore, P. P. (2002). A single motif responsible for ubiquitin recognition and monoubiquitination in endocytic proteins. *Nature* **416**, 451–455.

Pornillos, O., Alam, S. L., Rich, R. L., Myszka, D. G., Davis, D. R., and Sundquist, W. I. (2002). Structure and functional interactions of the Tsg101 UEV domain. *EMBO J.* **21**, 2397–2406.

Prag, G., Misra, S., Jones, E. A., Ghirlando, R., Davies, B. A., Horazdovsky, B. F., and Hurley, J. H. (2003). Mechanism of ubiquitin recognition by the CUE domain of Vps9p. *Cell* **113**, 609–620.

Rajesh, S., Sakamoto, T., Iwamoto-Sugai, M., Shibata, T., Kohno, T., and Ito, Y. (1999). Ubiquitin binding interface mapping on yeast ubiquitin hydrolase by NMR chemical shift perturbation. *Biochemistry* **38**, 9242–9253.

Russell, N. S., and Wilkinson, K. D. (2004). Identification of a novel 29-linked polyubiquitin binding protein, Ufd3, using polyubiquitin chain analogues. *Biochemistry* **43,** 4844–4854.

Ryu, K. S., Lee, K. J., Bae, S. H., Kim, B. K., Kim, K. A., and Choi, B. S. (2003). Binding surface mapping of intra- and interdomain interactions among hHR23B, ubiquitin, and polyubiquitin binding site 2 of S5a. *J. Biol. Chem.* **278,** 36621–36627.

Sakata, E., Yamaguchi, Y., Kurimoto, E., Kikuchi, J., Yokoyama, S., Yamada, S., Kawahara, H., Yokosawa, H., Hattori, N., Mizuno, Y., Tanaka, K., and Kato, K. (2003). Parkin binds the Rpn10 subunit of 26S proteasomes through its ubiquitin-like domain. *EMBO Rep.* **4,** 301–306.

Schnell, J. D., and Hicke, L. (2003). Non-traditional functions of ubiquitin and ubiquitinbinding proteins. *J. Biol. Chem.* **278,** 35857–35860.

Scott, P. M., Bilodeau, P. S., Zhdankina, O., Winistorfer, S. C., Hauglund, M. J., Allaman, M. M., Kearney, W. R., Robertson, A. D., Boman, A. L., and Piper, R. C. (2004). GGA proteins bind ubiquitin to facilitate sorting at the trans-Golgi network. *Nat. Cell Biol.* **6,** 252–259.

Seibenhener, M. L., Babu, J. R., Geetha, T., Wong, H. C., Krishna, N. R., and Wooten, M. W. (2004). Sequestosome 1/p62 is a polyubiquitin chain binding protein involved in ubiquitin proteasome degradation. *Mol. Cell. Biol.* **24,** 8055–8068.

Shekhtman, A., and Cowburn, D. (2002). A ubiquitin-interacting motif from Hrs binds to and occludes the ubiquitin surface necessary for polyubiquitination in monoubiquitinated proteins. *Biochem. Biophys. Res. Commun.* **296,** 1222–1227.

Shiba, Y., Katoh, Y., Shiba, T., Yoshino, K., Takatsu, H., Kobayashi, H., Shin, H. W., Wakatsuki, S., and Nakayama, K. (2004). GAT (GGA and Tom1) Domain responsible for ubiquitin binding and ubiquitination. *J. Biol. Chem.* **279,** 7105–7111.

Shih, S. C., Katzmann, K. J., Schnell, J. D., Sutanto, M., Emr, S. C., and Hicke, L. H. (2002). Epsins and Vps27/Hrs contain ubiquitin-binding domains that function in receptor endocytosis. *Nat. Cell Biol.* **4,** 389–393.

Shih, S. C., Prag, G., Francis, S. A., Sutanto, M. A., Hurley, J. H., and Hicke, L. (2003). A ubiquitin-binding motif required for intramolecular ubiquitylation, the CUE domain. *EMBO J.* **22,** 1273–1281.

Sloper-Mould, K. E., Pickart, C. M., and Hicke, L. (2001). Distinct functional surface regions on ubiquitin. *J. Biol. Chem.* **276,** 30483–30489.

Steen, H., and Mann, M. (2004). The abc's (and xyz's) of peptide sequencing. *Nat. Rev. Mol. Cell. Biol.* **5,** 699–711.

Swanson, K. A., Kang, R. S., Stamenova, S. D., Hicke, L., and Radhakrishnan, I. (2003). Solution structure of Vps27 UIM-ubiquitin complex important for endosomal sorting and receptor downregulation. *EMBO J.* **22,** 4597–4606.

van Nuland, N. A., Kroon, G. J., Dijkstra, K., Wolters, G. K., Scheek, R. M., and Robillard, G. T. (1993). The NMR determination of the IIA(mtl) binding site on HPr of the *Escherichia coli* phosphoenol pyruvate-dependent phosphotransferase system. *FEBS Lett.* **315,** 11–15.

Varadan, R., Assfalg, M., Haririnia, A., Raasi, S., Pickart, C., and Fushman, D. (2004). Solution conformation of Lys63-linked di-ubiquitin chain provides clues to functional diversity of polyubiquitin signaling. *J. Biol. Chem.* **279,** 7055–7063.

Varadan, R., Walker, O., Pickart, C., and Fushman, D. (2002). Structural properties of polyubiquitin chains in solution. *J. Mol. Biol.* **324,** 637–647.

Walters, K. J., Lech, P. J., Goh, A. M., Wang, Q., and Howley, P. M. (2003). DNA-repair protein hHR23a alters its protein structure upon binding proteasomal subunit S5a. *Proc. Natl. Acad. Sci. USA* **100,** 12694–12699.

Wang, B., Alam, S. L., Meyer, H. H., Payne, M., Stemmler, T. L., Davis, D. R., and Sundquist, W. I. (2003). Structure and ubiquitin interactions of the conserved zinc finger domain of Npl4. *J. Biol. Chem.* **278,** 20225–20234.

White, M. (1996). The yeast two-hybrid system: Forward and reverse. *Proc. Natl. Acad. Sci. USA* **93,** 10001–10003.

Wilkinson, K. D., Laleli-Sahin, E., Urbauer, J., Larsen, C. N., Shih, G. H., Haas, A. L., Walsh, S. T., and Wand, A. J. (1999). The binding site for UCH-L3 on ubiquitin: Mutagenesis and NMR studies on the complex between ubiquitin and UCH-L3. *J. Mol. Biol.* **291,** 1067–1077.

Yamakami, M., Yoshimori, T., and Yokosawa, H. (2003). Tom1, a VHS domain-containing protein, interacts with tollip, ubiquitin, and clathrin. *J. Biol. Chem.* **278,** 52865–52872.

Young, P., Deveraux, Q., Beal, R. E., Pickart, C. M., and Rechsteiner, M. (1998). Characterization of two polyubiquitin binding sites in the 26 S protease subunit 5a. *J. Biol. Chem.* **273,** 5461–5467.

Yuan, X., Simpson, P., McKeown, C., Kondo, H., Uchiyama, K., Wallis, R., Dreveny, I., Keetch, C., Zhang, X., Robinson, C., Freemont, P., and Matthews, S. (2004). Structure, dynamics and interactions of p47, a major adaptor of the AAA ATPase, p97. *EMBO J.* **23,** 1463–1473.

Zhu, G., Zhai, P., He, X., Terzyan, S., Zhang, R., Joachimiak, A., Tang, J., and Zhang, X. C. (2003). Crystal structure of the human GGA1 GAT domain. *Biochemistry* **42,** 6392–6399.

[10] Analysis of Ubiquitin Chain-Binding Proteins by Two-Hybrid Methods

By JENNIFER APODACA, JUNGMI AHN, IKJIN KIM, and HAI RAO

Abstract

Ubiquitin (Ub) regulates important cellular processes through covalent attachment to its substrates. Distinct fates are bestowed on multi-Ub chains linked through different lysine residues. Ub contains seven conserved lysines, all of which could be used for multi-Ub chain formation. K29 and K48 are the signals for proteasome-mediated proteolysis. Multi-Ub chains linked through K63 have nonproteolytic functions. Studies of Ub-binding factors are likely the key to understanding diverse functions of the Ub molecule. Yeast two-hybrid assay can be a powerful approach to dissect the interaction between Ub and its binding proteins and also the function of these Ub-chain binding proteins *in vivo*.

Introduction

Ub, an abundant 76-residue protein, is highly conserved among eukaryotes. Ubiquitylation—the covalent conjugation of Ub to lysine residues on other intracellular proteins—regulates a myriad of cellular processes, including cell cycle progression, signal transduction, DNA repair, and inflammation (Pickart and Cohen, 2004; Schwartz and Hochstrasser, 2003). In addition to its classical role in proteolysis, Ub is also emerging as a nonproteolytic signal for protein transport and processing. The fate of a substrate depends on the number of Ub moieties conjugated, as well as the lysine linkage of Ub-Ub conjugation (Pickart and Cohen, 2004; Weissman, 2001). Three types of Ub chains have known functions. K29 and K48 are the major sites used in the formation of Ub chains that target substrates for proteasome-mediated degradation. Ub chains linked through K63 have nonproteolytic functions (Weissman, 2001). One of the most challenging questions is to understand the mechanisms underlying diverse functions of the Ub molecule. Proteins that recognize specific Ub chain linkages are suspected to play a key role in determining the fate of ubiquitinated substrates (Chen and Madura, 2002; Rao and Sastry, 2002; Schnell and Hicke, 2003; Verma *et al.*, 2004).

Recently, a number of Ub-binding proteins have been isolated by various molecular and biochemical approaches (see Chapter 9 and Schnell and Hicke, 2003). These proteins regulate a broad range of cellular processes such as proteolysis, signal transduction, transcription, and endocytosis. Specific sequence motifs responsible for the Ub-association, including UBA, UIM, CUE, GAT, UEV, NZF, and PAZ, have been identified (Schnell and Hicke, 2003). The binding of these motifs could have been specifically to mono-Ub, multi-Ub chains, or ubiquitinated substrates. Do these proteins/domains prefer Ub-chains with specific linkages? Yeast two-hybrid assay, a powerful approach for protein–protein interaction studies, can be used to distinguish between these possibilities. We used this method to characterize the nature of the interaction between Ub and the UBA domain (Rao and Sastry, 2002). A major advantage of two-hybrid assay is that it detects *in vivo* interaction. Other biochemical binding assays often use Ub-chains with defined length and/or proteins purified from foreign hosts (*e.g.*, *Escherichia coli*, baculovirus), which may be different from *in vivo* settings. By use of a set of Ub derivatives, we found that the UBA motif specifically binds K48-linked Ub conjugates but not K63-linked Ub species (Rao and Sastry, 2002). Importantly, these results were confirmed by other biochemical binding assays (Bertolaet *et al.*, 2001; Raasi and Pickart, 2003; Rao and Sastry, 2002) and thus validated the two-hybrid approach as an effective means to characterize Ub-chain binding activities *in vivo*. How do these Ub-binding proteins achieve their substrate specificity *in vivo*? An array-based two-hybrid screen is used to identify proteins that facilitate the functioning of Ub-binding proteins (*e.g.*, Dsk2). The two-hybrid strategy described here could be used to characterize other Ub-binding molecules.

Materials

Strains, Media, And Plasmids

Yeast cultures were grown in rich (YPD) or in synthetic media containing standard ingredients. *S. cerevisiae* PJ69-4A (*MATa gal4Δ gal80Δ GAL2-ADE2 LYS2::GAL1-HIS3 met::GAL7-lacZ*) and PJ69–4α, which has the opposite mating type, were used for yeast two-hybrid assays (James *et al.*, 1996; Uetz *et al.*, 2000). Positive interactions can be identified by the growth on –his medium or –ade medium. The Ub-UBA interaction can activate the expression of the *HIS3* reporter gene but not the more stringent *ADE2* reporter (James *et al.*, 1996). pGBD and pGAD vectors, which contain Gal4 DNA-binding domain and activation domain, respectively, were used (James *et al.*, 1996). The use of this Gal4-based system is critical, because the involvement of specific lysine residues for the interactions between Ub- and wild-type UBA–containing proteins (*e.g.*, Rad23, Ddi1) was not easily detected by a lexA-based two-hybrid method (Bertolaet *et al.*, 2001), likely because of different sensitivities of the two systems.

Method and Results

Characterization of the Interactions Between Ub and UBA, a Ub-Binding Domain, by Two-Hybrid Assays

A genome-wide two-hybrid screen identified yeast Rad23 as a Ub-binding protein (Uetz *et al.*, 2000). The UBA (Ub associated) domains of Rad23 are sufficient for association with Ub (Bertolaet *et al.*, 2001; Chen and Madura, 2002; Rao and Sastry, 2002; Wilkinson *et al.*, 2001). To understand the basis of the Ub-UBA interactions, we constructed two-hybrid fusions containing mutations in Ub. Specifically, we made a series of mutations in the Ub coding region that would affect its ability to form various conjugates. The design of these constructs was based on the different roles of amino acid residues in Ub. Ub chain formation requires the C-terminal Gly76 of Ub to attach to a lysine residue of another Ub molecule or a substrate. Ub contains seven lysines, all of which can be used for multi-Ub chain assembly (Fig. 1A). Ub chains assembled through K29 and K48 are involved in proteolysis, whereas K63-based chains have nonproteolytic functions. We, therefore, tested the ability of the UBA domain to bind to several Ub variants, including Ub-V76, which contained a Gly to Val mutation that eliminated its ability to conjugate to other proteins, although it could still be an acceptor of other Ub molecules. Other constructs were Ub mutants containing various single or multiple

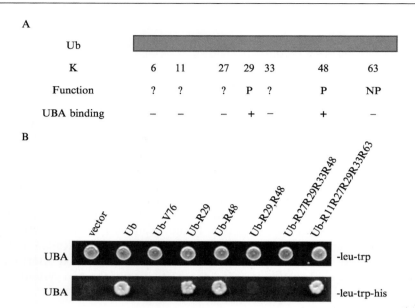

FIG. 1. Two-hybrid analysis of interactions between Ub and the UBA domain of Rad23 *in vivo*. (A) Different roles for different ubiquitin linkages. Ub has seven lysines, which could be used for multi-Ub chain assembly. Lys29- and Lys48-linked chains target proteins to the proteasome, and Lys63-linked chains have nonproteolytic functions. (B) Two-hybrid interactions between Ub and the UBA domain. Yeast cells were co-transformed with plasmids encoding GAL4 DNA-binding domain fused to the second UBA domain of Rad23 indicated on the left with the GAL4 activation domain fused to Ub and its derivatives indicated above each panel. The mutated residues in Ub are indicated. Growth on –his plates is indicative of protein-protein interactions.

Lys to Arg mutations, which block Ub conjugation to those lysine residues (Fig. 1).

1. The yeast strain PJ69–4A, harboring the bait plasmid expressing Gal4 DNA-binding domain fused to the UBA domain, was transformed with the plasmids containing the GAL4 activation domain fused to Ub derivatives. The transformants were spread onto synthetic plates lacking leucine and tryptophan (SD-leu-trp) and grown for 2 days at 30°.

2. A single colony from each transformation plate was inoculated in 3-ml SD-leu-trp medium and grown overnight at 30°. These cells were diluted in 3 ml fresh SD-leu-trp medium to $OD_{600} \sim 0.15$ and grown for ~8 h at 30°. The final OD_{600} should be approximately 0.6–0.8. This step ensures that yeast cells are growing actively and synchronously, which leads to more consistent results.

3. All cells were adjusted to the same OD_{600}; 100 μl cells were transferred to a U-bottom 96-well plate and subsequently spotted onto SD-leu-trp and SD-leu-trp-his plates with a microplate replicator. Growth on SD-leu-trp, which could be seen in 2 days, served as a loading control to ascertain that roughly equal amount of cells were used for each sample.

4. We incubated the SD-leu-trp-his plates for 4–6 days at 30°. Growth on the SD-leu-trp-his plate is indicative of positive protein–protein interactions.

Using this strategy, we found that the Ub interaction with the UBA motif required either K48 or K29, but not K63 (Fig. 1B). Because these interactions with Ub required Gly76, which is essential for substrate conjugation, and specific lysine residues that are important for Ub chain formation, it suggested that the UBA-containing proteins likely interact with multiubiquitylated substrates containing certain Ub linkages and that UBA proteins might be important for proteolysis, which was confirmed by follow-up genetic and biochemical experiments (Chen and Madura, 2002; Rao and Sastry, 2002; Wilkinson et al., 2001). Therefore, the two-hybrid assay described here could be used to study the regulators for multi-Ub chain formation and maintenance in vivo.

An Array-Based Method to Identify Partners of Dsk2, a UBA-Containing Protein

The traditional use of the two-hybrid method is to identify novel protein–protein interactions. High-throughput array-based strategies have been developed to conduct genome-wide two-hybrid screens in yeast (Uetz et al., 2000). Nearly all S. cerevisiae proteins have been linked to the Gal4-activation domain (Uetz et al., 2000) and are freely available from the Yeast Resource Center (University of Washington, Seattle). For most laboratories, at the moment, it is still formidable to perform such large-scale screens routinely. However, if one suspects that the protein of his or her interest may interact with a particular class of proteins, one could easily make a small array expressing these proteins linked to the Gal4-activation domain and perform a two-hybrid screening in 2 weeks. This array-based screen is quick and systematic. Importantly, the identity of the genes in each well is known. In addition, this deliberated screen can identify the interactions that are biologically relevant but have low affinity. One problem frequently associated with two-hybrid assays is the selection of the reporters for the screening (James et al., 1996). For example, *HIS3* is a very sensitive reporter but often leads to many false-positive results in the screens. *ADE2* is a much more stringent reporter, but its use often misses the weak but authentic interactions.

We used this strategy to identify the Ub-protein ligases (E3) associated with Dsk2, another Rad23-like Ub-binding protein in yeast. The homologs of Rad23 and Dsk2 were shown to bind several E3 enzymes including Mdm2 (Brignone et al., 2004), APC (Seeger et al., 2003), and E6AP (Kleijnen et al., 2000). We reasoned that yeast Dsk2 might also bind E3s. There are ~60 yeast proteins that contain the recognizable, known Ub-ligase motifs (e.g., RING finger, HECT domain, U box). Most of them have been linked to the Gal4-activation domain and transformed into PJ69-4A strain, which can be obtained from the Yeast Resource Center.

1. These prey strains were inoculated in 100 μl YPD medium in a 96-well plate and incubated at room temperature for 2 days.

2. The bait plasmid expressing the Gal4 DNA-binding domain; Dsk2 fusion was transformed to PJ69-4 strain. Then, the colonies were grown overnight in SD-trp medium. Yeast cells were diluted in fresh medium and grown to $OD_{600} \sim 0.6$; 80 μl of these cells were mixed with each of the prey strains described previously in the 96-well plate.

3. The matings were carried out at room temperature for 24 h. To select for diploid cells containing both bait and prey plasmids, cells were transferred to SD-leu-trp plates with a microplate replicator.

The mating mix can be directly transferred to SD-leu-trp-his plates to select for protein–protein interactions. However, we found that that approach led to a higher rate of false-positive results.

4. Diploid cells were picked and grown in 100 μl YPD medium in 96-well plates at room temperature for 24 h. Equal amounts of cells were spotted on SD-leu-trp plate for control and SD-leu-trp-his+2 mM 3-aminotriazole (3-AT) for interaction screening; 3-AT is a competitive inhibitor of the His3 protein.

Using this array-based method, we found that Dsk2 interacts with Ufd2, a E3/E4 enzyme (Fig. 2). This interaction was later confirmed by

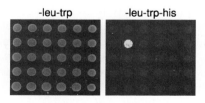

FIG. 2. Screen for the Dsk2-binding E3s using protein array. A positive interaction between Ufd2 and Dsk2. A yeast strain expressing Dsk2 was mated to the array of transformants carrying Ub-protein ligases, the diploids were transferred to SD-leu-trp, and SD-leu-trp-his+2 mM AT plates.

co-immunoprecipitations (data not shown). Because the Dsk2–Ufd2 interaction is weak and only activates the transcription of *HIS3* but not *ADE2*, a traditional two-hybrid screen would have missed the Dsk2–Ufd2 interaction. A similar strategy was used successfully to identify proteins that associate with the proteasome (Cagney *et al.*, 2001).

Summary

The two-hybrid method can be adapted to dissect the interactions between Ub-binding proteins and Ub or its regulators. The two-hybrid method has several useful features. The assay is highly sensitive. The system is easy and inexpensive to set up. The interactions detected occur *in vivo*. However, the two-hybrid system is an indirect assay, because it relies on transcriptional activation of reporter genes. It is important to confirm the two-hybrid results by other molecular, functional, or biochemical approaches, such as co-immunoprecipitation and co-purification.

Acknowledgments

We are grateful to S. Fields, T. Ito, and the Yeast Resource Center for strains and plasmids. This work was supported by grants to H. R. from San Antonio Cancer Institute, San Antonio Area Foundation, UT Health Science Center HHMI Research Resources Program, and UT Health Science Center Institutional Research Grant.

References

Bertolaet, B. L., Clarke, D. J., Wolff, M., Watson, M. H., Henze, M., Divita, G., and Reed, S. I. (2001). UBA domains of DNA damage-inducible proteins interact with ubiquitin. *Nat. Struct. Biol.* **8,** 417–422.
Brignone, C., Bradley, K. E., Kisselev, A. F., and Grossman, S. R. (2004). A post-ubiquitination role for MDM2 and hHR23A in the p53 degradation pathway. *Oncogene* **23,** 4121–4129.
Cagney, G., Uetz, P., and Fields, S. (2001). Two-hybrid analysis of the *Saccharomyces cerevisiae* 26S proteasome. *Physiol. Genomics* **7,** 27–34.
Chen, L., and Madura, K. (2002). Rad23 promotes the targeting of proteolytic substrates to the proteasome. *Mol. Cell. Biol.* **22,** 4902–4913.
James, P., Halladay, J., and Craig, E. A. (1996). Genomic libraries and a host strain designed for highly efficient two-hybrid selection in yeast. *Genetics* **144,** 1425–1436.
Kleijnen, M. F., Shih, A. H., Zhou, P., Kumar, S., Soccio, R. E., Kedersha, N. L., Gill, G., and Howley, P. M. (2000). The hPLIC proteins may provide a link between the ubiquitination machinery and the proteasome. *Mol. Cell* **6,** 409–419.
Pickart, C. M., and Cohen, R. E. (2004). Proteasomes and their kin: Proteases in the machine age. *Nat. Rev. Mol. Cell Biol.* **5,** 177–187.
Raasi, S., and Pickart, C. M. (2003). Rad23 ubiquitin-associated domains (UBA) inhibit 26 S proteasome-catalyzed proteolysis by sequestering lysine 48-linked polyubiquitin chains. *J. Biol. Chem.* **278,** 8951–8959.

Rao, H., and Sastry, A. (2002). Recognition of specific ubiquitin conjugates is important for the proteolytic functions of the ubiquitin-associated domain proteins Dsk2 and Rad23. *J. Biol. Chem.* **277,** 11691–11695.

Schnell, J. D., and Hicke, L. (2003). Non-traditional functions of ubiquitin and ubiquitin-binding proteins. *J. Biol. Chem.* **278,** 35857–35860.

Schwartz, D. C., and Hochstrasser, M. (2003). A superfamily of protein tags: Ubiquitin, SUMO and related modifiers. *Trends Biochem. Sci.* **28,** 321–328.

Seeger, M., Hartmann-Petersen, R., Wilkinson, C. R., Wallace, M., Samejima, I., Taylor, M. S., and Gordon, C. (2003). Interaction of the anaphase-promoting complex/cyclosome and proteasome protein complexes with multiubiquitin chain-binding proteins. *J. Biol. Chem.* **278,** 16791–16796.

Uetz, P., Giot, L., Cagney, G., Mansfield, T. A., Judson, R. S., Knight, J. R., Lockshon, D., Narayan, V., Srinivasan, M., Pochart, P., Qureshi-Emili, A., Li, Y., Godwin, B., Conover, D., Kalbfleisch, T., Vijayadamodar, G., Yang, M. J., Johnston, M., Fields, S., and Rothberg, J. M. (2000). A comprehensive analysis of protein-protein interactions in *Saccharomyces cerevisiae. Nature* **403,** 623–627.

Verma, R., Oania, R., Graumann, J., and Deshaies, R. (2004). Multiubiquitin chain receptors define a layer of substrate selectivity in the ubiquitin-proteasome system. *Cell* **118,** 99–110.

Weissman, A. M. (2001). Themes and variations on ubiquitylation. *Nat. Rev. Mol. Cell Biol.* **2,** 169–178.

Wilkinson, C. R. M., Seeger, M., Hartmann-Petersen, R., Stone, M., Wallace, M., Semple, C., and Gordon, C. (2001). Proteins containing the UBA domain are able to bind to multi-ubiquitin chains. *Nat. Cell Biol.* **3,** 939–943.

[11] Quantifying Protein–Protein Interactions in the Ubiquitin Pathway by Surface Plasmon Resonance

By RASMUS HARTMANN-PETERSEN and COLIN GORDON

Abstract

The commercial availability of instruments, such as Biacore, that are capable of monitoring surface plasmon resonance (SPR) has greatly simplified the quantification of protein–protein interactions. Already, this technique has been used for some studies of the ubiquitin-proteasome system. Here we discuss some of the problems and pitfalls that researchers should be aware of when using SPR analyses for studies of the ubiquitin-proteasome system.

Introduction

Intracellular protein degradation in eukaryotic cells plays an important role in a series of cellular and molecular functions, including the turnover of bulk proteins, cell cycle control, DNA repair, antigen presentation, vesicle transport, and in the regulation of signal transduction (Glickman and Ciechanover, 2002).

The ubiquitin-proteasome system constitutes the major intracellular proteolytic pathway (Glickman and Ciechanover, 2002). Since its discovery, the ubiquitin-proteasome system has been studied in great detail. However, some major challenges remain to be overcome before we may understand intracellular proteolysis in full. These include analyzing the specificity and regulation of the different components of the system, and perhaps, even more importantly, analyzing the interplay between these protein factors. To this end, the exploration of novel protein–protein interactions will be paramount. However, to fully comprehend the significance of pairs of interacting proteins, we will need to determine not only their cellular abundance but also quantify the strength of the protein–protein interactions.

The commercial availability of instruments, such as Biacore (Karlsson and Larsson, 2004), that are capable of monitoring surface plasmon resonance (SPR) has greatly simplified the quantification of protein–protein interactions. Already this technique has been put to use for some studies of the ubiquitin-proteasome system (Asher *et al.*, 2003; Garrus *et al.*, 2001; Kozlov *et al.*, 2004; Raasi and Pickart, 2003; Wilkinson *et al.*, 2001). Here we discuss some of the problems and pitfalls that researchers should be aware of when using SPR analyses for studies of the ubiquitin-proteasome system. We provide a general protocol, but for each pair of interacting proteins, it will most likely be necessary, often empirically, to optimize the assay conditions. For more specialized information regarding SPR analyses, we refer to recent methodological articles and reviews (Chinowsky *et al.*, 2003; Karlsson and Larsson, 2004; McDonnell, 2001; Medaglia and Fisher, 2001; Myszka, 2000; Nedelkov and Nelson, 2003).

Quantifying protein–protein interactions

The major experimental approaches used for the quantification of interactions between proteins and other macromolecules include equilibrium dialysis, analytical ultracentrifugation, isothermal titration calorimetry, and SPR. Equilibrium dialysis requires that there is a significant size difference between the proteins or protein complexes in question. Analytical ultracentrifugation provides evidence for the aggregation state of a protein, and isothermal titration calorimetry provides a set of thermodynamic parameters, including the equilibrium constant, enthalpy, and entropy changes, as well as revealing the stoichiometry of the reaction. Unfortunately, both analytical ultracentrifugation and isothermal titration calorimetry require fairly large amounts (milligrams) of pure protein. The strength of SPR compared with these techniques is its ability to obtain kinetic data, while requiring only microgram amounts of protein. Also SPR measurements are quite rapid, and results can often be obtained within a few days.

The drawbacks of SPR analysis are that the binding assays are performed in an unphysiological manner where one binding partner (the ligand) is attached to a solid support and other (the analyte) is in solution. Also SPR requires that the analytes must have a sufficient molecular mass (typically >2000 kDa). This is rarely a problem for studies of protein–protein interactions but may prove troublesome for analyses of interactions between smaller molecules like peptides, glycans, and synthetic inhibitors.

The kinetic data obtained by SPR are the association rate constant k_a and the dissociation rate constant k_d, which describe the interaction of two molecules binding to each other. The association rate constant is expressed by the Arrhenius equation: $k_a = A\exp(-\frac{E_a}{RT})$, in which the preexponential factor A is a statistical parameter describing the collision frequency, E_a is the activation energy, R is the gas constant, and T is the absolute temperature. The rate constant of dissociation k_d is given by an analogous equation.

From k_a and k_d, the affinity constant K_a or its reciprocal value, the dissociation constant K_d, can easily be calculated ($K_d = k_d/k_a$). As mentioned, knowledge of the dissociation constant of a protein complex is important, but more so compared with the dissociation constants of related protein complexes.

More elaborate SPR analyses may also yield a complete thermodynamic description of the molecular interaction. By performing SPR analyses at different temperatures, it is possible to determine the activation energies, and from these the enthalpy change ($\Delta H°$) of the interaction can be calculated because: $\Delta H° = E_a(\text{association}) - E_a(\text{dissociation})$. The changes in entropy ($\Delta S°$) and free energy ($\Delta G°$) of the interaction are determined from: $\Delta G° = -RT\ln K_a = \Delta H° - T\Delta S°$.

Surface Plasmon Resonance

Several SPR instruments are commercially available. These range from simple systems to more expensive and integrated robotic instruments. The most notable providers include Texas Instruments, XanTec, BioTul, and Biacore. We are mostly familiar with the Biacore system, but because the basic function and output of the different systems are largely comparable, the general experimental outlines described here should be similar for the other systems.

We shall not attempt to give a comprehensive account of the physics behind SPR analyses here. For a more technical account of these aspects of the SPR technique, we recommend reading Biacore SPR Technology Notes (available from http://www.biacore.com), which is readily understandable and does not require that the reader have a substantial background knowledge of physics. However, in brief, SPR is an optical phenomenon that occurs under specific conditions when light interacts with a metal. This resonance depends on the refractive index of the environment close to the metal surface and allows changes in refractive index to be monitored in real time. If molecular interactions, either binding or dissociation events, occur in this environment, it will cause a corresponding change in refractive index leading to an angular change of the reflected light, which is detected as a change in the SPR signal (Fig. 1). The arbitrary resonance unit (RU) is defined as 1/3000 of a degree in the Biacore instrument. When an experiment is finished, these RUs are used for fitting to certain binding models so that rate constants and K_d can be determined.

The major components of a Biacore SPR system are:

- The sensor chip to which the ligand is attached.
- A microfluidics system to deliver the analyte to ligand on the sensor chip.
- A data collection system with accompanying software for data analyses.

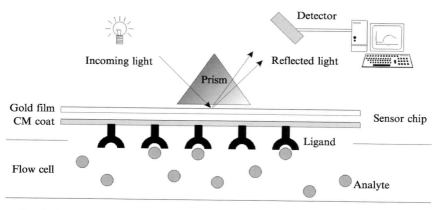

FIG. 1. Basic design of an SPR instrument. As analyte passes through the flow cell, it may interact with the immobilized ligand on the sensor surface. This leads to an increase of mass on the sensor surface that is proportional to a shift in the angle of the reflected light. This angular change is monitored by a detector and displayed as a sensorgram on the screen of an attached computer.

After the ligand has been immobilized to the chip surface, the actual binding experiments can begin.

The output from the SPR apparatus is a plot of SPR response vs. time called a sensorgram. This sensorgram (Fig. 2) is presented on the screen of an attached computer and is generated in real time.

Before the injection of the sample, running buffer is pumped across the surface of the sensor chip, and a baseline is recorded. Then samples of analyte are delivered to the flow cells in the sensor chip by the microfluidics system, which ensures an accurate and reproducible sample delivery and a low sample consumption. The Biacore instrument will then monitor changes in refractive index arising from binding events on the chip surface. Once a steady-state level has been reached, the sample delivery is terminated, and running buffer again flows over the chip, and the dissociation of the ligand–analyte complex is monitored (Fig. 2).

Although it seems to be straightforward, SPR analysis compared with qualitative methods of studying protein–protein interactions such as yeast two-hybrid analysis and co-precipitation, requires significant amount of optimization before the actual experiments can be run and the data then analyzed. For instance, the binding conditions must be optimized and the surface regeneration conditions determined.

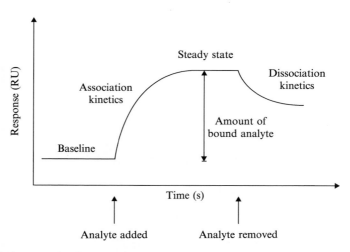

FIG. 2. An example of a sensorgram. The output from the Biacore instrument is a plot of SPR response *vs* time, called a sensorgram. This sensorgram is presented on the screen of an attached computer and is generated in real time. See text for details.

Preparation of Proteins for SPR

The Ligand

Unless one is using a capture system (see later), where the ligand can be purified directly on the sensor chip, it is important that the ligand is purified before immobilization. The presence of impurities in the ligand preparation can complicate results by introducing nonspecific binding. We recommend that the purity and homogeneity of the ligand are verified by electrophoresis before immobilization.

The concentration of the ligand is not critical, typically a 1 mg/ml solution is sufficient for conjugation.

Now one must also consider which chemistry to use for the ligand immobilization. We have generally been successful using amine coupling to the sensor chip; therefore, Tris buffers or other buffering agents containing primary amino groups cannot be used. Thus, after the ligand has been purified, we generally desalt the preparation on a column, equilibrated with HEPES-buffered saline (HBS) (10 mM HEPES, pH 7.4, 150 mM NaCl, 3 mM EDTA, 0.005 % (v/v) Tween-20). To protect the protein from oxidation 1–5 mM DTT can also be included, as long as one will not be using thiol chemistry for immobilization. Generally, we find that the viscosity of buffers containing glycerol or sucrose leads to unnecessary noise, and we, therefore, try to avoid them.

To spare the quite sensitive Biacore microfluidity system from potential clogs of insoluble protein, we recommend clearing the ligand preparation by either centrifugation (18,000g for at least 30 min) or filtration (0.2 μm pore size). Finally, to ensure an even flow, the sample should be degassed in a vacuum desiccator before conjugation.

The Analyte

If SPR is just used as a qualitative system for the discovery of protein–protein interactions, there is no need for a highly purified protein preparation. However, again we recommend centrifuging or filtering the preparation to remove aggregated material and thereby spare the microfluidity system from potential clogs of insoluble protein.

In the case of quantitative analyses, these are performed using purified protein of an accurately known concentration. The analyte normally needs to be quite highly concentrated; typically, at least a 1 mg/ml stock is required. However, if the interaction is weak, a more concentrated stock solution may be necessary. As for the buffering system, we generally use HBS. However, if the protein–protein interaction was previously analyzed by, for example, co-precipitation, then a buffer composition similar to that used for these experiments should work. However, at least 100 mM NaCl should be included to minimize unspecific electrostatic interactions with the sensor surface. Low concentrations of detergent can also help prevent analyte adsorption, and again we tend to omit viscous reagents such as glycerol from the buffers. To avoid potential artefacts arising from the buffer system, we generally use identical sample and running buffers. To ensure an even flow, all samples and buffers should be degassed in a vacuum desiccator before use.

When the samples are injected, any change in refractive index is detected. Such changes can be caused by either a binding event or by differences in the refractive index of the sample and the running buffer. As explained later, to circumvent buffer artefacts, the true ligand-binding signal is obtained by subtracting the signal from a control or reference flow cell.

Immobilizing the Ligand to the Sensor Chip

The Biacore sensor chip consists of a 1-cm^2 glass surface coated with gold within a plastic case. For most purposes, the ligand can be directly conjugated to a layer of carboxymethyl (CM) dextran on the sensor chip (Fig. 1). The carboxymethyl dextran–coated surface also provides a hydrophilic environment for the interactions to occur in. The standard Biacore sensor chip is the CM-5 chip, which allows for ligand immobilization by either native primary amine ($-NH_2$), thiol ($-SH$), aldehyde ($-CHO$), or carboxyl ($-COOH$) groups to its carboxymethyl dextran surface. Each sensor chip has four different flow cells to which ligands can be attached, so for instance three may be used for varying the ligand concentration while the fourth can be left blank or used for a control (with immobilized dummy ligand). If one requires a high throughput, then different ligands can be immobilized in each channel.

In some cases, it may be useful to capture the ligand with another molecule, because this will result in a known and uniform orientation of the ligand on the chip surface. One may prepare these oneself by, for instance, conjugating a specific antibody to the chip surface and then using this antibody to capture the ligand before analysis of the ligand–analyte interaction. Having the ligand as a histidine-tagged fusion protein can be quite useful for such purposes, because chips can be purchased with nitrilotriacetic acid (NTA) already coupled to the carboxymethyl dextran. Sensor chips coated with streptavidin for capturing biotinylated molecules are also available.

Choosing which of the interacting proteins to use as ligand can be crucial. In cases in which the interaction of several proteins with a common ligand is to be assessed, it is normally best to immobilize the common ligand, because this will save time, cut down on the amount of chips needed, and allow for easy data comparison.

Generally, large molecules are easier to conjugate to the chip surface than smaller molecules. However, to generate a significant mass difference on the sensor chip, it may be preferable to select the smallest molecule as ligand.

For ubiquitin-binding assays, monoubiquitin can be especially troublesome to conjugate, therefore glutathione S-transferase (GST) or histidine-tagged ubiquitin fusion proteins may be preferable either for cross-linking or capture in such assays. Also, we find that most of the commercially available nonrecombinant ubiquitin contains impurities of ubiquitin-chains and is, therefore, not ideally suited for SPR analysis. In our hands, K48-linked tetraubiquitin chains (available from Affiniti Research Products Ltd.) are readily conjugated by amine cross-linking to the CM-5 chip.

In general, we do not recommend attaching protein complexes to the solid support. So, for instance, if one is studying a protein interacting with a protein complex, we suggest immobilizing the free protein and leaving the complex in solution rather than the reverse experiment.

Before the actual immobilization, it is possible to run a few tests to determine under which conditions conjugation will occur most efficiently. We usually try varying the ligand concentration and pH. Generally, we find that conjugation occurs most readily at around pH 5.0. The operational software that accompanies the Biacore system contains a so-called pH-scout program to accommodate determination of the optimal pH for immobilization.

If only a limited amount of ligand is available, it is possible to perform the immobilization at a low flow rate (5 μl/min) to minimize ligand consumption.

After the ligand has been immobilized, the remaining sites on the chip surface are blocked, with, for instance, ethanolamine for amine coupling.

Usually the immobilized ligand is stable for several months when stored in HBS at 4° and can be used for hundreds of runs.

Standard Amine-Conjugation Protocol

1. Once the CM-5 chip has been docked in the Biacore, the surface should be primed with degassed HBS (10 mM HEPES, pH 7.4, 150 mM NaCl, 3 mM EDTA, 0.005% (v/v) Tween-20).
2. Prepare tubes containing 100 μl of:
    - 50 μg/ml ligand diluted in 10 mM Na-acetate with pH 5.0 or as determined using the pH scout program.
    - 50 mM NHS (*N*-hydroxysuccinimide).

- 200 mM EDC (N-ethyl-N'-dimethylamino propyl-carbodiimide).
- 1 M ethanolamine, pH 8.5.
- 20 mM HCl.

3. Place the tubes together with one empty tube (which will be used for mixing) in the Biacore sample rack according to the Immobilisation Wizard software.
4. Initiate conjugation with the Immobilisation Wizard software. We recommend a flow rate between 5 and 10 μl/min.
5. The robotics will now mix the NHS and EDC solutions and activate the surface for amine cross-linking. Then the ligand will be immobilized and the remaining sites blocked with ethanolamine.
6. Watch the immobilization level on the computer monitor. The final level should ideally be >5000 RU.

Steps 5 and 6 are performed manually on Biacore systems without robotic sampling handling.

Troubleshooting

If the ligand is not readily immobilized and an appropriate response level cannot be obtained, we suggest trying the following:

- Use fresh NHS and EDC stocks.
- Increase the concentration of the ligand.
- Decrease the flow rate.
- Change the pH of the ligand solution.
- Reverse the ligand/analyte relationship.

Activity Determination and Chip Regeneration

Once the ligand has been immobilized to an appropriate level, we recommend that the binding activity of the ligand is checked and that conditions for chip regeneration are optimized. The actions are easily accessed through the Biacore control software.

Activity Determination and Surface Regeneration Protocol

1. Prime the surface with degassed HBS (10 mM HEPES, pH 7.4, 150 mM NaCl, 3 mM EDTA, 0.005 % (v/v) Tween-20).
2. Inject 10 μl or more of highly concentrated (1–100 μM) analyte. We use flow rates between 10 and 30 μl/min and 1-min injections. Now the first indication of a ligand-analyte interaction should be evident from the on-screen sensorgram.

3. Once the analyte injection has ended, follow the dissociation of the complex for about 30 min. If the dissociation occurs very slowly and a baseline response has not been attained after 30 min, regeneration (steps 4–5) will be needed.
4. Try regeneration conditions (see below) starting with the mildest and only moving to successively more harsh conditions if needed. We normally use a contact time of 1 min or less for regeneration.
5. Once suitable surface regeneration conditions have been found, repeat steps 1–3 to ensure that the ligand is still active.

Regeneration Conditions

Between experiments, it may be necessary to wash off bound analyte to obtain a baseline response. For this purpose, one of a number of somewhat harsh conditions is used, and it is, therefore, important to check whether the ligand is still active after these treatments.

As mentioned, we recommend starting with the mildest conditions and only moving to successively more harsh conditions if needed. First, try regeneration at low pH, then at high pH or with high salt concentrations. We have never experimented with detergents for surface regeneration. Another parameter that may be useful to vary is the contact time. We suggest trying the following conditions using a contact time of 1 min:

- Low pH ($\leq$20 mM HCl)
- High pH ($\leq$100 mM NaOH)
- High salt (1–4 M of NaCl or MgCl$_2$)

Kinetic Analysis and Data Interpretation

Once the activity of the immobilized ligand has been verified and the surface regeneration conditions (if required) have been optimized, the kinetic analyses may begin.

The basic setup of these experiments is similar to that described for the activity determination in the preceding. However, to obtain kinetic information, the ligand-analyte interaction must be monitored over a range of different analyte concentrations. The analyte is diluted serially in running buffer to concentrations of typically 1 nM–100 μM. We normally use a flow rate of 10–30 μl/min and about 1-min injections and follow the dissociation for about 30 min. The software that accompanies the Biacore instrument is user-friendly and will guide you stepwise through the experimental setup.

For the highest analyte concentration (approximately 100 μM), the association and dissociation should be readily apparent from the sensorgram, because it should be evident whether the surface is saturated. If it is not, higher analyte concentrations might be needed.

After the sensorgrams for a variety of different analyte concentrations have been recorded, the accompanying software will align the different sensorgrams according to time and allow you to adjust the baseline to zero and subtract a blank/control sensorgram. Then it is often a good idea to exclude some data sets. Very high concentrations in which the surface is quickly saturated will not yield useful association kinetics. Also, low concentrations in which no significant ligand-analyte association is observed should be excluded. We generally only include data for about five different analyte concentrations in the final analysis.

The computer software will allow you to perform curve fitting of the sensorgrams to a selection of different binding models. Iteratively,

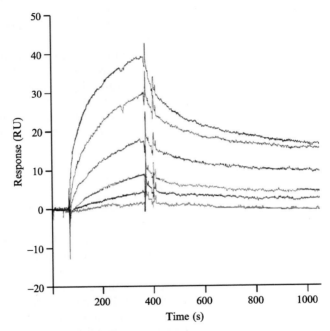

Fig. 3. Actual sensorgrams of Rhp23 interacting with Mts4. Most ubiquitin-like domain proteins interact directly with 26S proteasomes (Hartmann-Petersen and Gordon, 2004). One example is Rhp23/Rad23, which interacts with the Mts4/Rpn1 subunit (Seeger *et al.*, 2003; Wilkinson *et al.*, 2001). We analyzed the Rhp23–Mts4 interaction by SPR analysis on a Biacore 3000 instrument. Purified Mts4 was immobilized on a CM-5 chip by amine cross-linking, then increasing amounts of purified Rhp23 were injected, and the SPR response recorded. The running buffer was HBS. The sensorgrams related to Rhp23 binding alone is shown and was obtained by subtracting the signal from the reference surface from the signal that was obtained from the Mts4 surface. After curve fitting, the K_d for the Rhp23-Mts4 complex was determined to 149 nM.

using nonlinear least-squares regression analysis, the software will minimize the differences between the calculated and observed sets of data and determine the values for k_a and k_d that best describe the interaction.

We shall not attempt to give any detailed description of the Biacore data interpretation software here. However, a quick way to determine whether the obtained data are reliable is by inspecting the validity of the curve fitting by comparing observed and calculated sensorgrams to ensure that there are no major discrepancies between the graphs.

We have included an example of an SPR analysis of the interaction between the fission yeast proteins Rhp23 and Mts4 (Fig. 3).

Troubleshooting

In general, problems tend to fall into two different categories; either there is nonspecific binding of the analyte to the reference surface, or no observable response despite knowledge from other experiments (e.g., coprecipitation) that the proteins really do interact. In the following, we briefly give some clues that may help deal with such difficulties.

Nonspecific Binding

- Try different reference surfaces (i.e., blank, blocked, or dummy protein).
- Increase salt concentration in the analyte/running buffer.
- Switch to a different buffer (e.g., Tris/HCl or phosphate).
- Change the sensor surface to one with a different charge or hydrophilicity.
- Reverse the ligand/analyte relationship.

No Observable Response

- Use more concentrated analyte solution.
- Try changing the level of immobilized ligand.
- Change the composition of the analyte/running buffer.
- Change the sensor surface to one with a different charge or hydrophilicity.
- Reverse the ligand/analyte relationship.

Conclusions

The use of SPR is a powerful tool for measuring reaction kinetics and equilibrium constants. Obtaining such detailed binding data is especially useful when studying families of closely related or multifunctional proteins such as ubiquitin-like proteins (Jentsch and Pyrowolakis, 2000) and ubiquitin itself (Weissman, 2001) but also for studies of other proteins involved in the ubiquitin/proteasome system in which the activity of a protein is often directed to various cellular processes depending on its interaction with specific co-factors.

Acknowledgments

We are grateful to Douglas Stuart for assistance with figure preparation. We also thank Prof. Nick Hastie, Dr. Bent W. Sigurskjold, and Grete Koch for helpful discussions and comments on the manuscript. Our work is supported by funds from the Medical Research Council and the Wellcome Trust.

References

Asher, C., Sinha, I., and Garty, H. (2003). Characterization of the interactions between Nedd4-2, ENaC, and sgk-1 using surface plasmon resonance. *Biochim. Biophys. Acta* **1612,** 59–64.

Chinowsky, T. M., Quinn, J. G., Bartholomew, D. U., Kaiser, R., and Elkind, J. L. (2003). Performance of the Spreeta 2000 integrated surface plasmon resonance affinity sensor. *Sensors Actuators B* **6954,** 1–9.

Garrus, J. E., von Schwedler, U. K., Pornillos, O. W., Morham, S. G., Zavitz, K. H., Wang, H. E., Wettstein, D. A., Stray, K. M., Cote, M., Rich, R. L., Myszka, D. G., and Sundquist, W. I. (2001). Tsg101 and the vacuolar protein sorting pathway are essential for HIV-1 budding. *Cell* **107,** 55–65.

Glickman, M. H., and Ciechanover, A. (2002). The ubiquitin-proteasome proteolytic pathway: Destruction for the sake of construction. *Physiol. Rev.* **82,** 373–428.

Hartmann-Petersen, R., and Gordon, C. (2004). Integral UBL domain proteins: A family of proteasome interacting proteins. *Semin. Cell Dev. Biol.* **15,** 247–259.

Jentsch, S., and Pyrowolakis, G. (2000). Ubiquitin and its kin: How close are the family ties? *Trends Cell Biol.* **10,** 335–342.

Karlsson, R., and Larsson, A. (2004). Affinity measurement using surface plasmon resonance. *Methods Mol. Biol.* **248,** 389–415.

Kozlov, G., De Crescenzo, G., Lim, N. S., Siddiqui, N., Fantus, D., Kahvejian, A., Trempe, J. F., Elias, D., Ekiel, I., Sonenberg, N., O'Connor-McCourt, M., and Gehring, K. (2004). Structural basis of ligand recognition by PABC, a highly specific peptide-binding domain found in poly(A)-binding protein and a HECT ubiquitin ligase. *EMBO J.* **23,** 272–281.

McDonnell, J. M. (2001). Surface plasmon resonance: Towards an understanding of the mechanisms of biological molecular recognition. *Curr. Opin. Chem. Biol.* **5,** 572–577.

Medaglia, M. V., and Fisher, R. J. (2001). Analysis of interacting proteins with surface plasmon resonance spectroscopy using BIAcore. *In* "Molecular Cloning, a Laboratory Manual" (J. Sambrook and D. W. Russell, eds.), 3rd Ed., pp. 18.96–18.114. CSH Laboratory Press, New York.

Myszka, D. G. (2000). Kinetic, equilibrium, and thermodynamic analysis of macromolecular interactions with BIACORE. *Methods Enzymol.* **323,** 325–340.

Nedelkov, D., and Nelson, R. W. (2003). Surface plasmon resonance mass spectrometry: Recent progress and outlooks. *Trends Biotechnol.* **21,** 301–305.

Raasi, S., and Pickart, C. M. (2003). Rad23 ubiquitin-associated domains (UBA) inhibit 26 S proteasome-catalyzed proteolysis by sequestering lysine 48-linked polyubiquitin chains. *J. Biol. Chem.* **278**, 8951–8959.

Seeger, M., Hartmann-Petersen, R., Wilkinson, C. R., Wallace, M., Samejima, I., Taylor, M. S., and Gordon, C. (2003). Interaction of the anaphase-promoting complex/cyclosome and proteasome protein complexes with multiubiquitin chain-binding proteins. *J. Biol. Chem.* **278**, 16791–16796.

Weissman, A. M. (2001). Themes and variations on ubiquitylation. *Nat. Rev. Mol. Cell. Biol.* **2**, 169–178.

Wilkinson, C. R., Seeger, M., Hartmann-Petersen, R., Stone, M., Wallace, M., Semple, C., and Gordon, C. (2001). Proteins containing the UBA domain are able to bind to multiubiquitin chains. *Nat. Cell Biol.* **3**, 939–943.

[12] Using NMR Spectroscopy to Monitor Ubiquitin Chain Conformation and Interactions with Ubiquitin-Binding Domains

By RANJANI VARADAN, MICHAEL ASSFALG, and DAVID FUSHMAN

Abstract

Polyubiquitin (polyUb) chains function as signaling molecules that mediate a diverse set of cellular events. The outcome of polyubiquitination depends on the specific linkage between Ub moieties in the chain, and differently linked chains function as distinct intracellular signals. Although an increasing number of Ub-binding proteins that transmit the regulatory information conferred by (poly)ubiquitination have been identified, the molecular mechanisms of linkage-specific signaling and recognition still remain to be understood. Knowledge of the chain structure is expected to provide insights into the basis of diversity in polyUb signaling. Here we describe several NMR approaches aimed at determining the physiological conformation of polyUb and characterization of the chains' interactions with ubiquitin-binding proteins.

Introduction

Polyubiquitin chains (polyUb) function as signaling molecules in the regulation of a host of cellular processes, ranging from progression through the cell cycle, to transcriptional activation, antigen processing, and vesicular trafficking of proteins [reviewed in Aguilar and Wendland (2003); Hershko and Ciechanover (1998); McCracken and Brodsky (2003); Muratani and Tansey (2003); Osley (2004); and Pickart (2001)]. Very remarkably, conjugation of substrates to polyUb chains of different linkages commits the target protein to distinct fates in the cell [e.g., see Pickart and Fushman (2004)]. Despite an increasing wealth of information on the cellular processes regulated by polyubiquitination and the identification of numerous ubiquitin-binding proteins that tie the polyUb signal to downstream events, the molecular basis of diversity in Ub-mediated signaling remains puzzling.

Because the only chemical difference between the various polyUb is in the specific lysine residue involved in the chain's extension, one could expect that different linkages bestow unique conformational features to different polyUb chains, and knowledge of the chain conformation could provide clues to functional diversity of polyUb. Structure characterization of these chains, however, is not straightforward, as emphasized by the discrepancy between two crystal structures of Lys48-linked Ub_4 (Cook et al., 1994; Phillips et al., 2000) that likely reflects the inherent conformational flexibility of the chain. Although they provide valuable snapshots of the chain, it is unclear whether any of these structures represents the physiologically relevant conformation of polyUb. This exemplifies a general challenge for structure characterization of multidomain systems, like polyUb, by the conventional high-resolution methods, X-ray crystallography and NMR. Generally applicable to complexes formed by tight contacts between the individual, well-structured components of the macromolecule, these methods have more limited applicability in those cases in which there are weaker interactions between the components. Crystal structures of such complexes might be biased by packing forces comparable to the interdomain interactions, whereas the precision and accuracy of the conventional NMR structural approaches are necessarily limited by the available short-range interproton contacts (NOEs) between the domains and by interdomain flexibility rendering the already scarce NOE information uninterpretable. Alternative methods are, therefore, required to address the physiological conformation of these chains. This chapter describes several NMR approaches to characterization of the structure of polyUb chains and their interactions with ubiquitin-binding domains. We illustrate these methods by applying them to Lys48- and Lys63-linked Ub_2 constructs and highlight various challenges associated with such studies.

Segmental Isotope Labeling of PolyUb Chains

The first challenge for NMR studies of polyUb stems from the fact that this is a homopolymer (i.e., all Ub units in the chain are expected to be spectroscopically equivalent). To discriminate among NMR signals from various Ub units, we used segmental isotope labeling strategy, in which a single Ub domain at a given position in the chain was isotope labeled

(Varadan et al., 2002, 2004). Such chains were synthesized from recombinant Ub molecules (unlabeled and ^{15}N-labeled) using the method of controlled chain assembly described in Piotrowski et al. (1997). For example, Ub$_2$ chains ^{15}N-labeled at the proximal Ub were synthesized using ^{15}N-labeled D77 Ub and unlabeled K48C Ub (for Lys48-linked Ub$_2$) or K63C Ub (for Lys63-linked Ub$_2$) (Fig. 1). (Throughout this chapter, "proximal" and "distal" refer to the position in the chain with respect to a possible substrate.) Use of the D77 and K48C (or K63C) mutants instead of wt-Ub allows controlled synthesis of Lys48- (or Lys63-) linked Ub chains of desired length. ^{15}N-labeled Ub$_4$ chains can then be assembled from unlabeled Ub$_2$ and Ub$_2$ ^{15}N-labeled at proximal or distal ends as required. For example, as outlined in Fig. 1, Ub$_4$-1 and Ub$_4$-3 chains (isotope labeled at Ub units 1 or 3, counting from the proximal end) can be assembled from unlabeled and proximal-^{15}N-labeled Ub$_2$ molecules. Before their use in the synthesis of Ub$_4$ chains, the Ub$_2$s are deblocked to allow for chain extension. The deblocking was performed by alkylation of Cys48 using ethyleneimine or by enzymatic cleavage of the Asp77 by Ub C-terminal hydrolase. This segmental labeling strategy allows us to monitor precisely changes in each Ub unit on chain formation and ligand binding.

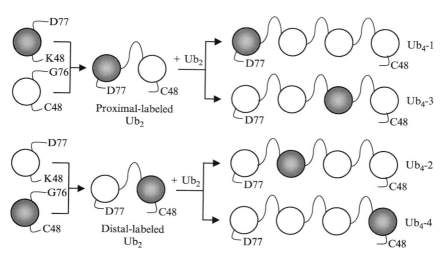

FIG. 1. Design and synthesis of segmentally isotope-labeled Lys48-linked polyUb chains. Open and shaded circles represent unlabeled and isotope-labeled Ub, respectively. Lys48-linked chains were assembled from D77 and K48C Ub mutants using enzymes E1 and E2-25K (Varadan et al., 2002). The use of these mutants instead of wt Ub allowed controlled synthesis of Lys48-linked polyUb chains of desired length and with desired position of the isotope-labeled unit. Lys63-linked chains were synthesized using a similar strategy with D77 and K63C Ub constructs (Varadan et al., 2004).

Methods for Structure Characterization of PolyUb Chains in Solution

As mentioned earlier, direct structure determination by conventional NMR methods relying on short-range interproton contacts (NOEs) also presents a challenge in the case of weakly coupled and, hence, inherently flexible multidomain systems, like polyUb. Thus, no interdomain NOEs could be reliably detected between the side-chain and amide hydrogens in Lys48-linked Ub_2 at pH6.8, most likely because of exchange broadening of the amide signals from the interface residues. In this situation, long-range orientational or distance constraints and interface mapping by chemical shift perturbations (CSP) become indispensable sources of information about the chain conformation.

Chemical Shift Perturbation Mapping of the Interdomain Interface in PolyUb

CSP mapping can be used as the first step in characterizing the conformation of polyUb to (1) verify the presence of the correct Ub–Ub linkage and (2) to identify the interfaces between Ub units in each chain. A similar approach is widely used for mapping protein surfaces involved in ligand binding; only in this case we look for differences in the fingerprint spectra (^1H–^{15}N HSQC) between Ub_2 or Ub_4 and monoUb. To quantify the perturbations, we use the combined amide CSP, $\Delta\delta = [(\Delta\delta_H)^2 + (\Delta\delta_N/5)^2]^{1/2}$, where $\Delta\delta_H$ and $\Delta\delta_N$ are the chemical shift differences for ^1H and ^{15}N, respectively, between polyUb and monoUb, determined for each NH group from the ^{15}N–^1H HSQC spectra of the corresponding constructs. Because chemical shifts are sensitive to changes in the electronic environment of the nucleus under observation, the residues exhibiting significant CSPs are most likely to be involved in the formation of the interdomain interface or in the Ub–Ub linkage.

The application of this method is illustrated in Fig. 2, which shows the observed CSPs in Lys48- and Lys63-linked Ub_2 compared with the corresponding monoUb constructs. Perturbations in the C-terminus (residues 74–76) of the distal Ub, and in residues adjacent to Lys48 (or Lys63) in the proximal Ub are not unexpected and confirm the formation of the Gly76–Lys48 (or Gly76–Lys63) isopeptide bond. Significant perturbations observed in the other backbone amides in Lys48-linked Ub_2 at pH6.8 indicate the presence of a well-defined Ub–Ub interface (Fig. 2C, D). These perturbations cluster around Leu8, Ile44, and Val70, suggesting that the interface is formed by the hydrophobic patches on both Ub units. A similar analysis showed well-defined CSPs in Lys48-linked Ub_4 (Varadan *et al.*, 2002), suggesting that this chain could be structured in solution. In

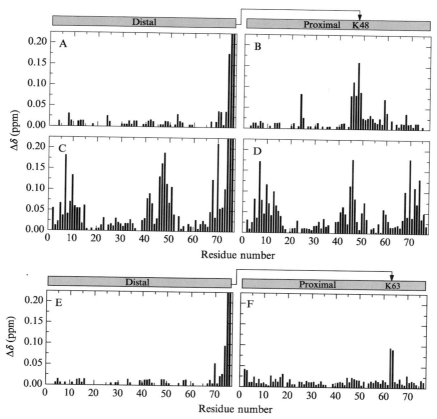

Fig. 2. Chemical shift perturbations in the backbone amides in Ub$_2$ versus the same residues in monoUb. Shown are data for Lys48-linked Ub$_2$ at (A, B) pH4.5, (C, D) pH6.8, and (E, F) for Lys63-linked Ub$_2$ at pH6.8. The left and right columns correspond to the distal and proximal Ub, respectively. The interdomain linkage is indicated on top of panels A, B and E, F.

contrast, no systematic perturbations except for the linker-adjacent sites were observed in Lys63-linked Ub$_2$ at the same pH (Fig. 2E, F), as well as in Lys48-linked Ub$_2$ under acidic conditions (pH 4.5, Fig. 2A, B), implying the absence of a defined interdomain contact in these two cases.

Use of Orientational Constraints for Determining the Interdomain Orientation in Di-Ubiquitin

In the absence of NOE constraints, the chain conformation in solution can be characterized using orientation-dependent approaches as outlined in Fushman *et al.* (2004). The interdomain orientation is determined on the basis of the overall rotational diffusion tensor (derived from ^{15}N spin-relaxation parameters) or the alignment tensor (from residual dipolar couplings [RDCs]) of Ub_2 measured for both Ub domains. If the two domains in Ub_2 reorient together in solution as a single molecule, these individual tensors represent the property of the whole molecule "seen" by the individual domains. Therefore, a rigid-body rotation of the domains, such that their tensors' axes become collinear, will properly orient them with respect to each other. A detailed description of the method along with the analysis of its accuracy and precision can be found in Fushman *et al.* (2004).

Figure 3 shows the structural properties of Lys48- and Lys63-linked Ub_2 obtained using this method. In agreement with the CSP data, these results show that under physiological conditions Lys48- and Lys63-linked Ub_2 adopt distinct conformations. The interdomain orientations derived independently from the rotational diffusion and alignment tensors agree well with each other and with the chemical shift perturbations (Fig. 2) (Varadan *et al.*, 2002, 2004). The good agreement between the results of these physically independent methods confirms the accuracy of the derived Ub_2 conformations.

Here are some issues to be aware of when using the domain orientation approach.

1. The advantage of this approach (over the NOE-based one) is that it relies on long-range orientational information about bond orientation within each domain. Its application, thus, requires the knowledge of the three-dimensional structures of the individual domains. Domain structures in the monomeric state can be used for domain alignment purposes, provided these are not perturbed when the multidomain system is formed. The latter needs to be verified for every application. For example, in our studies of Ub_2s, this was confirmed by the excellent agreement between the experimental and back-calculated RDCs (Fushman *et al.*, 2004).

2. The underlying concept of the method requires a certain correlation in the tumbling or orientation between the domains, so that they behave as a single entity. This could be a result of specific interactions between the domains (Lys48-linked Ub_2 at pH6.8) or could also be caused by steric restrictions resulting from short or relatively rigid interdomain linker (Lys48-linked Ub_2 at pH4.5 or Lys63-linked Ub_2). It is obvious that the domain orientation approach is not applicable to those multidomain systems where individual domains reorient independently (i.e., behave like "independent beads on a flexible string"). Therefore, the applicability of the approach to a particular protein system requires validation. The molecular weight dependence of ^{15}N T_1 (see later) and of ^{1}H T_2 (Varadan

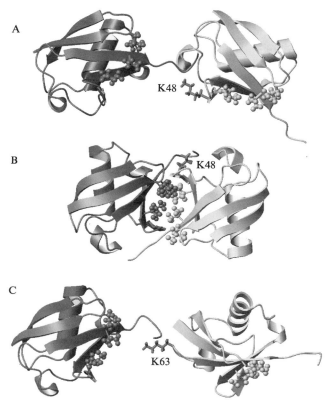

FIG. 3. Solution conformations of Lys48-linked Ub_2 at (A) pH 4.5, (B) pH 6.8, and (C) of Lys63-linked Ub_2 at pH 6.8. The ribbons are colored blue and green for the distal and proximal domains, respectively; the side chains of the hydrophobic patch residues Leu8-Ile44-Val70 are shown in ball-and-stick representation. Also shown in red is the side chain of the linkage lysine. Shown in (B) is the NMR structure of Lys48-linked Ub_2 (PDB entry 2BGF) obtained by domain docking on the basis of a combination of RDCs and CSP mapping data (van Dijk et al., 2005). (A and C) represent interdomain orientations (the positioning of the domains is arbitrary) obtained by aligning the rotational diffusion tensors reported by both domains (A) or the alignment tensors derived from RDCs (C). The ^{15}N relaxation measurements used the standard protocols [e.g., Fushman et al. (1997)]; the RDCs were measured in the liquid-crystalline phase of n-alkyl-poly(ethyleneglycol)/n-hexanol mixtures as detailed in (Varadan et al., 2002, 2004). The diffusion and alignment tensors were derived from the experimental data using in-house computer programs ROTDIF (Walker et al., 2004) and ALTENS (Varadan et al., 2002). (See color insert.)

et al., 2002) observed in Ub, Ub_2, and Ub_4 follows the Stokes–Einstein equation for a tumbling molecule, and hence serves as such a validation.

3. In reality, components of a multidomain system are rarely (if at all) locked in one particular conformation, so some interdomain mobility is always present. It is, therefore, important to bear in mind that this method provides a time-averaged orientation of the domains, which depends on the rates and the amplitudes of interdomain motions in a molecule. This effect will depend on a particular mechanism of interdomain dynamics and on the characteristic time scales of these motions that could vary from one molecular system to another. Because of the differences in the characteristic time windows between the RDC and relaxation measurements, the interdomain orientation obtained by these two methods could be somewhat different.

4. It is important to bear in mind that the domain orientation approach alone provides no distance information, hence it does not allow proper positioning of the domains with respect to each other in three-dimensional space. To obtain accurate structural properties, the orientational data have to be combined with contact- or distance-related information. The latter could be in the form of a contact map (from CSPs) or as interatomic distance constraints obtained from NOE measurements and/or from relaxation enhancement or pseudo-contact shifts caused by paramagnetic agents. Protein docking methods (Clore and Schwieters, 2003; Dominguez et al., 2003) provide a promising tool for modeling the structures of multidomain proteins (and protein–ligand complexes) on the basis of a combination of such restraints. For example, the Lys48-linked Ub_2 structure shown in Fig. 2B was obtained by protein docking combining the RDCs (domain orientation) with CSP data (van Dijk et al., 2005).

Site-Specific Spin Labeling as a Source of Long-Distance Information

The use of paramagnetic agents (e.g., spin labels) provides an additional source of long-range distance information for NMR studies of proteins. An unpaired electron of the spin label causes paramagnetic relaxation rate enhancement, ΔR_{2para}, resulting in broadening of resonances (hence, signal attenuation) for those nuclei that are close in space to the spin label. This effect is distance dependent; hence the observed signal attenuation can be used to derive structural information. For these purposes, ΔR_{2para} can be determined for each NH group from the ratio of signals in ^1H–^{15}N HSQC spectra recorded with the spin-label in the oxidized (paramagnetic) and reduced states (Jain et al., 2001): $\Delta R_{2para} = R_{2ox} - R_{2red} = \ln(I_{red}/I_{ox})/t$, where t is the total experimental time when the amide proton magnetization is in the transverse plane and undergoing paramagnetic relaxation. $\Delta R_{2\text{para}}$ is related to the distance r between the nucleus under observation and the unpaired electron as (Kosen, 1989):

$$\Delta R_{2\text{para}} = (1/20)\gamma_H^2 g_e^2 \beta_e^2 [4\tau_c + 3\tau_c/(1+\omega_H^2 \tau_c^2)]/r^6, \qquad (1)$$

where τ_c is the rotational correlation time of the molecule, γ_H and ω_H are the ^{1}H gyromagnetic ratio and resonance frequency, g_e is the electronic g-factor, and β_e is the Bohr magneton. An almost three orders of magnitude stronger gyromagnetic ratio ($g_e \beta_e/\hbar = 658\gamma_H$), of the electron spin makes it possible to obtain distance information of much longer range (up to ~25 Å) than that available through NOEs (up to ~5 Å).

For our studies, we used a paramagnetic spin label (1-oxyl-2,2,5,5-tetramethyl-3-pyrroline-3-methyl)methanesulfonate (MTSL) that is highly specific for sulfhydryl groups (Berliner et al., 1982), covalently attached to a single Cys residue in K48C Ub. MTSL was added to a protein sample from a concentrated stock in acetonitrile at a molar ratio of 3:1 MTSL/protein, and the mixture was incubated at room temperature for at least 1 h. Excess MTSL reagent was removed by dialysis into NMR buffer. To quantify the signal attenuations caused by MTSL, ^{1}H–^{15}N HSQC experiments with the spin-labeled protein were performed twice: first with MTSL in the oxidized (paramagnetic) state and then repeated after the spin-label was reduced with a threefold excess ascorbic acid, added in small volumes from a concentrated solution. The latter experiment also allows one to monitor any effects of the MTSL attachment on the structure of Ub or on its ligand-binding properties.

The attachment of MTSL to the distal Ub in Lys48-linked Ub$_2$ resulted in signal attenuations in the proximal domain, thus verifying the closed conformation of this chain (Fushman et al., 2004). In contrast, no attenuation was observed in the proximal Ub in Lys63-linked Ub$_2$ when MTSL was attached to Cys63 in the distal domain (K63C Ub), in agreement with the open conformation of this chain (Fig. 3C) (Varadan et al., 2004). The use of MTSL for structure analysis of the monoUb/UBA complex is discussed in detail in the next section.

It is worth mentioning that the use of paramagnetic ions (e.g., lanthanides) is a promising and versatile method for obtaining long-range distance and orientational information for these purposes, and various strategies of attaching them to proteins have recently been developed (Donaldson et al., 2001; Ikegami et al., 2004; Wohnert et al., 2003).

From Chain Conformation to Function: NMR Studies of the Interactions with UBA Domains

To relate the observed differences in the chain conformations to the functional diversity of polyUb, it is necessary to understand how these alternative conformations are differentially recognized. Studies of the interaction of polyUb chains with Ub-binding proteins can help address this question. NMR is particularly suited for studies of weak interactions ($K_d \sim 1$ mM), which are difficult to analyze by other methods, like ITC or SPR, that work well for stronger binding. NMR also provides a unique opportunity to monitor changes in virtually every group in a protein in the course of ligand titration. As an illustration, here we focus on the interaction of the UBA2 domain from hHR23A with monomeric Ub and Lys63-linked Ub$_2$.

Challenges for Structure Characterization of Ub Complexes with Ligands

Isolated UBA domains from hHR23A bind monomeric Ub in a 1:1 complex ($K_d \sim 300$–600 μM) that is held together predominantly by hydrophobic interactions, as inferred from the CSP mapping (Mueller *et al.*, 2004; Ryu *et al.*, 2003; Wang *et al.*, 2003). In the absence of NOEs or orientational constraints, the structure of the UBA/Ub complex was modeled with CSP-based protein docking (Mueller *et al.*, 2004), resulting in distinct orientations for UBA1 and UBA2 domains on Ub surface. These orientations also differ from the UBA/Ub structures (Mueller *et al.*, 2004) modeled after the CUE/Ub complex. These findings suggest that the hydrophobic surface of Ub could be promiscuous in that different ligands could bind Ub in different orientations. Thus, the results of computer modeling based on homology or ambiguous (CSP) constraints should be treated with caution, and additional experimental data reflecting the specificity of interaction with a particular ligand might be necessary to verify the predictions.

Use of Site-Specific Spin-Labeling Provides Insights into the Structure of UBA2/MonoUb Complex

Here we illustrate the use of MTSL to verify the correct model for the UBA2/monoUb complex. MTSL was attached to Cys48 in K48C Ub. According to our CSP and titration data, neither K48C mutation nor the spin labeling affected the UBA2–Ub interaction. The presence of paramagnetic MTSL resulted in signal attenuation in several amides in both Ub and UBA2, thus confirming the formation of the Ub/UBA complex. The pattern of attenuations observed in both binding partners was mapped (Fig. 4) onto the proposed models (Mueller *et al.*, 2004) of the complex. Although the location of the attenuated sites in the presence of the spin label is generally consistent with both proposed structures, the docked model seems to be in better qualitative agreement with the experiment.

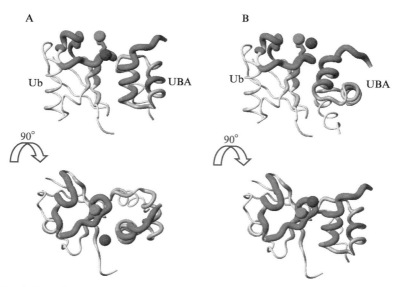

FIG. 4. Experimental verification of the existing models for the monoUb/UBA2 complex using site-directed spin labeling. (A) represents the Ub/UBA2 structure modeled by the Ub/CUE complex (Kang et al., 2003), whereas the docked structure (Mueller et al., 2004) is shown in (B). The bottom row shows views of the same structures from the top. The ribbons represent the backbone of Ub and UBA2; the atom coordinates are from Mueller et al. (2004). The ribbon width increases proportionally to the observed paramagnetic line broadening (hence closer distance to the spin label) and is color-coded by this distance as red (closest, <17 Å), orange (17–26 Å), and yellow (>26 Å). The spheres represent the reconstructed positions of the spin label as "seen" by Ub (green) and by UBA2 (blue). Also shown is the side chain of Cys48 (Ub), where the spin label was covalently attached. The coordinates of the spin label were obtained using a three-dimensional search algorithm aimed at satisfying all available amide–MTSL distance constraints. (See color insert.)

To quantify these observations, signal attenuations were converted into distance constraints between the amides and the unpaired electron (using Eq. 1), which were then used to reconstruct the position of the spin label. As control, the calculations based on signal attenuations observed in (spin-labeled) Ub positioned the unpaired electron at a distance of 4.75 Å from C_β of Cys48 (the site of MTSL attachment), in good agreement with its expected location. The position of the same spin label was also reconstructed from signal attenuations measured in UBA. The agreement between these "images" of the spin label "seen" by Ub and UBA is better for the docked model (Fig. 4B). Although this agreement is not perfect, the 5.6 Å distance between the spin-label locations reported by the two proteins (versus 10 Å for the homology-modeled structure, Fig. 4A) could reflect the imprecision in the MTSL positioning because of its inherent flexibility, the on–off dynamics in the complex, and the experimental errors. It is also possible that the relative orientation of the proteins in the docked model slightly deviates from their actual orientation. Note that the derived distances between amides and MTSL could, in principle, be included as distance constraints in further refinement of the Ub/UBA structure. It is not clear, however, to what extent the precision and accuracy of the resulting structure could be affected by the uncertainty in the position of the spin label because of its internal mobility.

Assessing the Stoichiometry of Binding Using Spin Relaxation Measurements

The extended conformation of Lys63-linked Ub_2 (Fig. 3C) makes hydrophobic patches on both Ub domains accessible for ligand binding, and CSP mapping suggests that both Ub domains in this chain can bind UBA independently and in the mode similar to the UBA2–monoUb interaction (Varadan *et al.*, 2004). Accurate characterization of the Ub_2-UBA interaction requires knowledge of its stoichiometry, which can be inferred from the size of the resulting complex.

Nuclear spin relaxation senses the rate of molecular tumbling in solution and, therefore, provides a means to measure the size of a macromolecule, which could complement the more conventional methods such as analytical ultracentrifugation and cross-linking/native gel-electrophoresis. To illustrate this, the ^{15}N longitudinal relaxation time (T_1) measured in the free monoUb, Ub_2, and Ub_4 constructs follows the theoretical dependence of ^{15}N T_1 on the size of the molecule (Fig. 5A) that can be used as a molecular weight "ruler." For example, an average ^{15}N T_1 of 548 ± 42 ms for UBA2 in the presence of monoUb corresponds to 13 kDa, in agreement with the expected molecular weight (14.5 kDa) of the UBA2/monoUb complex. The molecular weight of the Lys63-linked Ub_2/UBA2 complex was estimated from ^{15}N T_1 measured in Ub_2 in the presence of a saturating amount of UBA2 (UBA2/Ub_2, molar ratio 4:1). An average ^{15}N T_1 value of 956 ± 7 ms corresponds, according to the calibration curve, to an apparent molecular weight of 27.5 kDa. This number falls between the values expected for a 2:1 (29 kDa) and 1:1 (23 kDa) UBA2/Ub_2 complexes and, therefore, indicates that up to two UBA domains can be bound per Lys63-linked Ub_2 chain. Supporting this conclusion, the average ^1H transverse relaxation time T_2 at high UBA2/Ub_2 molar ratios (6:1 and 10:1) was in the range of 14–16 ms, in good agreement with 15 ms expected for a 2:1 UBA2/Ub_2 complex and shorter than 19 ms for a 1:1 complex. The overall tumbling time (hence the size) of the complex can also be assessed

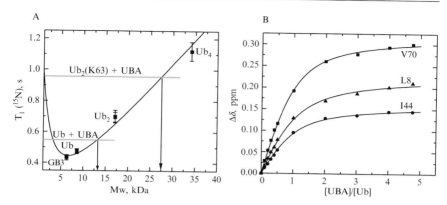

FIG. 5. NMR-based analysis of the (A) stoichiometry and (B) affinity of Ub_2-UBA2 binding. (A) illustrates the use of the molecular weight dependence of ^{15}N longitudinal relaxation time as a molecular weight calibration "ruler." The line represents the theoretical dependence, and the symbols correspond to experimentally measured data for various proteins, as indicated. The gray horizontal lines represent the average ^{15}N T_1 levels for UBA2/monoUb and UBA2/Ub$_2$(Lys63) complexes, the vertical arrows point to the molecular weight estimate from the calibration curve. The ^{15}N T_1 values were measured using standard methods [e.g., Fushman et al. (1997)]. The ^1H T_2 and ^{15}N T_1 (in Ub$_4$) were derived from envelope comparisons for a series of 1D spectra. (B) Representative binding curves for hydrophobic residues in the distal domain of Lys63-linked Ub$_2$ on titration of UBA2. The fitting curves correspond to a model assuming two independent identical UBA-binding sites on Ub$_2$ and allowing up to two UBA molecules bound per chain. The microscopic K_d values are 0.34 mM (Leu8), 0.27 mM (Ile44), and 0.23 mM (Val70). Both K_d and ΔB_{max} were treated as fitting parameters in a least-squares analysis using in-house computer program KDFIT.

from ^{15}N T_2 measurements, although these data should be treated with caution, because some sites might be affected by conformational exchange contributions.

Estimation of the Binding Affinity for the Lys63-Ub$_2$/UBA2 Complex

Knowledge of the stoichiometry of binding then allowed determination of the dissociation constant from the NMR titration data (Fig. 5B) by monitoring the CSPs in Ub$_2$ in the presence of increasing amounts of UBA2 and vice versa. As with titrations performed by other biophysical techniques (ITC, fluorescence, etc.), the working protein concentrations should be kept above the K_d, and small volumes of ligand (UBA2) should be added from a concentrated (5–10 mM) stock solution. Depending on the kinetics of the binding event in relation to the NMR time scale, the observed signals can be in the "fast," "intermediate," or "slow" exchange regimens [e.g., Zuiderweg (2002)]. In the limit of slow exchange, the equilibrium constant can be determined from the ratio of signals corresponding to the "free" and "bound" states of the protein, provided these signals are observed simultaneously.

In the fast exchange regime, when the signals gradually shift on titration, the dissociation constant can be obtained by fitting the observed titration curves (Fig. 4B) to the corresponding equations describing CSPs as a function of the ligand and protein concentrations. The observed CSP can be written as $\Delta\delta = \Delta\delta_{max}\, p_B$, where $\Delta\delta_{max}$ is the chemical shift difference between the free and ligand-bound state and p_B is the relative population of the bound state of the protein (Ub or UBA) under the observation. The equations relating p_B to the K_d and the total molar concentrations of the protein and ligand for some relevant stoichiometry models can be found in Varadan *et al.* (2004). Note that, because $\Delta\delta_{max}$ is not known *a priori*, it should be treated as a fitting parameter, together with K_d. Therefore, it is essential for a robust fit of the observed titration curves that ligand titration be carried out until CSPs saturate.

The data analysis is illustrated in Fig. 5B. The shape of the observed titration curves for both Ub domains in Lys63-linked Ub_2 suggests that UBA2 binding to each domain is an independent, noncooperative event. Fitting the data to a model assuming two independent identical UBA-binding sites on Ub_2 results in microscopic dissociation constants of 0.28 ± 0.1 mM for the distal Ub data, and 0.18 ± 0.08 mM for the proximal Ub, which are comparable to the K_d values for monoUb–UBA interaction. These numbers are in good agreement with the dissociation constant (0.21 ± 0.1 mM) derived from the titration curves observed for UBA2 on the addition of Ub_2.

Note that the use of an adequate binding model is essential for accurate K_d analysis, hence the importance of an independent assessment of the stoichiometry (see previous section). Selecting the right model solely on the basis of quality of fit of titration data might not be straightforward. For example, fitting the same titration data to a model assuming two binding sites but allowing only one ligand bound per chain resulted in only a slight increase in the residuals of fit, whereas the K_d values (1.2–1.6 mM) were well above those for monoUb. A comparison of the binding data obtained for both binding partners (e.g., Ub_2 and UBA) could also help determine the right binding model.

Acknowledgments

We thank Cecile Pickart for plasmid constructs used in these studies and for many insightful discussions and Juli Feigon for providing atom coordinates for the UBA/monoUb structure models. This work was supported by NIH grant R01 GM65334.

References

Aguilar, R. C., and Wendland, B. (2003). Ubiquitin: Not just for proteasomes anymore. *Curr. Opin. Cell Biol.* **15,** 184–190.

Berliner, L. J., Grunwald, J., Hankovszky, H. O., and Hideg, K. (1982). A novel reversible thiol-specific spin label: Papain active site labeling and inhibition. *Anal. Biochem.* **119,** 450–455.

Clore, G. M., and Schwieters, C. D. (2003). Docking of protein-protein complexes on the basis of highly ambiguous intermolecular distance restraints derived from 1H/15N chemical shift mapping and backbone 15N-1H residual dipolar couplings using conjoined rigid body/torsion angle dynamics. *J. Am. Chem. Soc.* **125,** 2902–2912.

Cook, W. J., Jeffrey, L. C., Kasperek, E., and Pickart, C. M. (1994). Structure of tetraubiquitin shows how multiubiquitin chains can be formed. *J. Mol. Biol.* **236,** 601–609.

Dominguez, C., Boelens, R., and Bonvin, A. M. (2003). HADDOCK: A protein-protein docking approach based on biochemical or biophysical information. *J. Am. Chem. Soc.* **125,** 1731–1737.

Donaldson, L. W., Skrynnikov, N. R., Choy, W. Y., Muhandiram, D. R., Sarkar, B., Forman-Kay, J. D., and Kay, L. E. (2001). Structural characterization of proteins with an attached ATCUN motif by paramagnetic relaxation enhancement NMR spectroscopy. *J. Am. Chem. Soc.* **123,** 9843–9847.

Fushman, D., Cahill, S., and Cowburn, D. (1997). The main-chain dynamics of the dynamin pleckstrin homology (PH) domain in solution: Analysis of 15N relaxation with monomer/dimer equilibration. *J. Mol. Biol.* **266,** 173–194.

Fushman, D., Varadan, R., Assfalg, M., and Walker, O. (2004). Determining domain orientation in macromolecules by using spin-relaxation and residual dipolar coupling measurements. *Progr. NMR Spectrosc.* **44,** 189–214.

Hershko, A., and Ciechanover, A. (1998). The ubiquitin system. *Annu. Rev. Biochem.* **67,** 425–480.

Ikegami, T., Verdier, L., Sakhaii, P., Grimme, S., Pescatore, B., Saxena, K., Fiebig, K. M., and Griesinger, C. (2004). Novel techniques for weak alignment of proteins in solution using chemical tags coordinating lanthanide ions. *J. Biomol. NMR.* **29,** 339–349.

Jain, N. U., Venot, A., Umemoto, K., Leffler, H., and Prestegard, J. H. (2001). Distance mapping of protein-binding sites using spin-labeled oligosaccharide ligands. *Protein Sci.* **10,** 2393–2400.

Kang, R. S., Daniels, C. M., Francis, S. A., Shih, S. C., Salerno, W. J., Hicke, L., and Radhakrishnan, I. (2003). Solution structure of a CUE-ubiquitin complex reveals a conserved mode of ubiquitin binding. *Cell* **113,** 621–630.

Kosen, P. A. (1989). Spin labeling of proteins. *Methods Enzymol.* **177,** 86–121.

McCracken, A. A., and Brodsky, J. L. (2003). Evolving questions and paradigm shifts in endoplasmic-reticulum-associated degradation (ERAD). *Bioessays* **25,** 868–877.

Mueller, T. D., Kamionka, M., and Feigon, J. (2004). Specificity of the interaction between ubiquitin-associated domains and ubiquitin. *J. Biol. Chem.* **279,** 11926–11936.

Muratani, M., and Tansey, W. P. (2003). How the ubiquitin-proteasome system controls transcription. *Nat. Rev. Mol. Cell. Biol.* **4,** 192–201.

Osley, M. A. (2004). H2B ubiquitylation: The end is in sight. *Biochim. Biophys. Acta* **1677,** 74–78.

Phillips, C. L., Thrower, J., Pickart, C. M., and Hill, C. P. (2000). Structure of a new crystal form of tetraubiquitin. *Acta Cryst. D.* **57,** 341–344.

Pickart, C. M. (2001). Ubiquitin enters the new millennium. *Mol. Cell.* **8,** 499–504.

Pickart, C. M., and Fushman, D. (2004). Polyubiquitin chains: Polymeric protein signals. *Curr. Opin. Chem. Biol.* **8,** 610–616.

Piotrowski, J., Beal, R., Hoffmann, L., Wilkinson, K. D., Cohen, R. E., and Pickart, C. M. (1997). Inhibition of the 26S proteasome by polyubiquitin chains synthesized to have defined lengths. *J. Biol. Chem.* **272,** 23712–23721.

Ryu, K. S., Lee, K. J., Bae, S. H., Kim, B. K., Kim, K. A., and Choi, B. S. (2003). Binding surface mapping of intra and inter domain interactions among hHR23B, ubiquitin and poly ubiquitin binding site 2 of S5a. *J. Biol. Chem.* **278,** 36621–36627.

van Dijk, A. D. J., Fushman, D., and Bonvin, A. M. (2005). Various strategies of using residual dipolar couplings in NMR-driven protein docking: Application to Lys48-linked di-ubiquitin and validation against ^{15}N-relaxation data. *Proteins* **60,** 367–381.

Varadan, R., Assfalg, M., Haririnia, A., Raasi, S., Pickart, C., and Fushman, D. (2004). Solution conformation of Lys63-linked di-ubiquitin chain provides clues to functional diversity of polyubiquitin signaling. *J. Biol. Chem.* **279,** 7055–7063.

Varadan, R., Walker, O., Pickart, C., and Fushman, D. (2002). Structural properties of polyubiquitin chains in solution. *J. Mol. Biol.* **324,** 637–647.

Walker, O., Varadan, R., and Fushman, D. (2004). Efficient and accurate determination of the overall rotational diffusion tensor of a molecule from 15N relaxation data using computer program ROTDIF. *J. Magn. Reson.* **168,** 336–345.

Wang, Q., Goh, A. M., Howley, P. M., and Walters, K. J. (2003). Ubiquitin recognition by the DNA repair protein hHR23a. *Biochemistry* **42,** 13529–13535.

Wohnert, J., Franz, K. J., Nitz, M., Imperiali, B., and Schwalbe, H. (2003). Protein alignment by a coexpressed lanthanide-binding tag for the measurement of residual dipolar couplings. *J. Am. Chem. Soc.* **125,** 13338–13339.

Zuiderweg, E. R. (2002). Mapping protein-protein interactions in solution by NMR spectroscopy. *Biochemistry* **41,** 1–7.

[13] Analysis of Ubiquitin-Dependent Protein Sorting Within the Endocytic Pathway in *Saccharomyces cerevisiae*

By DAVID J. KATZMANN and BEVERLY WENDLAND

Abstract

The plasma membrane protein composition of a eukaryotic cell is maintained in part through the removal of transmembrane proteins by endocytosis and delivery to the lysosome (or vacuole in yeast) for degradation. The endocytic and biosynthetic pathways converge at endosomes, where related sorting events occur for proteins arriving from either pathway before their lysosomal delivery. *Saccharomyces cerevisiae* has proven to be an excellent model organism for the study of fundamental cellular processes, and this complex process is no exception. The powerful genetics available in the yeast system have facilitated the identification of a large number of factors that drive protein sorting throughout the endocytic pathway. It is clear that ubiquitin plays a critical role in targeting cargoes into this degradative pathway and that this signal is recognized by a series of adaptor proteins between the cell surface and lysosome that are responsible for directing the cargo for degradation. Here we provide detailed protocols for studying the fate of cargo proteins within the endosomal system, as well as the role of putative ubiquitin-binding proteins.

Introduction

Ubiquitin is a highly conserved 76 amino acid protein that can be covalently attached posttranslationally to itself or other proteins. Recent advances in our understanding of transmembrane protein trafficking itineraries that are directed by ubiquitination have revealed the importance of this modification in determining protein sorting at a number of subcellular locations, including the cell surface, endosomes, and Golgi. In all cases, the end result seems to be the same: targeting of a transmembrane cargo protein into the multivesicular body (MVB) pathway and its delivery to the hydrolytic lumen of the vacuole/lysosome (reviewed in Hicke and Dunn [2003] and Katzmann *et al.* [2002]). In contrast to proteasomal substrates, which are typically targeted to the proteasome as a result of receiving a polyubiquitin chain, targeting transmembrane proteins to the MVB pathway seems to be the result of mono- or di-ubiquitin substrate modification. In addition, there seem to be multiple adapters at each of these subcellular locations that are responsible for recognizing the ubiquitin modification on cargo proteins and directing their sorting (Traub, 2003; Wendland, 2002). These *trans*-acting factors use a variety of ubiquitin-interacting motifs, which typically interact with ubiquitin at relatively low affinity (low micromolar range). This has led to the belief that *in vivo*, these ubiquitin interactions are stabilized by additional protein–protein or protein–lipid interactions in the context of selection at specific subcellular compartments. Proteins can be targeted into the MVB pathway either as endocytic cargo or as biosynthetic MVB cargo (without transiting through the cell surface). Here we highlight several methods that have been useful for studying the ubiquitination status of cargoes destined for degradation within the vacuole/lysosome and interactions with sorting machinery. Although these methods are specifically tailored for studies in yeast, the general concepts are applicable for any system.

When considering ubiquitin-dependent sorting of transmembrane cargo within the endosomal system, the terminal consequence is delivery into the lumen of the vacuole/lysosome. The result of this delivery to the lumen is either degradation, as in the case of cell surface proteins, or activation, as in the vacuolar hydrolase carboxypeptidase S (CPS) (Spormann *et al.*, 1992). A prediction for cargo that enters the MVB pathway, therefore, is vacuolar protease-dependent degradation or maturation of the cargo protein of interest. Delivery of cargo proteins to the vacuole lumen can be assayed in several ways. One is simply to construct a green fluorescent protein (GFP) chimera to the cargo of interest and to visualize cells expressing this chimera by means of fluorescent microscopy (for example, see Odorizzi *et al.* [1998]). Even in cells with normal vacuolar protease activity, GFP is resistant to degradation and will continue to fluoresce within the vacuole for a period of time that typically facilitates visualization of the vacuole lumen. However, strains deficient for vacuolar protease activity have also been used to stabilize the intravacuolar GFP signal that is the consequence of cargo–GFP delivery into the MVB pathway (for example, see Urbanowski and Piper [2001]). This microscopic, steady-state analysis reveals a predominantly vacuolar lumen signal for cargoes of the MVB pathway, including biosynthetic cargoes such as CPS and Phm5 (Odorizzi *et al.*, 1998; Reggiori and Pelham, 2001) and even cell surface proteins such as Ste2, Ste3, Ste6, and Gap1 (for examples, see Davies *et al.* [2003]; Losko *et al.* [2001]; Odorizzi *et al.* [1998]; and Scott *et al.* [2004]). This analysis can be further refined by the use of mutants defective for the function of the MVB pathway, such as the class E *vps* mutants. If MVB function is required for the delivery of the cargo to the vacuolar lumen, analysis in a class E *vps* mutant will result in its missorting to the limiting membrane of the vacuole, as well as appearing in the exaggerated endosomal structure apparent in these mutants (the class E compartment) (Odorizzi *et al.*, 1998; Raymond *et al.*, 1992). Similarly, in animal cells, dominant negative forms of class E Vps proteins or endocytic machinery, as well as siRNA, have been used to demonstrate a role for the MVB pathway in the delivery of cargo proteins to the lumen of the lysosome (for example, see Bache *et al.* [2003] and Bishop *et al.* [2002]). In this case, cargoes can be seen to accumulate in aberrant endosomes with a concurrent delay in lysosomal processing.

A second method involves assessing the turnover of cargo proteins within the vacuole either by metabolic pulse-chase immunoprecipitation or by "cycloheximide shut-off" of protein synthesis, followed with Western blot analysis of the cargo under investigation. By comparing the rates of protein turnover or maturation in wild-type versus vacuolar protease-deficient strains, it is possible to assign a role for the vacuole in this process.

Methodology

Measurement of Protein Turnover Rate

Pulse-Chase Immunoprecipitation (Denatured). To study ubiquitinated proteins, the ubiquitin modification must be preserved during the experimental manipulations. De-ubiquitinating enzymes (DUBs) are notoriously difficult to inactivate, and if special precautions are not taken, they can remove the ubiquitin modification on the protein under study. Inactivation of the DUBs can be done in at least two ways: the addition of N-ethylmaleimide (NEM) to alkylate the active site cysteine within DUBs and/or the complete denaturation of the sample. The latter is typically accomplished by adding trichloroacetic acid (TCA) to a final concentration of 10% to precipitate proteins out of solution, followed by removal of TCA and resuspension in an aqueous buffer that will allow subsequent manipulations. Although this method offers the advantage of dramatically stabilizing ubiquitinated intermediates, many polytopic membrane proteins are difficult (or impossible) to resolubilize after being precipitated with TCA (see review by Kaiser and colleagues). For this reason, it is important to understand how your cargo protein behaves after TCA precipitation before using this method. When using NEM in buffers to prevent DUB activity, it is critical to add the appropriate amount of solid NEM immediately before using the solution. In the following protocol, wild-type and vacuolar protease-mutant strains expressing the protein of interest should be used in parallel. Additional permutations can also be used, such as the overexpression of K48R ubiquitin, use of trafficking mutants, and mutation of candidate ubiquitin acceptor lysine residues within the substrate.

Buffers/Materials

- Urea cracking buffer: 6 M urea, 1% SDS, 50 mM Tris, pH 7.5, 1 mM EDTA.
- Tween-20 IP buffer: 50 mM Tris, pH 7.5, 150 mM NaCl, 0.5% Tween-20, 0.1 mM EDTA.
- Tween-20 urea IP buffer: 100 mM Tris, pH 7.5, 200 mM NaCl, 2 M urea.
- TBS: 50 mM Tris, pH 7.5, 150 mM NaCl.
- IP sample buffer: 20% glycerol, 10% BME, 6% SDS, 125 mM Tris, pH 6.8, 0.1% bromophenol blue.
- Protein-A sepharose buffer: 10 mM Tris, pH 7.5, 1 mg/ml BSA (protease free), 1 mM NaN$_3$.
- 50X Cys, Met chase: 250 mM methionine, 50 mM cysteine, 10% yeast extract.
- ^{35}S Cys/Met mix.

- Yeast minimal media (SM): 0.0067% yeast nitrogen base, 2% glucose, appropriate amino acids.
- Acid-washed 0.1-mm glass beads (after soaking overnight in nitric acid, neutralize with 0.5 M Tris, pH 8.0, wash extensively with water until pH is same as starting water, then wash with methanol and acetone before baking in a drying oven).
- Protein-A sepharose (or equivalent).

1. A starter yeast culture is grown overnight in minimal medium (SM). This culture will be used the following day to start a culture of the desired number of cells for labeling. Using the overnight culture, start the culture(s) for the experiment (in minimal medium lacking methionine and cysteine) at approximately 0.2 OD_{600}/ml, and continue growing until the culture reaches mid-log phase growth (0.5–0.6 OD_{600}/ml). *The final volume of the culture is determined by the number of OD_{600} units needed per time point and the number of time points. For example, 2 OD_{600} units per time point and five time points would mean that a total of 10 OD_{600} units are required. Because it is important to harvest cells during mid log phase growth, approximately 0.5 OD_{600}/ml culture of at least 20 ml would be inoculated. *One OD_{600} unit of yeast corresponds to approximately 1×10^7 cells, with an approximate protein yield of 70 μg.

2. Harvest the cells by centrifugation at 1000g for 5 min. Resuspend cell pellet to the desired number of OD_{600} units per milliliter in SM minus methionine and cysteine. For example, if 2 OD_{600} of cells per time point are desired, resuspend cells at 2 OD_{600} units per milliliter of medium.

3. Add ^{35}S Cys/Met to a concentration 10–20 μCi/OD_{600} unit of cells (typically on the order of 1–2 μl/OD_{600} unit). If the protein of interest is particularly rare, one can increase the amount of labeled Cys/Met by fivefold. *For optimal reproducibility, it is critical to stagger the start times for the addition of label. Choose an interval in the range of 30–60 sec between the addition of label to each culture tube for start of incorporation. This permits adequate time for sample processing between time points to ensure equivalency between cultures.

4. Incubate cells in a shaking water bath at 30° (or desired temperature) for 10 min; this is the labeling period. Add chase solution (final concentration of 5 mM methionine, 1 mM cysteine, and 0.2% yeast extract), vortex, remove 1 ml of culture (this is the "time zero" point) and transfer to 1.5 ml Eppendorf tube containing 100 μl of 100% TCA, vortex and place on ice. Repeat addition of chase to subsequent cultures, remembering to observe the appropriate staggered time interval. Return culture to shaking bath and continue incubation, removing and processing samples at the desired time points as was done for time zero. Leave

all samples on ice until the last time point is collected, allowing an additional 5 min on ice for the last sample.

5. Centrifuge samples in a Microfuge (approximately 12,000g) for 2 min, then aspirate supernatant. *It is critical to collect the supernatants of all steps, because there will be significant amounts of radiolabel present. This is readily accomplished by using a trap in the aspirator line; the collected liquid is subsequently disposed of appropriately.

6. Wash the pellets twice with 1 ml of cold ($-20°$) acetone to remove the TCA. This step is greatly facilitated by the use of a bath sonicator. One at a time, place the bottom of the tube with the pellet into the bath sonicator until the pellet is completely broken apart in the acetone. Centrifuge samples as before (2 min, 12,000g), remove supernatant by aspirating, and repeat the acetone wash. *It is critical to remove all of the TCA; otherwise, the TCA will alter the pH at later steps and prevent appropriate lysis and solubilization.

7. Dry the acetone-washed pellets in a rotary evaporator/concentrator ("SpeedVac") for approximately 5 min, making certain that all acetone is removed before proceeding. Alternately, the pellets can be dried on the benchtop by leaving the lids open and allowing them to evaporate for at least 30 min.

8. Resuspend each pellet in 100 μl urea cracking buffer plus 5 mM NEM; this step can also be facilitated by using the bath sonicator to help break up the pellet.

9. Add an equal volume of acid-washed glass beads (0.1-mm diameter), and mix using a vortex mixer on its highest setting for 1 min. *Transfer of the glass beads to the Eppendorf tubes can be greatly facilitated by constructing a 100-μl scoop. With a razor blade, cut the bottom off an Eppendorf tube at the 100-μl mark and epoxy it to a wooden applicator.

10. Heat samples at 65° for 4 min. Repeat the vortexing and heating steps.

11. Add exactly 1 ml of Tween-20 IP buffer plus 5 mM NEM. Invert to mix, then pellet the insoluble material by centrifugation (10 min, 12,000g). Transfer 950 μl to a new Eppendorf tube, being careful to avoid the pelleted material at the bottom of the tube. Remove a small amount of the cleared lysate (5 μl) and determine label incorporation by scintillation counting (typically 1–5 $\times$ 10^7 cpms/OD$_{600}$ unit).

12. Add 10 μl of 100 mg/ml protease-free BSA, the desired antibody, and incubate at 4° overnight by rocking or end-over-end mixing.

13. Briefly spin the tubes to bring the radioactive liquid down from the lid. Add 100 μl of protein A-sepharose (0.4 g protein-A sepharose swollen overnight in 11.2 ml protein-A sepharose buffer). Incubate at 4° with rocking or end-over-end mixing for 1–2 h. *Some primary antibodies will require the use of protein-G sepharose.

14. Pellet the protein A-sepharose beads in a Microfuge 1 min at 12,000g. Aspirate the supernatant, being careful to avoid the beads at the bottom of the tube.

15. Wash beads twice with 1 ml cold (4°) Tween-20 urea buffer, 1× with 1 ml Tween-20 IP buffer, and once with 1 ml TBS. For each wash, add approximately 1 ml of the appropriate wash buffer, close the tube, invert the tube several times, and pellet and aspirate as in step 14. *Note that some antibodies may require more or less stringent washing conditions.

16. Dry beads completely in a SpeedVac. Add 50–100 μl IP sample buffer, heat at 75° for 5 min to elute antigen, vortex, spin briefly in Microfuge to pellet beads. The sample is ready for SDS-PAGE and visualization by autoradiography or phosphor imaging. Depending on the subsequent application, it may be desirable to use signal-enhancing fluorogenic substrates such as sodium salicylate together with film. If the desired application is phosphor imaging and quantitation, then these signal-enhancing compounds will interfere with the function of the imaging screen and should not be used.

Although this technique will allow one to conclude that the protein of interest is degraded by the vacuolar system, it does not in any way address the role of ubiquitin in this process. Obviously, demonstration of ubiquitin-dependent sorting of cargo within the endosomal sorting system requires demonstration of ubiquitination of the cargo protein. The transient and labile nature of this modification can make detection of these ubiquitinated intermediates difficult. However, these intermediates can be stabilized through the use of mutants that are defective for the transport of cargo downstream of the ubiquitination step but upstream of the MVB sorting step/de-ubiquitination (see Fig. 1). For example, visualization of ubiquitinated forms of numerous cell surface proteins in yeast has been facilitated by the use of mutants defective for the internalization step of endocytosis (e.g., *end3*, *sla2/end4*.) (reviewed in Shaw *et al.* [2001]). Similarly, biosynthetic MVB cargoes that enter the MVB pathway without transiting through the cell surface can be stabilized by use of mutants that are defective for the delivery of transport vesicles to the late endosome (e.g., *pep12*) or the MVB sorting step itself (e.g., the class E Vps proteins, aka ESCRT complex components) (e.g., see Katzmann *et al.* [2001]). Demonstration of such stabilization likewise helps to localize the spatial and temporal role of ubiquitin in directing cargo sorting (as opposed to during folding/quality control earlier within the secretory pathway, for instance).

By use of the preceding procedure, it has been possible to visualize the appearance and disappearance of ubiquitinated intermediates, such as for the biosynthetic MVB cargo protein CPS (Katzmann *et al.*, 2004). At time

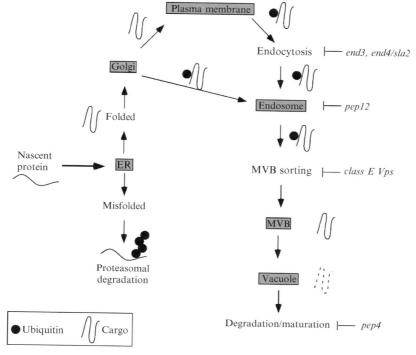

FIG. 1. Ubiquitin-dependent sorting events within the secretary and endocytic pathways. Improperly folded proteins within early secretory compartments (e.g., ER) can be retrotranslocated, polyubiquitinated, and degraded by the proteasome. Properly folded proteins are exported to the Golgi, where an additional sorting step can direct cargo to the cell surface, or into the MVB pathway (in a ubiquitin-dependent manner). Cell surface proteins can also be targeted into this degradative pathway through ubiquitin modification. Mutations that could be used to stabilize ubiquitinated MVB cargo at a number of steps are indicated.

zero during a pulse-chase experiment, it can be seen that a fraction of the protein exists in a higher molecular weight form, consistent with an ubiquitinated species. In wild-type cells, this form rapidly disappears, but in cells that fail to deliver MVB cargoes to the MVB sorting step (e.g., *pep12*) this form persists. To confirm that this represents an ubiquitinated form, the pulse-chase protocol can be modified in two ways. Overexpression of epitope-tagged ubiquitin, combined with the previous procedure is one way. Proteins receiving ubiquitin modification will incorporate both endogenous ubiquitin and epitope-tagged ubiquitin with an increased mass; this results in the conversion of ubiquitinated species to doublets, readily apparent compared with immunoprecipitated material from cells not expressing epitope-tagged ubiquitin. The second, and most convincing, demonstration is to perform immunoprecipitation of the protein of interest, followed by anti-ubiquitin Western blotting (or visualization of the epitope tag on ubiquitin by Western blotting). By simply TCA precipitating nonradiolabeled cells and performing the immunoprecipitation for the protein of interest, followed by antiubiquitin (or anti-epitope tag) Western blotting, it is possible to analyze not only the ubiquitination status of the protein but also the spatial location within the cell wherein this modification occurs (through the use of trafficking mutants).

Working with antiubiquitin antibodies is not as straightforward as many other antibodies. Most commercially available antiubiquitin antibodies (including the Zymed monoclonal that we frequently use) seem to work only when ubiquitin has been completely denatured after its transfer to nitrocellulose or PVDF. After transfer, sandwich the membrane with filter paper and place in a small tray such that it is weighted down under a small amount of water (an empty disposable pipet tip box with 200–500 g in it works well) and autoclave 30 min, liquid cycle. After the liquid cycle is complete, remove water from tray and return wet membrane, between wet filter paper sheets, to autoclave for an additional 10 min exposure under dry cycle. The membrane is now ready for blocking and antibody decoration.

Cycloheximide Chase

If it is known that the cargo protein under study will not resolubilize after TCA precipitation, it is possible to use native immunoprecipitations in the presence of NEM. However, it may be desirable to avoid the use of radioisotopes altogether. In such a case, it is still possible to follow protein turnover through the cessation of new protein synthesis by means of cycloheximide treatment and Western blotting.

Buffers and Solutions

- Yeast synthetic medium (SM) plus appropriate amino acids.
- 10% yeast extract: 10 g yeast extract dissolved in 100 ml water, autoclave, and keep sterile.
- Cycloheximide stock: 1 mg/ml in ethanol.
- 2× stop solution: 20 mM NaF, 20 mM NaN$_3$.
- 10× stop solution: 100 mM NaF, 100 mM NaN$_3$.
- 1× Laemmli urea sample buffer (LUSB) (5 ml): 1.1 ml H$_2$O, 400 μl 5× LSB, 720 mg urea, 50 μl β-mercaptoethanol.
- 5× LSB: 6.25 ml 0.5 M Tris-HCl, pH 6.8, 1 g SDS, 10 ml 50% glycerol, 0.5 ml 1% bromophenol blue, water to 50 ml.

1. Grow overnight yeast cultures in SM + 0.2% yeast extract (YE)+ amino acids (complete or selective, at 1×).
2. In the morning, dilute the cultures in SM + 0.2% YE + amino acids, as described in the preceding metabolic pulse-chase protocol. You will need 2 OD_{600}/time point (usually 12 OD_{600} units, for six time points). *1 OD_{600}/time point has also worked well in our experience.
3. Harvest the cells in a 50-ml disposable sterile conical screw cap tube and wash once in SM + amino acids. Resuspend cell pellets in SM + amino acids at 0.5 ml/time point (e.g., 3.0 ml/strain for six time points). Use a little extra volume for resuspending to avoid running out of cells at the last time point (e.g., 3.2 ml rather than 3.0 ml). *Alternate: if cells are sickly, or if the time course will extend out to 2 h or longer (i.e., if cell health is an issue during the course of the assay), consider this for better results: resuspend the cells after the wash in SM + 2× amino acids + 2× YE (0.4%) at 1.0 ml/time point (e.g., 6.0 ml/strain for six time points).
4. Place in a shaking water bath at the appropriate temperature for 5–10 min.
5. Using the same staggered time of addition approach described for metabolic pulse-chase in the previous protocol, add cycloheximide (CHX) to 5 μg/ml final (stock is 200×) and briefly vortex.
6. Immediately collect a sample for the zero time point and place into prechilled tubes on ice containing azide/fluoride (N_3/F) stop solution. After the addition of the cells, the stop solution should be 10 mM NaN_3 and 10 mM NaF. If collecting 0.5 ml/time point, use a 2× stock of N_3/F aliquoted at 0.5 ml and prechilled. If collecting 1.0 ml/time point, make a 10× stock and add 111 μl of this to each tube and prechill. Place the tube with the rest of the cells into a shaking water bath at the appropriate assay temperature to begin the chase period.
7. Continue to collect time points into chilled Eppendorf tubes containing stop solution as appropriate. A good starting series is 0 min, 5 min, 10 min, 20 min, 40 min, 90 min.
8. After all samples have been collected, pellet cells in centrifuge for 1 min at 13,000g at 4° and aspirate supernatant. Wash cells with 1× ice-cold PBS. Aspirate supernatant well. (This wash step is optional.)
9. To each cell pellet, add 50 μl LUSB and 200 μl glass beads. Close the tube, taking care that no glass beads are near the top of the tube where the lid inserts; trapped beads can prevent the lid from sealing and result in leakage.
10. Mix the cell/glass bead suspension on a vortex mixer for 90 sec and heat for 5 min. (*For soluble proteins, 95–100° should be used for the heating step. Transmembrane proteins are prone to aggregation and should be heated at lower temperatures; *as a general guideline* no higher than 37° is best for proteins with 12 or more transmembrane domains, up

to 70° can be used for other transmembrane proteins.) The vortex and heat steps can be repeated for optimal cell lysis if desired.

11. Add 150 μl LUSB to each tube, pulse spin (20 sec, ~5000g), and transfer the supernatant to a new tube. (A good way to do this transfer is to open the lid, pierce the bottom of the tube with a 22- to 25-gauge needle, and then squeeze the liquid through with a rubber Pasteur pipet bulb sealed over the tube opening.)

12. Pellet insoluble material by centrifugation at 13,000g for 5 min, transfer supernatant to a new Microfuge tube if desired. Store at −20°.

13. Run 5–10 μl of the sample on an acrylamide gel.

14. Determine the status of your protein by Western blotting with the appropriate antiserum.

Ubiquitin Binding by Candidate Sorting Machinery

Another aspect of ubiquitin-dependent protein sorting within the endosomal system is the ability of *trans*-acting machinery to bind to the ubiquitin signal on cargoes (or on other machinery components). A variety of methods have been used to demonstrate a noncovalent interaction between ubiquitin and a protein of interest. These methods range from fairly standard methods, such as yeast two-hybrid, *in vitro* "pull-downs" or *in vivo* co-purifications, to more quantitative biophysical methods, such as surface plasmon resonance, that are suited to detect very transient low-affinity interactions, or even the mapping of specific residues involved in making this contact, such as by NMR. Here we focus on two approaches that may be more approachable by most laboratories having standard equipment. Although *in vitro* "pull-down" assays using ubiquitin as bait may be a useful first step, the approaches highlighted here offer the advantage of detecting *in vivo* interactions. It should be noted that successful demonstration of an interaction with ubiquitin might only represent a first step. Determining the significance of this interaction is critical and must be addressed phenotypically, as well as biochemically.

Co-Purification of Ubiquitinated Cargo with Ubiquitin-Binding Machinery

Protein-A Chimera-Facilitated Affinity Purification

Buffers/Materials

- Softening buffer: 0.1 M Tris, pH 9.4/10 mM dithiothreitol (DTT).
- Spheroplasting media: 0.0067% yeast nitrogen base, 2% glucose, 1× amino acids, 1 M sorbitol, 20 mM Tris, pH 7.5.

- Phosphate-buffered saline: this is one suitable lysis buffer that can be used with or without the addition of detergent (as described below).
- IgG-sepharose.
- Zymo lyase 100 T.
- Elution buffer: 0.5 M sodium acetate, pH 3.4.
- Protease inhibitors: for example, 1× "Complete" from Roche

Softening/Spheroplasting

1. Harvest 10–20 OD_{600} cells expressing the protein A fusion protein of interest. *May need to scale this up on the basis of the abundance of the proteins being studied. Always include a negative control strain as well (e.g., a protein-A fusion) to some irrelevant protein, or protein-A alone.

2. Resuspend at 5 OD_{600}/ml in softening buffer, incubate 10 min at room temp. *Add DTT to softening buffer fresh, from frozen stock.

3. Pellet cells at 1000g for 5 min, aspirate supernatant, and resuspend pellet in spheroplasting media at 5 OD_{600}/ml. *Starting OD_{600} should be 5.0; this is important for measuring spheroplasting efficiency (add 10 μl cells to 990 μl water before addition of Zymolyase and expect a reading of ~0.05 OD_{600}).

4. Add zymolyase 100T: make a 10 mg/ml stock in 1 M sorbitol, add 1 μl of this stock per OD_{600} unit of cells.

5. Spheroplast for 20–30 min at 26–30°, check lysis to make sure it is greater than 90%. To check spheroplasting, add 10 μl spheroplasting reaction to 990 μl water and expect this OD_{600} measurement to be less than 0.005. *Spheroplasts are relatively fragile and should be handled more gently than intact cells.

6. Harvest cells by pelleting at 500g for 5 min, remove all supernate by aspirating (it is important to remove all the Zymolyase enzyme), and place pellet on ice.

Lysis/Clearing. This step can be performed in several ways. If it is not clear whether the interaction is occurring in the cytoplasm or on a membrane, the lysis can be performed in a low concentration of detergent (e.g., 0.1–1% of Tween-20, Triton X-100, LDAO, or digitonin). If the interaction occurs on membranes, it is possible to generate a membrane pellet that can subsequently be solubilized. If the protein of interest exists in both soluble and membrane fractions, it is also possible to purify it from both membrane and soluble fractions to test which fraction interacts with ubiquitinated proteins.

1. Resuspend spheroplasts in lysis buffer (e.g., PBS with 5 mM NEM plus protease inhibitors) at 5 OD_{600}/ml.

2. Using a glass Dounce homogenizer with a tight piston, subject each sample to 20 up/down strokes (try to avoid making bubbles). Use a small volume (1–2 ml) Dounce with a clearance of 0.013 mm, available from Wheaton or Kontes.
3. Clear the lysate, depending on application:

- for S100/P100 fractions, spin at 100,000g for 30 min to generate membrane and supernatant fractions. After the spin, transfer supernatant to new Eppendorf tube, solubilize pellet in lysis buffer plus 0.1–1.0% detergent on ice for 10 min, and reclear in a refrigerated Microfuge (4°, 10 min, 120,000g). Adjust S100 fraction to the same detergent concentration before using as input.
- For lysis in the presence of detergent, simply clear at 36,000g for 20 min after lysis.

Isolation of Protein A Fusion Chimera and Potential Interaction Partners

1. Use approximately 20 μl of IgG-sepharose per isolation reaction (this is in large excess but is a volume that can be worked with to have a visible pellet). Prepare the beads in batch and subsequently split between lysates. For two isolations, wash approximately 45 μl IgG-sepharose three times with 1 ml elution buffer to remove free IgG. After low pH washes, equilibrate beads in lysis buffer and divide into two tubes. Pellet the beads, remove the remaining buffer, and place on ice.
2. Add cleared lysate to the washed IgG-sepharose pellet.
3. Incubate at 4°, rotating end-over-end for 30–60 min.
4. Spin down beads (12,000g, 1 min), wash three times with lysis buffer (containing detergent, if appropriate), once with lysis buffer minus detergent, once with 0.5× lysis buffer minus detergent. *Obviously, the stringency of washing will have an effect on the recovery of interacting proteins and may be altered accordingly.
5. Elute sample with 1 ml 4° elution buffer on ice for 10 min or boil in sample buffer (so long as the IgG bands do not migrate at a similar size as the eluted material).

Troubleshooting this protocol will be greatly facilitated by keeping samples from the various steps performed. When loading the samples for SDS-PAGE, try to load equivalent OD_{600} units. That is, if you load 0.1 OD_{600} of the "start" material (e.g., the lysate generated before fractionation), then load 0.1 OD_{600} units of the "input" (e.g., the cleared lysates before their incubation with IgG-sepharose) and "unbound" (e.g., the supernatant after incubation of extracts with IgG-sepharose). For "bound" material (whether low pH eluted or SDS-sample buffer boiled), load an equivalent OD_{600}, as well as 5–10× more (because interactions may be weak). After gel electrophoresis, samples can be analyzed by protein stain (e.g., Coomassie or silver) and/or Western blotting.

Binding Experiments to Test for Ubiquitin-Dependent Recruitment to Membranes

The following approaches can be used to detect an ubiquitin-dependent membrane interaction of a *trans*-acting trafficking factor or for a quantitative assessment of the contribution of a candidate ubiquitin-binding domain to ubiquitin-dependent membrane interactions. Ubiquitinated membrane proteins, or ubiquitinated proteins that associate with membranes, can in some cases present a target that more resembles "*in vivo*" partners for a protein under study than does pure ubiquitin bound to beads. Variations of this experiment include using either ^{35}S-labeled protein from an *in vitro* transcription/translation reaction or recombinant protein as the "ligand" for binding to the membranes.

Preparation of Membranes

These membranes will be used as the "substrate" in the subsequent binding assays.

Buffers/Solutions

- Softening solution: 0.1 M Tris, pH 9.4, 10 mM DTT, 5 mM NEM.
- 2× YPD: 20 g yeast extract, 40 g peptone to 1 liter of water, autoclave in 100-ml aliquots.
- Spheroplasting solution: Mix equal volumes of 2× YPD and 2 M sorbitol; add protease inhibitor cocktail just before use.
- Zymolyase stock: 10 mg/ml Zymolyase 100T in PBS, 1 M sorbitol (100-μl aliquots can be stored at $-20°$).
- Spheroplast washing solution: 1× PBS, 1 M sorbitol, 5 mM NEM.
- HEPES/KOAc lysis buffer: 0.2 M sorbitol, 50 mM KOAc, 2 mM EDTA, 20 mM HEPES, pH 6.8, 50 mM NEM, protease inhibitor cocktail.
- Phosphatase inhibitors: 3 mM NaF, 0.1mM EDTA, 0.6 mM sodium-orthovanadate, 0.2 μM FK-506, 0.1 mM AEBSF.
- Protease inhibitor cocktail: e.g., 1× "Complete^{TM}" from Roche.

1. Grow overnight yeast cultures in YPD or selective medium at appropriate temperature. Dilute in the morning to 0.3 OD_{600}/ml. Cells can be either wild-type, RL120 [sole source of Ub is His6-Ub; (Ling *et al.*, 2000)], or strain of your choice harboring a plasmid encoding epitope-tagged Ub.

2. Harvest 20–50 OD_{600} cells/strain by pelleting at 1000g for 5 min at 25° and decant supernatant. For all subsequent steps, resuspend at 10 OD_{600}/ml (unless otherwise noted).

3. Soften cell walls by resuspending in softening solution. Incubate on the bench (no shaking necessary) for 10 min at room temperature. Harvest in centrifuge 1000g, aspirate off supernatant.

4. Generate spheroplasts in 1× YPD/1 M sorbitol + protease/phosphatase inhibitors. Before adding zymolyase, remove 10 μl cells, dilute in 990 μl ddH_2O, and read the OD_{600}. This is a reference point to monitor the spheroplasting process. Add 1 μl zymolyase stock per OD_{600} cells. Incubate at 30° (or lower, if working with a temperature-sensitive mutant strain) for 15–30 min with gentle agitation. Test spheroplasting efficiency as described previously.

5. Harvest the cells by centrifuging at 300g, 5 min, and aspirate spheroplasting solution. Gently wash the pellet in spheroplast washing solution. To resuspend spheroplasts efficiently, add 1 ml solution with a 1-ml pipette and gently pipet up/down to resuspend the pellet. Then add the rest of the volume to the final concentration of 10 OD_{600}/ml. Harvest the spheroplasts and aspirate the spherowashing solution.

6. Resuspend the spheroplast pellet in HEPES/KOAc lysis buffer + inhibitors and transfer to an ice-cold Dounce homogenizer. Homogenize with 10 strokes (10 up and 10 down) using the tightest-fitting pestle. *To wash homogenizer between samples, rinse with H_2O, EtOH, H_2O, then a small volume of lysis buffer and aspirate dry. Do not use detergent, because residual detergent can dissolve cell membranes.

7. To isolate the heavy membranes (including the plasma membrane), transfer 1-ml aliquots (10 OD_{600} equivalent) of each sample to Eppendorf tubes and centrifuge at 300g for 5 min at 4°. Carefully remove the tubes from the centrifuge, because the pellets are very soft. Remove most of the supernatant using a pipetter fitted with a 1-ml plastic tip. Wash each pellet by resuspension in 1 ml HEPES/KOAc lysis buffer and centrifugation as before. Remove all of the supernatant with a pipet. The pellets can be stored at $-80°$ or used immediately.

The membrane pellets that are now prepared are used as the "substrate" in the following binding assays.

Recombinant Protein Binding Assays

In this protocol, wild-type and mutant recombinant proteins are assayed for ubiquitin-dependent binding to membranes. The recombinant proteins can either be purified from bacteria or generated as radiolabeled species from an *in vitro* transcription/translation system. The latter is appropriate for a quantitative analysis of the contribution of ubiquitin-binding domains to the binding. For instance, once ubiquitin dependence of membrane recruitment has been established for your protein, mutant versions with deletions or substitutions of suspected ubiquitin-binding domains can be synthesized as radiolabeled proteins and tested quantitatively for binding. The behavior of the mutants is compared with the behavior of the wild-type protein ± free ubiquitin; the addition of free ubiquitin as a competitor allows one to determine the contribution of ubiquitin-dependent interactions in the membrane recruitment assay. In the following, we describe two variations of the binding experiment, depending on whether purified recombinant proteins or *in vitro* transcription/translation products are being used.

Buffers/Solutions

- TBS: 10 mM Tris, pH 8.0, 150 mM NaCl.
- TBS/BSA: 1× TBS plus 0.1% BSA.
- TBST: TBS + 0.1% Tween-20.
- TBST + 10 mM CaCl$_2$.

Purified Recombinant Protein Binding

1. Purify your recombinant protein using standard techniques, and exchange into TBS buffer using a desalting column. Mix the protein ± 10 mM ubiquitin with end-over-end mixing overnight at 4°. *100 μl of sample is used for each binding reaction. Our recombinant proteins typically range from 1–10 μM.

2. Prepare the membrane pellets for the binding experiment: to each tube containing 10 OD$_{600}$ equivalents of membrane, add 1 ml TBS + 0.1% BSA and triturate by passing through a 25-gauge needle five times. Each tube contains sufficient membranes for 10 binding reactions. *Insulin, gelatin, or BSA can be used as a blocking protein, depending on the size of the protein you are studying; choose one that migrates at a different mass from your proteins of interest.

3. Combine 100 μl of protein, 100 μl membranes, and 100 μl TBS/BSA buffer. Incubate with end-over-end mixing for 1 h at 4°.

4. Add 1 ml cold TBST buffer, centrifuge 6 min at 21,000g at 4°. Remove all but 100 μl of the supernatant. Repeat this wash twice more. For each wash, pipet up and down with a mechanical pipetter fitted with a 200-μl tip to disrupt the pellet.

5. Wash the membrane pellet with TBST + 10 mM CaCl$_2$, and remove all supernatant after centrifugation. *Ca^{2+} is not included until the final wash buffer, because it aggregates the membranes and interferes with efficient pellet disruption that facilitates the washing steps.

6. Add 15 µl protein sample buffer, heat for 5 min. Resolve samples on appropriate percentage of SDS-PAGE and analyze by Western blotting to detect the presence or absence of the recombinant protein. **It is generally best to detect the recombinant protein using an antiserum recognizing an epitope- or affinity-tag on the recombinant protein to distinguish it from endogenous yeast proteins that may be present in the starting sample.

In Vitro *Transcription/Translation Product Binding*

1. Prepare your labeling protein for the binding experiment using ^{35}S-methionine and cysteine in an *in vitro* transcription/translation system. Preincubate ±10 mM ubiquitin overnight at 4° with end-over-end mixing.

2. Prepare the membranes for the binding experiment as described previously.

3. Add 3–10 × 10^6 cpm of ^{35}S-labeled proteins (±Ub) to 100 µl membranes and 100 µl TBS/BSA buffer to a final volume of 300 µl. Incubate with end-over-end mixing for 1 h at room temperature or 4°. *Perform each reaction in triplicate.

4. Wash the membranes as described previously (twice in TBST, once in TBST + 10 mM CaCl$_2$), using care to collect and dispose properly of the radioactive supernatants.

5. Resuspend the pellet in 200 µl TBST and transfer to a scintillation vial. Add scintillation fluid and assess the amount of labeled protein that has bound to the membranes by counting the radioactivity associated with the pellets; average the counts for each condition. *A good control is to compare the recruitment of your protein ± Ub with the recruitment of a mutant version of your protein in which the candidate ubiquitin-binding motif has been altered or deleted.

Crosslinking to Identify Candidate Partner Proteins in the Membranes. This approach is used to enrich for protein complexes that can form between ubiquitin-binding proteins and ubiquitinated membrane cargo proteins. Interactions uncovered through this method should subsequently be verified by an *in vivo* approach.

Buffers/Solutions

- 10× DTSSP: 15 mM, prepared in water immediately before use.
- Quenching solution: 100 mM Tris base, 2% SDS, 2 mM AEBSF.
- TBS: 10 mM Tris, pH 8.0, 150 mM NaCl.
- Bead-blocking solution: TBS + 1% BSA + 0.1% SDS.
- TBST: TBS + 0.1% Tween-20.

1. Follow the purified recombinant protein binding experiment described previously, using a recombinant protein that has an affinity tag such as His_6, up to step number 4, leaving the bound membranes in 100-μl volume after three washes. *These washes are included to remove the unbound free Ub before addition of the cross-linker. If the free Ub is not removed, it will quench the cross-linker and yield results that artifactually appear to be specific. It is best to work quickly to minimize dissociation of complexes that have formed. If you have many samples, wash the pellets in batches of no more than three at a time.

2. Add 10 μl 10× DTSSP and pipet up and down (using a 200-μl pipet tip, followed by a narrow-bore disposable pipet tip such as a gel-loading tip) to resuspend pellet. Incubate on ice 2 h with occasional agitation or tube "flicking" to mix samples.

3. Quench cross-linker with 110 μl quenching solution, incubate 30 min at room temperature with end-over-end or rocking to mix.

4. Move samples to 4° for overnight incubation with end-over-end mixing to solubilize membranes.

5. Pellet insoluble material 15 min at 21,000g at 4°.

6. Transfer 150 μl supernatant to a fresh tube and add 1.25 ml TBS to dilute the SDS.

7. Add 100 μl affinity resin (e.g., Ni^{2+} or Co^{2+} beads for His_6-tagged protein; GSH beads for GST-tagged protein, etc.) and incubate 1 h at 4° with end-over-end mixing. *Affinity resin is preblocked by overnight incubation in TBS + 1% BSA + 0.1% SDS. In the morning, the beads are washed three times with TBS (no BSA or SDS).

8. Harvest beads 3 min at 21,000g at 4°. Wash beads three times with 1 ml cold TBST per wash. After the final wash, remove as much supernatant as possible with a pipet.

9. Dry the beads either in a SpeedVac or by removing all traces of liquid using a Hamilton syringe (the bore of the needle is small enough to exclude most beads).

10. Add 20 μl Laemmli protein sample buffer + reducing agent (βME or DTT), heat 5 min, and resolve on appropriate percentage SDS-PAGE. Analyze presence of candidate partner proteins by Western blotting.

**If the membranes were prepared from cells expressing an epitope-tagged Ub, the blot can be probed with antibodies against the epitope to reveal possible Ub-conjugated cargo that was crosslinked to your recombinant protein.

Conclusions

Protein ubiquitination has received tremendous attention in recent times, uncovering previously unappreciated roles for this modification. Within the endosomal system, ubiquitin plays a critical role in directing protein cargo into the MVB pathway; both at the level of tagging these cargos but, apparently, also at the level of regulating the machinery that directs this process. The protocols presented here can facilitate the ability to detect ubiquitin-modified forms of proteins or to bind to ubiquitin and/or ubiquitinated partners. Further studies are nearly always required to establish the *in vivo* relevance or physiological significance of such modifications or interactions. The protocols presented here were selected because they can easily be altered to address *in vivo* significance, allowing for a more informed interpretation of the results.

Acknowledgments

The authors wish to acknowledge Dr. Bruce Horazdovsky for critical reading of the manuscript, Rubén Claudio Aguilar for developing the *in vitro* binding protocols, and Drs. Greg Payne and Jim Howard for advice on cycloheximide chase experiments.

References

Bache, K. G., Brech, A., Mehlum, A., and Stenmark, H. (2003). Hrs regulates multivesicular body formation via ESCRT recruitment to endosomes. *J. Cell Biol.* **162,** 435–442.

Bishop, N., Horman, A., and Woodman, P. (2002). Mammalian class E vps proteins recognize ubiquitin and act in the removal of endosomal protein-ubiquitin conjugates. *J. Cell Biol.* **157,** 91–101.

Davies, B. A., Topp, J. D., Sfeir, A. J., Katzmann, D. J., Carney, D. S., Tall, G. G., Friedberg, A. S., Deng, L., Chen, Z., and Horazdovsky, B. F. (2003). Vps9p CUE domain ubiquitin binding is required for efficient endocytic protein traffic. *J. Biol. Chem.* **278,** 19826–19833.

Hicke, L., and Dunn, R. (2003). Regulation of membrane protein transport by ubiquitin and ubiquitin-binding proteins. *Annu. Rev. Cell Dev. Biol.* **19,** 141–172.

Katzmann, D. J., Babst, M., and Emr, S. D. (2001). Ubiquitin-dependent sorting into the multivesicular body pathway requires the function of a conserved endosomal protein sorting complex, ESCRT-I. *Cell* **106,** 145–155.

Katzmann, D. J., Odorizzi, G., and Emr, S. D. (2002). Receptor downregulation and multivesicular-body sorting. *Nat. Rev. Mol. Cell. Biol.* **3,** 893–905.

Katzmann, D. J., Sarkar, S., Chu, T., Audhya, A., and Emr, S. D. (2004). Multivesicular body sorting: Ubiquitin ligase rsp5 is required for the modification and sorting of carboxypeptidase s. *Mol. Biol. Cell* **15,** 468–480.

Ling, R., Colon, E., Dahmus, M. E., and Callis, J. (2000). Histidine-tagged ubiquitin substitutes for wild-type ubiquitin in *Saccharomyces cerevisiae* and facilitates isolation and identification of *in vivo* substrates of the ubiquitin pathway. *Anal. Biochem.* **282,** 54–64.

Losko, S., Kopp, F., Kranz, A., and Kolling, R. (2001). Uptake of the ATP-binding cassette (ABC) Transporter Ste6 into the yeast vacuole is blocked in the doa4 mutant. *Mol. Biol. Cell* **12,** 1047–1059.

Odorizzi, G., Babst, M., and Emr, S. D. (1998). Fab1p PtdIns(3)P 5-kinase function essential for protein sorting in the multivesicular body. *Cell* **95,** 847–858.

Raymond, C. K., Howald-Stevenson, I., Vater, C. A., and Stevens, T. H. (1992). Morphological classification of the yeast vacuolar protein sorting mutants: Evidence for a prevacuolar compartment in class E *vps* mutants. *Mol. Biol. Cell* **3,** 1389–1402.

Reggiori, F., and Pelham, H. R. (2001). Sorting of proteins into multivesicular bodies: Ubiquitin-dependent and -independent targeting. *EMBO J.* **20,** 5176–5186.

Scott, P. M., Bilodeau, P. S., Zhdankina, O., Winistorfer, S. C., Hauglund, M. J., Allaman, M. M., Kearney, W. R., Robertson, A. D., Boman, A. L., and Piper, R. C. (2004). GGA proteins bind ubiquitin to facilitate sorting at the trans-Golgi network. *Nat. Cell Biol.* **6,** 252–259.

Shaw, J. D., Cummings, K. B., Huyer, G., Michaelis, S., and Wendland, B. (2001). Yeast as a model system for studying endocytosis. *Exp. Cell Res.* **271,** 1–9.

Spormann, D. O., Heim, J., and Wolf, D. H. (1992). Biogenesis of the yeast vacuole (lysosome). The precursor forms of the soluble hydrolase carboxypeptidase yscS are associated with the vacuolar membrane. *J. Biol. Chem.* **267,** 8021–8029.

Traub, L. M. (2003). Sorting it out: AP-2 and alternate clathrin adaptors in endocytic cargo selection. *J. Cell Biol.* **163,** 203–208.

Urbanowski, J. L., and Piper, R. C. (2001). Ubiquitin sorts proteins into the intralumenal degradative compartment of the late-endosome/vacuole. *Traffic* **2,** 622–630.

Wendland, B. (2002). Epsins: Adaptors in endocytosis? *Nat. Rev. Mol. Cell. Biol.* **3,** 971–977.

Section III

Methods to Study the Proteasome

[14] Preparation of Ubiquitinated Substrates by the PY Motif-Insertion Method for Monitoring 26S Proteasome Activity

By Y. SAEKI, E. ISONO, and A. TOH-E

Abstract

For analysis of the mechanism of the 26S proteasome–mediated protein degradation *in vitro*, the preparation of well-defined substrate, the ubiquitinated proteins, of the 26S proteasome is inevitable. However, no method has been available to ubiquitinate a given protein. Here, we propose a relatively simple method for preparation of the ubiquitinated substrates using HECT-type ubiquitin ligase Rsp5, termed the PY motif-insertion method. The principle of this method is that the PY motif, known as the Rsp5-binding motif, is inserted into protein to be ubiquitinated by Rsp5. In this communication, we describe that Sic1 was successfully ubiquitinated by the PY motif-insertion method and demonstrate that Sic1 thus ubiquitinated was degraded by the purified yeast 26S proteasome.

Introduction

In eukaryotic cells, the ubiquitin-proteasome system regulates various cellular processes (Hershko and Ciechanover, 1998; Pickart, 2001). In this pathway, substrate proteins are polyubiquitinated by E1/E2/E3 enzymes, and thus a formed polyubiquitin chain is recognized by the 26S proteasome, and the substrate portion is degraded in an ATP-dependent manner. The substrate recognition by the ubiquitin ligase (E3) is the critical step for the substrate specificity of the protein ubiquitination. Studies of this step revealed that this recognition signal usually requires modification, such as phosphorylation, oxidation, or glycosylation. For example, polyubiquitination of Sic1, a CDK inhibitor in *S. cerevisiae*, requires multiple phosphorylations by CDK for recognition by the E3 complex SCF^{Cdc4}. Deshaies and his colleagues reconstituted this SCF-catalyzed Sic1 ubiquitination system *in vitro* and applied polyubiquitinated Sic1 thus produced to the biochemical study of the 26S proteasome assay (Verma *et al.*, 2001). But the procedure for the preparation of polyubiquitinated Sic1 is technically difficult, because the reconstitution of this ubiquitination system requires CDK and the SCF^{Cdc4} complex. Also, the several components must be produced by

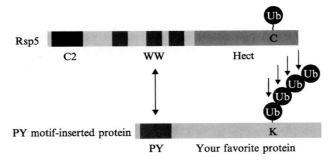

FIG. 1. Schematic illustration of the PY motif-insertion method. The PY motif-inserted proteins are recognized by Rsp5 by means of its WW domain; then, Rsp5 ubiquitinates the PY motif-inserted protein. Ub, ubiquitin; K, lysine; Hect, homologous to the E6-AP carboxyl terminus.

the insect expression system. Obviously, a less tedious method for polyubiquitination of proteins should be developed.

We searched for an E3 enzyme that does not require posttranslational modification of its substrate for ubiquitination. Among the E3 recognition motifs, the PY motif is known as a binding site of the Rsp5/Nedd4 family by way of their WW domain without any modification (Sudol and Hunter, 2000). The PY motif consists of a short sequence, Pro-Pro-X-Tyr, and directly binds with the WW domain *in vitro*. We assumed that the PY motif-inserted proteins might bind with Rsp5 and be ubiquitinated by it (Fig. 1). To test this, we chose a well-characterized substrate Sic1 and introduced the PY motif into the N-terminal region of Sic1, designated as Sic1^{PY}. As expected, Sic1^{PY} was efficiently polyubiquitinated by Rsp5. Importantly, all components of this system, including Rsp5, can be easily expressed in *Escherichia coli* and purified from bacterial lysates.

In Vitro Polyubiquitination of the PY Motif-Inserted Protein

Purification of the PY Motif-Inserted Sic1

Sic1^{PY} was constructed as follows. The CCACCGCCGTAT sequence encoding Pro-Pro-Pro-Ser was inserted between +18 and +25 bp of the *SIC1* open-reading frame by polymerase chain reaction. The adenine residue of ATG corresponding to the putative translation initiation codon was defined as the +1 nucleotide.

Materials and Reagents. pET21-Sic1, pET21-Sic1^{PY}, pET21-Sic1K36^{PY}, *E. coli* Rosetta (DE3) competent cells (Novagen), protease inhibitor cocktail (EDTA free, Roche), TALON resin (Clontech). MicroSpin Empty columns (Amersham).

Solutions and Buffers. LB medium (supplemented with 50 μg/ml ampicillin and 24 μg/ml chloramphenicol); 1 M isopropyl thio-β-galactopyranoside (IPTG); lysis buffer (50 mM Na-phosphate, pH 7.0, 300 mM NaCl, 10% glycerol, 1 mM β-mercaptoethanol); wash buffer (50 mM Na-phosphate, pH 7.0, 300 mM NaCl, 10% glycerol, 1 mM β-mercaptoethanol, 0.2% Triton X-100); elution buffer (50 mM Na-phosphate, pH 7.0, 300 mM NaCl, 10% glycerol, 150 mM imidazole); buffer A (50 mM Tris-HCl, pH 7.5, 100 mM NaCl, 10% glycerol, 1 mM DTT).

Procedure

1. Using standard methods, grow 200 ml of *E. coli* Rosetta (DE3) transformed with respective plasmids in LB medium supplemented with 50 μg/ml ampicillin and 24 μg/ml chloramphenicol to OD_{600} of 0.7.
2. Cells are cooled to 30° and induced with 0.5 mM IPTG for 3 h at 30°.
3. Cells are harvested by centrifugation (3000g, 10 min) and stored at −80° until use.
4. Cells are suspended with 15 ml of ice-cold lysis buffer supplemented with 1× protease inhibitor cocktail and lysed by sonication.
5. Triton X-100 is added to 0.2%, and then incubated on ice for 10 min.
6. The lysate is cleared by centrifugation (15,000g for 30 min), and the supernatant is recovered to 15-ml tube.
7. Add 200 μl of the TALON resin that has been preequilibrated with lysis buffer to the supernatant. Rotate for 1–2 h at 4°.
8. Collect the TALON resin beads by centrifugation (3000 rpm for 2 min, TOMY 1500 or its equivalent) and wash them three times with 10 ml of wash buffer (spin the tube at 3000 rpm for 2 min and discard the supernatant).
9. Transfer the beads to a MicroSpin Empty column. Wash twice with 400 μl of lysis buffer.
10. Add 300 μl of elution buffer to the column. Cap the column and rotate for 20 min. Recover the eluted materials by centrifugation (5000 rpm for 1 min).
11. Dialyze the eluate against 500 ml of buffer A for 12 h at 4°.
12. Analyze the protein by SDS-PAGE. Scan the CBB-stained gel by the LAS3000 system (Fuji) and quantify the protein band by Image Gauge software (Fuji).
13. Dilute the proteins to appropriate concentration and divide into small aliquots. Store at −80°.

Comment. Usually we obtained 2 mg of T7-Sic1-His$_6$, 2 mg of T7-Sic1^{PY}-His$_6$, and 0.2 mg of T7-Sic1K36^{PY}-His$_6$ from 200-ml culture (Fig. 2A).

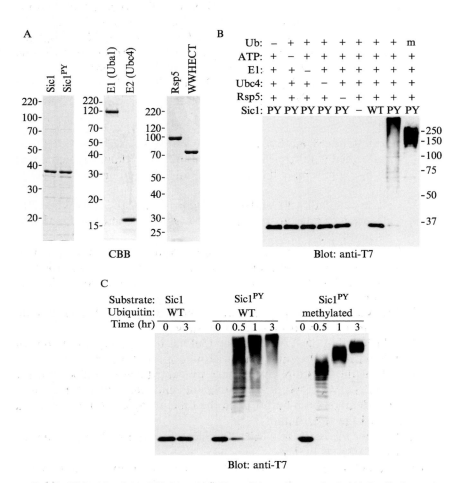

FIG. 2. Ubiquitination of Sic1 by the PY motif-insertion method. (A) Purified proteins used in this study. Proteins (1 μg each) were analyzed by SDS-polyacrylamide gel electrophoresis followed by Coomassie staining. (B) Ubiquitination assay using the purified enzymes and Sic1^{PY}. Reaction mixtures containing indicated materials were incubated for 3 h at 25°. Then, the reactions were terminated by the addition of SDS-loading buffer and subjected to SDS-PAGE, followed by Western blotting with anti-T7 antibody. Ub, ubiquitin; m, methylated ubiquitin; PY, PY motif-inserted Sic1; WT, wild-type Sic1. (C) Kinetics of Sic1^{PY} ubiquitination by Rsp5. The reactions were terminated at indicated times and analyzed as in (B).

Purification of Rsp5

Rsp5 has a multidomain topology: an N-terminal C2 domain, three WW domains, and a C-terminal HECT domain. Because the C2 domain functions as a membrane localization signal (Dunn *et al.*, 2004), this domain might be unnecessary for ubiquitin ligase activity. We tested the ubiquitination activity of an Rsp5 derivative that had been constructed by deleting the N-terminal 220 amino acids from the wild-type Rsp5, designated as WWHECT, and found that the WWHECT had ubiquitination activity as high as the wild-type Rsp5.

Materials and Reagents. pGEX6P1-Rsp5, pGEX6P1-WWHECT, *E. coli* Rosetta (DE3) competent cells (Novagen), glutathione sepharose 4B (Amersham). MicroSpin Empty columns (Amersham), PreScission protease (Amersham).

Solutions and Buffers. LB medium (supplemented with 50 μg/ml ampicillin and 24 μg/ml chloramphenicol), 1 M isopropyl thio-β-galactopyranoside (IPTG), buffer A (50 mM Tris-HCl, pH 7.5, 100 mM NaCl, and 10% glycerol).

Procedure

1. Using standard methods, grow 200 ml of *E. coli* Rosetta (DE3) transformed with respective plasmids in LB medium supplemented with 50 μg/ml ampicillin and 24 μg/ml chloramphenicol to OD$_{600}$ of 0.5.
2. Cool cells to 20° and induce with 0.2 mM IPTG for 15 h at 20°.
3. Harvest cells by centrifugation (3000g, 10 min) and store at −80° until use.
4. Suspend cells with 15 ml of ice-cold buffer A and lyse by sonication.
5. Add Triton X-100 to 0.2% followed by incubation on ice for 10 min.
6. Clear the lysate by centrifugation (15,000g for 30 min) and recover the supernatant to a 15-ml tube.
7. Add 200 μl of the glutathione sepharose that has been preequilibrated with buffer A to the supernatant. Rotate for 1 h at 4°.
8. Collect agarose beads by centrifugation at 300 rpm for 2 min and wash them three times with 10 ml of buffer A supplemented with 0.2% Triton X-100 (Spin the tube at 3000 rpm for 2 min and discard the supernatant).
9. Transfer the beads to a MicroSpin Empty column. Wash twice with 400 μl of buffer A (spin the column at 5000 rpm for 30 sec).
10. Add 200 μl of protease buffer (8 μl of PreScission protease in buffer A). Cap the column and rotate for 12 h at 4°.
11. Recover the eluted materials by centrifugation (5000 rpm for 1 min).

12. Analyze the protein by SDS-PAGE. Scan the CBB-stained gel by the LAS3000 system (Fuji) and quantify the protein band by Image Gauge software (Fuji).
13. Dilute the proteins to appropriate concentration and divide into small aliquots. Store at $-80°$.

Comments. Usually we obtained 600 μg of Rsp5 and 400 μg of WWHECT from 200 ml culture (Fig. 2A). The GST tag does not seem to affect the enzymatic activity of both Rsp5 and WWHECT. The purified Rsp5 tends to be easily aggregated at high concentration (above 1 mg/ml); this probably might be due to a property of the N-terminal C2 domain, because WWHECT does not have such a tendency.

Preparation of Polyubiquitinated Sic1^{PY}

Materials and Reagents. Five milligrams per milliliter bovine ubiquitin dissolved in buffer A (Sigma), 5 mg/ml methylated-ubiquitin dissolved in buffer A (Calbiochem), 400 ng/μl T7-Sic1-His$_6$ (Sic1), 400 ng/μl T7-Sic1^{PY}-His$_6$ (Sic1^{PY}), 450 ng/μl Uba1 (E1) (we obtained Uba1 by affinity purification from the yeast $UBA1^{3FLAG}$ strain [YYS41] [Fig. 2A] and stored it at $-80°$ [Saeki et al., 2004]), 1 μg/μl Ubc4 (E2) (we purified recombinant Ubc4 protein expressed in *E. coli* [Fig. 2A] and stored it at $-80°$ [Saeki et al., 2004]), 920 ng/μl Rsp5 or 680 ng/μl WWHECT (E3), HRP-conjugated mouse anti-T7 antibody (Novagen).

Solutions and Buffers. Buffer A (50 mM Tris-HCl, pH 7.5, 100 mM NaCl, and 10% glycerol), 5 × ATP solution (10 mM ATP, pH 7.5, 50 mM MgCl$_2$, 5 mM DTT in buffer A).

Procedure. The ubiquitination reaction was typically performed for 3 h at 25° in a 20-μl mixture containing buffer A, 2 mM ATP, 5 mM MgCl$_2$, 2 pmol Uba1, 60 pmol Ubc4, 10 pmol Rsp5 or WWHECT, 10 pmol Sic1^{PY}, and 1.2 nmol ubiquitin.

1. Prepare mix 1 consisting of the following reagents: 4.5 μl of buffer A, 2 μl of 5 × ATP solution, 0.5 μl of Uba1, 1 μl of Ubc4, 2 μl of ubiquitin. Preincubate for 5–10 min at 25°.
2. During the preincubation, prepare mix 2 consisting of the following reagents: 6 μl of buffer A, 2 μl of 5 × ATP solution, 1 μl of Rsp5 or WWHECT, 1 μl of Sic1^{PY}.
3. Combine mix 1 and mix 2. Incubate for 3–12 h at 25°.

Comments. The ubiquitination reaction was monitored by Western blotting with anti-T7 antibody (1:2000 dilution) using 2–4 μl of the reaction mixture. Products can be stored at 4° (for 2 days) or $-80°$. Usually we prepared several controls, ubiquitin-omitted reactions, using wild-type Sic1 instead of Sic1^{PY} and using methylated-ubiquitin instead of ubiquitin.

Isolation of the Yeast 26S Proteasome from S. cerevisiae

Our laboratory uses the following procedure for the preparation of the 26S proteasome by the affinity purification with a slight modification described by Verma *et al.* (2000). An integral lid subunit Rpn11 was tagged with the PreScission protease site and 3× FLAG epitope tag at its C terminus by chromosomal homologous recombination. The epitope tagging to Rpn11 does not seem to affect the structure and enzymatic activity of the 26S proteasome (Sone *et al.*, 2004).

Materials and Reagents

S. cerevisiae $RPN11^{3FLAG}$ strain (YYS40, *MATa rpn11::RPN11^{3FLAG}-HIS3 leu2 his3 ura3 trp1 ade2 can1 ssd1*), M2-agarose (Sigma), 3× FLAG peptide (Sigma), PreScission protease (Amersham), glutathione sepharose 4B (Amersham), MicroSpin Empty columns (Amersham), PreScission protease (Amersham), Multibeads shocker (MB501, YASUI KIKAI Corp.) glass beads (0.5-mm diameter).

Solutions and Buffers

Yeast nutrient-rich medium (YPD, 1% yeast extract, 2% polypepton, 2% glucose, 400 μg/ml adenine, 20 μg/ml uracil), buffer A (50 mM Tris-HCl, pH 7.5, 100 mM NaCl, and 10% glycerol), 50 × ATP regeneration system (50 × ARS, 0.5 mg/ml creatine phosphokinase, 0.5 M creatine phosphate dissolved in buffer A), buffer A' (buffer A supplemented with 2 mM ATP, 5 mM MgCl$_2$, 1 mM DTT), buffer A'' (buffer A supplemented with 4 mM ATP, 10 mM MgCl$_2$, 1 mM DTT, 2 × ARS).

Procedure

1. Grow the $RPN11^{3FLAG}$ strain in 1 l of YPD to mid-log phase.
2. Harvest cells by centrifugation (3000g, 10 min) and wash twice with 400 ml of dH$_2$O and once with 50 ml of buffer A. Adjust packed cell weight to 4 g and store the cells in a 50-ml conical tube at $-80°$ until use.
3. Suspend cells in 8 ml of buffer A'' Add glass beads to 70% of total volume. Lyse with a multibead shocker with following parameters: vortexing for 20 sec at 2500 rpm followed by chilling for 40 sec, total time 360 sec. Check the efficiency of disruption by microscopy. Usually, 85–95% of the cells are disrupted.

4. Pass the lysate through a 30-ml syringe attached with a 18-gauge needle to remove glass beads. Add 8 ml of buffer A' with 1 × ARS.
5. Clear the lysate by centrifugation (20,000g for 30 min) and recover the supernatant to a 15-ml tube.
6. Add 100 µl of M2-agarose to the supernatant. Rotate for 2 h at 4°.
7. Wash twice with 10 ml of buffer A', once with 5 ml of buffer A' containing 0.2% Triton X-100, and again once with 5 ml of buffer A'.
8. Transfer the beads to a MicroSpin Empty column. Wash twice with 400 µl buffer A' (centrifuge the column at 5000 rpm for 30 sec).
9. Add 300 µl of protease buffer (20 units of PreScission protease in buffer A). Cap the column and rotate for 12 h at 4°.
10. Recover the eluted materials by centrifugation (5000 rpm for 1 min): the eluted material should contain the tag-removed 26S proteasome and PreScission protease.
11. Add 20 µl of the glutathione sepharose preequilibrated with buffer A' to remove the PreScission protease that is GST-fused protein. Rotate for 2 h at 4°. Recover the flow through fraction by centrifugation (5000 rpm for 1 min).
12. Divide into small aliquots and store at −80°.
13. Analyze 10 µl of the purified protein by SDS-PAGE. Scan the CBB-stained gel by LAS3000 system (Fuji) and quantify the protein band by Image Gauge software (Fuji).

Alternately, 26S proteasome can be eluted by 3× FLAG peptide from the beads after step 8.

9' Incubate with 300 µl of 3× FLAG peptide (100 µg/ml) in buffer A' for 15 min at 25°.
10' Recover the eluted materials by centrifugation (5000 rpm for 1 min): the eluted material contains the 26S proteasome with tag and 3× FLAG peptide.
11' Dialyze against 100 ml of buffer A' overnight at 4°.

Comments

The 26S proteasomes obtained by either elution method seem to have an equal degradation activity at least against the ubiquitinated Sic1^{PY}. Also, 3× FLAG peptide does not affect the degradation activity. To quantify the 26S proteasome, we usually analyzed the protein band corresponding to Rpn11 and calculated the mol concentration of the 26S proteasome. To confirm the purified 26S proteasome does not contain excess free 19S RP, we evaluated the ratio of Pre10, a CP subunit, and Rpn12. If the value of Rpn12/Pre10 rate is near 1, then the condition of the

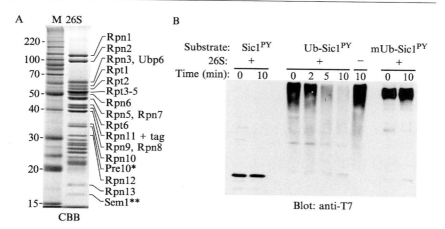

FIG. 3. *In vitro* degradation assay using the purified yeast 26S proteasome. (A) SDS-PAGE analysis of the affinity-purified 26S proteasome from $RPN11^{3FLAG}$ strain (YYS40). 3× FLAG peptide eluate was incubated with SDS-loading buffer at 37° for 1 h, then subjected to 12.5% SDS-PAGE and stained by CBB. The protein bands corresponding to all 19S subunits and Pre10 (*), a CP subunit, were indicated. **, When heating with SDS-loading buffer, the protein band of Sem1 was detected at a position of 19 kDa (confirmed by Western blotting with anti-Sem1 antibody). M, molecular mass standards (Invitrogen). (B) Degradation of polyubiquitinated $Sic1^{PY}$ and the 26S proteasome. The polyubiquitinated $Sic1^{PY}$ (Ub-$Sic1^{PY}$) (200 nM) was incubated with the 26S proteasome (100 nM) or mock purified sample for indicated times. Then, degradation of the ubiquitinated $Sic1^{PY}$ was analyzed by Western blotting with anti-T7 antibody. As a control, nonubiquitinated ($Sic1^{PY}$) or monoubiquitinated $Sic1^{PY}$ (mUb-$Sic1^{PY}$) produced by using methylated ubiquitin was used.

purified 26S proteasome is a doubly capped form. We usually obtained 300 pmol of the 26S proteasome (about 600 μg) in this procedure (Fig. 3A).

In Vitro Degradation Assay of the Polyubiquitinated Proteins by the Purified 26S Proteasome from *S. cerevisiae*

Materials and Reagents

1 μM 26S proteasome, 0.5 μM polyubiquitinated $Sic1^{PY}$, 0.5 μM $Sic1^{PY}$ (the ubiquitination reaction without ubiquitin).

Solutions and Buffers

5 × ATP solution (10 mM ATP, pH 7.5, 50 mM MgCl$_2$, 5 mM DTT in buffer A), buffer A (50 mM Tris-HCl, pH 7.5, 100 mM NaCl, and 10% glycerol).

Procedure

Degradation assay was typically performed in 10-μl mixtures containing buffer A, 2 mM ATP, 5 mM MgCl$_2$, 1 mM DTT, 2 pmol ubiquitinated-Sic1^{PY}, 1 pmol purified 26S proteasome per lane.

1. Mix the following reagents, 3 μl of buffer A, 2 μl of 5 × ATP solution, 1 μl of 26S proteasome. Preincubate for 1–5 min at 25°.
2. Add 4 μl of ubiquitinated Sic1^{PY} and mix gently.
3. Incubate for appropriate time (2–10 min) at 25°. Terminate the reaction by adding 5 μl of 3 × SDS-loading buffer and boil for 5 min.
4. Subject 5–15 μl of each sample to 10% SDS-PAGE. Transfer the proteins onto PVDF membrane completely using semidry system (1 h with a current of 4 mA/cm^2, 240 mA per mini-gel). Blot with anti-T7 antibody (1:2000 dilution). Analyze by the LAS3000 system (Fuji) and quantify the signals by Image Gauge software (Fuji) (Fig. 3B).

Comments

Because the preparation of the sample at time point 0 is technically difficult, we denatured the 26S proteasome by the addition of the SDS-loading buffer before mixing with the ubiquitinated Sic1^{PY}. To analyze the amount of the ubiquitinated Sic1 quantitatively, the exposure should be stopped before the signals are saturated. Apparently, the signals of ubiquitinated Sic1^{PY} are stronger than that of an unmodified one because of the difference in the antibody's accessibility. Note that Rsp5 is also highly polyubiquitinated itself, and the autoubiquitinated Rsp5 seem to bind with the 26S proteasome.

Discussion

We described a convenient method for *in vitro* ubiquitination designated the PY insertion method and demonstrated that polyubiquitinated Sic1^{PY} was degraded by the purified 26S proteasome. The insertion of the PY motif into the target proteins was essential for ubiquitination by Rsp5 or WWHECT (Fig. 2B and unpublished data). When we used methylated ubiquitin, monoubiquitinated Sic1^{PY} was detected at 150–250 kDa (Fig. 2B and C). Because Sic1 has 20 lysines, these data indicate that almost all lysines of Sic1 are ubiquitinated by long incubation. The polyubiquitinated Sic1^{PY} were rapidly degraded by the 26S proteasome, whereas the unmodified Sic1^{PY} and multiple monoubiquitinated Sic1^{PY} were not (Fig. 3). It

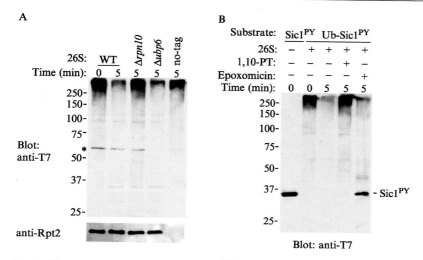

FIG. 4. Characterization of ubiquitinated Sic1^{PY}. (A) Degradation assay using proteasome mutant. Purified 26S proteasomes from the $\Delta rpn10$ or $\Delta ubp6$ mutants were incubated with ubiquitinated Sic1^{PY} for 5 min. The presence of an equal amount of the mutant 26S proteasomes was confirmed by Western blotting with anti-Rpt2 antibody (AFFINIT research) and Coomassie staining (data not shown). *, Sometimes, anti-T7 antibody seems to cross-react with Ubp6. (B) Degradation assay in the presence of inhibitors. 26S proteasomes were preincubated with 1% DMSO, 1 mM 1, 10-phenanthroline (1,10-PT), or 100 μM epoxomicin for 20 min at 25°. Then, the degradation assay was initiated by the addition of ubiquitinated Sic1^{PY}. The deubiquitinated Sic1^{PY} (recovery was ~36% input) was detected in the epoxomicin treatment.

has been thought that the K48-linked chain of at least four Ubs in length is a signal for the proteasome-mediated degradation (Thrower et al., 2000). So, we thought at first that Rsp5 might catalyze the formation of the K48-linked Ub chain, although Rsp5 is known to catalyze the K63-linked Ub chain in vivo (Galan and Haguenauer-Tsapis, 1997). To analyze which types of Ub chains are formed by Rsp5, we tested the ubiquitination assay using mutant Ubs with a single K and Sic1K36, Sic1 with a single K at a position of K36 residue. We found that Rsp5 is able to catalyze the K11-, K33-, K48-, and K63-Ub chains of at least two to more Ubs (data not shown). This suggests that the ubiquitinated Sic1^{PY} described here contains several types of the Ub-linkage. This unexpected result raises the possibility that the ubiquitinated Sic1^{PY} produced by the PY motif-insertion method is somewhat different from the SCF-catalyzed one. So, we carefully characterized the ubiquitinated Sic1^{PY}: (1) Rpn10 and Rad23, known as the ubiquitin receptors of the 26S proteasome (Verma et al., 2004), were able to bind efficiently with the Rsp5-catalyzed ubiquitinated Sic1^{PY} (data not shown); (2) the 26S proteasome from the deletion mutant of *RPN10* was not able to degrade the ubiquitinated Sic1^{PY} (Fig. 4A) as reported previously in the case of ubiquitinated Sic1 (Verma *et al.*, 2004); (3) the deubiquitination of the ubiquitinated Sic1^{PY} by the 26S proteasome seems to be facilitated by Rpn11 but not by Ubp6 (Fig. 4); and (4) the ubiquitinated Sic1^{PY} was not degraded but deubiquitinated by the 26S proteasome when treated with epoxomicin, a strong proteasome inhibitor (Fig. 4B), as reported previously (Verma *et al.*, 2002). Thus, the behavior of the Rsp5-catalyzed ubiquitinated Sic1^{PY} seems to be the same as that of the SCF-catalyzed one so far tested.

The advantage of the PY motif-insertion method is that any protein can be used for substrate to be ubiquitinated. We have already succeeded in the ubiquitination of several proteins, including green fluorescent protein, by using this method. Our preliminary data suggest that the 26S proteasome cannot degrade all polyubiquitinated proteins. The degradability of each protein is likely to depend on its structural property itself as reviewed in Pickart and Cohen (2004). As mentioned previously, the PY motif-insertion method is a novel and useful tool for analyzing the mechanism of the ubiquitin-dependent proteolysis.

Acknowledgments

We thank Dr. R. Deshaies for providing the pET-Sic1K36 plasmid, Dr. J. Jantti for anti-Sem1 antibody, and Dr. K. Tanaka for pGEX-UbK0, UbK29, K48, and K63 plasmids. This work was supported by a grant-in-aid for scientific research from the Ministry of Education, Science, Sports, Culture, and Technology of Japan (MEXT). Y. S. is a recipient of Research Fellowship of Science for Young Scientists (JSPS).

References

Dunn, R., Klos, D. A., Adler, A. S., and Hicke, L. (2004). The C2 domain of the Rsp5 ubiquitin ligase binds membrane phosphoinositides and directs ubiquitination of endosomal cargo. *J. Cell Biol.* **165,** 135–144.

Galan, J. M., and Haguenauer-Tsapis, R. (1997). Ubiquitin lys63 is involved in ubiquitination of a yeast plasma membrane protein. *EMBO J.* **16,** 5847–5854.

Hershko, A., and Ciechanover, A. (1998). The ubiquitin system. *Annu. Rev. Biochem.* **67,** 425–479.

Pickart, C. M. (2001). Mechanisms underlying ubiquitination. *Annu. Rev. Biochem.* **70,** 503–533.

Pickart, C. M., and Cohen, R. E. (2004). Proteasomes and their kin: proteases in the machine age. *Nat. Rev. Mol. Cell Biol.* **5,** 177–187.

Saeki, Y., Tayama, Y., Toh-e, A., and Yokosawa, H. (2004). Definitive evidence for Ufd2-catalyzed elongation of the ubiquitin chain through Lys48 linkage. *Biochem. Biophys. Res. Commun.* **320,** 840–845.

Sone, T., Saeki, Y., Toh-e, A., and Yokosawa, H. (2004). Sem1p is a novel subunit of the 26 S proteasome from *Saccharomyces cerevisiae. J. Biol. Chem.* **279,** 28807–28816.

Sudol, M., and Hunter, T. (2000). NeW wrinkles for an old domain. *Cell* **103,** 1001–1004.

Thrower, J. S., Hoffman, L., Rechsteiner, M., and Pickart, C. M. (2000). Recognition of the polyubiquitin proteolytic signal. *EMBO J.* **19,** 94–102.

Verma, R., Aravind, L., Oania, R., McDonald, W. H., Yates, J. R., 3rd, Koonin, E. V., and Deshaies, R. J. (2002). Role of Rpn11 metalloprotease in deubiquitination and degradation by the 26S proteasome. *Science* **298,** 611–615.

Verma, R., Chen, S., Feldman, R., Schieltz, D., Yates, J., Dohmen, J., and Deshaies, R. J. (2000). Proteasomal proteomics: identification of nucleotide-sensitive proteasome-interacting proteins by mass spectrometric analysis of affinity-purified proteasomes. *Mol. Biol. Cell* **11,** 3425–3439.

Verma, R., McDonald, H., Yates, J. R., 3rd, and Deshaies, R. J. (2001). Selective degradation of ubiquitinated Sic1 by purified 26S proteasome yields active S phase cyclin-Cdk. *Mol. Cell* **8,** 439–448.

Verma, R., Oania, R., Graumann, J., and Deshaies, R. J. (2004). Multiubiquitin chain receptors define a layer of substrate selectivity in the ubiquitin-proteasome system. *Cell* **118,** 99–110.

[15] Large- and Small-Scale Purification of Mammalian 26S Proteasomes

By Yuko Hirano, Shigeo Murata, and Keiji Tanaka

Abstract

The 26S proteasome is an ATP-dependent protease known to collaborate with ubiquitin, whose polymerization acts as a marker for regulated and enforced destruction of unnecessary proteins in eukaryotic cells. It is an unusually large multi-subunit protein complex, consisting of a central catalytic machine (called the *20S proteasome* or *CP/core particle*) and two terminal regulatory subcomplexes, termed *PA700* or *RP/regulatory particle,* that are attached to both ends of the central portion in opposite orientations to form an enzymatically active proteasome. To date, proteolysis driven by the ubiquitin-proteasome system has been shown to be involved in a diverse array of biologically important processes, such as the cell cycle, immune response, signaling cascades, and developmental programs; and the field continues to expand rapidly. Whereas the proteasome complex has been highly conserved during evolution because of its fundamental roles in cells, it has also acquired considerable diversity in multicellular organisms, particularly in mammals, such as immunoproteasomes, PA28, S5b, and various alternative splicing forms of S5a (Rpm 10). However, the details of the ultimate pathophysiological roles of mammalian proteasomes have remained elusive. This article focuses on methods for assay and purification of 26S proteasomes from mammalian cells and tissues.

Introduction

The 26S proteasome is a protein-destroying apparatus capable of degrading a variety of cellular proteins in a rapid and timely fashion. Most, if not all, substrates are modified by ubiquitin before their degradation by the 26S proteasome. The covalent attachment of multiple ubiquitins on target proteins is catalyzed by a multienzyme cascade, consisting of the E1 (Ub-activating), E2 (Ub-conjugating), and E3 (Ub-ligating) enzymes (Hershko and Ciechanover, 1998; Pickart, 2001). The resulting polyubiquitin chain serves as a signal for trapping the target protein, and, consequently, the substrate is destroyed after proteolytic attack by the 26S proteasome (Baumeister *et al.*, 1998; Coux *et al.*, 1996). The 26S proteasome is a dumbbell-shaped particle, consisting of a centrally located, cylindrical 20S proteasome (alias core particle, CP) that functions as a catalytic machine and two large terminal PA700 modules (alias 19S complex, or regulatory particle, RP) attached to the 20S core particle in opposite orientations.

The 20S proteasome/CP is a complex with a sedimentation coefficient of 20S and a molecular mass of approximately 750 kDa (see a model of Fig. 1). It is a barrel-like particle formed by the axial stacking of four rings made up of two outer α-rings and two inner β-rings, which are each made up of seven structurally similar α- and β-subunits, respectively, being associated in the order of $\alpha_{1-7}\beta_{1-7}\beta_{1-7}\alpha_{1-7}$. The overall architectures of the highly ordered structures of yeast (*Saccharomyces cerevisiae*) and mammalian (bovine) 20S proteasomes are indistinguishable, as demonstrated by x-ray crystallography (Groll *et al.*, 1997; Unno *et al.*, 2002). Three of the β-type subunits of each inner ring have catalytically active threonine residues at their N-terminus, all of which show N-terminal nucleophile (Ntn) hydrolase activity, indicating that the proteasome is a novel threonine protease, differing from the known protease families categorized as seryl-, thiol-, carboxyl-, and metalloproteases. The catalytic $\beta1$, $\beta2$, and $\beta5$ subunits correspond to caspase-like/PGPH (peptidylglutamyl-peptide hydrolyzing), trypsin-like, and chymotrypsin-like activities, respectively, which are capable of cleaving peptide bonds at the C-terminal side of acidic, basic, and hydrophobic amino acid residues, respectively. Two copies of these three active sites face the interior of the cylinder and reside in a chamber formed by the centers of the abutting β rings.

PA700/RP contains approximately 20 distinct subunits of 25–110 kDa, which can be classified into two subgroups: a subgroup of six ATPases,

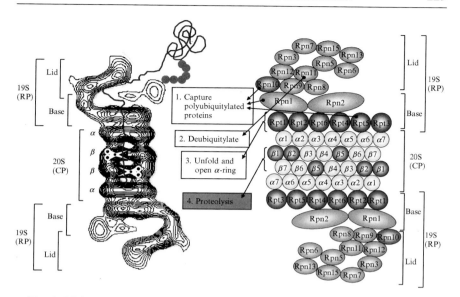

Fig. 1. Molecular organization of 26S proteasomes. (Left panel) Averaged image of the 26S proteasome complex of rat based on electron micrographs. The α and β rings of the 20S proteasome are indicated. Photograph kindly provided by W. Baumeister. (Right panel) Schematic drawing of the subunit structure. CP, core particle (alias 20S proteasome); RP, 19S regulatory particle (alias PA700) consisting of the base and lid subcomplexes; Rpn, RP non-ATPase; Rpt, RP triple–ATPase. Note that relative positions of 19S subunits have not been established. (See color insert.)

numbered from Rpt1 to Rpt6 (i.e., <u>R</u>P <u>t</u>riple ATPases 1–6), that are structurally similar and have been highly conserved during evolution, and a subgroup of more than 15 heterogeneous subunits, numbered from Rpn1 to Rpn15 (i.e., <u>R</u>P <u>n</u>on-ATPases 1–15), that are structurally unrelated to the members of the ATPase family (Tanaka *et al.*, 2005). The PA700/RP structurally consists of two subcomplexes, known as "base" and "lid," which, in the 26S proteasome, correspond to the portions of PA700 proximal and distal, respectively, to the 20S proteasome (Glickman *et al.*, 1998). The base is made up of six ATPases (Rpt1–Rpt6) and two large regulatory components Rpn1 and Rpn2, functioning as presumptive receptor(s) of ubiquitin-like proteins, and the lid contains multiple non-ATPase subunits (Rpn3–Rpn15). The base-complex is thought to bind in an ATP-dependent manner to the outer α-ring of the central 20S proteasome. The six ATPases in this base-complex are assembled into one ring complex. The main role of the ATPase ring is to supply energy continuously for the degradation of target proteins. In fact, the metabolic energy liberated by ATP consumption is probably used for unfolding target proteins, gate opening of the 20S proteasome, and substrate translocation so that they can penetrate into the channel formed by the α- and β-rings of the 20S proteasome (Ogura and Tanaka, 2003). On the other hand, the lid-complex is thought to be involved in the recognition of polyubiquitylated target proteins, deubiquitylation of substrates for reutilization of ubiquitin, and physical interactions with various other proteins that influence proteasome activity. The details of molecular bases for functions of individual subunits, however, are largely unknown to date.

Assay of Proteasome Activity

Assay of Suc-LLVY-MCA Degrading Activity

The 26S proteasome is incubated at 37° for 10 min in 50 mM Tris-HCl buffer (pH 8.5) containing 1 mM dithiothreitol (DTT) and a 0.1 mM concentration of a fluorogenic substrate, the synthetic peptide succinyl-Leu-Leu-Val-Tyr-4-methyl-coumaryl-7-amide (Suc-LLVY-MCA) (Peptide Institute). This substrate is added to the assay mixture at a final concentration of DMSO of 1% (v/v). The reaction was stopped by adding 10% SDS at final concentration of 1%, and the reaction product is measured fluorometrically (excitation 380 nm, emission 460 nm).

The 26S proteasome can be visualized on electrophoretic gels as a Suc-LLVY-MCA-degrading enzyme. Samples are subjected to nondenaturing polyacrylamide gel electrophoresis (PAGE) at 4° before the gels are overlaid with 0.1 mM Suc-LLVY-MCA for 10 min at room temperature. Fluorescence was then detected under ultraviolet light.

Assay of Polyubiquitylated ^{125}I-Lysozyme Degrading Activity

Preparation of polyubiquitylated ^{125}I-lysozyme can be prepared by using purified E1, E2, and E3 enzymes, as described previously (Fujimuro et al., 1994; Tamura et al., 1991). For assay of degradation of polyubiquitylated ^{125}I-lysozyme, samples of ^{125}I-lysozyme-ubiquitin conjugates (5000–10,000 cpm) are incubated at 37° for 15–60 min in a total volume of 100 μl of reaction mixture containing 50 mM Tris-HCl buffer (pH 8.5) with 5 mM MgCl$_2$, 2 mM ATP, an ATP-regeneration system (10 μg of creatine phosphokinase and 10 mM phosphocreatine), 1 mM DTT, and a suitable amount of the 26S proteasome. After the reaction is stopped by adding SDS-PAGE sample buffer, the proteins are subjected to

SDS-PAGE and autoradiographed. The gels are dried and exposed to x-ray film at $-70°$ with an intensifying screen. For measuring the degradation of ^{125}I-lysozyme-ubiquitin conjugates into acid-soluble fragments by the 26S proteasome, the reaction is terminated by addition of 575 μl of 10% trichloroacetic acid (TCA) with 125 μl of 4% bovine serum albumin (BSA) as a carrier, and the radioactivity recovered in the acid-soluble fraction after centrifugation is determined in a γ-counter.

Assay of Polyubiquitylated Sic1 Degrading Activity

Deshaies and his colleagues devised an *in vitro* assay method of 26S proteasomes using polyubiquitylated Sic1, a CDK inhibitor in the budding yeast, as a substrate. Polyubiquitylation of Sic1 phosphorylated by CDK is catalyzed by E1, E2 (Cdc34), and E3 (SCF^{Cdc4}). The details of the methods were described previously (Verma and Deshaies, 2005; Verma et al., 2001).

Saeki *et al.* devised an improved method by preparing PY motif-inserted Sic1 (Sic1^{PY}) that is effectively polyubiquitylated by Rsp5 E3-ligase and rapidly degraded by 26S proteasomes in an ATP-dependent fashion (for details, see Chapter 14 [Saeki et al., 2005]). It is of note that all components used in this assay system can be easily expressed and purified using bacterial cells.

Assay of ^{35}S-ODC Degrading Activity

For quantitative and sensitive measurement of ATP-dependent proteolysis activity of mammalian proteasomes *in vitro*, ornithine decarboxylase (ODC) is a useful substrate. ODC is the best-known natural substrate of the proteasome whose recognition and degradation are independent of ubiquitylation (Murakami *et al.*, 1992). Antizyme (AZ), an ODC inhibitory protein that is needed for this *in vitro* degradation assay, is prepared as a recombinant protein (Murakami *et al.*, 1999). Rat AZ cDNA Z1 is expressed in *Escherichia coli*, and an extract of the *E. coli* (800 mg protein) is applied to a monoclonal anti-AZ antibody (HZ-2E9)-AffiGel 10 column (1 ml); the column is washed with 25 mM Tris-HCl buffer (pH 7.5) containing 1 mM EDTA, 1 mM DTT, and 0.01% Tween 80, supplemented with 4 M NaCl. AZ is eluted with 4 ml of 3 M MgCl$_2$, and the eluate is dialyzed against the same buffer. ^{35}S-labeled ODC is produced by an *in vitro* translation system using rabbit reticulocyte lysate containing rat ODC mRNA, ^{35}S-labeled methionine, and ^{35}S-labeled cysteine (Du Pont NEN). The reaction is applied to a monoclonal anti-ODC antibody (HO101)-AffiGel 10 column (0.15 ml). The procedures for wash and elution are the same as AZ purification.

The degradation of the recombinant ^{35}S-labeled-ODC (2000–3000 cpm) is assayed in the presence of ATP, an ATP-regenerating system, and AZ (Murakami *et al.*, 1999). After incubation for 60 min at 37°, the amount of TCA-soluble radioactivity in the reaction mixture is measured, and the activity is expressed as a percent of total ODC added.

Comments for Assays

1. Suc-LLVY-MCA (i.e., a substrate of chymotrypsin-like activity) is recommended as a sensitive substrate. Various other fluorogenic peptides, such as Boc (*t*-Butyloxycarbonyl)-Leu-Arg-Arg-MCA and Z (benzyloxycarbonyl)-Leu-Leu-Glu-MCA for monitoring trypsin-like and caspase-like/PGPH (peptidylglutamyl-peptide hydrolyzing) activity, respectively, are suitable for measurement of 20S and 26S proteasomal activity, because proteasomes show broad substrate specificity. The hydrolytic activities toward various fluorogenic substrates are determined by measuring the fluorescence of groups liberated from these peptides. Latent 20S proteasomes can be activated in various ways. We recommend the use of SDS at low concentrations of 0.02–0.08% for activation of Suc-LLVY-MCA breakdown; the optimal concentration depends on the enzyme source and the protein concentration used. The fluorogenic peptide (e.g., Suc-LLVY-MCA) can be used for assay of the 26S proteasome, because it is active without any treatment unlike the latent 20S proteasome. MCA (4-methyl-coumaryl-7-amide) is used as a reference compound for analysis with peptidyl-MCAs.

2. Various fluorogenic peptides are suitable for measurement of 20S and 26S proteasomal activity, but note that all of them are not specific substrates for these proteasomes. For specific assay, ATP-dependent degradation of polyubiquitinated ^{125}I-lysozyme, or Sic1/Sic1^{PY} should be measured, although such assay is not easy, because three kinds of enzymes, E1, E2, and E3, must be purified for *in vitro* preparation of ubiquitinated substrates. Therefore, for quantitative and sensitive measurement of ATP-dependent proteolysis activity of mammalian proteasomes *in vitro*, ODC is a useful substrate. Note that AZ is not present in lower organisms such as yeasts, and thus this assay is not fit for proteasomes isolated from these cells.

3. The purification of the 26S proteasome is monitored by measuring ATPase activity at later steps of its purification, because the 26S proteasome has intrinsic ATPase activity (Ugai *et al.*, 1993). Note that this assay is not sensitive and cannot be used in crude extracts because of the existence of numerous other ATPases in cells.

Large-Scale Purification of 20S and 26S Proteasomes from Rat Liver

Purification Procedure of 20S Proteasomes

Step 1. Homogenize 200–400 g samples of rat liver in 3 vol of 25 mM Tris-HCl buffer (pH 7.5) containing 1 mM DTT and 0.25 M sucrose in a Potter-Elvehjem homogenizer. Centrifuge the homogenate for 1 h at 70,100g, and use the resulting supernatant as the crude extract.

Step 2. Add glycerol at a final concentration of 20% to the crude extract. Then mix the extract with 500 g of Q-Sepharose (Amersham) that has been equilibrated with buffer A (25 mM Tris-HCl [pH 7.5] containing 1 mM DTT [or 10 mM 2-mercaptoethanol] and 20% glycerol). Wash the Q-Sepharose with the buffer A on a Büchner funnel and transfer to a column (5 × 60 cm). Wash the column with buffer A and elute the material with 2 liters of a linear gradient of 0–0.8 M NaCl in buffer A, and measure the activity of proteasomes using Suc-LLVY-MCA as a substrate.

Step 3. Pool fractions containing 20S proteasomes from the Q-Sepharose column and add 50% polyethylene glycol 6000 (Sigma) (adjust to pH 7.4) to a final concentration of 15% with gentle stirring. After 15 min, centrifuge the mixture at 10,000g for 20 min, dissolve the resulting pellet in a minimum volume (approximately 50 ml) of buffer A, and centrifuge at 20,000g for 10 min to remove insoluble material.

Step 4. Fractionate the material precipitated with polyethylene glycol on a Bio-Gel A-1.5m column (5 × 90 cm) in buffer A. Collect fractions of 10 ml and assay their proteasome activity. Pool fractions of 20S proteasomes.

Step 5. Apply the active fractions from the Bio-Gel A-1.5m (Bio-Rad) column directly to a column of hydroxylapatite equilibrated with buffer B (10 mM phosphate buffer [pH 6.8] containing 1 mM DTT and 20% glycerol). Wash the column with the same buffer and elute the material with 400 ml of a linear gradient of 10–300 mM phosphate. Collect fractions of 4 ml. 20S proteasomes are eluted with approximately 150 mM phosphate.

Step 6. Combine the active fractions from the hydroxylapatite (Bio-Rad), dialyze against buffer A, and apply to a column of heparin-Sepharose CL-6B (Amersham) equilibrated with buffer A. Wash the column with the same buffer until the absorbance of the eluate at 280 nm returns to baseline. Then elute with 200 ml of a linear gradient of 0–0.4 M NaCl in buffer A, and collect fractions of 2 ml. 20 S proteasomes are eluted with approximately 75 mM NaCl.

Step 7. Pool the fractions with high proteasomal activity, dialyze against buffer A, and concentrate to about 5 mg/ml protein by ultrafiltration in an Amicon cell with a PM-10 membrane (Millipore). The enzyme can be stored at −80° for at least 2–3 years. The SDS-PAGE analysis of purified enzyme revealed that it consists of a set of proteins, displaying the molecular weights of 20–32 kDa (see left panel of Fig. 2).

Purification Procedure of 26 Proteasomes

Step 1. Homogenize 200–400-g samples of rat liver in 3 vol of 25 mM Tris-HCl buffer (pH 7.5) containing 1 mM DTT, 2 mM ATP, and 0.25 M sucrose in a Potter-Elvehjem homogenizer. Centrifuge the homogenate for 1 h at 70,100g and use the resulting supernatant as the starting material.

Step 2. Recentrifuge the crude supernatant for 5 h at 70,100g to obtain 26S proteasomes, which precipitate almost completely. Dissolve the precipitate in a suitable volume (40–50 ml) of buffer C (buffer A containing 0.5 mM ATP) and centrifuge at 20,000g for 30 min to remove insoluble material.

Step 3. Apply samples of the preparation from step 2 to a Bio-Gel A-1.5m column (5 × 90 cm) in buffer C. Collect fractions of 10 ml and assay the 26S proteasome activity in the fractions. Pool fractions of 26S proteasomes.

Step 4. Add ATP at a final concentration of 5 mM to the pooled fractions of 26S proteasomes from the Bio-Gel A-1.5m column. Apply the sample directly to a hydroxylapatite column with a 50-ml bed volume that has been equilibrated with buffer D (buffer B containing 5 mM ATP). Recover the 26S proteasome in the flow-through fraction, because they do

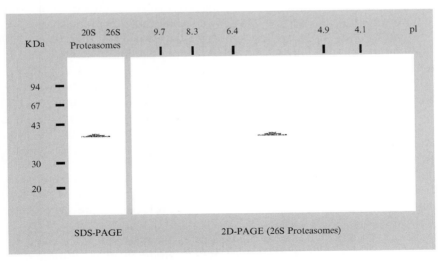

FIG. 2. Electrophoretic analyses of 20S and 26S proteasomes from rat liver. (Left panel) SDS-PAGE pattern of purified 20S and 26S proteasomes. (Right panel) 2D-PAGE pattern of purified 26S proteasomes. Proteins were stained with Coomassie Brilliant Blue (CBB).

not associate with this column in the presence of 5 mM ATP. Approximately 70% of the proteins, including free 20S proteasomes, bind to the hydroxylapatite resin.

Step 5. Apply the flow-through fraction from the hydroxylapatite column to a Q-Sepharose column that has been equilibrated with buffer C without ATP and washed with 1 bed volume of buffer C. Wash the column with 5 bed volumes of buffer C, and elute the adsorbed materials with 300 ml of a linear gradient of 0–0.8 M NaCl in buffer C. Collect fractions of 3.0 ml. Proteins with ability to degrade Suc-LLVY-MCA with or without 0.05% SDS are eluted with approximately 0.4 M NaCl as a single symmetrical peak. ATPase activity and the ATP-dependent activity necessary to degrade ^{125}I-lysozyme-Ub conjugates are observed at the same position as the peptidase activity and are eluted as superimposable symmetrical peaks, which suggests a specific association of ATPase with the 26S proteasome complex. Collect the protein in fractions exhibiting high activity.

Step 6. Concentrate the 26S proteasome fraction obtained by Q-Sepharose chromatography to 2.0 mg/ml by ultrafiltration with an Amicon PM-30 membrane, and subject samples of 2.0 mg of protein to 10–40% glycerol density-gradient centrifugation (30 ml in buffer C containing 2 mM ATP). Centrifuge for 22 h at 82,200g in a SW rotor, and collect fractions of 1 ml from the bottom of the centrifuge tube. A single major peak of peptidase activity, measured in the absence of SDS, is eluted around fraction 15, but when the activity is assayed with 0.05% SDS, another small peak is observed around fraction 20. The latter peak corresponds to the elution position of 20S proteasomes. ATPase activity is observed at the same position as peptidase activity. Activity for ATP-dependent degradation of ^{125}I-lysozyme-Ub conjugates is also observed as a single symmetrical peak, coinciding in position with the ATPase and peptidase activities in the absence of SDS. No significant ^{125}I-lysozyme-Ub conjugate degrading activity is detected in fractions of 20S proteasomes. Pool fractions 12–16 and store at −80°. Two-dimensional (2D) PAGE revealed that the purified enzyme consists of a set of approximately 40 proteins displaying the molecular weights of 20–110 kDa and isoelectric points (pIs) of 3–10 (see right panel of Fig. 2).

Small-Scale Purification of 26S Proteasomes

Conventional Chromatographic Purification of Nuclear 26S Proteasomes

Preparation of Nuclear Extracts. The nuclei from rat liver were prepared as described previously (Tanaka *et al.*, 1989).

Step 1. Homogenize animal tissues (mouse or rat) (50 g) in 4 volumes (200 ml) of 50 mM Tris-HCl (pH 8.0) buffer containing 1 mM DTT, 15 mM KCl, 1 mM EDTA, 5 % glycerol, 2.2 M sucrose, and Complete protease inhibitor cocktail (Roche Molecular Biochemical). The resulting homogenates are layered on a cushion of 50 mM Tris-HCl (pH 8.0) buffer containing 1 mM DTT, 15 mM KCl, 1 mM EDTA, 10% glycerol, and 2 M sucrose occupying one third the volume of centrifuge tubes and are centrifuged at 83,000g for 60 min in aSW rotor to pellet the nuclei.

Step 2. Disrupt the isolated nuclei by sonication in 50 mM Tris-HCl (pH 8.0) buffer containing 1 mM DTT, 2 mM ATP, and Complete protease inhibitor cocktail. The nuclear extracts were obtained by centrifugation at 10,000g for 20 min as the resulting supernatants (approximately 40 mg). The purity of the nuclear extracts should be examined by Western blot analysis. Histone H1, a marker of nucleus (detected with antibodies from Upstate Biotechnology), but not LDH, a marker of cytosol (detected with antibodies from Abcam), should be detected in the nuclear extracts without obvious cross-contamination.

Purification of Nuclear 26S Proteasomes

Step 1. Load the nuclear extracts on a RESOURCE Q column (Amersham Biosciences) equilibrated with buffer E (50 mM Tris-HCl [pH 8.0] buffer containing 1 mM DTT, 2 mM ATP, and 10% glycerol), wash the column with buffer E, and elute bound proteins with a gradient of 0–0.8 M NaCl in buffer E. Pool the fractions with Suc-LLVY-MCA degrading activity. 26S proteasomes are eluted with 450–500 mM NaCl.

Step 2. Add ATP at a final concentration of 5 mM to the pooled fractions of 26S proteasomes from RESOURCE Q column. Load the fractions on a Hydroxylapatite column (Bio-Rad) equilibrated with buffer D. Recover 26S proteasomes in the flow-through fractions. (Check Suc-LLVY-MCA degrading activity).

Step 3. Load the flow-through fractions on a Mono Q column (Amersham Biosciences) equilibrated with buffer E, wash the column with buffer E, and elute bound proteins with a gradient of 0–0.8 M NaCl in buffer E (0.5 ml/fraction). Pool the fraction exhibiting peak activity and the adjacent fractions. 26S proteasomes are eluted with 450–500 mM NaCl. This step helps to concentrate 26S proteasomes for the next step.

Step 4. Subject the pooled fractions (approximately 2.0 mg protein/ 1.5 ml) to 10–40% glycerol density-gradient centrifugation (30 ml in buffer F [50 mM Tris-HCl {pH 8.0} buffer containing 1 mM DTT and 2 mM ATP]). Centrifuge for 22 h at 82,200g in SW28 (Beckman) or P28S (HITACHI) rotor, collect fractions of 1 ml from the top of the centrifuge tube, and check

Suc-LLVY-MCA degrading activity. A single major peak of peptidase activity, measured in the absence of SDS, corresponds to 26S proteasomes sedimented around fraction 20 (approximately 0.1 mg protein). Pool fractions with high Suc-LLVY-MCA degrading activity and store at −80°.

Affinity Purification

Conventional biochemical techniques for purification of 26S proteasomes use chromatographic columns as described previously. During the purification steps, 26S proteasomes are exposed to high ionic strength buffers, which cause dissociation of proteins bound to proteasomes transiently or with low affinity. In yeast, tagging of certain subunits of 26S proteasomes that are driven by their own promoters and purification by the tag in milder conditions has enabled identification of many novel proteasome-interacting proteins (PIPs). Mammalian proteasomes are expected to have a more complicated network and it is essential to clarify mammalian PIPs to fully understand the roles of proteasomes. To solve this problem, we developed an ES cell line that has one allele of the human Rpn11 gene tagged with a C-terminal flag epitope (Rpn11$^{FLAG/+}$ ES cells) by a homologous recombination technique. The method for establishing the ES cell line will be described elsewhere.

Procedure

Step 1. Grow Rpn11$^{FLAG/+}$ ES cells on six 10-cm dishes on which mitomycin C–treated murine embryonic fibroblasts were laid.

Step 2. Collect cells using an appropriate scraper with PBS in a conical tube, centrifuge at 1500g for 10 min. Wash cells once more with PBS.

Step 3. The cell pellet was resuspended in 6 ml of buffer G (20 mM HEPES-NaOH [pH 7.5], 0.2% NP-40, 2 mM ATP, 1 mM DTT) by gentle pipetting and placed on ice for 10 min.

Step 4. Centrifuge at 10,000g for 10 min to remove cell debris.

Step 5. To preclear the lysate, pass the lysate through a column packed with 0.5 ml (bed volume) of Sepharose CL-4B (Sigma).

Step 6. Apply the flow-through onto the column packed with 50 μl (bed volume) of M2-agarose (Sigma). Pass the flow-thorough through the column five times.

Step 7. Wash the column 10 times with 5 ml of buffer G supplemented with 50 mM NaCl.

Step 8. Incubate the column with 50 μl of FLAG peptide (Sigma; dissolved at 100 μg/ml in buffer G) on ice for 3 min.

Step 9. Recover the eluted proteins by centrifugation at 1000 rpm for 1 min.

Step 10. Repeat step 8 and step 9 three more times, and collect all the eluted materials in one tube. We usually obtained about 60 μg of 26S proteasome in this procedure. The 2D PAGE pattern of 26S proteasomes purified by this method is shown in Fig. 3.

Discussion

Proteasomes have been purified from a variety of eukaryotic cells by many investigators. Many purification methods have been reported, but no special techniques are necessary, because 20S proteasomes are very stable and abundant in cells, constituting 0.5–1.0% of the total cellular proteins. The procedures used for purification of 20S proteasomes obviously differ, depending on whether they are small or large operations. For their isolation from small amounts of biological materials, such as cultured cells, 10–40% glycerol density gradient centrifugation is very effective. 20S proteasomes are present in a latent form in cells and can be isolated in this form in the presence of 20% glycerol. For their isolation in high yield, a key point is to keep them in their latent form, because their activation results in autolytic loss of a certain subunit(s) and marked reduction of enzymatic activities, particularly their hydrolysis of various proteins. Accordingly, all buffers used contain 10–20% glycerol as a stabilizer. Furthermore, a reducing agent is required, because 20S proteasomes precipitate in its absence.

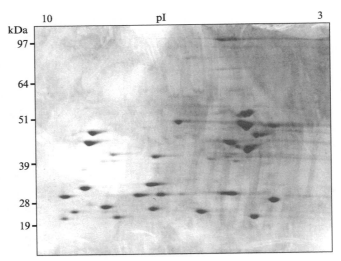

FIG. 3. Two-dimensional PAGE pattern of 26S proteasomes purified from Rpn11$^{FLAG/+}$ ES cells. Proteins were stained with CBB.

All purification procedures are performed at 4°, but operations in a high-performance liquid chromatography (HPLC) apparatus can be carried out within a few hours at room temperature.

For purification of the 26S proteasome, ATP (0.5 mM or 2 mM) together with 20% glycerol and 1 mM DTT should be added to all solutions used, because they strongly stabilize the 26S proteasome complex: the purified enzyme is stable during storage at $-70°$ for at least 6 months in the presence of 2 mM ATP and 20% glycerol. Chromatographic steps that require high salt concentrations or extremes of pH should be avoided, because these operations may result in dissociation of the 26S complex into its constituents.

References

Baumeister, W., Walz, J., Zuhl, F., and Seemuller, E. (1998). The proteasome: Paradigm of a self-compartmentalizing protease. *Cell* **92**, 367–380.

Coux, O., Tanaka, K., and Goldberg, A. L. (1996). Structure and functions of the 20S and 26S proteasomes. *Annu. Rev. Biochem.* **65**, 801–847.

Fujimuro, M., Sawada, H., and Yokosawa, H. (1994). Production and characterization of monoclonal antibodies specific to multi-ubiquitin chains of polyubiquitinated proteins. *FEBS Lett.* **349**, 173–180.

Glickman, M. H., Rubin, D. M., Coux, O., Wefes, I., Pfeifer, G., Cjeka, Z., Baumeister, W., Fried, V. A., and Finley, D. (1998). A subcomplex of the proteasome regulatory particle required for ubiquitin-conjugate degradation and related to the COP9-signalosome and eIF3. *Cell* **94**, 615–623.

Groll, M., Ditzel, L., Lowe, J., Stock, D., Bochtler, M., Bartunik, H. D., and Huber, R. (1997). Structure of 20S proteasome from yeast at 2.4 A resolution. *Nature* **386**, 463–471.

Hershko, A., and Ciechanover, A. (1998). The ubiquitin system. *Annu. Rev. Biochem.* **67**, 425–479.

Murakami, Y., Matsufuji, S., Hayashi, S. I., Tanahashi, N., and Tanaka, K. (1999). ATP-dependent inactivation and sequestration of ornithine decarboxylase by the 26S proteasome are prerequisites for degradation. *Mol. Cell Biol.* **19**, 7216–7227.

Murakami, Y., Matsufuji, S., Kameji, T., Hayashi, S., Igarashi, K., Tamura, T., Tanaka, K., and Ichihara, A. (1992). Ornithine decarboxylase is degraded by the 26S proteasome without ubiquitination. *Nature* **360**, 597–599.

Ogura, T., and Tanaka, K. (2003). Dissecting various ATP-dependent steps involved in proteasomal degradation. *Mol. Cell* **11**, 3–5.

Pickart, C. M. (2001). Ubiquitin enters the new millennium. *Mol. Cell* **8**, 499–504.

Saeki, Y., Isono, E., and Toh-e, A. (2005). Preparation of ubiquitinated substrates by the PY motif-insertion method for monitoring 26S proteasome activity. *Methods Enzymol.* **399**, 215–227.

Tamura, T., Tanaka, K., Tanahashi, N., and Ichihara, A. (1991). Improved method for preparation of ubiquitin-ligated lysozyme as substrate of ATP-dependent proteolysis. *FEBS Lett.* **292**, 154–158.

Tanaka, K., Kumatori, A., Ii, K., and Ichihara, A. (1989). Direct evidence for nuclear and cytoplasmic colocalization of proteasomes (multiprotease complexes) in liver. *J. Cell Physiol.* **139**, 34–41.

Tanaka, K., Yashiroda, H., and Murata, S. (2005). Ubiquitin and diversity of the proteasome system. *In* "Protein Degradation" (R. J. Mayer, A. Ciechanover, and M. Rechsteiner, eds.). Wiley-VCH Verlag, Weinheim (in press).

Ugai, S., Tamura, T., Tanahashi, N., Takai, S., Komi, N., Chung, C. H., Tanaka, K., and Ichihara, A. (1993). Purification and characterization of the 26S proteasome complex catalyzing ATP-dependent breakdown of ubiquitin-ligated proteins from rat liver. *J. Biochem. (Tokyo)* **113,** 754–768.

Unno, M., Mizushima, T., Morimoto, Y., Tomisugi, Y., Tanaka, K., Yasuoka, N., and Tsukihara, T. (2002). The structure of the mammalian 20S proteasome at 2.75 A resolution. *Structure (Camb).* **10,** 609–618.

Verma, R., and Deshaies, R. J. (2005). Assaying degradation and deubiquitination of a ubiquitinated substrate by purified 26S proteasomes. *Methods Enzymol.* **398,** 391–399.

Verma, R., McDonald, H., Yates, J. R., 3rd, and Deshaies, R. J. (2001). Selective degradation of ubiquitinated Sic1 by purified 26S proteasome yields active S phase cyclin-Cdk. *Mol. Cell* **8,** 439–448.

Section IV

Identification and Characterization of Substrates and Ubiquitin Ligases

[16] Is This Protein Ubiquitinated?

By PETER KAISER and CHRISTIAN TAGWERKER

Abstract

Covalent modification of proteins with ubiquitin plays an important role in a wide array of cellular processes (Hershko and Ciechanover, 1998; Pickart, 2004). For this reason an increasing number of investigators in diverse research fields are confronted with the question whether their favorite proteins are ubiquitinated. Experiments to demonstrate covalent modification with ubiquitin *in vivo* can be quite challenging because of low steady-state levels of the ubiquitinated forms caused by degradation by the 26S proteasome and/or highly active deubiquitinating enzymes (Dubs) that remove the ubiquitin units (Pickart and Cohen, 2004; Wilkinson and Hochstrasser, 1998). Several different methods to determine whether a particular protein is ubiquitinated have been developed (Beers and Callis, 1993; Ellison and Hochstrasser, 1991; Hochstrasser *et al.*, 1991; Treier *et al.*, 1994). Some of these assays were described in detail by Laney and Hochstrasser (2002). This chapter is focused on one experimental approach using expression of hexahistidine-tagged ubiquitin. It can be applied to most situations in which one suspects ubiquitination of a particular protein and has been successfully used in various organisms (Kaiser *et al.*, 2000; Treier *et al.*, 1994).

Identification of Ubiquitinated Proteins by Expression of 6xHis-Ubiquitin

The basic strategy is outlined in Fig. 1. Amino-terminal hexahistidine-tagged ubiquitin (6xHis-Ubi) is expressed in cells. 6xHis-Ubi is efficiently conjugated to ubiquitinated proteins, which allows purification by Ni^{2+}-chelate affinity chromatography. This purified pool of ubiquitinated proteins can be analyzed for the presence of the protein of interest by immunoblotting. This strategy has several advantages over other approaches, mainly because lysate preparation and purification are done under highly denaturing conditions (8 M urea or 6 M guanidinium), which limits Dub activity and, therefore, preserves the ubiquitination status during the entire procedure. In addition, expression of variants of 6xHis-Ubi can be helpful. For example, 6xHis-Ubi^{G76A} protects from

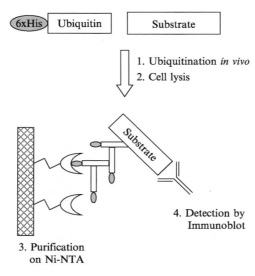

FIG. 1. Schematic representation of the steps involved in analyzing the ubiquitination status of proteins.

deubiquitination *in vivo* and can thus increase the fraction of the ubiquitinated form of the protein of interest (Hodgins *et al.*, 1992). Similarly, the chain terminator 6xHis-Ubi^{K48R} can shorten K48-linked ubiquitin chains so that proteasome targeting is reduced, which increases the likelihood of detection of the ubiquitinated forms of the protein of interest (Willems *et al.*, 1996). Likewise, treatment of cells with proteasome inhibitors (e.g., MG132, Calbiochem, San Diego, CA) or the use of conditional proteasome mutants (e.g., the yeast *pre1-1 pre4-1* double mutant) (Hilt *et al.*, 1993) increases the steady-state levels of the ubiquitinated forms of many proteins (Fig. 2A). If proteasome inhibitors are used in yeast, it is important to use a *pdr5Δ* mutant, because it is more permeable to these drugs (Fleming *et al.*, 2002).

Plasmids for overexpression of 6xHis-Ubi in yeast (Flick *et al.*, 2004; Ling *et al.*, 2000) and mammalian cells (Kaiser, *unpublished*; Treier *et al.*, 1994) are available and can be easily generated for other organisms. It is important to carry out control experiments using cells expressing untagged ubiquitin instead of 6xHis-Ubi to verify the specificity of the assay. The following protocol describes the procedure for budding yeast using a 6xHis-Ubi construct under control of the inducible *CUP1* promoter, but it can be readily adapted to other expression constructs or other organisms.

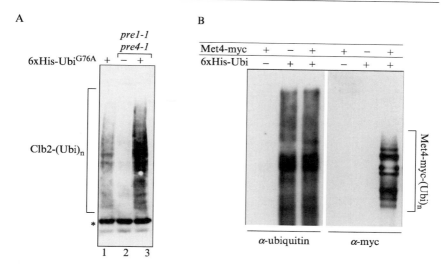

Fig. 2. (A) Effects of proteasome mutants on ubiquitin conjugates: Wild-type cells (lane 1) or cells carrying temperature-sensitive alleles of the 20S proteasome subunits *PRE1* and *PRE4* (lanes 3) expressing 6xHis-Ubi^{G76A} were analyzed. As a control, cells expressing untagged Ubi^{G76A} (lane 2) were processed in parallel. All cells expressed a HA3-tagged version of the cyclin Clb2. Cells were grown at 37° for 4 h to inactivate proteasome function and processed as described in this chapter. Ubiquitinated Clb2 was detected by immunoblotting with a monoclonal anti-HA antibody (clone 16B12, Covance, Denver, PA). The asterisk indicates a cross-reacting protein that copurified on Ni-NTA. Inactivation of proteasome function significantly increased the fraction of ubiquitinated Clb2 (compare lanes 1 and 3). (B) Analysis of ubiquitination of the yeast transcription factor Met4: Cells expressing 6xHis-Ubi ("+") or untagged Ubi ("−") and myc-epitope tagged Met4 as indicated were analyzed as described in this chapter. The immunoblot shown was analyzed with antibodies against ubiquitin (Zymed Laboratories, San Francisco, CA) and a monoclonal anti-myc antibody (clone 9E10, Covance, Denver, PA).

Preparation of Cell Lysates

1. Grow 50 ml yeast cells expressing the 6xHis-Ubi to an $A_{600} = 0.3$, add CuSO$_4$ to a final concentration of 100 μM to induce expression of *CUP1*-6xHis-Ubi and continue incubation for 4 h at 30°. Collect cells by centrifugation (2000*g*, 3 min, 4°) or filtration. Expression conditions are the same for cells grown in minimal or rich media. However, we noticed that *CUP1*-6xHis–Ubi constructs express higher levels when cells are grown in synthetic complete media (SC) (Guthrie and Fink, 1991) compared with rich media (YEPD). Therefore, we recommend the use of SC media to increase 6xHis-ubiquitin levels.

2. Wash the cell pellet in 1 ml lysis buffer (50 mM sodium-phosphate buffer, pH 8.0, 8 M urea, 300 mM NaCl, 0.5% Triton X-100, 1 mM PMSF, *optional:* 10 mM N-ethylmaleimide [NEM, Sigma, St. Louis, MO]) and transfer to a 1.5-ml tube. Because of dissociation of urea, the pH of the lysis buffer should be readjusted immediately before use with HCl or NaOH as needed. If NEM is included, prepare a 0.5 M or 1 M solution in 100% ethanol immediately before it is added, because it is unstable. NEM alkylates sulfhydryl groups and can, therefore, block Dub activity.

3. Spin cells for 10 sec at 14,000g, remove the supernatant, and quickly freeze the pellet and store at $-80°$.

4. Break the cells with glass beads (0.5-mm size, Biosspec Products Inc., Bartlesville, OK) in 400 μl lysis buffer. Very reliable results are achieved when cells are broken with a FastPrep FP120 (Qbiogene, Carlsbad, CA) at setting 4.5 for 120 sec at 4°. Alternatively, cells can be vortexed with glass beads for 5 min at 4°. To separate the lysate from the glass beads, invert the tube, and with a 21-gauge needle poke a hole at the bottom of the tube. Insert the tube with the hole into a fresh tube and spin carefully for 30 sec at 800g. The lysate will pass through the hole in the upper tube and is collected in the lower tube, whereas the glass beads remain in the upper tube.

5. Add 600 μl lysis buffer, vortex, and centrifuge the lysate to remove cell debris (30 min, 16,000g, 20°). Transfer the clarified supernatant to a fresh tube. One can expect to obtain between 1.5 and 2.5 mg of total protein in the supernatant.

Purification of the Ubiquitinated Protein of Interest

All steps are carried out at room temperature. We do not observe any increase in proteolysis or deubiquitination at room temperature as long as highly denaturing conditions are maintained.

6. Take 100 μl of Ni-NTA agarose (50% slurry, 50-μl bead volume) (Qiagen, Germantown, MD) for each sample, let the beads settle by gravity or centrifuge 2 min at 100g, and remove the supernatant by aspiration with a 27-gauge needle. Wash beads once with 300 μl lysis buffer.

7. Add the Ni-NTA agarose to the lysate and bind overnight at room temperature with gentle mixing.

8. Let the beads settle by gravity for about 5 min and carefully remove the supernatant by aspiration with a 27-gauge needle. Wash the beads twice with 1 ml lysis buffer for 1 min and three times with 1 ml wash buffer (identical to lysis buffer but pH 6.3 and 10 mM imidazole). Adjust the pH of the wash buffer immediately before use. Between the wash steps, let the beads settle by gravity or centrifuge 2 min at 100g and remove the supernatant by aspiration with a 27-gauge needle.

9. Elute twice by incubation for 5 min with 100 µl elution buffer (50 mM sodium-phosphate buffer, 100 mM Tris-HCl, pH 4.3, 8 M urea, 50 mM NaCl, 2% SDS, 10 mM EDTA). Adjust pH immediately before use with HCl or NaOH as needed.

10. Pool the two eluted fractions and add 200 µl 2× PAGE loading buffer (2×: 4% SDS, 125 mM Tris-HCl, pH 6.8, 20% glycerol, 200 mM DTT, 0.002% bromophenol blue) and analyze by Western blotting with an antibody against the protein of interest. Do not boil samples containing urea, because this induces carbamylation, which can interfere with antibody binding. Depending on the protein being analyzed, immunoblots show either smears of immunoreactivity (Fig. 2A) or discrete bands that form a ladder (Fig. 2B). The eluted fractions can also be analyzed with antiubiquitin antibodies (Zymed Laboratories, San Francisco, CA) to verify successful purification of ubiquitinated proteins (Fig. 2B).

Concluding Remarks

Generally, only a small fraction of the protein of interest is in the ubiquitinated form, and the detected signals are relatively weak and often appear as high molecular weight smears. To ensure that the detected signals reflect antibody reactivity with the protein of interest and are not caused by antibody artifacts, one should include controls to confirm the identity of the protein detected in the immunoblot. For experiments with epitope-tagged proteins, a strain lacking the epitope tag can serve as an effective control (Fig. 2B). When specific antibodies directed against the protein of interest are used, a strain carrying a deletion, or in case of essential genes a temperature-sensitive allele, of this protein of interest should be included as a control. The use of epitope tags on target proteins can lead to artifacts. In some cases, addition of tags can induce ubiquitination and proteasomal degradation (Schauber et al., 1998). It is very difficult to recognize such effects of epitope tags, particularly when no phenotype is associated with the tagging. If available, the use of specific antibodies against the protein of interest prevents this problem. Overexpression of the protein of interest should be avoided if possible, because it can lead to artifactual ubiquitination. This is most likely caused by ubiquitination of a misfolded fraction of the overexpressed protein.

Acknowledgments

We thank Rob Steele for critical comments on the manuscript. P. Kaiser acknowledges support from NIH, the California Breast Cancer Research Program, and the Cancer Research Coordinating Committee. C. Tagwerker is a DOC-fellow of the Austrian Academy of Sciences.

References

Beers, E. P., and Callis, J. (1993). Utility of polyhistidine-tagged ubiquitin in the purification of ubiquitin-protein conjugates and as an affinity ligand for the purification of ubiquitin-specific hydrolases. *J. Biol. Chem.* **268,** 21645–21649.

Ellison, M. J., and Hochstrasser, M. (1991). Epitope-tagged ubiquitin. *J. Biol. Chem.* **266,** 21150–21157.

Fleming, J. A., Lightcap, E. S., Sadis, S., Thoroddsen, V., Bulawa, C. E., and Blackman, R. K. (2002). Complementary whole-genome technologies reveal the cellular response to proteasome inhibition by PS-341. *Proc. Natl. Acad. Sci. USA* **99,** 1461–1466.

Flick, K., Ouni, I., Wohlschlegel, J. A., Capati, C., McDonald, W. H., Yates, J. R., and Kaiser, P. (2004). Proteolysis-independent regulation of the transcription factor Met4 by a single Lys 48-linked ubiquitin chain. *Nat. Cell Biol.* **6,** 634–641.

Guthrie, C., and Fink, G. R. (1991). "Guide to Yeast Genetics and Molecular Biology." Academic Press Inc., San Diego.

Hershko, A., and Ciechanover, A. (1998). The ubiquitin system. *Annu. Rev. Biochem.* **67,** 425–479.

Hilt, W., Enenkel, C., Gruhler, A., Singer, T., and Wolf, D. H. (1993). The PRE4 gene codes for a subunit of the yeast proteasome necessary for peptidylglutamyl-peptide-hydrolyzing activity. Mutations link the proteasome to stress- and ubiquitin-dependent proteolysis. *J. Biol. Chem.* **268,** 3479–3486.

Hodgins, R. R., Ellison, K. S., and Ellison, M. J. (1992). Expression of a ubiquitin derivative that conjugates to protein irreversibly produces phenotypes consistent with a ubiquitin deficiency. *J. Biol. Chem.* **267,** 8807–8812.

Hochstrasser, M., Ellison, M. J., Chau, V., and Varshavsky, A. (1991). The short lived MATα2transcriptional regulator is ubiquitinated *in vivo*. *Proc. Natl. Acad. Sci. USA* **88,** 4606–4610.

Kaiser, P., Flick, K., Wittenberg, C., and Reed, S. I. (2000). Regulation of transcription by ubiquitination without proteolysis: Cdc34/SCF(Met30)-mediated inactivation of the transcription factor Met4. *Cell* **102,** 303–314.

Laney, J. D., and Hochstrasser, M. (2002). Assaying protein ubiquitination in *Saccharomyces cerevisiae*. In "Methods in Enzymology" (C. Guthrie and G. R. Fink, eds.), Vol. 351, pp. 248–257. Academic Press, New York.

Ling, R., Colon, E., Dahmus, M. E., and Callis, J. (2000). Histidine-tagged ubiquitin substitutes for wild-type ubiquitin in *Saccharomyces cerevisiae* and facilitates isolation and identification of *in vivo* substrates of the ubiquitin pathway. *Anal. Biochem.* **282,** 54–64.

Pickart, C. M. (2004). Back to the future with ubiquitin. *Cell* **116,** 181–190.

Pickart, C. M., and Cohen, R. E. (2004). Proteasomes and their kin: Proteases in the machine age. *Nat. Rev. Mol. Cell Biol.* **5,** 177–187.

Schauber, C., Chen, L., Tongaonkar, P., Vega, I., Lambertson, D., Potts, W., and Madura, K. (1998). Rad23 links DNA repair to the ubiquitin/proteasome pathway. *Nature* **391,** 715–718.

Treier, M., Staszewski, L. M., and Bohman, D. (1994). Ubiquitin-dependent c-Jun degradation *in vivo* is mediated by the d domain. *Cell* **78,** 787–798.

Wilkinson, K. D., and Hochstrasser, M. (1998). The deubiquitinating enzymes. In "Ubiquitin and the Biology of the Cell" (J. M. Peters, J. R. Harris, and D. Finley, eds.), pp. 99–125. Plenum Press, New York.

Willems, A. R., Lanker, S., Patton, E. E., Craig, K. L., Nason, T. F., Mathias, N., Kobayashi, R., Wittenberg, C., and Tyers, M. (1996). Cdc53 targets phosphorylated G1 cyclins for degradation by the ubiquitin proteolytic pathway. *Cell* **86,** 453–463.

[17] Experimental Tests to Definitively Determine Ubiquitylation of a Substrate

By JOANNA BLOOM and MICHELE PAGANO

Abstract

Ubiquitin-mediated proteolysis is a major pathway of protein degradation that regulates numerous cellular processes. An understanding of the circumstances that contribute to the ubiquitylation of a specific protein can yield vast insight into its regulation. This article examines multiple procedures that explain whether a protein is ubiquitylated and suggests methods to investigate the factors that specifically target the substrate for ubiquitylation, as well as the site of ubiquitin conjugation.

Introduction

It is becoming increasingly evident that most cellular proteins are regulated by ubiquitylation. The attachment of a ubiquitin chain to a target protein typically targets it for degradation by the multicatalytic protease, the 26S proteasome reviewed in Hershko and Ciechanover (1998). Other cellular systems exploit the ubiquitin pathway for nondegradative pathways, including internalization of membrane receptors and histone modification (Pickart, 2004). As a result, the ability to assess ubiquitylation of a protein allows one to examine the kinetics of protein degradation or other cellular processes that use ubiquitin ligation. This chapter analyzes methods for establishing whether a protein is ubiquitylated both *in vivo* and *in vitro*, as well as protocols to determine whether ubiquitylation occurs on a lysine residue or the N-terminal α-amino group. The methods described are not limited to proteins that are modified with K48-linked ubiquitin chains, which are associated with proteasomal degradation, but are also useful for K63-linked chains, which are associated with endocytosis, and K11- or K29-linked chains, whose functions are less well characterized (Pickart, 2001). In addition, with the appropriate modifications, these methods can be used to investigate whether a protein is covalently conjugated to ubiquitin-like proteins, such as SUMO and NEDD8 (Cope and Deshaies 2004; Pan *et al.*, 2004; Seeler and Dejean 2003).

Confirmation of Substrate Ubiquitylation In Vivo

General Considerations

A clear way to establish whether a protein is a substrate of the ubiquitin system is to evaluate its ubiquitylation status *in vivo*. This can be achieved by various methods, which involve direct Western blot of a total cell lysate containing the protein of interest, as an easy way to initially detect the presence of ubiquitylated species, and then ultimately purification of the ubiquitylated protein followed by Western blot analysis to identify the ubiquitylated species. The methods described in the following make use of an overexpressed protein to analyze ubiquitylation. This helps to ascertain the appropriate conditions to observe ubiquitylation; however, to conclusively establish that a protein is subject to the ubiquitin pathway, ubiquitylated forms of the endogenous protein should be seen. Therefore, once it appears that a protein is ubiquitylated after its overexpression, the techniques described should be adjusted to observe the endogenous version.

Several factors must be taken into consideration when performing these types of experiments. For one, many ubiquitylated proteins are rapidly turned over by proteasomes, making this transient event difficult to detect. It is useful to treat cells with an inhibitor of the proteasome before analysis to preserve the ubiquitylated forms. Generally, treating cells with the proteasome inhibitor MG132 at a final concentration of 5–10 μM for 2–4 h is sufficient to block the catalytic activity of the proteasome without resulting in toxicity for the cells. An additional factor to consider is the presence of deubiquitylating enzymes (isopeptidases) in the cell extract, which can remove the ubiquitin molecules from the protein of interest, thereby preventing detection of the ubiquitylated species. These problems can be avoided by preparing the extraction buffer with N-ethylmaleimide (NEM), which blocks the critical cysteine residue present in the active site of most deubiquitylating enzymes. Another important aspect is the conditions under which a protein is ubiquitylated and degraded. For example, in the case of IκBα, cytokine treatment is required to induce efficient ubiquitylation, whereas in untreated cells IκBα remains stable (Yaron *et al.*, 1998). In the case of Skp2 and Cks1, ubiquitylation is limited to the early G1 phase of the cell cycle, rendering ubiquitylated forms difficult to detect if asynchronous cells are used (Bashir *et al.*, 2004; Wei *et al.*, 2004). Therefore, it is critical that the conditions that promote substrate degradation be established if ubiquitylated forms are to be detected. Another factor to consider is the abundance of the protein of interest. In some cases, the ubiquitylation of an endogenous protein can be assessed, whereas in others, overexpression of the target protein is required. Last, but not least, it is necessary to denature the extract before purification of the target protein to demonstrate that the protein of interest is itself ubiquitylated and not merely binding to additional copurified proteins that are ubiquitylated.

Materials and Reagents. 293T cells from American Type Culture Collection, vectors encoding His-tagged ubiquitin, and the protein of interest for expression in mammalian cells. MG132 (Boston Biochem), Tris-HCL, sodium chloride, Triton X-100 (Roche), EDTA, sodium fluoride, dithiothreitol (DTT), sodium vanadate, phenyl-methyl sulfonyl fluoride (PMSF), leupeptin, soybean trypsin inhibitor, L-1 Chlor-3-(4-tosylamido)-4 phenyl-2-butanon (TPCK), L-1 chlor-3-(4-tosylamido)-7-amino-2- heptanon-hydrochloride (TLCK), aprotinin, N-ethylmaleimide (NEM), sodium dodecyl sulfate (SDS), nickel-NTA agarose (Qiagen), guanidine-HCl, and beta-mercaptoethanol.

Buffers

LYSIS BUFFER. 50 mM Tris-HCl, pH 7.4, 0.25 M NaCl, 0.1% Triton X-100, 1 mM EDTA, 50 mM NaF, 1 mM DTT, 0.1 mM Na$_3$VO$_4$. The following protease inhibitors should also be added to prevent protein degradation: 0.1 mM PMSF (from an 10-mg/ml stock solution dissolved in isopropyl alcohol and stored at $-20°$), 1 μg/ml leupeptin (from an 10-mg/ml stock solution dissolved in isopropyl alcohol and stored at $-20°$), 10 μg/ml soybean trypsin inhibitor (from a 10-mg/ml stock solution dissolved in PBS and stored at $-20°$), 10 μg/ml TPCK (from a 10-mg/ml stock solution dissolved in ethanol and stored at $-20°$), 5 μg/ml TLCK (from a 5-mg/ml stock solution dissolved in PBS and stored at $-20°$), and 1 μg/ml aprotinin (from a 1-mg/ml stock solution dissolved in PBS and stored at $-20°$).

WESTERN BLOT DENATURING BUFFER. 6 M guanidine-HCl, 20 mM Tris-HCl, pH 7.5, 5 mM β-mercaptoethanol, 1 mM PMSF (from an 18-mg/ml stock solution dissolved in isopropyl alcohol and stored at $-20°$).

Analysis of Ubiquitylated Proteins by Direct Western Blot

An initial way to determine whether a protein of interest is ubiquitylated is by direct immunoblot analysis. If the protein is ubiquitylated, antibodies to the protein should recognize bands that migrate slower than the potential substrate. If these bands represent Lys48-linked ubiquitylated species (which target substrates to the proteasome), their intensity should increase when the cells are pretreated with proteasome inhibitors. Direct immunoblot analysis can also be applied to determine whether inhibiting polyubiquitin chain elongation prevents the appearance or changes the pattern of bands that migrate slower than the potential substrate. This can be verified by expressing a ubiquitin "lysineless" mutant in which all of the lysines are mutated to arginines [Ub(K0)]. Ub(K0) terminates ubiquitin chains, and its overexpression stabilizes ubiquitylated proteins by blocking polyubiquitylation, thus preventing their recognition by 26S proteasomes. Because ubiquitin is a very abundant protein, high levels of Ub(K0) must be expressed to overcome the effects of endogenous ubiquitin. In fact, the Ub(K0) mutant has a dose-dependent effect on the polyubiquitylation and steady-state levels of substrates of the ubiquitin system (Bloom *et al.*, 2003). If a protein is ubiquitylated, coexpressing Ub(K0) should alter the pattern of slower migrating bands (i.e., yield less ubiquitylated forms and/or shorter chains) observed by direct Western blot analysis. A final note: for obvious reasons, direct immunoblot analysis can only be performed using antibodies with little or no cross-reactivity with unspecific proteins contained in the total cell extract.

Protocol

1. Cotransfect 293T cells (in duplicate) with a vector encoding the protein of interest in the absence or presence of a vector encoding Ub(K0). Use as much Ub(KO) DNA as possible encoding the protein of interest to compete with endogenous ubiquitin. Transfections can be carried out in 293T cells using the standard calcium phosphate method (Rolfe, 1995).

2. Approximately 24 h after transfection, add MG132 at a final concentration of 5–10 μM (from a 10-mM stock solution dissolved in DMSO and stored at $-20°$) to one set of transfected cells such that experimental conditions include $+/-$ Ub(K0) and $+/-$ MG132.

3. After incubation with MG132 for 2–4 h, wash the cells two times with PBS and then prepare protein extracts by resuspending the cell pellet in 3–5 volumes of lysis buffer. In addition, the lysis buffer should include 2 mM NEM to block the activity of isopeptidases and deubiquitylating enzymes. NEM should be freshly dissolved at a stock concentration of 100 mM.

4. Resolve proteins on an SDS-PAGE and transfer to a nitrocellulose membrane.

5. Analyze the pattern of bands in the absence or presence of Ub(K0) and MG132 by immunoblotting with an antibody to the protein of interest.

Isolation of Histidine-Tagged Ubiquitylated Proteins

If slower migrating bands are observed in the presence of MG132 and this pattern changes with overexpression of Ub(K0), it is likely that the protein of interest is ubiquitylated. This can be more formally assessed by affinity purification of the bulk of ubiquitylated proteins followed by Western blot analysis of the protein of interest. Overexpression of ubiquitin with a histidine (His) tag allows for the purification of ubiquitylated proteins using nickel agarose (Treier et al., 1994). Purified ubiquitinylated proteins are then resolved by SDS-PAGE and probed with an antibody to the protein of interest. If the protein is, indeed, conjugated to ubiqutin, a ladder of bands with sizes corresponding to ubiquitylated forms of the protein should be observed (Fig. 1). Tags other than His, such as HA or Myc, are useful for marking the putative substrate, because they are easily detectable by Western blot analysis using commercial antibodies. Moreover, the ability to recognize ubiquitylated forms with antibodies against different substrate epitopes (i.e., using antibodies to the protein as well as to the peptide tag) is convincing evidence that the potential substrate is ubiquitylated. This experiment can also be performed with His-tagged Ub

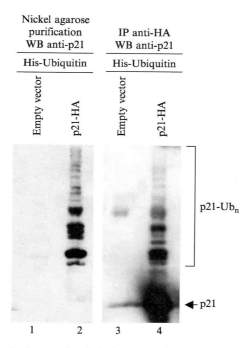

FIG. 1. Analysis of substrate ubiquitylation *in vivo* (using p21 as an example). 293T cells were transfected with His-tagged ubiquitin in the presence of an empty vector (lanes 1 and 3) or p21 tagged at the C-terminus with HA (lanes 2 and 4). Cell extracts were subjected to nickel agarose purification (lanes 1 and 2) or immunoprecipitated with a polyclonal antibody to HA (lanes 3 and 4). Immunoprecipitates were analyzed by immunoblotting with an antibody to p21.

(K0) substituted for His-ubiquitin to determine whether the pattern of chains is altered when polyubiquitylation is inhibited.

Protocolr

1. Cotransfect 293T cells with a vector encoding the protein of interest and a vector encoding His-ubiquitin. Use four to five times the amount of DNA encoding His-ubiquitin (or His-Ub[K0]) as the DNA encoding the protein of interest to compete with endogenous ubiquitin. Transfections can be carried out in 293T cells using the standard calcium phosphate method.

2. Twenty-four hours after transfection, add MG132 at a final concentration of 5–10 μM (from a 10-mM stock solution dissolved in DMSO and stored at $-20°$) to the transfected cells.

3. After incubation with MG132 for 2–4 h, wash the cells two times with PBS and then prepare extract by resuspending the cell pellet in 3–5 volumes of lysis buffer. In addition, the lysis buffer should include 2 mM NEM to block the activity of isopeptidases and deubiquitylating enzymes. NEM should be freshly dissolved at a stock concentration of 100 mM.

4. Add 50 μl of a 1:1 slurry of nickel-NTA agarose prewashed two times with lysis buffer to the cell lysates diluted with lysis buffer to 0.5–1 ml. Rotate for 2 h at 4°.

5. Wash the nickel-NTA agarose four times with lysis buffer and then resuspend the dry beads in 25 μl 2× sample buffer.

6. Resolve proteins by SDS-PAGE and transfer to a nitrocellulose membrane.

7. Analyze the pattern of bands by immunoblotting with an antibody to the protein of interest.

Analysis of Ubiquitin Chains after Immunoprecipitation of a Target Protein

Ubiquitylation can be confirmed by performing a complementary experiment, that is, purifying the target protein and detecting ubiquitylated species by Western blot analysis with an antibody to ubiquitin. If an epitope tag is added to ubiquitin for detection, HA or Myc is recommended rather than a His tag, because commercial antibodies to the His tag are generally not effective for Western blot analysis. In this case, overexpression of both the target protein and ubiquitin by transient transfection will greatly increase the ability to detect these higher molecular weight forms. The purified protein is then subjected to SDS-PAGE and probed with an antibody specific for ubiquitin chains. Before blotting, the membrane should be denatured to better expose the ubiquitin epitopes. If the protein is, indeed, conjugated to ubiquitin, a higher molecular weight ladder or smear of bands should be observed (Fig. 1). This experiment can also be performed with Ub(K0) substituted for wild-type ubiquitin to determine whether the pattern of chains is altered when polyubiquitylation is inhibited.

Protocol

1. Cell transfection, MG132 treatment, and extract preparation can be performed according to steps 1–3 in the protocol for "Method for the isolation of histidine-tagged ubiquitylated proteins" (above).
2. Denature the extract by adding 1% SDS to the lysate in a final volume of 0.1 ml and boiling for 10 min. To quench and dilute the SDS, lysates should then be incubated with 0.1 ml of 10% Triton X-100 and 0.8 ml of lysis buffer on ice for 30 min.
3. Add antibody to the protein of interest. Rotate for 2 h at 4°.
4. Add 50 μl of a 1:1 slurry of protein G-agarose or protein A-agarose (depending on whether a monoclonal or polyclonal antibody is used) prewashed two times with lysis buffer. Rotate for 30 min at 4°.
5. Wash the protein G- or protein A-agarose pellets four times with lysis buffer and then resuspend the dry beads in 25 μl 2× sample buffer.
6. Resolve immunoprecipitated proteins by SDS-PAGE and transfer to a nitrocellulose membrane.
7. Denature nitrocellulose membrane by incubating in Western blot denaturing buffer for 30 min at 4° or by autoclaving.
8. Wash membrane several times with PBS before Western blotting with an antibody to ubiquitin, such as monoclonal anti-ubiquitin from Zymed Laboratories (catalogue #13–1600).

In Vitro Ubiquitylation

General Considerations

In vitro ubiquitylation of a substrate is a useful method to confirm the ubiquitylation observed *in vivo*. Moreover, *in vitro* ubiquitylation can be a convenient tool to dissect the pathway by which a particular protein is ubiquitylated. Because posttranslational modifications and/or binding partners may be essential for the ubiquitylation to occur, this *in vitro* system allows one to determine which factors must be added for efficient ubiquitylation. Once the components of the pathway that ubiquitylates a protein have been established, a purified system containing only recombinant enzymes, substrate, ubiquitin, and an ATP-regenerating system can be used for *in vitro* ubiquitylation. Such a system will demonstrate the specificity of the reaction. This assay can be performed using either an *in vitro* translated protein or a purified protein as a substrate. Rabbit reticulocyte lysate, which is used for *in vitro* protein translation, is an established source of ubiquitylating enzymes (Hershko, 1988). As a result, some ubiquitylation, either specific or nonspecific, can occur. An ATP-regenerating system is included in the reaction to sustain the activity of ubiquitylating enzymes. It is important to note, however, that ubiquitylation of the putative substrate will often require posttranslational modifications that do not occur in rabbit reticulocyte lysate or may require a specific ubiquitin ligase or the binding to other factors that are not present in the lysate. In such instances, additional purified proteins can be added to the reaction to allow ubiquitylation to proceed *in vitro*. For instance, some background ubiquitylation is observed for *in vitro* translated p27. Specific *in vitro* ubiquitylation of p27 is observed only with the addition of suitable extract; extracts from cells in S phase or G2/M generate abundant ubiquitylated forms of *in vitro* translated p27, whereas extract from cells in G1 does not cause ubiquitylation greater than the background (Montagnoli *et al.*, 1999). If cyclin E/Cdk2, the binding partners or p27, and Skp2 and Cks1, components of the ligase that ubiquitylate p27, are added as purified proteins to the G1 extract, specific ubiquitylation of p27 can now be observed (Carrano *et al.*, 1999; Ganoth *et al.*, 2001; Montagnoli *et al.*, 1999). Moreover, when S/G2/M extract is immunodepleted of Cdks (Montagnoli *et al.*, 1999) or Skp2 (Carrano *et al.*, 1999), the ubiquitylation of *in vitro* translated p27 returns to background levels. Once the specific ubiquitin ligase and the other factors necessary for p27 ubiquitin ligation were identified, *in vitro* ubiquitylation of purified recombinant p27 could be achieved using only purified proteins (Ganoth *et al.*, 2001).

Although rabbit reticulocyte lysate also contains 26S proteasomes, the rate of ubiquitylation exceeds the rate of degradation in this system because of "clogging" of the endogenous proteasomes by hemin contained in the reticulocyte lysate. This allows for the accumulation of ubiquitylated forms of the substrate, such that proteasome inhibitors are not required in the reaction. Proteasomes may be further inhibited by adding adenosine-5′-(γ-thio)-triphosphate (ATP-γ-S), a nonhydrolyzable ATP analog, which does not affect the activity of ubiquitylating enzymes but cannot be used by proteasomes (Pagano *et al.*, 1995). Alternately, proteasomes can be removed from rabbit reticulocyte lysate before ubiquitylation reactions by ultracentrifugation at 100,000g for 6 h in the presence of 5 mM MgCl$_2$ (Pagano *et al.*, 1995). Ubiquitin aldehyde should be added to prevent the deubiquitylating activity of endogenous isopeptidases.

Methylated ubiquitin, which is chemically modified to block all of its free amino groups, is a useful tool for *in vitro* ubiquitylation reactions, because it prevents the formation of polyubiquitin chains. Methylated ubiquitin will alter the pattern of ubiquitylation resulting in the appearance of lower molecular weight ubiquitylated species. Because these lower molecular weight ubiquitylated forms are not efficiently recognized by proteasomes, substrates that are conjugated with methylated ubiquitin display increased stability.

Materials and Reagents. TNT-coupled *in vitro* transcription/translation kit (Promega), Tris-HCl, magnesium chloride, ATP, phosphocreatine, creatine phosphokinase, ubiquitin aldehyde (Boston Biochem), and methylated ubiquitin (Boston Biochem).

Buffers

5× PROTEIN BREAKDOWN MIX (PBDM). 0.25 M Tris-HCl, pH 7.5, 25 mM MgCl$_2$, 10 mM ATP, 50 mM phosphocreatine, and 17.5 U/ml creatine phosphokinase.

Ubiquitylation of a ^{35}S-Labeled Substrate In Vitro

In vitro reactions using an *in vitro* translated substrate that is labeled with ^{35}S-methionine is a common method to measure the kinetics of ubiquitylation. When the substrate is incubated with an ATP-regenerating system, enzymes that are present in the rabbit reticulocyte lysate could give some background ubiquitylation of the target protein. Addition of crude extracts or purified proteins can yield specific ubiquitylated forms of the substrate. Ubiquitylated forms of the radiolabeled protein can then be observed by resolving the proteins on an SDS gel and subsequent autoradiography. If the substrate is ubiquitylated, a high molecular weight ladder of bands will appear, and if methylated ubiquitin is included in the reaction, lower molecular weight bands corresponding to mono-, di-, and tri-ubiquitylated species should become apparent.

Protocol

1. *In vitro* translate the putative substrate using TNT-coupled *in vitro* transcription/translation kit (Promega) in the presence of ^{35}S-methionine.
2. Prepare a reaction mix in a final volume of 10 μl containing 0.5 μl *in vitro* translated [^{35}S] substrate, 1 μM ubiquitin aldehyde (from a 50-mg/ml stock solution dissolved in water and stored at $-20°$), and 2 μl of freshly prepared 5× PBDM. The mix can include an extract that has been prepared using a nitrogen-disruption bomb (Montagnoli *et al.*, 1999), which preserves the activity of ubiquitinylating enzymes. This extract can either be depleted

of proteins required for ubiquitylation or mock depleted to demonstrate that the depleted factor(s) are essential for efficient ubiquitylation. Alternately, purified proteins that are necessary for or responsible for the ubiquitylation of the substrate can be added to the extract. Ten micrograms of methylated ubiquitin (from a 10-mg/ml stock solution dissolved in water and stored at $-20°$) may also be included in the reaction mix.

3. Incubate reaction mixture at $37°$ for 30 min and then stop the reaction by the addition of $2\times$ sample buffer.

4. Resolve proteins by SDS-PAGE followed by autoradiography.

Ubiquitylation of a Purified Recombinant Substrate In Vitro

If methylated ubiquitin changes the pattern of bands of a radiolabeled substrate, the protein is likely ubiquitylated *in vitro;* however, a purified recombinant substrate is useful to confirm that the high molecular weight bands observed after an *in vitro* ubiquitylation reaction do, indeed, correspond to ubiquitylated forms of the protein of interest and to test whether the system is complete. Here, the ubiquitylation of a purified substrate is carried out by the same procedure that is used for the *in vitro* translated protein. However, after the proteins are resolved by SDS-PAGE, they are immunoblotted with an antibody to the protein of interest. If a ladder of high molecular weight bands that is dependent on the supplementation of ubiquitin (and obviously on temperature and time of incubation) is observed, and if lower molecular weight species are observed if methylated ubiquitin is included in the reaction, the protein is a substrate for ubiquitylation *in vitro*.

Protocol

1. Purify substrate from *Escherichia coli* or baculovirus-infected insect cells.

2. Prepare a reaction mix in a final volume of 10 μl containing 20 ng purified substrate, 1 μM ubiquitin aldehyde (from a 50-mg/ml stock solution dissolved in water and stored at $-20°$), 1 μl of rabbit reticulocyte lysate (as a source of ubiquitylating enzymes), and 2 μl of freshly prepared $5\times$ PBDM. The mix can include extract that has been prepared using a nitrogen-disruption bomb (Montagnoli *et al.*, 1999), which can either be depleted of proteins required for ubiquitylation or mock depleted. Alternately, purified proteins are necessary for or responsible for ubiquitylation of the substrate. Ten micrograms of methylated ubiquitin (from a 10-mg/ml stock solution dissolved in water and stored at $-20°$) may also be included in the reaction mix.

3. Incubate reaction mixture at 37° for different times (e.g., 0, 30, 60, and 90 mins) and then stop the reaction by the addition of 2× sample buffer.

4. Resolve proteins by SDS-PAGE followed by Western blot analysis with an antibody to the putative substrate.

Additional Methods to Show that a Substrate Is Ubiquitylated

General Considerations

There are additional assays that can be used to provide indirect evidence that a protein is ubiquitylated, two of which are described in the following. These methods rely on blocking the activity of the ubiquitin system, either by inactivating the ubiquitin-activating enzyme or by overexpressing a ubiquitin mutant that terminates ubiquitin chains, and then determining the steady-state levels of the protein of interest. Therefore, these assays can be used to demonstrate that inhibition of the ubiquitin system results in increased levels of a putative substrate. However, these methods do not eliminate the possibility that the accumulation of the protein of interest results from indirect effects (e.g., blocking the ubiquitylation of an upstream regulator). As a result, these assays can be useful in preliminary tests of the effect of ubiquitylation on a substrate or as supporting evidence that a substrate is ubiquitylated.

Materials and Reagents. 293T cells from American Type Culture Collection cell line that is temperature sensitive for the ubiquitin-activating enzyme (tsBN75) (Nishimoto *et al.*, 1980), vectors encoding wild-type ubiquitin, ubiquitin with all of its lysines mutated to arginines, and the protein of interest for expression in mammalian cells, cycloheximide (Sigma), Tris-HCL, sodium chloride, Triton X-100 (Roche), EDTA, sodium fluoride, DTT, PMSF, leupeptin, soybean trypsin inhibitor, TPCK, TLCK, aprotinin.

Buffers

LYSIS BUFFER. 50 mM Tris-HCl, pH 7.4, 0.25 M NaCl, 0.1% Triton X-100, 1 mM EDTA, 50 mM NaF, 1 mM DTT, 0.1 mM Na$_3$VO$_4$. The following protease inhibitors should also be added to prevent protein degradation: 0.1 mM PMSF (from an 18-mg/ml stock solution dissolved in isopropyl alcohol and stored at $-20°$), 1 μg/ml leupeptin (from an 18-mg/ml stock solution dissolved in isopropyl alcohol and stored at $-20°$), 10 μg/ml soybean trypsin inhibitor (from a 10-mg/ml stock solution dissolved in PBS and stored at $-20°$), 10 μg/ml TPCK (from a 10-mg/ml stock solution dissolved in ethanol and stored at $-20°$), 5 μg/ml TLCK (from a 5-mg/ml stock solution dissolved in PBS and stored at $-20°$), and 1 μg/ml aprotinin (from a 1-mg/ml stock solution dissolved in PBS and stored at $-20°$).

Use of a Cell Line That Is Temperature-Sensitive for the Ubiquitin-Activating Enzyme

The murine cell line, tsBN75, allows for the inactivation of the ubiquitin-activating enzyme (E1) at high temperatures (Nishimoto et al., 1980). At the permissive temperature (34°), the E1 enzyme is functional, and substrates are appropriately ubiquitylated and degraded. At the nonpermissive temperature (40°), the ubiquitin-activating enzyme is much less active, and all substrates of the ubiquitin pathway accumulate. This cell line allows for the comparison of steady-state levels and half-lives of a protein of interest when the E1 is functional versus when the E1 is less active. If levels of a particular protein are increased at the nonpermissive temperature, that protein is a direct or indirect downstream target of the ubiquitin system.

Protocol

1. Grow duplicate plates of tsBN75 cells at the permissive temperature (34°).
2. Switch one plate of tsBN57 cells to the restrictive temperature (40°).
3. Twenty-four hours after incubation at the permissive or nonpermissive temperature, wash the cells two times with PBS and then prepare extracts by resuspending the cell pellets in 3–5 volumes of lysis buffer. In addition to measuring steady-state levels, a cycloheximide time-course (Bashir et al., 2004) or a pulse-chase analysis (Pagano et al., 1995) may be used at this step to measure the half-life of the protein at the permissive or nonpermissive temperature.
3. Resolve proteins from two experimental conditions by SDS-PAGE and transfer to a nitrocellulose membrane.
4. Analyze the levels of the substrate at the permissive or nonpermissive temperature by immunoblotting with an antibody to the protein of interest.

Analysis of Substrate Half-Life in the Presence of Ub(K0)

Ub(K0), as discussed previously, terminates ubiquitin chains, similar to what is observed with methylated ubiquitin *in vitro*. In addition to its use in determining whether inhibition of polyubiquitin chain formation alters the pattern of ubiquitylation of a substrate, Ub(K0) can be used to determine whether a putative substrate is stabilized in its presence. Proteins that are regulated by ubiquitylation should display an increased half-life in the presence of Ub(K0).

Protocol

1. Cotransfect 293T cells with a vector encoding the protein of interest and a vector encoding wild-type ubiquitin or a vector encoding Ub(K0).

Use four to five times the amount of DNA encoding wild-type ubiquitin or Ub(K0) as the DNA encoding the protein of interest to compete with endogenous ubiquitin. Transfections can be carried out in 293T cells using the standard calcium phosphate method.

2. Twenty-four hours after transfection, wash the cells two times with PBS and then prepare extract by resuspending the cell pellet in 3–5 volumes of lysis buffer. In addition to measuring steady-state levels, a cycloheximide time-course (Bashir *et al.*, 2004) or a pulse-chase analysis (Pagano *et al.*, 1995) may be used at this step to measure the half-life of the protein in the presence of wild-type ubiquitin versus Ub(K0).

3. Resolve proteins from two experimental conditions on an SDS gel and transfer to a nitrocellulose membrane.

4. Analyze the levels of the substrate in the presence of wild-type ubiquitin versus Ub(K0) by immunoblotting with an antibody to the protein of interest.

Assays To Determine Whether Ubiquitylation Occurs on an Internal Lysine or the N-Terminal Methionine

General Considerations

A number of proteins that lack any lysine residues have been shown to be ubiquitylated (Aviel *et al.*, 2000; Ben-Saadon *et al.*, 2004; Bloom *et al.*, 2003; Breitschopf *et al.*, 1998; Coulombe *et al.*, 2004; Ikeda *et al.*, 2002; Kuo *et al.*, 2004; Reinstein *et al.*, 2000). In these cases, ubiquitylation is thought to occur on the N-terminal methionine. In fact, mass spectrometric analysis has been used to directly demonstrate ubiquitin modifications on N-terminal residues. This technique has proven that ubiquitin is fused directly to the N-terminus of HPV E7-58 (Ben-Saadon *et al.*, 2004) and p21 (Coulombe *et al.*, 2004), as well as to demonstrate the specific lysine residue that is modified (Moren *et al.*, 2003). If the protein of interest naturally lacks lysines or is engineered to lack lysines but is still sensitive to proteasome inhibition, the protein is a potential candidate for N-terminal ubiquitylation. Although mass spectrometry is the ideal tool to identify the N-terminus of a protein as a site for ubiquitylation, several assays can be used to determine whether modification of the N-terminal residue blocks protein degradation, which would imply that ubiquitylation occurs on the N-terminal methionine. Here, we describe two methods to assess N-terminal ubiquitylation, although additional methods have been described, including modification of the N-terminal residue by acetylation, which should preclude N-terminal ubiquitylation (Kuo *et al.*, 2004) or fusion of a cleavable tag to the N-terminus of a protein to determine if the cleaved tag is ubiquitylated *in vivo* (Bloom *et al.*, 2003; Kuo *et al.*, 2004).

Materials and Reagents. 293T cells from American Type Culture Collection, cycloheximide (Sigma), Tris-HCL, sodium chloride, Triton X-100 (Roche), EDTA, sodium fluoride, Dithiothreitol (DTT), sodium vanadate, phenyl-methyl sulfonyl fluoride (PMSF), leupeptin, soybean trypsin inhibitor, L-1 chlor-3-(4-tosylamido)-4 phenyl-2-butanon (TPCK), L-1 chlor-3-(4-tosylamido)-7-amino-2-heptanon-hydrochloride (TLCK), aprotinin, TNT coupled *in vitro* transcription/translation kit (Promega), Tris-HCl, magnesium chloride, ATP, phosphocreatine, creatine phosphokinase, and ubiquitin aldehyde (Boston Biochem).

Buffers

LYSIS BUFFER. 50 mM Tris-HCl, pH 7.4, 0.25 M NaCl, 0.1% Triton X-100, 1 mM EDTA, 50 mM NaF, 1 mM DTT, 0.1 mM Na$_3$VO$_4$. The following protease inhibitors should also be added to prevent protein degradation: 0.1 mM PMSF (from an 18-mg/ml stock solution dissolved in isopropyl alcohol and stored at $-20°$), 1 μg/ml leupeptin (from an 18-mg/ml stock solution dissolved in isopropyl alcohol and stored at $-20°$), 10 μg/ml soybean trypsin inhibitor (from a 10-mg/ml stock solution dissolved in PBS and stored at $-20°$), 10 μg/ml TPCK (from a 10-mg/ml stock solution dissolved in ethanol and stored at $-20°$), 5 μg/ml TLCK (from a 5-mg/ml stock solution dissolved in PBS and stored at $-20°$), and 1 μg/ml aprotinin (from a 1-mg/ml stock solution dissolved in PBS and stored at $-20°$).

5 × PROTEIN BREAKDOWN MIX (PBDM). 0.25 M Tris-HCl, pH7.5, 25 mM MgCl$_2$, 10 mM ATP, 50 mM phosphocreatine, and 17.5 U/ml creatine phosphokinase.

Modification of the N-terminal methionine with a 6×-myc tag in vivo. Several groups have shown that the addition of 6×-myc tag to the N-terminus of proteins that are N-terminally ubiquitylated blocks their degradation by proteasomes *in vivo* (Aviel *et al.*, 2000; Bloom *et al.*, 2003; Breitschopf *et al.*, 1998; Coulombe *et al.*, 2004; Reinstein *et al.*, 2000). This 78-amino acid tag is thought to function by blocking the recognition site for ubiquitin ligases, blocking ubiquitin, or both. Once cells have been transfected with an untagged version of the protein of interest or a 6×-myc tagged version, the steady-state levels of the protein can be determined by direct Western blot analysis. If the potential substrate is more stable with 6×-myc tag, the N-terminus is a likely site for ubiquitylation. This can be more accurately assayed using a cycloheximide time-course to measure the half-lives of untagged and 6×-myc tagged versions of the protein.

Protocol

1. Transfect 293T cells with a vector encoding an untagged version of the protein of interest or a 6×-myc tagged version of the protein of

interest. Transfections can be carried out in 293T cells using the standard calcium phosphate method.
2. Twenty-four hours after transfection, add 100 μg/ml cycloheximide to each plate (from a stock of 100 mg/ml dissolved in 100% ethanol) and incubate cells for various time points.
3. Wash the cells two times with PBS and then prepare extract by resuspending the cell pellet in 3–5 volumes of lysis buffer.
4. Resolve proteins from various experimental conditions by SDS-PAGE and transfer to a nitrocellulose membrane.
5. Analyze the levels of the untagged or 6×-myc tagged substrate by immunoblotting with an antibody to the protein of interest.

Modification of the N-Terminal Methionine In Vitro

The N-terminus of a recombinant protein can be chemically modified to assay whether ubiquitin ligation occurs on the N-terminal methionine. The free amino group of the N-terminal methionine (α-NH2) can be modified by

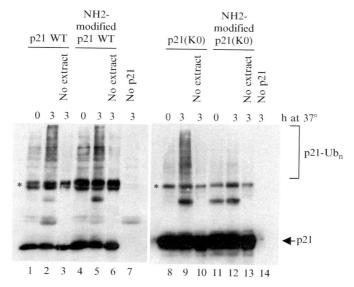

FIG. 2. Modification of the N-terminus of a substrate missing lysine residues blocks its ubiquitylation *in vitro*. *In vitro* ubiquitylation of wild-type p21 (lanes 1–3), carbamylated wild-type p21 (NH$_2$-modified p21 WT; lanes 4–6), lysineless p21 (lanes 8–10), or carbamylated lysineless p21 (NH$_2$-modified p21 ΔK; lanes 11–13) was carried out using a rabbit reticulocyte lysate (lanes 1–2, 4–5, 8–9, 11–12) for the indicated intervals of times. Lanes 3, 6, 10, and 13 contain no lysate, whereas lanes 7 and 14 contain no p21. Unmodified and ubiquitylated forms of p21 were detected with an antibody to p21. The asterisks indicate nonspecific bands present in the p21 prep.

carbamylation, which does not affect the ε-amino groups of internal lysines (Breitschopf et al., 1998). Unmodified and carbamylated forms of the recombinant protein can be subjected to an *in vitro* ubiquitylation assay to determine whether blocking the N-terminal residue prevents ubiquitylation (Fig. 2).

Protocol

1. Purify substrate from *E. coli* or baculovirus-infected insect cells.
2. Dissolve 200 μg of purified protein in 0.2 M potassium phosphate pH 6.0/6 M urea/50 mM potassium cyanate and incubate at 37° for 8 h.
3. Stop reaction by the addition of Gly-Gly to a final concentration of 150 mM. Adjust the pH to 8.1 with 30% (w/v) K_2HPO_4, pH 11. Incubate at 37° for an additional hour to remove carbamyl groups from non-amine groups in the protein.
4. Remove reagents by dialysis against water overnight.
5. Prepare a reaction mix in a final volume of 10 μl containing 20 ng purified substrate (unmodified or carbamylated), 1 μM ubiquitin aldehyde, 1 μl of rabbit reticulocyte lysate (as a source of ubiquitylating enzymes), and 2 μl of 5× PBDM. Ten micrograms of methylated ubiquitin may also be included in the reaction mix.
6. Incubate reaction mixture at 37° for 30 min and then stop the reaction by the addition of 2× sample buffer.
7. Resolve proteins by SDS-PAGE followed by Western blot analysis with an antibody to the putative substrate.

Acknowledgments

We thank A. Hershko for critically reading this manuscript. J. B. is supported by funds from The Rockefeller University's *Woman & Science* Fellowship program. MP is supported by grants from the NIH (R01-CA76584 and R01-GM57597).

References

Aviel, S., Winberg, G., Massucci, M., and Ciechanover, A. (2000). Degradation of the Epstein-Barr virus latent membrane protein 1 (LMP1) by the ubiquitin-proteasome pathway. Targeting via ubiquitination of the N-terminal residue. *J. Biol. Chem.* **275**, 23491–23499.

Bashir, T., Dorrello, N. V., Amador, V., Guardavaccaro, D., and Pagano, M. (2004). Control of the SCF(Skp2-Cks1) ubiquitin ligase by the APC/C(Cdh1) ubiquitin ligase. *Nature* **428**, 190–193.

Ben-Saadon, R., Fajerman, I., Ziv, T., Hellman, U., Schwartz, A. L., and Ciechanover, A. (2004). The tumor suppressor protein p16INK4a and the human papillomavirus oncoprotein E7-58 are naturally occurring lysine-less proteins that are degraded by the ubiquitin system: Direct evidence for ubiquitination at the N-terminal residue. *J. Biol. Chem.* **279**, 41414–41421.

Bloom, J., Amador, V., Bartolini, F., DeMartino, G., and Pagano, M. (2003). Proteasome-mediated degradation of p21 via N-terminal ubiquitinylation. *Cell* **115**, 71–82.

Breitschopf, K., Bengal, E., Ziv, T., Admon, A., and Ciechanover, A. (1998). A novel site for ubiquitination: The N-terminal residue, and not internal lysines of MyoD, is essential for conjugation and degradation of the protein. *EMBO J.* **17**, 5964–5973.

Carrano, A. C., Eytan, E., Hershko, A., and Pagano, M. (1999). SKP2 is required for ubiquitin-mediated degradation of the CDK inhibitor p27. *Nat. Cell Biol.* **1**, 193–199.

Cope, G. A., and Deshaies, R. J. (2004). COP9 signalosome: A multifunctional regulator of SCF and other cullin-based ubiquitin ligases. *Cell* **114**, 663–671.

Coulombe, P., Rodier, G., Bonneil, E., Thibault, P., and Meloche, S. (2004). N-terminal ubiquitination of extracellular signal-regulated kinase 3 and p21 directs their degradation by the proteasome. *Mol. Cell Biol.* **24**, 6140–6150.

Ganoth, D., Bornstein, G., Ko, T. K., Larsen, B., Tyers, M., Pagano, M., and Hershko, A. (2001). The cell-cycle regulatory protein Cks1 is required for SCF(Skp2)-mediated ubiquitinylation of p27. *Nat. Cell Biol.* **3**, 321–324.

Hershko, A. (1988). Ubiquitin-mediated protein degradation. *J. Biol. Chem.* **263**, 15237–15240.

Hershko, A., and Ciechanover, A. (1998). The ubiquitin system. *Annu. Rev. Biochem.* **67**, 425–479.

Ikeda, M., Ikeda, A., and Longnecker, R. (2002). Lysine-independent ubiquitination of Epstein-Barr virus LMP2A. *Virology* **300**, 153–159.

Kuo, M. L., den Besten, W., Bertwistle, D., Roussel, M. F., and Sherr, C. J. (2004). N-terminal polyubiquitination and degradation of the Arf tumor suppressor. *Genes Dev.* **18**, 1862–1874.

Montagnoli, A., Fiore, F., Eytan, E., Carrano, A. C., Draetta, G. F., Hershko, A., and Pagano, M. (1999). Ubiquitination of p27 is regulated by Cdk-dependent phosphorylation and trimeric complex formation. *Genes Dev.* **13**, 1181–1189.

Moren, A., Hellman, U., Inada, Y., Imamura, T., Heldin, C. H., and Moustakas, A. (2003). Differential ubiquitination defines the functional status of the tumor suppressor Smad4. *J. Biol. Chem.* **278**, 33571–33582.

Nishimoto, T., Takahashi, T., and Basilico, C. (1980). A temperature-sensitive mutation affecting S-phase progression can lead to accumulation of cells with a G2 DNA content. *Somatic Cell Genet.* **6**, 465–476.

Pagano, M., Tam, S. W., Theodoras, A. M., Beer-Romero, P., Del Sal, S., Chau, V., Yew, P. R., Draetta, G. F., and Rolfe, M. (1995). Role of the ubiquitin-proteasome pathway in regulating abundance of the cyclin-dependent kinase inhibitor p27. *Science* **269**, 682–685.

Pan, Z. Q., Kentsis, A., Dias, D. C., Yamoah, K., and Wu, K. (2004). Nedd8 on cullin: Building an expressway to protein destruction. *Oncogene* **23**, 1985–1997.

Pickart, C. M. (2001). Mechanisms underlying ubiquitination. *Annu. Rev. Biochem.* **70**, 503–533.

Pickart, C. M. (2004). Back to the future with ubiquitin. *Cell* **116**, 181–190.

Reinstein, E., Scheffner, M., Oren, M., Ciechanover, A., and Schwartz, A. (2000). Degradation of the E7 human papillomavirus oncoprotein by the ubiquitin-proteasome system: Targeting via ubiquitination of the N-terminal residue. *Oncogene* **19**, 5944–5950.

Rolfe, M. (1995). Stable and transient transfections in cell cycle studies. In "Cell cycle: Materials and methods" (M. Pagano, ed.), pp. 87–92. Springer-Verlag, New York.

Seeler, J. S., and Dejean, A. (2003). Nuclear and unclear functions of SUMO. *Nat. Rev. Mol. Cell Biol.* **4**, 690–699.

Treier, M., Staszewski, L. M., and Bohmann, D. (1994). Ubiquitin-dependent c-Jun degradation *in vivo* is mediated by the delta domain. *Cell* **78**, 787–798.

Wei, W., Ayad, N. G., Wan, Y., Zhang, G. J., Kirschner, M. W., and Kaelin, W. G., Jr. (2004). Degradation of the SCF component Skp2 in cell-cycle phase G1 by the anaphase-promoting complex. *Nature* **428,** 194–198.

Yaron, A., Hatzubai, A., Davis, M., Lavon, I., Amit, S., Manning, A. M., Andersen, J. S., Mann, M., Mercurio, F., and Ben-Neriah, Y. (1998). Identification of the receptor component of the IkappaBalpha-ubiquitin ligase. *Nature* **396,** 590–594.

[18] Identification of Ubiquitination Sites and Determination of Ubiquitin-Chain Architectures by Mass Spectrometry

By PETER KAISER and JAMES WOHLSCHLEGEL

Abstract

The identification of protein modification sites is an important step toward understanding the biological role of covalent modifications. For example, the mapping of phosphorylation sites and analyses of phosphorylation site mutants have tremendously contributed to our knowledge of different cellular processes. Given the diverse functions of ubiquitination, similar studies with ubiquitin attachment site mutants are becoming increasingly important in understanding the molecular roles of ubiquitination. Relatively few studies to date have mapped ubiquitination sites, and in almost all cases the identification of the acceptor lysines were based on indirect evidence (Petroski and Deshaies, 2003; Scherer *et al.*, 1995); that is, mutation of particular lysines to arginines blocked ubiquitination. Direct evidence for ubiquitin attachment sites has been obtained by mapping of hydroxylamine-derived peptides from ubiquitinated proteins (Chau *et al.*, 1989); however, these experiments can be very challenging. Recent advances in protein mass spectrometry have enabled ubiquitinated lysine residues to be identified directly, thereby providing more convincing evidence for the exact location of the modification (Flick *et al.*, 2004; Peng *et al.*, 2003). In addition to mapping attachment sites, mass spectrometry can also be used to determine the type of ubiquitin chain linkage (Flick *et al.*, 2004; Peng *et al.*, 2003). *In vivo* evidence for the covalent attachment of the carboxyl terminus of one ubiquitin molecule to lysine residues in several locations in a different ubiquitin molecule demonstrates the complexity of ubiquitin biology (Peng *et al.*, 2003). These different ubiquitin chain topologies can dramatically affect the molecular function of ubiquitin chains (Hoege *et al.*, 2002; Spence *et al.*, 1995), and, hence, the mass spectrometric determination of the ubiquitin chain architecture can provide important insight into the mechanisms of ubiquitin function.

This chapter describes mass spectrometric approaches for identifying ubiquitin acceptor lysines on target proteins and analyzing the ubiquitin chain topology.

Identification of Ubiquitin Attachment Sites

The basic strategy is outlined in Fig. 1. The protein of interest is purified, and peptides generated by digestion with trypsin and other proteases are analyzed by mass spectrometry. Trypsin cuts after the carboxyl-terminal arginine in ubiquitin and thus removes all but the two terminal glycine residues of ubiquitin from the ubiquitinated peptide. The 114-Dalton mass increase caused by the addition of the two glycine

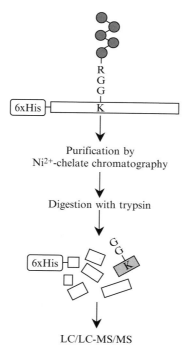

FIG. 1. Identification of ubiquitin acceptor lysines: The protein of interest is purified, and peptides generated by digestion with trypsin and other proteases are analyzed by mass spectrometry. The 114-Dalton mass increase caused by the addition of the two glycine residues to the target lysine can be monitored by mass spectrometry and is diagnostic for the ubiquitination of that residue.

residues to the target lysine can be monitored by mass spectrometry and is diagnostic for the ubiquitination of that residue (Peng et al., 2003). The critical step in this procedure is the preservation of the ubiquitin modification during the purification process. Although this is less of a problem for the analysis of proteins that were ubiquitinated in a partially purified or reconstituted *in vitro* system, the highly active deubiquitinating enzymes (Dubs) present in whole cell lysates are a major hurdle in the identification of *in vivo* ubiquitination sites (Pickart and Cohen, 2004; Wilkinson and Hochstrasser, 1998). We have found that purification of the protein of interest under highly denaturing conditions prevents Dub activity and largely preserves the ubiquitin modification. We therefore recommend using a 6xHis-tagged version of the protein of interest, which can be purified in the presence of 6 M guanidinium or 8 M urea. Dub activity can also be prevented by the highly specific inhibitor ubiquitin aldehyde (Affinity Research Products Ltd., Devon, UK). However, the relatively high price of this reagent makes it less attractive for the large amounts of cell lysates required for this analysis. Alternately, Dubs have a cysteine residue in their active site that is essential for their catalytic activity and can be blocked by the sulfhydryl alkylating agent N-ethylmaleimide (NEM) (Sigma, St. Louis, MO). Although we have found that NEM was not necessary when cell lysates were prepared in 8 M urea- or 6 M guanidinium-based buffers, NEM can be quite effective when purification under less denaturing conditions is required. It is important to note that the treatment of cell lysates with NEM will lead to the alkylation of solvent-exposed, reduced cysteine residues in the sample, which must be accounted for when analyzing the mass spectrometric data.

The protocol we used to determine the ubiquitination site on the yeast transcription factor Met4 is outlined in the following (Flick et al., 2004).

Preparation of Cell Lysates

If possible, we suggest using a Pep4 (vacuolar proteinase A)–deficient yeast strain. *pep4* mutants exhibit a strong reduction of protease activity during lysate preparation.

1. Grow 2 liters of yeast cells expressing the 6xHis-tagged protein of interest to an $A_{600} = 1.5$ and collect cells by centrifugation (2000g, 3 min, 4°) or filtration.

2. Wash cells with 0.025 vol. (50 ml) lysis buffer (50 mM sodium-phosphate buffer, pH 8.0, 8 M urea, 300 mM NaCl, 0.5% Triton X-100, 1 mM PMSF, *optional:* 10 mM N-ethylmaleimide [NEM]). Because of dissociation of urea, the pH of the lysis buffer should be readjusted immediately before use. If NEM is included, prepare a 0.5 M or 1 M

solution in 100% ethanol immediately before it is added, because it is unstable.

3. Spin cells as before and resuspend the pellet in one pellet volume of lysis buffer (~8 ml). Pipette droplets into liquid nitrogen and store at −80°.

4. Break cells by grinding in liquid nitrogen. We use a powered mortar grinder (Retsch RM 100, Newtown, PA), but manual grinding is possible as well. Cell lysis can be monitored microscopically, and grinding should be continued until >90% of the cells are broken. The ground cell powder can be stored at −80°. Cell lysis by grinding in liquid nitrogen leads to lysates with maximum preservation of the ubiquitination status; however, other methods such as glass bead lysis (Peng et al., 2003) and chemical lysis with NaOH work as well (Johnson and Blobel, 1999; Wohlschlegel et al., 2004).

Purification of the Ubiquitinated Protein of Interest

All steps are carried out at room temperature. We do not observe any increase in proteolysis or deubiquitination at room temperature as long as highly denaturing conditions are maintained. It is important to wear gloves at all times during this procedure to minimize the contamination of the sample with keratin, which can hinder analysis by mass spectrometry.

5. Add an equal vol. (~16 ml) of lysis buffer to the cell powder and thaw at room temperature under continuous mixing.

6. Spin lysate for 20 min at 25,000g and transfer supernatant to a fresh tube. If the lysate is viscous, use a syringe and pass it once through an 18-gauge needle and twice through a 21-gauge needle. Save 200 μl of the lysate for analysis. We usually obtain around 180 mg of total protein.

7. Take 2 ml of Ni-NTA agarose (50% slurry, 1-ml bead volume) (Qiagen, Germantown, MD), let the beads settle by gravity or centrifuge 2 min at 100g. Remove the supernatant and wash once with 3 ml lysis buffer.

8. Add the Ni-NTA agarose to the lysate and bind overnight at room temperature with gentle mixing.

9. Let beads settle by gravity for about 5 min and carefully remove the supernatant. Keep the supernatant in case a second binding step is necessary. Wash the beads with 10 ml lysis buffer for 1 min, remove wash, and resuspend the beads in 5 ml lysis buffer. Pour the slurry into an appropriate chromatography column (e.g., Poly-Prep column, Bio-Rad, Hercules, CA). Wash once with an additional 5 ml lysis buffer and twice with 5 ml wash buffer (identical to lysis buffer but pH 6.3). Readjust the pH of the buffer immediately before use.

10. Elute six times with 500 μl elution buffer (same as lysis buffer but pH 4.3). Make sure to adjust pH immediately before use.

11. Analyze the samples by immunoblotting. Take 10 μl of each, the lysate from step 6, the supernatant from step 9, and the six eluted fractions from step 10. Mix with 10 μl 2× PAGE loading buffer (2×: 4% SDS, 125 mM Tris-HCl, pH 6.8, 20% glycerol, 200 mM DTT, 0.002% bromophenol blue). Separate samples by SDS-PAGE and analyze by standard immunoblotting. Do not boil samples containing urea, because it induces carbamylation. A specific antibody for the RGS6xHis epitope is available (Qiagen, Germantown, MD). Thus, it is advisable to fuse the protein of interest to the RGS6xHis tag, which combines Ni-NTA affinity and antibody detection. Comparison of total lysate and supernatant indicates efficiency of binding of your favorite protein to Ni-NTA agarose. For unknown reasons, binding to Ni-NTA agarose is sometimes inefficient (<50%). In this case, rebind the supernatant from step 9 to Ni-NTA agarose by following steps 7 to 11. Rebinding is usually very efficient.

Mass Spectrometric Analysis

12. Pool all fractions containing the protein of interest. Add 100% w/v trichloroacetic acid (TCA) to the sample to a final concentration of 20%. Incubate on ice for 30 min, centrifuge at >10,000g for 20 min, and decant the supernatant. Wash the pellet with 500 μl of ice-cold acetone and repeat centrifugation and decanting steps. Dry the pellet using vacuum centrifugation.

13. Resuspend the sample pellet in resolubilization buffer (100 mM Tris-HCl, pH 8.5, 8 M urea). Add Tris(2-carboxyethyl)-phosphine (Pierce, Rockford, IL) to a final concentration of 3 mM and incubate at room temperature for 30 min to reduce protein disulfide bonds. Alkylate the resulting free thiols by the addition of iodoacetamide to a final concentration of 20 mM.

14. Digest the reduced and alkylated protein sample by the serial addition of lys-C and trypsin proteases. First, add Lys-C (Roche, Indianapolis, IN) at an estimated enzyme-to-substrate ratio of 1:100 (wt/wt) and incubate at 37° for 4 h. Then, dilute the sample to a final urea concentration of ~2 M using 100 mM Tris, pH 8.5. Add $CaCl_2$ to a concentration of 1 mM and modified trypsin (Promega, Madison, WI) to an estimated enzyme-to-substrate ration of 1:50. Incubate the reaction for 12–24 h at 37°, quench by adding 90% formic acid to a 4% final concentration, and then store at −20° until analysis.

15. Pressure load peptide digests onto a 100-μm inner diameter fused-silica capillary column with a 5-μm pulled tip and packed with 7 cm of 5 μm Aqua C18 resin (Phenomenex, Ventura, CA) followed by 3 cm of 5 μm Partisphere strong cation exchange resin (Whatman, Clifton, NJ)

and finally with 3 cm of 5 μm Aqua C18 resin (Phenomenex, Ventura, CA).

16. After loading, place the column inline with an Agilent 1100 quaternary HPLC (Agilent Technologies, Palo Alto, CA) and fractionate the peptide mixture using a six-step separation. Step 1 should consist of a 100-min gradient beginning at 100% buffer A (95% water, 5% acetonitrile, 0.1% formic acid) and ending at 100% buffer B (20% water, 80% acetonitrile, 0.1% formic acid). Steps 2 through 6 should each be composed of a 3-min pulse of increasing percentages of buffer C (500 mM ammonium acetate, 5% acetonitrile, 0.1% formic acid) followed by a 100-min gradient from 10–40% buffer B. Percentages for buffer C should be 10%, 25%, 50%, 75%, and 100% for steps 2 through 6, respectively. For the final step, the 100-min gradient from 10–40% buffer B should be followed by a 20-min gradient from 40% buffer B to 100% buffer B. Operate the HPLC pump at a flow rate of 100 μl/min and then split the flow to obtain a final flow rate of 200 nl/min through the capillary column.

17. Electrospray the peptides eluting from the column directly into an LCQ-Deca mass spectrometer (ThermoFinnigan, Palo Alto, CA) by applying a 2.4 kV spray voltage. Data acquisition should be performed in a data-dependent fashion, with one cycle consisting of the collection of one full mass spectrum (400–1400 m/z) followed by three MS/MS spectra (35% normalized collision energy) corresponding to the three most abundant ions from the full mass scan. This data acquisition cycle should be repeated throughout the multidimensional separation.

18. MS/MS spectra are then analyzed using a series of software algorithms. First, analyze the spectra with the 2to3 program, which assigns the appropriate charge state to multiply charged peptide spectra and removes poor-quality spectra (Sadygov *et al.*, 2002). Then, search the spectra against a protein database from the appropriate organism using the SEQUEST algorithm (Eng *et al.*, 1994). This search should be set up to consider a possible modification of +114 on lysines, which corresponds to the two glycine residues from ubiquitin that remain attached to a target lysine after trypsinization. After the database search is completed, use the program DTASelect to filter peptide identifications and manually validate spectra that correspond to putative ubiquitin attachment sites (Tabb *et al.*, 2002).

Possible Modifications

The ubiquitinated fraction of the protein of interest can be increased in several ways. If the ubiquitinated protein is degraded by the proteasome, inhibition of proteasome activity can be very effective. This can be achieved by using a yeast strain carrying temperature-sensitive alleles of proteasome subunits. For example, the temperature-sensitive *pre1-1 pre4-1* double mutant works well (Hilt *et al.*, 1993). Alternately, proteasome inhibitors can be used (e.g., MG132, Calbiochem, San Diego, CA). For experiments in yeast, it is important to use a Δ*pdr5* mutant, because they are more permeable for proteasome inhibitors (Fleming *et al.*, 2002). For both 26S substrates and other ubiquitinated proteins, the ubiquitinated fraction can be increased by overexpression of a mutant form of ubiquitin that has the carboxyl-terminal glycine residue replaced with alanine (Ubi^{G76A}). Ubi^{G76A} does not affect ubiquitin chain formation but protects from deubiquitination (Hodgins *et al.*, 1992). If Ubi^{G76A} is used, the database search algorithms should consider a mass increase of $+128$ instead of $+114$ for ubiquitinated lysines to account for the glycine to alanine change.

Although the mass spectrometry experiments described in this section have been optimized for the instrumentation present in our laboratory, the overall approach is very general. Any mass spectrometer capable of acquiring tandem mass spectra (MDS Sciex QSTAR, Micromass Q-Tof, etc), as well any algorithm able to perform differential modification searches (Pep_probe, Mascot, etc), can also be used. Similarly, although we routinely perform multidimensional separations of our peptide mixtures, simple mixtures can be effectively analyzed using single dimension chromatography.

Analysis of Ubiquitin Chain Architecture

Although a direct approach comparing yields of tryptic peptides of free ubiquitin and isolated polyubiquitin chains has been used successfully to determine a ubiquitin chain topology (Chau *et al.*, 1989), until recently, ubiquitin chain architectures have been almost exclusively determined by indirect approaches that use ubiquitin mutants carrying lysine to arginine mutations (Finley *et al.*, 1994; Spence *et al.*, 1995). For example, expression of Ubi^{K48R} shortens ubiquitin chains that are linked through lysine 48 on ubiquitin but not other ubiquitin chains, and this chain shortening can be assayed by immunoblotting (Banerjee *et al.*, 1993; Flick *et al.*, 2004). This approach has potential problems in that shortening of a ubiquitin chain on a particular protein might be an indirect result of shortened K48-linked chains on other proteins (e.g., stabilization of an ubiquitin ligase inhibitor). The mass spectrometric analysis we describe here provides direct insight into the ubiquitin chain architecture without affecting the physiology of the cell.

Because an ubiquitin chain linkage is just a specific example of an ubiquitin attachment site in which ubiquitin is attached to a second ubiquitin molecule instead of a cellular substrate, the method for identifying these linkages is conceptually identical to the procedure outlined previously. The primary difference in this approach is the requirement that the ubiquitinated form of the protein of interest be the only ubiquitinated protein present in the analysis. Depending on the stringency of the purification, it is possible for the purified ubiquitinated protein of interest to be contaminated with other cellular factors containing different ubiquitin chains. Because these proteins are both proteolytically digested before analysis, when an ubiquitin chain linkage is identified in the mixture, it is not possible to determine from which ubiquitinated protein it was generated. To avoid this potential pitfall, we prefer to identify ubiquitin chain linkages from highly purified ubiquitinated proteins excised from gels whenever possible. The basic strategy is outlined in Fig. 2. To achieve the required purity, Ni-chelate chromatography is followed by separation by SDS-PAGE. Mass spectrometric analysis is performed after in-gel digest of the ubiquitinated forms of the protein of interest. Although this analysis

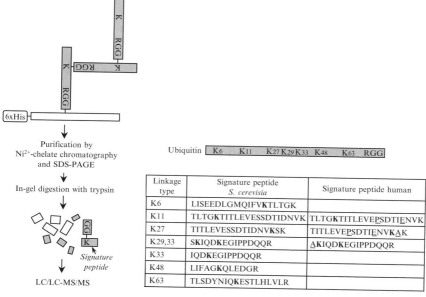

FIG. 2. Determination of the ubiquitin chain topology: To achieve the required purity, the partially purified protein of interest is separated by SDS-PAGE, and the ubiquitinated forms are excised, digested with trypsin, and analyzed by mass spectrometry. Depending on the linkage type, two glycine residues are attached to one of the signature peptides listed in the table, which is diagnostic for the linkage type.

Linkage type	Signature peptide *S. cerevisiae*	Signature peptide human
K6	LISEEDLGMQIFVKTLTGK	
K11	TLTGKTITLEVESSDTIDNVK	TLTGKTITLEVEPSDTIENVK
K27	TITLEVESSDTIDNVKSK	TITLEVEPSDTIENVKAK
K29,33	SKIQDKEGIPPDQQR	AKIQDKEGIPPDQQR
K33	IQDKEGIPPDQQR	
K48	LIFAGKQLEDGR	
K63	TLSDYNIQKESTLHLVLR	

can be successfully performed on femtomole levels of ubiquitinated protein, higher protein quantities (~5–10 pmol) greatly facilitate the identification of ubiquitin chains.

The protocol we used to determine the topology of the ubiquitin chain that is attached to the yeast transcription factor Met4 is outlined in the following (Flick *et al.*, 2004).

Preparation of Cell Lysates

Follow steps 1 to 4 as described previously. Depending on abundance of the ubiquitinated form of the protein of interest, it might be necessary to increase the culture volume to obtain enough material. Alternately, the fraction of the ubiquitinated form can be increased by proteasome inhibition and/or expression of Ubi^{G76A} as mentioned previously.

Purification of the Ubiquitinated Protein of Interest

Follow steps 5 to 11 of the preceding section.

12. To concentrate the sample so it can be loaded in one or two wells of the polyacrylamide gel, all fractions containing the protein of interest are pooled, the pH is adjusted to 8.0 with NaOH, and the sample is incubated with 100 μl Ni-NTA agarose (of a 50% slurry) at room temperature overnight. The amount of Ni-NTA agarose used in this second purification step is much lower than that added to the total lysate (step 7). For unclear reasons purification of 6xHis-tagged proteins from total yeast extracts requires unproportionately large amounts of Ni-NTA agarose, but once the sample is partially purified, the Ni-NTA agarose needed is an approximate reflection of the binding capacity (5–10 mg/ml). Sample concentration can also be achieved by TCA precipitation. However, we noticed that resolubilization of ubiquitinated proteins is sometimes very inefficient. In addition, the second Ni-chelate chromatography further purifies the protein of interest.

13. Let beads settle by gravity for about 5 min and carefully remove the supernatant by aspiration with a 27-gauge needle. Wash the beads with 1 ml lysis buffer for 1 min, remove wash, and resuspend the beads in 1 ml lysis buffer. Pour the slurry into an appropriate chromatography column (e.g., Bio-Spin Chromatography Columns, Bio-Rad, Hercules, CA). Wash once with an additional 0.5 ml lysis buffer and three times with 0.5 ml wash buffer (identical to lysis buffer but pH 6.3). Readjust the pH of the buffer immediately before use.

14. Elute 15 25-μl fractions with elution buffer (50 mM sodium-phosphate buffer, 100 mM Tris-HCl, 8 M urea, 50 mM NaCl, 2% SDS, 10 mM EDTA, pH 4.3). Make sure to adjust pH immediately before use.

15. Add 25-μl 2× PAGE loading buffer (step 11) to each fraction and analyze 1–5 μl by immunoblotting (see step 11). Pool the peak fractions and separate by SDS-PAGE.

16. Stain the gel with Coomassie brilliant blue and excise band(s) corresponding to the ubiquitinated protein of interest. Other protein staining methods such as silver and SYPRO Ruby (Molecular Probes, Eugene, OR) are also compatible with this approach.

17. Perform an in-gel digest of the protein using trypsin (Taylor *et al.*, 2000). First, cut gel slices into 1-mm^3 cubes. Wash gel pieces for 20 min with 100 mM ammonium bicarbonate and remove the supernatant. Reduce proteins by covering gel pieces with 3 mM dithiothreitol/100 mM ammonium bicarbonate and incubating at room temperature for 30 min. Add iodoacetamide to a final concentration of 6 mM and incubate at room temperature in the dark for 20 min. Discard the supernatant and dehydrate the gel pieces by incubating the slice in 50% acetonitrile/100 mM ammonium bicarbonate for 20 min. Dry cubes by vacuum centrifugation and re-swell. Dry the pieces with 0.2 μg of modified trypsin dissolved in 25 mM ammonium bicarbonate (add just enough to cover the gel pieces). Incubate overnight at 37°. Extract peptides with 100 μl of 60% acetonitrile/0.1% TFA at room temperature for 20 min. Lyophilize the supernatants and reconstitute the peptides in 20 μl of 5% formic acid.

18. Pressure load peptide digest onto a 100-μm inner diameter fused-silica capillary column with a 5-μm pulled tip and packed with 12 cm of 5-μm Aqua C18 resin (Phenomenex, Ventura, CA). After loading, place the column inline with an Agilent 1100 quaternary HPLC and fractionate the peptide mixture using a 120-min gradient beginning at 100% buffer A (95% water, 5% acetonitrile, 0.1% formic acid) and ending at 45% buffer B (20% water, 80% acetonitrile, 0.1% formic acid). Operate the HPLC pump at a flow rate of 100 μl/min, and then split the flow to obtain a final flow rate of 200 nl/min through the capillary column.

19. Electrospray eluting peptides into the mass spectrometer as described previously. Collect data by sequentially isolating and fragmenting (35% normalized collision energy) seven different 3 *m/z* windows (730.9, 819.2, 926.9, 990.1, 1046.8, 1122.6, and 1189.4) corresponding to branched-chain peptides that are diagnostic for specific ubiquitin chain linkages (Lys 48, Lys 33, Lys 29, Lys 6, Lys 27, Lys 63, and Lys 11, respectively). As opposed to the data-dependent acquisition strategy described in the previous section, sequentially monitoring specific m/z windows in this fashion ensures that mass ranges corresponding to all peptides of interest are continuously sampled and increases the sensitivity and dynamic range of the analysis. Analyze fragmentation spectra using SEQUEST as described previously.

Possible Modifications

Although we describe an approach to identify ubiquitin branched-chain peptides from their tandem mass spectra, these peptides can also be effectively observed in protein mixtures using selected reaction monitoring (SRM) strategies. SRM approaches can provide several benefits including increased sensitivity and the ability to quantify the absolute amount of the ubiquitinated peptide when coupled with stable isotope dilution strategies.

Acknowledgments

P. Kaiser acknowledges support from NIH, the California Breast Cancer Research Program, and the Cancer Research Coordinating Committee. J. Wohlschlegel is supported by a postdoctoral fellowship from the American Cancer Society.

References

Banerjee, A., Gregori, L., Xu, Y., and Chau, V. (1993). The bacterially expressed yeast CDC34 gene product can undergo autoubiquitination to form a multiubiquitin chain linked protein. *J. Biol. Chem.* **268,** 5668–5675.

Chau, V., Tobias, J. W., Bachmair, A., Marriott, D., Ecker, D. J., Gonda, D. K., and Varshavsky, A. (1989). A multiubiquitin chain is confined to specific lysine in a targeted short-lived protein. *Science* **243,** 1576–1583.

Eng, J., McCormack, A., and Yates, J. (1994). An approach to correlate tandem mass spectral data of peptides with amino acid sequences in a protein database. *J. Am. Soc. Mass Spectrom* **5,** 976–989.

Finley, D., Sadis, S., Monia, B. P., Boucher, P., Ecker, D. J., Crooke, S. T., and Chau, V. (1994). Inhibition of proteolysis and cell cycle progression in a multiubiquitination-deficient yeast mutant. *Mol. Cell. Biol.* **14,** 5501–5509.

Fleming, J. A., Lightcap, E. S., Sadis, S., Thoroddsen, V., Bulawa, C. E., and Blackman, R. K. (2002). Complementary whole-genome technologies reveal the cellular response to proteasome inhibition by PS-341. *Proc. Natl. Acad. Sci. USA* **99,** 1461–1466.

Flick, K., Ouni, I., Wohlschlegel, J. A., Capati, C., McDonald, W. H., Yates, J. R., and Kaiser, P. (2004). Proteolysis-independent regulation of the transcription factor Met4 by a single Lys 48-linked ubiquitin chain. *Nat. Cell Biol.* **6,** 634–641.

Hilt, W., Enenkel, C., Gruhler, A., Singer, T., and Wolf, D. H. (1993). The PRE4 gene codes for a subunit of the yeast proteasome necessary for peptidylglutamyl-peptide-hydrolyzing activity. Mutations link the proteasome to stress- and ubiquitin-dependent proteolysis. *J. Biol. Chem.* **268,** 3479–3486.

Hodgins, R. R., Ellison, K. S., and Ellison, M. J. (1992). Expression of a ubiquitin derivative that conjugates to protein irreversibly produces phenotypes consistent with a ubiquitin deficiency. *J. Biol. Chem.* **267,** 8807–8812.

Hoege, C., Pfander, B., Moldovan, G. L., Pyrowolakis, G., and Jentsch, S. (2002). RAD6-dependent DNA repair is linked to modification of PCNA by ubiquitin and SUMO. *Nature* **419,** 135–141.

Johnson, E. S., and Blobel, G. (1999). Cell cycle-regulated attachment of the ubiquitin-related protein SUMO to the yeast septins. *J. Cell Biol.* **147,** 981–994.

Peng, J., Schwartz, D., Elias, J. E., Thoreen, C. C., Cheng, D., Marsischky, G., Roelofs, J., Finley, D., and Gygi, S. P. (2003). A proteomics approach to understanding protein ubiquitination. *Nat. Biotechnol.* **21,** 921–926.

Petroski, M. D., and Deshaies, R. J. (2003). Context of multiubiquitin chain attachment influences the rate of Sic1 degradation. *Mol. Cell.* **11,** 1435–1444.

Pickart, C. M., and Cohen, R. E. (2004). Proteasomes and their kin: Proteases in the machine age. *Nat. Rev. Mol. Cell Biol.* **5,** 177–187.

Sadygov, R. G., Eng, J., Durr, E., Saraf, A., McDonald, H., MacCoss, M. J., and Yates, J. R., 3rd (2002). Code developments to improve the efficiency of automated MS/MS spectra interpretation. *J. Proteome Res.* **1,** 211–215.

Scherer, D. C., Brockman, J. A., Chen, Z., Maniatis, T., and Ballard, D. W. (1995). Signal-induced degradation of I kappa B alpha requires site-specific ubiquitination. *Proc. Natl. Acad. Sci. USA* **92,** 11259–11263.

Spence, J., Sadis, S., Haas, A. L., and Finley, D. (1995). A ubiquitin mutant with specific defects in DNA repair and multiubiquitination. *Mol. Cell. Biol.* **15,** 1265–1273.

Tabb, D. L., McDonald, W. H., and Yates, J. R., 3rd (2002). DTASelect and Contrast: Tools for assembling and comparing protein identifications from shotgun proteomics. *J. Proteome Res.* **1,** 21–26.

Taylor, R. S., Wu, C. C., Hays, L. G., Eng, J. K., Yates, J. R., 3rd, and Howell, K. E. (2000). Proteomics of rat liver Golgi complex: Minor proteins are identified through sequential fractionation. *Electrophoresis* **21,** 3441–3459.

Wilkinson, K. D., and Hochstrasser, M. (1998). The deubiquitinating enzymes. *In* "Ubiquitin and the Biology of the Cell" (J. M. Peters, J. R. Harris, and D. Finley, eds.), pp. 99–125. Plenum Press, New York.

Wohlschlegel, J. A., Johnson, E. S., Reed, S. I., and Yates, J. R., III (2004). Global analysis of protein sumoylation in *Saccharomyces cerevisiae*. *J. Biol. Chem.* **279,** 45662–45668.

[19] Mapping of Ubiquitination Sites on Target Proteins

By Shigetsugu Hatakeyama, Masaki Matsumoto, and Keiichi I. Nakayama

Abstract

Although the identification of ubiquitin-conjugated lysine residues on target proteins is extremely significant, to date it is generally quite difficult to identify ubiquitinated sites by usual mutation analysis. More recently, the technology of mass spectrometry is answering these difficult questions. In this chapter, we introduce the method of purification of ubiquitinated target proteins using affinity chromatography with anti-polyubiquitin antibody and the identification of ubiquitinated lysine residues on target proteins using mass spectrometry. Using these techniques, we can obtain comprehensive information about ubiquitinated proteins in various cells and tissues.

Introduction

Ubiquitin is a highly conserved 76-amino acid polypeptide present in all eukaryotes. Individual ubiquitin molecules are linked to substrates through isopeptide bonds between the carboxyl-terminal glycine of ubiquitin and ε-amino groups of lysine residues in the substrate. Additional ubiquitin monomers are added to substrate-bound ubiquitin moieties in a sequential manner, resulting in the formation of branched polyubiquitin chains through linkage between lysine-48 of the last ubiquitin in the chain and the carboxyl terminus of the new ubiquitin molecule. Protein ubiquitination is mediated by a multienzyme cascade and involves the formation by ubiquitin of thiol esters with at least two, and, in most instances, three, distinct types of enzyme: E1 (ubiquitin-activating enzyme), E2 (ubiquitin-conjugating enzyme), and E3 (ubiquitin-protein ligase). E3 enzymes catalyze the final step in the ubiquitination pathway, the formation of a stable isopeptide linkage between the carboxyl-terminal glycine of ubiquitin, and the ε-amino group of a lysine residue in the target protein (Hershko, 1983). Recently, E4 enzymes, such as UFD2 and CHIP, have been identified as ubiquitin chain assembly factors that catalyze polyubiquitin chain formation (Imai et al., 2002; Koegl et al., 1999).

A few reports have identified ubiquitin-conjugated lysine residues on target protein such as IκBα, whose lysine-21 and -22 are ubiquitinated (Baldi et al., 1996; Rodriguez et al., 1996). Recently, it has been shown that the conjugation of MyoD and p21^{Cip1} occurs by means of a novel modification involving the attachment of ubiquitin to the N-terminal residue, suggesting that the ubiquitination reaction may not require the lysine residues in target proteins (Bloom et al., 2003; Breitschopf et al., 1998). Notwithstanding these examples, analyzing the conjugation sites of ubiquitin on target protein is typically quite complicated.

Still more formidable is the analysis of the polyubiquitin chain. The formation of a polyubiquitin chain by linkage of the carboxyl terminus of a new ubiquitin molecule to lysine-48 of the last ubiquitin moiety of the chain is thought to mark a protein for proteolysis by the 26S proteasome. However, recent observations indicate that polyubiquitin chains are also assembled through conjugation to lysine residues of ubiquitin other than lysine-48, and the resulting chains seem to function in distinct biological processes. A polyubiquitin chain composed of lysine-63–linked ubiquitin moieties is implicated in DNA repair (Spence et al., 1995), the cellular response to stress (Arnason and Ellison, 1994), inheritance of mitochondrial DNA (Fisk and Yaffe, 1999), endocytosis of certain plasma membrane proteins (Galan and Haguenauer-Tsapis, 1997), and ribosomal function (Spence et al., 2000). TRAF6, a RING-finger type E3, in conjunction with the E2 Ubc13 and the Ubc-like protein Uev1A, targets lysine-63 of ubiquitin and plays an important role in the phosphorylation of IκB in the NF-κB signaling pathway (Deng et al., 2000). UFD2a, a mammalian homolog of yeast Ufd2, was found to conjugate ubiquitin not only to lysine residues of ubiquitin at positions 29, 48, or 63 but also to lysines at other positions (Hatakeyama et al., 2001). It is thus possible that polyubiquitination formation of heterogeneous or multiply branched structures on target proteins results in the biological functions distinct from the provision of a marker for proteolysis.

Although protein ubiquitination has been the subject of many studies, few comprehensive analyses of proteins that are ubiquitinated during normal biological processes have been performed. Peng et al. recently described a method for enriching ubiquitinated proteins and, with this approach, identified 72 ubiquitin-protein conjugates in yeast (Peng et al., 2003). Most of the identified proteins, however, were integral membrane proteins and not short-lived regulators, probably because these investigators did not seem to take steps, such as the use of a chemical proteasome inhibitor, to stabilize such labile proteins.

Although the identification of ubiquitin-conjugated lysine residues is extremely significant, to date it is generally quite difficult to identify ubiquitinated sites on target proteins by mutation analysis. More recently, the technology of mass spectrometry is answering these difficult questions. In this chapter, we introduce the method of purification of ubiquitinated target proteins and the analysis of ubiquitinated lysine residues using mass spectrometry.

Affinity Purification of Ubiquitinated Proteins

With regard to the analysis for tyrosine-phosphorylated proteins, many biochemical approaches using antibodies specifically recognizing tyrosine phosphorylation on proteins have previously been reported. A similar method is likely to be available for ubiquitination. However, because free or unconjugated ubiquitins are far more abundant in the cell than the amount of conjugated ubiquitins on target proteins, it is difficult to concentrate only on the ubiquitin-conjugated proteins by affinity purification using anti-ubiquitin antibody. It has been previously reported that one of the anti-ubiquitin monoclonal antibodies, FK2, can selectively recognize the polyubiquitin chains conjugated on target proteins by isopeptide bond but not free ubiquitins (Fujimuro et al., 1994). Therefore, we tried to use an affinity column conjugated with the FK2 antibody to enrich the ubiquitinated proteins. Furthermore, FK2 antibody can recognize polyubiquitin chains even in denatured condition, including 4 M urea (Matsumoto et al., in preparation). Such denatured conditions can interfere with the association of ubiquitins with ubiquitin-interacting proteins. Thus, the FK2 antibody can collect both ubiquitinated proteins and ubiquitin-interacting proteins in native state as Urp-N (*u*biquitin-*r*elated *p*roteome only under the *n*ative condition), but only ubiquitinated proteins in denatured condition as Urp-D (*u*biquitin-*r*elated *p*roteome under the *d*enaturing condition). Then they were digested with trypsin, and the resulting peptides were analyzed by two-dimensional liquid chromatography and tandem mass spectrometry (Fig. 1). As described previously, ubiquitins are linked to substrates through isopeptide bonds between the carboxyl-terminal glycine of ubiquitin and ε-amino groups of lysine residues in the substrate. Because the carboxyl-terminal sequence of ubiquitin has -Arg-Gly-Gly (-R-G-G), ubiquitinated proteins generate peptides with Gly-Gly on lysine residues by trypsin treatment. Such peptides with Gly-Gly are distinguishable from peptides without Gly-Gly residues by MS/MS analysis using mass spectrometry (Fig. 2). Eventually, mass spectrometry is likely to lead to the identification of the ubiquitinated sites of target proteins. Our results demonstrate the potential of proteomics analysis of protein ubiquitination to provide important insight into the regulation of protein stability and other ubiquitin-related cellular functions.

Experimental Procedures

Reagents

Cross-Linker. 50 mM dimethyl pimelimidate (Pierce, Rockford, IL) in 100 mM triethanolamine-HCl (pH 8.3).

Buffer A. 50 mM Tris-HCl (pH 7.4), 300 mM NaCl, 0.5% Triton X-100, aprotinin (10 µg/ml), leupeptin (10 µg/ml), 1 mM PMSF, 400 µM Na$_3$VO$_4$, 400 µM EDTA, 10 mM NaF, and 10 mM sodium pyrophosphate.

Generation of Anti-Ubiquitin-Conjugated Column

FK2 monoclonal antibody against polyubiquitin chain can be purchased from Nippon Bio Test Lab. (Tokyo, Japan) or MBL (Nagoya, Japan). To prepare an immunoaffinity column, we added 2 mg of FK2 mouse mAb to ubiquitin in 1 ml of PBS to 1 ml of protein A–Sepharose (Amersham-Pharmacia Biotech, Little Chalfront, UK) and rotated the mixture for 2 h

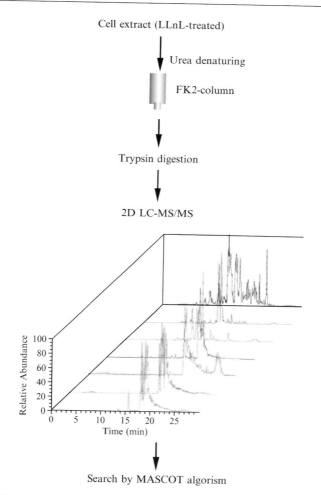

FIG. 1. (A) Strategy for analysis of ubiquitin-related proteome from mammalian cells. Cell extracts were loaded onto the FK2-column. After extensive washing, bound proteins were eluted with 100 mM glycine-HCl and concentrated. Then, proteins were digested with trypsin, and the resulting peptides were separated by strong cation exchange chromatography. Each fraction was analyzed by LC-MS/MS system and followed by MASCOT™ search.

at 4°. The resin was then washed with 100 mM triethanolamine-HCl (pH 8.3), and the antibody was cross-linked to the resin by incubation for 4 h at 4° with 50 mM dimethyl pimelimidate in 100 mM triethanolamine-HCl (pH 8.3). The reaction was terminated by incubation of the resin for 2 h

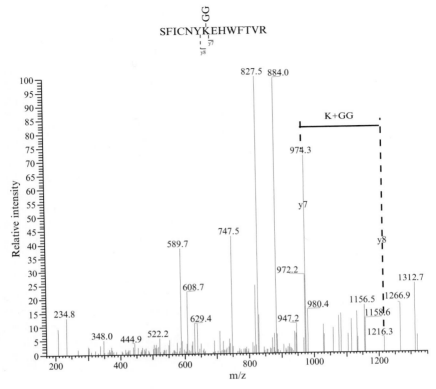

Fig. 2. An example of MS/MS spectrum with ubiquitin-signature. MS/MS spectrum of ubiquitinated peptide from ataxin-3 (SFICNYKEHWFTVR) was represented. Note the gap between y-8 and y-7 ion corresponding to GG-conjugated lysine (K+114).

at room temperature with 0.1 M Tris-HCl (pH 7.4). Unconjugated antibody was removed from the resin by washing with 100 mM glycine-HCl (pH 2.8), after which the resin was equilibrated with PBS.

Cell Culture with Proteasome Inhibitor

HEK293T cells were cultured under an atmosphere of 5% CO_2 at 37° in DMEM (Roche, Indianapolis, IN) supplemented with 10% FBS (Invitrogen, Groningen, The Netherlands) and antibiotics. The accumulation of ubiquitinated proteins in the cells was promoted by culturing in the presence of 20 μM leucyl-leucyl-norleucinal (LLnL) (Roche) for 8 h. The cells (5×10^8) were then harvested, lysed in 10 ml of buffer A, and centrifuged at 16,000g for 10 min at 4°.

Affinity Chromatography in Denaturing Condition

Under the denaturing condition, urea was added to the supernatant of the cell lysate to a final concentration of 8 M, and the mixture was incubated for 30 min at room temperature, cooled to 4°, and combined with an equal volume of buffer A before application to the FK2 column after its equilibration with buffer A containing 4 M urea. After washing the column with buffer A containing 4 M urea and then with PBS, bound proteins were eluted with two volumes of 100 mM glycine-HCl (pH 2.8). The eluates from the column were concentrated by precipitation with trichloroacetic acid or by chromatography as described in the following.

Consequently, 10–20 μg of proteins were usually recovered from 200 mg of cellular proteins. We can estimate the enrichment of ubiquitinated proteins by Western blot analysis with FK2. Thus, ubiquitinated proteins/peptides are enriched to approximately 1000 times as much as the concentration of ubiquitinated proteins in cell extracts.

In-Gel Digestion of Purified Proteins

Proteins purified with the FK2 column were concentrated as described previously, fractionated by SDS-PAGE, and stained with Coomassie blue G250. Protein bands were excised from the stained gel and were washed and destained with acetonitrile. After reduction with 10 mM DTT, proteins were alkylated with 55 mM iodoacetamide and then digested for 16 h at 37° with sequence-grade trypsin (Promega, Madison, WI). The resulting peptides were sequentially extracted from the gel with 0.1% trifluoroacetic acid (TFA) in 2% acetonitrile, 0.1% TFA in 33% acetonitrile, and 0.1% TFA in 70% acetonitrile. The combined extracts were evaporated, and the residue was dissolved with 0.1% TFA in 2% acetonitrile.

In-Solution Digestion of Purified Proteins

Proteins purified with the FK2 column were concentrated by chromatography on a Tskgel Phenyl 5PW RP column (Tosoh, Japan). After elution from the column with 0.1% TFA in 70% acetonitrile, proteins were dried by centrifugation under vacuum and dissolved in a solution containing 100 mM Tris-HCl (pH 8.0), 6 M urea, and 0.05% SDS. They were then reduced with 2 mM DTT, alkylated with 10 mM iodoacetamide, and digested with trypsin (1 μg/ml) for 16 h at 37°. The resulting peptides were applied to a cation-exchange column (1.0 × 50 mm, Polysulfoethyl A; Poly LC, Columbia, MD) that had been equilibrated with a solution containing 0.2% formic acid and 25% acetonitrile. The peptides were eluted with a gradient of 0 to 350 mM ammonium acetate in the column equilibration buffer. Eighteen fractions (100 μl) were collected and evaporated, and the dried peptides were dissolved with 0.1% TFA in 2% acetonitrile.

MS and Database Searching

Peptides were subjected to chromatography on a C_{18} column (0.2 mm × 5 cm, L-column) with an HPLC system (Magic 2002) (Michrom Bio-Resources, Auburn, CA) that was directly coupled to an ion-trap mass spectrometer (LCQ-Deca, Finnigan) (Thermo Electron Corp. San Jose, CA) equipped with a nano-LC electrospray ionization source (AMR). Collision-induced dissociation (CID) spectra were acquired automatically in the data-dependent scan mode with the dynamic exclusion option. Uninterpretable CID spectra were compared with the International Protein Index (IPI, European Bioinformatics Institute) with the use of MASCOT algorithm. Assigned high-scoring peptide sequences were processed by in-house software (data integration and mining system, or DIMS) and were manually confirmed by comparison with the corresponding CID spectra.

Comments and Critical Points

When the yield of ubiquitinated proteins by affinity chromatography is low, the temperature in denaturation by urea may be high. Thus, it is recommended that samples be incubated with urea at exactly 25°. Because the incubation at higher temperature may cause the ubiquitin chain to completely denature, FK2 may not effectively collect ubiquitinated proteins. When we actually incubated with urea at high temperature (e.g., 50°), we could not recover few ubiquitinated proteins. As another reason, de-ubiquitinating enzyme activity may interfere with the purification of the intact ubiquitinated proteins. It is better to add alkylating reagents such as iodoacetamide (10 mM) or N-ethylmaleimide (10 mM) into cell lysates to inhibit the activity of deubiquitinating enzymes.

By these procedure, we usually identify approximately 200–300 ubiquitinated proteins from a single experiment using 100-mg cell extract. Although ubiquitinated ubiquitin peptide such as Gly-Gly-Lys48, Gly-Gly-Lys6, and Gly-Gly-Lys63 can be easily identified, it is difficult to identify Gly-Gly-peptides on substrate proteins (usually 5–10 peptides per one experiment). Repetitive experiments may increase the number of identified proteins and facilitate identification of low abundant proteins, as well as detection of peptides with ubiquitin signature.

Summary

In this chapter, we introduced the ubiquitin-related proteome analysis using affinity chromatography with anti-polyubiquitin antibody to enrich ubiquitinated proteins and showed the possibility of the identification of ubiquitinated lysine residues on target proteins. We were able to distinguish ubiquitinated proteins from the contamination of ubiquitin binding proteins by using the high reactivity of the FK2 antibody even in denatured condition with 8 M urea and effectively yield the real ubiquitinated proteins by the high specificity of FK2 antibody. Although several ubiquitinated sites of proteins were identified by our methods, not all ubiquitinated proteins may be identified. Furthermore, the results should be confirmed through other methods. Recently, affinity chromatography using a metal chelating column that uses whole cell extracts from yeast expressing hexa-histidine-tagged ubiquitin has been reported (Peng *et al.*, 2003). Moreover, ER-associated ubiquitinated proteins in yeast cells were analyzed using a similar strategy (Hitchcock *et al.*, 2003). Thus, affinity purification of hexa-histidine-tagged ubiquitin is likely to be a powerful tool to explore ubiquitin-related proteome in yeast. These studies showed the many ubiquitinated lysines on target proteins. But ubiquitin-related proteome analysis using mammalian cells is quite difficult, because the amount of protein is much more abundant than with yeast. However, it is possible to identify the ubiquitinated sites even in mammalian cells with new advanced procedures in mass spectrometry or biochemical techniques.

Generally, ubiquitinated proteins are not clearly separated by electrophoresis because of a uniform number of ubiquitin chains. Therefore, it is difficult to quantify ubiquitinated proteins by classical electrophoresis-based methods, such as SDS-PAGE and 2D-PAGE. Recently, quantitative analysis methods have been established by using stable-isotope labeling and mass spectrometry (Veenstra *et al.*, 2000). In these strategies, proteins are quantified at the level of peptides derived from corresponding protein. Therefore, these methods have an advantage for the quantification of ubiquitinated proteins and may provide functional information about ubiquitin-related proteome. For example, the effects of proteasome inhibitor on ubiquitinated proteins could be verified by such methods. Such information may yield important new insight into the regulation of cellular structure and function.

References

Arnason, T., and Ellison, M. J. (1994). Stress resistance in *Saccharomyces cerevisiae* is strongly correlated with assembly of a novel type of multiubiquitin chain. *Mol. Cell. Biol.* **14,** 7876–7883.

Baldi, L., Brown, K., Franzoso, G., and Siebenlist, U. (1996). Critical role for lysines 21 and 22 in signal-induced, ubiquitin-mediated proteolysis of I kappa B-alpha. *J. Biol. Chem.* **271,** 376–379.

Bloom, J., Amador, V., Bartolini, F., DeMartino, G., and Pagano, M. (2003). Proteasome-mediated degradation of p21 via N-terminal ubiquitinylation. *Cell* **115,** 71–82.

Breitschopf, K., Bengal, E., Ziv, T., Admon, A., and Ciechanover, A. (1998). A novel site for ubiquitination: The N-terminal residue, and not internal lysines of MyoD, is essential for conjugation and degradation of the protein. *EMBO J.* **17,** 5964–5973.

Deng, L., Wang, C., Spencer, E., Yang, L., Braun, A., You, J., Slaughter, C., Pickart, C. M., and Chen, Z. J. (2000). Activation of the IkappaB kinase complex by TRAF6 requires a dimeric ubiquitin-conjugating enzyme complex and a unique polyubiquitin chain. *Cell* **103,** 351–361.

Fisk, H. A., and Yaffe, M. P. (1999). A role for ubiquitination in mitochondrial inheritance in *Saccharomyces cerevisiae*. *J. Cell Biol.* **145,** 1199–1208.

Fujimuro, M., Sawada, H., and Yokosawa, H. (1994). Production and characterization of monoclonal antibodies specific to multi-ubiquitin chains of polyubiquitinated proteins. *FEBS Lett.* **349,** 173–180.

Galan, J. M., and Haguenauer-Tsapis, R. (1997). Ubiquitin lys63 is involved in ubiquitination of a yeast plasma membrane protein. *EMBO J.* **16,** 5847–5854.

Hatakeyama, S., Yada, M., Matsumoto, M., Ishida, N., and Nakayama, K. I. (2001). U box proteins as a new family of ubiquitin-protein ligases. *J. Biol. Chem.* **276,** 33111–33120.

Hershko, A. (1983). Ubiquitin: Roles in protein modification and breakdown. *Cell* **34,** 11–22.

Hitchcock, A. L., Auld, K., Gygi, S. P., and Silver, P. A. (2003). A subset of membrane-associated proteins is ubiquitinated in response to mutations in the endoplasmic reticulum degradation machinery. *Proc. Natl. Acad. Sci. USA* **100,** 12735–12740.

Imai, Y., Soda, M., Hatakeyama, S., Akagi, T., Hashikawa, T., Nakayama, K., and Takahashi, R. (2002). CHIP is associated with Parkin, a gene responsible for familial Parkinson's disease, and enhances its ubiquitin ligase activity. *Mol. Cell.* **10,** 55–67.

Koegl, M., Hoppe, T., Schlenker, S., Ulrich, H. D., Mayer, T. U., and Jentsch, S. (1999). A novel ubiquitination factor, E4, is involved in multiubiquitin chain assembly. *Cell* **96,** 635–644.

Peng, J., Schwartz, D., Elias, J. E., Thoreen, C. C., Cheng, D., Marsischky, G., Roelofs, J., Finley, D., and Gygi, S. P. (2003). A proteomics approach to understanding protein ubiquitination. *Nat. Biotech.* **21,** 921–926.

Rodriguez, M. S., Wright, J., Thompson, J., Thomas, D., Baleux, F., Virelizier, J. L., Hay, R. T., and Arenzana-Seisdedos, F. (1996). Identification of lysine residues required for signal-induced ubiquitination and degradation of I kappa B-alpha *in vivo.*. *Oncogene* **12,** 2425–2435.

Spence, J., Gali, R. R., Dittmar, G., Sherman, F., Karin, M., and Finley, D. (2000). Cell cycle-regulated modification of the ribosome by a variant multiubiquitin chain. *Cell* **102,** 67–76.

Spence, J., Sadis, S., Haas, A. L., and Finley, D. (1995). A ubiquitin mutant with specific defects in DNA repair and multiubiquitination. *Mol. Cell. Biol.* **15,** 1265–1273.

Veenstra, T. D., Martinovic, S., Anderson, G. A., Pasa-Tolic, L., and Smith, R. D. (2000). Proteome analysis using selective incorporation of isotopically labeled amino acids. *J. Am. Soc. Mass Spectrom.* **11,** 78–82.

[20] Identification of Substrates for F-Box Proteins

By JIANPING JIN, XIAOLU L. ANG,
TAKAHIRO SHIROGANE, and J. WADE HARPER

Abstract

F-box proteins serve as specificity factors for a family of ubiquitin protein ligases composed of Skp1, Cul1, and Rbx1. In SCF complexes, Cul1 serves as a scaffold for assembly of the catalytic components composed of Rbx1 and a ubiquitin-conjugating enzyme and the specificity module composed of Skp1 and an F-box protein. F-box proteins interact with Skp1 through the F-box motif and with ubiquitination substrates through C-terminal protein interaction domains such as WD40 repeats. The human genome contains ~68 F-box proteins, which fall into three major classes: Fbws containing WD40 repeats, Fbls containing leucine-rich repeats, and Fbxs containing other types of domains. Most often, F-box proteins interact with their targets in a phosphorylation-dependent manner. The interaction of F-box proteins with substrates typically involves a phosphodegron, a small peptide motif containing specific phosphorylation events whose sequence is complementary to the F-box protein. The identification of substrates of F-box proteins is frequently a challenge because of the relatively weak affinity of substrates for the requisite F-box protein. Here we describe approaches for the identification of substrates of F-box proteins. Approaches include stabilization of ubiquitination targets by Cul1-dominant negatives, the use of shRNA hairpins to disrupt F-box protein expression, and the use of collections of F-box proteins as biochemical reagents to identify interacting proteins that may be substrates. In addition, we describe approaches for the use of immobilized phosphopeptides to identify F-box proteins that recognize particular phosphodegrons.

Introduction

F-box proteins function as specificity factors for a family of ubiquitin ligases that use Cul1 as a scaffold (reviewed in Deshaies, 1999; Koepp et al., 1999; Patton et al., 1998a). These complexes, referred to as *SCF ubiquitin ligases,* contain four major components (Feldman et al., 1997; Patton et al., 1998b; Skowyra et al., 1997). The C-terminus of Cul1 binds the Ring-finger protein Rbx1 (also called Roc1 and Hrt1) to form the core E3 ubiquitin ligase, which binds and activates the E2 ubiquitin conjugating enzyme (Seol et al., 1999; Skowyra et al., 1999; Zheng et al., 2002) (Fig. 1). Cul1 is

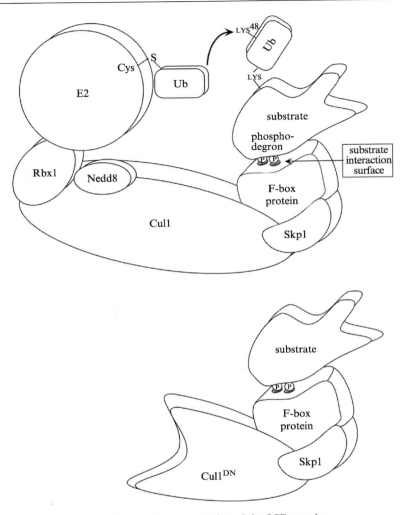

FIG. 1. Schematic representation of the SCF complex.

also modified by the ubiquitin-like protein Nedd8 (Deshaies, 1999). This modification greatly stimulates ubiquitin ligase activity. The specificity component is composed of Skp1 and a member of the F-box family of proteins (Bai *et al.*, 1999). Skp1 binds to the N-terminus of Cul1 and also interacts with the F-box motif. F-box proteins also contain additional protein interaction domains that bind substrates. In many cases, the interaction of an ubiquitylation substrate with the F-box protein is phosphorylation dependent.

Currently, 68 F-box–containing genes have been identified in the human genome (Jin et al., 2004). However, only a small fraction of these have been examined biochemically or genetically. Most frequently, individual F-box proteins interact with multiple ubiquitylation substrates. For example, β-TRCP (Fbw1) functions in the ubiquitylation of β-catenin (Latres et al., 1999; Winston et al., 1999), IκBα (Spencer et al., 1999; Winston et al., 1999; Yaron et al., 1998), Emi1 (Margottin-Goguet et al., 2003), and Cdc25A (Busino et al., 2003; Jin et al., 2003). In contrast, Fbw7/Sel-10 functions in the ubiquitylation of cyclin E, c-myc, notch-1, and c-jun (Hubbard et al., 1997; Koepp et al., 2001; Strohmaier et al., 2001; Welcker et al., 2004; Yada et al., 2004). Given that there is a large number of F-box proteins in eukaryotic genomes, a major emphasis of current research in the field concerns the identification of targets of ubiquitin ligases in general and SCF complexes in particular. In this chapter, we describe several approaches that can be used to identify F-box proteins that control the turnover of proteins whose degradation is known to be controlled in a phosphorylation-dependent manner.

Overview of Approaches

Several approaches have been used to identify F-box protein involved in the degradation of particular substrates. Typically, the identification of relevant F-box proteins has relied on prior knowledge concerning turnover of a particular ubiquitylated protein. In most cases, the ubiquitylation protein is known to be rapidly turned over in cells, either constitutively or in response to particular signals. Frequently, protein kinases important for turnover are known, because they are the identity of phosphorylation sites within the protein that are important for turnover. Phosphorylation of ubiquitylation targets by relevant kinases leads to the formation of a phosphorylated motif, referred to as a "phosphodegron," which is capable of specifically interacting with the appropriate F-box protein (Koepp et al., 2001; Nash et al., 2001; Winston et al., 1999). To date, two crystal structures of phosphodegrons bound to F-box proteins have been reported. The β-catenin phosphodegron (DpSGIHpS) binds to the surface of the WD40 repeats in β-TRCP in a highly specific manner by means of clusters of arginine residues (Wu et al., 2003). A distinct phosphodegron derived from cyclin E (LPpTPP) has been crystallized with yeast Cdc4, which is the apparent isolog of the human F-box protein Fbw7 (Koepp et al., 2001; Strohmaier et al., 2001). In this case, a single phosphothreonine provides the major interactions required for recognition by Cdc4 (Orlicky et al., 2003).

Before attempting to identify F-box proteins for particular ubiquitylation targets, it is frequently useful to determine whether the SCF is involved using genetic approaches. In situations in which the ubiquitylation target is known to be properly degraded in yeast, it is possible to use temperature-sensitive mutants in core SCF components to examine whether turnover *in vivo* requires the SCF. Temperature-sensitive mutations in *skp1*, the E2 ubiquitin–conjugating enzyme *cdc34*, or *cdc53* (the budding yeast Cul1 isolog) have been used for this purpose (Koepp *et al.*, 2001; Strohmaier *et al.*, 2001). Alternately, overexpression of cullin-dominant negative proteins can be used to examine whether turnover of a particular protein involves a specific cullin family member. Thus far, this approach has been used for targets of both Cul1 (Donzelli *et al.*, 2002; Jin *et al.*, 2003) and Cul3 (Cullinan *et al.*, in press).

Once the involvement of Cul1 has been established, biochemical methods can be used to screen for F-box proteins that interact with the ubiquitylation target in a manner that depends on either the protein kinase of interest or particular phosphorylation sites that are critical for turnover (Jin *et al.*, 2003; Koepp *et al.*, 2001; Winston *et al.*, 1999). Such experiments frequently make use of a collection of F-box protein expression constructs that allow production of these proteins either by *in vitro* transcription-translation or *in vivo* by transfection. Although screening approaches have been useful, direct purification of ubiquitylation targets has also been used to identify relevant F-box proteins, as in the case of the identification of β-TRCP as the ubiquitin ligase for IκBα (Yaron *et al.*, 1998). However, the general utility of such an approach is currently unknown. For example, during the systematic affinity purification of proteins from budding yeast using affinity purification approaches, few known substrates of SCF complexes were purified as proteins associated with F-box proteins (Ho *et al.*, 2002). This may reflect either the relatively weak interactions that F-box proteins display toward their targets or the labile nature of the substrate *in vivo* once complexed with the F-box protein. In rare cases, ubiquitylation targets have been identified using two-hybrid systems. For example, the ubiquitylation target Cdc6 was found to interact with the F-box protein Cdc4 in the two-hybrid system (Drury *et al.*, 1997). Likewise, the target of the Met30 F-box protein, Met4, was captured in the two-hybrid system (Thomas *et al.*, 1995). The utility of the two hybrid system for the identification of multicellular F-box protein–substrate pairs may be limited by the absence of appropriate kinases in yeast that are required to generate relevant phosphodegrons in the substrate. Once candidate F-box proteins are identified, loss-of-function experiments that use RNAi coupled with reconstitution of ubiquitin ligase reactions are required to verify the involvement of particular F-box proteins in turnover of the substrate of interest.

Use of Dominant-Negative Cul1 to Identify SCF Targets

A truncated cDNA version of Cul1—containing the first 452 amino acid residues followed by a TGA stop codon—is cloned into a mammalian expression vector with a strong promoter (e.g., pcDNA3 with a CMV promoter; Jin et al., 2003). This truncated protein acts as a dominant-negative (DN) version of Cul1, because even though it has lost its ability to interact with Rbx1 and the E2 Ub–conjugating enzyme, it retains its interaction with Skp1 and, indirectly, the F-box proteins (Fig. 1). Accordingly, overexpression of Cul1^{DN} ultimately sequesters the F-box proteins, depleting them from the cellular pool and blocking the Cul1-based Ub-pathway from ubiquitylating its targets. In such a situation, an SCF target would be expected to be stabilized. To date, this approach has been used with several substrates, including cyclin E (Jin et al., 2003), p27 (Jin et al., 2003), and Cdc25A (Donzelli et al., 2002; Jin et al., 2003). Described below is a protocol investigating the DNAdamage–induced turnover of Cdc25A, as well as the steady-state levels of cyclin E and p27 (Jin et al., 2003). It is frequently advantageous to initially examine the steady-state level of the protein in question in the presence and absence of overexpressed Cul1^{DN}, because increased accumulation at steady state provides an indication that the stability has been altered. This can be done by examining the total levels of the protein in question using immunoblotting of whole cell extracts. This should be followed by a direct examination of the stability of the protein in question, using either pulse-chase or cycloheximide (CHX) chase approaches. In many cases, protein degradation through the SCF pathway is activated by particular stimuli. Thus, it is also possible to examine the time-course of protein turnover after the appropriate stimuli in the presence and absence of Cul1^{DN} expression. Each of these approaches will be described below. The use of these approaches requires that the candidate ubiquitylation target be detectable by immunoblotting or by immunoprecipitating from ^{35}S-methionine–labeled cells using available antibodies. If suitable antibodies are not available, it is also possible to use an analogous approach with transiently transfected plasmids encoding epitope-tagged versions of the ubiquitylation target of interest. Although the specific examples used concern Cul1, similar experiments can be performed with other cullins, including Cul3 (Cullinan et al., 2004).

Transfection

1. Seed HEK293T cells in 6-well plates (~1 million cells/well).
2. Approximately 24 h after seeding the cells, transfect 3 μg of pcDNA3-Cul1^{DN} construct or 3 μg of empty vector pcDNA3 into appropriate wells

using lipofectamine 2000 or an analogous transfection reagent. At the time of transfection, cells should be ~70–80% confluent. Typically, this quantity of plasmid produces sufficient quantities of Cul1^{DN} to block much of the SCF-dependent ubiquitylation in the transfected cells (Jin et al., 2003). However, when analyzing the bulk cell population, it is important to remember that the transfection efficiency will determine the overall extent of stabilization of SCF targets. If cotransfection of an epitope-tagged version of the candidate target is required, it is imperative that the quantity of Cul1^{DN} expression plasmid be in excess of that used for the candidate target protein. Typically, a 5:1 ratio will suffice.

Steady-State Protein Accumulation

1. Forty-eight hours after transfection, cells are washed with 1 ml of ice-cold PBS, collected in Eppendorf tubes, and centrifuged for 1 min at 2500g in a Microfuge.
2. Resuspend cell pellets in NETN lysis buffer (50 mM Tris-HCl, pH 7.5; 100 mM NaCl; 0.5% NP40; 1 mM EDTA) with protease inhibitors, and incubate lysates on ice for 10 min.
3. Clear debris and membranes from cell lysates by centrifugation at 20,000g for 10 min at 4°. Transfer supernatant (lysates) to new tubes.
4. Quantitate the total protein concentration in each tube using Bio-Rad Protein Assay reagent (Cat# 500-0006) or equivalent as described by the supplier.
5. Forty micrograms of total protein is then subjected to SDS-PAGE before transfer to nitrocellulose and immunoblotting. Protein detection is then accomplished using relevant antibodies. In the example shown in Fig. 2, blots were probed with anti-Cul1 to detect Cul1^{DN} with several SCF targets (cyclin E, p27, and Cdc25A). In addition, blots should be stripped in 0.1 M glycine (pH 3) for 5 min and re-probed with an appropriate loading control to verify equal protein loading (e.g., anti-tubulin). As shown in Fig. 2A, cyclin E, p27, and Cdc25A protein levels all accumulated in response to overexpression of Cul1^{DN}, whereas the levels of Cdc25C (which is not an SCF substrate) were unchanged.

^{35}S-Methionine Pulse Chase

Optimally, the effects of Cul1^{DN} on turnover of a particular protein should be examined by pulse-chase analysis. If immunoprecipitating antibodies are available, this can be accomplished by metabolic labeling using ^{35}S-methionine. With some proteins, the methionine content is relatively low, making metabolic labeling difficult. In such cases, CHX-chase

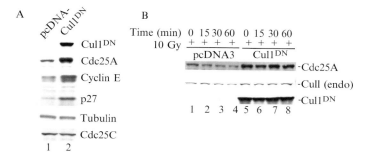

FIG. 2. Use of Cul1 dominant-negative proteins (Cul1^{DN}) to assess the involvement of the SCF pathway in controlling the degradation of proteins of interest. (A) 293T cells were transfected with 3 μg of pcDNA3-Cul1^{DN} or empty vector. After 48 h, cell extracts were subjected to immunoblotting with the indicated antibodies. (B) Effect of Cul1^{DN} on DNA damage-dependent degradation of Cdc25A in 293T cells. 293T cells were transfected with 3 μg of pcDNA3-Cul1^{DN} or empty vector. After 48 h, cells were subjected to γ-irradiation (10 Gy). Cells were harvested at the indicated times and extracts then subjected to immunoblotting with the indicated antibodies. Reprinted by permission from Cold Spring Harbor Press.

experiments may also be performed (see below). Because each protein has a particular turnover rate, preliminary experiments need to be performed to determine the approximate half-life. Once a half-life measurement is obtained, sufficient dishes for several time points that span the half-life should be used.

1. Forty-eight hours after transfection, cells are washed twice with PBS and subsequently incubated in culture medium lacking methionine for 30 min to 1 h followed by addition of 100 μCi of ^{35}S-methionine (1000 Ci/mmole)/ml of medium for 30 min. It is critical that the medium being used both before and during the labeling be free of exogenous methionine, and this can be achieved by using dialyzed serum.

2. After completion of the labeling period, the methionine-containing medium is removed and the cells supplemented with fresh media containing 1 mM methionine. Cells are harvested at 0 h (time of addition of fresh methionine-containing medium) and at time points spanning the half-life, and lysed in buffers containing detergents (e.g., RIPA buffer). The use of RIPA buffer facilitates more efficient immunoprecipitation of the target protein.

3. Equal quantities of cell extracts are then used for immunoprecipitation using specific antibodies. Immune complexes are then separated by SDS-PAGE and dried gels subjected to autoradiography or phosphoimager analysis. The later method is advantageous, because it allows relative

levels of labeled proteins to be quantitated more accurately. The time at which the quantity of metabolically labeled protein is reduced to 50% of that found at 0 time is considered the half-life of the protein. The expectation is that if the SCF complex is involved in degradation of the target protein, the half-life will be substantially extended in the presence of Cul1^{DN}.

Cycloheximide Time Course

For some proteins that are difficult to metabolically label or for which immunoprecipitating antibodies are not available, a CHX time course coupled with immunoblotting can be used to estimate the half-life. CHX is an inhibitor of *de novo* translation, and its presence results in a reduction in the abundance of proteins that are intrinsically unstable.

CHX is applied 48 h after transfection by replacing the media on the cells with fresh media containing CHX at a final concentration of 25 μg/ml. Again, the number of time points required will depend on the half-life of the protein but should contain times spanning the half-life.

Cells are harvested at appropriate time points by titration in PBS on ice. Cells are lysed as described previously for measurement of steady-state protein levels. Cell extracts are then subjected to SDS-PAGE and immunoblotting using specific antibodies, and the blots are re-probed with a loading control antibody (e.g., tubulin) to demonstrate equal protein levels. Proteins that are degraded through the SCF pathway are expected to have an extended half-life in the presence of Cul1^{DN} compared with cells transfected with a control plasmid. Previous studies have used this type of analysis to demonstrate a role for Cul1 in the degradation of Cdc25A (Donzelli *et al.*, 2002).

Stimulus-Induced Turnover

In some cases, degradation of particular proteins is triggered by cellular signals. One example of this is DNA damage–dependent degradation of Cdc25A. Cdc25A is normally unstable during S-phase, but its turnover is increased in response to DNA damage. Evidence of the involvement of the SCF in this process was deduced using Cul1^{DN} to block DNA damage dependent turnover as described in the following.

1. Forty-eight hours after transfection with either pcDNA3 or pcDNA3-Cul1^{DN}, cells were treated with γ-irradiation (10Gy).

2. Cells were washed with PBS and lysed in NETN buffer at 0, 15, 30, and 60 min after irradiation.

3. Clear cell extracts were quantitated as described previously and subjected to SDS-PAGE followed by immunoblotting for Cdc25A. As shown in Fig. 2, γ-irradiation leads to decreased steady-state levels of Cdc25A over time. However, expression of Cul1^{DN} leads to increased levels of Cdc25A at the 0 h time point, and its levels were not decreased in response to DNA damage, implicating the SCF in DNA damage–inducible Cdc25A degradation. Immunoblotting with anti-Cul1 antibodies revealed equal quantities of proteins during the time course, as well as demonstrating constant expression of Cul1^{DN} (Fig. 2B).

Screening F-Box Proteins for Interaction with Substrates *In Vivo*

Having validated the involvement of the SCF pathway in degradation of the protein of interest using Cul1^{DN}, it is then possible to begin a search for relevant specificity components (i.e., F-box proteins). Because F-box proteins bind tightly and specifically to their targets, protein interaction approaches can be used to identify candidate F-box proteins. Several F-box proteins have been shown to bind tightly enough to their targets to be immunoprecipitated as a complex from extracts of mammalian cells. Although association between F-box proteins and substrates has been detected at endogenous protein levels (Jin *et al.*, 2003), most frequently association is demonstrated with overexpressed proteins. The ability to form stable complexes when overexpressed may reflect the fact that the proteins are present at high levels, possibly overwhelming the downstream degradation machinery. In cases in which overexpression of the F-box protein of interest leads to degradation of the protein of interest, it is possible to overexpress Cul1^{DN} as well, thereby uncoupling binding of the target to the F-box protein from the ubiquitylation process. In the following section, we describe approaches for identifying F-box proteins involved in degradation of particular proteins (i.e., Cdc25A) and approaches for verifying the use of the identified F-box protein under normal physiological conditions.

Cloning and Expression of Mammalian F-Box Proteins

The identification of relevant F-box proteins through protein interaction approaches requires the availability of expression vectors for these proteins. Recent studies have led to the identification of 68 human F-box proteins and their corresponding mouse orthologs (Jin *et al.*, 2004). These proteins fall into three main classes: those containing WD40 repeats (Fbws), those containing leucine rich repeats (Fbls), and those containing other classes of functional domains (Fbxs). To facilitate the identification of substrates for these proteins, we have cloned and expressed a number of these proteins in mammalian cells (Jin *et al.*, 2003; Koepp *et al.*, 2001). Open reading frames for β-TRCP1(Fbw1), β-TRCP2 (Fbw11), Fbl3, Fbw8, Fbw4, Fbl6, Fbl2, Fbw7, Fbl12, Fbl16, Fbw10, Fbl7, Fbx28, Fbx30, Fbl8, Fbx16, and Fbl18 were sub-cloned into pUNI50 (Liu *et al.*, 2000). pUNI50 represents the entry vector in the Universal Plasmid Subcloning (UPS) system (Liu *et al.*, 2000). In this system, open reading frames in pUNI can be placed under different expression contexts using *in vitro* Cre recombinase mediated sub-cloning. To create mammalian expression vectors for the F-box proteins, we used Cre-mediated fusion reactions using pUNI50 plasmids and pCMV-Myc-lox. Cre-mediated recombination places the Myc-tagged F-box open reading frame under control of the CMV promoter. As shown in Fig. 3, expression from these plasmids is readily detectable in extracts of HEK293T cells after transient transfection (see following for experimental details). These plasmids provide a resource for identifying F-box proteins that interact with particular targets.

To identify F-box proteins that interact with a particular target, co-transfection assays coupled with binding assays are performed. In the example shown, pCMV-GST-Cdc25A or pCMV-GST as control is cotransfected with vectors expressing a particular F-box protein. Subsequently, GST-pull-down assays followed by immunoblotting are used to determine whether one or more F-box proteins stably assemble with the degradation target Cdc25A. Other candidate F-box proteins may be substituted in this assay for Cdc25A.

Transfection

1. Seed HEK293T cells in 12-well plates (half million cells/well).
2. When cells are 70–80% confluent, cells are transfected with 1 μg of each of 23 pCMV-Myc-F-box protein expression plasmids and 1 μg of either pCMV-GST or pCMV-Cdc25A using lipofectamine 2000 (Invitrogen).

Binding Assay of F-Box Proteins with Your Protein of Interest

1. Forty-eight hours after transfection, cells are harvested and lysed with NP-40 lysis buffer (50 mM Tris-HCl, pH 7.5, 150 mM NaCl, 10 mM NaF, 0.5% NP-40) containing 10 mM β-glycerophosphate, 10 mM p-nitrophenyl phosphate (NPP), and 100 nM okadaic acid, protease inhibitor tablets (Roche).
2. Cleared lysates are transferred to new tubes and equal quantities of protein incubated with 5 μl of GSH-Sepharose (Pharmacia) for 1 h at 4°.

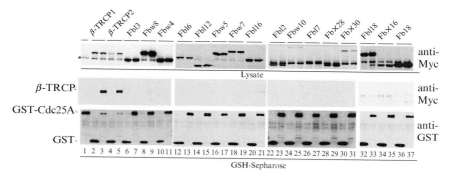

Fig. 3. Screening a panel of F-box proteins for association with Cdc25A reveals specific association with β-TRCP1 and β-TRCP2. Vectors expressing the indicated Myc-tagged F-box proteins (1 μg) were cotransfected into 293T cells with either pCMV-GST or pCMV-GST-Cdc25A (1 μg) and association with GST-Cdc25A determined 48 h later after purification using GSH-Sepharose. Extracts or GSH-Sepharose–associated proteins were examined by immunoblotting. Names of F-box proteins conforms to the new nomenclature recently adopted by the Human Genome Nomenclature Committee (Jin et al., 2004). Reprinted by permission from Cold Spring Harbor Press.

3. Beads are then washed three times with 1 ml of NP-40 lysis buffer, and the bound proteins are analyzed by SDS-PAGE and immunoblotting. In parallel, 40 μg of crude extract is subjected to SDS-PAGE to determine the expression of each of the transfected F-box proteins. Blots are probed first with anti-myc tag (9E10) antibodies to determine the identity of any associated F-box proteins. Subsequently, blots are stripped and reprobed with anti-GST antibodies to demonstrate expression of GST and GST-Cdc25A. With this approach, it was demonstrated that Cdc25A specifically associates with β-TRCP1 and β-TRCP2 but did not associate with a large number of other F-box proteins (Jin et al., 2003) (Fig. 3).

This approach can be modified, including the use of alternative epitope tags on the target protein of interest. Once a candidate F-box protein is identified, it is frequently possible to demonstrate that the endogenous target and the endogenous F-box protein form a complex by coimmunoprecipitation, possibly assisted by the use of proteasome inhibitors.

Use of RNAi to Validate the *In Vivo* Requirement of Particular F-Box Proteins in Targeted Degradation

Once candidate F-box proteins are identified, it is critical that further genetic and biochemical evidence of functional interactions be obtained. Of particular interest is the reconstitution of ubiquitin ligase activity toward the substrate of interest, as described later. In addition, loss-of-function experiments are required to demonstrate a requirement for a particular F-box protein in degrading the protein of interest. An alternative to RNAi is the use of dominant-negative F-box proteins, although this approach has certain limitations, including the possibility of indirect effects. In the following section, we describe assays for examining the stability of Cdc25A in the presence and absence of siRNA against the β-TRCP1 and β-TRCP2 F-box proteins.

Two approaches can be used to generate RNAi toward a particular F-box protein. In one case, synthetic oligonucleotides can be synthesized and tested for their ability to deplete candidate F-box protein by immunoblotting or its mRNA by RT-PCR approaches. Several approaches are available for designing such siRNAs. The alternative is to deliver the siRNA by means of an expression plasmid expressing short hairpin RNAs (shRNAs). Several strategies have been put forward. In the strategy described below, plasmid vectors expressing an shRNA capable of depleting both β-TRCP1 and β-TRCP2 are used to determine whether either of these F-box proteins is required for degradation of Cdc25A. Other studies have used similar approaches to identify F-box proteins controlling the abundance of cyclin E, c-myc, c-jun, and other SCF targets. Although it is possible to examine the steady-state level of the ubiquitylation target, definitive evidence of a role for the candidate F-box protein in turnover of the target requires that the half-life be measured using pulse-chase or CHX-based approaches.

1. After suitable synthetic or vector-driven siRNA sequences have been identified, transfection experiments can be performed. As with the Cul1^{DN}, the precise parameters for transfection will depend on the intrinsic stability of the protein of interest. In the case of Cdc25A, the apparent half-life in the presence of CHX is $\sim$30 min. Thus, four time points representing 0, 30, 60, and 90 min are selected. HEK293T cells are seeded at 8 million cells in one 10-cm dish per shRNA vector to be examined. When cells reach 70–80% confluence, cells are transfected with 20 μg of either pSUPER-β-TRCP or pSUPER-GFP as a negative control using lipofectamine 2000. The pSUPER-β-TRCP plasmid targets a conserved sequence found in both β-TRCP1 and β-TRCP2 and is therefore effective at depleting both mRNAs.

2. Twenty-four hours after transfection, cells of each 10-cm dish are split equally into four 6-cm dishes.

3. Forty-eight hours after transfection, the media is removed and replaced with media containing CHX (25 μg/ml).

4. Cells are harvested at 0, 30, 60, and 90 min after addition of CHX and extracts prepared as described previously. Forty micrograms of extract is separated by SDS-PAGE and immunoblotted using antibodies against

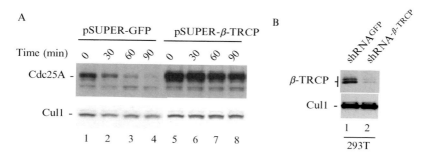

Fig. 4. (A) Stabilization of Cdc25A by depletion of β-TRCP. 293T cells (2-cm dish) were transfected with 5 μg of pSUPER-β-TRCP of pSUPER-GFP shRNA vectors, and 48 h after transfection, cells were treated with CHX. Cell extracts were generated at the indicated time and analyzed by SDS-PAGE and immunoblotting with anti-Cdc25A and anti-Cul1 antibodies as a loading control. (B) Analysis of β-TRCP expression in cells expressing β-TRCP RNAi.

Cdc25A. Blots are stripped and reprobed with anti-Cul1 as a loading control. In the absence of β-TRCP depletion, the half-life of Cdc25A is ~30 min (Fig. 4A, lanes 1–4). In the presence of β-TRCP deletion, the half-life is increased to >90 min, implicating β-TRCP in Cdc25A degradation *in vivo* (lanes 5–8). To validate the effectiveness of RNAi, cell extracts should be blotted for the F-box protein of interest (Fig. 4B). Depending on turnover rates, the time and number of time points may be adjusted.

Use of Synthetic Phosphodegrons to Identify F-Box Proteins for Particular Substrates

The term "degron" is used to refer to a minimal peptide sequence required to recruit an ubiquitylation substrate to its cognate E3 and was first used in the context of the N-end Rule Pathway (Dohmen *et al.*, 1994). SCF ubiquitin ligases often interact with short phosphopeptide motifs referred to as "phosphodegrons." The crystal structure of the β-TRCP1-Skp1 and β-catenin phosphodegron complex has demonstrated that the phosphoserine, aspartic acid, and hydrophobic residues in the β-catenin phosphodegron are directly recognized by specific residues in the WD40 domain of β-TRCP1 (Wu *et al.*, 2003). Interactions between properly phosphorylated SCF targets and their cognate F-box proteins are highly specific. For example, the phosphodegrons of IκB, β-catenin, and Cdc25A bind tightly with β-TRCP but do not bind other WD40-repeat–containing F-box proteins including Fbw2, Fbw7, or Fbw8 (Jin *et al.*, 2003; Winston *et al.*, 1999). Conversely, the phosphodegrons derived from cyclin E and c-Myc are recognized by Fbw7 but not by β-TRCP, Fbw2, Fbw5, or Fbw8 (Koepp et al., 2001; Welcker et al., 2004). This specificity, coupled with the fact that F-box proteins frequently interact with small contiguous peptide elements within the ubiquitylation substrate, provides a simple approach for the identification of F-box proteins that interact with known phosphodegron. Immobilized peptide corresponding to known phosphodegron can be used to search for candidate F-box proteins, and in this context, corresponding unphosphorylated peptides can be used as negative controls. The source of F-box proteins could either be ^{35}S-methionine–labeled *in vitro* translation products, or, alternately, extracts from cells ectopically expressing epitope-tagged F-box proteins could also be used, in which case immunoblotting is used to detect relevant F-box proteins. Described below is the general protocol to conjugate peptides to agarose beads, to make radiolabeled F-box proteins in reticulocyte lysate, and to perform binding assays.

Immobilization of Phosphopeptides on Agarose Beads

1. Dissolve 1 mg of peptide in 450 μl of water, then add 50 μl of 1 M Mops (pH 7.5).
2. Wash Affigel-10 (Bio-RAD, Cat# 153-6099) sequentially with one volume of water and 0.1 M Mops (pH 7.5). Finally, make 50% slurry in 0.1 M Mops (pH 7.5).
3. Mix 1 ml of Affigel-10 (50% slurry) with 1 mg of peptide. Rotate gently at 4° for 4 h.
4. Wash the beads with 1 ml of 0.1 M Tris-HCl (pH 7.5) once to remove the bulk of unconjugated peptide. The beads are then incubated with 1 ml of 0.1 M Tris-HCl (pH 7.5) at 4° overnight.
5. Centrifuge at 1000g for 2 min and aspirate the supernatant.
6. Wash the beads three times with 1ml of 0.1 M Mops (pH 7.5).
7. Resuspend the beads in 0.5 ml of 0.1 M Mops (pH 7.5) and make it as 50% slurry (peptide concentration at 1 mg/ml). The peptide slurry can be stored at 80°.

To determine the efficiency of coupling, 5-μl aliquots of peptides before and after coupling can be taken and subsequently subjected to either absorbance measurements at 280 nm in cases in which peptides contain aromatic amino acids, or alternately, peptides can be subjected to analytical high-performance liquid chromatography on C8 or C18 columns. The fractional reduction of peptide in the postcoupling supernatant relative to the starting mixture provides a measure of coupling efficiency. In most cases, coupling is greater than 90%.

In Vitro *Translation of F-Box Proteins in Reticulocyte Lysate System*

The ^{35}S-methionine–labeled F-box proteins are made by T_NT T7–coupled reticulocyte lysate system (Promega). Each reaction contains 2 μg CMV/T7-Myc-F-box expression vector, 25 μl of T_NT rabbit reticulocyte lysate, 2 μl of T_NT reaction buffer, 1 μl of T_NT amino acid mixture minus methionine, 5 μl of ^{35}S-methionine, 2 μl of ribonuclease inhibitor (20 units, Invitrogen), 1 μl of T_NT T7 RNA polymerase in a total volume of 50 μl. The reactions are carried out at 30° for 90 min.

Phosphodegron Binding Assay

In most cases, a panel of *in vitro* translated F-box proteins will be used in combination with both phosphorylated and unphosphorylated peptides. For each binding reaction:

1. Take 10 μl of 50% peptide slurry and wash with 1 ml of ice-cold NP-40 lysis buffer. Carefully remove the supernatant.
2. Resuspend the beads in 100 μl of ice-cold NP-40 lysis buffer.
3. Add 5 μl of [S^{35}]-methionine radiolabeled F-box proteins. Before removing the aliquot, the *in vitro* translation product should be centrifuged at 20,000g for 5 min.
4. Rock gently at 4° for 1 h.
5. Wash the resin three times each with 1 ml of NP-40 lysis buffer.
6. Samples are separated by SDS-PAGE and bound proteins visualized by autoradiography of dried gels. Eight percent polyacrylamide gels are appropriate for most applications. To aid in determining the extent of binding, 1 μl (20% of input) *in vitro* translation product is subjected to electrophoresis alongside relevant bead samples. The extent of interaction observed in such assays will vary with the F-box proteins being examined and with the nature of the phosphodegron. Typically, the interaction between a phosphodegron and a relevant F-box protein may result in retention of 10–50% of the input *in vitro* translation product, whereas less than 1% binding will be seen between unphosphorylated peptides and the relevant F-box protein. We have found this assay to be highly selective, with little nonspecific association seen with a variety of F-box proteins (Koepp *et al.*, 2001; Winston *et al.*, 1999).

Reconstitution of Ubiquitin Ligase Activity

Several approaches have been used to demonstrate specific ubiquitylation of proteins by candidate SCF complexes. One approach involves reconstitution of ubiquitin ligase activity in reticulocyte lysate. We and others have found that F-box proteins synthesized by *in vitro* translation will assemble with Skp1/Cul1/Rbx1 present in the reticulocyte extract to form active E3s (Jin et al., 2003, Welcker et al., 2004; Wu et al., 2003). Interestingly, although neddylated Cul1 represents a small fraction of Cul1 in the reticulocyte extract (<5%), SCF complexes purified from reticulocyte extracts using epitope tags on the F-box protein contain >30% neddylated Cul1. Thus, it appears that F-box proteins either assemble preferentially with neddylated Cul1 or that assembly of an F-box protein with the Skp1/Cul1/Rbx1 complex promotes neddylation of Cul1 (Jin and Harper, unpublished). A major advantage of this approach is that reticulocyte lysates may contain one or more essential factors required for the ubiquitin reaction. Alternately, it is possible to purify active SCF complexes after expression of relevant components in either insect cells using recombinant baculoviruses (Feldman et al., 1997; Ganoth et al., 2001; Skowyra et al., 1997) or in mammalian cells after transfection of the F-box protein (Strohmaier et al., 2001). A major advantage of the use of insect cells is that large quantities of purified materials can be obtained, allowing quantitative analysis of reaction mechanisms. One limitation of this approach, however, is that in some cases, the extent of Cul1 neddylation is relative low (<5%), leading to complexes with relatively low specific activity. This problem can be alleviated by *in vitro* neddylation of purified SCF complexes, as described in the following.

Reconstitution of SCF$^{\beta\text{-}TRCP}$ Ubiquitin-Ligase Activity for the Cdc25A Protein Phosphatase in Reticulocyte Lysate

Ubiquitylation reactions for several F-box protein/substrate pairs have been developed in reticulocyte extracts including IκB and Cdc25A with β-TRCP, cyclin E and c-Myc with Fbw7, and p27 with Skp2 (Jin et al., 2003; Koepp et al., 2001; Welcker et al., 2004; Wu et al., 2003). In the following section, we describe a method for reconstituting Chk1-dependent ubiquitylation of Cdc25A by SCF$^{\beta\text{-}TRCP}$ in reticulocyte extracts (Jin et al., 2003).

Preparation of Radiolabeled hCdc25A and Unlabeled β-TRCP in Reticulocyte Lysate. *In vitro* translation reactions are performed as described previously separately for substrate and F-box protein in separate reactions. The substrate is synthesized in the presence of ^{35}S-methionine to yield a radiolabeled product, whereas the F-box protein is synthesized in the presence of 1 mM methionine (instead of ^{35}S-methionine), yielding an unlabeled F-box protein. The presence of the F-box protein should be confirmed by immunoblotting or by SDS-PAGE and autoradiography before performing *in vitro* ubiquitylation reactions. In principle, the amount of substrate (e.g., Cdc25A) included in the ubiquitylation reaction should be such that it is readily detectable on autoradiography for 2 h.

Phosphorylation of hCdc25A by GST-hChk1. Cdc25A ubiquitylation by SCF$^{\beta\text{-TRCP}}$ requires that it be phosphorylated by Chk1. This is achieved using GST-Chk1 purified from insect cells (Jin *et al.*, 2003). The kinase reaction includes 16 μl of ^{35}S-methionine labeled hCdc25A, 50 mM Tris-HCl (pH 7.5), 10 mM MgCl$_2$, 1 mM DTT, 800 ng GST-Chk1 (human), and 200 μM ATP in a total volume of 20 μl. The reactions are performed at 30° for 30 min. To assess phosphorylation, 1-μl aliquots with and without Chk1 can be subjected to SDS-PAGE. Chk1 phosphorylation leads to decreased mobility of Cdc25A (Jin *et al.*, 2003).

Ubiquitylation Reactions. To allow assembly of SCF-substrate complexes, 2.5 μl of phosphorylated Cdc25A is preincubated with 2.3 μl of unlabeled β-TRCP at 30° for 10 min. Then, 5.2 μl of the ubiquitylation reaction mixture containing other reagents was added before incubation at 30° for 60 min. Each ubiquitylation reaction contains 250 ng of ubiquitin-activating enzyme (E1), 250 ng of His-UbcH5a, 5 μg of ubiquitin (Sigma), 0.2 μl of 100 mM ATP, 0.5 μl of 20 μM ubiquitin aldehyde, 1 μl of 10× *u*biquitylation *r*eaction *b*uffer (10× URB: 500 mM Tris-HCl [pH 7.5], 50 mM KCl, 50 mM NaF, 50 mM MgCl$_2$, and 5 mM DTT), 1 μl of 10× Energy Regeneration Mix (10× EM: 200 mM creatine phosphate, and 2 μg/μl creatine phosphokinase), 0.5 μl of 20× protease inhibitor (one pellet of protease inhibitor cocktail is dissolved in 1 ml water, Roche, Cat#1873580), and 0.5 μl of 0.5 mM LLnL. The reactions are stopped by addition of 10 μl of 2× SDS-PAGE sample buffer. Reaction products are then resolved by 4–12% SDS-PAGE gradient gel and dried gels subjected to autoradiography.

The presence of high molecular weight conjugates is suggestive of ubiquitylation. However, because reticulocyte lysates represent a crude extract system, it is critical that controls be performed to demonstrate that the formation of high-molecular weight conjugates is dependent on both ubiquitin and on the added F-box protein. The former can be achieved by addition of methyl ubiquitin, which leads to a collapse in the ubiquitin conjugates because of the inability of methyl ubiquitin to form polyubiquitin conjugates. To demonstrate F-box protein specificity, a panel of F-box proteins can be used in analogous ubiquitylation reactions. It should also be noted that reticulocyte extracts contain E1 enzyme, and, in some cases, it is not necessary to supplement reactions with recombinant E1 (see Passmore *et al.*, 2003).

Reconstitution of SCF^{Fbw7} Ubiquitin Ligase Activity by Purified SCF^{Fbw7} Complex from Recombinant Baculovirus-Infected Insect Cells

An alternative to the use of reticulocyte extracts is the use of SCF complexes purified from insect cells. When coupled with *in vitro* neddylation of Cul1, highly active complexes can be produced. Previous genetic and biochemical studies have implicated Fbw7 in the degradation of cyclin E, in response to phosphorylation by Cdk2 and/or GSK3β (Koepp *et al.*, 2001; Moberg *et al.*, 2001; Strohmaier *et al.*, 2001; Welcker *et al.*, 2003). However, cyclin E ubiquitylation has not been accomplished using biochemically purified components. The following protocol describes the reconstitution of cyclin E ubiquitylation by SCF$^{Fbw7\alpha}$ using insect cell–derived SCF complexes.

Purification of SCF$^{Fbw7\alpha}$ Complex from Baculovirus-Infected SF9 Cells

1. Seed 3×10^7 SF9 cells with 25 ml insect culture media in 150 mm × 25 mm tissue culture dish.

2. One hour later, infect the cells with a mixture of recombinant baculoviruses expressing Flag-Fbw7α, His$_6$-Cul1, His$_6$-Skp1, and His$_6$-Myc-Rbx1 individually. Optimal expression requires the use of high-titer viruses. Untagged Cul1, Skp1, and Rbx1 can be substituted for the epitope-tagged versions.

3. Forty hours after infection, cells are collected by gentle disruption and centrifuged at 1000g for 2 min.

4. Cells are washed once with ice-cold PBS.

5. Cells are then frozen at $-80°$ for at least 30 min to facilitate lysis.

6. Cells are resuspended in 2.5 ml of ice-cold NETN lysis buffer with protease inhibitors cocktail (per 150 × 25 mm culture dish), and incubated on ice for 10 min.

7. Lysates are cleared by centrifugation at 20,000g for 30 min at 4°.

8. The supernatant is transferred to a 15-ml conical tube, and 30 μl of ANTI-FLAG M2 Agarose Affinity Gel (Sigma Cat# A-2220) (50% slurry) is added.

9. The mixture is rocked gently at 4° for 1 h.

10. The beads are collected by brief centrifugation at 2000g and subsequently washed three times with 2.5 ml of ice-cold NETN lysis buffer.

11. The beads are then washed with 1 ml of TBS (50 mM Tris-HCl, pH 7.5, 200 mM NaCl).

12. Beads are collected by centrifuge at 1000g for 1 min and the supernatant aspirated.

13. The purified SCF complex can be used for ubiquitylation reactions while bound to beads (Skowyra *et al.*, 1997) or can be eluted. For elution, one bed volume of 3× Flag elution solution (500 μg/ml 3× Flag peptide in TBS) is added to the resin.

14. The sample is incubated with gently shaking for 30 min at 4°.

15. Centrifuge the resin for 10 sec at 20,000g. Transfer the supernatants to a new Eppendorf tube, being careful to exclude the resin.

16. Repeat steps 13–15 once.
17. Combine the eluted samples and transfer them into Mini Dialysis Unit (Pierce Cat# 69550).
18. Dialyze against 200 ml of dialysis buffer (50 mM Tris-HCl, pH 7.5, 200 mM NaCl, 50% [v/v] Glycerol) containing 1 mM phenylmethylsulfonyl fluoride (PMSF) for 2 h at 4°.
19. Repeat the dialysis with fresh dialysis buffer for another 2 h.
20. Samples are collected and stored at −20°.
21. To examine the purity of SCF complexes, SDS-PAGE is performed. Typically, 3 μl of eluate should allow detection of proteins by Coomassie staining (approximately 0.5 μg of SCF components). An example of SCF$^{Fbw7\alpha}$ purified using this method is shown in Fig. 5A.

Reconstitution of Cyclin E Ubiquitylation. Cyclin E degradation is known to require phosphorylation on Thr-380 (Clurman et al., 1996; Won and Reed, 1996). Previous studies have demonstrated that cyclin E/CDK2 (E/K2) purified from SF9 insect cells (Connell-Crowley et al., 1997) are appropriately phosphorylated for ubiquitylation by Fbw7 (Koepp et al., 2001). The E/K2 complex was incubated with an equal amount of SCF^{Fbw7} at 30° for 10 min before addition of ubiquitylation components. The ubiquitylation reaction contains 250 ng His$_6$-E1, 250 ng His$_6$-UbcH3, 5 μg ubiquitin, 2 mM ATP, 1 μM ubiquitin aldehyde, 1× URB, 1× EM, 6 nM

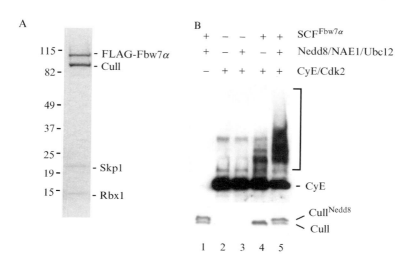

FIG. 5. *In vitro* ubiquitylation. (A) Recombinant SCF$^{Fbw7\alpha}$ complexes purified from insect cells. (B) Reconstitution of cyclin E and Cdc25A ubiquitylation. The indicated components were incubated together with E1, Ubc3, ubiquitin, and ATP before analysis by SDS-PAGE and immunoblotting to determine the extent of cyclin E ubiquitylation and Cul1 neddylation.

E/K2, 6 nM SCF^{Fbw7}, 12 nM Nedd8, 6 nM UbcH12, and 6 nM Nedd8-activating enzyme (NAE1) in a total volume of 10 μl and was performed at 30° for 60 minutes. Reactions were terminated by adding 10 μl of 2× SDS-PAGE sample buffer, separated by 4–12% SDS-PAGE gradient gel and detected by Western blot using an antibody against cyclin E (Santa Cruz, C-19). The extent of neddylation can be determined by immunoblotting with antibodies against Cul1.

Neddylation is essential for the function of SCF complexes in multicellular eukaryotes (Ohh et al., 2002; Osaka et al., 2000). Unlike the reticulocyte lysate system, Cul1 present in SCF^{Fbw7} complexes purified from insect cells largely lacks the nedd8 modification (Fig. 5B, lane 4). Addition of components required for Cul1 neddylation (Nedd8, NAE1, and Ubc12) leads to a dramatic decrease in the electrophoretic mobility of Cul1, consistent with the linkage of a single Nedd8 molecule ($\sim$8 kDa). In the example shown, >50% of the Cul1 was converted to the neddylated form (Fig. 5B, compare lanes 4 and 5), and the extent of this conversion can be increased to >90% by longer incubation times or by the use of higher quantities of neddylation components. Importantly, little cyclin E ubiquitylation is observed in the absence of Cul1 neddylation (lane 4). However, in the presence of Cul1-neddylation, efficient cyclin E ubiquitylation by SCF$^{Fbw7\alpha}$ is observed (lane 5).

Conclusions

Available data indicate the existence of a large number of F-box proteins, which presumably ubiquitylate an even larger number of cellular proteins. To date, substrates for only a small fraction of F-box proteins have been identified. A further understanding of the role of the SCF pathway in protein homeostasis will require a detailed understanding of the proteins whose stability is regulated by F-box proteins. In this chapter, we have described several biochemical approaches that should facilitate the identification of additional substrates of the SCF. Analogous approaches may be useful in the identification of substrates of other classes of cullin-based ubiquitin ligases, including the Cul3/BTB protein E3s (Furukawa et al., 2003; Geyer et al., 2003; Pintard et al., 2003; Xu et al., 2003) and the Cul2/ BC-box protein E3s (reviewed in Deshaies, 1999).

Acknowledgments

This work is supported by NIH grant AG11085 and by the Department of Defense (DAMD17-01-1-0135) to J. W. H., and by the Department of Defense (DAMD17-02-1-0284) to J. J.

References

Bai, C., Sen, P., Hofmann, K., Ma, L., Goebl, M., Harper, J. W., and Elledge, S. J. (1999). Skp1 connects cell cycle regulators to the ubiquitin proteolysis machinery through a novel motif, the F-box. *Cell* **86**, 263–274.

Busino, L., Donzelli, M., Chiesa, M., Guardavaccaro, D., Ganoth, D., Dorrello, N. V., Hershko, A., Pagano, M., and Draetta, G. F. (2003). Degradation of Cdc25A by beta-TrCP during S phase and in response to DNA damage. *Nature* **426**, 87–91.

Clurman, B. E., Sheaff, R. J., Thress, K., Groudine, M., and Roberts, J. M. (1996). Turnover of cyclin E by the ubiquitin-proteasome pathway is regulated by cdk2 binding and cyclin phosphorylation. *Genes Dev.* **10**, 1979–1990.

Connell-Crowley, L., Harper, J. W., and Goodrich, D. W. (1997). Cyclin D1/Cdk4 regulates retinoblastoma protein-mediated cell cycle arrest by site-specific phosphorylation. *Mol. Biol. Cell.* **8**, 287–301.

Cullinan, S., Gordon, J. D., Jin, J., Harper, J. W., and Diehl, A. (2004). The Keap1-BTB protein is an adaptor that bridges Nrf2 to a Cul3-based E3 ligase: Oxidative stress sensing by a Cul3-Keap1 ligase. *Mol. Cell. Biol.* **24**, 8477–8486.

Deshaies, R. J. (1999). SCF and cullin/ring H2-based ubiquitin ligases. *Annu. Rev. Cell Dev. Biol.* **15**, 435–467.

Dohmen, R. J., Wu, P., and Varshavsky, A. (1994). Heat-inducible degron: A method for constructing temperature-sensitive mutants. *Science* **263**, 1273–1276.

Donzelli, M., Squatrito, M., Ganoth, D., Hershko, A., Pagano, M., and Draetta, G. F. (2002). Dual mode of degradation of Cdc25 A phosphatase. *EMBO J.* **21**, 4875–4884.

Drury, L. S., Perkins, G., and Diffley, J. F. (1997). The Cdc4/34/53 pathway targets Cdc6p for proteolysis in budding yeast. *EMBO J.* **16**, 5966–5976.

Feldman, R. M., Correll, C. C., Kaplan, K. B., and Deshaies, R. J. (1997). A complex of Cdc4p, Skp1p, and Cdc53p/cullin catalyzes ubiquitylation of the phosphorylated CDK inhibitor Sic1p. *Cell* **91**, 209–219.

Furukawa, M., He, Y. J., Borchers, C., and Xiong, Y. (2003). Targeting of protein ubiquitylation by BTB-Cullin 3-Roc1 ubiquitin ligases. *Nat. Cell Biol.* **5**, 1001–1007.

Geyer, R., Wee, S., Anderson, S., Yates, J., and Wolf, D. A. (2003). BTB/POZ domain proteins are putative substrate adaptors for cullin 3 ubiquitin ligases. *Mol. Cell.* **12**, 783–790.

Ganoth, D., Bornstein, G., Ko, T. K., Larsen, B., Tyers, M., Pagano, M., and Hershko, A. (2001). The cell-cycle regulatory protein Cks1 is required for SCF(Skp2)-mediated ubiquitinylation of p27. *Nat. Cell Biol.* **3**, 321–324.

Ho, Y., Gruhler, A., Heilbut, A., Bader, G. D., Moore, L., Adams, S. L., Millar, A., Taylor, P., Bennett, K., Boutilier, K., Yang, L., Wolting, C., Donaldson, I., Schandorff, S., Shewnarane, J., Vo, M., Taggart, J., Goudreault, M., Muskat, B., Alfarano, C., Dewar, D., Lin, Z., Michalickova, K., Willems, A. R., Sassi, H., Nielsen, P. A., Rasmussen, K. J., Andersen, J. R., Johansen, L. E., Hansen, L. H., Jespersen, H., Podtelejnikov, A., Nielsen, E., Crawford, J., Poulsen, V., Sorensen, B. D., Matthiesen, J., Hendrickson, R. C., Gleeson, F., Pawson, T., Moran, M. F., Durocher, D., Mann, M., Hogue, C. W., Figeys, D., and Tyers, M. (2002). Systematic identification of protein complexes in *Saccharomyces cerevisiae* by mass spectrometry. *Nature* **415**, 180–183.

Hubbard, E. J., Wu, G., Kitajewski, J., and Greenwald, I. (1997). sel-10, a negative regulator of lin-12 activity in *Caenorhabditis elegans*, encodes a member of the CDC4 family of proteins. *Genes Dev.* **11**, 3182–3193.

Jin, J., Shirogane, T., Xu, L., Nalepa, G., Qin, J., Elledge, S. J., and Harper, J. W. (2003). SCF1-beta-TRCP links Chk1 signaling to degradation of the Cdc25A protein phosphatase. *Genes Dev.* **17,** 3062–3074.

Jin, J. Cardoza,T., Lovering, R. C., Elledge, S. J., Pagano, M., and Harper, J. W. (2004). Systemic analysis and nomenclature of mammalian F-box proteins. *Genes Dev.* **18,** 2573–2580.

Koepp, D. M., Harper, J. W., and Elledge, S. J. (1999). How the cyclin became a cyclin: Regulated proteolysis in the cell cycle. *Cell* **97,** 431–434.

Koepp, D. M., Schaefer, L. K., Ye, X., Keyomarsi, K., Chu, C., Harper, J. W., and Elledge, S. J. (2001). Phosphorylation-dependent ubiquitylation of cyclin E by SCF^{Fbw7} ubiquitin ligase. *Science* **294,** 173–177.

Latres, E., Chiaur, D. S., and Pagano, M. (1999). The human F box protein β-TrCP associates with the Cul1/Skp1 complex and regulates the stability of β-catenin. *Oncogene* **18,** 849–854.

Liu, Q., Li, M. Z., Liu, D., and Elledge, S. J. (2000). Rapid construction of recombinant DNA by the univector plasmid-fusion system. *Methods Enzymol.* **328,** 530–549.

Margottin-Goguet, F., Hsu, J. Y., Loktev, A., Hsieh, H. M., Reimann, J. D., and Jackson, P. K. (2003). Prophase destruction of Emi1 by the SCF(betaTrCP/Slimb) ubiquitin ligase activates the anaphase promoting complex to allow progression beyond prometaphase. *Dev. Cell.* **4,** 813–826.

Nash, P., Tang, X., Orlicky, S., Chen, Q., Gertler, F. B., Mendenhall, M. D., Sicheri, F., Pawson, T., and Tyers, M. (2001). Multisite phosphorylation of a Cdk inhibitor sets a threshold for the onset of DNA replication. *Nature* **414,** 514–521.

Ohh, M., Kim, W. Y., Moslehi, J. J., Chen, Y., Chau, V., Read, M. A., and Kaelin, W. G., Jr. (2002). An intact NEDD8 pathway is required for Cullin-dependent ubiquitylation in mammalian cells. *EMBO Rep.* **3,** 177–182.

Orlicky, S., Tang, X., Willems, A. R., Tyers, M., and Sicheri, F. (2003). Structural basis for phosphodependent substrate selection and orientation by the SCF^{Cdc4} ubiquitin ligase. *Cell* **112,** 243–256.

Osaka, F., Saeki, M., Katayama, S., Aida, N., Toh-e, A., Kominami, K., Toda, T., Suzuki, T., Chiba, T., Tanaka, K., and Kato, S. (2000). Covalent modifier NEDD8 is essential for SCF ubiquitin-ligase in fission yeast. *EMBO J.* **19,** 3475–3484.

Passmore, L. A., McCormack, E. A., Au, S. W., Paul, A., Willison, K. R., Harper, J. W., and Barford, D. (2003). Doc1 mediates the activity of the anaphase-promoting complex by contributing to substrate recognition. *EMBO J.* **22,** 786–796.

Patton, E. E., Willems, A. R., and Tyers, M. (1998a). Combinatorial control in ubiquitin-dependent proteolysis: Don't Skp the F-box hypothesis. *Trends Genet.* **14,** 236–243.

Patton, E. E., Willems, A. R., Sa, D., Kuras, L., Thomas, D., Craig, K. L., and Tyers, M. (1998b). Cdc53 is a scaffold protein for multiple Cdc34/Skp1/F-box protein complexes that regulate cell division and methionine biosynthesis in yeast. *Genes Dev.* **12,** 692–705.

Pintard, L., Willis, J. H., Willems, A., Johnson, J. L., Srayko, M., Kurz, T., Glaser, S., Mains, P. E., Tyers, M., Bowerman, B., and Peter, M. (2003). The BTB protein MEL-26 is a substrate-specific adaptor of the CUL-3 ubiquitin-ligase. *Nature* **425,** 311–316.

Seol, J. H., Feldman, R. M., Zachariae, W., Shevchenko, A., Correll, C. C., Lyapina, S., Chi, Y., Galova, M., Claypool, J., Sandmeyer, S., Nasmyth, K., Shevchenko, A., and Deshaies, R. J. (1999). Cdc53/cullin and the essential Hrt1 RING-H2 subunit of SCF define a ubiquitin ligase module that activates the E2 enzyme Cdc34. *Genes Dev.* **13,** 1614–1626.

Skowyra, D., Craig, K. L., Tyers, M., Elledge, S. J., and Harper, J. W. (1997). F-box proteins are receptors that recruit phosphorylated substrates to the SCF ubiquitin-ligase complex. *Cell* **91,** 209–219.

Skowyra, D., Koepp, D. M., Kamura, T., Conrad, M. N., Conaway, R. C., Conaway, J. W., Elledge, S. J., and Harper, J. W. (1999). Reconstitution of G1 cyclin ubiquitylation with complexes containing SCFGrr1 and Rbx1. *Science* **284,** 662–665.

Spencer, E., Jiang, J., and Chen, Z. J. (1999). Signal-induced ubiquitylation of IκBα by the F-box protein Slimb/β-TrCP. *Genes Dev.* **13,** 284–294.

Strohmaier, H., Spruck, C. H., Kaiser, P., Won, K.-A., Sangfelt, O., and Reed, S. I. (2001). Human F-box protein hCdc4 targets cyclin E for proteolysis and is mutated in a breast cancer cell line. *Nature* **413,** 316–322.

Thomas, D., Kuras, L., Barbey, R., Cherest, H., Blaiseau, P. L., and Surdin-Kerjan, Y. (1995). Met30p, a yeast transcriptional inhibitor that responds to S-adenosylmethionine, is an essential protein with WD40 repeats. *Mol. Cell. Biol.* **15,** 6526–6534.

Welcker, M., Singer, J., Loeb, K. R., Grim, J., Bloecher, A., Gurien-West, M., Clurman, B. E., and Roberts, J. M. (2003). Multisite phosphorylation by Cdk2 and GSK3 controls cyclin E degradation. *Mol. Cell.* **12,** 381–392.

Welcker, M., Orian, A., Jin, J., Grim, J. A., Harper, J. W., Eisenman, R. N., and Clurman, B. E. (2004). The Fbw7 tumor suppressor regulates glycogen synthase kinase 3 phosphorylation-dependent c-Myc protein degradation. *Proc. Natl. Acad. Sci. USA* **101,** 9085–9090.

Winston, J. T., Strack, P., Beer-Romero, P., Chu, C. Y., Elledge, S. J., and Harper, J. W. (1999). The SCF$^{\beta\text{-TrCP}}$-ubiquitin ligase complex associates specifically with phosphorylated destruction motifs in IκBα and β-catenin and stimulates IκBα ubiquitylation *in vitro*. *Genes Dev.* **13,** 270–283.

Won, K. A., and Reed, S. I. (1996). Activation of cyclin E/CDK2 is coupled to site-specific autophosphorylation and ubiquitin-dependent degradation of cyclin E. *EMBO J.* **15,** 4182–4193.

Wu, G., Xu, G., Schulman, B. A., Jeffrey, P. D., Harper, J. W., and Pavletich, N. P. (2003). Structure of a beta-TrCP1-Skp1-beta-catenin complex: Destruction motif binding and lysine specificity of the SCF(beta-TrCP1) ubiquitin ligase. *Mol. Cell.* **11,** 1445–1456.

Xu, L., Wei, Y., Reboul, J., Vaglio, P., Shin, T. H., Vidal, M., Elledge, S. J., and Harper, J. W. (2003). BTB proteins are substrate-specific adaptors in an SCF-like modular ubiquitin ligase containing CUL-3. *Nature* **425,** 316–321.

Yada, M., Hatakeyama, S., Kamura, T., Nishiyama, M., Tsunematsu, R., Imaki, H., Ishida, N., Okumura, F., Nakayama, K., and Nakayama, K. I. (2004). Phosphorylation-dependent degradation of c-Myc is mediated by the F-box protein Fbw7. *EMBO J.* **23,** 2116–2125.

Yaron, A., Hatzubai, A., Davis, M., Lavon, I., Amit, S., Manning, A. M., Andersen, J. S., Mann, M., Mercurio, F., and Ben-Neriah, Y. (1998). Identification of the receptor component of the IκBα-ubiquitin ligase. *Nature* **396,** 590–594.

Zheng, N., Schulman, B. A., Song, L., Miller, J. J., Jeffrey, P. D., Wang, P., Chu, C., Koepp, D. M., Elledge, S. J., Pagano, M., Conaway, R. C., Conaway, J. W., Harper, J. W., and Pavletich, N. P. (2002). Structure of the Cul1-Rbx1-Skp1- Fbox Skp SCF ubiquitin ligase. *Nature* **416,** 703–709.

[21] Fusion-Based Strategies to Identify Genes Involved in Degradation of a Specific Substrate

By RANDOLPH Y. HAMPTON

Abstract

Fusion proteins have been used in many instances to allow genetic screening for genes required for the degradation of a specific substrate. This straightforward, yet powerful, approach can be applied in many circumstances to facilitate gene characterization and discovery. Some general principles are discussed and then several successful uses of these tactics are described in detail.

General Principles of Using Fusion Reporters to Discover Degradation Genes

Because protein degradation is highly processive and evolved to destroy a wide variety of proteins, the addition of fusion partners will often allow normal degradation of the resulting protein. A fusion gene added to a degradation substrate can render an otherwise tedious or infeasible genetic screen facile and practical. In this chapter, some general principles for using this technique will be discussed, and then several detailed examples will be described. The examples are all from studies using yeast. It is hoped that the combination of general concepts and detailed examples will allow the largest number of applications.

Reporter Design: Pathway Fidelity and Degrons

There can be many ways for a protein in a particular compartment to undergo ubiquitin-mediated degradation, and altering a protein by fusion addition could send it down a distinct pathway from the one designated for study. It is important to confirm that the engineered reporter fusion undergoes degradation by the pathway of interest. If some mutants deficient in the degradation pathway already exist, testing the behavior of the reporter protein in those mutants will confirm the fidelity of the fusion. If no mutants are available, characterization of the fusion's fidelity by other means can be useful. In studying the degradation of yeast Hmg2p isozyme of HMG-CoA reductase (HMGR), a GFP fusion reporter was tested both for stabilization by mutants and normal physiological regulation of HMGR stability.

Sometimes loss of degradation pathway information can be desirable, if the resulting fusion simplifies a genetic analysis. This is the case for the ground-breaking studies of α2 repressor protein degradation (Matα2) in yeast. This very rapidly degraded protein undergoes ubiquitination by two distinct pathways, one using the E2s Ubc6/Ubc7 and the E3 ligase Doa10p, and the other by Ubc4/Ubc5 and a still-unknown ligase (Chen *et al.*, 1993). Entry into these two degradation pathways is mediated by separate portions of the protein, called "degrons," with the N-terminal 67 amino acids, called "deg1," mediating only the Ubc6/7 branch (Hochstrasser and Varshavsky, 1990). The genetic analysis of Matα2 degradation was facilitated by the use of only the deg1 portion of the entire Matα2 protein fused to reporters, because this tactic effectively isolated that single branch of degradation for study.

The degron concept is useful in the design of reporter fusions. A degron is defined as a discrete, transferable region of a protein that is necessary and sufficient for the ubiquitin-mediated degradation of the protein in which it resides (e.g., Dohmen *et al.* [1994]). deg1 above is an example of this idea. However, discrete degrons do not always mediate selective protein degradation. There are also cases in which the information for pathway entry involves a large portion of a protein, including sequence or structural elements that are far removed from each other. The entire multi-spanning N-terminal domain (approximately half the protein) of Hmg2p is required for its regulated degradation by the HRD pathway (Gardner and Hampton, 1999; see later). Degradation reporters for Hmg2p must include this complete N-terminal domain (Cronin and Hampton, 1999). When a protein of interest has other activities, it is also useful to know whether these must be included in the reporter. For example, Hmg2p has a discrete C-terminal catalytic domain that is responsible for an essential step of the sterol pathway but not required for regulated degradation. Thus, another decision that goes into making reporters for Hmg2p is whether to leave the C-terminal catalytic region intact or to replace the C-terminal catalytic region with the reporter. In general, an understanding of the protein regions that are and are not needed for physiologically relevant degradation is an important aid in designing the most effective reporter fusions.

Mode and Stability of Expression

Most genetic screens for alterations in degradation score alterations in protein level. For the simple case of a protein that is synthesized at a rate that is independent of the protein's levels, alterations in the protein degradation rate will cause the same proportional change in the steady-state level. Thus, the difference between the mutant and wild-type steady-state concentration of a reporter will often be proportional to the change in degradation rates for the two cases. If variability in the expression of the reporter is on the order of the difference caused by the desired mutations, there will be more difficulty in finding the desired mutants among false positives that arise because of expression level variation. For example, in yeast, expression plasmids can be either of one of two types of autonomously replicating plasmids, called ARS/CEN or 2 micron (YCp or YEp, respectively), or they can be integrating plasmids (YIp) that must be incorporated into the yeast genome to be replicated (Botstein and Fink, 1988). Autonomously replicating plasmids can vary in copy number over severalfold, whereas an integrating plasmid has a copy number that is preserved because of its presence as a true part of the chromosomal genome. Although our experience is in yeast, the principle of limiting changes in expression that could enhance false positives is applicable to any biological system. In mammalian cell lines, transfected reporters are often present in multiple arrays. Alterations in expression because of changes in reporter gene copy number, unequal recombination, or gene amplification can result in inheritable alterations in steady-state level that are not due to changes in degradation. In general, it is desirable to use the most stable source of expression possible, especially if the difference between the stable and degraded phenotypes is in the range of only fourfold to fivefold.

Another choice in the design of reporter fusions is the choice of promoter to drive the expression of the heterologous gene. Constitutive promoters that are not subject to regulation by signaling pathways limit the number of undesired ways that the steady-state levels can be affected in a screen. In truth, no promoter is completely free from communication with the cell, but those that drive always-needed housekeeping genes can limit the number of spurious phenotypes in a genetic analysis. We have had much success with the yeast GAPDH promoter (TDH3) (Schena et al., 1991). The strength of the promoter used to drive the reporter fusion can also affect some studies, because there are degradation pathways that can be overwhelmed by overly strong production of a pathway substrate (e.g., HMG-CoA reductase [HMGR] in mammals; Sever et al. [2003]). Again, the best strategy is to test the behavior of an engineered reporter plasmid before applying it to full-blown screen.

Phenotypes of Degradation

When using a fusion protein to evaluate degradation, the degradation phenotype must be amenable with the throughput demanded by the genetics. Either the steady-state level of the protein is used as a gauge, or the degradation of the fusion is directly determined. Direct assays of degradation are more difficult to use in screening, but such approaches have been employed successfully. This method of scoring includes subjecting samples from a master plate to conditions in which protein synthesis is curtailed and subsequent analysis of the levels of the degradation substrate under study after a degradation period (e.g., Knop *et al.* [1996]).

Once a screening phenotype has been decided, it is a good idea to test the screen for its ability to detect desired mutants. This is best done by using previously available mutants with the new assay/screen. With the ubiquitin proteasome pathway, one can often use more downstream mutants, such as one of many proteasome alleles (e.g., cim3–1, Hiller *et al.* [1996]; hrd2-1/rpn1, Hampton *et al.* [1996a]), which are still viable but cause a general stabilization of many ubiquitinated substrates. Alternately, proteasome inhibitors such as lactacystin or MG132 can be used to test the involvement of the pathway. One can also use mutant versions of the reporter protein that are resistant to the degradation pathway under study. For example, sometimes changing key amino acids can stabilize a protein under study (Gardner and Hampton, 1999). Creating the analogous fusion reporter with one or several stabilizing mutations provided a reliable alternative to test the feasibility of a screening strategy. We have used such "in cis phenocopies" in a variety of approaches, although these tools are luxuries of previous detailed analyses of a substrate under study. Alternately, if a reporter undergoes physiological regulation of its degradation, the stabilizing conditions can be used to determine whether a screen can locate degradation-deficient candidates. For example, plating yeast cells expressing the regulated HMGR reporter Hmg2p-GFP on a small dose of lovastatin causes physiological stabilization that can be scored by an optical colony screen for GFP (see later) (Cronin *et al.*, 2000).

Eliminating Fusion Reporter Mutants in a Screen

Sometimes mutations in the reporter protein that arise in the screen can phenocopy the desired mutants. When possible, the best strategy is to test for plasmid independence of the degradation phenotype. In yeast screens, each candidate mutant is cured of the original reporter-expressing plasmid, retransformed with a fresh sample of the same plasmid, and then scored for the desired degradation phenotype. In cases in which the plasmid cannot be removed easily, such as in larger eukaryotic cell lines, or when the reporter gene is replacing the endogenous allele without flanking homologous DNA, the independence of the degradation phenotype can be rescored by adding a different reporter of the same protein that can be detected while being coexpressed with the original reporter. Alternately when feasible, the native protein from which the reporter is derived can be evaluated if it is also coexpressed in the candidate mutants, using a method that allows unambiguous detection of this second substrate.

A variation of this idea is to use two distinct reporter proteins coexpressed in the parent strain or line. The screen would entail scoring separate phenotypes for each reporter; candidates will show both mutant phenotypes. The likelihood of a *cis* mutation in each independently expressed reporter protein causing the desired phenotype is much lower, thus biasing the screen toward the desired genomic mutations. Use of this approach is detailed later in the COD screen, but the idea is generally applicable if independent reporters can be developed.

Sources of Mutants

Along with the classical (and still very useful) randomly mutagenized cells, complete collections of viable null mutations are now available for *Saccharomyces cerevisiae* and undoubtedly other organism collections are on the way. Screening fusions for degradation phenotypes with these organized collections is a nice complement to traditional screening for mutants and has the further advantage of immediate identification of involved genes. The principal hurdle in this sort of analysis is to generate a strain or line of the null array with the desired reporter. In yeast, some creative applications of robotic mating and mating-type–specific selection have been applied to automate this task (Tong *et al.*, 2001).

Directed Genetic Screens

Available DNA genomic libraries, cDNA expression libraries, or more recently siRNA libraries for use in *C. elegans* (Kamath *et al.*, 2003), *Drosophila* S2 cells (Kiger *et al.*, 2003), or mammals (Paddison *et al.*, 2004), can also by analyzed with these approaches. In the case of protein-encoding DNA, candidates would be proteins that alter the degradation pathway by being overabundant, either as the native protein or some alteration or truncation that exists in the library. In the case of siRNA-expressing libraries, the phenotypically interesting candidate is presumed to encode a gene required for the wild-type phenotype, such that lowering its levels by siRNA will cause the alteration in degradation that is being screened for. One example of this approach, the high copy screen with a yeast 2micron library, is given in detail, but the principle is the same for many variations of this idea. In all cases, the fusion reporter-expressing wild-type strain is produced by transforming the reporter plasmid into the desired cell line or organism, and the resulting reporter strain is screened with the library to evaluate degradation effects.

Detailed Examples of Screens and Selections

The following examples are from the Hochstrasser laboratory (Yale University, New Haven, CT) or our own work. They are all yeast screens; together they include most of the ideas stated in the preceding general considerations. Standard yeast techniques can be found in a previous volume of this series (Guthrie et al., 2004) or other sources (Burke et al., 2000), including the many-linked *Saccharomyces* Genome Database (www.yeastgenome.org).

The DOA Pathway: Screens and Selections with Reporter Enzymes

The rapidly degraded yeast alpha two repressor (Matα2) is involved in the control of mating-type specific genes in yeast. A yeast genetic analysis of Matα2 degradation was launched early in the studies of this substrate; the resulting genes are collectively known as DOA (*D*egradation *O*f *A*lpha2) (Hochstrasser and Varshavsky, 1990). Degradation of this protein is complex, being mediated by separate regions at the N- and C-terminus. The N-terminal degron is known as deg1 (aa 1–67). deg1 is necessary and sufficient for Ubc6/Ubc7-dependent degradation mediated by the Doa10p ubiquitin ligase (Chen et al., 1993; Swanson et al., 2001). As mentioned previously, use of only deg1 as the fusion partner isolates only this branch of Matα2 degradation for analysis.

The First DOA Screen: deg1-lacZ

The reporter fusion called deg1-lacZ has the first 67 amino acids of α2 fused to the entire β-galactosidase protein and is expressed by inclusion of the natural Matα2 promoter in the plasmid. The fusion has enzyme activity and a very short half-life imparted by the presence of the deg1 degron. The yeast vector used was the low-copy ARS/CEN plasmid YCp50, which is selected in yeast with the *URA3* gene on the vector. Although this plasmid can have between one and four stable copies in yeast, this variability did not have an impact on the success of the screen. The wild-type parent strain included a *ura3-52* mutation, allowing for continued selection for the deg1-lacZ reporter plasmid by growth on uracil-minus medium. The strain also was of the alpha mating type to allow expression of the reporter fusion from the Matα2 promoter included in the expression plasmid. This parental reporter strain harbored very low levels of β-galactosidase because of the continued deg1-dependent degradation of the reporter plasmid.

To find *doa* mutants, colonies deficient for deg1-lacZ degradation were screened for increased levels of β-galactosidase (Hochstrasser and

Varshavsky, 1990). Wild-type cells were mutagenized with ethyl methane sulfonate to ~20% survival. A typical procedure involves growing ~10^8 cells to stationary phase in YPD medium, washing the cells by centrifugation, and resuspending in 1–2 ml 0.1 M sodium phosphate buffer (pH 7). Fifty microliters of EMS is added, and the cells are incubated at 30° for a time calibrated to give the desired level of killing (usually an hour). Cells are resuspended and washed twice with 200 μl 5% sodium thiosulphate ($Na_2S_2O_3$, which neutralizes the EMS) and then stored refrigerated in sodium phosphate buffer. Some investigators let cells divide once or twice in liquid so that mutagenized S/G2/M cells do not give rise to chimeric colonies composed of wt and mutant cells. (*Caution*: EMS is a powerful mutagen; all equipment that contacts EMS should be thiosulfate treated). Mutagenized cells were then plated onto uracil-deficient agar plates containing the chromogenic substrate X-gal, which is hydrolyzed to a blue product by β-galactosidase. The X-gal plates have a final concentration of X-gal of 80 μg/ml, and the medium is phosphate buffered at pH 7.0 as required for the indicator (7 g Na_2HPO_6, 3 gNaH_2PO_4 in 100 ml, pH 7, per final liter of medium). The X-gal is added after the sterilized agar is allowed to cool to ~50° because of its thermolability. Many companies that sell two-hybrid interaction systems (like Clonetech MATCHMAKER) have detailed descriptions of media formulations and use, as do many investigator's, web sites (e.g., Dr. Russel Finely (Wayne St. University, Detroit MI) (http://proteome.wayne.edu/Interactiontrap.html).

Colonies were plated to a density that allows visual scoring, approximately 200–300 per plate, ~40,000 colonies total. After 3–5 days, the plates were evaluated for colonies with elevated blue color. The candidate mutants were then isolated, colony purified on selective medium, and individually tested for a true-breeding phenotype. Next, candidates were subjected to individual pulse-chase analysis to ascertain whether the heightened steady-state levels of β-galactosidase activity were, indeed, due to slow degradation of the reporter. The candidates were then analyzed using yeast classical and molecular genetics.

A DOA Selection: Use of an Enzyme Reporter to Complement an Auxotrophy

When deg1 is fused to the *URA3* gene product Ura3p, the resulting enzyme is rapidly degraded in a *DOA* gene–dependent manner. Consequently, wild-type cells that express this reporter have very low levels of the deg1-Ura3p protein, whereas doa mutants have elevated levels of the fusion gene. The difference in levels of the degraded and stabilized report was sufficient to use complementation of uracil auxotrophy as a doa phenotype. That, is, wild-type *ura3-52* cells expressing the deg1-Ura3p fusion cannot grow on uracil-deficient medium, whereas doa mutants can (Chen *et al.*, 1993). Thus, plating a large number of the parent strain on uracil-deficient medium imposed a selection for doa mutants, allowing a second independent approach for discovering DOA genes. In fact, the selection revealed the ubiquitin ligase, Doa10p, which is responsible for ubiquitination of deg1-bearing proteins (Swanson *et al.*, 2001).

The DOA uracil auxotrophy selection was performed using wild-type Ura-cells (with the non-reverting *ura3-52* mutation) harboring an ARS/CEN plasmid expressing the deg1-URA3 fusion gene. This expression plasmid was marked with *LEU2*, thus allowing selection for the plasmid that is independent of the uracil auxotrophy needed to distinguish mutant from wild type. Cells were mutagenized with EMS at a level of $\sim$70% killing. A total of 3×10^6 mutagenized cells were plated on 80 uracil-deficient agar plates and incubated at 30°. The first 960 colonies to appear on the selection plates were collected in 96-well dishes and retested for maintained uracil prototrophy. The candidates were tested for complementation group by mating with a number of previously known doa mutants. This is done by replica plating candidate mutants onto a lawn of test nulls of opposite mating time, with auxotrophies such that only the diploids will grow on the replica-plated medium. The resulting diploids were tested for growth on uracil-minus medium by streaking on separate plates. More than 600 of the candidates were alleles of the previously known *DOA2 (UBC6)*, one of the ubiquitin E2s involved in the DOA pathway. More than 300 others were alleles of *DOA10* that encodes the ligase.

In comparing the two DOA isolation strategies, the deg1-lacZ method is a screen, in which every colony is queried for the desired phenotype (blue color in this case), and the deg1-Ura3p is a selection, in which only the mutant candidates were allowed to grow. Both have strengths and weaknesses. Selections allow use of far more individual genomes (in this case more than 10^6 cells), whereas screens typically allow analysis of tens of thousands of colonies. In the DOA uracil auxotrophic selection, the authors isolated two very rare mutants (one each in almost 1000 candidates) that would not have turned up in the lacZ screen. However, screens have the advantage that poorly growing colonies are included in the scoring, whereas in a selection there is a greater chance that these could be missed.

The HRD Pathway: Use of Optical Proteins for Genetics

HMG-CoA reductase (HMGR) is an essential early enzyme of the sterol synthetic pathway. This ER-anchored essential enzyme undergoes regulated, ubiquitin-mediated degradation (Hampton, 2002). In yeast, the

Hmg2p isozyme is subjected to this regulation. When sterol pathway activity is high, the Hmg2p protein is degraded rapidly. When sterol pathway activity is slowed, as when the cells are treated with the HMGR-inhibitor lovastatin, Hmg2p degradation is slowed (Hampton and Rine, 1994). The first yeast genetic analysis of the pathway revealed that this process was ubiquitin-mediated, because the *HRD2* gene encodes a proteasome subunit, and *HRD1* encodes an ER-membrane bound ubiquitin ligase that recognizes Hmg2p and numerous other substrates (Hampton, 2002). The first *HRD* genes were isolated using lovastatin killing to select for cells with elevated levels of Hmg2p because of slow degradation (with some modifications to allow the selection to work with an acceptable background). Subsequently, the availability of GFP allowed construction of an Hmg2p-GFP reporter fusion, in which the catalytic C-terminal domain of Hmg2p was replaced with GFP (Hampton *et al.*, 1996b). The resulting protein consisted of the large N-terminal multispanning membrane domain (525 amino acids), which is required for both ER localization and regulated HRD-dependent degradation, followed by the GFP reporter. The cloning and properties of this fusion reporter, and its use in the study of degradation, has been extensively described in an earlier volume of this series (Cronin and Hampton, 1999). In all cases, Hmg2p-GFP is expressed in cells that also have active HMGR, because this is an essential enzyme, and the optical reporter has no catalytic domain. Hmg2p-GFP has been used in a variety of genetic screens, two of which will be described below.

Regulation of the HRD Pathway: Isolation of *COD1* by Two-Protein Screening

Hmg2p and Hmg2p-GFP undergo regulated degradation in yeast. When the sterol pathway is slowed with HMGR inhibitor lovastatin, the signals for degradation decrease, and the Hmg2p-GFP reporter protein is stabilized. Hmg2-GFP was used to find *cod* mutants (*CO*trol of hmgr *D*egradation) that continue to degrade Hmg2p-GFP even when the signals for degradation were low (Cronin *et al.*, 2000). Specifically, the desired mutants remain dark when plated on a low dose of lovastatin that normally causes stabilization of the fluorescent reporter and brightening of the cells. Because many uninteresting mutations could make the cells dark (poor expression of Hmg2p-GFP, poor permeability to lovastatin, increased metabolism of the drug, mutations in the *HMG2-GFP* reporter itself), we included another reporter protein, a functional copy of the Hmg2p enzyme with a myc tag (1myc-Hmg2p) expressed from an integrated plasmid at a locus distinct from the optical reporter protein. Although this is not technically a fusion protein, its use as a second reporter in the same cells as a built-in secondary screen is instructive and can be applied to many other circumstances in which plasmid-based reporters with distinct phenotypes are available.

We scored for poor regulation of the catalytically active 1myc-Hmg2p by toxicity of lovastatin, seen at much higher doses than those used to test the regulation of Hmg2p-GFP. As the sole active HMGR in the cells, 1myc-Hmg2p activity is essential for cell growth, so at sufficiently high doses of lovastatin, the cells will die. In cells in which lovastatin induces stabilization of Hmg2p, the elevated levels of the Hmg2p blunt the toxicity of the lovastatin, shifting the killing curve of lovastatin to the right compared with strains that cannot slow the degradation of Hmg2p. In other words, more lovastatin is required to kill cells that can stabilize 1myc-Hmg2p (wild-type) than needed to kill cells that cannot stabilize 1myc-Hmg2p (cod mutants). We confirmed this idea by using engineered variants of Hmg2p with sequence changes that removed lovastatin-induced stabilization. The increase in sensitivity that accompanies loss of regulation is about threefold to fourfold, so that 200 μg/ml lovastatin on agar plates will kill the *cod* mutants but not the wild-type cells, thus providing an independent phenotype for poor Hmg2p regulation. Because the dose used to kill the cells is much higher than the concentration that first causes the regulatory response (12.5 μg/ml), the use of low-dose lovastatin to first score the optical phenotype of the coexpressed Hmg2p-GFP does not affect the growth of the cod mutants.

The Two-Gene COD Screen

The wild-type parent strain for the cod screen coexpressed Hmg2p-GFP and 1myc-Hmg2p from separate integrating plasmids with the TDH3 promoter. The 1myc-Hmg2p expression plasmid was integrated at the native locus and was maintained in cells by virtue of its being the only form of the essential HMGR activity in the cells. The Hmg2p-GFP expression vector was integrated at the *ura3-52* locus and was maintained by complementation of uracil prototrophy. We confirmed that each reporter underwent HRD-dependent regulated degradation when the two were coexpressed before launching the screen. This can be important, because there are cases in which degradation pathways can be saturated by an overabundance of a specific substrate.

The COD screen was performed by plating EMS mutagenized cells ($\sim$30% survival; see earlier) onto agar plates supplemented with the appropriate nutrients and 12.5 μg/ml lovastatin to a colony density of approximately 250 colonies per plate. After 2–3 days of growth, the plates were examined for fluorescent colonies, using a plate-based assay developed for this purpose. This technique has been detailed in a previous volume of this series (Cronin and Hampton, 1999). A narrow bandpass filter with a maximum wavelength of 488 nm was custom-made by Omega Optical, Inc. (Brattleboro, VT; www.omegafilters.com; cost ~$200). The filter was designed to exclude both lower and higher wavelengths that normally come from a bright white light. The filter had the dimensions of a photographic slide (50 mm × 50 mm), so that it could be put into a Kodak slide projector (now readily available in the surplus sites of many universities), providing an intense blue light field for inspection of multiple plates. The blue-illuminated plates are examined through a Kodak No. 12 Wratten filter placed on a pair of laboratory goggles. The colony assay was calibrated both with hrd mutants and with strains expressing variants of Hmg2p-GFP that do not respond to lovastatin to ensure that the screen would distinguish the desired mutants. Once established, this assay is quite facile; 300,000 colonies were examined for lovastatin-induced GFP fluorescence in the cod screen. Colonies with low fluorescence were isolated on no-lovastatin plates. These unresponsive candidates were then tested for growth onto agar plates with 200 μg/ml lovastatin, a dose that will kill hypersensitive cod mutants but not the wild-type parent cells. Candidates that were both unresponsive to lovastatin in the low-dose optical screen and hypersensitive to lovastatin in high-dose toxicity assay were then checked directly for altered regulation of Hmg2p-GFP by *in vivo* flow cytometry and biochemical analysis. Finally, successful candidates were analyzed by classical and molecular genetic means. From this, the *cod1-1* mutant and 39 independent alleles of the same gene, were isolated.

The use of these two independent reporters automatically rules out a large number of distinct false positives. For example, if a candidate were sufficiently impermeable to lovastatin, the optical screen would score as a mutant (unresponsive to lovastatin), but the secondary screen would not show increased sensitivity to the drug at higher doses. Furthermore, the use of two separate Hmg2p-based reporters strongly decreases the isolation of mutations in the Hmg2p-GFP reporter itself, because these would not affect the regulation of the independent 1myc-Hmg2p reporter. With the plethora of reporters available, it is often possible to devise a two-reporter screen that obviates many of the typical concerns of classical genetic screening.

Use of the Optical Screen to Find Genes That Block Hmg2p Degradation

Another powerful approach in many organisms is to test expression libraries for the ability to cause phenotypes of interest. We used a wild-type strain expressing Hmg2p-GFP from the *TDH3* promoter to screen for yeast genes that disrupt the degradation of Hmg2-GFP at high doses (20–40 copies per cell) (22, 23). The recipient strain expressed Hmg2p-GFP and (from an independent locus) high levels of the soluble catalytic domain of Hmg2p. This source of HMGR activity provided a high level of the cellular degradation signal to ensure rapid degradation of the Hmg2p-GFP and, consequently, low colony fluorescence in the parent strain. The low wild-type fluorescence in the parent strain allowed more sensitive detection of candidates with increased fluorescence because of degradation-inhibiting plasmids.

The source DNA was a publicly available 2 μ yeast genomic library with a *LEU2* prototrophy marker. These plasmids are harbored in yeast cells at a level of 20–40 copies of a single plasmid per cell. The screen is performed on leucine-deficient and uracil-deficient medium to allow continuous selection for the library plasmid and the Hmg2p-GFP reporter plasmid, respectively. Cells were transformed with library DNA and spread on the leucine-deficient agar plates to give 250 colonies per plate (actual number used depends on investigator eyesight). The plates were allowed to grow for 2–3 days and examined for colony fluorescence. Because a typical genomic fragment is represented approximately once every 1000–2000 times in a genomic library, it is good to examine at least 5–10 times this number to improve the possibility of finding a desired clone. Colonies that have elevated fluorescence were isolated, regrown on selective medium, and, if still bright, grown in liquid medium and tested by flow cytometry for direct effects on the degradation rate of the Hmg2p-GFP reporter.

The plasmid DNA was next isolated in bacteria from interesting candidates and purified. Candidate plasmids were analyzed by restriction analysis, and each unique plasmid was tested for phenotypic fidelity by transformation into the parent strain. Successful candidates were analyzed by sequencing, and the candidate coding regions were then individually analyzed by subcloning and retesting in the parent strain. In this way, we isolated both a dominant-negative truncated version of *HRD1* and a yeast homolog of INSIG (Flury *et al.*, 2005), a protein involved in regulated degradation of HMGR in mammals (Sever *et al.*, 2003)

Acknowledgments

The author wishes to thank members of the Hampton laboratory for continuous intellectual and inspirational support and for developing the techniques and putting them to practice. No number of Turkey-off-the-Bird sandwiches (Elijah's Delicatessen, La Jolla, CA) can repay their creativity and high spirit. This work was supported by NIH (NIDDK) grant #GM51996-06, and an AHA Established Investigator Award.

References

Botstein, D., and Fink, G. R. (1988). Yeast: An experimental organism for modern biology. *Science* **240**, 1439–1443.

Burke, D., Dawson, and Stearns, T. (2000). "Methods in Yeast Genetics." Cold Spring Harbor Laboratory Press, Cold Spring Harbor.

Chen, P., Johnson, P., Sommer, T., Jentsch, S., and Hochstrasser, M. (1993). Multiple ubiquitin-conjugating enzymes participate in the *in vivo* degradation of the yeast MAT alpha 2 repressor. *Cell* **74**, 357–369.

Cronin, S., and Hampton, R. Y. (1999). Measuring protein degradation with GFP. *Methods Enzymol.* **302**, 58–73.

Cronin, S. R., Khoury, A., Ferry, D. K., and Hampton, R. Y. (2000). Regulation of HMG-CoA reductase degradation requires the P-type ATPase Cod1p/Spf1p. *J. Cell Biol.* **148**, 915–924.

Dohmen, R. J., Wu, P., and Varshavsky, A. (1994). Heat-inducible degron: A method for constructing temperature-sensitive mutants. *Science* **263**, 1273–1276.

Flury, I., Garza, R., Shearer, A., Cronin, S., Rosen, J., and Hampton, R. Y. (2005). INSIG: A broadly conserved transmembrane chaperone for start-sensing domain (SSD) proteins. Manuscript submitted.

Gardner, R. G., and Hampton, R. Y. (1999). A 'distributed degron' allows regulated entry into the ER degradation pathway. *EMBO J.* **18**, 5994–6004.

Guthrie, C., Fink, G., Abelson, J., and Simon, M. (eds.), (2004). *In* "A Guide to Yeast Genetics and Molecular Biology; Part A."

Hampton, R. Y. (2002). Proteolysis and sterol regulation. *Annu. Rev. Cell Dev. Biol.* **18**, 345–378.

Hampton, R. Y., Gardner, R. G., and Rine, J. (1996a). Role of 26S proteasome and HRD genes in the degradation of 3-hydroxy-3-methylglutaryl-CoA reductase, an integral endoplasmic reticulum membrane protein. *Mol. Biol. Cell* **7**, 2029–2044.

Hampton, R. Y., Koning, A., Wright, R., and Rine, J. (1996b). *In vivo* examination of membrane protein localization and degradation with green fluorescent protein. *Proc. Natl. Acad. Sci. USA* **93**, 828–833.

Hampton, R. Y., and Rine, J. (1994). Regulated degradation of HMG-CoA reductase, an integral membrane protein of the endoplasmic reticulum, in yeast. *J. Cell Biol.* **125**, 299–312.

Hiller, M. M., Finger, A., Schweiger, M., and Wolf, D. H. (1996). ER degradation of a misfolded luminal protein by the cytosolic ubiquitin-proteasome pathway. *Science* **273**, 1725–1728.

Hochstrasser, M., and Varshavsky, A. (1990). *In vivo* degradation of a transcriptional regulator: The yeast alpha 2 repressor. *Cell* **61**, 697–708.

Kamath, R. S., Fraser, A. G., Dong, Y., Poulin, G., Durbin, R., Gotta, M., Kanapin, A., Le Bot, N., Moreno, S., Sohrmann, M., Welchman, D. P., Zipperlen, P., and Ahringer, J. (2003). Systematic functional analysis of the *Caenorhabditis elegans* genome using RNAi. *Nature* **421**, 231–237.

Kiger, A. A., Baum, B., Jones, S., Jones, M. R., Coulson, A., Echeverri, C., and Perrimon, N. (2003). A functional genomic analysis of cell morphology using RNA interference. *J. Biol.* **2**, 27.

Knop, M., Finger, A., Braun, T., Hellmuth, K., and Wolf, D. H. (1996). Der1, a novel protein specifically required for endoplasmic reticulum degradation in yeast. *EMBO J.* **15**, 753–763.

Paddison, P. J., Silva, J. M., Conklin, D. S., Schlabach, M., Li, M., Aruleba, S., Balija, V., O'Shaughnessy, A., Gnoj, L., Scobie, K., Chang, K., Westbrook, T., Cleary, M.,

Sachidanandam, R., McCombie, W. R., Elledge, S. J., and Hannon, G. J. (2004). *Nature* **428,** 427–431.

Schena, M., Picard, D., and Yamamoto, K. R. (1991). Vectors for constitutive and inducible gene expression in yeast. *Methods Enzymol.* **194,** 389–398.

Sever, N., Song, B. L., Yabe, D., Goldstein, J. L., Brown, M. S., and DeBose-Boyd, R. A. (2003). Insig-dependent ubiquitination and degradation of mammalian 3-hydroxy-3-methylglutaryl-CoA reductase stimulated by sterols and geranylgeraniol. *J. Biol. Chem.* **278,** 52479–52490.

Swanson, R., Locher, M., and Hochstrasser, M. (2001). A conserved ubiquitin ligase of the nuclear envelope/endoplasmic reticulum that functions in both ER-associated and Matalpha2 repressor degradation. *Genes Dev.* **15,** 2660–2674.

Tong, A. H., Evangelista, M., Parsons, A. B., Xu, H., Bader, G. D., Page, N., Robinson, M., Raghibizadeh, S., Hogue, C. W., Bussey, H., Andrews, B., Tyers, M., and Boone, C. (2001). Systematic genetic analysis with ordered arrays of yeast deletion mutants. *Science* **294,** 2364–2368.

[22] Bi-substrate Kinetic Analysis of an E3-Ligase–Dependent Ubiquitylation Reaction

By David C. Swinney, Michael J. Rose, Amy Y. Mak, Ina Lee, Liliana Scarafia, and Yi-Zheng Xu

Abstract

Little is known about the kinetic mechanism of E3 ubiquitin ligases. This work describes basic methodology to investigate the kinetic mechanism of E3 ubiquitin ligases. The method used steady state, bi-substrate kinetic analysis of an E3 ligase–catalyzed monoubiquitylation reaction using ubiquitin-conjugated E2 (E2ub) and a mutant IκBα as substrates to evaluate whether the E3-catalyzed ubiquitin transfer from E2ub to protein substrate was sequential, meaning both substrates bound before products leaving, or ping pong, meaning that ubiquitin-conjugated E2 would bind, transfer ubiquitin to the E3, and debind before binding of protein substrate. The method requires the E3 reaction to be rate limiting and at steady state. This was accomplished through optimization of the conditions to ensure that the E3-dependent transfer of ubiquitin from E2ub to substrate was rate limiting. We observed a sequential bi-substrate E3-dependent ubiquitylation reaction on using E2UBCH7 and IκBαSS32/36EE (IκBαee as substrates and a partially purified Jurkat cell lysate as a source for the E3 ligase activity). The sequential bi-substrate kinetic mechanism is consistent with the formation of a ternary complex among E2UBCH7, IκBαSS32/36EE, and E3 before the transfer of ubiquitin from E2UBCH7 to IκBαSS32/36EE. The described method should be of use to characterize the kinetic mechanism of other E3 ligase–catalyzed ubiquitylation reactions.

Introduction

Ubiquitylation of a protein substrate is a complex process involving at least three different classes of enzymes—E1 activating enzymes, E2 conjugating enzymes, and E3 ligases—that can conjugate ubiquitin to lysine side chains of a protein substrate. An additional level of complexity is added when the ubiquitylation enzymes add multiple ubiquitin molecules to a substrate to form polyubiquitin chains. Studies into the mechanism of enzyme reactions generally use a combination of structural and kinetic analysis. Understanding of the ubiquitylation system has been facilitated by much structural work (reviewed in Passmore and Barford, 2004). However, studies of the kinetic mechanism of the ubiquitylation enzymology have been limited.

The ubiquitylation reaction is traditionally described as a type of bucket brigade in which ubiquitin is activated by E1 and transferred by means of E2 and E3 to lysine side chains of a protein substrate (Haas and Seipmann, 1997; Hershko and Ciechanover, 1998; Wilkinson, 2000). This model of the ubiquitylation reaction, although helpful to conceptually understand the complex ubiquitylation enzymology, is challenging to reconstitute experimentally for investigation of kinetic mechanisms. We approached the E3-ubiquitylation reaction as a steady-state bi-substrate reaction: the two substrates were E2ub and the protein to be conjugated with ubiquitin, and the two products were E2 and the ubiquitin-conjugated protein substrate. Bi-substrate kinetic analysis can be used to determine whether E3 enzymes follow a sequential mechanism, in which catalysis only proceeds after binding of both E2ub and protein substrate, or a ping-pong mechanism, in which catalysis and product release occur, leaving a covalent enzyme ubiquitin intermediate before the binding and reaction with the second substrate (Fig. 1) (Segel, 1975). Because enzyme-catalyzed ubiquitin group transfer reactions can use ubiquitin-enzyme intermediates (ping-pong) or direct transfer (sequential), bi-substrate steady-state analyses can differentiate among the possible mechanisms.

Meaningful kinetic data are facilitated by kinetic isolation of rate-limiting steps; only slowest steps will generally be kinetically detectable. The ubiquitylation process requires multiple kinetic transitions between ubiquitin activation by E1 and final transfer of ubiquitin to the protein substrate; the purpose of this work was to develop a method to isolate and measure the kinetics associated with the E3 ligation component of

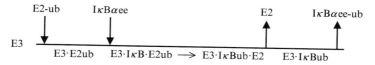

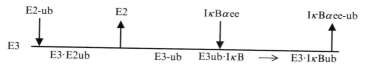

FIG. 1. Schematic of BiBi kinetic mechanisms. The term "sequential" is used to denote systems in which all substrates must bind to the enzyme before any product is released. A random system is sequential even though both substrates bind randomly. An ordered system in which one substrate must bind before the other is also sequential, because no product is released before formation of the ternary complex (E3:IκB:E2ub). Ping-pong systems, in which a product is released between substrate additions, are nonsequential.

ubiquitylation. The complication for the ubiquitin system is that one of the substrates (E2ub) must be "kinetically" generated but must "not" be kinetically detected. This is in contrast to a classic bi-substrate system that has one enzyme and two substrate molecules at essentially a static concentration. For examples of classic bi-substrate systems, see the work of Burke *et al.* (1998) on the IκB kinase (IKK) and the work of Leskovac *et al.* (2004) on yeast alcohol dehydrogenase.

The system we investigated used a mutant form IκBα, serines 32 and 36 were mutated to glutamic acid (IκBαee), as one substrate and ubiquitin-conjugated E2UBCH7 as the second substrate. The E3 activity was provided by a partially purified Jurkat cell lysate. The method requires the E3 reaction to be rate limiting and at steady state. This was accomplished through optimization of the conditions to ensure that the E3-dependent transfer of ubiquitin from E2ub to IκBαee was rate limiting. Saturated steady-state levels of E2-ub were attained by ensuring that the activation and transfer of ubiquitin to E2UBCH7 by the E1 activating enzyme was not rate limiting. Another complicating kinetic factor is the potential for polyubiquitylation. The approach we used to minimize the contribution of polyubiquitylation to the kinetics was to limit the reaction to monoubiquitylation through the use of an excess of nonradioactive reductively methylated ubiquitin in the reaction. Reductive methylation of lysine side chains prevents ubiquitin from forming polyubiquitin chains with other ubiquitin molecules (Hersko and Heller, 1985). Methylated ubiquitin can form an E1-catalyzed active thioester at the C-terminus, allowing the molecule to be transferred to the lysines of substrate proteins (monoubiquitylation). Under this protocol, only the addition of a single ubiquitin molecule to the substrate is detectable.

The E3 ligase activity investigated in this study has been shown to be important for NF-κB activation stimulated by TNFα (Swinney et al., 2002). The exact molecular species that catalyzes the IκBαee ubiquitylation has yet to be established. Further properties of the system are described in a previous report (Swinney et al., 2002).

Materials and Methods

Materials

The E1, E2UBCH7, and E3 ligase enzyme preparations were prepared as previously described (Swinney et al., 2002). E1 was purified from rabbit liver, recombinant E2UBCH7 was purified from transfected *E. coli*, and the E3-dependent IκBαee-ubiquitylation activity was fractionated from TNFα-activated Jurkat cells. The recombinant substrate IκBαee with an N-terminal His$_6$ tag and serines 32 and 36 mutated to glutamic acid was also purified from transfected *E. coli* as previously described. Concentrations of E2 and IκBαee were calculated assuming 100% purity; SDS-acrylamide gels showed purity to be greater than 95% (data not shown). ^{125}I-ubiquitin was obtained from Amersham (Piscataway, NJ), and methyl-ubiquitin was obtained from Boston Biochem (Boston, MA). All other reagents were purchased from Sigma Chemical Co. (St Louis, MO).

IκBαee Ubiquitylation Assay Procedure

Ubiquitylation was assayed in 10 mM Hepes buffer, pH 7.5, containing 1 mM ATP, 5 mM MgCl$_2$, ^{125}I-ubiquitin, 0.2-1.0 μg E1, 0.5–3.0 μg E2 UBCH7, 0.5–2.0 μg E3, and 0.2–5.0 μM IκBαee in 50 μl total volume. Typically, E1, E2, and E3 were preincubated with ubiquitin, ATP, and MgCl$_2$ for 20 min at 37°. IκBαee was added to the preincubated mixture to initiate the reaction. The reaction was stopped after 1 h by adding 50 μl of 25 mM DTT to cleave thioester bonds. After 5 min, the reaction was diluted with 1 ml of binding buffer (5 mM imidazole, 0.5 M NaCl, 20 mM Tris-HCl, pH 7.9) and a mixture of urea and iodoacetamide to a final concentration of 5.3 M and 50 mM, respectively. A suspension of Ni-NTA beads (Qiagen Inc., Valencia, CA) (100 μl) in binding buffer was added to the sample and shaken for 1 h at room temperature. The supernatant was removed after centrifugation, and the NTA beads were washed (4×) with 1 ml of 1% NP-40 in 60 mM imidazole, 0.5 M NaCl, 20 mM Tris-HCl, pH 7.9. The ^{125}I-ubiquitin–conjugated IκBαee product was eluted with 200 μl of 1 M imidazole, 0.5 MNaCl, 20 mM Tris-HCl, pH 7.9, and the amount of ^{125}I was determined in the gamma counter. Typically, incubations were run in duplicate, and nonspecific binding activity was determined in incubations without substrate.

Results and Discussion

Assay Optimization

Optimization of the conditions to ensure that the E3-dependent transfer of ubiquitin from E2ub to IκBαee was rate limiting presents unique challenges because of the large number of reagents that are required for activity. The results of the optimization are described in the following. The overall strategy was to determine the time of reaction and concentrations of ATP, E1, ub/meub, E2, and E3 required for formation of IκBαee conjugates.

1. Specific binding: A nickel-bead pull-down format was used to separate conjugated from free ^{125}I-ubiquitin. N-terminal His$_6$-tagged IκBαee bound to the beads was washed and quantified as described in "Materials and Methods." To minimize the detection of nonspecific ubiquitin conjugates, the enzymatic reactions were stopped with 25 mM DTT, 5.3 M urea, and 50 mM iodoacetamide. Control incubation without substrate (minus IκBαee) was also included at each reaction condition to determine the nonspecific binding to the beads. The nonspecific bound radioactivity was then subtracted from the total bound radioactivity to calculate the amount of ^{125}I ubiquitin specifically bound to IκBαee.

2. Progress curves: The reaction was optimized for incubation time (Fig. 2). Progress curves were run with and without ubiquitin and ATP preincubation with E1, E2, and E3. A time lag of 5–10 min was observed when all the enzymes were not preincubated with ubiquitin and ATP, presumably because of the time required for the system to reach steady state. For all further studies, a 20-min preincubation was added to ensure that the system was at steady state. After preincubation, the reaction was observed to be linear for at least 1 h (Fig. 2).

3. Linearity with E3 concentration: The reactions were also shown to be linear with E3 concentrations up to 10 μg/0.05 ml. The reactions were run at saturating concentrations of ATP, E1, E2, ub, and IκBαee. E1, E2, E3, ATP, and ubiquitin were preincubated for 20 min, and the reaction was initiated with IκBαee (data not shown).

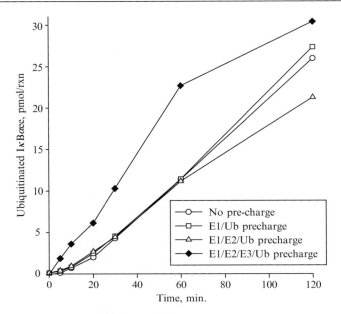

FIG. 2. Progress curves for IκBαee ubiquitylation with and without preincubation of ubiquitylation enzymes. E1 (3 pmol), E2 (3 μg), and/or E3 (10 μg) were preincubated for 20 min with ATP and 5 μM ^{125}I-ubiquitin, and the reaction was then initiated with the substrate IκBαee (5 μM). All preincubation reactions were carried out at 37°. Reactions with no preincubations were initiated with ATP. At the appropriate time, the reactions were stopped with a 25 mM DTT, followed 5 min later with 5.3 M urea, 59 mM iodoacetamide (final) solution, and ubiquitin conjugated IκBαee separated from free ubiquitin with Ni beads as described in "Materials and Methods."

4. Methylubiquitin saturation curves: Methylubiquitin plays two roles in these experiments. Primarily, it prevents the formation of polyubiquitin chains, and secondarily, it provides an alternative substrate for E1 and E2. For the system to operate such that E3 is rate limiting, the rate of formation of the E1 and E2 ubiquitin conjugates must not be rate limiting. Accordingly the methylubiquitin/ubiquitin concentration must be sufficient not to limit the rate of formation of the E1 and E2 ubiquitin intermediates. In the methylubiquitin experiments, a tracer concentration of ^{125}I-ubiquitin was used with increasing concentrations of unlabeled methylubiquitin. Saturation curves were adjusted to represent the total amount of ubiquitin conjugated (ubiquitin plus methylubiquitin), assuming the rate of ubiquitin and methylubiquitin activation and conjugation were identical. By use of this assumption, methylubiquitin displayed saturable

kinetics under the conditions of these experiments with an apparent Km of 1.5 ± 0.06 μM (Fig. 3, right panel); 4 μM methylubiquitin was used in bi-substrate kinetic experiments. Saturation curves with ubiquitin alone gave similar Km values (data not shown).

5. Optimization of E1: A saturable effect of E1 concentration on IκBαee activity was assumed to reflect the attainment of steady-state levels of ubiquitin conjugated E2. (This assumes that the rate of formation of E2ub by reaction of E1ub and E2 is faster than the E3-limiting reaction of E2ub with IκBαee.) Optimization for steady-state E2ub concentrations required ensuring that the steady-state levels of E2ub were rapidly achieved and maintained. We first preincubated the E1, E2UBCH7, and the E3 prep for 20 min with ATP and an ubiquitin/methylubiquitin mix to load them with ubiquitin/methylubiquitin. The reaction was then initiated with the substrate IκBαee. The dependence of the reaction on E1 concentration is shown in the left panel of Fig. 3. E1 displayed saturable kinetics under the conditions of these experiments with an apparent Km of 0.15 ± 0.06 μg/ml; 0.5 g/ml of E1 was used in further experiments.

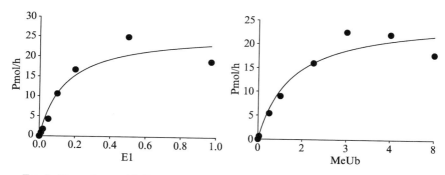

FIG. 3. Dependence of IκBαee-ubiquitylation on E1 and methylubiquitin. The left panel shows the saturation of IκBαee ubiquitylation with increasing E1 concentration. E1, E2 (2 μg), and E3 (10 μg) were preincubated for 20 min with ATP and methylubiquitin/ubiquitin mixture, and the reaction was initiated with the addition of substrate IκBαee (5 μM). After 60 min, the reaction was stopped with 25 mM DTT followed 5 min later with 5.3 M urea and 59 mM iodoacetamide (final). Ubiquitin-conjugated IκBαee separated from free ubiquitin with Ni beads as described in the "Materials and Methods" section. The final ubiquitin concentration of 4 μM was composed of 200 pmol unlabeled methylubiquitin and 0.075 μCi, 0.0375 pmol labeled ^{125}I-ubiquitin. The right side of the figure shows the effect of titrating methylubiquitin on the IκBαee ubiquitylation activity. Methylubiquitin was added in concentrations from 50 nM to 8 μM (2.5–400 pmol/0.05 ml reaction), and [125]I -labeled ubiquitin was held constant (0.075 μCi, 0.0375 pmol per reaction). The specific activity was calculated assuming methylubiquitin and ubiquitin contributed equally to the formation of IκBαee–ubiquitin conjugates.

6. Optimization of IκBαee and E2UBCH7 concentrations: Before running the bi-substrate experiments, preliminary studies were run to determine the appropriate concentration range of IκBαee and E2UBCH7. In the preliminary experiments, one substrate was fixed at a saturating concentration, whereas the other was varied. For the bi-substrate kinetic experiments, concentrations were chosen to cover a range of at least five times below to five times above the substrate's apparent Km.

Bi-substrate Kinetic Analysis

Bi-substrate kinetic analyses were carried out to distinguish between a possible sequential or ping-pong kinetic mechanism. Initial velocities were determined as a function of both substrates using a 6 × 7 matrix of substrate concentrations (IκBαee vs E2), ranging from approximately five times above to five below the respective apparent Km; E2UBCH7 concentrations ranging from 0.05–2.5 μM and IκBαee concentrations from 0.2–5 μM. The use of a matrix approach allows collection of all the data required to distinguish between a sequential versus ping-pong mechanism in a single experiment. The results presented here describe a single experiment that is representative of three separate experiments. Initial velocities were measured and saturation plots generated (Fig. 4). The velocity vs. IκBαee curves were saturable at all E2UBCH7 concentrations as were the

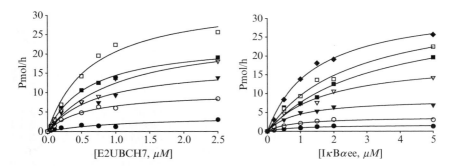

Fig. 4. Saturation curves for the substrates E2UBCH7 and IκBαee. The left panel shows saturation curves for E2UBCH7 at fixed concentrations of IκBαee and the right panel saturation curves for IκBαee at fixed E2UBCH7 concentrations. E1 (0.5 μg), E2, and E3 (10 μg) were preincubated for 20 min with ATP and methylubiquitin/ubiquitin mixture and the reaction then initiated with the substrate IκBαee. After 60 min, the reaction was stopped with 25 mM DTT followed 5 min later with 5.3 M urea and 59 mM iodoacetamide (final), and the ubiquitin-conjugated IκBαee was separated from free ubiquitin with Ni beads as described in the "Materials and Methods" section. The final ubiquitin concentration of 4 μM was composed of 200 pmol unlabeled methylubiquitin and 0.075 μCi, 0.0375 pmol labeled ^{125}I-ubiquitin.

TABLE I
KINETIC PARAMETERS FOR E2UBCH7ub INTERACTIONS DEPENDENT ON IκBαee

[IκBαee, μM]	Km, app, μM	Vmax
0.2	1.18 (0.72)	4.04 (1.22)
0.5	0.54 (0.08)	10.1 (0.60)
1.0	0.75 (0.18)	17.2 (1.81)
1.5	1.08 (0.26)	25.9 (3.00)
2.0	0.63 (0.06)	23.5 (0.90)
5.0	0.66 (0.14)	34.1 (2.88)

Kinetic constants were determined by fitting to a one-site saturation curve using SigmaPlot software program. Vmax is expressed as pmol/h/0.01 mg E3; parentheses indicate the standard error.

TABLE II
KINETIC PARAMETERS FOR IκBαee INTERACTIONS DEPENDENT ON E2UBCH7ub

[E2UBCH7-ub, μM]	Km, app, μM	Vmax
0.05	0.81 (0.28)	1.68 (0.20)
0.1	0.75 (0.21)	3.78 (0.35)
0.2	0.88 (0.20)	8.54 (0.69)
0.5	1.75 (0.34)	19.1 (1.69)
0.75	3.16 (0.83)	31.7 (4.51)
1.0	2.57 (0.47)	33.8 (3.14)
2.5	1.42 (0.15)	33.2 (1.53)

Kinetic constants were determined by fitting to a one-site saturation curve using SigmaPlot software program. Vmax is expressed as pmol/h/0.01 mg E3; parentheses indicate the standard error.

velocity vs. E2UBCH7 curves saturable at all IκBαee concentrations. The apparent Km values and Vmax for each curve were calculated by fitting to a single site saturation kinetic model (Tables I and II). Whereas the Vmax of IκBαee monoubiquitylation increased with addition of both substrates, the Km values were not significantly changed, suggesting that the interactions of the two substrates with the E3 enzyme were independent of each other. The mean Km value for IκBαee and E2UBCH7ub binding were 1.62 ± 0.38 and 0.81 ± 0.12 μM, respectively.

Double-reciprocal plots of 1/v vs. 1/[substrate] were also generated. The plot of 1/v versus 1/[IκBαee] (Fig. 5) shows a series of lines that intersects to the left of the 1/v axis and on the 1/[IκBαee] axis in a noncompetitive manner. Similarly, the plot of 1/v vs. 1/[E2UBCH7]

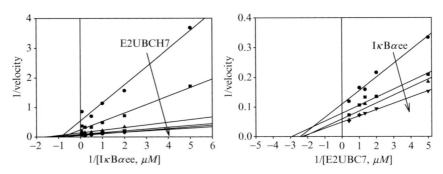

FIG. 5. Lineweaver–Burk plots of substrate-dependent monoubiquitylation. Replot of the data in Fig. 4.

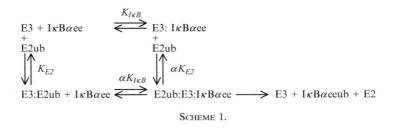

SCHEME 1.

(Fig. 5) also shows a series of lines that intersect to the left of the 1/v axis and on the 1/[E2UBCH7] axis in a noncompetitive manner. The noncompetitive nature of each interaction is consistent with the formation of a ternary complex by way of a sequential mechanism (Segel, 1975). The sequential mechanism requires that before any catalytic step, both ubiquitylated E2UBCH7 and IκBαee bind to form a ternary complex with the E3 enzyme. The intersection of the plots on the horizontal axis suggests that the binding of one substrate has no effect on the binding of the other substrate. Although the data clearly indicate the requirement for a ternary complex, the order of substrate addition and product release cannot be established from these experiments alone. Additional experiments with competitive inhibitors are required to confirm the kinetic order of the reaction (Burke et al., 1998; Leskovac et al., 2004; Segel, 1975).

The kinetic model for a random sequential reaction is defined in Scheme 1, where $K_{I\kappa B}$ and $\alpha K_{I\kappa B}$ are the dissociation constants for IκBαee in the absence and presence, respectively, of E2ub in the active site; K_{E2} and αK_{E2} are the dissociation constants for E2ub in the absence and presence, respectively, of IκBαee in the active site. A value of $\alpha = 1$

indicates that the equilibrium dissociation constants are not influenced by binding of the alternative substrate. The substrates potentially can bind randomly or in an ordered manner. The preceding scheme represents random binding. The results of these studies are consistent with sequential random binding but are by themselves not conclusive.

The observations of a sequential mechanism for this ubiquitylation system as opposed to a ping pong suggests the two substrates bind simultaneously to the E3 at structurally distinct binding sites. It is consistent with a mechanism of ubiquitin transfer not requiring an E3–ubiquitin complex. If a ping-pong mechanism had been observed, it would have suggested the existence of a distinct E3ub complex (Fig. 1). It will be of interest to evaluate whether ping-pong kinetic behavior is observed with HECT E3s that utilize a thioubiquitin complex as part of their catalytic mechanism.

The goal of this work was to provide a method to evaluate the kinetic mechanism of E3-dependent ubiquitylation reactions. This method can be of use to classify the kinetic mechanism of the E3 ligase reactions, determine kinetic constants, and evaluate the mechanism and affinity of E3 inhibitors.

References

Burke, J. R., Miller, K. R., Wood, M. K., and Meyers, C. A. (1998). The multisubunit IκB kinase complex shows random sequential kinetics and is activated by the c-terminal domain of IκBα. *J. Biol. Chem.* **273**, 12041–12046.

Haas, A. L., and Seipmann, T. J. (1997). Pathways of ubiquitin conjugation. *FASEB J.* **11**, 1257–1268.

Hershko, A., and Heller, H. (1985). Occurrence of a polyubiquitin structure in ubiquitin-protein conjugate. *Biochem. Biophys. Res. Comm.* **128**, 1079–1085.

Hershko, A., and Ciechanover, A. (1998). The ubiquitin system. *Annu. Rev. Biochem.* **67**, 425–479.

Leskovac, V., Trivic, S., Pericin, D., and Kandrac, J. (2004). A general method for the analysis of random bisubstrate enzyme mechanisms. *J. Ind. Microbiol. Biotechnol.* **31**, 155–160.

Passmore, L. A., and Barford, D. (2004). Getting into position: The catalytic mechanisms of protein ubiquitylation. *Biochem. J.* **379**, 513–525.

Segel, I. H. (1975)."Enzyme Kinetics, Behaviour and Analysis of Rapid Equilibrium and Steady-State Enzyme Systems" John Wiley & Sons Inc., New York.

Swinney, D. C., Xu, Y.-Z., Scarafia, L., Lee, I., Mak, A. Y., Gan, Q.-F., Ramesha, C. S., Mulkins, M. A., Dunn, J., So, O.-Y., Biegel, T. A., Dinh, M., Volkel, P., Barnett, J., Dalrymple, S. A., Lee, S., and Huber, M. (2002). A small molecule ubiquitination inhibitor blocks NF-κB-dependent cytokine expression in cells and rats. *J. Biol. Chem.* **277**, 23573–23581.

Wilkinson, K. D. (2000). Ubiquitination and deubiquitination: Targeting of proteins for degradation by the proteasome. *Semin. Cell Dev. Biol.* **11**, 141–148.

[23] Screening of Tissue Microarrays for Ubiquitin Proteasome System Components in Tumors

By Norman L. Lehman, Matt van de Rijn, and Peter K. Jackson

Abstract

The turnover of key proteins that mediate development, cellular proliferation, and a host of essential biological processes is controlled by the ubiquitin proteasome system (UPS). In several well-studied examples, notably in the cell cycle, regulatory proteins that control ubiquitin-dependent destruction are themselves substrates of the UPS, creating a multilayered system to ensure precise and dynamic control of protein stability. UPS regulators controlled at the level of protein stability—including the F-box protein Skp2 and the VHL protein (substrate adapter proteins for multicomponent E3 ubiquitin ligases)— seem to be misregulated in tumors. In these cases, especially, measuring levels of critical regulatory and target proteins will often present a more biologically meaningful picture than examining relative mRNA levels, which do not always reflect corresponding protein levels. Tissue microarrays (TMAs) allow simultaneous screening of large numbers of tumors for expression of specific proteins by immunohistochemical staining of a single microscope slide prepared from a TMA paraffin block. Replicate slides prepared from the same block can be immunostained for multiple proteins functioning in a related pathway, and a semiquantitative protein expression profile for a given subset of UPS pathway components, or other subsets of proteins of interest, can be assembled. Protein expression profiles of individual tumors or tissue types can be compared and visualized by hierarchical clustering methods. These expression profiles may be used as screening tools to investigate the relative abundance of components of a biochemical pathway in tumors or other tissues. TMAs have an exciting future as tools for basic research, diagnostic pathology, and drug targeting. In this article, we provide an introduction to the use of TMAs to study the expression of UPS component proteins and substrates in tumors by immunohistochemistry.

Introduction

The ubiquitin proteasome system (UPS) is an important regulatory system controlling diverse biological processes. The UPS participates in the control of cell cycle progression, circadian clocks, the coordination of growth regulation and development, and membrane receptor turnover, to name just a few examples. Multiple classes of E3 ubiquitin ligases catalyze the covalent attachment of activated 8.0-kDa ubiquitin proteins to substrate protein lysine residues to form polyubiquitin chains (Jackson *et al.*, 2002). In some cases, polyubiquitin chain formation seems to be facilitated by other proteins known as E4s (Koegl *et al.*, 1999). Polyubiquitin chains may bind the 26S proteasome directly or be recognized by UBA/UBL domain containing factors that direct the target protein to the proteasome for destruction (Buchberger, 2002). The E3 ubiquitin ligases—which include possibly more than 300 single polypeptide E3 ubiquitin ligases, but also several large families of multicomponent complexes with dozens of adaptor proteins—recognize a range of substrates determined by a variety of specific primary amino acid sequences within the substrate known as *degrons*. Specific posttranslational modifications of degron sites, such as serine phosphorylation or prolyl hydroxylation, are required for E3 binding in some instances (Kaelin, 2002).

In 1991, Glotzer *et al.* discovered that the precisely timed, ubiquitin-dependent destruction of cyclins is responsible for their oscillatory expression that drives the cell cycle. In the interim, investigators have shown that a multilayered cross talk of ubiquitin-dependent proteolysis exists between ubiquitin ligase substrates and components of different ubiquitin ligase complexes and regulatory proteins critical to growth and development. These developments have prompted several groups to study various aspects of the UPS in tumors.

Examples of UPS Misregulation in Tumors

The importance of the UPS in human malignancies is illustrated by the following three examples. The protein β-catenin is an important effector of the Wnt signaling pathway that is known to be involved in both normal and cancer cell proliferation. β-catenin combines with TCF/LEF transcription factors to activate transcription of critical genes controlling growth and development. Elevated β-catenin is characteristic of both familial and sporadic forms of colon cancer (Morin *et al.*, 1997). In each case, increased β-catenin is largely due to its abnormally increased protein stability. β-TrCP is one of several F-box proteins that act as substrate adapters for a multimeric E3 ubiquitin ligase known as the SCF (for Skp-Cullin-F-box) (Deshaies, 1999). Ubiquitination of β-catenin requires its binding to β-TrCP in conjunction with the SCF ubiquitin ligase. β-TrCP binding requires prior phosphorylation of two serine residues within the β-TrCP recognition motif (DSGxxS degron) of β-catenin by glycogen synthase kinase 3β (GSK-3β). The adenomatous polyposis Coli (APC) gene is necessary for GSK-3β to interact with β-catenin for this phosphorylation event to occur. Germline mutations of the APC gene are found in familial polyposis coli and Gardner familial colon cancer syndromes, and somatic mutations are present in most sporadic colon cancers (Morin *et al.*, 1997). β-catenin mutations that abolish the β-TrCP recognition motif phosphorylation sites have been identified in sporadic colon cancers lacking APC mutations.

In another example, the von Hippel–Lindau syndrome gene (VHL) is the substrate recognition subunit of a multicomponent ubiquitin ligase similar to the SCF that mediates ubiquitin-dependent degradation of hypoxia inducible factor alpha (HIF-α), a positive regulator of transcription of vascular endothelial growth factor (VEGF) and other factors (Kaelin, 2002). Germline and somatic mutations of VHL result in elevated VEGF levels in both von Hippel–Lindau syndrome–associated and sporadic hemangioblastomas and renal cell cancers. *Here,* the modification of HIF-α by prolyl hydroxylation in response to high oxygen tension forms the essential determinant in the HIF-α degron for recognition by the VHL protein. Loss of this pathway in VHL and renal cell cancers leads to hypervascularization and growth of the tumors.

The anaphase-promoting complex/cyclosome (APC/C) ubiquitin ligase is responsible for the ubiquitin-dependent destruction of cyclin A and other key regulators of S phase and early mitosis (Harper *et al.*, 2002). Many of these S phase and mitotic regulators, including Skp2, Polo-like kinase-1, Aurora A kinase, and Securin are overexpressed in tumors (Gstaiger *et al.*, 2001; Knecht *et al.*, 1999; Tanaka *et al.*, 1999; Zou *et al.*, 1999) and when misexpressed can lead to genomic instability.

The regulation of the APC/C is complex and involves inhibition by a SCF ubiquitin ligase substrate known as Emi1. Emi1 is a direct inhibitor of the APC/C that functions from late G1 to mitotic prophase to allow accumulation of cyclins A and B and other APC/C substrates (Hsu *et al.*, 2002). Thus, Emi1 is a critical regulator of APC/C E3 ligase activity. Emi1 must be ubiquitinated by the SCF and destroyed in early mitosis for normal cell cycle progression to continue beyond prometaphase (Margottin-Goguet *et al.*, 2003). Emi1 is thus a key mediator of crosstalk between the two major E3 ubiquitin ligases—the APC and SCF—that globally regulate cell cycle progression.

In summary, many of the substrates, components, and regulators of the SCF and APC/C ubiquitin ligases are critical to normal cell proliferation and mitosis. Misregulation of these important destruction pathways leads to the unregulated- or hyper-accumulation of substrates, with the consequence that alterations in the levels of substrates alter the timing of important cellular events. The effects can lead to alterations in growth or genomic instability, ultimately leading to tumor progression.

At this point, only a few of the UPS pathways likely affected in pathological states are known. Moreover, even when one pathway has been implicated in a specific cancer, a more comprehensive survey is rarely performed. Instead, research groups have focused on individual tumors and typically publish expression studies over many years, with little sense of how these results compare with results from other neoplasms. We believe that the use of TMAs allows broader surveys of tumors and also improves the uniformity of data by enabling a better sense of the relative importance of each marker protein in different tumors.

Methods to Study UPS Components in Tissues

Several different methods are available to screen tumors for expression of specific gene products. Many depend on quantification of mRNA levels in tumors rather than protein levels. These methods include RT-PCR, dot blots, *in situ* nucleic acid hybridization, and various DNA microarray technologies. These techniques offer some advantages, including highly accurate quantification of mRNA species in the case of RT-PCR (Schneider *et al.*, 2004), and high-throughput (the ability to compare very large numbers of gene products in several samples simultaneously) in the case of DNA microarrays.

Measurement of levels of an mRNA species as a quantitative index of gene expression is based on the assumption that mRNA levels of a given gene product are proportional to protein levels. However, in tumors and other tissues, this is often not the case (Chen *et al.*, 2002). Any specific protein present in any given cellular compartment is subject to various translational and posttranslational regulatory mechanisms. These include different conformational states; various covalent modifications such as glycosylation, phosphorylation, sumoylation, or ubiquitination; and transport to specific cellular compartments such as the cytoplasm, cell membrane, or nucleus. Many of these posttranslational events will affect not only the function but also the stability of the protein.

Because the expression of components and targets of the UPS are frequently regulated at the level of protein stability, it is often more biologically pertinent to determine levels of protein expression rather than RNA message. This can be accomplished, for example, by examining tumor extracts for protein levels by Western blotting, enzyme-linked immunosorbent assay (ELISA), protein microarrays, or various proteomic techniques. One advantage of some of these methods is the ability to simultaneously differentiate proteins with and without posttranslational modifications such as phosphorylation (Espina *et al.*, 2003). Most of these technologies are labor intensive and depend on the availability of fresh or fresh frozen tumor samples from which to prepare extracts.

Tissue Microarrays

A practical method for evaluating protein levels in tissue samples that does not rely on fresh or fresh frozen material is the use of tissue microarrays (TMAs). Conventional sections from formalin-fixed paraffin-embedded material have been used for many years for immunohistochemistry to determine protein expression in human tissue samples. A major advancement in this technology occurred when Kononen et al. (1998) published their method to combine hundreds of small tissue cores obtained from hundreds of specimens in a single new paraffin block (TMA). In TMA blocks, the cores are placed in orderly rows and columns, thus forming an array of tissue samples where the identity of each sample is know by its location on the array (Fig. 1).

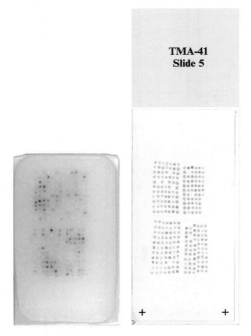

FIG. 1. TMA Construction. A finished TMA paraffin block and corresponding H&E–stained TMA slide. (See color insert.)

Thin sections prepared from a TMA block and adhered to standard microscope glass slides can be immunostained by conventional immunoperoxidase or other methods to screen for the expression of a specific protein of interest. TMAs can also be used for other detection methods such as *in situ* hybridization for RNA message detection. This is particularly helpful when good antibodies for immunohistochemical staining are not available, because *in situ* probes can be readily designed from published gene sequences.

Advantages of TMAs

The great utility of TMAs lies in the ability to simultaneously screen numerous tumors or other tissue samples through immunostaining a single slide. Most tumor samples consist of a mixture of malignant cells and reactive and nonmalignant stroma. With Western blots and DNA and protein microarrays, which involve homogenation of tissue samples, it is impossible to distinguish which part of the sample is generating the positive signal. TMAs thus have an advantage over methods involving sample homogenation, because one can readily distinguish which component of the specimen is immunopositive (Fig. 2). Specific characteristics of the tumor, such as whether the tissue displays various forms of dysplasia or the presence or absence of tumor invasiveness, can allow for a more detailed analysis of the biological correlates of protein expression. Last, unlike for DNA and protein microarrays, TMAs do not require normalization of the primary data to adjust for variability in sample total RNA or protein concentrations (Espina *et al.*, 2003; Quackenbush, 2001).

Availability of TMAs

TMAs are typically made with formalin-fixed, paraffin-embedded tissue from human patients. This is tissue that has been surgically removed from a patient for purposes of diagnosis in the course of receiving medical care or tissue removed from a deceased patient at autopsy. In both cases, samples of both diseased and nondiseased tissue are often removed. After the samples have been used for pathological diagnosis, the remaining paraffin-embedded material is stored in hospital pathology department archives. These archives constitute a valuable source of material for the study of human disease. Use of patient tissue for research requires prior institutional review board (IRB) approval and often patient or family consent. A tissue pathologist may be helpful for the logistics of obtaining appropriate human paraffin-embedded material to construct a TMA and to select appropriate representative areas of the sample (see later).

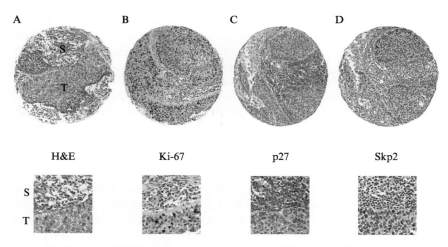

FIG. 2. Example of H&E–stained and immunostained TMA cores. The first micrograph (A) is an H&E–stained core section of a squamous cell carcinoma demonstrating areas of tumor (T) and surrounding stroma (S). The latter consists mostly of chronic inflammatory cells in this case. The micrograph below each core is an enlargement of the stomal/tumor interface. Ki-67 is a general marker of cellular proliferation. The second core (B) shows that the tumor cells are immunoreactive for Ki-67, whereas the benign inflammatory cells within the stroma are immunonegative. p27 is a cyclin-dependent kinase inhibitor most often significantly expressed in benign nonproliferating cells. The third core (C) shows p27 immunopositivity of the stromal inflammatory cells. The tumor cells are p27 immunonegative. The last core (D) is immunostained for Skp2, the SCF ubiquitin ligase adapter subunit responsible for binding p27 and initiating the ubiquitin-dependent degradation of p27. Skp2 has been shown to be overexpressed in cancers correlating with low p27 levels and cell cycle deregulation. Note that the immunostaining pattern is the reverse of that of p27; the benign inflammatory cells are Skp2 immunonegative, whereas the tumor cells are Skp2 immunopositive. The immunostains are counterstained with a blue stain (Mayer's hematoxylin) so that individual cells can be visualized. The arrangement of tumor and stromal cells appears somewhat different in each core, because the corresponding TMA sections are from different levels (depths) of the TMA block. Original magnifications of the cores are 100×. (See color insert.)

TMAs can also be made using cell lines or tissues from experimental animals to support other *in vitro* or animal model data (Hoos and Cordon-Cardo, 2001). Here, specific controls, such as sham treatment groups, may be included in the array to improve the interpretability of TMA results. Alternatives to constructing a TMA include collaboration with an investigator who routinely constructs TMAs or purchasing commercial TMA slides of which several choices are currently available. Commercial sources of TMAs include Zymed Laboratories (South San Francisco, CA), NxGen BioSciences (San Diego, CA), US Biomax (Rockville, MD), BioCat

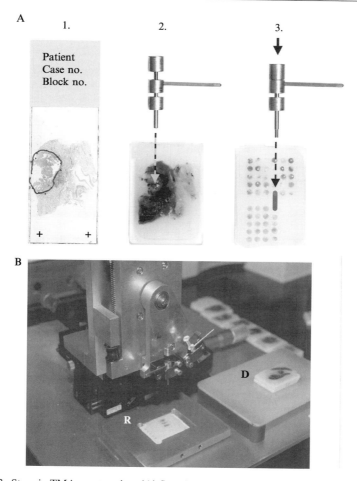

Fig. 3. Steps in TMA construction. (A) Step 1, A suitable area of tumor is identified on an H&E–stained microscope slide and circled. Step 2, A core is then punched from the corresponding area of the paraffin block used to make the slide (donor block). Step 3, Next the core is extruded into a prepunched hole in a new block (recipient block). Specimen cores are placed in orderly rows and columns in this manner. The block is then used to cut thin sections, which are adhered onto new microscope slides as depicted in Fig. 1. The punches shown are from the Beecher manual arrayer. (B) To facilitate block construction, we use the Beecher Manual Arrayer. Here, a core is taken from the donor block (D) and repositioned in the recipient block (R). (See color insert.)

(Heidelberg, Germany), Cytomyx (Cambridge, UK), Chemicon International (Temecula, CA), LifeSpan Biosciences (Seattle, WA), American Research Products (Belmont, MA), and others. Some of these companies also offer custom TMA manufacturing.

TMA Construction

Choosing a Method of Construction

TMAs are constructed by use of a punch device to obtain a core of paraffin-embedded tissue from a *donor* paraffin block and placing that core in a new, *recipient* paraffin block in which receiving holes were previously punched (Fig. 3A, B). The diameter of the core can vary, but most researchers use 0.6-mm diameter punches. This process is repeated until cores of all the desired samples to be studied are in place in the new block. The block is then heated to anneal the cores within the block. Once the block has been created, 4- to 5-μm paraffin sections are obtained with a standard tissue microtome and annealed onto electrostatically charged glass microscope slides. Before immunostaining, the slides should be baked at 60 to 65° for at least 1 h to firmly anneal the section to the slide. These slides now contain a thin cross-section of each of the cores represented in the TMA and are ready for immunostaining. Typically, 100 or more such slides can be prepared from a TMA block.

TMA block construction can be performed completely manually with a simple dermatological skin biopsy punch; disposable punches work well (Jensen, 2004), or with the aid of *manual tissue arrayer* (Fig. 3B) or *automated tissue arrayer* machines currently available from Beecher Instruments (Sun Prairie, WI) and Chemicon International (Temecula, CA). Manual machines consist of a base to support the punch device plus an area to firmly hold the recipient or both the donor and recipient blocks. The punch device contains a mechanism to precisely control the horizontal X and Y coordinates, such that recipient block holes can be precisely positioned and the donor cores can be precisely inserted into the recipient block holes. Automated machines provide nearly complete automation of this process and high-throughput construction of identical TMA blocks (Beecher Instruments). They also allow for maximum use of the donor block for construction of multiple arrays in some instances.

A low-tech and inexpensive alternative to punching holes in a recipient block, which is suitable for making small arrays, is to array large cores (e.g., 2 mm in diameter) onto double-stick tape within a paraffin block mold. The mold is then filled with paraffin to form the TMA block around the cores (Jensen, 2004).

Planning the Array

Careful planning is the first step to quality TMA construction. The size of the sample cores, the total number of samples, and whether single or duplicate cores of each sample are used are all important questions to consider in planning a TMA. Typically a paraffin-embedded tumor will contain both neoplastic and nonneoplastic tissue (e.g., normal tissue, fibrosis, hemorrhage, or inflammation). It is, therefore, necessary to preselect an area of the donor block that represents the type of tissue that one wants represented on the TMA. This is accomplished by examining an H&E-stained section from the donor block and circling a suitable area of the specimen with a marker. The marked slide is then matched with the donor block to identify the corresponding area of the specimen within the block.

An important parameter in the design of a TMA is deciding on a sample core size. Smaller cores enable more samples to be incorporated in a TMA block (e.g., 500 or more samples can be arrayed in a standard paraffin block using 0.6-mm diameter cores). Larger cores provide a larger sample of each specimen but decrease the number of specimens that can be incorporated into the TMA. Cores are available in 0.6-, 1.0-, 1.5-, and 2.0-mm diameters for the manual arrayer from Beecher.

TMA cores can be arranged in one continuous grid or divided into two or more sectors to aid in the scoring process (Liu *et al.*, 2002). We usually place several control tissues in the upper left corner of the array to allow for easy orientation of samples. We typically use skeletal muscle as the unique corner marker, but any easily recognized tissue such as liver or placenta could be used. Standard control cores such as a series of normal tissues or common tumors may also be used to aid in the evaluation of staining performance between different TMAs.

Antibodies

Not all antibodies that work well for other applications (Western blotting, immunoprecipitation, or immunofluorescence) are suitable as primary antibodies in immunohistochemistry. Monoclonal antibodies are often ideal for reproducibility and specificity (to obtain a clean background) but may not always work well on paraffin-embedded sections. Polyclonal antibodies provide reactivity to multiple protein epitopes, therefore providing higher avidity, but this may also result in higher nonspecific background staining, particularly spurious staining of cellular nuclei. Many commercially supplied antibodies are tested for their suitability for immunohistochemistry, and the manufacturer may even provide a suggested antibody dilution range or staining protocol, particularly for antibodies marketed for use in clinical diagnostic pathology.

Optimization of immunohistochemistry procedures and conditions (including antibody dilution, antigen retrieval, washing, and blocking of nonspecific binding) are similar to those used on conventional sections, and these methods will not be reviewed in detail here. Several protocols are available in texts (Carson, 1997; Harlow and Lane, 1999), online (www.protocolsonline.org, www.ihc.com, www.ihcworld.com, www.antibodyresource.com), and from commercial antibody manufacturers. When testing an antibody not previously used for immunohistochemistry, blocking of the primary antibody with antigen by preincubation is desirable to demonstrate specific staining. Another method of antibody validation that has only recently been possible with the advent of RNA interference is the use of cultured cells known to express the protein of interest transfected with control oligonucleotides or their targeted siRNA oligonucleotides as positive and negative controls (Lehman and Jackson, unpublished data). Such cells can be grown on microscope slides or coverslips and fixed with methanol or acetone and immunostained or grown in culture dishes, harvested, pelleted, and processed for standard paraffin-embedding (Hoos and Cordon-Cardo, 2001).

Potential Pitfalls of Using TMAs

Visual interpretation of immunohistochemical staining is semiquantitative at best, even when well controlled. In some situations, biologically significant differences in protein levels may not be detectable by standard immunoperoxidase techniques. A recent study by Camp *et al*. (2002) demonstrates great promise for much more quantitative analysis of TMA data using immunofluorescence-based detection of TMA immunoreactivity.

Another general problem is misinterpretation of nonspecific staining as positive immunoreactivity when using a poorly validated primary antibody. Unfortunately, many commercially produced research-grade antibodies are not reliable in regard to antigen specificity. As a minimum, examining the relative monospecificity of the antibody by Western blotting is recommended. Further validation by antigen blocking or protein knockdown by RNA interference as described above is desirable. Given that most TMAs are a precious resource, blocking experiments or antibody titrations can be performed on less precious samples such as sections from common tumors, lymph nodes, or tonsil.

Subcellular localization of antibody reactivity may be different in paraffin-embedded tissues compared with methanol, acetone, or paraformaldehyde-fixed frozen sections or cultured cells. Standard paraffin-embedded clinical samples are processed in automated tissue processors that use several heated stages of fixation from formalin to several fixation and dehydration steps in increasing concentrations of ethanol, followed by extraction of the ethanol in xylene and paraffin impregnation. Formalin causes extensive protein–protein and protein–nucleic acid crosslinking, which may mask antigens and result in altered immunoreactivity in the case of some antibodies. The effects of antigen retrieval techniques or heat and/or alcohol fixation in standard tissue processing can extract or alter the tertiary structure of many proteins, rendering them undetectable by immunohistochemistry or only detectable in other compartments such as the cytoplasm. Inclusion of acetic acid or zinc in fixative solutions tends to preserve nuclear structure and nucleoprotein antigenicity (Carson, 1997). Examples of nuclear proteins that are frequently only detected in the cytoplasm of formalin-fixed paraffin-embedded cells by immunohistochemistry include cyclin D1 and β-catenin (Anton et al., 2000; Iwaya et al., 2003; Ninomiya et al., 2000; Tut et al., 2001).

Optimization of tissue fixation and/or antigen retrieval techniques for individual antibodies or construction of TMAs from frozen tissues (cryoarrays) may circumvent the preceding problems in some cases (Carson 1997; Hoos and Cordon-Cardo, 2001). Nevertheless, many cell biologists may need to retune their expectations for specific protein localizations or levels they expect from studies in frozen tissue sections or from cultured cells.

Data Analysis

Scoring Primary Data

The various methods used at Stanford for data analysis of TMAs have been reviewed in detail (Liu et al., 2002). The following is a basic summary for getting started. The first step in collecting TMA data is scoring the immunostained slide. Several approaches may be appropriate for a given set of experiments depending on the scope of the study. We use a three-tiered scoring method. Specimens that are negative for immunostaining, usually defined as less than 3–5% of cells showing immunoreactivity, are assigned a score of 0; weakly positive specimens, usually defined as 5–30% of cells being immunoreactive, are scored 2; and strong positivity, with >30% of cells immunoreactive are scored 3. Examples where immunoreactivity cannot be determined because tumor tissue (or other tissue of interest) is absent in the section, where immunoreactivity is equivocal, or where other technical problems seem are invariably encountered. Such technical problems may include uneven application of primary or secondary antibodies, faulty washing steps, or loss of the tissue core section from the slide. These cores are essentially unscored, but are assigned a "score" of 1 for bookkeeping purposes.

The degree of immunopositivity could be further characterized as weak, moderately positive, strongly positive, very strongly positive, etc., but in most cases, two tiers of positivity as judged by a human observer is fairly reproducible for standard immunohistochemistry, whereas additional subdivision may introduce more noise in the data because of both intraobserver and interobserver variability and is also more time consuming.

Fluorescently labeled secondary antibodies provide a more useful signal dynamic range and are therefore more amenable to quantification than standard immunohistochemistry chromogens (e.g., 3,3-diaminobenzidine [DAB], commonly referred to as brown stain; Camp et al., 2003). Automated scoring of TMAs with image analysis systems, particularly with fluorescently labeled antibodies, is also possible but probably best suited to only very well-characterized primary antibodies. In the future, we can imagine these more quantitative methods of TMA signal detection being routinely applied.

Data Archiving

Careful scoring of TMA data is a critical factor for building a meaningful protein expression profile. Review of immunohistochemical or *in situ* hybridization data on TMAs is essential, because interpretation of positive reactivity remains somewhat subjective, and the original data often need to be revisited. More importantly, crucial biological insights may be missing during early phases of a TMA project. For example, during the course of study, it may become apparent that the expression of a specific protein of interest occurs in the stromal cells that surround tumor cells rather than in the tumor cells themselves. Thus, reevaluation of TMA slides may be an ongoing process during the course of some studies. When one needs to compare different stains on different sections of a TMA, it can become laborious to find core sections from the same specimen among the hundreds of other cores on the TMA. Many groups now generate digital images of TMA cores to allow for rapid review of staining results and to facilitate sharing of results between different groups. In addition, as specific antibody probes are improved or replaced, the need to examine the staining of existing data requires that archives of essential data be readily available for review.

For some purposes, the shelf life of the primary data slides will suffice. But to ensure long-term durability and rapid remote access to primary immunohistochemistry data, an electronic archiving system is essential. At Stanford we use a semiautomated image collection system (Bliss system, Bacus Laboratories, Lombard, IL; *http://bacuslabs.com*) for archiving digital images. Immunostained tissue cores are imaged and stored as 400-kb compressed JPEG files. To date, archived images of more than 90,000 immunostained tissue cores have been collected.

Stainfinder™ is a web-based program (freely available at http://genome-www.stanford.edu/TMA) that links all protein marker data for a single sample core to the corresponding recorded core images. Digital images are presented in a thumbnail format that can be enlarged by selecting the name of the thumbnail. This allows for quick verification of the score given to a particular immunostained core sample by comparing the color representing the score in the TreeView™ document (see later) to the corresponding digital image. This feature allows reevaluation of original staining scores and easy comparison of multiple immunostains performed on the same core. In addition, variables such as nuclear versus membrane staining and the presence of benign inflammation or stromal cell immunoreactivity can most easily be accessed by reexamining the digital images instead of reexamining multiple cores on multiple TMA slides with a microscope.

Using the Stainfinder™ software, it is possible to rapidly compare interobserver or interlaboratory variations in results using the same TMA or antibody. Digital image collections are also easily shown over the web. Thus, rather than showing only a few examples of staining results in the published manuscript, primary data consisting of thousands of digital images can be made available to the scientific community (http://microarray-pubs.stanford.edu/tma_portal/index.shtml).

Data Analysis Software and Methods

TMA scoring data can be entered into database spreadsheets (e.g., Microsoft Excel™ files), where it can be readily used for statistical analysis using the database software or specialized statistical software packages. At Stanford University we use an Excel macro subroutine (TMA-deconvoluter program, freely available at http://genome-www.stanford.edu/TMA/decon-manop.shtml) to convert raw TMA scores into scaled values centered around zero (e.g., a raw score of 0 is converted to -2, 2 to 1, and 3 to 2), and to match these converted scores with the identifying information for the sample (e.g., sample #100 colon adenocarcinoma). This information is outputted as a simple text file that can be used for standard types of statistical analysis or hierarchical clustering.

Data for many different markers can be displayed and compared in clusters of tissues with similar antibody reactivity across several markers using Michael Eisen's Cluster™ and TreeView™ programs or other software (Eisen *et al.*, 1998; Liu *et al.*, 2002). The Eisen software is well known to groups examining RNA profiles by DNA microarray analysis and is freely downloadable at http://rana.lbl.gov/EisenSoftware.htm. The original Eisen laboratory software is only for PC computers; however, similar software has been developed for Macintosh™ computers (de Hoon *et al.*, 2004) and is available at http://bonsai.ims.u-tokyo.ac.jp/~mdehoon/software/cluster/ index.html.

Clustering programs group biological samples according to similarities and differences in biomarker expression. The resultant clustergrams are often displayed using graphics programs, such as TreeView™, which produce a heat map, in which different colors and shades of color are used to indicate relative expression levels (Fig. 4). Conventionally, green is used to represent low expression and red for high expression. Linkage relationships between biomarkers are shown by dendograms similar to those used to show evolutionary relationships between animal species. Linkages between samples of like marker expression (subclusters) are shown in the same way. Our current method of display of TMA data differs from DNA microarray data in that biomarkers are placed on the horizontal axis, and sample identifying data are arranged on the vertical axis; the opposite of the convention for DNA microarrays (Fig. 4).

Hierarchical clustering programs use straightforward algorithms to group single sample expression profiles into groups, which are further grouped in a hierarchical tree fashion (Quackenbush, 2001). They offer several options for performing clustering in different ways. Two commonly used algorithms are average-linkage clustering and complete-linkage clustering. Average-linkage analysis links sample groups based on average relatedness of expression of pairs of markers or average distances between clusters. Complete-linkage analysis uses worst-case relatedness; distances between clusters are determined by the greatest distance between the expression of pairs of members. In many cases, these two approaches yield similar clustergrams; however, in some cases, average-linkage clustering may reveal subtle trends not apparent by complete-linkage clustering. On the other hand, complete-linkage clustering may sometimes yield more clearly demarcated cluster groups.

In standard unsupervised hierarchical clustering, the samples and markers are ordered into subclusters based solely on the results of the clustering algorithm with equal weighting of all markers or samples. In the simplest form of weighted clustering, clustering of similar expression profiles can be subdivided into subclusters focused on the expression of a single parameter gene. This is accomplished by increasing the numerical weight value for that protein marker (normally set to 1.0 for equal weighting) in the text file before clustering. For example, one could use this technique to divide a group of tumor samples into Her-2 immunopositive and Her-2 immunonegative groups. All the Her-2 immunopositive tumors are displayed together as are the negative tumors, thus allowing direct comparison of the expression profiles of other genes within these two groups. The profile of protein marker expression within each sample remains the same.

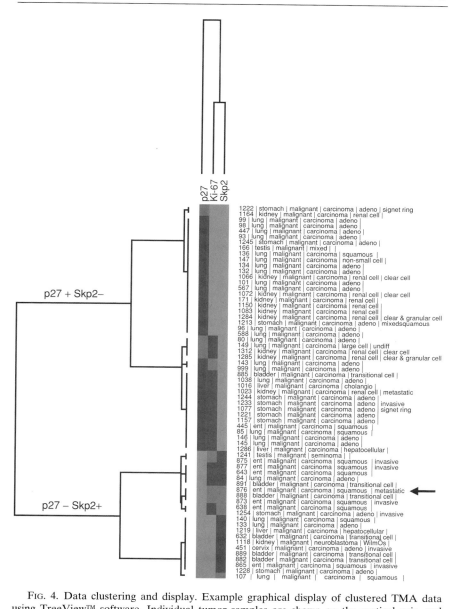

FIG. 4. Data clustering and display. Example graphical display of clustered TMA data using TreeView™ software. Individual tumor samples are shown on the vertical axis, and protein markers (p27, Ki-67, and Skp2) are shown on the horizontal axis above. Green indicates no expression; dark red, low to intermediate expression (3 to <30% of tumor cells

More complex forms of directed clustering (known as supervised clustering) are also possible, for example, clustering a group of query genes associated with a known phenotype or clustering based on specific tissue type subgroups or clinical outcomes (Quackenbush, 2001). One should be cautious, however, of the possible overinterpretation of clustering algorithm results in the analysis of TMA data (F. D. Hsu *et al.*, 2002). These methods were developed for the analysis of very large data sets based on mRNA expression detection that has a much larger dynamic range than standard immunohistochemical staining. Therefore, overinterpretation of clustering patterns of TMA data may be misleading in some cases. Using appropriate statistical analysis of the data may help avoid this potential pitfall.

Utility of TMA Analysis

Clustering can be used to suggest biologically significant expression profiles. For example, a poor prognosis in breast cancer has been linked to Her-2 overexpression and ER immunonegativity (Wright *et al.*, 1989). Systematic cDNA microarray expression studies have confirmed this result and identified other markers defining larger sets of genes that predict poor outcome with greater statistical confidence than two or three markers alone (Sørlie *et al.*, 2001; van 't Veer *et al.*, 2002). Using hierarchical clustering, Sørlie *et al.* (2001) identified five subtypes of breast cancer that correlated with survival differences. Whereas clustering algorithms were initially designed for the analysis of extremely large data sets, as mentioned above (e.g., hierarchical distance-based clustering for gene arrays), they can also be applied to much smaller datasets derived from TMA studies. In a study using a breast cancer TMA, Makretsov *et al.* (2004) found that breast cancer prognosis could be predicted with a high degree of statistical significance by analysis of only 11 proteins.

One can also use TMAs and hierarchical clustering to interrogate expression of UPS components or substrates in a particular biochemical pathway (e.g., SCF or APC/C substrates) and regulatory components in tumors (Lehman and Jackson, unpublished data) (Fig. 5). Because the UPS pathway regulates the accumulation of proteins, the use of *in situ*

---

immunopositive); bright red, high expression (>30% of tumor cells immunopositive). The tumors are separated into p27-positive/Skp2-negative and p27-negative/Skp2-positive groups and then further subdivided based on Ki67 expression. Ki67 and Skp2 cluster more closely together than with p27 as indicated by the shorter branches linking the two proteins. The arrow indicates the tumor (#876) shown in Fig. 2. (See color insert.)

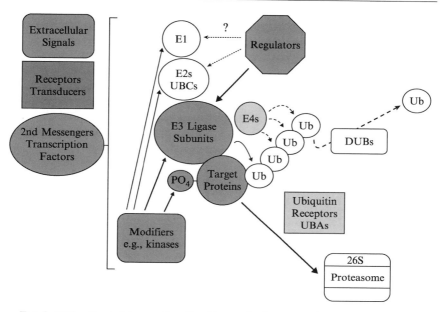

Fig. 5. UPS pathway interrogation. Cytokines and other extracellular ligands may activate second messengers or transcription factors that direct the transcription of UPS components, inases, prolyl hydrolases, sumoylation enzymes, or other modifiers of UPS components or target proteins. E3 ligase regulators (shaded medium gray) are all good candidates for study by TMAs. E4 enzymes that promote the formation of polyubiquitin chains and UBA domain–containing proteins involved in the delivery of ubiquitinated proteins to the proteasome (shaded light gray) may be important regulators of ubiquitin-dependent proteolysis. Other UPS components such as E2 ubiquitin–conjugating enzymes (UBCs), deubiquitinating enzymes (DUBs), and ubiquitin monomers (Ub) are shown unshaded.

immunodetection in tissues is especially appropriate. Generally, decreased expression of a particular ubiquitin ligase or adapter subunit (such as Skp2) would cause an increase in the levels of its substrates (e.g., $p27^{Kip1}$), such as shown in Figs. 2 and 4. A useful internal control for Skp2 staining is the presence of p27 in the surrounding stromal tissue lacking obvious Skp2 accumulation. Within the same tissue, the effect of Skp2 expression and p27 loss causes cells to enter S phase, which is reflected by staining for the proliferation marker Ki-67 (Gerdes et al., 1984).

The expression of multiple components of the UPS system and pathways outside the UPS that regulate UPS proteins may be examined using TMAs. Extracellular signals from cytokines and other ligands may activate second messengers or transcription factors that affect the transcription of UPS components. Kinases, prolyl hydrolases, sumoylation enzymes, or other posttranslational modifiers may affect the activity or stability of UPS components such as E1 and E2 enzymes, E3 ligase subunits, or target proteins. Signal transduction pathways, modifier enzymes, many E3 ligase subunits, and many target proteins are well characterized, and antibodies against them are available. The modification state of substrates can sometimes be determined by TMAs, for example, by using phospho-specific antibodies.

Some E3 ligase regulators are also well described (JY Hsu et al., 2002). Deubiquitinating enzymes (DUBs), E4 enzymes that promote the formation of polyubiquitin chains, and UBA/UBL domain–containing proteins involved in the interaction of ubiquitinated proteins with the proteasome are not as well studied but may be found to be important regulators of ubiquitin-dependent proteolysis that may warrant comparative expression studies.

Protein expression patterns within tissues elucidated by TMA analysis can provide clues to dominantly active parts of a pathway or deregulation of a subpathway. For example, high Skp2 expression in tumors with low p27 expression might indicate the dominance or activation of the ubiquitination of p27 by the SCF^{Skp2} E3 ligase in tumors. At the same time, high expression of both proteins may indicate deregulation of p27 degradation as triggered by Skp2. Last, abnormally high or low expression of UPS substrates with abnormally high or low expression of regulators may suggest misregulation of the pathway.

Summary

Tissue microarrays are a powerful research tool for investigating physiological processes in animal and clinical samples. TMAs circumvent many disadvantages of technologies for measuring gene expression. Because of recent increased commercial availability of several premade TMAs, many can be obtained at relatively modest cost. They are also easily constructed with a modest investment of time and funds for specialized research applications or at medical centers when tissues from specific patient populations are to be studied. TMAs are useful for both interrogating hypothesis-driven research and as screening tools for discovery.

Although fewer gene products are screened in TMAs compared with DNA microarrays, TMAs measure the expression of proteins, which are the ultimate effectors of biological processes. Protein expression often does not correlate with RNA message expression measured by DNA microarrays. TMAs can be used alone or as confirmatory adjuncts to

RNA message-based screening. TMAs are, therefore, extremely useful to verify the biological significance of message-based methods. Because TMAs measure protein expression, they are particularly suited for studying dynamic substrates and components of the ubiquitin proteasomal system *in situ*. In regard to the UPS and other biological pathways, TMAs have a promising future as adjunctive and discovery tools for basic research, in screening for new clinical diagnostic and prognostic indicators, and in drug targeting.

Acknowledgments

The authors are grateful to Dr. Rob Tibshirani, Dr. Jorge Torres, and Kelli Montgomery for helpful suggestions, and Caroline Tudor for photographic assistance. NL Lehman is supported by K08 NS45077 from the National Institute of Neurological Disorders and Stroke (NINDS).

References

Anton, R. C., Coffey, D. M., Gondo, M. M., Stephenson, M. A., Brown, R. W., and Cagle, P. T. (2000). The expression of cyclins D1 and E in predicting short-term survival in squamous cell carcinoma of the lung. *Mod. Pathol.* **13,** 1167–1172.
Buchberger, A. (2002). From UBA to UBX: New words in the ubiquitin vocabulary. *Trends Cell Biol.* **12,** 216–221.
Camp, R. L., Chung, G. G., and Rimm, D. L. (2002). Automated subcellular localization and quantification of protein expression in tissue microarrays. *Nat. Med.* **8,** 1323–1327.
Carson, F. L. (1997). "Histotechnology. A Self-Instructional Text." 2nd Edition, ASCP Press, Chicago.
Chen, G., Gharib, T. G., Huang, C. C., Taylor, J. M., Misek, D. E., Kardia, S. L., Giordano, T. J., Iannettoni, M. D., Orringer, M. B., Hanash, S. M., and Beer, D. G. (2002). Discordant protein and mRNA expression in lung adenocarcinomas. *Mol. Cell Proteomics* **1,** 304–313.
de Hoon, M. J. L., Imoto, S., Nolan, J., and Miyano, S. (2004). Open source clustering software. *Bioinformatics* **20,** 1453–1454.
Deshaies, R. J. (1999). SCF and Cullin/Ring H2-based ubiquitin ligases. *Annu. Rev. Cell Dev. Biol.* **15,** 435–467.
Eisen, M. B., Spellman, P. T., Brown, P. O., and Botstein, D. (1998). Cluster analysis and display of genome-wide expression patterns. *Proc. Natl. Acad. Sci. USA* **95,** 14863–14868.
Espina, V., Mehta, A. I., Winters, M. E., Calvert, V., Wulfkuhle, J., Petricoin, E. F. 3rd, and Liotta, L. A. (2003). Protein microarrays: Molecular profiling technologies for clinical specimens. *Proteomics* **3,** 2091–2100.
Gerdes, J., Lemke, H., Baisch, H., Wacker, H. H., Schwab, U., and Stein, H. (1984). Cell cycle analysis of a cell proliferation-associated human nuclear antigen defined by the monoclonal antibody Ki-67. *J. Immunol.* **133,** 1710–1715.
Glotzer, M., Murray, A. W., and Kirschner, M. W. (1991). Cyclin is degraded by the ubiquitin pathway. *Nature* **349,** 132–138.
Gstaiger, M., Jordan, R., Lim, M., Catzavelos, C., Mestan, J., Slingerland, J., and Krek, W. (2001). Skp2 is oncogenic and overexpressed in human cancers. *Proc. Natl. Acad. Sci. USA* **98,** 5043–5048.

Harlow, E., and Lane, D. (1999). "Using Antibodies. A Laboratory Manual." Cold Spring Harbor Laboratory Press, Cold Spring Harbor.

Harper, J. W., Burton, J. L., and Solomon, M. J. (2002). The anaphase-promoting complex: It's not just for mitosis any more. *Genes Dev.* **16,** 2179–2206.

Hoos, A., and Cordon-Cardo, C. (2001). Tissue microarray profiling of cancer specimens and cell lines: Opportunities and limitations. *Lab. Invest.* **81,** 1331–1338.

Hsu, F. D., Nielsen, T. O., Alkushi, A., Dupuis, B., Huntsman, D., Liu, C. L., van de Rijn, M., and Gilks, C. B. (2002). Tissue microarrays are an effective quality assurance tool for diagnostic immunohistochemistry. *Mod. Pathol.* **15,** 1374–1380.

Hsu, J. Y., Reimann, J. D., Sorensen, C. S., Lukas, J., and Jackson, P. K. (2002). E2F-dependent accumulation of hEmi1 regulates S phase entry by inhibiting APC(Cdh1). *Nat. Cell Biol.* **4,** 358–366.

Iwaya, K., Ogawa, H., Kuroda, M., Izumi, M., Ishida, T., and Mukai, K. (2003). Cytoplasmic and/or nuclear staining of beta-catenin is associated with lung metastasis. *Clin. Exp. Metastasis.* **20,** 525–529.

Jackson, P. K., Eldridge, A. G., Freed, E., Furstenthal, L., Hsu, J. Y., Kaiser, B. K., and Reimann, J. D. (2000). The lore of the RINGs: Substrate recognition and catalysis by ubiquitin ligases. *Trends Cell Biol.* **10,** 429–439.

Jensen, T. (2004). Tissue microarrays–done inexpensively. *LabLeader,* Thermo Electron Corp. Publication, Spring, 9 (http://www.thermo.com/com/cda/article/general/1,517,00.html).

Kaelin, W. G. (2002). Molecular basis of the VHL hereditary cancer syndrome. *Nat. Rev. Cancer* **2,** 673–682.

Knecht, R., Elez, R., Oechler, M., Solbach, C., von Ilberg, C., and Strebhardt, K. (1999). Prognostic significance of polo-like kinase (PLK) expression in squamous cell carcinomas of the head and neck. *Cancer Res.* **59,** 2794–2797.

Koegl, M., Hoppe, T., Schlenker, S., Ulrich, H. D., Mayer, T. U., and Jentsch, S. (1999). A novel ubiquitination factor, E4, is involved in multiubiquitin chain assembly. *Cell* **96,** 635–644.

Kononen, J., Bubendorf, L., Kallioniemi, A., Barlund, M., Schraml, P., Leighton, S., Torhorst, J., Mihatsch, M. J., Sauter, G., and Kallioniemi, O. P. (1998). Tissue microarrays for high-throughput molecular profiling of tumor specimens. *Nat. Med.* **4,** 844–847.

Liu, C. L., Prapong, W., Natkunam, Y., Alizadeh, A., Montgomery, K., Gilks, C. B., and van de Rijn, M. (2002). Software tools for high-throughput analysis and archiving of immunohistochemistry staining data obtained with tissue microarrays. *Am. J. Path.* **161,** 1557–1565.

Makretsov, N. A., Huntsman, D. G., Nielsen, T. O., Yorida, E., Peacock, M., Cheang, M. C., Dunn, S. E., Hayes, M., van de Rijn, M., Bajdik, C., and Gilks, C. B. (2004). Hierarchical clustering analysis of tissue microarray immunostaining data identifies prognostically significant groups of breast carcinoma. *Clin. Cancer Res.* **10,** 6143–6151.

Margottin-Goguet, F., Hsu, J. Y., Loktev, A., Hsieh, H. M., Reimann, J. D., and Jackson, P. K. (2003). Prophase destruction of Emi1 by the SCF(betaTrCP/Slimb) ubiquitin ligase activates the anaphase promoting complex to allow progression beyond prometaphase. *Dev. Cell.* **4,** 813–826.

Morin, P. J., Sparks, A. B., Korinek, V., Barker, N., Clevers, H., Vogelstein, B., and Kinzler, K. W. (1997). Activation of beta-catenin-Tcf signaling in colon cancer by mutations in beta-catenin or APC. *Science* **27,** 1787–1790.

Ninomiya, I., Endo, Y., Fushida, S., Sasagawa, T., Miyashita, T., Fujimura, T., Nishimura, G., Tani, T., Hashimoto, T., Yagi, M., Shimizu, K., Ohta, T., Yonemura, Y., Inoue, M., Sasaki, T.,

and Miwa, K. (2000). Alteration of β-catenin expression in esophageal squamous-cell carcinoma. *Int. J. Cancer* **85,** 757–761.
Quackenbush, J. (2001). Computational analysis of microarray data. *Nat. Rev. Genet.* **2,** 418–427.
Schneider, S., Uchida, K., Salonga, D., Yochim, J. M., Danenberg, K. D., and Danenberg, P. V. (2004). Quantitative determination of p16 gene expression by RT-PCR. *Methods Mol. Biol.* **281,** 91–103.
Sørlie, T., Perou, C. M., Tibshirani, R., Aas, T., Geisler, S., Johnsen, H., Hastie, T., Eisen, M. B., van de Rijn, M., Jeffrey, S. S., Thorsen, T., Quist, H., Matese, J. C., Brown, P. O., Botstein, D., Eystein Lonning, P., and Borresen-Dale, A. L. (2001). Gene expression patterns of breast carcinomas distinguish tumor subclasses with clinical implications. *Proc. Natl. Acad. Sci. USA* **98,** 10869–10874.
Tanaka, T., Kimura, M., Matsunaga, K., Fukada, D., Mori, H., and Okan, Y. (1999). Centrosomal kinase AIK1 is overexpressed in invasive ductal carcinoma of the breast. *Cancer Res.* **59,** 2041–2044.
Tut, V. M., Braithwaite, K. L., Angus, B., Neal, D. E., Lunec, J., and Mellon, J. K. (2001). Cyclin D1 expression in transitional cell carcinoma of the bladder: Correlation with p53, waf1, pRb and Ki67. *Br. J. Cancer* **84,** 270–275.
van de Rijn, M., and Gilks, C. B. (2004). Applications of microarrays to histopathology. *Histopathology* **44,** 97–108.
van 't Veer, L. J., Dai, H., van de Vijver, M. J., He, Y. D., Hart, A. A., Mao, M., Peterse, H. L., van der Kooy, K., Marton, M. J., Witteveen, AT, Schreiber, G. J., Kerkhoven, R. M., Roberts, C., Linsley, P. S., Bernards, R., and Friend, S. H. (2002). Gene expression profiling predicts clinical outcome of breast cancer. *Nature* **415,** 530–536.
Wright, C., Angus, B., Nicholson, S., Sainsbury, J. R., Cairns, J., Gullick, W. J., Kelly, P., Harris, A. L., and Horne, C. H. (1989). Expression of c-erbB-2 oncoprotein: A prognostic indicator in human breast cancer. *Cancer Res.* **49,** 2087–2090.
Zou, H., McGarry, T. J., Bernal, T., and Kirschner, M. W. (1999). Identification of a vertebrate sister-chromatid separation inhibitor involved in transformation and tumorigenesis. *Science* **285,** 418–422.

[24] Structure-Based Approaches to Create New E2–E3 Enzyme Pairs

By G. Sebastiaan Winkler and H. Th. Marc Timmers

Abstract

The study of ubiquitin-conjugating enzymes (E2) and ubiquitin–protein ligases (E3) is complicated by the fact that a relatively limited number of E2 proteins interacts with a large number of E3 enzymes. Many E3 enzymes contain a RING domain. Based on structural and biochemical analysis of the complex between UbcH5b and the CNOT4 RING finger, we describe a rationale to design new E2–E3 enzyme pairs with altered specificity. In such enzyme pairs, the E2 and E3 proteins are each mutated so that they do not interact with their wild-type partner. However, a functional enzyme pair is reconstituted when both E2 and E3 mutants are combined. Such altered-specificity enzyme pairs may be valuable to study the physiological significance of particular E2–E3 interactions.

Introduction

An extensive protein machinery is devoted to the multiple covalent attachment of ubiquitin to protein substrates (Hershko and Ciechanover, 1998; Pickart, 2001; Weissman, 2001). Although only one ubiquitin-activating enzyme (E1) is present in humans, more than 25 ubiquitin-conjugating enzymes (E2) have been identified. These proteins contain the conserved core UBC domain encompassing approximately 150 amino acids. The UBC domain is structurally well characterized and contains an N-terminal α-helix followed by a four-stranded anti-parallel β-sheet and three α-helices (see Fig. 1A, C; Pickart, 2001; VanDemark and Hill, 2002). E3 proteins usually contain either a HECT or RING domain (Huibregtse et al., 1995; Joazeiro and Weissman, 2000). Based on the number of polypeptides identified with these domains, the RING finger E3 enzymes comprise the largest group of proteins involved in ubiquitylation; more than 350 RING finger proteins have been identified in man. Usually, the RING domain consists of approximately 70 amino acids. Typical features are the presence of two Zn^{2+} ions that are coordinated by eight cysteine and histidine residues in a cross-brace manner and the presence of a hydrophobic cluster. RING domain proteins can act as single subunit E3 enzymes or as part of multi-subunit E3 complexes (VanDemark and Hill, 2002).

Although the UBC domain is highly conserved, each RING E3 enzyme interacts functionally and physically with only few E2 enzymes. How this specificity is obtained and how it determines substrate specificity is an important question. Insight into this comes from the structure of the complex between the RING domain c-Cbl protein and UbcH7 (Zheng et al., 2000). Specific residues in two loop regions, L1 and L2, of UbcH7 make contacts with residues in a hydrophobic pocket of the c-Cbl RING domain that are coordinated by the Zn^{2+} ions. Furthermore, the N-terminal α-helix of UbcH7 is involved in additional contacts with a region of c-Cbl outside the RING domain. Further understanding is provided by the analysis of the RING finger protein CNOT4, a component of the human Ccr4-Not transcriptional regulatory complex (Albert et al., 2000; Collart and Timmers, 2004). A structure of CNOT4 bound to UbcH5b was

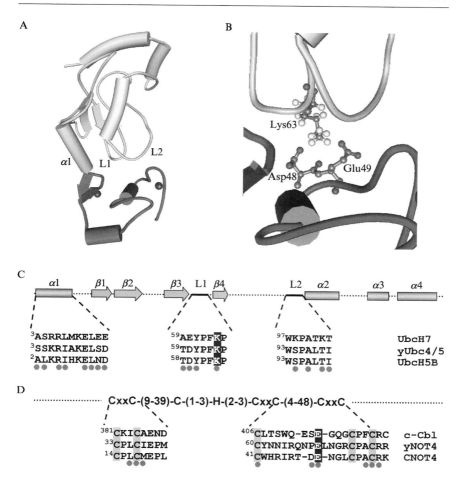

FIG. 1. (A) Overview of the UbcH5b–NOT4 RING structure (PDB accession number 1UR6). UbcH5b is shown in yellow, the CNOT4 RING domain in red, and the Zn^{2+} ions in gray. Indicated are the three UbcH5b regions making contact with the CNOT4 RING. (B) Detail of the UbcH5b–CNOT4 RING structure. The UbcH5b residue Lys63 interacts with acidic residues Asp48 and Glu49 of the CNOT4 RING domain. (A) and (B) were generated using WebLab Viewer Lite (Molecular Simulations Inc). Note that not all structures in the ensemble of five best solutions contain β-sheets in the CNOT4 RING domain. (C) Schematic diagram of the UBC domain. Regions important for the interaction with E3 enzymes are expanded. Shown are amino acid sequences of UbcH5b, yeast (y) Ubc4 and Ubc5, and UbcH7. Circles indicate UbcH5b residues involved in the interaction with the CNOT4 RING domain as identified by chemical-shift perturbation experiments (combined chemical shift differences >0.1 ppm) (Dominguez et al., 2004). The conserved loop L1 residue corresponding to UbcH5b Lys63 is highlighted. (D) Schematic diagram of the RING domain. Expanded are

proposed by combining NMR titration, mutagenesis, and docking methods (Fig. 1A; Dominguez et al., 2004). The overall structure is similar to that of the UbcH7–c-Cbl complex. However, in addition to hydrophobic interactions mediated by UbcH5b residues in loops L1 and L2, the N-terminal α-helix of UbcH5b binds directly to the CNOT4 RING domain by means of a network of hydrogen bonds and salt bridges (Dominguez et al., 2004). Although these structures do not fully explain the selective interaction between RING fingers and E2 enzymes, they provide substantial information about the interaction between these proteins, highlighting the importance of particular regions on the interfaces of both enzymes.

The hierarchical nature of the ubiquitin system and the relative number of E2 and E3 enzymes identified implies that a given E2 protein interacts with several E3 enzymes. This complicates and restricts the study of particular E2–E3 pairs. These problems can potentially be overcome by using new E2–E3 enzyme pairs with altered specificities. This can be achieved by creating hybrids containing parts of different E2 enzymes (Martinez-Noel et al., 2001), but a more specific manner is to generate E2–E3 enzyme pairs by mutation of a critical residue and subsequent screening for compensatory mutations in the enzyme partner based on the structure of E2–E3 complexes. Using the latter strategy, an altered-specificity UbcH5b-CNOT4 enzyme pair was designed (Winkler et al., 2004).

Experimental Rationale

Homology-Based Approach: Ubc4/5–E3 Interaction

During the characterization of the binding interface of the CNOT4 RING domain and UbcH5b, we identified acidic residues (amino acids Glu49 and Asp48) of CNOT4 that interact with a basic amino acid (Lys63) of UbcH5b (Fig. 1B; Albert et al., 2002; Dominguez et al., 2004; Winkler et al., 2004). Although charge-reversal mutations on either the CNOT4 or UbcH5b surface disrupted binding, reciprocal substitution of the charged amino acids re-created a functional enzyme pair.

---

amino acid sequences of CNOT4, yeast Not4, and c-Cbl. Circles indicate CNOT4 residues involved in the interaction with UbcH5b as identified by chemical-shift perturbation experiments (combined chemical shift differences >0.05 ppm) (Albert et al., 2002). Indicated in gray are cysteine residues involved in Zn^{2+}-coordination. The conserved acidic residue corresponding to CNOT4 Glu49 is highlighted. (See color insert.)

A similar strategy may be used for other E2–E3 enzymes. In particular, regions in the UBC domain are well delineated (Fig. 1C). For example, loop L1 residue Lys63 of UbcH5b is highly conserved in members of the Ubc4/5 sub-group of E2 enzymes (Fig. 1C). Also, residue Glu49 of CNOT4 is conserved in orthologs from yeast to man and positioned approximately four residues N-terminal from the seventh Zn^{2+}-chelating residue (Cys53 in CNOT4; Fig. 1D). Furthermore, several demonstrated and potential RING-type E3 ligases contain acidic residues at positions corresponding to Glu49 of CNOT4 (Fig. 1D, Winkler et al., 2004). Because some of these E3 enzymes may also interact with E2 enzymes of the Ubc4/5 subgroup in a manner like CNOT4–UbcH5b, it may be expected that similar amino acid substitutions result in new E2–E3 enzyme pairs. Indeed, based on this assumption, the orthologous altered-specificity enzyme pairs consisting of yeast Ubc4/Ubc5–Not4 were designed (Fig. 1D; Winkler et al., 2004).

In some cases, it may be difficult to identify the RING finger residues, because the loop region linking the sixth and seventh Zn^{2+}-coordinating residues is highly variable (Fig. 1D). In those circumstances, it may be necessary to scan a limited region by mutational analysis. Alternately, additional structural data may be acquired (see below).

This approach can be extended further to other E3 RING enzymes that interact with the Ubc4/5 subgroup. We noted that the amino acid substitutions D48A and E49A of CNOT4 allow binding to UbcH5b with similar affinity compared with wild type (Albert et al., 2002). Also, we noted that the K63E substitution in UbcH5b affected the ubiquitylation activity in conjunction with another RING-type E3 enzyme (unpublished observations). Therefore, charge-reversal mutations in E2 enzymes corresponding to K63E of UbcH5b may be compensated for by introducing basic amino acid alterations at positions corresponding to Glu49 of CNOT4 in E3 enzymes that do not contain an acidic patch. By scanning a limited region, a compensatory mutation may be identified in the E3 (Fig. 1D).

Homology-Based Approach: Interactions Involving E2 Enzymes Not Belonging to the Ubc4/5 Subgroup

Some RING-type E3 enzymes will not interact with E2 enzymes of the Ubc4/5 subgroup. Two approaches may be envisaged to design altered-specificity enzyme pairs in these cases. First, an artificial acidic–basic amino acid pair analogous to UbcH5b–CNOT4 may be introduced. Because the structure of the UBC domain is highly conserved, modeling the E2 of interest may identify the position at which the acidic residue should be introduced with relative confidence. The compensatory mutations in the E3 enzyme may require scanning a limited region, as described above.

A second approach may constitute to design a second independent mutation to disrupt the E2–E3 interaction. In particular, residues in the N-terminal α-helix and loops L1 and L2 of the UBC domain are good targets for mutagenesis (Fig. 1C). By subsequent screening of a mutagenized library of the corresponding partner (e.g., using a yeast two-hybrid system), compensatory mutations may be isolated (Crispino et al., 1999; Mak and Parker, 2001).

NMR Chemical-Shift Perturbation-Based Approach

An alternative strategy is based on the mapping of the interaction surfaces of specific E2–E3 enzymes. NMR chemical-shift perturbation mapping is well suited to study protein–protein interactions, because the chemical shifts of amide protons are highly sensitive to the binding of interaction partners or conformational changes (Zuiderweg, 2002). Although the observed chemical shifts do not reveal direct contacts *per se*, it is an excellent tool to direct mutational analysis.

In case backbone resonance assignments of one of the proteins of interest is available (e.g., of Ube2b or UbcH5b), chemical shift perturbation mapping may be obtained relatively quickly by recording successive NMR spectra of ^{15}N-labeled protein in the presence of increasing amounts of unlabeled binding partner. As an example, residues of UbcH5b and CNOT4 with chemical shift perturbations are indicated (Fig. 1C, D) and include Lys63 of UbcH5b and both Asp48 and Glu49 of CNOT4. In addition, when both interaction surfaces have been mapped, high-quality models of the E2–E3 complex can be obtained using molecular docking approaches (e.g., using the HADDOCK package; Dominguez et al., 2003; 2004), which can be further used to direct mutational analysis.

Methods

After the design of amino acid substitutions, several assays can be used to analyze the interaction between E2 and E3 enzymes. As an example, we provide protocols for the expression and purification of UbcH5b and several other E2 enzymes. These proteins can be used to study the binding between these enzymes and RING proteins using glutathione-S-transferase (GST)-pull down. In addition, a protocol is provided for a quantitative yeast two-hybrid analysis, which is, in particular, suitable for screening collections of mutants. Finally, a protocol is provided describing *in vitro* ubiquitylation reactions. It is of particular importance to complement physical interaction studies with an activity assay, because binding between E2 and E3 enzymes does not necessarily result in a productive enzyme pair (Brzovic et al., 2003).

Expression and Purification of E2 Enzymes

We routinely express and purify a panel of E2 enzymes as GST-fusion proteins. Plasmids used in our studies were described previously (Albert *et al.*, 2002; Winkler *et al.*, 2004). Expression of GST–E2 fusion proteins is induced by addition of 0.4 mM isopropyl-β-D-thiogalactopyranoside (IPTG) to *E. coli* BL21(DE3) cells grown to mid-log phase (OD$_{600}$, 0.6–0.7) and further growth for 3 h at 30° or room temperature. Cells are collected by centrifugation and lysed by repeated freeze-thawing in lysis buffer (50 mM Tris-HCl, pH 8.0, 300 mM KCl, 20% sucrose, 2 mM EDTA, 0.1% Triton X-100, 1 mM dithiothreitol [DTT], and protease inhibitors) containing freshly added 250 μg/ml lysozyme. Lysis is carried out in 40 ml buffer per liter culture volume and repeated until a very viscous solution is obtained. Lysates are sonicated (three times 15 sec with 15-sec intervals; Dr. Hielscher Ultraschallprozessor UP200S using a 6-mm diameter midsized horn) on ice to fragment high-molecular weight DNA and cleared by centrifugation (60 min at 112,000g or 25,000 rpm in a Beckman SW28 rotor). The soluble lysate is subsequently applied at a flow rate of less than 10 column volumes/h onto a glutathione-agarose column (2.5-ml bed volume per liter culture volume using a 1-cm diameter column) equilibrated in five column volumes lysis buffer. After washing with wash buffer (50 mM Tris-HCl, pH 7.5, 50 mM KCl, 2.5 mM MgCl2, 0.5 mM EDTA, 0.25 mM DTT, and 0.5 mM PMSF), the column is eluted in the same buffer containing 20 mM reduced glutathione. After extensive dialysis versus wash buffer, the proteins can be used for GST-pull down analysis. Routinely, we obtain between 6 and 20 mg GST–E2 enzyme per liter induced culture.

To remove the GST tag, 3 units thrombin (Sigma) per mg GST fusion protein is added directly to the purified and dialyzed preparation. After incubation for 3 h at 37°, thrombin is inactivated by addition of 0.5 mM PMSF. It is advised to verify the activity of the protease in a small test reaction, because some activity of thrombin can be lost during storage. A second glutathione–agarose column (2.5-ml bed volume per liter culture volume) is used after thrombin treatment to remove free GST and undigested GST–E2. Alternatively, GST proteins and free E2 enzymes are separated by gel filtration (see below).

Expression and Purification of Labeled E2 Enzymes for NMR Analysis

For NMR analysis, ^{15}N-labeled proteins are obtained by growing bacteria in minimal medium (6 g/l Na$_2$HPO$_4 \cdot$2H$_2$O, 3 g/l KH$_2$PO$_4$, 0.5 g/l NaCl, 25 mg/l MgSO$_4$, 0.29 mg/l CaCl$_2$, 1.0 μg/l FeSO$_4 \cdot$7 H$_2$O, 0.27 mg/l ZnCl$_2$, 5.0 mg/l thiamine, 4 g/l glucose) containing ^{15}N-labeled NH$_4$Cl (0.5 g/l) as the single source of nitrogen. ^{15}N-labeled NH$_4$Cl is a relatively inexpensive compound that can be obtained from several vendors. Freshly transformed *E. coli* BL21(DE3) carrying the desired expression plasmid is replated onto minimal medium containing 50 μg/ml ampicillin and grown for 2 days at 37°. A single colony is used to inoculate a preculture that is grown for 2 days at 37° with vigorous shaking. This culture is diluted 1:2500 (0.4 ml culture/l medium) and grown for about 18 h at 37° until an OD_{600} of 0.7–0.75 is reached.

Induction of protein expression (at 30°) by IPTG and subsequent purification of GST-UbcH5b by glutathione affinity-purification, thrombin cleavage, and gel filtration is carried out as described above with few modifications. After cell lysis and loading of the cleared soluble lysate onto glutathione–agarose, the column is further washed with 10 column volumes WB buffer (20 mM K-PO$_4$, pH 7.0, 150 mM KCl, and protease inhibitors) and eluted in buffer WB containing 10 mM reduced glutathione. The peak fractions containing GST–UbcH5b are identified by SDS-PAGE followed by Coomassie staining. After treatment with thrombin, 0.5 mM PMSF is added. Thrombin is then removed by batch binding to benzamidine-agarose (Sigma). A stirred Amicon ultrafiltration cell with a 3-kDa cutoff filter (Millipore) is used to concentrate the sample to less than 4.0 ml. Subsequently, a Superdex-75 16/60 gel filtration column (AP Biotech) is run in buffer WB to separate GST and UbcH5b collecting 4-ml fractions. Peak fractions are identified by SDS-PAGE and Coomassie staining, pooled, and again concentrated using a stirred cell to about 1 mM or 14 mg/ml protein. About 10 mg ^{15}N-labeled GST-UbcH5b or 2 mg labeled UbcH5b is normally obtained per liter of induced culture. Purified UbcH5b is stored at −80° until further use. At this stage, repeated freeze-thaw cycles of the concentrated protein preparation should be avoided, but the purified protein is stable for several months at 4°.

Site-Directed Mutagenesis

Mutations in the open reading frame of fusion genes can be introduced by site-directed mutagenesis using standard overlap-PCR techniques or using the Quick-change protocol (Stratagene). No obvious changes in expression levels or purification properties were observed for any of the mutant proteins used.

Analysis of Binding Using GST Pull-Down Analysis

GST–E2 fusion proteins are expressed as described above. Although we used purified GST–E2 enzymes in some cases, proteins present in cleared bacterial lysates can be bound directly to glutathione–agarose beads. The concentrations of GST fusion proteins present in the cleared lysates are determined by SDS-PAGE analysis followed by Coomassie staining and immunoblotting using anti-GST antibodies (Santa Cruz, B-14) and purified GST–UbcH5b protein with a known concentration as a standard. Routinely, 10 μg GST–E2 enzyme is immobilized onto 20 μl glutathione–agarose (Sigma) by binding 1–2 h or overnight at 4°. After washing with binding buffer (50 mM K-PO$_4$, pH 6.6, 50 mM KCl, 0.1% Nonidet P-40, 10 μM ZnCl$_2$, 1 mM DTT, and protease inhibitors), 5 μg purified CNOT4 RING domain (Hanzawa et al., 2001) is allowed to bind while mixing for 1 h at 4° in binding buffer (500 μl). After washing in binding buffer (3 × 500 μl), samples are boiled and subjected to 15% SDS-PAGE. Binding of CNOT4 is determined by immunoblotting using anti-CNOT4 mouse monoclonal antibody 19A12 (Albert et al., 2002), whereas the recovery of GST–E2 proteins is assessed by immunoblotting using anti-GST antibodies. Binding to GST protein without a fusion partner is used to assess background binding in these assays. In addition, GST–Ubc9, the specialized E2 enzyme for the ubiquitin-like polypeptide SUMO, can be included as a negative control. In these experiments, typically 5–10% CNOT4 is recovered. We have noticed that recovery of CNOT4 is strongly dependent on the pH of the binding and washing buffers.

Analysis of Binding Using Yeast Two-Hybrid Analysis

We use the yeast two-hybrid system based on fusions with the bacterial LexA DNA-binding domain (driven by a constitutive ADH1 promoter) and the galactose-inducible B42 acidic region that acts as an activation domain in yeast (Estojak et al., 1995; commercially available as the Matchmaker LexA system, Clontech). Growth, selection, and transformation by LiAc of yeast strain EGY48 was carried out using standard protocols (Burke et al., 2000).

Appropriate expression of the LexA and B42 fusions is verified by immunoblotting. For this purpose, single colonies are used to inoculate selective medium (synthetic complete medium containing 2% galactose and 1% sucrose as the sole carbon sources) and grown overnight at 30° with vigorous shaking. A 1-ml aliquot is transferred to a microtube and spun 10 sec at maximum speed to pellet the yeast cells. After removal of the medium, cells are washed in 100 μl sterile water and spun briefly. Subsequently, cells are resuspended in 100 μl 0.2 M NaOH and incubated for 5 min at room temperature. Finally, cells are spun and resuspended in 100 μl SDS-sample buffer and placed 5 min at 100°. After removal of insoluble material by centrifugation (5 min, maximum speed), 5 μl is routinely used for immunoblotting (Kushnirov, 2000). LexA and B42 proteins can be detected using mouse monoclonal antibodies 2–12 (Santa Cruz) and 12CA5 (Roche), respectively.

To evaluate binding between the LexA and B42 proteins in yeast strain EGY48, growth on medium lacking leucine can be carried out. In addition, when the reporter plasmid pSH18-34 is cotransformed, binding can be assayed by β-galactosidase assays. Single colonies from the primary transformants are resuspended in 100 μl sterile water. Subsequently, small drops (2–3 μl) are spotted onto plates containing 5-bromo-4-chloro-3-indolyl β-D-galactopyranoside (X-Gal, synthetic complete medium lacking the appropriate amino acids containing 2% galactose as the sole carbon source, 0.1 M K-PO$_4$, pH 7.0, and 40 μg/ml X-gal) and incubated at 30° for 1–3 days. However, X-gal staining is notoriously nonquantitative, and, therefore, a chemiluminescent β-galactosidase assay is preferred to compare interactions between various mutants. To this end, yeast cells are grown in 3 ml selective medium (synthetic complete medium containing 2% galactose and 1% sucrose as the sole carbon sources) overnight with vigorous shaking. The cells are harvested by centrifugation in a Microfuge, washed once with 1 ml sterile water, and disrupted with Zirconia/silica beads using a mini beadbeater (Biospec Products, 1 min maximum speed at room temperature) in 100 μl yeast lysis buffer (100 mM Tris-HCl, pH 8.0, 20% (v/v) glycerol, 1 mM β-mercaptoethanol, 0.5% sodium dodecyl sulfate, and protease inhibitors). After removal of insoluble material by centrifugation, β-galactosidase activities are determined using the Galacto-Light Plus chemiluminescent reporter assay (Tropix). Usually, 0.5 μl yeast lysate is assayed in 70 μl 1× reaction buffer (Tropix) and incubated for 30 min at room temperature in darkness. After addition of accelerator solution (100 μl, Tropix), β-galactosidase activity is determined by counting luminescence (Berthold Lumat LB 9507). Activities are normalized to total protein content as determined by a Bradford protein assay using 2–3 μl yeast lysate and 1 ml protein assay reagent (BioRad).

In Vitro *Ubiquitylation*

A ubiquitylation assay is used to complement physical interactions with an assay of enzyme function. In addition to the E2 and E3 enzymes that are purified as described above, ubiquitin and the ubiquitin-activating enzyme E1 are required. Both can be acquired from several commercial sources or expressed in *E. coli* or insect cells. Ubiquitin can easily be obtained after expression as a GST fusion protein. Purification of untagged ubiquitin after thrombin treatment is carried out essentially according to the protocol provided above for the purification of GST–UbcH5b. By "leave-out" reactions, we observed that some GST–E2 enzymes, including GST–UbcH5b, display strong polyubiquitylation activity in the absence of the appropriate E3 enzyme. Therefore, it is also important that E2 enzymes free of the GST moiety are used in functional assays that address E2–E3 interactions.

Ubiquitylation reactions (20 μl) typically contain 50 ng rabbit E1 enzyme (Boston Biochem), 150 ng UbcH5b, 250 ng purified CNOT4, and 500 ng ubiquitin in buffer containing 50 mM Tris-HCl, pH 7.5, 50 mM KCl, 2.5 mM MgCl$_2$, 0.5 mM EDTA, 0.25 mM DTT, and 2 mM ATP. In some cases, we included an ATP-regenerating system (10 mM creatine phosphate and 10 units creatine phosphokinase, Calbiochem), but it is not an essential component in our experience. After 90 min incubation at 30°, reactions are stopped by addition of SDS-sample buffer and subjected to 10% SDS-PAGE. Polyubiquitylation is easily detected by immunoblotting using mouse monoclonal antibody P4D1 raised against ubiquitin (Santa Cruz). Auto-ubiquitylation is detected using mouse monoclonal antibody 19A12 recognizing CNOT4 (Albert *et al.*, 2002; Winkler *et al.*, 2004).

Conclusion

Altered-specificity mutants have been widely used to study the significance of protein–protein interactions in eukaryotic systems (see, for example, Crispino *et al.*, 1999; Mak and Parker, 2001; Tansey and Herr, 1997). Here, a structure-based rationale is provided to generate new E2–E3 enzyme pairs. Although the procedure described is limited in that it is based on RING-type E3 enzymes, it may be valuable for the design of new E2–E3 enzyme pairs to dissect the physiological roles of specific E2–E3 enzymes involved in ubiquitin conjugation.

Acknowledgments

We thank Drs. Rolf Boelens and Cyril Dominguez for continued collaboration and stimulating discussions. F. M. A. van Schaik and Y. I. A. Legtenberg are acknowledged for expert technical assistance. This work was supported by grants from The Netherlands Organisation for Scientific Research (NWO-MW Pioneer 900-98-142 and NWO-CW 700-50-634) and by The Netherlands Centre for Proteomics.

References

Albert, T. K., Hanzawa, H., Legtenberg, Y. I. A., de Ruwe, M. J., van den Heuvel, F. A. J., Collart, M. A., Boelens, R., and Timmers, H. T. M. (2002). Identification of a ubiquitin-protein ligase subunit within the CCR4-NOT transcription repressor complex. *EMBO J.* **21**, 355–364.

Albert, T. K., Lemaire, M., van Berkum, N. L., Gentz, R., Collart, M. A., and Timmers, H. T. M. (2000). Isolation and characterization of human orthologs of yeast CCR4-NOT complex subunits. *Nucleic Acids Res.* **28**, 809–817.

Brzovic, P. S., Keeffe, J. R., Nishikawa, H., Miyamoto, K., Fox, D., 3rd, Fukuda, M., Ohta, T., and Klevit, R. (2003). Binding and recognition in the assembly of an active BRCA1/BARD1 ubiquitin-ligase complex. *Proc. Natl. Acad. Sci. USA* **100,** 5646–5651.

Burke, D., Dawson, D., and Stearns, T. (2000). Methods in yeast genetics: A Cold Spring Harbor Laboratory course manual. Cold Spring Harbor Laboratory Press, Cold Spring Harbor.

Collart, M. A., and Timmers, H. T. (2004). The eukaryotic Ccr4-not complex: A regulatory platform integrating mRNA metabolism with cellular signaling pathways? *Prog. Nucleic Acid Res. Mol. Biol.* **77,** 289–322.

Crispino, J. D., Lodish, M. B., MacKay, J. P., and Orkin, S. H. (1999). Use of altered specificity mutants to probe a specific protein-protein interaction in differentiation: The GATA-1:FOG complex. *Mol. Cell* **3,** 219–228.

Dominguez, C., Boelens, R., and Bonvin, A. M. (2003). HADDOCK: A protein-protein docking approach based on biochemical or biophysical information. *J. Am. Chem. Soc.* **125,** 1731–1737.

Dominguez, C., Bonvin, A. M., Winkler, G. S., van Schaik, F. M., Timmers, H. T., and Boelens, R. (2004). Structural model of the UbcH5B/CNOT4 complex revealed by combining NMR, mutagenesis, and docking approaches. *Structure (Camb)* **12,** 633–644.

Estojak, J., Brent, R., and Golemis, E. A. (1995). Correlation of two-hybrid affinity data with *in vitro* measurements. *Mol. Cell. Biol.* **15,** 5820–5829.

Hanzawa, H., de Ruwe, M. J., Albert, T. K., van der Vliet, P. C., Timmers, H. T. M., and Boelens, R. (2001). The structure of the C_4C_4 RING finger of human NOT4 reveals features distinct from those of C_3HC_4 RING fingers. *J. Biol. Chem.* **276,** 10185–10190.

Hershko, A., and Ciechanover, A. (1998). The ubiquitin system. *Annu. Rev. Biochem* **67,** 425–479.

Huibregtse, J. M., Scheffner, M., Beaudenon, S., and Howley, P. M. (1995). A family of proteins structurally and functionally related to the E6-AP ubiquitin-protein ligase. *Proc. Natl. Acad. Sci. USA* **92,** 2563–2567.

Joazeiro, C. A. P., and Weissman, A. M. (2000). RING finger proteins: Mediators of ubiquitin ligase activity. *Cell* **102,** 549–552.

Kushnirov, V. V. (2000). Rapid and reliable protein extraction from yeast. *Yeast* **16,** 857–860.

Mak, H. Y., and Parker, M. G. (2001). Use of suppressor mutants to probe the function of estrogen receptor-p160 coactivator interactions. *Mol. Cell Biol.* **21,** 4379–4390.

Martinez-Noel, G., Muller, U., and Harbers, K. (2001). Identification of molecular determinants required for interaction of ubiquitin-conjugating enzymes and RING finger proteins. *Eur. J. Biochem.* **268,** 5912–5919.

Pickart, C. M. (2001). Mechanisms underlying ubiquitination. *Annu. Rev. Biochem.* **70,** 503–533.

Tansey, W. P., and Herr, W. (1997). Selective use of TBP and TFIIB revealed by a TATA-TBP-TFIIB array with altered specificity. *Science* **275,** 829–831.

VanDemark, A. P., and Hill, C. P. (2002). Structural basis of ubiquitylation. *Curr. Opion. Struct. Biol.* **12,** 822–830.

Weissman, A. M. (2001). Themes and variations on ubiquitylation. *Nat. Rev. Mol. Cell Biol.* **2,** 169–178.

Winkler, G. S., Albert, T. K., Dominguez, C., Legtenberg, Y. I., Boelens, R., and Timmers, H. T. (2004). An altered-specificity ubiquitin-conjugating enzyme/ubiquitin-protein ligase pair. *J. Mol. Biol.* **337,** 157–165.

Zheng, N., Wang, P., Jeffrey, P. D., and Pavletich, N. P. (2000). Structure of a c-Cbl-UbcH7 complex: RING domain function of ubiquitin-protein ligases. *Cell* **102,** 533–539.

Zuiderweg, E. R. P. (2002). Mapping protein-protein interactions in solution by NMR spectroscopy. *Biochemistry* **41,** 1–7.

[25] Proteomic Analysis of Ubiquitin Conjugates in Yeast

By JUNMIN PENG and DONGMEI CHENG

Abstract

Although the list of proteins modified by ubiquitination has been growing rapidly, a reliable method to biochemically isolate and identify *in vivo* ubiquitinated substrates is needed. Here we describe a proteomic approach to enrich, identify, and validate ubiquitinated proteins from *Saccharomyces cerevisiae* on a large scale. To facilitate the purification of ubiquitinated proteins, all four ubiquitin genes were knocked out, and a plasmid coding N-terminal His-tagged ubiquitin was introduced in a yeast strain. Ubiquitinated proteins from the strain were purified by metal chelation chromatography under denaturing condition to minimize co-isolation of interacting proteins. Purified proteins were further analyzed by highly sensitive mass spectrometry to determine their identities. The ubiquitination sites in the proteins could be determined in many cases according to the mass shift caused by the modification. Moreover, the polyubiquitin chain topology could be detected by the same method as well. To confirm the genuineness of the identified ubiquitin conjugates, several strategies were developed: (1) to subtract proteins detected in the control experiment (metal chelation purification from wild-type yeast strain); (2) to remove proteins that did not show the increase of apparent molecular weight because of ubiquitination; (3) to accept proteins with identified ubiquitination sites; and (4) to confirm the state of ubiquitination of individual protein by immunoprecipitation and Western blotting analysis. The method provides a generic approach for biochemical characterization of ubiquitinated proteins and can be extended to analyze targets of ubiquitin-like molecules.

Introduction

The conserved ubiquitin (Ub) molecule, a 76 amino acid polypeptide, is covalently attached to other cellular proteins by the coordinated activities of several classes of enzymes: ubiquitin-activating enzyme (E1), ubiquitin-conjugating enzymes (E2s), and ubiquitin ligases (E3s) (Weissman, 2001), and this modification is reversible by the action of deubiquitinating enzymes (Wilkinson, 2000). Generally, the carboxyl terminus of a Ub molecule forms an isopeptide bond with the ε-amino group of a lysine residue in protein substrates (Fig. 1A). It seems that many substrates are modified by polyubiquitin chains in which additional Ub molecules are conjugated to lysine residues of the first Ub (Fig. 1B). All seven lysine residues in ubiquitin itself have been suggested in chain assembly to generate different types of polyUb linkages (Pickart, 2004). The consequence of ubiquitination of given substrates is dependent on the length and linkage of added ubiquitin chains. For example, Lys48 polyUb chains are the principal signal for targeting substrates for degradation by proteasome (Chau *et al.*, 1989), and efficient targeting requires a chain with at least four ubiquitin molecules (Thrower *et al.*, 2000). By contrast, polyUb chains linked through Lys63 have been shown to act in endocytosis (Haglund *et al.*, 2003), DNA repair (Hoege *et al.*, 2002), kinase activation (Deng *et al.*, 2000), and translational control (Spence *et al.*, 2000). Similarly, monoubiquitination of proteins functions as a sorting signal in a variety of membrane trafficking events (Hicke and Dunn, 2003). Thus, identifying ubiquitinated

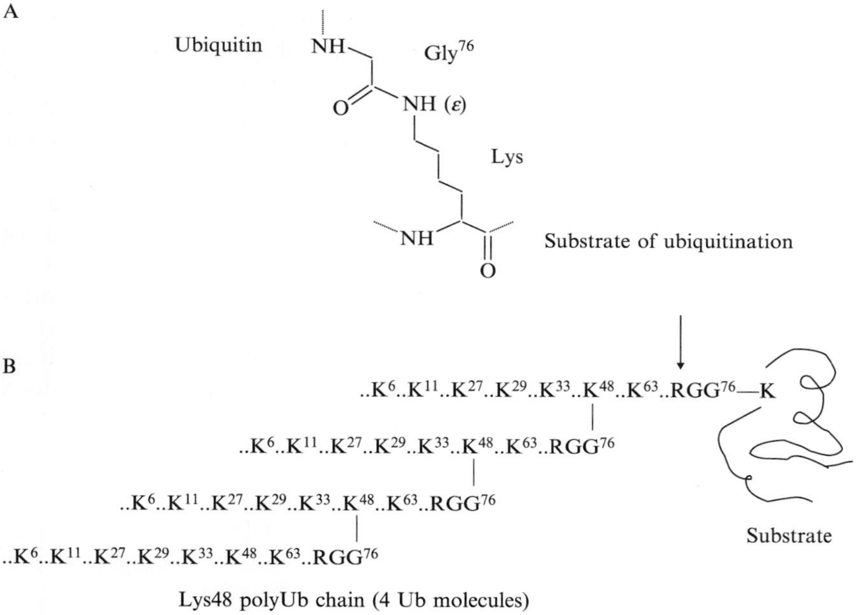

FIG. 1. (A) The isopeptide (amide) bond between the carboxyl group of ubiquitin C-terminal Gly residue and the amino group of the Lys side chain from a substrate. Gly76 of ubiquitin is on the top left. (B) Representative Lys48-linked polyUb chain on a target protein. All lysine residues of ubiquitin are shown. The arrow indicates a tryptic site in the Ub sequence.

conjugates and measuring the length and linkage of Ub chains can be highly instructive for understanding the function of ubiquitination.

Biochemical characterization of endogenous ubiquitinated species has been difficult until the development of protein sequencing technology by mass spectrometry with femtomolar or even sub-femtomolar sensitivity (Aebersold and Mann, 2003; Yates, 2004), because the abundance of these proteins is often low in cells. Ubiquitinated conjugates can be enriched by a number of affinity approaches, including ubiquitin antibodies, ubiquitin-binding proteins, and epitope-tagged ubiquitin (e.g., FLAG, hemagglutinin [HA]-tag, *myc*-tag, and six-His). Although the yield and efficiency of these approaches varies dramatically, His-tag is well suited for the purpose of biochemical purification (Peng *et al.*, 2003b). His-tag is the only available tag compatible with fully denaturing conditions, under which protein–protein interaction is almost eliminated and contaminants are greatly reduced. In addition, like other tags, His-tag at the N-terminus has not been reported to interfere with the physiological function of ubiquitin. The isolated Ub-conjugates are typically analyzed by liquid chromatography coupled with tandem mass spectrometry (LC-MS/MS) to determine protein identities in a complex mixture. The protein sample is digested with a protease (e.g., trypsin), and the resulting peptides are fractionated by liquid chromatography (LC). These peptides are ionized and transferred into a mass spectrometer, where they are further separated based on mass-to-charge ratio (m/z). It should be mentioned that tandem mass spectrometry (MS/MS) itself is a high-resolution separation tool, because it can physically isolate a peptide ion according to its m/z value despite the presence of many other co-eluting peptides. The isolated peptide is then fragmented to generate a specific MS/MS spectrum containing its sequence information. Bioinformatics software, such as SEQUEST (Eng *et al.*, 1994), Mascot (Perkins *et al.*, 1999), can be used to match the experimental MS/MS spectra with theoretical (computer-produced) spectra of predicted peptides from a database, leading to the identification of peptides/proteins sequences. More details of the LC-MS/MS method are reviewed in other tutorial papers (Peng and Gygi, 2001; Steen and Mann, 2004).

In addition, amino acid sites modified posttranslationally can be determined by mass spectrometry (Mann and Jensen, 2003), because any modification changes the mass of target amino acid residues. We and others developed a strategy to detect ubiquitination sites in modified proteins (Marotti *et al.*, 2002; Peng and Gygi, 2001; Peng *et al.*, 2003b). After trypsin digestion, the original ubiquitin molecule is trimmed to a dipeptide (glycine–glycine) remnant that adds a mass of 114.1 Da on the affected lysine residue (Fig. 1B). This modification leads to a unique MS/MS spectrum that can be matched by database-searching algorithms. In addition, the modified peptides carry a missed proteolytic cleavage, because trypsin digestion cannot occur at the modified lysine sites. The method is equally effective in detecting the lysine residues in ubiquitin itself for polyUb chain assembly.

In this article, we discuss the methodological details of how to isolate and identify unknown Ub-conjugates from the engineered yeast strain and how to reduce false-positive proteins in biochemical purification and in mass spectrometric analysis. The cautions for data interpretation will also be highlighted.

Methods

Purification of Ubiquitinated Proteins from Yeast

Growth and Harvest of the Yeast Strains. Two yeast strains were recovered from frozen stocks: SUB592 (Spence *et al.*, 1995, 2000) (also named JSY171 with deletion of all ubiquitin genes and supplement of a plasmid expressing His-*myc*–tagged ubiquitin (named His-Ub for simplicity) under the control of CUP1 promoter) and the control strain SUB280 (similar to SUB592 except introduction of a wild-type ubiquitin plasmid). The strains were inoculated first into a small volume (e.g., 20 ml) of YPD medium (2% peptone, 1% yeast extract, and 2% dextrose) and grown overnight at 30° and then transferred into liters of YPD medium. The cells were grown to log phase of OD_{610} 1–1.5. (Optional step: the ubiquitin expression could be further induced by upregulating CUP1 promoter with 0.1 mM $CuSO_4$ for 2 h, which might increase the abundance of some Ub-conjugates in the yeast.) The cells were centrifuged at 2000g for 10 min at 4° and washed twice with ice-cold water. Approximately 10 mg (wet weight) of cells was obtained from 1 liter of culture of OD_{610} 1.5. The cells were lysed in denaturing buffer A (10 mM Tris, pH 8.0, 0.1 M NaH_2PO_4, 8 M urea, and 10 mM β-mercaptoethanol) using glass beads with buffer/sample ratio of 2:1 (v/v). The volume of yeast cells was also considered, and extra solid urea was added to maintain the final concentration at 8 M. The total cell lysates were clarified by centrifugation at 70,000g for 30 min. The protein concentration of the samples was measured by Bio-Rad Protein Assay, and a typical concentration of the lysates was ~5mg/ml. After lysis, all steps were performed at 4° unless specified.

Isolation of Ubiquitinated Proteins by Metal Chelation Chromatography. We performed a small-scale experiment first to test the yeast lysates and to optimize the ratio between total proteins applied and the size of a Nickel-NTA–agarose column (capacity of ~5 mg/ml beads, Qiagen, Chatsworth, CA). A small amount of resins (e.g., 10 μl) was incubated with a titrated

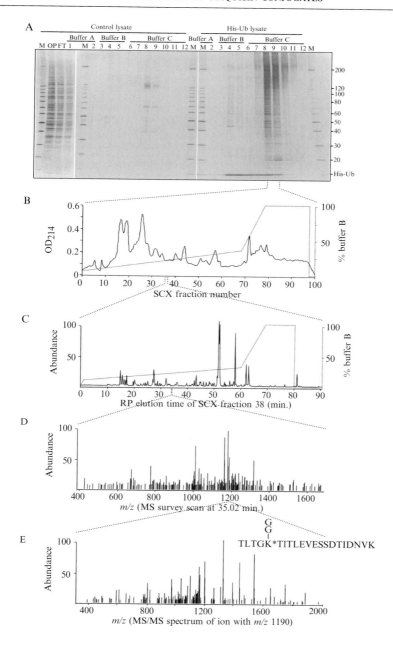

amount of total yeast proteins (e.g., 0.1–5 mg). The efficiency of binding was tested in a Western blotting analysis by monitoring His-*Myc*–tagged Ub-conjugates remaining in the supernatant. Based on the result, we loaded ~100 mg of yeast proteins on a 0.5-ml Nickel-NTA column in the following large-scale purification experiment. TALON™ metal affinity resin (BD Biosciences Clontech, CA) was also tested and shown to be much less effective in binding Ub-conjugates under the condition.

In the large-scale purification, a Nickel-NTA column was equilibrated with denaturing buffer A, loaded with total yeast lysate as the onput, and reloaded once with the flowthrough. The column was then washed sequentially with 30 V (bed volume) of buffer A twice (fraction 1 and 2), 1 V of buffer B three times (10 mM Tris, pH 6.3, 0.1 M NaH$_2$PO$_4$, 8 M urea, fraction 3–5), and eluted with 0.5 V of buffer C seven times (10 mM Tris, pH 4.5, 0.1 M NaH$_2$PO$_4$, 8 M urea, fraction 6–12). All fractions were examined on a 6–16% SDS gel and silver staining (Fig. 2A). Much less proteins were found in fraction 3, 6, and 7, because the elution of proteins on the column was delayed because of the presence of dead volume (~1 bed volume). According to Fig. 2A, extensive wash with a denaturing buffer at pH 8.0 decreased the background proteins to an undetected level (compare fraction 1 to fraction 2). A more stringent wash at pH 6.3 stripped some proteins from the column including His-Ub monomer (see fraction 3–5), suggesting that some monoubiquitinated proteins could be partially eluted at this step. Finally, the column was eluted by lowering the pH to 4.5. In contrast to only several major protein bands visible from the control lysate, many proteins were recovered from the His-Ub lysate,

---

FIG. 2. Isolation and identification of His-tagged yeast Ub-conjugates. (A) Nickel chromatographic fractions on a silver-stained SDS gel. Total cell lysates from control strain or His-Ub strain were loaded on Nickel columns in parallel. After loading, the columns were washed sequentially with 30 V (bed volume) of buffer A (pH 8.0) twice (fraction 1–2), 1 V of buffer B three times (pH 6.3, fraction 3–5), and eluted with 0.5 V of buffer C seven times (pH 4.5, fraction 6–12). The loading amount for each fraction was as follows: 0.0005% of the onput (OP), flowthrough (FT), and fraction 1 of the control lysate; 0.02% of fraction 2, 0.5% of fraction 3–5, and 1% of fraction 6–12. A molecular weight marker (M) was shown on either side of the gel (10-kDa ladder of 10–120 kDa and 200 kDa, Invitrogen). The monomer form of His-tagged ubiquitin was indicated. (B) Elution profile of strong cation exchange (SCX) chromatography with UV detection (OD$_{214}$) and fraction collection every minute. The onput was pooled from fraction 8 and 9 in (A). The elution buffer gradient is indicated by the dotted line. (C) A representative elution profile of reverse-phase (RP) LC-MS/MS. The onput was fraction 38 in (B). (D) MS survey scan at the point of 35.02 min during the RP chromatography in (C). Several major ions are shown with many background ions in the range of 400–1700 m/z. (E) MS/MS scan of the precursor ion 1190 m/z in (D), which led to the identification of a ubiquitinated peptide by database searching.

[25] PROTEOMIC ANALYSIS OF UBIQUITIN CONJUGATES 373

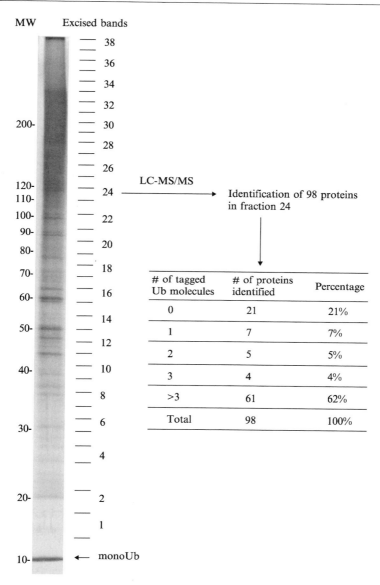

FIG. 3. Analysis of ubiquitinated proteins by 1D gel-LC-MS/MS strategy. A small fraction (~ 0.5%) of pooled Ub-conjugates and molecular weight marker (m) were run on a SDS gel followed by silver-staining as indicated. The remaining samples (~0.2 mg) were resolved on another SDS gel and stained with Coomassie blue G250 (data not shown). The sample lane was cut into 38 pieces that were subjected to trypsin digestion and LC-MS/MS analysis. The

yielding ~0.2 mg proteins from 100 mg yeast lysate. The staining pattern of the enriched proteins was highly similar to that in Western analysis using ubiquitin antibodies, as shown by the high molecular weight smear (in fraction 8–9). The recovery of Ub-conjugates was estimated to be approximately 30% with ~500-fold enrichment.

Identification of Enriched Ubiquitinated Proteins by Mass Spectrometry

We have used two strategies to identify ubiquitinated proteins in a complex mixture: multidimensional liquid chromatography-tandem mass spectrometry (LC/LC-MS/MS) and one-dimensional SDS gel (1D gel) coupled with LC-MS/MS. In the protocol of LC/LC-MS/MS, Ub-conjugate mixture is directly digested in a solution. The resulting peptide mixture is prefractionated by strong cation exchange (SCX) chromatography. Every SCX fraction is further separated by reverse phase (RP) chromatography and analyzed by a coupled mass spectrometer (Fig. 2). Direct digestion of Ub-conjugates in a mixture provides some advantages to increase *sensitivity*, because Ub-conjugates derived from a single protein are often heterogeneous because of variable lengths of ubiquitin chains, but they generate the same set of proteolytic fragments. By contrast, the 1D gel-LC-MS/MS approach helps obtain the *apparent molecular weight* information of the proteins identified (Fig. 3), because the proteins are first resolved on an SDS gel, and the entire gel lane is excised into pieces according to the molecular weight marker. The proteins in each of the gel pieces are digested and analyzed by reverse-phase LC-MS/MS.

Identification of Ubiquitinated Proteins by LC/LC-MS/MS. Fractions containing highly concentrated Ub-conjugates (fraction 8 and 9 in Fig. 2A, ~0.2 mg) were pooled and adjusted the pH to 7.5. The Cys residues in proteins were reduced with 10 mM DTT at 37° for 1 h and then alkylated with 50 mM iodoacetamide at room temperature in dark for 30 min. The step helps increase the recovery of Cys-containing peptides. The treated sample was dialyzed to switch buffer to 5 mM Tris-HCl, pH 7.5, and 1 M urea. Trypsin (Promega, Madison, WI) was added to a final protease/protein ratio of 1:20 (w/w) for digestion overnight at 37°.

To run the strong cation exchange column, the sample was supplemented with acetonitrile to 25%, acidified by trifluoroacetic acid (TFA) to

---

gel excision pattern is shown according to the marker. Ninety-eight proteins have been found in fraction 24 (MW between 110–120 kDa). The number of ubiquitin molecules attached to the proteins has been estimated according to the apparent MW on the gel and calculated MW from protein sequences.

0.5%, and loaded on a column (2.1 mm × 20 cm Polysulfoethyl A, Poly LC Inc., Columbia, MD). The column was then washed with 10 V of solvent A (5 mM phosphate buffer, 25% acetonitrile, pH 3.0) and eluted in a 5–20% gradient of solvent B (5 mM phosphate buffer, 25% acetonitrile, pH 3.0, 350 mM KCl) in 70 min followed by a 20–100% solvent B gradient in 10 min with the flow rate at 0.2 ml/min. Fractions were collected every minute (Fig. 2B).

After lowering the volume and concentration of acetonitrile by evaporation, each fraction was analyzed individually by reverse-phase LC-MS/MS using 75 μm id × 12 cm self-packed fused-silica C18 capillary columns. Peptides were eluted during a 60-min 10–30% gradient (solvent A: 0.4% acetic acid, 0.005% heptafluorobutyric acid (HFBA), 5% acetonitrile; solvent B: 0.4% acetic acid, 0.005% HFBA, 95% acetonitrile; flow rate: ~300 nl/min). Eluted peptides were ionized under high voltage (1.8–2.0 kV), detected in a MS survey scan from 400–1700 m/z with two μscans followed by three data-dependent MS/MS scans (three μscans each, isolation width 3 m/z, 35% normalized collision energy, dynamic range 3 min) in a completely automated fashion on an LCQ-DECA ion trap mass spectrometer (Thermo Finnigan, San Jose, CA; Fig. 2).

MS/MS spectra were searched against the yeast protein database supplemented with the sequence of recombinant His-*myc*-ubiquitin using SEQUEST algorithm (Eng *et al.*, 1994). Modifications were permitted to allow for detection of the following (mass shift shown in Daltons): oxidized methionines (+16), carboxymethylated cysteine (+57), ubiquitinated lysine (+114), and phosphorylated serine, threonine, and tyrosine (+80). The result was filtered according to the established criteria (Peng *et al.*, 2003a; Washburn *et al.*, 2001). We also manually examined all phosphorylated and/or ubiquitinated peptides and the peptides from proteins identified by two or less peptides.

Analysis of Ubiquitinated Proteins by 1D Gel-LC-MS/MS. The pooled Ub-conjugates (~0.2 mg) were first concentrated by acetone precipitation. Four times the sample volume of cold acetone (−20°) was added and incubated at −20° for 1 h. The precipitated proteins were centrifuged at 15,000g for 10 min, and dried by careful evaporation. The proteins were dissolved and reduced in SDS loading buffer (10 mM Tris-HCl, pH 8.0, 4% Ficoll, 2% SDS, 0.02% bromophenol blue, and 10 mM DTT) at 95° for 5 min, and then alkylated with the addition of iodoacetamide to 50 mM and incubation in dark at room temperature for 30 min. The sample and molecular weight markers were loaded on a 6–12% SDS gel (~12 cm in length) and run slowly for 4 h. The gel was stained with Coomassie G-250, soaked for at least 3 h to reduce residual SDS inside, and excised into gel pieces according to the flanking markers. A total of 38 gel pieces were cut to achieve reasonable molecular weight resolution in the entire gel lane (Fig. 3).

Proteins in the gel pieces were trypsinized with an in-gel digestion protocol as previously reported (Shevchenko et al., 1996). The gel pieces were cut into 1 mm^3, destained with a buffer (50% acetonitrile and 50% 50 mM ammonium bicarbonate), and soaked into 100% acetonitrile for 10 min and vacuum dried. Trypsin was absorbed into the gel by incubating the dried pieces with trypsin solution (12.5 ng/μl in 50 mM ammonium bicarbonate) for 45 min at 4°. Proteins were then digested into peptides overnight at 37°. The peptides were collected from the supernatant after spinning for 1 min, and the peptides remaining in the gel pieces were extracted three times with a buffer containing 5% formic acid and 50% acetonitrile. All extracted peptides were pooled and vacuum dried. Finally, the peptide sample from each gel piece was analyzed by reverse-phase LC-MS/MS as described previously. A total of ∼700 proteins and ∼50 ubiquitination sites were identified in the gel (Peng J., unpublished data).

Validation of the Identified Ubiquitinated Proteins

It is important to verify whether the proteins identified are really conjugated by ubiquitin or are simply present in the samples as contaminants. Although the absolute amount of contaminants can be controlled at a very low level using highly stringent conditions, such as denaturing condition for Nickel affinity chromatography, it is still possible that a large number of contaminants are co-isolated at low abundance. Such contaminants could be prominent when mass spectrometry is sensitive enough to detect them. To address this issue, we have developed several strategies to differentiate the *bona fide* Ub-conjugates from potential co-purified proteins.

Subtraction of Proteins That Were Purified by the Nickel Column from Wild-Type Yeast Lysate. The most common contaminants in Nickel affinity purification are His-rich and/or highly abundant proteins. We consider that His-rich proteins generally contain at least three consecutive histidine residues. Moreover, the protein abundance level can be reasonably indexed by the codon bias, a measure of the propensity of a gene to use only a subset of the 61 potential codons to produce its amino acids (Bennetzen and Hall, 1982, Gygi et al., 2002). As a general rule, a codon bias value <0.1 is indicative of low protein expression levels in cells, whereas a value >0.1 would reflect highly abundant proteins. More than one-half of the genes in yeast have codon bias values <0.1 and are thus thought to be expressed at low abundance (Gygi et al., 2002). In the control experiment, a total of 48 proteins was isolated (Fig. 2A) and identified by the approach of

LC/LC-MS/MS. Eighteen proteins (~38%) were rich in histidine residues, and the remaining were abundant proteins, such as metabolic enzymes and cytoskeletal components (Peng et al., 2003b). Although we could not rule out the possibility that some His-rich proteins or highly abundant proteins were genuinely modified by ubiquitin, we discarded these proteins from the protein list generated from the His-Ub sample. Even so, we still considered the list was preliminary, because it was possible that the His-Ub sample contained additional false-positive proteins that had substantial affinity for already bound proteins to the Nickel column in the presence of 8 M urea.

Removal of the Proteins That Did Not Show an Increase of Apparent Molecular Weight as a Result of Ubiquitination. During the 1D gel-LC-MS/MS analysis, the apparent molecular weight information obtained in the 1D SDS gel can be used to reconstitute a virtual Western blotting picture for each identified protein (Peng J., unpublished data). When compared with the molecular weight of its unmodified form that is calculated from its amino acid sequence, one can estimate the length of ubiquitin chain on each protein, because ubiquitination of proteins causes a dramatic shift in molecular weight (~10 kDa for monoubiquitination, ~20 kDa for diubiquitination, and so on). On the basis of this principle, we analyzed the proteins identified in the 24th gel piece (Fig. 3) and found that 21 proteins (21%) of 98 identified proteins did not show a molecular weight increase of 10 kDa, indicating that they were not likely to be ubiquitinated. The molecular weight information is complexed by unusual gel mobility of irregular-shaped proteins and other posttranslational modifications (e.g., phosphorylation, glycosylation, and lipidation). But phosphorylation of a protein does not typically cause gel mobility shift comparable to ubiquitination. In addition, it may be helpful to pretreat the sample with specific enzymes to trim some other modifications. Nevertheless, the potential ubiquitinated proteins that are changed in size will be further examined by the methods that follow.

Confirmation of Ubiquitinated Proteins by Identified Ubiquitination Sites. Performing a database search with a mass shift of di-glycine tag (114.1 Da) on lysine led to the identification of ubiquitination sites in many proteins (>100) (Peng et al., 2003b). The presence of di-glycine tag prevents C-terminal tryptic cleavage of ubiquitinated lysines. In some cases, incomplete trypsinization may lead to a longer tag of Leu-Arg-Gly-Gly (383.5 Da) on the modified peptides. It should be mentioned that a ubiquitin-like protein Rub1 produces indistinguishable signature peptides from ubiquitin after trypsin digestion (Fig. 4A). It is essential to separate proteins modified by ubiquitin or Rub1 by prefractionation. Similarly, the method can be used to define the lysine residues used for polyUb chain

A

UB_YEAST	LSDYNIQKESTLHLVLRLR*GG*
UB_HUMAN	LSDYNIQKESTLHLVLRLR*GG*
RUB1/NEDD8_YEAST	VTDAHLVEGMQLHLVLTLR*GG*
NEDD8_HUMAN	AADYKILGGSVLHLVLALR*GG*
ISG15/UCRPII_HUMAN	LGEYGLKPLSTVFMNLRLR*GG*
HUB1_YEAST	K*DHISLEDYEVHDQTNLELYY*
HUB1/UBL5_HUMAN	K*DHVSLGDYEIHDGMNLELYY*
SMT3_YEAST	PEDLDMEDNDIIEAHRE*QIGG*
SUMO-1_HUMAN	PK*ELGMEEEDVIEVYQEQTGG*
APG12_YEAST	WMQFK*TNDELIVSYCASVAFG*
APG12_HUMAN	YECFGSDGKLVLHYCK*SQAWG*
APG8_YEAST	QEHKDK*DGFLYVTYSGENTFG*

B

Signature Peptides of Various PolyUb Linkages:

Lys6 of wild type Ub: MQIFVK⁶TLTGK
                           |
                           GG

Lys6 of 6xHis-myc-Ub: LISEEDLGMQIFVK⁶TLTGK
                                 |
                                 GG

Lys11: TLTGK¹¹TITLEVESSDTIDNVK
            |
            GG

Lys27: TITLEVESSDTIDNVK²⁷SK
                       |
                       GG

Lys29: SK²⁹IQDK
         |
         GG

Lys33: IQDK³³EGIPPDQQR
          |
          GG

Lys48: LIFAGK⁴⁸QLEDGR
             |
             GG

Lys63: TLSDYNIQK⁶³ESTLHLVLR
                |
                GG

FIG. 4. (A) The C-terminal amino acid sequences of some proteins in the ubiquitin family. Only the mature forms of the proteins are shown, and the derived tags after trypsin digestion are italic and underlined. Some large tags may be reduced by using other proteases with different substrate specificity. (B) The list of ubiquitinated signature peptides of yeast polyUb chains. Detection of these peptides by mass spectrometry helps resolve the polyUb chain topologies. Mammalian ubiquitin sequence differs at three residues: S19P, D24E, and S28A, which causes the change of four signature peptides (Lys6, Lys11, Lys27, and Lys29).

elongation (Fig. 4B). All seven lysine residues in ubiquitin itself have been suggested to be involved with a relative abundance order of Lys48 > Lys63 and Lys11 > Lys6, Lys27, Lys29, and Lys33 in yeast (Peng *et al.*, 2003b). One technical challenge is that complete modification site mapping requires almost 100% coverage of peptides "sequenced" by tandem mass spectrometry. In our initial large-scale analysis, only a fraction of ubiquitination sites could be detected; because the technique for peptide sequencing (LC-MS/MS) was a "shotgun" approach, we were not able to sequence every peptide from the proteins and could not identify ubiquitinated sites for all ubiquitinated species. The problem can be alleviated by preenrichment of ubiquitinated signature peptides using antibodies that recognize the di-glycine tag (Dr. Steve P. Gygi, personal communication, July 2004).

Validation of Ubiquitinated Proteins Individually Using Western Blotting Experiment. It will be more conclusive to assess the state of protein ubiquitination by an independent approach once the proteins are identified. The protein (or tagged protein) of interest is expressed in cells, immunoprecipitated, resolved on a SDS gel, and immunoblotted. Ub-conjugates are evidenced as high molecular weight species detected by antibodies that react with ubiquitin or differentially tagged ubiquitin. High stringent conditions (e.g., SDS) should be used to minimize coimmunoprecipitation of other ubiquitinated proteins. One should be cautious to interpret the data if the tagged proteins are overexpressed in the cells. More details on characterizing known Ub-conjugates can be found in parallel chapters in this book.

Future Applications

The method described can be used not only to determine the identity of ubiquitinated substrates but also to characterize ubiquitination sites and the length and topology of ubiquitin chains of target proteins, although many techniques need further refinement. We anticipate that this method can serve as a general tool for dissecting the role of ubiquitination in other eukaryotic cells. However, the population of His-rich proteins is much larger in mammalian cells and could be potentially purified as contaminants. A tandem affinity tag may be required for the enrichment of ubiquitinated proteins. Finally, this method can also be adapted to the study of other ubiquitin-like proteins such as SUMO (Li *et al.*, 2004; Panse *et al.*, 2004; Vertegaal *et al.*, 2004; Wohlschlegel *et al.*, 2004; Zhou *et al.*, 2004), ISG15, and Apg12.

Acknowledgments

We thank Drs. D. Finley, J. Roelofs, and G. Marsischky for providing the yeast strains and verifying identified Ub-conjugates by immunoprecipitation, Drs. T. Yao and R. Cohen for sharing ubiquitin proteins to initiate the projects, and C. C. Thoreen, J. E. Elias, D. Schwartz, J. Eng, and D. M. Duong for bioinformatics support. We thank Drs. A. Goldberg, D. Finley, R. Cohen, D. Moazed, S. Gerber, and S. P. Gygi for encouraging discussions. We also thank

Dr. S. P. Gygi for his generous support. This work was partially supported by Jane Coffin Childs Memorial Fund for Medical Research, the start-up funds from Emory University, and NIH grant DK069580 to J.P.

References

Aebersold, R., and Mann, M. (2003). Mass spectrometry-based proteomics. *Nature* **422**, 198–207.

Bennetzen, J. L., and Hall, B. D. (1982). Codon selection in yeast. *J. Biol. Chem.* **257**, 3026–3031.

Chau, V., Tobias, J. W., Bachmair, A., Marriott, D., Ecker, D. J., Gonda, D. K., and Varshavsky, A. (1989). A multiubiquitin chain is confined to specific lysine in a targeted short-lived protein. *Science* **243**, 1576–1583.

Deng, L., Wang, C., Spencer, E., Yang, L., Braun, A., You, J., Slaughter, C., Pickart, C., and Chen, Z. J. (2000). Activation of the IkappaB kinase complex by TRAF6 requires a dimeric ubiquitin-conjugating enzyme complex and a unique polyubiquitin chain. *Cell* **103**, 351–361.

Eng, J., McCormack, A. L., and Yates, J. R., 3rd. (1994). An approach to correlate tandem mass spectral data of peptides with amino acid sequences in a protein database. *J. Am. Soc. Mass Spectrom.* **5**, 976–989.

Gygi, S. P., Rist, B., Griffin, T. J., Eng, J., and Aebersold, R. (2002). Proteome analysis of low abundance proteins using multidimensional chromatography and isotope coded affinity tags. *J. Proteome Res.* **1**, 47–54.

Haglund, K., Di Fiore, P. P., and Dikic, I. (2003). Distinct monoubiquitin signals in receptor endocytosis. *Trends Biochem. Sci.* **28**, 598–603.

Hicke, L., and Dunn, R. (2003). Regulation of membrane protein transport by ubiquitin and ubiquitin-binding proteins. *Annu. Rev. Cell Dev. Biol.* **19**, 141–172.

Hoege, C., Pfander, B., Moldovan, G. L., Pyrowolakis, G., and Jentsch, S. (2002). RAD6-dependent DNA repair is linked to modification of PCNA by ubiquitin and SUMO. *Nature* **419**, 135–141.

Li, T., Evdokimov, E., Shen, R. F., Chao, C. C., Tekle, E., Wang, T., Stadtman, E. R., Yang, D. C., and Chock, P. B. (2004). Sumoylation of heterogeneous nuclear ribonucleoproteins, zinc finger proteins, and nuclear pore complex proteins: A proteomic analysis. *Proc. Natl. Acad. Sci. USA* **101**, 8551–8556.

Mann, M., and Jensen, O. N. (2003). Proteomic analysis of post-translational modifications. *Nat. Biotechnol.* **21**, 255–261.

Marotti, L. A., Jr., Newitt, R., Wang, Y., Aebersold, R., and Dohlman, H. G. (2002). Direct identification of a G protein ubiquitination site by mass spectrometry. *Biochemistry* **41**, 5067–5074.

Panse, V. G., Hardeland, U., Werner, T., Kuster, B., and Hurt, E. (2004). A proteome-wide approach identifies sumolyated substrate proteins in yeast. *J. Biol. Chem.* **279**, 41346–41351.

Peng, J., Elias, J. E., Thoreen, C. C., Licklider, L. J., and Gygi, S. P. (2003a). Evaluation of multidimensional chromatography coupled with tandem mass spectrometry (LC/LC-MS/MS) for large-scale protein analysis: The yeast proteome. *J. Proteome Res.* **2**, 43–50.

Peng, J., and Gygi, S. P. (2001). Proteomics: The move to mixtures. *J. Mass Spectrom.* **36**, 1083–1091.

Peng, J., Schwartz, D., Elias, J. E., Thoreen, C. C., Cheng, D., Marsischky, G., Roelofs, J., Finley, D., and Gygi, S. P. (2003b). A proteomics approach to understanding protein ubiquitination. *Nat. Biotechnol.* **21**, 921–926.

Perkins, D. N., Pappin, D. J., Creasy, D. M., and Cottrell, J. S. (1999). Probability-based protein identification by searching sequence databases using mass spectrometry data. *Electrophoresis* **20,** 3551–3567.

Pickart, C. M. (2004). Back to the future with ubiquitin. *Cell* **116,** 181–190.

Shevchenko, A., Wilm, M., Vorm, O., and Mann, M. (1996). Mass spectrometric sequencing of proteins silver-stained polyacrylamide gels. *Anal. Chem.* **68,** 850–858.

Spence, J., Gali, R. R., Dittmar, G., Sherman, F., Karin, M., and Finley, D. (2000). Cell cycle-regulated modification of the ribosome by a variant multiubiquitin chain. *Cell* **102,** 67–76.

Spence, J., Sadis, S., Haas, A. L., and Finley, D. (1995). A ubiquitin mutant with specific defects in DNA repair and multiubiquitination. *Mol. Cell Biol.* **15,** 1265–1273.

Steen, H., and Mann, M. (2004). The abc's (and xyz's) of peptide sequencing. *Nat. Rev. Mol. Cell Biol.* **5,** 699–711.

Thrower, J. S., Hoffman, L., Rechsteiner, M., and Pickart, C. M. (2000). Recognition of the polyubiquitin proteolytic signal. *EMBO J.* **19,** 94–102.

Vertegaal, A. C., Ogg, S. C., Jaffray, E., Rodriguez, M. S., Hay, R. T., Andersen, J. S., Mann, M., and Lamond, A. I. (2004). A proteomic study of SUMO-2 target proteins. *J. Biol. Chem.* **279,** 33791–33798.

Washburn, M. P., Wolters, D., and Yates, J. R., 3rd. (2001). Large-scale analysis of the yeast proteome by multidimensional protein identification technology. *Nat. Biotechnol.* **19,** 242–247.

Weissman, A. M. (2001). Themes and variations on ubiquitylation. *Nat. Rev. Mol. Cell Biol.* **2,** 169–178.

Wilkinson, K. D. (2000). Ubiquitination and deubiquitination: Targeting of proteins for degradation by the proteasome. *Semin. Cell Dev. Biol.* **11,** 141–148.

Wohlschlegel, J. A., Johnson, E. S., Reed, S. I., and Yates, J. R., 3rd. (2004). Global analysis of protein sumoylation in *Saccharomyces cerevisiae*. *J. Biol. Chem.* **279,** 45662–45668.

Yates, J. R., 3rd. (2004). Mass spectral analysis in proteomics. *Annu. Rev. Biophys. Biomol. Struct.* **33,** 297–316.

Zhou, W., Ryan, J. J., and Zhou, H. (2004). Global analyses of sumoylated proteins in *Saccharomyces cerevisiae*. Induction of protein sumoylation by cellular stresses. *J. Biol. Chem.* **279,** 32262–32268.

Section V

Genome and Proteome-wide Approaches to Identify Substrates and Enzymes

[26] Two-Step Affinity Purification of Multiubiquitylated Proteins from *Saccharomyces cerevisiae*

By THIBAULT MAYOR and RAYMOND J. DESHAIES

Abstract

In budding yeast and higher eukaryotic genomes, there are, respectively, 50 and up to 400 or more distinct genes that encode for ubiquitin-ligases, and ~15–90 genes that encode for ubiquitin isopeptidases (TM and RJD, Semple *et al.*, 2003). This puts ubiquitylation on par with phosphorylation as the most common reversible posttranslational modifications in eukaryotic cells. A key challenge that has met with limited success to date is to identify the proteins that are the substrates for this large collection of enzymes. To begin to address this daunting challenge, we sought to identify ubiquitylated proteins that are potential substrates of the 26S proteasome. Here, we describe a two-step affinity purification protocol that uses a budding yeast strain that expresses hexahistidine-tagged ubiquitin. In the first step, native cell lysate was chromatographed on a UBA domain-containing matrix that binds preferentially to K48-linked multiubiquitin chains. Free ubiquitin and presumably monoubiquitylated proteins did not bind this column, whereas proteins that are potential substrates of the proteasome were enriched. In the second step, UBA domain–binding proteins were subjected to immobilized metal ion affinity chromatography (IMAC) under denaturing conditions on magnetic nickel beads, resulting in >3000-fold enrichment of ubiquitin conjugates relative to crude cell extract.

Yeast Strain Design

A ubiquitin gene modified to encode an amino-terminal hexahistidine tag (H6-ubiquitin) was integrated into the *TRP1* locus of *Saccharomyces cerevisiae* W303 strain [*MATalpha, can1-100, his3-11,-15, trp1-1, ura3-1, ade2-1, leu2-3, -112*]. RDB1848 was created by placing H6-ubiquitin coding sequences between the GPD constitutive promoter and PGK terminator sequences of the modified yeast expression vector pG-1 (Schena *et al.*, 1991) lacking the 2μ region (excised with EcoRI). Two primers (5′ GCGGATCCATGAGAGGTAGTCATCATCACCATCATCACGG TGGTATGCAGATTTTCGTCAAGACT 3′ and 5′ GAGCTCGAGAC-CACCTCTTAGCCTTAGCAC 3′) were used to amplify by PCR the yeast

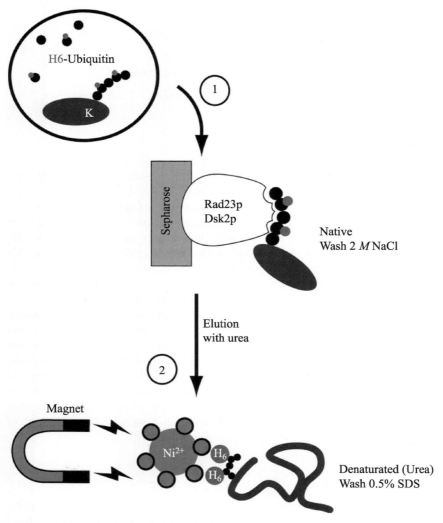

FIG. 1. Flow diagram for the two-step purification of multiubiquitin-conjugates. Yeast cells expressing a hexahistidine-tagged ubiquitin are lysed in a nondenaturing buffer, and proteins conjugated to multiubiquitin chains are purified using UBA domain–containing proteins (1). Proteins bound to the resin are washed with 2 M NaCl and eluted with urea. Denatured proteins are then mixed with magnetic nickel beads. Stringent washing conditions (up to 0.5% SDS) are used to wash away contaminants (2).

ubiquitin gene (first ubiquitin of the *UBI4* locus). The PCR fragment was digested with *Bam*HI and *Xho*I and ligated into the modified pG-1 (digested with *Bam*HI and *Sal*I). RDB1848 was then digested with *Eco*RV (which cleaves within *TRP1* sequences on the plasmid) and transformed into W303. Expression of H6-ubiquitin in RJD2997 was confirmed by Western blotting and corresponded to a 1:1 ratio with the unmodified ubiquitin (see Fig. 2B, first lane). The control strain RJD2998 was obtained by integrating the empty pG-1 vector into the *TRP1* locus.

Immobilization of Multiubiquitin Chain-Binding
  Proteins on Sepharose

We used recombinant Rad23p and Dsk2p proteins for the first step in our affinity purification scheme. Both proteins contain UBA domains that bind K48-linked ubiquitin chains with greater affinity than monoubiquitin or K63-linked chains (Raasi *et al.*, 2004; Wilkinson *et al.*, 2001). The glutathione S-transferase (GST) fusions of Dsk2 (pGEX-KG) and Rad23 (pGEX-6P1) were generous gifts from H. Kobayashi and H. Yokosawa, respectively. Fusion proteins were induced in BL21(DE3)/pLysS for 4 h at 30° with 1 mM IPTG and purified in sodium phosphate/Triton X-100 lysis buffer with Sepharose 4B resin following the manufacturer's instructions (Amersham Biosciences). Typically, 1 liter of induced bacterial culture yielded 10 to 15 mg of recombinant protein. Purified proteins were then dialyzed into 100 mM NaHCO$_3$, pH8.3, 0.5 M NaCl and coupled to CNBr-activated Sepharose 4B resin (Amersham Biosciences, ~10 mg of GST-Dsk2p and ~20 mg of GST-Rad23 were coupled each separately to 1.5 ml of resin). The recombinant proteins coupled to the resin could be stored for several months at 4° in a 50% slurry with 100 mM TrisHCl, pH8.0, 0.5 M NaCl, 0.02% NaN$_3$.

Two-Step Purification of Conjugates Bearing a Multiubiquitin Chain

1. The protocol described here is for 1 liter of yeast culture. Grow the cells at 25° in YPD (1% yeast extract, 2% peptone, 2% dextrose, optional 25 μg/ml ampicillin) until the OD$_{600}$ reaches 1–1.5. Collect the cells in a 1-liter centrifuge bottle (Nalgene) using the H-6000A/HBB-6 rotor in a RC-3B Sorvall centrifuge run at 5000 rpm for 10 min. Wash the cells first with 200 ml ice-cold TBS (150 mM NaCl, 50 mM TrisHCl, pH 7.5), and then with 100 ml ice-cold TBS, 1 mM 1,10-phenanthroline (Sigma, P9375), 10 mM iodoacetamide (Sigma, I-1149, alternatively use 5 mM NEM, Sigma, E-3878). The latter two compounds are added to inhibit zinc- and cysteine-dependent ubiquitin isopeptidases. Iodoacetamide is preferred

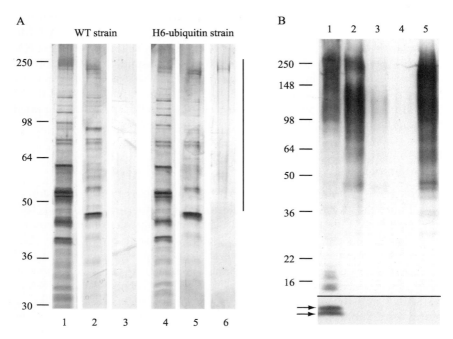

FIG. 2. (A) Silver staining of the two-step purification. Purifications were performed using the WT control strain (lanes 1–3) or the H6-ubiquitin expressing strain (lanes 4–6). Aliquots of total cell extract (lanes 1, 4), proteins eluted after the first step of the purification (lanes 2, 5), and proteins bound to the nickel magnetic beads (lanes 3, 6) were separated onto a 4–20% Tris-glycine PAGE. The fraction of each sample loaded was 4×10^{-7} (lanes 1, 4), 1/600 (lanes 2, 5), and 1/60 (lanes 3, 6). Lane 1 corresponds to 1 μg of loaded proteins. All the lanes were processed in the same way. (B) Immunoblotting of the two-step purification. Aliquot of total cell extract (lane 1), elution after the first step (lane 2), material that did not bind to the magnetic beads (lane 3), proteins washed by UB buffer containing 0.5% SDS (lane 4), and purified proteins after the second step of the purification (lane 5) were separated on a 4–20% Tris-glycine SDS-polyacrylamide gel, transferred to a nitrocellulose membrane, and blotted with an antiubiquitin antibody (mAb1510, ICN). The lower part of the gel corresponds to a lower exposure to allow visualization of distinct monoubiquitin and H6-ubiquitin bands. Lane 1 corresponds to a 10% input, lanes 4 and 5–5x more concentrated samples as compared with lanes 2 and 3. Note that only 25–30% of the purified ubiquitin conjugates were released from the magnetic beads in SDS-PAGE sample buffer.

over NEM if the sample is to be analyzed by mass spectometry. Freeze the cell pellet in liquid N_2 and store at $-80°$.

2. Thaw the cell pellet on ice for 5 min and add 14 ml of cold lysis buffer (50 mM sodium phosphate, pH 8.0, 300 mM NaCl, 0.5% Triton X-100, 0.5 mM AEBSF (MP Biochemicals, 193503), 5 μg/ml aprotinin,

5 μg/ml chymostatin, 5 μg/ml leupeptin, 1 μg/ml pepstatin A, 1 mM 1,10-phenanthroline, 10 mM iodoacetamide (or 5 mM NEM)). Lyse cells using a One Shot Cell Disrupter (Constant Systems) by applying 30,000 psi (foaming might occur). Collect the lysate as rapidly as possible and centrifuge at 4° in a Sorvall SS34 rotor for 20 min at 14,000 rpm. Alternately, cells can be lysed using the previously described mortar and pestle grinding method with liquid nitrogen (Verma *et al.*, 2000). Perform all the subsequent steps at 4°.

3. After clearing, collect the supernatant (typically 150–200 mg of proteins) in a 15-ml polypropylene conical tube and add 0.5 mg of each GST-Rad23 and GST-Dsk2 coupled to Sepharose (preequilibrated with lysis buffer). Mix gently for 90 min at 4° using a rotating wheel.

4. Centrifuge to pellet the resin. Use low velocity when centrifuging to prevent crushing of the resin (typically 1000 rpm for 5 min in a clinical centrifuge). Remove the buffer and wash the resin with 14 ml lysis buffer. Centrifuge, and then remove the buffer and add 14 ml of 50 mM sodium phosphate, pH 8.0, 2 M NaCl (avoid storing the 2 M NaCl solution directly on ice or alternately use a 1 M NaCl solution to prevent salt precipitation). Mix for 15 min using a rotating wheel. Centrifuge, remove the buffer, and perform one wash with 14 ml of 50 mM sodium phosphate, pH 8.0, 2 M NaCl, and two more washes with 14 ml of 50 mM sodium phosphate, 300 mM NaCl, 0.1% Triton X-100. If necessary, the washes can be performed in a gravity-flow column (e.g., Econo-Pac column, Bio-Rad).

5. The following steps are performed at room temperature. Elute the multiubiquitin conjugates by adding 2 resin-volumes of urea buffer (UB; 8 M urea, 100 mM NaH$_2$PO$_4$, 10 mM TrisHCl, pH 8.0). Mix well, centrifuge, and collect the supernatant. Repeat the procedure once and mix together the two elution fractions in an Eppendorf tube. Centrifuge for 2 min at 10,000 rpm in a Microfuge to pellet the residual resin and collect the eluate.

6. Add imidazole to the eluate to a final concentration of 20 mM. Add 25 μl of a magnetic nickel bead slurry (Promega, V8565) prewashed in the urea buffer (the actual bead volume should be around 7–8 μl). Mix gently for 60 min on a rotating wheel. Adapt the size of the tube (0.5, 1.5, or 2 ml) to the eluate volume to minimize loss of the nickel beads on the tube surface.

7. Wash the beads three times with 1 eluate-volume of UB. For the washes, place the tube in a magnetic rack for 10 sec, remove the supernatant, remove the tube from the rack, and mix the beads with a fresh aliquot of wash buffer.

8. If nonspecific background is a major concern, the wash protocol described here can be substituted for step 7. Remove the supernatant and

mix the beads with 1 eluate-volume of UB supplemented with 0.5% SDS. Incubate the beads for 15 min with gentle mixing and remove buffer. For mass spectrometry analysis, the SDS is extracted with Triton X-100. Wash the beads briefly with 1 eluate-volume of urea buffer with 0.5% Triton X-100, remove buffer, and add again 1 eluate-volume of urea buffer with 0.5% Triton X-100. Incubate for 15 min while mixing gently. Remove the buffer and repeat the same procedure with plain UB to remove the Triton. This detergent wash step may decrease the yield of the purification up to 20% but will greatly reduce the background.

9. After the washes, the proteins can be eluted with 500 mM imidazole or in sample buffer for PAGE analysis. For mass spectrometry analysis, we directly performed tryptic digests with the proteins bound to the beads (Mayor *et al.*, 2005, see following).

Comments

We performed the two-step purification in parallel with a strain expressing H6-ubiquitin and a control strain that only expresses endogenous ubiquitin. Although a similar amount of proteins was recovered from both strains after the first step, no signal was detected in the control strain preparation on silver staining after the second step (Fig. 2A). By contrast, when the strain that expresses H6-ubiquitin was used, silver staining revealed a continuous spread of proteins ranging from 50–250 kDa. This is expected, because a collection of proteins of different molecular weights conjugated to multiubiquitin chains of variable lengths should produce a spread rather than a pattern composed of discrete bands.

In the first step of the purification, approximately 15% of the high molecular weight ubiquitin conjugates in the cell were recovered (Fig. 2B). It is important to note that ubiquitin conjugates were enriched, whereas the highly abundant monoubiquitin was not recovered. When starting with 150 mg of protein, we estimated that 20–50 μg of protein were eluted after the first step of purification, and some 3–5 μg remained bound to the nickel beads after the IMAC step (representing about 10% of ubiquitin conjugates in the cell). This represented a 3000- to 5000-fold enrichment for ubiquitin conjugates.

Our method provides an effective approach to purify conjugates that bear a multiubiquitin chain and can be adapted to study other related pathways. For example, purification can be biased in favor of monoubiquitylated proteins by using CUE or UIM domains instead of the UBA domain in the first step (Donaldson *et al.*, 2003; Kang *et al.*, 2003; Shih *et al.*, 2002). A related method for the purification of ubiquitylated proteins has been described by Gygi and colleagues (Peng *et al.*, 2003; see Chapter 25 in this volume). Our approach differs from that reported by Peng *et al.* in that their method uses a single-step purification of ubiquitylated proteins by IMAC, starting with a strain that contains H6-ubiquitin as the only source of ubiquitin. By contrast, our method uses a strain that expresses normal ubiquitin in addition to H6-ubiquitin and involves two consecutive affinity purification steps. The relative benefit of using a strain in which the tagged ubiquitin is the only source of ubiquitin, or is expressed in addition to normal ubiquitin, is not clear. On the one hand, our approach leads to an approximately two-fold overproduction of ubiquitin, which could potentially cause artifacts. On the other hand, in our experience tagged forms of ubiquitin are clearly less active than natural ubiquitin when the two are compared in reconstitution systems, and thus there may be less overall perturbation to the ubiquitin-proteasome system (UPS) if the tagged ubiquitin is diluted by the presence of natural ubiquitin (Rati Verma and RJD, personal communications). Indeed, we did not observe any increase of cadmium chloride (3 mM) sensitivity after integration of H6-ubiquitin (data not shown). By contrast, we believe that the inclusion of a purification step before IMAC is important, because even under denaturing conditions a significant fraction (0.5–1%) of total yeast extract proteins bind to Ni^{2+}–NTA resin. By using UBA domain affinity chromatography in tandem with IMAC, we were able to greatly enrich for multiubiquitin chain conjugates (>3000-fold).

When we analyzed the ubiquitylated proteins purified from 6 liters of culture by multidimensional chromatography-mass spectrometry (MudPIT or LC/LC-MS/MS), we routinely identified 150–200 proteins (Mayor *et al.*, 2005). Subsequent validation analyses enabled us to confirm the presence of a multiubiquitin chain on 7 proteins from a pool of 10 tested candidates expressed at their endogenous level. This strongly suggests that most of the purified proteins are *bona fide* substrates of the UPS. Using this purification method, we were also able to identify ubiquitylated proteins that specifically accumulated when part of the UPS system was impaired (i.e., deletion of the proteasome substrate receptor Rpn10). From this effort, we identified the cell cycle regulator Sic1 and the transcriptional activator Gcn4 as being candidate ligands for Rpn10. The fact that these two UPS targets are known to be present at low abundance underscores the power of our two-step affinity purification method (Mayor *et al.*, 2005).

Acknowledgments

We thank K. R. Yamamoto, H. Kobayashi, and H. Yokosawa for providing reagents, laboratory members for helpful discussions and support, and G. Kleiger for a critical reading of the manuscript. This project was supported by the Swiss National Science Foundation and EMBO fellowships.

References

Donaldson, K. M., Yin, H., Gekakis, N., Supek, F., and Joazeiro, C. A. (2003). Ubiquitin signals protein trafficking via interaction with a novel ubiquitin binding domain in the membrane fusion regulator, Vps9p. *Curr. Biol.* **13**, 258–262.

Kang, R. S., Daniels, C. M., Francis, S. A., Shih, S. C., Salerno, W. J., Hicke, L., and Radhakrishnan, I. (2003). Solution structure of a CUE-ubiquitin complex reveals a conserved mode of ubiquitin binding. *Cell.* **113**, 621–630.

Mayor, T., Lipford, J. R., Graumann, J., Smith, G. T., and Deshaies, R. J. (2005). Analysis of polyubiquitin conjugates reveals that the Rpn10 substrate receptor contributes to the turnover of multiple proteasome targets. *Mol. Cell. Proteomics* **4**, 741–751.

Peng, J., Schwartz, D., Elias, J. E., Thoreen, C. C., Cheng, D., Marsischky, G., Roelofs, J., Finley, D., and Gygi, S. P. (2003). A proteomics approach to understanding protein ubiquitination. *Nat. Biotechnol.* **21**, 921–926.

Raasi, S., Orlov, I., Fleming, K. G., and Pickart, C. M. (2004). Binding of polyubiquitin chains to ubiquitin-associated (UBA) domains of HHR23A. *J. Mol. Biol.* **341**, 1367–1379.

Schena, M., Picard, D., and Yamamoto, K. R. (1991). Vector for constitutive and inducible gene expression in yeast. "Method in Enzymology." Vol. 194, pp. 389–398. Academic Press, San Diego.

Semple, C. A. M., RIKEN GER Group and GSL Members (2003). The comparative proteomics of ubiquitination in mouse. *Genome Res.* **13**, 1389–1394.

Shih, S. C., Katzmann, D. J., Schnell, J. D., Sutanto, M., Emr, S. D., and Hicke, L. (2002). Epsins and Vps27p/Hrs contain ubiquitin-binding domains that function in receptor endocytosis. *Nat. Cell Biol.* **4**, 389–393.

Verma, R., Chen, S., Feldman, R., Schieltz, D., Yates, J., Dohmen, J., and Deshaies, R. J. (2000). Proteasomal proteomics: Identification of nucleotide-sensitive proteasome-interacting proteins by mass spectrometric analysis of affinity-purified proteasomes. *Mol. Biol. Cell.* **11**, 3425–3439.

Wilkinson, C. R., Seeger, M., Hartmann-Petersen, R., Stone, M., Wallace, M., Semple, C., and Gordon, C. (2001). Proteins containing the UBA domain are able to bind to multiubiquitin chains. *Nat. Cell Biol.* **3**, 939–943.

[27] Identification of SUMO–Protein Conjugates

By Meik Sacher, Boris Pfander, and Stefan Jentsch

Abstract

Modification of proteins by covalent attachment of ubiquitin and the ubiquitin-like modifier SUMO are widespread regulatory events of all eukaryotic cells. SUMOylation has received much attention, because several identified targets play prominent roles, in particular, in cell signaling, gene expression, and DNA repair. Notably, only a very small fraction of a substrate is usually SUMOylated at steady-state levels, which could be because modification is reversible and transient. Because of the low level of modification, SUMOylated proteins are often overlooked or sometimes misinterpreted as a less important fraction of a protein pool. Here we discuss procedures that can circumvent identification problems and describe methods for their verification.

Introduction

Posttranslational modification of proteins is an important means to alter their function (e.g., their activity or intracellular localization). A particularly striking example is the modification of proteins by the covalent attachment of the small protein ubiquitin. In addition to its well-known function to earmark proteins for degradation by the 26S proteasome, modification by ubiquitin ("ubiquitylation") serves a variety of different functions (e.g., in protein sorting, transcriptional activation, or DNA repair; for a review see Pickart [2001]). A prerequisite for proteasomal degradation is usually the modification of the protein by a specific multiubiquitin chain ("multiubiquitylation"), in which several ubiquitin moieties are linked through isopeptide bonds via lysine (K) residue 48 of ubiquitin (other lysine residues, e.g., K29, are less frequently used). In contrast, modification of proteins by a single ubiquitin moiety ("monoubiquitylation") or by K63-linked multiubiquitin chains mediates nonproteolytic functions (Pickart, 2000). Recent studies suggest that the different types of ubiquitin modifications recruit distinct ubiquitin-binding factors, which in turn seem to dictate the different functions of ubiquitin.

Modification of proteins by covalent attachment of another protein is more common than initially assumed. Since 1997, several "ubiquitin-like modifiers" (UBLs) have been discovered, which share different degrees of sequence identity with ubiquitin (for a review see Jentsch and Pyrowolakis [2000]). Among these, the modifier SUMO ("small ubiquitin-like modifier") has probably received the widest attention (for a review on SUMO, see Johnson [2004]; Melchior [2000]; and Müller et al. [2001]). This interest stems from the fact that modification by SUMO ("SUMOylation") is rather prevalent (only surpassed by ubiquitylation) and that several known SUMOylation targets are proteins of singular interest (e.g., p53, IκB, PCNA). SUMO is approximately 18% identical in sequence to ubiquitin and highly conserved across species. In the budding yeast, *Saccharomyces cerevisiae*, SUMO (Smt3) is essential for viability. In the fission yeast, *Schizosaccharomyces pombe*, SUMO (Pmt3) is not essential, but deletion mutants exhibit severe growth defects. Both budding and fission yeasts have only one form of SUMO. In contrast, humans have four isoforms:

SUMO-1, the highly similar proteins SUMO-2 and SUMO-3, and SUMO-4. These isoforms differ mainly in their amino (N)-terminal extensions, but whether they mediate different functions is currently unclear. SUMOylation occurs both in the cytosol and the nucleus. However, most identified SUMO targets are nuclear proteins, and, indeed, several nuclear events are modulated by SUMOylation. These include intranuclear protein sorting (e.g., localization to PML bodies), transcriptional silencing, and DNA repair. In a few cases, in particular if the same lysine residue of a target protein can be alternatively modified by either ubiquitin or SUMO, SUMOylation may compete with ubiquitin-dependent function. For example, in IκB, SUMOylation seems to inhibit ubiquitin-dependent degradation (Desterro *et al.*, 1998), whereas in PCNA, SUMOylation and ubiquitylation mediate alternative S-phase–dependent functions (Hoege *et al.*, 2002). How SUMO mediates the different functions is currently unknown. It is believed that at least in some cases, specific SUMO-conjugate binding proteins may be recruited (analogous to the ubiquitin system; see earlier) that mediate the SUMO-dependent functions.

Analogous to ubiquitin, the primary translation product of SUMO is a precursor, which is processed to the mature form by a SUMO-specific protease (Ulp1 in *S. cerevisiae*; Li and Hochstrasser, 1999). This reaction exposes a di-glycine sequence at SUMO's carboxy (C)-terminus through which SUMO is attached to ε-amino groups of lysine residues of target proteins. Conjugation of mature SUMO to target proteins proceeds by an enzymatic reaction that resembles ubiquitylation (Johnson, 2004). It is first activated in an ATP-dependent reaction by an (E1) activating enzyme (Aos1/Uba2 heterodimer in *S. cerevisiae*), which forms a thioester-linked complex between a cysteine residue of itself and SUMO's C-terminus. Activated SUMO is then transferred on a cysteine residue of the (E2) enzyme Ubc9, which resembles ubiquitin-conjugating enzymes in sequence. Ubc9 usually binds SUMOylation targets directly; however, the transfer of SUMO to the target protein is often stimulated by so-called SUMO ligases. Three unrelated types of SUMO ligases have been reported: proteins of the Siz/PIAS family, RanBP2, and PC2 (Johnson, 2004). In most cases, SUMO is attached to target proteins as a single molecule, albeit sometimes simultaneously at several residues of the substrate (Hoege *et al.*, 2002). Short multi-SUMO chains have also been reported, but the functional significance of this modification is currently unclear (Bylebyl *et al.*, 2003). However, there is so far no indication that mono-SUMOylation and multi-SUMOylation serve different functions (in contrast to ubiquitylation; see earlier).

SUMOylation is a reversible modification. The removal of SUMO from conjugates is catalyzed by specific de-SUMOylation enzymes. In *S. cerevisiae*, two of these enzymes have been described, Ulp1 and Ulp2, with Ulp1 serving additionally as a SUMO-precursor processing enzyme (Li and Hochstrasser, 1999, 2000; Schwienhorst et al., 2000). Contrasting ubiquitin, the pool of free SUMO appears rather small, suggesting that de-SUMOylation enzymes may play an important regulatory role.

Properties of SUMO Conjugates

Studies of SUMO conjugates should take a number of typical characteristics of SUMOylation into consideration. Because of these features, a casual experimental analysis can lead to misinterpretations or even a failure of detection. In the following we will compile and discuss the most relevant aspects.

First, and most strikingly, usually only a very small fraction of a SUMO substrate is modified by SUMO at steady state. This small fraction should not be misinterpreted as a minor, less important pool, but probably it rather reflects the reversible nature of the SUMO modification system. SUMOylation is often highly regulated and therefore restricted to very specific cellular events. For example, SUMOylation of the DNA-polymerase processivity factor PCNA in yeast occurs exclusively in S-phase and disappears at later stages of the cell cycle (probably by deconjugation) (Hoege et al., 2002). Moreover, only a fraction of yeast PCNA that is actively coupled to replication seems to be modified by SUMO.

Second, a SUMO target is sometimes modified at more than one lysine residue of the protein (Hoege et al., 2002). This can lead to multiple bands in an SDS electrophoresis gel, which even further reduces the detectable levels of modified protein species. A "ladder" of SUMO–protein conjugates in a gel may, therefore, not necessarily be due to a multi-SUMOylation of the substrate (see earlier), but rather could be a consequence of mono-SUMOylation at several sites. Importantly, experimental removal of one acceptor lysine residue of a substrate can lead to an induction of SUMOylation at another lysine residue. Moreover, as in the case for yeast PCNA, the protein may be alternatively modified by ubiquitin (Hoege et al., 2002). However, a protein modified simultaneously by ubiquitin and SUMO has not yet been described, and mixed SUMO-ubiquitin chains have so far not been reported to occur *in vivo*.

Third, the mere absence of so-called consensus sites for SUMOylation in a protein does not formally exclude the possibility for substrate SUMOylation. This consensus sequence, ΨKXE (Ψ being an aliphatic residue, often L, I, V), has been found in many SUMOylated proteins, and it was, therefore, postulated to be a criterion for SUMOylation (Johnson, 2004). Indeed, structural studies have shown that this sequence makes direct contacts with the SUMO-conjugation enzyme Ubc9, which transfers SUMO onto the lysine residue within this sequence (Bernier-Villamor et al., 2002). It is important to emphasize, however, that several proteins have been identified recently that are modified at non-consensus sites as well. For example, yeast PCNA is modified at two lysine residues, K127 and K164, and only one of them (K127) lies within the consensus sequence (Hoege et al., 2002). As a matter of fact, a larger fraction of PCNA is modified at the non-consensus site residue K164 compared with K127. An obvious corollary of this finding is that experimental attempts in identifying SUMOylation sites in a given protein should not be restricted a priori to consensus sites.

Fourth, and for an experimental analysis most crucial, SUMO modifications are extremely rapidly cleaved by the activity of de-SUMOylation enzymes (e.g., Ulp1) after cell disruption. Therefore, special care must be undertaken to avoid deconjugation during preparation. To demonstrate to what extent the detectable SUMO-conjugate pattern can vary, we tested different lysis conditions and strains (Fig. 1). We used either wild-type (WT) cells (lanes 1, 3, and 5) or the *ulp1-ts* strain (lanes 2, 4, and 6), which is partially defective in de-SUMOylation (Li and Hochstrasser, 1999). Cells were lysed under native conditions by glass bead lysis in the absence (lanes 3 and 4) or presence (lanes 5 and 6) of NEM and were incubated on ice for 2 h, or they were lysed under denaturing conditions (lanes 1 and 2). SUMO conjugates are visualized by immunoblotting using an anti-SUMO antibody. Evidently, SUMOylation of cellular proteins is best preserved under denaturing conditions (see later) in the *ulp1-ts* background (lane 2). To maintain SUMOylation under native conditions, it seems absolutely necessary to prevent de-SUMOylation by either taking advantage of the *ulp1-ts* background (lane 4), by applying NEM that inhibits Ulp activity (lane 5), or by a combination of both means (lane 6). Otherwise, virtually all SUMO conjugates will be de-SUMOylated on lysis (lane 3). It is reasonable to assume that this extensive de-SUMOylation is due to a release of Ulp1 from the nuclear pore to which its activity is usually confined (Schwienhorst et al., 2000; Takahashi et al., 2000). To illustrate the preceding findings for a known SUMO substrate, we show an anti-PCNA blot of the identical samples.

Procedures

The following describes a large-scale two-step affinity approach to purify SUMO-conjugates that makes use of expressed His-tagged SUMO (^{HIS}SUMO). In the first step of the protocol, ^{HIS}SUMO and its conjugates are pulled down with Ni-NTA agarose. This purification works and should

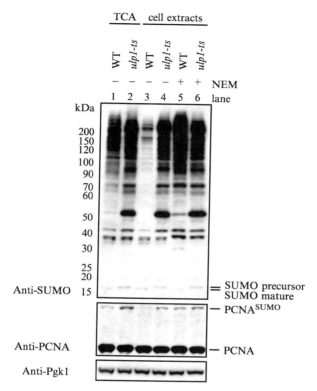

FIG. 1. Comparison of lysis conditions and strain backgrounds to maintain SUMO-protein conjugates. Cells were lysed under denaturing conditions (TCA, lanes 1 and 2) or under native conditions by glass bead lysis in the absence (lanes 3 and 4) or presence (lanes 5 and 6) of 10 mM NEM and were incubated for 2 h on ice. SUMO and SUMO conjugates are visualized by immunoblotting using an anti-SUMO antibody. PCNA SUMOylation is shown in an anti-PCNA Western blot. The anti-Pgk1 Western blot serves as a loading control. For each condition, either the WT strain (lanes 1, 3, and 5) or the *ulp1-ts* strain (lanes 2, 4, and 6) was used, which is partially deficient in de-SUMOylation. SUMO conjugates can be preserved under native conditions by either adding NEM or by using the *ulp1-ts* strain.

be performed under denaturing conditions to prevent de-SUMOylation and proteolysis (furthermore, proteins are quantitatively precipitated and solubilized). In a second step, after elution and exchange of buffers, the ^{HIS}SUMO-conjugates are collected by immunoprecipitation (IP) using SUMO-specific antibodies (Fig. 2A).

For the purification of SUMO-protein conjugates from *S. cerevisiae* (Hoege *et al.*, 2002), we used a mutant strain that carried a temperature-sensitive mutation in *ULP1* (*ulp1-333;* Li and Hochstrasser, 1999) that was

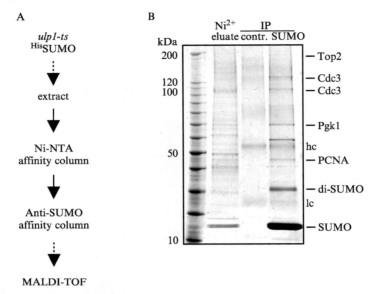

FIG. 2. Purification of SUMO-protein conjugates. (A) Scheme of a two-step strategy for purifying SUMOylated yeast proteins. From an *ulp1-ts* strain that expresses His-tagged SUMO, an extract is made under denaturing conditions. The SUMO–protein conjugates are first purified by a Ni-NTA affinity chromatography and subsequently under native conditions by an anti-SUMO IP. The purified proteins are separated on a SDS polyacrylamide gel and investigated by mass spectrometry. (B) Pattern of SUMO-protein conjugates visualized by Coomassie blue staining. Shown are either an eluate of the Ni-NTA column (step 1) or the material of the anti-SUMO or control IP (step 2). Proteins that are specifically enriched by the anti-SUMO IP were analyzed by mass spectrometry.

transformed with a linearized integrative plasmid expressing 6His-tagged SUMO from the *ADH1*-promoter. Cells were grown at 23° in YPD, and 6 liters of a culture of OD 1 were harvested, washed with water, pelleted by centrifugation, frozen in liquid nitrogen, and stored at −80°. The frozen pellet was resuspended in cold lysis buffer (1.85 M NaOH, 7.5% β-mercaptoethanol) to a final volume of 150 ml (about 1 ml of lysis buffer per 30–40 ODs). After 20 min on ice, an equal volume of cold 55% trichloroacetic acid (TCA) was added and kept on ice for another 20 min. Proteins were pelleted by centrifugation at 2500g for 20 min (4°), and the pellet was washed once with cold water. Subsequently, the pellet was thoroughly resuspended in 150 ml of buffer A plus 0.05% Tween 20 (buffer A: 6 M guanidine-HCl, 100 mM NaH$_2$PO$_4$, 10 mM Tris·Cl; pH adjusted to pH 8.0 with NaOH). After transfer into centrifuge tubes, the suspensions were shaken for 1 h at 180 rpm at room temperature for resolubilization. Insoluble material was removed by centrifugation (20 min, 4°, 15,000g), and the clear supernatant was transferred into a fresh tube. After adding imidazole to a final concentration of 20 mM, 2 ml of Ni-NTA agarose (Qiagen) was added to the tubes for the protein pull-down, and the sample was incubated on a rotating wheel at 4° overnight. The material was then poured into a column, washed with 20 bed volumes of buffer A plus 0.05% Tween 20, and subsequently with 100 bed volumes of buffer C plus 0.05% Tween 20 (buffer C: 8 M urea, 100 mM NaH$_2$PO$_4$, 10 mM Tris/HCl; pH adjusted to pH 6.3 with HCl). Proteins bound to the Ni-NTA column were eluted in four subsequent steps with 1 ml of buffer C plus 200 mM imidazole each, and the eluates were pooled. One milliliter of the sample that serves as the input control for the IP was concentrated by TCA precipitation, dissolved in 25 μl HU buffer plus 10 μl water, and denatured for 10 min at 65° (HU buffer: 8 M urea; 5% SDS; 200 mM Tris/HCl, pH 6.8; 1.5% DTT). The rest of the sample (3 ml) was distributed into one fraction for IP with anti-SUMO antibodies and another for a control IP. For IP, RIPA buffer (50 mM Tris/HCl, pH 7.4; 150 mM NaCl; 1% Triton-X-100; 0.1% SDS; 0.5% DOC) was added to the volume of 40 ml, and either anti-SUMO antibodies or unspecific IgGs coupled to ProteinA magnetic beads (Dynal) were added. These had been prepared by coupling 60 μg of the respective IgGs to 1 ml of the beads with dimethyl pimelimidate (DMP). The IPs were incubated in the presence of 10 mM NEM and 0.2 mM PMSF on a rotating wheel at 4° overnight. Subsequently, by using a magnetic rack, the beads were washed four times with cold RIPA buffer containing 10 mM NEM and once with 50 mM Tris/HCl, pH 8.0. If mass spectrometric analysis of the purified proteins is desired, ultra pure and filtered buffers should be used in all subsequent steps. To elute the proteins from the beads, the samples were incubated in two subsequent steps each time with 50 μl 1% SDS at 65°. After thorough drying of the pooled fractions in a vacuum centrifuge, the proteins were dissolved in 25 μl HU buffer plus 10 μl water and denatured for 10 mins at 65°. The three protein samples (Ni-NTA eluate, anti-SUMO IP, control IP) were separated on a 4–12% Bis/Tris NuPAGE gel (Invitrogen) using MOPS buffer. The proteins were visualized by staining with Coomassie R-250. Bands appearing solely in the anti-SUMO IP were sliced out and investigated by MALDI TOF mass spectrometric (MS) analysis. Among the SUMO conjugates identified by this procedure we found PCNA (Hoege et al., 2002), Top2 (Bachant et al., 2002), the septin Cdc3 (Johnson and Blobel, 1999), and Pgk1 (Zhou et al., 2004), which may serve as useful positive controls in further studies.

Verification

Independent protocols should be used to verify the identified potential SUMO targets. If antibodies against the potential targets are not available, expression of epitope-tagged versions of the proteins (if the tags do not disturb the functionality of the protein) should be used. Because usually only a small fraction of a given target protein is modified by SUMO, one might be able to see an up-shifted band only in long exposures of Western blots of whole cell extracts. Although free (nonconjugated) SUMO runs with an apparent molecular mass of roughly 12 kDa in SDS-polyacrylamide gels, singly SUMO modified species run usually about 15–20 kDa larger than the unmodified protein, most likely because branched peptides migrate abnormally slow in gels. To verify that the protein band represents the SUMO modified protein, two strategies might be used.

First, an N-terminal–tagged version of SUMO (His-tagged or myc-tagged) can be used to increase the size of the SUMO modifier. As a result, an upshift of the modification band can be detected in SDS gels. For the experiment shown in Fig. 3A, we used a SUMO (*smt3*) deletion strain in which we overexpressed either N-terminally 6His-tagged or 3myc-tagged SUMO from an integrative plasmid under the control of the *GAL1-10* promoter. Both SUMO variants are able to complement the *smt3* knock out and form conjugates with PCNA, however, to different degrees.

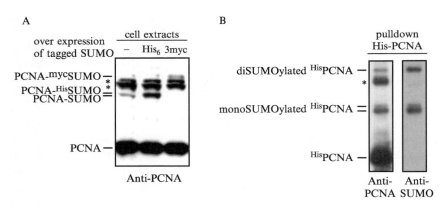

FIG. 3. Strategies for verifying SUMO modifications. (A) Modified PCNA forms detected by anti-PCNA Western blots. Overexpression of either 6His- or 3myc-tagged SUMO result in the appearance of an up-shifted (and sometimes enhanced) PCNA-SUMO protein band that reflects the additional molar mass of the tag. The asterisk denotes a cross-reacting protein. (B) Pull-down of 6His-tagged PCNA using denaturing conditions that preserve SUMOylation and impede precipitation of complexes. Shown are anti-PCNA and anti-SUMO Western blots of the purified material.

Second, the potential SUMO conjugates can be verified by pull-downs (or IPs) of either the potential target protein or of SUMO, followed by Western blotting against the respective other protein. To minimize de-SUMOylation on lysis, Ni-NTA pull-downs of His-tagged proteins using denaturing lysis are preferred over IPs under native conditions. Furthermore, the denaturing conditions prevent complex formation and thus rule out the possibility that the identified protein is a noncovalent interactor of a SUMO conjugate. To exclude the possibility that the observed SUMOylation is due to overexpression of the target or of SUMO, the tagged proteins should be ideally expressed from the endogenous promoter. In Fig. 3B, we show an example of endogenously expressed His-tagged PCNA for which we demonstrate SUMOylation by Ni-NTA pull-down followed by Western blotting using anti-SUMO antibodies.

For this experiment we used 200 ml of a cell culture ($OD_{600} = 1$) and prepared the samples essentially as described previously. After lysis in 6 ml lysis buffer (see "Procedures") and incubation for 15 min on ice, proteins were precipitated by addition of an equal volume of 55% TCA and further incubation for 10 min on ice. Centrifugation (15 min; 3500 rpm; 4°) yielded a protein pellet that was washed twice with acetone ($-20°$) and resuspended in 12 ml buffer A plus 0.05% Tween-20 (see "Procedures"). Proteins were solubilized by shaking for 1 h at room temperature, and insoluble material was removed by centrifugation (20 min; 13000 rpm; 4°). To the supernatant, imidazole was added to 20 mM to reduce unspecific binding to the beads. Fifty microliters of Ni-NTA magnetic beads (Qiagen) was added, and the mixture was incubated at 4° on a tubing roller overnight. Using a magnetic rack, the beads were washed three times with buffer A plus 0.05% Tween-20 and 2 mM imidazole and five times with buffer C plus 0.05% Tween-20 (see "Procedures") before elution with 30 μl 1% SDS for 10 min at 65°. The eluate was dried using a vacuum centrifuge and resuspended in 10 μl water and 15 μl HU buffer (see "Procedures").

Other Purification Procedures

Recently, several reports described proteomic approaches for the identification of new SUMO conjugates from mammalian cells (Li *et al.*, 2004; Vertegaal *et al.*, 2004; Zhao *et al.*, 2004) and budding yeast (Panse *et al.*, 2004; Wohlschlegel *et al.*, 2004; Zhou *et al.*, 2004). The three studies of mammalian SUMO conjugates used one-step purifications with 6His-, HA-, or biotin-tags, respectively. The proteins were identified by MS or tandem MS.

In the following, we will briefly summarize the different approaches that have been applied for yeast. Zhou *et al.* (2004) used a two-step strategy starting with 2000 OD of cells, which expressed 6His-FLAG-tagged SUMO. After denaturing lysis in 8 M urea, the SUMO conjugates were first purified with a Ni-NTA resin. The second step involved anti-FLAG affinity chromatography. In contrast, Panse *et al.* (2004) performed a one-step purification using ProtA-tagged SUMO. The authors used a native lysate from 7000 OD of spheroplasts, in which de-SUMOylation was inhibited by addition of iodoacetamide (an alkylating agent that irreversibly alkylates cysteine and histidine residues). Subsequently, the lysate was incubated with IgG-Sepharose beads to bind SUMO conjugates. Also, Wohlschlegel *et al.* (2004) performed a one-step purification of SUMO conjugates. In this study 10,000 OD of cells expressing 8His-tagged SUMO was lysed under denaturing conditions with NaOH and β-mercaptoethanol, followed by TCA precipitation. After resuspension with 6 M guanidine-HCl, SUMO substrates were purified by means of Ni-NTA agarose.

A substantial advance of the protocols used in the aforementioned studies is the use of tandem liquid chromatography MS. This method is not only highly sensitive, but it also bears the potential of identifying the acceptor lysine residues for SUMOylation. For example, Zhou *et al.* (2004) performed a three-step purification of Sod1, one of their identified SUMO substrates. The substrate was labeled with a 6His-FLAG double tag, and cells expressed HA-tagged SUMO. The modified protein was then purified sequentially by a Ni-NTA column, an anti-HA matrix, and finally by anti-FLAG agarose. For the MS analysis, the protein was fragmented by trypsin digestion. Treatment by this enzyme leaves the SUMO-derived peptide EQIGG at the acceptor lysine residue of the substrate. This extra mass when found added to the mass of a given peptide is, therefore, a clear indication for a SUMOylation site within the peptide sequence. By using liquid tandem MS of the purified material, Zhou *et al.* (2004) identified two target lysine residues in the Sod1 protein, K19, and K70. Furthermore, by using a similar protocol, the authors were also able to identify the acceptor lysine residues for SUMOylation in several other proteins and confirmed the known target residues for PCNA (Hoege *et al.*, 2002). The alternative strategy for the identification of acceptor lysines by systematic elimination of SUMOylation consensus sites has its clear limitations, because SUMOylation frequently occurs at other sites as well (see earlier).

An interesting perspective of the aforementioned proteomic approaches is the possibility to identify ideally the complete set of SUMOylated proteins in a given cell. As noted in the introduction, the pattern of SUMOylated proteins can change dramatically upon extracellular or intracellular signals. Therefore, systematic studies of whole cell changes of

SUMOylation activities are expected to reveal the full dimension of the regulatory potential of the SUMO modification system.

Acknowledgments

M. S. and B. P. contributed equally to this work. S. J. is supported by the Max Planck Society, Deutsche Forschungsgemeinschaft, Deutsche Krebshilfe, and Fonds der chemischen Industrie.

References

Bachant, J., Alcasabas, A., Blat, Y., Kleckner, N., and Elledge, S. J. (2002). The SUMO-1 isopeptidase Smt4 is linked to centromeric cohesion through SUMO-1 modification of DNA topoisomerase II. *Mol. Cell.* **9,** 1169–1182.

Bernier-Villamor, V., Sampson, D. A., Matunis, M. J., and Lima, C. D. (2002). Structural basis for E2-mediated SUMO conjugation revealed by a complex between ubiquitin-conjugating enzyme Ubc9 and RanGAP1. *Cell.* **108,** 345–356.

Bylebyl, G. R., Belichenko, I., and Johnson, E. S. (2003). The SUMO isopeptidase Ulp2 prevents accumulation of SUMO chains in yeast. *J. Biol. Chem.* **278,** 44113–44120.

Desterro, J. M., Rodriguez, M. S., and Hay, R. T. (1998). SUMO-1 modification of Ikappa-Balpha inhibits NF-kappaB activation. *Mol. Cell.* **2,** 233–239.

Hoege, C., Pfander, B., Moldovan, G. L., Pyrowolakis, G., and Jentsch, S. (2002). RAD6-dependent DNA repair is linked to modification of PCNA by ubiquitin and SUMO. *Nature* **419,** 135–141.

Jentsch, S., and Pyrowolakis, G. (2000). Ubiquitin and its kin: How close are the family ties? *Trends Cell Biol.* **10,** 335–342.

Johnson, E. S. (2004). Protein modification by SUMO. *Annu. Rev. Biochem.* **73,** 355–382.

Johnson, E. S., and Blobel, G. (1999). Cell cycle-regulated attachment of the ubiquitin-related protein SUMO to the yeast septins. *J. Cell Biol.* **147,** 981–994.

Li, S. J., and Hochstrasser, M. (1999). A new protease required for cell-cycle progression in yeast. *Nature* **398,** 246–251.

Li, S. J., and Hochstrasser, M. (2000). The yeast ULP2 (SMT4) gene encodes a novel protease specific for the ubiquitin-like Smt3 protein. *Mol. Cell Biol.* **20,** 2367–2377.

Li, T., Evdokimov, E., Shen, R. F., Chao, C. C., Tekle, E., Wang, T., Stadtman, E. R., Yang, D. C., and Chock, P. B. (2004). Sumoylation of heterogeneous nuclear ribonucleoproteins, zinc finger proteins, and nuclear pore complex proteins: A proteomic analysis. *Proc. Natl. Acad. Sci. USA* **101,** 8551–8556.

Melchior, F. (2000). SUMO–nonclassical ubiquitin. *Annu. Rev. Cell Dev. Biol.* **16,** 591–626.

Müller, S., Hoege, C., Pyrowolakis, G., and Jentsch, S. (2001). SUMO, ubiquitin's mysterious cousin. *Nat. Rev. Mol. Cell Biol.* **2,** 202–210.

Panse, V. G., Hardeland, U., Werner, T., Kuster, B., and Hurt, E. (2004). A proteome-wide approach identifies sumoylated substrate proteins in yeast. *J. Biol. Chem.* **279,** 41346–41351.

Pickart, C. (2000). Ubiquitin in chains. *Trends Biochem. Sci.* **25,** 544–548.

Pickart, C. M. (2001). Mechanisms underlying ubiquitination. *Annu. Rev. Biochem.* **70,** 503–533.

Schwienhorst, I., Johnson, E. S., and Dohmen, R. J. (2000). SUMO conjugation and deconjugation. *Mol. Gen. Genet.* **263,** 771–786.

Takahashi, Y., Mizoi, J., Toh, E. A., and Kikuchi, Y. (2000). Yeast Ulp1, an Smt3-specific protease, associates with nucleoporins. *J. Biochem. (Tokyo)* **128,** 723–725.

Vertegaal, A. C., Ogg, S. C., Jaffray, E., Rodriguez, M. S., Hay, R. T., Andersen, J. S., Mann, M., and Lamond, A. I. (2004). A proteomic study of SUMO-2 target proteins. *J. Biol. Chem.* **279,** 33791–33798.

Wohlschlegel, J. A., Johnson, E. S., Reed, S. I., and Yates, J. R., 3rd. (2004). Global analysis of protein sumoylation in *Saccharomyces cerevisiae*. *J. Biol. Chem.* **279,** 45662–45668.

Zhao, Y., Kwon, S. W., Anselmo, A., Kaur, K., and White, M. A. (2004). Broad spectrum identification of cellular small ubiquitin-related modifier (SUMO) substrate proteins. *J. Biol. Chem.* **279,** 20999–21002.

Zhou, W., Ryan, J. J., and Zhou, H. (2004). Global analyses of sumoylated proteins in *Saccharomyces cerevisiae*. Induction of protein sumoylation by cellular stresses. *J. Biol. Chem.* **279,** 32262–32268.

[28] Identification of Ubiquitin Ligase Substrates by *In Vitro* Expression Cloning

By NAGI G. AYAD, SUSANNAH RANKIN, DANNY OOI, MICHAEL RAPE, and MARC W. KIRSCHNER

Abstract

The number of identified E3 ubiquitin ligases has dramatically increased in recent years. However, the substrates targeted for degradation by these particular ligases have not been easily identified. One reason for the inability of matching substrates and ligases is the finding that E3 recognition elements in substrates are often poorly defined. This minimizes the likelihood that bioinformatic approaches will lead to the identification of E3 substrates. For example, the multi-subunit complex the anaphase promoting complex (APC) is an E3 that recognizes destruction boxes (RXXLXXXXD/N/E) or KEN motifs within substrates (Glotzer *et al.*, 1991; Pfleger and Kirschner, 2000). However, many proteins that contain either a potential destruction or a KEN motif are not recognized by the APC *in vitro* or *in vivo*, suggesting that there are other, less well-defined characteristics of substrates that contribute to their ability to serve as APC substrates (Ayad, Rankin, and Kirschner, unpublished observations). Aside from bioinformatic approaches of identifying APC substrates, several groups have also attempted to use affinity techniques to discover novel APC substrates. This has not been widely successful, because many APC substrates are not abundant. Also, as is the case with many ligase-substrate interactions, the affinity of substrates for the APC is likely to be very low. All these considerations have motivated a search for other techniques to assist in identifying substrates of this particular E3 ligase. Here, we describe the use of *in vitro* expression cloning to identify novel APC substrates.

In Vitro Expression Cloning

One technique that has led to the discovery of several APC substrates is *in vitro expression cloning* (IVEC; reviewed in Lustig *et al.*, 1997). IVEC has been used successfully in our laboratory to identify substrates of both kinases and proteases that are activated in a cell cycle–dependent manner (McGarry and Kirschner, 1998; Stukenberg *et al.*, 1997; Zou *et al.*, 1999). In this approach, small pools of cDNAs (50–300 clones each) are translated in an *in vitro* expression reaction in the presence of ^{35}S methionine and cysteine, and the radiolabeled reaction products are incubated in cellular extracts containing the kinase or degradation activity of interest. As described later, we generally have used *Xenopus* embryonic extracts, which can be manipulated to enter particular cell-cycle states. More recently, we have used extracts of somatic cells synchronized in particular stages of the cell cycle or treated to activate particular signaling pathways (Wan *et al.*, 2001). In all cases, the ability to compare the same protein pools incubated in control extracts greatly reduces the likelihood of identifying inherently unstable proteins (in the case of degradation screens) or constitutively phosphorylated proteins (in screens for kinase substrates). In addition, identifying positive clones from within small pools is generally straightforward and has been described in detail elsewhere (Lustig *et al.*, 1997).

We have typically used *Xenopus* cDNA libraries originally created to assist in identifying components of developmental pathways when carrying out IVEC screens. These libraries were constructed by cloning cDNAs from distinct developmental stages, such as egg, blastula, and neurula, into an expression vector, pCS2$^+$ (Turner and Weintraub, 1994), which allows *in vitro* transcription under the control of commercially available polymerases (e.g., SP6 from Promega). We have also used cDNAs from adult mammalian brain (Promega) to search for kinase and protease substrates. Because the kinase assays were already described in a prior *Methods in Enzymology* chapter (Lustig *et al.*, 1997), we will mainly concentrate on the degradation screens that have led to the identification of several mitotic and G1 substrates of the APC (Fig. 1).

Identification of Mitotic APC Substrates Using *Xenopus* Egg Extracts

The APC must be activated to target substrates for degradation. Activation requires phosphorylation of core APC subunits and association with either one of two proteins, Cdc20 or Cdh1 (Peters, 2002), which confer

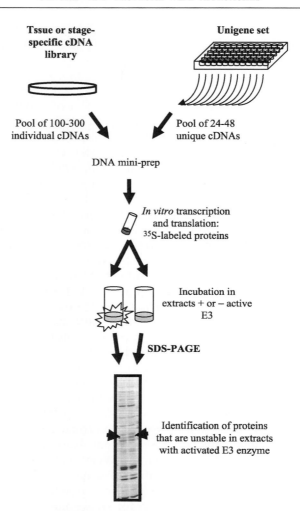

FIG. 1. Schematic illustration of *in vitro* expression cloning (IVEC). Small pools of *in vitro* transcribed and translated radiolabeled proteins are incubated in extracts that have been treated to activate the E3 enzyme of interest. The proteins that are specifically ubiquitinated by the activated E3 are degraded in the extract and can be identified when the pools are run on SDS-PAGE gels adjacent to the same pool incubated in control extracts. See text for details. (See color insert.)

substrate specificity to the complex. In the early *Xenopus* embryo, there is little or no Cdh1 expression and no apparent Cdh1-dependent APC activity. The APC^{CDC20} form of the APC is predominant (Lorca *et al.*, 1998) and only active during mitosis (not interphase), because Cdc20 association with the APC is increased in the presence of high mitotic kinase activity (Kotani et al., 1999). These observations provided the rationale for conducting a screen designed to identify proteins degraded when the APC^{Cdc20} is active (mitosis) relative to when it was inactive (interphase).

Xenopus Egg Extract Preparation

Interphase extracts (Murray, 1991) are prepared from activated *Xenopus* eggs presoaked in freshly made cycloheximide (Sigma; 100 mg/ml stock in dimethyl sulfoxide is diluted to a final concentration of 100 μg/ml) to prevent translation of cyclin B from endogenous RNA, which would eventually drive the extract into mitosis. The eggs can be released from a metaphase arrest either using ionophore (A23187; Sigma) at 200 ng/ml final concentration or by using an electrical activation chamber as described in Murray (1991). Extracts are supplemented either with sucrose (300 mM final concentration) or glycerol (4% final concentration), stored frozen in liquid nitrogen, and thawed at room temperature or on ice before use. A cocktail (1:1:1 ratio) of energy mix (150 mM creatine phosphate, 20 mM ATP, pH 7.4, 2 mM EGTA, pH 7.7, and 20 mM MgCl$_2$, stored at $-20°$), cycloheximide (0.1 μg/ml), and 0.1 μg/ml ubiquitin is added to the extracts (1 μl cocktail:21 μl of preactivated extract) immediately before use. To generate mitotic extracts, interphase egg extracts are incubated with a nondegradable form of cyclin B to drive the extract into mitosis (to yield mitotic extracts) or a buffer control (interphase extracts). The extracts contain all of the accessory activities, including ubiquitin, E1 and E2 enzymes, and proteasomes, required for the efficient ubiquitination and degradation of E3 substrates. Extracts driven into mitosis by the addition of recombinant nondegradable cyclin contain activated APC capable of initiating degradation of several mitotic proteins.

Preparation of Substrate Pools and Degradation Assays

Pools of radiolabeled *in vitro* transcribed and translated cDNAs are prepared. The cDNA libraries are maintained as frozen glycerol stocks, each stock containing a mixture of bacterial strains representing 100–300 discrete cDNA clones. These stocks are used to inoculate liquid bacterial cultures, and small-scale (1–3 ml culture) plasmid preparations from each pool are isolated using a Qiagen spin Miniprep kit (50 μl total per pool). Subsequently, 1 μl of each pool (0.2–0.5 μg/μl DNA) is incubated in a TNT-coupled transcription–translation reaction containing ^{35}S methionine (6 μl total reaction volume, 1 μl DNA:5 μl TnT mix containing rabbit reticulocyte lysate from Promega; we now routinely use trans-^{35}S-label from ICN, Cat#510064, 1175Ci/mmol) at 30° for 1.5 h (we have recently found that TNT Gold™ from Promega provides optimal translation). A small volume (1.0 μl) from each pool of *in vitro* translated ^{35}S-labeled proteins is aliquoted into two wells of a 96-well microtiter plate on ice (Costar Thermowell 96-well plate); to one is added 5 μl of the mitotic extract; to the other, 5 μl of the control interphase extract. Once all of the reaction mixes are assembled, the plate is sealed (Microseal A film, MSA-5001; MJ research), moved to room temperature, and both experimental and control reactions are allowed to proceed for 1 h. The reactions are then stopped by the addition of 40 μl of SDS-containing sample buffer (125 mM Tris-Cl, pH 6.8, 2% (w/v) sodium dodecyl sulfate, and 10% (w/v) glycerol) supplemented with 1 mM DTT. The samples are then heated to 95° for 5 min in a PCR machine (MJ Research PTC-100 programmable thermocycler), and 5 μl of each stopped reaction mixture resolved by SDS-PAGE. Pools containing ^{35}S-labeled proteins present in interphase extracts but missing in mitotic extracts are easily identified. An example of such a degradation screen is shown in Fig. 1. The same pool incubated in control extract is resolved in an adjacent lane of the SDS gel to facilitate identification of positive clones. We routinely use gradient gels (most often 10–15% acrylamide to allow the resolution of proteins of a wide variety of molecular weights).

Three novel APC substrates were isolated from an identical mitotic APC screen: the DNA replication inhibitor geminin, the separase inhibitor securin, and the chromokinesin Xkid (Funabiki and Murray, 2000; McGarry and Kirschner, 1998; Zou *et al.*, 1999). All three proteins contain destruction boxes and were subsequently confirmed to be *bona fide* APC substrates *in vitro* and *in vivo*. Interestingly, in the case of securin, *in vitro* expression cloning was the only successful means of identifying this protein, despite our efforts to purify it using conventional chromatography. This illustrates the advantage of *in vitro* expression cloning over standard biochemical techniques when substrates are not abundant or are only transiently expressed during the cell cycle (Zou *et al.*, 1999).

In Vitro *Transcription/Translation Mix*	In Vitro *Transcription/Translation Reaction*
200 μl Reticulocyte lysate from Promega	5 μl *in vitro* transcription/translation mix
16 μl TNT buffer from Promega	1 μl Miniprep DNA from pool
32 1 ^{35}S-methionine-cysteine trans-label ICN (1175 Ci/mmol)	30°, 1.5 h
8 μl methionine-cysteine free amino acid mix from Promega	
8 μl Rnasin from Promega	
8 μl SP6, T7, or T3 enzyme	
128 μl dh20	
400 μl total	

Identification of G1 APC Substrates Using *Xenopus* Egg Extracts

There is no apparent G1 phase in the cell cycle in the early *Xenopus* embryo. Work from our laboratory has indicated that the APC present in interphase *Xenopus* egg extracts is inactive but can be activated by the addition of recombinant or *in vitro* translated Cdh1, the G1 activator of the APC (Fang et al., 1998; Pfleger and Kirschner, 2000). This finding facilitated our search for proteins specifically targeted for degradation by the G1 form of the APC. We searched for proteins that are unstable in extracts containing Cdh1-activated APC using a strategy similar to the one used to identify substrates of the mitotically activated form of the APC. How this specific screen was performed is described in the following.

Activation of the G1 Form of the APC Complex

Recombinant Cdh1 is produced in SF9 cells and added to concentrated interphase extracts, prepared as described previously, to activate the endogenous APC. In our case, the ratio of interphase egg extracts to recombinant Cdh1 was 20:1 and yielded a 0.4 μM final concentration of Cdh1 in the extract. The Cdh1 or buffer-supplemented extract is incubated at room temperature for 20 min (to allow APC activation) and then returned to ice to assemble the degradation reactions. Again, small pools of *in vitro* transcribed and translated proteins are aliquoted into wells of a microtiter plate and gently mixed separately with the two extracts. The degradation assays are initiated by shifting the plates to room temperature and stopped after 60 min by the addition of SDS sample buffer. *In vitro* translated ^{35}S-labeled Cdc20 is used as a positive control, because Cdc20 is degraded specifically when APC is activated by Cdh1 in *Xenopus* egg extracts.

We were able to isolate two novel APC^{CDH1} substrates using this screening protocol, both of which contained KEN motifs. The first substrate, Tome-1, regulates the G2-M transition in both *Xenopus* egg extracts and somatic cells (Ayad et al., 2003; Ayad and Kirschner, in preparation) and is a *bona fide in vitro* substrate of APC^{CDH1}; mutation of the KEN sequence in Tome-1 inhibits its degradation in Cdh1 supplemented *Xenopus* egg extracts. Tome-1 is also degraded by means of the APC *in vivo*, because its levels are lowest during G1 (Ayad et al., 2003). The second novel APC^{CDH1} substrate identified is also degraded during G1 and will be described in detail elsewhere (Rankin et al., manuscript in preparation). A third APC^{CDH1} substrate has been identified using a modification of this screening method in which a unigene set of *Drosophila* cDNAs was used instead of pools of *Xenopus* cDNAs (Fig. 2; Ooi and Kirschner, unpublished observations).

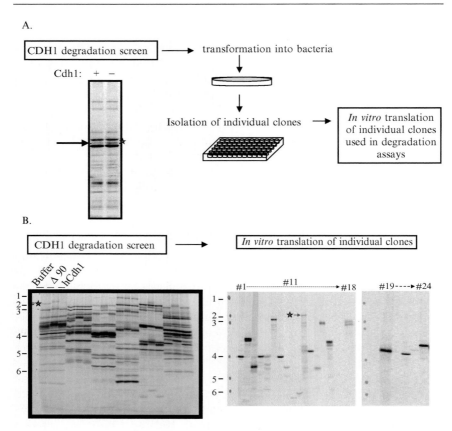

FIG. 2. Sib selection versus isolating positive clones from unigene set. (A) SDS-PAGE of pools of *in vitro* translated proteins incubated in *Xenopus* egg extracts supplemented with Cdh1 or buffer (+ or −, respectively). Once the desired pool is identified, a sib selection procedure is performed to isolate the single clone of interest labeled with star. (B) Alternately, if a unigene set of cDNAs is used, sib selection is not necessary because the single clone can be identified simply by matching the molecular weight of *in vitro* translation products by SDS-PAGE. An example of this technique is shown for a protein degraded in *Xenopus* egg extracts supplemented with Cdh1 and stable in extracts supplemented with buffer or a nondegradable version of cyclin B (Δ90). The band of interest is labeled with a star and has a slightly lower electrophoretic mobility than the #2 molecular weight standard. (See color insert.)

Identification of APC Substrates Using Somatic Cell Extracts

We have found *Xenopus* egg extracts invaluable for studying the early embryonic cell cycle and identifying novel APC substrates. However, we realize that many investigators do not have access to *Xenopus laevis* but would like to use *in vitro* expression cloning to identify E3 ligase substrates. Those investigators who have access to tissue culture cells can also perform *in vitro* expression cloning screens. Recently, we have developed highly concentrated somatic cell extracts that recapitulate *in vitro* degradation of various APC substrates. Using extracts from mink lung epithelial cells or HeLa cells, we have demonstrated that Sno-N and Skp-2 are APC substrates *in vitro*, respectively (Wan et al., 2001; Wei et al., 2004). Recently, we have been able to recapitulate degradation of APC substrates using frozen extracts as described later (when freezing extracts, it is imperative to do so as quickly as possible using liquid nitrogen and to store the extract at $-80°$).

Preparation of Somatic Cell Extracts

To prepare concentrated Hela cell extracts from mitotic or G1 cells, 1 to 2 liters of cells grown in suspension are first synchronized with a thymidine/nocodazole block. Cells are grown at 37° in 5% CO_2 in DMEM supplemented with 10% fetal calf serum and 100 μg/ml each penicillin and streptomycin. Thymidine is added to a final concentration of 2 mM from a 100× sterile stock solution for 24 h to asynchronous Hela cells. The cells are released from this block by washing with medium twice and returning them to the tissue culture incubator for 3 h (thymidine release). Nocodazole is then added (330 nM final concentration) for 11 h (maximum is 12 h; less is better). Hela S3 cells are generally grown in spinner flasks (Bellco μ-carrier 1000 ml spinner flasks; cells are inoculated at 2×10^5 cells/ml; final concentration at harvesting is 5×10^5 cells/ml). To prepare extracts, cells are pelleted (1000 RPM, 290g for 5 min in Sorvall RC3C), and washed three times with PBS. Subsequently, half of the cells are saved on ice (nocodazole-arrested cells for the preparation of mitotic extracts) and the remainder resuspended in culture medium and returned to the tissue culture incubator for 4 h. After 4 h, these cells are pelleted again and washed three times with PBS. These are now a source of G1 cells for the preparation of extracts. As much PBS as possible is removed from both pellets (the nocodazole-released G1 cells and the nocodazole-arrested cells left on ice for 4 h), and the cells are resuspended in swelling buffer (Swelling buffer = 20 mM Hepes, pH 7.7, 5 mM $MgCl_2$, 5 mM KCl, 1 mM DTT, ATP regenerating system, and protease inhibitor tablet from Pierce, EDTA-free Complete tablets; 0.75 ml buffer:1 ml packed cells; typical volumes were 2–3 ml of packed cell volume). The samples are frozen and thawed twice (Liquid Nitrogen/30° water bath), then passed through a 20.5-gauge needle twice (1-ml syringes are used; we ice syringes and needles beforehand). The resulting lysates are spun at 5000 RPM

(2655g) for 5 min in the cold room in an Eppendorf centrifuge. The supernatants are then collected and spun for an additional 30 min (at top speed, 20,000g, 4°, Eppendorf). Finally, the supernatants from the mitotic and G1 cell extracts are carefully isolated (avoiding the top lipid-containing layer) and used in degradation assays that are assembled on ice. These degradation assays are assembled similarly to those detailed previously using *Xenopus* egg extracts. One microliter of *in vitro* translated ^{35}S-labeled substrate is mixed with 20 μl extract (mixed thoroughly but not too vigorously, avoiding frothing) and 1 μl of a cocktail of energy mix, cycloheximide, and ubiquitin (1:1:1 ratio as above) on ice in a 96-well microtiter plate. The reactions are started by shifting the plate to room temperature. Four microliters of each of the reactions are added to 10 μl sample buffer at various time points (usually 60–120 min) to stop the reaction. The stopped reactions are heated to 95° and separated by SDS-PAGE. Using this protocol, we have identified several proteins that are degraded in G1 extracts but not mitotic extracts. The cell cycle role of these proteins is currently being investigated (Rape and Kirschner, unpublished observations).

Practical Considerations

We used 10–15% gradient gels, because they optimally separated proteins in the size range in which we were interested (we rarely found very high molecular weight proteins in our pools); we routinely resolved 30–40 bands (from a 100 cDNA pool). After fixing these gels in a 5% methanol, 7.5% glacial acetic acid solution, and drying them, we exposed them to film for 1 week (we routinely exposed six gels to the same large film), and then proceeded to search for protein bands that were specifically absent in reactions containing Cdh1 (Figs. 1 and 2). Film gave higher resolution than phosphorimaging, and this was necessary in several cases. This initial phase of the screen in which approximately 1300 pools of cDNAs (mostly from neurula stage library, some from egg and blastula stages as well) lasted approximately one and a half months, including exposure to film. Routinely, one person could easily perform 50 *in vitro* transcription-translation and degradation reactions per day, as well as resolve these reactions by SDS-PAGE. After potential APC^{CDH1} substrates within a pool were identified, the degradation reactions were repeated with the positive pool to ensure reproducibility. Beyond this short initial screening phase, the sib selection phase lasted approximately 1 week. Sib selection has been described previously and involves subdividing a pool containing a cDNA encoding a putative substrate until a single clone can be attained (Lustig *et al.*, 1997). Once a single clone was isolated, control reactions were performed to determine whether substrate degradation could be competed with a known APC substrate such as cyclin B. Finally, the construct encoding the positive substrate was retransformed into bacteria, and then the cDNA was sequenced to identify destruction boxes or KEN sequences.

As mentioned, we have begun to use the *Drosophila* unigene set in addition to other cDNA libraries as a source of cDNAs. The use of unigene sets will ultimately greatly simplify the analysis. First, in libraries that have not been normalized, many rare cDNAs are likely not to be represented, whereas others will appear multiple times. Therefore, in the past, we have identified the same substrates several times, although we failed to identify several previously characterized substrates. The second limitation involved the sib selection procedure. In several cases, we were unable to sib select cDNAs encoding large molecular weight proteins, perhaps because of the toxicity of these clones in bacteria, which might result in their underrepresentation in pools of bacterial transformants. Because the *Drosophila* unigene set is an arrayed set of individual cDNAs, we are able to identify positive clones without sib selection (Fig. 2). We are currently duplicating this technology to create pools of human and mouse cDNAs derived from unigene sets of cDNAs, which should obviate the need to perform sib selection.

Conclusions

The protocols we have discussed were mainly created to search for novel APC substrates. However, they can be easily adapted to identify substrates of other E3s. This can be accomplished by identifying conditions in which the particular E3 is active or inactive, perhaps by isolating cells from different phases of the cell cycle or distinct developmental stages. One major advantage of using the IVEC strategy for the identification of E3 substrates is that it can be done in complex cellular extracts. This might be particularly important if the signals or activities that target particular substrates for degradation might not be well characterized. For example, many SCF complex substrates must be phosphorylated to be recognized by the relevant F-box protein. Identification of the specific kinase might not be necessary if extracts can be prepared from cells isolated under the appropriate conditions. In terms of developmental control of proteolysis, we found that the homeobox transcriptional repressor Xom is degraded during a specific stage of *Xenopus* development by means of an SCF complex containing β-TRCP (Zhu and Kirschner, 2002). It is likely that other degradation reactions are similarly controlled during development, thereby providing the basis for various *in vitro* expression-cloning screens. It may also be possible to include dominant-negative versions of a particular E3 into an extract capable of supporting degradation and search for proteins that are stabilized in the presence of the dominant negative. Finally, we had the advantage of having some known APC substrates to use as internal controls. Therefore, if any substrates are known for an E3, they can be used to judge the robustness of the *in vitro* expression cloning screen.

References

Ayad, N. G., Rankin, S., Murakami, M., Jebanathirajah, J., Gygi, S., and Kirschner, M. W. (2003). Tome-1, a trigger of mitotic entry, is degraded during G1 via the APC. *Cell* **113,** 101–113.

Fang, G., Yu, H., and Kirschner, M. W. (1998). Direct binding of CDC20 protein family members activates the anaphase-promoting complex in mitosis and G1. *Mol. Cell.* **2,** 163–171.

Funabiki, H., and Murray, A. W. (2000). The *Xenopus* chromokinesin Xkid is essential for metaphase chromosome alignment and must be degraded to allow anaphase chromosome movement. *Cell* **102,** 411–424.

Glotzer, M., Murray, A. W., and Kirschner, M. W. (1991). Cyclin is degraded by the ubiquitin pathway. *Nature* **349,** 132–138.

Kotani, S., Tanaka, H., Yasuda, H., and Todokoro, K. (1999). Regulation of APC activity by phosphorylation and regulatory factors. *J. Cell Biol.* **146,** 791–800.

Lorca, T., Castro, A., Martinez, A. M., Vigneron, S., Morin, N., Sigrist, S., Lehner, C., Doree, M., and Labbe, J. C. (1998). Fizzy is required for activation of the APC/cyclosome in *Xenopus* egg extracts. *EMBO J.* **17,** 3565–3575.

Lustig, K. D., Stukenberg, P. T., McGarry, T. J., King, R. W., Cryns, V. L., Mead, P. E., Zon, L. I., Yuan, J., and Kirschner, M. W. (1997). Small pool expression screening: Identification of genes involved in cell cycle control, apoptosis, and early development. *Methods Enzymol.* **283,** 83–99.

McGarry, T. J., and Kirschner, M. W. (1998). Geminin, an inhibitor of DNA replication, is degraded during mitosis. *Cell* **93,** 1043–1053.

Murray, A. W. (1991). Cell cycle extracts. *Methods Cell Biol.* **36,** 581–605.

Peters, J. M. (2002). The anaphase-promoting complex: Proteolysis in mitosis and beyond. *Mol. Cell.* **9,** 931–943.

Pfleger, C. M., and Kirschner, M. W. (2000). The KEN box: An APC recognition signal distinct from the D box targeted by Cdh1. *Genes Dev.* **14,** 655–665.

Stukenberg, P. T., Lustig, K. D., McGarry, T. J., King, R. W., Kuang, J., and Kirschner, M. W. (1997). Systematic identification of mitotic phosphoproteins. *Curr. Biol.* **7,** 338–348.

Turner, D. L., and Weintraub, H. (1994). Expression of achaete-scute homolog 3 in Xenopus embryos converts ectodermal cells to a neural fate. *Genes Dev.* **8,** 1434–1447.

Wan, Y., Liu, X., and Kirschner, M. W. (2001). The Anaphase-promoting complex mediates TGF-beta signaling by targeting SnoN for destruction. *Mol. Cell.* **8,** 1027–1039.

Wei, W., Ayad, N. G., Wan, Y., Zhang, G. J., Kirschner, M. W., and Kaelin, W. G., Jr. (2004). Degradation of the SCF component Skp2 in cell-cycle phase G1 by the anaphase-promoting complex. *Nature* **428,** 194–198.

Zhu, Z., and Kirschner, M. (2002). Regulated proteolysis of Xom mediates dorsoventral pattern formation during early *Xenopus* development. *Dev. Cell.* **3,** 557–568.

Zou, H., McGarry, T. J., Bernal, T., and Kirschner, M. W. (1999). Identification of a vertebrate sister-chromatid separation inhibitor involved in transformation and tumorigenesis. *Science* **285,** 418–422.

[29] In Vitro Screening for Substrates of the N-End Rule–Dependent Ubiquitylation

By Ilia V. Davydov, John H. Kenten, Yassamin J. Safiran, Stefanie Nelson, Ryan Swenerton, Pankaj Oberoi, and Hans A. Biebuyck

Abstract

We describe a systematic, high-throughput approach to the discovery of protein substrates of ubiquitylation. This method uses a library of cDNAs in combination with a reticulocyte lysate–based, transcription–translation system that acts as both an excellent means for high-throughput protein expression and a source of ubiquitylation enzymes. Ubiquitylation of newly expressed proteins occurs in this milieu from the action of any one of a number of E3 ligases that are present in the lysate. Specific detection of ubiquitylated proteins is carried out using electrochemiluminescence-based assays in conjunction with a multiplexing scheme that provides replicate measurements of the ubiquitylated products and two controls in each well of a microtiter plate. We used this approach to identify putative substrates of the N-end rule–dependent ubiquitylation (mediated by the UBR family of ubiquitin ligases), a system already well known to have high endogenous activity in reticulocyte lysates. We screened a library of ~18,000 cDNA clones, one clone per well, by expressing them in reticulocyte lysate and measuring the extent of modification. We selected ~500 proteins that showed significant ubiquitylation. This set of modified proteins was redacted to ~60 potential substrates of the N-end rule pathway in a secondary screen that involved looking for inhibition of ubiquitylation in reticulocyte lysates supplemented with specific inhibitors of the N-end rule ubiquitylation. We think our system provides a general approach that can be extended to the identification of substrates of other E3 ligases.

Introduction

Systematic analysis of the functional interactions in the human proteome is a formidable challenge facing biologists in the post-genomic era. Sequencing of the human genome has, in particular, provided insight into the complexity of the ubiquitylation of proteins. We now know that there are hundreds of proteins potentially involved in different aspects of the conjugation and release of ubiquitin (Ub) and ubiquitin-like proteins, including ubiquitin-activating (E1) and ubiquitin-conjugating (E2) enzymes, ubiquitin ligases (E3), and deubiquitylating enzymes (DUBs) (Semple *et al.*, 2003; Wong *et al.*, 2003). Many of these proteins appear in loci of disease (Ciechanover, 2003; Pagano and Benmaamar, 2003; Sun, 2003), providing ample motivation for characterizing the biological pathways influenced by these proteins and studying the role of these proteins in disease.

Functional genomic technologies that enable comprehensive and high-throughput analysis of various protein activities are becoming more prevalent as researchers negotiate the vast stores of information derived from various sequencing efforts. Large-scale isolation of complex mixtures of ubiquitin conjugates followed by proteolytic digestion and peptide analysis using mass spectrometry, for example, yields valuable information on the nature of ubiquitylation substrates and the identity of sites of ubiquitylation (Gururaja *et al.*, 2003; Peng *et al.*, 2003). Another approach involves *in vitro* expression screening of pooled cDNAs with electrophoresis-based detection of proteins based on their biochemical properties, such as, for example, their ubiquitin-dependent degradation under specific conditions (Davydov and Varshavsky, 2000; Lustig *et al.*, 1997). Here, we demonstrate an alternative high-throughput screening strategy for identification of protein substrates of a specific ubiquitylation activity. The approach uses an *in vitro* expression system for generating, in parallel, large libraries of individual proteins. The individual proteins are exposed to the ubiquitylation activity, and a high-throughput ubiquitylation assay is used to rapidly identify those proteins that are modified by the ubiquitylation activity. In this report, we describe the use of this approach to identify substrates of the N-end rule–dependent ubiquitylation.

Our screening strategy required a simple and reliable protein expression system and exploited the availability of highly curated libraries of cDNA clones. The latter are available from many sources and feature complete or partially sequenced genes in vectors convenient for the expression of proteins. Commercially available cell lysate systems provide an attractive means of *in vitro* protein expression amenable to high-throughput automation. We used a commercial transcription and translation (TnT) system that is based on a rabbit reticulocyte lysate containing a high level of protein translation machinery. The lysate is combined with bacterial polymerases to drive transcription of cDNA. The lysate is also supplemented with tRNA to increase the yield of protein. The use of tRNAs with modified amino acids provides a convenient approach for the introduction of various labels at the protein level that allow the specific interrogation of the cDNA product.

The TnT-type *in vitro* expression systems are capable of making proteins and introducing posttranslational modifications that are normally found *in vivo*, like phosphorylation, myristoylation, prenylation, arginylation, and ubiquitylation. Hershko and others demonstrated that reticulocyte lysates from rabbits contain an active ubiquitin system (i.e., Ub-activating and -conjugating enzymes, and Ub ligases). In particular, rabbit reticulocyte) lysates have been shown to contain the ubiquitin E3 ligase UBR1 (E3α) that recognizes substrates for ubiquitylation through the N-end rule pathway (Hershko *et al.*, 2000).

The N-end rule pathway targets proteins for ubiquitylation and proteasomal degradation based on the nature of their N-terminal residue (Varshavsky, 2003). In mammals, the ubiquitin ligases UBR1 and UBR2 recognize primary destabilizing residues of two types, basic residues Arg, Lys, and His ("type 1") and bulky hydrophobic residues Phe, Leu, Tyr, Trp, and Ile ("type 2") (Kwon *et al.*, 2003). So-called tertiary destabilizing residues Asn and Gln are deamidated by two different N-terminal amidases, yielding the secondary destabilizing residues Asp and Glu, which in turn are arginylated by Arg-transferase (Varshavsky, 2003). The tertiary destabilizing residue Cys serves as a substrate of Arg-transferase after enzymatic oxidation (Kwon *et al.*, 2002). In addition, several known N-end rule substrates result from proteolytic cleavage of the original full-length protein to yield a fragment having a destabilizing amino acid residue at its N-terminus (Davydov and Varshavsky, 2000; deGroot *et al.*, 1991; Ditzel *et al.*, 2003; Hamilton *et al.*, 2003; Rao *et al.*, 2001). In one specific example, RGS4, RGS5, and RGS16 proteins become N-end rule substrates in reticulocyte lysates after successive, N-terminal modifications: (1) removal of Met-1 by Met-aminopeptidases (a proteolytic cleavage event), (2) oxidation of Cys-2 into cysteic acid, and (3) conjugation of Arg to the N-terminus (Davydov and Varshavsky, 2000; Kwon *et al.*, 2002). Ubiquitylation and degradation of type 1 N-end rule substrates like RGS4, RGS5, and RGS16 is inhibited *in vitro* by the dipeptide Arg-Ala (Davydov and Varshavsky, 2000), whereas ubiquitylation and degradation of type 2 N-end rule substrates can be inhibited by the dipeptide Trp-Ala (Gonda *et al.*, 1989). In either case, this inhibition presumably occurs by competition with the peptide for the respective substrate binding site on UBR1 or UBR2.

In this chapter, we describe a method to identify protein substrates of ubiquitylation in reticulocyte lysates using a library of individually characterized cDNAs as our starting basis for an "unbiased" *in vitro* survey. We applied this method specifically to the identification of potential substrates of the N-end rule pathway. We demonstrate its success by the blinded recovery of known substrates of the N-end rule and the discovery of a number of potentially novel ones.

Principles of the Method

Our method is based on *in vitro* protein expression in a reticulocyte lysate system. Test proteins are synthesized from individual cDNA templates using a coupled transcription-translation system derived from reticulocyte lysates fortified with SP6 (or T7) RNA polymerase. The TnT reactions are further supplemented with biotinylated Lys-tRNA that incorporates biotinylated lysine residues into the nascent proteins in place of natural amino acid. The level of biotinylation achieved with this method is expected to be low (estimated to be <1 label per protein based on a streptavidin binding assay, data not shown) because of competition with the unlabeled lysine in the lysates. We favored this approach to protein labeling compared with epitope tagging, because the former better preserves native protein structures and allows use of unmodified cDNAs. Nonetheless, detrimental effects of biotin labels on protein structure remain a possibility.

Ubiquitylation of newly synthesized proteins occurs in the same reaction mixtures and is driven by the ubiquitylation system present in reticulocyte lysates (see "Introduction"). Ubiquitylated cDNA products will carry both ubiquitin groups and biotin groups (from the modified lysine-tRNA). Aliquots of the reaction mixtures are transferred into multiwell plates for quantitative detection of these proteins.

Ubiquitylated proteins are measured using an electrochemiluminescence-based binding assay format. The principle of detection is outlined in Fig. 1. The measurement is based on the use of a label (Sulfo-TAG, a highly water-soluble derivative of ruthenium *tris*-bipyridine or RuBpy) that emits light when oxidized at an electrode in an appropriate chemical environment. The electrochemiluminescence measurements are carried out in specially designed microtiter plates having screen-printed carbon electrodes that are used as both a capture phase for solid-phase binding assays and as a source of electrical energy for inducing electrochemiluminescence. Multiplexed measurements can be carried out on plates that have microarrays of binding reagents patterned on the electrodes in each well. These microarray patterns are defined by a patterned dielectric layer deposited on the working electrode. The dielectric layer has a pattern of holes that define exposed "spots" on the working electrode (as shown in Fig. 1). Binding reagents are immobilized by passive adsorption from solutions printed on the spots using a nanoliter dispenser. An imaging plate reader adapted to apply electrical potential to the electrodes and image the resulting electrochemiluminescence is used to measure and quantify the electrochemiluminescence generated at each spot (see www.mesoscale.com for more details on this technology).

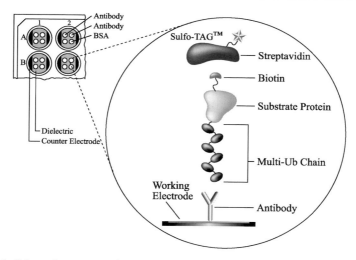

FIG. 1. Schematic representation of the capture and detection of ubiquitylated proteins in Multi-Spot plates. Ubiquitylated protein is captured on the surface of the two spots coated with the antibody against ubiquitylated proteins. Sulfo-TAG–labeled streptavidin is recruited to the surface through its interaction with biotinylated Lys residues in the ubiquitylated protein.

This study used plates having four spots (Fig. 1) in each well. Two of the spots were coated with an antibody against ubiquitylated proteins and were used to provide replicate measurements of ubiquitylated proteins in a sample. Another spot was coated with bovine serum albumin (BSA) and provided a background control. The fourth spot was an additional control based on antibodies for a different posttranslational modification. Ubiquitylated proteins are specifically recruited to the surface of the antiubiquitin antibody-coated spots. RuBpy-labeled streptavidin binds to the biotin labels, creating a high concentration of RuBpy on the surface of these spots *only* where those wells have newly synthesized, ubiquitylated protein. Electrochemiluminescent signals from each spot are then measured and quantitated on an appropriate reader.

We found that the presence of a BSA-coated spot in the well helps to exclude false-positive results caused by proteins that nonspecifically bind to the plate surface. We encountered such proteins in our primary screen, although their frequency (<10% of all hits) was not high enough to prohibit a simplified primary screening strategy that omitted this control. Similarly, the primary screen might be carried out in a "single-spot" plate without the control spots, where false-positive clones would be eliminated during the secondary screening.

One important variable that influences the success of ubiquitylation screens in TnT systems is the incubation time of the reaction. Available evidence indicates that accumulation of small proteins ($\leq$40 kDa) largely stops after 30 min of the TnT reaction at 30° (Davydov and Varshavsky, 2000), but larger proteins may potentially continue to accumulate over a longer period of time. Ubiquitylation rates can also vary greatly depending on the substrate. Ubiquitylated proteins are also degraded by proteasomes present in reticulocyte lysates. The presence of a proteasome inhibitor, such as MG-132, in the reactions can slow this degradation considerably but will not totally prevent degradation of very unstable proteins, especially for long reaction times. We decided to limit the reticulocyte lysate reactions to 45 min during the primary screen, a reaction time that provided sufficient levels of a wide assortment of proteins having a range of molecular weights while minimizing their degradation by proteasomes (data not shown).

We measured ubiquitylation of RGS4 as a known N-end rule substrate (Davydov and Varshavsky, 2000) to validate the electrochemiluminescent assay and compared its signals to that for a mutant version of the protein, RGS4$_{C2V}$, known to be poorly ubiquitylated. Lysates expressing wild-type RGS4 produced substantially higher signals in the assay than the lysates programmed with an empty vector or the lysates expressing the mutant RGS4$_{C2V}$ (Fig. 2A). The excellent signal-to-background and low noise of these measurements further corroborated the sensitivity and robustness of the detection scheme and recommended its practice for a comprehensive screen. In agreement with the published data (Davydov and Varshavsky, 2000), RGS4 ubiquitylation was inhibited by the type 1 N-end rule inhibitor Arg-β-Ala, but not by a type 2 dipeptide Trp-Ala (Fig. 2B). The presence of the proteasome inhibitor MG-132 in the reaction was important to achieve maximum assay signal (Fig. 2B) by minimizing degradation of ubiquitylated proteins by proteasomes. *In tota*, these data demonstrated that the electrochemiluminescent assay was capable of detecting N-end rule–dependent ubiquitylation of a known substrate in an efficient and practical manner and suggested a secondary screening method based on use of an inhibitor.

In addition to the assay format presented in Fig. 1, we also developed and tested two alternative detection strategies. In the first, ubiquitylated proteins were captured in streptavidin-coated plates and detected with sulfo-TAG–labeled antibodies against ubiquitylated proteins ("reverse sandwich"). In the second, sulfo-TAG–labeled ubiquitin (Ub) was added to reticulocyte lysates and conjugated to proteins. The proteins were captured on streptavidin-coated plates, and the attached sulfo-TAG-labeled Ubs were then detected directly. Although both of these formats demonstrated

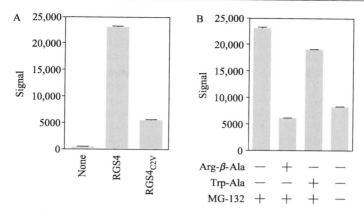

FIG. 2. N-end rule–dependent ubiquitylation of RGS4 detected in an electrochemiluminescence assay. (A) An empty plasmid vector, a plasmid encoding wild type RGS4, or a mutant RGS4$_{C2V}$ were used for protein expression, and ubiquitylation of the synthesized proteins was measured in electrochemiluminescence assay. (B) Ubiquitylation of the wild-type RGS4 was measured in the absence or in the presence of the indicated N-end rule inhibitors (Arg-β-Ala, or Trp-Ala), or a proteasome inhibitor MG-132.

specific detection of ubiquitylated RGS4 and were thus plausible alternatives, the primary assay format shown in Fig. 1 was more sensitive and provided the highest signal-to-background ratios.

Materials

cDNA Library

The cDNA library (Meso Scale Discovery, MSD, Gaithersburg, MD) is composed of mouse and human cDNA clones under the control of a bacteriophage RNA polymerase promoter. Individual clones were stored in separate wells of 96-well plates. The cDNA library consists of 60,000 full-length clones representing 22,000 unique genes (according to Unigene, www.ncbi.nlm.nih.gov/entrez/query.fcgi?db=unigene) based on the analysis of the 5′ sequences of the cDNAs. A study of the protein expression from 250 random clones from the library showed an average molecular weight ∼55 kDa, with ∼70% of these clones expressing evident levels of protein having approximately the right mass.

Reagents

1. Streptavidin labeled with MSD Sulfo-TAG (MSD)
2. TnT Quick Coupled transcription-translation system (Promega, Madison, WI)

3. Biotinylated lysine-tRNA (Transcend tRNA, Promega)
4. MG-132 (Calbiochem, La Jolla, CA)
5. Arg-β-Ala (Sigma, St. Louis, MO)
6. Trp-Ala (Sigma)
7. Bestatin (Sigma)
8. Binding buffer (MSD)
9. Assay buffer T (MSD)

MULTI-ARRAY Plates for Electrochemiluminescence Measurements

Electrochemiluminescence measurements are carried out using specially designed multi-well plates having integrated carbon ink electrodes (MULTI-ARRAY or Multi-Spot plates, MSD). A dielectric layer patterned over the working electrode in each well exposes four regions or "spots" on the working electrode. The exposed working electrode in two of the spots is coated with a mouse monoclonal antibody that is specific for ubiquitylated proteins (MSD). Another spot is coated with BSA and used as a negative control.

Electrochemiluminescence Reader

Electrochemiluminescence is induced and measured in the Multi-Spot plates using a SECTOR Imager 6000 reader (MSD).

High-Throughput Automation

1. Biomek FX pipetting station (Beckman Coulter, Fullerton, CA)
2. Plate washer (Bio-Tek Instruments, Winooski, VT)
3. Multidrop 384 (Labsystems, Vantaa, Finland)

Protocols

Electrochemiluminescence Assay for Ubiquitylation

Proteins are produced from individual cDNAs in transcription–translation reaction mixtures containing rabbit reticulocyte lysate and SP6 (or T7) RNA polymerase (TnT SP6, or T7 Quick Mix, Promega), further supplemented by 20 μM methionine, 8–30 μg/ml plasmid DNA, 20 μg/ml biotinylated Lys-tRNA, and 50 μM of the proteasome inhibitor MG-132 in a total volume of 12.5 μl/well in polypropylene 96-well plates. The reactions proceed for 20–60 minutes at 30°. One microliter of each reaction mixture is mixed with 50 μl binding buffer (MSD) with 1 μg/ml sulfo-TAG–labeled streptavidin in the well of the Multi-Spot plates. The plate is incubated on a tabletop shaker for 1 h. During this time, newly synthesized proteins containing both a biotin label and polyubiquitin bind to both immobilized antibody and sulfo-TAG–labeled streptavidin resulting in the accumulation of the electrochemiluminescent labels on the surface (Fig. 1). Thereafter, the plate is washed three times with 20 mM Tris-HCl, 0.004% Triton X-100 followed by the addition of 150 μl of a buffered solution containing ECL coreactant tripropylamine (Assay buffer T, MSD) into each well. ECL signals from labels bound to the electrode surfaces are measured using a SECTOR Imager 6000 instrument (MSD).

Screen for Substrates of the N-End Rule Pathway

To determine whether the protein produced from a specific clone is ubiquitylated through the N-end rule pathway, the reticulocyte lysate–mediated ubiquitylation reactions are repeated in the presence of 1 mM Arg-β-Ala (an inhibitor of ubiquitylation of type 1 N-end rule substrates), 1 mM Trp-Ala (an inhibitor of ubiquitylation of type 2 N-end rule substrates), or a mixture of 1 mM Arg-β-Ala and 1 mM Trp-Ala. These reactions also have 0.15 mM bestatin, an aminopeptidase inhibitor that is added to prevent degradation of the dipeptides.

Analysis of Protein Ubiquitylation by Immunoprecipitation-Western Blot

Proteins are synthesized in reticulocyte lysates as described in the previous section in 50-μl reactions. One microliter of the reaction mixture is removed for analysis of ubiquitylated proteins (as earlier), and 4 μl is added to an SDS sample buffer for detection of the total nascent protein by Western blot. The remaining 45 μl is mixed with 1 ml immunoprecipitation buffer (IP buffer: 20 mM Tris-HCl, pH 7.4, 150 mM NaCl, 20 mM EDTA, 1% Triton X-100, and a cocktail of protease inhibitors [Roche Applied Sciences, Indianapolis, IN]) and 20 μl of a suspension of Protein A/G beads (Santa Cruz Biotechnology, CA), prewashed in IP buffer, and tumbled for 30 min at 4°. The mixture is centrifuged in a microcentrifuge for 5 min at 1500g, and the supernatant is transferred into a new tube, and mixed with 10 μg of the antibody against ubiquitylated proteins (MSD). The contents of the tube are additionally tumbled for 1 h at 4°, then 20 μl of a new suspension of Protein A/G beads is added, and the mixing continues as before for an additional 2 h. The Protein A/G beads are washed three times with IP buffer containing 0.1% SDS, pelleted by centrifugation, and resuspended in 20 μl SDS sample buffer.

The samples of "total proteins" and "immunoprecipitated proteins" are subjected to PAGE on 4–20% SDS Tris-Gly gels (Invitrogen, Carlsbad, CA) followed by Western blot analysis with streptavidin conjugated to horseradish peroxidase (Amersham Biosciences, Piscataway, NJ) and enhanced chemiluminescence detection (Amersham Biosciences) according to the manufacturer's protocols.

Analysis of Protein Cleavage and Degradation by Denaturing Gels

Proteins are synthesized and ubiquitylated in reticulocyte lysates as described previously except that biotinylated Lys-tRNA is substituted with [^{14}C]-lysine or [^{35}S]-methionine, and the MG-132 is omitted. The reactions are incubated at 30° for various times. Reactions are stopped by transferring aliquots into SDS sample buffer. These samples are heated for 5 min at 95° and fractionated by SDS-PAGE on 4–20% gradient polyacrylamide gels (Invitrogen). Gels are soaked in Amplify fluorographic reagent (Amersham Pharmacia Biotech), dried, and exposed to X-ray films.

Library Screening

Primary Screening for Ubiquitylated Proteins

On the basis of our protein expression, ubiquitylation, and detection system, we developed a protocol for high-throughput, primary screening using standard laboratory automation (Fig. 3). We screened a cDNA library of 18,000 individual clones (a subset of the total library of ~60,000 clones) stored in 190, 96-well plates to identify proteins that showed statistically elevated levels of ubiquitylation in reticulocyte lysates (see later). Figure 4 shows data for the primary screen. We picked 528 ubiquitylated proteins for secondary screening using several criteria to sort the database of measurements. First, the lowest 50% of signals on each plate were assumed to be background. Second, we filtered the set to select for clones having signal-to-background ratios >10, or those having signals higher than 30 times the standard deviation of the background. Using these criteria, we identified (in blinded fashion) known N-end rule substrates RGS4, RGS5, RGS16, as well as clones encoding natural ubiquitin fusions. This recovery validated our method and its appropriateness for the potential discovery of substrates of ubiquitylation.

We chose several random strong hits from the screen as well as RGS5, RGS4, and a few non-hits for immunoprecipitation–Western blot analysis to confirm ubiquitylation of the selected proteins by an independent method. Ubiquitylated proteins were immunoprecipitated with an antibody recognizing conjugated Ub, followed by Western blot with

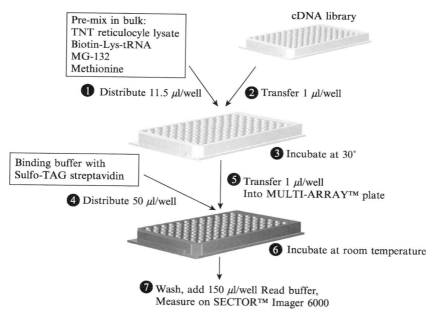

FIG. 3. Flowchart of the screen for substrates of ubiquitylation.

streptavidin-HRP to visualize multi-Ub ladders (see "Protocols" for the detailed procedure). The data in Fig. 4B–D demonstrates the correlation between the levels of ubiquitylation revealed by conventional gels and measured by the electrochemiluminescence, further confirming that proteins selected as hits in the screen were truly ubiquitylated.

The signals detected in the ubiquitylation assay are directly proportional to the degree of ubiquitylation of the expressed protein and also are a function of the level of accumulation of this protein in the reaction. Indeed, we observed that many of the strongest hits in the primary screen were both well-expressed and highly ubiquitylated proteins (Fig. 4B–D), although several proteins with relatively low expression levels were recovered as well (e.g., RGS4 and RGS5). In contrast, many (but certainly not all) of the "non-hits" were characterized by relatively low (but still clearly detectable) accumulation levels.

One potential source of false-positive proteins in the screen would be represented by those library proteins that have the ability to tightly bind to endogenous ubiquitylated proteins from reticulocyte lysates, and thus would generate signals in the electrochemiluminescent assay.

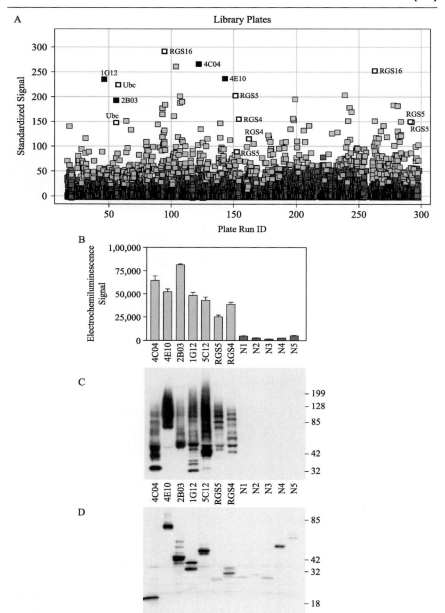

However, unlike true-positive proteins, these false-positive proteins would not generate multi-Ub ladders in immunoprecipitation–Western blot analysis as described in "Protocols" and thus can be eliminated in secondary screening. In our random survey of positive clones by immunoprecipitation–Western blot analysis, we have not seen any false-positive proteins of this type, which may serve as an indication that they are not very abundant.

Screen for N-End Rule Substrates

We further refined the ubiquitylated hits from the primary screen into a set of putative N-end rule substrates using a mixture of Arg-β-Ala and Trp-Ala to modulate the ubiquitylation reaction by the UBR family of E3 ligases present in the reticulocyte lysate. Type 1 and type 2 N-end rule substrates are expected to have lower levels of ubiquitylation in the presence of the dipeptides. Hits having no N-end rule participation should have no significant changes in levels of ubiquitylation. The assay was run in duplicate at 20-, 40-, and 60-min time points. The ratio of the signal in the presence of dipeptides to the signal in the absence of dipeptides was used to identify clones selected for further investigation. We selected 60 clones that showed significantly suppressed ubiquitylation in the presence of the dipeptides (Fig. 5A). We further selected additional clones based on kinetic profiles that showed significant signal suppression at some of the three time points (Fig. 5B). In addition, we picked 8 clones as negative controls (no response to inhibitors) and 24 clones that showed some enhancement of ubiquitylation with the addition of inhibitors.

Potential N-end rule substrates from the secondary screen were grouped into type 1 or type 2 substrates on the basis of their individual response to Arg-β-Ala (type 1 inhibitor) and Trp-Ala (type 2 inhibitor). Inhibition of ubiquitylation of selected N-end rule substrates by the dipeptides was also characterized by immunoprecipitation–Western blot analysis. The data from the electrochemiluminescent assay was found to be in agreement with the immunoprecipitation–Western blot images (Fig. 6).

Fig. 4. Primary screen of 18,000 cDNA clones for ubiquitylated proteins. (A) The primary screen involved 270 96-well plates, including 35 control plates, 45 plates run in duplicate, and the remainder of the library plates. Electrochemiluminescence signals from the selected positive and negative clones (B) were compared with the gel images from immunoprecipitation–Western blot (C), and a Western blot for total labeled proteins (D).

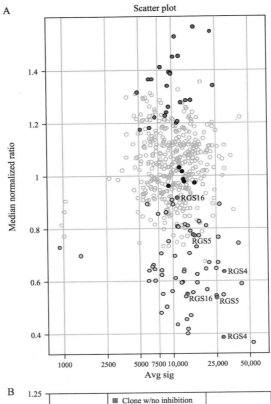
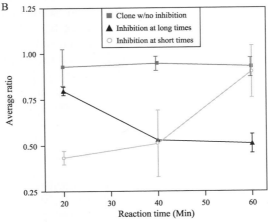

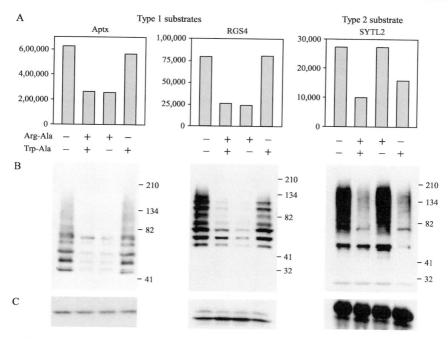

FIG. 6. Characterization of the N-end rule substrates from the screen. Comparison of signals from selected clones measured by electrochemiluminescence (A), immunoprecipitation and Western blot (B), and Western blot for total labeled proteins (C).

Further Characterization of the Hits

N-end rule substrates typically require an initial processing event resulting in generation of a new N-terminus with a destabilizing amino acid residue, and this initial processing by proteolytic or peptidolytic cleavage may also be followed by further modifications (see "Introduction"). In some cases, proteolytic cleavage and subsequent reduction of the protein size can be detected in denaturing gels. Several hits identified in our N-end rule screen revealed multiple bands in their SDS gels with patterns consistent with the proteolytic processing of the nascent full-length protein (see

FIG. 5. Screen for N-end rule substrates. (A) Response of ubiquitylated proteins to inhibitors of the N-end rule pathway showing the degree of inhibition. (B) Differential kinetics of inhibition at different time points exhibited by certain clones. Selection of clones should take into account these kinetic effects to select clones for further study.

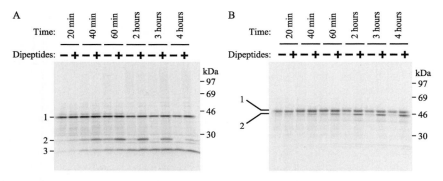

FIG. 7. Time course of posttranslational processing of putative N-end rule substrates in reticulocyte lysates. Putative N-end rule substrates were synthesized in the presence of [^{35}S]-methionine and ubiquitylated in reticulocyte lysates either in the absence or in the presence of a mixture of Arg-β-Ala and Trp-Ala dipeptide inhibitors. Reactions were stopped at different reaction times and separated on SDS-PAGE followed by autoradiography. (A) Full-length aprataxin (Aptx, band 1) is cleaved into two fragments (bands 2 and 3). Band 2 likely represents a C-terminal fragment of the original protein with a type 1 destabilizing residue at its N-terminus. (B) Synaptotagmin-like protein 2 (SYTL2, band 1) is probably proteolytically trimmed from the N-terminus to generate a shorter polypeptide (band 2) with a type 2 residue at the N-terminus that is metabolically stabilized by dipeptides.

Fig. 7 and the accompanying legend for specific examples). In those cases, one of the shorter protein fragments visible on SDS gel appeared metabolically unstable and was quickly degraded in reticulocyte lysates (MG-132 was omitted in these reactions to facilitate proteasomal degradation), but this fragment was stabilized in the presence of the N-end rule inhibitors (Fig. 7). These studies provided further validation of the N-end rule hits identified from the *in vitro* screen. To understand the specific physiological roles of N-end rule–dependent ubiquitylation of these putative new substrates, functional *in vivo* studies will need to be carried out.

Other Potential Screening Approaches

In general, systematic identification of protein substrates for a specific ubiquitylation pathway or a given E3 presents considerable methodological challenges. Technical feasibility of our screening strategy for identification of the N-end rule substrates was determined by (1) the availability of an efficient *in vitro* protein expression system that, at the same time, contained all the functional components of the N-end rule pathway; (2) the availability of N-end rule inhibitors; and (3) the property of the reticulocyte lysates to introduce posttranslational modifications necessary to generate N-end rule substrates (such as limited proteolytic cleavage, N-terminal arginylation).

The method should be applicable to other pathways within the system of ubiquitylation that satisfy these criteria. In addition, with some modifications, the method could potentially overcome these limitations. For example, small molecular weight E3 inhibitors are currently unavailable for most E3s and could be replaced with recombinant E3 fragments possessing dominant-negative activities. If the E3 of interest is not present endogenously in reticulocyte lysates, it could be generated and tested for activity separately (either as a recombinant protein or partially purified from an active cell lysate) and added (in some cases together with the matching E2) to the reticulocyte lysate expression system. With these modifications of the method, one should be able to expand it to screening for substrates of a wider range of different E3s.

Conclusions

We developed a facile, proteome-wide method for investigation of protein posttranslational modifications. The technique takes advantage of a unique cDNA library, high-throughput protein production, and multiplexed electrochemiluminescence measurements. This approach relies on the ability of reticulocyte lysate expression systems to efficiently produce proteins and support many of the posttranslational protein modifications seen *in vivo*, including protein ubiquitylation. We used our method to screen for protein substrates of the N-end rule–dependent ubiquitylation from a library of ~18,000 individual mouse and human cDNA clones. We showed blinded recovery of the known substrates of this pathway from the library, as well as identification of several previously unknown putative substrates. The newly recovered substrates were further validated and confirmed by independent, more traditional, and lower throughput methods. Our approach could be generally applied to identify and characterize novel N-end rule substrates that are processed in reticulocyte lysates from other libraries of cDNA clones from human, mouse, and other higher organisms. The method could also potentially be modified and expanded to screen for substrates of other ubiquitylation pathways and specific E3s.

Acknowledgments

We thank Steven Cheng, Paul Grulich, Nisar Pampori, Jon Reeves, and Laura Schaefer for their help and support, and George Sigal for the comments on the manuscript.

References

Ciechanover, A. (2003). The ubiquitin proteolytic system and pathogenesis of human diseases: A novel platform for mechanism-based drug targeting. *Biochem. Soc. Trans.* **31**, 474–481.

Davydov, I. V., and Varshavsky, A. (2000). RGS4 is arginylated and degraded by the N-end rule pathway *in vitro*. *J. Biol. Chem.* **275,** 22931–22941.

Ditzel, M., Wilson, R., Tenev, T., Zachariou, A., Paul, A., Deas, E., and Meier, P. (2003). Degradation of DIAP1 by the N-end rule pathway is essential for regulating apoptosis. *Nat. Cell Biol.* **5,** 467–473.

deGroot, R. J., Rümenapf, T., Kuhn, R. J., and Strauss, J. H. (1991). Sindbis virus RNA polymerase is degraded by the N-end rule pathway. *Proc. Natl. Acad. Sci. USA* **88,** 8967–8971.

Gonda, D. K., Bachmair, A., Wünning, I., Tobias, J. W., Lane, W. S., and Varshavsky, A. (1989). Universality and structure of the N-end rule. *J. Biol. Chem.* **264,** 16700–16712.

Gururaja, T., Li, W., Noble, W. S., Payan, D. G., and Anderson, D. C. (2003). Multiple functional categories of proteins identified in an *in vitro* cellular ubiquitin affinity extract using shotgun peptide sequencing. *J. Proteome Res.* **2,** 394–404.

Hamilton, M. H., Cook, L. A., McRackan, T. R., Schey, K. L., and Hildebrandt, J. D. (2003). γ2 subunit of G protein heterotrimer is an N-end rule ubiquitylation substrate. *Proc. Natl. Acad. Sci. USA* **100,** 5081–5086.

Hershko, A., Ciechanover, A., and Varshavsky, A. (2000). The ubiquitin system. *Nat. Med.* **6,** 1073–1081.

Lustig, K. D., Stukenberg, P. T., McGarry, T. J., King, R. W., Cryns, V. L., Mead, P. E., Zon, L. I., Yuan, J., and Kirschner, M. W. (1997). Small pool expression screening: Identification of genes involved in cell cycle control, apoptosis, and early development. *Methods Enzymol.* **283,** 83–99.

Kwon, Y. T., Kashina, A. S., Davydov, I. V., Hu, G. G., An, J. Y., Seo, J. W., Du, F., and Varshavsky, A. (2002). An essential role of N-terminal arginylation in cardiovascular development. *Science* **297,** 96–99.

Kwon, Y. T., Xia, Z., An, J. Y., Tasaki, T., Davydov, I. V., Seo, J. W., Sheng, J., Xie, Y., and Varshavsky, A. (2003). Female lethality and apoptosis of spermatocytes in mice lacking the UBR2 ubiquitin ligase of the N-end rule pathway. *Mol. Cell. Biol.* **23,** 8255–8277.

Pagano, M., and Benmaamar, R. (2003). When protein destruction runs amok, malignancy is on the loose. *Cancer Cell.* **4,** 251–256.

Peng, J., Schwartz, D., Elias, J. E., Thoreen, C. C., Cheng, D., Marsischky, G., Roelofs., J., Finley, D., and Gygi, S. P. (2003). A proteomics approach to understanding protein ubiquitination. *Nat. Biotechnol.* **21,** 921–926.

Rao, H., Uhlmann, F., Nasmyth, K., and Varshavsky, A. (2001). Degradation of a cohesion subunit by the N-end rule pathway is essential for chromosome stability. *Nature* **410,** 955–959.

Semple, C. A. M., RIKEN GER group, and GSL members (2003). The comparative proteomics of ubiquitination in mouse. *Genome Res.* **13,** 1389–1394.

Sun, Y. (2003). Targeting E3 ubiquitin ligases for cancer therapy. *Cancer Biol. Ther.* **2,** 623–629.

Varshavsky, A. (2003). The N-end rule and regulation of apoptosis. *Nat. Cell. Biol.* **5,** 373–376.

Wong, B. R., Parlati, F., Qu, K., Demo, S., Pray, T., Huang, J., Payan, D. G., and Bennett, M. K. (2003). Drug discovery in the ubiquitin regulatory pathway. *Drug Discovery Today* **8,** 746–754.

[30] Genome-Wide Surveys for Phosphorylation-Dependent Substrates of SCF Ubiquitin Ligases

By XIAOJING TANG, STEPHEN ORLICKY, QINGQUAN LIU, ANDREW WILLEMS, FRANK SICHERI, and MIKE TYERS

Abstract

The SCF (Skp1–Cullin–F-box) family of ubiquitin ligases target numerous substrates for ubiquitin-dependent proteolysis, including cell cycle regulators, transcription factors, and signal transducers. Substrates are recruited to an invariant core SCF complex through one of a large family of substrate-specific adapter subunits called F-box proteins, each of which binds multiple specific substrates, often in a phosphorylation-dependent manner. The identification of substrates for SCF complexes has proven difficult, especially given the requirement of often complex phosphorylation events for substrate recognition. The archetype for such interactions is the binding of the yeast F-box protein Cdc4 to its various substrates by means of multiple motifs that weakly match an optimal consensus called the Cdc4 phosphodegron (CPD), which is phosphorylated by cyclin-dependent kinases (CDKs) and possibly other kinases. Provided phosphodegron recognition motifs and/or the targeting kinases for SCF substrates are delineated, it is possible to use genome-wide methods to identify new substrates. Here we describe two methods for the systematic retrieval of SCF substrates based on membrane arrays of synthetic phosphopeptides and on genome-wide kinase substrate profiles. In the first approach, which identifies substrates with strong matches to the CPD, a search of the predicted yeast proteome with the optimal CPD motif identified ~1100 matches. A phosphopeptide membrane array corresponding to each of these sequences is then probed with recombinant Cdc4, thereby identifying potential substrates. In the second approach, which identifies substrates that lack strong CPD motifs, a genome-wide set of recombinant CDK substrates is phosphorylated and directly assayed for binding to Cdc4. The proteins corresponding to these hits from each approach can then be subjected to the more stringent criteria of phosphorylation-dependent binding to Cdc4, ubiquitination by SCF^{Cdc4} *in vitro*, and Cdc4-dependent protein instability *in vivo*. Both methods have identified novel substrates of Cdc4 and may, in principle, be used to identify numerous new substrates of other SCF and SCF-like complexes from yeast to humans.

Introduction

The ubiquitin–proteasome system mediates the rapid and selective degradation of numerous regulatory proteins, including key cell-cycle regulators, transcription factors, tumor suppressor proteins, oncoproteins, membrane receptors, among many others (Hershko and Ciechanover, 1998). The formation of ubiquitin conjugates is catalyzed by the now famous enzymatic cascade E1 → E2 → E3, which activates ubiquitin in an ATP-dependent manner, then serially transfers ubiquitin for conjugation as an isopeptide to an acceptor lysine residue on the substrate (Hershko et al., 1983). The spatial and temporal specificity of ubiquitin conjugation is dictated by the many hundreds of E3 enzymes, also referred to as ubiquitin ligases, which selectively bind to one or more substrates. E3 enzymes fall into two general categories, the HECT domain class, which forms a catalytically essential thioester with ubiquitin, and the RING domain class, which seems to simply juxtapose the substrate next to the E2 catalytic center (reviewed in Pickart [2001]). Substrate interactions may be controlled at the level of E3 subunit expression or by posttranslational modification of either the E3 or the substrate. Given that substrates are usually of low abundance and rapidly degraded on ubiquitination, the assignment of substrates to their cognate E3 enzymes has proven to be an ongoing challenge in the dissection of the ubiquitin system. Despite much effort, the substrate profile for any E3 enzyme has yet to be defined at a proteome-wide level.

The SCF (Skp1–Cullin–F-box) family of E3 enzymes are multi-subunit complexes first discovered through genetic requirements for cell cycle progression in budding yeast (reviewed in Patton et al. [1998b]). The modular nature of SCF architecture is conserved from yeast to humans (reviewed in Deshaies [1999] and Willems et al. [2004]). SCF complexes are constructed from an invariant core platform composed of the Skp1 linker protein, the Cdc53/Cul1 scaffold protein, and the Rbx1/Roc1/Hrt1 RING domain protein, and a suite of variable substrate-specific adaptors called F-box proteins, which recruit substrates for ubiquitination by an associated E2 enzyme (Fig. 1). F-box proteins directly bind to Skp1 by means of a conserved 40-amino acid motif, termed the F-box, whereas substrate recognition is mediated by protein–protein interaction domains, such as WD40 repeat or leucine-rich repeat (LRR) domains (Bai et al., 1996; Feldman et al., 1997; Patton et al., 1998a; Skowyra et al., 1997). The E2 enzyme is recruited to the complex by Rbx1 (Seol et al., 1999; Skowyra et al., 1999). The assembly of SCF complexes is dynamically regulated through F-box protein turnover and complex interactions of auxiliary binding partners and modifications (reviewed in Willems et al. [2004]).

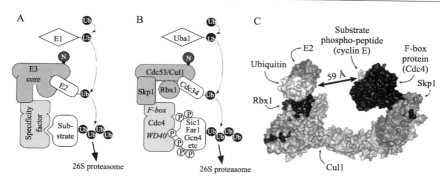

Fig. 1. SCF ubiquitin ligase architecture. (A) Generic schematic of cullin-based ubiquitin ligases. N indicates site for modification by Nedd8/Rub1, which is essential for activity of all SCF complexes, except in budding yeast, where it is entirely dispensable. (B) Schematic of SCF^{Cdc4} complex. P indicates substrate phosphorylation. (C) Space-filling representation of modeled holo-SCF^{Cdc4} complex drawn with Mac PyMol using coordinates from Orlicky et al. (2003).

The human SCF^{Skp2} complex forms an elongated cradlelike structure, with Skp1 and the F-box protein bound at the N-terminal domain of the Cdc53/Cul1 scaffold and Rbx1 at the C-terminal domain (Zheng et al., 2002). As in all E3 structures determined to date, a considerable gap of ~60 Å exists between the substrate binding site and the E2 catalytic site, presumably to accommodate the large ubiquitin moiety and to facilitate ubiquitin polymer formation. The modular nature of the SCF complex has allowed the structures of two other F-box proteins, Cdc4 and β-TrCP, to be modeled on the holo-complex (Orlicky et al., 2003; Wu et al., 2003), as shown in Fig. 1.

Substrate recognition by E3 substrates is typically mediated by one or more short primary sequence elements, termed degrons, or phosphodegrons in the case of phosphorylation-dependent interactions. Most, but not all, known SCF substrates must be phosphorylated to be bound by their cognate F-box protein. Phosphorylation-dependent recognition underlies many cell-cycle–regulated degradation events, such as elimination of phosphorylated CDK inhibitors and phosphorylated G1 cyclins by the F-box proteins Cdc4 and Grr1, respectively. Similarly, in metazoans, including humans, phosphorylation targets the CDK inhibitor p27^{Kip1} and cyclin E to the F-box proteins Skp2 and Cdc4, respectively, whereas phosphorylated forms of the protooncogenic transcription factor β-catenin and the antiinflammatory factor IκBα are recognized by the F-box protein β-TrCP. A representative list of known SCF substrates and phosphodegron degron sequences is compiled in Table I. As an ever-increasing number of SCF

TABLE I
Characterized Phosphodegrons

F-box protein	Substrate	Sequence	Consensus	References
β-TRCP	HIV Vpu	$_{51}$DpSGNEpS$_{56}$	DpSGX$_{2-4}$pS	Margottin et al., 1998
	β-Catenin	$_{32}$DpSGIHpS$_{37}$		Hart et al., 1999
	IκBα	$_{31}$DpSGLDpS$_{36}$		Yaron et al., 1998
	Emi1	$_{144}$DpSGYSpS$_{149}$		Margottin-Goguet et al., 2003
	Wee1	$_{48}$pSTGEDpSAFQEPDpS$_{60}$		Watanabe et al., 2004
	NFκB1/p105	$_{117}$EEGFGpSSpSPVKpS$_{127}$		
	ATF4	$_{926}$DpSGVETpS$_{932}$		Lang et al., 2003
	CDC25A	$_{218}$DpSGICMpS$_{224}$		Lassot et al., 2001
	Smad4	$_{81}$DpSGFCLDpS$_{88}$		
Cdc4		$_{142}$DLpSGLTLQpS$_{150}$	[IL][ILP]pTP[KR]	Wan et al., 2004
	Sic1	$_{43}$PVpTPSTTK$_{50}$		Nash et al., 2001
	Far1	$_{61}$PIpSPPPSLK$_{69}$		Nash et al., 2001
	Gcn4	$_{163}$PVpTPVLED$_{170}$		Meimoun et al., 2000
	Cdc6	$_{37}$DVpTPESpSPEKL$_{47}$		Perkins et al., 2001
		$_{366}$PLpTPTTpSPVK$_{375}$		
Fbw7/Cdc4	Tec1	$_{271}$LLpTPITAS$_{278}$		Chou et al., 2004
	Cyclin E	$_{378}$LLpTPPQSG$_{385}$		Koepp et al., 2001
	Myc	$_{56}$LPpTPPLpSPSRR$_{66}$		Yada et al., 2004
	Jun	$_{61}$LTpSPDVGLLK$_{70}$		Nateri et al., 2004
		$_{71}$LApSPELER$_{78}$		
		$_{89}$TTpTPpTQFLCPK$_{99}$		
	SV40 LT	$_{699}$PPpTPPPEPET$_{678}$		Welcker and Orian, 2004
	Notch	CDK8 sites	Not defined	Fryer et al., 2004
Skp2	p27/Kip	$_{185}$EQpTPKKPG$_{192}$	Not defined	Tsvetkov et al., 1999
Grr1	Cln2	CDK sites	Not defined	Berset et al., 2002
	Mth1	CK2 sites		Spielewoy et al., 2004

pathways are linked to human disease, methods to systematically discover relevant F-box protein substrates have become of paramount importance.

The archetypal SCF^{Cdc4} complex of budding yeast targets a number of important regulatory proteins for degradation, including the CDK inhibitors Sic1 and Far1, the replication protein Cdc6, and the transcription factor Gcn4 (Willems *et al.*, 2004). Of these substrates, Sic1 is arguably the most critical, because its elimination is essential for progression through G1 phase, such that defects in SCF^{Cdc4} function are manifest as a permanent G1 phase arrest. At the end of G1 phase, Sic1 is heavily phosphorylated by G1 cyclin CDK activity (Cln1/2-Cdc28), which enables Cdc4 to recruit Sic1 to the SCF core complex for ubiquitination by the E2 enzyme Cdc34 (Verma *et al.*, 1997). The mechanics of Sic1–Cdc4 interaction are understood in considerable detail. In contrast to conventional phosphorylation-dependent binding, such as the interaction of SH2 domains with phosphotyrosine epitopes (Pawson, 1995), at least six of the nine consensus CDK sites on Sic1 must be phosphorylated for Cdc4 binding to occur (Nash *et al.*, 2001). Although none of the individual nine phosphodegrons bind Cdc4 with appreciable affinity, a heterologous phosphopeptide derived from human cyclin E (centered around Thr380, a site that is necessary for cyclin E degradation in yeast and human cells) binds Cdc4 with high affinity (Kd ~1 μM). Although a single copy of the high-affinity cyclin E site is sufficient to target Sic1 for elimination, it seems that the nine natural sites establish an ultrasensitive response that buffers against kinase fluctuation in G1 phase, which would otherwise result in premature elimination of Sic1 (Nash *et al.*, 2001). Exploration of the sequence space for high-affinity binding of short peptides to Cdc4, beginning with the high-affinity cyclin E phosphopeptide, yielded an optimal consensus binding site termed the Cdc4 phosphodegron (CPD) of sequence I/L-I/L/P-pT-P-{K/R}$_4$, where {} indicates disfavored residues (Fig. 2). The CPD consensus suggested that the reason individual phosphoepitopes in Sic1 bind with such low affinity is the presence of consensus mismatches, most notably the presence of basic residues favored by targeting CDK kinases (Nash *et al.*, 2001). Importantly, the anti-selection of basic residues in the CPD would preclude detection of the optimum motif by peptide library methods that rely on positive selection, because the suppression of single disallowed residues cannot be readily detected (Songyang and Cantley, 1995). Structural studies of a high-affinity CPD peptide bound to Cdc4 suggests that electrostatic repulsion between a conserved basic patch on the surface of Cdc4 and the basic residues natural Sic1 sites accounts for this anti-selection effect (Orlicky *et al.*, 2003). The mechanism whereby multiple weak sites combine to generate high-affinity binding is not fully elaborated but may involve the interaction of multiple weak CPD sites in

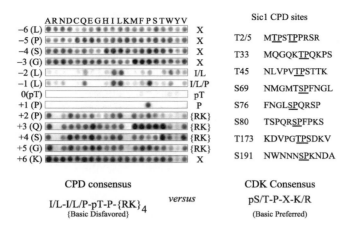

Fig. 2. Delineation of the optimal Cdc4 phosphodegron by Spot peptide membrane array. Data reproduced from Nash et al. (2001) with copyright permission from Nature Publishing Group (http://www.nature.com/). The optimal CPD and CDK consensus sequences differ primarily in that the former disfavors whereas the latter favors C-terminal basic residues. The nine CDK phosphorylation sites and flanking sequences of Sic1 comprise only very weak matches to the CPD, such that at least six of the nine sites must be phosphorylated for efficient recognition by Cdc4. Mismatches are shown in outline font.

rapid equilibrium with the single peptide–binding pocket on Cdc4 (Klein et al., 2003). Finally, although Cln–Cdc28 activity is the dominant kinase that targets Sic1, contributions to Sic1 regulation from the related CDK enzyme Pho85 and casein kinase 2 have also been reported (Coccetti et al., 2004; Nishizawa et al., 1998).

Characterized F-box proteins bind individual phosphodegrons over a range of affinities (see Table I for examples and references). In a number of cases, synthetic phosphopeptides that correspond to substrate sequences bind with moderate to high affinity to their cognate F-box protein. In one of the best-defined examples of a phosphodegron, diphosphorylated peptides derived from β-catenin, HIV Vpu, IκBα, Wee1, Cdc25A, and other substrates bind avidly to β-TrCP, whereas monophosphorylated or unphosphorylated peptides do not detectably interact. All of the known β-TrCP phosphodegrons match the consensus motif DpSGX$_{2-4}$pS. In another example, a phosphopeptide centered on the critical Thr187 site of p27 is able to capture Skp2 from solution, although a precise phosphodegron consensus for Skp2 has not yet been reported. The metazoan homolog of Cdc4, variously termed Fbw7, hCdc4, Ago, and SEL-10, seems to recognize phosphodegrons that closely match the yeast CPD consensus sequence in substrates such as cyclin E, Myc, and Jun. In most of the preceding cases, the natural sequence essentially defines the optimal phosphodegron. Even within known Cdc4 substrates in yeast, the affinities of individual phosphodegrons vary widely (Table I). In Sic1, none of the natural sites interact appreciably with Cdc4, at least within the detection limits of a fluorescence polarization assay (Nash *et al.*, 2001). In contrast, Far1 contains one site that binds with moderate affinity ($\sim$3 μM), and Gcn4 contains a site that binds with high affinity ($<$1 μM). Even in these instances, additional weak CPD sites contribute to recognition and destabilization *in vivo* (Chi *et al.*, 2001; Henchoz *et al.*, 1997; Meimoun *et al.*, 2000). Human Cdc4 also seems to bind its substrates through a combination of dominant and weak sites phosphorylated by both CDK and GSK-3 kinases (Strohmaier *et al.*, 2001; Welcker *et al.*, 2003, 2004). Finally, nonphosphorylated epitopes may also contribute to high-affinity binding, as in the case of the p27 recognition, which depends in part on Cks1 as a co-factor (Ganoth *et al.*, 2001; Spruck *et al.*, 2001). It is clear that substrate recognition by F-box proteins is highly tunable, in terms of the number, nature, and spacing of phosphodegrons, and in terms of the targeting kinases.

Overall Strategy

Most F-box proteins bind to their substrates through phosphodegrons that exhibit a range of affinities. Despite the often facile nature of these interactions, the lability and/or low abundance of SCF substrates, and the presence of ubiquitous cellular phosphatases that counteract the targeting kinases, conspire to make substrate identification an arduous task. In addition to genetic identification and educated guesses, direct purification of substrates has on occasion yielded E3 components, as in the case of the G1 cyclin Cln2 in yeast (Willems *et al.*, 1996) and β-TrCP in human cells (Yaron *et al.*, 1998). Systematic mass spectrometric and two-hybrid screens have not, however, yielded F-box protein–substrate interactions (Ho *et al.*, 2002; Uetz *et al.*, 2000). The nature of the Cdc4-substrate interaction poses a particularly difficult problem in substrate identification in that high-affinity recognition may often demand quantitative phosphorylation at a number of sites, which can be difficult to achieve both *in vitro* and *in vivo* (Nash *et al.*, 2001). Once in hand, though, phosphodegrons and/or targeting kinases provide handles for systematic methods to survey the full repertoire of SCF substrates. This chapter describes two such approaches: (1) the application of Spot peptide membrane arrays to pan for high-affinity Cdc4 phosphodegrons and (2) direct surveys of full-length recombinant CDK substrates for Cdc4 interactions (Fig. 3). These approaches complement and extend the conventional genetic and biochemical methods initially used to identify SCF substrates.

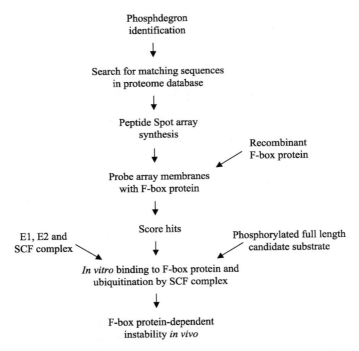

Fig. 3. Schema for genome-wide approaches to SCF substrate identification.

Genome-Wide Scans for Functional CPD Motifs

The development of Spot peptide synthesis technology has afforded the ability to synthesize and subsequently to screen large numbers of synthetic peptides, as well as other organic compounds, in an array format on a cellulose membrane support (Frank, 1992). Semiautomated peptide synthesis on a robotic workstation requires little experience in organic chemistry and has made the technique available to many research laboratories. Through synthesis of overlapping peptide series that span an entire protein sequence, the Spot peptide array has emerged as a powerful technology to identify phosphorylation sites, antibody-binding sites, and linear-sequence motifs that mediate protein–protein interactions (Frank, 1992). Previously, we used a Spots array, in which every residue of a high-affinity CPD derived from human cyclin E was systematically substituted with each of the 20 naturally occurring amino acids, to deduce an optimal CPD consensus motif (Fig. 2). To identify candidate Cdc4 substrates, we searched the predicted yeast proteome for matches to the CPD. To expand the scope of this in silico screen, weakly favored residues at the -1 and

−2 positions and basic residues at the +3 and +4 sites were allowed in the genome-wide motif search. That is, all matches to the sequence [ILQFPT-WYV]-[ILPQFWVA] -pT-P-{RK}-{RK}, where {} indicates any amino acid other than Arg or Lys, were scored as hits. Note that possible CPD sites that contain the less favored serine phosphorylation site (Nash *et al.*, 2001) were not included because of limitations in synthesis for this pilot scale study. The pattern match search function available at the *Saccharomyces* Genome Database (SGD; http://www.yeastgenome.org) was used to identify 1132 unique peptide matches in 960 proteins. A Spot array corresponding to each of these 1132 peptides, built as 11 mers centered on the threonine phosphorylation site, was synthesized on membranes and screened for Cdc4 interaction, as described in the following.

Spot-Synthesis of Peptide Arrays

Peptide arrays were constructed according to the Spot-synthesis method (Frank, 1992). Acid-hardened cellulose membranes prederivatized with polyethylene glycol (AbiMed, Langfield, Germany) were spotted with a grid of Fmoc β-alanine (Bachem) before peptide synthesis. Amino acids were spotted in a high density 24 × 16 format on 130 × 90-mm membranes using an AbiMed ASP422 robot. Activated Fmoc-amino acids were twice spotted at a volume of 0.2 μl for each coupling reaction. At the end of each synthesis cycle, 20 min was allowed for reaction completion. Amino acid stock preparation, activation, washing steps, Fmoc de-protection, and side chain de-protection were performed manually essentially according to the manufacturer's instructions.

Amino Acid Stock Solution Preparation. Fmoc-amino acids are supplied in cartridges of 0.5 mmol each in a dry and nonactivated form. To the dry stocks, add 1.5 ml of 0.5 M hydroxybenzotriazole in dry 1-methyl-2-pyrrolidine (NMP). Vortex briefly and rotate the cartridges for 0.5–1 h. Check each vial visually for complete dissolution. Nonactivated Fmoc-amino acid stock solutions can be stored at 4° for up to 1 week.

Activation of Fmoc-Amino Acids. On the day of synthesis, prepare an activator solution of 1.1 M diisopropyl carbodiimide in NMP. Activate Fmoc-amino acids by adding one part of the above activator solution to three parts of the above amino acid stock solution. The resulting 0.25 M activated Fmoc-amino acid solutions are stable for 1 day with the exception of arginine, which decomposes rapidly and, therefore, should be freshly made just before each synthesis cycle.

Array Synthesis. After each synthesis cycle, the membrane is transferred to a polypropylene box and processed as follows. *All steps should be carried out in a fume hood with appropriate protection.* Glass vessels should be used for all solution manipulations and waste disposal.

1. Capping: submerge membrane in capping solution (15 ml dimethylformamide (DMF) + 300 µl acetic anhydride) and incubate for 30 sec. Decant and repeat with fresh solution for 2 min. Note that although this step is recommended by the manufacturer to increase coupling efficiency at each cycle, we have found that synthesis efficiency is adequate without capping, so this step is optional.
2. DMF wash: 1 × 30 sec, 1 × 2 min.
3. Fmoc de-protection: mix 80 ml DMF with 20 ml piperidine, add 25 ml of the resultant solution to each membrane, and incubate for 10 min.
4. DMF wash: 1 × 30 sec, 4 × 2 min.
5. Ethanol wash: 1 × 30 sec, 2 × 2 min.
6. Methanol wash: 1 × 2 min, then air-dry the membrane.
7. For the next cycle of synthesis, mount the membranes in their previous position on the robot holder.

All the preceding steps are carried out with gentle agitation of membrane on a nutator. DMF, ethanol, and methanol are flammable solvents, and so never use hot air, such as from a hair dryer, to dry the membranes. If multiple membranes are used, each sheet must be processed in a separate polypropylene container. Note that the membranes should be differentially marked with pencil (only graphite pencil survives the chemical procedures). For subsequent detection of peptide interactions on the membrane, it is also important to mark the peptide grid so as to distinguish rows and columns. Mark the grid with pencil when peptide spots are still wet in the last cycle. After the final synthesis, treat the membranes following steps 2–6 above (i.e., skip the capping step). Let the membrane dry overnight in a vacuum desiccator.

Side Chain De-Protection. This step to remove blocking groups is performed with trifluoroacetic acid (TFA) and triisopropylsilane as a catalyst. Note that TFA is a strong acid that causes severe burns and lung damage if vapors are inhaled. Side chain de-protection and solution preparation must be carried out in a fume hood and reactions performed in a polypropylene box with a lid. *Do not mix DMF waste with TFA waste, because these reactants undergo a strongly exothermic reaction that can splash or boil over. Always wear lab coat, gloves, and eye protection.*

Prepare de-protection solution by mixing the following:

5 ml TFA
5 ml dichloromethane (DCM)
300 µl triisopropylsilane
200 µl water

Add the de-protection solution onto the membrane and let stand for 1 h, then wash the membrane as follows: 4 × with 20 ml DCM; 4 × 2 min with DMF; 2 × 2 min with ethanol; 1 × 2 min with methanol. Air-dry the membrane, which can then be probed immediately or stored dry at 4° for a short term or in a vacuum desiccator at −20° for several months.

Preparation of Recombinant Cdc4-Skp1 Complex

Expression and purification of recombinant Cdc4-Skp1 complex has been described previously (Orlicky et al., 2003). Briefly, ^{His6}Cdc4 and ^{GST}Skp1 were cloned into same expression vector (pPROEX HTa) and the proteins co-expressed in BL21 Codon Plus competent cells (Stratagene). The cell pellet was lysed in 50 mM HEPES, pH 7.5, 500 mM NaCl, 10% glycerol, and 5 mM imidazole with 2 mM PMSF using an EmulsiFlex-C5 High Pressure Homogenizer (Avestin, Inc.). The lysate was centrifuged at 48,000g for 30 min to remove cell debris and then filtered through a two-stage Whatman/0.45-μm PVDF syringe filter (Fisher). ^{His6}Cdc4-^{GST}Skp1 complex was affinity purified on a Ni^{2+} chelate column (HiTrap HP, Pharmacia), concentrated, and used without further purification. Sample purity was estimated at ∼65%. One liter of culture yielded ∼1 mg of ^{His6}Cdc4 in a 1:1 complex with ^{GST}Skp1, an amount sufficient to probe one 130 × 90 mm membrane. For other applications, we have found that a similar quantity of protein is required, for example, in probing arrays with isolated protein domains or active protein kinases.

Probing the Spot Peptide Array

In all procedures, the membrane arrays are handled and probed in the same manner as an immunoblot. To obviate potential nonspecific antibody interactions with certain peptides, it is desirable to detect protein interactions with a different antibody against the probe protein of interest. Signals obtained with both antibodies are likely to reflect specific protein–peptide interactions. The membrane can be stripped and reprobed up to three times with little reduction in signal intensity (Darji et al., 1996; Niebuhr et al., 1997). A generic procedure for most protein probes is as follows:

1. Immerse the membrane in ethanol before transfer into aqueous solution.
2. Transfer ethanol-wetted membrane into TBS/0.05% Tween-20 (TBST). Wash in TBST three times, block nonspecific binding at 4° overnight with 5% (w/v) skim milk powder in TBST in a sealed plastic pouch.

3. Incubate the membrane with 5 ml TBS containing 2 μM ^{His6}Cdc4-^{GST}Skp1 at 4° for 1.5 h. Save the ^{His6}Cdc4-^{GST}Skp1 solution, because it can be reused several times, for example, in probing multiple membranes.
4. Wash 3 × 10 min in TBS.
5. For detection, incubate the membrane with anti-Skp1 polyclonal antibody (Bai *et al.*, 1996) at 1:2000 dilution in 5% (w/v) skim milk powder in TBST at room temperature for 30 min. For detection of other proteins of interest, use antibodies at concentrations appropriate for immunoblot analysis.
6. Wash 3 × 10 min in TBST.
7. Add HRP-coupled anti-rabbit secondary antibody (Amersham) at 1:10000 dilution in 5% (w/v) skim milk powder in TBST.
8. Wash 3 × 10 min in TBST.
9. Detect with SuperSignal ECL reagent (Pierce).
10. To ascertain true positives, strip the membrane (see below), reblock, and reprobe with ^{His6}Cdc4-^{GST}Skp1 followed by second antibody against the primary protein probe. For the study reported here, a mouse monoclonal antibody (Qiagen) directed against tetrahistidine on Cdc4 was used at 1:1250 dilution.

Regenerating the Membrane. The membrane may be gently stripped for use with subsequent probes (Darji *et al.*, 1996). Each washing step is performed three times for 10 min at room temperature.

1. Wash the membrane with water.
2. Wash with stripping buffer A (8 M urea, 1% w/v SDS, 0.5% v/v β-mercaptoethanol).
3. Wash with stripping buffer B (10% (v/v) acetic acid, 50% (v/v) ethanol).
4. Wash with ethanol. If desired, verify complete elimination of the original signal, rinse in TBS, and re-develop with ECL reagent.
5. Rinse in TBST, then either reprobe immediately or rinse in methanol, air-dry, and store at $-20°$ for later use.

Synopsis of CPD Peptides That Bind Cdc4

Peptides that retrieved the Skp1-Cdc4 complex in both the anti-Skp1 and anti-tetrahistidine (Cdc4) blots were considered *bona fide* CPD candidates. Two representative peptide membranes yielded a range of signal intensities (Fig. 4). Signals were qualitatively classified as strong (strong signals with both antibodies), weak (weak signals with both antibodies or strong with one and weak with the other), or none (no

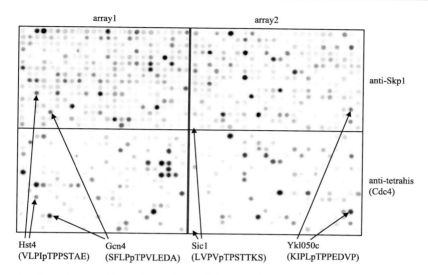

FIG. 4. Proteome-wide analysis of potential Cdc4 substrates by Spot peptide membrane array. Phosphopeptides corresponding to 1132 matches to the CPD consensus in the yeast proteome were synthesized on a membrane array, probed with recombinant ^{His6}Cdc4-^{GST}Skp1 complex, and then blotted with anti-Skp1 antibody (upper panels) and anti-tetrahistidine antibody (lower panels). Arrows indicate both positive and negative hits with corresponding phosphopeptide sequences.

signal with one or both antibodies). The 1132 original peptide array yielded ~50 strong hits and ~300 weak hits, several of which are annotated in Fig. 4. Among the strong hits was a peptide derived from the transcription factor Gcn4 that corresponds to a known high-affinity site (Table I). Notably, the best natural sequence match to the CPD consensus in Sic1, LVPVT$_{45}$PSTTKS, exhibited only background signals (Fig. 4), consistent with its weak K_d of >100 μM (Nash et al., 2001). A candidate CPD centered on Thr63 of Far1, another known multisite phosphorylation-dependent substrate, was also negative in this assay. Thus, as expected, the Spot membrane array method detects only high-affinity CPD interactions. An ancillary benefit of the peptide array approach is that a more sophisticated definition of the CPD may be built, which, for example, may disallow pairs of mismatched residues that in isolation would otherwise be permitted. Finally, the full repertoire of functional CPD sites in the yeast proteome might be further elaborated in more degenerate screens, for example, by allowing the somewhat more disfavored serine phosphorylation site in the motif search.

Secondary Tests of Candidate Cdc4 Substrates

Full-length proteins corresponding to the CPD peptide hits were then validated in secondary assays to establish *bona fide* Cdc4 substrates. Standard assays for SCF^{Cdc4} substrates include: (1) ability to bind recombinant Cdc4 *in vitro*; (2) *in vitro* ubiquitination by SCF^{Cdc4}; (3) *CDC4*-dependent protein instability *in vivo*. Details for each of these assays are described in the following.

Preparation of Recombinant Substrates

Recombinant proteins for *in vitro* assays are readily produced to analytical levels in yeast by expression as C-terminal FLAG epitope fusions from the inducible *GAL1* promoter using the Gateway cloning system (Invitrogen), as described previously (Ho *et al.*, 2002). This method has the advantage that, unlike bacterial expression systems, most proteins are expressed to some degree, and potential phosphorylation or other modifications that might augment recognition by Cdc4 are more likely to occur in a homologous overexpression system. Fortuitous overlap of strong hits in the CPD peptide array with a previously cloned set of genes (Ho *et al.*, 2002) enabled 35 candidate Cdc4 substrates to be tested in secondary assays. Yeast culture was performed according to standard methods (Kaiser *et al.*, 1994). Proteins were expressed in a *pep4* strain, which is deficient in vacuolar proteases, to minimize postlysis degradation. After overnight growth of a 100 ml culture in non-inducing raffinose (2% w/v) medium to a cell density of $\sim 1 \times 10^7$ cells per ml, protein expression was induced with galactose (2% w/v) for 1.5 h. Cell extracts were prepared by glass bead lysis and FLAG-tagged proteins immunopurified on 10 μl of anti-FLAG M2 monoclonal antibody resin (Sigma) as described (Ho *et al.*, 2002). Because of intrinsic differences in expression and stability for any given protein, yields ranged from 0.1 μg–2 μg, amounts easily sufficient for *in vitro* binding and ubiquitination assays.

To ensure maximal levels of substrate phosphorylation, purified proteins were phosphorylated by recombinant Cln2-Cdc28 kinase produced in insect cells (Feldman *et al.*, 1997; Skowyra *et al.*, 1997). Sf9 insect cells were coinfected with recombinant baculoviruses that express Cln2, GST-HA-Cdc28, and His_6-Cks1. After 3 days, cells were lysed in buffer 3 (50 mM Tris-Cl, pH 7.5, 100 mM NaCl, 5 mM NaF, 5 mM EDTA, 0.1% NP-40, 1 mM DTT, 100 μg/ml PMSF, 1 μg/ml aproteinin, 1 μg/ml leupeptin, 1 μg/ml pepstatin) and the kinase complex purified on glutathione-agarose resin (Sigma). Soluble Cln2-Cdc28 was eluted from the resin with 20 mM glutathione in a buffer containing 100 mM Tris-HCl, pH 7.0, 100 mM NaCl, 1 mM EDTA, 0.5 mM NaF, and 10% glycerol. *Ensure that the final pH of the solution is correct after glutathione is dissolved.* For substrate phosphorylation, elute half of the yeast protein preparation from the anti-FLAG M2 resin with 25 μl of FLAG peptide at 0.2 mg/ml in 50 mM HEPES, pH 7.5, 100 mM NaCl. Both immobilized and eluted protein fractions are phosphorylated with Cln2-Cdc28 kinase for *in vitro* binding and ubiquitination assays, respectively. To phosphorylate eluted proteins, 20 μl of eluted protein, 10 μl of 5× kinase buffer (250 mM Tris-HCl, pH 7.5, 50 mM MgCl$_2$, 5 mM ATP, 5 mM DTT), and 5 μl of Cln2-Cdc28 kinase purified on glutathione-agarose are incubated in a total volume of 50 μl for 1 h at 30°. To phosphorylate proteins bound to anti-FLAG resin, the bead slurry is mixed with 10 μl of 5× kinase buffer and 5 μl of eluted Cln2-Cdc28 kinase in a total volume of 50 μl for 1 h at 30°.

Capture of Substrate by Cdc4 In Vitro

A primary requirement for Cdc4 substrates is that they detectably interact with Cdc4 *in vitro*. This *in vitro* method is preferable to detection *in vivo* because of low substrate abundance, substrate instability, and the potent activity of phosphatases that remove targeting phosphate residues. To determine whether the phosphorylated substrate protein binds to Cdc4, 40 μl of the soluble substrate preparation is added to 50 μl of glutathione-sepharose slurry containing 0.2 μg Cdc4 (in the form of Skp1-Cdc4 complex; see previously) and incubated at 4° for 1h on a nutator. Blank glutathione-Sepharose may be used as a control for nonspecific binding to resin. The beads are washed three times in buffer 3, then solubilized in SDS-PAGE sample buffer. Cdc4 bound proteins and 5 μl of input proteins are analyzed by anti-FLAG immunoblot. As further controls, phosphorylation-dependent binding can be demonstrated by loss of binding on lambda protein phosphatase pretreatment of substrate and the reappearance of binding when dephosphorylated proteins are retreated with Cln2-Cdc28 kinase. Phosphorylation-dependent capture of a representative new Cdc4 substrate, Hst4, which is an NAD$^+$-dependent deacetylase of the Sir2 family, is shown in Fig. 5.

Ubiquitination of Substrate by SCF^{Cdc4} In Vitro

Substrate ubiquitination reaction may be carried out in a fully soluble format or with either substrate or E3 complex bound to resin. As described previously, phosphorylated substrate was bound to beads, such that the SCF^{Cdc4} complex must be prepared in soluble form. Each component of the recombinant ubiquitination reaction mixture was prepared as follows. The E1 enzyme (^{His6}Uba1) enzyme produced by overexpression in yeast was purified over a Ni^{2+} chelate column and step-gradient elution from a

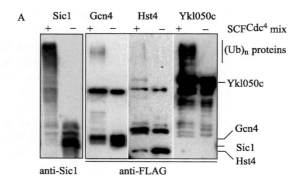
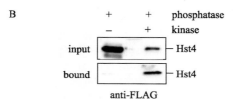

FIG. 5. In vitro ubiquitination and Cdc4 binding of selected proteins that contain functional CPD motifs. (A) Purified FLAG-tagged proteins were phosphorylated with Cln2-Cdc28 kinase then incubated with complete SCF^{Cdc4} ubiquitination mix containing yeast Uba1, Cdc34, SCF^{Cdc4}, ubiquitin, and ATP. Ubiquitination results in depletion of input phosphorylated substrate and accumulation of high M_h substrate–ubiquitin conjugates. (B) Phosphorylation-dependent Cdc4 binding of a representative substrate. FLAG-Hst4 was dephosphorylated with lambda phosphatase, then either phosphorylated or not by recombinant Cln2-Cdc28 kinase before capture on Cdc4-Skp1 resin.

DEAE-Sepharose column (Skowyra et al., 1997). The cognate E2 enzyme (^{His6}Cdc34) is produced in bacteria from an inducible pET16b expression vector and purified on Ni^{2+} chelate resin (Skowyra et al., 1997). The SCF^{Cdc4} complex is expressed in Sf9 insect cells by coinfection with baculoviruses that individually express ^{Myc3}Cdc53, Cdc4, Skp1, and ^{His6}Rbx1. The complex is captured from crude cell lysates on a Ni^{2+} chelate column and eluted with 100 mM imidazole in ubiquitination buffer (see later). The amount of active SCF^{Cdc4} complex in the preparation is estimated by the quantity of Cdc4, as judged by Coomassie Brilliant Blue stain of SDS-PAGE separated material.

For the ubiquitination reaction, equilibrate the phosphorylated substrate on beads with 1× ubiquitination buffer (50 mM Tris-HCl, pH 7.5, 2 mM ATP, 10 mM MgCl$_2$, 0.1 mM DTT), then add 10 μl of a ubiquitination mix that contains 0.2 μg His$_6$-E1, 0.4 μg His$_6$-Cdc34, 0.2 μg SCF^{Cdc4} (judged by Cdc4 abundance), 1 μg ubiquitin (Sigma), 1× ubiquitination buffer. Incubate the reaction at 30° for 1 h with occasional agitation. Reactions are terminated with 4 μl of 6× SDS-PAGE sample buffer and analyzed by anti-FLAG immunoblot. Productive ubiquitination reactions result in depletion of the input phosphorylated substrates and accumulation of high M_r substrate–ubiquitin conjugates. Representative SCF^{Cdc4} ubiquitination reactions for a Sic1 positive control and three substrates identified in the CPD peptide array screen, the transcription factor Gcn4, the NAD$^+$-dependent deacetylase Hst4, and an uncharacterized protein Ykl050c are shown in Fig. 5.

CDC4-Dependent Protein Instability In Vivo

A crucial test of whether or not a protein is a *bona fide* substrate of an E3 enzyme is stabilization of the substrate *in vivo* on inactivation of the E3 by a conditional mutation. The substrates that passed *in vitro* biochemical tests must thus be assessed for protein stability in a wild-type strain and in a temperature-sensitive *cdc4-1* strain. Cdc4-dependent stability of candidate substrates is assessed as follows:

1. Transform plasmids bearing desired ORF expressed under the *GAL1* promoter (see preceding) into wild-type and *cdc4-1* strain using standard procedures (Kaiser *et al.*, 1994).

2. Inoculate a starter culture in selective medium containing raffinose (2% w/v) as carbon source and grow at 25° overnight. Dilute the culture 1 to 50 in 50 ml of raffinose medium, grow until culture density reaches ~0.8 × 10^7 cells per ml, induce protein expression by adding galactose (2% w/v) for 30 min, then shift the cultures to a 37° shaking water bath for another 30 min.

3. Repress expression by adding glucose (2% w/v) and harvest 5 ml of culture on addition of glucose (0 min) and at 20, 40, 60, and 80 min thereafter. For each time point, cells are rapidly pelleted, resuspended in 300 μl of cold 20% TCA, transferred to a 2-ml Eppendorf tube, and snap frozen in liquid N$_2$.

4. To protect proteins from degradation by proteases during cell lysis, total protein is extracted directly in TCA (Kuras *et al.*, 2002). Add ~300 μl volume of acid-washed glass beads to the frozen TCA pellet, vortex 4 × 30 sec in cold room, and chill the tubes on ice for 1 min after each vortex cycle. Transfer the broken cell lysate to a fresh tube with a cutoff 200-μl tip. Wash the beads with 200 μl 5% TCA and combine with original lysate.

5. Spin the crude cell lysate at 4° for 10 min at 14,000 rpm in a Microfuge. Resuspend the white protein pellet in 30 μl buffer 3 and add 7 μl 6× SDS-PAGE loading buffer. The acidic solution will turn the dye yellow. Neutralize by adding 1 M NaOH with a P20 pipetman while vortexing until solution just turns blue (1–3 μl of 1 M NaOH is usually sufficient).

6. Boil samples for 5 min, spin briefly, and load 10 μl of supernatant onto an SDS-PAGE gel for immunoblot analysis. A representative experiment demonstrating *CDC4*-dependent instability of Hst4 on *GAL1* promoter repression is shown in Fig. 6.

As an alternative to the preceding promoter shut-off regimen, protein stability in wild-type and *cdc4-1* cells can be crudely estimated by inhibition of translation with cycloheximide, followed by immunoblot analysis of epitope-tagged protein expressed from the endogenous locus, or if available, a polyclonal antibody against native protein (Schneider *et al.*, 1998). A more rigorous estimate of protein stability requires a [^{35}S]-methionine/cysteine pulse-chase protocol, again using either epitope-tagged alleles or polyclonal antibody to the native protein (Patton *et al.*, 1998a).

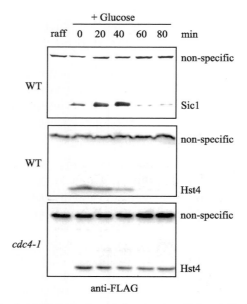

FIG. 6. Cdc4-dependent substrate degradation *in vivo*. Sic1 and Hst4 were conditionally expressed from the *GAL1* promoter in the indicated strains and protein abundance assessed by immunoblot at the indicated times after promoter repression.

Synopsis of Secondary Tests for Cdc4 Substrates

In the preceding studies, the *in vitro* criteria of substrate binding to recombinant Cdc4 and substrate ubiquitination by recombinant SCF^{Cdc4} indicate that 12 of 35 randomly chosen hits from the Spot peptide array are indeed *bona fide* Cdc4 substrates (Fig. 5). Furthermore, one of the substrates tested *in vivo*, Hst4, is rapidly degraded in wild-type cells and completely stabilized in the *cdc4-1* temperature-sensitive strain (Fig. 6). Hst4 is one of four homologs of the Sir2 NAD$^+$-dependent deacetylase, which is implicated in gene silencing in yeast (Smith *et al.*, 2000). It is possible that Hst4 instability limits its activity and prevents silencing at ectopic sites in the genome. Hits in the Spots peptide array that fail to meet secondary *in vitro* and *in vivo* substrate criteria are likely to contain sites that are altogether inaccessible for phosphorylation in the context of tertiary protein structure, sites that are only inefficiently phosphorylated by the chosen recombinant kinase, and/or are localized in different subcellular compartments than Cdc4, which is found predominantly in the nucleus (Blondel *et al.*, 2000). Most of the ~350 candidate substrates from the Spots peptide array screen remain to be characterized by the preceding assays, but if the initial fraction of true positives holds, we estimate that ~100 new Cdc4 substrates may be identified by this method.

Multisite CDK Substrates as Potential SCF^{Cdc4} Substrates

A drawback of the Spot peptide array method is that by definition it can only identify substrates with one or more high-affinity phosphodegrons, such as in Gcn4 or cyclin E (Nash *et al.*, 2001). Substrates that interact with Cdc4 through multiple low-affinity phosphodegrons, such as Sic1, were predictably not recovered on the array. However, substrates of this nature may be uncovered in a focused candidate approach by directly screening proteins that contain multiple kinase phosphorylation sites. The prevalence of minimal sites for proline-directed kinases (S/T-P), preferred CDK sites (S/T-P-X-K/R), and CPD sites in the yeast proteome is shown in Fig. 7. The number of candidate Cdc4 substrates based on the criterion of containing multiple CDK consensus sites is substantial, because in the 2004 annotation of the yeast genome, 402 proteins contain two or more consensus S/T-P-X-K/R motifs (Fig. 7). This number of potential targeting sites is significantly larger if the minimal S/T-P phosphorylation site motif is used. Recently, a set of some ~200 CDK substrates has been identified by direct Cdc28 phosphorylation of a candidate substrate set (Ubersax *et al.*, 2003). In this approach, each candidate protein was produced as a GST fusion protein in yeast, purified, and then phosphorylated in crude yeast lysate bearing an

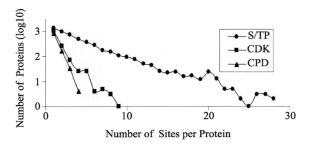

FIG. 7. Occurrence of minimal consensus S/TP sites, CDK sites, and CPD sites in the yeast proteome. Site occurrence was calculated with the pattern match algorithm available at http://www.yeastgenome.org/.

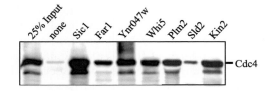

	p-score	S/TP	CDK	CPD	spot score
Sic1	3.7	9	3	1	0
Far1	2.8	15	4	2	0
Plm2	3.9	14	4	2	0
Kin2	3.6	11	1	2	0
Ynr047w	3.6	16	3	0	N/A
Whi5	2	12	4	0	N/A
Sld2	3	11	5	0	N/A

FIG. 8. *In vitro* binding of CDK substrates by Cdc4. Candidate substrates with high *p* scores as defined by Ubersax *et al.* (2003), but that did not score (0) or were not tested (N/A) in the CPD peptide array, were purified as immobilized FLAG-tagged proteins from yeast, phosphorylated with Cln2-Cdc28, and used to capture ^{His6}Cdc4-^{GST}Skp1 complex from solution. Cdc4 was detected by immunoblot with anti-tetrahistidine antibody (Qiagen).

analog-sensitive version of Cdc28, which is able to use a modified form of ATP that is refractory to all natural kinases (Ubersax *et al.*, 2003). By the combined criterion of site density, protein abundance, and [^{32}P]-phosphate incorporation, *p* value scores were established that suggested at least 181 likely *in vivo* substrates for Cdc28. To survey whether these Cdc28 substrates may also be substrates for Cdc4, we selected an initial set of 94 substrates enriched for multiple CDK sites but that lacked optimal CPD sites. These proteins were produced as FLAG fusions using preexisting Gateway clones (Ho *et al.*, 2002), phosphorylated by Cln2-Cdc28, and assessed for their ability to capture Cdc4 onto anti-FLAG resin, essentially as described previously. As positive controls, the known Cdc4 substrates Sic1 and Far1, which score highly as CDK substrates, were effectively captured by Cdc4 (Fig. 8). By the criterion of Cdc4 binding, 23 additional candidate substrates that did not contain consensus matches to the optimal CPD, or that were negative in the CPD peptide arrays, were recovered by this method. The capture of Cdc4 by a representative set of these candidates is shown in Fig. 8. Whether or not these proteins exhibit Cdc4-dependent instability *in vivo* or serve as ubiquitination substrates *in vitro* remains to be determined. However, if the initial hit rate in this candidate approach holds, we estimate that approximately 50 new Cdc4 substrates may be uncovered. The CDK substrate candidate approach thus represents an efficient means to capture Cdc4 substrates as targeted by CDK-dependent phosphorylation.

Conclusions

Spot peptide array technology has emerged as a powerful yet facile means for identification and characterization of protein interaction motifs. Application of this method enables a rapid survey of short degrons that target substrates to E3 enzymes (Varshavsky, 1991). In the example described previously, delineation of the optimal CPD consensus followed by synthesis of all potential matches in the yeast proteome identified many dozens, potentially even hundreds, of candidate Cdc4 substrates. In parallel, the straightforward, but somewhat more labor intensive, approach of directly screening potential substrates on the basis of multiple CDK sites effectively captures additional substrates that lack overt matches to the CPD. All told, this broad repertoire of new substrates significantly enlarges the scope of potential biological processes under Cdc4 control. These methods should be widely applicable once degrons for other F-box proteins, other SCF-like complexes, and, indeed, E3 enzymes in general are identified (Willems *et al.*, 2004). For example, the di-phosphorylated degron recognized by β-TrCP, $DpSGX_{2-4}pS$, has been defined in many substrates (Table 1) and, thus, is ideally suited for the Spots peptide array approach. Analogous stringent criteria for mammalian SCF substrates could be readily applied to candidate hits from Spots peptide array screens (see Chapter 20).

The two direct approaches to E3 substrate identification described here complement a number of genome-wide methods that have been deployed in other studies. For example, it is now possible to identify with some reliability ubiquitinated species on a proteomic scale by mass spectrometry based on capture of histidine-tagged ubiquitin derivatives (Peng et al., 2003) and on the signature di-glycine modification of lysine in substrate peptide fragmentation patterns (Cooper et al., 2004). Differential analysis of wild-type versus E3 mutant strains or tissue culture cell lines by these methods should in principle identify candidate E3 substrates. Cognate substrates may also in principle be directly captured by recombinant protein interaction domains of F-box proteins and identified by mass spectrometry, although the requirements for substrate phosphorylation and the presence of highly active phosphatases in cell lysates has rendered such efforts problematic to date. In a genetic approach, overexpression of substrates in E3 mutant strains often results in lethality, as for example with *FAR1* or *SIC1* overexpression in *cdc4* strains (Henchoz et al., 1997). Systematic overexpression of all yeast ORFs using genetic array technology may thus identify new substrates without the need for difficult biochemical manipulations (Tong et al., 2001). The combination of all these methods should elaborate the full corpus of SCF substrates and open a host of new questions on issues ranging from the signals that trigger substrate degradation to the evident competitive flux of myriad substrates through an E3 pathway.

Acknowledgments

We thank Danielle Dewar for technical assistance, Steve Elledge for anti-Skp1 antibody, and David Morgan, Jeff Ubersax, Mart Loog, and Wade Harper for providing unpublished data and helpful discussions. MT and FS are supported by grants from the National Cancer Institute of Canada and the Canadian Institutes of Health Research.

References

Bai, C., Sen, P., Hofmann, K., Ma, L., Goebl, M., Harper, J. W., and Elledge, S. J. (1996). SKP1 connects cell cycle regulators to the ubiquitin proteolysis machinery through a novel motif, the F-box. *Cell* **86,** 263–274.

Berset, C., Griac, P., Tempel, R., La Rue, J., Wittenberg, C., and Lanker, S. (2002). Transferable domain in the G_1 cyclin Cln2 sufficient to switch degradation of Sic1 from the E3 ubiquitin ligase SCF^{Cdc4} to SCF^{Grr1}. *Mol. Cell Biol.* **22,** 4463–4476.

Blondel, M., Galan, J. M., Chi, Y., Lafourcade, C., Longaretti, C., Deshaies, R. J., and Peter, M. (2000). Nuclear-specific degradation of Far1 is controlled by the localization of the F-box protein Cdc4. *EMBO J.* **19,** 6085–6097.

Chi, Y., Huddleston, M. J., Zhang, X., Young, R. A., Annan, R. S., Carr, S. A., and Deshaies, R. J. (2001). Negative regulation of Gcn4 and Msn2 transcription factors by Srb10 cyclin-dependent kinase. *Genes Dev.* **15,** 1078–1092.

Chou, S., Huang, L., and Liu, H. (2004). Fus3-regulated Tec1 degradation through SCF^{Cdc4} determines MAPK signaling specificity during mating in yeast. *Cell* **119,** 981–990.

Coccetti, P., Rossi, R. L., Sternieri, F., Porro, D., Russo, G. L., di Fonzo, A., Magni, F., Vanoni, M., and Alberghina, L. (2004). Mutations of the CK2 phosphorylation site of Sic1 affect cell size and S-Cdk kinase activity in *Saccharomyces cerevisiae. Mol. Microbiol.* **51**, 447–460.

Cooper, H. J., Heath, J. K., Jaffray, E., Hay, R. T., Lam, T. T., and Marshall, A. G. (2004). Identification of sites of ubiquitination in proteins: A Fourier transform ion cyclotron resonance mass spectrometry approach. *Anal. Chem.* **76**, 6982–6988.

Darji, A., Niebuhr, K., Hense, M., Wehland, J., Chakraborty, T., and Weiss, S. (1996). Neutralizing monoclonal antibodies against listeriolysin: Mapping of epitopes involved in pore formation. *Infect. Immun.* **64**, 2356–2358.

Deshaies, R. J. (1999). SCF and Cullin/Ring H2-based ubiquitin ligases. *Annu. Rev. Cell Dev. Biol.* **15**, 435–467.

Feldman, R. M., Correll, C. C., Kaplan, K. B., and Deshaies, R. J. (1997). A complex of Cdc4p, Skp1p, and Cdc53p/cullin catalyzes ubiquitination of the phosphorylated CDK inhibitor Sic1p. *Cell* **91**, 221–230.

Frank, R. (1992). Spot-synthesis: An easy technique for positionally addressable, parallel chemical synthesis on a membrane support. *Tetrahedron.* **48**, 9217–9232.

Fryer, C. J., White, J. B., and Jones, K. A. (2004). Mastermind recruits CycC:CDK8 to phosphorylate the Notch ICD and coordinate activation with turnover. *Mol. Cell.* **16**, 509–520.

Ganoth, D., Bornstein, G., Ko, T. K., Larsen, B., Tyers, M., Pagano, M., and Hershko, A. (2001). The cell-cycle regulatory protein Cks1 is required for SCF^{Skp2}-mediated ubiquitinylation of p27. *Nat. Cell Biol.* **3**, 321–324.

Hart, M., Concordet, J. P., Lassot, I., Albert, I., del los Santos, R., Durand, H., Perret, C., Rubinfeld, B., Margottin, F., Benarous, R., and Polakis, P. (1999). The F-box protein b-TrCP associates with phosphorylated b-catenin and regulates its activity in the cell. *Curr. Biol.* **9**, 207–210.

Henchoz, S., Chi, Y., Catarin, B., Herskowitz, I., Deshaies, R. J., and Peter, M. (1997). Phosphorylation and ubiquitin-dependent degradation of the cyclin-dependent kinase inhibitor Far1p in budding yeast. *Genes Dev.* **11**, 3046–3060.

Hershko, A., and Ciechanover, A. (1998). The ubiquitin system. *Annu. Rev. Biochem.* **67**, 425–479.

Hershko, A., Heller, H., Elias, S., and Ciechanover, A. (1983). Components of ubiquitin-protein ligase system. Resolution, affinity purification, and role in protein breakdown. *J. Biol. Chem.* **258**, 8206–8214.

Ho, Y., Gruhler, A., Heilbut, A., Bader, G. D., Moore, L., Adams, S. L., Millar, A., Taylor, P., Bennett, K., Boutilier, K., *et al.* (2002). Systematic identification of protein complexes in *Saccharomyces cerevisiae* by mass spectrometry. *Nature* **415**, 180–183.

Kaiser, C. S., Michaelis, S., and Mitchell, A. (1994). "Methods in yeast genetics." Cold Spring Harbor Laboratory Press, Cold Spring Harbor, NY.

Klein, P., Pawson, T., and Tyers, M. (2003). Mathematical modeling suggests cooperative interactions between a disordered polyvalent ligand and a single receptor site. *Curr. Biol.* **13**, 1669–1678.

Koepp, D. M., Schaefer, L. K., Ye, X., Keyomarsi, K., Chu, C., Harper, J. W., and Elledge, S. J. (2001). Phosphorylation-dependent ubiquitination of cyclin E by the SCF^{Fbw7} ubiquitin ligase. *Science* **294**, 173–177.

Kuras, L., Rouillon, A., Lee, T., Barbey, R., Tyers, M., and Thomas, D. (2002). Dual regulation of the Met4 transcription factor by ubiquitin-dependent degradation and inhibition of promoter recruitment. *Mol. Cell.* **10**, 69–80.

Lang, V., Janzen, J., Fischer, G. Z., Soneji, Y., Beinke, S., Salmeron, A., Allen, H., Hay, R. T., Ben-Neriah, Y., and Ley, S. C. (2003). b-TrCP-mediated proteolysis of NF-kB1 p105 requires phosphorylation of p105 serines 927 and 932. *Mol. Cell Biol.* **23,** 402–413.

Lassot, I., Segeral, E., Berlioz-Torrent, C., Durand, H., Groussin, L., Hai, T., Benarous, R., and Margottin-Goguet, F. (2001). ATF4 degradation relies on a phosphorylation-dependent interaction with the SCF$^{b\text{-}TrCP}$ ubiquitin ligase. *Mol. Cell Biol.* **21,** 2192–2202.

Margottin, F., Bour, S. P., Durand, H., Selig, L., Benichou, S., Richard, V., Thomas, D., Strebel, K., and Benarous, R. (1998). A novel human WD protein, hb-TrCP, that interacts with HIV-1 Vpu connects CD4 to the ER degradation pathway through an F-box motif. *Mol. Cell.* **1,** 565–574.

Margottin-Goguet, F., Hsu, J. Y., Loktev, A., Hsieh, H. M., Reimann, J. D., and Jackson, P. K. (2003). Prophase destruction of Emi1 by the SCF$^{b\text{-}TrCP/Slimb}$ Ubiquitin ligase activates the anaphase promoting complex to allow progression beyond prometaphase. *Dev. Cell.* **4,** 813–826.

Meimoun, A., Holtzman, T., Weissman, Z., McBride, H. J., Stillman, D. J., Fink, G. R., and Kornitzer, D. (2000). Degradation of the transcription factor Gcn4 requires the kinase Pho85 and the SCF^{CDC4} ubiquitin-ligase complex. *Mol. Biol. Cell.* **11,** 915–927.

Nash, P., Tang, X., Orlicky, S., Chen, Q., Gertler, F. B., Mendenhall, M. D., Sicheri, F., Pawson, T., and Tyers, M. (2001). Multisite phosphorylation of a CDK inhibitor sets a threshold for the onset of DNA replication. *Nature* **414,** 514–521.

Nateri, A. S., Riera-Sans, L., Da Costa, C., and Behrens, A. (2004). The ubiquitin ligase SCF^{Fbw7} antagonizes apoptotic JNK signaling. *Science* **303,** 1374–1378.

Niebuhr, K., Ebel, F., Frank, R., Reinhard, M., Domann, E., Carl, U. D., Walter, U., Gertler, F. B., Wehland, J., and Chakraborty, T. (1997). A novel proline-rich motif present in ActA of *Listeria monocytogenes* and cytoskeletal proteins is the ligand for the EVH1 domain, a protein module present in the Ena/VASP family. *EMBO J.* **16,** 5433–5444.

Nishizawa, M., Kawasumi, M., Fujino, M., and Toh-e, A. (1998). Phosphorylation of Sic1, a cyclin-dependent kinase (Cdk) inhibitor, by Cdk including Pho85 kinase is required for its prompt degradation. *Mol. Biol. Cell.* **9,** 2393–2405.

Orlicky, S., Tang, X., Willems, A., Tyers, M., and Sicheri, F. (2003). Structural basis for phosphodependent substrate selection and orientation by the SCF^{Cdc4} ubiquitin ligase. *Cell* **112,** 243–256.

Patton, E. E., Willems, A. R., Sa, D., Kuras, L., Thomas, D., Craig, K. L., and Tyers, M. (1998a). Cdc53 is a scaffold protein for multiple Cdc34/Skp1/F-box protein complexes that regulate cell division and methionine biosynthesis in yeast. *Genes Dev.* **12,** 692–705.

Patton, E. E., Willems, A. R., and Tyers, M. (1998b). Combinatorial control in ubiquitin-dependent proteolysis: don't Skp the F-box hypothesis. *Trends Genet.* **14,** 236–243.

Pawson, T. (1995). Protein modules and signalling networks. *Nature* **373,** 573–580.

Peng, J., Schwartz, D., Elias, J. E., Thoreen, C. C., Cheng, D., Marsischky, G., Roelofs, J., Finley, D., and Gygi, S. P. (2003). A proteomics approach to understanding protein ubiquitination. *Nat. Biotechnol.* **21,** 921–926.

Perkins, G., Drury, L. S., and Diffley, J. F. (2001). Separate SCF^{CDC4} recognition elements target Cdc6 for proteolysis in S phase and mitosis. *EMBO J.* **20,** 4836–4845.

Pickart, C. M. (2001). Mechanisms underlying ubiquitination. *Annu. Rev. Biochem.* **70,** 503–533.

Schneider, B. L., Patton, E. E., Lanker, S., Mendenhall, M., Wittenberg, C., Futcher, B., and Tyers, M. (1998). Yeast G1 cyclins are unstable in G1 phase. *Nature* **395,** 86–89.

Seol, J. H., Feldman, R. M., Zachariae, W., Shevchenko, A., Correll, C. C., Lyapina, S., Chi, Y., Galova, M., Claypool, J., Sandmeyer, S., Nasmyth, K., and Deshaies, R. J. (1999). Cdc53/

cullin and the essential Hrt1 RING-H2 subunit of SCF define a ubiquitin ligase module that activates the E2 enzyme Cdc34. *Genes Dev.* **13,** 1614–1626.

Skowyra, D., Craig, K. L., Tyers, M., Elledge, S. J., and Harper, J. W. (1997). F-box proteins are receptors that recruit phosphorylated substrates to the SCF ubiquitin-ligase complex. *Cell* **91,** 209–219.

Skowyra, D., Koepp, D. M., Kamura, T., Conrad, M. N., Conaway, R. C., Conaway, J. W., Elledge, S. J., and Harper, J. W. (1999). Reconstitution of G1 cyclin ubiquitination with complexes containing SCF^{Grr1} and Rbx1. *Science* **284,** 662–665.

Smith, J. S., Brachmann, C. B., Celic, I., Kenna, M. A., Muhammad, S., Starai, V. J., Avalos, J. L., Escalante-Semerena, J. C., Grubmeyer, C., Wolberger, C., and Boeke, J. D. (2000). A phylogenetically conserved NAD$^+$-dependent protein deacetylase activity in the Sir2 protein family. *Proc. Natl. Acad. Sci. USA.* **97,** 6658–6663.

Songyang, Z., and Cantley, L. C. (1995). SH2 domain specificity determination using oriented phosphopeptide library. *Methods Enzymol.* **254,** 523–535.

Spielewoy, N., Flick, K., Kalashnikova, T. I., Walker, J. R., and Wittenberg, C. (2004). Regulation and recognition of SCFGrr1 targets in the glucose and amino acid signaling pathways. *Mol. Cell Biol.* **24,** 8994–9005.

Spruck, C., Strohmaier, H., Watson, M., Smith, A. P., Ryan, A., Krek, T. W., and Reed, S. I. (2001). A CDK-independent function of mammalian Cks1: Targeting of SCF^{Skp2} to the CDK inhibitor p27^{Kip1}. *Mol. Cell.* **7,** 639–650.

Strohmaier, H., Spruck, C. H., Kaiser, P., Won, K. A., Sangfelt, O., and Reed, S. I. (2001). Human F-box protein hCdc4 targets cyclin E for proteolysis and is mutated in a breast cancer cell line. *Nature* **413,** 316–322.

Tong, A. H., Evangelista, M., Parsons, A. B., Xu, H., Bader, G. D., Page, N., Robinson, M., Raghibizadeh, S., Hogue, C. W., Bussey, H., Andrews, B., Tyers, M., and Boone, C. (2001). Systematic genetic analysis with ordered arrays of yeast deletion mutants. *Science* **294,** 2364–2368.

Tsvetkov, L. M., Yeh, K. H., Lee, S. J., Sun, H., and Zhang, H. (1999). p27^{Kip1} ubiquitination and degradation is regulated by the SCF^{Skp2} complex through phosphorylated Thr187 in p27. *Curr. Biol.* **9,** 661–664.

Ubersax, J. A., Woodbury, E. L., Quang, P. N., Paraz, M., Blethrow, J. D., Shah, K., Shokat, K. M., and Morgan, D. O. (2003). Targets of the cyclin-dependent kinase Cdk1. *Nature* **425,** 859–864.

Uetz, P., Giot, L., Cagney, G., Mansfield, T. A., Judson, R. S., Knight, J. R., Lockshon, D., Narayan, V., Srinivasan, M., Pochart, P., Qureshi-Emili, A., Li, Y., Godwin, B., Conover, D., Kalbfleisch, T., Vijayadamodar, G., Yang, M., Johnston, M, Fields,S., and Rothberg, J. M. (2000). A comprehensive analysis of protein-protein interactions in *Saccharomyces cerevisiae*. *Nature* **403,** 623–627.

Varshavsky, A. (1991). Naming a targeting signal. *Cell* **64,** 13–15.

Verma, R., Annan, R. S., Huddleston, M. J., Carr, S. A., Reynard, G., and Deshaies, R. J. (1997). Phosphorylation of Sic1p by G1 Cdk required for its degradation and entry into S phase. *Science* **278,** 455–460.

Wan, M., Tang, Y., Tytler, E. M., Lu, C., Jin, B., Vickers, S. M., Yang, L., Shi, X., and Cao, X. (2004). Smad4 protein stability is regulated by ubiquitin ligase SCF$^{b\text{-}TrCP1}$. *J. Biol. Chem.* **279,** 14484–14487.

Watanabe, N., Arai, H., Nishihara, Y., Taniguchi, M., Hunter, T., and Osada, H. (2004). M-phase kinases induce phospho-dependent ubiquitination of somatic Wee1 by SCF$^{b\text{-}TrCP}$. *Proc. Natl. Acad. Sci. USA* **101,** 4419–4424.

Welcker, M., Orian, A., Jin, J., Grim, J. A., Harper, J. W., Eisenman, R. N., and Clurman, B. E. (2004). The Fbw7 tumor suppressor regulates glycogen synthase kinase 3 phosphorylation-dependent c-Myc protein degradation. *Proc. Natl. Acad. Sci. USA* **101,** 9085–9090.

Welcker, M., Singer, J., Loeb, K. R., Grim, J., Bloecher, A., Gurien-West, M., Clurman, B. E., and Roberts, J. M. (2003). Multisite phosphorylation by Cdk2 and GSK3 controls cyclin E degradation. *Mol. Cell.* **12,** 381–392.

Willems, A. R., Lanker, S., Patton, E. E., Craig, K. L., Nason, T. F., Mathias, N., Kobayashi, R., Wittenberg, C., and Tyers, M. (1996). Cdc53 targets phosphorylated G1 cyclins for degradation by the ubiquitin proteolytic pathway. *Cell* **86,** 453–463.

Willems, A. R., Schwab, M., and Tyers, M. (2004). A hitchhiker's guide to the cullin ubiquitin ligases: SCF and its kin. *Biochim. Biophys. Acta* **1695,** 133–170.

Wu, G., Xu, G., Schulman, B. A., Jeffrey, P. D., Harper, J. W., and Pavletich, N. P. (2003). Structure of a b-TrCP1-Skp1-b-catenin complex: Destruction motif binding and lysine specificity of the $SCF^{b-TrCP1}$ ubiquitin ligase. *Mol. Cell.* **11,** 1445–1456.

Yada, M., Hatakeyama, S., Kamura, T., Nishiyama, M., Tsunematsu, R., Imaki, H., Ishida, N., Okumura, F., Nakayama, K., and Nakayama, K. I. (2004). Phosphorylation-dependent degradation of c-Myc is mediated by the F-box protein Fbw7. *EMBO J.* **23,** 2116–2125.

Yaron, A., Hatzubai, A., Davis, M., Lavon, I., Amit, S., Manning, A. M., Andersen, J. S., Mann, M., Mercurio, F., and Ben-Neriah, Y. (1998). Identification of the receptor component of the IkBa-ubiquitin ligase. *Nature* **396,** 590–594.

Zheng, N., Schulman, B. A., Song, L., Miller, J. J., Jeffrey, P. D., Wang, P., Chu, C., Koepp, D. M., Elledge, S. J., Pagano, M., *et al.* (2002). Structure of the Cul1-Rbx1-Skp1-Skp2 SCF ubiquitin ligase complex. *Nature* **416,** 703–709.

Further Reading

Welcker, M., and Clurman, B. E. (2005). The SV40 large T antigen contains a decoy phosphodegron that mediates its interactions with Fbw7/hCdc4. *J. Biol. Chem.* **280,** 7654–7658.

[31] Yeast Genomics in the Elucidation of Endoplasmic Reticulum (ER) Quality Control and Associated Protein Degradation (ERQD)

By ANTJE SCHÄFER and DIETER H. WOLF

Abstract

The endoplasmic reticulum (ER) is the eukaryotic organelle where most secreted proteins enter the secretory pathway. They enter this organelle in an unfolded state and are folded by a highly active folding machinery to reach their native state. The ER contains an efficient protein quality control system, which recognizes malfolded and orphan proteins and targets them for elimination by a mechanism called ER-associated degradation (ERAD). Both processes are tightly linked, and they will be abbreviated as ERQD (ER quality control and associated degradation). Because ERQD is highly conserved from yeast to man, the easy amenability of yeast to genetic and molecular biological studies combined with the knowledge of its genome and proteome makes it a preferred organism to study such "housekeeping" functions of eukaryotic cells. New genomic and proteomic methods have led to new experimental concepts. Genome-wide screens using genomic deletion libraries led to the identification of genes involved in the processes in question. Using such a genome-wide approach, we devise a sensitive growth test for selection of yeast mutants defective in ERQD. A chimeric protein (CTL*) was generated consisting of the ER luminal, N-glycosylated CPY* protein fused to a transmembrane domain and cytoplasmic 3-isopropylmalate dehydrogenase, the Leu2 protein. In addition, the nonglycosylated ER-membrane–located ERQD substrate Sec61-2p was fused to Leu2p (Sec61-2-L*). Cells carrying a *LEU2* deletion can only grow on medium lacking leucine when the chimeric protein CTL* or Sec61-2-L* is not degraded. Thus, only mutant cells defective in an ERQD component can grow. A genome-wide screen can be performed by transforming the CTL* or Sec61-2-L* coding DNA into the ~5000 individual deletion mutants of the EUROSCARF yeast library. Examples for new components required for ERQD found by this method are the mannose-6-phosphate receptor domain protein Yos9p and the ubiquitin domain proteins Dsk2p and Rad23p.

Introduction

The endoplasmic reticulum (ER) is the central organelle in eukaryotic cells responsible for membrane crossing, folding, and Golgi delivery of secretory proteins (Haigh and Johnson, 2002). Because secretory proteins have to arrive at their site of action in a properly folded, native state, the ER contains a highly active folding machinery. Interestingly, ~30% of all newly synthesized proteins are thought to be malfolded and degraded by the proteasome (Schubert *et al.*, 2000). In the ER, a specific quality control system checks the proper folding of proteins (Ellgaard and Helenius, 2003). The ER prevents aggregation of malfolded or orphan proteins by use of chaperones and upregulation of the unfolded protein response (UPR) (Friedländer *et al.*, 2000; Travers *et al.*, 2000). Closely connected is an elimination process of these proteins termed ER degradation or ER-associated degradation (ERAD). Elimination by means of ERAD is achieved after recognition of malfolded proteins in the ER followed by their retrotranslocation to the cytosol, their ubiquitination, and proteasomal degradation (Kostova and Wolf, 2002). Because of the close linkage of ER quality control and ERAD and with respect to the gene products expected from the mutant screen, the abbreviation of both processes as ERQD is introduced. A hyperactive or hypoactive ERQD process underlies many diseases in humans such as cystic fibrosis, Parkinson's, and Creutzfeld–Jakob disease or leads to the escape of viruses from immune recognition (Kostova and Wolf, 2002; Rutishauser and Spiess, 2002). Malfolded soluble and membrane proteins were found to require a set of basic components to proceed through the ERQD pathway. This basic machinery found until now is composed of (1) ER-luminal lectin Htm1/Mnl1, for recognizing and preventing malfolded N-glycosylated proteins from secretion; (2) the ubiquitination machinery consisting of the ubiquitin activating enzymes E1, the ubiquitin conjugating enzymes E2 (Ubc1, Ubc6, Ubc7), and the ubiquitin ligases E3 (Der3/Hrd1 and/or Doa10) for polyubiquitination of proteins during the retrotranslocation process; (3) the AAA–ATPase complex Cdc48-Ufd1-Npl4p for the release of polyubiquitinated proteins from the ER membrane; and (4) the 26S proteasome, which finally degrades the misfolded proteins (Kostova and Wolf, 2003). Additional components may vary, depending on the diversity of malfolded substrates, and different results indicate the existence of more than one pathway to achieve the final degradation by means of the proteasome (Hirsch *et al.*, 2004). So far, the limited knowledge of ERQD leaves many questions open. Therefore, the identification of new components involved in this process is still a major focus of interest.

Because ERQD is highly conserved from yeast to man (Ellgaard and Helenius, 2003; Kostova and Wolf, 2003), the availability of various yeast mutants and the elegant genetic and molecular biological approaches make this organism one of the preferred eukaryotic model systems in the life sciences. The classical method of screening for ERQD mutants in yeast was based on EMS (ethyl-methane sulfonate) mutagenesis (Cronin et al., 2000; Finger et al., 1993; Knop et al., 1996; Lawrence, 2002). This method generates random mutations in the yeast genome, and the mutated cells have to be screened for stabilization of an ERQD substrate. This must be followed by the identification of the respective gene and its encoded protein product. In this chapter, a different genetic approach is presented, which makes use of an available yeast deletion library and a growth test for selection of yeast mutants defective in ERQD. This approach allows a direct assignment of the gene and the respective protein to the ERQD process.

Strategy

For identification of new components involved in ERQD, an available yeast mutant library is used for a genome-wide screen, which involves approximately 5000 mutant strains each deleted in a single nonessential gene (EUROSCARF, Frankfurt, Germany). Use of this genomic library is possible because cells defective in ERQD can tolerate defects in this process as long as the unfolded protein response (UPR) is intact (Friedländer et al., 2000; Travers et al., 2000). A plasmid encoding an ERQD substrate fused to a marker protein is transformed into all 5000 mutants. In case of wild-type degradation of this chimeric protein, the cells are not able to grow on medium lacking the complementing nutrient. In contrast, disturbance of the ERQD pathway prevents the degradation of the ERQD substrate carrying the marker protein fusion and allows cell growth under selective conditions.

Procedures

Choice and Generation of the Substrate

To screen a deletion library for yeast mutants defective in ERQD, a known, characterized ERQD substrate is chosen that allows mutant selection.

The mutated carboxypeptidase yscY (Gly^{255}-Arg^{255}; CPY*) is a classical, characterized ERQD model substrate in *Saccharomyces cerevisiae*. It is a soluble, malfolded protein carrying four N-linked carbohydrate chains. It is completely translocated into the ER and rapidly degraded (Hiller *et al.*, 1996; Plemper *et al.*, 1999b). Also, the related derivatives CT* and CTG* (Fig. 1) have been established as membrane-bound substrates to elucidate components of the ERQD process of membrane proteins (Taxis *et al.*, 2003). Both CT* and CTG* consist of luminal CPY* and the last transmembrane domain of the ATP-binding cassette transporter Pdr5p. In the case of CTG*, a cytosol-located GFP domain is added to the transmembrane domain. Soluble CPY* and membrane-anchored CT* and CTG* (Fig. 1) require the cytosolic trimeric Cdc48 complex for degradation in addition to the ubiquitination and proteasomal degradation machinery. Furthermore, the elimination of CTG* depends on the cytosolic chaperones of the Hsp70, Hsp40, and Hsp104 classes (Taxis *et al.*, 2003). To identify new components involved in the degradation of malfolded ER proteins as found for CTG*, a chimeric protein was constructed that instead of the GFP-protein contained the selectable marker protein Leu2p (3-isopropylmalate dehydrogenase) in the cytosol yielding CTL* (Fig. 1).

Conventional molecular biological techniques were used to generate plasmid pRS316 expressing CTL* under control of the *GAL4*-promotor, which makes low expression under noninducing conditions possible (Lawrence, 2002; Sherman, 2002). Low expression of CTL* was required to allow the observation of well-defined growth differences (Buschhorn *et al.*, 2004; Medicherla *et al.*, 2004). A plasmid expressing the mutated polytopic ER membrane protein Sec61-2p carboxyterminally fused to Leu2p (Sec61-2-L*) (Biederer *et al.*, 1996; Plemper *et al.*, 1999a) was similarly constructed as a second substrate (Fig.1).

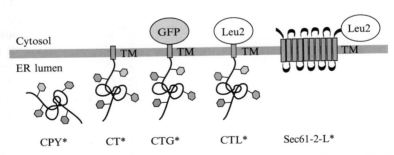

FIG. 1. Schematic representation of the ERQD substrates based on CPY* and Sec61-2p. CPY*, a mutated form of carboxypeptidase yscY, is a soluble, malfolded ERQD substrate. The modified versions consist of luminal CPY*, the last transmembrane domain of Pdr5p (CT*), and additional fusions of cytosolic domains to CT* as are GFP (CTG*) or Leu2p (CTL*). The membrane protein Sec61-2-L* is a chimeric derivative of the ERQD substrate Sec61-2p fused to Leu2p, which is degraded when cells are shifted to 38°. (From Buschhorn *et al.*, 2004.)

Choice of the Gene Deletion Library

The set of approximately 5000 single-gene deletion clones generated in the EUROFAN II project using the BY4743 strain background *MATa/MATα; his3Δ1/his3Δ1; leu2Δ/leu2Δmet15Δ0/MET15; LYS2/lys2Δ0; ura3Δ0/ura3Δ0* (EUROSCARF, University of Frankfurt, Germany) was used for the screen. BY4743 is derived from strain S288C and has been used in the sequencing project of the *S. cerevisiae* genome. For minimizing unwanted mutational effects during the screen, the diploid strain library of approximately 5000 *S. cerevisiae* strains each homozygously deleted for a single nonessential gene was taken. The strains are deficient in the activity of 3-isopropylmalate-dehydrogenase (Leu2p), the enzyme chosen as cytosolic domain for creation of the ERQD substrates CTL* and Sec61-2-L*. The library is contained in YPD medium in 96-multiwell microtiter plates. For continuous use of the library, it is advisable to prepare a $-80°$ stock in similar multiwell plates containing YPD and 15% glycerol.

Transformation of the Library Strains and Screening

Each deletion strain is precultured at $30°$ overnight to stationary phase in sterile 24-well plates with lids containing 0.6 ml of YPD liquid medium in each well. After inoculation of 60 μl of the preculture into new 24-well plates containing 0.6 ml YPD, the strains are regrown at $30°$ for 2 h to logarithmic phase (A_{600} of 0.8–1.0/ml). The use of multichannel-pipettes makes the handling efficient. After growth, cells are sedimented in the plates by centrifugation for 5–10 min. The liquid supernatant is removed, and the sedimented cells are resuspended in sterile, deionized water followed by an additional centrifugation step. Thereafter, each strain is transformed with 2 μg of plasmid DNA that encodes the ERQD substrate protein (e.g., CTL* or Sec61-2-L*) together with 120-μg carrier DNA using the standard heat shock lithium acetate method (Gietz and Woods, 2002). Afterwards, 600 μl selection medium is added into each well, and cells are grown while shaking for 14–16 h at $30°$. For selection of positive transformants containing the ERQD substrate encoding plasmid, 50 μl of each transformed strain is dropped on synthetic dropout medium plates without uracil (approximately 12 different strains per plate). The plates are incubated for 2–3 days at $30°$. For subsequent selection of mutants defective in ERQD, each growing strain is streaked on synthetic dropout medium without uracil and leucine (six mutants plus two control strains per plate). As controls, ERQD wild-type and ERQD-defective strains (i.e., $\Delta der3/hrd1$ [Deak and Wolf, 2001; Gardner *et al.*, 2000; Plemper *et al.*, 1999a]) transformed with CTL*– or Sec61-2L*–expressing plasmids were used (Fig. 2).

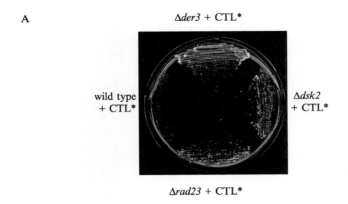

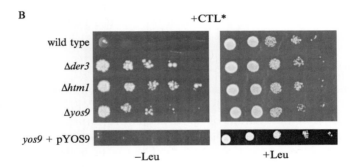

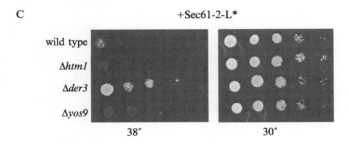

Fig. 2. The genomic screen using cell growth as readout identifies components of ERQD. (A) BY4743 wild-type (WT) and mutant cells were plated on CM medium without leucine. Plates were incubated for 2–3 days at 30°. Uracil was omitted to select for cells carrying the plasmid expressing CTL*. The leucine auxotrophic WT strain expressing CTL* cannot grow in the absence of leucine. The leucine deficiency is complemented by CTL* in the ERAD-defective strain Δ*der3* and in strains carrying the deletions in *DSK2* and *RAD23* (from Medieherla *et al.*, 2004.) (B) Isogenic wild-type (WT) and mutant cells expressing CTL* were plated in serial dilutions on CM medium with and without leucine. Plates were incubated for

Growth at 30° will identify cells in which the degradation of the substrate CTL* or after growth at 38° Sec61-2-L* is disturbed. The growing yeast strains are retested by replating on dropout medium without uracil and leucine.

Results

The growth phenotypes of ~5000 transformed individual deletion mutants of the yeast library expressing CTL* or Sec61-2-L* were screened. For mutants showing reproducible growth in the absence of leucine, the EUROSCARF strain list revealed the identity of the deleted corresponding gene. For further information, yeast genomic databases were consulted. Proof of the reliability of the method was provided by the fact that most of the genes identified in the screen had previously been found to be involved in ERQD (Table I). New mutants found by this method exhibited strong complementation because of defective substrate degradation. They fell into both categories of ERQD, being defective in ER quality control and in ERAD (Buschhorn et al., 2004; Medicherla et al., 2004). Biochemical proof for participation in ERQD came from

TABLE I
Known ERAD Components Found in the Screen

ORF name	Gene name	Function
YOL013C	HRD1/DER3	Ubiquitin-protein ligase
YLR207W	HRD3	Der3p interacting protein
YHR204W	HTM1/MNL1	ER-luminal lectin
YNR030W	ECM39/ALG12	N-glycan synthesis
YJR131W	MNS1	ER α-mannosidase I
YMR022W	UBC7	Ubiquitin-conjugating enzyme
YEL031W	SPF1/COD1	ER calcium pump

2–4 days at 30°. The leucine auxotrophic WT strain fails to grow on CM medium lacking leucine, whereas the ERQD-defective strains Δder3 and Δhtm1 permit growth. Also, deletion of YOS9 allows growth of the leucine auxotrophic strain. As control, wild-type phenotype is observed when Yos9p is expressed from plasmid pYOS9 together with CTL* in Δyos9 cells. (C) Wild-type and mutant cells expressing the conditional nonglycosylated ERAD substrate Sec61-2-L* were plated as described in (B) and incubated at 30° (control) and 38° for 4–5 days. Lethality because of leucine deficiency is complemented only in the Δder3 cells, indicating that Yos9p is not involved in the degradation of nonglycosylated substrates (From Buschhorn et al., 2004.)

cycloheximide decay or pulse-chase analyses after the degradation kinetics of well known ERQD substrates as are CPY* or CTG* (Taxis et al., 2003).

Yeast mutants have also proven to be highly valuable tools in the elucidation of components of the ERQD system for mammalian proteins known to underlie this process. The best-studied example is the degradation of the cystic fibrosis transmembrane conductance regulator (CFTR), which causes the severe disease cystic fibrosis in humans (Kostova and Wolf, 2002; Rutishauser and Spiess, 2002). The disease is directly linked to the ERQD process (Jensen et al., 1995; Kerem et al., 1989; Ward et al., 1995). This autosomal recessive disorder is mostly caused by the $\Delta F508$ mutation in this polytopic membrane ATP-binding cassette transporter, leading to rapid ER-associated degradation of the mutated protein. Also the wild-type CFTR protein is very unstable and is degraded to approximately 70–80% (Jensen et al., 1995; Ward et al., 1995). It has been shown that CFTR expressed in yeast activates the ER quality control and is degraded. Yeast mutants were able to unravel a multitude of components required for CFTR degradation in this organism (Gnann et al., 2004; Kiser et al., 2001; Zhang et al., 2001). Mutants identified by genomic screens will most likely continue to uncover new players of the ERQD machinery of this important protein. They may also lead to the identification of mammalian homologs of these processes by cross-complementation with genes of mammalian counterparts (Gnann et al., 2004). This makes the discovery of new players or functional tests of mammalian ERQD components possible.

References

Biederer, T., Volkwein, C., and Sommer, T. (1996). Degradation of subunits of the Sec61p complex, an integral component of the ER membrane, by the ubiquitin-proteasome pathway. *EMBO J.* **15,** 2069–2076.

Buschhorn, B. A., Kostova, Z., Medicherla, B., and Wolf, D. H. (2004). A genome-wide screen identifies Yos9p as essential for ER-associated degradation of glycoproteins. *FEBS Lett.* **577,** 422–426.

Cronin, S. R., Khoury, A., Ferry, D. K., and Hampton, R. Y. (2000). Regulation of HMG-CoA reductase degradation requires the P-type ATPase Cod1p/Spf1p. *J. Cell Biol.* **148,** 915–924.

Deak, P. M., and Wolf, D. H. (2001). Membrane topology and function of Der3/Hrd1p as a ubiquitin-protein ligase (E3) involved in endoplasmic reticulum degradation. *J. Biol. Chem.* **276,** 10663–10669.

Ellgaard, L., and Helenius, A. (2003). Quality control in the endoplasmic reticulum. *Nat. Rev. Mol. Cell Biol.* **4,** 181–191.

Finger, A., Knop, M., and Wolf, D. H. (1993). Analysis of two mutated vacuolar proteins reveals a degradation pathway in the endoplasmatic reticulum or a related compartment of yeast. *Eur. J. Biochem.* **218,** 565–574.

Friedländer, R., Jarosch, E., Urban, J., Volkwein, C., and Sommer, T. (2000). A regulatory link between ER-associated protein degradation and the unfolded-protein response. *Nat. Cell Biol.* **2**, 379–384.

Gardner, R. G., Swarbrick, G. M., Bays, N. W., Cronin, S. R., Wilhovsky, S., Seelig, L., Kim, C., and Hampton, R. Y. (2000). Endoplasmic reticulum degradation requires lumen to cytosol signaling. Transmembrane control of Hrd1p by Hrd3p. *J. Cell Biol.* **151**, 69–82.

Gietz, R. D., and Woods, R. A. (2002). Transformation of yeast by lithium acetate/single-stranded carrier DNA/polyethylene glycol method. *In* "Guide in Yeast Genetics and Molecular Biology" (C. Guthrie and G. Fink, eds.), Vol. 350, pp. 87–96. Academic Press, San Diego, CA.

Gnann, A., Riordan, J. R., and Wolf, D. H. (2004). Cystic fibrosis transmembrane conductance regulator degradation depends on the lectins Htm1p/EDEM and the Cdc48 protein complex in yeast. *Mol. Biol. Cell.* **15**, 4125–4135.

Haigh, N. G., and Johnson, A. E. (2002). Protein sorting at the membrane of the endoplasmic reticulum. *In* "Protein Targeting, Transport and Translocation" (R. E. Dalbey and G. von Heijne, eds.), pp. 74–106. Academic Press, London-New York.

Hiller, M. M., Finger, A., Schweiger, M., and Wolf, D. H. (1996). ER degradation of a misfolded luminal protein by the cytosolic ubiquitin-proteasome pathway. *Science* **273**, 1725–1728.

Hirsch, C., Jarosch, E., Sommer, T., and Wolf, D. H. (2004). Endoplasmic reticulum associated protein degradation—one model fits all? *Biochim. Biophys. Acta Mol. Cell Res.* **1695**, 215–223.

Jensen, T. J., Loo, M. A., Pind, S., Williams, D. B., Goldberg, A. L., and Riordan, J. R. (1995). Multiple proteolytic systems, including the proteasome, contribute to CFTR processing. *Cell* **83**, 129–135.

Kerem, B., Rommens, J. M., Buchanan, J. A., Markiewicz, D., Cox, T. K., Chakravarti, A., Buchwald, M., and Tsui, L. C. (1989). Identification of the cystic fibrosis gene: genetic analysis. *Science* **245**, 1073–1080.

Kiser, G. L., Gentzsch, M., Kloser, A. K., Balzi, E., Wolf, D. H., Goffeau, A., and Riordan, J. R. (2001). Expression and degradation of the cystic fibrosis transmembrane conductance regulator in *Saccharomyces cerevisiae. Arch. Biochem. Biophys.* **390**, 195–205.

Knop, M., Finger, A., Braun, T., Hellmuth, K., and Wolf, D. H. (1996). Der1, a novel protein specifically required for endoplasmic reticulum degradation in yeast. *EMBO J.* **15**, 753–763.

Kostova, Z., and Wolf, D. H. (2002). Protein quality control in the export pathway: the endoplasmic reticulum and its cytoplasmic proteasome connection. *In* "Protein Targeting, Transport and Translocation" (R. E. Dalbey and G. von Heijne, eds.), pp. 180–213. Academic Press, London-New York.

Kostova, Z., and Wolf, D. H. (2003). For whom the bell tolls: protein quality control of the endoplasmic reticulum and the ubiquitin-proteasome connection. *EMBO J.* **22**, 2309–2317.

Lawrence, C. W. (2002). Classical mutagenesis techniques. *In* "Guide in Yeast Genetics and Molecular Biology" (C. Guthrie and G. Fink, eds.), Vol. 350, pp. 189–199. Academic Press, San Diego, CA.

Medicherla, B., Kostova, Z., Schaefer, A., and Wolf, D. H. (2004). A genomic screen identifies Dsk2p and Rad23p as essential components of ER-associated degradation. *EMBO Rep.* **5**, 692–697.

Plemper, R. K., Bordallo, J., Deak, P. M., Taxis, C., Hitt, R., and Wolf, D. H. (1999a). Genetic interactions of Hrd3p and Der3p/Hrd1p with Sec61p suggest a retro-translocation complex mediating protein transport for ER degradation. *J. Cell Sci.* **112**, 4123–4134.

Plemper, R. K., Deak, P. M., Otto, R. T., and Wolf, D. H. (1999b). Re-entering the translocon from the lumenal side of the endoplasmic reticulum. Studies on mutated carboxypeptidase yscY species. *FEBS Lett.* **443,** 241–245.

Rutishauser, J., and Spiess, M. (2002). Endoplasmic reticulum storage diseases. *Swiss Med. Wkly.* **132,** 211–222.

Schubert, U., Anton, L. C., Gibbs, J., Norbury, C. C., Yewdell, J. W., and Bennink, J. R. (2000). Rapid degradation of a large fraction of newly synthesized proteins by proteasomes. *Nature* **404,** 770–774.

Sherman, F. (2002). Getting started with yeast. *In* "Guide in Yeast Genetics and Molecular Biology" (C. Guthrie and G. Fink, eds.), Vol. 350, pp. 3–41. Academic Press, San Diego, CA.

Taxis, C., Hitt, R., Park, S. H., Deak, P. M., Kostova, Z., and Wolf, D. H. (2003). Use of modular substrates demonstrates mechanistic diversity and reveals differences in chaperone requirement of ERAD. *J. Biol. Chem.* **278,** 35903–35913.

Travers, K. J., Patil, C. K., Wodicka, L., Lockhart, D. J., Weissman, J. S., and Walter, P. (2000). Functional and genomic analyses reveal an essential coordination between the unfolded protein response and ER-associated degradation. *Cell* **101,** 249–258.

Ward, C. L., Omura, S., and Kopito, R. R. (1995). Degradation of CFTR by the ubiquitin-proteasome pathway. *Cell* **83,** 121–127.

Zhang, Y., Nijbroek, G., Sullivan, M. L., McCracken, A. A., Watkins, S. C., Michaelis, S., and Brodsky, J. L. (2001). Hsp70 molecular chaperone facilitates endoplasmic reticulum-associated protein degradation of cystic fibrosis transmembrane conductance regulator in yeast. *Mol. Biol. Cell* **12,** 1303–1314.

[32] Mechanism-Based Proteomics Tools Based on Ubiquitin and Ubiquitin-Like Proteins: Synthesis of Active Site-Directed Probes

By HUIB OVAA, PAUL J. GALARDY, and HIDDE L. PLOEGH

Abstract

The families of ubiquitin and ubiquitin-like modifiers are involved in the regulation of many biochemical pathways. The steady-state level of polypeptide modification with these molecules depends on the opposing activity of conjugating and deconjugating enzymes. Here we describe the generation of mechanism-dependent active site-directed probes that target the large family of ubiquitin/ubiquitin-like isopeptidases. To maintain substrate specificity for these enzymes, we have based the development of these probes on full-length sequences of ubiquitin and several ubiquitin-like molecules. For their construction, this approach necessitates the use of a combination of organic synthesis and expressed protein ligation. These probes have been used in the isolation and identification of active isopeptidases from crude cell extracts and have been instrumental in the discovery of novel ubiquitin and ubiquitin-like deconjugating enzymes. These probes may be generated with or without an epitope tag that enables activity profiling of proteases from cell extracts. In addition, we have used a ubiquitin-based probe in the structural analysis of the ligand-bound form of the enzyme UCH-L3 by x-ray crystallography. Together, these probes greatly facilitate the study of ubiquitin and ubiquitin-like proteases.

Conjugation and Removal of Ubiquitin and Ubiquitin-Like Proteins

Ubiquitin (Ub) is a 76-residue polypeptide that is attached to substrate proteins to regulate protein half-life or function. Conjugation of Ub to target proteins occurs through the sequential action of a single Ub-activating enzyme (E1), many Ub-conjugating enzymes (E2s), and a multitude of Ub ligases (E3s). Removal of Ub from conjugates occurs through the action of a large group of Ub-specific proteases (USPs) that aid in the regulation of Ub-dependent processes. A small group of ubiquitin-like proteins (UBLs) share similarity in sequence, structure, or enzymology to Ub. Ub and all UBLs are ligated by means of their C-terminal residues to the ε-amino group of a lysine residue of another protein, forming an isopeptide bond. Occasionally, the Ub/UBL is ligated to the N-terminal amino group of the substrate (Glickman and Ciechanover, 2002; Glickman and Maytal, 2002). Distinct from all other UBL proteins, Apg8 and its mammalian homologs are conjugated to lipids rather than proteins (Ichimura et al., 2000).

A Ub/UBL-specific monomeric or dimeric E1-activating enzyme or enzyme complex initiates Ub/UBL conjugation. Activation of the C-terminus of the Ub/UBL protein, by means of an ATP-dependent adenylate formation, charges the E1 active site cysteine residue with the Ub/UBL protein cargo. The activated E1–Ub/UBL donor then transfers its load onto the active site cysteine residue of one of approximately 50 human E2-conjugating enzymes. Ub and certain UBL molecules are then ligated to their substrates with substrate specificity conferred by a large family of E3 ligases (Hershko and Ciechanover, 1998; Jentsch, 1992). Conjugation factors (E2s, E3s) may have overlapping specificity toward different Ub/UBLs, as exemplified by Mdm2. Mdm2 (Harper, 2004; Xirodimas et al., 2004) has been shown to conjugate both Ub and Nedd8 onto the tumor suppressor p53.

All E1s, E2s, and HECT-domain E3 ligases rely on an active site cysteine residue and hence can be probed chemically. In principle, several Ub/UBL conjugation factors can be forced to react *in vitro* with appropriate Ub/UBL-based probes described here, providing information on reactivity and cross-reactivity and affording potentially novel tools for crystallography and biochemistry.

Reversal of Ub/UBL Modification

Similar to the manner in which kinases are counterbalanced by the action of phosphatases, the action of Ub/UBL-specific ligases is counterbalanced by the action of Ub/UBL-specific proteases. These proteases play an important regulatory role by determining the conjugation status of substrate proteins (Wilkinson, 2000; Wing, 2003). In some cases, removal of Ub from a targeted protein results in its rescue from proteasomal degradation. The removal of monoubiquitin groups and UBL proteins plays a more discrete role in the function of many biochemical pathways, ranging from receptor internalization and vesicular sorting to transcriptional regulation and DNA repair pathways. Perhaps the most basic role of Ub/UBL-specific proteases is the processing of precursor translation products. Most Ub/UBLs are expressed with C-terminal extensions or, in the case of Ub, as linear fusion protein repeats that require the action of Ub/UBL-specific proteases to liberate the active monomer. Not surprisingly, a variety of USPs plays important roles in the development of disease, because they have been shown to act as oncoproteins or tumor suppressor proteins (Borodovsky *et al.*, 2002; Li *et al.*, 2002; Yao and Cohen, 2002). More than one Ub/UBL-specific protease exists for many Ub/UBL modifiers. Indeed, more than 100 genes encode potential Ub/UbL-deconjugating enzymes, raising the question of their biological role. Current evidence suggests that Ub and UBL isopeptidases are highly regulated enzymes with tissue-specific expression patterns and restricted substrate specificities.

Most enzymes have been assigned to the UB/UBL protease families by sequence alignment alone, with direct evidence of catalytic activity lacking in many cases. With few exceptions, detailed information on substrate specificities of Ub- and UBL-specific proteases is not available, although some Ub- and UBL-specific proteases have been shown to possess overlapping substrate specificity (Hemelaar *et al.*, 2004). By specifically forming covalent adducts with Ub/UBL-specific proteases, the tools described here thus provide information on the specificity and activity of these proteases. The generation of tools that allows enzyme identification and confirmation of catalytic activity should greatly facilitate the study of these enzymes. We here describe the synthesis of Ub/UBL-based probes (Borodovsky *et al.*, 2001, 2002) using a combination of chemical synthesis and expressed protein ligation. For detailed descriptions of the intein-based approach to protein expression and purification, the reader is referred to Chapter 3. For details on the use of Ub/UBL-based probes, please refer to the accompanying chapter in this volume.

USP and UBL-Specific Protease Families

The ubiquitin-specific proteases (USPs) or deubiquitinating enzymes (DUBs) form the largest class of Ub/UBL-specific proteases described to date. These enzymes can be subdivided into different families based on sequence alignment and structural motifs. At present, USPs are classified as ubiquitin-specific processing proteases (UBP), ubiquitin carboxyterminal hydrolases (UCH), ovarian tumor (OTU) (Borodovsky et al., 2002)–domain containing proteases, and JAMM family metalloproteases (Borodovsky et al., 2002; Verma et al., 2002; Yao and Cohen, 2002). In addition to a single Ub-specific metalloprotease that resides in the 19S cap of the proteasome, a similar JAMM motif is found within the COP9 signalosome and is responsible for the cleavage of Nedd8 from cullins, components of SCF-type E3 ligases. The Nedd8 modification of cullins is required to arm this E3 for efficient Ub transfer. Proteins with both USP and E3 ligase activities residing in different domains exist as well. An example of the latter family is the protein A20 that contains both a functional OTU protease domain and zinc-finger domains with independent ligase activity (Evans et al., 2004; Wertz et al., 2004). Excluding the JAMM family of metalloproteases, all known USPs and UBL-deconjugating enzymes are characterized by the presence of a highly nucleophilic cysteine residue in the catalytic sites. This nucleophilicity makes the Ub/UBL proteases chemically distinguishable from the conjugation machinery. Whereas adducts with conjugating enzymes can be generated only when combining purified ligase and probe at high concentrations, Ub/UBL-specific proteases can be targeted in a highly chemoselective manner in cell lysates at relatively low probe concentrations. To date, no spurious cross-reactivities have been detected; all polypeptides retrieved by means of reactivity with a Ub/UBL-specific probe have proven to be true Ub/UBL-specific proteases. Other proteases, such as lysosomal proteases, do not participate in the reaction.

Probe Design

The crystal structure of the UBP USP7 (HAUSP) in the free state and in complex with the inhibitor ubiquitin aldehyde (Ubal) demonstrates that USP7 makes contacts with a large surface area of Ub and that a large conformational change occurs on substrate binding (Hu et al., 2002). The interactions between UCH enzymes and Ub have been revealed in the crystal structures of human UCH-L3 in complex with UbVME and yeast Yuh1 in complex with Ub aldehyde (Ubal). Although the interacting surface area of Ub with UCH enzymes is less than that observed for HAUSP, critical contacts exist between residues in the N-terminus of Ub and both UCH-L3 and Yuh1. It is, therefore, expected that the entire ubiquitin sequence is needed to obtain significant substrate specificity toward designer probes, especially for UBPs. For this reason, we chose an expressed protein ligation approach toward the synthesis of ubiquitin probes as depicted in Fig. 1. Expression of an appropriately tagged Ub/UBL sequence in fusion with an intein and a chitin-binding domain allows easy purification of the protein over a chitin column. Cleavage of the intein using an excess of MesNa then releases the tagged Ub/UBL thioester. The Ub/UBL sequence is engineered to lack the conserved C-terminal glycine residue. Incorporation of a warhead now introduces a reactive group in the exact position or in close proximity from the position (the Gly76 carbonyl) where a Ub/UBL modification or polymer is normally deconjugated from fusions. As warheads, both Michael acceptors (vinyl sulfones, α,β-unsaturated esters) and alkylhalides were chosen (Fig. 2).

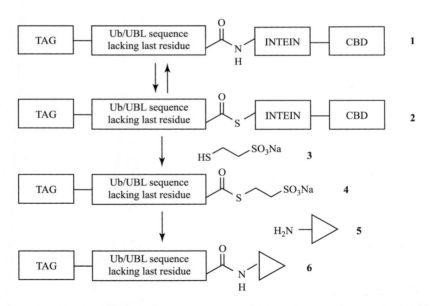

FIG. 1. Probe synthesis by expressed protein ligation. The desired, tagged, Ub/UBL sequence is expressed as an intein chitin–binding domain fusion protein. MesNa (**3**)-mediated cleavage of the intein releases the tagged Ub/UBL as a homogeneous thioester that can be directly equipped with a warhead by ligation with (**5**) to give the Ub/UBL-based probe (**6**).

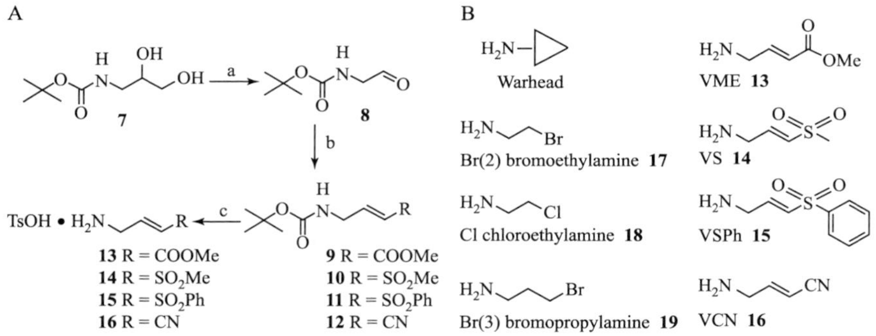

FIG. 2. Synthesis of Michael acceptors. (A) NaIO$_4$ H$_2$O/CH$_2$Cl$_2$. (b) Ph$_3$=CHC(=O)OMe (**9**) or Ph$_3$=CHCN (**10**), THF, 0°. (EtO)$_2$P(=O)CH$_2$SO$_2$Me (**11**) or (EtO)$_2$P(=O)CH$_2$SO$_2$Ph (**12**), NaH, THF 0°. C. pTsOH, Et$_2$O, 0°. (B) Structures of different cysteine-reactive elements that have been incorporated into HAUb-based probes.

Materials and Methods

HPLC-grade organic solvents *N,N*-dimethyl formamide (DMF) and dichloromethane (American Bioanalytical), *n*-hexane, and ethyl acetate (Fisher) were used as received. Tetrahydrofuran was obtained from Acros and distilled over LiAlH$_4$ under a nitrogen atmosphere before use. 2-Chloroethylamine hydrochloride and 2-bromoethylamine hydrobromide were purchased from Acros. β-(D)-isopropyl thiogalactoside was from Fluka. 3-bromopropylamine hydrobromide, methyl triphenylphosphoranylidene acetate, and triphenylphosphoranylidene acetonitrile were purchased from Aldrich. Methanesulfonylmethyl-phosphonic acid diethyl ester and benzenesulfonylmethyl-phosphonic acid diethyl ester were synthesized according to literature procedures. Slide-a-lyzer™ dialysis membranes were from Pierce. NMR spectra were recorded on a Varian 200- or 500-MHz spectrometer and mass spectra were recorded on an electrospray LCZ or LCT LC-MS instrument (LC HP1100 Hewlett Packert, MS Micromass, UK) equipped with a Waters DeltaPak C4 (3.9 × 150 mm) column.

Plasmid Construction

pTYB-(HA)Ub plasmid omitting the C-terminal codon was constructed by cloning the sequence of human Ub (lacking Gly76) into the pTYB2 vector (New England Biolabs) to generate an in-frame fusion with the intein and chitin-binding domain (CBD). A hemagglutinin (HA) epitope tag can be introduced at this point by inserting an oligonucleotide cassette into the NdeI site at the 5′ end of the Ub sequence. Similar plasmids were obtained by cloning into the pTYB1 vector (New England Biolabs) for Nedd8, FLAGNedd8, ISG15, HAISG15, UCRP, HASUMO1, Fau, URM-1, HUB1, FAT10, GATE-16, MAP1-LC3, GABARAP, Apg8L, and Apg12. All constructs generated were sequence verified.

Synthesis of Ub/UBL-Based Probes

The strategy takes advantage of an intein-mediated amide to thioester transthioesterification reaction. This thioester intermediate can be used to generate an Ub/UBL MesNa adduct. Reaction of the MesNa thioester with different warheads affords the desired probes. Ub is depicted as an example.

Synthesis of $HAUb_{75}$–MESNa

HAUb-intein-chitin binding domain fusion protein was expressed in *E. coli* BL21(DE3) in 2 liters LB medium and induced for 2 h at an optical density of 0.6 at 600 nm with 0.5 mM IPTG at 30°. Cell pellets were lysed by French press (2000 psi) in 50 mM HEPES, pH 6.5, 100 mM NaOAc, supplied with PMSF or complete protease inhibitor cocktail (Roche). The cell extract was clarified and protein was absorbed onto a chitin bead (New England Biolabs) column followed by on-column cleavage by transthioesterification with 50 mM MesNa at 37° for 2 days. At this point, Ub_{75}MesNa or $HAUb_{75}$-MesNa can be purified by MonoS (Pharmacia) cation exchange chromatography using a gradient of 0–500 mM NaCl in 50 mM NaOAc, pH 4.5. The MesNa adducts ($HAUb_{75}$MesNa or Ub_{75}MesNa) elute under these conditions at 170 mM NaCl. Identical methods apply for the generation of UBL–MesNa polypeptide thioesters that were used without additional purification. FAT10, HUB1, and Apg12 proved insoluble under all conditions tested.

Synthesis and Purification of Cysteine-Reactive Groups for Chemical Ligation to $HAUb_{75}$MESNa or Ub_{75}MESNa

N-tert-butyloxycarbonyl (Boc) protected glycinal was obtained by sodium periodate–mediated oxidative diol cleavage of *N*-Boc-1-amino-2,3-propanediol and directly used in the next step. Thus obtained, N-*Boc*-glycinal was reacted according to standard literature procedures with either Wittig ylides (methyl triphenyl phosporanylidene acetate for VME, triphenylphosporanylidene acetonitrile for VCN) at room temperature, or with appropriate Horner-Wadsworth-Emmons reagents (methanesulfonylmethyl-phosphonic acid diethyl ester, sodium hydride for HAUbVS, benzenesulfonylmethyl-phosphonic acid diethyl ester, sodium hydride for HAUbVSPh) in THF at 0°. In all cases, yields of ∼80% were obtained of the *N*-Boc protected compounds as *Z/E* mixtures. The *E*-isomer was in all cases the major isomer and could be purified by column chromatography on silica gel with hexane/ethyl acetate gradients. (*E*)-BocGlyVS can be crystallized selectively from the *Z*-isomer in hexane/ethyl acetate (1/1 v/v). BocGlyVsPh (**11**): ^{1}H-NMR (200 MHz, CDCl$_3$): δ 1.40 (s, 9H), 3.95 (m, 2H), 4.71 (bs, 1H), 6.44 (m, 1H), 6.95 (m, 1H), 7.58 (m, 3H), 7.89 (m, 2H). BocGlyVCN (**12**): ^{1}H-NMR (200 MHz, CDCl$_3$): δ 1.45 (s, 9H), 3.92 (m, 2H), 4.72 (bs, 1H), 5.52 (m, 1H), 6.72 (m, 1H). BocGlyVME (**9**): ^{1}H-NMR (200 MHz, CDCl$_3$): δ 1.44 (s, 9H), 3.72 (s, 3H), 3.89 (m, 2H), 4.82 (m, 2H), 5.92 (m, 1H)., 6.88 (m, 1H). Boc-groups were removed by adding dry toluenesulfonic acid (three equivalents, no stirring, room temperature overnight) in diethylether on which the deprotected amines crystallize as *para*-toluene sulfonic acid salts. Salts were filtered off, washed with cold ether, and collected. Residual ether was removed under high vacuum, and salts were stored at −20° in a desiccator.

Synthesis and Purification of HAUb-Derived Active Site Thiol-Reactive Probes

HAUbCl (**18**), *HAUbBr2* (**17**), *HAUbBr3* (**19**). To a solution of HAUb$_{75}$–MESNa (2–5 mg/ml) in column buffer (500 μl), was added subsequently: 0.2 mmol of the desired haloalkylamine haloacid salt (BrCH$_2$CH$_2$NH$_2$.HBr, BrCH$_2$CH$_2$ CH$_2$NH$_2$.HBr, ClCH$_2$CH$_2$NH$_2$.HCl) and 100 μl of 2.0 *M* aqueous NaOH, and the mixture was immediately vortexed. After 10 min at room temperature, 100 μl of 2.0 *M* aqueous HCl was added, and the solution was dialyzed immediately against 50 m*M* NaOAc, pH 4.5, in a 3.5-ml Pierce Slide-a-lyzer cassette (3500 MWCO) for 1 h with stirring at room temperature. The resulting product (>90% conversion estimated from LC-MS) was divided into aliquots and stored at −80°. Significant deterioration is observed within this probe series on prolonged storage, especially for HAUbBr2.

HAU$_B$VME (**13**), HAU$_B$VS (**14**), HAU$_B$VCN (**16**). To a solution of HAUb$_{75}$–MESNa (2–5 mg/ml, 500 μl) was added subsequently: 0.125 mmol of the desired Michael acceptor as *para*-toluene sulfonic acid salt followed by 75 μl of 2 *M* *N*-hydroxy succinimide and 125 μl 2 *M* NaOH. The mixture was incubated at 37° for 2 h, and reaction progress was monitored by LC-MS to give the desired products. Conversions were 50–70% in all cases. The reaction mixture was neutralized by the addition of 125 μl of 2 **M** HCl and dialyzed against 50 m*M* NaOAc, pH 4.5, in a 3.5-ml Pierce Slide-a-lyzer cassette (3500 MWCO) overnight at + 4°.

HAU$_B$VsPh (**15**). To a solution of HAUb$_{75}$–MESNa (4–5 mg/ml, 500 μl) was added subsequently: a solution of glycine vinyl phenyl sulfone tosic acid salt (0.2 mmol, 46 mg) in 250 μl of DMF followed by 50 μl of 1 *M*

DMAP in DMF and 100 µl of 2 M aqueous NaOH. Reaction progress was monitored by LC-MS, and after 45 min 100 µl of 2 M aqueous HCl and 2 ml of 50 mM NaOAc pH 4.5 was added. A clear precipitate formed and was removed by centrifugation. The cleared solution was filtered and dialyzed overnight at 4° against the same buffer. The product was filtered, and concentrated to approximately 500 µl.

Purification

All synthetically modified HA-tagged ubiquitin derivatives were purified, except HAUbBr2, which proved to be too unstable for purification. HAUbBr2 was used directly after dialysis (synthesis yield was >90%). Other Ub-derived probes were purified to 95% purity using a Pharmacia SMART system MonoS 1.6/5 column with a linear gradient from 0–30% B; 50 mM NaOAc, pH 4.5 (buffer A), 50 mM NaOAc, pH 4.5, 1 M NaCl (buffer B). Hydrolyzed HAUbMesNa elutes at 15% B, and products (HAUbVS and HAUbVME elute at 20%B. HAUbVSPh gave a different elution profile, presumably because of the hydrophobicity of its C-terminus. All HAUb-derived probes were analyzed by LC-MS (ESI) using a C4 RP-HPLC column with a 0–80% gradient over 20 min in a 0.1% formic acid/acetonitrile buffer system. Purified probes can be aliquoted, snap-frozen, and stored at −80°. HAUbBr2 slowly deteriorates on prolonged storage at −80°. We have not observed any deterioration for HAUbVME or HAUbVS after prolonged storage (>1 y) at −80°. All UBL probes were used without further purification.

Analytical Data

Observed and predicted molecular weights of HAUb derivatives were determined by LC/ESI-MS: Multicharged species observed for each HAUb-derived probe are given (species listed all contain the N-terminal methionine residue, which is frequently partially removed or processed): Ub-MESNa: found (calculated); $[M + 9H]^{9+}$ 960 (960); $[M + 8H]^{8+}$ 1079 (1080); $[M + 7H]^{7+}$ 1233 (1234); $[M + 6H]^{6+}$ 1439 (1440). HAUb-MESNa: found (calculated) $[M + 11H]^{11+}$ 936 (936); $[M + 10H]^{10+}$ 1029 (1029); $[M + 9H]^{9+}$ 1143 (1144); $[M + 8H]^{8+}$ 1286 (1287); $[M + 7H]^{7+}$ 1469 (1470). HAUb: Found (calculated) $[M + 11H]^{11+}$ 929 (928); $[M + 10H]^{10+}$ 1022 (1021); $[M + 9H]^{9+}$ 1135 (1134); $[M + 8H]^{8+}$ 1277 (1276); $[M + 7H]^{7+}$ 1459 (1458). HAUbVS: found (calculated); $[M + 11H]^{11+}$ 935 (935); $[M + 10H]^{10+}$ 1027 (1029); $[M + 9H]^{9+}$ 1142 (1143); $[M + 8H]^{8+}$ 1286 (1286); $[M + 7H]^{7+}$ 1468 (1469). HAUbVME: found (calculated) $[M + 11H]^{11+}$ 933 (933); $[M + 9H]^{10+}$ 1027 (1027); $[M + 9H]^{9+}$ 1140 (1141); $[M + 8H]^{8+}$ 1283 (1283); $[M + 7H]^{7+}$ 1466 (1466). HAUbVCN: found (calculated):

[M + 11H]$^{11+}$ 932 (930); [M + 10H]$^{10+}$ 1025 (1023); [M + 9H]$^{9+}$ 1137 (1137); [M + 8H]$^{8+}$ 1280 (1279). HAUbCl: Found (calculated): [M + 11H]$^{11+}$ 930 (930); [M + 10H]$^{10+}$ 1023 (1023); [M + 9H]$^{9+}$ 1337 (1137); [M + 8H]$^{8+}$ 1279 (1279). HAUbBr2: found (calculated); [M + 11H]$^{11+}$ 934 (934); [M + 10H]$^{10+}$ 1027 (1027); [M + 9H]$^{9+}$ 1141 (1142); [M + 8H]$^{8+}$ 1284 (1284); [M + 7H]$^{7+}$ 1467 (1468). HAUbBr3: found (calculated): [M + 11H]$^{11+}$ 935 (936); [M + 10H]$^{10+}$ 1028 (1029); [M + 9H]$^{9+}$ 1286 (1286); [M + 8H]$^{8+}$ 1469 (1469). HAUbVsPh: found (Calculated): [M + 11H]$^{11+}$ 940 (941); [M + 10H]$^{10+}$ 1034 (1035); [M + 9H]$^{9+}$ 1149 (1150); [M + 8H]$^{8+}$ 1292 (1293); [M + 7H]$^{7+}$ 1477 (1478).

Radioiodination of Ub/UBL Probes

Forty micrograms of UBL-VS was iodinated in 50 mM phosphate buffer (pH 7.5) containing 1 mCi of Na^{125}I, using Iodogen as a catalyst (iodogen-coated Eppendorf). The reaction was allowed to proceed for 30 min on ice and was quenched with 0.1 mg of tyrosine per milliliter. Hen egg lysozyme (1 mg/ml) was added as a carrier protein for the subsequent purification over a Sephadex G-25 (Pharmacia) spin column. Iodinated UBL-probes were stored at $-80°$.

References

Borodovsky, A., Kessler, B. M., Casagrande, R., Overkleeft, H. S., Wilkinson, K. D., and Ploegh, H. L. (2001). A novel active site-directed probe specific for deubiquitylating enzymes reveals proteasome association of USP14. *EMBO J.* **20,** 5187–5196.
Borodovsky, A., Ovaa, H., Kolli, N., Gan-Erdene, T., Wilkinson, K. D., Ploegh, H. L., and Kessler, B. M. (2002). Chemistry-based functional proteomics reveals novel members of the deubiquitinating enzyme family. *Chem. Biol.* **9,** 1149–1159.
Evans, P. C., Ovaa, H., Hamon, M., Kilshaw, P. J., Hamm, S., Bauer, S., Ploegh, H. L., and Smith, T. S. (2004). Zinc-finger protein A20, a regulator of inflammation and cell survival, has de-ubiquitinating activity. *Biochem. J.* **378,** 727–734.
Glickman, M. H., and Ciechanover, A. (2002). The ubiquitin-proteasome proteolytic pathway: Destruction for the sake of construction. *Physiol. Rev.* **82,** 373–428.
Glickman, M. H., and Maytal, V. (2002). Regulating the 26S proteasome. *Curr. Top. Microbiol. Immunol.* **268,** 43–72.
Harper, J. W. (2004). Neddylating the guardian; Mdm2 catalyzed conjugation of Nedd8 to p53. *Cell* **118,** 2–4.
Hemelaar, J., Borodovsky, A., Kessler, B. M., Reverter, D., Cook, J., Kolli, N., Gan-Erdene, T., Wilkinson, K. D., Gill, G., Lima, C. D., Ploegh, H. L., and Ovaa, H. (2004). Specific and covalent targeting of conjugating and deconjugating enzymes of ubiquitin-like proteins. *Mol. Cell. Biol.* **24,** 84–95.
Hershko, A., and Ciechanover, A. (1998). The ubiquitin system. *Annu. Rev. Biochem.* **67,** 425–479.

Hu, M., Li, P., Li, M., Li, W., Yao, T., Wu, J. W., Gu, W., Cohen, R. E., and Shi, Y. (2002). Crystal structure of a UBP-family deubiquitinating enzyme in isolation and in complex with ubiquitin aldehyde. *Cell* **111,** 1041–1054.

Ichimura, Y., Kirisako, T., Takao, T., Satomi, Y., Shimonishi, Y., Ishihara, N., Mizushima, N., Tanida, I., Kominami, E., Ohsumi, M., Noda, T., and Ohsumi, Y. (2000). A ubiquitin-like system mediates protein lipidation. *Nature* **408,** 488–492.

Jentsch, S. (1992). The ubiquitin-conjugation system. *Annu. Rev. Genet.* **26,** 179–207.

Li, M., Chen, D., Shiloh, A., Luo, J., Nikolaev, A. Y., Qin, J., and Gu, W. (2002). Deubiquitination of p53 by HAUSP is an important pathway for p53 stabilization. *Nature* **416,** 648–653.

Verma, R., Aravind, L., Oania, R., McDonald, W. H., Yates, J. R., 3rd, Koonin, E. V., and Deshaies, R. J. (2002). Role of Rpn11 metalloprotease in deubiquitination and degradation by the 26S proteasome. *Science* **298,** 611–615.

Wertz, I. E., O'Rourke, K. M., Zhou, H., Eby, M., Aravind, L., Seshagiri, S., Wu, P., Wiesmann, C., Baker, R., Boone, D. L., Ma, A., Koonin, E. V., and Dixit, V. M. (2004). De-ubiquitination and ubiquitin ligase domains of A20 downregulate NF-kappaB signalling. *Nature* **430,** 694–699.

Wilkinson, K. D. (2000). Ubiquitination and deubiquitination: Targeting of proteins for degradation by the proteasome. *Semin. Cell Dev. Biol.* **11,** 141–148.

Wing, S. S. (2003). Deubiquitinating enzymes–the importance of driving in reverse along the ubiquitin-proteasome pathway. *Int. J. Biochem. Cell Biol.* **35,** 590–605.

Xirodimas, D. P., Saville, M. K., Bourdon, J. C., Hay, R. T., and Lane, D. P. (2004). Mdm2-mediated NEDD8 conjugation of p53 inhibits its transcriptional activity. *Cell* **118,** 83–97.

Yao, T., and Cohen, R. E. (2002). A cryptic protease couples deubiquitination and degradation by the proteasome. *Nature* **419,** 403–407.

Section VI

Real Time/Non-Invasive Technologies

[33] Application and Analysis of the GFP^u Family of Ubiquitin-Proteasome System Reporters

By Neil F. Bence, Eric J. Bennett, and Ron R. Kopito

Abstract

The relevance of the ubiquitin proteasome system (UPS) to disease and fundamental cellular processes has generated a demand for methods to monitor the activity of this system in living cells and organisms. Here we describe the GFP^u family of UPS reporters. These reporters are constitutively degraded, ubiquitin-dependent proteasome substrates that can be used to monitor UPS function in the living cell. The GFP^u reporter family consists of three variants that can report on global, nuclear, and cytoplasmic UPS function. This article focuses on the properties and design of these reporters and highlights appropriate techniques and applications for their use.

Introduction

Reporter Assays of UPS Function

Three predominant methods are used to assess UPS function in cells, tissues, or whole animals. Fluorogenic peptidase assays are widely used and permit rapid assessment of 20S or 26S proteasome catalytic activity. Because these reporters are cell impermeant, they are suitable only for assessment of proteasome activity in cell lysates or tissue homogenates. Moreover, because cleavage of these fluorogenic peptides does not require ATP-dependent unfolding, a rate-limiting step in proteolysis (Thrower *et al.*, 2000), or ubiquitin conjugation, they more closely resemble the products of the UPS than its substrates. Another alternative is to monitor the levels of a short-lived endogenous substrate protein by immunoblotting or immunofluorescence microscopy. Signal-to-noise ratios can plague immunofluorescence analysis, and quantitative immunoblotting is complicated by the altered mobility of ubiquitylated species on SDS-PAGE gels. These antibody-based detection methods can be circumvented through the use of reporter protein fusions, but caution must be exercised to avoid dominant negative effects of overexpressed fusion proteins.

The use of luciferase or GFP fusions to a UPS-specific degron, like GFP^u, overcomes many of the aforementioned limitations (Neefjes and

Dantuma, 2004). They do not mimic or participate in any known cellular signaling or metabolic pathways, do not require cell or tissue disruption, and are true polypeptide substrates that require protein unfolding and ubiquitylation (with the exception of ornithine decarboxylase [ODC]–based reporters) before degradation. Exploitation of enzymatic or fluorescent detection helps to reduce the level of expression required to achieve a high signal-to-noise ratio.

UPS reporters now use a targeting element of the ubiquitin-independent substrate ODC (Li *et al.*, 1998), an uncleavable ubiquitin fusion (UFD) (Dantuma *et al.*, 2000), a cleavable ubiquitin fusion that permits the creation of an N-end rule substrate (Dantuma *et al.*, 2000), and the CL1 degron (Bence *et al.*, 2001; Gilon *et al.*, 1998). The major difference among the non-ODC degrons is the ubiquitin conjugation pathway traversed en route to the proteasome. These reporters are all constitutively degraded and are widely applicable UPS monitoring reagents.

GFP^u and the GFP^u Family of UPS Reporters

GFP^u is a UPS reporter that relies on a 16 amino acid degron (CL1) fused to the carboxyl terminus of GFP (Fig. 1A,B). The CL1 degron was first identified in a yeast screen for sequences that destabilize β-galactosidase in a Ubc6- and Ubc7-dependent manner (Gilon *et al.*, 1998). Structural predictions indicate that the peptide sequence can form an amphipathic helix, perhaps mimicking the endogenous Ubc6/7-dependent degron of the yeast protein Matα (Gilon *et al.*, 2000). Although GFP^u is degraded in mammalian cells in an ubiquitin-dependent manner by the proteasome, it is not known whether GFP^u degradation is Ubc6/7 dependent. The failure of a dominant-negative Ubc6 homolog to influence GFP^u degradation implies at the very least that Ubc6 is not the sole E2 enzyme responsible for GFP^u degradation (Lenk *et al.*, 2002). Interestingly, to promote GFP^u accumulation with pharmacological proteasome inhibitors, the chymotryptic activity of the proteasome must be attenuated in excess of 70%. This is consistent with the observations made for the ubiquitin fusion GFP reporter (Masucci and Dantuma, 2000) and may reflect either a robust cellular reservoir of proteasome activity or the need to also inhibit the tryptic or peptidyl-glutamyl peptide hydrolytic (PGPH) activities of the proteasome to impair proteolysis (most inhibitors show the highest affinity for the chymotryptic site and only affect the latter sites at high concentrations).

Supporting a role for ubiquitin in the degradation of GFP^u, the reporter coimmunoprecipitates with ubiquitin, and degradation is impaired by the expression of dominant-negative K48R ubiquitin (unpublished data). The CL1 degron contains an internal lysine that may act as an ubiquitin

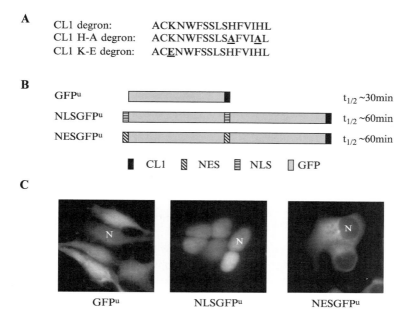

FIG. 1. The GFP^u family of UPS reporters and CL1 sequence variants. (A) The CL1 peptide sequence variants that are UPS degrons. (B) The architecture of the GFP^u, NLSGFP^u, and NESGFP^u reporter proteins. (C) Microscopy images demonstrating the subcellular localization of the GFP^u, NLSGFP^u, and NESGFP^u reporter proteins (N denotes a nucleus).

acceptor, but the mutation of the CL1 lysine to a glutamate (CL1 K-E) increases the half-life from 30 min to approximately 1 h but does not alter the eventual proteasomal fate of the reporter (both in yeast and mammalian cells). GFP itself contains a number of lysine residues on the surface of its β-barrel structure that may act as interchangeable ubiquitin acceptors similar to the multiple lysine residues on the yeast protein Sic1 (Petroski and Deshaies, 2003).

Additional CL1 amino acid changes also affect the properties of the degron. Mutation of the two histidine residues to alanine circumvents the Ubc6/7 enzymes and diverts ubiquitylation to an unidentified E2 pathway in yeast (Gilon et al., 2000). Our unpublished experiments have shown that the corresponding mutations in the GFP^u degron still yield an unstable UPS reporter protein in mammalian cells, although it is currently unknown whether these mutations alter the ubiquitylation pathway as in yeast. As more is understood about the degradation pathway of the CL1 degron and its sequence variants, it should be possible to further tailor reporters to the study of specific ubiquitylation pathways.

Recently, nuclear and cytoplasmic versions of the GFP^u reporters have been generated and characterized (Bennett *et al.*, 2005). Both reporters contain tandem GFP molecules (to exceed the ~60-kDa diffusional limit of the nuclear pore complex) with a CL1 degron and either a nuclear localization sequence (NLS) or a nuclear export sequence (NES) (Fig. 1B,C). The nuclear and cytoplasmic GFP^u reporters (NLSGFP^u and NESGFP^u) are useful for the study of localized UPS insults that may affect the nuclear or cytoplasmic UPS pools differently. The nuclear and cytoplasmic GFP^u reporters, in combination with the original GFP^u and its sequence variants, comprise a powerful set of tools for UPS research.

Application of the GFP^u Reporter System in Mammalian Cells

GFP^u experiments can be performed with either transient or stable expression of the reporter. Stable expression is preferred, because transient overexpression frequently results in reporter synthesis rates that saturate the capacity of the UPS, making the reporter insensitive to changes in UPS function. The GFP^u plasmid and a stable expressing clone in HEK293 cells (Bence *et al.*, 2001) are available through American Type Culture Collection (ATCC), #MBA-87 and #CRL-2794 respectively. GFP^u has been expressed successfully in a variety of cell lines and does not seem to exhibit toxicity when expressed at levels typically achieved through stable expression. Single-cell clones with moderate to low levels of detectable basal GFP^u expression should be selected to obtain a maximal dynamic range of reporter response and to minimize the demands of GFP^u degradation on cellular UPS capacity. It should be noted that all cell-based protocols in this article assume the use of a stable GFP^u clonal cell line.

Analysis of GFP^u UPS Reporters Using Flow Cytometry and Pharmacological Proteasome Inhibitors

The simplest way to quantify GFP^u levels is to analyze GFP fluorescence intensity in live cells by means of flow cytometry. Clonal stable cell lines are optimal, because they exhibit a narrow distribution of basal intensities yielding uniform histograms that are easy to quantify and permit the detection of small changes in population fluorescence (Fig. 2A).

Protocol 1

1. Plate out $>2 \times 10^6$ GFP^u cells in a 6-cm or greater dish. Throughout cell maintenance and experimental treatments (for example, the addition

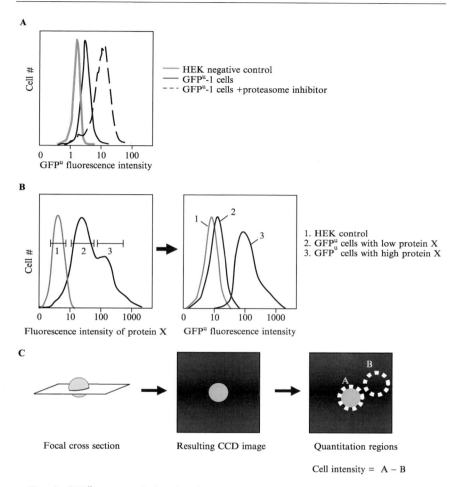

FIG. 2. GFP^u assay analysis using flow cytometry and microscopic image analysis. (A) Example of flow cytometry histograms of control cells, GFP^u cells under basal conditions, and GFP^u cells treated with a proteasome inhibitor, resulting in a total shift of the histogram. (B) Analyzing GFP^u levels relative to the levels of a protein (protein X) requires a two-step process. A histogram of protein X intensity is first generated and subpopulations of the histogram, regions 1, 2, and 3, are used to generate GFP^u intensity histograms. This example represents data in which the increased expression of protein X initiates UPS inhibition as shown by the substantial shift of GFP^u intensity in the GFP^u histogram generated from population 3. (C) Schematic of the use of spherical cells for quantitative microscopy analysis. Equatorial images generate round image intensity regions that are easily quantified by use of image analysis software and simple mathematics to generate a representative GFP^u cell intensity.

of proteasome inhibitors), maintain cells at subconfluent densities (40–60% is optimal).

2. Cells may be treated with and without proteasome inhibitors (i.e., 5 μM MG132 up to 12 h). To achieve uniform distribution of drug, only media containing freshly prediluted and mixed compound should be added to cells. The addition of small volumes directly to medium in the culture dish results in poor mixing and broadens flow cytometry intensity histograms because of uneven compound distribution.

3. At appropriate time points, wash cells with PBS and remove adherent cells with trypsin-EDTA or PBS + 10 mM EDTA. Resuspend pellet in a minimum of 2 ml PBS in a 5-ml FACS tube.

4. Pellet cells at 200g for 5 min in a swinging bucket centrifuge and wash cells in ice-cold PBS by gently resuspending the pellet and recentrifuging cells.

5. Add propidium iodide to a concentration of 10 μg/ml and analyze cells by flow cytometry. Generate a GFP intensity histogram by gating on the live, propidium iodide negative, cell population.

6. After proteasome inhibition, GFP^u cells will exhibit higher overall fluorescence, resulting in shifts of the GFP intensity histograms (Fig. 2A). Compare the mean fluorescence intensities of the treated versus control histograms to discern the presence or absence of proteasome inhibitor exposure.

Analysis of GFP^u Relative to a Second Protein Marker

The preceding protocol is optimal for pharmacological proteasome inhibition. Conditions such as transient transfection of a dominant-negative inhibitor of UPS function (i.e., K48R ubiquitin) will likely result in the shift of a subpopulation of GFP^u cells. This is because only a portion of the transfected GFP^u cells will express sufficient levels of the dominant-negative inhibitor to achieve UPS impairment. UPS function within such a subpopulation can be illuminated by applying a two-color flow cytometric analysis that compares the relative levels of GFP^u intensity to the levels of the expressed protein.

Protocol 2

1. Maintain cells at subconfluent densities throughout the experiment. This protocol requires a large number of washes and centrifugation steps, and, as a result, a large number of cells (minimum of $\sim 10 \times 10^6$ cells per sample) should be used to compensate for cell loss. Transfect GFP^u cells using a lipid or calcium-phosphate–based method with the gene

of choice and a control vector. Process samples 24 to 72 h after transfection.

2. Remove adherent cells with trypsin-EDTA or PBS + 10 mM EDTA if trypsin is undesirable. Pellet cells at 200g for 5 min in a 5-ml FACS tube in a swinging bucket rotor. Resuspend cells in ice-cold PBS and place on ice for 5 min to promote rounding of the cells. Pellet at 200g for 5 min.

3. Resuspend and incubate cell pellet in 1 ml 4% p-formaldehyde in PBS at room temperature for 15 min, swirling tube every few minutes to gently maintain cells in suspension. Methanol or other alcohol-based fixation protocols destroy the GFP chromophore and should be avoided.

4. Permeabilize cells by resuspending them in 1 ml cold PBS + 2% BSA (buffer A) with 0.1% Triton X-100. Lipid extraction occurs rapidly and should be performed on ice with an incubation time of less than 5 min to maintain sample integrity. Cells should be pelleted and washed in >2 ml buffer A. The protocol for antibody standing must be optimized for each antibody. For reference, we typically incubate with the 1° antibody for 30 min at RT in buffer A. Wash two times with >2 ml buffer A. Incubate with 2° antibody in buffer A for 30 min. Finish by washing cells in buffer A three times. The 2° antibody fluorophore conjugate choice should match the excitation and emission parameters of the microscope or flow cytometer to be used in each experiment. To prevent signal contamination that artificially elevates GFP^u signals, fluorophores with minimal spectral overlap and longer emission wavelengths than GFP are recommended (alexa 594 or texas red conjugates for microscopy and phycoerythrin (PE) or allophycocyanin (APC) for two-color flow cytometry.

5. Analyze the cells by flow cytometry (do not use propidium iodide on fixed cells) or use the microscopy assay. If using flow cytometry, determine whether the levels of GFP^u correlate with the levels of the stained protein in a manner that may indicate UPS dysfunction. This can be done by gating on cells with low and high levels of the stained protein and then generating a GFP histogram for each of the subpopulations. The mean GFP fluorescence of each subpopulation can be determined and compared (Fig. 2B).

Quantitative Microscopy of the GFP^u UPS Reporter

GFP^u quantification using an epifluorescence microscope with a CCD (charge coupled device) camera and image analysis software is an accurate way to make GFP^u measurements. Although two-color flow cytometry allows for rapid comparison of GFP^u intensity versus the total cell intensity of a second protein on an individual cell basis, this type of analysis is not always optimal. For example, if a dominant-negative inhibitor exhibited both nuclear and cytoplasmic localization phenotypes with the nuclear phenotype associated with UPS dysfunction, a flow cytometer would be unable to make this important distinction. A microscopy assay permits the researcher to select for such target phenotypes. When performed properly, this method yields quantitative data of high quality.

Protocol 3: Microscope Setup. The critical components of the assay are a quality microscope objective, a CCD camera, and image capture and analysis software. Because the initial intensity of the UPS reporter line is low, high numerical aperture (NA) oil objectives (NA ~1.2–1.4) should be used. Objective power should be at least 40× to yield a final image in which individual cells are represented by a sufficient number of pixels to aid quantification. Before performing experiments, the uniformity of field illumination and the linearity of the CCD camera should be determined and accounted for.

1. Check for even-field illumination using all filter sets that will be used for imaging. Seal a solution containing the fluorophore to be used in the study (e.g., fluorescent 2° antibody conjugate) between a slide and a coverslip. Image the solution with a CCD camera at a variety of exposure times. Use image analysis software to survey individual pixel intensity values throughout the image. The presence of regions of overly high or low intensity in the imaging field is typically the result of an imbalance in field illumination. Refer to the appropriate literature concerning excitation lamp alignment and manipulate the lamp and optics until an even field illumination can be achieved. If this fails to correct the uniformity issue, the problem may result from non-uniformity in the light path. Clean or replace any components of the light path (e.g., objectives, prisms, and filters) that exhibit discoloration or defects.

2. Check the linearity of the CCD camera. This can be done with a fluorescent object (i.e., a fixed cell mounted with an anti-fade reagent) with a fluorescence intensity similar to that expected in your experiments. Select a number of objects and create a series of images at exposure times that increase by a factor of 2 (10 mS, 20 mS, 40 mS, 80 mS etc.) Quantify the object fluorescence intensity (as described later) and plot exposure time versus normalized object intensity to determine the linear range of the camera (usually 2 or more orders of magnitude). Refer to this information when performing experiments to ensure that imaged object intensities are within the linear range of CCD sensitivity.

Protocol 4: Collecting Image Sets and Image Quantification. We have devised a method for quantifying single-cell GFP^u intensities that is both simple and accurate. Cells are imaged in spherical form at the equatorial plane, yielding circular cross-sections that are nearly uniform in overall area from cell to cell (Fig. 2C). This process helps to reduce variation in the data because of differential cell spreading.

1. Trypsinize adherent cells (or lifted in PBS + 10 mM EDTA if trypsin is undesirable) to promote rounding. Fix cells and, if necessary, further process samples using steps 1 through 4 of protocol 2. The inclusion of a UV excitable DNA dye (i.e., DAPI, bisbenzimide) should be avoided because of potential bleed through of the blue emission into the green channel and the damaging effects of ultraviolet light on GFP fluorescence.

2. Once cells are trypsinized into spheres, fixed, and, if necessary, stained for a second protein (as in protocol 2), cells can be resuspended in mounting media with an anti-photobleaching reagent. Place small drops of this suspension onto slides, cover with coverslips, and seal with fingernail polish.

3. Scan coverslips for the individual cell of interest on a microscope (e.g., through the red channel for the staining of a second protein to minimize photobleaching of GFP^u) and then image the cell in all channels for which the cell contains fluorophores of interest, including the GFP channel for GFP^u. Images should be captured with an exposure that results in image intensities within the linear range of the CCD camera with the highest grayscale bit rate.

4. Images should be collected and analyzed with a suitable software program. To analyze, two identical circular regions are drawn around the cell of interest (A) and a blank region (B) of the field (Fig. 2C). A total integrated pixel intensity of each region should be acquired, and the final cell intensity should be calculated as the integrated intensity in region A minus the integrated pixel intensity of region B. Large data sets of >100, preferably 400 or more, are needed to obtain statistically significant data. Compiled cell intensities can be binned to generate fluorescence histograms, with mean and standard deviation calculations used for sample comparison.

Conclusion

The GFP^u family of UPS reporters permits the assessment of UPS function in the nuclear, cytoplasmic, or global cellular compartments. Their fluorescence readout, dependence on ubiquitylation, and rapid response to proteasome inhibition make them simple and reliable tools. In this article we have described the basic flow cytometry and epifluorescence microscopy methods that can be applied to the study of UPS dysfunction arising from pharmacological agents, proteotoxic stress, mutations, or other UPS insults. With proper use, they represent a valuable asset to the UPS researcher.

References

Bence, N. F., Sampat, R. M., and Kopito, R. R. (2001). Impairment of the ubiquitin-proteasome system by protein aggregation. *Science* **292,** 1552–1555.
Bennett, E. J., Bence, N. F., Rajadas, J., and Kopito, R. R. (2005). Global impairment of the ubiquitin-proteasome system by nuclear or cytoplasmic protein aggregates precedes inclusion body formation. *Mol. Cell* **17,** 351–365.
Dantuma, N. P., Lindsten, K., Glas, R., Jellne, M., and Masucci, M. G. (2000). Short-lived green fluorescent proteins for quantifying ubiquitin/proteasome-dependent proteolysis in living cells. *Nat. Biotechnol.* **18,** 538–543.
Gilon, T., Chomsky, O., and Kulka, R. G. (1998). Degradation signals for ubiquitin system proteolysis in *Saccharomyces cerevisiae*. *EMBO J.* **17,** 2759–2766.
Gilon, T., Chomsky, O., and Kulka, R. G. (2000). Degradation signals recognized by the Ubc6p-Ubc7p ubiquitin-conjugating enzyme pair. *Mol. Cell Biol.* **20,** 7214–7219.
Lenk, U., Yu, H., Walter, J., Gelman, M. S., Hartmann, E., Kopito, R. R., and Sommer, T. (2002). A role for mammalian Ubc6 homologues in ER-associated protein degradation. *J. Cell Sci.* **115,** 3007–3014.
Li, X., Zhao, X., Fang, Y., Jiang, X., Duong, T., Fan, C., Huang, C. C., and Kain, S. R. (1998). Generation of destabilized green fluorescent protein as a transcription reporter. *J. Biol. Chem.* **273,** 34970–34975.
Masucci, M. G., and Dantuma, N. P. (2000). Reply to 'ubiquitin/proteasome system'. *Nat. Biotechnol.* **18,** 807.
Neefjes, J., and Dantuma, N. P. (2004). Fluorescent probes for proteolysis: Tools for drug discovery. *Nat. Rev. Drug Discov.* **3,** 58–69.
Petroski, M. D., and Deshaies, R. J. (2003). Context of multiubiquitin chain attachment influences the rate of Sic1 degradation. *Mol. Cell* **11,** 1435–1444.
Thrower, J. S., Hoffman, L., Rechsteiner, M., and Pickart, C. M. (2000). Recognition of the polyubiquitin proteolytic signal. *EMBO J.* **19,** 94–102.

[34] Monitoring of Ubiquitin-Dependent Proteolysis with Green Fluorescent Protein Substrates

By Victoria Menéndez-Benito, Stijn Heessen, and Nico P. Dantuma

Abstract

A reliable and robust means of evaluating the functional status of ubiquitin-dependent proteolysis in living cells is to follow the turnover of readily detectable reporter substrates. During the past few years, several reporter substrates have been generated by use of the green fluorescent protein (GFP), which is converted for this purpose from a normally very stable protein into a short-lived substrate of the ubiquitin/proteasome system. These short-lived substrates are valuable tools providing researchers with unique information about the absence or presence of blockades in this system in living cells. We have recently generated the first transgenic mouse model for monitoring the ubiquitin/proteasome system based on the ubiquitous expression of a GFP-based proteasome substrate. Together these models can be used to study ubiquitin-dependent degradation in health and disease and for the identification of small synthetic compounds or proteins capable of modifying the activity of the system. In this chapter, we describe the basic principles of GFP-based reporter substrates, their strengths and weaknesses, and a number of protocols that can be used to study the ubiquitin/proteasome system in yeast, cell lines, and transgenic mice.

Introduction

Cellular proteases have throughout the years been appreciated as therapeutic targets largely owing to their well-defined and easily assessable enzymatic activities, which enable high-throughput screens for compounds with inhibitory activity. These assays are often based on small fluorogenic peptide substrates that emit fluorescence on cleavage at a specific target sequence. Similar fluorogenic peptide substrates are available for the three distinct proteolytic activities of the proteasome: the chymotrypsin-like, the trypsin-like, and the post-glutamyl peptidyl hydrolyzing activities. Although these fluorogenic substrates elicit detailed information about the proteolytic activities of the proteasome and have been helpful in the identification of active-site inhibitors, they do not truly reflect the biologically relevant activity of this proteolytic machinery. Moreover, the proteolytic activity of the proteasome is only one of the many events within the ubiquitin/proteasome system that eventually lead to destruction of proteins (Baumeister *et al.*, 1998; Hershko and Ciechanover, 1998).

Ubiquitination, recruitment, unfolding, and translocation of substrates are all equally important for ubiquitin/proteasome–dependent proteolysis. Disturbance in any of these events may seriously affect the handling of substrates by the ubiquitin/proteasome system. Assessing the functional status of the ubiquitin/proteasome system is an additional challenge that requires alternative tools. Several studies have highlighted a possible link between a number of human diseases and the ubiquitin/proteasome system (Ciechanover and Brundin, 2003; Hernandez *et al.*, 2004; Lindsten and Dantuma, 2003), making such tools of particular interest in drug development. Moreover, some proteasome inhibitors have anticancer activity (Meng *et al.*, 1999a). Indeed, the first drug based on this concept has been recently approved for treatment of patients with multiple myeloma (Adams, 2004).

One means of following the functionality of the ubiquitin/proteasome system is to evaluate the turnover of one of its many natural substrates. Although this can provide us with valuable information, this approach has a number of important shortcomings. First and foremost, degradation of most substrates of the ubiquitin/proteasome system is regulated. Hence, delayed turnover of a particular substrate is not necessarily caused by functional impairment of ubiquitin/proteasome–dependent proteolysis but may reflect functional stabilization. For example, if under certain conditions, degradation of the tumor suppressor p53 is shown to be inhibited, this may not necessarily be caused by a blockade of the ubiquitin/proteasome system but rather be a consequence of the fact that the cells are undergoing p53-dependent apoptosis (Scheffner, 1998). Second, monitoring turnover of natural proteasome substrates requires either pulse-chase metabolic labeling or administration of protein synthesis inhibitors, followed by biochemical analysis of the turnover of the protein of interest, both of which are laborious procedures that may themselves affect the physiology of the cell. Third, these assays cannot be used for real-time monitoring of ubiquitin-dependent proteolysis. Finally, it is virtually impossible to use these methods to analyze the ubiquitin/proteasome system in individual cells.

The discovery of the green fluorescent protein (GFP) of the jellyfish *Aequorea victoria* (Chalfie *et al.*, 1994), the more recent identification of fluorescent homologs of *Anthozoa* corals (Matz *et al.*, 1999), and the generation of a collection of spectral variants (van Roessel and Brand, 2002) has opened up the possibility to follow the fate of designed reporter substrates in living cells. We and others have used GFP and its relatives to develop authentic fluorescent protein substrates of the ubiquitin/proteasome system that allow real-time monitoring of ubiquitin-dependent degradation (Neefjes and Dantuma, 2004). GFP has several characteristics that identify it as an excellent candidate protein for the development of reporter substrates (Tsien, 1998). First, the formation of its chromophore is an autocatalytic event that does not require any other proteins or cofactors. Second, GFP is a very stable protein that in its native conformation is not an efficient substrate of the ubiquitin/proteasome system or other cellular proteases. Third, loss of structural integrity of the fluorescent protein results in a dramatic loss of fluorescence. Thus, even though the proteasome generates peptide fragments of up to 20–30 amino acids (Kisselev *et al.*, 1999), the chance that any fluorescent GFP fragments will remain after processive degradation by the proteasome is slim. One should be aware that unfolding of GFP-based reporter substrates, which must precede the actual proteasomal degradation, might be sufficient for loss of fluorescence. However, it is unlikely that this will occur in the absence of proteolytic cleavage, because unfolded GFP rapidly refolds under physiological conditions (Tsien, 1998). Indeed, loss of GFP fluorescence as a consequence of unfolding has only been successfully measured *in vitro* in the presence of a chaperone trap that freezes the protein in its unfolded conformation (Hoskins *et al.*, 2000; Weber-Ban *et al.*, 1999).

The different fluorescent protein reporters for the ubiquitin/proteasome system have been designed according to a general principle (Neefjes and Dantuma, 2004). Either by introducing a degradation signal into GFP or fusing GFP with a proteasome substrate, fluorescent protein reporters can be generated that are destroyed shortly after synthesis by ubiquitin-dependent degradation. Degradation signals are domains, small motifs, or aberrant structures that are recognized by ubiquitin ligases (Laney and Hochstrasser, 1999). Subsequently, ubiquitin ligases provide those proteins that contain degradation signals with polyubiquitin chains, ultimately leading to the destruction of the ubiquitinated protein by the proteasome (Pickart, 2001). The choice of degradation signal or fusion protein used in the reporters is crucial, because it will not only determine the half-life of the protein but also whether ubiquitination is regulated by internal or external conditions. For reporter substrates that are designed to give information on the functional status of the ubiquitin/proteasome system, a prerequisite is that degradation occurs constitutively. Fusions of GFP with proteasome substrates whose degradation is tightly regulated, such as IkBα (Li *et al.*, 1999), may be very useful for studying degradation of that particular substrate but will be not very informative for questions relating to functional impairment of the ubiquitin/proteasome system. Fortunately, there are a number of degradation signals that are targeted for constitutive degradation and are active across species. These degradation signals have been successfully used for the generation of reporter substrates (see later).

Cells and transgenic mice expressing GFP-based proteasome substrates typically contain low levels of the fluorescent protein, which are often undetectable by native fluorescence or immunostaining (Dantuma *et al.*, 2000b; Heessen *et al.*, 2003; Lindsten *et al.*, 2003). A number of membrane-permeable compounds can block the ubiquitin/proteasome system by inhibiting the proteolytic activities of the proteasome, which can be used in both cell lines and mice. The most commonly used proteasome inhibitors are lactacystin (Fenteany *et al.*, 1995), peptide aldehydes (such as MG132 and PSI) (Jensen *et al.*, 1995), epoxomicin (Meng *et al.*, 1999b), peptide boronic acids (such as MG262 and bortozemib) (Adams, 2004), and peptide vinyl sulfones (Bogyo *et al.*, 1997). Treatment of cells expressing reporter substrates with proteasome inhibitors results in a dramatic increase in the fluorescence intensity that can be readily quantified (Dantuma et al., 2000b). Accumulation of the reporter is predictive, and a steep rise in reporter levels occurring just hours after administration of inhibitor correlates with induction of cell cycle arrest and cell death after 24–48 h (Dantuma et al., 2000b). Intraperitoneal administration of proteasome inhibitors to transgenic mice ubiquitously expressing a GFP-based proteasome substrate causes an increase in reporter levels in those tissues affected by the inhibitors (Lindsten et al., 2003). Besides the application of GFP-based substrates for studies on the ubiquitin/proteasome system, it has been shown that shortening the half-life of GFP by introduction of a degradation signal also makes them useful as transcriptional reporters (Li et al., 1998). Because of the low steady-state levels and improved signal-to-noise ratio, small changes in promoter activity can be detected with greater sensitivity.

An important difference between analyses of the ubiquitin/proteasome system using reporter substrates compared with assays that probe into specific events in this pathway (ubiquitination, deubiquitination, proteolysis, etc.) lies in the fact that the reporters give information about the functionality of the total system but leave uncertainty about what causes the blockade. Experimental data obtained with reporter substrates basically tells the researcher whether cells or tissues can or cannot deal with the vast amount of endogenous ubiquitinated substrates. How the cells degrade their ubiquitinated substrates or, in the case that they do not, why they accumulate the substrates, cannot directly be answered and requires further investigation. On the other hand, with assays that focus on an isolated process, it is impossible to predict whether aberrations in that particular event will have an overall effect on the ubiquitin/proteasome system. Cells contain an enormous amount of excessive proteasome activity, and inhibition of up to 80% of the chymotrypsin-like activity is required for functional impairment of the system to occur (Bence et al., 2001; Dantuma et al., 2000b). The postglutamyl peptidyl hydrolyzing activity of the proteasome seems to be even more redundant for a functional ubiquitin/proteasome system in mammalian cells (Myung et al., 2001). Thus, at best, only educated guesses can be made as to whether a certain level of inhibition of the individual sites of the proteasome will cause a general impairment. A striking example is that under specific conditions, degradation of reporter substrates is influenced by an inhibitor of tripeptidyl peptidase (TPP)-II (Kessler et al., 2003), a peptidase that plays a crucial role in degrading the peptide fragments generated by the proteasome (Reits et al., 2004). Hence, assays based on reporter substrates may not reveal reductions in proteasome activity that, although substantial, fail to block ubiquitin-dependent proteolysis or may respond to compounds that affect other proteases than the proteasome. This, however, does not mean that the assays are less sensitive or less specific. Rather these examples show that assays using reporter substrates represent truly functional assays that summarize the complexity of the system in a simple readout, namely, whether the cells are capable of clearing proteins targeted for ubiquitin-dependent degradation. It is this type of information that is highly relevant if we talk about functional impairment of the ubiquitin/proteasome system in diseases (Ciechanover and Brundin, 2003; Hernandez et al., 2004; Lindsten and Dantuma, 2003) or for the development of proteasome inhibitors as therapeutic drugs (Adams, 2004).

GFP Reporters for Ubiquitin-Dependent Degradation

We generated a set of GFP-based substrates of the ubiquitin/proteasome system by introduction of constitutively active degradation signals into GFP (Dantuma et al., 2000b). For this purpose, we adapted the ubiquitin fusion strategy developed by Varshavsky and coworkers that allows the half-life of the protein of interest to be changed from stable to an N-end rule substrate or a ubiquitin-fusion degradation (UFD) substrate by single amino acid substitutions (Johnson et al., 1992; Varshavsky, 1996). This is accomplished by expressing GFP fused to an N-terminal ubiquitin moiety. N-terminal ubiquitin moieties are rapidly cleaved from precursors by endogenous deubiquitination enzymes (DUBs) (Chung and Baek, 1999). Thus, shortly after synthesis, the ubiquitin moiety is cleaved from the ubiquitin–GFP (Ub-GFP) fusion product. The half-life of the GFP cleavage product can be regulated by varying the amino terminal residue of the GFP fragment generated by this cleavage. The nature of the N terminal amino acid is a major determinant of ubiquitination, a phenomenon known as the N-end rule (Varshavsky, 1996). If the fusion is expressed as Ub-methionine-GFP (Ub-M-GFP), cleavage by DUBs will generate M-GFP, which is a long-lived protein. However, in the case of Ub–arginine-GFP (Ub-R-GFP), the resulting R-GFP will be recognized by a specific ubiquitin ligase, provided with a ubiquitin chain in close proximity to its N-terminus and rapidly degraded by the proteasome (Fig. 1). Exploitation of the N-end rule has been shown to be a reliable and robust tool for targeting proteins for proteasomal degradation. Conversely, cleavage of the N-terminal ubiquitin can be strongly impeded by substituting the last amino acid of the ubiquitin moiety from a glycine to valine (Johnson et al., 1992). In Ub^{G76V}-GFP, the N-terminal ubiquitin forms the anchor for polyubiquitin chains that will target the complete fusion product for

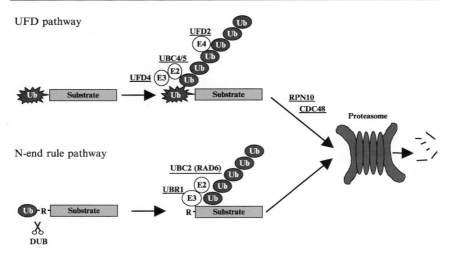

FIG. 1. GFP reporters for the ubiquitin/proteasome system. Schematic drawing of the UFD and N-end rule GFP reporter substrates and some of the different ubiquitination enzymes and accessory proteins involved in each pathway. UFD substrates are ubiquitinated within the N-terminal ubiquitin moiety, whereas N-end rule substrates before ubiquitination on the substrate itself require cleavage of the N-terminal ubiquitin by isopeptidases. Cdc48 and Rpn10 are required to target ubiquitinated UFD substrates to the proteasome.

degradation (Fig. 1). This is known as the UFD pathway. An interesting observation about UFD substrates is that both Lys^{29} and Lys^{48} polyubiquitin chains are engaged in the targeting process (Johnson et al., 1992; Lindsten et al., 2002). The ubiquitination pathways for the N-end rule and UFD pathways have been mapped in detail and shown not to overlap with different E2s, E3s, E4s, chaperones, and ubiquitin-binding subunits involved in proteasomal targeting of each of the substrates (Fig. 1) (Bartel et al., 1990; Johnson et al., 1992; Koegl et al., 1999; van Nocker et al., 1996). More recently, others have developed similar reporters, such as GFP-CL1 (also known as GFPu) (Bence et al., 2001), based on the introduction of a small 16 amino acid–long motif known as the CL1 degradation signal (Gilon et al., 1998, 2000), and a GFP-MHC class I fusion that can be targeted by means of a viral protein for endoplasmic reticulum associated degradation by the ubiquitin/proteasome system (Kessler et al., 2001).

GFP-based reporters based on the ubiquitin-fusion technology, and in particular the Ub-R-GFP and Ub^{G76V}-GFP, have been shown to be valuable tools for the evaluation of ubiquitin/proteasome–dependent proteolysis in yeast, cell lines, and transgenic animals (Neefjes and Dantuma, 2004).

In the next few sections, we will describe some of the protocols that have been instrumental in our work with Ub-GFP reporter substrates.

Monitoring Ubiquitin-Dependent Degradation in Budding Yeast

The budding yeast *Saccharomyces cerevisiae* has been extensively used to study ubiquitin-dependent proteasomal degradation (Wolf, 2004). The key components and molecular mechanisms in ubiquitin-dependent proteolysis are conserved from yeast to mammals (Finley *et al.*, 1998). In yeast, genetic manipulations can easily be combined with biochemical and cell biology analyses, making it an ideal model organism to learn about the individual roles of proteins in intracellular proteolysis. Many mutant strains deficient for proteins involved in the ubiquitin/proteasome system have been generated and characterized extensively. Furthermore, the N-end rule and UFD degradation pathways were originally discovered in yeast and reporter substrates based on enzymatically active proteins carrying such degradation signals have been used in genome-wide screens for proteins involved in these pathways (Bartel *et al.*, 1990; Johnson *et al.*, 1992). GFP-based reporter substrates allow the analysis of protein stability directly with fluorimetric tools, such as fluorimetry and flow cytometry. The GFP reporters have been used for functional analysis of the N-end rule and UFD pathways in yeast mutants (Bays *et al.*, 2001; Heessen *et al.*, 2003). Notably, proteasome inhibitors, which are commonly used for studies in metazoan cells, cannot be used in most yeast strains, because they do not pass the cell wall unless specific mutant strains are used (Lee and Goldberg, 1996). In the following section, we describe a protocol for determining steady-state levels of GFP reporters in yeast by flow cytometric analysis.

Protocol for Flow Cytometric Analysis of Yeast-Expressing GFP Reporters

1. Obtain yeast expressing the GFP reporter substrate by transformation of yeast with an episomal or centromeric plasmid expressing the GFP reporter substrate or by genomic integration at an appropriate locus. Include yeast transformed with an empty plasmid and Ub-M-GFP expression plasmid as negative and positive controls, respectively. Optionally, yeast mutants deficient for the N-end rule or UFD pathway can be included. The *ubr1*Δ (Bartel *et al.*, 1990) and the *ubc4/ubc5*Δ (Johnson *et al.*, 1992) double mutant are advisable for the N-end rule and UFD substrates, respectively (Heessen *et al.*, 2003).

2. Inoculate colonies from each transformation in selective liquid medium. For each wild-type or mutant strain to be tested, include

transformants with empty plasmid, reporter-encoding plasmid (e.g., Ub-R-GFP or Ub^{G76V}-GFP) and Ub-M-GFP-encoding plasmid.

3. If the reporters are expressed from a GAL1 promoter, optimal expression can be obtained by successively growing dense overnight cultures of the transformants in selective medium with 2% glucose, 2% raffinose, and, finally, 2% galactose as sole carbon source.

4. Dilute a dense overnight culture in selective medium with 2% galactose to an OD600 of 0.1–0.2 and grow to log phase.

5. Take 100 µl of log-phase yeast (OD600 0.5–1.0) and dilute 1:5 in medium.

6. Analyze 1×10^4 cells with a flow-automated cell sorter (FACSort; Becton-Dickinson, San Jose, CA) with a standard FL1 filter ($\lambda_{excitation}$ 488 nm, $\lambda_{emission}$ 515 nm).

7. Analyze the raw data with CellQuest software by making a histogram plot of the FL1 channel and performing histogram statistics. Most informative are the mean fluorescence intensities of the total yeast populations.

Episomal 2-µm–based yeast expression plasmids have a high and variable copy number (~20–40 per cell) and give, when the stable Ub-M-GFP is expressed, a broad fluorescence peak, ranging from background fluorescence to an approximate 100-fold increase over background. More homogenous expression can be obtained by using low-copy number plasmids, such as CEN-based or integrative plasmids (Bays et al., 2001) (NPD, unpublished observations). Fluorescence intensities of yeast expressing the destabilized Ub-R-GFP or Ub^{G76V}-GFP reporters from 2-µm–based plasmids are around the background levels observed with vector-transformed yeast. Expressing these fluorescent reporters in yeast strains lacking essential enzymes in the N-end rule (for Ub-R-GFP) or UFD pathway (for Ub^{G76V}-GFP) increases the fluorescence to levels similar to that of Ub-M-GFP expressing yeast, showing that such mutant strains can be used to evaluate the correct targeting of the reporters.

Monitoring Ubiquitin-Dependent Degradation in Cell Lines

GFP-based reporter substrates have been used to study the effect of proteins or compounds on the functionality of the ubiquitin/proteasome system (Kessler et al., 2003; Lindsten et al., 2002; Lundgren et al., 2003; Myung et al., 2001). Cell lines stably expressing Ub-R-GFP or Ub^{G76V}-GFP typically contain very low levels of the reporter protein that may be difficult to distinguish from background fluorescence and to detect in Western blot analysis. Blockage of the ubiquitin/proteasome system by proteasome inhibitor treatment can provoke up to a 500-fold increase in fluorescence intensities of reporter-expressing cells. Ub-R-GFP and Ub^{G76V}-GFP reporter cell lines have been generated with human cells (Dantuma et al., 2000b), murine cells (Kessler et al., 2003), and insect cells (Lundgren et al., 2003). In the following section, we describe a protocol for generation of cell lines stably expressing GFP reporters.

Protocol for the Generation of Cell Lines Stably Expressing GFP Reporters

1. Culture cells in 100-mm dishes.
2. Transfect cells with Ub-R-GFP or Ub^{G76V}-GFP with the transfection method of choice.
3. At 48 h after transfection, add 0.5 to 2 mg/ml G418 (Sigma) to complete culture medium. The optimal concentration depends on the sensitivity of the cell line and has to be empirically determined by growing the parental cells in the presence of G418. The optimal concentration is the lowest concentration that kills untransfected parental cells in 4 to 6 days.
4. Grow the cells in the presence of G418 until colonies of G418-resistant cells have a radius of 2 to 5 mm. With most cell lines this will take 23 weeks.
5. Check the colonies for fluorescence with an inverted fluorescence microscope. Most colonies should be nonfluorescent because of high turnover of the GFP reporter. There may be occasionally GFP positive colonies, but these are unlikely to behave as functional GFP reporter clones.
6. Culture the colonies for 6 to 8 h in complete medium with 5 μM proteasome inhibitor MG132 (Affiniti, UK). Clones expressing functional GFP reporter substrates will accumulate GFP during proteasome inhibitor treatment and emit fluorescence. This normally occurs only in a small percentage of the clones.
7. Check frequently during the 6- to 8-h incubation for the appearance of fluorescent clones. Because proteasome inhibitors are toxic to cells and will eventually induce apoptosis, it is important to keep the proteasome-inhibitor treatment as short as possible to minimize damage to the clones. The responsive clones have to survive the treatment, as they will be used for establishing reporter cell lines. Therefore, continue as soon as dim fluorescent clones appear. Other inhibitors can be used, but an advantage of MG132 is its reversible nature, making it possible to wash away after the treatment. Most other inhibitors will cause irreversible inhibition of proteasomes (lactacystin, epoxomicin, vinyl sulfone) or will have a prolonged effect because of slow off-rates (MG262). These proteasome inhibitors can be obtained from Affiniti (UK).

8. Identify fluorescent clones exhibiting a homogenous fluorescence throughout the clone and mark them on the bottom of the plate.

9. Wash the cells three times with complete medium without inhibitor, leaving a small volume of complete medium on the cells after the final wash.

10. Collect the marked clones from the plate by careful scraping with the tip of a micropipette the clone from the plate and simultaneously collecting with mild suction the dissociated cells. Avoid contamination of the clone with surrounding cells. If necessary, clones can be isolated with the aid of an inverted light microscope. Transfer the clones to individual wells on a 24-well plate containing the appropriate growth medium and G418.

11. When 70% confluence is reached, split the cells over three wells on a 24-well plate.

12. When the triplet wells reach 70% confluence, incubate one well from each clone with 10 μM MG132 in complete medium for 10-16 h (other inhibitors can also be used at this point, such as 10 μM lactacystin or z-Leu-Leu-Leu-vinyl sulfone (ZLVS) or 500 nM epoxomicin).

13. Using flow cytometry, compare the fluorescence intensity of the inhibitor-treated well with one of the two untreated wells (Fig. 2). The third well is kept to maintain the clone.

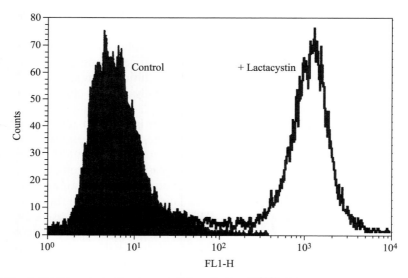

FIG. 2. FACS analysis of reporter cell line. Stable Ub^{G76V}-GFP HeLa cells were treated for 16 h with 10 μM lactacystin or left untreated. Cells were harvested and directly analyzed by flow cytometry for GFP fluorescence (FL1 channel).

14. Identify the clones that accumulate the GFP reporter in response to proteasome inhibition and expand the cultures in the presence of G418 to maintain positive selection for the reporter plasmid.

15. In some cases, it may be necessary to sort the clones several times to obtain a homogenous response. To eliminate cells that are fluorescent in the absence of inhibitor, harvest untreated cells and collect the GFP^{negative} or GFP^{low} cells by FACS. If the clone contains a substantial population of unresponsive cells, treat the cells for 6-8 h with 5 μM MG132, wash away the MG132, and immediately collect the GFP^{high} population by FACS. The combination of inhibitor treatment and FACS sorting may cause considerable cell death, but most cell lines tolerate this short treatment rather well and will survive.

An alternative strategy is to make a polyclonal cell line by collecting the total pool of inhibitor-responsive cells in successive FACS sorting rounds. Both cells that are fluorescent in the absence of proteasome inhibitor and cells that remain nonfluorescent in the presence of proteasome inhibitor must be eliminated.

If cells do not tolerate the 6- to 8-h incubation with MG132, other inhibitors can be tried. Alternately, clones can be picked without inhibitor treatment and tested afterward for inhibitor responsiveness. It is advisable to check the sensitivity of the cell lines to proteasome inhibitors before the transfection procedure.

Identification of Stabilization Signals

Stabilization signals are defined as *cis*-acting, transferable domains that can counteract degradation signals and thereby protect proteins from proteasomal degradation (Dantuma and Masucci, 2002). Stabilization signals have been identified in viral proteins and pathogenic proteins (Levitskaya *et al.*, 1997; Verhoef *et al.*, 2002). It has been postulated that cellular proteins may use stabilization signals to regulate their turnover (Dantuma and Masucci, 2002; Heessen *et al.*, 2005). *Cis*-inhibitory effects of domains on proteasomal degradation have been studied with reporter constructs in yeast and mammalian cells (Dantuma *et al.*, 2000a; Heessen *et al.*, 2003; Michalik and Van Broeckhoven, 2004; Verhoef *et al.*, 2002).

By comparing the effect of insertion of domains on the steady-state levels of reporters, putative stabilization signals (PSS) can be identified and characterized. An increase in the steady-state levels of the reporter substrate indicates that the domain of interest delays turnover, which can be confirmed by determining the half-life of the reporter without and with the PSS by either promoter shutoff experiments or pulse-chase analysis. One should also check whether the domain of interest causes a general impairment of the ubiquitin/proteasome system, which may indirectly block degradation of the reporter substrate. Authentic stabilizations signals should selectively protect the protein that carries the domain (analogous to degradation signals). This can be determined by coexpressing an unmodified GFP reporter with the GFP reporter fused to the PSS and determine the turnover of both GFP reporters simultaneously. In the following section, we describe protocols for analyzing the *cis*-inhibitory effect of domains on proteasomal degradation in yeast and mammalian cells.

Protocol for Analysis of Stabilization Signals in Yeast

1. PCR amplify DNA encoding the PSS and preferably introduce a 5′ flanking sequence with an *Ssp*B1 restriction site, and 3′ flanking sequence with a stop codon directly followed by a *Not*1 restriction site. Check that there are no *Ssp*B1 or *Not*1 sites present in the target sequence.

2. Digest the PCR product with *Ssp*B1 (Boehringer-Mannheim) and *Not*1 (Invitrogen), and clone the PSS fragment into the unique *Ssp*B1 and *Not*1 restriction sites flanking the stop-codon of the original Ub-GFP fusion open reading frame. It may be necessary to perform the first cloning step in the mammalian Ub-GFP plasmids (Dantuma *et al.*, 2000b), which are based on the commercial EGFP-N1 vector (CLONTECH Laboratories), followed by subcloning of the Ub-GFP-PSS into a yeast vector, because *Ssp*B1 and *Not*1 are unique in the mammalian expression vector. Alternately, the Ub-GFP-PSS fusion can be made directly in yeast by *in vivo* recombination.

3. Transform yeast with the following constructs: (1) empty vector, (2) Ub-M-GFP (positive control), (3) Ub-R-GFP, (4) Ub^{G76V}-GFP, (5) Ub-R-GFP-PSS, and (6) Ub^{G76V}-GFP-PSS. Preferably transform both wild type yeast and the N-end rule or UFD mutants.

4. Analyze the fluorescence intensities of each of the yeast strains by flow cytometry as described previously. Calculate the absolute mean fluorescence values by subtracting the mean fluorescence intensity of yeast transformed with empty vector from the mean fluorescence intensity of each strain. Then, calculate the relative steady-state levels of the protein by comparison with the Ub-M-GFP control, which should be set at 100%. Relative values for the reporter substrates should be close to zero. If the relative value is significantly higher in the presence of PSS, the domain may have a protective effect on proteasome degradation.

Protocol for Analysis of Stabilization Signals in Mammalian Cells

1. Clone the PSS in the reporter substrates as described previously.

2. Transiently transfect cells in a 12-well plate with (1) empty vector, (2) stable Ub-M-GFP (positive control), (3) Ub-R-GFP, (4) Ub^{G76V}-GFP, (5) Ub-R-GFP-PSS, and (6) Ub^{G76V}-GFP-PSS. Transfect six wells for each construct (for triplicate controls and inhibitor treatments).

3. After 48 h, add to three wells from each construct proteasome inhibitor (10 μM MG132, lactacystin or ZLVS, or 500 nM epoxomicin or MG-262) in complete media and incubate for 10 h. Also change media in three control wells (include same concentration of inhibitor solvent).

4. Harvest cells in both the untreated wells and the inhibitor-treated wells for each construct and analyze the fluorescence of the cells by flow cytometry. Determine the percentage and mean fluorescent intensity of the GFP-positive cells in each sample. Calculate the ratio between GFP-positive cells in the absence and presence of inhibitor. For a stable protein (Ub-M-GFP), the ratio should be approximately 1. For the reporter substrates, this ratio should be $\ll 1$. If the ratio of reporter-PSS is significantly higher than the ratio of the unmodified reporter and/or the mean fluorescent intensity of the reporter-PSS is more intense, this is indicative of a protective effect by the PSS.

Monitoring Ubiquitin-Dependent Degradation in Transgenic Mice

Reporter mice based on the expression of GFP or luciferase reporter proteins have been generated for various purposes (Maggi et al., 2004). Most importantly, the use of reporter mice simplifies the study of pharmacological effects *in vivo* and enables the efficacy of biological systems to be analyzed in the context of complex pathologies. A xenotransplantation model for the ubiquitin/proteasome system has been generated, whereby nude mice were injected with human cells expressing a luciferase-based proteasome substrate (Luker et al., 2003). Although this model allows noninvasive imaging of the ubiquitin/proteasome system in living mice, it can only be used for analysis of the transplanted tumor cells. Off-target effects will be missed with this model, because the surrounding tissue does not express the reporter, nor can this model reveal aberrations in the ubiquitin/proteasome system in mouse models of diseases. We have developed a transgenic mouse model of the ubiquitin/proteasome system based on ubiquitous expression of the Ub^{G76V}-GFP reporter substrate (Lindsten et al., 2003). These mice can be used for studying the effect of proteasome inhibitors on normal and malignant tissue. Moreover, these mice can be used to address the heavily debated question of whether impairment of the ubiquitin/proteasome system plays a role in those diseases that are characterized by the accumulation of misfolded proteins (Hernandez et al., 2004). The strategy used in the generation of this mouse model and the protocols that can be used to investigate the ubiquitin/proteasome system using this model are outlined in the following. The animal experiments described in this section have been approved by the Ethical Committee in Stockholm (Ethical permission number N-36/00).

Generation of the Ub^{G76V}-GFP Reporter Mice

The Ub^{G76V}GFP reporter was placed under the control of the cytomegalovirus immediate early enhancer and the chicken β-actin promoter and introduced into the pronucleus of fertilized CBA × C57BL/6 F1 oocytes. This promoter had previously been used to generate transgenic mice expressing unmodified GFP (Okabe et al., 1997). Transgenic mice expressing the unmodified GFP, having the same genetic background as the reporter mice, are available from Jackson Laboratory (Jackson code: C57BL/6-Tg(ACTbEGFP)1Osb/J) and may be used as a positive control in studies with the Ub^{G76V}-GFP reporter mice. Two transgenic Ub^{G76V}-GFP founders were obtained giving rise to the two congenic strains presently used: Ub^{G76V}-GFP/1 and Ub^{G76V}-GFP/2, which are both maintained in a C57BL/6 background. There are no apparent differences between these two reporter strains. Both strains express the reporter transcript in all analyzed tissues: cerebrum, cerebellum, heart, lungs, spleen, pancreas, stomach, intestines, ovaries, and testis. Mice carrying Ub^{G76V}-GFP are viable, fertile, and indistinguishable in appearance from their nontransgenic littermates. The tissues of Ub^{G76V}-GFP transgenics do not have the characteristic yellowish glow, which is readily observed even under normal light in the aforementioned GFP transgenic strain.

Breeding of Ub^{G76V}-GFP Reporter Mice

The mouse colony is maintained by successive back-crosses of hemizygous male or female Ub^{G76V}GFP mice with C57BL/6 mice. In general, these crosses produce normal litter sizes and a Mendelian inheritance pattern. A recurrent problem in our hands with the Ub^{G76V}-GFP/2 strain is that mothers frequently eat the litter directly after birth, reducing the litter sizes and making this strain more difficult to breed.

Protocol for Genotyping Ub^{G76V}-GFP Reporter Mice

Two precautions in the design of the PCR protocol for genotyping have been made to minimize the risk of false positives because of contamination and amplification of endogenous genes. First, the PCR product that is generated is derived from the chicken β-actin promoter and the 5' part of the Ub^{G76V}-GFP open reading frame. Second, the antisense primer anneals at the boundary of the ubiquitin/GFP fusion having its 5' end at the G76V substitution in the mutant ubiquitin. These precautions make it almost impossible to obtain PCR products from other GFP fusions that may be used in the laboratory or from the highly conserved ubiquitin genes present in the mouse genome.

1. Prepare genomic DNA from tail biopsy specimens taken from litters of Ub^{G76V}-GFP/1 × C57BL/6 or Ub^{G76V}-GFP/2 × C57BL/6 at weaning according to standard procedure (Hogan *et al.*, 1994).
2. Make a 20-μl PCR mix with 2 μl genomic DNA template (approximately 0.5 μg total DNA) with the sense primer 'Transgene 5-1': CCT ACA GCT CCT GGG CAA CGT, and the antisense primer 'UbG76V-2': TCG ACC AAG CTT CCC CAC CAC. PCR mix: 2 μl 10× PCR buffer, 2 μl 2.5 m*M* dNTP, 1 μl 10 pmol/μl Transgene 5-1 primer, 1 μl 10 pmol/μl UbG76V-2 primer, 2 μl template, 12 μl distilled water, 1 unit Taq DNA polymerase (Invitrogen). It is recommended to include as a positive control a sample from a previously identified transgenic mouse.
3. Amplify the target DNA with step 1, 2 min 97°; step 2, 30 sec, 95°; step 3, 30 sec 60°; step 4, 1 min, 72°; repeat steps 2–4 30 times; step 5, 7 min 72° on a PCT200 (MJ Research, Reno, NV) or comparable PCR machine.
4. Run PCR product on a 1% agarose gel.
5. Mice that carry the Ub^{G76V}-GFP transgene can be identified by the presence of a 423 base pair PCR product. No PCR product will be obtained from samples of non-transgenic mice.

Protocol for Administration of Proteasome Inhibitors to Ub^{G76V}-GFP Mice

Although the Ub^{G76V}-GFP transcript is present in all analyzed tissues, the protein product is generally undetectable because of the intended rapid turnover by the proteasome. To get acquainted with the model and the reporter levels that can be expected when the ubiquitin/proteasome system is impaired, it may be helpful to first administer moderate levels of the proteasome inhibitors MG262 or epoxomicin and evaluate the effect on the reporter. This treatment should cause accumulation of the Ub^{G76V}-GFP reporter in the liver within 20 h (Fig. 3). The same protocol can be used to test the effect of a compound of interest on ubiquitin-dependent degradation.

1. Weigh Ub^{G76V}-GFP mice. Prepare a solution of 200 μl of 60% DMSO/40% PBS with 1 μmol/kg body weight MG262 or epoxomicin (Affiniti, UK). MG262 and epoxomicin can be stored as stock solution of 1 m*M* in DMSO at –20°.
2. Inject the mice intraperitoneally with the inhibitor solution. As controls, injection of Ub^{G76V}-GFP mice with control solution (60% DMSO/40% PBS) and injection of non-transgenic littermates with inhibitor or control solution may be considered.
3. Kill the mice 20 h after injection and prepare liver and optionally other tissues for microscopic examination. Both inhibitors induce accumulation

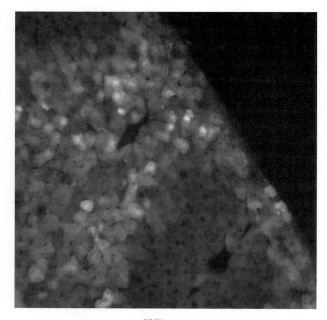

FIG. 3. Liver of a MG262-treated Ub^{G76V}-GFP/1 transgenic mouse. Fluorescence micrograph of native Ub^{G76V}-GFP reporter in a liver section from a Ub^{G76V}-GFP/1 transgenic mouse treated with MG262 (1 μmol/kg body weight). The picture was captured with an LEITZ-BMRB fluorescence microscope (Leica) provided with a Hamamatsu cooled CCD camera.

primarily in the cells surrounding the portal vein, with MG262 giving a more pronounced accumulation (Fig. 2). The Ub^{G76V}-GFP can be detected by its native GFP fluorescence, and the signal can be further enhanced by immunofluorescence or immunohistochemistry.

A fivefold higher concentration of MG262 inhibitor gives accumulation of reporter in multiple tissues (liver, intestine, pancreas, spleen, kidney, lung) but should be used with caution because of the considerable toxicity for the mice.

Protocols for Detection of the Reporter in Cryosections from Ub^{G76V}GFP Mice

The GFP fluorescence can be detected either directly by its native fluorescence or the signal can be enhanced by immunofluorescence and immunohistochemical methods (Bowman *et al.*, 2005). Some protocols for tissue processing and GFP imaging by immunohistochemistry and immunofluorescence are described in the following.

Protocol For Tissue Processing

1. Anesthetize mice with an overdose of isoflurane (4.4% isoflurane in oxygen) and transcardially perfuse with 50 ml of PBS. Alternately, tissue can be processed without perfusion, although this may cause an increase in background fluorescence.
2. Excise the tissue and fix by immersion in 4% paraformaldehyde (PFA) in PBS (pH 7.4) for 16 h at 4°. Fixation with PFA preserves both fluorescent properties and antigenicity of GFP.
3. Immerse the tissues in a graded series of sucrose in PBS at 4° as follows: 3 h in 7% sucrose, 3 h in 14% sucrose, and 16 h in 25% sucrose.
4. Embed the tissue in Tissue-Tek OCT compound (Sakura Finetek USA), freeze with dry-ice. Samples can be stored at –70°.

Protocol for Native Ub^{G76V}-GFP Imaging

1. Prepare 10-μm sections with a cryotome and fix the section on the object glass (SuperFrost Plus, Menzel-Glaser, Germany) with 4% PFA for 20 min at room temperature. Wash the sections 3 × 5 min with PBS to remove excessive fixative.
2. Permeabilize the tissue with 0.2% Triton X-100 in PBS for 5 min at room temperature. Wash the sections 3 × 5 min with PBS and counterstain the sections with Hoechst 33258 (2 μg/ml in distilled water) for 15 min at room temperature in the dark. Wash 3 × 5 min with PBS and mount the sections with DABCO mounting solution. (*DABCO stock*: 2.5 g DABCO, 9 ml glycerol, 1 ml distilled water. *DABCO mounting solution*: 30 ml glycerol, 1 ml 1 M Tris (pH 8.0), 4 ml 25% DABCO stock, 5 ml distilled water).

Protocol for Immunofluorescent Staining of Ub^{G76V}-GFP

1. Prepare and fix the sections as described previously.
2. Add blocking solution (normal goat serum diluted 1:50 in PBS + 0.2% Triton X-100) and block for 30 min at room temperature.
3. Remove blocking solution and add primary rabbit polyclonal anti-GFP antibody (Molecular Probes Europe, Leiden, The Netherlands) diluted 1:1000 in blocking solution. Incubate for 16 h at 4°.
4. Wash 3 × 5 min with PBS and incubate with secondary Alexa fluor 488-conjugated anti-rabbit antibody (Molecular Probes Europe) diluted 1:1000 in blocking solution for 1 h at room temperature in the dark.
5. Wash 3 × 5 min with PBS, counterstain the sections with Hoechst 33258 (2 μg/ml in distilled water) and mount with DABCO mounting solution.

Protocol for Immunohistochemical Staining of Ub^{G76V}-GFP

1. Prepare and fix the sections as described previously.
2. Quench endogenous peroxidase by immersion of sections in 3% H_2O_2 for 10 min, followed by 3 × 5 min washes with PBS at room temperature.
3. Block and incubate with primary anti-GFP antibody as described previously.
4. Wash 3 × 5 min with PBS and incubate with secondary biotinylated anti-rabbit IgG antibody (Vector Laboratories) 1:1000 in blocking solution for 30 min at room temperature.
5. Incubate the sections with avidin/biotinylated peroxidase complex (ABC kit, Vector Laboratories) for 30 min at room temperature.
6. Develop with diaminobenzidine (DAB) solution (Vector Laboratories). A dark brown staining will appear. When an appropriate level of staining is achieved, stop the reaction by washing with PBS and counterstain with hematoxylin. Too long DAB incubation times will cause background staining in the negative control.
7. Dehydrate by sequentially immersing the sections in 75%, 80%, 95%, and 100% ethanol. Clear the sections in xylene (Sigma) and mount with Entellan mounting solution (Merck).

Monitoring the Ubiquitin/Proteasome System in Primary Cells from Ub^{G76V}-GFP Mice. The Ub^{G76V}GFP transgenic mice may serve as a source of primary cells expressing the reporter. These primary cultures can be used to quantify the sensitivity of different cell types to proteasome inhibitors and to pathological conditions that affect the ubiquitin/proteasome system. We have established primary fibroblasts, cardiomyocytes, hepatocytes, and neurons from Ub^{G76V}GFP mice and shown that the reporter is functionally expressed in these cell types (Lindsten *et al.*, 2003).

Acknowledgments

We thank Christa Maynard, Lisette Verhoef, and Kristina Lindsten for critical reading of the manuscript and helpful suggestions. This work was supported by grants awarded by the Swedish Research Council, Swedish Cancer Society, the High Q Foundation, the Wallenberg Foundation, and the Karolinska Institute.

References

Adams, J. (2004). The development of proteasome inhibitors as anticancer drugs. *Cancer Cell* **5,** 417–421.
Bartel, B., Wunning, I., and Varshavsky, A. (1990). The recognition component of the N-end rule pathway. *EMBO J.* **9,** 3179–3189.

Baumeister, W., Walz, J., Zuhl, F., and Seemuller, E. (1998). The proteasome: Paradigm of a self-compartmentalizing protease. *Cell* **92**, 367–380.
Bays, N. W., Wilhovsky, S. K., Goradia, A., Hodgkiss-Harlow, K., and Hampton, R. Y. (2001). HRD4/NPL4 is required for the proteasomal processing of ubiquitinated ER proteins. *Mol. Biol. Cell.* **12**, 4114–4128.
Bence, N. F., Sampat, R. M., and Kopito, R. R. (2001). Impairment of the ubiquitin-proteasome system by protein aggregation. *Science* **292**, 1552–1555.
Bogyo, M., McMaster, J. S., Gaczynska, M., Tortorella, D., Goldberg, A. L., and Ploegh, H. (1997). Covalent modification of the active site threonine of proteasomal beta subunits and the *Escherichia coli* homolog HslV by a new class of inhibitors. *Proc. Natl. Acad. Sci. USA* **94**, 6629–6634.
Bowman, A. B., Yong, S. Y., Dantuma, N. P., and Zoghbi, H. Y. (2005). Polyglutamine neuropathology occurs in the absence of detectable proteasome impairment and inversely correlates with nuclear inclusions. *Hum. Mol. Genet.* **14**, 679–691.
Chalfie, M., Tu, Y., Euskirchen, G., Ward, W. W., and Prasher, D. C. (1994). Green fluorescent protein as a marker for gene expression. *Science* **263**, 802–805.
Chung, C. H., and Baek, S. H. (1999). Deubiquitinating enzymes: Their diversity and emerging roles. *Bichem. Biophys. Res. Commun.* **266**, 633–640.
Ciechanover, A., and Brundin, P. (2003). The ubiquitin proteasome system in neurodegenerative diseases: Sometimes the chicken, sometimes the egg. *Neuron* **40**, 427–446.
Dantuma, N. P., Heessen, S., Lindsten, K., Jellne, M., and Masucci, M. G. (2000a). Inhibition of proteasomal degradation by the Gly-Ala repeat of Epstein-Barr virus is influenced by the length of the repeat and the strength of the degradation signal. *Proc. Natl. Acad. Sci. USA* **97**, 8381–8385.
Dantuma, N. P., Lindsten, K., Glas, R., Jellne, M., and Masucci, M. G. (2000b). Short-lived green fluorescent proteins for quantifying ubiquitin/proteasome-dependent proteolysis in living cells. *Nat. Biotechnol.* **18**, 538–543.
Dantuma, N. P., and Masucci, M. G. (2002). Stabilization signals: A novel regulatory mechanism in the ubiquitin/proteasome system. *FEBS Lett.* **529**, 22–26.
Fenteany, G., Standaert, R. F., Lane, W. S., Choi, S., Corey, E. J., and Schreiber, S. L. (1995). Inhibition of proteasome activities and subunit-specific amino-terminal threonine modification by lactacystin. *Science* **268**, 726–731.
Finley, D., Tanaka, K., Mann, C., Feldmann, H., Hochstrasser, M., Vierstra, R., Johnston, S., Hampton, R., Haber, J., McCusker, J., Silver, P., Frontali, L., Thorsness, P., Varshavsky, A., Byers, B., Madura, K., Reed, S. I., Wolf, D., Jentsch, S., Sommer, T., Baumeister, W., Goldberg, A., Fried, V., Rubin, D. M., Toh-e, A., *et al.* (1998). Unified nomenclature for subunits of the *Saccharomyces cerevisiae* proteasome regulatory particle. *Trends Biochem. Sci.* **23**, 244–245.
Gilon, T., Chomsky, O., and Kulka, R. G. (1998). Degradation signals for ubiquitin system proteolysis in *Saccharomyces cerevisiae*. *EMBO J.* **17**, 2759–2766.
Gilon, T., Chomsky, O., and Kulka, R. G. (2000). Degradation signals recognized by the Ubc6p-Ubc7p ubiquitin-conjugating enzyme pair. *Mol. Cell. Biol.* **20**, 7214–7219.
Heessen, S., Dantuma, N. P., Tessarz, P., Jellne, M., and Masucci, M. G. (2003). Inhibition of ubiquitin/proteasome-dependent proteolysis in *Saccharomyces cerevisiae* by a Gly-Ala repeat. *FEBS Lett.* **555**, 397–404.
Heessen, S., Masucci, M. G., and Dantuma, N. P. (2005). The UBA2 domain functions as an intrinsic stabilization signal that protects Rad23 from proteasomal degradation. *Mol. Cell* **18**, 225–235.
Hernandez, F., Diaz-Hernandez, M., Avila, J., and Lucas, J. J. (2004). Testing the ubiquitin-proteasome hypothesis of neurodegeneration *in vivo*. *Trends Neurosci.* **27**, 66–69.

Hershko, A., and Ciechanover, A. (1998). The ubiquitin system. *Annu. Rev. Biochem.* **67**, 425–479.

Hogan, B., Beddington, R., Costantini, F., and Lacy, E. (1994). "Manipulating the Mouse Embryo: A Laboratory Manual." Cold Spring Harbor Laboratory Press, New York.

Hoskins, J. R., Singh, S. K., Maurizi, M. R., and Wickner, S. (2000). Protein binding and unfolding by the chaperone ClpA and degradation by the protease ClpAP. *Proc. Natl. Acad. Sci. USA* **97**, 8892–8897.

Jensen, T. J., Loo, M. A., Pind, S., Williams, D. B., Goldberg, A. L., and Riordan, J. R. (1995). Multiple proteolytic systems, including the proteasome, contribute to CFTR processing. *Cell* **83**, 129–135.

Johnson, E. S., Bartel, B., Seufert, W., and Varshavsky, A. (1992). Ubiquitin as a degradation signal. *EMBO J.* **11**, 497–505.

Kessler, B., Hong, X., Petrovic, J., Borodovsky, A., Dantuma, N. P., Bogyo, M., Overkleeft, H. S., Ploegh, H., and Glas, R. (2003). Pathways accessory to proteasomal proteolysis are less efficient in major histocompatibility complex class I antigen production. *J. Biol. Chem.* **278**, 10013–10021.

Kessler, B. M., Tortorella, D., Altun, M., Kisselev, A. F., Fiebiger, E., Hekking, B. G., Ploegh, H. L., and Overkleeft, H. S. (2001). Extended peptide-based inhibitors efficiently target the proteasome and reveal overlapping specificities of the catalytic beta-subunits. *Chem. Biol.* **8**, 913–929.

Kisselev, A. F., Akopian, T. N., Woo, K. M., and Goldberg, A. L. (1999). The sizes of peptides generated from protein by mammalian 26 and 20S proteasomes. Implications for understanding the degradative mechanism and antigen presentation. *J. Biol. Chem.* **274**, 3363–3371.

Koegl, M., Hoppe, T., Schlenker, S., Ulrich, H. D., Mayer, T. U., and Jentsch, S. (1999). A novel ubiquitination factor, E4, is involved in multiubiquitin chain assembly. *Cell* **96**, 635–644.

Laney, J., and Hochstrasser, M. (1999). Substrate targeting in the ubiquitin system. *Cell* **97**, 427–430.

Lee, D. H., and Goldberg, A. L. (1996). Selective inhibitors of the proteasome-dependent and vacuolar pathways of protein degradation in *Saccharomyces cerevisiae*. *J. Biol. Chem.* **271**, 27280–27284.

Levitskaya, J., Sharipo, A., Leonchiks, A., Ciechanover, A., and Masucci, M. G. (1997). Inhibition of ubiquitin/proteasome-dependent protein degradation by the Gly-Ala repeat domain of the Epstein-Barr virus nuclear antigen 1. *Proc. Natl. Acad. Sci. USA* **94**, 12616–12621.

Li, X., Fang, Y., Zhao, X., Jiang, X., Duong, T., and Kain, S. R. (1999). Characterization of NFκB activation by detection of green fluorescent protein-tagged IκB degradation in living cells. *J. Biol. Chem.* **274**, 21244–21250.

Li, X., Zhao, X., Fang, Y., Jiang, X., Duong, T., Fan, C., Huang, C. C., and Kain, S. R. (1998). Generation of destabilized green fluorescent protein as a transcription reporter. *J. Biol. Chem.* **273**, 34970–34975.

Lindsten, K., and Dantuma, N. P. (2003). Monitoring the ubiquitin/proteasome system in conformational diseases. *Ageing Res. Rev.* **2**, 433–449.

Lindsten, K., de Vrij, F. M., Verhoef, L. G., Fischer, D. F., van Leeuwen, F. W., Hol, E. M., Masucci, M. G., and Dantuma, N. P. (2002). Mutant ubiquitin found in neurodegenerative disorders is a ubiquitin fusion degradation substrate that blocks proteasomal degradation. *J. Cell Biol.* **157**, 417–427.

Lindsten, K., Menendez-Benito, V., Masucci, M. G., and Dantuma, N. P. (2003). A transgenic mouse model of the ubiquitin/proteasome system. *Nat. Biotechnol.* **21**, 897–902.

Luker, G. D., Pica, C. M., Song, J., Luker, K. E., and Piwnica-Worms, D. (2003). Imaging 26S proteasome activity and inhibition in living mice. *Nat. Med.* **9**, 969–973.

Lundgren, J., Masson, P., Realini, C. A., and Young, P. (2003). Use of RNA interference and complementation to study the function of the Drosophila and human 26S proteasome subunit S13. *Mol. Cell. Biol.* **23,** 5320–5330.

Maggi, A., Ottobrini, L., Biserni, A., Lucignani, G., and Ciana, P. (2004). Techniques: Reporter mice - a new way to look at drug action. *Trends Pharmacol. Sci.* **25,** 337–342.

Matz, M. V., Fradkov, A. F., Labas, Y. A., Savitsky, A. P., Zaraisky, A. G., Markelov, M. L., and Lukyanov, S. A. (1999). Fluorescent proteins from nonbioluminescent Anthozoa species. *Nat. Biotechnol.* **17,** 969–973.

Meng, L., Kwok, B. H., Sin, N., and Crews, C. M. (1999a). Eponemycin exerts its antitumor effect through the inhibition of proteasome function. *Cancer Res.* **59,** 2798–2801.

Meng, L., Mohan, R., Kwok, B. H., Elofsson, M., Sin, N., and Crews, C. M. (1999b). Epoxomicin, a potent and selective proteasome inhibitor, exhibits *in vivo* antiinflammatory activity. *Proc. Natl. Acad. Sci. USA* **96,** 10403–10408.

Michalik, A., and Van Broeckhoven, C. (2004). Proteasome degrades soluble expanded polyglutamine completely and efficiently. *Neurobiol. Dis.* **16,** 202–211.

Myung, J., Kim, K. B., Lindsten, K., Dantuma, N. P., and Crews, C. M. (2001). Lack of proteasome active site allostery as revealed by subunit-specific inhibitors. *Mol. Cell* **7,** 411–420.

Neefjes, J., and Dantuma, N. P. (2004). Fluorescent probes for proteolysis: Tools for drug discovery. *Nat. Rev. Drug Discov.* **3,** 58–69.

Okabe, M., Ikawa, M., Kominami, K., Nakanishi, T., and Nishimune, Y. (1997). 'Green mice' as a source of ubiquitous green cells. *FEBS Lett.* **407,** 313–319.

Pickart, C. M. (2001). Mechanisms underlying ubiquitination. *Annu. Rev. Biochem.* **70,** 503–533.

Reits, E., Neijssen, J., Herberts, C., Benckhuijsen, W., Janssen, L., Drijfhout, J. W., and Neefjes, J. (2004). A major role for TPPII in trimming proteasomal degradation products for MHC class I antigen presentation. *Immunity* **20,** 495–506.

Scheffner, M. (1998). Ubiquitin, E6-AP, and their role in p53 inactivation. *Pharmacol. Ther.* **78,** 129–139.

Tsien, R. Y. (1998). The green fluorescent protein. *Annu. Rev. Biochem.* **67,** 509–544.

van Nocker, S., Sadis, S., Rubin, D. M., Glickman, M., Fu, H., Coux, O., Wefes, I., Finley, D., and Vierstra, R. D. (1996). The multiubiquitin-chain-binding protein Mcb1 is a component of the 26S proteasome in *Saccharomyces cerevisiae* and plays a nonessential, substrate-specific role in protein turnover. *Mol. Cell. Biol.* **16,** 6020–6028.

van Roessel, P., and Brand, A. H. (2002). Imaging into the future: visualizing gene expression and protein interactions with fluorescent proteins. *Nat. Cell Biol.* **4,** E15–E20.

Varshavsky, A. (1996). The N-end rule: Functions, mysteries, uses. *Proc. Natl. Acad. Sci. USA* **93,** 12142–12149.

Verhoef, L. G., Lindsten, K., Masucci, M. G., and Dantuma, N. P. (2002). Aggregate formation inhibits proteasomal degradation of polyglutamine proteins. *Hum. Mol. Genet.* **11,** 2689–2700.

Weber-Ban, E. U., Reid, B. G., Miranker, A. D., and Horwich, A. L. (1999). Global unfolding of a substrate protein by the Hsp100 chaperone ClpA. *Nature (London)* **401,** 90–93.

Wolf, D. H. (2004). From lysosome to proteasome: The power of yeast in the dissection of proteinase function in cellular regulation and waste disposal. *Cell. Mol. Life Sci.* **61,** 1601–1614.

[35] Monitoring Proteasome Activity *In Cellulo* and in Living Animals by Bioluminescent Imaging: Technical Considerations for Design and Use of Genetically Encoded Reporters

By SHIMON GROSS and DAVID PIWNICA-WORMS

Abstract

The ubiquitin-proteasome pathway is the central mediator of regulated proteolysis, instrumental for switching on and off a variety of signaling cascades. Deregulation of proteasomal activity or improper substrate recognition and processing by the ubiquitin-proteasome machinery may lead to cancer, stroke, chronic inflammation, and neurodegenerative diseases. Quantifying total and substrate-specific proteasome activity in intact cells and living animals would enable analysis *in vivo* of proteasomal regulation and facilitate the screening and validation of potential modulators of the proteasome or its substrates. We discuss examples of tetra-ubiquitin or IκBα fused to firefly luciferase as genetically encoded reporters for monitoring total and IκBα-specific proteasomal activity by bioluminescence imaging. Such technology enables repetitive, temporally resolved, and regionally targeted assessment of proteasomal activity *in vivo*.

Preface

The 26S proteasome degrades proteins that control essential signaling pathways, including proteins crucial to cell cycle regulation, programmed cell death, and inflammation (Voorhees *et al.*, 2003). Several important proteins that are regulated by the proteasome include inhibitor of nuclear factor κB (IκB) (Karin and Ben-Neriah, 2000), β-catenin (Aberle *et al.*, 1997), tumor suppressor p53 (Moll and Petrenko, 2003), cyclin-dependent kinase inhibitors p21 (Blagosklonny *et al.*, 1996; Bloom and Pagano, 2004) and p27 (Pagano *et al.*, 1995), hypoxia-inducible transcription factor HIF-1α (Maxwell *et al.*, 1999), proapoptotic protein Bax (Li and Dou, 2000) and, to some extent, epidermal growth factor receptor ErbB1 (Levkowitz *et al.*, 1998).

Given the myriad of proteasome substrates, it is not surprising that inhibition of proteasome activity has been explored as a therapeutic strategy for cancer (Voorhees *et al.*, 2003), stroke (Wojcik and Di Napoli, 2004), and a variety of inflammatory and autoimmune diseases (Elliott *et al.*, 2003;

Voorhees et al., 2003). This can be accomplished by direct inhibition of the proteasome or modulation of regulatory pathways that are upstream to proteasomal degradation (i.e., by inhibiting an E3-ligase or a kinase that phosphorylates and thereby renders its target a substrate for ubiquitin ligation and proteasomal degradation). In fact, one of these drug candidates, the proteasome inhibitor bortezomib (Velcade®), was recently approved by the FDA for treatment of multiple myeloma (Bross et al., 2004). However, the need for appropriate molecular reporters for high-throughput screening (HTS) assays in vitro and animal models for target validation and translational studies in vivo impede the discovery of novel therapeutics that modulate proteasome action or analyses of substrates.

This chapter describes methods for detecting and quantifying changes in proteasome activity in vivo using genetically encoded reporters and remote detection devices. Such technology enables repetitive and noninvasive assessment of the regulation of proteasomal activity and the potency and pharmacodynamics of proteasome inhibitors or modulators of proteasome targets.

The reader is provided with a brief introduction to in vivo molecular imaging, followed by two examples of genetically encoded, fusion reporters used for imaging activation and inhibition of signaling pathways associated with proteasome degradation: (1) a tetra-ubiquitinated firefly luciferase (Ub-FL) for imaging total proteasome activity and (2) IκBα-firefly luciferase (IκBα-FL) for imaging ligand-induced degradation of the proteasome substrate IκBα. Finally, technical considerations for designing suitable functional reporters and imaging intact cells and animals with such reporters are discussed.

Introduction to Molecular Imaging

Molecular imaging is broadly defined as the characterization and measurement of biological processes in living animals, model systems, and humans at the cellular and molecular levels using remote imaging detectors. With refined genomic maps of human, mouse, and many pathogens completed, genetic information is expected to lead to new medical therapies, diagnostics, and, ultimately, cures previously not imagined. In the post-genomic era, in which functionality will be added to this vast array of genetic information, opportunity exists for imaging to play a significant role in basic and translational research and in clinical care for patients.

Molecular imaging is focused on noninvasive, repetitive monitoring of gene expression in vivo. The target genes can be either endogenous or exogenous genes. Reporter strategies enable direct imaging of exogenous gene expression and indirect imaging of endogenous gene expression by the use of endogenous promoters driving reporter constructs. Another fundamental advantage of any reporter gene/reporter probe system is that once validated, the reporter gene can theoretically be cloned into an appropriate vector, and any gene of interest can be interrogated with the same validated reporter probe. A variety of reporter genes have been introduced and validated for different imaging modalities, including various luciferases (i.e., firefly luciferase, *Renilla* luciferase) for bioluminescence imaging (BLI) (Bhaumik and Gambhir, 2001; Contag *et al.*, 1997), herpes simplex virus-1 thymidine kinase (HSV-1 TK) for positron emission tomography (PET) (Hospers *et al.*, 2000; Luker *et al.*, 2002; Tjuvajev *et al.*, 1995), various fluorescent proteins for fluorescence microscopic and macroscopic imaging (Yang *et al.*, 2000), transferrin receptor (ETR) for magnetic resonance imaging (MRI) (Weissleder *et al.*, 2000) and a variety of receptors or transporters (i.e., somatostatin receptor type 2 [Chaudhuri *et al.*, 2001], dopamine receptor type 2 [Maclaren *et al.*, 1999], and NaI symporter for single photon emission computerized tomography (SPECT) and PET imaging [Sharma *et al.*, 2002]). Each of these reporter genes can be used for investigating a wide variety of biological processes, such as (1) transcriptional regulation pathways by introducing the reporter gene downstream of one or more copies of a promoter or a regulated DNA element of interest (i.e., a transgenic mouse expressing FL under the control of an NF-kB–responsive element [Carlsen *et al.*, 2002]); (2) monitoring tumor burden or cell tracking by expressing (constitutively or conditionally) the reporter gene in cells of interest (i.e., tracking hematopoietic stem cell engraftment or tumor cell migration by bioluminescent imaging [Rettig *et al.*, 2004; Wang *et al.*, 2003]); (3) assessing pretranscriptional and posttranscriptional regulation of gene expression by fusing in-frame the reporter gene to a protein of interest (i.e., fusing p27 to FL and thereby assessing activation and pharmacological inhibition of its upstream kinase Cdk2 [Zhang *et al.*, 2004]); and (4) monitoring protein–protein interactions by using split reporters fused to binding partners of interest (Luker and Piwnica-Worms, 2004; Luker *et al.*, 2004; Ozawa *et al.*, 2001; Paulmurugan *et al.*, 2002, 2004) or by fusing the binding partners of interest to an appropriate set of DNA-binding and activator domains to drive the transcription of the imaging reporter gene ("two-hybrid system") (Luker *et al.*, 2002, 2003b).

When establishing strategies to monitor proteasome activity in live cells and intact organisms, two major approaches should be considered: (1) the use of exogenously delivered probes that are activated or deactivated by proteasomal processing (Reinheckel *et al.*, 1998) or (2) genetically encoded reporters that serve as proteasome substrates (specific or nonspecific), and, thus, the signals produced by such reporters correspond to the level of activity of the proteasome toward those substrates (i.e., Ub-FL [Luker et al., 2003a] and Ub-GFP [Dantuma et al., 2000] for monitoring total proteasome activity and IκBα-FL for monitoring IκBα-specific proteasomal activity as will be described later).

Although theoretically the luciferase moiety can be interchanged with any of the reporter genes mentioned previously to be used by different imaging modalities, BLI provides a simple and robust method to repetitively monitor proteasome action in intact cells and animals. Another key advantage of BLI is that no light is produced or is detectable until the substrate/enzyme interaction of a D-luciferin and luciferase occurs. Consequently, background luminescence levels in most animals are very low, enabling sensitive, high signal-to-noise analyses *in vivo*. Therefore, this chapter will focus on technical aspects of designing and imaging genetically encoded, luciferase-fused reporters for monitoring proteasome action *per se* or degradation of a direct substrate.

Imaging Total Proteasome Activity with Tetra-Ubiquitinated Luciferase

To directly assay total proteasome activity in intact cells and living animals, one can engineer an ubiquitin–luciferase bioluminescence imaging reporter by fusing the N-terminus of firefly luciferase to four copies of a mutant ubiquitin (Ub^{G76V}). The tetra-ubiquitin fusion degradation motif has been shown to significantly destabilize heterologous proteins in cultured cells (Stack et al., 2000), whereas the glycine-to-valine substitution at the C-terminus of ubiquitin limits cleavage by ubiquitin hydrolases (Johnson et al., 1992; Stack et al., 2000). The plasmid pGL-3 basic (Promega) containing codon-optimized firefly (*Photinus Pyralis*) luciferase (FL) is digested with *Hind*III and *Xba*I to remove FL, which is ligated to *Hind*III and *Not*I sites (blunted) in EGFP-N1 (Clontech). EGFP is excised from the vector during cloning, producing a construct that expresses FL from a CMV promoter and a *Neo* cassette from an SV40 promoter. Four tandem copies of Ub^{G76V} are produced as described previously (Stack et al., 2000) and fused in frame with the N-terminus of FL, using *Nhe*I and *Hind*III sites. The resultant ubiquitin–luciferase fusion construct has been designated Ub-FL (Fig. 1A). An unfused firefly luciferase–expressing construct (FL, Fig. 1A) sharing the same backbone as Ub-FL is also engineered to serve as a nonspecific control.

To stably express Ub-FL or FL in HeLa cells, 35,000 cells are seeded per well in 24-well plates for transient transfections with Fugene-6 (Roche). Clonal cell lines are selected and maintained in 500 μg/ml G418. Note that clonal selection of noninducible luciferase–expressing cells directly from

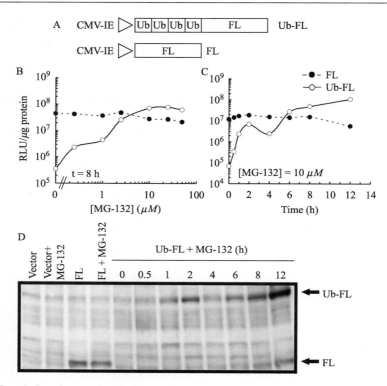

Fig. 1. Imaging total proteasome activity and inhibition *in vitro*: (A) Schematic representation of the constructs used for this study. Bioluminescence from Ub-FL cells shows concentration- (B) and time-dependent (C) increases in response to inhibition of the proteasome by MG-132. (D) Western blot of Ub-FL cells treated with MG-132 (10 μM) for various periods of time. Lysates of vector only and FL-expressing cells under baseline conditions and after 12 h of incubation with MG-132 are also shown. (Modified from Luker *et al.*, 2003a).

plates under steady-state conditions, before propagation and expansion of G418 resistant colonies, can be directly achieved by use of a cooled charged couple device (CCD) camera such as the IVIS 100 system (Xenogen, Alameda, CA, see later). This saves the time, effort, and the expense of expanding resistant colonies to assay for expression of the desired reporter phenotype by conventional methods. However, such a strategy cannot be applied to selecting stable clones expressing inducible or unstable luciferase reporters, because total photon flux is often limiting.

To assay luciferase activity in tissue culture, stably expressing reporter cells are seeded at 50,000 cells/well in 24-well plates and incubated with compounds that reversibly (e.g., MG-132; Sigma, St. Louis, MO and bortezomib; Millennium Pharmaceuticals, Cambridge, MA) or irreversibly (e.g., lactacystin; Sigma) inhibit the proteasome (Kisselev and Goldberg, 2001), or vehicle controls at the doses or times required for individual experiments. Luciferase activity in cell lysates is determined with a luciferase assay kit (Promega) using a standard plate luminometer and normalized to micrograms of cell protein as measured by BCA assay (Pierce). Luciferase activity can also be directly measured from intact cells using the IVIS bioluminescence imaging system. However, data should be normalized to protein content, cell number, or to signal from a second unfused reporter (i.e., *Renilla* luciferase [RL] or β-Gal).

Under baseline conditions, this Ub-FL fusion reporter has approximately 150-fold less bioluminescence than unfused FL transfectants (Fig. 1B,C). Proteasome inhibition has been shown to produce concentration-dependent and time-dependent increases in bioluminescence from Ub-FL (see Fig. 1B,C for MG-132) with no significant effect on FL bioluminescence.

To confirm that detected changes in bioluminescence truly reflect changes in relative levels of Ub-FL or FL, luciferase levels should be analyzed by Western blotting. One hundred-microgram aliquots of total cell lysates are separated, blotted, and probed with a polyclonal anti-luciferase primary antibody (Promega). Immune complexes are detected by enhanced chemiluminescence (Amersham Biosciences) with a donkey anti-goat secondary antibody coupled to horseradish peroxidase (Santa Cruz Biotechnology). Note that anti-luciferase antibodies presently available on the market possess relatively high nonspecific cross-reactivity. It is, therefore, crucial to dilute these antibodies in RIA-grade BSA (1% in TBS/T) instead of the commonly used dry milk powder. Western blot analysis of lysates prepared from Ub-FL cells treated with MG-132 for increasing periods of time have shown increases in Ub-FL levels and a small amount of unfused FL that generally parallel changes in bioluminescence (compare Figs. 1C and 1D).

To monitor proteasome function *in vivo* with bioluminescence imaging, tumor xenografts are formed by injecting $8–10 \times 10^6$ Ub-FL stably expressing HeLa cells subcutaneously near the right forelimb of 15–20 g, male *NCr nu/nu* nude mice (Taconic). HeLa cells stably expressing unfused FL ($8–10 \times 10^6$ cells) are also injected subcutaneously near the left forelimb. These control xenografts enable monitoring of any global, nonspecific effects of injected drugs or biologics on tumor perfusion, substrate availability, or reporter expression levels. A third tumor xenograft composed of HeLa cells containing the vector only is generated close to the right hind limb, and the bioluminescence of this tumor is considered as the background (noise) level. Note that cells are cultured without G418 for 1 day before implantation into the mice. Imaging studies are started 4 days after inoculation, at which time ~4-mm tumors are palpable. Bortezomib is administered by tail vein injection, whereas MG-132 and lactacystin are delivered intraperitoneally (i.p.). Mice are injected i.p. with 150 μg/g D-luciferin (Xenogen) in PBS, and bioluminescence imaging with a CCD camera (IVIS, Xenogen) is initiated 10 min after injection. After acquiring a gray-scale photograph, a 60-sec bioluminescent image is obtained, using a 15-cm FOV, binning (resolution) factor of 8, f/stop of 1, and an open filter. Signal intensities from regions of interest (ROI) are defined manually using Living-Image software (Xenogen), and data are expressed as photon flux (photons/sec/cm^2/steradian). Background photon flux is defined from a ROI drawn over the vector control HeLa cell tumor, and these data are subtracted from flux values quantified in FL and Ub-FL tumors. We generally express data as a ratio of photon flux from the Ub-FL tumor relative to FL tumor in each mouse to account for animal-to-animal variations in delivery and pharmacokinetics of the D-luciferin substrate. Failure to include both FL and vector-only tumors in the *same* animal as Ub-FL tumors results in larger standard errors in the *in vivo* analysis arising from differences not only in individual substrate pharmacokinetics but also perfusion, blood pressure, and anesthetic levels. Careful inclusion of appropriate controls enables more refined analyses and improved sensitivity for detecting significant but smaller differences in biological signals.

With this approach, we detect robust bioluminescence from FL xenografts, whereas light from size-matched Ub-FL tumors is either undetectable or barely detectable above background (Fig. 2, left image in each panel). The mean ratios of Ub-FL to FL under baseline conditions are typically 0.01 to 0.04, showing that bioluminescence imaging can monitor steady-state proteasome activity in living mice. When mice are injected with bortezomib (0.1, 0.5, or 1.0 μg/g BW) and monitored for bioluminescence 4 h later, we can observe dose-dependent increases in luciferase activity only in Ub-FL tumors (Fig. 2, right image in each panel). Bortezomib (1.0 μg/g BW) increases photon flux from Ub-FL tumors by two orders of magnitude, resulting in Ub-FL to FL ratios of 0.8 to 1.0 without affecting bioluminescence from FL tumors, whereas treatment with vehicle alone does not significantly affect this ratio. Similar results have been obtained with MG-132 and lactacystin.

Moreover, repetitive imaging enables *in vivo* pharmacodynamics to be observed in living animals. For example, one dose of bortezomib revealed that proteasome function in the Ub-FL tumor xenografts is blocked within 30 min and returned to nearly baseline by 46 h (Luker *et al.,* 2003a). However, after a 2-week regimen of bortezomib, imaging of target tumors shows significantly enhanced proteasome inhibition that no longer returns

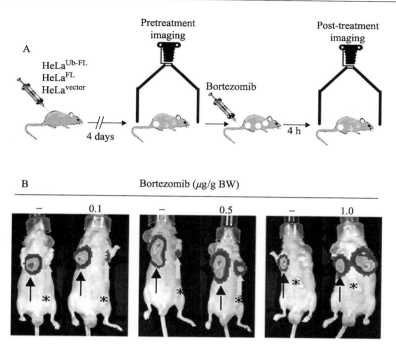

FIG. 2. *In vivo* bioluminescence imaging of Ub-FL monitors proteasome function and inhibition in living mice. (A) Schematic representation of the experimental timeline. (B) Mice bearing size-matched tumors were imaged one day before (−) and 4 h after tail vein injection of the indicated doses of bortezomib. FL, Ub-FL, and vector control tumors are denoted by black arrows, yellow arrows, and asterisks, respectively (Modified from Luker *et al.*, 2003a). (See color insert.)

to baseline (Luker *et al.*, 2003a). These types of data may be critical in designing clinical dosing regimens of a variety of drugs.

Overall, these results validate Ub-FL as an *in vivo* reporter for 26S proteasome function in a therapeutic target tissue. The Ub-FL reporter enables repetitive tissue-specific analysis of 26S proteasome activity *in vivo* and should facilitate development and validation of proteasome inhibitors in mouse models and in investigations of the ubiquitin-proteasome pathway in disease pathogenesis. However, when imaging drug-induced stabilization and accumulation of Ub-FL, functional readout is coupled to transcription and translation of the reporter. Consequently, the temporal changes in bioluminescence do not necessarily accurately reflect relevant kinetic aspects of drug pharmacodynamics. Therefore, Ub-FL is appropriate for assaying proteasomal degradation and for comparing the potency of different proteasome inhibitors *in vitro* and *in vivo*, but extracting

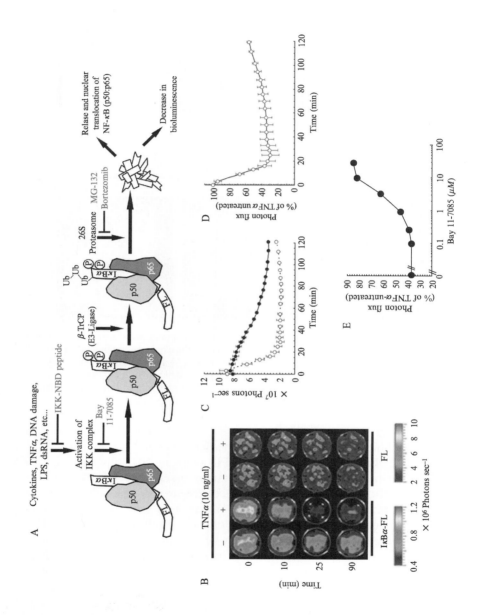

meaningful rate constants for drug action or upstream regulators *in vitro* or temporally resolving rapid pharmacodynamic data *in vivo* from such data sets is impractical. Hence, the next section is focused on strategies to overcome this problem by introducing methods for uncoupling the bioluminescence readout from reporter transcription/translation, thereby enabling real-time imaging of proteasomal substrates.

Imaging Proteasome-Dependent Degradation of IκBα

The transcription factor NF-κB is a key regulator of cellular activation, proliferation, and apoptosis. Defects in the NF-κB pathway contribute to cancer, neurodegenerative diseases, rheumatoid arthritis, asthma, inflammatory bowel disease, and atherogenesis, thus this pathway is an important target for drug development. In resting cells, inactive NF-κB dimers reside in the cytoplasm, bound to members of the IκB family, thereby masking NF-κB nuclear localization signals and preventing nuclear uptake and DNA-binding of NF-κB (Ghosh and Karin, 2002; Karin *et al.*, 2002; Li and Verma, 2002). Activation of NF-κB depends on induced degradation of IκBs in a phosphoserine-dependent manner (S32/S36) (Karin and Ben-Neriah, 2000). This stimulus-induced phosphorylation is executed by a large upstream kinase complex (IKK). Phosphorylation of IκBα renders it a substrate for rapid polyubiquitination (on lysines 21/22) by a specific E3-ligase (SCFβ-TrCP) (Karin and Ben-Neriah, 2000) and targeted degradation by the 26S proteasome (Karin and Ben-Neriah, 2000) (Fig. 3A).

FIG. 3. Validation of IκBα-FL functionality *in vitro*. (A) Schematic representation of stimulus-induced degradation of IκBα-FL. In resting cells, IκBα-FL binds NF-κB (p50:p65 dimer). On stimulation (e.g., TNFα binding to its cognate receptor), the upstream kinase complex (IKK) is activated and in turn phosphorylates IκBα-FL. This double phosphorylation renders IκBα-FL a substrate for the specific E3-ligase (β-TrCP) that polyubiquitinates IκBα-FL. Polyubiquitinated IκBα-FL is then selectively degraded by the 26S proteasome, producing a decrease in bioluminescence. NF-κB then freely translocates to the nucleus to promote κB-dependent gene transcription. Molecular targets of various NF-κB modulators are shown in red: IKK-NBD peptide specifically interferes with activation of the IKK complex. Bay 11-7085 inhibits the kinase activity of IKK, and MG-132 and bortezomib inhibit the 26S proteasome. (B) Bioluminescence imaging of HeLa$^{I\kappa B\alpha\text{-}FL}$ (left) and HeLa^{FL} (right) cells before and at the indicated time points after addition of TNFα (10 ng/ml) or vehicle (PBS). Images show color-coded maps of photon flux superimposed on black-and-white photographs of the assay plates. (C) Changes in raw bioluminescence plotted as a function of time after addition of TNFα ($\circ$) or vehicle ($\bullet$) to HeLa$^{I\kappa B\alpha\text{-}FL}$ cells. (D) TNFα-induced net degradation and resynthesis of IκBα-FL over time calculated from the photon flux ratio of treated HeLa$^{I\kappa B\alpha\text{-}FL}$ cells over unstimulated control values. (E) Concentration-dependent inhibition of TNFα-induced degradation of IκBα-FL by the IKK inhibitor Bay 11-7085. (See color insert.)

Thus, to monitor ligand-induced, proteasome-dependent degradation of IκBα in intact cells by bioluminescence imaging, an IκBα-FL fusion reporter can be engineered. The codon-optimized firefly (*Photinus pyralis*) luciferase (FL) is amplified from the pGL-3 control plasmid (Promega, Madison, WI) with the primers: 5'ATGGATCCTCGGGAGGTTCCAGTTCTGGAGGGCTAGCCGAA GACGCCAAAAACATAAAGAAAGGCCCG-3' and 5'-TAGCGG CCGCT TACACGGCGATCTTTCCGCCCTTCTTGGC-3', introducing a sequence encoding for an 11-mer polypeptide linker (SSGGSSSGGLA) (Robinson and Sauer, 1998) upstream to FL to minimize any potential steric hindrance imposed by FL on IκBα interactions with NF-κB dimers through its six ankyrin repeats (Jacobs and Harrison, 1998), with IKK and β-TrCP through the regulatory N-terminus (Karin and Ben-Neriah, 2000) or with CK-II through the PEST domain (Barroga *et al.*, 1995). The resulting PCR product is blunt-end cloned to a blunt II-TOPO pCR4 vector (Invitrogen, Carlsbad, CA), sequenced, cut with *BamHI/NotI* and ligated into the pIκBα-EGFP plasmid (BD-Bioscience, San Jose, CA), thereby replacing EGFP.

The same strategy can be used to clone pFL control plasmid encoding for unfused FL, but instead, FL is amplified with the primers: 5'-ATCCGCGGATGGAAGACGCCAAAAACATAAAGAAAGGC-3' and 5'-TAGCGGCCGCTTACACGGCGATCTTTCCGCCCTTCTTGG C-3' and ligated into the *SacII/NotI* sites of the pIκBα-EGFP plasmid, thereby replacing IκBα-EGFP.

To stably express IκBα-FL or FL in HeLa cells, typically 10^5 cells are seeded in 10-cm dishes for transfections (1 μg DNA/10^5 cells) with Fugene-6 (Roche, Basel, Switzerland). Stable clonal cell lines expressing IκBα-FL (HeLa$^{I\kappa B\alpha\text{-}FL}$) or unfused FL (HeLa^{FLu}) are selected in 1 mg/ml G418.

To monitor ligand-induced degradation of IκBα in intact cells and to analyze the time-dependent and concentration-dependent effects of various proteasome or IKK inhibitors on this process in real-time, stable HeLa$^{I\kappa B\alpha\text{-}FL}$ or HeLa^{FL} cells (10^5 cells/well) are cultured in 24-well plates. Twenty-four hours later, cells are washed twice with colorless MEBSS buffer (mM: 4.0 HEPES, 5.4 KCl, 144 NaCl, 0.8 Na$_2$HPO$_4$, 1.2 CaCl$_2$, 0.8 MgSO$_4$, and 5.6 D-glucose, pH 7.4), and incubated (1 h) in the same buffer supplemented with 10% FBS and 50 μg/ml D-luciferin in the absence or presence of increasing concentrations of proteasome inhibitors such as MG-132 or bortezomib or IKK inhibitors such as Bay 11-7085 (Calbiochem, San Diego, CA) or IKK-NBD peptide (Biomol, Plymouth, PA). An upstream ligand activator of the pathway, such as TNFα, is then added to a final concentration of 10 ng/ml. Bioluminescence imaging is performed before and at the indicated time points after addition of TNFα using the

IVIS 100 imaging system (Xenogen, Alameda, CA). Typical acquisition parameters are: exposure time, 10 sec; binning, 4; no filter; f/stop, 1; FOV, 10 cm. This assay should preferably be performed in black-coated plates (*In Vitro* Systems GmbH, Gottingen, Germany) to minimize "well-to-well" cross-contamination arising from photon reflection and refraction.

On addition of TNFα (10 ng/ml) to HeLa$^{I\kappa B\alpha\text{-}FL}$ cells (in the absence of IKK or proteasome inhibitors), a rapid decrease in raw bioluminescence signal is observed compared with untreated cells (Fig. 3B,C). In HeLa^{FL} control cells, while overall signal slowly declines, no significant difference in bioluminescence over time is identified between TNFα-treated and untreated cells (Fig. 3B, two right columns). Note that D-luciferin is added only once before initiation of imaging. Therefore, the kinetic pattern of IκBα degradation overrides the uptake and consumption profile of D-luciferin (Fig. 3C). By normalizing the bioluminescence at any given time point to that of a TNFα-untreated control, one may monitor the net degradation effect, regardless of the uptake and consumption levels of D-luciferin, and data should, therefore, be expressed as the normalized (% of initial) ratio of the photon counts of TNFα-treated over untreated cells. Such an analysis reveals a rapid decrease in normalized bioluminescence in HeLa$^{I\kappa B\alpha\text{-}FL}$ cells (reaching a minimum value of $\sim$30% of initial, 20 min after addition of TNFα), which slowly rebounds up to 60% at 120 min after addition of TNFα (Fig. 3D). Signal rebounding can be attributed to resynthesis of IκBα-FL and is in a good agreement with the previously reported ligand-induced stabilization of newly synthesized endogenous IκBα (Place *et al.*, 2001).

To eliminate the need for normalization to an untreated control, one can fuse IkBα to green click-beetle luciferase (CBG, Promega) and coexpress (preferably, on the same construct) an unfused red click-beetle luciferase (CBR, Promega). Both luciferases use the same substrate (D-luciferin) but emit light with different spectral outputs (peak emissions are 542 and 610 nm for CBG and CBR, respectively). By use of appropriate optical filters with the IVIS system (a "green" 500–570-nm bandpass filter and a "red" >650-nm longpass filter) and acquiring two images (one for each filter) for any given time point, one can define ROIs and plot the "green to red" ratio over time. This type of analysis produces essentially similar results compared with normalization to untreated controls for analysis of cultured cells (unpublished data).

To confirm that the changes in bioluminescence truly reflect changes in relative levels of IκBα-FL and that these changes correspond to changes in levels of native IκBα, whole-cell lysates of TNFα-treated HeLa$^{I\kappa B\alpha\text{-}FL}$ cells (100 μg) can be analyzed by Western blot using a rabbit polyclonal antibody against human IκBα (SCBT, Santa Cruz, CA). This analysis reveals a dramatic drop in the levels of IκBα-FL and endogenous IκBα on addition of TNFα (by 20 min) that slowly recovers (data not shown), recapitulating the kinetic pattern of changes in bioluminescence (Fig. 3D).

Preincubation of the cells (1 h) with increasing concentrations of IKK or proteasome inhibitors reveals that all the examined agents (mentioned earlier) inhibit the TNFα-induced degradation of IκBα-FL in a time- and concentration-responsive manner (data not shown). To determine the apparent IC$_{50}$ values for these agents *in cellulo*, the normalized bioluminescence is expressed as a function of drug concentration at the 20-min time point (maximal degradation as shown in Fig. 3D). The apparent IC$_{50}$ value calculated for the IKK inhibitor Bay 11-7085 using such analysis is 2.3 μM (Fig. 3E). This value is in a good agreement with previously published IC$_{50}$ values for Bay 11-7085 derived from a different assay system (Pierce *et al.*, 1997).

IκBα-FL can generate functional readouts reflecting native IκBα status noninvasively in living animals. For example, one can use the reporter to study acute liver inflammation rapidly induced (within minutes) by a single systemic administration of LPS (Streetz *et al.*, 2003). To generate an appropriate model organism, hepatocytes can be transiently transfected *in vivo* by I.V. injection of naked DNA plasmids diluted in a sufficiently high volume (1 ml/10 g BW) (Liu *et al.*, 1999). For example, mice (BALB/C, 6-week-old males, typically 3–4 per group) are rapidly (5–7 sec) injected with plasmids encoding IκBα-FL (3 μg) or FL (3 μg) using a 3-ml syringe fitted with a 27-gauge needle. A construct encoding for *Renilla* luciferase (RL, BioSignal Packard, Montreal, Canada, 1 μg) is coinjected into all mice as a control reporter of transfection efficiency. Eighteen hours later, mice are anesthetized (isoflurane) and imaged for liver RL and FL expression (after I.V. injection of coelenterazine or i.p. injection of D-luciferin [4 and 150 μg/g BW, respectively]) using the IVIS 100 imaging system (exposure time, 2–30 sec; binning, 4; no filter; f/stop, 1; FOV, 10 cm). Mice are then allowed to recover for 5 h to allow D-luciferin clearance from the circulation. LPS (1 μg/g BW, I.V.) or vehicle (PBS) is administered, and a second posttreatment imaging round is performed (as earlier) 1 h later (see Fig. 4A for a graphic description). By use of analysis software, regions of interest (ROI) are defined manually over the liver for determining signal intensities (SI) and data are expressed as (FL/RL)$_{posttreatment}$/(FL/RL)$_{pretreatment}$.

By use of this technique, a fivefold decrease in IκBα-FL hepatic bioluminescence (as normalized to RL bioluminescence) has been recorded as soon as 1 h after LPS injection (Fig. 4B). Mice injected with vehicle only or LPS-treated mice expressing unfused FL did not exhibit a significant change in RL-normalized FL bioluminescence (data not shown).

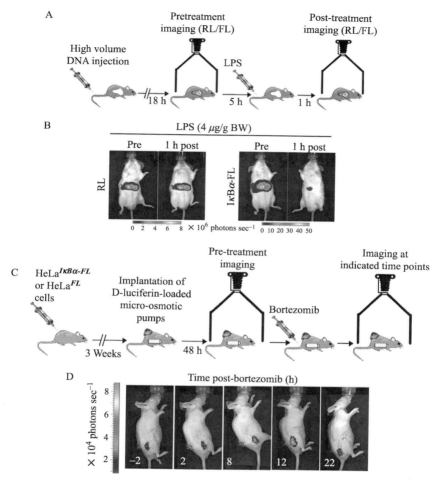

Fig. 4. Imaging proteasomal-dependent IκBα degradation in living mice. (A, B) Imaging pharmacological modulation of LPS-induced IκBα degradation. (A) Schematic representation of the experimental timeline. (B) Representative bioluminescence images of RL (left two panels) and IκBα-FL (right two panels) taken before or 1 h after LPS stimulation. All images correspond to an individual mouse. (C, D) Real-time imaging of IκBα accumulation in tumors of bortezomib-treated mice. (C) Schematic representation of the experimental timeline. (D) Representative bioluminescence images of a HeLa$^{IκBα-FL}$ tumor-bearing mouse, taken at the indicated time points (h) before and after i.p. administration of bortezomib (1 μg/g BW). Note that D-luciferin was continuously delivered by a subcutaneously implanted microosmotic pump. (See color insert.)

As another example, IκBα stabilization and accumulation in a tumor model can be imaged after administration of a single dose of the proteasome inhibitor bortezomib. Herein, HeLa$^{I\kappa B\alpha\text{-}FL}$ cells (2×10^6) are injected subcutaneously onto the back of 6-week-old male NCr *nu/nu* nude mice (Taconic Farms, Germantown, NY). The utility of reporters such as IκBα-FL can be further enhanced through use of an implanted microosmotic pump for persistent and constant delivery of the bioluminescent substrate D-luciferin, thereby providing a stable baseline for each animal to serve as its own control and eliminating temporal constrains of i.p. bolus reinjections of D-luciferin. When tumors reach a diameter of $\sim$5 mm (typically $\sim$3 weeks), D-luciferin (50 mg/ml)–loaded microosmotic pumps (Alzet Model 1007D; Durect, Cupertino, CA) are surgically implanted (s. c.) in the dorsal neck fat pad. Mice are then allowed to recover for 48 h and are then imaged for tumor bioluminescence using the IVIS 100 imaging system (exposure time, 180 sec; binning, 4; no filter; f/stop, 1; FOV, 10 cm) at various time points before and after (typically -4 to $+48$ h) administration of bortezomib (1 μg/g BW) or vehicle alone (for a graphic description see Fig. 4C). ROIs are defined manually over the tumor and the thorax to determine tumor and background signal intensities, respectively. Data should be expressed as tumor-to-background ratios (fold of initial). By use of this technique, a time-dependent increase in tumor bioluminescence has been recorded with bortezomib, peaking $\sim$12 h after drug administration at $\sim$3.5-fold over pretreatment values, followed by a gradual decrease over the subsequent 32 h after treatment to levels of vehicle-treated mice (see Fig. 4D for representative images taken from a single mouse). Thus, the implanted pump enables continuous real-time molecular imaging of reporter activity throughout the time course of a multiday experiment, while simultaneously allowing rapid temporally resolved analysis of the short time constants characterizing drug action.

Conclusions

The emerging notion that proteasomal degradation regulates a wide array of signaling pathways involved in cell proliferation and differentiation, apoptosis, innate and acquired immunity, and cancer propels the need for experimental systems to measure total and substrate-specific proteasomal activity both in intact cells and in living animals. In this chapter, we presented several techniques to meet this requirement through innovative use of genetically encoded fusion reporters. Technical considerations that should be carefully addressed when designing such reporters include: (1) the need for imaging total vs. substrate-specific proteasomal activity,

(2) potential requirement for signals with high temporal resolution (i.e., minutes vs. hours for IκBα-FL and Ub-FL, respectively) as opposed to high dynamic range (i.e., threefold to fourfold vs. orders of magnitude change in the bioluminescence of IκBα-FL and Ub-FL, respectively), (3) appropriate controls and normalization techniques to generate physiologically meaningful data sets, and (4) implementation of independent assays for validation of bioluminescence data (i.e., Western blot analysis).

Overall, such fusion reporters enable target-specific analysis of the ubiquitin-proteasome pathway *in vitro* and *in vivo* and should facilitate development and validation of new therapeutics and temporal and regional investigations of signaling pathways associated with proteasomal degradation.

Acknowledgments

The authors thank colleagues in the Washington University Molecular Imaging Center for their help and insights in completing work described herein. Supported by NIH grant P50 CA94056.

References

Aberle, H., Bauer, A., Stappert, J., Kispert, A., and Kemler, R. (1997). Beta-catenin is a target for the ubiquitin-proteasome pathway. *EMBO J.* **16**, 3797–3804.

Barroga, C. F., Stevenson, J. K., Schwarz, E. M., and Verma, I. M. (1995). Constitutive phosphorylation of I kappa B alpha by casein kinase II. *Proc. Natl. Acad. Sci. USA* **92**, 7637–7641.

Bhaumik, S., and Gambhir, S. (2001). Optical imaging of *Renilla* luciferase reporter gene expression in living mice. *Proc. Natl. Acad. Sci. USA* **99**, 377–382.

Blagosklonny, M. V., Wu, G. S., Omura, S., and el-Deiry, W. S. (1996). Proteasome-dependent regulation of p21WAF1/CIP1 expression. *Biochem. Biophys. Res. Commun.* **227**, 564–569.

Bloom, J., and Pagano, M. (2004). To be or not to be ubiquitinated? *Cell Cycle* **3**, 138–140.

Bross, P. F., Kane, R., Farrell, A. T., Abraham, S., Benson, K., Brower, M. E., Bradley, S., Gobburu, J. V., Goheer, A., Lee, S. L., Leighton, J., Liang, C. Y., Lostritto, R. T., McGuinn, W. D., Morse, D. E., Rahman, A., Rosario, L. A., Verbois, S. L., Williams, G., Wang, Y. C., and Pazdur, R. (2004). Approval summary for bortezomib for injection in the treatment of multiple myeloma. *Clin. Cancer Res.* **10**, 3954–3964.

Carlsen, H., Moskaug, J., Fromm, S., and Blomhoff, R. (2002). In vivo imaging of NF-κB activity. *J. Immunol.* **168**, 1441–1446.

Chaudhuri, T., Rogers, B., Buchsbaum, D., Mountz, J., and Zinn, K. (2001). A noninvasive reporter system to image adenoviral-mediated gene transfer to ovarian cancer xenografts. *Gynecol. Oncol.* **83**, 432–438.

Contag, C., Spilman, S., Contag, P., Oshiro, M., Eames, B., Dennery, P., Stevenson, D., and Benaron, D. (1997). Visualizing gene expression in living mammals using a bioluminescent reporter. *Photochem. Photobiol.* **66**, 523–531.

Dantuma, N. P., Lindsten, K., Glas, R., Jellne, M., and Masucci, M. G. (2000). Short-lived green fluorescent proteins for quantifying ubiquitin/proteasome-dependent proteolysis in living cells. *Nat. Biotechnol.* **18,** 538–543.

Elliott, P. J., Zollner, T. M., and Boehncke, W. H. (2003). Proteasome inhibition: A new antiinflammatory strategy. *J. Mol. Med.* **81,** 235–245.

Ghosh, S., and Karin, M. (2002). Missing pieces in the NF-kappaB puzzle. *Cell* **109**(Suppl.), S81–S96.

Hospers, G., Calogero, A., van Waarde, A., Doze, P., Vaalburg, W., Mulder, N., and de Vries, E. (2000). Monitoring of herpes simplex virus thymidine kinase enzyme activity using positron emission tomography. *Cancer Res.* **60,** 1488–1491.

Jacobs, M. D., and Harrison, S. C. (1998). Structure of an IkappaBalpha/NF-kappaB complex. *Cell* **95,** 749–758.

Johnson, E., Bartel, B., Seufert, W., and Varshavsky, A. (1992). Ubiquitin as a degradation signal. *EMBO J.* **11,** 497–505.

Karin, M., and Ben-Neriah, Y. (2000). Phosphorylation meets ubiquitination: The control of NF-[kappa]B activity. *Annu. Rev. Immunol.* **18,** 621–663.

Karin, M., Cao, Y., Greten, F. R., and Li, Z. W. (2002). NF-kappaB in cancer: From innocent bystander to major culprit. *Nat. Rev. Cancer.* **2,** 301–310.

Kisselev, A., and Goldberg, A. (2001). Proteasome inhibitors: From research tools to drug candidates. *Chem. Biol.* **8,** 739–758.

Levkowitz, G., Waterman, H., Zamir, E., Kam, Z., Oved, S., Langdon, W. Y., Beguinot, L., Geiger, B., and Yarden, Y. (1998). c-Cbl/Sli-1 regulates endocytic sorting and ubiquitination of the epidermal growth factor receptor. *Genes Dev.* **12,** 3663–3674.

Li, B., and Dou, Q. P. (2000). Bax degradation by the ubiquitin/proteasome-dependent pathway: Involvement in tumor survival and progression. *Proc. Natl. Acad. Sci. USA* **97,** 3850–3855.

Li, Q., and Verma, I. M. (2002). NF-kappaB regulation in the immune system. *Nat. Rev. Immunol.* **2,** 725–734.

Liu, F., Song, Y., and Liu, D. (1999). Hydrodynamics-based transfection in animals by systemic administration of plasmid DNA. *Gene Ther.* **6,** 1258–1266.

Luker, G., Pica, C., Song, J., Luker, K., and Piwnica-Worms, D. (2003a). Imaging 26S proteasome activity and inhibition in living mice. *Nat. Med.* **9,** 969–973.

Luker, G., Sharma, V., Pica, C., Dahlheimer, J., Li, W., Ochesky, J., Ryan, C., Piwnica-Worms, H., and Piwnica-Worms, D. (2002). Noninvasive imaging of protein-protein interactions in living animals. *Proc. Natl. Acad. Sci. USA* **99,** 6961–6966.

Luker, G., Sharma, V., Pica, C., Prior, J., Li, W., and Piwnica-Worms, D. (2003b). Molecular imaging of protein-protein interactions: Controlled expression of p53 and large T antigen fusion proteins *in vivo*. *Cancer Res.* **63,** 1780–1788.

Luker, K., and Piwnica-Worms, D. (2004). Optimizing luciferase protein fragment complementation for bioluminescent imaging of protein-protein interactions in live cells and animals. *Methods Enzymol.* **385,** 349–360.

Luker, K. E., Smith, M. C., Luker, G. D., Gammon, S. T., Piwnica-Worms, H., and Piwnica-Worms, D. (2004). Kinetics of regulated protein-protein interactions revealed with firefly luciferase complementation imaging in cells and living animals. *Proc. Natl. Acad. Sci. USA* **101,** 12288–12293.

Maclaren, D., Gambhir, S., Satyamurthy, N., Barrio, J., Sharfstein, S., Totykuni, T., Wu, L., Berk, A., Cherry, S., Phelps, M., and Herschman, H. (1999). Repetitive, non-invasive imaging of the dopamine D2 receptor as a reporter gene in living animals. *Gene Ther.* **6,** 785–791.

Maxwell, P. H., Wiesener, M. S., Chang, G. W., Clifford, S. C., Vaux, E. C., Cockman, M. E., Wykoff, C. C., Pugh, C. W., Maher, E. R., and Ratcliffe, P. J. (1999). The tumour suppressor protein VHL targets hypoxia-inducible factors for oxygen-dependent proteolysis. *Nature* **399**, 271–275.

Moll, U. M., and Petrenko, O. (2003). The MDM2-p53 interaction. *Mol. Cancer Res.* **1**, 1001–1008.

Ozawa, T., Kaihara, A., Sato, M., Tachihara, K., and Umezawa. (2001). Split luciferase as an optical probe for detecting protein-protein interactions in mammalian cells based on protein splicing. *Anal. Chem.* **73**, 2516–2521.

Pagano, M., Tam, S. W., Theodoras, A. M., Beer-Romero, P., Del Sal, G., Chau, V., Yew, P. R., Draetta, G. F., and Rolfe, M. (1995). Role of the ubiquitin-proteasome pathway in regulating abundance of the cyclin-dependent kinase inhibitor p27. *Science* **269**, 682–685.

Paulmurugan, R., Massoud, T., Huang, J., and Gambhir, S. (2004). Molecular imaging of drug-modulated protein-protein interactions in living subjects. *Cancer Res.* **64**, 2113–2119.

Paulmurugan, R., Umezawa, Y., and Gambhir, S. S. (2002). Noninvasive imaging of protein-protein interactions in living subjects by using reporter protein complementation and reconstitution strategies. *Proc. Natl. Acad. Sci. USA* **99**, 15608–15613.

Pierce, J. W., Schoenleber, R., Jesmok, G., Best, J., Moore, S. A., Collins, T., and Gerritsen, M. E. (1997). Novel inhibitors of cytokine-induced IkappaBalpha phosphorylation and endothelial cell adhesion molecule expression show anti-inflammatory effects *in vivo*. *J. Biol. Chem.* **272**, 21096–21103.

Place, R. F., Haspeslagh, D., Hubbard, A. K., and Giardina, C. (2001). Cytokine-induced stabilization of newly synthesized I(kappa)B-alpha. *Biochem. Biophys. Res. Commun.* **283**, 813–820.

Reinheckel, T., Sitte, N., Ullrich, O., Kuckelkorn, U., Davies, K. J., and Grune, T. (1998). Comparative resistance of the 20S and 26S proteasome to oxidative stress. *Biochem. J.* **335** (Pt 3), 637–642.

Rettig, M. P., Ritchey, J. K., Prior, J. L., Haug, J. S., Piwnica-Worms, D., and DiPersio, J. F. (2004). Kinetics of *in vivo* elimination of suicide gene-expressing T cells affects engraftment, graft-versus-host disease, and graft-versus-leukemia after allogeneic bone marrow transplantation. *J. Immunol.* **173**, 3620–3630.

Robinson, C. R., and Sauer, R. T. (1998). Optimizing the stability of single-chain proteins by linker length and composition mutagenesis. *Proc. Natl. Acad. Sci. USA* **95**, 5929–5934.

Sharma, V., Luker, G., and Piwnica-Worms, D. (2002). Molecular imaging of gene expression and protein function *in vivo* with PET and SPECT. *J. Magn. Reson. Imaging* **16**, 336–351.

Stack, J., Whitney, M., Rodems, S., and Pollok, B. (2000). A ubiquitin-based tagging system for controlled modulation of protein stability. *Nat. Biotechnol.* **18**, 1298–1302.

Streetz, K. L., Wustefeld, T., Klein, C., Kallen, K. J., Tronche, F., Betz, U. A., Schutz, G., Manns, M. P., Muller, W., and Trautwein, C. (2003). Lack of gp130 expression in hepatocytes promotes liver injury. *Gastroenterology* **125**, 532–543.

Tjuvajev, J. G., Stockhammer, G., Desai, R., Uehara, H., Watanabe, K., Gansbacher, B., and Blasberg, R. G. (1995). Imaging the expression of transfected genes *in vivo*. *Cancer Res.* **55**, 6126–6132.

Voorhees, P. M., Dees, E. C., O'Neil, B., and Orlowski, R. Z. (2003). The proteasome as a target for cancer therapy. *Clin. Cancer Res.* **9**, 6316–6325.

Wang, X., Rosol, M., Ge, S., Peterson, D., McNamara, G., Pollack, H., Kohn, D. B., Nelson, M. D., and Crooks, G. M. (2003). Dynamic tracking of human hematopoietic stem cell engraftment using *in vivo* bioluminescence imaging. *Blood* **102**, 3478–3482.

Weissleder, R., Moore, A., Mahmood, U., Bhorade, R., Benveniste, H., Chicocca, E., and Basilion, J. (2000). In vivo magnetic resonance imaging of transgene expression. *Nat. Med.* **6,** 351–355.

Wojcik, C., and Di Napoli, M. (2004). Ubiquitin-proteasome system and proteasome inhibition: New strategies in stroke therapy. *Stroke* **35,** 1506–1518.

Yang, M., Baranov, E., Jiang, P., Sun, F.-X., Li, X.-M., Li, L., Hasegawa, S., Bouvet, M., Al-Tuwaijri, M., Chishima, T., Shimada, H., Moossa, A., Penman, S., and Hoffman, R. (2000). Whole-body optical imaging of green fluorescent protein-expressing tumors and metastases. *Proc. Natl. Acad. Sci. USA* **97,** 1206–1211.

Zhang, G. J., Safran, M., Wei, W., Sorensen, E., Lassota, P., Zhelev, N., Neuberg, D. S., Shapiro, G., and Kaelin, W. G., Jr. (2004). Bioluminescent imaging of Cdk2 inhibition in vivo. *Nat. Med.* **10,** 643–648.

[36] Bioluminescent Imaging of Ubiquitin Ligase Activity: Measuring Cdk2 Activity *In Vivo* Through Changes in p27 Turnover

By GUO-JUN ZHANG and WILLIAM G. KAELIN, JR.

Abstract

Optical imaging of reporter molecules such as firefly luciferase has become a popular method of tracking and visualizing cells in living animals. Many biological processes involve ubiquitin ligases, which target specific proteins for destruction under specific sets of conditions. Importantly, the motifs recognized by different ubiquitin ligases are often modular and can be used to target foreign proteins for destruction in cis. We recently fused the Cdk inhibitor p27, which is polyubiquitylated by a Skp2-containing ubiquitin ligase if phosphorylated by cdk2 to firefly luciferase. The resulting fusion protein, p27-Luc, was induced by cdk2 inhibitors in living cells grown in culture or in nude mice. This article describes protocols for validation of p27-Luc in cell culture using siRNA against cdk2 (or its partner cyclin A) and for imaging cells producing p27-Luc grown in transparent hollow fibers after treatment with cdk2 inhibitory drugs *in vivo*. These approaches should be generalizable to other ubiquitin–ligase substrate pairs.

Introduction

The development of sensitive imaging-based assays to monitor molecular and cellular processes in animal models *in vivo* should greatly enhance the study of normal physiology and disease states. Molecular imaging holds particular promise in the area of drug discovery, where it might provide rapid indications of drug action (Rudin and Weissleder, 2003). Three different noninvasive imaging technologies suitable for molecular imaging have been developed over the past two decades (1) MR imaging; (2) nuclear imaging (quantitative autoradiography, gamma camera, and PET); and (3) optical imaging (Blasberg, 2003). Optical imaging is based on quantitative or qualitative changes in light emission by fluorescent or bioluminescent probes, which may be protein based.

Early optical imaging studies were performed using cells in which the expression of the bioluminescent reporter (typically wild-type firefly luciferase) or fluorescent reporter (typically GFP) was driven by a constitutive promoter such as the CMV early promoter. In these early studies, signal strength was primarily a function of the number of viable cells harboring the reporter. To gain additional information, it is also possible to place a protein such as luciferase or GFP under the control of a promoter that interrogates a particular pathway or process of interest. For example, a luciferase cDNA under the control of p53-responsive promoter was used to monitor the status of p53 protein (Wang and El-Deiry, 2003), whereas a luciferase cDNA driven by the regulatory elements of the heme oxygenase promoter was used to evaluate heme oxygenase-1 transcription (Contag and Stevenson, 2001). Alternately, bioluminescent or fluorescent proteins can themselves be reengineered to respond to different signals. For example, this approach has been used to make reporters that respond to proteasome inhibition or to phosphorylation by particular kinases or cleavage by specific proteases (Jenkins *et al.*, 1990; Luker *et al.*, 2003; Sala-Newby and Campbell, 1991, 1992; Zhang *et al.*, 2002). These efforts have been enhanced by a growing understanding of the targeting motifs recognized by enzymes within their respective substrates.

The Cdk inhibitor p27 is polyubiquitylated in a cell-cycle–dependent manner by an SCF complex containing Skp1, Cul1, and Skp2 (Shirane *et al.*, 1999). p27 is phosphorylated on Thr187 by cyclin E/Cdk2 or cyclin A/Cdk2 in late G1 and S-phase, and the recognition of p27 by SCF^{Skp2} is enhanced after Thr187 phosphorylation (Montagnoli *et al.*, 1999; Tsvetkov *et al.*, 1999). Taking advantage of Cdk2-dependent p27 phosphorylation, we developed a reporter, p27-luciferase fusion protein (p27Luc), for monitoring Cdk2 activity. As predicted, p27Luc levels increased after blocking Cdk2 activity with inhibitory proteins, peptides, or siRNA in cell-based assays (Fig. 1 and data not shown) (Zhang *et al.*, 2004).

As a segue to monitoring p27Luc *in vivo*, we first introduced this reporter into mammalian cancer cells that were then grown in transparent semipermeable hollow fibers *in vitro* and *in vivo*. Hollow fiber assays have become popular in the pharmaceutical industry because they provide a rapid preliminary evaluation of *in vivo* therapeutic efficacy. Hollow fibers

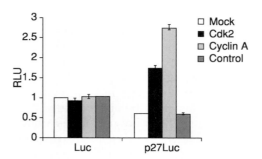

FIG. 1. Validation of p27Luc as a reporter of Cdk2 activity. HeLa cells stably transfected to produce Luc or p27Luc were transfected with indicated siRNAs. Forty-eight hours after treatment, luciferase activity was assayed. Relative luciferase unit (RLU) was corrected by protein concentration and normalized to the mock-treated HeLa-Luc cells. Error bars indicate ± 1 SD.

containing viable cells represent packages that can be implanted into animals (typically subcutaneously or in the peritoneal cavity), exposed to drugs *in vivo*, and retrieved for further analysis *ex vivo* (Hollingshead *et al.*, 1995). For example, such assays have been used to test whether potential anticancer drugs can affect cancer cell viability *in vivo*. An additional advantage of hollow fiber assays is that they are amenable to multiplexing—up to six fibers, each containing a different cell line, can be implanted in the same mouse and assayed simultaneously.

We, therefore, adapted the use of hollow fibers to ask whether p27Luc was responsive to cdk2 inhibitors *in vivo*. Cells expressing p27Luc or wild-type luciferase (Luc) were grown in hollow fibers, which were then maintained in cell culture media in multiwell plates or implanted in nude mice. In both environments, exposure to cdk2 inhibitors led to increased light emission by the p27Luc cells, but not the Luc cells (Figs. 2 and 3). These results heralded similar findings obtained when these cells were allowed to form tumors in nude mice after subcutaneous administration in conventional xenograft assays (Zhang *et al.*, 2004).

In this chapter, we describe the methods that were used to validate p27Luc in cell culture experiments using siRNA directed against cdk2 (or its partner cyclin A) and the imaging of p27Luc in hollow fibers grown *in vitro* or *in vivo*.

Generation of Plasmid Expressing p27-Luciferase Fusion Protein

To make a plasmid encoding full-length p27 fused to firefly luciferase, we first amplified a p27 cDNA with primers (forward: 5'-GC GC AAGCTTGCCACCATGTCAAACGTGCGAGTGTCTAAC-3' reverse:

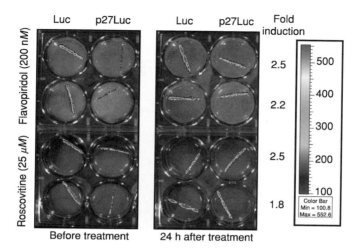

FIG. 2. Bioluminescent imaging of hollow fibers *in vitro*. Hollow fibers were filled with polyclonal U2OS cells producing Luc or p27Luc and cultivated in 6-well plates. Bioluminescent images were acquired before and after treatment with 200 nM flavopiridol or 25 μM roscovitine using Xenogen IVIS™ imaging system. Fold induction was calculated as p27Luc/Luc$_{posttreatment}$ ÷ p27Luc/Luc$_{pretreatment}$. (See color insert.)

5′-GCGCCCATGGTCGTTTGACGTCTTCTGAGGCCAGG - 3′ that introduced a 5′ *Hind*III site and a 3′ *Nco*I site. The resulting PCR product was then introduced into pGL3-control (Promega) cut with these two enzymes using standard methods. Note that *Nco*I cuts pGL3-control at a unique site located immediately 5′ of the firefly luciferase cDNA, and the *Nco*I site within the 3′ PCR amplimer was designed so that the p27 open reading frame (ORF) would be in-frame with the luciferase ORF.

Generation of Cell Lines Stably Expressing Luciferase (Luc) or p27Luc

Materials

- 0.05% trypsin-EDTA
- HeLa cells, U2OS cells
- PcDNA3 (Invitrogen)
- PGL3-p27Luc and pGL3-control
- P100 tissue culture dishes
- DMEM + 10% FBS + penicillin/streptomycin

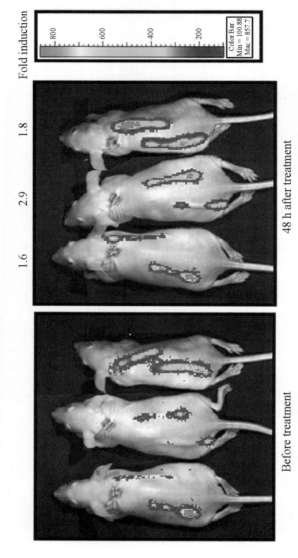

FIG. 3. Bioluminescent imaging of hollow fibers *in vivo*. Hollow fibers filled with polyclonal U2OS cells producing Luc (left flank) or p27Luc (right flank) were implanted subcutaneously into nude mice. Seven days later, baseline bioluminescent images were acquired using Xenogen IVIS™ imaging system (left). After two doses of flavopiridol (5 mg/kg, once a day by i.p.), repeat images were obtained. Fold induction was calculated as p27Luc/Luc$_{posttreatment}$ ÷ p27Luc/Luc$_{pretreatment}$. (See color insert.)

- LIPOFECTAMINE reagent (Invitrogen)
- PLUS reagent (Invitrogen)
- Opti-MEMI (Invitrogen)
- 15-ml Falcon polypropylene conical tube
- 1.5-ml Eppendorf tube
- 37° incubator with 5% CO_2
- 37° incubator with 10% CO_2
- G418 100 mg/ml
- EBC buffer (50 mM Tris, pH = 7.5, 150 mM NaCl, 0.5% NP40)
- Complete mini (protease inhibitor cocktail tablets) (Roche)
- PBS (phosphate-buffered saline; 20 mM sodium phosphate, 150 mM NaCl, pH = 7.4)
- Trans-Blot® 0.45-μm nitrocellulose membrane (Bio-Rad)
- TBS (Tris-buffered saline; Tris 10 mM pH = 8.0, 150 mM NaCl)
- TBS-T (0.1% Tween20 in TBS)
- Dry milk (grocery store)
- Anti-luciferase antibody (rabbit polyclonal) (Sigma, cat No., L-0159)
- HRP-conjugated rabbit IgG (Pierce)
- ECL detection reagent (Pierce)

Procedures

1. The day before transfection, trypsinize and resuspend the HeLa or U2OS cells in DMEM + 10%FBS + penicillin/streptomycin at $\sim 2 \times 10^6$ (HeLa) or 3×10^6 (U2OS) cells/ml. Plate 1 ml of the suspended cells per P100 tissue culture dish and add fresh media to a total of 10 ml. The goal here is to have cells 70% confluent on the day of transfection.

2. Premix plasmid DNA with the PLUS Reagent: Dilute 0.5 μg pcDNA3 (for G418 selection) and 5 μg of pGL3-control (Luc) or 5 μg of pGL3-p27Luc (p27Luc) plasmid DNA into 750 μl Opti-MEMI medium in 15-ml Falcon tube. Immediately add 20 μl PLUS Reagent to this diluted DNA, gently vortex, and incubate for 15 min at room temperature (RT).

3. Set up two Eppendorf tubes with each tube containing 30 μl of LIPOFECTAMINE reagent diluted into 750 μl Opti-MEMI medium. Gently vortex. (*Note:* When performing multiple transfections, it is useful to make a master tube containing 30 μl $\times$ N of LIPOFECTAMINE reagent diluted into 750 μl $\times$ N Opti-MEMI medium where "N" = the total number of transfections being performed).

4. Transfer the diluted LIPOFECTAMINE Reagent (from step 3) into the 15 Falcon tube containing the precomplexed DNA (from step 2). Gently vortex and incubate for 15 min at RT.

5. Replace culture medium on the cells with 5 ml Opti-MEMI.

6. Using a glass Pasteur pipet, add DNA-PLUS-LIPOFECTAMINE complexes (from step 4) dropwise to the culture dish. Manually shake the dishes gently to mix complexes into medium. Incubate for 4 h in a 37° incubator containing 5% CO_2.

7. Four hours later, replace the transfection medium with 10 ml DMEM+ 10% FBS + penicillin/streptomycin and return culture dishes to 37° incubator containing 10% CO_2.

8. 24 h after transfection, replace medium with 10 ml fresh medium containing 1 mg/ml G418. In parallel, replace the medium of a plate of untransfected cells with the same medium to serve as a control for G418-induced killing.

9. Change G418 containing medium every 2–3 days until all the cells in the untransfected control dishes (step 8) are dead. Small colonies containing 50–100 cells should be visible in the transfected plates.

10. Collect and pool G418-resistant cells for each reporter by trypsinization. Maintain them in DMEM+ 10% FBS + penicillin/streptomycin as polyclonal cell lines (we continue to supplement the media with G418 for early passages).

11. Harvest cells for luciferase assay (see later under "Validation of p27Luc Reporter by siRNA against Cyclin A or Cdk2") or for anti-luciferase Western blot assay (see steps 12–20).

12. For Western blots assay, lyse cells in EBC buffer supplemented with Complete Mini (1 tablet/ml) (for example, 200 μl per well for cells grown in 6-well plates). Aspirate media, wash cells once with PBS, and rock with EBC plus inhibitors for 15 min at 4°. Transfer lysates to prechilled 1.5-ml Eppendorf tubes and centrifuge at 14,000 rpm for 10 min at 4°. Save supernatants.

13. Load 20 μg total cell extracts on 10% SDS-PAGE mini-gel and separate proteins by electrophoresis at 130 V for 1.5 h.

14. Transfer proteins onto 0.45-μm nitrocellulose membrane at 100 V for 2 h on ice.

15. Block the membrane with TBST containing 5% dry milk for 30 min at RT.

16. Discard blocking cocktail and incubate membrane with anti-luciferase (1:10,000 in TBS-T with 4% dry milk) overnight at 4°.

17. Wash membrane three times with TBS-T at RT (10 min per wash).

18. Incubate membrane with HRP-conjugated rabbit IgG (1:5,000 in TBS-T with 4% dry milk) for 1 h at RT.

19. Wash membrane three times with TBS-T RT (10 min per wash).

20. Develop with ECL detection reagent according to manufacturer's protocol.

21. Freeze cells expressing Luc or p27Luc for later use.

Validation of p27Luc Reporter by siRNA against Cyclin A or Cdk2

Materials

- 0.05% Trypsin-EDTA
- 6-well tissue culture plates
- siRNA duplexes (Dharmicon): Cdk2, 5'-GGUGGUGGCGCUUA AGAAAdTdT-3'; cyclin A, 5'-GGCAGCGCCCGUCCAA-CAAdTdT-3'; control, 5'-CAACCUGC CGCGACGGAAdTdT-3'. (*Note:* shown are sense sequences consisting of 19-bp targeted sequence followed by dTdT)
- 1× Universal buffer (supplied by Dharmacon)
- Opti-MEMI (Invitrogen)
- Oligofectamine (Invitrogen)
- RNase-free 1.5-ml microcentrifuge tubes (Ambion)
- DMEM + 10% FBS (without antibiotics)
- 37° incubator with 5% CO_2
- DMEM + 30% FBS
- 37° incubator with 10% CO_2
- PBS
- Shaker (Labnet)
- Bio-Rad protein assay–dye concentrate reagent (Bio-Rad)
- Luciferase assay kit (Promega)
- BD Monolight™ 3010 cuvettes (BD Biosciences)
- Lumat LB 9507 luminometer (Berthold Technologies)

Procedure

1. 24 h before transfection, plate 5×10^4 HeLa cells expressing Luc or p27Luc into 6-well plates in DMEM + 10% FBS (without antibiotics) in total 2 ml media/well. Goal is to have cells ~30% confluent at the time of transfection. (*Note:* all media from this point should be antibiotic-free or transfection efficiency will be lower).

2. Resuspend siRNAs in 1× universal buffer at final concentration of 20 μM.

3. Add 5 μl of Cdk2 or cyclin A or control siRNA to 180 μl Opti-MEMI in an RNase-free 1.5-ml microcentrifuge tube. Gently vortex and incubate at RT for 10 min.

4. Mix 4 μl of Oligofectamine and 11 μl Opti-MEMI in separate RNase-free 1.5-ml microcentrifuge tube. Gently vortex and incubate at RT for 10 min. (*Note:* it is useful to make a master mix at this step when performing multiple transfections. Mix 4 μl of Oligofectamine × N and

11 µl Opti-MEMI × N in separate RNase-free 1.5-ml microcentrifuge tube where "N" is number of transfections.)

5. Add 15 µl of Oligofectamine/Opti-MEMI to the siRNA/Opti-MEMi mixtures from step 3. Gently mix by inverting the tubes 4–6 times. (*Note:* do not vortex or transfection efficiency will be lowered.) Incubate at RT for 20 min.

6. Replace cell culture media with fresh 0.8 ml Opti-MEMI.

7. Pipet the siRNA mixture from step 5 up and down twice and add dropwise to cells using a P200 Pipetman. Manually shake the plates 4–6 times to evenly distribute the transfection mix.

8. Incubate for 4 h in a 37° incubator containing 5% CO_2.

9. Add 500 µl DMEM + 30%FBS (without antibiotics) and transfer plates to a 10% CO_2 incubator.

10. 48 h after transfection, remove the transfection media and wash the cells once with PBS. Lyse the cells by rocking with 200 µl passive lysis buffer on a shaker for 10 min at RT. Collect lysates using Pipetman and transfer to a fresh Eppendorf tube.

11. Measure protein concentration by Bradford method using Bio-Rad protein assay reagent.

12. Dissolve the contents of 1 vial of luciferase assay substrate into 1 vial of luciferase assay buffer provided in the luciferase assay kit. Keep at RT until use.

13. Mix 10 µl of cell lysates and 100 µl of luciferase assay solution (from step 12) in 3010 cuvettes. (*Note:* unused luciferase assay solution can be aliquoted and stored at –20° for 1 month provided it is brought back to RT at the time of use).

14. Measure luciferase activity using a Lumat LB 9507 luminometer and 5-sec measurement duration. (*Note:* Lysate prepared from parental HeLa cells (no luciferase cDNA) should be used as a negative control to control for background in these assays).

15. Normalize the luciferase activities using the protein concentration determined in step 11 (Fig. 1).

Note: This procedure was used to validate p27Luc as an indicator of Cdk2. We have also carried out validation studies in which cells were transiently (rather than stably) transfected to produce p27-Luc. For example, we showed that p27-Luc was induced by cotransfection with a plasmid encoding a dominant-negative version of cdk2 (Zhang *et al.*, 2004). A theoretical risk with this approach is that high-level production of ubiquitination targets (in this case, p27) might out-titrate certain proteins required for their destruction. For example, we have encountered this problem when studying the turnover of p53 and HIF (unpublished data).

Preparation of Hollow Fibers

Materials

- 70% EtOH
- Gloves (non-sterile)
- Polyvinylidene fluoride (PVDF) hollow fibers (50-kDa MW cut off) with inner diameter of 1.0 mm and outer diameter 1.2 mm (Spectrum Medical Inc.) (Fig. 4A)
- Scissors (Fig. 4B, B) (Fisher Scientific)
- Autoclave pans (46 × 15 × 6.7 cm, Nalgene)
- Ruler

Procedure

1. Clean work surface in tissue culture hood with 70% EtOH.
2. Wearing gloves, remove the coil of fibers from the bag in which they come and place them on the work surface.
3. Cut the fibers into 30-cm-long pieces using scissors.
4. Open the ends of the cut fibers by squeezing gently at the points of the crimp created by scissors.
5. Soak fibers for at least 72 h in 70% EtOH before use (*Note:* use a tray or pan that is large enough to avoid kinking the fibers).
6. Fill an autoclave pan with 500 ml of sterile distilled water.
7. Transfer the desired number of fibers to be autoclaved from the alcohol bath into the autoclave pan.
8. Autoclave 20 min.
9. Place autoclave pan containing sterile fibers into plastic bag and place in cold room until use. Seal bag to prevent evaporation/contamination.

Important: The fibers should not be squeezed at any time. Cells grow poorly in fibers that have kinks.

Encapsulation of U2OS Cells Expressing Luc or p27Luc in Hollow Fibers

Materials

- Smooth-jawed needle holder (Fig. 4B, D)
- Scissors (Fig. 4B, B)
- Forceps (Fig. 4B, E)
- Perforated stainless steel work surface (Fig. 4A)

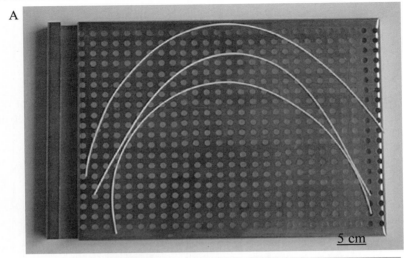
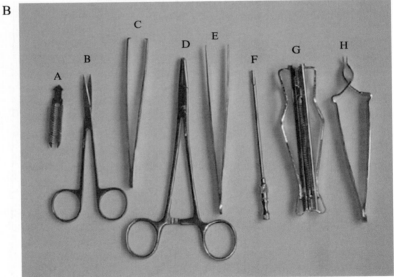

Fig. 4. Instruments used for implantation of hollow fibers. (A) Perforated stainless work surface and hollow fibers cut into 30 cm. Line indicates 5 cm in scale. (B) Autoclip (A); scissors (B); tissue forceps (C); needle holder (D); forceps (E); 11-gauge trocar (F); Autoclip applier (G); Autoclip remover (H).

- Sterile field drapes (Johnson and Johnson)
- Gloves (sterile, single wrapped)
- 3-ml syringes (1 per cell line) (Becton Dickinson)
- 20-ml syringe (1) (Becton Dickinson)
- 20-gauge needles (1 per cell line + 1)
- 50-ml conical polystyrene centrifuge tubes (1 per cell line)
- Suspended U2OS-p27Luc and U2OS-Luc cells ($\sim 10^6$ cells/ml and viability >95% as determined by trypan blue staining)
- Bunsen burner
- Six well plates
- 500-ml bottle of prechilled (4°) DMEM + 10% FBS media + penicillin/streptomycin

Note: The cells should be freshly trypsinized, resuspended, and counted immediately before transfer to hollow fibers. It is probably advisable to use cells that are subconfluent and proliferating at the time of harvest.

Procedure

1. Wrap and autoclave the stainless steel work surface and all metal instruments.

2. Remove autoclave pan containing hollow fibers from plastic bag. Wipe exterior of pan with 70% EtOH and transfer to tissue culture hood.

3. Transfer media, wrapped work surface, and instruments, along with sterile wrapped syringes, needles, centrifuge tubes, and sterile field drapes to tissue culture hood.

4. Add DMEM + 10% FBS media + penicillin/streptomycin to two 6-well plates (4 ml media/well). Label one plate "Luc" and the other "p27Luc." Put plates into 37° incubator to prewarm the media.

5. Don sterile gloves and spray the gloves with 70% ethanol.

6. Carefully move gloved hands into the hood without touching anything else. (*Note:* If you need to remove your gloved hand from the hood from this point on, repeat step 5).

7. Unwrap sterile drapes and fold in half. Then lay two of the folded drapes on the tissue culture hood surface such that each drape overlaps slightly with the other and covers the area immediately in front of you and extending to the back of the hood.

8. Unwrap the perforated stainless steel work surface and place on sterile drapes. Do the same with the wrapped metal instruments.

9. Unwrap syringes, remove caps, and attach needles. Place on sterile drapes.

10. Fill the 20-ml syringe with chilled media and use it to wet down the center of the perforated stainless steel work surface (the length of the wet area should exceed the length of a hollow fiber). *Note:* the work surface must remain hydrated or the quality of the fibers will be compromised.

11. Remove a fiber from the pan and place it on the wet perforated tray. Using the 20-ml syringe, rinse the outside of the fiber and then flush the inside. In total should use about 3–4 ml of media per wash.

12. Slowly fill a 3-ml syringe with 2 ml of U2OS-p27Luc suspension ($\sim 10^6$ cells/ml).

13. Hold the 3-ml syringe in one hand and gently grasp one end of the fiber between the thumb and index finger of your other hand.

14. Insert the needle of the syringe into the end of the fiber and inject the cell suspension into the fiber, making sure not to introduce air bubbles, so that the fiber is full (typically requires about 0.5–1 ml).

15. With your free hand, heat the tip of the needle holders in the Bunsen burner and use the hot needle holder to seal the fiber just beyond the point of needle insertion. Lightly clamp down on the fiber and then gently slide needle holder along the length of the fiber so that the seal is approximately 4 mm. *Note:* Don't heat the needle holder too much in this or subsequent steps or the fiber will break. Heat the needle holder initially for ~ 8 sec and for 4–5 seconds for step 16.

16. Release the sealed end of the fiber, reheat the needle holders, and seal the opposite end of the fiber.

17. If desired, can gently agitate the remaining cells in the 3-ml syringe and repeat steps 13–16. (*Note:* if making more than two fibers can discard the remaining cells in the 3-ml syringe in a waste tube and repeat steps 12–16).

18. During all of these procedures make sure filled fibers remain wet using 20-ml syringe if needed to rewet the surface.

19. Heat the needle holder in the Bunsen burner and seal the filled fibers approximately every 1.5 cm (seal width ~ 4 mm) to create a "sausage link" appearance.

20. Using scissors, cut the hollow fibers into 1.5-cm sections by cutting across the middle of the seals.

21. Line the fibers up parallel to each other at the bottom left-hand corner of the perforated work surface. The seals of each fiber can be used for handling purposes (*Note:* use the forceps to handle the ends of the fibers from this point onward).

22. Transfer fibers to 6-well plates containing DMEM + 10% FBS + penicillin/streptomycin (4 ml/well). At this point can place up to 14 fibers per well. Place in 37° incubator containing 10% CO_2.

23. Clean the perforated stainless steel work surface using 70% EtOH and sterile gauze.
24. Repeat steps 12–22 to make U2OS-Luc hollow fibers.
25. Allow cells to equilibrate at least 12 h before use in subsequent assays.

Note: If serum proteins build up on the needle holders, use the scissors to remove them. Always handle the fibers gently and do not squeeze or bend them at any point.

Bioluminescent Imaging of Hollow Fibers *In Vitro*

Materials

- D-luciferin (7.5 mg/ml) (Xenogen)
- Two 6-well tissue culture plates
- Forceps (Fig. 4B, E)
- DMEM + 10% FBS + penicillin/streptomycin
- Xenogen IVIS™ imaging system (Xenogen)
- Flavopiridol: stock at 10 mM in DMSO (National Cancer Institute)
- Roscovitine: stock at 10 mM in DMSO (Cyclacel Ltd, UK)

Procedure

1. Wear sterile gloves and clean the tissue culture hood work surface with 70% EtOH.
2. Add DMEM + 10% FBS + penicillin/streptomycin to two 6-well plates (2 ml media/well).
3. Bring plates with hollow fibers prepared in the section "Encapsulation of U2OS Cells Expressing Luc or p27Luc in Hollow Fibers" into the tissue culture hood.
4. Place one fiber (U2OS-p27Luc or U2OS-Luc) per well into each well of 6-well plate using forceps (grab fiber by end). (*Note:* make sure fiber is completely submerged and not floating.)
5. Add 14 μl of D-Luciferin (7.5 mg/ml) to each well (final concentration of 50 μg/ml). Mix well by manually shaking the plates 6–8 times.
6. Start Living Image 2.02 program from the Windows start menu. (*Note:* this and subsequent steps are based on the instructions provided by Xenogen.)
7. Click the Initialize button in the Camera Control panel.
8. Put 1 or 2 plates (maximum, 2 plates) into imaging dark box of the Xenogen system.
9. Adjust the field of view (FOV) by raising the shelf with choosing B mode (height = 15 cm) of FOV control in the Camera Control panel.

10. Ensure that the Auto exposure box for the Photographic image is chosen in the Camera Control panel.

11. Get photographic image by clicking the Photographic box and clicking the Acquire button in the Camera Control panel. If necessary, adjust the position of samples.

12. Acquire the luminescent image: Click the Overlay box in the Camera Control panel and set the exposure time to 2 min and set Binning to 2. Click the Acquire button in the Camera Control panel. If necessary, adjust the settings based on image quality.

13. After the exposure is complete, the overlaid image will be displayed. Under the Living Image menu item, choose "Save imaging Image Data..." to save images in the hard drive or other devices.

14. Acquire the luminescent images every 5–10 min up to 30 min (data shown in Fig. 2 are images obtained after 10 min).

15. 5 h after imaging, replace cell culture media with fresh media containing flavopiridol or roscovitine to a final concentration of 200 nM or 25 μM respectively.

16. Repeat steps 5–14 every 24 h. Data shown in Fig. 2 are 24 h after treatment.

Implantation of Hollow Fibers into Nude Mice

Materials

- Scissors (Fig. 4B, B)
- Forceps (Fig. 4B, E)
- Tissue forceps (Fig. 4B, C)
- 11-gauge trocar (Popper and Sons Inc.) (Fig. 4B, F)
- Autoclip kit with 5 sets of clips (Becton Dickinson) (Autoclip, Fig. 4B, A; Autoclip applier, Fig. 4B, G; Autoclip remover, Fig. 4B, H)
- Sterile field drapes (Johnson & Johnson)
- Heating pad
- Ketamine 100 mg/ml (Phoenix Pharmaceuticals)
- Xylazine 20 mg/ml (Phoenix Pharmaceuticals)
- Sterile gloves
- 10% Clorox
- Novalsan® Solution (Fort Dodge Animal Health)
- 250-ml beaker (autoclaved)
- Sterile, distilled water

- Female athymic–nu (Ncr nu/nu) at approximately 25 g each (Taconic)
- Glass vials with stoppers (2 ml, Wheaton)

Procedure

1. Wrap and autoclave all metal instruments.
2. Transfer wrapped sterile instruments, sterile syringes, and sterile drapes into a laminar flow hood.
3. Mix ketamine, xylazine, and sterile distilled water 0.7:0.3:4.0 ml, respectively, in glass vial (will need 0.2 ml per mouse).
4. Spray down the anesthesia glass vial with 70% EtOH and place it in the laminar flow hood.
5. Place plates containing the hollow fibers into the hood.
6. Spray down the heating pad with 0.58% $NaClO_4$ (10% commercial Clorox) and place it in the laminar flow hood.
7. Add 150 ml sterile, distilled water, and 50 ml of Novalsan® into 250-ml beaker.
8. Transfer beaker into laminar flow hood. Transfer 1 cage containing 5 nude mice into hood.
9. Wear sterile gloves. Spray gloves with 0.58% $NaClO_4$ (10% commercial Clorox) followed by 70% EtOH.
10. Place the sterile drape over the heating pad.
11. Place the plate containing the fibers on the sterile drape covering the heating pad.
12. Unwrap the sterilized instruments and the 1-ml insulin syringe in the hood and drop them onto the sterile drape. (*Note:* if at any point it is necessary to leave the hood, reglove as in step 9.)
13. Administer 0.2 ml of ketamine/xylazine per mouse by intraperitoneal injection using the 1-ml syringe.
14. Return mice to cage. Wait (typically 2–3 min) until mice are sedated and do not respond to mild pain (pedal reflex). They should remain anesthetized for ~30 min.
15. Place one anesthetized mouse in prone position. For right-handed investigators, step 17 is easier if the mouse is oriented with tail on the left and head on the right.
16. Apply a few drops Novalsan® at the nape of the neck at the planned incision site.
17. Hold the skin at the nape of the neck with tissue forceps with left hand, make a small skin incision (about 5 mm) perpendicular to the long axis of the mouse, with scissors with right hand.

18. Turn the mouse so that its head is facing toward you.

19. Using the regular forceps, place a hollow fiber filled with U2OS-p27Luc cells into the end of the 11-gauge trocar with the plunger pulled halfway back.

20. Insert the tip of the trocar into the incision and slide it back toward the tail end and to one side of the centerline running from head to tail (see Fig. 5A).

21. Insert the fibers as you are retracting the trocar by depressing the plunger.

22. Repeat steps 19–21 for a U2OS-Luc cell fiber, and place it on the side opposite the fiber that was already inserted.

23. Use an Autoclip applier to close the wound with an Autoclip (Fig. 5B).

24. Place each implanted mouse to the side on the drape-covered heating pad until they begin to recover from the anesthesia.

25. Put mice back into their regular cages.

Note: Autoclips can be removed with Autoclip remover if hollow fibers need to be removed or repositioned.

Bioluminescent Imaging of Hollow Fibers *In Vivo*

Materials

- D-luciferin (7.5 mg/ml)
- Ketamine 100 mg/ml (Phoenix Pharmaceuticals)

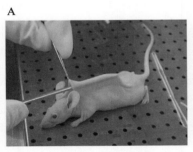
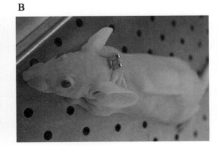

Incision and implantation of fiber   Closure of incision

FIG. 5. Insertion of hollow fiber into nude mouse. (A) A 5-mm incision was made at the nape of the neck perpendicular to the mouse long axis. Shown is insertion of 11-gauge trocar containing a hollow fiber. (B) Mouse with hollow fibers (one on each side) after placement of Autoclip.

- Xylazine 20 mg/ml (Phoenix Pharmaceuticals)
- Heating pads
- Sterile field drapes (Johnson & Johnson)
- 1-ml syringe
- 27-gauge needle (need 1 per mouse)
- Xenogen IVIS™ imaging system
- Isoflurane (Isoflo) (100%, Abbott Laboratories)
- Flavopiridol: stock at 1 mg/ml in DMSO (National Cancer Institute)

Procedure

1. Mix 760 μl of D-Luciferin (50 mg/kg), 170 μl of ketamine (140 mg/kg), and 70 μl of xylazine (12 mg/kg) in glass vial (need ~220 μl) per mouse).
2. Transfer sterile drapes and 1-ml syringes to hood.
3. Place one cage of mice implanted with hollow fibers containing U2OS-p27Luc or U2OS-Luc in the hood (see protocol entitled "Implantation of Hollow Fibers into Nude Mice").
4. Spray down the heating pad with 0.58% $NaClO_4$ (10% commercial Clorox) and place it in the laminar flow hood.
5. Wear sterile gloves. Spray gloves with 0.58% $NaClO_4$ (10% commercial Clorox) and then with 70% EtOH.
6. Place the sterile drape over the heating pad.
7. Unwrap the 1-ml syringe in the hood and drop it onto the sterile drape. Attach 27-gauge needle. Draw up ~1 ml of anesthesia mix.
8. Administer 8.8 ml/kg of anesthesia mixture from step 1 by intraperitoneal injection using the 1-ml insulin syringe. Change needles between mice.
9. Start Living Image 2.02 program from the Windows start menu. (*Note:* this and subsequent steps are based on the instructions provided by Xenogen.)
10. Click the Initialize button in the Camera Control panel.
11. Put up to 4 mice into dark box of Xenogen imaging system.
12. Adjust FOV by raising the shelf and choosing C mode (height = 20 cm) the FOV control in the Camera Control panel. If necessary, adjust the height.
13. Ensure that the Auto exposure box for the Photographic image is chosen in the Camera Control panel.
14. Get photographic image by clicking the Photographic box and clicking the Acquire button in the Camera Control panel. If necessary, adjust the position of samples.

15. Acquire the luminescent image: Check the Overlay box in the Camera Control panel and set the exposure time as 5 min, Binning as 4 for the luminescent image. Click the Acquire button in the Camera Control panel. If necessary, change the setting.

16. After the exposure is complete, the overlaid image will be displayed. Under the Living Image menu item, choose "Save imaging Image Data..." to save images in the hard drive or other devices.

17. Acquire the luminescent images every 10 min up to 30 min (data shown in Fig. 3 are images obtained after 20 min.)

18. Place each mouse on the drape-covered heating pad until they begin to recover from the anesthesia.

19. Put mice back into their reguiar cages.

20. 5 h after pretreatment imaging, anesthetize mice by the inhalation of 4% Isoflo in an anesthesia chamber until each mouse is unconscious.

21. Administer flavopiridol 5 mg/kg by intraperitoneal injection using 1-ml insulin syringe. (*Note:* Flavopiridol should be given once a day.)

22. Place each mouse to the side on the drape-covered heating pad until they begin to recover from the anesthesia.

23. Repeat steps 1–19 once a day. Images shown in Fig. 3 were obtained after 2 days.

References

Blasberg, R. G. (2003). Molecular imaging and cancer. *Mol. Cancer Ther.* **2**, 335–343.
Contag, C. H., and Stevenson, D. K. (2001). *In vivo* patterns of heme oxygenase-1 transcription. *J. Perinatol* **21**(Suppl 1), S119–124; discussion S25–27.
Hollingshead, M. G., Alley, M. C., Camalier, R. F., Abbott, B. J., Mayo, J. G., Malspeis, L., and Grever, M. R. (1995). *In vivo* cultivation of tumor cells in hollow fibers. *Life Sci.* **57**, 131–141.
Jenkins, T. M., Sala-Newby, G., and Campbell, A. K. (1990). Measurement of protein phosphorylation by covalent modification of firefly luciferase. *Biochem. Soc. Trans.* **18**, 463–464.
Luker, G. D., Pica, C. M., Song, J., Luker, K. E., and Piwnica-Worms, D. (2003). Imaging 26S proteasome activity and inhibition in living mice. *Nat. Med.* **9**, 969–973.
Montagnoli, A., Fiore, F., Eytan, E., Carrano, A. C., Draetta, G. F., Hershko, A., and Pagano, M. (1999). Ubiquitination of p27 is regulated by Cdk-dependent phosphorylation and trimeric complex formation. *Genes Dev.* **13**, 1181–1189.
Rudin, M., and Weissleder, R. (2003). Molecular imaging in drug discovery and development. *Nat. Rev. Drug Discov.* **2**, 123–131.
Sala-Newby, G., and Campbell, A. K. (1992). Engineering firefly luciferase as an indicator of cyclic AMP-dependent protein kinase in living cells. *FEBS Lett.* **307**, 241–244.
Sala-Newby, G. B., and Campbell, A. K. (1991). Engineering a bioluminescent indicator for cyclic AMP-dependent protein kinase. *Biochem. J.* **279**(Pt 3), 727–732.

Shirane, M., Harumiya, Y., Ishida, N., Hirai, A., Miyamoto, C., Hatakeyama, S., Nakayama, K., and Kitagawa, M. (1999). Down-regulation of p27(Kip1) by two mechanisms, ubiquitin-mediated degradation and proteolytic processing. *J. Biol. Chem.* **274**, 13886–13893.

Tsvetkov, L. M., Yeh, K. H., Lee, S. J., Sun, H., and Zhang, H. (1999). p27(Kip1) ubiquitination and degradation is regulated by the SCF(Skp2) complex through phosphorylated Thr187 in p27. *Curr. Biol.* **9**, 661–664.

Wang, W., and El-Deiry, W. S. (2003). Bioluminescent molecular imaging of endogenous and exogenous p53-mediated transcription *in vitro* and *in vivo* using an HCT116 human colon carcinoma xenograft model. *Cancer Biol. Ther.* **2**, 196–202.

Zhang, J., Campbell, R. E., Ting, A. Y., and Tsien, R. Y. (2002). Creating new fluorescent probes for cell biology. *Nat. Rev. Mol. Cell Biol.* **3**, 906–918.

Zhang, G. J., Safran, M., Wei, W., Sorensen, E., Lassota, P., Zhelev, N., Neuberg, D. S., Shapiro, G., and Kaelin, W. G., Jr. (2004). Bioluminescent imaging of Cdk2 inhibition *in vivo*. *Nat. Med.* **10**, 643–648.

[37] Monitoring the Distribution and Dynamics of Proteasomes in Living Cells

By TOM A. M. GROOTHUIS and ERIC A. J. REITS

Abstract

The proteasome is a large protease complex present in the cytoplasm and the nucleus of eukaryotic cells. This chapter describes how proteasomes in living cells can be visualized using fluorescently tagged subunits. The use of noninvasive fluorescent tags like the green fluorescent protein enables visualization of various subunits of the ubiquitin-proteasome system and prevents possible artefacts like disruption by microinjection or altered fluorescence distribution caused by fixation. Once quantitative incorporation of tagged subunits into proteasomes is ensured, the distribution of proteasome complexes can be visualized *in vivo*. In addition, different bleaching techniques can be applied to study the dynamics of proteasomes within the cell. Finally, we describe how proteasomes can be recruited to particular sites of degradation during various cellular conditions like aggregate formation and virus infection.

Introduction

The proteasome is a large, multicatalytic protease complex present in both the nucleus and the cytoplasm of all eukaryotic cells and is responsible for the turnover of most intracellular proteins. These proteins are continuously being generated and may have a half-life up to several days, but many proteins live only for minutes. Short-lived proteins include key regulatory molecules involved in cell cycle but also newly synthesized, but defective, proteins that are rapidly degraded by the proteasome (Yewdell *et al.*, 2003). Protein turnover not only generates new building blocks for protein synthesis but also prevents accumulation of potentially hazardous unfolded and damaged proteins.

The 26S proteasome complex consists of a 20S catalytic core containing the proteolytic subunits and a 19S regulatory complex that is involved in the unfolding and deubiquitination of proteins targeted for degradation. The 20S core has a cylindrical structure made up of four rings containing each seven subunits. The outer rings contain α-subunits, whereas the inner two rings consist of β-subunits. Three constitutively expressed β-subunits, delta (also known as Y or β1), MB1 (X or β5), and Z (β2), contain proteolytic active sites facing the inner chamber of the 20S core (reviewed by Groll and Huber [2003] and Voges *et al.* [1999]). These subunits are generated with N-terminal extensions that are cleaved off during proteasome assembly, and the new N-termini are now formed by a threonine residue that forms the catalytic active site attacking the peptide bond of the unfolded protein entering the 20S barrel. After interferon-γ stimulation, these subunits are replaced by LMP2 (low-molecular mass polypeptide 2, also known as β1i), LMP7 (β5i), and MECL-1 (multicatalytic endopeptidase complex-like-1, β2i), respectively (Eleuteri *et al.*, 1997). The resulting "immunoproteasome" has an altered peptidase activity that affects the site of polypeptide cleavage but not the overall rate of degradation. As a result, peptides with different N- and C-termini are generated that may be favored by MHC class I molecules for subsequent presentation to the immune system (Cerundolo *et al.*, 1995; Gaczynska *et al.*, 1994; Kloetzel, 2001).

To monitor the distribution and dynamics of proteasomes in living cells, subunits of either the 19S cap or the 20S core can be tagged with fluorescent proteins like the green fluorescent protein (GFP). GFP has become the most widely used fluorescent marker in cell biology, because it can be tagged to many proteins and expressed as a fusion protein in the living cell (reviewed by Miyawaki *et al.* [2003] and Zhang *et al.* [2002]). Therefore, GFP has to be cloned in frame with the gene encoding the protein of interest, and the resulting fusion-protein can be expressed in cells on transfection. Ideally, the fusion with GFP should not affect either function or localization of the tagged protein, and because of the compact size of GFP, many proteins localized in different organelles can be successfully tagged. Fluorescence will reflect subcellular localization (as determined by targeting domains within the tagged protein) but also the level of gene expression. GFP is relatively resistant to photobleaching, and fluorescence can be followed for long periods. However, at high laser power, GFP can be irreversibly photobleached, which enables the measurement of fluorescence recovery because of diffusion. The dynamics are an important indicator for the mobility of the tagged complex within specific cell compartments like the nucleus but also movement between different compartments. In addition, protein interactions and conformational changes can be measured with bleaching protocols.

Tagging the Proteasome with GFP

To visualize the proteasome in living cells, we tagged the β-subunit LMP2 with GFP (Reits *et al.*, 1997). In principle, any subunit could be tagged, as long as the fusion-protein is quantitatively incorporated in the 20S complex. We chose the LMP2 subunit, because antibodies raised against the C-terminus of this subunit could detect intact proteasomes (Patel *et al.*, 1994). This suggested that the C-terminus was exposed at the outside of the 20S barrel and that tagging the LMP2 C-terminus with GFP should not interfere with incorporation. The full-length LMP2 cDNA was used to generate a LMP2 fragment with the final stop codon replaced for the first codons of GFP. Similarly, a GFP fragment was created by PCR with the last codons of LMP2 (without the stop codon) at the N-terminus. The two fragments would serve as primers for the complementary strands, creating a LMP2–GFP fusion protein that was cloned into pcDNA3 (Invitrogen). This two-step approach of creating fusion proteins is not necessary anymore, because there is now a family of pEGFP vectors containing EGFP in front of or behind a multiple cloning site, resulting in a tagged protein of interest with EGFP at the N- or C-terminal, respectively (Clontech). The E stands for enhanced, and the EGFP protein is the result of a chromophore mutation that produces fluorescence 35 times more fluorescent than wild-type GFP. These vectors make it much easier to generate a fusion-protein, because one can now generate an LMP2 fragment without stop codon by PCR and directly clone it in frame with EGFP.

When tagging a particular subunit of a larger complex with a fluorophore, it has to be ensured that the observed fluorescence is derived from the entire complex instead of nonincorporated fusion-proteins (i.e., proteasome complexes instead of free LMP2-GFP subunits). A number of ways exist to test whether most (if not all) tagged subunits are incorporated.

1. Immunoprecipitate the complex with an antibody recognizing a different subunit of the complex and determine whether nonincorporated fusion-proteins are remaining in the lysate. In our case, we made lysates of cells stably transfected with LMP2-GFP. The lysates were subjected to

three rounds of immunoprecipitation with the antibody MCP21 that recognizes the proteasome α-subunit HC3 (α2). This should remove all LMP2-GFP subunits incorporated into complete proteasomes. Western blotting with anti-GFP antisera showed that LMP2-GFP was coimmunoprecipitated with MCP21 and that no LMP2-GFP could be detected in total lysate after the three rounds with MCP21, showing that LMP2–GFP was incorporated quantitatively into proteasomes.

2. When both nonincorporated subunits and large complexes are present in the same cellular compartment, sucrose density centrifugation can be used to separate the two pools. Cells stably transfected with either GFP or LMP2-GFP were washed two times with phosphate-buffered saline (PBS), harvested by scraping, and diluted in 1 ml cold homogenization buffer (HB, 0.25 mM sucrose, 10 mM TEA [triethanolamine], 10 mM acetic acid, and 0.5 mM MgCl$_2$) and homogenized with a ball-bearing homogenizer. Nuclei and debris were removed by low-speed centrifugation (2.500 rpm), and the resulting cytosolic fraction was layered on top of a stepwise sucrose gradient containing 1.5-ml layers of 10/15/20/25/30/35/40% sucrose prepared in HB. The samples were centrifuged in a Beckman SW 40 Ti for 24 h at 31.000 rpm at 4° (no brake). Fractions of 0.5 ml were taken from the top, and proteins were recovered by 10% TCA (trichloroacetic acid) precipitation before analysis by 12% SDS-PAGE and Western blotting. If LMP2-GFP was not incorporated into the proteasome, it should be in the same fraction as other low molecular weight proteins such as free GFP, because it has the same density. However, LMP2-GFP was only detected in a different fraction also containing other 20S proteasomal subunits, indicating that it is incorporated quantitatively in the 20S barrel.

3. Measure the diffusion rate of the fluorescent subunit in living cells. As will be shown later in more detail, the diffusion rate of the fusion-protein can be measured using fluorescence recovery after photobleaching (FRAP). When a small region in the cell is selectively photobleached by high laser power, recovery of fluorescence will occur because of diffusing of nonbleached molecules from the environment into that region. The recovery time is related to the size of the molecule, because a large complex will diffuse slower than a small molecule (the diffusion rate is mainly determined by the radius of the complex; Arrio-Dupont *et al.* [1996]; Gribbon and Hardingham [1998]). The diffusion rate of LMP2-GFP is relatively slow compared with free GFP or small GFP fusion-proteins. In addition, the recovery curve shows a single-exponential recovery of the fluorescent signal, suggesting that there is only one pool of LMP2-GFP that is present in a large complex.

4. Measure diffusion of the fluorescent subunit between two compartments. When studying a relatively small fusion-protein like LMP2-GFP (~52 kDa), the molecular size is often small enough to allow diffusion through the nuclear pore (which has a size limit of approximately 60 kDa). When no specific localization signals are present, fluorescence will be observed in both the nucleus and the cytoplasm, and loss of fluorescence in one compartment because of bleaching will result in loss of fluorescence in the other compartment (because both fluorescent and bleached molecules simply diffuse through the nuclear pore in both directions). As will be shown later, no rapid diffusion of LMP2-GFP between the nuclear and cytoplasmic compartment was observed, suggesting that it is incorporated in a large complex that is unable to pass the nuclear pore by diffusion.

Incorporation of the tagged subunit into the 20S complex should not affect the function of the proteasome. To determine any effect of incorporation of LMP2-GFP on the catalytic activity of the proteasome, the activity of proteasomes isolated with either the antibody MCP21 (against the α-subunit HC-3) or anti-GFP (to immunoprecipitate proteasomes containing LMP2-GFP) antisera was assayed. This can be done by measuring the chymotrypsin-like and the trypsin-like proteasomal activities using small fluorogenic peptide substrates. Therefore, approximately 2×10^7 cells were washed three times with PBS, and pellets were lysed in a hypotonic buffer (10 mM TEA (triethanolamine), 10 mM acetic acid, 1 mM EDTA, 0.25 M sucrose, pH 7.4), homogenized with a ball-bearing homogenizer and subsequently kept at 4°. Equal amounts of supernatant (corresponding to 300 μg protein) were used per immunoprecipitation. Overnight preclearing with normal rabbit serum coupled to Protein A beads was followed by another round of Protein A beads for 2 h. Immunoprecipitations were done for 1.5 h using MCP21, GFP antisera, or TAP1 antisera (TAP1 as a negative control). After 5× washing, the pellets were split into equal portions and assayed against 100 μM of the peptide substrates z-Ala-Ala-Arg-AMC and z-Gly-Gly-Leu-AMC (both in DMSO carrier) (Novabiochem and Bachem, Switzerland) in a total volume of 200 μl hypotonic buffer at 37°. Aliquots of 10 μl in duplicate were quenched into 1 ml ethanol after 8 h of incubation and the fluorescence of the free AMC ($\lambda_{excitation} = 370$ nm, $\lambda_{emission} = 460$ nm) measured on a spectrofluorometer. We observed similar levels of activity from proteasomes isolated with either MCP21 or anti-GFP antibodies in the LMP2-GFP expressing cell line, whereas the parental cell line lacking LMP2-GFP did not show any proteolytic activity after precipitating with anti-GFP antibody. Apparently, incorporation of GFP-tagged LMP2 into proteasomes has little effect on the chymotryptic and tryptic activities.

Distribution of Proteasomes in Living Cells

Mammalian cells can be transiently or stably transfected with cDNA encoding the fusion protein to visualize the distribution of fluorescence. Transient transfection will lead to high levels of fluorescence within 24–48 h, but most cell lines will lose fluorescence after 3 days. In addition, very high levels of fusion proteins may lead to only partial incorporation into complexes, making it problematic to perform reliable biochemistry experiments. By placing the transfected cells on selection (determined by the resistance genes present in the vector backbone), either single colonies stably transfected with the introduced gene can be picked and grown separately, or stably transfected cells expressing low levels of fluorescence can be isolated by FACS sort. Again, the selection of low-fluorescent cells should favor quantitative incorporation of the expressed fusion-proteins.

The intracellular distribution of GFP-tagged proteasomes in living cells can be determined with most inverted fluorescence microscopes. Green fluorescence is observed in both the cytoplasm and the nucleus of the transfected cells, but proteasomes are excluded from nucleoli and membrane-rich areas like the nuclear envelope and the ER/Golgi region (Fig. 1A). The GFP-tag enables visualization of protein distribution using noninvasive techniques and prevents the need to fix cells to stain the protein of interest with fluorescent antibodies. The latter may also lead to misinterpretation, because fixing cells with cold methanol for 5 min changes the pattern of fluorescence dramatically (Fig. 1B, C). Methanol treatment precipitates the GFP-tagged proteasomes to membrane-rich areas like the ER, leading to the false interpretation that proteasomes

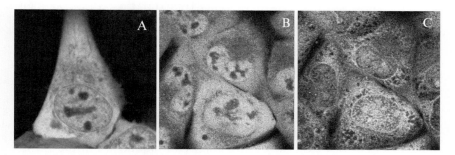

FIG. 1. Distribution of GFP-tagged proteasomes. Stably transfected Mel JuSo (human melanoma cells) expressing LMP2-GFP and analyzed by confocal microscopy show fluorescence in both the cytoplasm and the nucleus, but not in nucleoli or membranes like the ER or the nuclear envelope (A). The distribution of fluorescent proteasomes is dramatically changed when living cells (B) are fixed with methanol for 5 min (C).

may dock to the transporter associated with antigen processing (TAP) to facilitate peptide transport into the ER lumen and subsequent MHC class I loading. A better approach to stain cells expressing GFP-tagged molecules with antibodies is to fix cells with 3.7% formaldehyde for 10 min, followed by 0.1% Triton X-100 treatment to make the fixed interior accessible for antibodies.

Dynamics of Proteasomes in Living Cells

Many researchers use GFP only to show the localization of their protein of interest in living cells. Using inverted fluorescence microscopy, one can easily visualize the intracellular distribution and also changes in distribution over time. It is, however, not possible to show kinetics like protein mobility within the cell and binding to other components. Most research facilities now have access to a confocal laser scanning laser microscope (CSLM) equipped with photobleaching protocols, which makes it possible to determine the mobility of fluorescent proteins within and between different cellular compartments. To show movement of proteins, a small region in the cell can be photobleached (using a high-power focused laser beam), resulting in depletion of fluorescence in that selected region. The fluorescent molecules are irreversibly photobleached but will diffuse out of the region, while at the same time surrounding fluorescent molecules will diffuse into the bleached area. Using low laser power, fluorescence recovery within this region can be followed in time, and both the rate and extent of recovery can be quantified (reviewed by Lippincott-Schwartz et al. [2001] and Reits and Neefjes [2001]).

To determine whether proteasomes containing LMP2-GFP are freely moving through the cell or mainly maintained in large, immobile complexes, the mobility is determined using a photobleaching technique called fluorescence recovery after photobleaching (FRAP) (Axelrod et al., 1976). A time-lapse protocol can be made in which first an image is made of the entire cell before bleaching, followed by photobleaching a small region in the cell briefly at 100% laser power, immediately followed by time-lapse imaging the entire cell for an extended period (until no further recovery is observed). It is important to use an identical setting of laser intensity when imaging the cell before and after bleaching to determine parameters like mobile fraction and loss of fluorescence because of imaging. Imaging at low laser power can still lead to some loss of fluorescence, which can be checked by measuring the fluorescence in a nearby cell (which is not photobleached at high laser power) during the FRAP assay. If loss of fluorescence in nearby cells is observed, the recovery rates in the photobleached cell should be corrected. Continuing imaging after photobleaching is critical, because fluorescence recovery will start immediately after bleaching.

When, for example, a region in the cytoplasm of LMP2-GFP expressing cells is bleached, the mobile fraction and the rate of recovery can be determined (Fig. 2). The fluorescence in the bleached area (indicated by the white circle) will drop from the initial fluorescence F_i to F_0 (just after bleaching). Fluorescence will recover in time by diffusion until the fluorescence has recovered to a plateau phase (F_∞). The mobile fraction R can be calculated by comparing the fluorescence in the bleached region after full recovery (F_∞) with the fluorescence before bleaching (F_i) and just after bleaching: $R = (F_\infty - F_0)/(F_i - F_0)$. R will be low when many proteins are restricted in their movement (e.g., because of interaction with immobile

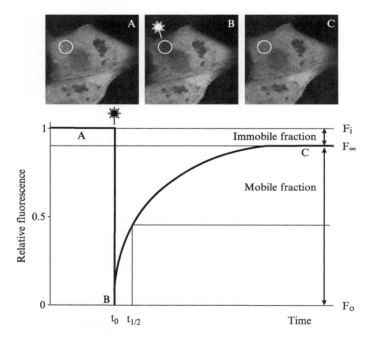

FIG. 2. Measuring the dynamics of proteasomes using fluorescence recovery on photobleaching (FRAP). To determine the mobility of fluorescent proteasomes, a small region in the cytoplasm can be bleached (indicated by the white circle in A). Fluorescence will recover in the bleached region because of rapid diffusion of proteasomes within the cytoplasm (compare B and C, at t_0 the region is bleached). When the recovery is plotted, two parameters can be obtained: the diffusion time $t_{1/2}$, indicating the time at which half of the fluorescence has recovered, and the mobile fraction. F_i is the initial fluorescence, F_0 the fluorescence just after bleaching, and F is the fluorescence after full recovery.

molecules or structures). The second parameter is the diffusion time $t_{1/2}$, which indicates the time at which half of the fluorescence has recovered. For a freely diffusing molecule, the diffusion time is mainly determined by the radius of the molecules but also by collision with other molecules and hindrance by immobile obstacles like cytoskeletal filaments. In the case of LMP2-GFP, fluorescence recovery is fast, and almost no immobile fraction is observed, suggesting that proteasomes move unrestricted through the cytoplasm. The mobility is independent of energy, because ATP depletion has no effect on the mobility of tagged proteasomes (using 0.05% NaAz and 50 μM 2-deoxyglucose for 30 min to deplete intracellular ATP levels), indicating that proteasomes move by diffusion and not by active transport. Similarly, when a region in the nucleus is bleached, fluorescence recovery in the nucleus is similar to the cytoplasm. Apparently, proteasomes diffuse rapidly within both compartments, thereby colliding and interacting with proteins targeted for degradation.

In addition, one can photobleach a region for a longer period (or multiple times) to determine fluorescence loss in photobleaching (FLIP). All fluorescent molecules diffusing through that region will be photobleached, and other cellular compartments connected to the bleached area will lose fluorescence in time (Cole *et al.*, 1996; Phair and Misteli, 2000). Similar to the FRAP assay, time-lapse imaging a number of cells can be combined with repeated photobleaching of a small region, and recovery of fluorescence between each photobleaching period can be monitored. At the same time, fluorescence levels will remain unaffected in those areas of the cell that are not connected with the bleached region (Fig. 3A). For example, free GFP will be present in both the cytoplasm and the nucleus, and repeated photobleaching of the cytoplasm will decrease fluorescence in both the cytoplasm and the nucleus because of diffusion of GFP between both compartments. However, the size of the proteasome will prevent passive diffusion through the nuclear pore, and repeated photobleaching of the cytoplasm will lead to loss of fluorescence in the cytoplasm but not the nucleus. Similarly, extended photobleaching of the nucleus will result in loss of fluorescence in the nucleus but hardly in the cytoplasm (Fig. 3B). Note that the cytoplasm has lost some fluorescence; this is due to the composition of the cell where the nucleus is surrounded by the cytoplasm, and the photobleaching laser beam going "through" the nucleus will also hit the small volume cytoplasm above and below the nucleus. Although there is no immediate exchange of proteasomes between the two compartments, slow recovery of fluorescence in the nucleus can be observed after photobleaching the nucleus and monitoring fluorescence levels over an extended period (hours instead of minutes), which is due to the presence of nuclear localization signals in the proteasome. In

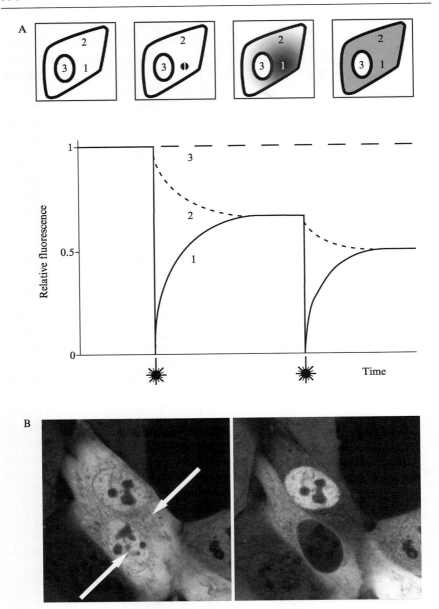

Fig. 3. Measuring the dynamics of proteasomes using fluorescence loss in photobleaching (FLIP). To determine the exchange of fluorescent proteasomes between different cellular compartments, a region in the cytoplasm can be repeatedly bleached (region 1), and fluorescence levels within the same compartment (2) or a different compartment (3) can be

addition, proteasomes can redistribute during mitosis when the nuclear envelope has disassembled.

Conclusions and Perspectives

We have used the LMP2 subunit of the 20S proteasome, but successful tagging of subunits of the 26S proteasome is not limited to LMP2 alone. Other parts of the 26S proteasome, like α-subunits of the 20S core but also subunits of the 19S cap, can similarly be tagged with GFP. Microscopic analysis revealed that both the 20S and 19S subunits of the yeast proteasome are accumulated mainly in the inner side of the nuclear envelope and endoplasmic reticulum (ER) network (Enenkel *et al.*, 1998; Russell *et al.*, 1999; Wilkinson *et al.*, 1998), which is different from eukaryotic cells. Several subunits of the 19S cap are also shown to be involved in other nonproteolytic processes like nucleotide excision repair (NER) and transcription elongation (Ferdous *et al.*, 2001). Because most of the research on 19S subunits has been done in yeast, and because of differences between proteasomes in yeast and eukaryotic cells, there is still a long road ahead for research on the 19S cap in eukaryotic cells. If subunits of the cap might indeed be involved in transcription elongation or DNA repair, some of the fluorescence recovery techniques described in this chapter might be very useful. Sites of transcription or repair might bind the subunit(s) for extended periods, which can be visualized with both FRAP and FLIP analysis.

In addition to components of the 26 proteasome, ubiquitin can also be tagged with GFP. For example, tagging GFP to the N-terminus of ubiquitin will lead to a completely functional ubiquitin molecule that will be involved in various cellular processes like polyubiquitination of proteins targeted for degradation but also endocytosis (Qian *et al.*, 2002; Varshavsky, 2000). However, when GFP is tagged to the C-terminus of ubiquitin, GFP will be removed by deubiquitination enzymes (DUBs), resulting in two independent proteins, of which the free GFP can be used as a transfection marker. Visualization of ubiquitin, 20S, and 19S subunits can be done simultaneously with two or more fluorochromes, because the use of FRAP and FLIP is not limited to GFP alone. Depending on the available laser excitation lines, one can also use other variants of GFP, like CFP (cyan)

---

monitored (A). When there is diffusion within compartments but no exchange between different compartments, region 2 will drop to the same fluorescence level as the bleached region (on recovery), whereas region 3 will be unaffected. When either a region in the cytoplasm or the nucleus of two cells is bleached (indicated by the arrow in B), those compartments will lose fluorescence but not the other compartments in the same cells (C).

and YFP (yellow), or the recently optimized mRFP (a red fluorescent coral reef protein). A inverted variant of FRAP can be achieved with the photoactivatable form of GFP (PA-GFP). This variant can be activated with an UV laser on which the GFP fluorochromes are converted to the anionic state, resulting in a 100-fold fluorescence increase after excitation at 488 nm (Patterson and Lippincott-Schwartz, 2002). The local increase in fluorescence will redistribute throughout the cell similar to bleached molecules in a classical FRAP experiment.

Tagging the proteasome with GFP will supply information both about the distribution and the dynamics of the complex at a given time point, and changes can be used as a readout system to measure cellular responses to various stress conditions like aggregation, infection, and starvation. The proteasome is recruited to large perinuclear structures containing misfolded and ubiquitinated proteins, called aggresomes (Garcia-Mata et al., 1999; Johnston et al., 1998; Wigley et al., 1999). Aggresomes are thought to be specialized sites for degradation where proteasomes, heat shock proteins, and other degradation-specific proteins are present and may be a solution to prevent the formation of dangerous aggregates throughout the cell. Subsequent clearance of the aggresome is, however, a process that is not fully understood, but it has been suggested that the aggresome can be removed from the cell by autophagy, a process used by cells to degrade among others mitochondria and peroxisomes. The structures are engulfed into autophagosomes and fuse with lysosomes leading to degradation by lysosomal hydrolases (Garcia-Mata et al., 2002). Recently, Perrin et al. (2004) showed that the ubiquitin-proteasome system could also be recruited to intracellular bacteria like *Salmonella typhimurium* that may escape from the phagosomes into the cytoplasm of macrophages. Recruitment of proteasomes to aggresomes and internalized bacteria may be very specific reactions of cytoplasmic proteasomes that are normally freely diffusing in the cytoplasm.

There are also examples of nuclear accumulations of proteasomes. For example, elevated expression levels of the nuclear transcription factor c-Myc give rise to accumulation of c-Myc and proteasomes in nucleoli, structures normally devoid of proteasomes (Arabi et al., 2003). Myc is normally a substrate for rapid turnover by the proteasome system, but expression of elevated levels of c-myc (or proteasome inhibition) leads to accumulation of c-myc within nucleoli. Proteasomes are also re-localized on expression of a viral nuclear protein to PODs (protomyelocytic leukemia gene product Oncogenic Product; Anton et al. [1999]), and components of the 26S proteasome are sequestered in discrete foci in the nucleus of infected cells on herpes simplex virus type 1 (HSV-1) infection (Burch and Weller, 2004).

Different groups have, however, observed different proteasome re-localization patterns, which may be due to the use of different fixation protocols and antibodies. In addition, expression of proteins from introduced plasmids is often under a CMV promoter, and differences in (over) expression may easily lead to different results. Therefore, the functionality and the incorporation of the introduced protein should be checked, which will at least ensure that the protein is not completely disturbing the behavior of the cell. In addition, diffusion coefficients and mobile fractions may vary between different cell lines because of differences in viscosity and architecture. Differences in confocal setups (like applied time-lapse protocols and variations in temperature) will also affect the measured mobility of soluble and membrane-associated molecules. When bleaching experiments are performed in a carefully controlled temperature stage, live cell imaging with many of its interesting aspects may then lead to a model system that is close to the *in vivo* situation and will be a great opportunity to visualize the dynamics of components of the ubiquitin-proteasome pathway. It might also be a tool to visualize whether proteasomes found in intracellular structures are still able to diffuse out of these structures. A quick FRAP and/or FLIP experiment might give the answer already.

Acknowledgment

We thank Jacques Neefjes for critically reading the manuscript.

References

Anton, L. C., Schubert, U., Bacik, I., Princiotta, M. F., Wearsch, P. A., Gibbs, J., Day, P. M., Realini, C., Rechsteiner, M. C., Bennink, J. R., and Yewdell, J. W. (1999). Intracellular localization of proteasomal degradation of a viral antigen. *J. Cell Biol.* **146**, 113–124.

Arabi, A., Rustum, C., Hallberg, E., and Wright, A. P. (2003). Accumulation of c-Myc and proteasomes at the nucleoli of cells containing elevated c-Myc protein levels. *J. Cell Sci.* **116**, 1707–1717.

Arrio-Dupont, M., Cribier, S., Foucault, G., Devaux, P. F., and d'Albis, A. (1996). Diffusion of fluorescently labeled macromolecules in cultured muscle cells. *Biophys. J.* **70**, 2327–2332.

Axelrod, D., Koppel, D. E., Schlessinger, J., Elson, E., and Webb, W. W. (1976). Mobility measurement by analysis of fluorescence photobleaching recovery kinetics. *Biophys. J.* **16**, 1055–1069.

Burch, A. D., and Weller, S. K. (2004). Nuclear sequestration of cellular chaperone and proteasomal machinery during herpes simplex virus type 1 infection. *J. Virol.* **78**, 7175–7185.

Cerundolo, V., Kelly, A., Elliott, T., Trowsdale, J., and Townsend, A. (1995). Genes encoded in the major histocompatibility complex affecting the generation of peptides for TAP transport. *Eur. J. Immunol.* **25**, 554–562.

Cole, N. B., Smith, C. L., Sciaky, N., Terasaki, M., Edidin, M., and Lippincott-Schwartz, J. (1996). Diffusional mobility of Golgi proteins in membranes of living cells. *Science* **273**, 797–801.

Eleuteri, A. M., Kohanski, R. A., Cardozo, C., and Orlowski, M. (1997). Bovine spleen multicatalytic proteinase complex (proteasome). Replacement of X, Y, and Z subunits by LMP7, LMP2, and MECL1 and changes in properties and specificity. *J. Biol. Chem.* **272**, 11824–11831.

Enenkel, C., Lehmann, A., and Kloetzel, P. M. (1998). Subcellular distribution of proteasomes implicates a major location of protein degradation in the nuclear envelope-ER network in yeast. *EMBO J.* **17**, 6144–6154.

Ferdous, A., Gonzalez, F., Sun, L., Kodadek, T., and Johnston, S. A. (2001). The 19S regulatory particle of the proteasome is required for efficient transcription elongation by RNA polymerase II. *Mol. Cell.* **7**, 981–991.

Gaczynska, M., Rock, K. L., Spies, T., and Goldberg, A. L. (1994). Peptidase activities of proteasomes are differentially regulated by the major histocompatibility complex-encoded genes for LMP2 and LMP7. *Proc. Natl. Acad. Sci. USA* **91**, 9213–9217.

Garcia-Mata, R., Bebok, Z., Sorscher, E. J., and Sztul, E. S. (1999). Characterization and dynamics of aggresome formation by a cytosolic GFP-chimera. *J. Cell Biol.* **146**, 1239–1254.

Garcia-Mata, R., Gao, Y. S., and Sztul, E. (2002). Hassles with taking out the garbage: Aggravating aggresomes. *Traffic* **3**, 388–396.

Gribbon, P., and Hardingham, T. E. (1998). Macromolecular diffusion of biological polymers measured by confocal fluorescence recovery after photobleaching. *Biophys. J.* **75**, 1032–1039.

Groll, M., and Huber, R. (2003). Substrate access and processing by the 20S proteasome core particle. *Int. J. Biochem. Cell Biol.* **35**, 606–616.

Johnston, J. A., Ward, C. L., and Kopito, R. R. (1998). Aggresomes: A cellular response to misfolded proteins. *J. Cell Biol.* **143**, 1883–1898.

Kloetzel, P. M. (2001). Antigen processing by the proteasome. *Nat. Rev. Mol. Cell Biol.* **2**, 179–187.

Lippincott-Schwartz, J., Snapp, E., and Kenworthy, A. (2001). Studying protein dynamics in living cells. *Nat. Rev. Mol. Cell Biol.* **2**, 444–456.

Miyawaki, A., Sawano, A., and Kogure, T. (2003). Lighting up cells: Labelling proteins with fluorophores. *Nat. Cell Biol.* Suppl. S1–7.

Patel, S. D., Monaco, J. J., and McDevitt, H. O. (1994). Delineation of the subunit composition of human proteasomes using antisera against the major histocompatibility complex-encoded LMP2 and LMP7 subunits. *Proc. Natl. Acad. Sci. USA* **91**, 296–300.

Patterson, G. H., and Lippincott-Schwartz, J. (2002). A photoactivatable GFP for selective photolabeling of proteins and cells. *Science* **297**, 1873–1877.

Perrin, A. J., Jiang, X., Birmingham, C. L., So, N. S., and Brumell, J. H. (2004). Recognition of bacteria in the cytosol of Mammalian cells by the ubiquitin system. *Curr. Biol.* **14**, 806–811.

Phair, R. D., and Misteli, T. (2000). High mobility of proteins in the mammalian cell nucleus. *Nature* **404**, 604–609.

Qian, S. B., Ott, D. E., Schubert, U., Bennink, J. R., and Yewdell, J. W. (2002). Fusion proteins with COOH-terminal ubiquitin are stable and maintain dual functionality *in vivo*. *J. Biol. Chem.* **277**, 38818–38826.

Reits, E. A., Benham, A. M., Plougastel, B., Neefjes, J., and Trowsdale, J. (1997). Dynamics of proteasome distribution in living cells. *EMBO J.* **16**, 6087–6094.

Reits, E. A., and Neefjes, J. J. (2001). From fixed to FRAP: Measuring protein mobility and activity in living cells. *Nat. Cell Biol.* **3**, E145–47.

Russell, S. J., Steger, K. A., and Johnston, S. A. (1999). Subcellular localization, stoichiometry, and protein levels of 26 S proteasome subunits in yeast. *J. Biol. Chem.* **274,** 21943–21952.

Varshavsky, A. (2000). Ubiquitin fusion technique and its descendants. *Methods Enzymol.* **327,** 578–593.

Voges, D., Zwickl, P., and Baumeister, W. (1999). The 26S proteasome: A molecular machine designed for controlled proteolysis. *Annu. Rev. Biochem.* **68,** 1015–1068.

Wigley, W. C., Fabunmi, R. P., Lee, M. G., Marino, C. R., Muallem, S., DeMartino, G. N., and Thomas, P. J. (1999). Dynamic association of proteasomal machinery with the centrosome. *J. Cell Biol.* **145,** 481–490.

Wilkinson, C. R., Wallace, M., Morphew, M., Perry, P., Allshire, R., Javerzat, J. P., McIntosh, J. R., and Gordon, C. (1998). Localization of the 26S proteasome during mitosis and meiosis in fission yeast. *EMBO J.* **17,** 6465–6476.

Yewdell, J. W., Reits, E., and Neefjes, J. (2003). Making sense of mass destruction: Quantitating MHC class I antigen presentation. *Nat. Rev. Immunol.* **3,** 952–961.

Zhang, J., Campbell, R. E., Ting, A. Y., and Tsien, R. Y. (2002). Creating new fluorescent probes for cell biology. *Nat. Rev. Mol. Cell Biol.* **3,** 906–918.

Section VII

Small Molecule Inhibitors

[38] Identifying Small Molecule Inhibitors of the Ubiquitin-Proteasome Pathway in *Xenopus* Egg Extracts

By ADRIAN SALIC and RANDALL W. KING

Abstract

Small molecule inhibitors of the proteasome have been crucial for dissecting the mechanism of proteasome-dependent protein degradation and identifying substrates of the ubiquitin-proteasome system (UPS). To identify small molecules that block ubiquitin-dependent protein degradation through other mechanisms, we have developed pathway-based screening approaches in *Xenopus* egg extracts. The regulated degradation of UPS substrates can be reconstituted in these extracts, providing an excellent system in which to perform forward chemical genetic screens. The ability to manipulate extracts biochemically and to compare the activity of small molecules across different assays facilitates the identification of potential target proteins. Here we describe methods for identifying inhibitors of the proteolytic pathways that regulate cell cycle progression and Wnt signaling in *Xenopus* extracts.

Introduction

Small molecule inhibitors of the proteasome have been essential for understanding the mechanism of proteasome-dependent degradation and identifying substrates and functional roles of the ubiquitin-proteasome pathway (Kisselev and Goldberg, 2001). The recent approval of a proteasome inhibitor for the treatment of multiple myeloma also highlights the importance of this pathway as a target for the development of new cancer therapies (Adams, 2004). Despite the utility of proteasome inhibitors, the biochemical complexity of ubiquitin metabolism suggests that small molecules that inhibit other steps in the pathway, such as ubiquitin chain formation or removal, will also be valuable, both as tools and as therapeutics (Pray *et al.*, 2002; Robinson and Ardley, 2004). Recently, a targeted screen has identified a class of small molecules (nutlins) that block the interaction between MDM2 and p53, thereby preventing the ability of MDM2 to catalyze p53 ubiquitination (Vassilev *et al.*, 2004). Inhibitors of ubiquitin C-terminal hydrolases have also been discovered through high-throughput screening (Liu *et al.*, 2003). Although such targeted approaches with purified proteins have been successful, it remains difficult to know which steps in the ubiquitin-proteasome pathway are most amenable to small molecule inhibition with the chemical libraries that are available today.

Chemical Genetics

Forward chemical genetics provides an opportunity to identify small molecules that inhibit a biochemical pathway without making assumptions about which step is likely to be most sensitive to inhibition (Lokey, 2003; Mayer, 2003). In addition to identifying new inhibitors, the approach can provide tools that illuminate new components of a pathway or identify unexpected steps in the pathway that are sensitive to inhibition by small molecules. We have taken a forward chemical genetic approach to identify small molecules that inhibit the UPS in *Xenopus* egg extracts. These screens, combined with reconstituted biochemical assays, led to the identification of ubistatins, small molecules that inhibit ubiquitin-dependent degradation by binding to the ubiquitin chain (Verma *et al.*, 2004).

Xenopus egg extracts provide a convenient system for performing chemical genetic screens for inhibitors of the UPS. Complex biochemical pathways can be reconstituted in these extracts, enabling many potential targets to be screened simultaneously. Unlike cell-based assays, compounds do not need to be membrane-permeable to be active. Extracts can be fractionated and biochemically manipulated to facilitate target identification, which is often the rate-limiting step in forward chemical genetics (Tochtrop and King, 2004). The ability to generate large quantities of extracts facilitates high-throughput screening of large chemical libraries (Verma *et al.*, 2004). Several screens have been performed in *Xenopus* extracts, including screens for inhibitors of actin assembly (Peterson *et al.*, 2001, 2004), spindle assembly (Wignall *et al.*, 2004), and cell cycle progression (Verma *et al.*, 2004). Targets of active molecules have been identified either by affinity purification (Wignall *et al.*, 2004) or biochemical reconstitution and candidate testing (Peterson *et al.*, 2004; Verma *et al.*, 2004).

Here we describe the application of chemical genetic methods to identify small molecule inhibitors of cyclin B degradation or β-catenin degradation. Both of these substrates are degraded by ubiquitin-dependent proteolysis, yet are targeted for ubiquitination by distinct ubiquitin ligases. Comparison of the activity of molecules in these two assays provides a convenient method for identifying potential targets.

Studies of Cell Cycle Progression and Cyclin B Proteolysis in Xenopus *Extracts. Xenopus* egg extracts have been especially useful for understanding the mechanism and regulation of cyclin B proteolysis during the cell cycle. In this system, anaphase onset and exit from mitosis require the activation of a multisubunit ubiquitin ligase, the anaphase-promoting complex/cyclosome (APC/C), which is responsible for ubiquitinating cyclin B and targeting it for destruction (Peters, 2002). The APC/C can work in concert with one of two different E2 enzymes, UbcH5 or UbcH10 (Aristarkhov *et al.*, 1996; Yu *et al.*, 1996). The APC/C is activated by mitotic phosphorylation catalyzed by cyclin B/cdc2 and also polo kinase and requires the participation of an activator protein called Fizzy or Cdc20 (Peters, 2002). Cyclin degradation requires the presence of a 9-amino acid sequence in its N-terminal domain, called the destruction box, which targets the protein to the APC/C (Glotzer *et al.*, 1991; King *et al.*, 1996). Deletion of the destruction box, or the N-terminal domain from cyclin B, results in a stable protein that cannot be recognized or ubiquitinated by the APC/C. However, this nondegradable protein remains capable of binding and activating Cdc2 and can be used to generate stably arrested mitotic extracts in which APC/C is constitutively activated (Glotzer *et al.*, 1991).

Xenopus extracts are an ideal system to screen for compounds that directly target the core cell-cycle machinery, because egg extracts lack the checkpoint pathways that normally respond to DNA damage (Dasso and Newport, 1990) or spindle damage (Minshull *et al.*, 1994) unless exogenous nuclei are added. Because small molecule libraries typically contain a large number of compounds that can perturb DNA replication or microtubule function (Mayer *et al.*, 1999), this is a great benefit, because these compounds do not inhibit cell cycle progression in *Xenopus* extracts as they would in mammalian cells.

Studies of the Wnt Pathway and β-Catenin Proteolysis in Xenopus *Extracts.* The Wnt pathway, which takes its name from the secreted Wnt signaling proteins, is one of the most ancient signaling pathways in metazoans, conserved from Hydra to humans (Logan and Nusse, 2004; Moon *et al.*, 2004). It is used repeatedly during embryonic development in many different contexts, regulating cell fate and tissue specification in embryos from the unicellular stage to late organogenesis. Wnt signaling is also important in cancer (Polakis, 2000), because most colon cancers express increased levels of β-catenin. A few dozen genes are involved in Wnt signaling, but a core module transduces extracellular Wnt signals to control the rate of degradation of β-catenin. Normally β-catenin is unstable in the absence of Wnt stimulation, because of a futile cycle of synthesis and ubiquitin-dependent degradation. To be targeted for degradation, β-catenin must be first phosphorylated on conserved serine and threonine residues clustered close to the N-terminus of the protein. A casein kinase 1 (CK1) site is phosphorylated first ("priming"), which then triggers phosphorylation by glycogen synthase kinase 3 (GSK3) (Amit *et al.*, 2002; Liu *et al.*, 2002). These phosphorylations occur in a large, multi-subunit complex built around two scaffold proteins: axin and the adenomatous polyposis coli protein (APC, not to be confused with the anaphase-promoting complex that degrades cyclin), both required for β-catenin degradation in *Xenopus* extracts (Salic *et al.*, 2000). Axin concentration is limiting for β-catenin phosphorylation and degradation in extracts. Once phosphorylated, β-catenin is recognized by the F-box protein β-TRCP (Hart *et al.*, 1999; Kitagawa *et al.*, 1999; Liu *et al.*, 1999; Winston *et al.*, 1999) and polyubiquitinated by the SCF complex, followed by proteasomal degradation. Although the exact mechanism is still unclear, a Wnt signal inhibits β-catenin phosphorylation and degradation, resulting in β-catenin accumulation and transcriptional activation of target genes. For more details, the reader is referred to several recent reviews (Logan and Nusse, 2004; Moon *et al.*, 2004; Seidensticker and Behrens, 2000).

The Wnt pathway is active in early frog development, where it plays critical roles in axis formation (Heasman *et al.*, 1994). Early embryos respond dramatically to perturbations of Wnt signaling: Wnt stimulation results in embryos with exaggerated dorsal structures, whereas Wnt inhibition generates embryos with expanded ventral structures (Heasman *et al.*, 1994; McMahon and Moon, 1989; Sokol *et al.*, 1991). We have taken advantage of the responsiveness of the egg cytoplasm to Wnt signaling and reconstituted the cytoplasmic steps of Wnt signaling in a physiological, unsimplified context (Salic *et al.*, 2000). The rate of β-catenin degradation in *Xenopus* extracts is very similar to that in embryos. β-Catenin degradation in extracts requires axin, GSK3, APC, and β-TRCP, and is inhibited by Dishevelled, consistent with the genetics of Wnt signaling. Experiments in egg extracts facilitated the biochemical dissection of the mechanism by which Dishevelled signals (Salic *et al.*, 2000) and uncovered a role for Tcf3 in β-catenin turnover (Lee *et al.*, 2001). More recently, the ability to manipulate Wnt pathway components and to make precise biochemical measurements of Wnt signaling in extracts was used to develop a mathematical model of Wnt signaling. The model accurately describes signal propagation through the Wnt pathway and has predicted several interesting features of Wnt signaling (Lee *et al.*, 2003).

Preparation of Extracts and Reporter Proteins

Preparation of Reporter Proteins

Most analyses of ubiquitin-dependent protein degradation in *Xenopus* extracts have detected endogenous substrates by immunoblotting or relied on addition of exogenous radiolabeled substrates. To develop reporter proteins suitable for high-throughput screening, we generated fusions of cyclin B or β-catenin to firefly luciferase (Deluca, 1976; Gould and Subramani, 1988). These fusion proteins permit simple determination of reporter protein level using a well-established luciferase assay that can be easily adapted to a high-throughput screening format. For small-scale screens, we have found expression of reporter proteins in reticulocyte lysate to be a convenient approach. To construct a cyclin-luciferase fusion protein (cyc-luc) for expression in reticulocyte lysate, the N-terminal sequence of *Xenopus laevis* cyclin B1, including amino acids 2–97, was amplified by PCR, digested with *BstE*II, and ligated into the pSP-lucNF expression vector (Promega). The fusion protein was expressed by coupled *in vitro* transcription and translation in reticulocyte lysate using the SP6-TNT Coupled Reticulocyte Lysate System (Promega), flash frozen in liquid nitrogen, and stored at $-80°$ until the time of use. The parental pSP-lucNF vector was used to express unmodified luciferase as a stable control protein. A similar approach was used to generate a luciferase β-catenin fusion protein (Salic *et al.*, 2000).

To express higher amounts of reporter proteins for large-scale screens, expression in *Escherichia coli* or baculovirus can be performed. A vector for expression of cyclin-luciferase in *E. coli* (pET cyc-luc) was constructed, and the *E. coli*–expressed protein was found to behave identically in all assays to the protein expressed in reticulocyte lysate. To generate this reporter protein, pSP cyc-luc was digested with *Hind*III and *Xho*I. The resulting 1949-bp fragment containing the cyclin B1-luciferase sequence was ligated into the pET 28b expression vector (Novagen) containing an N-terminal hexahistidine tag for protein purification. To express this fusion protein, 1 liter of LB containing *E. coli* strain BL21(DE3) was grown at 37° to an OD600 of 0.6. Expression was induced for 3 h with 1 mM IPTG. The cells were pelleted and lysed and protein purified by Ni-NTA batch purification under native protein conditions (Qiagen). This procedure typically yields approximately 500 μg of protein per liter of culture.

For large-scale screens of β-catenin degradation, we expressed luciferase-β-catenin through baculovirus-mediated expression in Sf9 cells (Salic *et al.*, 2000). A hexahistidine-tagged luciferase-β-catenin fusion was built in the pFastBac vector, and a baculovirus was generated using the Bac-to-Bac system (Invitrogen). The recombinant protein was purified from insect cells in high yield by standard Ni-NTA affinity chromatography. Yields were typically 10 mg per liter of cultured Sf9 cells. After dialysis against the desired buffer (XB, see later) the protein was concentrated to 1 mg/ml and aliquots flash frozen in liquid nitrogen. Although the protein remains soluble at this concentration, we have noted that precipitation occurs at higher concentrations.

Preparation of Proteins to Stimulate Protein Degradation

An important feature of the *Xenopus* system that makes it especially valuable for small molecule screening is that proteolysis can be specifically stimulated by addition of critical regulatory proteins to the extract. For example, in studies of cyclin degradation, we can activate APC/C in interphase extracts in one of two ways. The extracts can be induced to enter mitosis by addition of nondegradable cyclin B, which activates Cdc2 and stimulates mitotic phosphorylation, resulting in APC/C activation. Alternately, APC/C activity can be stimulated by adding recombinant Cdh1 to interphase extracts, which can induce cyclin proteolysis in the absence of mitotic phosphorylation (Pfleger and Kirschner, 2000). The ability to stimulate APC/C-dependent destruction by two different mechanisms provides a useful way for characterizing the mechanism of action of inhibitors discovered in the screen (Verma *et al.*, 2004).

To express nondegradable cyclin B, we generated a fusion of the maltose-binding protein to *Xenopus* cyclin B lacking its N-terminal 90 amino acids (MBP-Δ90). This protein can be expressed in a soluble form and purified according to standard procedures (New England Biolabs). We have found that MBP-Δ90 preparations are more reproducible than inclusion body preparations that use untagged sea urchin cyclin Δ90 (Glotzer *et al.*, 1991) and yield 2–3 mg/liter of culture. To express Cdh1, we use his-tagged human protein expressed in baculovirus (Pfleger and Kirschner, 2000). We have found this protein difficult to purify to homogeneity, with yields less than 1 mg/liter of culture, but even in impure form it is capable of stimulating cyclin proteolysis in interphase extracts.

In a similar manner, β-catenin degradation in *Xenopus* extracts can be significantly accelerated by supplementing the extract with recombinant axin and/or GSK3. For small-scale experiments, supplementing the extracts with axin expressed by *in vitro* translation in reticulocyte lysates works well. For large-scale screens, MBP-tagged full-length mouse axin can be purified in soluble and active form from either bacteria or Sf9 cells. For GSK3, we obtain the active his-tagged *Xenopus* protein by expression in insect cells. Alternately, the protein can be expressed and purified from bacteria as an MBP fusion.

Preparation of Extracts

Xenopus egg extracts are prepared from eggs laid overnight according to the protocol of Murray (1991) with several modifications.

A. Solutions

Extract buffer (XB): 100 mM KCl, 0.1 mM CaCl$_2$, 1 mM MgCl$_2$, 10 mM potassium HEPES (pH 7.7), 50 mM sucrose.

MMR (1X): 100 m*M* NaCl, 2 m*M* KCl, 1 m*M* MgCl$_2$, 2 m*M* CaCl$_2$, 0.1 m*M* EDTA, 5 m*M* HEPES (pH 7.8). We typically prepare a 25× solution, pH to 7.8, with NaOH and dilute to 1× just before use.

Energy mix: For a 20× stock, prepare a solution of 150 m*M* creatine phosphate, 20 m*M* ATP, 2 m*M* EGTA, and 20 m*M* MgCl$_2$ in water and adjust pH to 7.7.

Dejellying solution: Dissolve 3% w/v cysteine HCl (Sigma C-7880) in water. Titrate to pH 7.8 with NaOH. Prepare within 1 h of use.

Protease inhibitors: For a 1000× stock, prepare a mixture of leupeptin (Calbiochem NC9267778), chymostatin (MP Biomedical 15284550), and pepstatin (MP Biomedical 19536825) dissolved to a final concentration of 10 mg/ml each in DMSO. Store in aliquots at −20°.

Cytochalasin B: For a 1000× stock, dissolve cytochalasin B (MP Biomedical 19511910) at 10 mg/ml in DMSO and store in aliquots at −20°.

Calcium ionophore: For a 5000× stock, dissolve A23187, free acid form (Calbiochem 100105) at 10 mg/ml in DMSO and store in aliquots at −20°.

Cycloheximide: For a 100× stock, dissolve at 10 mg/ml in water and store in aliquots at −20°.

Pregnant mare serum gonadotropin (PMSG): 100 U/ml PMSG (Calbiochem, 367222) made up in water and stored at −20°.

Human chorionic gonadotropin (hCG): 500 U/ml hCG (Sigma CG-5) made up in water and stored at 4°.

B. Induction of Ovulation

Frogs are primed with 50 U of PMSG on day 1 and 25 U of PMSG on day 3. Ovulation can be induced by administration of 250 U of hCG on days 5–12. After injection of hCG, frogs are placed in separate 6-liter buckets containing 2 liters of 1× MMR. Frogs are allowed to lay eggs overnight (12–18 h) at 18°.

C. Preparation of Extracts

1. Frogs are removed from containers and eggs examined. If more than 10% of the eggs are white in color, have an abnormal morphology, or are bound together in strings, all of the eggs laid by the frog are discarded.

2. The remaining eggs are pooled and washed three times in 1× MMR (prechilled to 16°) to remove debris. Excess buffer is removed, and the eggs are incubated with three volumes of 3% cysteine to remove the jelly coat. Eggs should be gently swirled during the dejellying procedure, which should take approximately 5 min. Dejellying is complete when the eggs pack as dense spheres.

3. Wash eggs thoroughly in 1× MMR until the buffer remains clear, typically 5 times. Remove dead eggs (white or puffy in appearance) with a

Pasteur pipette that has been modified such that the mouth opening is wide enough to accommodate the eggs without lysing them.

4. To prepare interphase extracts, dilute calcium ionophore to 2 μg/mL in 1× MMR with rapid mixing, and immediately incubate eggs with a twofold volume excess of the diluted ionophore solution. Continue incubation until cortical contraction is observed (contraction of the pigmented area of the egg into a smaller circle), approximately 5 min. For β-catenin degradation, egg activation with ionophore is not necessary.

5. After egg activation, eggs are washed three times in XB (prechilled to 16°) and then into XB containing 1 × protease inhibitors (10 μg/ml each). For interphase extracts, the eggs are allowed to incubate for 25 min after activation. During this period, dead eggs can be removed with a pipette.

6. After the incubation period, eggs are transferred to centrifuge tubes that have been prechilled on ice. For a large preparation involving more than 25 ml of eggs, we use 50-ml centrifuge tubes (Nalgene 3110-0500) that contain 2 ml of XB with 100 μg/ml cytochalasin B and 10 μg/ml protease inhibitors. Mix cytochalasin rapidly with buffer to prevent precipitation. Eggs should be transferred to tubes in a minimum of buffer. For smaller-scale preparations, eggs can be transferred to 1.5-ml microcentrifuge tubes that contain 200 μl of buffer with diluted cytochalasin and protease inhibitor. Allow eggs to settle and remove excess buffer with a pipette.

7. Spin tubes at low speed (for 50-ml tubes, we use a Sorvall HB-6 rotor at 860 rpm) for 1 min to pack the eggs. For small preparations, a brief spin at 600 rpm in a refrigerated Microfuge is sufficient to pack the eggs. Aspirate excess buffer.

8. For interphase extracts for studies of cyclin proteolysis, we crush the eggs by spinning for 15 min at 4° in an HB-6 rotor at 12,000 rpm (23,000g) or at 14,000 rpm in a Microfuge (21,000g) for small preparations. The cytoplasmic layer (the middle layer between the yellow lipid on top and dark yolk at the bottom) is then removed by needle puncture and aspiration. Heating the needle in a flame facilitates puncture of thick-walled tubes. The extracts are placed on ice, cytochalasin and protease inhibitors are each added to 10 μg/ml, and cycloheximide is added to 100 μg/ml, and the extracts mixed thoroughly by pipetting. The extracts are then spun a second time under the same conditions, and the cytoplasmic layer is harvested as before. The extract is then supplemented with 1× energy mix and 4% glycerol, mixed well, and aliquots snap frozen in liquid nitrogen and stored at −80°. If the eggs are good quality, we generally obtain 1–2 ml of extract from the eggs laid overnight by one PMSG-primed frog.

9. For preparation of extracts to study β-catenin degradation, eggs are packed as previously but crushed by spinning for 5 min at 21,000g in a refrigerated Microfuge. For larger-scale extract preparations, eggs can be

spun at 21,000g in either an HB-6 (swinging bucket) or an SS-34 (fixed angle) rotor. The middle layer of cytoplasm is removed by using a P-1000 Pipetman (Rainin) to pierce the top lipid layer. The lipid that sticks to the sides of the pipette tip is removed with a Kimwipe before expelling the contents into a fresh chilled tube. The crude extract is supplemented with cytochalasin B (10 μg/ml), mixed well, and then spun again for 5 min at 21,000g. The cytoplasmic layer is again removed and spun a third time under identical settings. The final extract should be a clear yellow-gold color. Protease inhibitors (10 μg/ml) and energy mix are added, and aliquots are snap frozen in liquid nitrogen and stored at $-80°$. We did not find it necessary to add cryoprotectants (sucrose or glycerol) to the extracts before freezing for studies of β-catenin degradation. A good batch of extract retains its ability to degrade cyclin or β-catenin for at least 1 y, and even 2- to 3-year-old extracts can degrade β-catenin if supplemented with axin.

Assessment of Extract Quality

Extracts can vary significantly in their ability to degrade cyclin B or β-catenin. For reproducible cyclin B proteolysis, we have found it essential to adhere rigorously to the priming protocol used to induce ovulation. Although frogs can be induced to ovulate without prior PMSG priming, we have found that extracts prepared from eggs laid from unprimed frogs often do not degrade cyclin B efficiently. Also, we have found that efficient egg activation with calcium ionophore is essential for extracts to reproducibly degrade cyclin B.

Earlier protocols (Murray, 1991) use lower-speed centrifugation (10,000 rpm or 16,000g) for crushing the eggs. We have noted that such lower-speed extracts are often plagued by caspase activation that can obscure *bona-fide* ubiquitin-dependent β-catenin degradation (Fig. 1; A. Salic and E. Lee, unpublished). β-Catenin is a very good substrate for caspases (Brancolini *et al.*, 1997), and proteolysis of β-catenin by caspases generates cleavage products that are no longer degraded in the Wnt pathway. The same pattern of β-catenin cleavage is seen when extracts that do not have high caspase activity are supplemented with dATP (10 mM) or with purified cytochrome C, two known apoptotic triggers (Kluck *et al.*, 1997; Liu *et al.*, 1996). Also, caspase 3 is cleaved and activated in a similar manner in these *Xenopus* extracts (Fig. 1), further supporting the idea that β-catenin clipping in these extracts is due to caspase activation. We found that a more clarified extract produced by higher-speed centrifugation has negligible caspase activity, thus allowing robust reconstitution of β-catenin degradation without caspase interference. We speculate this is due to reduced amounts of mitochondria, which are required for triggering apoptosis in *Xenopus* extracts.

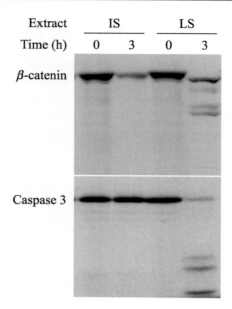

FIG. 1. Intermediate speed extracts are devoid of caspase activity. Extracts were prepared using low-speed centrifugation (LS, 16,000g) or intermediate-speed centrifugation (IS, 21,000g). Extracts were incubated with [^{35}S]-labeled β-catenin or procaspase 3 for 3 h. Aliquots were removed at the indicated times and processed by SDS-PAGE and radiography. In low-speed extracts (LS), pro-caspase 3 is activated, and β-catenin is cleaved, but the cleavage products remain stable. In intermediate speed extracts (IS), pro-caspase 3 is not activated, β-catenin is degraded by ubiquitin-dependent proteolysis.

Before embarking on a large-scale screen, we recommend making a large batch of extract and testing it for activity before initiating the screen. Extracts that generate β-catenin cleavage products indicative of caspase activity should be discarded. If the β-catenin degradation rate is too slow (half-life longer than 1–1.5 h), an extract batch can usually be "rescued" for screening by supplementing it with recombinant axin.

High-Throughput Screening

Screening for Inhibitors of Cyclin Proteolysis

We have performed small molecule screening in 384-well plates, which use about 5–10 μl of extract per well, and also in 1536-well plates, which use 2 μl of extract per well. Here we describe detailed procedures for screening in 384-well plates, but methods for screening in 1536-well plates are available in other articles (Verma et al., 2004; Walling et al., 2001).

1. Interphase extracts are thawed rapidly and diluted to a final concentration of 75% in extract buffer (XB) just before assay. Extracts are kept on ice and supplemented with 1 mg/ml bovine ubiquitin (Sigma), 10 μg/ml MBP-cyclin BΔ90, and cyclin-luciferase (cyc-luc) reporter protein. If using an *in vitro* translated reporter protein, the reticulocyte lysate is mixed directly into the *Xenopus* extract at a dilution of 1:200. If bacterially expressed cyc-luc is used instead, it is added to a final concentration of 0.1 μg/ml. It is important to mix all components well.

2. Extract is then distributed to individual wells of chilled 384-well plates. We use white, low-volume 384-well plates, such as Cliniplates (Thermo Labsystems, 95040010). Extract can be distributed using a multichannel pipetter or a plate dispenser such as the Multidrop Dispenser (Labsystems).

3. Compound libraries are stored as 5 mg/ml stocks in DMSO in 384-well polypropylene plates. For high-throughput screening, compounds are transferred to plates using an array of stainless steel pins attached to a robotic arm (Walling et al., 2001). The amount of compound transferred is determined by pin size and the speed at which the pin array is removed from the compound stock solution. We use pin arrays that transfer approximately 100–200 nl of compound per well to yield a screening concentration of 50–100 μg/ml. High compound concentrations are used because of the high protein (30–50 mg/ml) and lipid concentration of *Xenopus* extracts, resulting in a large fraction of added compound to be nonspecifically sequestered by protein binding or partitioning into membrane compartments. The contents of the wells are mixed using an orbital plate shaker followed by a brief spin in a tabletop centrifuge equipped with microplate carriers (Sorvall Legend RT). For more accurate transfer of compounds for retesting and determination of dose–response, compounds are first diluted at varying concentrations in DMSO. Compounds are then diluted 10-fold in XB and mixed well. The compound stocks are then diluted 10-fold again in extracts, adding 1 μl of compound solution to 9 μl of extract that has already been dispensed into the plate. The contents of each well are mixed with a multichannel pipette.

4. Plates are then warmed to room temperature, allowing the extracts to proceed into mitosis, activating the APC/C, and initiating cyclin degradation. After 60–75 minutes, 30 μl of luciferin reagent (20 mM tricine, pH 7.8, 470 μM D-luciferin [Molecular Probes]), 270 μM coenzyme A, 0.1 mM EDTA, 33 mM DTT, and 530 μM ATP) is added using a multidrop

dispenser (Labsystems). Luminescence is then measured on a plate reader, such as the Analyst (Molecular Devices).

5. For each screen we also prepare interphase extracts that contain the cyclin-luciferase reporter protein but lack MBP-cyclin BΔ90. These extracts remain in interphase and thus do not degrade the cyclin-luciferase reporter protein. We calculate percent inhibition as $100*(T-M)/(I-M)$, where T equals the luminescence value for the test compound in mitotic extract, M equals the value in a mitotic extract lacking inhibitor, and I equals the value in an interphase extract lacking inhibitor.

6. To characterize inhibitors identified in the screen, the preceding protocol can be modified in several ways. First, extracts can be preactivated to enter mitosis before addition of compound and reporter protein. In this case, interphase extracts are mixed with MBP-cyclin BΔ90 and allowed to incubate for 40–60 minutes at room temperature. The extract is then chilled on ice and mixed with cyc-luc. The extract is distributed to chilled plates, and compounds are then added as described previously. The plates are allowed to warm to room temperature, incubated for 30–60 min, and luciferin reagent is added and luminescence measured. In this assay, only compounds that directly interfere with the cyclin degradation machinery remain active, whereas compounds that act by blocking the transition from interphase into mitosis lose activity. Alternately, interphase extracts can be stimulated to degrade cyclin B by addition of the APC/C activator protein Cdh1. This activation step does not require mitotic phosphorylation of the APC/C, and, therefore, compounds that remain active in this assay are likely to function by direct inhibition of the cyclin proteolysis machinery rather than by blocking the transition from interphase to mitosis.

Screening for Inhibitors of β-Catenin Proteolysis

1. Extracts are thawed rapidly and placed on ice. If desired, the extract can be diluted with cold XB. Extracts are supplemented with bovine ubiquitin (0.3 mg/ml), protease inhibitors, and energy mix. Axin and GSK3 are added to 20–50 nM and 200 nM, respectively. β-Catenin-luciferase is then added at a concentration of 50 nM, and the extract is mixed thoroughly. The extract is then distributed to chilled 384-well plates as described for the cyclin degradation assay. Compounds are transferred and the plates mixed as described previously.

2. The plates are then incubated at room temperature in a closed, humidified chamber (such as a large Tupperware container) for 3 h. Luciferin solution is then added and the assay read as described previously.

Characterization of Active Compounds

Comparison of the activity of compounds across a variety of assays in *Xenopus* extracts has proved to be a useful method for determining the mechanism of action of inhibitors (Verma *et al.*, 2004). Figure 2 illustrates how known inhibitors of cell cycle progression or the Wnt pathway affect degradation of the reporter proteins in the two different assays described previously. Figure 3 demonstrates how comparison of compounds in multiple assays can be used to characterize the activity of unknown compounds discovered through high-throughput screening. Using the cyclin-luciferase assay in which degradation was stimulated from interphase extracts with nondegradable cyclin B, we screened more than 100,000 compounds to identify 22 inhibitors (Verma *et al.*, 2004). These compounds were subsequently retested in the four assays described previously. Figure 3 shows the activity of 12 of these inhibitors in the four assays. The compounds were tested in assays in which extracts were first allowed to enter mitosis before compound addition or were stimulated to degrade cyclin B-luciferase by Cdh1 addition. Finally, the compounds were tested in a β-catenin-luciferase degradation assay. As shown in Fig. 3, compounds have unique patterns of activity across each of these four assays, allowing compounds to be grouped into different functional classes. For example, compounds in class I blocked degradation in interphase but not mitotic extracts, suggesting they acted by inhibiting the transition from interphase to mitosis. A subset of these compounds (class IB) also inhibited degradation in the β-catenin assay, which may be due to inhibition of kinases such as GSK3, CK1, and cyclin B/cdc2 kinase. Other compounds, such as those in Class IIA, seem to be specific inhibitors of the cyclin proteolysis machinery, because they did not inhibit β-catenin degradation. Compounds in Class IIB inhibited both β-catenin and cyclin degradation, suggesting they inhibited a common component of the ubiquitin-proteasome pathway. Two of these compounds (C59 and C92) were identified as ubistatins, compounds that inhibit proteasome-dependent degradation by binding to the ubiquitin chain (Verma *et al.*, 2004).

Surprisingly, this screening method did not identify small molecules that directly inhibit the peptidase activity of the proteasome. We have found that inhibitors such as MG132 that target proteasome peptidase activity scored only weakly in our assay (Verma *et al.*, 2004) when we measure luciferase activity. However, immunoblotting of the reporter protein indicated that proteasome inhibitors indeed stabilize cyclin-luciferase fusion proteins in *Xenopus* extracts. One potential explanation for this finding is that the luciferase reporter may become unfolded by the proteasome when the peptidase activity is blocked, rendering the reporter protein

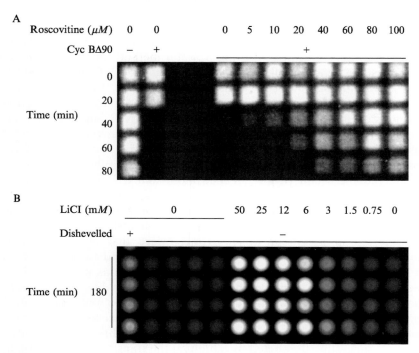

FIG. 2. Known inhibitors block the regulated degradation of cyc-luciferase and luciferase–β-catenin reporter proteins in *Xenopus* extracts. (A) Roscovitine, a small molecule inhibitor of Cdc2 kinase, blocks cyclin degradation by preventing mitotic entry. Interphase extracts containing the cyc-luciferase reporter protein were stimulated to enter mitosis by addition of nondegradable cyclin B (cyclin BΔ90), in the presence or absence of varying concentrations of roscovitine. At the indicated time points, aliquots were snap frozen. At the end of the experiment, samples were thawed and processed for luminescence imaging in a 384-well plate. The reporter protein is stable in interphase extracts that lack cyclin BΔ90, or in the presence of roscovitine, which blocks mitotic entry by inhibiting Cdc2 kinase activation. (B) Luciferase–β-catenin degradation by axin-supplemented extracts in a 384-well plate. Addition of recombinant mouse Dishevelled1 (1 μM) or lithium chloride (LiCl), a GSK3 inhibitor, block the degradation of the reporter protein in a dose-dependent manner.

inactive. Compounds such as ubistatins, which block recruitment of ubiquitinated proteins to the proteasome, stabilize luciferase reporter proteins in an active, folded conformation.

A variety of secondary assays can be performed to characterize the mechanism of inhibition. Extracts can be probed with antibodies that monitor the phosphorylation status of extract components in the presence of inhibitor. For example, CDC27, a subunit of the APC/C, undergoes a mitosis-specific change in mobility on SDS-PAGE that is due to mitotic

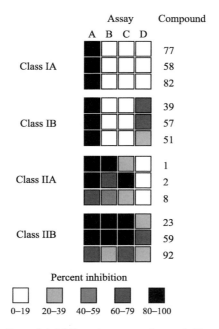

FIG. 3. Characterization of inhibitors by comparing activities in multiple assays in *Xenopus* extracts. More than 100,000 compounds were screened for the ability to block cyc-luciferase turnover in interphase extracts stimulated to enter mitosis with nondegradable cyclin B (Verma *et al.*, 2004). The most active compounds were then retested at 200 μM concentration in several different assays. (A) Activity of compounds in the original assay. (B) Extracts were allowed to enter mitosis before addition of compound and reporter protein. (C) Interphase extracts were stimulated to degrade cyc-luciferase by addition of Cdh1. (D) Determination of activity in the luciferase–β-catenin assay.

phosphorylation (King *et al.*, 1995; Peters *et al.*, 1996). Analysis of class I and II inhibitors showed that class I compounds blocked CDC27 phosphorylation, consistent with inhibition of mitotic entry, whereas class II compounds resulted in mitotic arrest with sustained CDC27 phosphorylation, consistent with specific inhibition of the cyclin proteolysis machinery (N. Peters and R. King, unpublished data). Compounds can be tested directly in a variety of reconstituted biochemical assays for direct effects on kinase inhibition, APC/C-dependent ubiquitination (King *et al.*, 1995), or proteasome-dependent degradation (Verma *et al.*, 2004).

Similar approaches can be used to further study compounds that inhibit β-catenin degradation. Phospho-specific antibodies are available to assay the status of β-catenin phosphorylation by GSK3 or CK1 (Amit *et al.*, 2002;

Liu et al., 2002). To test whether a compound inhibits GSK3 or CK1, β-catenin can be first probed with phospho-specific antibodies directly in the extract. If inhibition of specific phosphorylation is detected, the compound(s) can be tested with *in vitro* kinase assays with purified components. Axin stimulates β-catenin phosphorylation by both GSK3 and CK1; it would, thus, be best to perform the kinase assay using purified kinase, axin, and β-catenin, rather than just assaying the phosphorylation of a peptide substrate by the kinase. This more complex kinase assay might also identify inhibitors that interfere with the kinase-axin or axin–β-catenin interactions that would be missed in a simple peptide phosphorylation assay. Protein–protein interactions in the β-catenin degradation complex are regulated by phosphorylation, such as the binding of β-catenin to APC that is stimulated by APC phosphorylation. To test whether a compound inhibits a protein–protein interaction, gel filtration or density gradient centrifugation can be performed to determine whether the size of the known complex is affected by the small molecule. Another interaction required for β-catenin degradation is recognition of phosphorylated β-catenin by the F-box protein β-TRCP. Compounds that stabilize β-catenin by blocking this interaction could be identified by performing binding assays between β-TRCP on beads and recombinant β-catenin phosphorylated *in vitro* with axin, CK1, and GSK3. Extracts also allow quick "epistasis" experiments to narrow down the level at which a given inhibitor acts. If a compound inhibits β-catenin degradation by blocking its recognition by β-TRCP, adding more axin and/or GSK3 would not be expected to rescue the effect; if, however a compound inhibits β-catenin degradation upstream of or at the level of the degradation complex, increasing axin levels will likely reverse the effect of the compound.

Ubiquitin conjugates of cyclin B and β-catenin are ultimately degraded by the proteasome. Therefore, small molecules that inhibit protein degradation downstream of substrate ubiquitination are likely to score in both the luciferase–β-catenin and cyclin-luciferase screens. The activity of such molecules can be characterized in peptidase assays that monitor catalytic activity of the 20S proteasome core or in reconstituted assays using ubiquitinated proteins and purified 26S proteasomes (Verma et al., 2004).

Acknowledgments

A. S. was a Damon-Runyon postdoctoral fellow and is currently supported through a fellowship from the Charles King Trust of the Medical Foundation. R. W. K. is a Damon Runyon Scholar, and work in the laboratory is funded by grants from the NIH (CA78048, GM66492), the McKenzie Family Foundation, and the Harvard-Armenise Foundation.

References

Adams, J. (2004). The development of proteasome inhibitors as anticancer drugs. *Cancer Cell.* **5**, 417–421.

Amit, S., Hatzubai, A., Birman, Y., Andersen, J. S., Ben-Shushan, E., Mann, M., Ben-Neriah, Y., and Alkalay, I. (2002). Axin-mediated CKI phosphorylation of beta-catenin at Ser 45: A molecular switch for the Wnt pathway. *Genes Dev.* **16**, 1066–1076.

Aristarkhov, A., Eytan, E., Moghe, A., Admon, A., Hershko, A., and Ruderman, J. V. (1996). E2-C, a cyclin-selective ubiquitin carrier protein required for the destruction of mitotic cyclins. *Proc. Natl. Acad. Sci. USA* **93**, 4294–4299.

Brancolini, C., Lazarevic, D., Rodriguez, J., and Schneider, C. (1997). Dismantling cell-cell contacts during apoptosis is coupled to a caspase-dependent proteolytic cleavage of beta-catenin. *J. Cell Biol.* **139**, 759–771.

Dasso, M., and Newport, J. W. (1990). Completion of DNA replication is monitored by a feedback system that controls the initiation of mitosis *in vitro*: Studies in *Xenopus*. *Cell* **61**, 811–823.

Deluca, M. (1976). Firefly luciferase. *Adv. Enzymol. Relat. Areas Mol. Biol.* **44**, 37–68.

Glotzer, M., Murray, A. W., and Kirschner, M. W. (1991). Cyclin is degraded by the ubiquitin pathway. *Nature* **349**, 132–138.

Gould, S. J., and Subramani, S. (1988). Firefly luciferase as a tool in molecular and cell biology. *Anal. Biochem.* **175**, 5–13.

Hart, M., Concordet, J. P., Lassot, I., Albert, I., del los Santos, R., Durand, H., Perret, C., Rubinfeld, B., Margottin, F., Benarous, R., and Polakis, P. (1999). The F-box protein beta-TrCP associates with phosphorylated beta-catenin and regulates its activity in the cell. *Curr. Biol.* **9**, 207–210.

Heasman, J., Crawford, A., Goldstone, K., Garner-Hamrick, P., Gumbiner, B., McCrea, P., Kintner, C., Noro, C. Y., and Wylie, C. (1994). Overexpression of cadherins and underexpression of beta-catenin inhibit dorsal mesoderm induction in early *Xenopus* embryos. *Cell* **79**, 791–803.

King, R. W., Glotzer, M., and Kirschner, M. W. (1996). Mutagenic analysis of the destruction signal of mitotic cyclins and structural characterization of ubiquitinated intermediates. *Mol. Biol. Cell.* **7**, 1343–1357.

King, R. W., Peters, J. M., Tugendreich, S., Rolfe, M., Hieter, P., and Kirschner, M. W. (1995). A 20S complex containing CDC27 and CDC16 catalyzes the mitosis-specific conjugation of ubiquitin to cyclin B. *Cell* **81**, 279–288.

Kisselev, A. F., and Goldberg, A. L. (2001). Proteasome inhibitors: From research tools to drug candidates. *Chem. Biol.* **8**, 739–758.

Kitagawa, M., Hatakeyama, S., Shirane, M., Matsumoto, M., Ishida, N., Hattori, K., Nakamichi, I., Kikuchi, A., Nakayama, K., and Nakayama, K. (1999). An F-box protein, FWD1, mediates ubiquitin-dependent proteolysis of beta-catenin. *EMBO J.* **18**, 2401–2410.

Kluck, R. M., Bossy-Wetzel, E., Green, D. R., and Newmeyer, D. D. (1997). The release of cytochrome c from mitochondria: A primary site for Bcl-2 regulation of apoptosis. *Science* **275**, 1132–1136.

Lee, E., Salic, A., and Kirschner, M. W. (2001). Physiological regulation of [beta]-catenin stability by Tcf3 and CK1epsilon. *J. Cell Biol.* **154**, 983–993.

Lee, E., Salic, A., Kruger, R., Heinrich, R., and Kirschner, M. W. (2003). The roles of APC and axin derived from experimental and theoretical analysis of the Wnt pathway. *PLoS Biol.* **1**, E10.

Liu, C., Kato, Y., Zhang, Z., Do, V. M., Yankner, B. A., and He, X. (1999). beta-Trcp couples beta-catenin phosphorylation-degradation and regulates Xenopus axis formation. *Proc. Natl. Acad. Sci. USA* **96,** 6273–6278.

Liu, C., Li, Y., Semenov, M., Han, C., Baeg, G. H., Tan, Y., Zhang, Z., Lin, X., and He, X. (2002). Control of beta-catenin phosphorylation/degradation by a dual-kinase mechanism. *Cell* **108,** 837–847.

Liu, X., Kim, C. N., Yang, J., Jemmerson, R., and Wang, X. (1996). Induction of apoptotic program in cell-free extracts: Requirement for dATP and cytochrome c. *Cell* **86,** 147–157.

Liu, Y., Lashuel, H. A., Choi, S., Xing, X., Case, A., Ni, J., Yeh, L. A., Cuny, G. D., Stein, R. L., and Lansbury, P. T., Jr. (2003). Discovery of inhibitors that elucidate the role of UCH-L1 activity in the H1299 lung cancer cell line. *Chem. Biol.* **10,** 837–846.

Logan, C. Y., and Nusse, R. (2004). The Wnt signaling pathway in development and disease. *Annu. Rev. Cell. Dev. Biol.* **20,** 781–810.

Lokey, R. S. (2003). Forward chemical genetics: Progress and obstacles on the path to a new pharmacopoeia. *Curr. Opin. Chem. Biol.* **7,** 91–96.

Mayer, T. U. (2003). Chemical genetics: Tailoring tools for cell biology. *Trends Cell Biol.* **13,** 270–277.

Mayer, T. U., Kapoor, T. M., Haggarty, S. J., King, R. W., Schreiber, S. L., and Mitchison, T. J. (1999). Small molecule inhibitor of mitotic spindle bipolarity identified in a phenotype-based screen. *Science* **286,** 971–974.

McMahon, A. P., and Moon, R. T. (1989). Ectopic expression of the proto-oncogene int-1 in *Xenopus* embryos leads to duplication of the embryonic axis. *Cell* **58,** 1075–1084.

Minshull, J., Sun, H., Tonks, N. K., and Murray, A. W. (1994). A MAP kinase-dependent spindle assembly checkpoint in *Xenopus* egg extracts. *Cell* **79,** 475–486.

Moon, R. T., Kohn, A. D., De Ferrari, G. V., and Kaykas, A. (2004). WNT and beta-catenin signalling: Diseases and therapies. *Nat. Rev. Genet.* **5,** 691–701.

Murray, A. W. (1991). Cell cycle extracts. *Methods Cell Biol.* **36,** 581–605.

Peters, J. M. (2002). The anaphase-promoting complex: Proteolysis in mitosis and beyond. *Mol. Cell* **9,** 931–943.

Peters, J. M., King, R. W., Hoog, C., and Kirschner, M. W. (1996). Identification of BIME as a subunit of the anaphase-promoting complex. *Science* **274,** 1199–1201.

Peterson, J. R., Bickford, L. C., Morgan, D., Kim, A. S., Ouerfelli, O., Kirschner, M. W., and Rosen, M. K. (2004). Chemical inhibition of N-WASP by stabilization of a native autoinhibited conformation. *Nat. Struct. Mol. Biol.* **11,** 747–755.

Peterson, J. R., Lokey, R. S., Mitchison, T. J., and Kirschner, M. W. (2001). A chemical inhibitor of N-WASP reveals a new mechanism for targeting protein interactions. *Proc. Natl. Acad. Sci. USA* **98,** 10624–1069.

Pfleger, C. M., and Kirschner, M. W. (2000). The KEN box: An APC recognition signal distinct from the D box targeted by Cdh1. *Genes Dev.* **14,** 655–665.

Polakis, P. (2000). Wnt signaling and cancer. *Genes Dev.* **14,** 1837–1851.

Pray, T. R., Parlati, F., Huang, J., Wong, B. R., Payan, D. G., Bennett, M. K., Issakani, S. D., Molineaux, S., and Demo, S. D. (2002). Cell cycle regulatory E3 ubiquitin ligases as anticancer targets. *Drug Resist. Update.* **5,** 249–258.

Robinson, P. A., and Ardley, H. C. (2004). Ubiquitin-protein ligases—novel therapeutic targets? *Curr. Protein Pept. Sci.* **5,** 163–176.

Salic, A., Lee, E., Mayer, L., and Kirschner, M. W. (2000). Control of beta-catenin stability: Reconstitution of the cytoplasmic steps of the wnt pathway in *Xenopus* egg extracts. *Mol. Cell.* **5,** 523–532.

Seidensticker, M. J., and Behrens, J. (2000). Biochemical interactions in the wnt pathway. *Biochim. Biophys. Acta* **1495,** 168–182.

Sokol, S., Christian, J. L., Moon, R. T., and Melton, D. A. (1991). Injected Wnt RNA induces a complete body axis in *Xenopus* embryos. *Cell* **67,** 741–752.
Tochtrop, G. P., and King, R. W. (2004). Target identification strategies in chemical genetics. *Comb. Chem. High Throughput Screen.* **7,** 677–688.
Vassilev, L. T., Vu, B. T., Graves, B., Carvajal, D., Podlaski, F., Filipovic, Z., Kong, N., Kammlott, U., Lukacs, C., Klein, C., Fotouhi, N., and Liu, E. A. (2004). In vivo activation of the p53 pathway by small-molecule antagonists of MDM2. *Science* **303,** 844–848.
Verma, R., Peters, N. R., D'Onofrio, M., Tochtrop, G. P., Sakamoto, K. M., Varadan, R., Zhang, M., Coffino, P., Fushman, D., Deshaies, R. J., and King, R. W. (2004). Ubistatins inhibit proteasome-dependent degradation by binding the ubiquitin chain. *Science* **306,** 117–120.
Walling, L. A., Peters, N. R., Horn, E. J., and King, R. W. (2001). New technologies for chemical genetics. *J. Cell. Biochem.* **S37,** 7–12.
Wignall, S. M., Gray, N. S., Chang, Y. T., Juarez, L., Jacob, R., Burlingame, A., Schultz, P. G., and Heald, R. (2004). Identification of a novel protein regulating microtubule stability through a chemical approach. *Chem. Biol.* **11,** 135–146.
Winston, J. T., Strack, P., Beer-Romero, P., Chu, C. Y., Elledge, S. J., and Harper, J. W. (1999). The SCF-β-TRCP-ubiquitin ligase complex associates specifically with phosphorylated destruction motifs in IκB-α and β-catenin and stimulates IκB-α ubiquitination in vitro. *Genes Dev.* **13,** 270–283.
Yu, H., King, R. W., Peters, J. M., and Kirschner, M. W. (1996). Identification of a novel ubiquitin-conjugating enzyme involved in mitotic cyclin degradation. *Curr. Biol.* **6,** 455–466.

[39] Development and Characterization of Proteasome Inhibitors

By KYUNG BO KIM, FABIANA N. FONSECA, and CRAIG M. CREWS

Abstract

Although many proteasome inhibitors have been either synthesized or identified from natural sources, the development of more sophisticated, selective proteasome inhibitors is important for a detailed understanding of proteasome function. We have found that antitumor natural product epoxomicin and eponemycin, both of which are linear peptides containing a α,β-epoxyketone pharmacophore, target proteasome for their antitumor activity. Structural studies of the proteasome–epoxomicin complex revealed that the unique specificity of the natural product toward proteasome is due to the α,β-epoxyketone pharmacophore, which forms an unusual six-membered morpholino ring with the amino terminal catalytic Thr-1 of the 20S proteasome. Thus, we believe that a facile synthetic approach for α,β-epoxyketone linear peptides provides a unique opportunity to develop proteasome inhibitors with novel activities. In this chapter, we discuss the detailed synthetic procedure of the α',β'-epoxyketone natural product epoxomicin and its derivatives.

Introduction

Studies into the chemical synthesis of proteasome inhibitors were initiated during the early 1990s after standard serine/cysteine protease inhibitors were shown to inhibit the 20S proteasome (Figueiredo-Pereira et al., 1994; Orlowski, 1990). Early synthetic efforts were largely focused on the modification of the amino acid sequence of the serine/cysteine protease inhibitors. For example, MG115 and MG132, which were developed by Rock and colleagues (1994) and have been widely used in proteasome biology, are tripeptide aldehydes that share similar peptide backbones and aldehyde pharmacophores with known protease inhibitors such as calpain inhibitor 1 (Fig. 1). Other known peptide-based protease inhibitors possessing different pharmacophores such as vinylketones (Bogyo et al., 1997, 1998) and boronates (Adams, 2002; Adams et al., 1998; Iqbal et al., 1996) have also been developed as proteasome inhibitors (Fig. 1). The major advantage of these peptide backbone–based proteasome inhibitors is their ease of preparation and derivatization, potentially providing an easy access for the development of proteasome inhibitors with novel activities. Although these peptide inhibitors, in general, are cell-permeable potent inhibitors of the 20S proteasome and are still widely used in the study of the role of proteasome in many cellular processes, the cross-reactivity with other proteases remains a major concern for their use as

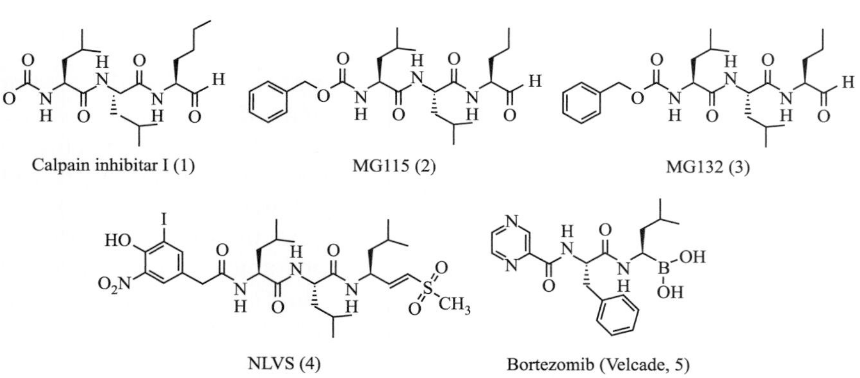

FIG. 1. Proteasome inhibitors derived from other known peptide-based protease inhibitors possessing common pharmacophores.

molecular probes in dissecting complex signaling pathways (Kim and Crews, 2003; Kisselev and Goldberg, 2001; Myung et al., 2001a).

In addition to synthetic approaches, natural products have also provided both peptide and nonpeptide proteasome inhibitors (Kim and Crews, 2003). Examples include epoxomicin (Hanada et al., 1992), eponemycin (Sugawara et al., 1990), lactacystin (Fenteany et al., 1995), TMC-95s (Koguchi et al., 2000a), phepropeptins (Sekizawa et al., 2001), and epigallocatechin-3-gallate (ECGC) (Nam et al., 2001). Among these, epoxomicin and eponemycin are members of a growing family of α',β'-epoxyketone natural products having a linear peptide structure (Fig. 2) (Koguchi et al., 1999, 2000b,c; Sugawara et al., 1990; Tsuchiya et al., 1997). We reported the first total synthesis of epoxomicin (Sin et al., 1999), isolated from an unidentified actinomycete strain No.Q996–17, and showed that epoxomicin potently inhibits the 20S proteasome (Meng et al., 1999b) using an affinity

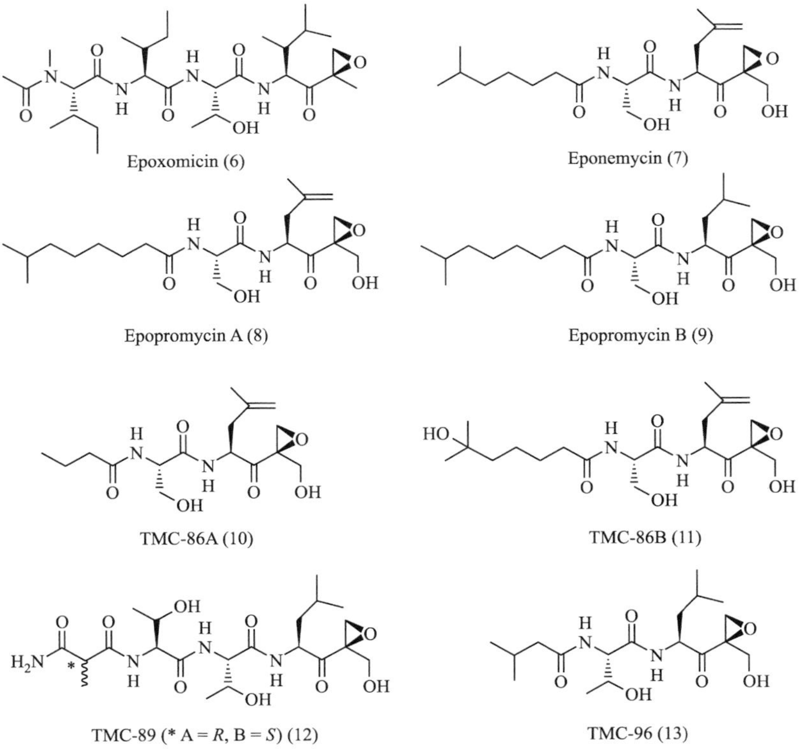

FIG. 2. α',β'-Epoxyketone-containing proteasome inhibitors from natural sources.

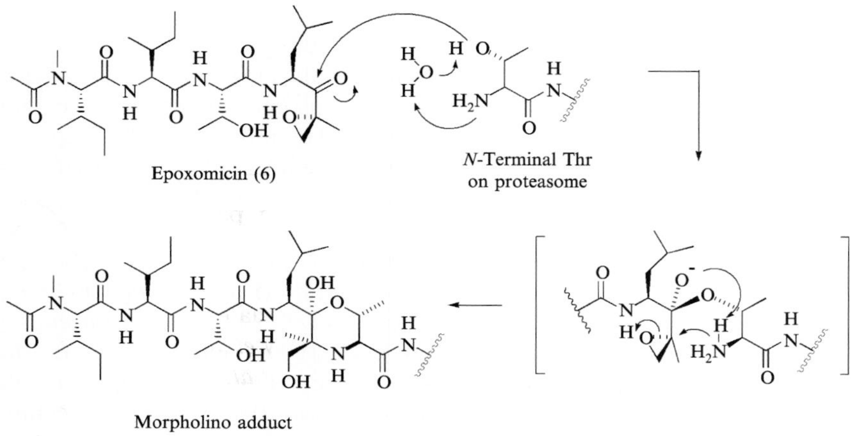

FIG. 3. The mechanism of proteasome inhibition by epoxomicin is proposed on the basis of the x-ray structure of yeast 20S proteasome–epoxomicin complex. It is postulated that the unique specificity of epoxomicin is due to the formation of an unusual six-membered morpholino ring between Thr-1 of the catalytic subunit of 20S proteasome and the α',β'-epoxyketone pharmacophore of epoxomicin.

reagent–labeled epoxomicin (Meng et al., 1999a; Sin et al., 1998). Interestingly, antitumor natural products epoxomicin and eponemycin are shown to be specific for the proteasome despite their common peptide backbone that resembles structures of known serine/cysteine protease inhibitors (Meng et al., 1999b; Sin et al., 1999). Structural studies of the yeast 20S proteasome complexed with epoxomicin revealed that the unique specificity of epoxomicin is due to the formation of an unusual six-membered morpholino ring between the amino terminal catalytic Thr-1 of the 20S proteasome and the α',β'-epoxyketone pharmacophore of epoxomicin (Fig. 3) (Groll et al., 2000). The facile synthetic strategy of α',β'-epoxyketone linear peptides developed through the total synthesis of epoxomicin and the unique specificity of its α',β'-epoxyketone pharmacophore for proteasome have prompted the development of proteasome inhibitors possessing higher potency or novel inhibitory specificities, such as YU101 (Elofsson et al., 1999) and YU102 (Myung et al., 2001b) (Fig. 4), respectively.

In this chapter, we discuss the synthesis and characterization of this important class of proteasome inhibitors, focusing on the α',β'-epoxyketone natural product epoxomicin and its derivatives. It should be noted that the synthetic strategy of epoxomicin described in this chapter can be applied easily to the development of proteasome inhibitors with novel activities.

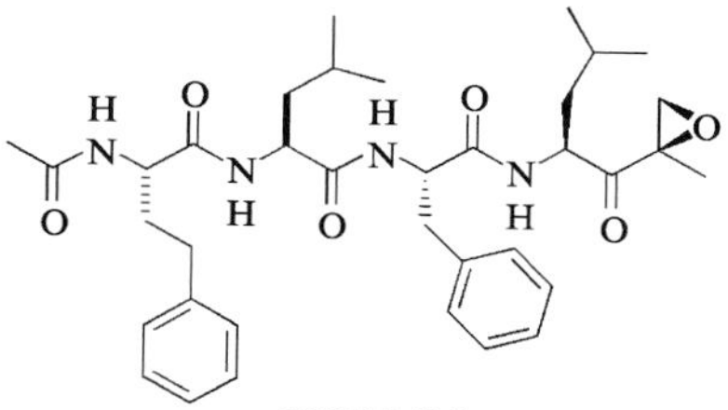

YU101 (14) 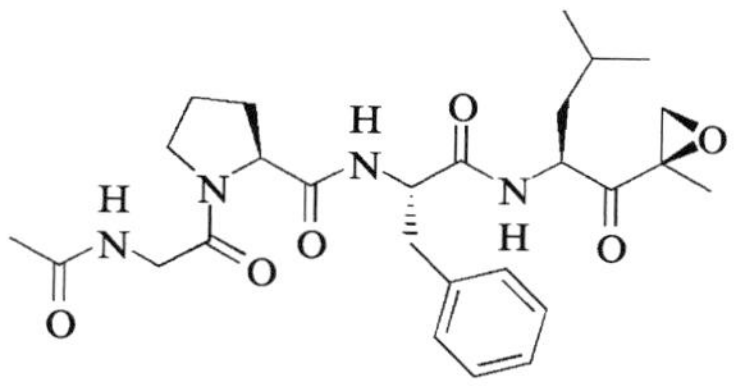   YU102 (15)

Fig. 4. Synthetic α′,β′-epoxyketone proteasome inhibitors designed to target a certain proteasomal proteolytic activity/subunit with a high degree of specificity. YU101 is a chymotrypsin-like activity (CT-L)-selective inhibitor, whereas YU102 is shown to be specific for caspase-like activity.

Synthesis Strategy of α′,β′-Epoxyketone Peptide Inhibitors: Total Synthesis of Epoxomicin

General

There are two main strategies for the chemical synthesis of linear peptides (Fields, 1997; Lloyd-Williams *et al.*, 1997). First, chain elongation in linear synthesis is carried out by repetitive N^α-amino group deprotection and protected amino acid coupling steps. Alternately, convergent synthesis involves the synthesis and coupling of protected peptide segments. Both strategies can be carried out in solution or on a solid support, although solution and solid-phase methods can coexist in convergent synthesis. For the synthesis of epoxomicin, chain elongation on solid support may be a challenging task because of the C-terminus epoxyketone pharmacophore, which lacks a handle for resin attachment. Solution phase chain elongation was also ruled out because of the possibility of decomposition of epoxyketone group during repetitive coupling and deprotection reactions. On the basis of these considerations, a convergent approach was chosen for the total synthesis of epoxomicin (Fig. 5).

The convergent approach in the synthesis of epoxomicin involves: (1) solution-phase synthesis of the right-hand fragment, (2) solution or solid-phase synthesis of left-hand fragment, (3) solution-phase assembly of complete peptide backbone of epoxomicin, and (4) purification and characterization of the coupled product. After the final coupling, HPLC purification is carried out to separate epoxomicin from the stereoisomeric counterpart generated during the final coupling reaction. Finally, the enzymatic inhibitory activity of epoxomicin was measured using purified bovine proteasome.

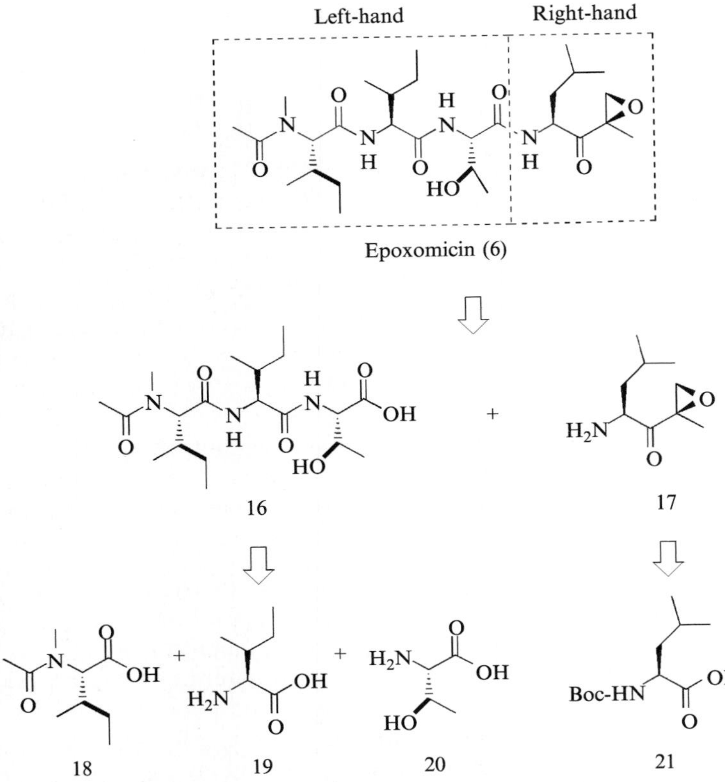

FIG. 5. A convergent approach for the total synthesis of epoxomicin.

Synthesis of the Right-Hand Fragment of Epoxomicin

There are two alternative approaches to introduce an epoxy group in the course of synthesis of epoxomicin. First, the epoxidation reaction can be performed after complete assembly of the peptide backbone of epoxomicin, and the resulting desired product is purified from the mixture of epoxomicin and *epi*-epoxomicin by HPLC (Schmidt and Schmidt, 1994) (Fig. 6A). The second approach relies on the preparation and isolation of stereochemically defined α',β'-epoxyketone leucine (Bennacer *et al.*, 2003; Dobler, 2001; Hoshi *et al.*, 1993; Iwabuchi *et al.*, 2001; Sin *et al.*, 1998), which is then coupled to the tripeptide left-hand fragment to yield the complete backbone of epoxomicin (Fig. 6B). Because the α',β'-epoxyketone leucine having the same configuration as that of epoxomicin can be

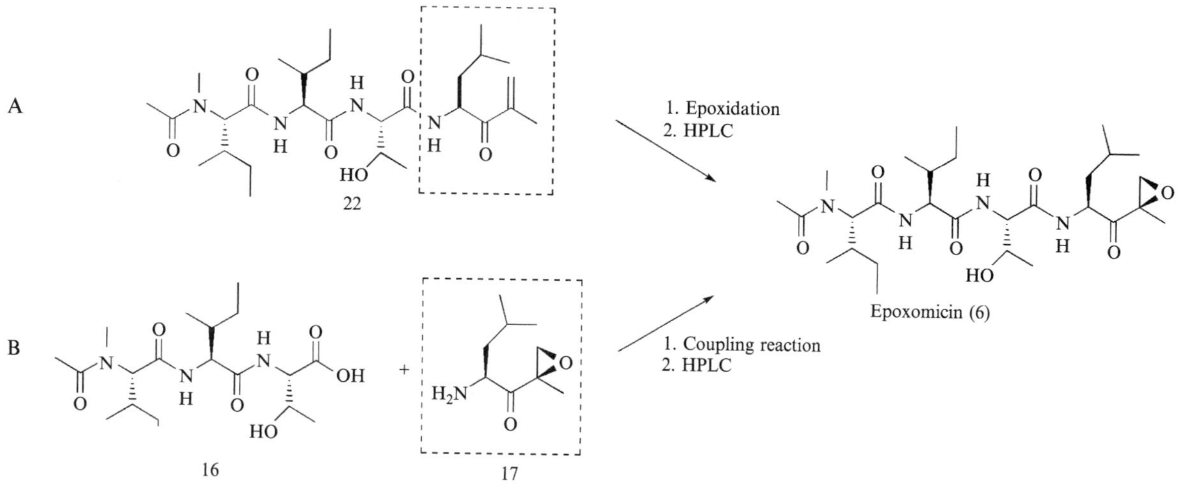

FIG. 6. Two potential strategies for the introduction of a α',β'-epoxyketone group.

easily chromatographically purified from its stereoisomeric counterpart, the second approach was initially chosen to prepare the right-hand fragment of epoxomicin (Fig. 7).

As the first step of the right-hand fragment synthesis, α-*tert*-butyloxycarbonyl (Boc)-leucine (Boc-Leu-OH) was coupled to *N*-methoxy-*N*-methylamine with *O*-benzotriazo-1-yl-*N*,*N*,*N'*, *N'*-tetramethyluronium hexafluorophosphate (HBTU) and 1-hydroxybenzotriazole (HOBt) to yield the Boc-leucine Weinreb amide 23. Readily preparable Weinreb amides (*N*-methoxy-*N*-methylamides) (Nahm and Weinreb, 1981) are known to couple in good yields with Grignard and organolithium reagents to produce ketones, and to be reduced with hydrides to afford aldehydes. Fluoren-9-ylmethoxycarbonyl-leucine (Fmoc-Leu-OH), however, cannot be used, because the Fmoc group is unstable under Grignard or organolithium reaction conditions, which are used to introduce the α',β'-unsaturated ketone 24 in the following step.

Reaction of α-Boc-leucine-Weinreb amide 23 with commercially available propen-2-yl magnesium bromide (Grignard reagent) led to the formation of α-Boc-amino-α',β'-unsaturated ketone 24 without racemization at low temperature ($-78°$ to room temperature) in tetrahydrofuran (THF). The nucleophilic addition proceeds through stable metal-chelate intermediates that block over-addition. The resulting α-Boc-amino-α',β'-unsaturated ketone 24 was readily purified by flash column chromatography using hexanes-ethyl acetate system (10:1, v/v) as eluant. Subsequent epoxidation of α-Boc-leucine-α',β'-unsaturated ketone with alkaline hydrogen peroxide (Schmidt and Schmidt, 1994) in methanol yielded a mixture of epoxide stereoisomers; the ratio of 2-(*R*)-epoxide 25a to 2-(*S*)-epoxide 25b was 1.7. The two isomers of leucine epoxyketone were readily separated by

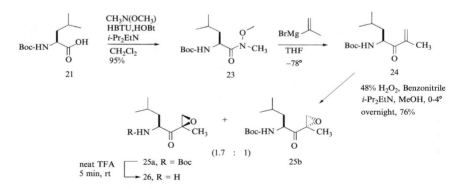

FIG. 7. Synthesis of the right-hand fragment (leucine epoxyketone).

flash column chromatography using hexanes-ethyl acetate (5:1, v/v) system. The isomer (2-(R)-epoxide) 25a, which migrates faster than the 2-(S)-epoxide 25b in thin-layer chromatography (TLC, hexanes-ethyl acetate = 5:1, v/v), was found to have the same configuration as that of epoxomicin (Sin et al., 1999). Finally, the Boc group of Boc-Leu-α',β'-epoxyketone 25a was deprotected by brief treatment (~5 min) with neat trifluoroacetic acid (TFA) without concomitant opening of the epoxide ring. Excess TFA was removed under high vacuum. The resulting TFA salt of leucine-α',β'-epoxyketone was used without further purification for the final coupling reaction with the left-hand fragment.

Synthesis of the Left-Hand Fragment

The tripeptide left-hand fragment of epoxomicin can be readily prepared either in solution or on a solid support (Sin et al., 1999). Given that peptide synthesis on a solid support has been a subject of many extensive reviews (Fields, 1997), in this chapter we discuss only the solution-phase synthesis.

In the first step (Fig. 8), Fmoc-isoleucine-OH was coupled to threonine benzylester 27 with HBTU and HOBt. Although there are many coupling reagents that may be used for peptide bond formation, HBTU/HOBt coupling system was found to be very effective for reducing loss of configuration at the carboxylic acid residue in this coupling reaction. The resulting dipeptide 28 was purified by flash column chromatography using hexanes-ethyl acetate system (1:1, v/v) as eluant. The solubility of the dipeptide 28 was so poor that it was readily precipitated during flash column chromatography, requiring a large volume of elution solvent to redissolve and collect the dipeptide. The protection of threonine hydroxyl group of the dipeptide 28 was then accomplished with *tert*-butyldiphenylsilylchloride (TBDPSCl) in methylene chloride (CH_2Cl_2) at room temperature. The same TBDPSCl-protection reaction, however, was less effective in tetrahydrofuran (THF). The advantages of using TBDPS protecting group compared with other commonly used protecting groups such as *tert*-butyldimethylsilyl (TBDMS) and trimethylsilyl (TMSCl) are: (1) enhanced solubility of TBDPS protected peptides in organic solvents facilitating the following solution-phase reactions and purification process; (2) better stability under acid and base conditions; (3) easy UV detection provided by the two phenyl rings of TBDPS group, thus aiding the HPLC purification of completely assembled TBDPS protected epoxomicin from the stereoisomeric counterpart generated during the final coupling of the left- and right-hand fragments.

The Fmoc group of TBDPS-protected dipeptide 29 was deprotected using standard protocol (20% piperidine in dimethylformamide, v/v)

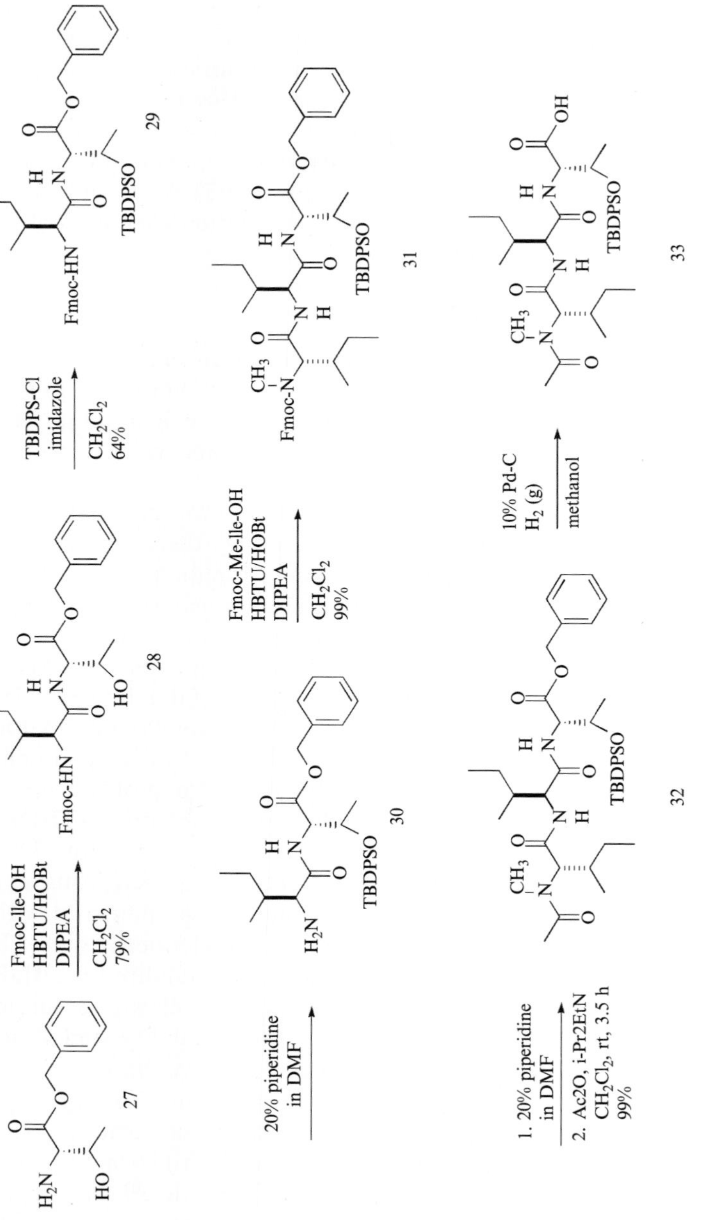

FIG. 8. Preparation of the tripeptide left-hand fragment.

(Fig. 8). The resulting Fmoc-deprotected dipeptide 30 was purified by flash column chromatography. In this type of Fmoc deprotection reaction, only one fourth or one fifth of the normal quantity of dried silica (SiO_2) that is used for routine column chromatography was packed for the purification of the Fmoc-deprotected product. Specifically, after directly loading the Fmoc-deprotected crude product into the silica gel column, fast-migrating Fmoc-adducts, which are shown as bright fluorescent spots under UV, were eluted away using a solvent system (hexanes-ethyl acetate, 1:1, v/v); under this elution condition, the Fmoc-deprotected product 30 was retained in the silica gel column. The retained deprotected dipeptide 30 was then eluted and collected using a different solvent system (CH_2Cl_2-MeOH, 9:1, v/v).

Next, Fmoc-N-methyl-isoleucine was coupled to the deprotected dipeptide 30 with HBTU/HOBt to yield the tripeptide 31, which was easily purified by flash column chromatography (hexanes-ethyl acetate, 1:1, v/v). After the Fmoc-protected tripeptide 31 was treated with 20% piperidine in DMF (v/v) for 20 min at room temperature, DMF and piperidine were removed under high vacuum. Without further purification, the crude product was mixed with excess acetic anhydride in pyridine for 1 h at room temperature. The resulting N-acetylated tripeptide 32 was purified by flash column chromatography (hexanes-ethyl acetate, 1:1, v/v). Finally, the C-terminus benzyl group of the left-hand fragment 32 was deprotected by catalytic hydrogenolysis mediated by 10% activated palladium-charcoal in methanol under hydrogen gas atmosphere. In this reaction, instead of performing hydrogenolysis under high pressure, a stream of hydrogen gas was directly applied to the reaction mixture providing hydrogen bubbles into the solution (Fig. 9). The benzyl protecting group was readily removed in nearly quantitative yield, and the reaction mixture was filtered through a coarse, Celite-packed, fritted-glass filter to afford the TBDPS-protected left-hand fragment 33, which was coupled to the right-hand fragment without further purification for the final coupling with the right-hand fragment.

Assembly of Complete Epoxomicin Peptide Backbone

The final coupling reaction between the right- and left-hand fragments of epoxomicin was performed with O-(7-azabenzotriazol-1-yl)-1,1,3,3-tetramethyluronium hexafluorophosphate (HATU) and 1-hydroxy-7-azabenzotriazol (HOAt) to give TBDPS protected epoxomicin 34b (Fig. 10). Unlike the coupling reactions performed during the synthesis of left-hand fragment, HATU was used for the final assembly of epoxomicin peptide backbone to enhance the efficiency of coupling reaction. HATU was developed by Carpino (Albericio and Carpino, 1997; Carpino, 1993) and

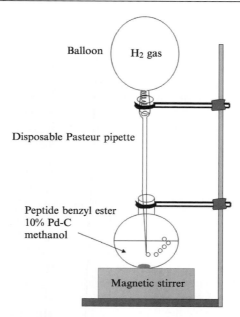

FIG. 9. Schematic of hydrogenation reaction, in which hydrogen gas is provided to the reaction mixture through a stream of hydrogen gas.

shown to be particularly effective with hindered couplings. After overnight stirring at room temperature, the resulting TBDPS protected epoxomicin 34b was initially purified by flash column chromatography (hexanes-ethyl acetate, 1:1, v/v) to verify the successful assembly of complete peptide backbone of epoxomicin by 1H NMR. 1H NMR spectra analysis showed the complete assembly of epoxomicin peptide backbone and the presence of minor stereoisomer 34a (5–15% of the assembled product), the formation of which during the amide bond formation is well documented (Carpino, 1993, 1997). It is assumed that the loss of configuration occurs at the carboxylic acid residue of P2 amino acid during the final coupling reaction (Fig. 10). It was difficult to purify the TBDPS protected epoxomicin 34b from the isomer 34a using standard column chromatography, but it was readily separated by normal-phase HPLC (hexanes-isopropanol, linear gradient, hexanes 100% to 50%). In normal-phase HPLC, the TBDPS-protected epoxomicin 34b displayed a longer retention time compared with its stereoisomeric counterpart 34a. Finally, treatment of the TBDPS protected epoxomicin with tetrabutylammonium fluoride (TBAF) gave epoxomicin (Fig. 10), which was purified by silica gel column chromatography (CH_2Cl_2-MeOH, 98:2, v/v), followed by reverse-phase HPLC.

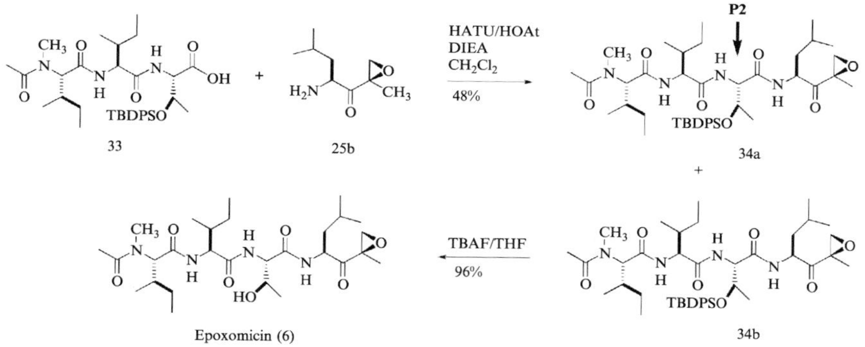

FIG. 10. The final assembly of epoxomicin peptide backbone and preparation of epoxomicin.

Purification of Epoxomicin by HPLC

Epoxomicin was purified by RP-HPLC (Rainin Dynamax) system composed of two solvent delivery pumps (Model SD200) and variable wavelength detector (Model UV-1) set at 214 nm. The column used was an YMC-Pack ODS-AM, 250 mm × 20 mm, 5-μm particles size and 120 Å pore size (Waters, Milford, MA) coupled to a guard column ODS-AM (10 mm × 10 mm) with same stationary phase specifications. The separation was carried out with a linear gradient of solvent A (water) into solvent B (MeOH): 65% of B to 80% of B over 90 min, run at 5.0 ml/min, room temperature. Epoxomicin was eluted at 49 min and the peak was collected as seven fractions of nearly 1 ml each.

Chemical Characterization of Epoxomicin

Analytical HPLC traces of collected fractions were carried out in a Waters Separation Module 2795 coupled to a Waters 2795 Photodiode Array Detector (set at 214 nm) and to a Micromass ZQ 2000 Electrospray Mass detector (cone voltage = +20V). A linear gradient of solvent A (water) in solvent B (MeOH), 30% of B to 90% of B in 30 min was run at 0.2 ml/min, at 25°. The analytical column used was an XTerra MS C18 (4.6 mm × 50 mm, 2.5-μm particle size, and 80 Å pore size).

Epoxomicin eluted at 19.8 min [observed m/z = 555.2 (M + H)$^+$, 577.6 (M + Na)$^+$]. Fractions showing purity ≥98% were grouped together and vacuum-dried with no heat until only water remained. Samples were then lyophilized. Epoxomicin was further characterized by ^{1}H NMR and ^{13}C NMR (Meng et al., 1999b; Sin et al., 1999).

Development Strategy of Proteasome Inhibitors with Novel Activities

General

The emergence of the ubiquitin-proteasome pathway as a major player in many important biological processes has prompted many synthetic efforts in proteasome inhibitor development (Kim and Crews, 2003; Myung et al., 2001a). In addition, systematic natural product screening has also provided a number of novel proteasome inhibitors (Kim and Crews, 2003). Thus far, most of these proteasome inhibitors have been shown to target the chymotrypsin-like (CT-L) activity. Although the CT-L activity of the proteasome has been shown to be largely responsible for the proteolytic function of the proteasome *in vivo* and *in vitro* (Myung et al., 2001b), the contribution of other major activities remains to be determined. Therefore, a novel class of inhibitors that target individual catalytic subunit/activity of proteasome may be required to dissect the role of each catalytic subunit.

Exposure of cells to stimuli such as IFN-γ, TNF-α, and LPS induces the synthesis of certain catalytic subunits (respectively, LMP2, MECL-1, and LMP7) that together are incorporated into alternative proteasome form (Kloetzel, 2001). This isoform, known as the immunoproteasome, has an enhanced capacity to generate peptides bearing hydrophobic and basic amino acids at their C-termini and a reduced capacity to produce peptides bearing acidic residues at their C-terminus (Kloetzel, 2001; Rock and Goldberg, 1999). Consequently, the spectrum of the produced peptides is shifted toward peptides that associate with MHC class I molecules with increased affinity (Fruh et al., 1994), implicating a major role of immunoproteasome in antigen presentation. Furthermore, it has been suggested that the immunoproteasome may be involved in some pathological processes, such as diabetes and autoimmune diseases (Casp et al., 2003). However, the exact role of the immunoproteasome remains unclear, caused in large part by the lack of appropriate molecular probes. Given that currently available proteasome inhibitors target both the constitutive proteasome and the immunoproteasome, development of immunoproteasome-specific inhibitors may be useful in the dissection of the role of immunoproteasomes.

Lessons from SAR Studies on the Natural Products Epoxomicin and Dihydroeponemycin

Although antitumor natural products epoxomicin (Meng et al., 1999b; Sin et al., 1999) and eponemycin (Meng et al., 1999a) have been shown to target proteasome, they markedly differ in proteasome subunit binding specificity and rates of proteasome inhibition. To understand such differences and develop proteasome inhibitors with novel activities, a combinatorial approach was taken to investigate the structure-activity relationship (SAR) of these compounds (Kim et al., 1999). Although both are members of the α',β'-epoxyketone linear peptide natural product family, there are some structural differences in their left-, central-, and right-hand fragments (Fig. 11) that may cause the different subunit binding specificity and inhibitory potency between two natural products. To this end, epoxomicin/dihydroeponemycin chimerae and their biotinylated counterparts were synthesized to correlate these structural features with proteasome inhibitory activity and their specific subunit labeling (Fig. 11). Kinetic data of epoxomicin/dihydroeponemycin chimerae for proteasome activities showed that α',β'-epoxyketone tetrapeptide inhibitors (compounds 35–37) generally display 300–500 fold greater inhibition for the CT-L activity compared with that of isooctanoic-containing tripeptide inhibitors (compounds 38–40) (Kim et al., 1999).

On the other hand, the α',β'-epoxyketone tetrapeptide inhibitors (compounds 35–37), regardless of changes in the central and right-hand fragments, displayed the same proteasome subunit binding pattern as epoxomicin, which predominantly binds LMP7/X and MECL1/Z (Meng et al., 1999b), whereas tripeptide inhibitors (compounds 38–40) that possess isooctanoic residue at the N-terminus displayed the same labeling pattern as that of dihydroeponemycin, which predominantly labels immunoproteasome subunit LMP2 and, to a lesser extent, LMP7/X (Meng et al., 1999a). Taken together, SAR studies on eponemycin/dihydroepoxomicin show that (1) α',β'-epoxyketone tetrapeptide inhibitors generally possess a higher inhibitory activity for the proteasome than tripeptide counterparts; (2) α',β'-epoxyketone tripeptide inhibitors containing the isooctanoic residue at the N-terminus, regardless of changes in the central and right-hand fragments, display greater specificity toward immunoproteasome over the constitutive proteasome. It is expected that the information on SAR studies will provide a basis for development of more potent proteasome inhibitors. In addition, the insights gained from SAR studies may facilitate the development of proteasome inhibitors with a high degree of immunoproteasome specificity.

Preparation of Epoxomicin/Dihydroeponemycin Chimerae

All the chimerae were prepared following the procedure used for the synthesis of epoxomicin with the exception of the right-hand fragment of dihydroeponemycin. The right-hand fragment of dihydroeponemycin was synthesized by a slight modification of a previously reported procedure (Hoshi et al., 1993; Schmidt and Schmidt, 1994). Recently, similar procedures for the preparation of the right-hand fragment have

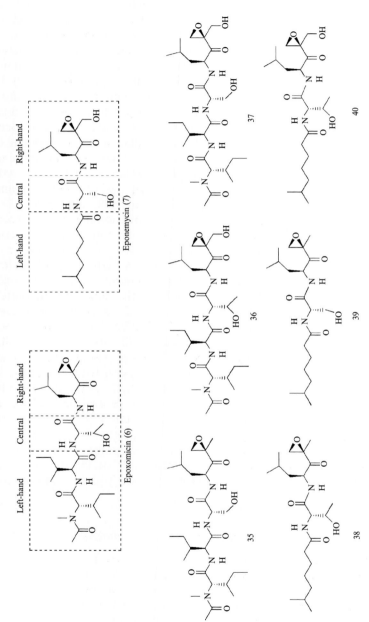

FIG. 11. Epoxomicin/eponemycin chimerae were prepared by a random combination of left- and right-hand and central fragments.

also been reported (Bennacer *et al.*, 2003; Dobler, 2001; Iwabuchi *et al.*, 2001). Given that a convergent approach that is applied for the preparation of epoxomicin/dihydroeponemycin chimerae is essentially the same as that described for epoxomicin, we only describe the total synthesis of dihydroeponemycin.

The synthesis of the right-hand fragment of dihydroeponemycin began with the addition of vinyl lithium derived from treatment of 2-bromo-1-hydroxy-2-propene (Hoshi *et al.*, 1993) with *t*-butyl lithium (Corey and Widiger, 1975) to leucine aldehyde 41, which afforded a 1:1 inseparable mixture of diol 42 (Fig. 12). The leucine aldehyde was prepared by the reduction of leucine Weinreb amide with LiAlH$_4$ in anhydrous ether. Selective protection of the primary alcohol of 42 with *t*-butyldimethylsilyl (TBS) group and a Swern oxidation of the remaining secondary alcohol yielded the α,β-unsaturated ketone 43. Epoxidation of 43 with hydrogen peroxide (Schmidt and Schmidt, 1994) afforded two epoxyketone isomers 44a and 44b as a 1.2:1 mixture that were readily separated by flash column chromatography using an elution system (hexanes-ethyl acetate = 5:1, v/v). The isomer (2-(*R*)-epoxide) 44a, which migrates faster than the 2-(*S*)-epoxide 44b in thin-layer chromatography (TLC), was found to have the same configuration as that of eponemycin epoxide. Final removal of the benzyloxycarbonyl (Z) protecting group of 44a through a catalytic hydrogenation reaction (Hoshi *et al.*, 1993) gave leucine epoxyketone 45 in nearly quantitative yield. The left-hand fragment was prepared by coupling of isooctanoic acid with serine benzylester using HBTU, followed by protection of the hydroxyl side chain of serine residue with TBDPS group and hydrogenolysis of the benzyl protecting group yielded TBDPS-protected left-hand fragment 49 (Fig. 13). The final coupling reaction between epoxyketone 45 and dipeptide 49 was performed with HATU, followed by

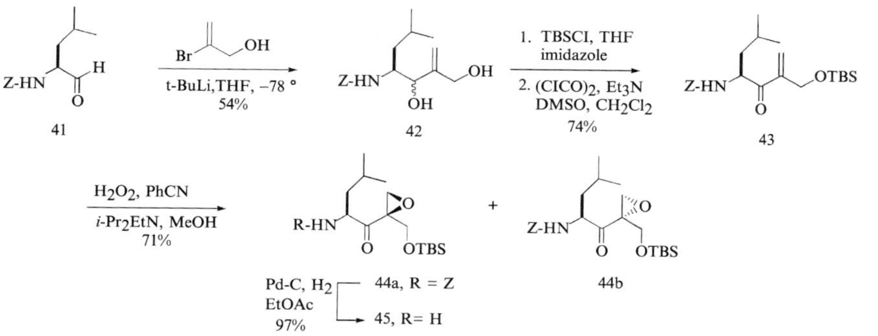

FIG. 12. Preparation of the epoxyketone residue of eponemycin.

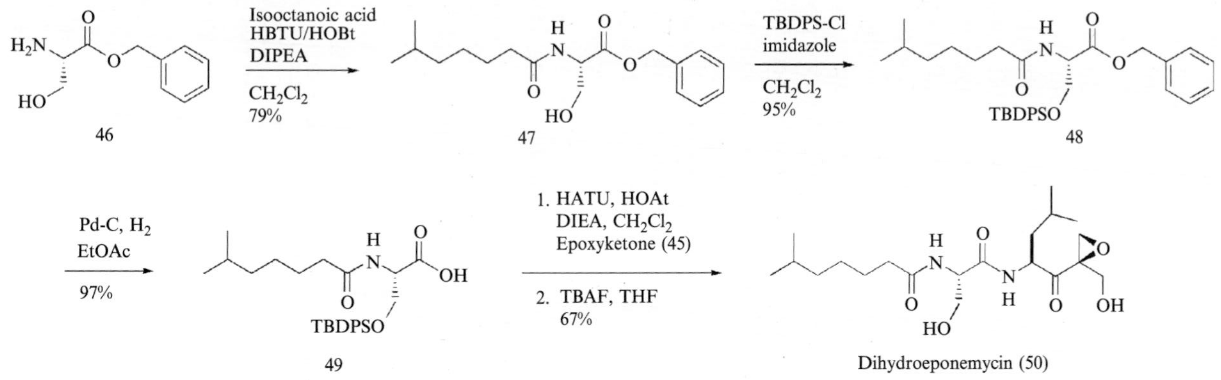

FIG. 13. The final assembly of eponemycin peptide backbone, which yielded dihydroeponemycin by TBDPS deprotection from serine side chain.

removal of the TBDPS group and normal-phase HPLC (hexanes-isopropanol, linear gradient, hexanes 100–50%) to yield dihydroeponemycin 50.

Development of Activity-Specific α',β'-Epoxyketone Proteasome Inhibitors

CT-L Activity-Specific α',β'-Epoxyketone Inhibitors

On the basis of the information gathered from SAR studies on epoxomicin/dihydroeponemycin, α',β'-epoxyketone tetrapeptide was chosen as a lead compound in the development of CT-L activity-specific proteasome inhibitors. To develop highly potent CT-L specific α',β'-epoxyketone proteasome inhibitors, a combinatorial positional scanning approach with a variety of hydrophobic amino acids, such as alanine, leucine, phenylalanine, homophenylalanine, 3-(1-naphthyl)-alanine, and p-benzoylphenylalanine, was applied to find the optimal amino acids at each position (Elofsson et al., 1999). After the synthetic strategy of epoxomicin described previously, peptide α',β'-epoxyketones were assembled from left-hand fragments, prepared by standard solid-phase synthesis, and the leucine α',β'-epoxyketone. The coupling reactions were carried out with HATU/HOAt and diisopropylethylamine (DIEA) in DMF. Unlike the final coupling in epoxomicin synthesis, in which CH_2Cl_2 was used as a solvent, DMF was used because of the poor solubility of the left-hand fragments. It seems that a slightly higher level of stereoisomer (5–30%) is formed in DMF conditions compared with CH_2Cl_2 environment. After reverse-phase HPLC purification, the resulting α',β'-epoxyketone tetrapeptides were tested for inhibition of three catalytic activities of the 20S proteasome. Once amino acids that display the best CT-L activity inhibition at each position were identified, the optimized inhibitor with optimized amino acids at each position was prepared, tested for CT-L activity inhibition to yield the most potent CT-L activity selective inhibitor (YU101, Fig. 4) to date (Elofsson et al., 1999).

Purification of YU101. Impure YU101 was purified by RP-HPLC (for HPLC system and column specifications see the purification procedure for epoxomicin). A linear gradient of solvent A into B (water and methanol, respectively) consisting of 70% of B to 85% of B in 90 min, run at 5 ml/min, room temperature. YU101, which was eluted at 62.5 min, was collected as seven fractions of nearly 2 ml each.

Chemical Characterization. LC-MS analysis of collected fractions were performed in a Waters Separation Module 2795 coupled to a Waters 2795 Photodiode Array Detector (set at 214 nm), and to a Micromass ZQ 2000

Electrospray Mass detector (cone voltage = +20V). A linear gradient of solvent A (water) in solvent B (MeOH), 5% of B to 98% of B in 30 min was run at 0.2 ml/min, 25°. The analytical column used was the same as the one described for epoxomicin.

YU101 was eluted at 10.3 min [observed $m/z = 635.5$ $(M + H)^+$, 658.4 $(M + Na)^+$]. Fractions showing purity $\geq 98\%$ were grouped together and vacuum-dried with no heat. YU101 was also confirmed by ^1H NMR.

Development of Caspase-Like Activity-Specific α',β'-Epoxyketone Peptide Inhibitors

α',β'-Epoxyketone-based caspase-like activity-specific proteasome inhibitors were also developed using a combinatorial positional scanning approach (Myung et al., 2001b). Initial efforts involved peptidyl α',β'-epoxyketones with glutamic acid at the P1 position. It was reasoned that the presence of Glu at the P1 position of peptide inhibitors alone would suffice to render higher selectivity toward the caspase-like activity. However, α',β'-epoxyketone peptide inhibitors with Glu at the P1 position did not display any significant selectivity for the caspase-like activity (Myung et al., 2001b). To screen an optimum P1 amino acid residue for the caspase-like activity inhibition, α',β'-epoxyketones with various natural and unnatural amino acid residues (i.e., Leu, Ala, Val, Phe, norleucine, cyclohexyl, and tert-butyl) in place of Glu at the P1 position were prepared, which were then coupled to dipeptides containing the serine at the C-terminus and isooctanoic group at the N-terminus. Among these, α',β'-epoxyketone tetrapeptide inhibitors with Leu at the P1 position provided the best caspase-like selectivity. In this regard, incorporation of leucine at the P1 position in an N-terminus protected peptidyl aldehyde proteasome inhibitor (i.e., Z-GPFL-CHO) has previously been shown to inhibit competitively the CASPASE-LIKE activity with a $\sim$13-fold selectivity over the CT-L activity (Vinitsky et al., 1994).

Interestingly, α',β'-epoxyketone inhibitors with Pro-Phe-Leu at the P3-P1 positions display no significant inhibition for the T-L activity, regardless of the nature of residues at the P4 position. Expanding on this finding, therefore, it was decided to derivatize the N-terminus of Pro-Phe-Leu-α',β'-epoxyketone inhibitors to explore the importance of length, steric bulk, and protecting group at the P4 position for the caspase-like activity inhibition. The results from these studies showed that the presence of a bulky aromatic protecting group in place of an acetyl group at the amino terminus provides a rather stronger inhibition toward the CT-L activity, thus making them less caspase-like selective. On the basis of these findings, several α',β'-epoxyketone peptides possessing a smaller amino-terminal group (i.e., acetyl) instead of a bigger group (i.e., benzyloxycarbonyl group) were prepared and tested for their inhibitory activity.

To this end, one compound (YU102, Fig. 4) showed the highest selectivity toward the caspase-like activity with ∼50-fold higher values of $k_{obs}/[I]$ for inhibition of the caspase-like activity than the CT-L activity (Myung et al., 2001b); 8 μM YU102 was found to inhibit only the caspase-like activity but not the CT-L activity at 8 μM concentration. Up to 90% of the caspase-like activity was found to be inhibited under the assay conditions. Against T-L activity, YU102 is a very poor inhibitor; even at concentrations of 100–150 μM, it displayed no significant inhibition of the T-L activity. Moreover, YU102 showed a time-dependent inhibition, indicating the irreversible modification of the catalytic Thr-1 of the proteasome. Therefore, a greater than 90% inhibition of the caspase-like activity under these conditions suggests that the caspase-like subunits were at least 90% saturated irreversibly with YU102. The CT-L activity of the 20S proteasome, however, was not affected. This results show that near quantitative occupancy of the caspase-like sites with YU102 did not trigger inhibition off the CT-L activity. YU102 was applied to probe the role of different catalytic subunits in cell-based protein degradation assays in living cells (Myung et al., 2001b), revealing that selective caspase-like activity inhibition is not sufficient to inhibit total protein degradation.

Enzyme Kinetic Assays

Purification of 20S Proteasome

20S proteasome was purified from bovine reticulocyte lysates by batch DE-52 binding, DEAE Sephacel chromatography, gel filtration on Sephacryl S-300, and chromatography on hydroxyapatite. This procedure is as previously described (Elofsson et al., 1999; Myung et al., 2001b), with the exception that bovine reticulocytes were used as a starting source.

To evaluate the rates of proteolytic inactivation by epoxomicin and YU101, $K_{association}(K_{obs}/[I])$, values were determined by use of fluorogenic peptide substrates over a range of inhibitor concentrations.

Epoxomicin and YU101 inhibition of the chymotrypsin-like catalytic activity of the 20S proteasome complex was determined as follows: Inhibitors solubilized in DMSO were mixed with a final concentration of 5 μM of the fluorogenic peptide substrate Suc-LLVY-AMC and assay buffer (20 mM Tris·HCl, pH 8.0, 0.5 mM EDTA/0.035% SDS) in a 96-well plate. Inhibitors concentrations were adjusted so that the final DMSO concentration would not exceed 1%. Hydrolysis was initiated by the addition of bovine red blood cell 20S proteasome to a final volume of 100 μl/well, and the reaction was followed by fluorescence (360 nm excitation/460 nm detection) using a multilable plate-reader Wallac Victor^2, set at 25°. Reactions were allowed to proceed for 50 min, and fluorescence data were collected every 10 sec. Fluorescence was quantified as arbitrary units, and progression curves were plotted for each reaction as a function of time. $K_{obs}/[I]$ values were obtained using KALEIDOGRAPH software by non-linear least-squares fit of the data to the following equation: fluorescence = $v_S t + [(v_O - v_S)/K_{obs}][1 - \exp(-K_{obs}t)]$, where v_O and v_S are the initial and final velocities respectively, and K_{obs} is the reaction rate constant. $K_{association} = K_{obs}/[I]$ ($M^{-1}s^{-1}$).

The range of inhibitors final concentrations tested were 40–100 nM for epoxomicin and 5–12 nM for YU101. Bovine erythrocyte 20S proteasome (2.5 mg/ml) was diluted 1:500.

YU101 most potently inhibits the chymotrypsin-like activity of the 20S proteasome with a $K_{association} = 310{,}000 \pm 19{,}000$ ($M^{-1}s^{-1}$), CV = 60%. The chymotrypsin-like inhibitory activity of epoxomicin was $K_{association} = 20{,}000 \pm 4{,}600$ ($M^{-1}s^{-1}$), CV = 24%.

For YU102, peptide-AMC (10 μM Suc-LLVY-AMC, 10 μM Z-LLE-AMC, or 20 μM Boc LRR-AMC) and 20S proteasome were added to assay buffer (20 mM Tris·HCl, pH 8.0, and 0.5 mM EDTA). For Suc-LLVY and Z-LLE-AMC assays, 0.035% (w/v) SDS was added to the assay buffer. After the steady state of hydrolysis for each substrate was established, an inhibitor was added to the assay buffer containing substrate and enzyme in a Dynex™ 96-well plate at room temperature. Release of fluorescent 7-amino-4-methylcoumarin (AMC) was measured using a Cytofluor spectrofluorometer with an excitation wavelength of 360 nm, and kinetic data were processed as described previously.

Summary

Given the complex proteolytic activities associated with the proteasome and poorly understood biological role of each catalytic subunit in many important signaling pathways, there are unmet needs for more sophisticated, selective proteasome inhibitors to dissect proteasome function. Here we describe strategies for developing α',β'-epoxyketone peptide-based proteasome inhibitors: (1) a general approach for the synthesis of α',β'-epoxyketone peptides that was developed through the total synthesis of epoxomicin; (2) SAR studies on epoxomicin/dihydroeponemycin, potentially shedding light on a means to design catalytic subunit- or immunoproteasome-specific α',β'-epoxyketone peptides; (3) development of highly potent, CT-L activity-specific α',β'-epoxyketone peptides; and (4) development of caspase-like activity-specific α',β'-epoxyketone inhibitors. Fortunately, all of the reactions for the synthesis of α',β'-epoxyketone peptides can readily be carried out in solutions (CH_2Cl_2 or DMF solvents) with good to excellent yields and easily repeatable. In conclusion, our studies have shown that derivatization of α',β'-epoxyketone peptide proteasome inhibitors at positions P1–P4 can be easily accomplished to provide novel proteasome-specific, subunit-selective small molecule inhibitors.

References

Adams, J. (2002). Development of the proteasome inhibitor PS-341. *Oncologist* **7**, 9–16.
Adams, J., Behnke, M., Chen, S., Cruickshank, A. A., Dick, L. R., Grenier, L., Klunder, J. M., Ma, Y. T., Plamondon, L., and Stein, R. L. (1998). Potent and selective inhibitors of the proteasome: Dipeptidyl boronic acids. *Bioorg. Med. Chem. Lett.* **8**, 333–338.
Albericio, F., and Carpino, L. A. (1997). Coupling reagents and activation. *Methods Enzymol.* **289**, 104–126.
Bennacer, B., Rivalle, C., and Grierson, D. (2003). A new route for the total synthesis of 6,7-dihydroeponemycin. *Eur. J. Org. Chem.* **23**, 4569–4574.
Bogyo, M., McMaster, J. S., Gaczynska, M., Tortorella, D., Goldberg, A. L., and Ploegh, H. (1997). Covalent modification of the active site threonine of proteasomal beta subunits and the *Escherichia coli* homolog HslV by a new class of inhibitors. *Proc. Natl. Acad. Sci. USA* **94**, 6629–6634.
Bogyo, M., Shin, S., McMaster, J. S., and Ploegh, H. L. (1998). Substrate binding and sequence preference of the proteasome revealed by active-site-directed affinity probes. *Chem. Biol.* **5**, 307–320.
Carpino, L. A. (1993). 1-Hydroxy-7-azabenzotriazole. An efficient peptide coupling additive. *J. Am. Chem. Soc.* **115**, 4397–4398.
Casp, C. B., She, J. X., and McCormack, W. T. (2003). Genes of the LMP/TAP cluster are associated with the human autoimmune disease vitiligo. *Genes Immun.* **4**, 492–499.
Corey, E. J., and Widiger, G. N. (1975). *J. Org. Chem.* **40**, 2975–2976.
Dobler, M. R. (2001). Total synthesis of (+)-epopromycin B and its analogues—studies on the inhibition of cellulose biosynthesis. *Tetrahedron Lett.* **42**, 215–218.
Elofsson, M., Splittgerber, U., Myung, J., Mohan, R., and Crews, C. M. (1999). Towards subunit-specific proteasome inhibitors: Synthesis and evaluation of peptide α',β'-epoxyketones. *Chem. Biol.* **6**, 811–822.
Fenteany, G., Standaert, R. F., Lane, W. S., Choi, S., Corey, E. J., and Schreiber, S. L. (1995). Inhibition of proteasome activities and subunit-specific amino-terminal threonine modification by lactacystin. *Science* **268**, 726–731.
Fields, G. B. (1997). "Methods in Enzymology." Academic Press, New York.
Figueiredo-Pereira, M. E., Banik, N., and Wilk, S. (1994). Comparison of the effect of calpain inhibitors on two extralysosomal proteinases: The multicatalytic proteinase complex and m-calpain. *J. Neurochem.* **62**, 1989–1994.
Fruh, K., Gossen, M., Wang, K., Bujard, H., Peterson, P. A., and Yang, Y. (1994). Displacement of housekeeping proteasome subunits by MHC-encoded LMPs: A newly discovered mechanism for modulating the multicatalytic proteinase complex. *EMBO J.* **13**, 3236–3244.
Groll, M., Kim, K. B., Kairies, N., Huber, R., and Crews, C. M. (2000). Crystal structure of epoxomicin: 20s proteasome reveals a molecular basis for selectivity of α',β'-epoxyketone proteasome inhibitors. *J. Am. Chem. Soc.* **122**, 1237–1238.
Hanada, M., Sugawara, K., Kaneta, K., Toda, S., Nishiyama, Y., Tomita, K., Yamamoto, H., Konishi, M., and Oki, T. (1992). Epoxomicin, a new antitumor agent of microbial origin. *J. Antibiot. (Tokyo).* **45**, 1746–1752.

Hoshi, H., Ohnuma, T., Aburaki, S., Konishi, M., and Oki, T. (1993). A total synthesis of 6,7-dihydroeponemycin and determination of stereochemistry of the epoxide ring. *Tetrahedron Lett.* **34**, 1047–1050.

Iqbal, M., Chatterjee, S., Kauer, J. C., Mallamo, J. P., Messina, P. A., Reiboldt, A., and Siman, R. (1996). Potent a-ketocarbonyl and boronic ester derived inhibitors of proteasome. *Bioorg. Med. Chem. Lett.* **6**, 287–290.

Iwabuchi, Y., Sugihara, T., Esumi, T., and Hatakeyama, S. (2001). An enantio- and stereocontrolled route to epopromycin B via cinchona alkaloid-catalyzed Baylis–Hillman reaction. *Tetrahedron Lett.* **42**, 7867–7871.

Kim, K. B., and Crews, C. M. (2003). Natural product and synthetic proteasome inhibitors. *In* "Cancer Drug Discovery and Development: Proteasome Inhibitors in Cancer Therapy" (J. Adams, ed.), pp. 47–63. Humana Press Inc., Totowa, NJ.

Kim, K. B., Myung, J., Sin, N., and Crews, C. M. (1999). Proteasome inhibition by the natural products epoxomicin and dihydroeponemycin: Insights into specificity and potency. *Bioorg. Med. Chem. Lett.* **9**, 3335–3340.

Kisselev, A. F., and Goldberg, A. L. (2001). Proteasome inhibitors: From research tools to drug candidates. *Chem. Biol.* **8**, 739–758.

Kloetzel, P. M. (2001). Antigen processing by the proteasome. *Nat. Rev. Mol. Cell Biol.* **2**, 179–187.

Koguchi, Y., Kohno, J., Nishio, M., Takahashi, K., Okuda, T., Ohnuki, T., and Komatsubara, S. (2000a). TMC-95A, B, C, and D, novel proteasome inhibitors produced by *Apiospora montagnei* Sacc. TC 1093. Taxonomy, production, isolation, and biological activities. *J. Antibiot. (Tokyo)* **53**, 105–109.

Koguchi, Y., Kohno, J., Suzuki, S., Nishio, M., Takahashi, K., Ohnuki, T., and Komatsubara, S. (1999). TMC-86A, B and TMC-96, new proteasome inhibitors from *Streptomyces* sp. TC 1084 and *Saccharothrix* sp. TC 1094. I. Taxonomy, fermentation, isolation, and biological activities. *J. Antibiot. (Tokyo)* **52**, 1069–1076.

Koguchi, Y., Kohno, J., Suzuki, S., Nishio, M., Takahashi, K., Ohnuki, T., and Komatsubara, S. (2000b). TMC-86A, B and TMC-96, new proteasome inhibitors from *Streptomyces* sp. TC 1084 and *Saccharothrix* sp. TC 1094. II. Physico-chemical properties and structure determination. *J. Antibiot (Tokyo)* **53**, 63–65.

Koguchi, Y., Nishio, M., Suzuki, S., Takahashi, K., Ohnuki, T., and Komatsubara, S. (2000c). TMC-89A and B, new proteasome inhibitors from *Streptomyces* sp. TC 1087. *J. Antibiot. (Tokyo)* **53**, 967–972.

Lloyd-Williams, P., Albericio, F., and Giralt, E. (1997). "Chemical Approaches to the Synthesis of Peptides and Proteins." CRC Press, Boca Raton, Florida.

Meng, L., Kwok, B. H., Sin, N., and Crews, C. M. (1999a). Eponemycin exerts its antitumor effect through the inhibition of proteasome function. *Cancer Res.* **59**, 2798–2801.

Meng, L., Mohan, R., Kwok, B. H., Elofsson, M., Sin, N., and Crews, C. M. (1999b). Epoxomicin, a potent and selective proteasome inhibitor, exhibits *in vivo* antiinflammatory activity. *Proc. Natl. Acad. Sci. USA* **96**, 10403–10408.

Myung, J., Kim, K. B., and Crews, C. M. (2001a). The ubiquitin-proteasome pathway and proteasome inhibitors. *Med. Res. Rev.* **21**, 245–273.

Myung, J., Kim, K. B., Lindsten, K., Dantuma, N. P., and Crews, C. M. (2001b). Lack of proteasome active site allostery as revealed by subunit-specific inhibitors. *Mol. Cell.* **7**, 411–420.

Nahm, S., and Weinreb, S. M. (1981). N-methoxy-N-methylamides as effective acylating agents. *Tetrahedron Lett.* **22**, 3815–3818.

Nam, S., Smith, D. M., and Dou, Q. P. (2001). Ester bond-containing tea polyphenols potently inhibit proteasome activity *in vitro* and *in vivo*. *J. Biol. Chem.* **276**, 13322–13330.

Orlowski, M. (1990). The multicatalytic proteinase complex, a major extralysosomal proteolytic system. *Biochemistry* **29,** 10289–10297.
Rock, K. L., and Goldberg, A. L. (1999). Degradation of cell proteins and the generation of MHC class I-presented peptides. *Annu. Rev. Immunol.* **17,** 739–779.
Rock, K. L., Gramm, C., Rothstein, L., Clark, K., Stein, R., Dick, L., Hwang, D., and Goldberg, A. L. (1994). Inhibitors of the proteasome block the degradation of most cell proteins and the generation of peptides presented on MHC class I molecules. *Cell* **78,** 761–771.
Schmidt, U., and Schmidt, U. (1994). The total synthesis of eponemycin. *Synthesis* **3,** 300–304.
Sekizawa, R., Momose, I., Kinoshita, N., Naganawa, H., Hamada, M., Muraoka, Y., Iinuma, H., and Takeuchi, T. (2001). Isolation and structural determination of phepeptins A, B, C, and D, new proteasome inhibitors, produced by *Streptomyces* sp. *J. Antibiot. (Tokyo)* **54,** 874–881.
Sin, N., Kim, K. B., Elofsson, M., Meng, L., Auth, H., Kwok, B. H., and Crews, C. M. (1999). Total synthesis of the potent proteasome inhibitor epoxomicin: A useful tool for understanding proteasome biology. *Bioorg. Med. Chem. Lett.* **9,** 2283–2288.
Sin, N., Meng, L., Auth, H., and Crews, C. M. (1998). Eponemycin analogues: Syntheses and use as probes of angiogenesis. *Bioorg. Med. Chem.* **6,** 1209–1217.
Sugawara, K., Hatori, M., Nishiyama, Y., Tomita, K., Kamei, H., Konishi, M., and Oki, T. (1990). Eponemycin, a new antibiotic active against B16 melanoma. I. Production, isolation, structure and biological activity. *J. Antibiot. (Tokyo)* **43,** 8–18.
Tsuchiya, K., Kobayashi, S., Nishikiori, T., Nakagawa, T., and Tatsuta, K. (1997). Epopromycins, novel cell wall synthesis inhibitors of plant protoplast produced by *Streptomyces* sp. NK04000. *J. Antibiot. (Tokyo)* **50,** 261–263.
Vinitsky, A., Cardozo, C., Sepp-Lorenzino, L., Michaud, C., and Orlowski, M. (1994). Inhibition of the proteolytic activity of the multicatalytic proteinase complex (proteasome) by substrate-related peptidyl aldehydes. *J. Biol. Chem.* **269,** 29860–29866.

[40] Screening for Selective Small Molecule Inhibitors of the Proteasome Using Activity-Based Probes

By MATTHEW BOGYO

Abstract

The proteasome's role in fundamental biological processes ranging from control of the cell cycle to production of peptides for display to immune cells has been uncovered with the help of small molecule inhibitors. Most of the commonly used inhibitors have been designed and synthesized by organic chemists or by Nature. To continue to develop new inhibitors and reagents for the proteasome, a rapid screening method is required that allows not only assessment of potency but also selectivity of inhibitors for each of the primary catalytic sites in the complex. This chapter outlines methods for the solid-phase synthesis of diverse peptide vinyl sulfone libraries and a rapid screen for potent and selective inhibitors that makes use of an active site label (Nazif and Bogyo, 2001). This assay can be performed with small quantities of total cellular extracts as a source of enzyme and can be used to rapidly screen virtually any potential inhibitor.

Introduction

The proteasome is a large multi-subunit protease complex that is responsible for the turnover of virtually all cytosolic proteins (Coux et al., 1996). It is a large barrel-shaped complex made up of multiple individual subunits that create a closed cavity in which controlled proteolysis takes place. The central core of the proteasome contains six catalytically active protease subunits with three distinct types of active subunits that process protein substrates after they enter the core chamber. As a result of this complex, multifaceted architecture, studies of the enzymatic processes that take place within the catalytic cavity have remained challenging.

To address some of the outstanding questions regarding substrate processing by the proteasome, a number of classes of small molecule inhibitors have been designed to target its multiple active sites (for reviews see Bogyo and Wang [2002] and Kisselev and Goldberg [2001]). In addition, the proteasome's fundamental role in crucial cellular processes has made it a target for the development of inhibitors with therapeutic potential (Adams and Kauffman, 2004). All these inhibitors have been valuable tools that have helped to explain the functional roles of the proteasome in basic biological processes of the cell.

A number of studies have used substrate-based methods to define the primary sequence specificity of proteases (Backes et al., 2000; Harris et al., 2000). These methods rely on libraries of peptide sequences that carry a masked fluorescent reporter that is released upon proteolytic processing. By scanning through a large number of libraries of sequences in which a single or multiple positions on the substrate are held constant, it is possible to begin to define optimal substrate binding sequences. One such study used this approach to examine the specificity of each of the primary catalytic centers of the proteasome (Harris et al., 2001). However, the results obtained for the proteasome are often difficult to interpret, because a single substrate can be processed by multiple active sites in the complex. Therefore, inhibitors that can be used to selectively inhibit individual active sites can be used in combination with substrate scanning approaches to begin to define substrate specificities of each of the three primary active sites. Furthermore, peptide-based inhibitors that form a covalent link to the active site threonine residue and that bind in a manner analogous to a protein substrate can be used to begin to directly define substrate recognition elements within individual active sites (Fig. 1; Nazif and Bogyo, 2001).

This chapter will outline methods used for the synthesis and screening of peptide-based covalent inhibitors of the proteasome. This process involves the synthesis of positional scanning libraries of peptides containing a reactive vinyl sulfone "warhead" group at their C-termini. These inhibitors form irreversible covalent adducts to the threonine nucleophile within the catalytically active β-subunits (Fig. 1B; Bogyo et al., 1997, 1998). Binding to each of the primary active sites can be monitored with a broad-spectrum inhibitor carrying a radioactive tag in a competition assay using SDS-PAGE as the readout (Fig. 2). This method allows screening of

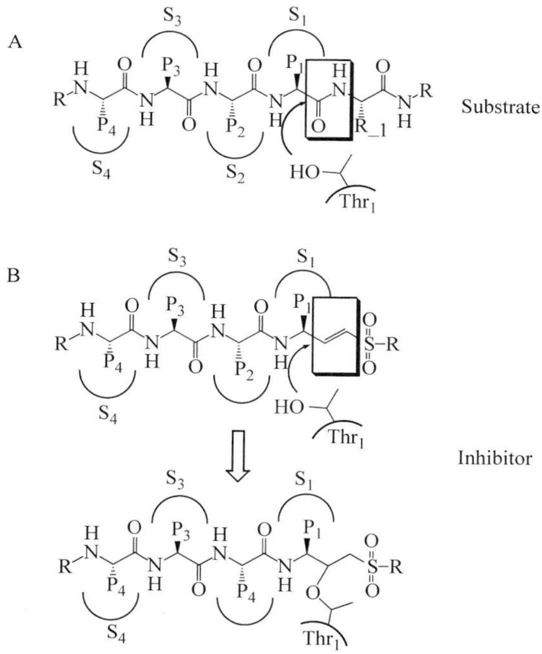

FIG. 1. Mode of substrate and inhibitor binding to the active site of a catalytically active β-subunit of the proteasome. (A) A substrate binds in the active site, making contacts with specificity pockets on both sides of the site of hydrolysis. The catalytic threonine-1 residue attacks the substrate using its hydroxyl nucleophile. The substrate binding pockets are termed the "S sites," and the corresponding residues on the substrate are the "P residues." (B) A peptide vinyl sulfone binds in the active site in a manner analogous to a substrate. This allows the primary amino acids to bind in the specificity pockets of the enzyme such that the reactive vinyl group is aligned for attack by the threonine-1 nucleophile. Reaction results in the formation of stable covalent ether bond between the inhibitor and the catalytic β-subunit.

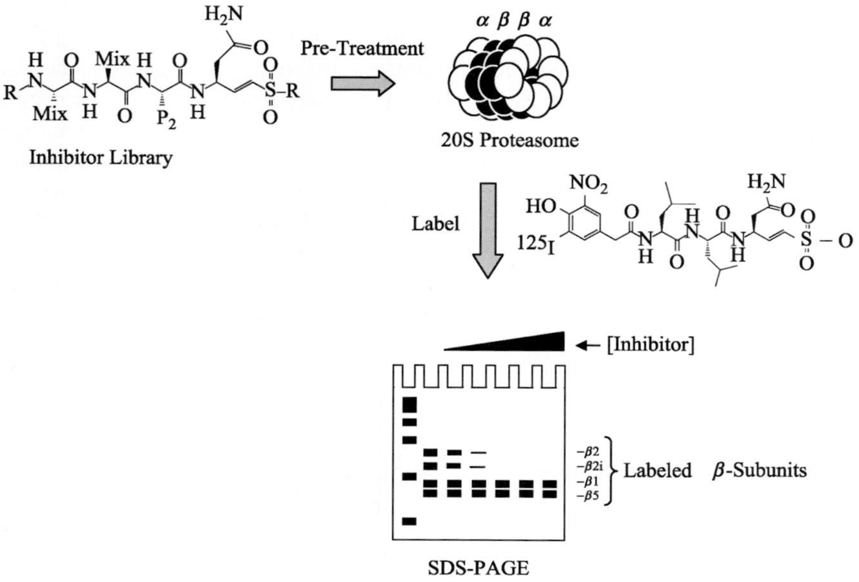

FIG. 2. Screening of inhibitor libraries using a competition assay with an activity-based probe. Small molecule libraries can be screened by treatment of purified proteasomes or crude cellular extracts followed by active site labeling using a radiolabeled probe that binds to each of the three primary active β-subunits. By increasing the concentration of the library, it is possible to assess the potency and selectivity of the inhibitor or inhibitor library by monitoring loss of labeling of each of the primary active sites by the radiolabeled probe. In this example, a mixture library with a constant P2 residue is added to purified proteasomes over a range of concentrations. This example library reduces labeling of primarily the $\beta2$ subunit, suggesting that this P2 element is driving selectivity for this active site in the complex. Typically, large numbers of libraries are screened at a single concentration to identify selective hits that can then be analyzed individually over a broad range of concentrations as shown in this example.

virtually any inhibitor or set of inhibitors and provides a direct assessment of subunit selectivity and overall potency. Furthermore, it can be performed with crude mixtures of proteins from whole cells or tissues, thereby circumventing the need for large quantities of purified enzyme.

Design of the Competition Assay Using a Selective Active Site Label

The most commonly used methods for screening of protease inhibitors rely on fluorogenic substrates that are processed by a protease to yield a fluorescent by-product. Such substrate assays can be monitored for a large number of samples with a fluorescence spectrophotometer. Although this assay provides a readout of overall kinetic parameters for a given inhibitor, for enzymes such as the proteasome that contain multiple active sites, it provides limited information about overall subunit selectivity of compounds. Thus, inhibition data are difficult to interpret in the absence of an exquisitely selective substrate for each of the primary active sites. Although it is clear that the proteasome's three primary active β-subunits each have different preferences for the amino acid found directly adjacent to the scissile amide bond (the P1 position), this selectivity is far from absolute (Harris *et al.*, 2001). Therefore, all the commonly used substrates remain only partially selective and can be cleaved at multiple active sites in the complex.

As an alternative to the substrate-based screening approach, selective small molecule labels can be used to directly monitor binding of an inhibitor to each of the primary active sites (Bogyo *et al.*, 1997, 1998; Nazif and Bogyo, 2001). These labeled probes bind covalently, labeling each of the primary catalytic subunits (Fig. 2). A simple SDS-PAGE analysis then allows labeled subunits to be resolved and monitored simultaneously. If a small molecule inhibitor is added before labeling, its binding to each of the active sites can be directly monitored as a loss of signal from the general probe. Furthermore, if the probe is sufficiently selective, it can be used to label the proteasome in crude cellular homogenates, thereby circumventing the need to obtain large quantities of purified enzyme.

Design and Synthesis of the Positional Scanning Libraries

The search for subunit-selective inhibitors begins with the synthesis of libraries of potential compounds. Because peptide vinyl sulfones have been shown to act as inhibitors of the proteasome by covalently modifying the active site threonine residues of the three catalytic β-subunits, they serve as an ideal starting point for library design (Bogyo *et al.*, 1997, 1998). Ideally, the synthesis of tetrapeptide vinyl sulfones containing all possible natural primary amino acid sequences would allow systematic evaluation of substrate specificity and potency. However, this would require extensive chemistry efforts. To access a significant diversity set with minimal synthetic chemistry efforts, a positional scanning method was selected to generate libraries of inhibitors (Nazif and Bogyo, 2001). This positional scanning method, originally designed for standard peptide libraries, is based on the synthesis of small pools (or sublibraries) of sequences in which a single amino acid position is held constant, whereas the others are varied (Fig 3A; Houghten *et al.*, (1991). The contribution of this constant amino acid to enzyme binding is measured for the mixture and then the constant position

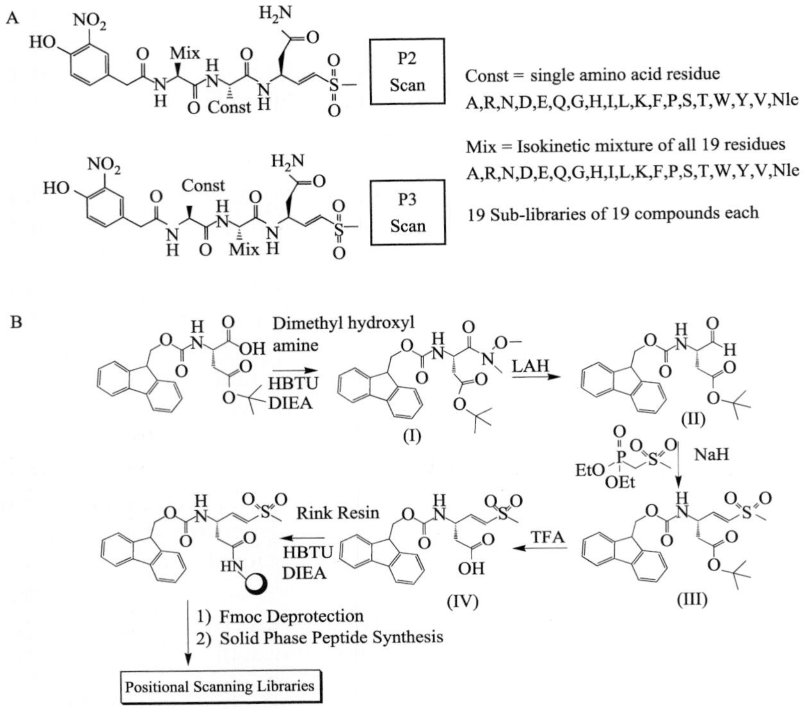

FIG. 3. Synthesis of positional scanning libraries of peptide vinyl sulfones. (A) Examples of P2 and P3 scanning libraries of peptide vinyl sulfones. In these libraries, an isokinetic mixture of all natural amino acids (see one letter codes) is coupled in the mix position, and a single residue from the same set is used in the constant position. Scanning of all residues in the constant positions results in two sets of 19 sublibraries made up of 19 members in each that can be used to determine the optimal binding residues for each fixed position. (B) Scheme for synthesis of the P1 vinyl sulfone building block and attachment to the Rink resin. Note the use of aspartic acid for coupling by means of the side chain acid to the Rink resin. After synthesis of the peptide backbone, the products are removed from the resin, thus generating a constant P1 asparagine residue.

is changed to the next residue. This process continues until the subsite on the inhibitor is scanned through all the desired amino acids. The constant position is then moved to the next subsite, and the scanning begins again. The end result is a set of binding data that provides a fingerprint of the optimal binding residues for each subsite on the inhibitor. If these binding interactions do not depend on cooperative binding effects with other positions, it is possible to design individual compounds with optimal potency and selectivity properties by combining the information from each site.

In this study, a set of positional scanning libraries of tetrapeptide vinyl sulfones was synthesized by use of solid-phase chemistries (Nazif and Bogyo, 2001). The P1 position on the inhibitors was used as the anchor to the resin and was, therefore, held constant as asparagine. This residue was found to be optimal for binding to all three catalytic β-subunits and could be generated by loading of a vinyl sulfone aspartic acid onto a Rink resin that results in conversion to asparagine after cleavage from the resin. The P2–P4 positions were scanned for each of the natural 20 amino acids (minus cysteine and plus norleucine in place of methionine) with an isokinetic mixture of the same 19 amino acids used in the other positions. The details of the synthesis are listed in the following and are shown in Fig. 3B.

Synthesis of Fmoc-Asp(OtBu)-Dimethyl Hydroxyl Amide (I)

Fmoc-Asp(OtBu)-OH (10 g, 24.3 mmol), HOBT (3.6 g, 26.7 mmol), and dicyclohexylcarbodiimide (DCC; 5.5 g, 26.7 mmol) were dissolved in DMF. After stirring 30 min at room temperature, a solid formed and was removed by filtration. N,O-dimethyl hydroxyl amine (2.84 g, 29.2 mmol) was added as a solid to the filtered reaction mixture along with triethyl amine (2.94 g, 29.2 mmol). The reaction was stirred for an additional 12 h and then concentrated by rotary evaporation of the solvent. The crude oil was then dissolved in ethyl acetate and extracted with three portions each of saturated sodium bicarbonate, 0.1 N HCl, and brine. The organic layer was dried over magnesium sulfate and concentrated to dryness. This crude preparation was used in subsequent reactions without further purification.

Synthesis of Fmoc-Asp(OtBu)-H (II)

Crude I (11.14 g, 24.5 mmol) was dissolved in anhydrous ethyl ether and stirred on ice under a positive argon flow. Lithium aluminum hydride (0.928 g, 24.5 mmol) was slowly added, leading to gas evolution. The reaction was stirred for an additional 20 min and then quenched by the addition of potassium hydrogen sulfate (6.67 g, 49 mmol). The quenched reaction was stirred on ice for 20 min and then at room temperature for an additional 30 min. The mixture was extracted three times with ethyl acetate, and the combined organics were washed with three portions of 0.1 M HCl, saturated sodium carbonate, and brine. The organic phase was concentrated to an oil. The product was purified by flash chromatography in hexane/ethyl acetate (2:1, v:v), yield (5.58 g, 58.3%).

Synthesis of Fmoc-Asp(OtBu)-Vinyl Sulfone (III)

Diethyl (methylthiomethyl) phosphonate was oxidized to the corresponding phosphonate sulfone with peracetic acid in aqueous dioxane as described (Bogyo *et al.*, 1998). Diethyl phosphonate sulfone (4.06 g, 17.66 mmol) was dissolved in anhydrous THF under argon. Sodium hydride (678 mg, 16.96 mmol) was added and the reaction stirred for 30 min at room temperature. Pure II (5.58 g, 14.1 mmol) was dissolved in anhydrous THF under argon and added to the stirring reaction by cannula. The reaction was then allowed to stir for 30 min, quenched with water, and the resulting aqueous phase washed three times with dichloromethane. The organic phases were combined, dried over $MgSO_4$, and evaporated to dryness. The crude oil was purified by flash chromatography over silica gel in hexane/ethyl acetate (1.5:1 v:v), yield (4.1 g, 8.7 mmol, 62%).

Synthesis of Fmoc-Asp-Vinyl Sulfone (IV)

Pure III (4.1 g, 8.7 mmol) was dissolved in methylene chloride (10 ml), and an equal volume of anhydrous trifluoroacetic acid was added. The reaction was stirred for 4 h and quenched by the addition of an excess volume of toluene. The reaction was concentrated by rotary evaporation and then resuspended in toluene. The reaction was evaporated to an oil, and the product was precipitated by addition of ethyl ether. The resulting white precipitate was collected by centrifugation and dried under vacuum, yield (2.95 g, 81%).

Coupling of Fmoc-Asp-Vinyl Sulfone (IV) to Rink Amide Resin

The pure product IV was coupled to Rink amide resin (0.8 mmol/g resin load) in DMF with standard coupling chemistry (DIC, HOBT), and the resin load was determined by absorbance of free Fmoc upon deprotection. The resin was dried under vacuum and used as the starting material for all library synthesis.

Synthesis of Positional Scanning and Single-Component Peptide Libraries

Fmoc-Asp-vinyl sulfone–loaded resin (20 mg/well 0.7 mmol/g) was weighed into 96 position filter blocks (FlexChem system; Robbins Scientific). Each library consisted of 19 sublibraries in which each of the natural amino acids (minus cysteine and methionine plus norleucine) was used at a designated constant position, and an isokinetic mixture of those same 19 amino acids was coupled in the variable position. The isokinetic mixture was created by use of a ratio of equivalents of amino acids on the basis of their reported coupling rates (Ostresh *et al.*, 1994). The total mixture was adjusted to 10-fold excess total amino acids over resin load. For constant positions, a single amino acid was coupled using 10-fold excess. Couplings were carried out using DIC and HOBT under standard conditions. The mixture libraries (P2 and P3) were amino terminally capped with 4-hydroxy-3-nitrophenyl acetic acid, and the single component P4 libraries were capped with acetic anhydride. Libraries and single components were cleaved from the resin by addition of 90% trifluoroacetic acid, 5% water, and 5% triisopropyl silane for 2 h. Cleavage solutions were collected and products precipitated by addition of cold diethyl ether. Solid products were isolated by centrifugation followed by lyophilization to yield the crude peptides that were dissolved in DMSO (50 mM stock) based on average weights for each mixture. Libraries and single compounds were stored at $-20°$ and further diluted to 5 mM stock plates for use in experiments.

Screening of Inhibitors

Initial inhibitor libraries were composed of mixtures of 361 peptides (two variable positions with 19 amino acids and two constant positions of a single amino acid). These mixtures were screened by use of the competition assay described earlier (Fig. 2) with an active site probe, ^{125}I-NIP-LLN-VS, that labels all three primary catalytic β-subunits. The competition assays were performed in crude cell extracts and with purified preparations of 20S proteasomes.

Screening of Libraries in Crude Extract and Purified 20S Proteasomes

In a typical experiment, NIH-3T3 or EL-4 lysates were prepared by continuous vortexing in the presence of glass beads (<104 microns; Sigma) in buffer A (50 mM Tris, pH 5.5, 1 mM DTT, 5 mM MgCl$_2$, 250 mM sucrose). Supernatants were centrifuged for 15,000g for 15 min at 4°, and the total protein concentration of the final supernatants (soluble) was determined by BCA protein quantification (Pierce). Lysates were diluted to 1 mg/ml in reaction buffer (RB) (50 mM Tris, pH 7.4, 5 mM MgCl$_2$, 2 mM DTT) for the assays. Lysates (100 μg total protein) or purified 20S proteasomes (1 μg per sample in RB with 0.01% SDS) were incubated with 1 μl from a 5-mM library stock (final concentration of 50 μM) for 30 min at room temperature. Iodinated inhibitors (diluted 1:10 or 1:5 in RB) were added to each reaction and incubation continued for an additional 90 min at room temperature. The reactions were quenched by addition of 4× SB to 1× followed by boiling for 5 min. All samples were separated on 12.5 % SDS-PAGE gels, and data were obtained by exposure of the gels to PhophorImager screens (Molecular Dynamics). Bands of activity were quantified using ImageQuant software (Molecular Dynamics) and ratios

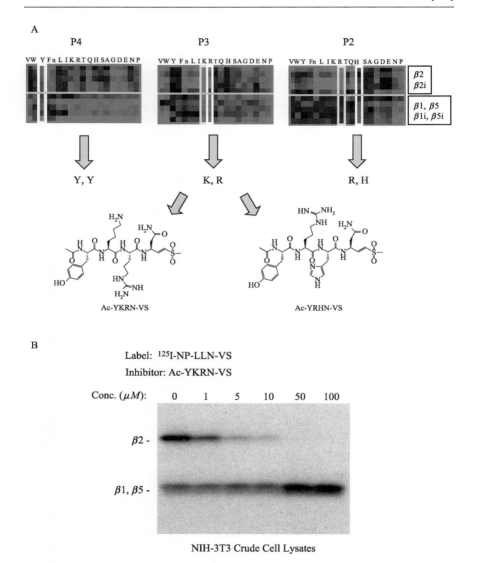

FIG. 4. Designing a subunit-selective inhibitor based on library inhibition data. (A) Data from P2, P3, and P4 library competition screens (see Fig. 2) are compiled and visualized using expression array analysis software (see text). The resulting graph groups binding data for the $\beta 2$ and $\beta 2i$ subunits at the top and binding data for the $\beta 1$, $\beta 1i$, $\beta 5$, $\beta 5i$ at the bottom. Red blocks indicate potent binders, whereas green blocks represent poor binders. The constant residues used in the inhibitors are listed across the top. Residues at each position that show

of each band's intensity relative to the corresponding control untreated sample were obtained.

Identification of Subunit Selective Inhibitors

The results from the screens were quantified and compiled as percent competition values relative to an untreated control. The resulting numerical data were plotted using standard microarray data processing and clustering programs (Eisen et al., 1998). The resulting clustergram allowed the data to be quickly analyzed to identify residues at each of the P2–P4 positions that showed selective binding to a given subunit (Fig. 4A). In the initial study, inhibitors that target the $\beta2$ or trypsin-like site were designed. These individual compounds were assessed for selectivity and potency using the same competition assay outlined previously (Fig. 4B). These results confirm that the positional scanning libraries could be used to identify multiple residues at several distinct subsites, which, when combined into a single compound, produced highly selective inhibitors. These results also confirm that optimal inhibitors can be designed by screening of compounds by competition assay in crude cellular homogenates as an enzyme source.

Expanding the Libraries of Inhibitors

Although the results from this specific example of a positional scanning library screen are promising and provided novel, $\beta2$ subunit-selective inhibitors, additional methods have been developed that should allow selective targeting of the other primary active sites and optimization of selective inhibitors. These methods make use of alternative solid-phase chemistries that allow synthesis of peptide vinyl sulfones with a range of variable P1 elements (Overkleeft et al., 2000; Wang and Yao, 2003; Wang et al., 2003). Thus, these methods are not limited to a constant P1 residue and can be used to further optimize subunit selective inhibitors. Furthermore, efforts using different reactive warhead groups have successfully targeted the $\beta1$ subunit of the proteasome (Elofsson et al., 1999). In addition, the general

---

selectivity for the $\beta2$ subunit are highlighted with white and yellow boxes. The resulting two inhibitors designed on the basis of these data are shown below. (B) A typical competition assay in which the $\beta2$-specific inhibitor Ac-YKRN-VS from (A) is used to treat crude NIH-3T3 lysates at a range of concentrations. Addition of the general label ^{125}I-NP-LLN-VS shows the selective binding of the inhibitor to the $\beta2$ subunit at concentrations as high as 100 μM. (See color insert.)

competition assay can be applied to any range of potential libraries, provided the number of compounds being analyzed is not excessively large.

Virtues and Limitations of the Competition-Based Screening Method

Although there have been several examples of successful uses of inhibitor screens with activity-based probes in crude proteomes (Greenbaum et al., 2002a,b; Leung et al., 2003), there remain both assets and limitations to this method. Perhaps the greatest virtue of this screen is that it allows rapid screening of inhibitors without the need to express and purify individual targets, something that is often challenging for proteolytic enzymes and multicomponent enzymes such as the proteasome. Furthermore, it allows compounds to be screened in a more physiologically relevant sample and provides information about binding to multiple proteins at once. Finally, the assay is easy to set up, does not require optimization for each target enzyme, and provides a direct readout of active site binding, thus eliminating false positives that result from nonspecific inhibition mechanisms. However, there are several significant limitations to this method. First, it is dependent on an analytical method that enables labeled proteins to be resolved from one another. The commonly used SDS-PAGE methods are somewhat low throughput and, therefore, do not allow for screening of extremely large libraries of compounds. This limiting factor could potentially be overcome by the adaptation of this method to high-throughput analytic separations such as capillary electrophoresis.

In addition, the positional scanning libraries have several caveats that need to be considered. These scanning approaches do not provide information regarding cooperative interactions at multiple sites, because these interactions are averaged out by the mixing of the variable residues. Thus, potentially interesting selectivity pairs are easily missed. This limitation can be overcome by generating libraries with multiple fixed positions. Such libraries have been successfully used to scan substrate specificity of the proteasome (Harris et al., 2001). Although the virtue of the inhibitor library scanning method is its ability to provide information regarding each of the active sites in the complex, the optimal inhibitor sequences do not necessarily match those of a true substrate. Therefore, care must be taken when translating inhibitor specificity to substrate specificity. However, comparison of the results from substrate scanning (Harris et al., 2001) to inhibitor scanning (Nazif and Bogyo, 2001) indicates a close correlation owing to the well-defined specificity pockets in the active sites that can accommodate both substrates and inhibitors using similar binding modes.

Despite the potential pitfalls and limitations, the results from this study and others that have made use of competition labeling assays strongly suggest that it is a valuable technique. It is likely that with advances in technology and an expansion of available activity-based probes, these types of screens will identify inhibitors for functional studies of a wide range of enzyme targets.

References

Adams, J., and Kauffman, M. (2004). Development of the proteasome inhibitor Velcade (Bortezomib). *Cancer Invest.* **22**, 304–311.

Backes, B. J., Harris, J. L., Leonetti, F., Craik, C. S., and Ellman, J. A. (2000). Synthesis of positional-scanning libraries of fluorogenic peptide substrates to define the extended substrate specificity of plasmin and thrombin. *Nat. Biotechnol.* **18**, 187–193.

Bogyo, M., McMaster, J. S., Gaczynska, M., Tortorella, D., Goldberg, A. L., and Ploegh, H. L. (1997). Covalent modification of the active site Thr of proteasomal β-subunits and the E. coli homolog HslV by a new class of inhibitors. *Proc. Natl. Acad. Sci. USA* **94**, 6629–6634.

Bogyo, M., Shin, S., McMaster, J. S., and Ploegh, H. (1998). Substrate binding and sequence preference of the proteasome revealed by active-site-directed affinity probes. *Chem. Biol.* **5**, 307–320.

Bogyo, M., and Wang, E. W. (2002). Proteasome inhibitors: Complex tools for a complex enzyme. *Curr. Top. Microbiol. Immunol.* **268**, 185–208.

Coux, O., Tanaka, K., and Goldberg, A. L. (1996). Structure and functions of the 20S and 26S proteasomes. *Annu. Rev. Biochem.* **65**, 801–847.

Eisen, M. B., Spellman, P. T., Brown, P. O., and Botstein, D. (1998). Cluster analysis and display of genome-wide expression patterns. *Proc. Natl. Acad. Sci. USA* **95**, 14863–1488.

Elofsson, M., Splittgerber, U., Myung, J., Mohan, R., and Crews, C. M. (1999). Towards subunit-specific proteasome inhibitors: Synthesis and evaluation of peptide alpha',beta'-epoxyketones. *Chem. Biol.* **6**, 811–822.

Greenbaum, D., Baruch, A., Hayrapetian, L., Darula, Z., Burlingame, A., Medzihradszky, K., and Bogyo, M. (2002a). Chemical approaches for functionally probing the proteome. *Mol. Cell Proteo.* **1**, 60–68.

Greenbaum, D. C., Baruch, A., Grainger, M., Bozdech, Z., Medzihradszky, K. F., Engel, J., DeRisi, J., Holder, A. A., and Bogyo, M. (2002b). A role for the protease falcipain 1 in host cell invasion by the human malaria parasite. *Science* **298**, 2002–2006.

Harris, J. L., Alper, P. B., Li, J., Rechsteiner, M., and Backes, B. J. (2001). Substrate specificity of the human proteasome. *Chem. Biol.* **8**, 1131–1141.

Harris, J. L., Backes, B. J., Leonetti, F., Mahrus, S., Ellman, J. A., and Craik, C. S. (2000). Rapid and general profiling of protease specificity by using combinatorial fluorogenic substrate libraries. *Proc. Natl. Acad. Sci. USA* **97**, 7754–7759.

Houghten, R. A., Pinilla, C., Blondelle, S. E., Appel, J. R., Dooley, C. T., and Cuervo, J. H. (1991). Generation and use of synthetic peptide combinatorial libraries for basic research and drug discovery. *Nature* **354**, 84–86.

Kisselev, A. F., and Goldberg, A. L. (2001). Proteasome inhibitors: From research tools to drug candidates. *Chem. Biol.* **8**, 739–758.

Leung, D., Hardouin, C., Boger, D. L., and Cravatt, B. F. (2003). Discovering potent and selective reversible inhibitors of enzymes in complex proteomes. *Nat. Biotechnol.* **21**, 687–691.

Nazif, T., and Bogyo, M. (2001). Global analysis of proteasomal substrate specificity using positional-scanning libraries of covalent inhibitors. *Proc. Natl. Acad. Sci. USA* **98,** 2967–2972.

Ostresh, J. M., Winkle, J. H., Hamashin, V. T., and Houghten, R. A. (1994). Peptide libraries: Determination of relative reaction rates of protected amino acids in competitive couplings. *Biopolymers* **34,** 1681–1689.

Overkleeft, H. S., Bos, P. R., Hekking, B. G., Gordon, E. J., Ploegh, H. L., and Kessler, B. M. (2000). Solid phase synthesis of peptide vinyl sulfone and peptide epoxyketone proteasome inhibitors. *Tetrahedron Lett.* **41,** 6005–6009.

Wang, G., Mahesh, U., Chen, G. Y., and Yao, S. Q. (2003). Solid-phase synthesis of peptide vinyl sulfones as potential inhibitors and activity-based probes of cysteine proteases. *Org. Lett.* **5,** 737–740.

Wang, G., and Yao, S. Q. (2003). Combinatorial synthesis of a small-molecule library based on the vinyl sulfone scaffold. *Org. Lett.* **5,** 4437–4440.

[41] Development of E3-Substrate (MDM2-p53)–Binding Inhibitors: Structural Aspects

By DAVID C. FRY, BRADFORD GRAVES, and LYUBOMIR T. VASSILEV

Abstract

Inhibition of E3 ligase–substrate binding is the most direct approach for blocking protein ubiquitylation and degradation. However, protein–protein interactions have proven to be difficult targets for discovery of small molecules that bind at the interface and modulate protein activity in a selective manner. Recently, we developed the first potent and selective small-molecule inhibitors of the binding between MDM2 E3 ligase and its substrate p53 (Vassilev *et al.*, 2004). This process was aided significantly by the acquisition and use of structural information. We describe herein how such information was obtained and used at various stages in the program. These applications included assessment of MDM2 as a target, evaluation of hits from high-throughput screening and the selection of lead molecules, and analysis of binding strategies used by the inhibitors as a basis for guiding studies of similar systems. These tools are likely to be useful in any attempt to find and develop druglike compounds that modulate the function of a protein-protein interaction.

Introduction

The tumor suppressor p53 is a transcription factor that plays a central role in the cell's defense against tumor development (Levine, 1997; Vogelstein *et al.*, 2000). In response to stress, p53 accumulates in cell nuclei by a posttranslational mechanism and induces multiple p53 target genes, leading to cell cycle arrest, apoptosis, or senescence. In unstressed cells, p53 is controlled by its negative regulator MDM2 (mouse double minute 2), which uses multiple mechanisms to repress p53 function (Freedman and Levine, 1999; Oliner et al., 1992). MDM2 binds p53 with high affinity and inhibits its transcriptional activity by concealing the transactivation domain of p53 (Oliner et al., 1993). Also, MDM2 serves as an E3 ubiquitin ligase that targets it for ubiquitin-dependent degradation in the proteasome (Haupt et al., 1997; Kubbutat et al., 1997; Midgley and Lane, 1997). Furthermore, MDM2 is involved in facilitating the nuclear export of p53. MDM2 and p53 form an autoregulatory feedback loop in which p53 exerts control over MDM2 transcription, whereas MDM2 modulates p53 activity and stability (Ashcroft and Vousden, 1999; Michael and Oren 2003; Picksley and Lane, 1993). This mechanism ensures that p53 is kept at a very low level in cycling cells to allow them to proliferate.

MDM2 has been found overexpressed in many human tumors and effectively disables p53 function (Momand et al., 1998). Therefore, stabilization and activation of p53 by antagonizing the p53–MDM2 interaction has been proposed as a novel strategy for therapy of tumors that have retained wild-type p53 (Lane, 1999). p53 has a very short half-life, and its cellular level is controlled almost exclusively by the rate of degradation. Therefore, inhibiting the E3 ligase activity of MDM2 should lead to stabilization of p53. Proof-of-concept experiments using inhibitors of MDM2 E3 ligase activity have shown that this approach can lead to accumulation of p53 and activation of p53-regulated genes (Lai et al., 2002). However, in addition to its role in p53 degradation, MDM2 inhibits directly the transcriptional activation by p53. Therefore, inhibitors of the E3 ligase activity of MDM2 that do not prevent its binding to p53 may not be able to completely restore p53 function. Antagonists of p53–MDM interaction that can bind specifically to MDM2 will release p53 from negative control and should stabilize and activate p53 to its full potential.

Recently, we reported the identification of the first potent and selective small-molecule inhibitors of the p53–MDM2 binding interaction, the nutlins (Vassilev et al., 2004). These compounds bind to the p53 site of MDM2 with high specificity, and the consequent release of p53 leads to its stabilization and activation of the p53 pathway in vitro and in vivo. Structural information obtained using NMR and x-ray crystallography was essential at every stage of development of the nutlins, from selecting MDM2 as a target to guiding the strategies used during lead identification and optimization. In this chapter, we describe some of the structure-based approaches used in the development of these inhibitors of protein–protein binding.

Structural Considerations Unique to Protein–Protein Interactions

For a small molecule to function effectively as an inhibitor of a protein–protein interaction it must overcome two key hurdles. (1) *Affinity:* Most protein–protein interaction surfaces are large and devoid of deep indentations. Affinity is attained by a multitude of weak interactions. It is difficult for a small molecule to effectively duplicate these interactions, because they are too numerous and too widely spaced. However, pioneering studies on the human growth hormone system showed that every situation is unique, and cases exist in which such inhibitors should be obtainable (Clackson and Wells, 1995). In the case of growth hormone and its receptor, it was shown by extensive mutagenesis and structural studies that a limited number of amino acids mediate nearly all of the key interactions that comprise the binding affinity. This subregion of the binding contact area was termed the "hot spot." Its dimensions were comparable to the size of a small organic molecule. (2) *Specificity:* In a typical protein–protein interaction, specificity is not attained by means of shape complementarity, the strategy normally used in a deep pocket, but rather by the requirement for a large number of weak interaction partners to be presented in a specific spatial arrangement. It is difficult for a small molecule to possess more than a subset of these specificity-encoding binding elements, and the molecule may, therefore, bind to other proteins whose interaction surface happens to contain a complementary subset. Furthermore, proteins can use protrusions that are harmlessly far from the interaction site with the desired partner protein but might bump into a secondary region of an undesired partner and thereby prevent binding, even though that partner has an accommodating primary interaction site. A small molecule would be unable to position such an "antibinding" element sufficiently far from its binding epitope.

Small-molecule inhibitors of protein–protein interactions have been developed, although such successes have been rare (reviewed in Toogood [2002], Berg [2003], and Arkin and Wells [2004]). In the search for these molecules, affinity was the main goal, and there was no concern toward specificity or any drive to ensure that the molecules be druglike in their other physical properties. In these respects, these molecules typically possess liabilities, such as charged groups, an excessive number of rotatable bonds, or too much peptidic character. It has been very rare to encounter a case in which a reported small molecule inhibitor of a protein–protein interaction has been druglike.

Structural Aspects of the p53–MDM2 Interaction

As discussed, any attempt to disrupt a protein–protein interaction is likely to encounter substantial difficulties at the molecular level. However, such a blanket characterization is not appropriate, and each case must be judged individually. Structural data are critical for making such a judgment in an informed manner. For the p53–MDM2 interaction, structural attributes were well characterized. The domain of MDM2 that binds to p53 was known to be reasonably small, contained within its first 120 N-terminal residues (Chen *et al.*, 1993). The region of p53 that participates in the interaction was known to also be quite small—a 15-residue peptide fragment had been shown to efficiently compete with the intact protein for binding (Picksley *et al.*, 1994). X-ray structures were available for p53 peptides in complex with the binding domain of MDM2 from two species, human (residues 17–125) and *Xenopus* (residues 13–119) (Kussie *et al.*, 1996). These structures revealed that the p53 peptide adopts a helical structure when bound and inserts three hydrophobic side chains into subpockets of the MDM2 site. Therefore, the dimensions of the binding interface appeared relatively compact, suggesting that a small organic molecule might be able to mimic enough of the critical interactions to act as a high-affinity inhibitor of the interaction.

To apply a more quantitative method for assessing the ability of a site to support binding of a small molecule, we used cavSearch (Stahl *et al.*, 2000), a computational program written in-house. CavSearch identifies and maps cavities in proteins using a grid-based measure of accessibility. The accessibility value is a unitless term that ranges from 1 (full accessibility to solvent) to 0 (completely buried within the protein). The parameters of the program have been adjusted using known examples to establish what constitutes a meaningful cavity. One output from the program is a graph of accessibility versus hydrophobic surface of the cavity (in square Angstroms). In essence, this indicates the size of the cavity at any given "depth," where depth is defined as distance a solvent molecule would have to traverse to achieve access. It is expected that an amenable binding site would be deep enough to sequester a ligand molecule away from substantial contact with solvent. To calibrate how such a graph would look for an amenable binding site, we examined known examples of such sites. We used Relibase to identify within the PDB a variety of ligands that appeared druglike (Oprea, 2000) and then extracted x-ray structures of these ligands in complex with their associated proteins. We edited the list down to one representative case per protein, for a total of 29 proteins. For each protein, a cavSearch analysis was performed. Graphs were made in which accessibility was plotted against hydrophobic surface area. We also examined, in the same manner, selected proteins for which it has appeared difficult to find druglike small molecule effectors that bind with high affinity (PDZ domain, SH2 domain, EH domain, and Ubc9). These were intended to represent nonamenable sites. The two types of sites turned out to be separately clustered. The amenable sites were not only larger overall but also possessed substantially more binding area at greater depths. For example, if one focuses on the accessibility value of 0.5, the amenable cavities are found to retain 200–600 sq Å of surface area, whereas for the nonamenable cavities, the area is less than 100 sq Å (as given for selected examples in Table I). Admittedly, the cavSearch analysis considers individual pockets, so a site with a second subpocket (such as the SH2 domain) may score poorly even though a compound that spanned both subpockets

TABLE I
COMPARATIVE ANALYSIS OF THE DIMENSIONS OF BINDING SITES USING THE PROGRAM CAVSEARCH[a]

CRABP	526.3
Glucokinase	521.5
FR-1	484.2
Thymidylate synthase	452.5
Lck kinase	408.7
DHODH	396.7
Neuraminidase	391.2
CDK2	379.8
ACHE	378.7
Stromelysin	372.5
Ricin	335.2
sPLA2	305.2
Beta lactamase	301.7
DHFR	285.5
Factor Xa	258.3
DHMPP kinase	257.3
HIV protease	239.8
Carbonic anhydrase	234.8
MDM2	120.8
PDZ domain	90.8
Ubc9	76.8
EH domain	54.7
SH2 domain	48.3

[a] Values represent available binding area at an accessibility value of 0.5. The table lists selected proteins from an analysis using the program cavSearch (Stahl et al., 1999). X-ray structures were obtained from the PDB, and for each, the dimensions of the primary binding site were graphed as hydrophobic surface area versus accessibility value. The accessibility value is a unitless term that ranges from 1 (full accessibility to solvent) to 0 (completely buried within the protein). The value of the area in square Angstroms at an accessibility value of 0.5 was taken from the graph. The first 18 proteins on the list are known to bind druglike small molecules. The bottom four proteins participate in protein–protein interactions and have appeared to be difficult targets for discovery of druglike inhibitors.

might be able to bind tightly. However, such a molecule is likely to be large and/or extended in conformation, and thereby not druglike.

After calibrating the cavSearch tool, we used it to assess the amenability of the MDM2 binding pocket. The profile for MDM2 fell in between those of the amenable and the nonamenable sites (see also Table I). This suggested that finding a druglike inhibitor of this system should be possible but would be challenging.

Hit Assessment by NMR

As typical of most modern drug discovery efforts, modulators of MDM2 were sought by performing a high-throughput functional screen, using a diverse library of synthetic chemicals. From the screen, a number of compounds were identified as active. NMR was used to assess the binding mechanism for each of the most interesting hits. There were 21 compounds identified as having sufficient potency and druglike characteristics to warrant this type of assessment. We pursued an NMR structure of the complex with MDM2 for each of these compounds.

The pursuit of NMR structures of ligands in complex with a protein historically has required acquisition and processing of at least five 2-D and 3-D NMR experiments, as exemplified for the case of MDM2 in Fry *et al.* (2004). This demands a protein sample that is stable over time at the high concentrations needed for NMR. Preliminary studies using a recombinant human MDM2 fragment (residues 1–118) indicated that the protein was not very stable in terms of behavior in an NMR tube and precipitated over time. MDM2 from *Xenopus leavis* (residues 13–119) was found to be much more stable. A hybrid version of the protein was ultimately created, designated "humanized-*Xenopus*-MDM2" ("hx-MDM2"), which consisted of residues 13–119 of the *Xenopus* sequence with three amino acid changes in the active site, corresponding to residues found in those positions in the human sequence—Ile50 to Leu; Pro92 to His; and Leu95 to Ile. The stability characteristics of hx-MDM2 were found to be intermediate between those of the human and *Xenopus* forms. Also, the binding attributes of hx-MDM2 were found to be about midway between those of the human and *Xenopus* forms. This was viewed as an optimal combination of stability and relevancy.

The first step in determining an NMR structure of a protein-ligand complex is to prepare the sample and acquire initial 1-D and 2-D spectra. During this phase, the behaviors of the hits from the high-throughput screening of MDM2 were found to cluster into distinct classes. For one class, attempts to prepare samples and acquire initial NMR spectra were prevented by low solubility of the compounds in aqueous buffer. Our protocol for testing the compounds was to first create a 63 mM stock solution in deuterated DMSO and then test the solubility of each compound by stepwise addition to buffer. For this test, 630 μl of MDM2 buffer (50 ml MES-d13, 150 ml KCl, 50 ml DTT-d10, 1.5 ml NaN$_3$, pH 7.0) was placed into a prerinsed 1.5-ml centrifuge tube. Into the sample was inserted a GH 4-mm combination electrode from a Radiometer PHM210 pH Meter. Sequential additions of 2 μl from the 63 mM stock solution were introduced to a upper limit of 5 aliquots, mixing immediately on each addition by quickly moving the centrifuge tube around so that the pH probe acted as a stationary stirring rod. Any changes in pH that occurred because of addition of the compound were monitored. The appearance of visible precipitate was used to determine the saturating concentration. A 1-D 1H NMR spectrum of the solution was acquired to ensure that the compound was present in solution and was not aggregated. Compounds that did not pass a solubility criterion of 0.2 mM were designated "insoluble."

Compounds found soluble to at least 0.2 mM were added to hx-MDM2. This was accomplished by placing 630 μl of an hx-MDM2 NMR sample into a prerinsed 1.5-ml centrifuge tube and sequentially adding compound in steps of 0.2 mM to an upper limit of 1 mM in a similar manner to that described previously (the preparation and composition of the NMR sample of hx-MDM2 is described in Fry *et al.* [2004]). If pH changes were detected, the pH was restored by adding microliter aliquots of 1 M NaOD or 1 M DCl with immediate mixing. Appearance of precipitate was checked visually, and additions were stopped if precipitate was observed. A 1-D 1H NMR spectrum of the sample was acquired. Depending on available time, other NMR experiments were acquired such as 2D 1H-15N-HSQC, 3D 1H-1H-15N-NOESY-HSQC, or 2D 13C/15N-filtered-1H-1H-NOESY (detailed conditions for acquisition of these spectra are described in Fry *et al.* [2004]). A second class of compound behavior became identifiable at this stage, designated "Causes protein to aggregate, unfold, etc." This behavior was revealed by a dramatic change in the character of the NMR spectra of MDM2 on addition of the compound. Such changes included the overall signal intensity becoming significantly reduced, the chemical shifts converting into a narrow range similar to that exhibited by unstructured polypeptide chains, or the line widths becoming exceedingly broad, or some combination of these. These spectral changes are consistent with the explanation that the compounds in this category bound to the protein and, thereby, triggered some degree of unfolding of its tertiary structure. It is expected that such unfolding can lead to exposure of opportunistic sites to which additional compound molecules can bind, further exacerbating the denaturation of the protein. Alternately, unnatural hydrophobic surface patches may be created, which cause aggregation of protein molecules. Either of these processes can ultimately lead to precipitation of the protein from the solution. Indeed, some amount of precipitation was visually observed in the sample after addition to the protein for each of the compounds in this class.

In a third category, designated "No binding detectable," compounds were found to stay in solution but to cause no change in the NMR parameters of the protein. Such a result indicated that the compounds were not binding to the protein. It has been consistently observed, even for weak binders in the millimolar range, that binding causes some observable change to the properties of the NMR resonances of the protein in terms of chemical shift positions or line widths. NMR, therefore, is one of the only techniques that can be a reliable indicator of lack of binding. These compounds must have been acting as false positives in the original assay.

Once the false positives and compounds with undesirable mechanisms were eliminated, structures could be pursued for the compounds that exhibited more conventional binding. Such binding was characterized by changes in NMR parameters for resonances belonging to a limited number of protein residues and the absence of visible precipitation. In some cases, the NMR parameters changed by only a very small amount. This indicated that the compound was probably not very deeply inserted into the body of the protein and was not making contact with much of the protein surface. The affinity of such compounds was expected to be low, and these were designated "Binds weakly." Pursuing detailed structures of such compounds bound to the protein was not expected to be very informative in terms of drug discovery, other than verifying that there was little potential for increasing affinity significantly.

Some compounds that produced perturbations to a limited number of residues exhibited changes in the NMR parameters that were more substantial in magnitude. These compounds were designated "Binds authentically." The pursuit of structures of these compounds in complex with hx-MDM2 was deemed valuable. In summary, the bar graph in Fig. 1 indicates number of compounds found for each of the various categories of hit behavior observed in the MDM2 project. For the compounds that bound authentically, a full structure of the complex between hx-MDM2 and the compound was pursued. An example of the acquisition and processing of NMR data required for determining such a structure is provided in Fry *et al.* (2004).

One feature of determining a structure by NMR is that the information is obtained steadily over time rather than as an all-or-none process. Accordingly, an approximate structure of the complex can be derived at

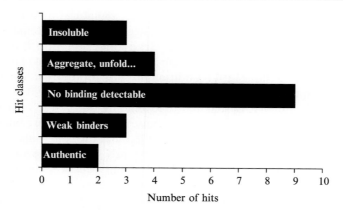

FIG. 1. Bar graph indicating the number of compounds found for each of the various categories of hit behavior observed in the MDM2 project.

an early stage, which then increases in resolution and accuracy (Fry and Emerson, 2000). Molecular modeling can be used to fill in gaps in knowledge at the early stages with guesses that are based on reasonable energetics and established interaction principles. At the beginning stages, we interpreted the NMR data obtained for hx-MDM2 by cataloging which resonances exhibited the largest changes in their linewidths, chemical shifts, and NOE patterns. For compounds belonging to each of the two chemical classes shown to bind authentically, the protein residues to which the most perturbed resonances belonged were found to be clustered in a 3-D sense within the structure of hx-MDM2. This information was used to derive initial approximations of these compounds bound to hx-MDM2.

Compounds of the imidazoline class were selected to serve as leads in a synthetic chemistry program aimed at improving potency. The emerging NMR-derived structures of these compounds docked into MDM2 were found to be useful during the drug design process. Ultimately, compounds were developed with potencies in the nanomolar range (Vassilev *et al.*, 2004).

X-Ray Crystallography

A critical aspect of the drug discovery process is the attainment of protein crystal structures complexed with key inhibitors. In the case of the MDM2 inhibitor program, such structures were obtained (for example, see Fig. 2A), and they verified the essential binding mode for the imidazolines that had been predicted on the basis of the NMR and molecular

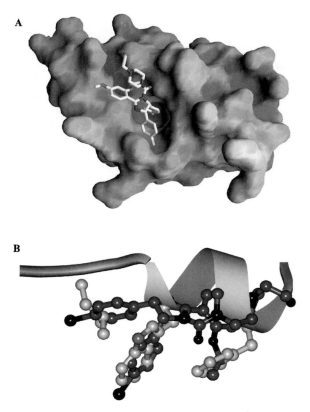

Fig. 2. Crystal structure of nutlin-2 bound to HDM2. (A) The structure of the p53-binding domain of HDM2 is depicted as a molecular surface with depressions in the surface colored as dark gray areas. Nutlin-2 is drawn as a stick model, with carbon atoms white and noncarbon atoms (nitrogen, oxygen, and bromine) various shades of gray. This image was drawn with the program GRASP (Nicholls, 1991). (B) Comparison of the ligand binding components of two structures with HDM2. The p53 peptide structure (Kussie, 1996) is represented as a gray ribbon for the main chain component, and the three critical side chains (Phe^{19}, Trp^{23}, and Leu^{26}) are shown explicitly with the atoms drawn as light gray spheres. The inhibitor structure (Vassilev, 2004) is drawn as a ball-and-stick model, with carbon atoms colored as dark gray spheres and noncarbon atoms (nitrogen, oxygen, and bromine) as black spheres. This image was drawn with program MOLSCRIPT (Kraulis, 1991).

modeling results. In addition, the structures confirmed that only one enantiomer (Fig. 2A) was bound (shown previously through binding assays with separated enantiomers) and defined the stereochemistry of the active enantiomer.

Another revelation from these structures was the manner in which the small molecule inhibitors mimicked the most important binding elements of the native ligand. The analysis shows that although the peptide is much larger, parts of it are not directly essential, and the functions of these parts can be duplicated in a more economical manner by the small inhibitor Historically, programs aimed at developing peptide mimetics have typically attempted to design a scaffold that would enable attachments that would overlay, in a 3-D sense, the C_α–C_β bonds of the critical amino acids. That was not an acceptable strategy in this case if we were to develop small molecule inhibitors with druglike properties. As can be seen in Fig. 2B, the imidazoline core is able to efficiently position groups that mimic the three crucial side chains of p53 (Phe^{19}, Trp^{23}, and Leu^{26}, as identified by Kussie et al. [1996]) but in a more direct fashion than would have been possible if we had forced the inhibitors to enter the subsites through the C_α–C_β vectors.

Acknowledgments

We acknowledge the contributions of S. Donald Emerson, Sung-Sau So, Frank Podlaski, Stefan Palme, Chao-Min Liu, and Binh T. Vu to the work described in this chapter. We thank Nader Fotouhi and David Heimbrook for critically reading the manuscript.

References

Arkin, M. R., and Wells, J. A. (2004). Small-molecule inhibitors of protein-protein interactions: Progressing towards the dream. *Nat. Rev. Drug Discov.* **3,** 301–317.

Ashcroft, M., and Vousden, K. H. (1999). Regulation of p53 stability. *Oncogene* **18,** 7637–7643.

Berg, T. (2003). Modulation of protein–protein interactions with small organic molecules. *Angew. Chem. Int. Ed. Engl.* **42,** 2462–2481.

Chen, J., Marechal, V., and Levine, A. J. (1993). Mapping of the p53 and mdm-2 interaction domains. *Mol. Cell. Biol.* **13,** 4107–4114.

Clackson, T., and Wells, J. A. (1995). A hot spot of binding energy in a hormone-receptor interface. *Science* **267,** 383–386.

Freedman, D. A., Wu, L., and Levine, A. J. (1999). Functions of the MDM2 oncoprotein. *Cell Mol. Life Sci.* **55,** 96–107.

Fry, D. C., and Emerson, S. D. (2000). Applications of biomolecular NMR to drug discovery. *Drug Des. Discov.* **17,** 13–33.

Fry, D. C., Emerson, S. D., Palme, S., Vu, B. T., Liu, C. M., and Podlaski, F. (2004). NMR structure of a complex between MDM2 and a small molecule inhibitor. *J. Biomol. NMR* **30,** 163–173.

Haupt, Y., Maya, R., Kazaz, A., and Oren, M. (1997). Mdm2 promotes the rapid degradation of p53. *Nature* **387,** 296–299.

Kraulis, P. J. (1991). MOLSCRIPT: A program to produce both detailed and schematic plots of protein structures. *J. Appl. Cryst.* **24,** 946–950.

Kubbutat, M. H., Jones, S. N., and Vousden, K. H. (1997). Regulation of p53 stability by Mdm2. *Nature* **387**, 299–303.

Kussie, P. H., Gorina, S., Marechal, V., Elenbaas, B., Moreau, J., Levine, A. J., and Pavletich, N. P. (1996). Structure of the MDM2 oncoprotein bound to the p53 tumor suppressor transactivation domain. *Science* **274**, 948–953.

Lai, Z., Yang, T., Kim, Y. B., Sielecki, T. M., Diamond, M. A., Strack, P., Rolfe, M., Caligiuri, M., Benfield, P. A., Auger, K. R., and Copeland, R. A. (2002). Differentiation of Hdm2-mediated p53 ubiquitination and Hdm2 autoubiquitination activity by small molecular weight inhibitors. *Proc. Natl. Acad. Sci. USA* **99**, 14734–14739.

Lane, D. P. (1999). Exploiting the p53 pathway for cancer diagnosis and therapy. *Br. J. Cancer* **80**(Suppl 1), 1–5.

Levine, A. J. (1997). p53, the cellular gatekeeper for growth and division. *Cell* **88**, 323–331.

Michael, D., and Oren, M. (2003). The p53-Mdm2 module and the ubiquitin system. *Semin. Cancer Biol.* **13**, 49–58.

Midgley, C. A., and Lane, D. P. (1997). p53 protein stability in tumour cells is not determined by mutation but is dependent on Mdm2 binding. *Oncogene* **15**, 1179–1189.

Momand, J., Jung, D., Wilczynski, S., and Niland, J. (1998). The MDM2 gene amplification database. *Nucleic Acids Res.* **26**, 3453–3459.

Nicholls, A., Sharp, K. A., and Honig, B. (1991). Protein folding and association: Insights from the interfacial and thermodynamic properties of hydrocarbons. *Proteins: Structure, Function Genetics* **11**, 281–296.

Oliner, J. D., Kinzler, K. W., Meltzer, P. S., George, D. L., and Vogelstein, B. (1992). Amplification of a gene encoding a p53-associated protein in human sarcomas. *Nature* **358**, 80–83.

Oliner, J. D., Pietenpol, J. A., Thiagalingam, S., Gyuris, J., Kinzler, K. W., and Vogelstein, B. (1993). Oncoprotein MDM2 conceals the activation domain of tumour suppressor p53. *Nature* **362**, 857–860.

Oprea, T. I. (2000). Property distribution of drug-related chemical databases. *J. Comp. Aided Mol. Design.* **14**, 251–264.

Picksley, S. M., and Lane, D. P. (1993). The p53-mdm2 autoregulatory feedback loop: A paradigm for the regulation of growth control by p53? *Bioessays* **15**, 689–690.

Picksley, S. M., Vojtesek, B., Sparks, A., and Lane, D. P. (1994). Immunochemical analysis of the interaction of p53 with MDM2—fine mapping of the MDM2 binding site on p53 using synthetic peptides. *Oncogene* **9**, 2523–2529.

Stahl, M., Bur, D., and Schneider, G. (2000). Mapping of protease active sites by surface-derived vectors. *J. Comp. Chem.* **20**, 336–347.

Toogood, P. L. (2002). Inhibition of protein-protein association by small molecules: Approaches and progress. *J. Med. Chem.* **45**, 1543–1558.

Vassilev, L. T., Vu, B. T., Graves, B., Carvajal, D., Podlaski, F., Filipovic, Z., Kong, N., Kammlott, U., Lukacs, C., Klein, C., Fotouhi, N., and Liu, E. A. (2004). In vivo activation of the p53 pathway by small-molecule antagonists of MDM2. *Science* **303**, 844–848.

Vogelstein, B., Lane, D., and Levine, A. J. (2000). Surfing the p53 network. *Nature* **408**, 307–310.

[42] Druggability of SCF Ubiquitin Ligase–Protein Interfaces

By TIMOTHY CARDOZO and RUBEN ABAGYAN

Abstract

The unique mechanism of the SCF ubiquitin ligase poses a challenge to drug discovery. A central enzymatic small molecule–binding active site is not evident in this multisubunit protein enzyme, as is the case with kinases or proteases. Instead, the SCF ligase seems to accomplish ubiquitylation through a series of cooperative movements dependent on the protein interfaces between its components and its substrate. Activity-modulating small molecules, therefore, need to interact with these protein interfaces. The three-dimensional structure of these interfaces may be the key asset in determining their suitability for small molecule binding. Computational tools and a systematic approach described in detail here can assess the "druggability" of an SCF ligase before the investment of effort in high-throughput screening (HTS), structure-based drug design (SBBD), or virtual library screening (VLS).

Introduction

Can Protein Interfaces Be Targeted for Drug Discovery?

Many, if not most, important biological interactions are interactions between biomolecular (e.g., protein) surfaces and *not* interactions between small molecules (drugs) and biomolecules. However, small molecules have overwhelming therapeutic advantages over biomolecules such as peptides and nucleic acids in humans because of their greater potential for bioavailability and oral delivery. Much prior effort has been dedicated to the discovery of small molecules that could disrupt protein interfaces. The fact that many point mutations disrupt protein interactions and the fact that a flexible amino acid side chain is equal to or smaller in size than many small molecule drugs suggests that many protein interfaces are susceptible to disruption by small chemical compounds. One of the prototype efforts in this regard however—the targeting of the SH2 domain interfaces—is generally regarded as a failure (Cochran, 2000). On the other hand, more recent drug discovery efforts targeted at protein interfaces have been successful (Berg *et al.*, 2002; Chen *et al.*, 2002; Lepourcelet *et al.*, 2004; Wang *et al.*, 2000). Notably, an inhibitor of the E3 ubiquitin ligase Mdm2, which targets the tumor suppressor p53 for degradation, was recently found

(Vassilev, 2004; Vassilev et al., 2004). In addition, some drugs already in clinical use, such as the vinca alkaloids and taxanes that noncatalytically alter microtubule interactions, modulate protein interfaces as their mechanism of action (Kavallaris et al., 2001). This body of evidence suggests what may, in retrospect, have been common sense: that all protein interfaces are not created equal and that at least some protein interfaces are susceptible to drug discovery ("druggable"). Methods that distinguish one interface from the other may be valuable in this regard.

Methods of this type include rational and nonrational approaches and equate the identification of a ligand-binding pocket with "druggability." Knowledge of the target can be used to infer the existence of a small molecule–binding pocket (rational approach), or binding assays can be used to directly establish it. Because the binding assays are redundant in both requirement and scale to high-throughput screening (HTS) assays themselves, only rational approaches can be efficient preliminary evaluation tests in the drug discovery process.

Rational approaches evaluate the presence of a drug binding area in a protein using three types of structural information: sequence information (there may be a motif for binding of a small molecule such as the ATP binding P-loop (Saraste et al., 1990), chemical/three-dimensional (3-D) structure of the ligand/substrate, or chemical/3-D structure of the receptor. Of these, the 3-D structure of the target receptor is the only rational approach that has a chance of being independent of ligand information, and, therefore, this is the only approach that can rule out the possibility that a suitable small molecule can bind to a receptor (negative predictive value), as well as being the approach that can predict the binding of a ligand whose structure is novel or not yet known (by rendering the surface pocket without knowledge of the ligand). Several algorithms have been developed to date that use only the 3-D structure of the receptor to identify ligand binding sites (An et al., 2004; Bliznyuk and Gready, 1998; Brady and Stouten, 2000; Dennis et al., 2002; Glick et al., 2002; Hendlich et al., 1997; Kortvelyesi et al., 2003; Laskowski, 1995; Levitt and Banaszak, 1992; Liang et al., 1998; Peters et al., 1996; Ruppert et al., 1997; Verdonk et al., 2001).

A leading surface pocket identification method (An et al., 2004) was used to analyze three "gold standards" of protein interface drug discovery (Fig. 1): the SH2 domain, which failed to accommodate a successful drug discovery screen (a negative control); and the mdm2 p53 binding domain and BCL-XL domains, for which published drug discovery efforts were successful (positive controls). The results show that the volume and location of surface pockets may help to predict drug discovery success: if a pocket is of comparable size to known drugs and is located in the vicinity of residues whose point mutation can disrupt the protein interface, compounds are likely to exist that bind to the pocket and also inhibit the

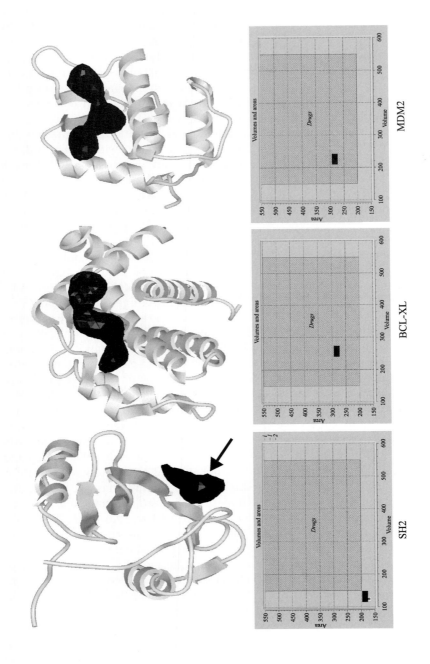

function associated with the surrounding protein interface. Interestingly, this analysis suggests a rationale for why SH2 inhibitors are elusive: although there is a pocket coinciding with interface point mutation residues, the pocket is too small to accommodate most compounds. At protein interfaces, then, not only does the critical possibility exist that compounds may not bind to the interface, but this analysis suggests that compound binding and compound antagonism of function may be uncoupled as well (if the pocket is not in the right location). These possibilities should be addressed at the initiation of drug discovery.

Drug Discovery Methods: High-Throughput Screening Versus SBDD and VLS

HTS tests thousands to millions of compounds directly in automated biochemical assays of activity. Although optimally sensitive, this type of screening is blind to the structure of the target protein and specific to varying degrees to the artificial conditions of the assay. HTS is also costly, and much of the expenditure occurs before potential success or failure can be evaluated.

Structure-based drug design (SBBD) methods are divided between those that incorporate an experimental structure determination step and those that are purely computational such as Virtual Library Screening (VLS) (Abagyan and Totrov, 2001). These methods are highly variable in cost and sensitivity but generally far less expensive than HTS for VLS. If sufficiently accurate, SBBD and VLS result in preliminary knowledge of the structural basis for inhibition by a drug, an important asset in lead optimization. HTS and SBBD/VLS are frequently used in concert to maximize the yield and quality of lead discovery (for review of HTS, VLS, and the complementarity of the two approaches, see Bajorath [2002]).

In the case of the SCF ubiquitin ligases, SBBD and VLS have additional advantages. Inhibitors discovered through HTS may be acting

---

FIG. 1. Receptor-defined pockets predict drug discovery success. The LigandPocket algorithm (An *et al.*, 2004) was used to identify and render (as black geometric objects) surface pockets in three query structures based only on the crystallographic coordinates of the protein (displayed in gray backbone ribbon depiction). The sizes of the rendered pockets shown are plotted as black rectangles on the grid plot in each panel: Volume (x-axis, $Å^3$) versus Area (y-axis, $Å^2$). The light gray hatched square overlays the region of the plot in which the volume/area of 15,000 known drugs in the structural database fall (between 100–500 $Å^{2/3}$ Area/Volume). Note that although a small pocket is found (arrow, left panel) in the Grb2 SH2 domain (PDB code 1jyr), it is smaller than that for known drugs (falls outside the hatched gray area). This interface proved resistant to drug discovery (see text). The pockets found on the structures of BCL-XL (PDB code 1r2d, middle panel) and MDM2 (PDB code 1ycr, left panel) are of appropriate size for binding druglike compounds (their size falls within the hatched gray area). Drug discovery targeted at these interfaces succeeded. In all three panels, published point mutations that disrupt the protein interface are in contact with the rendered pockets at these protein interfaces.

at the substrate-binding surface, at the E2 binding surface, or at allosteric sites throughout the enzyme, and it would not be trivial to determine which was the case. Where the inhibitors act will determine the specificity of the inhibitor: a compound competitively inhibiting the binding of the substrate should be specific for a particular F-box protein, whereas competitive inhibition of the binding of Skp1 or the E2 may inhibit many SCF ligases. In SBBD and VLS, the structural location of action is inherent.

Three-Dimensional Structures of SCF Ligases, Related Ligases, and Associated Factors

Three-dimensional structures of the target protein receptor may be available in the form of crystallographic structures, NMR structures, or models based either on sequence or low-resolution data (diffraction, electron microscopy). Engineering the interactions of a small molecule with a protein requires fine discrimination of interactions at the atomic level. Crystallographic structures are the "gold standard" for atomic level resolution, and NMR models may be no better than theoretical models (Abagyan and Totrov, 1997). Fortunately, the crystallographic structures of the Skp1-β-TrCP, Skp1-Skp2, and SCF^{Skp2} complexes have been reported (Schulman et al., 2000; Wu et al., 2003; Zheng et al., 2002b). Structures of several other F-box proteins and SCF ligase-associated proteins have also been resolved (Table I). The multiprotein SCF enzyme is a large C-shaped structure with a central cavity. The F-box protein binds the substrate on one side of the "C," and the E2–ubiquitin binds on the other side. Ligation then seems to takes place in the central cavity. The mechanism thus seems to operate through precise 3-D orientation and positioning of the substrate relative to the activated E2–ubiquitin complex (Cardozo and Pagano, 2004). Attachment of Nedd8, which modulates the activity of the enzyme, occurs near the E2 binding site (Wei and Deng, 2003). Cand1, a structurally unique HEAT-repeat inhibitory protein, binds both at the Skp1 and the E2 end of the central rod like cullin (Goldenberg et al., 2004; Zheng et al., 2002a). Several of the crystallographic structures include bound peptide fragments of the substrate for that respective ubiquitin ligase. In addition, almost all the structures were determined to a resolution of less than 3.0 Å. Several independent and high-quality crystallographic studies are, therefore, available that completely include four main interfaces of the SCF: the F-box protein–substrate interface, the F-box protein–Skp1 interface, the Skp1–cullin interface, and the cullin–Roc1/Rbx1 interface. The E2-binding interface and the Nedd8 attachment site are also available but not precisely defined at present. The general quality of the library of structural

TABLE I
STRUCTURAL BIOLOGY OF THE CULLIN-DEPENDENT E3 UBIQUITIN LIGASES (CDLS)

Structure (PDB code) (Reference)	Resolution (Å)	Topology
Skp1–Skp2 complex (1FQV, 1FS1, 1FS2) (Schulman et al., 2000)	2.8	
SCF–Skp2 complex (1LDD, 1LDK, 1LDN) (Zheng et al., 2002b)	3.1	
β-TrCP–Skp2 complex (1P22) (Wu et al., 2003)	3.0	

(continued)

TABLE I (continued)

Structure (PDB code) (Reference)	Resolution (Å)	Topology
SCF-CDC4 complex (1NEX) (Orlicky et al., 2003)	2.7	
Fbx2 (Fbs1) (1UMH)[a]	2.0	
Cand1-Cul1-Roc1 1U6G) (Goldenberg et al., 2004)	3.1	
VHL-HIF1(1LQB, 1LM8)[b]	2.0, 1.85	
VHL-ELC/B (1VC3)[c]	2.7	

Publicly available crystal structures of Cdls and Cdl-associated proteins. PDB, Protein Data Bank. SCF ubiquitin ligases are in bold letters, and the general topology of their F-box protein structures are shown in the right column, from the top row of the table to the bottom: the horseshoe-shaped leucine-rich repeat (LRR) domain of Skp2; the donut shaped WD-40 domains of β-TrCP and hCdc4/Fbw7; the lectin-like sugar binding domain of Fbx2.

[a] Mizushima, T., Hirao, T., Yoshida, Y., Lee, S. J., Chiba, T., Iwai, K., Yamaguchi, Y., Kato, K., Tsukihara, T., and Tanaka, K. (2004). Structural basis of sugar recognizing ubiquitin ligase. *Nat. Struct. Mol. Biol.* **11**, 365–370.

[b] Hon, W. C. Wilson, Mi. I., Harlos, K., Claridge, T. D., Schofield C. J., Pugh, C. W., Maxwell, P. H., Ratcliffe, P. J., Stuart, D. I., and Jones, E. Y. (2002). Structural basis for recognition of hydroxyproline in HIF-1 alpha by pVHL. *Nature* **417**, 975–978.

[c] Stebbins, C. E., Kaelin, W. G., Jr., and Pavletich, N. P. (1999). Structure of the VHL- ElonginC-ElonginB complex: Implications for VHL tumor suppressor function. *Science* **284**, 455–461.

information available to investigate the protein interfaces of the SCF complex is thus quite good and should continue to grow.

Using SCF Ligase Structures to Determine Whether to Target a Protein Interface for Drug Discovery

Local Quality of the Structural Data

Although the quality and quantity of global SCF ligase crystallographic information is quite good, drug discovery targeted at a particular site on the structure depends ultimately on the quality of the structural model at that particular site. Each atom of a 3-D protein model derived from crystallographic data has a certain reliability associated with it, which can be independent of the overall resolution of the structure (Chelvanayagam et al., 1994). The overall resolution of a structure may be excellent, but the individual atoms or protein fragments making up a small target surface are unreliable. In addition, artificial local conformations in the protein may be difficult to detect. In either case, modeling is required to improve that surface or pocket analysis or screening of that surface may give inaccurate results.

At least three errors or deviations may be inherent in solved structures that require special analyses for detection:

1. Backbone peptide flips: the mirror image of the true ϕ/φ configuration of a residue may fit the experimental electron density extremely well (Kleywegt and Jones, 1995).
2. High B factors (temperature factors): A measure of the reliability of the depicted atom positions in the crystallographic model (Kuriyan and Weis, 1991) usually correlates with disorder or flexibility of local protein segments. The correlate in NMR structures is local deviation between all models satisfying NMR restraints.
3. Crystal packing contact surfaces: contacts between different monomers in the crystal produce a local conformation of the atoms that is artificial for *in vivo* physiological conditions(Kossiakoff et al., 1992).

All the required data for detecting these errors is present in the publicly available Protein Data Bank (PDB) file (Sussman et al., 1998) for any given structure; however, common molecular modeling software packages may be required to expeditiously perform them. The command sequence provided here for illustration is from ICM (Abagyan et al., 1994) (Molsoft LLC, La Jolla CA), which has the largest set of tools specific for the tasks described in this chapter. Thus, the order of commands described hereafter, simply entered by the reader within the ICM Software package, will successfully execute the entire method described in this chapter.

Note, however, that the goal of each step in this method is clearly delineated so equivalent commands or subroutines in any molecular modeling package or programming language may be inferred or developed from the sequence of commands here. In addition, command line sequences are shown for clarity, but these functions are available from more user-friendly GUI pull-down menus in ICM and other molecular modeling packages.

The approach described here applies to any structurally resolved SCF ubiquitin ligase and, indeed, to any structurally resolved biomolecule with an entry in the PDB. To illustrate the method, we specifically targeted the substrate-binding interface of the F-box protein β-TrCP, which forms an SCF ligase that ubiquitylates β-catenin, cdc25A, and Emi1 and other substrates (Jin et al., 2004). The 3-D structure of β-TrCP in complex with Skp1 and a bis-phosphorylated β-catenin peptide was deposited in the PDB on July 8th, 2003.

Assessing Local Structural Reliability through Energy Strain

Potential energy scores (Hao and Scheraga, 1999) can be calculated from protein structural models and may detect steric clashes, exposed hydrophobic areas, and other physicochemical anomalies that are visually subtle or invisible in graphical views of the structure. These errors may be detected by analyzing the energy or force-field strain along the backbone (Maiorov and Abagyan, 1998) or less efficiently by careful (but tedious) inspection. Here, we download the PDB file for β-TrCP, and the energy strain along the backbone is automatically calculated and mapped graphically as an array onto the structure.

1. Reading files from the PDB and generating realistic 3-D virtual models (for this and all following commands, any PDB code for a structure may be substituted for "1p22"):
   [start ICM]
   Read pdb "1p22" (loads coordinates of the X-ray model of β-TrCP from the PDB)
   convertObject a_1p22. yes yes yes no (converts the PDB file to a 3-D virtual model in internal coordinates with added and optimized hydrogens)
2. Displaying energy strain along the backbone of a model:
   [start ICM]
   read pdb "1p22"
   convertObject a_1p22. yes yes yes no
   calcEnergyStrain a_1p22.b/A yes 7. (details of this subroutine available in the ICM language manual)

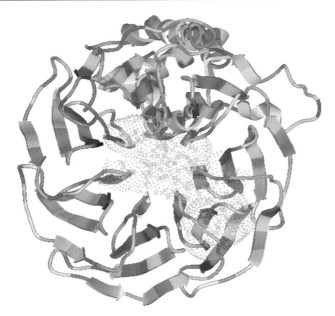

FIG. 2. Energy strain in the β-TrCP protein backbone. Face-on view of the WD-40 domain of β-TrCP. The protein backbone is displayed as a ribbon depiction with each amino acid position shaded according to energy strain score in the backbone: red = high energy; blue = low energy. The substrate-binding peptide in this complex is displayed as a green stippled cloud to identify the location of the substrate-binding interface of β- TrCP. (See color insert.)

This particular method colors strained residues red and unstrained residues blue with a spectrum of colors from red to blue corresponding to intermediate energy strain in between. The protein model is displayed in Fig. 2 with the backbone in ribbon depiction and the color spectrum chosen for black-and-white reproduction. The analysis indicates that the backbone in the region of the β-catenin binding pocket contains a few peptide flips and clashed side chains. Thus, an analysis for "druggability" of this specific surface should not be artifactual as a result of energy strain errors. In other targets in which problems are identified by this approach, they may be improved by means of well-established modeling approaches beyond the scope of this chapter (Cardozo et al., 1995).

Local Structural Reliability Inferred from Data in the Experimental Structure Determination

The second error, high B-factors (also known as temperature factors), may be interpreted as a score of the probability that the atom position observed in the model is in the true natural position of the atom. Very flexible distal side chain atoms may have high B-factors, because they are constantly in motion in the crystal solution conformation and their crystallographic electron density reflects this fact. Intrinsically disordered protein segments display the same behavior. The alternative conformations of such atoms to the initial one present in the rigid model constructed by the crystallographer should be explored by modeling lest these atoms preclude a pharmacophore from occupying the same space when, in fact, they are flexible and can easily accommodate the occupying group of an important ligand. B-factors for each atom are provided by the crystallographer in the PDB file. Here we color the atoms of the structure by B-factor, which reveals that all the atoms of the structure have relatively low B-factors. Thus, the substrate-binding surface is relatively stable and well resolved (Fig. 3).

3. Displaying B-factors associated with atoms in the model:
   [start ICM]
   read pdb "1p22"
   color a_//* Bfactor(a_//*) (maps the 1-D array of B-factors for each atom onto the array of the same length of atoms)

FIG. 3. B-factor mapping of β-TrCP structure. Face-on view of the WD-40 domain of β-TrCP. Left panel, All atoms are shown and colored according to B-factor (temperature factor): red = high B-factor, blue = low B-factor. Right panel, The protein backbone is displayed as a ribbon depiction with each amino acid position shaded according to B-factor for the entire residue: red = high B-factor, blue = low B-factor. The substrate-binding peptide in this complex is displayed as a green stippled cloud in both panels to identify the location of the substrate-binding interface of β-TrCP. The entire structure is different shades of blue, indicating high experimental local reliability throughout. (See color insert.)

The use of NMR structures is suboptimal as previously mentioned, but sometimes necessary. In NMR structures, the assessment of local reliability is more intuitive. When multiple models satisfying experimental NMR restraints are available, simply superimposing them and noting the areas of deviation between the models is sufficient. Extensive modeling is usually required, however, in cases in which the author submits only an averaged structure of all the models fitting NMR restraints. Averaged structures are likely only useful as a last resort.

Local Conformational Variability Caused by the Crystal

The x-ray diffraction crystal is a symmetrical array of protein molecules each positioned identically by contacts with each other. If one of the symmetrical crystal contacts in or between unit cells in the crystal occurs near or within the surface of interest, the observed structure may be an artifact of the crystal and not the native physiological structure. Under these conditions, this local conformation may be perfectly resolved and unstrained, but it is specific to the crystal and, therefore, artificial. The space group and crystallographic parameters $(a,b,c,\alpha,\beta,\gamma)$ of the crystal, which are either in the PDB file or in the original publication of the structure, may be used to apply the appropriate rotational and translational transformations and generate nearby symmetry-related molecules. These molecules may then be displayed, thereby visualizing the points of contact. Here, we used the crystallographic parameters of the β-TrCP structure to generate symmetry-related molecules, which shows that only the tail of the β-catenin substrate peptide is involved in crystal-packing contacts. The atoms of the substrate binding interface of β-TrCP and the bound portion of the β-catenin substrate peptide are unaffected by crystal packing, ruling out this source of artifact in this target surface (Fig. 4).

4. Generating symmetry-related molecules:
   ```
   [start ICM]
   read pdb "1p22"
   for i = 2,Nof(Symgroup(Symgroup(a_)))/12*2 (programming loop
       to generate symmetry-related molecules from the space group)
       copy a_1. "a" + I
       transform a_a$i. i-1
       display
       center
   endfor
   color ml a_*. (color each molecule differently)
   gcell = Grob("cell",Cell( ))
   ```

Fig. 4. Crystal packing analysis of β-TrCP substrate-binding interface. Depiction of the unit cell of the crystal used to solve the structure of β-TrCP. The index monomer is colored black, and the substrate peptide in the complex is displayed as green space-filling spheres to identify the substrate-binding surface. Symmetry-related molecules in the crystal packing around the index structure are shown in yellow ribbon depiction. Left panel, View of the entire assembly. Right panel, Close-up of the substrate- binding domain showing that only the tail of the substrate peptide (arrow) contacts a crystal neighbor, whereas the substrate-binding interface is free in solution (dashed circled). (See color insert.)

```
obl = Augment(Cell( ))
g1 = gcell + (−1)*obl[1:3,2]
g2 = g1 + obl[1:3,3] (generate the crystallographic cell)
display gcell g1 g2 (display the crystallographic cell)
```

Other Measures of Local Structural Quality

Assessment of local structural reliability in protein 3-D structural models is an area of active research (Cardozo et al., 2000). In addition, intuitive or experimental factors specific to any given structure, such as applied knowledge of the exact protein construct used for the crystallization experiment, may provide useful additional information for specific local areas of the structure. Thus, the techniques described here may be only the first of many in future years to assess the local reliability of the target structure for drug discovery. In general, at any given point in time, the literature should be searched for computational tools that may apply to any given target. One such example applicable to a protein structure on which the protein interaction surface is not known (protein not crystallographically resolved in complex with its partner) will soon be available (Fernandez-Recio et al., 2004). In addition, the investigator should attempt to apply any intuitive or experiential information at his or her disposal to the task of improving the local reliability of the structural model in the vicinity of the target surface before analysis of pockets and screening.

The Ultimate Goal: Identification of Surface Pockets as a Measure of Druggability

Global and local structural analysis procedures set the stage for methods that directly indicate that a particular surface of a structure may be amenable to drug discovery. One way to determine whether a particular interface is susceptible would be to use *in silico* docking and VLS procedures directly on the entire structure and analyze the fitness scores (goodness-of-fit) and binding locations of a large library of chemical compounds (the use of classic SBBD, with cycles of experimental structure determination, would not be efficient for this purpose). If the VLS effort results in many top hits being predicted to bind to the target protein interface and the fitness scores are reasonably high, the interface may be suitable for drug discovery. Many VLS and docking packages and chemical libraries are available to perform this procedure, with technological approaches differing primarily in the conformation generation of the ligand and the implementation of the atoms of the receptor (Abagyan and Totrov, 2001).

VLS is a form of drug discovery screening; however, a more concise and preliminary rational approach would be the identification of surface-solvent pockets that could accommodate drug-like ligands. One obvious pocket is that encompassing a ligand seen in the crystal structure, particularly a small molecule ligand (protein in complex with ATP, estrogen, etc.). This pocket is immediately a valid target for drug discovery, because it is already designed to bind a small chemical compound. Interestingly, pockets encompassing peptide ligands may be suitable as well. Indeed, the impetus for the successful recent discovery of an mdm2 inhibitor was the observation that the structure of Mdm2 occupied by a helical peptide of p53 represented a visible pocket that could accommodate a small drug when the helix was removed. In addition, point mutations disrupting the interaction were located near or in this pocket. Two SCF ligase structures to date were solved with bound substrate peptides, and in both cases, the studies demonstrated that point mutation of residues in contact with the substrate peptides could abolish function *in vitro* (Orlicky *et al.*, 2003; Wu *et al.*, 2003). Other drug discovery efforts targeted at protein interfaces have recently succeeded starting from this same bound-peptide starting point using *in silico* screening methods (Degterev *et al.*, 2001; Gruneberg *et al.*, 2002; Wang *et al.*, 2003). On the other hand, flexible peptides, especially charged moieties, may bind to shallow surfaces of a protein that may not be suitable for small molecule binding. In the case either that a peptide fragment in a crystallographic structure does not clearly occupy a suitably deep pocket or that no peptide or protein ligand is present, tools are needed to assess the suitability of the protein surface for drug design.

The algorithms that identify and render surface pockets, including the LigandPocket algorithm used in the Introduction section (Fig. 1), identify small molecule–binding pockets on the surface of 3-D protein structural models (An *et al.*, 2004; Bliznyuk and Gready, 1998; Brady and Stouten, 2000; Campbell *et al.*, 2003; Cavasotto *et al.*, 2003; Dennis *et al.*, 2002; Glick *et al.*, 2002; Hendlich *et al.*, 1997; Kortvelyesi *et al.*, 2003; Laskowski, 1995; Levitt and Banaszak, 1992; Liang *et al.*, 1998; Peters *et al.*, 1996; Ruppert *et al.*, 1997; Sotriffer and Klebe, 2002; Verdonk *et al.*, 2001). Many of these algorithms build a 3-D grid map of a binding potential in the virtual space of the structural model, and the position and size of the ligand-binding pocket are determined on the basis of the construction of equipotential surfaces (Totrov and Abagyan, 2001) along these maps. LigandPocket was trained on 17,126 small molecule–binding sites available in the PDB and identifies more than 90% of pockets of 100 Å^2–500 Å^2 in size. It is notable that most drugs in pharmacologic use today occupy a similar volume distribution. VLS docking tests indicate that the presence of a pocket identified by LigandPocket engenders a very high probability of identifying chemical compounds specific for that pocket in screening libraries (not shown). In addition, the presence of such pockets is almost always of functional significance. If not intended for a specific ligand, the pockets are usually found in hinge regions between protein domains and represent empty space required for domain movement in the protein. These, too, may be of functional significance, because their enhancement or removal (by binding of ligand) may bias the complex toward either an active or inactive form. Here we used LigandPocket on the β-TrCP substrate binding domain and interface for which we have just assessed the local structural reliability.

    5. Identifying solvent pockets on the surface of a 3-D virtual model:
       [start ICM]
       read pdb "1p22"
       convertObject a_1p22. yes yes yes no
       IcmPocketFinder a_1p22.b (details of this subroutine available in the ICM language manual)

The LigandPocket analysis of the β-TrCP substrate binding domain reveals a 480-Å^2 ligand-binding pocket in the central well of this WD-40 beta-propeller domain encompassing one of the phospate-binding surfaces (geometric object indicated with thick arrow in Fig. 5). The pocket overlaps

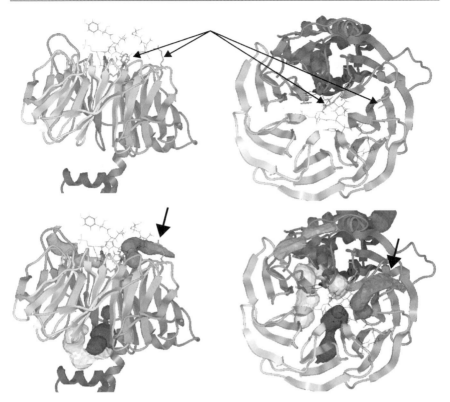

FIG. 5. Ligand pockets on the surface of β-TrCP. Side (left column) and face-on (right column) view of β-TrCP. The protein backbone is shown in ribbon depiction, the substrate peptide is displayed in thin wire depiction, and four residues known to abrogate ubiquitylation by of β-TrCP are displayed as stick depictions and indicated by the thin arrows. Ligand-binding pockets above the threshold size established by the plots shown in Fig. 1 are depicted as geometric objects. One ligand-binding pocket (red: thick arrows) is directly at the substrate-binding interface and in contact with the key functional residues. The method described in the text to perform this pocket analysis was also used for the analysis shown in Fig. 1. (See color insert.)

significantly with the substrates known natural ligand solved in the complex, is lined by point mutations that were shown to disrupt ubiquitylation (Wu *et al.*, 2003), and is of ideal size to accommodate a small compound of average size for a biologically active drug. This protein interface and this pocket seem, therefore, to be an attractive target for SCF$^{\beta\text{-TrCP}}$ drug discovery. Interestingly, four other pockets of suitable size are found on the structural surface of β-TrCP (Fig. 5), revealing additional targets for rational drug discovery and indicating that domain movements and allostery play a role in the mechanism of ubiquitylation by this F-box protein.

Summary

The available evidence indicates that protein interfaces may safely be targeted for drug discovery. The risk of doing so solely with nonrational approaches such as HTS may be prohibitive, as exemplified by the SH2 domain. A simple *rule* can determine which interfaces (and therefore which biomolecules) are "druggable": a suitable ligand-binding pocket must be detected with maximal confidence in a functionally sensitive location on the biomolecule. Specifically, here we have used the substrate-binding interface of β-TrCP as an example of how this determination is made for a protein interface of the SCF ubiquitin ligase. The example shows that the specific requirements for making this determination are good structural data for the target, assessments of the local reliability of the structural data on functionally sensitive surfaces, and tools for the identification and rendering of pockets on those surfaces. Several tools are already available to reduce operator dependence in this analysis for nonstructural biologists. We have demonstrated one such method here: (1) the exact conceptual steps in the method are delineated and (2) the exact command sequence to accomplish each step in one software package is provided. Improvements in each of the conceptual and technical steps of this method by means of ongoing computational research and by incorporation of forthcoming results of drug discovery efforts targeting the SCF ligase may improve the precision and utility of this approach to determining whether a biomolecular target is "druggable."

References

Abagyan, R., and Totrov, M. (2001). High-throughput docking for lead generation. *Curr. Opin. Chem. Biol.* **5,** 375–382.

Abagyan, R. A., and Totrov, M. M. (1997). Contact area difference (CAD): A robust measure to evaluate accuracy of protein models. *J. Mol. Biol.* **268,** 678–685.

Abagyan, R. A., Totrov, M. M., and Kuznetsov, D. A. (1994). ICM: A new method for protein modeling and design: Applications to docking and structure prediction from the distorted native conformation. *J. Comp. Chem.* **15,** 488–506.

An, J., Totrov, M., and Abagyan, R. A. (2004). Pocketome via comprehensive identification and classification of ligand binding envelopes. *Mol. Cell. Proteomics.* **4,** 752–761.

Bajorath, J. (2002). Integration of virtual and high-throughput screening. *Nat. Rev. Drug Discov.* **1,** 882–894.

Berg, T., Cohen, S. B., Desharnais, J., Sonderegger, C., Maslyar, D. J., Goldberg, J., Boger, D. L., and Vogt, P. K. (2002). Small-molecule antagonists of Myc/Max dimerization inhibit

Myc-induced transformation of chicken embryo fibroblasts. *Proc. Natl. Acad. Sci. USA* **99,** 3830–3835.

Bliznyuk, A. A., and Gready, J. E. (1998). Identification and energetic ranking of possible docking sites for pterin on dihydrofolate reductase. *J. Comput. Aided Mol. Des.* **12,** 325–333.

Brady, G. P., Jr., and Stouten, P. F. (2000). Fast prediction and visualization of protein binding pockets with PASS. *J. Comput. Aided Mol. Des.* **14,** 383–401.

Campbell, S. J., Gold, N. D., Jackson, R. M., and Westhead, D. R. (2003). Ligand binding: Functional site location, similarity and docking. *Curr. Opin. Struct. Biol.* **13,** 389–395.

Cardozo, T., Batalov, S., and Abagyan, R. (2000). Estimating local backbone structural deviation in homology models. *Comput. Chem.* **24,** 13–31.

Cardozo, T., and Pagano, M. (2004). The SCF ubiquitin ligase: Insights into a molecular machine. *Nat. Rev. Mol. Cell Biol.* **5,** 739–751.

Cardozo, T., Totrov, M., and Abagyan, R. (1995). Homology modeling by the ICM method. *Proteins* **23,** 403–414.

Cavasotto, C. N., Orry, A. J., and Abagyan, R. A. (2003). Structure- based identification of binding sites, native ligands and potential inhibitors for G-protein coupled receptors. *Proteins* **51,** 423–433.

Chelvanayagam, G., Roy, G., and Argos, P. (1994). Easy adaptation of protein structure to sequence. *Protein Eng.* **7,** 173–184.

Chen, J. K., Taipale, J., Young, K. E., Maiti, T., and Beachy, P. A. (2002). Small molecule modulation of smoothened activity. *Proc. Natl. Acad. Sci. USA* **99,** 14071–14076.

Cochran, A. G. (2000). Antagonists of protein-protein interactions. *Chem. Biol.* **7,** R85–R94.

Degterev, A., Lugovskoy, A., Cardone, M., Mulley, B., Wagner, G., Mitchison, T., and Yuan, J. (2001). Identification of small-molecule inhibitors of interaction between the BH3 domain and Bcl-xL. *Nat. Cell Biol.* **3,** 173–182.

Dennis, S., Kortvelyesi, T., and Vajda, S. (2002). Computational mapping identifies the binding sites of organic solvents on proteins. *Proc. Natl. Acad. Sci. USA* **99,** 4290–4295.

Fernandez-Recio, J., Totrov, M. M., Skorodumov, C., and Abagyan, R. A. (2004). Optimal docking area: A new method for predicting protein-protein interaction sites. *Proteins* **58,** 134–143.

Glick, M., Robinson, D. D., Grant, G. H., and Richards, W. G. (2002). Identification of ligand binding sites on proteins using a multi-scale approach. *J. Am. Chem. Soc.* **124,** 2337–2344.

Goldenberg, S. J., Cascio, T. C., Shumway, S. D., Garbutt, K. C., Liu, J., Xiong, Y., and Zheng, N. (2004). Structure of the Cand1-Cul1-Roc1 complex reveals regulatory mechanisms for the assembly of the multisubunit cullin-dependent ubiquitin ligases. *Cell.* **119,** 517–529.

Gruneberg, S., Stubbs, M. T., and Klebe, G. (2002). Successful virtual screening for novel inhibitors of human carbonic anhydrase: Strategy and experimental confirmation. *J. Med. Chem.* **45,** 3588–3602.

Hao, M. H., and Scheraga, H. A. (1999). Designing potential energy functions for protein folding. *Curr. Opin. Struct. Biol.* **9,** 184–188.

Hendlich, M., Rippmann, F., and Barnickel, G. (1997). LIGSITE: Automatic and efficient detection of potential small molecule-binding sites in proteins. *J. Mol. Graph. Model.* **15,** 359–363, 389.

Jin, J., Cardozo, T., Lovering, R. C., Elledge, S. J., Pagano, M., and Harper, J. W. (2004). Systematic analysis and nomenclature of mammalian F-box proteins. *Genes Dev.* **18,** 2573–2580.

Kavallaris, M., Tait, A. S., Walsh, B. J., He, L., Horwitz, S. B., Norris, M. D., and Haber, M. (2001). Multiple microtubule alterations are associated with Vinca alkaloid resistance in human leukemia cells. *Cancer Res.* **61,** 5803–5809.

Kleywegt, G. J., and Jones, T. A. (1995). Where freedom is given, liberties are taken. *Structure* **3,** 535–540.

Kortvelyesi, T., Silberstein, M., Dennis, S., and Vajda, S. (2003). Improved mapping of protein binding sites. *J. Comput. Aided Mol. Des.* **17,** 173–186.

Kossiakoff, A. A., Randal, M., Guenot, J., and Eigenbrot, C. (1992). Variability of conformations at crystal contacts in BPTI represent true low-energy structures: Correspondence among lattice packing and molecular dynamics structures. *Proteins* **14,** 65–74.

Kuriyan, J., and Weis, W. I. (1991). Rigid protein motion as a model for crystallographic temperature factors. *Proc. Natl. Acad. Sci. USA* **88,** 2773–2777.

Laskowski, R. A. (1995). SURFNET: A program for visualizing molecular surfaces, cavities, and intermolecular interactions. *J. Mol. Graph.* **13,** 307–308; 323–330.

Lepourcelet, M., Chen, Y. N., France, D. S., Wang, H., Crews, P., Petersen, F., Bruseo, C., Wood, A. W., and Shivdasani, R. A. (2004). Small- molecule antagonists of the oncogenic Tcf/beta-catenin protein complex. *Cancer Cell.* **5,** 91–102.

Levitt, D. G., and Banaszak, L. J. (1992). POCKET: A computer graphics method for identifying and displaying protein cavities and their surrounding amino acids. *J. Mol. Graph.* **10,** 229–234.

Liang, J., Edelsbrunner, H., and Woodward, C. (1998). Anatomy of protein pockets and cavities: Measurement of binding site geometry and implications for ligand design. *Protein Sci.* **7,** 1884–1897.

Maiorov, V., and Abagyan, R. (1998). Energy strain in three-dimensional protein structures. *Fold Des.* **3,** 259–269.

Orlicky, S., Tang, X., Willems, A., Tyers, M., and Sicheri, F. (2003). Structural basis for phosphodependent substrate selection and orientation by the SCFCdc4 ubiquitin ligase. *Cell.* **112,** 243–256.

Peters, K. P., Fauck, J., and Frommel, C. (1996). The automatic search for ligand binding sites in proteins of known three-dimensional structure using only geometric criteria. *J. Mol. Biol.* **256,** 201–213.

Ruppert, J., Welch, W., and Jain, A. N. (1997). Automatic identification and representation of protein binding sites for molecular docking. *Protein Sci.* **6,** 524–533.

Saraste, M., Sibbald, P. R., and Wittinghofer, A. (1990). The P- loop–a common motif in ATP- and GTP-binding proteins. *Trends Biochem. Sci.* **15,** 430–434.

Schulman, B. A., Carrano, A. C., Jeffrey, P. D., Bowen, Z., Kinnucan, E. R., Finnin, M. S., Elledge, S. J., Harper, J. W., Pagano, M., and Pavletich, N. P. (2000). Insights into SCF ubiquitin ligases from the structure of the Skp1-Skp2 complex. *Nature* **408,** 381–386.

Sotriffer, C., and Klebe, G. (2002). Identification and mapping of small-molecule binding sites in proteins: Computational tools for structure-based drug design. *Farmaco.* **57,** 243–251.

Sussman, J. L., Lin, D., Jiang, J., Manning, N. O., Prilusky, J., Ritter, O., and Abola, E. E. (1998). Protein Data Bank (PDB): Database of three-dimensional structural information of biological macromolecules. *Acta Crystallogr. Dev. Biol. Crystallogr.* **54,** 1078–1084.

Totrov, M., and Abagyan, R. (2001). Rapid boundary element solvation electrostatics calculations in folding simulations: Successful folding of a 23-residue peptide. *Biopolymers.* **60,** 124–133.

Vassilev, L. T. (2004). Small-molecule antagonists of p53-MDM2 binding: Research tools and potential therapeutics. *Cell Cycle* **3,** 419–421.

Vassilev, L. T., Vu, B. T., Graves, B., Carvajal, D., Podlaski, F., Filipovic, Z., Kong, N., Kammlott, U., Lukacs, C., Klein, C., Fotouhi, N., and Liu, E. A. (2004). *In vivo* activation of the p53 pathway by small-molecule antagonists of MDM2. *Science* **303,** 844–848.

Verdonk, M. L., Cole, J. C., Watson, P., Gillet, V., and Willett, P. (2001). SuperStar: Improved knowledge-based interaction fields for protein binding sites. *J. Mol. Biol.* **307,** 841–859.

Wang, J. L., Liu, D., Zhang, Z. J., Shan, S., Han, X., Srinivasula, S. M., Croce, C. M., Alnemri, E. S., and Huang, Z. (2000). Structure-based discovery of an organic compound that binds Bcl-2 protein and induces apoptosis of tumor cells. *Proc. Natl. Acad. Sci. USA* **97,** 7124–7129.

Wang, S., Yang, D., and Lippman, M. E. (2003). Targeting Bcl-2 and Bcl-XL with nonpeptidic small-molecule antagonists. *Semin. Oncol.* **30,** 133–142.

Wei, N., and Deng, X. W. (2003). The COP9 signalosome. *Annu. Rev. Cell Dev. Biol.* **19,** 261–286.

Wu, G., Xu, G., Schulman, B. A., Jeffrey, P. D., Harper, J. W., and Pavletich, N. P. (2003). Structure of a beta-TrCP1-Skp1-beta-catenin complex: Destruction motif binding and lysine specificity of the SCF(beta-TrCP1) ubiquitin ligase. *Mol. Cell.* **11,** 1445–1456.

Zheng, J., Yang, X., Harrell, J. M., Ryzhikov, S., Shim, E. H., Lykke-Andersen, K., Wei, N., Sun, H., Kobayashi, R., and Zhang, H. (2002a). CAND1 binds to unneddylated CUL1 and regulates the formation of SCF ubiquitin E3 ligase complex. *Mol. Cell.* **10,** 1519–1526.

Zheng, N., Schulman, B. A., Song, L., Miller, J. J., Jeffrey, P. D., Wang, P., Chu, C., Koepp, D. M., Elledge, S. J., Pagano, M., Conaway, R. C., Conaway, J. W., Harper, J. W., and Pavletich, N. P. (2002b). Structure of the Cul1-Rbx1-Skp1-F boxSkp2 SCF ubiquitin ligase complex. *Nature* **416,** 703–709.

[43] Overview of Approaches for Screening for Ubiquitin Ligase Inhibitors

By YI SUN

Abstract

E3 ubiquitin ligases are a large family of proteins that, together with ubiquitin-activating enzyme E1 and ubiquitin-conjugating enzyme E2, catalyze the ubiquitination of a variety of protein substrates for targeted degradation by means of the 26S proteasome. Because the turnover of many proteins involves targeted ubiquitination and degradation, E3 ubiquitin ligases, therefore, regulate almost every aspect of eukaryotic cellular functions or biological processes. Accumulated evidence in the past few years has suggested that a subset of E3 ubiquitin ligases that regulates the turnover of tumor suppressors and cell cycle regulators could be promising targets for mechanism-driven cancer drug discovery. Thus, it is highly desirable to optimize the methods of high-throughput screening (HTS) for specific inhibitors of these E3 ubiquitin ligases. Here I will give an overview of several approaches used for HTS for ubiquitin ligase inhibitors with a main focus on assay principles, applications, and the pros and cons of each approach. Experimental details for many of these assays can be found in other chapters in this volume.

Introduction

Ubiquitination of a protein substrate for targeted degradation involves multistep enzymatic reactions catalyzed by a cascade of enzymes, including ubiquitin-activating enzyme E1, ubiquitin-conjugating enzyme E2, and ubiquitin ligase E3. Ubiquitin is first activated by binding to E1 through a thioester bond between a cysteine residue at the active site of E1 and the C-terminal glycine (G76) of ubiquitin. Activated ubiquitin in the E1-ubiquitin complex is then transferred to E2, which also forms a thioester bond between its active-site cysteine residue and G76 of ubiquitin. Finally, ubiquitin is covalently attached to the protein substrate through an isopeptide bond between G76 of ubiquitin and the ε-amino group of an internal lysine residue of the protein substrate in a reaction catalyzed by E3 ubiquitin ligase. After the linkage of ubiquitin to a protein substrate, followed by multiple cycles of the same reaction, a polyubiquitin chain is formed in which the C-terminus of each ubiquitin unit is linked to a specific lysine residue (most commonly Lys[48]) of the previous ubiquitin. The polyubiquitinated substrates are then rapidly recognized and degraded by the 26S proteasome (Ciechanover, 1998). A typical ubiquitination reaction is illustrated in Fig. 1.

In addition to catalyzing substrate ubiquitination, many E3 ligases possess a feature of self-ubiquitination in the absence of a substrate, at least in *in vitro* experimental settings. Thus, depending on the need, recently developed high-throughput screening (HTS) assays for E3 ligase inhibitors are either (1) a substrate-dependent E3 reaction in which a protein substrate is included and polyubiquitination of the substrate is measured or (2) substrate-independent in which the reaction measures a self-polyubiquitination of E3 ligases. Following are a few approaches that have been developed for HTS in an effort to identify inhibitors of E3 ubiquitin ligases.

Screening Approaches

FRET Assay

Assay Principle. Fluorescence resonance energy transfer (FRET) or homogeneous time-resolved fluorescence (HTRF) is a widely used assay for HTS of E3 ubiquitin ligase inhibitors. This homogenous fluorescence

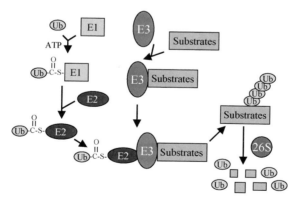

FIG. 1. The ubiquitination reaction. The ubiquitin is first attached to the E1 ubiquitin–activating enzyme in the presence of ATP. The activated ubiquitin is then transferred to E2 ubiquitin–conjugating enzyme. The E3 ubiquitin ligase recognizes a protein substrate, recruits the E2–ubiquitin complex, and catalyzes the ubiquitin transfer from the E2 to the substrate. Multiple cycles of these reactions leads to polyubiquitination of the substrate, which is recognized and degraded by the 26S proteasome.

assay involves individual labeling of two binding partners with two different fluorescent dyes: One serves as a fluorescence donor and the other as an acceptor. On binding, two fluorescent dyes are brought into close proximity, permitting energy transfer from the fluorescence donor to the acceptor that is measured as a ratio of fluorescence intensity with unique excitation and emission wavelengths. Traditional FRET assay has been modified to measure protein ubiquitination through the labeling of different components of ubiquitination reaction with fluorescent dyes.

Applications

P53 POLYUBIQUITINATION. The HTRF assay was set up to monitor polyubiquitination of wild-type p53 (Yabuki *et al.*, 1999). In this assay, europium chelate (Eu) and allophycocyanin (APC) that were attached to ubiquitin (Eu-Ub) and streptavidin (APC-SA) were used as the fluorescence donor and acceptor, respectively. The biotinylated p53 was ubiquitinated in the presence of E1, E2, E3, and Eu-Ub. Eu and APC were brought into close proximity through a biotin-SA binding to generate fluorescent signals that were measured with excitation of Eu^{3+} at 340 nm and time-resolved fluorescence at the emission wavelength of APC at 665 nm. The compounds presented in the reaction that caused a decrease in fluorescent signals were potential inhibitors of the ubiquitination reaction (Fig. 2).

UBIQUITIN TRANSFER FROM E2 TO E3. The FRET assay was also developed to screen for ubiquitination inhibitors that inhibited the transfer

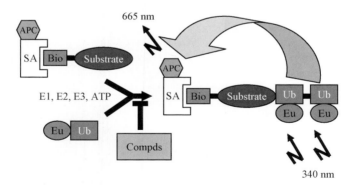

FIG. 2. FRET assay. Europium chelate (Eu)–labeled ubiquitin is attached to a biotin-labeled protein substrate in an ubiquitination reaction containing E1, E2, E3, and ATP. Fluorescent acceptor allophycocyanin (APC), which is attached to streptavidin (SA), is brought into close proximity to the fluorescent donor Eu through a SA–biotin binding, triggering energy transfer. Eu is excited at 340 nm, and the fluorescence of APC, generated when energy is transferred from Eu, is measured at 665 nm. The chemical compounds (Compds) that inhibit the ubiquitination reaction reduce the fluorescent signals.

of ubiquitin from an E2 to a HECT E3 (Boisclair *et al.*, 2000). In this assay, ubiquitin was prelabeled with biotin and transferred by ATP and E1 to E2 (bio-Ub-E2). Bio-Ub-E2 was preincubated with allophycocyanin-labeled streptavidin (APC-SA). GST-fused HECT E3 was preincubated with anti-GST antibody that was prelabeled with LANCE europium chelate (anti-GST Eu^{3+}-antibody). For the ubiquitination assay, prelabeled GST-E3 was mixed with prelabeled bio-Ub-E2 in the presence or absence of compounds. When E3 was ubiquitinated, Eu^{3+} and APC were brought into close proximity, permitting energy transfer between the two fluorescent labels, and fluorescent signals were detected at 340 nm/665nm. The compounds that inhibited ubiquitin transfer from E2 to E3 would decrease the fluorescent signals.

TRAF6 POLYUBIQUITINATION. A similar assay was recently developed for HTS to identify inhibitors of tumor necrosis factor receptor–associated factor 6 (TRAF6), a critical E3 ubiquitin ligase involved in inflammatory and immune response (Hong *et al.*, 2003). In this assay, ubiquitin was labeled with both Eu chelate and biotin. The ubiquitination reaction in the presence of E1, E2, and TRAF6 led to formation of a poly-Ub chain with randomly incorporated Eu and biotin labels. After termination of reaction and addition of SA-linked fluorescence acceptor APC, APC-Eu was brought into close proximity through a SA-biotin binding. FRET is then measured with an excitation at 340 nm and emission at 665 nm. Again, the compounds that decreased fluorescent signals would be the inhibitors of ubiquitination reaction.

Pros and Cons. This is a homogenous, robust, and cost-effective assay suitable for HTS. The assay is relatively precise, with low well-to-well variation. Another advantage is that the assay enabled the kinetics of the ubiquitin reaction to be monitored in real time. The biggest disadvantage is that the assay could be interfered by fluorescent compounds, although the fluorescence quenching was found to be minimal when the compound concentration was kept low (Boisclair *et al.*, 2000).

DELFIA Assay

Assay Principle. Dissociation-enhanced lanthanide fluoroimmunoassay (DELFIA) is a heterogeneous immunoassay modified from traditional ELISA (enzyme-linked immunosorbent assay). It is mainly used to measure the binding of protein–protein, protein–peptide, protein–DNA, and ligand-receptor. Typically, protein X coated into the plate is incubated with its GST-tagged binding partner (e.g., protein Y). After washing out unbound protein Y, the wells are incubated with anti-GST Eu^{3+}-antibody. After incubation and subsequent wash, the amount of bound Eu^{3+} is measured with excitation at 340 nm and fluorescence at 615 nm.

Application. DELFIA assay was modified and developed simultaneously with FRET assay to measure the transfer of ubiquitin from Ubc4 (an E2) to Rsc (a HECT E3) and used to screen for ubiquitination inhibitors (Boisclair et al., 2000). GST-Rsc was first incubated with biotin-Ub-UBC4. After transfer of biotin-Ub from E2 to E3, the biotin-Ub-Rsc-GST was immobilized to a streptavidin-coated plate. After incubation with anti-GST Eu^{3+} antibody, followed by washing, the amount of bound Eu^{3+} in each well was measured with excitation at 340 nm and fluorescence at 615 nm. The compound that inhibited ubiquitin transfer from E2 to E3 was detected with a decreased fluorescent signal (Fig. 3).

Pros and Cons. The assay in general has good precision with high sensitivity and is relatively easy to automate. Because the assay involves washing steps, it eliminates sample fluorescent interference during measurement and achieves a high signal/background ratio but has a higher well-to-well variation than FRET assay (Boisclair et al., 2000). Again, because of the involvement of time-consuming washing steps, it is not suitable for HTS, but rather it is to be used to confirm hits obtained by other HTS methods.

Electrochemiluminescence (ECL) for E3 Ubiquitin Ligases

Assay Principle. An electrochemiluminescence (ECL)-based high-throughput assay was developed for analyzing the self-ubiquitination of E3 ubiquitin ligase by Meso-Scale Discovery (formerly IGEN International, Inc.) using an M-SERIES™ M8 Analyzer workstation (Davydov, 2004). GST-fused E3 is immobilized to GSH-coated magnetic beads and self-polyubiquitinated in the presence of E1, E2, ubiquitin, and ATP.

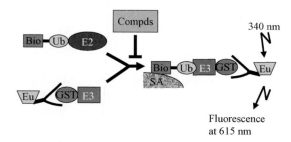

FIG. 3. DELFIA assay. Biotin-labeled ubiquitin is first transferred from E1 to E2 and then from E2 to GST-fused E3. Biotin-Ub-E3-GST is immobilized to SA-coated plate and recognized by Eu-labeled anti-GST antibody. After washing, fluorescence at 615 nm is detected after excitation of Eu at 340 nm. Compounds that inhibit ubiquitin transfer from E2 to E3 decrease fluorescent signals.

Polyubiquitin chain is recognized by ORI-TAG–labeled antiubiquitin antibody. In an M8 analyzer, the ORI-TAG label (an electrochemiluminescent compound) is excited at the surface of an electrode and emits light. After many excitation/emission cycles, the light signals in each ORI-TAG label are amplified and captured. The light signals are directly correlated to ubiquitin chain formation.

Application. The assay was initially designed for the measurement of Mdm2 autoubiquitination but could be easily optimized for many other E3s. The assay is conducted in a reconstituted system consisting of enzymes including E1, E2, and GST-fused E3 plus ATP and ubiquitin. GST-fused E3 is partially purified through binding to glutathione-coated beads followed by several washes. The ubiquitination reaction is initiated by adding the E1, E2, ATP, and ubiquitin to the beads with or without a compound. After incubation, magnetic bead–conjugated polyubiquitinated E3 is captured by a magnet built-in TRICORDER in a M8 analyzer that separates it from unbound labels to reduce background signals and improve assay sensitivity. After this separation, polyubiquitinated E3 is recognized by ORI-TAG–labeled antiubiquitin antibody, and the ECL-generated signal is detected (see Fig. 4).

Pros and Cons. The assay is simple, robust, and offers high signal/background ratios. The signal/background ratio, however, has a tendency for day-to-day variation. In addition, the M8 analyzer is quite expensive, and the routine maintenance of the machine will be costly.

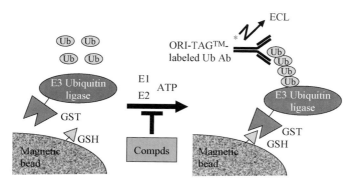

FIG. 4. Electrochemiluminescence assay. GST-E3 is first attached to glutathione-coated magnetic beads and partially purified through bead washing. The beads are then mixed with E1, E2, and ATP-containing buffer in the absence or presence of chemical compounds (Compds). The reaction is initiated by adding ubiquitin and stopped after 1 h incubation by adding EDTA. Polyubiquitinated E3 is detected by ORI-TAG™–labeled antibody against ubiquitin, and the ECL signal generated is captured in an M8 analyzer (IGEN International, Inc).

Scintillation Proximity Assay (SPA)

Assay Principle. The SPA is another binding-based assay. The assay is performed with radioactive labels that emit electrons with only a short range in water. Normally, one binding partner is immobilized to solid beads and impregnated with a scintillans, and the other is labeled with radioactive isotope such as ^{3}H or ^{125}I that emit electrons with the low energies required for SPA. On binding, which brought the isotope in proximity to a solid scintillator surface, the electron energy from a radioactive-labeled binding partner is transferred to the scintillator to produce photons detectable with a scintillation counter. Electrons emitted from labeled molecules not bound close to the surface dissipate their energy in the medium and are not detected, making separation of bound from free fraction unnecessary.

Application. A SPA assay using ^{125}I-labeled ubiquitin was developed to detect polyubiquitination of wild-type p53. ^{125}I-labeled ubiquitin was attached to biotin-labeled p53 in the presence of E1, E2 (UbcH4), and E3 (E6/E6AP). After termination of the ubiquitination reaction, polyubiquitinated p53 was captured by streptavidin-coated SPA beads, and the signals were quantified by a MicroBeta counter (Yabuki *et al.*, 1999). The compounds that caused a decrease of signals would be the inhibitors of the ubiquitination reaction (see Fig. 5).

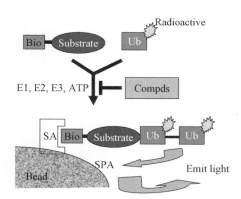

FIG. 5. SPA assay. Radioactive-labeled ubiquitin is attached to a biotin-labeled protein substrate in an ubiquitination reaction containing E1, E2, E3, and ATP. After termination of the reaction, streptavidin-labeled beads are added that bound to biotinylated substrate with radioactive ubiquitins attached. This biotin–SA binding makes the isotope close to a solid scintillator surface, allowing electron energy transfer from the isotope to the scintillator to produce photons to be detected with a scintillation counter.

Pros and Cons. The assay is homogeneous with a high sensitivity. It is highly generalizable and easy to automate. The biggest disadvantage is the use of radioactive materials.

Laboratory-Based In Vitro *and* In Vivo *Ubiquitination Assays*

Various laboratories have developed their favorite ubiquitination assays. They can be classified in general into test tube–based *in vitro* ubiquitination and cell-based *in vivo* ubiquitination assays. The *in vitro* reconstituted ubiquitination assay normally uses recombinant proteins of E1, E2, E3, and substrate, expressed in bacteria or baculovirus and purified through classic biochemical methodologies (Rolfe *et al.*, 1995). In some cases, the E3 source is provided by immunoprecipitates prepared from transiently cotransfected E3 components (Swaroop *et al.*, 2000). The cell-based *in vivo* ubiquitination assay normally uses endogenous sources of E1 and E2 in cells with other components of ubiquitin, E3, and substrate provided through a transient cotransfection. Transfected cells are typically lysed in a regular RIPA lysis buffer. Ubiquitinated substrate is first immunoprecipitated by antibodies against the substrate or substrate-tagged epitope and detected by immunoblotting by use of antibodies against ubiquitin or ubiquitin-tagged epitope. Alternately, cells in which His-tagged ubiquitin is transfected are lysed in 6 M guanidinium solution that inactivates deubiquitinase activity to prevent substrate deubiquitination during the sample preparation. Ubiquitinated substrate is partially purified through a nickel affinity column and eluted by imidazole, followed by Western blot analysis (Treier *et al.*, 1994; Xirodimas *et al.*, 2001). It is obvious that all these approaches are laboratory-based, low-throughput methods that are tedious and time consuming. Although they are not designed or suitable for large-scale drug screening, they can certainly be used for lead confirmation and specificity determination.

Challenge Versus Promise

Although a subset of E3 ubiquitin ligases seem to be good cancer targets (Pray *et al.*, 2002; Sun, 2003; Teodoro *et al.*, 2004) and few HTS assays have been developed and optimized for inhibitor screening as described previously, a major technical challenge is the assay complexity. Compared with a typical kinase HTS assay that requires kinase alone (autophosphorylation) or kinase plus a substrate (sometimes the substrates are synthetic artificial peptides), E3 ligase assay requires four to five proteins (ubiquitin, E1, E2, and E3, and sometimes a substrate). Except for the hits from assays designed to measure ubiquitin transfer from E2 to E3 that is more E3 specific, the hits identified from all other polyubiquitination assays described previously need to be deconvoluted as inhibitors of E3 versus E2 or even E1. Another challenge will be how to achieve the specificity among different E3s within the same class of E3 ligases, such as SCF with cullins and Rbx as the core ligase. Finally, even if a specific E3 inhibitor is identified, the investigator still faces the issue of compound selectivity between cancer and normal cells that determines a therapeutic window. Nevertheless, despite these challenges, the great promise of E3 ubiquitin ligase as a therapeutic target is drawing much attention in the pharmaceutical industry and academic laboratories for further target validation and actual drug screening and development.

References

Boisclair, M. D., McClure, C., Josiah, S., Glass, S., Bottomley, S., Kamerkar, S., and Hemmila, I. (2000). Development of a ubiquitin transfer assay for high throughput screening by fluorescence resonance energy transfer. *J. Biomol. Screen.* **5,** 319–328.

Ciechanover, A. (1998). The ubiquitin-proteasome pathway: On protein death and cell life. *EMBO J.* **17,** 7151–7160.

Davydov, I. V., Woods, D., Safiran, Y. J., Oberoi, P., Fearnhead, H. O., Fang, S., Jensen, J. P., Weissman, A. M., Kenten, J. H., and Vousden, K. H. (2004). Assay for ubiquitin ligase activity: High-throughput screen for inhibitors of HDM2. *J. Biomol. Screen.* **9,** 695–703.

Hong, C. A., Swearingen, E., Mallari, R., Gao, X., Cao, Z., North, A., Young, S. W., and Huang, S. G. (2003). Development of a high throughput time-resolved fluorescence resonance energy transfer assay for TRAF6 ubiquitin polymerization. *Assay Drug Dev. Technol.* **1,** 175–180.

Pray, T. R., Parlati, F., Huang, J., Wong, B. R., Payan, D. G., Bennett, M. K., Issakani, S. D., Molineaux, S., and Demo, S. D. (2002). Cell cycle regulatory E3 ubiquitin ligases as anticancer targets. *Drug Resist. Update* **5,** 249–258.

Rolfe, M., Beer-Romero, P., Glass, S., Eckstein, J., Berdo, I., Theodoras, A., Pagano, M., and Draetta, G. (1995). Reconstitution of p53-ubiquitinylation reactions from purified components: The role of human ubiquitin-conjugating enzyme UBC4 and E6-associated protein (E6AP). *Proc. Natl. Acad. Sci. USA* **92,** 3264–3268.

Sun, Y. (2003). Targeting E3 ubiquitin ligases for cancer therapy. *Cancer Biol. Therapy* **2,** 623–629.

Swaroop, M., Wang, Y., Miller, P., Duan, H., Jatkoe, T., Madore, S., and Sun, Y. (2000). Yeast homolog of human SAG/ROC2/Rbx2/Hrt2 is essential for cell growth, but not for germination: Chip profiling implicates its role in cell cycle regulation. *Oncogene* **19,** 2855–2866.

Teodoro, J. G., Heilman, D. W., Parker, A. E., and Green, M. R. (2004). The viral protein Apoptin associates with the anaphase-promoting complex to induce G2/M arrest and apoptosis in the absence of p53. *Genes Dev.* **18,** 1952–1957.

Treier, M., Staszewski, L. M., and Bohmann, D. (1994). Ubiquitin-dependent c-Jun degradation *in vivo* is mediated by the delta domain. *Cell* **78,** 787–798.

Xirodimas, D., Saville, M. K., Edling, C., Lane, D. P., and Lain, S. (2001). Different effects of p14ARF on the levels of ubiquitinated p53 and Mdm2 *in vivo. Oncogene* **20,** 4972–4983.

Yabuki, N., Watanabe, S., Kudoh, T., Nihira, S., and Miyamato, C. (1999). Application of homogeneous time-resolved fluorescence (HTRFTM) to monitor poly-ubiquitination of wild-type p53. *Comb. Chem. High Throughput Screen* **2,** 279–287.

[44] A Homogeneous FRET Assay System for Multiubiquitin Chain Assembly and Disassembly

By TARIKERE L. GURURAJA, TODD R. PRAY, RAYMOND LOWE, GUOQIANG DONG, JIANING HUANG, SARKIZ DANIEL-ISSAKANI, and DONALD G. PAYAN

Abstract

Ubiquitin (Ub, 76aa) is a small highly conserved protein present universally in eukaryotic cells. Covalent attachment of $(Ub)_n$ to target proteins is a well-known posttranslational modification that has been implicated in a wide array of cellular processes including cell biogenesis. Ubiquitin polymerization by the Ub activation-conjugation-ligation cascade and the reverse disassembly process catalyzed by Ub isopeptidases largely regulate substrate protein targeting to the 26S proteasome. Ub chains of four or more subunits attached by K48 isopeptide linkages have been shown to be necessary for the 26S proteasome association and subsequent degradation of protein molecules. To better understand this protein degradation event, it is important to develop Ub polymerization and depolymerization assays that monitor every reaction step involved in Ub attachment to, or detachment from, substrate protein molecules. In this chapter, we describe homogeneous, easy-to-use, nonradioactive, complementary continuous fluorescence assays capable of monitoring the kinetics of Ub chain formation by E3 Ub ligases, and their hydrolysis by isopeptidases, which rely on mixing a 1:1 population of fluorophore-labeled Ub molecules containing a FRET pair. The proximity of fluorescein (donor) and tetramethylrhodamine (acceptor) in Ub polymers results in fluorescein quenching on ligase-induced Ub chain assembly. Conversely, a dramatic enhancement of fluorescein emission was observed on Ub chain disassembly because of isopeptidase activity. These assays thus provide a valuable tool for monitoring Ub ligase and isopeptidase activities using authentic Ub monomers and polymers as substrates. Screening of a large number of small molecule compound libraries in a high-throughput fashion is achievable, warranting further optimization of these assays.

Introduction

Most regulated proteolysis in eukaryotes occurs by a mechanism in which conjugation to the conserved protein ubiquitin (Ub) targets substrates for degradation by the 26S proteasomes (Hershko and Ciechanover, 1998; Hochstrasser, 1996). Substrates of the Ub-proteasome pathway are marked for degradation by covalent ligation to Ub by means of an isopeptide bond between the C-terminus of Ub (G76) and a lysine residue of the target protein, which then acts as a signal for targeting the modified substrate to the proteasome (Pickart, 2001; VanDemark and Hill, 2002). This ubiquitination-mediated selective degradation of specific cellular proteins plays a pivotal role in regulating fundamental cellular processes, including metabolic homeostasis, protein quality control, transcription, translation, signal transduction, response to hypoxia, cell cycle progression, apoptosis, DNA repair, protein trafficking, and viral budding (Finley et al., 2004; Sun and Chen, 2004). Interestingly, the regulated assembly and disassembly of polymeric Ub chains on substrate proteins serves as an emerging paradigm of the reversible, covalent modification of polypeptide chains akin to phosphorylation/dephosphorylation. Recent interest in the ubiquitination/deubiquitination machinery has provided much insight into the protein families involved in these processes. The regulation of these enzymes and enzyme assemblies, critical to the proper proteasomal targeting of substrate molecules, has also been studied extensively, along with their potential catalytic mechanisms (Baek, 2003; Wilkinson, 2000). Aiding all of these studies is an emerging understanding of the structural motifs present in E1 Ub–activating, E2 Ub–conjugating, E3 Ub ligase, and Ub isopeptidase enzymes and their homologs.

Because either the stabilization or destabilization of regulatory proteins can alter nearly any cellular event, including cell cycle progression, metabolism, and signaling in general, ubiquitination/deubiquitination enzymes may be attractive targets for small molecule and/or other types of therapeutic intervention (Pray et al., 2002; Robinson, 2004; Sun, 2003; Wong et al., 2003). A unique feature of the ubiquitination/deubiquitination process is the polymeric nature of many Ub modifications. Although Ub and Ub-like (Ubl) modifiers are often stably added to substrate proteins as single subunits, often by use of attachment points other than Ub K48, proteasomal targeting seems to require the stepwise polymerization of Ub monomers in an outward growing chain originating on the substrate protein itself (Fang and Weismann, 2004). For this reason, it is important to develop polymerization-specific assays for Ub ligase activity. Similarly, because Ub isopeptidases bind and hydrolyze Ub monomers from one another by cleavage of the isopeptide bond between them, it is important to develop assays that monitor this event directly.

Homogeneous assays, in which all reagents remain in solution phase and no separation steps are required, are attractive for high-throughput screening (HTS) because of the reduction in assay complexity and duration. These assays are easily automated and have the ability to screen entire libraries of drug candidates. Because of these advantages over traditional heterogeneous assays such as ELISA, we sought to develop a fluorescence-based homogeneous assay for monitoring multi-Ub chain assembly and disassembly. Homogeneous assays based on fluorescence polarization are well established in the clinical laboratory and in drug discovery research by HTS strategies (Colbert and Childerstone, 1987; Owicki, 2000). Besides their precision and sensitivity, homogeneous fluorescence-based assays allow for real-time monitoring of the reaction, thus making them more versatile and flexible. This chapter presents the development of complementary Ub ligase and isopeptidase assays, using a simple approach that relies on fluorescence energy transfer (FRET) between two populations of fluorophore-labeled Ub molecules. The E3–Ub ligase assay is capable of monitoring the continuous formation of Ub polymers by the anaphase-promoting complex (APC) and other E3 ligases, based on the quenching of fluorescein–Ub (F-Ub) emission when in the proximity of tetramethylrhodamine-Ub (TAMRA-Ub designated as R-Ub) in Ub chains. Conversely, the cleavage and disassembly of these F-Ub/R-Ub polymers by the deubiquitinating enzyme, UbpM, and other Ub isopeptidases is readily detected by the enhancement of F-Ub emission on release from R-Ub chain neighbors. In addition, the production of these Ub reagents is facile and inexpensive; because native Ub contains no Cys residues, this strategy simply relies on the introduction of a Cys residue near the N-terminus of Ub and its concomitant labeling with commercially available sulfhydryl-reactive fluorophore moieties. The rapid, simple, inexpensive generation and purification of these and related fluorophore–Ub reagents should provide useful, general tools for the analysis of Ub ligase and isopeptidase activities for both basic research and identification of new drug candidates by means of inhibitor screening in a high-throughput format.

Experimental Procedures

Cloning, Expression, and Purification of Cys-Containing Ub. To facilitate fluorophore labeling of Ub, a Cys residue was inserted between the FLAG tag and Ub coding sequences. Ub does not contain any Cys residues, so FLAG-Cys-Ub will selectively incorporate thiol-reactive probes. Incorporation of Cys residue into the FLAG-ubiquitin sequence was accomplished by site-directed mutagenesis using the primer: 5′-CCCCCCAAGCTTTG CATGCAGATTTTCGTGAAGACCCTGACC-3′. The resulting ubiquitin fragment was subcloned into the pFlag-Mac Expression Vector (Sigma) as a HindIII-EcoRI fragment by PCR. The plasmid DNA was then transformed into BL21 DE3–competent *Escherichia coli* (Stratagene, Cat. No. 230132). Cultures were grown at 37 in Luria-Bertani (LB) medium containing ampicillin (100 μg/ml) and glucose (0.4%) with shaking until the OD_{600} reached approximately 0.6. At this point, IPTG (320 μM) was added, and the culture was allowed to grow for another 3 h before harvesting. The cells were harvested by centrifugation at 7700g, and the cell pellets were washed once with cold PBS before being resuspended in approximately 6 volumes of lysis buffer (20 mM Tris, 10% glycerol, 0.5 M NaCl, 2.5 mM EDTA, 1 mM TCEP plus complete EDTA-free protease inhibitor cocktail tablet, 1 tablet per 25 ml of resuspended cells, pH 8.0). The suspension was homogenized and sonicated (three times 30-sec pulsing mode). The nonionic detergent NP-40 was then added to a final concentration of 0.5%, and the tubes were rocked for 30 min at 4. After centrifugation at 11000 rpm for 25 to 30 min, the supernatant was loaded over an anti-FLAG M2 affinity resin media (VWR Scientific, Cat. No. IB13020) at a ratio of 15 ml of beads per liter of original culture. The resin was then washed with 10 bed volumes of lysis buffer. Elution of protein from the affinity column was accomplished with an acidic elution buffer (100 mM acetic acid, 10% glycerol, 200 mM NaCl, 2.5 mM EDTA, 0.1% NP-40, pH 3.5). Elutions were collected as 1-bed volume fractions directly into tubes containing $1/10^{th}$ volume of neutralization buffer (2 M Tris, 80 mM β-mercaptoethanol, pH 9.0) to neutralize the pH. Then each of the collected fractions was analyzed by SDS-PAGE and the appropriate fractions pooled based on their purity profiles and dialyzed against 400 volumes of a storage buffer (20 mM Tris, 10% glycerol, 200 mM NaCl, 2.5 mM EDTA, pH 8.0). Protein concentration was determined using a Micro-BCA protein estimation kit (Pierce) as per manufacturer's recommendations.

Production of Ubiquitin Pathway Enzymes E1, E2, and E3 (APC)

Rabbit E1 used in this study was obtained commercially (Affiniti Research Products, Exeter, UK), and the purity of the product was found to be greater than 90% using SDS-PAGE. For the production of E2, the open reading frame of E2 (UbcH5c) was amplified by PCR and cloned into the pGex-6p-1 *E. coli* expression vector (Amersham Pharmacia) as BglII–EcoRI fragments, with the N-terminus in frame fused to a GST-tag. For the detailed protocol used for the induction of protein expression in *E. coli* and its isolation and purification, please refer to the chapter by Huang *et al.* (2005) in this issue. The two subunits of E3 were expressed by co-infection in *Baculovirus* in the same SF-9 insect cells. For APC2/APC11,

APC2 was tagged using $(His)_6$ at the N-terminus and APC11 remained untagged. Although $(His)_6$-APC2 was cloned into the pFastBacHtb vector, the untagged APC11 was inserted into the pFastBac 1 vector before expression using the Bac-to-Bac *Bacoluvirus* expression system (Cat. No. 10359-016, Invitrogen). For the production and purification of E3 comprising human His-tagged APC2/APC11 refer to Huang *et al.* (Chapter 49, 2005).

Deubiquitinating Enzyme: UbpM Expression and Purification

The deubiquitinating enzyme UbpM was cloned into pFastBac HT vector and expressed as a fusion protein containing an affinity tag $(His)_6$ at the N-terminus using a Bac-to-Bac *Bacoluvirus* expression system (Cat. No. 10359-016, Invitrogen). Protein production of $(His)_6$-UbpM was essentially similar to that of $(His)_6$-APC2/APC11 as described previously, except that High-5 insect cells were used instead of SF9 cells. The starting High5 (Invitrogen) cell density was approximately 0.8×10^6 cells/ml. When the cell density reached 1.5×10^6 cells/ml, cells were infected with $(His)_6$-UbpM *Bacoluvirus* (MOI = 1–5). After viral infection, the cells are allowed to grow for 40–44 h, whereupon they were harvested (2500 g at 4°) and the pellet stored at $-80°$ before purification. Purification of $(His)_6$-UbpM was conducted at 4° unless otherwise specified after conventional IMAC procedure as recommended by Invitrogen, Inc.

Ub-Fluorophore Labeling and Purification

Purified FLAG-Cys-Ub protein was custom-labeled using either fluorescein-5-maleimide (F; Cat. No. F-150) or tetramethylrhodamine-6-maleimide (TAMRA/R; Cat. No. T-6028) obtained from Molecular Probes, Inc. (Eugene, OR). Labeling was carried out using approximately 800 μg of protein (100 μM/0.1 mmol) and 0.5 mmol of fluorophores (F or R) dissolved in PBS for 90 min at room temperature. A 10-fold molar excess of reducing agent, DTT, was added to quench any unincorporated maleimide fluorophores. After the reaction, unreacted fluorophore was separated from F- or R-incorporated Ub species by gel filtration over a Superdex 75 HR 10/30 column equilibrated in PBS containing 1 mM TCEP. Gel filtration was accomplished on an AKTA HPLC system (Amersham Biosciences) equipped with a variable UV detector. Purified fluorophore labeled Ub was snap-frozen in liquid nitrogen in PBS with 10% glycerol, and stored at $-80°$ until used.

Ubiquitin Polymerization Assay (APC Ligase Assay)

Reactions were performed in a final volume of 100 μl using fluorescence compatible black nonbinding surface (NBS) coated 96-well micro plates (Cat. No. 3650, Corning). The ligase buffer contained 62.5 mM Tris (pH 7.5) 6.25

mM MgCl$_2$, 0.75 mM DTT, 2.5 mM ATP, 2.5 mM NaF, and 100 ng of a 1:1 mixture of Flag-tagged F-Ub/R-Ub. The buffer solution was brought to a final volume of 80 μl with deionized milli-Q water. For assays directed toward identifying modulators of ubiquitin ligase activity, 10 μl of candidate modulator compound dissolved in DMSO was added to the solution. As a control, 10 μl of DMSO was used instead of candidate modulator. The ubiquitination reaction was triggered by the addition of 10 μl of enzyme mixture prepared in 20 mM Tris buffer (pH 7.5) containing 5% glycerol. The following amounts of each enzyme were used for this assay: 5 ng/well of E1; 25 ng/well of E2 (UbcH5c); and 100 ng/well of E3 (His-APC2/APC11). The reaction was then allowed to proceed at room temperature for 1 h. Reactions were either monitored continuously (for kinetics studies) or stopped by the addition of EDTA (10 mM) (for endpoint determinations). On polyubiquitination of APC, fluorescence quenching of fluorescein was monitored by excitation at 485 nm and emission at 520 nm. The 96-well assay plates were read with a Bio-Tek FL600 Fluorescence Microplate Reader equipped with a 485-nm laser for excitation. Formation of polyubiquitin chains was further confirmed by gel electrophoresis (SDS-PAGE) under nondenaturing conditions and by gel-filtration using a Hewlett Packard HP-1100 HPLC system (Agilent Technologies, Palo Alto, CA) equipped with a fluorescence detector.

Ubiquitin Depolymerization Assay (UbpM Isopeptidase Assay)

For conducting the depolymerization assay, a polymerization assay was first carried out as outlined previously and at the end of 1 h incubation, EDTA (10 mM) was added to stop the ligase reaction by sequestering Mg^{2+}. Purified UbpM was then added at varying concentrations to depolymerize APC polyubiquitin chains. Fluorescence enhancement was followed at the same wavelengths as for poly-Ub fluorescein quenching described previously. For quality control purposes, UbpM activity was also examined by use of an artificial fluorogenic substrate Ub-AMC, which contains AMC conjugated at the Ub C-terminal COOH group. Ubiquitin hydrolase activity of UbpM was measured using an assay buffer (50 mM HEPES, pH 7.8, 0.5 mM EDTA, 1 mM DTT, and 0.1 mg/ml BSA) containing 2 nM UbpM at 25° for 1 h. The reaction was started by the addition of 5 μl of 50 $\mu$$M$ ubiquitin COOH-terminal AMC (Ub-AMC; Boston Biochem) in DMSO. Reaction progress was monitored on a Bio-Tek FL600 Fluorescence Microplate Reader and was measured by the increase of fluorescence intensity at 460 nm ($\lambda_{ex} = 380$ nm) as a function of time, which is reflective of the cleavage of AMC from Ub-AMC.

Results and Discussion

Ub Transfer Assays

Although there are many reports in the literature regarding the development of various types of Ub transfer assays, these are mostly based on radiological and/or immunological methods; fluorescent-based assays have been explored to a very limited extent. Many assays reported are heterogeneous in nature and use either radioactive I^{125}-Ub and/or affinity-tagged Ub such as FLAG, $(His)_6$, biotin, etc. in both *in vivo* and *in vitro* experiments. In some instances, target protein substrates have been labeled with radioactive S^{35} *in vivo*, and ubiquitination is monitored by classical immunoprecipitation followed by Western analysis (Laney and Hochstrasser, 2002). Takada *et al.* (1995) have reported an immunoassay for the quantification of intracellular multiubiquitin chains, which uses a radioactive I^{125}-Ub followed by a sandwich ELISA. There is a growing trend, however, to avoid using radioactivity in screening applications because of problems with safety, disposal, and shelf life of the ligand reagents, and instead move toward labels that can generate less harmful radiation on demand. Fluorophores meet these criteria and typically are stable molecules that generate UV-VIS radiation only when excited by a primary source. At present, the most commonly used methods for assaying Ub transfer involve immunological detection of immobilized multi-Ub conjugates labeled with various affinity elements such as FLAG, $(His)_6$ and biotin, using their corresponding secondary reagents (Davydov *et al.*, 2004; Lai *et al.*, 2002). For instance, an enhanced chemiluminescence (ECL)–based high-throughput assay has been developed for analyzing self-ubiquitination of E3 ubiquitin ligase. The assay was initially designed for the measurement of Mdm2 autoubiquitination but could easily be optimized for many other E3s (Davydov *et al.*, 2004). Although these assays are known for their accuracy and reproducibility, they are time consuming and require multiple washing steps and secondary reagents for detection. In this context, fluorescent-based assays have a significant advantage over current immunological detection methods, because these assays offers direct detection of endpoints without using secondary reagents and minimize the number of steps involved in the assay.

Some fluorescence-based Ub attachment and cleavage assays have been developed previously to monitor Ub ligase and isopeptidase activities, respectively (Boisclair *et al.*, 2000; Dang *et al.*, 1998; Hong *et al.*, 2003; Wee *et al.*, 2000; Yabuki *et al.*, 1999). These ligase assays, although accurate and useful, have some shortcomings. At present, fluorescence-based Ub isopeptidase activity assays use fluorogenic leaving groups, such as AMC, which are chemically linked to the C-terminus of Ub (Dang et al., 1998). Although these assays are commonly used to study the kinetics of amide bond hydrolysis by protease enzymes, potentially critical features of Ub polymer substrate recognition may be lacking if Ub-AMC is used exclusively.

Design of FRET Assay

FRET is a well-established fluorescent technology (Clegg, 1996, Selvin, 1995; Szollosi et al., 1998) that is now becoming widely used in the drug discovery processes. The popularity of FRET for screening derives from the simplicity of the technique; separation and/or washing steps are avoided, and the method is truly homogeneous. Another great advantage of the homogeneous FRET method is its solution-phase nature, which completely eliminates unexpected artifacts arising from solid-phase methods, wherein one of the components of the assay system needs to be bound to a matrix and/or a solid surface. The principle of FRET is based on the ability of a higher energy donor (D) fluorophore to transfer energy directly to a lower energy acceptor (A) molecule, causing sensitized fluorescence of the acceptor molecule and simultaneous quenching of the donor fluorescence. The efficiency of energy transfer (E) is highly dependent on the distance (r) between the donor and acceptor chromophores, as described by the Forster equation; $E = R_o^6/(R_o^6 + r^6)$, where the Forster radius (R_o) is the distance at which the efficiency of energy transfer is 50% of the maximum and can be calculated from the spectral properties and relative orientation of each fluorescent molecule. Thus, FRET provides a very sensitive measure of small changes in intermolecular distances. In most cases, no FRET can be observed at distances greater than 100 Å, so the presence of FRET is a good indicator of close proximity, implying biologically meaningful protein–protein interactions. There are a number of donor/acceptor (D/A) pairs (Table I) having R_o (Å) values ranging from 30–60 that are commercially available (Haugland, 2002). Among them, fluorescein/tetramethylrhodamine and EDANS/DABCYL pairs having R_o (Å) value of 55 and 33, respectively, have been most widely used to develop FRET-based assays.

The strategy used for designing a homogenous assay for multiubiquitin chain assembly and disassembly is depicted in Fig. 1. This assay essentially uses the fluorescent-labeled reagents, Flag-Cys(F)-Ub and Flag-Cys(R)-Ub (R = TAMRA), as fluorescent donor and acceptor, respectively. On polyubiquitination of APC, the two fluorophores come into close proximity, generating the FRET signal. In the presence of an inhibitor of Ub ligation, a decrease in signal is observed, whereas on release of Ub chains, an

TABLE I
LIST OF COMMERCIALLY AVAILABLE DONOR/ACCEPTOR PAIRS ALONG WITH THEIR R_o VALUES*
COMMONLY USED IN FRET APPLICATIONS

Donor ($\lambda_{ex}/\lambda_{EM}$)	Acceptor ($_{ex}/\lambda_{em}$)	R_o (Å)	References
Fluorescein (494/518)	Tetramethylrhodamine (555/580)	49–56	Wu and Brand (1994)
IAEDANS (336/490)	FITC (494/528)	49	Wu and Brand (1994)
IAEDANS (336/490)	5-(Iodoacetamide) fluorescein (494/518)	49	Wu and Brand (1994)
Fluorescein (494/518)	Fluorescein (494/518)	44	Wu and Brand (1994)
EDANS (335/493)	DABCYL (453/none)	33	Wu and Brand (1994)
Tryptophan (280/348)	IAEDANS (336/490)	22	Lakowicz (1999)
Tryptophan (280/348)	Dansyl (335/518)	21–24	Lakowicz (1999)
Tryptophan (280/348)	Pyrene (340/376)	28	Lakowicz (1999)
Dansyl (335/518)	Fluorescein (494/518)	33–41	Fairclough and Canto (1978)
Naphthalene (336/490)	Dansyl (335/518)	22	Wu and Brand (1994)
Pyrene (340/376)	Coumarin (384/470)	39	Wu and Brand (1994)
B-Phycoerythrin (546/575)	Cys5 (650/670-700)	79	Wu and Brand (1994)

*The value may change under different experimental conditions.
λ Values are in nm. IAEDANS, 5-[(iodoacetyl)amino]naphthalene-1-sulfonic acid; FITC, fluorescein isothiocyanate; EDANS, 5-[(2-aminoethyl)amino]naphthalene-1-sulfonic acid; DABCYL, 4-[{4-(dimethylamino)phenyl}azo]benzoic acid.

enhancement in fluorescent signal of the donor occurs. The method used in this strategy involves the use of Cys-containing Ub and its direct labeling using fluorescent probes. The advantage of direct labeling is that the assay does not require the use of any secondary reagents such as antibodies and/ or affinity reagents to follow the reaction. There are a number of FRET-based assays reported in the literature in which indirect labeling methods have been used (Boisclair et al., 2000; Kane et al., 2000). For example, Boisclair et al. (2000) have developed a FRET assay that measures the transfer of ubiquitin from Ubc4 to the HECT protein Rsc 1083. Their strategy uses labeled secondary reagents such as anti-GST with Eu3+ and streptavidin with allophycocyanin. On ubiquitination, energy transfer occurs when Eu3+ chelate is brought into close proximity with allophycocyanin, which occurs when Ub labeled with biotin is transferred from Ubc4 to Rsc (which has a GST-tag). Compared with the DELFIA (dissociation-enhanced lanthanide fluoroimmunoassay) method, which was also developed by the same group, this FRET assay showed higher precision with minimum well-to-well variation. A potential drawback of indirect labeling is that a large molecular complex can form, resulting in inefficient FRET if the gap between the energy donor and acceptor is too great.

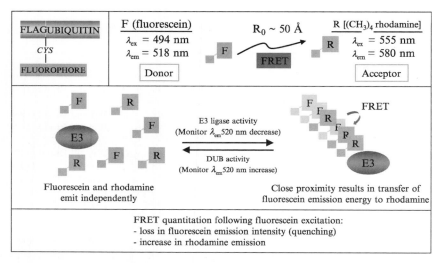

FIG. 1. Schematic representation of the FRET assay. On chain assembly, FRET occurs when the acceptor and the donor are brought into close proximity that is measured by fluorescent quenching. On the other hand, fluorescent emission decreases on Ub chain disassembly. The fluorescence emission at 520 nm excites the juxtaposed TAMRA, which emits fluorescence at 590 nm.

Production of Recombinant Proteins and Custom Labeling

All of the necessary proteins and enzymes, such as FLAG-Cys-Ub, GST-UbcH5c (E2), APC (E3), and UbpM required for the assay were produced in-house with the exception of rabbit-E1, which was procured commercially. For the purification of GST-tagged UbcH5c protein, glutathione Sepharose 4B media was used, whereas anti-FLAG M2 beads served as an affinity media for the purification of FLAG-Cys-Ub. Purification of $(His)_6$-tagged protein was carried out by IMAC chromatographic methods using Ni-NTA agarose beads. In most of the proteins, affinity tags were retained until the end of the purification steps, except in the case of GST-tagged UbcH5c (E2), in which the tags were removed subsequently using precision protease to yield untagged UbcH5c (E2). Quality control of the purified proteins was performed by SDS-PAGE (Fig. 2), with most proteins having purity greater than 90% (Fig. 2), and exhibited expected biological activities when used under *in vitro* biochemical assay conditions. In this study, we selected one of the most commonly used FRET pairs, fluorescein (F) and tetramethylrhodamine (TAMRA or R), which has a Ro value ranging from 49–56 Å. Fluorescein and TAMRA have excitation/

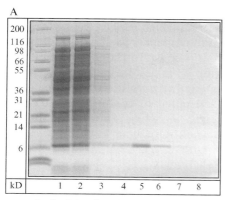
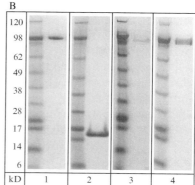

1 = load; 2 = flow through; 3 = wash;
4-8 = FLAG-Cys-Ub eluted fractions

1 = Rabbit E1; 2 = E2 (GST free UbcH5c);
3 = E3 (His-APC2/APC11); 4 = DUB (His-UbpM)

FIG. 2. (A) SDS-PAGE purification profile of recombinant FLAG-Cys-Ub protein column fractions collected during anti-FLAG column chromatography. Appropriate fractions were pooled on the basis of their purity and stored at −80°. (B) SDS-PAGE purification profiles of E1, E2, E3 (APC), as well as DUB (UbpM) proteins. Samples were run on 4–12% Tris-glycine gels under reducing conditions and were stained with Coomassie Blue R-250 protein staining reagent.

emission wavelengths of 494/518 and 555/580 nm, respectively. This broad separation in their excitation/emission wavelengths makes them a unique fluorescent donor and acceptor pair in FRET-based assay development. Using thiol-reactive maleimide derivatives of fluorescein (F) and tetramethylrhodamine (R), purified FLAG-Cys-Ub protein was custom labeled following the manufacturer's recommended protocol. Both the fluorophores were used in molar excess to ensure complete labeling of FLAG-Cys-Ub. After the conjugation, unreacted fluorophores were quenched using DTT, and final purification was accomplished by gel filtration.

Optimization and Utility of FRET Assay

Ub Polymerization Assay. Of two well-characterized E3 ligases, APC and SCF, we selected APC as an E3 ligase in the ubiquitination enzyme cascade for this study on the basis of its crucial role during mitosis. To simplify and reduce the number of components required to monitor the ligase activity, we used a substrate-independent APC autoubiquitination reaction as described in previous reports (Tang *et al.*, 2001). Tang *et al.* (2001) have shown that APC2 Cullin protein and APC11 RING protein, comprising minimal components of APC, are sufficient to catalyze ubiquitin polymerization in the absence of target protein substrate and that this activity is dependent on the inclusion of the proper E2 enzyme. It is important to note that this autoubiquitination assay itself cannot differentiate E3 inhibitors from E1 or E2 inhibitors, and a further deconvolution is certainly required for identifying specific E3 inhibitors.

For the purposes of assay development, reconstitution of a cell-free ubiquitination reaction was initially attempted to confirm the biological activities of the recombinant proteins, including fluorophore-labeled FLAG-Cys-Ub. As shown in Fig. 3A, multi-Ub chain assembly was successfully reconstructed *in vitro* using appropriate concentrations of E1/E2/E3 and a 1:1 mixture of FLAG-Cys(F)-Ub/FLAG-Cys(R)-Ub in the presence of ligase assay buffer containing ATP and $MgCl_2$. Because the ubiquitination reaction is dependent on the presence of each enzyme component, omission of either E1/E2 or E3 arrested the reaction (Fig. 3A). In addition to anti-FLAG Western analysis, multi-Ub chain formation was also analyzed and confirmed by gel-filtration using a superdex-75 HR column attached to an HPLC equipped with a fluorescent detector (Fig. 3B). Detection of F emission signal at 520 nm clearly showed

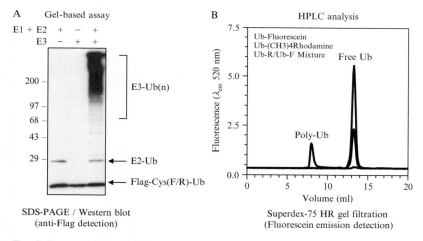

FIG. 3. Reconstitution of ubiquitination reaction using recombinant proteins (E1/E2/E3) and FLAG-Cys(F/R)-Ub in the presence of ATP and $MgCl_2$ *in vitro*. (A) Western analysis of the reaction mixture after 1 h incubation including appropriate controls carried out under nonreducing conditions over a 4–12% Bis-Tris (MES running buffer) NuPAGE gel system (Invitrogen). Blots were probed with anti-FLAG HRP–conjugated antibody (Sigma). (B) HPLC analysis of the reaction product showing the clear separation of multi-Ub species from their monomers. Peaks were monitored using a fluorescent detector set at fluorescein (F) emission wavelength ($\lambda_{em} = 520$ nm).

two distinct peaks comprising a higher molecular weight multi-Ub species and a low molecular weight unreacted free Ub monomer species.

To set up the homogenous FRET assay for multi-Ub chain assembly, we first optimized the concentrations of all enzyme components such as E1, E2, and E3 by titration with the ECL-based assay format developed in-house by Huang et al. (2005). The optimal concentrations of E1, E2, and E3 were found to be 5, 25, and 100 ng/well, respectively, in an assay volume of 100 μl (data not shown). The concentration of FLAG-Ub was kept constant at 100 ng per reaction. The details of assay optimization for the Ub-polymerization assay are available in Chapter 49. Identical assay conditions as those generated using the ECL-based heterogeneous assay were used in the development of the homogeneous FRET assay. The only difference was the use of a 1:1 mixture of FLAG-Cys(F)-Ub/FLAG-Cys(R)-Ub and a non-binding surface (NBS)-coated plate instead of a FLAG-Ub and a Ni-NTA–coated plate, respectively. Fig. 4A depicts the protocol used for the

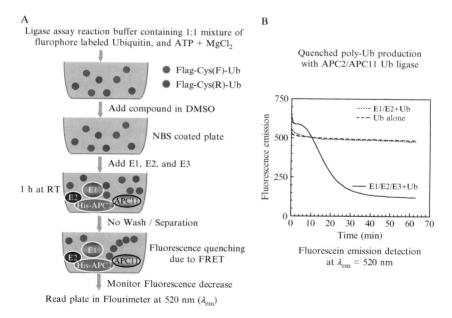

FIG. 4. Homogeneous FRET-based assay for multi-Ub chain assembly. (A) Schematic representation of the screening assay showing the stepwise addition of various components of the assay including test compounds in DMSO (inhibitors). (B) Plate-based assay (96-well) format for self-ubiquitination of E3 (APC) ubiquitin ligase. Formation of multi-Ub E3 (APC) was detected by quenched fluorescein emission signal at $\lambda_{em} = 520$ nm using a fluorescence plate reader.

homogeneous FRET assay to monitor multi-Ub chain assembly. The protocol is very similar to that used for ECL-based assay, but in this assay neither a washing nor an addition of secondary reagents was required to detect the signal. On multiubiquitination of E3 (APC), the close proximity of the two donor and acceptor fluorophores (F and R) results in a decrease in fluorescent intensity of F because of quenching, which is monitored at λ_{em} 520 nm using a fluorescent microplate reader (Fig. 4B). Elimination of either E1 + E2 or Ub did not quench the fluorescein emission signal, indicating the specificity of the Ub transfer assay. Although an average signal/background ratio of approximately 3–4 was obtained under the final assay conditions, this parameter can be fine tuned with further optimization. Signal/background ratios greater than 50 have been reported for some model FRET systems (Selvin et al., 1994), but ratios less than 10 are common for high-throughput screens developed for drug discovery (Earnshaw et al., 1999; Kolb et al., 1996).

Ub Depolymerization Assay. During the process of ubiquitin-mediated protein degradation in 26S proteasomes, ubiquitinated protein substrates of the proteasome undergo deubiquitination before breakdown by the proteasome (Pickart, 1997). Because a constant Ub concentration is crucial for normal cell function, internalized ubiquitinated plasma membrane proteins are probably deubiquitinated before their breakdown in the vacuole. Such deubiquitination is catalyzed by processing proteases called deubiquitinating enzymes or DUBs (Wilkinson, 1997). DUBs are a family of cysteine proteases constituting the largest known family in the Ub system. DUBs fall into two classes, the Ub C–terminal hydrolases and the Ub-specific processing proteases (UBPs). The former cleave primarily peptide bonds in poly-Ub precursor proteins or isopeptide bonds in small free Ub chains. UBPs, the larger class of DUBs, cleave isopeptide bonds between two UB residues or between Ub and another protein. From among various DUBs, we selected UbpM as a deubiquitinating enzyme to monitor the depolymerization event in this study. UbpM is known to get phosphorylated and then dephosphorylated at critical points in the mitotic cycle, and its properties and behavior suggest that it may play some role in regulating mitotic chromatin, possibly by deubiquitinating histones or other relevant substrates (Cai et al., 1999).

To the best of our knowledge, there is no assay that directly measures the isopeptidase activity of DUBs using authentic substrates containing polymeric Ub chains. This is partly due to the scarcity of true substrates that can be used for the measurement of its activity. This is our first report on the development of an assay to measure the DUB activity using a multi-Ub protein substrate. Although in previous reports a fluorogenic substrate Ub-AMC has been used to measure the Ub C-terminal hydrolase activity

(Dang et al., 1998), use of Ub-AMC as a substrate exclusively for monitoring UbpM activity has its own limitations. Absence of Ub polymeric chains in the substrate could lead to false-positive hits owing to a distorted enzyme–substrate complex during inhibitor screening. Ub-AMC, however, serves as an ideal indirect artificial substrate for optimizing some of the assay parameters of the UbpM assay. Figure 5A shows a schematic representation of the multi-Ub disassembly assay wherein poly-Ub APC serves as an authentic substrate for monitoring the UbpM activity. The multi-Ub disassembly assay was relatively simple to develop and run, because it involves just two more additional steps to the Ub-polymerization assay described previously. To measure UbpM activity, the ubiquitination of the APC reaction was stopped by adding EDTA at the end of 1 h incubation, followed by the addition of appropriate concentrations of UbpM to initiate the disassembly of multi-Ub chains. Disassembly of multi-Ub chains from APC results in an increase in fluorescent emission intensity of fluorescein

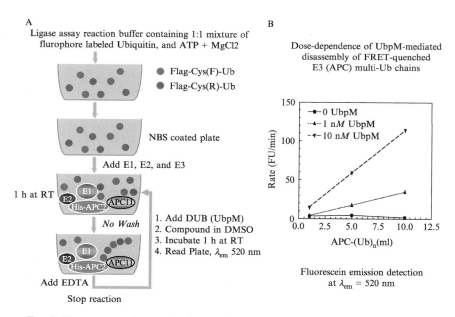

FIG. 5. Homogeneous FRET-based assay for multi-Ub chain disassembly. (A) Schematic representation of the screening assay showing the stepwise addition of various components of the assay including test compounds in DMSO (inhibitor). (B) Dose dependence of UbpM-mediated disassembly of FRET quenched E3 (APC) multi-Ub chains. Disassembly of multi-Ub E3 (APC) was detected by an increase in fluorescein emission signal at $\lambda_{em} = 520$ nm using a fluorescence plate reader.

(F), which was monitored at λ_{em} 520 nm using a fluorescent microplate reader (Fig. 5B). As shown in Fig. 5B, no increase in fluorescent intensity was observed in the reaction mixture lacking UbpM. Generation of fluorescent signal seems to be dose dependent and linear over the amount of APC-$(Ub)_n$ used in the assay.

For optimization of some of the assay conditions, we initially chose Ub-AMC as a substrate, because it is commercially available and saves assay time. As shown in Fig. 6A, UbpM responded well to Ub–AMC substrate, resulting in a steady-state cleavage of AMC from the Ub C-terminus. On cleavage of AMC from Ub-AMC, a clear increase in fluorescence intensity of AMC was noticed at λ_{em} 460 nm. The assay seems to be relatively intolerant to DMSO. As shown in Fig. 6A, 10% DMSO caused the loss of approximately 25–30% of the signal. Hence, when using this assay for compound screening, care must be taken to reduce the concentration of DMSO to less than 5% to nullify the background effect. The effects of DMSO in screening assays have been reported previously (Hong et al., 2003) for a TRAF6 FRET assay. In a separate experiment, we tested the reversible and irreversible enzyme activities of UbpM using free Ub and Ub-aldehyde (Ub-al) as inhibitors. As shown in Fig. 6B, in the presence of free Ub, UbpM showed a K_i of 1.9 μM, which was slightly higher than the enzymatic K_m. Similarly, on addition of Ub-al at various concentrations, UbpM underwent an irreversible modification because of covalent bond formation between the thiol group of UbpM and the aldehyde group of Ub-al, leading to a gradual loss in the activity (Fig. 6C). These data clearly suggest the potential application of this assay for screening isopeptidase inhibitors.

Pitfalls of FRET-Based Assays

Although when compared with heterogeneous assays the homogeneous FRET assay showed a higher precision with a far lesser well-to-well variation, it also had a lower signal/background ratio. Similar observations have been reported in previous FRET-based assays (Boisclair et al., 2000), and hence continuous efforts to address this issue are still needed. Another caveat inherent to all homogeneous high-throughput screens is the potential for test compounds to interfere with the assay signal. In the case of FRET assays, colored compounds can quench fluorescence by attenuating excitation of the donor fluorophore. Boisclair et al. (2000) have found that although some colored compounds showed substantial quenching at 100 μg/ml, quenching was insignificant at 1 μg/ml. Hence quenching of fluorescence was not a major problem when compounds were assayed at 10 μg/ml. However, caution should be exercised when screening

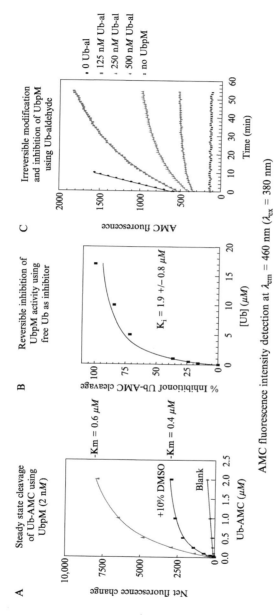

FIG. 6. Determination of UbpM activity using Ub-AMC as an alternative artificial substrate. (A) Steady-state cleavage of Ub-AMC using UbpM (2 nM). AMC cleavage from Ub C-terminus generates an increase in fluorescence intensity that was monitored at λ_{em} = 460 nm; (B) Reversible inhibition of UbpM activity using free Ub as inhibitor. Ub's affinity for proteases is in a useful range for screening and inhibition purposes (K_i seems to be slightly higher than enzymatic K_M). (C) Irreversible modification and inhibition of UbpM using Ub-aldehyde (Ub-al). UbpM/Ub-al adduct formation is diffusion controlled. Data analyzed using single exponential fit.

compounds at concentrations greater than 10 μg/ml, as, for example, when determining IC$_{50}$ measurements for hit compounds. In the case of compounds that interfere strongly with FRET assay measurements, it is preferable to corroborate the IC$_{50}$ using a companion heterogeneous assay method, such as ELISA or DELFIA.

Concluding Remarks

In summary, we have found that the homogeneous nature of the FRET assay offers significant advantages over conventional methods, which include DELFIA, radiological methods, and other colorimetric assay formats such as ELISA. In this assay format, one can monitor both ubiquitin polymerization and depolymerization reactions in the same reaction mixture simply by adding an additional reagent such as a DUB (UbpM) at the end of the polymerization reaction. Elimination of a number of washing steps and unnecessary use of secondary reagents during the detection step significantly shorten the overall assay time. Despite the need for further optimization of the assay, the current method has greater throughput capacity than currently available methods, and fine tuning of certain parameters should ease the automation of this assay. In general, the utility of the homogenous FRET assay, demonstrated here for multi-Ub chain assembly and disassembly, has the potential to improve the throughput and reliability of screening for new drug leads in cancer biology.

Acknowledgment

We thank Ms. Linette Fung for her help in preparing this manuscript.

References

Baek, K. H. (2003). Conjugation and deconjugation of ubiquitin regulating the destiny of proteins. *Exp. Mol. Med.* **35,** 1–7.
Boisclair, M. D., McClure, C., Josiah, S., Glass, S., Bottomley, S., Kamerkar, S., and Hemmila, I. (2000). Development of a ubiquitin transfer assay for high throughput screening by fluorescence resonance energy transfer. *J. Biomol. Screen.* **5,** 319–328.
Cai, S. Y., Babbitt, R. W., and Marchesi, V. T. (1999). A mutant deubiquitinating enzyme (Ubp-M) associates with mitotic chromosomes and blocks cell division. *Proc. Natl. Acad. Sci. USA* **96,** 2828–2833.
Clegg, R. M. (1996). Fluorescence resonance energy transfer (FRET). *In* "Fluorescence Imaging Spectroscopy and Microscopy" (X. F. Wang and B. Herman, eds.), pp. 179–252. John Wiley and Sons, New York.
Colbert, D. L., and Childerstone, M. (1987). Multiple drugs of abuse in urine detected with a single reagent and fluorescence polarization. *Clin. Chem.* **33,** 1921–1923.

Dang, L. C., Melandri, F. D., and Stein, R. L. (1998). Kinetic and mechanistic studies on the hydrolysis of ubiquitin C-terminal 7'-amido-4-methylcoumarin by deubiquitinating enzymes. *Biochemistry* **37,** 1868–1879.

Davydov, I. V., Woods, D., Safiran, Y. J., Oberoi, P., Fearnhead, H. O., Fang, S., Jensen, J. P., Weissman, A. M., Kenten, J. H., and Vousden, K. H. (2004). Assay for ubiquitin ligase activity: High-throughput screen for inhibitors of HDM2. *J. Biomol. Screen.* **9,** 695–703.

Earnshaw, D. L., Moore, K. J., Greenwood, C. J., Djaballah, H., Jurewicz, A. J., Murray, K. J., and Pope, A. J. (1999). Time-resolved fluorescence energy transfer DNA helicase assays for high throughput screening. *J. Biomol. Screen.* **4,** 239–248.

Fairclough, R. H., and Cantor, C. R. (1978). The use of singlet-singlet energy transfer to study macromolecular assemblies. *Methods Enzymol.* **48,** 347–379.

Fang, S., and Weissmann, A. M. (2004). Ubiquitin-proteasome system: A field guide to ubiquitylation. *Cell Mol. Life Sci.* **61,** 1546–1561.

Finley, D., Ciechanover, A., and Varshavsky, A. (2004). Ubiquitin as a central cellular regulator. *Cell* **116**(Suppl. 2), S29–S32, 2 p following S32.

Haugland, R. P. (2002). *In* "Handbook of Fluorescent Probes and Research Products." 9th Ed. Molecular Probes Inc., Eugene, OR.

Hershko, A., and Ciechanover, A. (1998). The ubiquitin system. *Annu. Rev. Biochem.* **67,** 425–479.

Hochstrasser, M. (1996). Ubiquitin-dependent protein degradation. *Annu. Rev. Genet.* **30,** 405–439.

Hong, C. A., Swearingen, E., Mallari, R., Gao, X., Cao, Z., North, A., Young, S. W., and Huang, S. G. (2003). Development of a high throughput time-resolved fluorescence resonance energy transfer assay for TRAF6 ubiquitin polymerization. *Assay Drug Dev. Technol.* **1,** 175–180.

Huang, J., Sheung, J., Dong, G., Coquilla, C., Daniel-Issakani, S., and Payan, D. G. (2005). High-throughput screening for inhibitors of ubiquitin ligase APC. *Methods Enzymol.* **399,** 739–753.

Kane, S. A., Fleener, C. A., Zhang, Y. S., Davis, L. J., Musselman, A. L., and Huang, P. S. (2000). Development of a binding assay for p53/HDM2 by using homogeneous time-resolved fluorescence. *Anal. Biochem.* **278,** 29–38.

Kolb, J. M., Yamanaka, G., and Manly, S. P. (1996). Use of a novel homogeneous fluorescent technology in high throughput screening. *J. Biomol. Screen.* **1,** 203–210.

Lai, Z., Yang, T., Kim, Y. B., Sielecki, T. M., Diamond, M. A., Strack, P., Rolfe, M., Caligiuri, M., Benfield, P. A., Auger, K. R., and Copeland, R. A. (2002). Differentiation of Hdm2-mediated p53 ubiquitination and Hdm2 autoubiquitination activity by small molecular weight inhibitors. *Proc. Natl. Acad. Sci. USA* **99,** 14734–14739.

Lakowicz, J. R. (1999). "Principles of Fluorescence Spectroscopy" Kluwer Academic/Plenum Publishers, New York.

Laney, J. D., and Hochstrasser, M. (2002). Assaying protein ubiquitination in *Saccharomyces cerevisiae*. *Methods Enzymol.* **351,** 248–257.

Owicki, J. C. (2000). Fluorescence polarization and anisotropy in high throughput screening: Perspectives and primer. *J. Biomol. Screen.* **5,** 297–306.

Pickart, C. M. (1997). Targeting substrates to the 26S proteasome. *FASEB J.* **11,** 1055–1066.

Pickart, C. M. (2001). Mechanisms underlying ubiquitination. *Annu. Rev. Biochem.* **70,** 505–533.

Pray, T. R., Parlati, F., Huang, J., Wong, B. R., Payan, D. G., Bennett, M. K., Issakani, S. D., Molineaux, S., and Demo, S. D. (2002). Cell cycle regulatory E3 ubiquitin ligases as anticancer targets. *Drug Resist. Update* **5,** 249–258.

Robinson, P. A. (2004). Ubiquitin-protein ligases-novel therapeutic targets? *Curr. Protein Pept. Sci.* **5,** 163–176.

Selvin, P. R. (1995). Flourescence resonance energy transfer. *Methods Enzymol* **246,** 300–334.
Selvin, P. R., Rana, T. M., and Hearst, J. E. (1994). Luminescence resonance energy transfer. *J. Am. Chem. Soc.* **116,** 6029.
Sun, Y. (2003). Targeting E3 ubiquitin ligases for cancer therapy. *Cancer Biol. Ther.* **2,** 623–629.
Sun, L., and Chen, Z. J. (2004). The novel functions of ubiquitination in signaling. *Curr. Opin. Cell Biol.* **16,** 119–126.
Szollosi, J., Damjanovich, S., and Matyus, L. (1998). Application of fluorescence resonance energy transfer in the clinical laboratory: Routine and research. *Cytometry* **34,** 159–179.
Takada, K., Nasu, H., Hibi, N., Tsukada, Y., Ohkawa, K., Fujimuro, M., Sawada, H., and Yokosawa, H. (1995). Immunoassay for the quantification of intracellular multi-ubiquitin chains. *Eur. J. Biochem.* **233,** 42–47.
Tang, Z., Li, B., Bharadwaj, R., Zhu, H., Ozkan, E., Hakala, K., Deisenhofer, J., and Yu, H. (2001). APC2 Cullin protein and APC11 RING protein comprise the minimal ubiquitin ligase module of the anaphase-promoting complex. *Mol. Biol. Cell* **12,** 3839–3851.
VanDemark, A. P., and Hill, C. P. (2002). Structural basis of ubiquitylation. *Curr. Opin. Struct. Biol.* **12,** 822–830.
Wee, K. E., Lai, Z., Auger, K. R., Ma, J., Horiuchi, K. Y., Dowling, R. L., Dougherty, C. S., Corman, J. I., Wynn, R., and Copeland, R. A. (2000). Steady-state kinetic analysis of human ubiquitin-activating enzyme (E1) using a fluorescently labeled ubiquitin substrate. *J. Protein Chem.* **19,** 489–498.
Wilkinson, K. D. (1997). Regulation of ubiquitin-dependent processes by deubiquitinating enzymes. *FASEB J.* **11,** 1245–1256.
Wilkinson, K. D. (2000). Ubiquitination and deubiquitination: Targeting of proteins for degradation by the proteasome. *Semin. Cell Dev. Biol.* **11,** 141–148.
Wong, B. R., Parlati, F., Qu, K., Demo, S., Pray, T., Huang, J., Payan, D. G., and Bennett, M. K. (2003). Drug discovery in the ubiquitin regulatory pathway. *Drug Discov. Today* **8,** 746–754.
Wu, P., and Brand, L. (1994). Resonance energy transfer: Methods and applications. *Anal. Biochem.* **218,** 1–13.
Yabuki, N., Watanabe, S., Kudoh, T., Nihira, S., and Miyamato, C. (1999). Application of homogeneous time-resolved fluorescence (HTRFTM) to monitor poly-ubiquitination of wild-type p53. *Comb. Chem. High Throughput Screen.* **2,** 279–287.

[45] Assays for High-Throughput Screening of E2 and E3 Ubiquitin Ligases

By JOHN H. KENTEN, ILIA V. DAVYDOV, YASSAMIN J. SAFIRAN, DAVID H. STEWART, PANKAJ OBEROI, and HANS A. BIEBUYCK

Abstract

We developed a series of assays for biochemical activities involving ubiquitin. These assays use electrochemiluminescence detection to measure the ubiquitylation of target proteins. To enable electrochemiluminescence detection, the target proteins were prepared as bacterially expressed fusion proteins and captured on the surface of specially designed microtiter plates having integrated electrodes. Ubiquitylation was quantitated directly, through the use of ubiquitin labeled with an electrochemiluminescent label, or indirectly, through the use of labeled antiubiquitin antibodies. Assays were carried out in both 96-well and 384-well plates. The success of the assay with this variety of formats allowed the selection of optimal work flows for specific applications on the basis of ease of use and overall reagent consumption and availability.

We used our ubiquitylation assays to measure the activities of E2 ubiquitin-conjugating enzymes and E3 class ubiquitin ligases. Signal/background ratios for many of our assays were greater than 50, significantly facilitating their conversion to high-throughput practice in a convenient manner. The speed, sensitivity, and convenience of the assay formats makes them well suited for comprehensive interrogations of libraries of compounds or genes in applications like drug and substrate discovery for ubiquitin ligases.

Introduction

Ubiquitylation, the attachment of the small protein ubiquitin to target proteins, typically proceeds in a multistep process involving attachment of a ubiquitin molecule to an E1 ligase (UBE1), transfer of the ubiquitin to an E2 ligase (UBC), and transfer of the ubiquitin to a target protein, this final transfer from the E2 ligase to the target being directed or catalyzed by an E3 ligase. Multiple ubiquitins can be transferred to a target resulting in the formation of a polyubiquitin chain attached to the target. The large number of human genes believed to encode UBCs (~50) and E3 ligases (~390) allows for the ubiquitylation of specific proteins or families of proteins to be independently controlled (Wiessman, 2001).

The emerging role of ubiquitylation as a key process for the control of numerous signaling pathways has sparked considerable research into this post-translational modification as a target for drug discovery (Weissman, 2001; Wong *et al.*, 2003). Interest intensified with the success of an inhibitor of the proteasome as an anticancer drug (Adams, 2003). A quick review of the recent literature combined with the number of potential ubiquitin E3 ligases in the genome suggests an exciting future. Given that the number of putative ubiquitin E3 ligases is similar to that for kinases in the mammalian genome, it is reasonable to assume that the role of E3 ligases in cell biology will also likely rival that of kinases in importance. Many problems remain toward forming a better understanding of ubiquitylation and its roles. Robust assay methods for investigating the function and activity of ubiquitin ligases pose outstanding challenges, particularly given the multiprotein character of many of the posttranslational events in these pathways.

Another complicating factor is that many E3 ligases function in ways that depart from classical biochemical paradigms presented by other systems of enzymes. In the case of the UBE1 (E1), UBC's (E2), and HECT domain E3 ligases, "classic" intermediate species are evident as expected in an enzymatic reaction. The large family of RING finger, E3 ligases have, by contrast, no such demonstrated chemistry and seem to act more as obligate scaffolds that bind and bring together the relevant E2 ubiquitin-conjugating enzyme and its ultimate target. The binding sites of these ligases tend to be large, presumably to allow the specific binding of large protein molecules, and do not have the recognizable catalytic active sites of classical enzymes. The nature of binding interactions with RING finger E3 ligases is not well understood, because only a small number of crystal structures documenting these interactions exist. In a few cases, however, it is clear that even single amino acid changes to the RING domain are sufficient to destroy the interaction with an E2 ubiquitin-conjugating enzyme, suggesting the possibility of intervention with small molecule inhibitors of the binding event. The important role of posttranslational modification in the control of E3 ligases, similarly, offers the possibility of actioning its activity through small molecule inhibitors of these modifications.

The typical reaction cycle for ubiquitylation of a substrate includes the substrate protein, ATP, ubiquitin, UBE1 (UBE1, 1085 aa), a UBC (UBC, ~50 human genes, usually between 140–250 aa), and an E3 (~390 human genes, typically ~150–1900 aa) (Pickart, 2004). In certain cases, UBCs are able to function like E3s and directly ubiquitylate substrates, as in the case of the unusually large BIRC6 (apollon, 4829 aa) protein (Hao et al., 2004). In a typical ubiquitylation process, ubiquitin is first activated for coupling by UBE1 using ATP to generate an UBE1-ubiquitin conjugate through a thioester bond. The bond occurs between the COOH terminus of ubiquitin and a Cys at the active site of UBE1. Activated ubiquitin transfers to the E2 ubiquitin-conjugating enzyme by means of a trans-thioesterification reaction that involves another Cys at the active site of the E2, but otherwise no additional energy. The generation of the activated E2s by UBE1 provides a pool of divergent UBCs that interact, at differing levels of specificity, with the great variety of E3 ligases to target specific proteins for ubiquitylation. Many E3 ligases show self- or auto-ubiquitylation in the presence of an activated E2, a property that has been used both in research and the development of assays, although its detailed mechanism and stoichiometries remain elusive (Lorick et al., 1999). The biological significance of self- or auto-ubiquitylation also requires elaboration, but the notion that autoregulation by this mechanism control levels of the ligase itself seems plausible and supported by observation (Li et al., 2004; Stommel and Wahl, 2004).

Together, the biochemical properties of ubiquitin ligases have slowed their selection as targets for drug discovery by pharmaceutical companies. Traditional methods have dominated the development of assays for the investigation of ubiquitylation to date: gel electrophoresis and Western blot performed with antibodies–enzyme conjugates or using radioactively labeled ubiquitin are the "gold standards." The literature shows more recent development of assays exploiting methods that allow the analysis of the state of ubiquitylation in multiple samples using simple tube-based formats amenable to discovery efforts like compound screening. In most cases, these assays reflect an abstraction of the pathway and measure the self- or auto-ubiquitylation of an E3 ligase and not the ubiquitylation of the E3 substrate (Lorick et al., 1999).

Published examples of simple, *in vitro*, tube-based assays used a variety of sources for reagents, including native and recombinant proteins having varying degrees of purity, and were largely built around fluorescence-based schemes of detection. One of the earliest references to potential *in vitro* methods described scintillation proximity assays (SPA), RIA, and ELISA-based formats using I^{125} labeled ubiquitin and antiubiquitin antibodies, although no data were shown (Holloway and Waddell, 1998). Yabuki *et al.* (1999) demonstrated a fluorescence resonant energy transfer (FRET) approach to the study of the formation of polyubiquitin on p53 using europium-labeled ubiquitin, allophycocyanin-labeled streptavidin, and biotinylated p53. After completion of the ubiquitylation reaction, the ubiquitylated p53 was detected by adding the labeled streptavidin and measuring the FRET signal. This group also compared the FRET assay to a SPA based on ^{125}I-labeled ubiquitin. Both assay formats were nonwashed (i.e., they required no separation of assay reagents from the ubiquitylated protein before its measurement). A variation of the FRET method used by Yabuki *et al.* was used by Hong *et al.* (2003) to develop an assay for TRAF6, a ringfinger E3 ubiquitin ligase. Hong *et al.* (2003) used a mixture of ubiquitin labeled with europium and ubiquitin labeled with biotin. The reaction was carried out with a mixture of the ubiquitylation enzymes UBE1, UBE2V1 (UeV1A), UBE2N (Ubc13), and TRAF6, allowing for the polyubiquitylation of TRAF6. The assay was stopped by the addition of EDTA (to inhibit the E1), and the addition of allophycocyanin-labeled streptavidin generated the final, detectable, species. A similar fluorescent-based assay was described by Boisclair *et al.* (2000) for ubiquitylation of the HECT domain E3 ligase UBE3C (Rsc). These fluorescent approaches typically have low signal-to-noise ratios and show enhanced sensitivity to interferences from fluorescent compounds present in most chemical libraries. Fluorescence was also used in the development of an assay for the ubiquitylation of IkBa, but in this assay, the europium-labeled ubiquitin conjugated to IkBa was detected directly after extensive washing, which helped to reduce the problems from fluorescent compounds (Swinney et al., 2002).

In a departure from the fluorescent-based approaches, Davydov et al. (2004) made use of electrochemiluminescence detection in a bead-based assay for MDM2. Although perhaps less well known than fluorescence, electrochemiluminescence-based methods have become important in the demanding arena of clinical diagnostics (Roberts and Roberts, 2004) and are increasingly used in life science and pharmaceutical research applications. These methods make use of electrochemiluminescent labels that emit light when oxidized at an electrode under appropriate chemical conditions (see Debad et al., 2004 for a more detailed explanation of the technique and its uses). The most widely used label is ruthenium *tris*-bypyridine, or RuBpy, because of its strong electrochemiluminescence signal, its chemical inertness, and the availability of a wide variety of derivatives that allow for facile incorporation of the labels into biomolecules using any one of a number of linking strategies (the most commonly used are derivatives containing N-hydroxysuccinimide esters). The increasing popularity of electrochemiluminescence-based assay methods is due to their attomole sensitivity, dynamic range, freedom from interference, and ease of use. An additional benefit is that light emission only occurs from labels present in close proximity to an electrode surface. This surface selectivity allows for electrochemiluminescence detection to be used with solid-phase assay formats having minimal or no wash steps (Debad et al., 2004). We advanced a number of *in vitro* ubiquitin ligase assays that combined the use of RuBpy-based labels with technologies developed at Meso-Scale Discovery for carrying out single-plex and array-based multiplex electrochemiluminescence measurements in a multi-well plate format (see www.meso-scale.com for a detailed description of the technology and examples of its use in a variety of applications). The following E2 and E3 ligase assays demonstrate the performance, simplicity, and speed that can be achieved using this approach.

Assay Principles

The electrochemiluminescence formats described here make use of either labeled ubiquitin or labeled antibodies to ubiquitin to measure the production of ubiquitylated target proteins by E2 ubiquitin-conjugating enzymes or E3 ligases. The label used is a highly water-soluble sulfonated derivative of RuBpy (MSD Sulfo-TAG™ label). The reaction products are measured by specifically capturing the products on a coated electrode surface, applying an electrical potential to the electrode to cause labels bound to the products to generate electrochemiluminescence, and measuring the emitted light to quantify the amount of captured product.

One approach to E3 ligase assays exploits their self-ubiquitylation, detected here using antiubiquitin antibodies. Figure 1 shows the basic scheme. These assays use recombinant E3 ligases expressed as the GST fusion protein in bacteria. The recombinant E3 ligase is combined with the appropriate amount (see later) of ATP, ubiquitin, UBE1, and the UBC of choice. The reaction can lead to ubiquitylation at multiple lysine ubiquitylation sites on the E3 ligase and the generation of polyubiquitin chains at one or more of these sites. If desired, a greater determinacy in the origin of signal can be conferred by use of a modified ubiquitin that allows only monomeric conjugation or by modifying the sequence of the E3 ligase to remove some of the ubiquitylation sites. The reaction products are detected by specifically capturing the E3 ligase on glutathione-coated

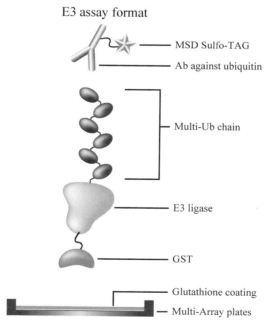

FIG. 1. Schematic drawing of the E3 assay format showing the final molecular complex formed at the end of a typical assay for E3 ligase activity. GST, glutathione S transferase; Ab, antibody; Ub, Ubiquitin.

electrodes (by means of glutathione–GST interactions) in the wells of specially designed multi-well plates having integrated electrodes, binding a labeled antiubiquitin antibody to ubiquitin on the captured E3 ligase, and measuring (by means of electrochemiluminescence) the amount of label on the electrode surface. The possibility remains that tight binding interactions between the E3 and E2 might cause spurious signals in the case in which the E3 is not ubiquitylated and the E2 remains charged. In our laboratory, this potential artifact is addressed by a combination of stringent washes designed to break up the binding interaction and direct measurement of the E2–E3 interaction.

We use either labeled ubiquitin or antibodies to ubiquitin to detect UBC–ubiquitin conjugates. Figure 2A and B shows these formats. Here, the UBC is a recombinant enzyme fused with a His_6 sequence at either its N or C terminus. The reaction takes place in a cocktail of reagents that includes ATP, MSD ubiquitin (either unlabeled or labeled with Sulfo-TAG depending on the format), and UBE1. Mixing in the UBC enzyme of interest leads to the formation of charged E2 with ubiquitin attached as a thioester. This complex is captured, through the His_6 tag on the recombinant E2, onto a multi-well plate having integrated electrodes coated with a

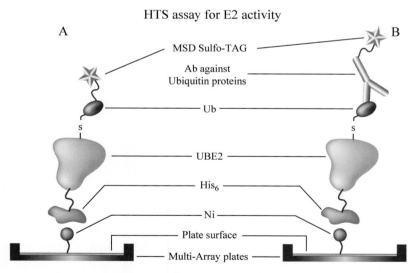

FIG. 2. Schema of two E2 ligase assay formats showing the final molecular complexes formed at the end of typical assays for UBC ligase activity. (A) Format based on using labeled ubiquitin. (B) Format based on using labeled antiubiquitin antibody. Ub, ubiquitin; UBC, ubiquitin conjugating enzyme of choice; and Ni-chelate plates.

nickel chelate. The complex is then detected by means of the labeled ubiquitin or after binding a labeled antiubiquitin antibody.

Materials and Instrumentation

The genes for the human ubiquitin E3 ligases XIAP, NEDD4, and MDM2 were obtained from Allan M. Wiessman; each was cloned to produce the N-terminal GST fusion protein (Fang et al., 2000; Lorick et al., 1999; Yang et al., 2000). We also cloned a number of ubiquitin E3 ligase genes using standard PCR methods from IMAGE clones to generate similar fusions in the pGEX-4T1 vector (Amersham Bioscience). The list of human ubiquitin E3 ligases we surveyed with the methods presented here include: TRIM32, RNF41, RNF11, SMURF1, RNF28 (MuRF1), HRD1, RNF25 (AO7), ZNF364, and TRIM25. The list of mouse ubiquitin E3 ligases include Rnf41, Rnf11, Smurf1, Hrd1, Rnf25 (AO7), and Trim25.

The E2 ubiquitin-conjugating enzymes; His_6-tagged UBE2D2 (UbcH5b), mutant UBE2D2-C85A, UBE2D1 (UbcH5a), and mutant UBE2D1-C85A were obtained from Allan M. Wiessman. We also cloned UBE2D2, the *Drosophila* E2 effete (eff), UBE2A (Rad6A), and UBE2B (Rad6B) both as C-terminal His_6-tagged and untagged sequences into pQE-60 (Qiagen). UBE1 enzyme (Meso-Scale Diagnostics, Gaithersburg, MD) was purified from rabbit reticulocyte lysate.

Electrochemiluminescence measurements were carried out using proprietary multi-well plates (Multi-Array™ plates) having integrated screen-printed electrodes that act as both a capture surface for solid-phase assays and as an energy source for electrochemiluminescence excitation. The electrodes were either coated with glutathione for binding GST fusion proteins or coated with a nickel chelate for binding $(His)_6$ fusion proteins. Electrochemiluminescence in the plates was induced and measured using an imaging plate reading system (Sector™ Imager 6000 reader). MSD Sulfo-TAG–labeled antiubiquitin antibody, MSD Sulfo-TAG–labeled ubiquitin, E2 and E3 Ubiquitin Reaction Buffers, Ubiquitin-Binding Buffer, MSD Read Buffer T, coated 96-well and 384-well Multi-Array plates, and the Sector Imager 6000 reader are products of Meso Scale Discovery (www.mesoscale.com).

Methods

Ubiquitin Ligase Expression

The various ubiquitin ligases were expressed in *Escherichia coli* following standard protocols and used as crude bacterial lysates. *In vitro* ligase reactions that used a cocktail of ATP, ubiquitin, UBE1, and a selection of

UBCs (see later) tested the recovery of active E3 ligases. The products of these reactions were run on SDS-PAGE gels to determine the level of ubiquitylation by observation of the parent and lower mobility bands using total protein staining. Western blot analysis was used to confirm that the lower mobility bands comprised ubiquitin. Typical results for the E3 ligases like MDM2, NEDD4, XIAP, RNF28 in combination with UBE2D2 appear in Fig. 3. The GST–E3 fusions produced strong bands with some degradation products before the ubiquitylation reaction (Fig. 3A, lane C). After they are subjected to the ubiquitylation reaction, gels of the GST–E3s show a continuum of the higher molecular weight species expected for the ubiquitylated product (Fig. 3A, lane U). The Western blot using an antiubiquitin antibody clearly showed that these higher molecular weight species were the consequence of the ubiquitylation reaction (Fig. 3B).

Ubiquitin E3 Ligase Assays

We developed and optimized a number of different protocols for measuring the E3 ligase self-ubiquitylation activity that suited different work flows, assay designs, and types of automation. These included changes to the number of wash steps, the orders of addition of reagent, and the preformation of the activated ubiquitin–UBC complex (charged UBC). The latter occurred by simply allowing the E1 and E2 to react in the absence of the E3. Typically, 30–60 min proved sufficient to get exhaustive

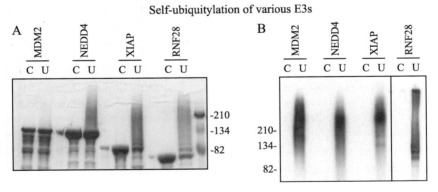

FIG. 3. Analysis of the self-ubiquitylation of MDM2, NEDD4, XIAP, and RNF28 using UBE2D2 as an E2. (A) SDS-PAGE of E3 ligases purified on glutathione sepharose. Lane C, control unreacted GST–E3 ligases, and lane U, GST–E3 ligases reacted in a ubiquitylation reaction (UBE1, UBC, ubiquitin, ATP). (B). Western blot analysis with antiubiquitin antibody of the gel from (A).

priming of the E2. Using precharged E2 provided a potential means of avoiding, in screens for inhibitors of E3 activity, artifacts because of inhibition of E1, E2, or other general components of the assay. The protocol that follows represents a first choice in the selection of a workflow for a screen aimed at finding inhibitors of E3 activity.

Optionally, glutathione-coated Multi-Array plates were blocked with bovine serum albumin and washed before use. We added 30 μl of the GST–E3 fusion protein to the wells of the plate and incubated for 60 min to prebind the GST–E3 to the surface (Fig. 1). We concurrently created a "master mix" of UBE1, UBE2D2, ubiquitin, and ATP and preincubated the mixture at room temperature for 30 min to precharge the UBE2D2. We then added 30 μl of the master mix and 1–2 μl of a potential inhibitor (in 50–100% DMSO) into the wells containing the GST–E3. The ubiquitylation reaction was run for 60 min to generate the multi-ubiquitylated GST–E3. Three washes with PBS effectively terminated the reaction. The ubiquitylated products were detected by the addition of 150 μl of a mixture of Ubiquitin Binding Buffer containing MSD Sulfo-Tag–labeled antiubiquitin antibody ($\sim$10 ng of antibody per well) and MSD Read buffer T (a buffer that provides the appropriate chemical environment for ECL generation). Plates were read using an SECTOR™ Imager 6000 instrument ($\sim$1 min per plate in 96-well or 384-well format). This workflow is illustrated in Fig. 4. We have also developed variations of this format that are configured for 384-well plates that eliminate wash steps and/or preincubate the E3 ligase with the potential inhibitors.

Assay Protocol in Detail.

1. Blocking of 96-well Multi Array glutathione plate (optional).

The Multi Array glutathione plates can be used directly, but blocking the plates with BSA may improve assay performance in certain cases.

    a. Add 200 μl of 5% BSA in PBS per well.
    b. Incubate 1 h at room temperature.
    c. Wash three times with PBS before use.

2. Preparation of buffers (120 tests).

E3 Reaction Buffer (120 tests): To 10 ml of E3 Ubiquitin Reaction Buffer add 20 μl of DDT (1 M), 50 μl of ATP (200 mM), and 0.01 g BSA.
Binding buffer (120 tests):
To 10 ml of 4$\times$ Ubiquitin Binding Buffer add 0.04 g BSA.

3. Preparation of E3 ligase.

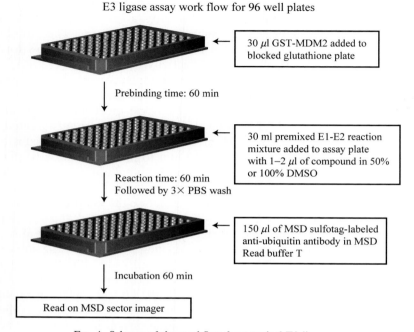

Fig. 4. Scheme of the workflow for a typical E3 ligase assay.

Titrations of the E3 ligases are required to establish the amount needed to generate acceptable signals (i.e., signal-to-backgrounds >10) without reaching saturation. The E3 ligase of choice is diluted into the reaction buffer to give the desired activity.

4. Prebind GST–E3 ligase.

   a. Add 30 μl of the diluted E3 ligase to the appropriate well of the 96-well glutathione plate.
   b. Incubate the plate at room temperature for 1 h.
   c. During the incubation, begin working on step 5.

5. Precharging UBE1–UBC mixture.

Combine the optimized amount of the following enzymes: UBE1 (purified) (between 0.1–0.003 μl) and UBC (lysate) (between 0.3–0.1 μl) in 4 ml of reaction buffer to make the UBE1–UBC mixture. At 30 min after starting the prebinding of GST-E3 (Step 4), add ubiquitin to the UBE1–UBC mixture for a final concentration of 1.5 μM to start the UBE1–UBC reactions.

Note: For a negative control, EDTA (25 mM final) can be added to a portion of the UBE1–UBC mixture before addition of ubiquitin; under these conditions, charging of UBC with ubiquitin does not occur.

6. Addition of test compounds and the precharged mixture from step 5.

   a. To all wells add 30 μl/well of the precharging mixture.
   b. Add 1–2 μl of 50–100% DMSO (during a compound library screening, the test compounds are dissolved in the DMSO).
   c. Incubate plate for 30–60 min.

Typically, at this step we first pipette up the UBE1–UBC mixture and then pipette up compound into the same pipette tip (an air gap separates the two aliquots) followed by dispensing of the entire volume into the Multi Array plate using a Biomek FX automated pipetting station.

7. Addition of the antibody.

   a. Wash plates three times with phosphate-buffered saline.
   b. Combine 4.5 ml of Binding Buffer (4×), 4.5 ml of Read Buffer T (4×), 9 ml of water, and 2.4 μl of MSD Sulfo-TAG–labeled antiubiquitin antibody (0.5 mg/ml).
   c. Add 150 μl/well of the above MSD Sulfo-Tag–Anti-Ub solution to all the reactions in 96-well plate.
   d. Incubate plates for 1 h at room temperature.

8. Read on the MSD Sector Imager. The plates are analyzed under the default instrument settings. The instrument returns a text file reporting the intensity of electrochemiluminescence in each well of the plate.

We evaluated and optimized a number of the key factors affecting the assay during the development for high-throughput screening. These factors included the selection of an active UBC, the time courses of reactions for both the precharging of the UBE1, UBC mixtures, and the E3 ligase reaction. Selection of the ideal UBC ubiquitin-conjugating enzyme typically would involve comparisons of the E3 ligase activity in combination with a range of selected UBCs. In many cases, a UBC ubiquitin-conjugating enzyme has been previously determined from some other biochemical or biological effort for a given E3 ligase so that we may only run a confirmatory gel before beginning our optimizations.

The assay response to a range of UBE1, UBC, E3 ligase, ubiquitin, ATP, and MSD antiubiquitin antibody concentrations are also studied. Figure 5 provides examples of these data that show a time course study for XIAP and an E3 ligase titration for Smurf1. There are many

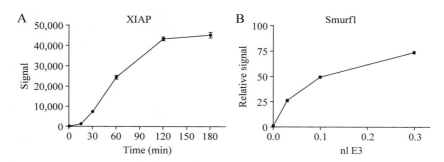

FIG. 5. Optimization of ubiquitin ligase assays. (A). Time course of XIAP self-ubiquitylation reaction. (B). Titration of Smurf1 lysate in a self-ubiquitylation reaction. These reactions were carried out using a ubiquitylation reaction mixture of UBE1, UBC, ubiquitin, and ATP at room temperature. Electrochemiluminescent signals were measured on a Sector Imager 6000.

considerations in looking for optima in such experiments. Principal among them is that the assay should demonstrate a linear relationship between its signal and the E3 amount in the reaction to detect the reduction of an E3 activity with a prospective inhibitor as sensitively as possible. Also, the time course of the reaction should demonstrate linear kinetics with increasing incubation time (i.e., none of the other reagents should be limiting in the generation of signal).

In summary, these investigations allowed us to select conditions that would optimize sensitivity of the E3 ligase to inhibition during a compound screen and help to minimize false positives because of inhibition of other assay components like UBE1 and UBC. Many additional factors are important for the development of robust compound screens, of course: *Inter alia* are the stability of the reactants and products at various steps and the tolerance of the reaction to DMSO, the solvent used in many cases for dissolving the compounds in small organic molecule libraries.

We applied our assay approach to a series of human E3 ligases, XIAP, Nedd4, MDM2, TRIM32, RNF41, RNF11, SMURF1, RNF28, HRD1, RNF25, ZNF364, and TRIM25, and the mouse ubiquitin E3 ligases, Rnf41, Rnf11, Smurf1, Hrd1, Rnf25, and Trim25. Each tested successfully, providing assays with performance well suited to HTS. We screened MDM2, RNF28, Nedd4, Hrd1, ZNF364, Smurf1, and XIAP against a collection of approximately 10,000 compounds. We used somewhat larger libraries (150,000–250,000 compounds including natural products) for Hrd1, ZNF364, Smurf1, and MDM2. The results from these screening efforts demonstrated that we had excellent assays that worked in a high-throughput mode with economic work flows. Figure 6 shows a sample of the data that we generated when screening RNF28 against a library of 10,000 compounds and illustrates the stability of the assay over a multi-day run. The data from this study fit a normal distribution and showed no systematic, or other, biases in our approach. This screen produced a hit rate of ~0.35% for compounds having 40% inhibition or greater. We also made use of multiple E3 ligases within a screening campaign to improve further our ability to eliminate compounds that were false positives by comparing their levels of inhibition for several different E3 ligases. This multiplexing of assays allowed us to improve further the selection of specific inhibitors for a given E3 ligase. An example of this approach is illustrated in Table I, where we measured the specificity of the inhibitors from a screen of RNF28 against MDM2, Nedd4, and XIAP. Each of these ligases used the same E2, UBE2D1. The data provide a clear illustration of both the pitfalls and benefits of screening for inhibitors of E3 ligases: there were a number of promiscuous inhibitors of all the E3 ligases tested (presumably these inhibitors target the E2 ligase), but by testing the compounds against a panel of E3 ligases, we were also able to clearly identify specific inhibitors of RNF28.

E2 Assays

We developed our assays for the activity of UBCs with the aim of complementing the recombinant E3 ligase–GST screens. We achieved this complementation in part by use of His_6-tagged versions of the various UBC ligases in combination with nickel-chelate coated Multi Array plates. The use of orthogonal affinity sequences, $(His)_6$ for UBCs, and GST for E3 ligases allows us to prepare different combinations of UBCs and E3 ligases and selectively capture and measure the ubiquitylation of either protein without interference from the other. We successfully investigated the activity of a number of different UBC ubiquitin-conjugating enzymes by use of this method, including UBE2A, UBE2B, UBE2C, UBE2D1, UBE2D2, and the *Drosophila* UBC ubiquitin-conjugating enzyme: eff.

A typical assay follows this basic outline:

The 96-well nickel-chelate Multi-Array plates are used as provided by the manufacturer without any additional treatments.

Preparation of Buffers and UBC Ubiquitin-Conjugating Enzyme

The E2 reaction buffer was prepared and supplemented with DTT (2 mM), ATP (1 mM), and MgCl$_2$ (5 mM). Binding buffer for the Sulfo-Tag–labeled antiubiquitin was prepared from PBS (pH 7.4) supplemented

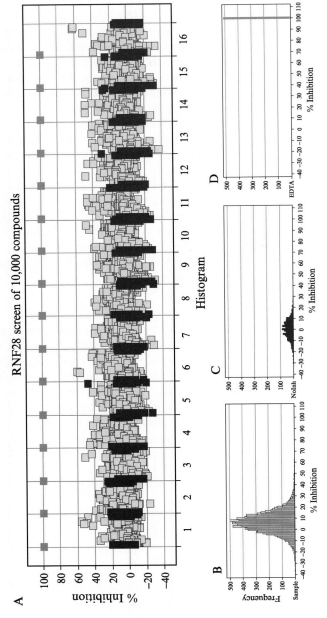

FIG. 6. Data sample from a 10,000-compound screen for inhibitors of RNF28. (A) This illustrates the stability of the screen over 3 days with 16, 384-well plates of compounds. (B) Histogram of the frequency distribution of the signals from the compounds. (C) Histogram of the frequency distribution of the DMSO control signals from the screen. (D) Histogram of the frequency distribution of the EDTA positive inhibitor.

TABLE I
SPECIFICITY OF COMPOUNDS IDENTIFIED FROM RNF28 SCREEN RELATIVE TO OTHER UBIQUITIN E3 LIGASES (IC$_{50}$ VALUES (μM) OF SELECTED COMPOUNDS FROM THE RNF28 SCREEN AGAINST DIFFERENT E3s)

Compound ID	RNF28	MDM2	XIAP	Nedd4
MSD00027753	4.53	>30	>30	>30
MSD00026777	6.00	>30	>30	>30
MSD00020203	8.66	>30	>30	>30
MSD00028848	11.19	>30	>30	>30
MSD00027000	13.80	>30	>30	>30
MSD00099999	15.38	>30	>30	>30
MSD00021445	17.76	>30	>30	>30
MSD00088888	23.58	>30	>30	>30
MSD00025033	0.14	0.63	>30	>30
MSD00012099	1.18	1.83	>30	>30
MSD00020131	1.48	1.25	1.16	>30
MSD00020007	2.83	10.25	8.92	>30
MSD00020218	4.03	8.45	13.62	>30
MSD00027842	1.43	1.71	7.05	13.04

with Tween-20 (0.2%) and bovine serum albumin (0.1%). Titrations of UBC ubiquitin-conjugating enzyme lysate will be needed for each UBC to establish the optimal amount to use per well. In the case of UBE2D2, we found that ~0.1 μl/well of lysate was optimal with our preparations. Also, the amounts of UBE1 used may vary with differing preparations of purified UBE1.

The assay can be performed in two of the following ways:

Direct Detection.

1. Add the following components to a 96-well nickel plate:
    a. 10 μl E2 reaction buffer containing 0.1 μl per reaction His-tagged recombinant E2 (i.e., UBE2D2)
    b. 10 μl E2 reaction buffer containing 0.03 μl per reaction of purified UBE1
    c. 1 μl of compound for testing in DMSO
    d. 20 μl E2 reaction buffer containing, 0.5 μM MSD Sulfo-TAG–labeled ubiquitin* to start the reaction.
2. Incubation for 2 h at room temperature.
3. Wash plates three times with phosphate-buffered saline.

*MSD Sulfo-Tag–labeled ubiquitin is labeled with ~3 Sulfo-TAG labels per ubiquitin.

4. Addition of 150 µl of MSD Read buffer T.
5. Read on MSD Sector Imager.

Indirect Method.
1. Add the following components to a 96-well nickel plate:
   a. 10 µl E2 reaction buffer containing 0.1 µl per reaction His-tagged recombinant E2 (i.e., UBE2D2)
   b. 10 µl E2 reaction buffer containing 0.03 µl per reaction of purified UBE1
   c. 1 µl of compound for testing in DMSO
   d. 20 µl E2 reaction buffer containing 0.5 µM ubiquitin to start the reaction
2. Incubation for 2 h at room temperature.
3. Wash plates three times with phosphate-buffered saline.
4. Addition of 50 ng of MSD Sulfo-Tag–labeled antibody in a 40-µl volume of binding buffer.
5. Incubation for 1 h at room temperature.
6. Wash plates three times with phosphate-buffered saline.
7. Add 150 µl of MSD Read buffer T.
8. Read on MSD Sector Imager.

Note: the second wash (step 6) can be omitted. In this case, the antibody should be added to 150 µl of binding buffer containing 1× MSD Read buffer T.

The direct approach with labeled ubiquitin provides a more rapid protocol and better workflow because of the elimination of extra steps but uses a modified ubiquitin that may alter the kinetics or substrate availability for different E2s (or E3 when these reactions are multiplexed). In practice, we found no such limitation for the E2s we surveyed. The only addition needed is 150 µl/well of Read Buffer T (1×) after the wash step followed by direct quantitation on a MSD Sector Imager instrument. Figure 7A and B highlights the results and demonstrates the large signal-to-background ratio seen with assays using ubiquitin labeled with MSD Sulfo-TAG for UBE2D2, UBE2D1, UBE2B, and the *Drosophila* E2 eff. The specificity of these assays was checked using reactions without UBE1 and mutant UBCs that had modified active site Cys (UBE2D2-C85A and UBE2D2-C85A). We also achieved excellent results in the case of the UBC ubiquitin ligase assay format using the labeled antibody. Figure 7C illustrates an optimization for UBE2D2 showing the strong signals possible in this format. The importance of titration of the various components in these reactions is evidenced by the changes in signal with concentration of

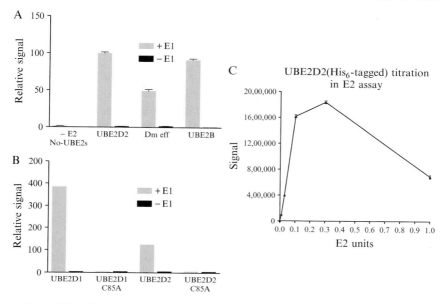

FIG. 7. Ubiquitin conjugation enzyme (E2) assays in two formats. (A) Comparison of the activities of UBE2D2 and UBE2B with the Drosophila homologue of UBE2D2, effete (eff). (B) Specificity study of UBE2D1 and UBE2D2 with the catalytically inactive mutants UBE2D1-C85A and UBE2D2-C85A. Electrochemiluminescent signals were detected on a Sector Imager 6000. Assays in both (A) and (B) made use of labeled ubiquitin. (C) Titration of UBE2D2 lysate using the assay format based on labeled antibody.

the E2 (Fig. 7C), where the increasing levels of UBE2D2 ultimately resulted in signal loss. Here, the most likely factor affecting the signal is the limited amount of UBE1 that allows competition of non-ubiquitylated UBE2D2 for the surface of the Ni-chelate coated Multi Array plates.

Applications for Ubiquitin Ligase Assays

The initial focus of much of our work on E2 and E3 ubiquitin ligases was the development of assays aimed at the discovery of inhibitors. We successfully used various protocols for screens of inhibitors of E2 and E3 ligase activities. This assay platform with its flexible protocols also provides an excellent starting point for the further investigation of aspects of the biochemistry and biology of ubiquitin. These assays can be adapted to studies of enzyme kinetics and biochemistry using purified components. The sensitivity and specificity of the electrochemiluminescent assay system also make it suitable for examining the ubiquitin ligase activities in cell and tissue lysates. Finally, the same basic techniques described here can be used to screen libraries of proteins for specific substrates of ubiquitin ligases or, alternately, for ligases that target specific substrates. This approach is illustrated in Chapter 29, which describes a screen for potential substrates of UBR1, an N-end rule E3 ligase (Varshavsky, 2003), using a library of putative protein substrates prepared and tested in cell lysates.

Note: All genes and proteins are referenced where possible using the Entrez Gene symbols, previously the LocusLink gene symbols: http://www.ncbi.nlm.nih.gov/entrez/query.fcgi?db=gene

References

Adams, J. (2003). Potential for proteasome inhibition in the treatment of cancer. *Drug Discov. Today* **8**, 307–315.

Boisclair, M. D., McClure, C., Josiah, S., Glass, S., Bottomley, S., Kamerkar, S., and Hemmila, I. (2000). Development of a ubiquitin transfer assay for high throughput screening by fluorescence resonance energy transfer. *J. Biomol. Screen.* **5**, 319–328.

Davydov, I. V., Woods, D., Safiran, Y. J., Oberoi, P., Fearnhead, H. O., Fang, S., Jensen, J. P., Weissman, A. M., Kenten, J. H., and Vousden, K. H. (2004). Assay for ubiquitin ligase activity: High throughput screen for inhibitors of hdm2. *J. Biomol. Screen.* **9**, 695–703.

Debad, J. D., Glezer, E. N., Leland, J. K., Sigal, G. B., and Wohlstadter, J. (2004). Clinical and biological applications of ECL. *In* "Electrogenerated Chemiluminescence" (A. J. Bard, ed.). Marcel Dekker Inc., New York.

Fang, S., Jensen, J. P., Ludwig, R. L., Vousden, K. H., and Weissman, A. M. (2000). Mdm2 is a RING finger-dependent ubiquitin protein ligase for itself and p53. *J. Biol. Chem.* **275**, 8945–8951.

Holloway, B. R., and Waddell, I. D. (1998). Assay for the identification of inhibitors of muscle protein degradation involving ubiquitin and human ubiquitin-carrier proteins. *UK Patent Application* GB 2320570A.

Hao, Y., Sekine, K., Kawabata, A., Nakamura, H., Ishioka, T., Ohata, H., Katayama, R., Hashimoto, C., Zhang, X., Noda, T., Tsuruo, T., and Naito, M. (2004). Apollon ubiquitinates SMAC and caspase-9, and has an essential cytoprotection function. *Nat. Cell Biol.* **6,** 849–860.

Hong, C. A., Swearingen, E., Mallari, R., Gao, X., Cao, Z., North, A., Young, S. W., and Huang, S. G. (2003). Development of a high throughput time-resolved fluorescence resonance energy transfer assay for TRAF6 ubiquitin polymerization. *Assay Drug Dev. Technol.* **1**, 175–180.

Li, M., Brooks, C. L., Kon, N., and Gu, W. (2004). A dynamic role of HAUSP in the p53-Mdm2 pathway. *Mol. Cell* **13**, 879–886.

Lorick, K. L., Jensen, J. P., Fang, S., Ong, A. M., Hatakeyama, S., and Weissman, A. M. (1999). RING fingers mediate ubiquitin-conjugating enzyme (E2)-dependent ubiquitination. *Proc. Natl. Acad. Sci. USA* **96**, 11364–11369.

Pickart, C. M. (2004). Back to the future with ubiquitin. *Cell* **116**, 181–190.

Roberts, R. F., and Roberts, W. L. (2004). Performance characteristics of five automated serum cortisol immunoassays. *Clin Biochem.* **37**, 489–493.

Stommel, J. M., and Wahl, G. M. (2004). Accelerated MDM2 auto-degradation induced by DNA-damage kinases is required for p53 activation. *EMBO J.* **23**, 1547–1556.

Swinney, D. C., Xu, Y. Z., Scarafia, L. E., Lee, I., Mak, A. Y., Gan, Q. F., Ramesha, C. S., Mulkins, M. A., Dunn, J., So, O. Y., Biegel, T., Dinh, M., Volkel, P., Barnett, J., Dalrymple, S. A., Lee, S., and Huber, M. (2002). A small molecule ubiquitination inhibitor blocks NF-kappa B-dependent cytokine expression in cells and rats. *J. Biol. Chem.* **277,** 23573–23581.

Varshavsky, A. (2003). The N-end rule and regulation of apoptosis. *Nat. Cell Biol.* **5,** 373–376.

Weissman, A. M. (2001). Themes and variations on ubiquitylation. *Nat. Rev. Mol. Cell Biol.* **2,** 169–178.

Wong, B. R., Parlati, F., Qu, K., Demo, S., Pray, T., Huang, J., Payan, D. G., and Bennett, M. K. (2003). Drug discovery in the ubiquitin regulatory pathway. *Drug Discov. Today* **8,** 746–754.

Yabuki, N., Watanabe, S., Kudoh, T., Nihira, S., and Miyamato, C. (1999). Application of homogeneous time-resolved fluorescence (HTRFTM) to monitor poly-ubiquitination of wild-type p53. *Comb. Chem. High Throughput Screen.* **2,** 279–287.

Yang, Y., Fang, S., Jensen, J. P., Weissman, A. M., and Ashwell, J. D. (2000). Ubiquitin protein ligase activity of IAPs and their degradation in proteasomes in response to apoptotic stimuli. *Science* **288,** 874–877.

[46] Quantitative Assays of Mdm2 Ubiquitin Ligase Activity and Other Ubiquitin-Utilizing Enzymes for Inhibitor Discovery

By KURT R. AUGER, ROBERT A. COPELAND, and ZHIHONG LAI

Abstract

Mdm2 is a negative regulator of p53 activity and functions as an E3 ubiquitin ligase of p53. Inhibition of mdm2 E3 ligase activity will block ubiquitination and subsequent proteasome-mediated degradation of p53, resulting in the stabilization of p53 protein that could lead to the restoration of its tumor-suppressor activity. This chapter describes quantitative biochemical assays for mdm2 E3 activity that can be applied to other ubiquitin-utilizing enzyme systems. Our unique assay format relies on the generation of labeled Ub–E2 conjugate that functions as a substrate for the E3 ligase enzyme. Reducing the E1-E2-E3 ubiquitin cascade to a single enzyme (E3) and bisubstrate (Ub-E2 and target protein) reaction makes it possible to carry out detailed biochemical characterization of the reaction mechanism, high-throughput screening to identify inhibitors of specific E3 ligases, and detailed characterization of the mode of inhibitor interactions with the target enzyme. In addition, preforming the Ub–E2 conjugate as an enzyme substrate for inhibitor screening minimizes interference from thiol-modifying compounds and from nucleotide analogs and other ATP-interfering compounds that might affect the E1 reaction. Using this type of format, we were able to identify small molecule inhibitors of mdm2 E3 ligase activity that are selective against E1 and other E3 ligases, including mdm2's own autoubiquitination activity. Detailed protocols on the labeling of Ub, the generation of Ub-E2, and the use of Ub–E2 in the E3 ligase reaction for inhibitor discovery and characterization are provided.

Introduction

The modification of specific target proteins by ubiquitin (Ub) has important regulatory functions in the cell. This small protein modification involves a cascade of enzymatic activities (Fig. 1) with increasing specificity conferred at each sequential step in the cascade. Target protein specificity is ultimately defined by a combination of ubiquitin-conjugating (Ubc or E2) and ubiquitin ligase (E3) enzymes. The increasing specificity is reflected by the size of each protein family. The E3 ligase family is very large (similar to the kinase family) and is often divided into two main groups, the HECT domain E3 ligases and the RING finger E3 ligases. A number of reviews discuss these enzymes in detail (Fang et al., 2003; Glickman and Ciechanover, 2002; Hershko and Ciechanover, 1998).

The p53 pathway is the most frequently mutated or altered pathway in the development of human cancer (Bourdon et al., 2003; Prives and Hall, 1999; Vogelstein et al., 2000). The p53 tumor suppressor protein is regulated by a number of modifications, including ubiquitination. Mdm2 (also known as hdm2 for the human form), a RING finger containing oncoprotein, has been shown to be a critical regulator of p53 activity and to function as an E3 ubiquitin ligase of p53 (Fang et al., 2000; Freedman et al., 1999; Honda et al., 1997; Lai et al., 2001). Elevated levels of mdm2 are associated with a number of human tumors that express wild-type p53. Hence, the E3 ligase activity of mdm2 is one mechanism that can abrogate the tumor-suppressive activity of p53. Inhibition of mdm2 E3 ligase activity would block ubiquitination and subsequent proteasome-mediated

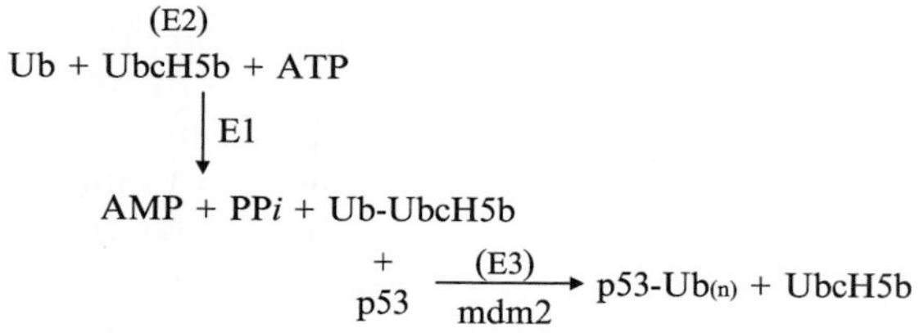

$$\text{Ub} + \underset{(E2)}{\text{UbcH5b}} + \text{ATP}$$
$$\downarrow \text{E1}$$
$$\text{AMP} + \text{PP}i + \text{Ub-UbcH5b}$$
$$+ \quad \xrightarrow[\text{mdm2}]{(E3)} \text{p53-Ub}_{(n)} + \text{UbcH5b}$$
$$\text{p53}$$

FIG. 1. Schematic representation of the ubiquitin cascade leading to the mdm2-mediated ubiquitination of p53.

degradation of p53, resulting in elevated protein levels and restoration of the tumor-suppressive activity. This has led us and other groups to identify small molecule inhibitors of mdm2 E3 ligase activity (Lai *et al.*, 2002). If successful, this could form the basis of a novel modality of cancer chemotherapy that could be extended to other therapeutic areas in which E3 ligase activity regulates the level and function of important cellular proteins.

To identify such inhibitors, we have established robust biochemical assays of mdm2 E3 ligase activity that are applicable to other ubiquitin-utilizing enzyme systems (Lai *et al.*, 2001, 2002; Wee *et al.*, 2000). In general, the unique assay format relies on the generation of a labeled Ub–E2 (*Ub-E2) conjugate that functions as a substrate for the E3 ligase enzyme. The ubiquitin acceptor protein constitutes the second substrate for the E3 ligase. Thus, the complex cascade illustrated in Fig. 1 is reduced to a simple, single-enzyme bisubstrate reaction for E3 ligases in which the transfer of labeled Ub from the Ub–E2 conjugate to the target protein is quantified. Because the E1-catalyzed Ub transfer to the E2 enzyme relies on active site Cys residues on both E1 and E2, this E1 reaction is highly sensitive to thiol-modifying agents. In contrast, the RING finger E3 ligases are much less sensitive to these agents (Lai *et al.*, 2001). Thus, screening for ligase inhibitors with the Ub–E2 conjugate minimizes interference from thiol-modifying compounds that are generally not pharmacologically tractable. In addition, the Ub–E2 conjugate assay format results in an enzymatic reaction independent of ATP. This eliminates the identification of nucleotide analogs and other ATP-interfering compounds that might inhibit the E1 reaction. To develop these assays, we relied on some of the biochemistry inherent to the ubiquitination cascade. In this chapter, we summarize the development of these assays and methods for generating the required protein reagents.

Covalent Labeling of Ub

A starting point for our assay design was the development of chemical methods to allow covalent modification of Ub with a convenient detection tag. We avoided radiolabeling strategies despite literature precedence for these approaches because of the significant safety and waste-disposal issues that arise from large-scale use of such reagents. Two additional criteria for the label were (1) the stoichiometry of incorporation could be controlled and analyzed to allow subsequent use of the labeled Ub as quantitatively as possible, and (2) label incorporation had to have minimal impact on the use of the labeled Ub by enzymes of the Ub cascade. These requirements were best met by controlled modification of lysine residues on Ub using succinimidyl ester chemistry. Succinimidyl esters of biotin, fluorescent molecules, chromogenic molecules, and radiolabeled groups are commercially available. We have made significant use of the fluorophore Oregon Green (OG). This labeling represents an illustrative example of chemistry for Ub modification by succinimidyl esters for specific and diverse label incorporation on Ub (Wee et al., 2000).

Ub (Sigma, St. Louis, MO) was dissolved in 100 mM NaHCO$_3$ (pH 8.0) to a concentration of 2.5 mg/ml as determined by the method of Lowry et al. (1951). The concentration of a Ub stock was verified by amino acid analysis and used to generate standard curves of Ub to quantify unknown Ub samples. OG-succinimidyl ester (Molecular Probes, Eugene, Oregon) was prepared as a stock solution of 20 mg/ml by dissolution in dimethyl sulfoxide (DMSO). The OG-succinimidyl ester solution was added to the Ub solution at a 1.5:1.0 molar ratio of label to Ub and allowed to mix in an amber vial for 90 min at room temperature. The reaction was stopped by addition of 0.15 M hydroxylamine (pH 8.5) and incubated at room temperature for 15 min. The labeled-Ub was separated from unreacted components by gel filtration chromatography on a disposable PD-10 desalting column (Pharmacia, Piscataway, NJ). Mass spectral analysis of the reaction product indicated a mixture of dilabeled (i.e., (OG)$_2$-Ub; 20% of product), monolabeled (OG-Ub; 50% of product), and unlabeled Ub (30% of product). The OG–Ub was then purified from the other reaction products by reverse-phase HPLC using a C18 column in conjunction with a 20–40% aqueous gradient of acetonitrile with 0.1% TFA. Each peak was analyzed by mass spectroscopy, and the peak containing exclusively OG-Ub was collected.

Ubiquitination of the target protein results in an isopeptide bond between a lysine residue on the target protein and the C-terminus of the Ub molecule (Glickman and Ciechanover, 2002). Subsequent poly-Ub chain formation is mediated by sequential isopeptide bonds between lysine residues on the Ub molecule and the C-terminus of the next Ub molecule in the growing chain. Most commonly, Lys48 of Ub is used in poly-Ub chain formation that leads to proteasome degradation. Lys29, 63, and other lysines of Ub have also been implicated in poly-Ub chain formation (Pickart, 2000, 2001). For this reason, it is critical to establish that the OG–Ub prepared represents a homogeneous population of OG-label on a specific lysine residue of Ub and that it does not involve Lys48 and other lysines that are critical for poly-Ub chain formation. This was established by amino acid sequencing of the labeled Ub population. We found that ∼95% of the OG–Ub represented a homogenous population in which the label was covalently attached to the ε-amino group of Lys6. The OG incorporation did not perturb the overall folding of Ub as was established by HSQC NMR spectroscopy (Wee et al., 2000).

Generation of the OG-Ub-E2 Conjugate

The use of OG–Ub as a substrate in the E1 enzymatic reaction results in OG-Ub-E2 product formation (Fig. 1). In our work toward an assay of mdm2, we used recombinant UbcH5b (also known as Ubc4 or Ubde2) as the E2. Wild-type E2 was expressed and purified from bacteria with standard chromatography techniques (Rolfe *et al.*, 1995; Wee *et al.*, 2000). Addition of either an N-terminal or C-terminal tag diminished activity of the E2 in subsequent activity assays. A recent report suggests that UbcH5b/c are the physiologically relevant E2s for mdm2's ligase activity (Saville *et al.*, 2004). We first established that OG–Ub was used as a substrate by recombinant human E1. E1 was expressed and purified from baculovirus-infected cells (Rolfe *et al.*, 1995; Wee *et al.*, 2000). Mixing varying micromolar concentrations of OG–Ub and E2 with 300 pM E1, we followed the formation of OG-Ub-E2 by measuring OG fluorescence (λ_{ex} = 492 nm, λ_{max} = 520 nm) after product separation was achieved by SDS-PAGE or by reverse-phase HPLC. The kinetic constants for OG–Ub utilization by human E1 are summarized in Table I. The incorporation of OG onto Lys6 increases the K_m relative to unlabeled Ub about 10-fold. Despite this reduction in apparent affinity, the OG–Ub is sufficiently used by E1 to form the OG-Ub-E2 conjugate. Having established that the OG–Ub was indeed used by E1, we optimized for large-scale production of OG-Ub-E2 for use in subsequent assays (Lai *et al.*, 2001).

Large-scale production of OG-Ub-E2 is performed as follows. A mixture composed of 0.5 μM E1, 32 units/ml adenylate kinase (Sigma) to remove AMP (a product inhibitor of the reaction) (Haas and Rose, 1982), 100 μM OG-Ub, and 50 μM E2 (UbcH5b) was prepared in 50 mM Tris (pH 7.5) at 37°. The E1 enzymatic reaction was initiated by addition of 2 mM MgCl$_2$/ATP, and the reaction was allowed to proceed at 37° for 10 min. The reaction was stopped by transferring the reaction vessel to ice and

TABLE I
KINETIC CONSTANTS FOR HUMAN E1-CATALYZED TRANSFER OF OG-Ub TO UbcH5b

Kinetic constant	Value
K_m, ATP	4.7 ± 1.0 μM
K_m, UbcH5b	1.9 ± 0.5 μM
K_i, UbcH5b	3.5 ± 1.7 μM
K_m, Ub	0.17 ± 0.03 μM
K_m, OG-Ub	2.0 ± 0.2 μM
k_{cat}	4.3 ± 1.2 min^{-1}

adding EDTA/NaCl (final concentrations of 10 and 400 mM, respectively). The volume of the reaction solution was reduced using an Amicon ultrafiltration cell equipped with a 3000-molecular weight cut-off filter and then loaded onto a Superdex 75 gel filtration column. The column was run with isocratic elution in 50 mM NaCl, 50 mM Tris (pH 7.5), 0.1 mM dithiothreitol. The OG-Ub-E2 eluted as the second peak as monitored by absorbance at 280 nm. Fractions containing OG-Ub-E2 were collected and stored as small volume aliquots at $-80°$. On average, 50–60% of input E2 can form OG-Ub-E2. We found that performing the E1 enzymatic reaction in large scale could generate two forms of OG-Ub-E2. One form was the expected product of the E1 reaction with OG-Ub linked to the active site cysteine of the E2 through a thioester bond as determined by its sensitivity to reducing agents. The second form of OG-Ub-E2 was apparently the product of trans-ubiquitination of OG-Ub-E2 on lysine residues. This second product was refractory to reducing agents and was not a viable substrate for subsequent E3 ligase reactions. Hence, the percentage of DTT reducible and nonreducible product is an important quality control parameter for batches of OG-Ub-E2. This was measured by subjecting a small sample of OG-Ub-E2 to 100 mM DTT and separating the resulting OG–Ub and E2 from the residual OG-Ub-E2 conjugate by reverse-phase HPLC on a C4 column. The products were quantified by the OG fluorescence and 214 nm absorbance of the peaks. Alternately, samples were resolved by SDS-PAGE and quantified with a FluorImager. Batches of OG-Ub-E2 to be used as substrate in subsequent E3 ligase reactions should display $\geq$80% reducible product. The OG-Ub-E2 produced in this manner was stable at $-80°$ for at least 2 months (Lai *et al.*, 2001).

Enzymatic Assays of Mdm2 E3 Ligase Activity

Having generated OG-Ub-E2, we used it as a substrate together with recombinant human p53 in assays of mdm2 catalytic activity. Human mdm2 was expressed as a GST fusion protein in the baculovirus expression system and purified by standard techniques (Lai *et al.*, 2001). The human p53 protein was also expressed by baculovirus but without a tag for purification. Standard ion exchange chromatography was used to purify the protein (Lai *et al.*, 2001). Catalytic amounts (1–5 nM) of mdm2 were mixed with varying micromolar concentrations of OG-Ub-E2 and p53 in 15 mM HEPES (pH 7.5), 5 mM NaCl, 10 mM octylglucoside at room temperature. Reactions were stopped at discrete time points by addition of 4× SDS-reducing sample buffer and heating to 95° for 5 min. The mdm2-catalyzed transfer of OG–Ub from E2 to p53 could be followed by fluorescence detection after separation of the proteins by SDS-PAGE. Figure 2 illus-

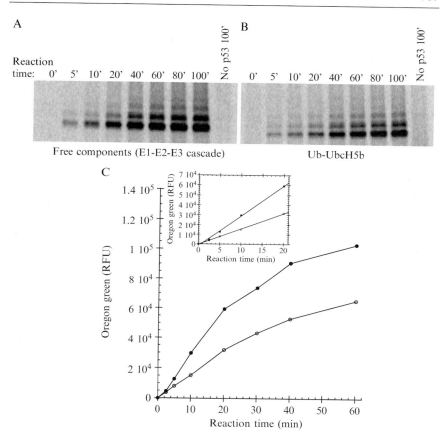

Fig. 2. Mdm2-mediated ubiquitination of p53. (A) Time course of OG–Ub incorporation onto p53 mediated by mdm2 with UbcH5b and E1. Purified recombinant proteins were incubated as indicated, analyzed by reducing SDS-PAGE, and scanned with a FluorImager. (B) Time course of p53 ubiquitination in a reaction with preconjugated OG-Ub-UbcH5b and mdm2. (C) Progress curves of OG–Ub incorporation onto p53-Ub_n. Data from three independent experiments were averaged for each. The filled circles represent the reaction with individual components, and the open circles represent the reaction using the preconjugated OG-Ub-UbcH5b. (Inset) Linear regression analysis of the data for the first 20 min of the reactions.

trates a typical time course for the formation of OG-Ub-p53 species catalyzed by mdm2 using both preformed OG-Ub-E2 conjugate and the free components of the E1-E2-E3 cascade. Both reactions produced a ladder of ubiquitinated p53 with similar kinetics, and the fluorescence intensity of the entire ladder of bands was quantified by fluorimaging. Plots of the integrated OG-fluorescence for these bands as a function of reaction time yielded linear progress curves up to 20–30 min, depending on the concentration of enzyme (mdm2) and substrates. At 1 μM of each substrate, reaction velocity was a linear function of mdm2 concentration up to 5 nM enzyme. These data indicate that the preformed OG-Ub-E2 can be used to simplify the reaction cascade to a single enzyme, bisubstrate reaction. For mdm2, a nonprocessive ligase that forms multiple mono-ubiquitinated products, our reaction is set up using equal molar p53 and OG-Ub-UbcH5b, because the K_m of both substrates were measured to be 1 μM. Setting up the reaction at balanced conditions (Copeland, 2005) for both substrates maximizes the chance to find inhibitors with diverse mechanisms of action with respect to both substrates. For a processive E3 ligase that forms poly-Ub chain products, the appropriate inhibitor screening conditions are dependent on the characteristic of the reaction (relative concentration of enzyme and each of the substrates, as well as K_m of substrates and the average number of Ub per substrate).

Distinguishing Polyubiquitination from Multiple Monoubiquitination

As previously mentioned, polyUb chain formation through linkage of Lys48 of Ub is a signal for proteasome-mediated degradation. PolyUb chains can also form through Lys29, 63, and other lysines of Ub, and these have been proposed to have other regulatory functions. Because a ladder of ubiquitinated p53 (up to 6–8 depending on conditions) was observed in our enzymatic reaction, we asked whether the ubiquitinated p53 were polyUb chains or multiple monoUb modifications on different lysine residues of p53. To address this, we used a K48R, a K63R, and a double mutant (K48R/K63R) of Ub in the E1-E2-E3 cascade reaction. The p53 products observed were the same. A methylated ubiquitin, in which all the lysine residues were blocked by reductive methylation, also worked equally well to form the pattern of p53-Ub$_n$ species. These results suggest that under our assay conditions, mdm2 catalyzes monoUb to multiple lysine residues on p53, rather than polyUb chain formation. Experiments of this type provide a means of distinguishing between target protein polyUb versus multiple monoUb modifications by any E3 ligase of interest.

Recently, additional factors have been described to facilitate the mdm2-mediated polyUb chain formation on p53 (Grossman *et al.*, 2003; Sui *et al.*, 2004); mdm2 has also been shown to promote polyUb of p53 at high ratios of mdm2 to p53 (Brooks *et al.*, 2004; Li *et al.*, 2003). These reports have extended our original observation of mdm2-catalyzed multiple monoUb of p53.

Use of the Mdm2 Assay for Mechanistic Studies

The simplification of the mdm2-catalyzed p53 ubiquitination reaction to a single enzyme, bisubstrate reaction, opens the way for detailed mechanistic studies of catalytic mechanism and inhibitor interactions (Lai *et al.*, 2001). Using the SDS-PAGE and FluorImager detection described previously, one has a great deal of flexibility with respect to changes in substrate concentrations, solution conditions, and inclusion of inhibitors. We have used this assay system to titrate the concentrations of both substrates, OG-Ub-E2 and p53, and determined the effects of different substrate concentrations on the initial velocity of the mdm2 reaction. The data generated in this way resulted in a series of parallel lines when plotted in double reciprocal fashion (Fig. 3). This is typically indicative of a double displacement (Ping Pong) reaction mechanism (Copeland, 2000). E2, a product of the mdm2 reaction, was analyzed for product inhibition. Titration of E2 into the mdm2/p53 assay demonstrated it to be a competitive inhibitor with respect to OG-Ub-E2 and noncompetitive with respect to p53 (Fig. 4). This pattern of product inhibition is inconsistent with a classical double-displacement reaction (Copeland, 2000; Segel, 1975) and led us to propose three potential reaction mechanisms for mdm2 (Lai *et al.*, 2001). A common feature of all three potential reaction mechanisms is the need for an enzyme isomerization step between initial substrate binding and final product release. This assay also afforded us a method for analyzing the concentration-dependent inhibition of mdm2 by small molecule inhibitors and a method for determining the mode of inhibition by these molecules as discussed in more detail in the following.

Assays for High-Throughput Screening for Mdm2 Inhibitors

The general scheme described previously is also amenable to development of assays for high-throughput screening (HTS) for inhibitors of E3 ligases. Using the same succinimidyl ester chemistry, alternate detection tags can be incorporated into Ub to facilitate HTS assays. For example, we and others have used this same chemical strategy to incorporate biotin onto Lys6 of Ub. This was then used to develop an assay for the E3 ligase SCF$^{\beta\text{-TRCP}}$ that catalyzes the ubiquitination of IκB (Lai *et al.*, 2002). After the E3 ligase reaction is allowed to proceed for a fixed period of time, the reaction is stopped (as described previously), and the mixture is transferred to a Protein G plate (Pierce) precoated with an antibody specific for the product, in this case anti-IκB antibody. In this way the product, biotin-Ub$_n$-IκB, and remaining substrate, IκB, are captured on the plate surface. After washing the plate with buffer, a solution of 0.1 μg/ml streptavidin-europium

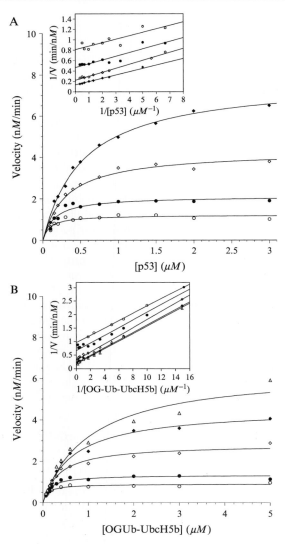

FIG. 3. Kinetic analysis of Ub transfer to p53. (A) Velocity of p53-Ub_n formation as a function of p53 concentration at several fixed concentrations of OG-Ub-UbcH5b. The lines represent the best global fitting of the entire data set to a Ping Pong mechanism using Grafit (OG-Ub-UbcH5b concentrations, ○, 0.25 μM; ●, 0.5 μM; ◇, 1.5 μM; ◆, 5 μM). (Inset) Double reciprocal replot of the same data. (B) Velocity of p53-Ub_n formation as a function of OG-Ub-UbcH5b concentration at several fixed concentrations of p53. The lines represent the best global fitting of the entire data set to a Ping Pong mechanism using Grafit (p53 concentrations: ○, 0.1 μM; ●, 0.15 μM; ◇, 0.25 μM; ◆, 0.5 μM; △, 1 μM). (Inset) Double reciprocal plot of the data.

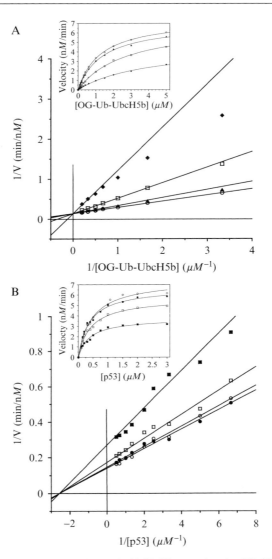

FIG. 4. Product inhibition of the mdm2/p53 Ub reaction by UbcH5b. (A) Double reciprocal plot of the initial velocity of p53-Ub$_n$ formation as a function of OG-Ub-UbcH5b concentration. Reactions were run at 1 μM p53 and different UbcH5b concentrations (○, 0 μM; ▲, 1 μM; □, 5 μM; ◆, 20 μM). (Inset) The untransformed data with lines depicting global fitting of the entire data set to a competitive inhibition model. (B) Double reciprocal plot of the initial velocity of p53-Ub$_n$ formation as a function of p53 concentration. Reactions were run at 5 μM OG-Ub-UbcH5b and different UbcH5b concentrations (○, 0 μM; ●, 1 μM; □, 5 μM; ■, 20 μM). (Inset) The untransformed data with lines depicting global fitting of the entire data set to a noncompetitive inhibition model.

(Wallac) is added and allowed to incubate with the immobilized protein for 40 min; this step differentiates between product and substrate. The solution is removed, the plate washed, and enhancement solution (Wallac) is added. The europium signal is then measured by time-resolved fluorescence spectroscopy.

A similar strategy was reported in a patent application by Issakani *et al.* (2004) to follow auto-ubiquitination (that is, the ability of the E3 ligase to "self-ubiquitinate") of the E3 ligase ROC1/Cul1. In this case, the recombinant ROC1 was produced with a polyhistidine tag, and the Ub was produced as a fusion protein containing a FLAG tag. After the reaction, the His-ROC1/Cul1 was captured on nickel plates, and product formation was detected using an anti-FLAG antibody. Also, Yabuki *et al.* (1999) reported the application of homogeneous time-resolved fluorescence (HTRF) for E6-E6AP–mediated polyUb of p53. Here the biotinylated p53 is ubiquitinated with europium cryptate–labeled ubiquitin and the addition of streptavidin-labeled allophycocyanin brings the donor and acceptor into close proximity to generate a fluorescence signal. Similarly, Boisclair *et al.* (2000) and Hong *et al.* (2003) have applied the HTRF format to monitor the ubiquitination of Rsc1083 (a HECT domain E3) and TRAF6 (a RING finger E3), respectively.

By methods like the ones described here, the E3 ligase–catalyzed ubiquitination of a target protein can be followed in 96-well and/or 384-well microplates to facilitate HTS for identification of E3 ligase inhibitors.

Identification and Characterization of Mdm2 Inhibitors

Screening compound libraries for inhibitors of mdm2-catalyzed p53 ubiquitination has resulted in identification of several distinct inhibitors. Compounds that reproducibly demonstrated significant inhibition of mdm2 activity at a fixed concentration of 10 μM from HTS were titrated against the enzyme in the preconjugated OG-Ub-E2 SDS-PAGE assay described previously. *Bona fide* inhibitors are expected to demonstrate a concentration-dependent inhibition of the enzyme that is quantified by the concentration of inhibitor required to reduce the enzymatic activity by 50% (IC_{50}). Compounds described to inhibit mdm2-catalyzed p53 ubiquitination had IC_{50} values ranging from 3–14 μM (Lai *et al.*, 2002). Using assays similar to those described previously, the selectivity of the compounds for inhibition of other Ub-utilizing enzymes was tested. We chose to test compounds against the following set of counter-screen enzymes: human, recombinant E1; Nedd-4; $SCF^{\beta\text{-TRCP}}$; and the auto-ubiquitination activity of mdm2 (similar to the mdm2/p53 assay but leaving out the p53).

The three compounds previously described were found to be selective inhibitors for mdm2-mediated ubiquitination of p53 after analysis in the preceding counter screens ($IC_{50} > 100\ \mu M$) (Lai et al., 2002).

To quantify the true affinity of an inhibitor toward an enzyme target, the phenomenological IC_{50} value is insufficient; instead, one must determine the equilibrium dissociation constant, K_i, for the enzyme-inhibitor complex. General methods for determination of K_i values for enzyme–inhibitor complexes have been described in this series and elsewhere (Copeland, 2000, 2005; Segel, 1975). The assay flexibility afforded by mdm2 assays using OG-Ub-E2 and the SDS-PAGE format has allowed us to determine the effects of simultaneous titration of each substrate and inhibitor on the initial velocity of the mdm2 enzymatic reaction (Lai et al., 2002). For example, the velocity as a function of OG-Ub-E2 concentration, at a fixed p53 concentration of 1 μM, was determined at concentrations of compound 3 (see Lai et al. [2002] for compound structure) of 0, 1, 3, and 10 μM. In a reciprocal manner, the velocity was also determined as a function of p53 concentration at a fixed OG-Ub-E2 concentration of 1 μM in the presence of inhibitor compound 3 at 0, 1, 3, and 10 μM concentrations. As illustrated in Fig. 5, the data thus generated yield a series of lines that intersect at the x-axis, beyond the y-axis, when plotted as double reciprocal plots of 1/velocity as a function of 1/[substrate] for either varied substrate of the mdm2 reaction. These data, and the corresponding fits of the untransformed data, indicate that compound 3 is noncompetitive with respect to both substrates of the enzymatic reaction and displays a K_i value of 3–4

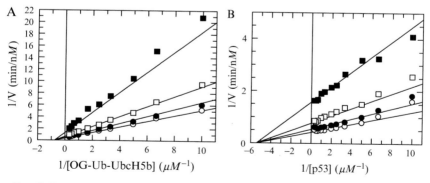

FIG. 5. Double reciprocal plots of mdm2 E3 ligase activity for p53 in the presence of compound 3 at 0 (○), 1 (●), 3 (□), and 10 (■) μM, (A) velocity as a function of OG-Ub-UbcH5b concentration at 1 μM p53 and (B) velocity as a function of p53 concentration at 1 μM OG-Ub-UbcH5b. The fitted lines represent global fitting of the untransformed data to a noncompetitive model.

μM. Similar data were obtained for compounds 1 and 2 (Lai *et al.*, 2002), and both of these compounds were also found to be noncompetitve with respect to both substrates.

In situations in which multiple inhibitors display noncompetitive inhibition of a target enzyme, a question that often arises is whether the compounds share a common binding site on the enzyme molecule. In the absence of structural data, this is a difficult question to resolve unambiguously. However, one can determine whether two compounds bind to a common target enzyme in a mutually exclusive manner by measuring the effects of combinations of the two inhibitors on the initial velocity of the enzymatic reaction. As first shown by Yonetani and Therorell (1964), the velocity in the presence of two inhibitors, I and J, is described by the following reciprocal equation:

$$\frac{1}{v_{ij}} = \frac{1}{v_0}\left(1 + \frac{[I]}{K_i} + \frac{[J]}{K_j} + \frac{[I][J]}{\alpha K_i K_j}\right) \quad (1)$$

where v_0 is the initial velocity in the absence of inhibitors; *[I]* and *[J]* are the molar concentrations of inhibitors I and J, respectively; K_i and K_j are the equilibrium dissociation constants for the enzyme–inhibitor complexes of I and J, respectively; and α is an interaction term that defines the degree to which binding of one inhibitor affects the affinity of the second inhibitor for the enzyme. When the two inhibitors bind in a mutually exclusive fashion to the target enzyme, the value of α is infinite, and the Dixon plots of 1/velocity as a function of concentration of one compound at varying concentrations of the second compound yield a series of parallel lines. This is the pattern found when various combinations of compounds 1, 2, and 3 were tested together as inhibitors of mdm2 (Fig. 6). Thus, in this example, all three compounds were found to be noncompetitive with respect to both substrates of the E3 ligase, and all three were found to bind to the enzyme in a mutually exclusive fashion. Although these data can be explained in a number of ways, the simplest interpretation is that the compounds bind in a common regulatory site on the mdm2 molecule.

Summary

Although the ubiquitination cascade is a complex series of enzymatic reactions, the specificity for target protein modification is conferred at the final step of the cascade, the E3 ligase reaction. Hence, inhibition of this step should provide the most selective inhibition of ubiquitination for specific target proteins. We have described methods by which a reductionist approach may be applied to the Ub cascade to study, in biochemical detail, the isolated E3 ligase enzymatic reaction. We have used the

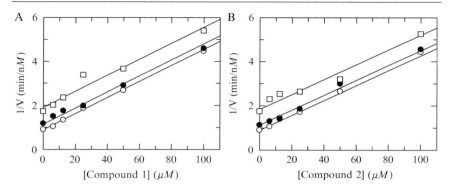

FIG. 6. Exclusivity studies of compounds 1–3. The fitted lines represent global fitting of the untransformed data to equation 1. (A) 1/V is plotted as a function of compound 1 concentration at 0 (○), 1 (●), and 4 (□) μM of compound 3. (B) 1/V is plotted as a function of compound 2 concentration at 0 (○), 1 (●), and 4 (□) μM of compound 3.

example of mdm2 from our work to illustrate this approach. Generation of labeled Ub–E2 thioester conjugate simplifies the cascade to an assay of E3 ligase as a single enzyme, bisubstrate reaction. This approach allows for detailed biochemical characterization of the reaction mechanism, for high-throughput screening to identify inhibitors of a specific E3 ligase, and for detailed characterization of the mode of inhibitor interactions with the target enzyme. The association of E3 ligase activity with particular human diseases, especially the association of mdm2 activity with a number of human cancers, suggests that selective inhibition of E3 ligase activity may constitute a novel paradigm for chemotherapeutic intervention in human medicine. The methods described here provide a solid biochemical foundation for the discovery of starting points toward this goal.

References

Boisclair, M. D., McClure, C., Josiah, S., Glass, S., Bottomley, S., Kamerkar, S., and Hemmila, I. (2000). Development of a ubiquitin transfer assay for high throughput screening by fluorescence resonance energy transfer. *J. Biomol. Screen.* **5**, 319–328.
Bourdon, J. C., Laurenzi, V. D., Melino, G., and Lane, D. (2003). p53: 25 years of research and more questions to answer. *Cell Death Differ.* **10**, 397–399.
Brooks, C. L., Li, M., and Gu, W. (2004). Monoubiquitination: The signal for p53 nuclear export? *Cell Cycle* **3**, 436–438.
Copeland, R. A. (2000). "Enzymes: A Practical Introduction to Structure, Mechanism, and Data Analysis." 2nd Ed. Wiley, New York.
Copeland, R. A. (2005). "Evaluation of Enzyme Inhibitors in Drug Discovery: A Guide for Medicinal Chemists and Pharmacologists." Wiley, New York.

Fang, S., Jensen, J. P., Ludwig, R. L., Vousden, K. H., and Weissman, A. M. (2000). Mdm2 is a RING finger-dependent ubiquitin protein ligase for itself and p53. *J. Biol. Chem.* **275,** 8945–8951.

Fang, S., Lorick, K. L., Jensen, J. P., and Weissman, A. M. (2003). RING finger ubiquitin protein ligases: Implications for tumorigenesis, metastasis and for molecular targets in cancer. *Semin. Cancer Biol.* **13,** 5–14.

Freedman, D. A., Wu, L., and Levine, A. J. (1999). Functions of the MDM2 oncoprotein. *Cell Mol. Life Sci.* **55,** 96–107.

Glickman, M. H., and Ciechanover, A. (2002). The ubiquitin-proteasome proteolytic pathway: Destruction for the sake of construction. *Physiol, Rev.* **82,** 373–428.

Grossman, S. R., Deato, M. E., Brignone, C., Chan, H. M., Kung, A. L., Tagami, H., Nakatani, Y., and Livingston, D. M. (2003). Polyubiquitination of p53 by a ubiquitin ligase activity of p300. *Science* **300,** 342–344.

Haas, A. L., and Rose, I. A. (1982). The mechanism of ubiquitin activating enzyme. *J. Biol. Chem.* **257,** 10329–10337.

Hershko, A., and Ciechanover, A. (1998). The ubiquitin system. *Annu. Rev. Biochem.* **67,** 425–479.

Honda, R., Tanaka, H., and Yasuda, H. (1997). Oncoprotein MDM2 is a ubiquitin ligase E3 for tumor suppressor p53. *FEBS Lett.* **420,** 25–27.

Hong, C. A., Swearingen, E., Mallari, R., Gao, X., Cao, Z., North, A., Young, S. W., and Huang, S. G. (2003). Development of a high throughput time-resolved fluorescence resonance energy transfer assay for TRAF6 ubiquitin polymerization. *Assay. Drug Dev. Technol.* **1,** 175–180.

Issakani, S. D., Huang, J., and Sheung, J. (2004). Ubiquitin ligase assay. US Patent 6740475.

Lai, Z., Ferry, K. V., Diamond, M. A., Wee, K. E., Kim, Y. B., Ma, J., Yang, T., Benfield, P. A., Copeland, R. A., and Auger, K. R. (2001). Human mdm2 mediates multiple monoubiquitination of p53 by a mechanism requiring enzyme isomerization. *J. Biol. Chem.* **276,** 31357–31367.

Lai, Z., Yang, T., Kim, Y. B., Sielecki, T. M., Diamond, M. A., Strack, P., Rolfe, M., Caligiuri, M., Benfield, P. A., Auger, K. R., and Copeland, R. A. (2002). Differentiation of Hdm2-mediated p53 ubiquitination and Hdm2 autoubiquitination activity by small molecular weight inhibitors. *Proc. Natl. Acad. Sci. USA* **99,** 14734–14739.

Li, M., Brooks, C. L., Wu-Baer, F., Chen, D., Baer, R., and Gu, W. (2003). Mono- versus polyubiquitination: Differential control of p53 fate by Mdm2. *Science* **302,** 1972–1975.

Lowry, O. H., Rosebough, N. J., Farr, A. L., and Randall, R. J. (1951). Protein measurement with the Folin phenol reagent. *J. Biol. Chem.* **193,** 265–275.

Pickart, C. M. (2000). Ubiquitin in chains. *Trends Biochem. Sci.* **25,** 544–548.

Pickart, C. M. (2001). Mechanisms underlying ubiquitination. *Annu. Rev. Biochem.* **70,** 503–533.

Prives, C., and Hall, P. A. (1999). The p53 pathway. *J. Pathol.* **187,** 112–126.

Rolfe, M., Beer-Romero, P., Glass, S., Eckstein, J., Berdo, I., Theodoras, A., Pagano, M., and Draetta, G. (1995). Reconstitution of p53-ubiquitinylation reactions from purified components: The role of human ubiquitin-conjugating enzyme UBC4 and E6-associated protein (E6AP). *Proc. Natl. Acad. Sci. USA* **92,** 3264–3268.

Saville, M. K., Sparks, A., Xirodimas, D. P., Wardrop, J., Stevenson, L. F., Bourdon, J. C., Woods, Y. L., and Lane, D. P. (2004). Regulation of p53 by the ubiquitin-conjugating enzymes UbcH5B/C *in vivo. J. Biol. Chem.* **279,** 42169–42180.

Segel, I. H. (1975). "Enzyme Kinetics: Behavior and Analysis of Rapid Equilibrium and Steady-State Enzyme Systems." Wiley, New York.

Sui, G., Affar, E. B., Shi, Y., Brignone, C., Wall, N. R., Yin, P., Donohoe, M., Luke, M. P., Calvo, D., Grossman, S. R., and Shi, Y. (2004). Yin Yang 1 is a negative regulator of p53. *Cell* **117,** 859–872.

Vogelstein, B., Lane, D., and Levine, A. J. (2000). Surfing the p53 network. *Nature* **408,** 307–310.

Wee, K. E., Lai, Z., Auger, K. R., Ma, J., Horiuchi, K. Y., Dowling, R. L., Dougherty, C. S., Corman, J. I., Wynn, R., and Copeland, R. A. (2000). Steady-state kinetic analysis of human ubiquitin-activating enzyme (E1) using a fluorescently labeled ubiquitin substrate. *J. Protein Chem.* **19,** 489–498.

Yabuki, N., Watanabe, S., Kudoh, T., Nihira, S., and Miyamato, C. (1999). Application of homogeneous time-resolved fluorescence (HTRFTM) to monitor poly-ubiquitination of wild-type p53. *Comb. Chem. High Throughput. Screen.* **2,** 279–287.

Yonetani, T., and Theorell, H. (1964). *Arch. Biochem. Biophys.* **106,** 243–255.

[47] High-Throughput Screening for Inhibitors of the Cks1–Skp2 Interaction

By KUO-SEN HUANG and LYUBOMIR T. VASSILEV

Abstract

The cyclin-dependent kinase inhibitor p27^{Kip1} is a critical cell cycle regulator frequently altered in human cancer. The cellular level of p27 is controlled by ubiquitin-dependent degradation mediated by the E3 ligase SCF^{Skp1}. Decreased p27 level in cancer cells has been associated with enhanced ubiquitin-dependent degradation and linked to poor prognosis. Therefore, restoration of p27 by inhibiting SCF^{Skp2} activity has been proposed as a novel therapeutic strategy. Recently, the small regulatory protein Cks1 has been found to bind Skp2 and dramatically increases the affinity of Skp2 to p27, thus facilitating its ubiquitylation and degradation. Here, we describe a high-throughput screening assay for inhibitors of the Cks1–Skp2 interaction. The assay measures the binding of recombinant human GST-Cks1 and His6-Skp2-Skp1 using a homogeneous time-resolved fluorescence format and permits a throughput in excess of 100,000 data points per day when implemented on the Zeiss uHTS system.

Introduction

The cyclin-dependent kinase (CDK) inhibitor p27^{Kip1} plays an important role in the regulation of multiple cellular processes, including proliferation, differentiation, and apoptosis (Morgan, 1995; Sherr and Roberts, 1999). It has been well documented that loss or decreased level of p27 in human tumors is associated with poor prognosis, and p27 has been accepted as an independent prognostic indicator in many major solid malignancies (Esposito et al., 1997; Slingerland and Pagano, 2000). The cellular level of p27 is regulated at the post-translation level by ubiquitin-dependent proteolysis (Carrano et al., 1999; Reed et al., 1994). Studies in colorectal, breast, stomach, and lung cancers have shown that the low level of p27 in cancer cells is due to enhanced degradation (Slingerland and Pagano, 2000; Tsihlias et al., 1999). Therefore, restoration of p27 in cancer cells by inhibiting its ubiquitin-dependent degradation could offer a novel approach to cancer therapy.

p27 is ubiquitylated by a multiprotein complex known as SCF^{Skp2} (E3 ubiquitin ligase for p27). The process is initiated with the phosphorylation of p27 on Thr187 by CDK2/cyclin E (Tsvetkov et al., 1999; Vlach et al., 1997). Phosphorylated p27 is recognized by Skp2, an F-box protein and a component of SCF that is specific for p27. Recent studies have demonstrated that the binding of the small nuclear protein Cks1 to Skp2 greatly increases its affinity to p27 and allows for efficient ubiquitylation of the protein (Ganoth et al., 2001; Spruck et al., 2001). It has been shown that Cks1 is critical for the ubiquitylation of p27 both *in vitro* and *in vivo* (Spruck et al., 2001). On the basis of these studies, Cks1 appears to function as an accessory factor of SCF^{Skp2} that can regulate the ubiquitylation of p27. According to this model, the disruption of Cks1–Skp2 protein complex should result in inhibition of p27 degradation. Recently, we have investigated the biochemical interactions between recombinant Cks1 and Skp2 *in vitro* and have shown that their binding constant ($K_d = 140$ nM) puts this protein–protein interaction in a tractable range for development of small-molecule inhibitors (Xu et al., 2002). Here, we describe the development of a high-throughput screening assay for Cks1–Skp2 inhibitors.

Recently, time-resolved fluorometric assays using Europium (Eu) chelates as fluorophores have been widely used for kinase, phosphatase, nuclear receptor, and protein binding assays (Hemmila, 1999; Mathis, 1999). These fluorophores exhibit an intense and long-lived fluorescence emission, making it possible to measure fluorescence after a time delay. The time-resolved fluorescence effectively reduces background emissions and makes the assay highly sensitive. Homogeneous time-resolved fluorescence (HTRF) assay format does not involve washing steps and permits one to measure the interactions between two molecules in solution under physiological conditions. It uses Eu chelate as a fluorophore donor and cross-linked allophycocyanin (APC) as an acceptor. When these two molecules are brought together by a specific binding interaction, the excited energy of the Eu-chelate is transferred by a non-radioactive resonance energy transfer mechanism to an acceptor within a short distance.

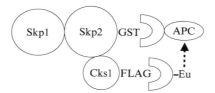

Fig. 1. Principal scheme of the Cks1–Skp2 binding assay. Arrow indicates fluorescence energy transfer between Eu Chelate and APC conjugated to antibodies specific for the FLAG-tag of Cks1 and the GST-tag of Skp2, respectively.

The long-lived acceptor emission at 665 nm minimizes the short-lived background fluorescence and substantially improves signal-to-noise ratio. Eu emission at 615 nm is not affected significantly upon energy transfer, so the ratio 665:615 nm is used to correct for fluorescence quenching.

To screen for inhibitors of the Cks1–Skp2 binding, we designed an HTRF assay in which APC-labeled anti-GST antibody and Eu-labeled anti-FLAG antibody are used to capture GST-Skp2/Skp1 and FLAG-Cks1, respectively (Fig. 1). On the basis of this assay, we developed a screening protocol using the Zeiss ultra high-throughput screening (uHTS) system and have achieved a daily throughput of more than 100,000 data points. However, the HTRF assay format described here can be implemented with slight modifications on any standard robotic screening system. Moreover, this is a generic assay format that is applicable to any protein–protein interaction that may represent a viable target for pharmacological intervention.

Methods

Materials

Europium-labeled anti-GST, anti-FLAG, and anti-His6 antibodies were purchased from Perkin-Elmer Life & Analytical Sciences. APC-labeled streptavidin, anti-GST, and anti-FLAG antibodies were from Prozyme (San Leandro, CA). Biotinylated p27 peptide phosphorylated on Thr[187] (Biotin-P-p27; biotin-Aca-SDG-SPN-AGS-VEQ-T[p]PK-KPG-LRR-RQT-CONH$_2$, where Aca is aminocaproic acid) and biotinylated unphosphorylated p27 peptide (biotin-p27; biotin-Aca-SDGSPNAGS-VEQTPKKPGLRRRQT-CONH$_2$) were synthesized at Hoffmann-La Roche Inc. Complete protease inhibitor cocktail was purchased from Roche Applied Science. All other chemicals were purchased from Sigma unless otherwise indicated in the text.

Expression and Purification of GST-Skp2/Skp1

We initially expressed GST-Skp2 in bacterial cells and found that the expression level was low because of rapid *in vivo* degradation of the protein. To overcome this problem, we coexpressed Skp2 and Skp1 from a dicistronic vector (Schulman *et al.*, 2000). This not only increased the yield of recombinant Skp2 but provided Skp2 bound to its natural partner in the SCF^{Skp2} complex. Expression plasmid for GST-Skp2 and Skp1 (pET/GST-Skp2-Skp1) was constructed by PCR amplification of the coding regions for Skp1 and Skp2 from pcDNA1/hSkp1 and pcDNA3/Skp2, respectively. Skp2 was fused in frame at the N-terminus with GST followed by a ribosomal binding site. These two DNA fragments were cloned into pET11 vector (Novagen, Madison, WI) in tandem with GST-Skp-2 upstream of the Skp1 coding sequence as a dicistronic expression unit. Plasmid DNA was transformed into BL-21(DE3) cells for protein expression. Cells from a 5-liter bacterial culture were collected by centrifugation, and pellets were resuspended in 200 ml of lysis buffer (20 mM Tris-HCl, pH 7.5; 150 mM NaCl, 1 mM DTT, 0.5% NP-40) containing 200 mg/ml lysozyme. The cell suspension was sonicated five times (1 min each) with a 2-min rest on ice between bursts. The homogenate was centrifuged at 10,000g for 20 min, and the supernatant was combined with 30 ml of glutathione Sepharose equilibrated with lysis buffer. The slurry was tumbled at 4° for 1 h. The resin was then washed with 10 volumes of lysis buffer followed by 20 volumes of lysis buffer minus NP-40. The protein complex was eluted by incubation of the resin with 5 volumes of 20 mM Tris-HCl, 150 mM NaCl, 1 mM DTT, 20 mM glutathione (pH 7.8) followed by vacuum filtration through a 0.2-μm PES membrane. The protein was concentrated using an Amicon Stir Cell (Amicon, Beverly, MA) with a 30,000 Da cut-off membrane. The samples were then dialyzed against 2 × 1 liter of 20 mM Tris-HCl, pH 7.5, 150 mM NaCl, 1 mM DTT, and 20% glycerol. Purified material shows two bands on SDS-PAGE with molecular weights corresponding to GST-Skp2 and Skp1.

Expression and Purification of FLAG-Cks1 and Cks-1

Escherichia coli containing FLAG-Cks1 expression plasmid was grown on LB media (3-liter fermentation) and induced with 0.6 mM isopropylthiogalactose (IPTG) for 3 h at 37°. Cells were then harvested by centrifugation at 6000g for 30 min. FLAG-Cks1 was purified as described previously (Arvai *et al.*, 1995). All procedures were carried out at 4°. *E. coli* cells (300 g wet weight) containing FLAG-Cks1 were lysed by sonication in 50 ml of 50 mM Tris-acetate, pH 7.5, 2 mM EDTA, 100 mM NaCl, 10%

glycerol, and complete protease inhibitor cocktail (Solution A). The homogenate was centrifuged at 10,000g for 20 min. To the supernatant was added polyethyleneamine to a final 0.65% to precipitate DNA. After centrifugation (Beckmann Ti45 rotor, 40,000 rpm), the supernatant containing FLAG-Cks1 was precipitated with an 80% ammonium sulfate cut, pelleted at 12,000g, and resuspended in solution A without protease inhibitors. The supernatant was fractionated on a Sephadex S-200 column equilibrated in solution A. Fractions containing FLAG-Cks1 were pooled. The purified material showed a major band (>95% purity) with a MW of 10,000 on SDS-PAGE. The identity of the purified protein was confirmed by mass spectroscopy and sequencing of N-terminal peptides. The procedures for expression and purification of non-tag Cks1 were identical to that of FLAG-Cks1.

HTRF Assay for Binding of FLAG-Cks1 and GST-Skp2/Skp1

Using the assay format described in Fig. 1, we studied the interaction between FLAG-Cks1 and GST-Skp2/Skp1 and determined their binding constant (Kd). When FLAG-Cks1 was titrated with increasing concentrations of GST-Skp2/Skp1, a concentration-dependent binding was observed (Fig. 2A). The binding data fitted well with the one-site binding equation with a Kd of 140 nM when analyzed by the GraphPad Prism software. As a background control, experiments were also carried out in the absence of GST-Skp2/Skp1 or FLAG-Cks1, and no binding signal was detected in either cases. To determine whether Skp1 plays a direct role in the binding to Cks1, we replaced GST-Skp2/Skp1 in the assay with His6-Skp1* and used Eu-labeled anti-His6 antibody and APC-labeled anti-FLAG antibody. No detectable binding between His6-Skp1 and Cks1 was observed. We also tested untagged Cks1 as a competitor in the assay (Fig. 2B), and the Ki was calculated to be 160 nM, implying that the FLAG tag has no significant effect on the affinity of Cks1 to GST-Skp2/Skp1. We also found that the binding is ionic strength dependent and pH sensitive, with optimal binding observed at pH 7.5 in a buffer containing 150–180 mM NaCl.

The Zeiss uHTS System

Figure 3 shows a diagram of a 5-module Zeiss uHTS system. Hotel module (HW1) contains a temperature, CO_2, and humidified control incubator capable of holding 189 microtiter plates. In addition, it has a plate lift

---

*His6-Skp1 was expressed in baculovirus and purified by Ni-chelate column chromatography (Xu et al., 2002).

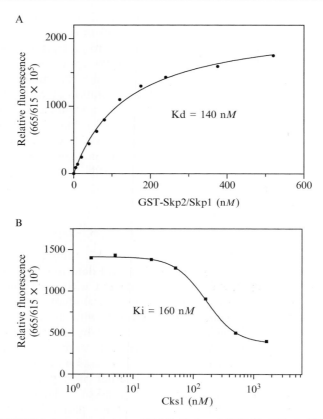

FIG. 2. HTRF binding assay between FLAG-Cks1 and GST-Skp2/Skp1. (A) Concentration-dependent binding. 20 μl/well of FLAG-Cks1 (final, 62.5 nM) was added to black flat-bottomed 384-well microtiter plates (Costar # 3710) followed by 20 μl/well of GST-Skp2/Skp1 (final 2.8–541 nM, 1.5-fold serial dilution) in assay buffer containing 50 mM Tris-HCl, pH 7.5, 150 mM NaCl, 0.05% (w/v) BSA, and 1 mM DTT. After mixing, plates were incubated at 37° with shaking for 1 h. Then, 10 μl/well of Eu-labeled anti-FLAG antibody (final, 0.32 μg/ml) and 10 μl/well of APC-labeled anti-GST antibody (final, 0.2–39 μg/ml, 1.5-fold serial dilution) were added in assay buffer. The mixture was shaken at room temperature for another 30 min and read on Victor 5 plate reader (PerkinElmer Life & Analytical Science) using excitation at 340-nm emission at 665 nm (50 μsec delay and 100 μsec acquisition window) and 615 nm (400 μsec delay and 400 μsec acquisition window). Because the donor Eu emission at 615 nm only decreased slightly on energy transfer, the ratio of emission at 665 nm to that at 615 nm is used to quantify the binding. Each data point was performed in duplicate. (B) Competition assay with Cks1. 10 μl/well of FLAG-Cks1 (final, 78 nM) and 10 μl/well of Cks1 (final, 0.69 nM–1.5 μM, three-fold serial dilution) in assay buffer were added to black flat-bottomed 384-well microtiter plates (Costar # 3710), followed by the addition of 20 μl of GST-Skp2/Skp1 (final, 44 nM). The mixture was incubated at 37° with shaking for 1 h. Then, 20 μl of premixed Eu-labeled anti-FLAG antibody (final, 0.32 μg/ml) and APC-labeled anti GST antibody (final, 6.9 μg/ml) were added, and the samples were shaken at room temperature for another 30 min and read as previously. Each data point was performed in quadruplicate.

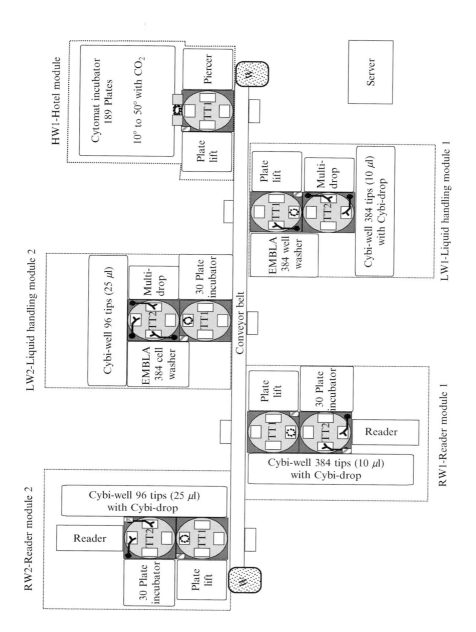

with 12 holders each capable of holding 24–28 different types of microtiter plates. It has a Piercer to punch holes on the compound plates, which are usually sealed with aluminum foil to prevent water moisture from entering DMSO compound solution. Each module contains 1 or 2 turntables for delivering plates into components. Liquid handling module 1 (LW1) contains a Cybi-well (384 tips) pipetter with Cybi-drop for handling small volume liquids (1–10 μl), a multidrop for dispensing large volumes (more than 10 μl), an EMBLA 384-well washer and a plate lift. Liquid handling module 2 (LW2) contains a CyBi-well (96 tips) pipetter, multidrop, EMBLA 384 cell washer (suitable for washing attached cells), and a 30-plate incubator with temperature control (25–50°) suitable for short-term incubations. Reader module 1 (RW1) contains a Cybi-well (384 tips) pipetter with Cybi-drop, a plate lift, a 30-plate incubator and a Zeiss multimode reader (capable of reading absorbance, fluorescence, luminescence, fluorescence polarization, but not time-resolved fluorescence*). Reader module 2 (RW2) contains the same components as that of RW1, except the Cybi-Well (96 tips) pipetter. Compound plates or assay plates are usually placed in the plate lift and picked up by a robot arm to a turntable. They can be delivered to individual components by turntables with pusher devices. Turntables are equipped with a flipper device to turn plates around for correct orientation and a vacuum-driven de-lidder device for removing and replacing lids. Plates are transported between modules by a conveyor belt.

---

* Zeiss recently introduced a new version of Multimode reader capable of reading time-resolved fluorescence.

FIG. 3. Diagram of the Zeiss uHTS modules. Each plate lift has 12 stackers, and each stacker is capable of holding 24 384-well microtiter plates (without lid) or 28 1536-well microtiter plates (BD Falcon, Low Base with lid). Cytomat incubator (Kendro Laboratory Products, CT) is capable of handling 189 microtiter plates with temperature (10–50°), CO_2, and humidity control. EMBLA washer and multidrop dispenser were manufactured by Skatron Instruments (Norway) and Labsystems (Finland), respectively. Cybi-well pipetter equipped with Cybi-drop dispenser was made by CyBio (Germany). Multimode reader with 96 channels is capable of reading absorbance, fluorescence, and luminance (Carl Zeiss, Germany). Microtiter plates can be transported between modules by the conveyor belt and delivered to individual components in the module by the turn tables (TT1 or TT2). They can also be disposed by the conveyor belt at both ends (shown as "W").

High-Throughput Screen for Cks1–Skp2 Binding Inhibitors Using the Zeiss uHTS Robot

A screening protocol was developed using the high-density 1536-well plate format and implemented on the Zeiss uHTS system. A flowchart of the assay protocol is shown in Fig. 4. Typically, polypropylene 384-well plates containing test compounds in columns 3–24 (2 μl per well, 1 mM in DMSO) were placed in the HW1 plate-lift, and polystyrene 1536-well assay plates were placed in LW1 plate-lift. An assay plate was lifted by a robot arm, placed on a turntable, and moved to the Cybi-well 384 pipetter in LW1. FLAG-Cks1 (3 μl/well, 3.0 μg/ml) in 20 mM Tris-HCl, pH 7.5, 180 mM NaCl, 1 mM DTT, and 0.075% BSA was then added through the Cybi-well 384 pipetter. Compound plates were then transported to the Cybi-drop in LW1, and 20 μl/well of 20 mM Tris-HCl, pH 7.5, 180 mM NaCl, 1 mM DTT was added by the Cybi-drop. Columns 1–2 contained assay buffer only, and columns 3–24 contained test compounds. Samples were mixed three times with the 384 pipetter, and 1.5 μl/well was transferred to the assay plate. Samples from four 384-well compound plates were added to one 1536-well plate. The assay plate was then moved to the Cybi-well 384 pipetter in RW1. GST-Skp2 (1.5 μl/well, 40 μg/ml) in 20 mM

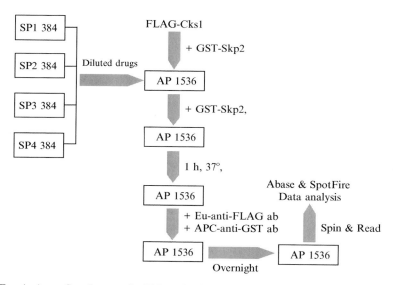

FIG. 4. Assay flowchart on the Zeisss uHTS system. SP1 384, SP2 384, SP3 384, and SP4 384 represent four compound plates. AP 1536 represents 1536-well assay plate. Abase, ActivityBase.

Tris-HCl, pH 7.5, 180 mM NaCl, 1 mM DTT, and 0.075% BSA was stored in "Smart Trough" (a divided reagent reservoir) and added to columns 5–48. Assay buffer was added to the top half of wells in columns 1–4 as a background. A premixed GST-Skp2 (40 μg/ml) and Cks1 (4.0 μg/ml) solution was added to the bottom half of wells in column 14 as an inhibitor control. The sample was then mixed three times using the 384 pipetter, and the plate was incubated at 37° for 30 min (I30 incubator in RW1).

For FLAG-Cks1/GST-Skp2 complex detection, a premixed Eu-labeled anti-FLAG antibody (3 nM) and APC-anti-GST antibody (100 nM) solution in 20 mM Tris-HCl, pH 7.5, 180 mM NaCl and 0.075% BSA was added to the assay plate (2 μl/well) through the Cybi-well pipetter in RW1, and the plate was incubated at 19° in a humidified incubator (I189 incubator) in HW1 for at least 1 h. The assay plates were centrifuged briefly to remove air bubbles before reading. To minimize evaporation of small assay volume (8 μl/well), the assay was carried out using plates with lids. Typically, less than 5% of volume evaporation was observed during the assay period. During overnight runs (12 h), plates were stored in a humidified incubator at 19°. In this case, approximately 15% volume reduction was observed. Nonetheless, the Z' factor (Zhang et al., 1999) was not affected significantly under those conditions. The Zeiss uHTS system has unique features in handling plates in a parallel process that increases assay throughput substantially. By use of the preceding process, we achieved a maximum throughput of 70 assay plates (1536-well) in 18 h, corresponding to 107,520 data points.

HTS Plate Reader

To shorten the reading time of 1536-well plates, we routinely used the ViewLux reader (PerkinElmer Life & Analytical Sciences) with the following settings: excitation 340 nm, emission 665 nm (50 μsec delay and 354 μsec acquisition window), and 615 nm (50 μsec delay and 354 μsec acquisition window). ViewLux reader is a CCD-based image reader and typically takes only 3 min to read a 1536-well plate. The ViewLux instrument was calibrated for the plate configuration using a blank 1536-well microtiter plate (BD Falcon, Cat # 353249) and for the donor spectrum cross-talk at 665 nm using 1 nM of Eu-labeled anti-FLAG antibody in the assay buffer. The output file contains a blank-corrected normalized ratio (Rn) calculated by the following formula:

$$Rn = [(A - B_a - C \times D)/(D - B_d)] \times (D_c - B_d)$$

Where A is the fluorescence intensity of the sample at 665 nm, D is the fluorescence intensity of the sample at 615 nm, B_a and B_d are plate backgrounds at 665 nm and 615 nm, respectively, D_c is the fluorescence intensity of 1 nM Eu-labeled anti-FLAG antibody in the assay buffer at 615 nm.

The cross-talk factor (C) is determined by the following formula:

$$C = (A_c - B_a)/(D_c - B_d)$$

Where A_c is the fluorescence intensity of 1 nM Eu-labeled anti-FLAG antibody in the assay buffer at 665 nm.

Conclusion

Protein–protein interactions represent a major class of potential drug targets, but their exploration has been limited. Assays that permit ultra high-throughput screening of large compound libraries at reasonable cost can balance the high risk associated with this target class. We have developed an HTRF-based assay platform for protein–protein interaction that allows ultra high-throughput screening in high-density formats. It is applicable to most protein–protein interaction that can tolerate epitope tags. The assay platform was tested by screening for inhibitors of the interaction between p27 ubiquitin ligase component Skp2 and Cks1 using the Zeiss uHTS system. Implementation of this screening assay to a diverse library of small molecules allowed us to identify several compounds with $IC_{50} < 20$ μM as leads for further optimization of potency and selectivity.

Acknowledgments

We thank Kui Xu, Andrew Schutt, Wei Chu, Chao-Min Liu, Charles Belunis, and David Weber for their contributions to this work.

References

Arvai, A. S., Bourne, Y., Williams, D., Reed, S. I., and Tainer, J. A. (1995). Crystallization and preliminary crystallographic study of human CksHs1: A cell cycle regulatory protein. *Proteins* **21**, 70–73.

Carrano, A. C., Eytan, E., Hershko, A., and Pagano, M. (1999). SKP2 is required for ubiquitin-mediated degradation of the CDK inhibitor p27. *Nat. Cell Biol.* **1**, 193–199.

Esposito, V., Baldi, A., De Luca, A., Groger, A. M., Loda, M, Giordano, G.G, Caputi, M., Baldi, F., Pagano, M., and Giordano, A. (1997). Prognostic role of the cyclin-dependent kinase inhibitor p27 in non-small cell lung cancer. *Cancer Res.* **57**, 3381–3385.

Ganoth, D., Bornstein, G., Ko, T. K., Larsen, B., Tyers, M., Pagano, M., and Hershko, A. (2001). The cell-cycle regulatory protein Cks1 is required for SCF^{Skp2}-mediated ubiquitinylation of p27. *Nat. Cell Biol.* **3**, 321–324.

Hemmila, I. (1999). LANCE™ Homogeneous Assay Platform for HTS. *J. Biomol. Screen* **4**, 303–308.

Mathis, G. (1999). HTRF (R) Technology. *J. Biomol. Screen* **4,** 309–314.
Morgan, D. O. (1995). Principles of CDK regulation. *Nature* **374,** 131–134.
Reed, S. I., Baillly, E., Dulic, V., Hengst, L., Resnitzky, D., and Slingerland, J. (1994). G1 control in mammalian cells. *J. Cell Sci.* **18,** 69–73.
Schulman, B. A., Carrano, A. C., Jeffrey, P. D., Bowen, Z., Kinnucan, E. R., Finnin, M. S., Elledge, S. J., Harper, J. W., Pagano, M., and Pavletich, N. P. (2000). Insights into SCF ubiquitin ligases from the structure of the Skp1–Skp2 complex. *Nature* **408,** 381–386.
Slingerland, J., and Pagano, M. (2000). Regulation of the cdk inhibitor p27 and its deregulation in cancer. *J. Cell Physiol.* **183,** 10–17.
Sherr, C. J., and Roberts, J. M. (1999). CDK inhibitors: Positive and negative regulators of G1-phase progression. *Genes Dev.* **13,** 1501–1512.
Spruck, C., Strohmaier, H., Watson, M., Smith, A. P., Ryan, A., Krek, T. W., and Reed, S. I. (2001). A CDK-independent function of mammalian Cks1: Targeting of SCF^{Skp2} to the CDK inhibitor p27Kip1. *Mol. Cell* **7,** 639–650.
Tsihlias, J., Kapusta, L., and Slingerland, J. (1999). The prognostic significance of altered cyclin-dependent kinase inhibitors in human cancer. *Ann. Rev. Med.* **50,** 401–423.
Tsvetkov, L. M., Yeh, K. H., Lee, S. J., Sun, H., and Zhang, H. (1999). $p27^{Kip1}$ ubiquitination and degradation is regulated by the SCF^{Skp2} complex through phosphorylated Thr187 in p27. *Curr. Biol.* **9,** 661–664.
Vlach, J., Hennecke, S., and Amati, B. (1997). Phosphorylation-dependent degradation of the cyclin-dependent kinase inhibitor p27. *EMBO J.* **16,** 5334–5344.
Xu, K., Belunis, C., Chu, W., Weber, D., Podlaski, F., Huang, K. S., Reed, S. I., and Vassilev, L. T. (2003). Protein–protein interactions involved in the recognition of p27 by E3 ubiquitin ligase. *Biochem. J.* **371,** 957–964.
Zhang, J.-H., Chung, T. D. Y., and Oldenburg, K. R. (1999). A simple statistical parameter for use in evaluation and validation of high throughput screening assays. *J. Biomol. Screen* **2,** 67–73.

[48] *In Vitro* SCF$^{\beta\text{-Trcp1}}$–Mediated IκBα Ubiquitination Assay for High-Throughput Screen

By Shuichan Xu, Palka Patel, Mahan Abbasian, David Giegel, Weilin Xie, Frank Mercurio, and Sarah Cox

Abstract

An increasing body of evidence indicates that constitutive activation of NF-κB contributes to tumorigenesis and inflammation. Ubiquitination and degradation of IκB plays an essential role in NF-κB activation. Here we describe an *in vitro* IκBα ubiquitination assay system in which purified E1, E2, SCF$^{\beta\text{-Trcp1}}$ E3, IκBα, IKK2, and Ub were used to generate ubiquitinated IκBα. The ubiquitination of IκBα is strictly dependent on its phosphorylation by IKK2, as well as the presence of E1, E2, E3, and Ub. The assay was adapted into 384-well plate format in which an antibody against IκBα was used to capture IκBα, and the biotinylated ubiquitin attached to IκBα was detected with europium (Eu)-labeled streptavidin. This assay can be used to discover inhibitors of IκBα ubiquitination. Such inhibitors would block NF-κB activation by stabilizing IκB levels in cells and thus provide a new therapeutic approach to NF-κB–related human diseases.

Introduction

NF-κB, a family of dimeric transcription factors, controls expression of a wide variety of genes (Karin and Ben-Neriah, 2000; Li and Verma, 2002). The products encoded by these genes regulate important cellular functions ranging from immune response to cell survival. NF-κB is activated by many different stimuli, including pathogens, proinflammatory cytokines such as IL-1β and TNFα, T- and B-cell mitogens, and stress signals. The activation mechanism of NF-κB has been well studied. In the resting state, NF-κB is bound to IκB proteins, and the complex remains in the cytoplasm in an inactive form. On stimulation, IκB is phosphorylated on Ser32 and Ser36 by the IKK complex, which is composed of IKKα, β, and γ (Karin and Ben-Neriah, 2000; Mercurio *et al.*, 1997). Subsequently, phosphorylated IκB is ubiquitinated and degraded by the 26S proteasome complex. The released NF-κB is transported into nucleus through its now exposed nuclear-localization sequence (NLS), where it turns on transcription of its target genes.

Among the IκB family members, IκBα is best characterized. The inducible degradation after ubiquitination of IκBα is the rate-limiting step in the activation of NF-κB. Ubiquitination of IκBα is mediated by the SCF$^{\beta\text{-Trcp1}}$ E3 complex, composed of Cul1, Roc1, Skp1, and β-Trcp1 (Amit and Ben-Neriah, 2003; Fuchs et al., 2004). X-ray structure studies indicate that the C-terminus of Cul1 interacts with Roc1 and presumably E2, whereas the N-terminus of Cul1 binds to Skp1 (Orlicky et al., 2003; Wu et al., 2003). β-Trcp associates with Skp1 by means of the F-box at the N-terminus and binds to IκBα through the WD40 repeats at the C-terminus. The destruction motif (DSGLDS) at the N-terminus of IκBα (that contains both phosphorylation sites) for IKK is the recognition site for the WD40 repeats of β-Trcp. Phosphorylation on both serines (underlined) is required for the interaction between IκBα and β-Trcp1 (Fuchs et al., 2004). On stimulation, IκBα is rapidly phosphorylated by the IKK complex. Subsequently, SCF$^{\beta\text{-Trcp}}$ binds to phosphorylated IκBα and facilitates ubiquitin transfer from E2 to IκBα. Ubiquitinated IκBα is then degraded by the 26S proteasome pathway. Here we describe a 384-well format microtiter plate assay to measure the ubiquitination of IκBα in a system reconstituted with purified E1, E2, SCF$^{\beta\text{-Trcp1}}$ complex, ubiquitin, and phosphorylated IκBα. Because constitutive activation of NF-κB is often associated with inflammatory diseases and cancer (Amit and Ben-Neriah, 2003; Baeuerle and Baichwal, 1997; Barnes and Adcock, 1997; Bharti and Aggarwal, 2002; Gilmore, 2003; Luque and Gelinas, 1997; Mayo and Baldwin, 2000), our assay can be used to screen for inhibitors of IκBα ubiquitination. Such inhibitors would block NF-κB activation by stabilizing IκB levels in cells and thus provide a new therapeutic approach to NF-κB–related human diseases.

Materials and Methods

Materials

Black Maxisorp plates (384 well) were purchased from Nalge Nunc (Rochester, NY), and 384-well conical polypropylene plates were purchased from E&K Scientific (Campbell, CA). All the anti-IκBα antibodies were purchased from Santa Cruz Biotechnology, Inc (Santa Cruz, CA) except 5A5, which was purchased from Cell Signaling Technology (Beverly, MA). Delfia™ Eu-Streptavidin and Enhancement solution were from Perkin Elmer (Boston, MA). Assay diluent was from BD Pharmingen (San Diego, CA). Ubiquitin was from Sigma (Saint Louis, MO). EZ-link™ Sulfo-NHS-LC-Biotin was from Pierce Biotechnology (Woburn, MA), and 2× SDS-loading buffer for protein PAGE was purchased from Invitrogen (San Diego, CA).

Preparation of Protein Components

His-E1 and His-UbcH3 were expressed in *Escherichia coli* and purified on Ni^{2+} chelate resin. The SCF$^{\beta\text{-Trcp1}}$ complex was affinity purified from Sf9 cells using glutathione agarose resin. GST-IκBα 1-54 and IKK2EE were expressed and purified from *E. coli* as described (Mercurio et al., 1997). All the preceding proteins were dialyzed into Ub dialysis buffer (30 mM Tris-HCl. pH 7.5, 20% glycerol, 1 mM DTT) and stored in small aliquots at −80°. Ubiquitin was labeled with biotin by incubating 100 mg of Ub (Sigma U6253) with 12.5 mg of EZ-link™ Sulfo-NHS-LC-Biotin in PBS on ice for 2 h. The stoichiometry of labeling was determined by LCMS. On average, each ubiquitin molecule contained 2–3 moieties of biotin. The biotinylated ubiquitin was dialyzed against 10 liters of 10 mM Hepes, pH8.0, and stored in small aliquots at −80°.

Preparation of Prephosphorylated IκBα (pIκBα)

Because phosphorylation of IκBα is required for its binding to SCF$^{\beta\text{-Trcp1}}$, prephosphorylated IκBα was prepared by incubating 0.1 mg/ml IKK2EE (1.1 μM) with 0.1 mg/ml (2.78 μM) GST-IκBα 1–54 at 37° for 2 h in kinase buffer (40 mM Tris-HCl, pH7.5, 10 mM MgCl$_2$, 1 mM DTT, 1 mM ATP). Aliquots were stored at −80°.

In Vitro IκBα Ubiquitination

Unless indicated specifically, pIκBα was ubiquitinated in a 15-μl reaction by incubation with E1, E2, E3, and Bio-Ub in UB buffer (40 mM Tris HCl, pH 7.5, 5 mM MgCl$_2$) containing 1 mM DTT and 0.5 mM ATP at RT for 1 h (Table I). For analysis of the ubiquitination of IκBα by SDS-PAGE,

TABLE I
ASSAY COMPONENTS

Components	Reagent name	Concentration
E1	His-E1	5 ng/μl (43 nM)
E2	His-UbcH3	150 ng/μl (5.6 μM)
E3 (SCF$^{\beta\text{-Trcp1}}$)	GST-Skp1/Flag-β-Trcp1/Roc1/His-Cul1	15 ng/μl (59 nM)
Substrate	pIκBα	0.5 ng/μl (14 nM)
Biotinylated ubiquitin	Bio-Ub	250 ng/μl (27.8 μM)
Tris-HCl, pH7.5		40 mM
MgCl$_2$		5 mM
DTT		1 mM
ATP		0.5 mM

the assay mixture was mixed with an equal volume of 2× SDS-loading buffer containing 100 mM DTT. For the plate assay, the assay mixture was diluted 1:3, 1:6, or 1:10 in stopping buffer (assay diluent containing 30 mM EDTA), and 20 μl of the diluted reaction was transferred to each well of a capture plate. For titration, each protein component was diluted in assay dilution buffer (30 mM Tris HCl, pH7.5, 1 mM DTT, 19.5% glycerol). The diluted protein was incubated with the rest of components (Table I), and the ubiquitination of pIκBα was measured as described in the following.

Western Blot

After SDS-PAGE, the proteins in the gel were transferred to a PVDF membrane and probed with either 5A5 or an anti-ubiquitin antibody. The signal on the membrane was detected and quantified on a LiCor imager using the manufacturer's software.

Plate Assay

Black Maxisorp plates (384-well) were coated with 22 μl of 2.5 μg/ml 5A5 in PBS at RT overnight. The plates were blocked with 100 μl/well of assay diluent (BD Pharmingen) at RT for 1–4 h and then incubated with 20 μl/well of diluted assay mixture. After incubation at RT for 1 h, the plates were washed six times with wash buffer (10 mM Tris-HCl, pH7.5, 0.05% [v/v] Tween 20). Eu-streptavidin (0.1 mg/ml) was diluted 1:250 in 25 mM Hepes, pH7.6, 1% BSA, 0.2% (v/v) Tween 20; 25 μl of Eu-streptavidin at a final concentration of 0.4 μg/ml was added to each well. After incubation at RT for 1 h, the plates were washed six times with wash buffer; 25 μl of Enhancement solution was added to each well, and time-resolved fluorescence was used to read each plate (excitation of 340 nm and emission of 620 nm) in a Victor II plate reader (PerkinElmer).

Results and Discussion

The principle of the assay is illustrated in Fig. 1A. E1 activates Ub in the presence of Mg^{2+} and ATP by forming a thioester bond between the C-terminus of Ub and an active cysteine on E1. Activated Ub is next transferred to an active site cysteine residue on E2, forming another thioester bond (Finley and Chau, 1991; Hershko and Ciechanover, 1998). In the presence of the $SCF^{\beta\text{-Trcp1}}$ E3 complex, the Ub on E2 is transferred to pIκBα and forms an isopeptide bond with the side chain of lysine residues on IκBα. Polyubiquitin chains are formed by isopeptide bonds between the C-terminus of one Ub and the side chain of lysine 48 of another Ub molecule. It is not yet known whether the polyubiquitin chains are formed on E2 or on substrate.

To develop a microtiter plate assay, Bio-Ub was used to generate polyubiquitinated IκBα. The ubiquitinated IκBα was captured with an anti-pIκBα antibody (5A5) coated on a plate and detected by Eu-labeled streptavidin that binds to the biotinylated ubiquitin. The protein components in the assay are shown in Fig. 1B.

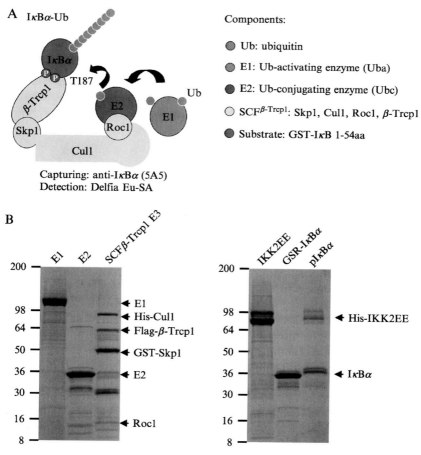

FIG. 1. (A) Schematic outline for the *in vitro* IκBα ubiquitination assay. (B) 5 μg of E1, E2, E3, IKK2EE, IκBα, and pIκBα were subjected to SDS-PAGE. The proteins in the gel were visualized by staining the gel with Coomassie blue.

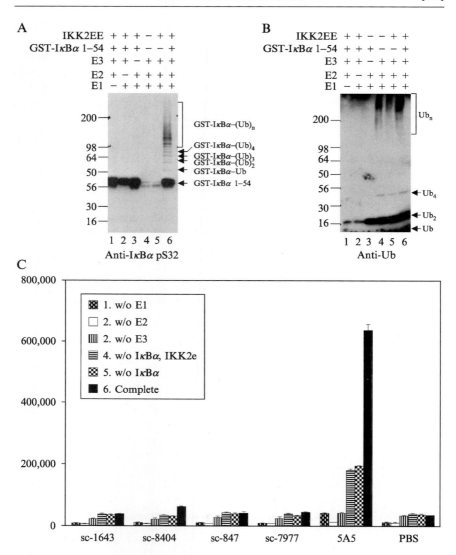

FIG. 2. *In vitro* ubiquitination of IκBα. GST-IκBα 1–54 was incubated with E1, E2, E3, or IKK2EE as indicated at RT for 1 h. The concentration of each component was as described in Table I. The reactions were stopped either with SDS-loading buffer (A and B) or by diluting 1:2.3 in assay diluent. (A) Western blot of pIκBα. Un-ubiquitinated and ubiquitinated pIκBα are indicated by the arrows. (B) Western blot of ubiquitin. Polyubiquitin chains were formed both in the absence and presence of IκBα. (C) Plate assay for ubiquitinated IκBα. The diluted reactions were incubated in wells coated with PBS alone or indicated antibodies diluted in PBS at 10 μg/ml. The ubiquitination was detected using Eu-SA. The value of the y-axis represents the fluorescence of Eu emission at 615 nm.

Reconstitution of the Ubiquitination of IκBα In Vitro

pIκBα was incubated with E1, E2, E3, and Bio-Ub in the presence of ATP and $MgCl_2$ (Fig. 2A). A ladder of IκBα with increasing molecular weights was observed. The increase in the molecular weight of the ladder was approximately at an increment of 10 kDa, close to the molecular weight of Bio-Ub (9.5 kDa). GST-IκBα exists as a doublet, probably because of degradation during purification. Interestingly, we observed that the doublet also appeared in the ladder, indicating that both forms of IκBα were ubiquitinated. The ubiquitination of pIκBα was strictly dependent on the presence of E1, E2, E3, and Bio-Ub. Figure 2B shows the total ubiquitination in the same reactions blotted with anti-ubiquitin antibody. Polyubiquitin was formed in the absence of IκBα, suggesting that other components in addition to pIκBα can be ubiquitinated in this system (Fig. 2B). However, no polyubiquitination was observed in the absence of E3. Interestingly, E1, E2, and E3 also led to formation of species corresponding in molecular weight to those of di- and tetra-ubiquitin. A species with molecular weight of tri-ubiquitin was not observed.

To measure IκBα ubiquitination specifically, a number of anti-IκBα antibodies were tested for their ability to capture ubiquitinated IκBα. The same reactions in Fig. 2A were incubated in a plate coated with the antibodies indicated, and ubiquitination of captured IκBα was detected

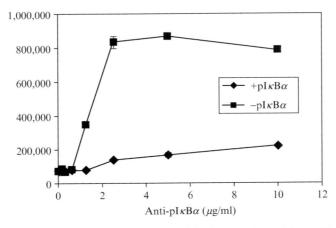

FIG. 3. Titration of capturing antibody. Ubiquitination reactions with or without pIκBα were performed and stopped by adding 40 μl of stopping buffer to 30 μl reaction; 20 μl of diluted reaction was transferred to a capturing plate coated with increasing amounts of 5A5. Ubiquitination was detected as described in Fig. 2.

by Eu-streptavidin. Antibody 5A5 gave a good signal for the complete reaction (Fig. 2C). Although the background for the controls without IκBα (4 and 5) was high in this initial test, the background was significantly reduced by including 30 mM EDTA in the stopping solution. A titration of the capture antibody (5A5) is shown in Fig. 3; 2.5 μg/ml was selected to coat the capture plate (Fig. 3).

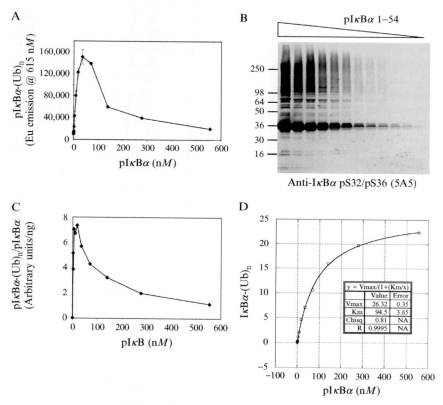

FIG. 4. Titration of pIκBα. Ubiquitination reactions were performed as described previously with increasing concentrations of pIκBα. Part of each reaction was diluted 1:3 with assay diluent containing 30 mM EDTA and transferred to a capture plate for detection of ubiquitination of pIκBα (A), while the rest of each reaction was used for an anti-pIκBα blot (B). The ubiquitinated IκBα, as judged by the ladder of pIκBα with increasing molecular weights in the gel, was scanned and quantified on the LiCor. The results were plotted (D). The ratio of ubiquitinated pIκBα (quantified by LiCor) over the total pIκBα in each reaction was plotted (C).

Optimization of pIκBα Concentration

The ubiquitination reactions were performed with increasing concentrations of pIκBα, whereas the other components were held at a fixed concentration. The ubiquitination of pIκBα was measured either in a plate capture assay (Fig. 4A) or by Western blot with 5A5 (Fig. 4B). The ubiquitinated IκBα detected by Western blot was quantified by the LiCor software (Fig. 4D). We found that the total ubiquitination of pIκBα increased as the concentration of pIκBα increased (Fig. 4B, D). After quantitation of the signal by LiCor (Fig. 4D), the apparent Km for pIκBα was found to be 94.5 nM. In the plate assay, the signal represents the amount of ubiquitinated pIκBα that is captured by the plate. Figure 4A shows that the signal increased as the concentration of pIκBα increased initially but decreased as the concentration of pIκBα further increased. A plot of the ratio of ubiquitinated pIκBα to total pIκBα gave a similar curve (Fig. 4C), suggesting that the decreased signal captured in the plate is likely due to competition between ubiquitinated and non-ubiquitinated pIκBα for antibody binding at the higher concentration of pIκBα. The binding capacity of 5A5 was estimated by transferring the reaction to a second plate after incubating it in the first plate. In Fig. 5, 80% of ubiquitinated pIκBα was captured in the first plate at a concentration of 17.4 nM (6.25 ng/μl) of pIκBα. At 34.8 nM (12.5 ng/μl) of pIκBα, only 50% of ubiquitinated

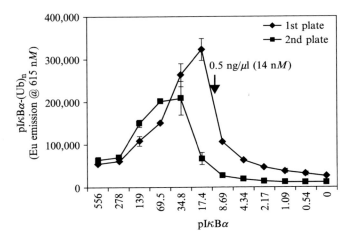

FIG. 5. Binding capacity of the capturing plate for pIκBα. pIκBα was ubiquitinated with an increasing amount of pIκBα and diluted 1:3 in assay diluent containing 30 mM EDTA. After incubating in the first plate for 1 h, pIκBα remaining in the supernatant was captured on a second plate coated with 5A5. At 0.625 ng/μl (17 nM), 80% IκBα bound to the well. Binding capacity of the well at 20 μl was calculated to be 3.3 ng (0.092 pMol) of pIκBα.

pIκBα was bound to the first plate. The binding capacity of the well was calculated to be 3.3 ng (0.092 pmol) of pIκBα in the 20-μl volume. Subsequent experiments were performed at 0.5 μg/ml (14 nM) of pIκBα in a 30-μl reaction. The reaction was stopped with 60 μl of stopping buffer, and 20 μl of the mixture was transferred to the capture plate.

Optimization of Other Parameters

Using of the substrate and capture condition described previously, the dependence of the signal on other components of the system was determined. Each component was titrated individually. The final concentrations chosen for E1 (43 nM), E2 (5.6 μM), and Bio-Ub (27.8 μM) were close to, or at, saturating. E3 (59 nM) was at a limiting concentration (Fig. 6). The apparent Km for Bio-Ub was measured as 3.73 μM (data not shown), close to the reported value for Oregon green (Og)–labeled Ub (Wee et al., 2000). The reaction was linear up to 2 h under these conditions (Fig. 7). A standard reaction time of 1 h was used to evaluate this assay for HTS. The assay performed well with Z′ between 0.5 and 0.7 and a signal to background ratio of 8–12 (Table II).

In summary, an *in vitro* system able to generate ubiquitinated IκBα has been developed. This system uses entirely purified recombinant

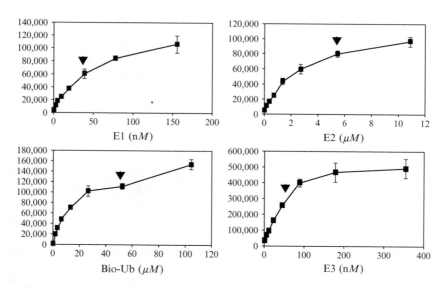

FIG. 6. Titration of E1, E2, E3, and Bio-Ub. Each component was titrated, while the other components were used at the indicated concentration (Table I).

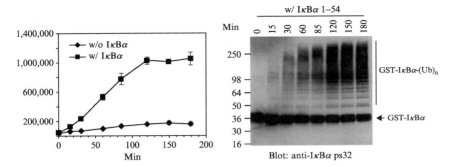

Fig. 7. Time dependence. Ubiquitination reactions were performed with or without pIκBα at RT. At the indicated time, 30 μl of each reaction was removed and mixed with 40 μl of assay diluent. The mixtures were kept on dry ice until all samples were collected. The samples were thawed, and ubiquitination of pIκBα was detected in a capture plate.

TABLE II
ASSAY STATISTICS (NEGATIVE, WITHOUT pIκBα)

	Positive ($n = 32$)	Negative ($n = 32$)
Average	60308	4496
STDEV	8460	384
Signal/background	13	
Z'	0.5	

components—E1, E2, SCF$^{\beta\text{-Trcp1}}$ E3, Bio-Ub, and GST-IκBα. Each of the components is essential for IκBα ubiquitination *in vitro*. This assay has been adapted to a robust 384-well plate assay detecting ubiquitination of IκBα by use of a specific antibody to capture IκBα and Eu-streptavidin to detect ubiquitination. This assay is specific, reliable, and suitable for HTS.

Acknowledgments

We thank Sholita Packer for assistance in obtaining the legal approval for publishing this chapter and Nathan Eller for helping with the preparation of the figures.

References

Amit, S., and Ben-Neriah, Y. (2003). NF-kappaB activation in cancer: A challenge for ubiquitination- and proteasome-based therapeutic approach. *Semin. Cancer Biol.* **13**, 15–28.

Baeuerle, P. A., and Baichwal, V. R. (1997). NF-kappa B as a frequent target for immunosuppressive and anti-inflammatory molecules. *Adv. Immunol.* **65**, 111–137.

Barnes, P. J., and Adcock, I. M. (1997). NF-kappa B: A pivotal role in asthma and a new target for therapy. *Trends Pharmacol. Sci.* **18,** 46–50.
Bharti, A. C., and Aggarwal, B. B. (2002). Nuclear factor-kappa B and cancer: Its role in prevention and therapy. *Biochem. Pharmacol.* **64,** 883–888.
Finley, D., and Chau, V. (1991). Ubiquitination. *Annu. Rev. Cell Biol.* **7,** 25–69.
Fuchs, S. Y., Spiegelman, V. S., and Kumar, K. G. (2004). The many faces of beta-TrCP E3 ubiquitin ligases: Reflections in the magic mirror of cancer. *Oncogene* **23,** 2028–2036.
Gilmore, T. D. (2003). The Rel/NF-kappa B/I kappa B signal transduction pathway and cancer. *Cancer Treat. Res.* **115,** 241–265.
Hershko, A., and Ciechanover, A. (1998). The ubiquitin system. *Annu. Rev. Biochem.* **67,** 425–479.
Karin, M., and Ben-Neriah, Y. (2000). Phosphorylation meets ubiquitination: The control of NF-[kappa]B activity. *Annu. Rev. Immunol.* **18,** 621–663.
Li, Q., and Verma, I. M. (2002). NF-kappaB regulation in the immune system. *Nat. Rev. Immunol.* **2,** 725–734.
Luque, I., and Gelinas, C. (1997). Rel/NF-kappa B and I kappa B factors in oncogenesis. *Semin. Cancer Biol.* **8,** 103–111.
Mayo, M. W., and Baldwin, A. S. (2000). The transcription factor NF-kappaB: control of oncogenesis and cancer therapy resistance. *Biochim. Biophys. Acta* **1470,** M55–62.
Mercurio, F., Zhu, H., Murray, B. W., Shevchenko, A., Bennett, B. L., Li, J., Young, D. B., Barbosa, M., Mann, M., Manning, A., and Rao, A. (1997). IKK-1 and IKK-2: Cytokine-activated IkappaB kinases essential for NF-kappaB activation. *Science* **278,** 860–866.
Orlicky, S., Tang, X., Willems, A., Tyers, M., and Sicheri, F. (2003). Structural basis for phosphodependent substrate selection and orientation by the SCFCdc4 ubiquitin ligase. *Cell* **112,** 243–256.
Wee, K. E., Lai, Z., Auger, K. R., Ma, J., Horiuchi, K. Y., Dowling, R. L., Dougherty, C. S., Corman, J. I., Wynn, R., and Copeland, R. A. (2000). Steady-state kinetic analysis of human ubiquitin-activating enzyme (E1) using a fluorescently labeled ubiquitin substrate. *J. Protein Chem.* **19,** 489–498.
Wu, G., Xu, G., Schulman, B. A., Jeffrey, P. D., Harper, J. W., and Pavletich, N. P. (2003). Structure of a beta-TrCP1-Skp1-beta-catenin complex: destruction motif binding and lysine specificity of the SCF(beta-TrCP1) ubiquitin ligase. *Mol. Cell* **11,** 1445–1456.

[49] High-Throughput Screening for Inhibitors of the E3 Ubiquitin Ligase APC

By JIANING HUANG, JULIE SHEUNG, GUOQIANG DONG, CHRISTINA COQUILLA, SARKIZ DANIEL-ISSAKANI, and DONALD G. PAYAN

Abstract

The anaphase-promoting complex (APC) is an E3 ubiquitin ligase that mediates the ubiquitination and degradation of the securin protein and mitotic cyclins, resulting in the regulation of the onset of sister-chromatid separation and mitotic exit. In an effort to identify novel therapeutic compounds that modulate cell proliferation and, therefore, have potential applications in oncology, a plate-based *in vitro* ubiquitination assay that uses recombinant purified E1, E2 (UbcH5c), E3 (APC11/APC2), and Flag-ubiquitin has been established and used to screen for small molecule inhibitors of APC E3 ligase activity. In this assay, APC2/APC11 is immobilized on the plate, and its E3 ligase activity (i.e., the incorporation of Flag-tagged polyubiquitin chain onto APC2/APC11 as a result of auto-ubiquitination) is detected with anti-Flag-horseradish peroxidase–conjugated antibody by monitoring the luminescence signal from the plate. Here we describe in detail the protocol for high-throughput screening of APC, including expression and purification of the individual proteins, assay development, and optimization. This assay has been validated in a 96-well plate format and successfully implemented to identify novel small molecule compounds that potently inhibit APC2/APC11 ligase activity.

Introduction

The anaphase-promoting complex/cyclosome (APC) is a cell cycle–regulated E3 ubiquitin ligase that controls both entry and exit from mitosis by directly promoting polyubiquitin chain formation on key mitotic cyclins and targeting them for destruction by the 26S proteasome. In addition to its role in degrading cyclin B and thereby inactivating cdk1, which is necessary for mitotic exit, APC is also required for the degradation of securin, resulting in the activation of separase for sister chromatin separation and spindle disassembly. A number of other important regulators of cell cycle and checkpoint activation, such as cyclin A, cdc6, and cdc20, have also been identified as APC substrates (Harper *et al.*, 2002; Jackson, 2004; Murray, 2004; Peters, 2002).

In the presence of the ubiquitin-activating enzyme E1, and certain ubiquitin-conjugating enzymes (E2s) such as Ubc4 and UbcH10, the APC complex is able to catalyze the attachment of ubiquitin to securin and mitotic cyclins *in vitro* (King *et al.*, 1995; Yu *et al.*, 1996). As a high molecular mass complex, APC is an unusual E3 ubiquitin ligase, with at least 11 core subunits identified in man. Sequence analysis has indicated that two of the subunits, APC2 and APC11, are homologous with proteins involved in other ubiquitination systems, including the Skp1-Cullin-F-box (SCF) pathway (Ohta *et al.*, 1999; Yu *et al.*, 1998). APC2 is a distant member of the cullin family, whereas APC11 contains a RING-H2 finger domain, a common motif shared with other RING finger E3 ligases that has been proposed to coordinate Zn^{2+} binding and is essential for ligase activity (Jackson *et al.*, 2000; Joazeiro and Weissman, 2000). In particular, it has been reported that the heterodimer of these two subunits represents the minimal ligase catalytic core of APC and is sufficient to promote polyubiquitin chain formation *in vitro* in the presence of E1 and E2 (Gmachl *et al.*, 2000; Tang *et al.*, 2001). Moreover, APC2/APC11 catalyzes its own auto-ubiquitination in an *in vitro* assay, a property that has been shown for many other ubiquitin ligases containing Ring finger domains (Chen *et al.*, 2002; Fang *et al.*, 2000).

The RING finger ubiquitin ligases represent a very large number of proteins that can be found intracellularly as members of supramolecular assemblies (e.g., APC, SCF) or monomers (e.g., Mdm2, Cbl, TRAF6) (Jackson *et al.*, 2000; Joazeiro and Weissman, 2000). Because of the importance of ubiquitination in cellular regulation, the ubiquitin ligases provide a class of novel targets for drug discovery (Pray *et al.*, 2002; Wong *et al.*, 2003). Several lines of evidence suggest that inhibition of the APC complex may provide a novel approach to anticancer therapy. First, mRNA levels of the APC11 subunit are found to be upregulated in a subset of tumor tissues relative to matched surrounding normal tissues, suggesting a selective dysregulation of this complex in tumors (Rigel, unpublished data). Second, overexpression of endogenous cellular APC inhibitory molecules such as Emi1 in mammalian cells inhibits cell cycle progression and results in cell death (Reimann *et al.*, 2001). Finally, RNA interference using siRNA to the catalytic subunits of APC11 and APC2 in HeLa cells leads to growth arrest and cell death (Rigel, unpublished data). The previous results taken together suggest that interfering with APC's ubiquitin ligase activity may have potent antitumor effects or may further sensitize tumor cells to existing chemotherapeutics.

Identification of such APC inhibitors requires a rapid, simple, and reliable ligase biochemical assay for small molecule screening. Traditional methods for measuring the formation of ubiquitin conjugation have relied primarily on radiolabeled ubiquitin and anti-ubiquitin immunoblots for detection; however, both methods are labor intensive and low throughput. In addition, neither method is sensitive enough to determine the potency of inhibitors within a dynamic range. Several assays for highthroughput drug screening of monomeric E3 ligase inhibitors have been reported that are based on time-resolved fluorescence technology (TR-FRET), including those targeting p53 ubiquitination by E6AP and TRAF6-mediated polyubiquitination (Hong *et al.*, 2003; Yabuki *et al.*, 1999). These cell-free E3-dependent HTS assays can be reconstituted with recombinant purified components or a partially purified fraction. However, the feasibility of screening the multicomponent E3s with these methods has yet to be reported. The unusual size and subunit complexity of the APC complex causes its conventional purification on a large scale to be difficult, which makes it challenging to develop HTS assays for the holoenzyme. To overcome this complexity, we have developed a substrate-independent ubiquitination assay for small molecule screening of APC inhibitors using the catalytic core subunits of APC2/APC11. The holo-APC activity is well supported by both Ubc4 and UbcH10 as E2s *in vitro*. However, Ubc4 seems to generate ubiquitin conjugates of higher molecular weight than those generated by UbcH10 (Tang *et al.*, 2001). We have observed a similar phenomenon in the plate assay. Reactions using UbcH5c (97% identical to Ubc4) as the E2 are able to generate a more robust signal compared with the assays using UbcH10, which also require a greater amount of E1 and E2 enzyme. Therefore, UbcH5c is used as the E2 ubiquitin-conjugating enzyme in the screen. This assay has been optimized and successfully used to identify novel small molecule compounds that potently inhibit APC ligase activity.

Expression and Purification of E1, UbcH5c (E2), APC2/APC11(E3), and Ubiquitin

E1 was expressed with an N-terminal 6×His tag fusion by means of recombinant baculoviral expression in insect cells (using the pFastBac-HT vector from Invitrogen). Active forms of UbcH5c with an N-terminal tag of GST (using the pGex-6P vector from Amersham Pharmacia Biotech) can be prepared relatively easily by recombinant expression in bacteria and the tag cleaved off after protein purification. Ubiquitin was fused to a Flag epitope tag (using the pFlag-Mac vector from Sigma) for immunodetection and expressed in *Escherichia coli*. The APC2/APC11 complex was expressed by coinfection of baculovirus of untagged APC11 (using the pFastBac vector from Invitrogen) and $(His)_6$-APC2 (using the pFastBac-HT vector) into insect cells, which allowed the active complex to be assembled *in vivo*. The propagation of *E. coli* and insect cell cultures, as well as the production, titer, and maintenance of recombinant baculoviruses, was performed according to the manufacturer's protocols (Amersham Pharmacia Biotech, Invitrogen).

All protein purification procedures were handled at 4°. The protocol for purification of GST-, 6×His-, and Flag-fusion proteins was based on the manufacturer's manual or published procedures and has been modified based on our experience. Protein concentration was analyzed by Bio-Rad reagent with BSA as a standard protein, and purity was verified by SDS-PAGE analysis and Coomassie Blue gel staining. The purity was generally greater than 90%. Purified proteins were aliquoted and stored at −80° after quick freezing by dry ice.

Purification of 6×His Fusion Proteins E1, and APC2/APC11

SF9 cell pellets from 1 liter fermentation cultures were resuspended in 100 ml lysis buffer (20 mM Tris-HCl, 15% glycerol, 0.5 M NaCl, 1 mM TCEP (Tris-2-carboxyethyl-phosphine), and 10 mM imidazole at pH 8.0) containing protease inhibitors (20 μg/ml PMSF, 2 μM leupeptin, and 1 μM pepstatin A). The non-ionic detergent NP-40 (10%) was added to a final concentration of 1.0 %, and cell lysis was conducted by gently mixing for 30 min at 4°. The lysate was sonicated by a Sonifier 450 cell disrupter for 30 sec three times, followed by centrifugation at 12,000 rpm twice for 40 min each. The supernatant was mixed together with 5 ml of Nickel-agarose beads (Sigma, Cat. No. P6611) that had been washed once with 20 ml water and twice with 20 ml wash buffer (20 mM Tris-HCl, 15% glycerol, 0.5 M NaCl, 1 mM TCEP, 0.1% NP-40, and 10 mM imidazole at pH 8.0). After 1 h of mixing, samples were centrifuged at 1500 rpm for 10 min. The pellet beads were washed twice with 20 ml wash buffer and then packed into a column. The column was washed with 4 volumes of the preceding wash buffer and eluted with 3 volumes of elution buffer (250 mM imidazole, 20 mM Tris-HCl, 15% glycerol, 50 mM NaCl, 1 mM TCEP, and 0.02% NP-40 at pH 8.0). The eluate was collected at 1 ml/tube. An aliquot of sample was run on SDS-PAGE to check the purity of the collected fractions. Further purification by Q-Sepharose chromatography and gel filtration chromatography was performed if the purity was not good enough (i.e., less than 90% pure). Otherwise, the purified fractions were pooled together and dialyzed against a dialysis buffer (20 mM Tris-HCl, 100 mM NaCl, 1.0 mM TCEP, 0.02% NP-40, 10% glycerol at pH 8.0) at 4° overnight. Figure 1 shows the

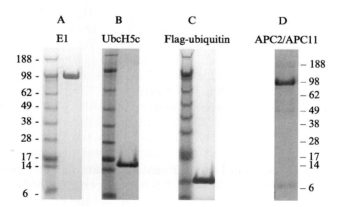

FIG. 1. SDS-PAGE analysis of the purified enzymes and proteins. (A) (His)$_6$-E1. (B) UbcH5c. (C) Flag-ubiquitin. (D) (His)$_6$-APC2/APC11.

SDS-PAGE analysis and Coomassie blue staining of purified $(His)_6$-E1 (yield of 5 mg/L) and $(His)_6$-APC2/APC11 (yield of 0.2 mg/L).

Purification of GST-UbcH5c

E. coli cell pellets from 1 liter fermentation cultures were resuspended in 180 ml lysis buffer (20 mM Tris-HCl, 10% glycerol, 0.5 M NaCl, 1 mM TCEP, and 2.5 mM EDTA at pH 8.0) containing protease inhibitors (20 μg/ml PMSF, 2 μM leupeptin, and 1 μM pepstatin A). After the suspension, 10% NP-40 was added to a final concentration of 1.0% with gentle mixing. The suspended cells were sonicated with a Sonifier 450 cell disrupter for 6 min at 4° with a 50% duty cycle, followed by centrifugation at 12,000 rpm at 4° for 40 min. The supernatants were pooled together and clarified by passing through a 0.2-μm microfiltrate membrane in a pellicon system. The clarified cell lysate was collected. Glutathione Sepharose 4 Fast Flow beads were packed into a XK50 column (5.0 cm × 7 cm), connected with an AKTAprimer purification system (Amersham Pharmacia Biotech). The column was balanced with 2 volumes of a wash buffer (20 mM Tris-HCl, 10% glycerol, 0.5 M NaCl, 0.5 mM TCEP, 0.1% NP-40, and 2.5 mM EDTA at pH 8.0), loaded with 4 volumes of cell lysate, and washed with 4 volumes of the wash buffer. Protein was eluted with 2 volumes of the elution buffer (20 mM glutathione, 20 mM Tris-HCl, 10% glycerol, 50 mM NaCl, 1 mM TCEP, 0.02% NP-40, and 2.5 mM EDTA at pH 8.0). The eluate was collected at 1 ml/tube. An aliquot was run on SDS-PAGE to check the purity of different fraction samples. To cleave the GST tag by PreScission Protease, purified fractions were pooled, and the pH was adjusted to 7.5 with 1 M acetic acid (pH 3.5). PreScission protease was added at 5 U/mg GST-UbcH5c. The protein mixture was loaded into dialysis tubes (10 K MWCO dialysis tube, Pierce) and dialyzed against the dialysis buffer (20 mM Tris-HCl, 50 mM NaCl, 1.0 mM TCEP, 0.5 mM EDTA, 0.02% NP-40, and 10% glycerol at pH 7.0) at 4°. The buffer was changed once during dialysis. After 24 h, a sample was taken and run on SDS-PAGE. If the GST was fully cut, dialysis was terminated; the protein was ready for further purification by a Q-Sepharose column. Figure 1 shows the SDS-PAGE analysis and Coomassie blue staining of purified UbcH5c (yield of 20 mg/L).

Purification of Flag Fusion Protein

E. coli cell pellets from 1 liter fermentation cultures were resuspended in 180 ml lysis buffer (20 mM Tris-HCl, 10% glycerol, 0.5 M NaCl, 1 mM TCEP, and 2.5 mM EDTA at pH 8.0) containing protease inhibitors (20 μg/ml PMSF, 2 μM leupeptin, and 1 μM pepstatin A). After resuspension, 10% NP-40 was added to a final concentration of 1.0% with gentle mixing. The suspended cells were sonicated by a Sonifier 450 cell disrupter for 6 min at 4° with a 50% duty cycle, followed by centrifugation at 12,000 rpm for 40 min. The supernatants were pooled together and further cleaned by passage through a 0.2-μm microfiltrate membrane in a pellicon system. The clarified lysate was collected. Flag-agarose beads (Sigma) were packed into a XK26 column (2.6 cm $\times$ 11 cm) and connected with an AKTA primer purification system (Amersham Pharmacia Biotech). The column was balanced with 2 volumes of a wash buffer (20 mM Tris-HCl, 10% glycerol, 0.5 M NaCl, 0.5 mM TCEP, 0.1% NP-40, and 2.5 mM EDTA at pH 8.0), loaded with 3 volumes of cell lysate, and washed with 4 volumes of the wash buffer. Protein was eluated with 2 volumes of the elution buffer (100 mM acetic acid, 10% glycerol, 100 mM NaCl, 1 mM TCEP, 0.02% NP-40, and 2.5 mM EDTA at pH 3.5). The eluate was collected in 1 ml/tube, which contained 0.08 ml of 2 M Tris-HCl at pH 9.0 for neutralization. An aliquot of each sample was run on SDS-PAGE to check the purity of collected fractions. The purified fractions were pooled together and dialyzed against a dialysis buffer (20 mM Tris-HCl, 50 mM NaCl, 1.0 mM TCEP, 0.5 mM EDTA, 0.02% NP 40, 10% glycerol at pH 8.0) at 4°. The dialysis buffer was changed three times. Fig. 1 shows the SDS-PAGE analysis and Coomassie blue staining of purified Flag-ubiquitin (yield of 10 mg/L).

High-Throughput Screening for APC Inhibitors

Standard *in vitro* ubiquitin ligase assay buffer:
50 mM Tris pH7.5
5 mM $MgCl_2$
0.6 mM DTT
5 μM ATP

In Vitro *APC2/APC11 Ubiquitination Assay Using Recombinant Proteins*

We first examined the activity of the recombinant proteins in a standard *in vitro* ubiquitination reaction. With some modifications, the amount of each protein used in the reaction was similar to that published by Ohta *et al.* (1999); 20 ng of E1, 200 ng of UbcH5c, 200 ng of APC2/APC11, and 500 ng of Flag-ubiquitin were mixed in the preceding ligase assay buffer to a final total volume of 30 μl. Reactions missing individual enzymes or ATP were included and served as negative controls. The ligase reaction mixtures were incubated at room temperature for 1 h, stopped by the addition of 10 μl of 4$\times$ SDS-PAGE sample buffer, and boiled for 5 min; 20 μl of each sample was analyzed on SDS-PAGE, followed by Western blotting with anti-Flag-HRP–conjugated antibody to detect the formation of poly-Flag-ubiquitin chains. As shown in Fig. 2A, the catalytic core subunit of APC2/APC11 was able to promote the formation of a high molecular weight ladder/smear on the blot in the presence of ATP, E1,

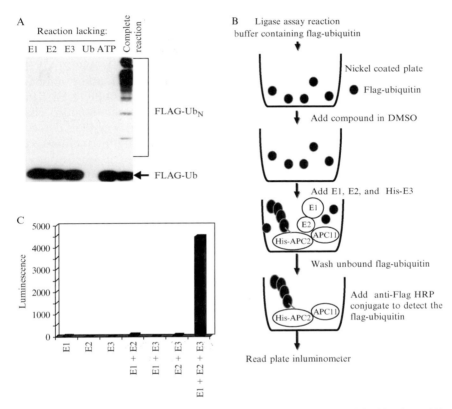

FIG. 2. *In vitro* ubiquitin ligase assay for APC2/APC11 auto-ubiquitination. (A) Reconstitution of the ubiquitination reaction using recombinant E1, UbcH5c, APC2/APC11, and Flag-ubiquitin. Reactions were performed at room temperature for 1 h followed by SDS-PAGE analysis and Western blot with an anti-Flag HRP–conjugated antibody. (B) Schematic representation of the APC2/APC11 plate assay for HTS. APC2/APC11 catalyzes its own auto-ubiquitination in the presence of ATP, E1, UbcH5c, and Flag-ubiquitin. During the 1-h incubation, His-APC2/APC11 was immobilized on the plate, while free Flag-ubiquitin and unbound enzymes were washed away. The poly-Flag-ubiquitin chains incorporated into the APC2/APC11 were detected with anti-FLAG HRP conjugate by monitoring th luminescence signal from the plate. (C) Results from the plate-based APC2/APC11 assay. A robust polyubiquitination signal as indicated by the strong luminescence activity was detected from the reaction containing all three of the necessary enzymes, E1, UbcH5c, and APC2/APC11.

and UbcH5c, an indication of ubiquitin polymerization. This polyubiquitin chain formation was not detected in the control reactions, in which either E1, UbcH5c, APC2/APC11, Flag-ubiquitin, or ATP were excluded.

APC2/APC11 Plate-Based Ubiquitination Assay

To make the assay feasible for high-throughput screening, we converted the SDS-PAGE/Western blot analysis into an ELISA plate–based format. The plate assay is illustrated in Fig. 2B. The reaction was performed in a Nickel coated-plate, which was used to immobilize $(His)_6$-APC2/APC11. After completion of the reaction, free ubiquitin and other unbound enzymes were washed away from the plate. The APC2/APC11 E3 ligase activity (i.e., the incorporation of Flag-tagged polyubiquitin chain onto APC2/APC11 as a result of auto-ubiquitination) was then detected with anti-Flag HRP–conjugated antibody by monitoring the luminescence signal from the plate.

In a typical 100-μl/well assay in a 96-well Nickel coated-plate, E1, UbcH5c, APC2/APC11, and Flag-ubiquitin were mixed in the standard ligase assay buffer and incubated at room temperature for 1 h. After incubation, the plate was washed with PBS three times. Then antibody mixture (100 μl/well) containing anti-Flag and HRP-conjugated anti-mouse IgG was added to the plate, and the incubation was continued for another hour at room temperature. Finally, the plate was washed with PBS three times, and luminol substrate was added to measure the luminescence activity. As can be seen in Fig. 2C, a robust polyubiquitination activity, as indicated by the strong luminescence signal, was detected in the wells containing all three of the necessary enzymes, E1, UbcH5c, and APC2/APC11. A very limited background signal was observed for the control reactions when any one of the enzymes was excluded.

Optimization of the APC2/APC11 Plate-Based Assay

For optimization of the plate assay, we first titrated E1 from a range of 1.25 ng/well to 40 ng/well. As shown in Fig. 3A, as little as 1.25 ng of E1 was able to give a strong signal for the assay, and starting at 10 ng, the reaction reached saturation. A similar titration of UbcH5c was done from 5 ng/well to 100 ng/well. The luminescence activity increased with higher UbcH5c concentration, and the reaction saturated at 50 ng UbcH5c (Fig. 3B). Figure 3C shows the APC2/APC11 titration from 25 ng/well to 400 ng/well. There was a dose-dependent increase in luminescence activity with increasing amounts of APC2/APC11. The reaction was linear up to 100 ng, and reached saturation at 200 ng. For screening, enzyme concentrations in the linear range are desirable. Therefore, 5 ng of E1, 20 ng of UbcH5c,

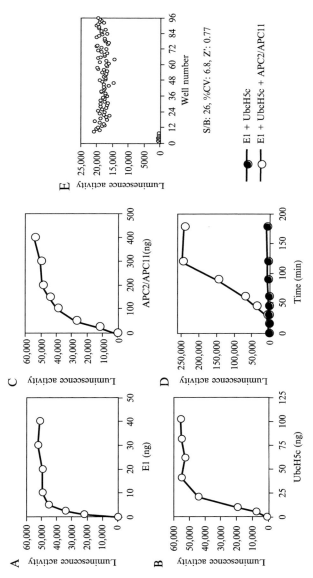

FIG. 3. Assay optimization. (A) Titration of E1 enzyme from 1.25 ng/well to 40 ng/well. (B) Titration of UbcH5c from 5 ng/well to 100 ng/well. (C) Titration of APC2/APC11 from 25 ng/well to 400 ng/well. (D) Time course for the assay. Reactions were incubated for various time points from 15 min to 3 h with 5 ng/well of E1, 20 ng/well of UbcH5c, and 80 ng/well of APC2/APC11. (E) Assay parameters.

and 80 ng of APC2/APC11 were used for the reaction (100 µl total volume). A time course experiment was also performed from 15 min to 3 h under the preceding conditions. Reactions were allowed to incubate for different lengths of time, after which the plates were washed with PBS and signal detected by an anti-Flag antibody. On the basis of the results of this experiment, an incubation time of 1 h at room temperature was selected, as it was sufficient to generate a strong polyubiquitination signal while staying within the range of reaction linearity (Fig. 3D). Because HTS is usually performed in the presence of diluted DMSO, DMSO tolerance of the ligase assay was also examined; 10% DMSO in the reaction had no significant effect on APC enzymatic activity (data not shown).

Standard Protocol for APC2/APC11 High-Throughput Screening

When screening, column 1 of the plate contained reactions with only E1 and UbcH5c in the ligase assay buffer as background controls, whereas columns 2–12 contained complete ligase reactions with E1, UbcH5c, and APC2/APC11. Compounds were added to wells in columns 2–11 to a final concentration of 10 μM in 10% DMSO, whereas column 1 (background control) and column 12 (signal control) contained only the 10% DMSO vehicle.

The final standard protocol for APC high-throughput screening was as follows: 96-well Nickel coated-plates (Pierce Cat. No. 15242) were blocked with 100 µl/well of 1% casein/PBS for 1 h at room temperature, and then washed three times with 200 µl/well of 1× PBS. 80 µl of the reaction buffer (62.5 mM Tris, pH 7.5, 6.25 mM MgCl2, 1.0 mM DTT, and 5 μM ATP) containing 100 ng of Flag-ubiquitin was added to each well, followed by the addition of 10 µl of 100 μM compound in 100% DMSO. Then the reaction was initiated by adding 10 µl of E1 (5 ng/well), E2 (UbcH5c, 20 ng/well), and E3 (APC2/APC11, 80 ng/well) in an enzyme dilution buffer of 20 mM Tris, pH 7.5, and 10% glycerol. The plates were shaken for 10 min on a plate shaker, and the reaction was incubated at room temperature for 1 h. After completion of the reaction, the plates were washed three times with 200 µl of 1× PBS. One hundred microliters of antibody mixture containing 1:30,000 anti-Flag (Sigma Cat. No. F-3165) and 1:150,000 anti-mouse IgG-HRP (Jackson Immunoresearch Cat. No. 115-035-146) in 1× PBS with 0.25% BSA was added to each well and incubated at room temperature for 1 h. Plates were washed with 200 µl/well of 1× PBS three times. For luminescence activity detection, 100 µl of Luminol substrate (Pierce Cat. No. 37070) was added to each well, and the plates were read using the luminometer (BMG Fluostar Optima).

We evaluated the assay quality in 96-well plates for the preceding conditions. Z' factor is a screening window coefficient which reflects both the assay signal dynamic range and the data variation associated with the signal measurements. It is a simple statistical characteristic to measure the overall quality of each HTS assay. A higher Z' (>0.5) indicates that the assay exhibits very little data variation and has a broad dynamic range. As shown in Fig. 3E, it is a sensitive assay with a signal-to-background ratio of 26-fold. The Z' factor of 0.77 indicates that this assay is suitable for high-throughput screening of a large chemical compound library (Zhang et al., 1999).

Identification of APC2/APC11 Inhibitors from HTS

HTS against APC has been completed at Rigel with a hit rate of 0.8%. The screen successfully identified novel small molecule compounds that potently inhibit APC2/APC11 ligase activity (Fig. 4A). The potency of the inhibitors (IC_{50}) was determined by compound dosing with concentrations ranging from 0.009–20 μM. Compounds R545 and R516 showed dose-dependent inhibition of APC2/APC11 activity in the plate assay, with IC_{50}s of 0.5 μM and 1.7 μM, respectively.

This "ELISA" plate–based assay is sensitive and relatively easy to perform; however, just as in other screenings, there may be false-positive hits. To

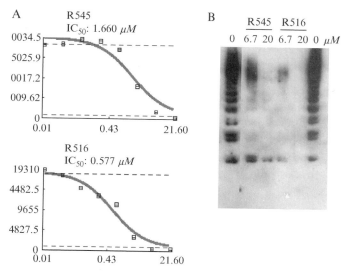

FIG. 4. Identification of APC2/APC11 E3 ligase inhibitors. (A) *In vitro* compound screening was performed in 96-well Nickel-coated plates. IC_{50}s for the compounds were determined by eight-point dose ranging from 0.009–20 μM. (B) Activity of the compounds was confirmed by the solution-based ligase assay using SDS-PAGE and Western blot analysis.

confirm a compound's inhibitory activity, we tested it in a solution-based *in vitro* ubiquitination assay using SDS-PAGE and Western blot analysis. This was done under the same conditions as those used for the plate assay in a total volume of 100 μl/reaction: 10 μl of compound in 100% DMSO was added to 80 μl of reaction buffer (62.5 mM Tris, pH 7.5, 6.25 mM MgCl2, 1.0 mM DTT, and 5 μM ATP) containing 100 ng of Flag-ubiquitin. The reaction was initiated by adding 10 μl of E1 (5 ng), E2 (UbcH5c, 20 ng), and E3 (APC2/APC11, 80 ng) in an enzyme dilution buffer of 20 mM Tris pH 7.5 and 10% glycerol. After a 1-h incubation at room temperature, the reaction was stopped by the addition of 33 μl of 4× SDS-PAGE sample buffer and boiled for 5 min; 20 μl of each sample was run on SDS-PAGE and Western blotted using an anti-Flag-HRP–conjugated antibody. Inhibition of APC2/APC11 activity was confirmed by a decrease in the polyubiquitination signal (i.e., the high molecular weight Flag-ubiquitin ladder/smear by Western blot) as shown in Fig. 4B. Although this assay is not as accurate or high throughput as the plate assay in determining an inhibitor's IC$_{50}$, it eliminates possible plate-binding effects (such as Ni–plate binding inhibition) and confirms a compound's activity in inhibiting APC2/APC11. This solution-based assay has been used to filter any false-positive hits resulting from the screen.

The APC ligase assay involves multiple enzymatic cascades of E1 ubiquitin charging, E2 (UbcH5c) ubiquitin conjugating, and E3 (APC2/APC11) ubiquitin polymerizing. Thus, APC2/APC11 inhibitors cannot be differentiated from inhibitors of E1 or UbcH5c on the basis of this screen. Secondary screens for hit deconvolution, such as an E2 thioester assay, in which only E1 and E2 are included in the reaction for E2-ubiquitin conjugation, can be used to filter out E1 and E2 inhibitors, including thiol-reactive compounds. Moreover, counter screens of other RING finger E3 ligases, such as the SCF complex, will allow for the identification of APC selective inhibitors.

Concluding Remarks

We have described detailed protocols of high-throughput screening for inhibitors of the ubiquitin ligase APC, including the recombinant expression and purification of individual enzymes and proteins used in the assay, as well as assay development and optimization. The assay has been used to identify small molecules that inhibit APC ligase activity.

Because the described *in vitro* assay contains only the ligase core subunit of APC2/APC11, secondary assays using native APC complex should be used to further evaluate hits resulting from the screen. Toward this end, endogenous APC complex can be immunoprecipitated from mitotic cell extracts with an anti-APC antibody, and then applied, in the presence of ATP, exogenous E1, E2, and Flag-ubiquitin, to the solution-based *in vitro* assay in lieu of recombinant APC2/APC11. Protocols for immunoprecipitation of APC from cell lysate and measurement of its activity have been previously described (King *et al.*, 1995; Yu *et al.*, 1996). The on-target effects of APC inhibitors can then be further examined by measuring substrate (i.e., cyclin B, securin) levels in cells after compound treatment.

Acknowledgments

We are grateful to Drs. Brian Wong and Tarikere L. Gururaja for critical reading of the manuscript, and to Ms. Linette Fung for her help in formatting the manuscript.

References

Chen, A., Kleiman, F. E., Manley, J. L., Ouchi, T., and Pan, Z.-Q. (2002). Autoubiquitination of the BRAC1.BARD1 RING ubiquitin ligase. *J. Biol. Chem.* **227,** 22085–22092.
Fang, S., Jensen, J. P., Ludwig, R. L., Vousden, K. H., and Weissman, A. M. (2000). Mdm2 is a RING finger-dependent ubiquitin protein ligase for itself and p53. *J. Biol. Chem.* **275,** 8945–8951.
Gmachl, M., Gieffers, C., Podtelejnikpv, A. V., Mann, M., and Peters, J.-M. (2000). The RING-H2 finger protein APC11 and the E2 enzyme UBC4 are sufficient to ubiquitinate substrates of the anaphase-promoting complex. *Proc. Natl. Acad. Sci. USA* **97,** 8973–8978.
Harper, J. W., Burton, J. L., and Solomon, M. J. (2002). The anaphase-promoting complex: It's not just for mitosis any more. *Genes Dev.* **16,** 2179–2206.
Hong, C. A., Swearingen, E., Mallari, R., Gao, X., Cao, Z., North, A., Young, S. W., and Huang, S.-G. (2003). Development of a high throughput time-resolved fluorescence resonance energy transfer assay for TRAF6 ubiquitin polymerization. *Assay Drug Dev. Technol.* **1,** 175–180.
Jackson, P. K. (2004). Linking tumor suppression, DNA damage and the anaphase-promoting complex. *Trends Cell Biol.* **14,** 331–334.
Jackson, P. K., Eldridge, A. G., Freed, E., Furstenthal, L., Hsu, J. Y., Kaiser, B. K., and Reimann, J. D. (2000). The lore of the RINGs: Substrate recognition and catalysis by ubiquitin ligases. *Trends Cell Biol.* **10,** 429–439.
Joazeiro, C. A., and Weissman, A. M. (2000). RING finger proteins: Mediators of ubiquitin ligase activity. *Cell* **102,** 549–552.
King, R. W., Peters, J. M., Tugendreich, S., Rolfe, M., Hieter, P., and Kirschner, M. W. (1995). A 20S complex containing CDC27 and CDC16 catalyzes the mitosis-specific conjugation of ubiquitin to cyclin B. *Cell* **81,** 279–288.
Murray, A. W. (2004). Recycling the cell cycle: Cyclins revisited. *Cell* **116,** 221–234.
Ohta, T., Michel, J. J., Schottelius, A. J., and Xiong, Y. (1999). ROC1, a homolog of APC11, represents a family of cullin partners with an associated ubiquitin ligase activity. *Mol. Cell* **3,** 535–541.
Peters, J.-M. (2002). The anaphase-promoting complex: Proteolysis in mitosis and beyond. *Mol. Cell* **9,** 931–943.

Pray, T. R., Parlati, F., Huang, J., Wong, B. R., Payan, D. G., Bennett, M. K., Issakani, S. D., Molineaux, S., and Semo, S. D. (2002). Cell cycle regulatory E3 ubiquitin ligases as anticancer targets. *Drug Resist. Updat.* **5,** 249–258.

Reimann, J. D., Freed, E., Hsu, J. Y., Kramer, E. R., Peters, J. M., and Jackson, P. K. (2001). Emi1 is a mitotic regulator that interacts with Cdc20 and inhibits the anaphase-promoting complex. *Cell* **105,** 645–655.

Tang, Z., Li, B., Bharadwaj, R., Zhu, H., Ozkan, E., Hakala, K., Deisenhofer, J., and Yu, H. (2001). APC2 cullin protein and APC11 RING protein comprise the minimal ligase module of the anaphase-promoting complex. *Mol. Biol. Cell* **12,** 3839–3851.

Wong, B. R., Parlati, F., Qu, K., Demo, S., Pray, T., Huang, J., Payan, D. G., and Bennett, M. K. (2003). Drug discovery in the ubiquitin regulatory pathway. *Drug Discov. Today* **8,** 746–754.

Yabuki, N., Watanabe, S., Kudoh, T., Nihira, S., and Miyamoto, C. (1999). Application of homogeneous time-resolved fluorescence (HTRF^{TM}) to monitor poly-ubiquitination of wild type p53. *Comb. Chem. High Throughput Screen.* **2,** 279–287.

Yu, H., Peters, J.-M., King, R. W., Page, A., Hieter, P., and Kirschner, M. W. (1998). Identification of a cullin homology region in a subunit of the anaphase-promoting complex. *Science* **279,** 1219–1223.

Yu, H., King, R. W., Peters, J.-M., and Kirschner, M. W. (1996). Identification of a novel ubiquitin-conjugating enzyme involved in mitotic cyclin degradation. *Curr. Biol.* **6,** 455–466.

Zhang, J. H., Chung, T. D., and Oldenburg, K. R. (1999). A simple statistical parameter for use in evaluation and validation of high throughput screening assays. *J. Biomol. Screen.* **4,** 67–73.

Section VIII

Generally Applicable Technologies

[50] The Split-Ubiquitin Sensor: Measuring Interactions and Conformational Alterations of Proteins *In Vivo*

By Christoph Reichel and Nils Johnsson

Abstract

The split-ubiquitin technique can monitor alterations in the conformation of proteins and can be used to detect the interaction between two proteins in living cells. The technique is based on unique features of ubiquitin, the enzymes of the ubiquitin pathway, and the reconstitution of a native-like ubiquitin from its N- and C-terminal fragments. By exploiting the reassociation of a protein from its defined fragments to monitor protein interactions, the split-ubiquitin assay served as the prototype of a still growing number of split-protein sensors to analyze protein function within living cells.

Introduction

Protein–protein interactions are at the center of almost all biological processes, and the identification and in-depth characterization of such interactions continues to be one of the main challenges in functional genomics and proteomics. The introduction of the yeast two-hybrid system by Fields and Song (1989) was a milestone in the analysis of protein interaction in living cells. However, a major limitation of this system is that the interaction to be studied has to be reproduced in the nucleus of the yeast. To overcome this restriction, the split-ubiquitin technique (split-Ub) was developed as a tool to detect and monitor a protein–protein interaction in a time-resolved manner at different locations in the living cell.

The Concept of the Split-Ubiquitin Sensor

Ubiquitin, a compact protein of only 76 residues, can be artificially separated into an N-terminal peptide of 35 (N_{ub}) and a C-terminal peptide of 41 amino acid residues (C_{ub}). Both ubiquitin halves will reassociate into a native-like ubiquitin once coexpressed in the same cell (Johnsson and Varshavsky, 1994). Proteins that are attached to the C-terminus of the intact ubiquitin are rapidly cleaved off *in vivo* by the ubiquitin-specific proteases (UBP). Because the UBPs identify features of the folded ubiquitin, only the reassembled N_{ub} and C_{ub}, but not the separated halves, are recognized as substrates. Consequently, the UBPs only effectively cleave off a reporter protein from the C-terminus of C_{ub}, once C_{ub} is in a complex with N_{ub} (Johnsson and Varshavsky, 1994) (Fig. 1). Importantly, the reassociation of N_{ub} and C_{ub} and the subsequent cleavage of a reporter protein (R) at the C-terminus of C_{ub} by the UBPs is slow and inefficient. Connecting N_{ub} and C_{ub} to interacting proteins forces them into close proximity and thereby accelerates the formation of the native-like ubiquitin. The resulting cleavage of the reporter provides an indirect measure of the interaction-induced proximity between N_{ub} and C_{ub} (Johnsson and Varshavsky, 1994).

In some applications, the spontaneous reconstitution of Ub might interfere with the robust detection of the interaction-driven N_{ub}/C_{ub} reassembly (Johnsson and Varshavsky, 1994). To suppress the spontaneous reassembly of the Ub halves, without eliminating the ability to reassociate when forced into close proximity, the isoleucines at position 3 and 13 of N_{ub} were replaced by amino acids bearing shorter side chains. These mutations very probably destabilize the fold of N_{ub} and, therefore, reduce

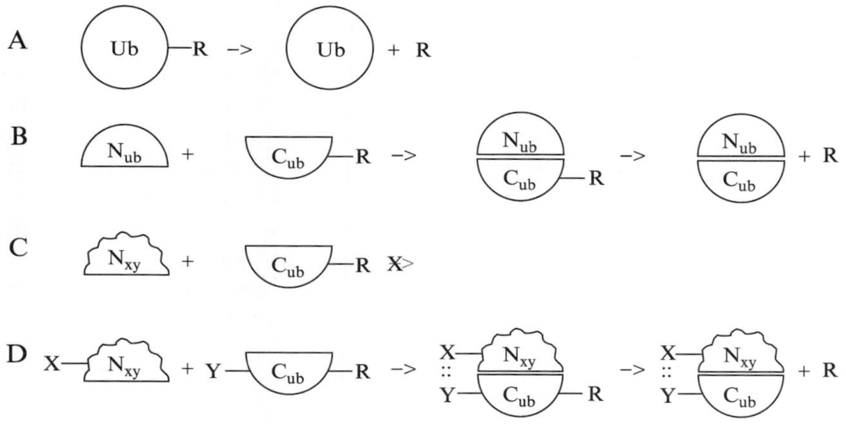

FIG. 1. Split-ubiquitin as a proximity sensor *in vivo*. (A) A newly formed ubiquitin (Ub) that was linked to a reporter protein (R) is recognized by the ubiquitin-specific proteases and the reporter is cleaved off, yielding the free reporter. (B) When the N- and C-terminal (N_{ub} and C_{ub}) subdomains of Ub are coexpressed as separate fragments, with C_{ub} still linked to the reporter, significant, albeit not complete, *in vivo* reconstitution of a native-like Ub takes place as indicated by the release of the reporter from the C_{ub}. (C) The *in vivo* reconstitution of Ub from its coexpressed fragments does not occur with mutants of N_{ub} (N_{xy}; indicated by the wavy line) that bear single- or double-residue replacements at position 3 and/or 13. (D) N_{xy} supports reconstitution of Ub if the two Ub fragments are linked to proteins X and Y that interact *in vivo*.

the reassembly rate between the two ubiquitin halves (Johnsson and Varshavsky, 1994). A collection of 12 N_{ub}-mutants with different affinities to C_{ub} was constructed in which N_{ub} (N_{ii}) bearing the two wild-type isoleucines at position 3 and 13 had the highest and N_{gg} with glycines instead of the two isoleucines had the lowest affinity to C_{ub} (Raquet et al., 2001). These mutations suppress the spontaneous reassociation to different degrees, and the cleavage of the reporter from C_{ub} becomes strictly dependent on the interaction between the two proteins that are linked to N_{ub} and to C_{ub} (Johnsson and Varshavsky, 1994) (Fig. 1).

The availability of N_{ub}s with different affinities to C_{ub} provides a certain flexibility toward the type of protein interaction that can be investigated with this assay. N_{ub}s with a low affinity to C_{ub} were especially used in monitoring the interactions of membrane-associated proteins and in detecting changes in the conformation of proteins (Raquet et al., 2001; Stagljar et al., 1998; Wittke et al., 1999).

Reporters for the Split-Ub Reconstitution

Any protein that is attached to the C-terminus of ubiquitin will be cleaved off by the UBPs. Once a measurable change in the property of the protein is associated with its release from the C-terminus of C_{ub}, the protein qualifies as a reporter for the split-Ub assay. Although the split-Ub system should work in all eukaryotic cells that contain ubiquitin and its enzymes, most of the reported applications were performed in yeast cells. This is reflected by the predominance of yeast-specific reporters that convert the interaction between N_{ub} and C_{ub} coupled proteins into a growth advantage or disadvantage of the yeast cells expressing this pair of fusion proteins. An exception is the use of the split-Ub system to detect the interaction between DNA binding proteins in mammalian cells (Rojo-Niersbach et al., 2000). Because the different reporter systems that were developed are not redundant but enable different applications of the split-Ub system, a short survey is given in the following.

Molecular Weight Reporter

When measuring the interaction between the two proteins X and Y, the most direct readout for this interaction in the split-Ub assay is the change in molecular weight of the Y-C_{ub}-R fusion and the simultaneous appearance of the cleaved R. On N_{ub}/C_{ub} reassociation, the reporter is detached, thereby changing the apparent molecular weight of Y-C_{ub}-R during SDS–polyacrylamide gel electrophoresis. Antibodies directed against either the protein Y or the reporter R can be used to monitor the N_{ub}-X/Y-C_{ub}-R reassociation after cell extraction and immunoblotting. Both antibodies detect the cleaved and the uncleaved fraction of the C_{ub} fusion and thereby allow to quantify of the extent of the N_{ub}/C_{ub} reassociation. The possibility to monitor the kinetics of cleavage by a pulse-chase experiment is a distinct advantage of this procedure (Johnsson and Varshavsky, 1994). This feature not only increases the sensitivity of the assay but also allows the measurement of subtle changes in the association rate of protein X and Y that may be induced by cellular signals or other changes in the environment of the proteins. We routinely use an epitope-tagged mouse dihydrofolate reductase (Dha) as R. Dha is monomeric and, as a mouse protein, is highly unlikely to interact with endogenous yeast proteins. The molecular weight readout is universal, because it depends exclusively on the presence of the UBPs.

Ura3p

Monitoring interactions by measuring changes in the molecular weight of the fusion proteins is time and labor intensive and, therefore, not suitable for the screening of new interaction partners. Ura3p is a yeast enzyme that catalyzes the decarboxylation of orotidine-5′-phosphate during the synthesis of uracil. Yeast cells lacking this activity cannot grow on media without uracil (SD-ura). Ura3p was first used as a reporter for the split-Ub assay to measure the interaction between a signal sequence bearing C_{ub}-Ura3p fusion protein and N_{ub}-Sec62p or N_{ub}-Sec61p, two members of the translocation apparatus of the endoplasmic reticulum (ER) membrane (Dünnwald *et al.*, 1999). The temporal interaction between the signal sequence bearing C_{ub}-Ura3p fusion and N_{ub}-Sec62p or N_{ub}-Sec61p during the translocation of the C_{ub}-fusion across the membrane of the ER resulted in the reassociation of N_{ub} and C_{ub}, the cleavage of Ura3p, and its release into the cytosol. N_{ub}-fusions not involved in translocation did not induce cleavage at the C-terminus of C_{ub}, thus enabling the unhindered escape of the signal sequence–bearing C_{ub}-Ura3 fusion into the lumen of the ER. Because Ura3p activity is normally required in the cytosol of the cell, interactions were easily scored by the growth of the N_{ub}/C_{ub} cotransformants on media lacking uracil (Dünnwald *et al.*, 1999; Wittke *et al.*, 2002). A prerequisite for the use of Ura3p is the rapid inactivation of the uncleaved Y-C_{ub}-Ura3p fusion protein in the absence of interaction. Beside translocation into the ER, this can be achieved by the import of the C_{ub}-Ura3p fusion protein into mitochondria, the peroxisome, or the rapid degradation of the uncleaved Y-C_{ub}-Ura3p fusion. Thus, depending on the mechanism of inactivation, the Ura3p reporter can be used to monitor a variety of other temporal protein interactions.

N-End–Rule Substrate RUra3p

To extend the use of Ura3p as reporter toward proteins that constantly reside in the cytosol, an arginine residue followed by a short peptide including acceptor lysines for ubiquitylation was inserted at the C_{ub}-Ura3p junction. After cleavage from C_{ub}, the arginine becomes the N-terminal amino acid of Ura3p (RUra3p) and, as a destabilizing residue according to the N-end–rule pathway of protein degradation, will induce the rapid destruction of RUra3p (Varshavsky, 1996; Wittke et al., 1999). Consequently, cells expressing a pair of interacting N_{ub}/C_{ub} fusion proteins will not grow on media lacking uracil, whereas cells expressing a pair of noninteracting proteins are uracil auxotrophs (Wittke et al., 1999) (Fig. 2).

Although the nongrowth of cells can be used as a sensitive readout for the interaction assays of selected pairs of proteins, it is problematic as a signal for split-Ub–based interaction screens. 5-fluoro-orotic acid (5-FOA) is a drug that is converted to the toxic compound 5-fluorouracil by Ura3p (Boeke et al., 1984). Because the interaction of the N_{ub}/C_{ub}-coupled proteins leads to the rapid degradation of the RUra3p moiety, supplying 5-FOA to the growth medium will revert the signal for a protein interaction in the RUra3p-based split–Ub growth assay. In cells expressing interacting N_{ub} and C_{ub} fusions, 5-FOA will no longer be converted to the toxic 5-fluorouracil because of the rapid degradation of RUra3p. These cells will survive on medium containing 5-FOA and uracil (5-FOA^r), whereas cells expressing a pair of noninteracting N_{ub}/C_{ub} fusion proteins will not grow (5-FOA^s).

By applying RUra3p as a reporter in yeast cells, one has to keep in mind that Ura3p is an endogenous yeast enzyme. N_{ub}-fusions of yeast proteins interacting with Ura3p may mimic the interaction with the protein Y of the Y-C_{ub}-RUra3p. Because these interactions are independent of Y, the origin of such "false positives" can easily be deduced. In addition, Ura3p has the capacity to form homodimers, which may affect the function of

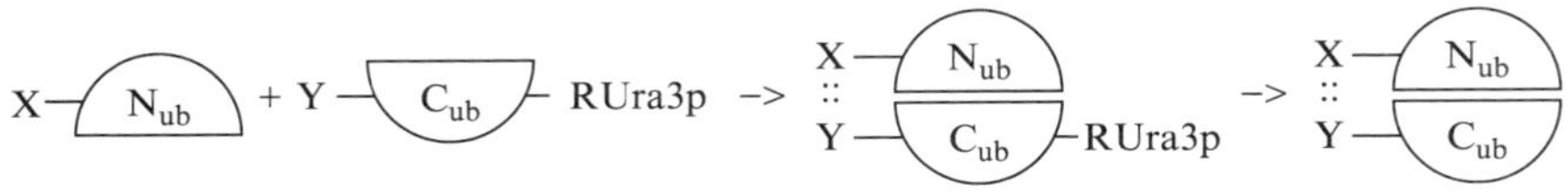

FIG. 2. The RUra3p reporter. N_{ub} and C_{ub} are linked to the interacting proteins X and Y. Reconstitution of Ub from its N- and C-terminal fragments leads to the release of RUra3p from C_{ub}. The newly displayed N-terminal arginine residue of RUra3p is recognized by the enzymes of the N-end–rule pathway, and RUra3p is rapidly degraded. Efficient Ub reconstitution leads to cells that are phenotypically ura- and 5-FOA resistant, whereas cells coexpressing N_{ub} and C_{ub} fusion proteins that do not interact remain ura+ and 5-FOA sensitive.

protein Y in the Y-C_{ub}-RUra3p fusion (Miller et al., 2000). However, on the basis of our experience with different Y-C_{ub}-RUra3p fusion proteins, the dimerization of Ura3p was not found to be a general problem. Examples for the application of the RUra3p reporter extend from nuclear proteins to membrane proteins of the ER, the peroxisome, or the mitochondrion (Ansari et al., 2002; Deslandes et al., 2003; Eckert and Johnsson, 2003; Gromöller and Lehming, 2000; Kim et al., 2002; Laser et al., 2000; Wittke et al., 1999, 2000).

The N-end rule–based degradation after cleavage from C_{ub} was also applied to other reporter proteins. One example is provided by the green fluorescent protein (GFP), where N_{ub}/C_{ub} reassociation leads to cleavage of RGFP from C_{ub} and its rapid degradation. Interaction of the N_{ub}/C_{ub} coupled proteins will, therefore, result in nonfluorescing cells, whereas cells expressing noninteracting N_{ub}/C_{ub} fusion proteins will stay fluorescent (Laser et al., 2000). Useful applications include interactions between proteins that are compartment specific. For example, Y-C_{ub}-RGFP fusion proteins that interact only in the nucleus but not in the cytosol will keep the nucleus dark but leave the cytosol green fluorescent, thereby reflecting the nucleus-specific interaction between the two respective fusion proteins.

Transcription-Based Readouts

In this configuration of the split-Ub assay, an artificial transcription factor (PLV, a fusion of Protein A, LexA, and VP16) was attached to the C-terminus of C_{ub} (Stagljar et al., 1998). On reassociation of N_{ub} and C_{ub}, PLV is cleaved off and will diffuse freely into the nucleus to activate transcription from promoters containing the matching lexA sites. Because the lexA sites control the expression of *HIS3* and β-galactosidase, interaction between N_{ub} and C_{ub} coupled fusion proteins will enable the cells to grow on medium lacking histidine and to convert colorless substrates of β-galactosidase into visible dyes. An important prerequisite for using the PLV reporter is that the uncleaved C_{ub} fusion protein does not move into the nucleus. Applications of the PLV reporter are, therefore, confined to C_{ub} fusion proteins that are strictly nonnuclear-like membrane proteins of the secretory pathway, the mitochondrion or the peroxisome. The transcriptional readout was successfully applied to detect and screen for new interaction partners of yeast and mammalian membrane proteins and was recently extensively reviewed (Fetchko and Stagljar, 2004, and references herein; Ludewig et al., 2003; Massaad and Herscovics; 2001; Obrdlik et al., 2004; Thaminy et al., 2003; Wang et al., 2003, 2004).

Measuring Interactions with the C_{ub}-RUra3p Reporter

To successfully monitor or detect new interactions with the RUra3p-based–split-Ub system we recommend the following steps:

1. Adjusting the concentration of the Y-C_{ub}-RUra3 fusion protein and selecting the appropriate N_{ub} mutant.
2. Introducing the N_{ub} library into the Y-C_{ub}-RUra3p expressing strain and selecting for interactors on plates containing 5-FOA.
3. Testing the 5-FOA resistance of the selected candidate strains for N_{ub} fusion dependency.
4. Isolating the N_{ub} fusion plasmids and testing their interactions against the original Y-C_{ub}-RUra3p and an unrelated Y*-C_{ub}-RUra3p.
5. Sequencing of the N_{ub} fusion candidates that preferentially interact with Y-C_{ub}-RUra3p.
6. Performing additional interaction tests including the split-Ub–based competition assay with the full-length X and Y proteins.

Adjusting the Concentration of the Y-C_{ub}-RUra3 Fusion Protein and Selecting the Appropriate N_{ub} Mutant

To perform a successful interaction assay, it is critical to first establish which N_{ub} mutant to use and at which level to express the C_{ub} fusion. When using the RUra3p reporter, we routinely use different promoters for the N_{ub} and C_{ub} fusion proteins. N_{ub} fusions can be expressed from the leaky P_{CUP1}-promoter usually under noninducing conditions or from the strong P_{ADH1}-promoter. The P_{MET17}- or the P_{CUP1}-promoters were alternately used to drive the expression of the Y-C_{ub}-RUra3p fusions. The level of expression can be adjusted by adding different concentrations of methionine or copper sulfate to the medium. Y-C_{ub}-RUra3p expression is tested by means of complementation of Ura3p deficiency on appropriate selective medium lacking uracil. The transformed strain should be strongly ura+ and 5-FOA^s. Y-C_{ub}-RUra3p constructs with weak or absent Ura3p activity are not suitable for interaction assays. Re-cloning of the Y-C_{ub}-RUra3p by changing fusion junctions or expressing subdomains of Y may rescue the Ura3p activity. Switching to high copy–number plasmids (2 μ) is an alternate strategy. The choice of the right mutant of N_{ub} depends on the type of interaction to be measured. For interaction assays, the wild-type N_{ub} and the mutants of N_{ub} carrying the alanine or the glycine in position 13 (N_{ia}, N_{ig}) are commonly used. Because a positive interaction signal requires the efficient N_{ub}/C_{ub} reassociation and subsequently the complete degradation of RUra3p, the wild-type N_{ub} has been used to successfully measure the interactions between cytosolic proteins, between membrane proteins and cytosolic proteins, and between membrane proteins of the peroxisome (Deslandes et al., 2003; Eckert and Johnsson, 2003; Kim et al., 2002; Laser et al., 2000). Measuring the interactions between membrane proteins of the ER always requires N_{ia} or even N_{ig} fusion proteins to obtain the required specificity (Stagljar et al., 1998; Wittke et al., 1999). Despite these guidelines, it is always recommended to first test the Y-C_{ub}-RUra3p fusion against different N_{ub}-fusion proteins that are known not to interact and that possess different cellular localizations. The most sensitive, yet still specific, conditions are achieved by using the lowest possible expression level of the Y-C_{ub}-RUra3p fusion and the N_{ub} mutant with the highest intrinsic affinity to C_{ub} that does not induce an unspecific interaction signal.

For the rapid degradation of the RUra3p reporter after the cleavage from C_{ub}, it is preferable to link the C_{ub}-RUra3p module to the C-terminus of the protein (i.e., Y-C_{ub}-RUra3). N_{ub} can be linked to either the N- or the C-terminus of the protein of interest to yield N_{ub}-X or X-N_{ub}. We have previously observed that linking N_{ub} to the N-terminus of a protein is the more sensitive configuration for the split-Ub assay (Eckert and Johnsson, 2003). A further important prerequisite for successfully screening interaction partners of a membrane protein is a reasonable prediction about the topology of the protein Y. The attachment site for the C_{ub}-RUra3p module has to point into the cytosol of the cell.

Protocol: Adjusting the Parameters for a Split-Ub Screen

- Transform into the yeast JD53 (Dohmen et al., 1995) the Y-C_{ub}-*RURA3* construct (i.e., coding for a membrane protein of the secretory pathway) under the control of the P_{MET17}-promoter together with N_{ii}-, N_{ia}-, N_{ig}-*SEC62* (marker for the ER); N_{ii}-, N_{ia}-, N_{ig}-*SED5* (marker for the early Golgi); N_{ii}-, N_{ia}-, N_{ig}-*SNC1* (marker for the late Golgi-plasma membrane); N_{ii}-, N_{ia}-, N_{ig}-*SSO1* (marker for the plasma membrane); N_{ii}-, N_{ia}-, N_{ig}-*GUK1* (marker for the cytosol). Provided that potential or known interaction partners of Y are available, they should be included as positive controls as N_{ii}-, N_{ia}-, N_{ig}-fusions. The expression of the N_{ub} fusions should be controlled by a different promoter than the P_{MET17}; for example, P_{CUP1}.
- Pick and restreak colonies on selective media.
- Inoculate in 2 ml of selective media. Let grow at 30° to an OD_{600} of 1.
- Spot 4 μl of each culture and dilutions of 1/10, 1/100, and 1/1000 on selective media that lack uracil or contain 5-FOA and uracil and in addition different concentrations of methionine (in steps from 0–10 mM).

- Let incubate for up to 5 days at 30°.
- Inspect growth and select the right concentrations of methionine and the right N_{ub} mutant for the interaction assay and the screen.

Introducing the N_{ub} Library into the Y-C_{ub}-RUra3p–Expressing Strain and Selecting for Interactors on Plates Containing 5-FOA

The introduction of plasmid DNA into yeast cells is based on the lithium acetate (LiAc) transformation protocol (Gietz and Schiestl, 1991; Gietz et al., 1992; Schiestl and Gietz, 1989). This method yields roughly 10^5 transformants/μg plasmid DNA. The efficiency drops by one order of magnitude when cotransforming two plasmids simultaneously. To increase transformation efficiency for library screens, a sequential transformation of the Y-C_{ub}-RURA3 construct and the library plasmids is, therefore, recommended. The exact amount of library plasmid DNA to be used in a transformation should be titrated for each library to obtain the optimal transformation efficiency. Raising the DNA concentrations might lead to problems in data interpretation, because many 5-FOA–resistant colonies will contain more than one N_{ub}-X_n–containing plasmid. We routinely use 4.5-μg library plasmid DNA per 0.5–1×10^9 cells.

Protocol: Genome-Wide Interaction Screen

- Inoculate a culture from a fresh plate with yeast containing the Y-C_{ub}-RURA3 construct, selecting for the plasmid (we prefer *HIS3* as selectable marker of the Y-C_{ub}-RURA3 plasmid) and Ura3p activity (SD-his-ura).
- To obtain a mid-log phase culture on the day of transformation, inoculate fresh medium with the overnight culture at OD_{600} ~0.2–0.25 and incubate at 30° with shaking to a final OD_{600} of ~0.6–0.8. This last step may be performed in YPD medium, but should be kept to 2–3 h to reduce the risk of Ura3p activity loss.
- Determine exact cell density before transformation. One OD_{600} corresponds roughly to 2×10^7 cells/ml. Plate 0.5–1×10^9 cells per large plate (22.5 cm × 22.5 cm). This corresponds to 0.5–1×10^6 cotransformants per plate in an experiment using the sequential transformation protocol.
- Prepare sample DNA for library transformation: per large plate, pipette 4.5 μg of library DNA, 100 μl carrier DNA (2 μg/μl), 0.5 ml cells (0.5–1×10^9) and 3 ml PEG/LiAc/TE solution into 15-ml sterile tubes; mix vigorously.
- Incubate for 30 min at 30° with gentle agitation.

- Heat shock samples for 20 min in 42° water bath; invert tubes several times to keep cells in suspension.
- Let samples adjust to room temperature.
- Adjust volume of each transformation to 5 ml with TE and plate transformation mixtures onto large SD-his-leu+5-FOA plates. Spreading of yeast cells is conveniently performed using 10–20 sterile glass beads with 5-mm diameter.
- Alternately, spin down transformation mix and resuspend cells carefully in 5 ml of SD-his-leu containing the tested concentrations of methionine and copper for the proper expression of the fusion proteins. Incubate at 30° for 1 h, spin down, and resuspend in 5 ml of TE. Spread cells as described on SD-his-ura + 5-FOA.
- Determine the transformation efficiency by plating appropriate dilutions of the transformation mixtures onto small plates, selecting for the presence of the two plasmids (i.e., SD-his-leu; 1:100 and 1:2000 dilutions work well).
- Incubate plates at 30° (it will take up to 3 days for the first colonies, but more colonies may appear during the course of the next 2 weeks).

Methods

YEAST STRAIN. JD53: *MATa his3-Δ200 leu2-3,112 lys2-801 trp1-Δ63 ura3-52*

PLASMIDS. Y-C_{ub}-RUra3p (Fig. 3): Several yeast shuttle plasmids were adapted to the split-Ub system. The plasmid carrying the *Y-C_{ub}-RURA3* fusion usually harbors a *HIS3* marker gene to allow selection for the plasmid in yeast. Stable, low copy–number maintenance of the plasmid is provided by a CEN locus. The plasmid contains an antibiotic resistance gene for the selection against kanamycin in *E. coli*. This plasmid should provide a different resistance than the N_{ub}-X_n plasmid to be able to discriminate between the plasmids when rescuing them from yeast.

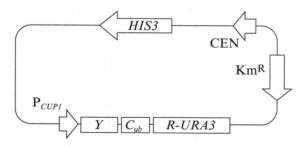

FIG. 3. Y-C_{ub}-RUra3p expressing shuttle vector. Km^R, kanamycin resistance marker; CEN locus for low copy number maintenance; P_{CUP1}, copper inducible promoter; HIS3, auxotrophic marker for selection in yeast.

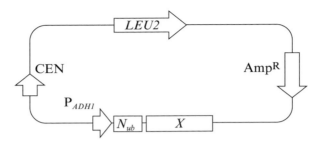

FIG. 4. N_{ub}-X–expressing shuttle vector. Amp^R, ampicillin resistance gene (β-lactamase); CEN, CEN locus for low copy number maintenance; P_{ADH1}, ADH1 promoter; LEU2, auxotrophic marker for selection in yeast.

N_{ub}-X (Fig. 4): The plasmid expressing a N_{ub}-X fusion protein may carry the *LEU2* gene and a CEN locus for low-copy maintenance in the yeast JD53. For propagation in *E. coli*, the plasmid contains the β-lactamase antibiotic resistance marker. By inserting a sequence between N_{ub} and X that codes for one of the common epitopes, the expression of the fusion protein can be monitored by Western blotting.

N_{ub} libraries: N_{ub} libraries were already constructed as translational fusions between N_{ii} or N_{ig} and various sources of genomic and cDNA (Laser *et al.*, 2000; Thaminy *et al.*, 2003; Wang *et al.*, 2003, 2004). For the construction of these libraries, polylinker sequences are inserted after the N_{ub} moiety; three individual plasmids are constructed with single nucleotide insertions between N_{ub} and the polylinker to reflect three different reading frames. The polylinker sequences are followed by three stop codons in the three different reading frames. The sources of DNA may be genomic DNA (e.g., from yeast) or cDNA from various tissues and organisms. The genomic DNA may be fragmented by partial digestion with multi-cutting restriction enzymes (e.g., *Tsp*509I) or by shearing followed by blunt end ligation into the vector. cDNA should be inserted using rare cutting enzymes (e.g., *Sal*I, *Not*I).

Materials for Transformation

LITHIUM ACETATE. 100 mM LiAc, 10 mM Tris-HCl, pH 9.5, 1 mM EDTA.

PEG/LiAc/TE. 40% PEG 3350, 100 mM LiAc, 10 mM Tris-HCl, pH 8, 1 mM EDTA

Preparation of Single-Stranded Carrier DNA (2 mg/ml)

- Dissolve 200 mg of salmon testes DNA (sodium salt) in 100 ml sterile TE buffer (10 mM Tris-HCl/1 mM EDTA, pH 8).

- Heat to approximately 50° while vigorously stirring.
- Aliquots are boiled 5–10 min, cooled in ice-water bath, and stored at −20°. Boiling and cooling may be repeated before transformation to increase transformation efficiency.

5-FOA Plates

- Prepare 2× concentrated Bacto Agar (40 g/l) in water and autoclave.
- Cool down to 50° in a water bath.
- Dissolve 0.5 g 5-FOA in appropriate 2× concentrated SD minimal medium (e.g., SD-leu-his); 5-FOA must not be heated to temperatures higher than 50°.
- Stir on heat plate set to 50° until FOA is dissolved; watch temperature closely.
- Filter sterilize.
- Pour 2× agar to 5-FOA mixture, mix and pour plates, pop bubbles with burner flame; 1000 ml is sufficient for three big plates (22.5 cm × 22.5 cm) or roughly 40 small plates (9 cm).

Testing the 5-FOA Resistance of the Selected Candidate Strains for N_{ub} Fusion Dependency

Before engaging in laborious downstream analysis of primary FOA-resistant clones; false-positives that will arise in every screen should be discarded to reduce effort and cost. Most false positives fall into two classes. The first class is composed of "non-Ura3p-plasmids" that either arise from mutations in the sequence of *RURA3* or from the unintended introduction of contaminating plasmids that contain the same auxotrophy marker as the Y-C_{ub}-*RURA3* plasmid. This first class of false positives gives rise to 5-FOA resistance independently of the expression of N_{ub}-X. Following a procedure of the classical two-hybrid screen, passive loss of the Nub-Xn plasmids under nonselective conditions is used to identify this class of false positives (Chien *et al.*, 1991). After passive loss of the N_{ub}-X_n plasmids, the yeasts are analyzed for their ability to support growth on minimal medium lacking uracil. Cells that do not express functional Y-C_{ub}-RUra3p and, therefore, fail to grow on medium lacking uracil are excluded from further analyses.

Avoiding Contaminating Plasmids. Contaminating plasmids can be a problem, especially for yeast laboratories that simultaneously work with different yeast plasmids and libraries. Two measures during the screen can reduce the source for this class of false positives. The first is to avoid laboratory space in which similar screens or work with yeast shuttle vectors are performed. The second is to use aerosol-free pipette tips when preparing DNA stocks to be used in library screens. Remember that you are actively selecting for the loss of the URA3 activity during the screen and that even traces of other standard yeast vectors that share the same auxotrophic marker gene can lead to this class of false positives.

Protocol: N_{ub}-X Dependency Analysis

- Inoculate all primary 5-FOA^r clones into 1 ml SD-his using 20 μl of cell suspension from fresh SD-his cultures and incubate for 3 days at 30°.
- Transfer 5 μl of each culture onto SD-his, SD-his-ura, and SD-leu plates and incubate for another 3 days.
- Homogenous growth of colonies is expected on SD-his plates, indicating that the Y-C_{ub}-$RURA3$ plasmid is present. Growth on SD-leu plates should be sparse, indicating efficient loss of the N_{ub}-X_n plasmids (Fig. 5). If growth on SD-leu is indistinguishable from the growth on SD-his, the experiment may have to be repeated with extended times in medium supporting the loss of the N_{ub}-X_n plasmid. However, even a slight reduction in cells that express N_{ub}-X_n will lead to the reversion of the interaction-dependent loss-of-URA3 phenotype. Thus, colonies that grow on SD-his-ura are indicative of a functional Y-C_{ub}-RUra3p plasmid and are likely to originate from primary 5-FOA^r clones harboring true interactors (Fig. 5). No growth on SD-his-ura despite a significant number of colonies on SD-his indicates a defective or absent Y-C_{ub}-RUra3p plasmid in the original 5-FOA^r clone.
- Discard clones that fail to give rise to cells that grow on SD-his-ura.
- Continue with those colonies that give rise to cells that are ura+ after the loss of N_{ub}-X_n.

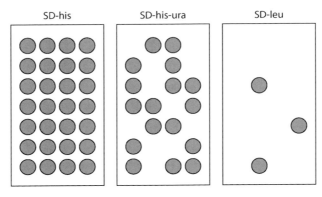

FIG. 5. N_{ub}-fusion dependency analysis. Primary 5-FOA–resistant clones that should express the original Y-C_{ub}-RUra3p and N_{ub}-X_n are cultured in nonselective media (SD-his) to allow loss of the N_{ub}-X_n plasmid. Phenotypes are then scored on plates selecting for the presence of the Y-C_{ub}-$RURA3$ plasmid (SD-his), for the activity of the RUra3p reporter (SD-his-ura), and for the presence of the N_{ub}-X_n plasmid (SD-leu). Growth on SD-his-ura indicates that the original 5-FOA resistance was dependent on the interaction between Y-C_{ub}-RUra3p and N_{ub}-X_n.

Isolating the N_{ub}-Fusion Plasmids and Testing Their Interactions Against the Original Y-C_{ub}-RUra3p and an Unrelated Y^*-C_{ub}-RUra3p

The second and more problematic class of false positives arises from N_{ub}-X_n fusions whose high local concentrations toward the Y-C_{ub}-RUra3p are not caused by direct interactions with Y. Such high local concentrations might, for example, be the result of an exceptionally strong expression level of X in combination with a relatively unspecific colocalization with Y.

N_{ub}-X_n plasmids are first isolated from primary 5-FOA^r clones and cotransformed with the original Y-C_{ub}-*RURA3* into JD53. The cells are tested for a FOA^r and an ura-phenotype as detailed under the *"Protocol: Adjusting the Parameters for a Split-Ub Screen."* Continue only with those N_{ub}-X_n plasmids that confer FOA^r and significantly reduce the growth of the cells on SD-ura. This already eliminates several classes of false-positives: (1) clones that are picked from background colonies, (2) contaminating *HIS3*-containing plasmids, (3) plasmids that were rescued from yeast containing more than one N_{ub}-X_n plasmid, and (4) N_{ub}-X_n fusions that eliminate the toxic action of 5-FOA by other means (transporters, pumps).

Those N_{ub}-X_n fusions that still interact with Y-C_{ub}-RUra3p in this assay can be tested for their interaction with an unrelated Y^*-C_{ub}-RUra3p. Again, care has to be taken for selecting the right Y^*-C_{ub}-RUra3p. An ideal candidate displays roughly the same expression level and cellular localization as Y-C_{ub}-RUra3p but should be involved in a completely different activity than Y. N_{ub}-X fusions that interact with Y-C_{ub}-RUra3p much better than with Y^*-C_{ub}-RUra3p are chosen as the first candidates for a further detailed characterization.

Protocol: Plasmid Isolation from Yeast

- N_{ub}-X_n plasmids are isolated from 5 ml yeast cultures grown in SD-leu medium for 1–2 days to high cell density.
- Yeast pellets are resuspended in Tris buffer (50 mM Tris-HCl, 10 mM EDTA, 10 μg/ml RNase A, pH 8), supplemented with zymolyase-100T (Sigma-Aldrich, USA) at a final concentration of 0.4 mg/ml, and incubated for 60 min at 37° to digest cell walls.
- Plasmid isolation is continued by adding two volumes of alkaline lysis buffer following the standard alkaline lysis protocol for the isolation of plasmid DNA from *E. coli* (Birnboim and Doty, 1979). When high throughput is required, vacuum-based plasmid preparation kits from various suppliers may be used instead. Follow manufacturer's instructions continuing with alkaline lysis buffer.
- Transform a 10% aliquot of the plasmid preparation into *E. coli* by standard procedure. If yields of plasmid preparations are low, which

may be the case when using CEN plasmids, plasmid DNA may have to be precipitated and resuspended in a smaller volume.

Sequencing of the N_{ub}-Fusion Candidates That Preferentially Interact with Y-C_{ub}-RUra3p

Revealing the identity of X_n requires the isolation of the N_{ub}-X_n plasmid followed by sequencing of the insert (see "Isolating the N_{ub}-Fusion Plasmids and Testing Their Interactions Against the Original Y-C_{ub}-RUra3p and an Unrelated Y*-C_{ub}-RUra3p"). Primers that anneal at the end of the N_{ub} sequence will not only allow the identification of X but will also clarify whether X is in frame with N_{ub}.

Performing Additional Interaction Tests Including the Split-Ub–Based Competition Assay with the Full-Length X and Y Proteins

Different methods exist to verify an interaction that was originally defined by means of a split-Ub–based interaction screen. Coimmunoprecipitation, pull-downs, or fluorescence energy transfer assays (to name a few) are independent of the split-Ub technique and can be used to confirm the interaction. All these techniques have their disadvantages, and a negative outcome does not necessarily prove this interaction as nonexisting. Colocalization of or a functional connection between X and Y will further increase the confidence in the split-Ub–based interaction result.

A test that confirms the specificity of the found interaction and still operates within the realm of the split-Ub technique is the successful competition between the unmodified X or Y and the N_{ub}- or C_{ub}-modified X and Y for their corresponding binding partner (see "Reverse Split-Ubiquitin").

Further Applications

The application of the split-Ub sensor is not restricted to measuring the interactions between two proteins. The technique can also be used to detect conformational alterations in proteins, to determine the influence of mutations or compounds on a certain interaction, or to map the binding sites of a given complex by a split-Ub–based competition assay.

Reverse Split-Ubiquitin

Depending on the composition of the medium, interactions between two N_{ub}- and C_{ub}-coupled proteins can be scored with the RUra3p reporter as growth or nongrowth of the cells expressing the N_{ub}/C_{ub} fusion proteins. Cells that express a pair of interacting N_{ub}/C_{ub} fusions will not grow on medium lacking uracil. Conditions that interfere with this interaction will lead to less cleavage of the RUra3p reporter and consequently to the growth of the cells on medium lacking uracil. This feature of the assay has been used to test the specificity of the interaction between N_{ig}-Sec62p and Sec63-C_{ub}-RUra3p. Expressing native Sec62p from the strong P_{GAL1}-promoter displaces the N_{ig}-Sec62p from its complex with Sec63-C_{ub}-RUra3p. Less RUra3p is cleaved off, and the cells grow on galactose-containing medium lacking uracil (SG-ura) (Wittke et al., 1999). The same assay can further be used to map the sites of interaction in a given complex. Expressing different fragments of Sec62p under the P_{GAL1}-promoter and scoring the growth of the N_{ig}-Sec62p/Sec63-C_{ub}-RUra3–expressing cells on SG-ura revealed two distinct sites on Sec62p that are needed for the strong binding to Sec63p (Wittke et al., 2000).

A further variation of this assay should allow screening compound libraries for small molecules interfering with the interaction of a given N_{ub}-X/Y-C_{ub}-RUra3p pair.

Protocol: Competition Assay

- Transform cells expressing Y-C_{ub}-RUra3p with a N_{ub}-X construct bearing the N_{ub} mutant that has the lowest possible affinity to C_{ub} for monitoring this interaction. Adjust the promoters of Y-C_{ub}-$RURA3$ and N_{ub}-X to obtain the lowest possible expression levels that still give a robust interaction signal.
- Transform these cells with a plasmid carrying either Y or X (for testing the putative interactors of a screen Y is desirable) under the control of the strong P_{GAL1}-promoter. Use an unrelated protein under the control of the P_{GAL1}-promoter as a reference.
- Inoculate the triple transformants on minimal media containing either glucose or galactose as carbon source. Plate 10,000 cells on selective medium with and without uracil and either glucose or galactose.
- Count colonies on all plates after incubation for 3–7 days at 30°. Compare numbers from cells overexpressing the protein of interest with the number of cells expressing the reference protein. Make sure that both cell types grow equally well on the plates containing uracil.
- A significant increase in the number of surviving colonies on SG-ura of the cells expressing either X or Y from the P_{GAL1}-promoter confirms the specificity of the split-ubiquitin measured interaction.

Split-Ub–based Conformational Sensor

Changes in the conformation of proteins are frequently the basis for the modulation of their activities. Monitoring these changes in living cells is, therefore, similarly important for understanding their function as measuring their interactions with other proteins or ligands.

Structural analysis revealed for most proteins a fixed distance between the N-terminus and the C-terminus. Attaching N_{ub} and C_{ub} to the N- and C-terminus of the same polypeptide allows the measurement of intramolecular N_{ub} and C_{ub} reassociation by quantifying the ratio of cleaved to uncleaved fusion protein (Fig. 6). This ratio is defined by the affinity of N_{ub} to C_{ub}, and by the nature of the polypeptide, separating N_{ub} from C_{ub}. Two features of the inserted protein dominate the efficiency of the N_{ub}/C_{ub} reassociation: the position of the N- and C-termini relative to each other in the folded conformation and the rigidity of the structure. Mutations or conditions that favor the unfolded state, or that induce a conformation that will alter the time-averaged distance between the N- and the C-terminus of the protein, will influence the ratio of cleaved to uncleaved N_{ub}-Y-C_{ub}-R (Raquet et al., 2001).

The balance between the folded and the unfolded state of a protein is sensitively adjusted. Attaching N_{ub} and C_{ub} to the termini of a protein will, therefore, disturb this balance. To monitor alterations in the conformation of a protein, a N_{ub}/C_{ub}-pair has to be selected that does not disturb the balance between the folded and the unfolded state too much, yet still displays sufficient affinity to monitor changes in the average distance between the N- and C-termini of the protein (Raquet et al., 2001). The detection of both cleaved and uncleaved N_{ub}-Y-C_{ub}-Dha in a steady-state analysis is a good first indication. On the basis of experience with different proteins, a N_{vg}-Y-C_{ub}-Dha fusion protein provides a good starting point for the analysis (Raquet et al., 2001). Once a change in the ratio of cleaved to uncleaved fusion protein is measured during the experiment, different N_{ub} mutants can be used to maximize this effect and to increase the sensitivity of the assay. Calculating the ratio of cleaved to uncleaved fusion protein after cell extraction and immunoanalysis of the separated proteins is the

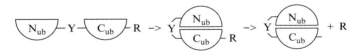

FIG. 6. The split-Ub conformational sensor. The protein Y is sandwiched between N_{ub} and C_{ub}. The efficiency of the reconstitution of Ub from N_{ub} and C_{ub} depends on the spatial arrangement of the N- and C-terminus in the structure of Y and the rigidity of the structure. Mutations or conditions that alter the conformational properties of Y and that change the time average distance between its N- and C-terminus will influence the efficiency of the N_{ub} and C_{ub} reassociation and thereby the ratio of cleaved to uncleaved N_{ub}-Y-C_{ub}-R fusion protein.

most robust parameter for this assay. A molecular weight readout of the split-Ub assay is, therefore, recommended. However, it is important to control that the sum of cleaved and uncleaved fusion protein stays roughly constant under the different conditions. Otherwise, it is difficult to exclude that changes in this ratio might result from changes in the degradation rates of the uncleaved fusion. Measuring the interaction between N_{ub} (i.e., N_{vg})-Y and Y-C_{ub}-R can exclude that the observed changes in intramolecular N_{ub}/C_{ub}-reassociation are not caused by an altered aggregation or dissociation behavior of the protein Y. Examples for the successful application of the split-Ub conformational sensor are the detection of the mutation-induced changes in the conformation of the proteins p53, Sec62p, guanylate kinase, or Fk506 binding protein, and monitoring the alteration in the conformation of the γ-subunit of the trimeric G-receptor on binding to its β-subunit (Dues *et al.*, 2001; Johnsson, 2002; Raquet *et al.*, 2001).

Protocol: Detecting a Conformational Change in Protein Y on Binding Protein X

- Construct N_{vg}-*Y*-C_{ub}-*Dha* under the control of the P_{CUP1}-promoter and the unmodified gene *X* under the control of the P_{GAL1}-promoter. Transform yeast cells with P_{CUP1}-N_{vg}-*Y*-C_{ub}-*Dha* and P_{GAL1}-*X* or P_{CUP1}-N_{vg}-*Y*-C_{ub}-*Dha* and an empty control plasmid.
- Inoculate 10 ml of selective media containing galactose or glucose as carbon source.
- Let the cells grow to an OD_{600} of 0.8 and induce the expression of N_{vg}-Y-C_{ub}-Dha by adding copper sulfate to 100 μM.
- Let cells grow for an additional hour. Spin down, dissolve cell pellet in 100 μl of twofold Laemmli sample buffer.
- Boil immediately for 5 min. Transfer tubes into liquid nitrogen. Repeat four times.
- Load 20 μl on a SDS–polyacrylamide gel. Transfer the separated proteins on nitrocellulose to detect N_{vg}-Y-C_{ub}-Dha and the free Dha with an anti-ha antibody and a peroxidase coupled secondary antibody.
- Quantify the amount of cleaved and uncleaved fusion protein. Calculate the ratio and the sum of the cleaved and uncleaved fusion proteins.
- Compare the ratios and the sums of N_{vg}-Y-C_{ub}-Dha and Dha from extracts derived from the cells that were either grown in glucose or in galactose. A change in the conformation of Y is indicated once the ratio, but not the sum, of cleaved and uncleaved N_{vg}-Y-C_{ub}-Dha is clearly different in those cells that contain P_{GAL1}-*X* and that were grown in galactose.

Acknowledgments

We thank past and present members of both groups for their contributions and especially Dr. Dirnberger for introducing the plasmid dependency analysis.

References

Ansari, A. Z., Koh, S. S., Zaman, Z., Bongards, C., Lehming, N., Young, R. A., and Ptashne, M. (2002). Transcriptional activating regions target a cyclin-dependent kinase. *Proc. Natl. Acad. Sci. USA* **99,** 14706–14709.

Birnboim, H. C., and Doty, J. (1979). A rapid alkaline extraction procedure for screening recombinant plasmid DNA. *Nucleic Acids Res.* **7,** 1513–1523.

Boeke, J. D., LaCroute, F., and Fink, G. R. (1984). A positive selection for mutants lacking orotidine-5′-phosphate decarboxylase activity in yeast: 5-fluoro-orotic acid resistance. *Mol. Gen. Genet.* **197,** 345–346.

Chien, C. T., Bartel, P. L., Sternglanz, R., and Fields, S. (1991). The two-hybrid system: A method to identify and clone genes for proteins that interact with a protein of interest. *Proc. Natl. Acad. Sci. USA* **88,** 9578–9582.

Deslandes, L., Olivier, J., Peeters, N., Feng, D. X., Khounlotham, M., Boucher, C., Somssich, I., Genin, S., and Marco, Y. (2003). Physical interaction between RRS1-R, a protein conferring resistance to bacterial wilt, and PopP2, a type III effector targeted to the plant nucleus. *Proc. Natl. Acad. Sci. USA* **100,** 8024–8029.

Dohmen, R. J., Stappen, R., McGrath, J. P., Forrova, H., Kolarov, J., Goffeau, A., and Varshavsky, A. (1995). An essential yeast gene encoding a homolog of ubiquitin activating enzyme. *J. Biol. Chem.* **270,** 18099–18109.

Dues, G., Müller, S., and Johnsson, N. (2001). Detection of a conformational change in G gamma upon binding G beta in living cells. *FEBS Lett.* **505,** 75–80.

Dünnwald, M., Varshavsky, A., and Johnsson, N. (1999). Detection of transient *in vivo* interactions between substrate and transporter during protein translocation into the endoplasmic reticulum. *Mol. Biol. Cell* **10,** 329–344.

Eckert, J. H., and Johnsson, N. (2003). Pex10p links the ubiquitin conjugating enzyme Pex4p to the protein import machinery of the peroxisome. *J. Cell Sci.* **116,** 3623–3634.

Fetchko, M., and Stagljar, I. (2004). Application of the split-ubiquitin membrane yeast two-hybrid system to investigate membrane protein interactions. *Methods Enzymol.* **32,** 349–362.

Fields, S., and Song, O. K. (1989). A novel genetic system to detect protein-protein interactions. *Nature* **340,** 245–246.

Gietz, D., Jean, A. S., Woods, R. A., and Schiestl, R. H. (1992). Improved method for high efficiency transformation of intact yeast cells. *Nucleic Acids Res.* **20,** 1425.

Gietz, R. D., and Schiestl, R. H. (1991). Applications of high efficiency lithium acetate transformation of intact yeast cells using single-stranded nucleic acids as carrier. *Yeast* **7,** 253–264.

Gromöller, A., and Lehming, N. (2000). Srb7p is a physical and physiological target of Tup1p. *EMBO J.* **19,** 6845–6852.

Johnsson, N. (2002). A split-ubiquitin-based assay detects the influence of mutations on the conformational stability of the p53 DNA binding domain *in vivo*. *FEBS Lett.* **531,** 259–264.

Johnsson, N., and Varshavsky, A. (1994). Split ubiquitin as a sensor of protein interactions *in vivo*. *Proc. Natl. Acad. Sci. USA* **91,** 10340–10344.

Kim, M. C., Panstruga, R., Elliott, C., Muller, J., Devoto, A., Yoon, H. W., Park, H. C., Cho, M. J., and Schulze-Lefert, P. (2002). Calmodulin interacts with MLO protein to regulate defence against mildew in barley. *Nature* **416,** 447–451.

Laser, H., Bongards, C., Schueller, J., Heck, S., Johnsson, N., and Lehming, N. (2000). A new screen for protein interactions reveals that the *Saccharomyces cerevisiae* high mobility group proteins Nhp6A/B are involved in the regulation of the GAL1 promoter. *Proc. Natl. Acad. Sci. USA* **97,** 13732–13737.

Ludewig, U., Wilken, S., Wu, B., Jost, W., Obrdlik, P., El Bakkoury, M., Marini, A. M., Andre, B., Hamacher, T., Boles, E., von Wiren, N., and Frommer, W. B. (2003). Homo- and hetero-oligomerization of ammonium transporter-1 NH4 uniporters. *J. Biol. Chem.* **278,** 45603–45610.

Massaad, M. J., and Herscovics, A. (2001). Interaction of the endoplasmic reticulum alpha 1,2-mannosidase Mns1p with Rer1p using the split-ubiquitin system. *J. Cell Sci.* **114,** 4629–4635.

Miller, B. G., Hassell, A. M., Wolfenden, R., Milburn, M. V., and Short, S. A. (2000). Anatomy of a proficient enzyme: The structure of orotidine 5′-monophosphate decarboxylase in the presence and absence of a potential transition state analog. *Proc. Natl. Acad. Sci. USA* **97,** 2011–2016.

Obrdlik, P., El-Bakkoury, M., Hamacher, T., Cappellaro, C., Vilarino, C., Fleischer, C., Ellerbrok, H., Kamuzinzi, R., Ledent, V., Blaudez, D., Sanders, D., Revuelta, J. L., Boles, E., Andre, B., and Frommer, W. B. (2004). K^+ channel interactions detected by a genetic system optimized for systematic studies of membrane protein interactions. *Proc. Natl. Acad. Sci. USA* **101,** 12242–12247.

Raquet, X., Eckert, J. H., Müller, S., and Johnsson, N. (2001). Detection of altered protein conformations in living cells. *J. Mol. Biol.* **305,** 927–938.

Rojo-Niersbach, E., Morley, D., Heck, S., and Lehming, N. (2000). A new method for the selection of protein interactions in mammalian cells. *Biochem. J.* **348,** 585–590.

Schiestl, R. H., and Gietz, R. D. (1989). High efficiency transformation of intact yeast cells using single stranded nucleic acids as a carrier. *Curr. Genet.* **16,** 339–346.

Stagljar, I., Korostensky, C., Johnsson, N., and Te Heesen, S. (1998). A genetic system based on split-ubiquitin for the analysis of interactions between membrane proteins *in vivo*. *Proc. Natl. Acad. Sci. USA* **95,** 5187–5192.

Thaminy, S., Auerbach, D., Arnoldo, A., and Stagljar, I. (2003). Identification of novel ErbB3-interacting factors using the split-ubiquitin membrane yeast two-hybrid system. *Genome Res.* **13,** 1744–1753.

Varshavsky, A. (1996). The N-end rule: Functions, mysteries, uses. *Proc. Natl. Acad. Sci. USA* **93,** 12142–12149.

Wang, B., Nguyen, M., Breckenridge, D. G., Stojanovic, M., Clemons, P. A., Kuppig, S., and Shore, G. C. (2003). Uncleaved BAP31 in association with A4 protein at the endoplasmic reticulum is an inhibitor of Fas-initiated release of cytochrome c from mitochondria. *J. Biol. Chem.* **278,** 14461–14468.

Wang, B., Pelletier, J., Massaad, M. J., Herscovics, A., and Shore, G. C. (2004). The yeast split-ubiquitin membrane protein two-hybrid screen identifies BAP31 as a regulator of the turnover of endoplasmic reticulum-associated protein tyrosine phosphatase-like B. *Mol. Cell. Biol.* **24,** 2767–2778.

Wittke, S., Lewke, N., Müller, S., and Johnsson, N. (1999). Probing the molecular environment of membrane proteins *in vivo*. *Mol. Biol. Cell* **10,** 2519–2530.

Wittke, S., Dünnwald, M., and Johnsson, N. (2000). Sec62p, a component of the endoplasmic reticulum protein translocation machinery, contains multiple binding sites for the Sec complex. *Mol. Biol. Cell* **11,** 3859–3871.

Wittke, S., Dünnwald, M., Albertsen, M., and Johnsson, N. (2002). Recognition of a subset of signal sequences by Ssh1p, a Sec61p-related protein in the membrane of endoplasmic reticulum of yeast *Saccharomyces cerevisiae*. *Mol. Biol. Cell* **13,** 2223–2232.

[51] Ubiquitin Fusion Technique and Related Methods

By ALEXANDER VARSHAVSKY

Abstract

The ubiquitin fusion technique, developed in 1986, is still the method of choice for producing a desired N-terminal residue in a protein of interest *in vivo*. This technique is also used as a tool for protein expression. Over the past two decades, several otherwise unrelated methods were invented that have in common the use of ubiquitin fusions as a component of design. I describe the original ubiquitin fusion technique, its current applications, and other methods that use the properties of ubiquitin fusions.

Introduction

The ubiquitin (Ub) fusion technique was invented through experiments in which a segment of DNA encoding the 76-residue Ub was joined, in-frame, to DNA encoding *Escherichia coli* β-galactosidase (βgal) (Bachmair *et al.*, 1986; Varshavsky, 1996b, 2000). When the resulting protein fusion was expressed in the yeast *Saccharomyces cerevisiae* and detected by radiolabeling and immunoprecipitation with anti-βgal antibody, only the moiety of βgal was observed, even if pulse-labeling was close to the time (1–2 min) required for translation of the Ub-βgal's open reading frame (ORF). It was found that the Ub moiety of the fusion was rapidly cleaved off after the last residue of Ub (Fig. 1) (Bachmair *et al.*, 1986). The proteases involved are called deubiquitylating enzymes (DUBs) (Amerik and Hochstrasser, 2004; Baker, 1996; Gilchrist *et al.*, 1997; Hemelaar *et al.*, 2004; Pickart and Cohen, 2004; Verma *et al.*, 2002; Wilkinson, 2000; Wilkinson and Hochstrasser, 1998). A mammalian genome encodes at least 80 distinct DUBs that are specific for the Ub moiety. The *in vivo* cleavage of a Ub fusion at the Ub-polypeptide junction is largely cotranslational (Johnsson and Varshavsky, 1994b; Turner and Varshavsky, 2000).

A note on terminology: ubiquitin whose C-terminal (Gly-76) carboxyl group is covalently linked to another compound is called the *ubiquityl* moiety, with derivative terms *ubiquitylation* and *ubiquitylated*. The acronym Ub refers to both free ubiquitin and the ubiquityl moiety. This nomenclature (Varshavsky, 1997; Webb, 1992), which brings Ub-related terms in line with standard chemical terminology, has been adopted by most Ub researchers. Shorthand for "degradation signal" is "degron"

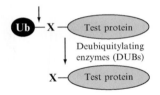

FIG. 1. The ubiquitin fusion technique. Linear fusions of Ub to other proteins are cleaved after the last residue of Ub by deubiquitylating enzymes (DUBs) (see the main text) (Bachmair et al., 1986; Varshavsky, 1996b).

(Dohmen et al., 1994; Gardner and Hampton, 1999; Varshavsky, 1991). Through the use of prefixes, subscripts, or superscripts, this acronym can be employed to denote, in a uniform and succinct way, different types of degradation signals. For example, "N-degron" denotes one class of degradation signals recognized by the N-end–rule pathway, specifically those in which an essential determinant is a substrate's destabilizing N-terminal residue (Bachmair and Varshavsky, 1989; Rao et al., 2001; Suzuki and Varshavsky, 1999; Varshavsky, 1996b).

One physiological function of DUB-mediated cleavage reactions (Fig. 1) is the excision of Ub from its natural DNA-encoded precursors, either linear poly-Ub (Finley et al., 1987) or Ub fusions to specific ribosomal proteins (Finley et al., 1989; Redman and Rechsteiner, 1989). Many DUBs that process linear Ub fusions can also cleave Ub off its branched, posttranslationally formed conjugates, in which Ub is conjugated either to itself, as in a branched poly-Ub chain, or to other proteins. A Ub–protein conjugate usually is composed of a single poly-Ub chain covalently linked to an internal Lys residue of a substrate protein. The ubiquitylated substrate is recognized (in part through its poly-Ub chain) and processively degraded by the 26S proteasome, an ATP-dependent multisubunit protease (Baumeister et al., 1998; Rechsteiner and Hill, 2005). For reviews of the Ub system, see Fang and Weissman (2004); Hershko et al. (2000); Hicke and Dunn (2003); Petroski and Deshaies (2005); and Pickart (2004).

The finding of a rapid *in vivo* deubiquitylation of Ub fusions (Fig. 1) led to the discovery of N-end rule, a relation between the *in vivo* half-life of a protein and the identity of its N-terminal residue (Fig. 2) (Bachmair et al., 1986). First, it was shown that the cleavage of a Ub-X-polypeptide after the last residue of Ub takes place regardless of the identity of a junctional residue X, proline (Pro) being the single exception. By allowing a bypass of "normal" N-terminal processing of a newly formed protein, this finding yielded an *in vivo* method for placing different residues at the N-termini of

Residue X	Half-life of X-βgal	
	E. coli	S. cerevisiae
Arg	2 min	2 min
Lys	2 min	3 min
Phe	2 min	3 min
Leu	2 min	3 min
Trp	2 min	3 min
Tyr	2 min	10 min
His	>10 h	3 min
Ile	>10 h	30 min
Asp	>10 h	3 min
Glu	>10 h	30 min
Asn	>10 h	3 min
Gln	>10 h	10 min
Cys	>10 h	>30 h
Ala	>10 h	>30 h
Ser	>10 h	>30 h
Thr	>10 h	>30 h
Gly	>10 h	>30 h
Val	>10 h	>30 h
Pro	>10 h	>30 h
Met	>10 h	>30 h

FIG. 2. The N-end rule of the yeast *S. cerevisiae*. Specific residues at the N-terminus of a test protein such as βgal are produced using the Ub fusion technique (Fig. 1 and the main text). The *in vivo* half-lives of the corresponding X-βgal proteins are indicated on the right. Stabilizing N-terminal residues (Met, Gly, Ala, Ser, Thr, Cys, Val, and Pro) are not recognized by UBR1, the E3 Ub ligase of the N-end–rule pathway. The N-end rule of mammalian cells is similar but contains fewer stabilizing residues (Kwon *et al.*, 2002, 2003).

otherwise identical proteins. It was found that the *in vivo* half-lives of resulting test proteins were strongly dependent on the identities of their N-terminal residues, a relation referred to as the N-end rule (Fig. 2) (Bachmair *et al.*, 1986). The underlying, universally present N-end rule pathway has a variety of functions; their list continues to expand (Du *et al.*, 2002; Kwon *et al.*, 2002, 2003; Rao *et al.*, 2001; Turner *et al.*, 2000; Varshavsky, 1996b, 2003; Yin *et al.*, 2004).

The Ub fusion technique (Figs. 1 and 2) remains the method of choice for producing, *in vivo*, a desired N-terminal residue in a protein of interest. The requirement for a "technique" to do so stems from a constraint imposed by the genetic code. All nascent proteins bear N-terminal Met (formyl-Met in prokaryotes). The known methionine aminopeptidases (MetAPs) that remove N-terminal Met would do so if, and only if, a residue at position 2, to be made N-terminal after cleavage, is small enough (Bradshaw et al., 1998; Varshavsky, 1996b). Specifically, MetAPs do not remove N-terminal Met if it is followed by any of the 12 destabilizing residues in the yeast-type N-end rule (Fig. 2). The exception, in metazoans, is Cys, whose side chain is small enough to allow cleavage of the Met-Cys bond by MetAPs. N-terminal Cys is a destabilizing residue in mammals and (apparently) other multicellular eukaryotes, but a stabilizing residue in fungi such as *S. cerevisiae* (Gonda et al., 1989; Kwon et al., 2002). (All destabilizing residues, including Cys, can be made N-terminal through cleavages by other intracellular proteases, such as separases, caspases, and calpains, which act, in this capacity, as upstream components of the N-end rule pathway.) The Ub-specific DUB proteases are free of constraints imposed by the preceding property of MetAPs, except when the residue X of a Ub-X polypeptide is Pro, in which case the cleavage still takes place but at a much lower rate (Bachmair et al., 1986; Johnson et al., 1992, 1995). However, there also exists a DUB that can efficiently cleave at the Ub-Pro junction (Gilchrist et al., 1997).

Ub fusions can be deubiquitylated *in vitro* as well (Baker, 1996; Catanzariti et al., 2004; Gonda et al., 1989). High activity and specificity of DUBs make them reagents of choice for applications that involve, for example, the removal of affinity tags from overexpressed and purified proteins. A particularly efficacious version of the Ub fusion technique for high-level production and easy purification of recombinant proteins expressed in *E. coli* was described by R. Baker and colleagues (Fig. 3) (Baker et al., 2005; Catanzariti et al., 2004).

Yet another advantage of the Ub fusion technique stems from the finding that expression of a protein as a Ub fusion can dramatically augment protein's yield (Baker et al., 1994; Butt et al., 1989; Ecker et al., 1989; Mak et al., 1989). The yield-enhancement effect of Ub was observed with short peptides as well (Pilon et al., 1997; Yoo et al., 1989). This and other applications of Ub fusions are described in the following, with references to original articles and specific constructs.

Production and Uses of N-Degrons

An N-degron is composed of a protein's destabilizing N-terminal residue and an internal Lys residue (Bachmair and Varshavsky, 1989; Hill et al., 1993; Suzuki and Varshavsky, 1999; Varshavsky, 1996b). The lysine determinant is the site of formation of a substrate-linked poly-Ub chain (Chau et al., 1989; Pickart and Fushman, 2004). One way to produce an N-degron in a protein of interest is to express the protein as a Ub fusion whose junctional residue, which becomes N-terminal on removal of the Ub

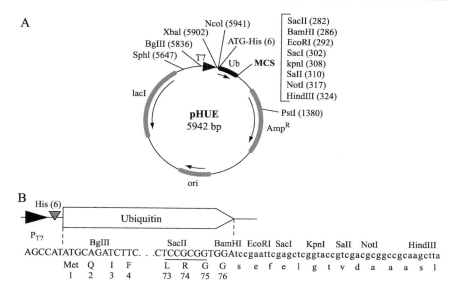

FIG. 3. The ubiquitin fusion-based expression-purification technique of R. Baker and colleagues (Catanzariti et al., 2004). (A) Plasmid map of pHUE, a histidine-tagged ubiquitin expression vector. It shows the Ub-coding region (black box), the T7 polymerase promoter (black triangle), and other relevant regions (shaded boxes). Arrows indicate the direction of transcription. Restriction enzyme recognition sites within the multiple cloning site (MCS) are listed, and other useful recognition sites are also shown, all of them unique, except BglII; locations are cited relative to the start codon upstream of the His-tag, ATG = 1. (His)6, polyhistidine tag; Amp^r, β-lactamase gene; ori, colE1 origin of replication; lacI, Lac repressor gene. (B) DNA and encoded protein sequence of the 5' and 3' ends of the Ub-coding region, showing the engineered SacII site (underlined) within Leu-73, Arg-74, and Gly-75, and the 3' polylinker. Restriction sites and encoded amino acid residues are shown above and under the DNA sequence, respectively. A Ub fusion expressed using pHUE can be deubiquitylated by the histidine-tagged DUB enzyme USP2 (Baker et al., 2005; Catanzariti et al., 2004).

moiety, is destabilizing (Fig. 2). An appropriately positioned internal Lys residue (or residues) is the second essential determinant of N-degron. Many natural proteins lack such "targetable" lysines, and therefore would remain long-lived even if their N-terminal residue were replaced by a destabilizing one. One way to bypass this difficulty is to link a protein of interest to a portable N-degron that contains both a destabilizing N-terminal residue (produced through a Ub fusion) and a requisite Lys residue(s). The earliest portable N-degron of this kind is still among the most efficacious known (Fig. 4B) (Bachmair and Varshavsky, 1989) and can be improved through insertion of additional Lys residues (Suzuki and Varshavsky, 1999). A screen in the sequence space of two amino acids,

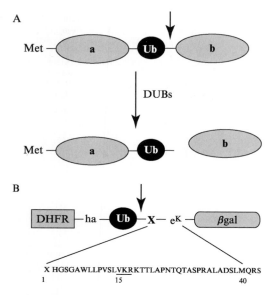

Fig. 4. The ubiquitin-protein-reference (UPR) technique. (A) A tripartite fusion containing **a** the reference protein moiety whose C-terminus is linked, by means of a spacer peptide, to the Ub moiety. The C-terminus of Ub is linked to **b**, a protein of interest (Lévy et al., 1996; Suzuki and Varshavsky, 1999). In vivo, this tripartite fusion is cotranslationally cleaved (Turner and Varshavsky, 2000) by deubiquitylating enzymes (DUBs) at the Ub-**b** junction, yielding equimolar amounts of the unmodified protein **b** and **a**-Ub, the reference protein **a** bearing a C-terminal Ub moiety. If **a**-Ub is long-lived, determining the ratio of **a**-Ub to **b** as a function of time or at steady-state yields, respectively, the in vivo decay curve or the relative metabolic stability of protein **b**. (B) Example of a specific UPR-type Ub fusion (Suzuki and Varshavsky, 1999). This fusion contains the following elements: DHFRha, a mouse dihydrofolate reductase (DHFR) moiety extended at the C-terminus by a sequence containing the hemagglutinin-derived ha epitope; the Ub moiety (more specifically, the Ub^{R48} moiety bearing the Lys → Arg alteration at position 48); a 40-residue, E. coli Lac repressor–derived sequence, termed e^K (extension [*e*] containing lysines [*K*]) and shown below in single-letter abbreviations for amino acids; a variable residue X between Ub and e^K; the E. coli βgal moiety lacking the first 24 residues of wild-type βgal. A short arrow in (A) and (B) indicates the site of in vivo cleavage by DUBs.

Lys and Asn, has shown that certain sequences containing exclusively lysines and asparagines can also function in vivo as strong N-degrons (Suzuki and Varshavsky, 1999). At least some natural N-degrons, such as the one in a separase-produced fragment of SCC1, a subunit of S. cerevisiae cohesin, are highly efficacious as well (Rao et al., 2001) and are also likely to be portable. The modularity and portability of N-degrons make possible a variety of applications whose common feature is the conferring of a constitutive or conditional metabolic instability on a protein of interest.

N-Degron and Reporter Proteins

A change in the physiological state of a cell that is preceded or followed by the induction or repression of specific genes can be monitored through the use of promoter fusions to a variety of protein reporters, such as, for example, βgal, β-glucuronidase, luciferase, and green fluorescent protein (GFP). A long-lived reporter is useful for detecting the induction of genes but is less suitable for monitoring either a rapid repression or a temporal pattern that involves an up- or down-regulation of a gene product of interest. A short-lived reporter is required in such settings. N-degron–containing X-βgal proteins of the original N-end rule study (Fig. 2) (Bachmair et al., 1986) were the first such reporters. Over the past two decades, other reporters, including the ones mentioned previously, were metabolically destabilized by extending them with either a portable N-degron or a "nonremovable" Ub moiety (Dantuma et al., 2000; Deichsel et al., 1999; Paz et al., 1999; Worley et al., 1998). The latter is targeted by a distinct Ub-dependent proteolytic pathway called the UFD pathway (Ub-fusion-degradation) (Bachmair et al., 1986; Johnson et al., 1992, 1995; Koegl et al., 1999). Metabolically unstable reporters are particularly useful in settings in which the reporter's concentration must reflect a recent level of gene activity. Portable N-degrons were also used to destabilize specific protein antigens, thereby enhancing presentation of their peptides to the immune system (Tobery and Siliciano, 1999; Townsend et al., 1988).

N-Degron and Conditional Mutants

Conditional mutants based on N-degrons are described in detail in Chapter 52. A frequent problem with conditional phenotypes is their leakiness (i.e., unacceptably high residual activity of either a temperature-sensitive (ts) protein at nonpermissive temperature or a gene of interest in the "off" state of its promoter). Another problem is "phenotypic lag," which often occurs between the imposition of nonpermissive conditions and the emergence of a relevant null phenotype. Phenotypic lag tends to be longer with proteins that are required in catalytic rather than stoichiometric amounts.

In one application of Ub fusions and the N-end–rule pathway to the problem of phenotypic lag, a constitutive N-degron (produced as a Ub fusion) was linked to a protein expressed from an inducible promoter (Park *et al.*, 1992). This otherwise useful method is constrained by the necessity of using a heterologous promoter and by a constitutively short half-life of a target protein, whose levels may, therefore, be suboptimal under permissive conditions. An alternative approach is to link an N-degron to a normally long-lived protein in a strain where the N-end–rule pathway can be induced or repressed. Such strains have been constructed with *S. cerevisiae* (Ghislain *et al.*, 1996; Moqtaderi *et al.*, 1996) but can also be designed for other organisms, including mammalian cells. The metabolic stabilities, and hence also the levels of N-degron-bearing proteins in a cell with an inducible N-end rule pathway, are either normal or very low, depending on whether UBR1, the pathway's E3 Ub ligase, is absent or present (Ghislain *et al.*, 1996; Moqtaderi *et al.*, 1996). These conditional mutants can be constructed with any cytosolic or nuclear protein whose function tolerates an N-terminal extension.

Yet another design is a portable N-degron that is inactive at low (permissive) temperature but becomes active at high (nonpermissive) temperature. Such an N-degron was constructed, using the Ub fusion technique, in the context of a specific *ts* allele of the 20-kDa mouse dihydrofolate reductase (DHFR) bearing N-terminal Arg, a strongly destabilizing residue (Dohmen and Varshavsky, 2005; Dohmen *et al.*, 1994). Linking this DHFR-based, heat-activated N-degron to proteins of interest yielded a new class of *ts* mutants, called *td* (temperature-activated degron). The *td* method does not require an often unsuccessful search for a nonleaky *ts* mutation in a gene of interest. If a protein can tolerate N-terminal extensions, the corresponding *td* fusion is likely to be functionally unperturbed at permissive temperature. (By contrast, a low activity of a *ts* protein at permissive temperature is a frequent problem with conventional *ts* mutants.) The *td* method eliminates or reduces phenotypic lag, because the activation of N-degron results in rapid disappearance of a *td* protein. Yet another advantage of the *td* technique is the possibility of using two sets of conditions: a *td* protein-expressing strain at permissive versus nonpermissive temperature or, alternately, the same strain versus a congenic strain lacking the N-end–rule pathway, with both strains at nonpermissive temperature (Dohmen *et al.*, 1994). This powerful internal control, provided in the *td* technique by two alternative sets of permissive/nonpermissive conditions, is unavailable with conventional *ts* mutants. Since 1994, a number of laboratories described successful uses of the *td* technique to construct *ts* (*td*) alleles of specific proteins (Amon, 1997; Aparicio, 2003; Caponigro and Parker, 1995; Hardy, 1996; Kanemaki *et al.*, 2003; Kesti *et al.*, 2004; Labib *et al.*, 2000; Valasek *et al.*, 2003; Wang *et al.*, 2004; Wolf *et al.*, 1998).

N-Degron and Conditional Toxins

A major limitation of current pharmacological strategies stems from the absence of drugs that are specific, in a predetermined manner, for two or more independent molecular targets. For reasons discussed elsewhere (Varshavsky, 1995, 1998), it is desirable to have a therapeutic agent that possesses a multitarget, combinatorial selectivity, which requires the presence of two or more predetermined targets in a cell and simultaneously the *absence* of one or more targets for the drug to exert its effect. Note that simply combining two or more "conventional" drugs against different targets in a multidrug regimen will not yield a multitarget selectivity, because two drugs together would perturb not only cells containing both targets but also cells containing either one of the targets.

A strategy for designing protein-based reagents that are sensitive to the presence or absence of more than one target at the same time was proposed a decade ago (Varshavsky, 1995). One key feature of such reagents is their ability to use codominance, the property characteristic of many signals in proteins, including degrons and nuclear localization signals (NLSs). Codominance, in this context, refers to the ability of two or more signals in the same molecule to function independently and not to interfere with each other. The critical feature of a degron-based multitarget reagent is that its intrinsic toxicity (or another intended property) is the same in all cells, whereas its *in vivo* half-life, and, consequently, its steady-state level and overall toxicity, depends on the cell's protein composition, specifically on the presence of multiple "target" proteins that have been chosen to define the profile of a cell to be eliminated or otherwise modified (Varshavsky, 1995). A related but different design involves a toxic protein made short-lived (and therefore relatively nontoxic) by the presence of a degradation signal such as N-degron, produced using the Ub fusion technique. For example, if a cleavage site for a specific viral or nonviral protease is placed between the fusion's toxic moiety and the N-degron, the fusion would be cleaved only in cell containing the relevant protease. As a result, the toxic moiety of the fusion would become long-lived, and, therefore, more toxic, only in a target cell population, for example, virus-infected cells (Falnes and Olsnes, 1998; Falnes *et al.*, 1999; Varshavsky, 1996a). Analogous Ub fusion–based approaches can use a different degron, for example, the one recognized by the UFD (Ub-fusion-degradation) pathway (Tcherniuk *et al.*, 2004). The interference/codominance (IC) concept and the ideas about protein-size multitarget reagents have been extended to small (<1 kDa) multitarget compounds (Varshavsky, 1998).

Overexpression of Proteins as Ubiquitin Fusions

A major application of the Ub fusion technique is its use to augment the yields of recombinant proteins (Baker et al., 1994; Butt et al., 1989; Ecker et al., 1989; Mak et al., 1989; Pilon et al., 1997). See Fig. 3 for a particularly effective version of the Ub fusion technique for high-level expression and purification of recombinant proteins expressed in E. coli, by R. Baker and colleagues (Baker et al., 2005; Catanzariti et al., 2004).

The yield-enhancing effect of Ub was observed not only with eukaryotic cells (where the Ub moiety is present in a nascent fusion but not in its mature counterpart) but also in prokaryotes, which lack the Ub system, including DUBs, and, therefore, retain the Ub moiety in a translated fusion. (E. coli transformed with a plasmid expressing the S. cerevisiae DUB UBP1 acquires the ability to deubiquitylate Ub fusions [Tobias et al., 1991].) The effect of N-terminal Ub moiety of a fusion on the fusion's yield is likely to stem from rapid folding of the nascent Ub moiety, whose presence at the N-terminus of an emerging protein may thereby either partially protect, in ways that are not understood, a still unfolded part of the fusion from proteolytic attack, facilitate its folding, or both. The Ub-mediated increase in total yield is often accompanied by an increase in solubility of overexpressed protein. In this regard, the effect of Ub is analogous to that of several other proteins, such as thioredoxin and maltose-binding protein (MBP) (Kapust and Waugh, 1999). When these moieties are cotranslationally linked to a protein of interest, they often increase its yield and solubility. A model of the underlying mechanism suggested for MBP (Kapust and Waugh, 1999) may also be relevant to the effect of Ub moiety. Specifically, a partially unfolded nascent protein is presumed to weakly interact with the nearby (upstream) MBP moiety, thereby transiently precluding intermolecular self-interactions that can result in irreversible aggregation before the protein has had the time to attain its mature conformation.

The first engineered Ub fusions used pUB23-X, a family of high-copy plasmids that expressed Ub-X-βgal proteins containing different junctional residues (X) in S. cerevisiae from a galactose-inducible, glucose-repressible promoter (Bachmair and Varshavsky, 1989; Bachmair et al., 1986). Subsequent designs facilitated the construction of ORFs encoding Ub-X-polypeptide fusions by introducing a SacII (SstII) site within the codons for the last three residues of Ub moiety (Baker et al., 1994). In this cloning scheme, an ORF of interest is amplified using PCR and a primer in which the 5' extension encodes the last three residues of Ub. Another cloning route uses double-stranded oligonucleotides with SacII cohesive ends that are used to join DNA fragments (Baker et al., 1994). The expression of a resulting Ub-X-polypeptide fusion in a eukaryotic cell (or in a prokaryotic cell that contains the *S. cerevisiae* UBP1 DUB) yields an X-polypeptide bearing a predetermined N-terminal residue X (Figs. 1 and 2).

In their natural milieu, proteins of biotechnological or pharmacological interest are often products of secretory pathways and, therefore, are cleaved by signal peptidase on their entrance into the endoplasmic reticulum (ER). This cleavage frequently yields destabilizing residues at the N-termini of these proteins. When the same proteins are overexpressed in the cytosol of a heterologous bacterial or eukaryotic host, their N-terminal Met tends to be retained, because MAPs cannot cleave off N-terminal Met if it is followed by a destabilizing residue (see earlier). It is in these, quite frequent, cases that the expression of a protein as a Ub-X-protein fusion can attain two aims at once: producing a protein of interest bearing the desired N-terminal residue (Fig. 2) and also, quite often, increasing protein's yield compared with an otherwise identical expression of Ub-lacking protein (Baker, 1996).

There are numerous examples of Ub-mediated increases in the yield and solubility of overexpressed proteins. For instance, a conventional heterologous expression of the *Streptomyces* tyrosinase in *E. coli* yielded inactive enzyme, whereas expression of tyrosinase as a Ub fusion resulted in an abundant and active enzyme (Han *et al.*, 1994). Another example was a high yield, in *E. coli*, of the soluble human collagenase catalytic domain as a Ub fusion (Gehring *et al.*, 1995). In contrast, the expression of the same protein in the absence of N-terminal Ub moiety resulted in low yield and insoluble product. A 60-fold increase in the yield of the human pi class glutathione transferase GSTP1 was observed on the addition of a Ub-coding sequence to the *GSTP1* ORF (Baker *et al.*, 1994). A strong increase of protein yield in *E. coli* was reported using a combination of the T7 RNA polymerase promoter system and Ub fusions (Koken *et al.*, 1993). Several other examples of the Ub fusion approach to protein overexpression (Coggan *et al.*, 1995; Mak *et al.*, 1989; Rian *et al.*, 1993; Sabin *et al.*, 1989) are described in an earlier review (Baker, 1996). More recently, Vierstra and colleagues applied the Ub fusion technique to augment protein expression in transgenic plants (Hondred *et al.*, 1999).

Ubiquitin-Assisted Analysis of Protein Translocation across Membranes

A method, developed in 1994 and called UTA (u̲biquitin t̲ranslocation a̲ssay), uses Ub as an *in vivo* kinetic probe in the context of signal sequence-bearing Ub fusions (Johnsson and Varshavsky, 1994b). After emerging from ribosomes in the cytosol, a protein may remain in the cytosol or may be transferred to compartments separated from the cytosolic space by membranes. With a few exceptions, noncytosolic proteins begin journeys to their respective compartments by crossing membranes that enclose intracellular organelles such as the ER and mitochondria in eukaryotes or the periplasmic space in bacteria. Amino acid sequences that enable a protein to cross the membrane of a compartment are often located at the protein's N-terminus. These "signal" sequences (Walter et al., 1984) are targeted by translocation pathways specific for each compartment. The translocation of a protein across a compartment's membrane can start before the protein's synthesis is completed, resulting in docking of the still translating ribosome at the transmembrane channel. The UTA technique takes advantage of rapid (largely cotranslational) cleavage of a Ub fusion to examine temporal aspects of protein transport across the ER membrane in living cells (Johnsson and Varshavsky, 1994b). Specifically, if a Ub fusion that has been engineered to bear an N-terminal signal sequence (SS) upstream of the Ub moiety is cleaved in the cytosol by DUBs, the fusion's reporter moiety would fail to be translocated into the ER. Conversely, if a nascent SS mediates the docking of a translating ribosome at the transmembrane channel rapidly enough, or if the fusion's Ub moiety is located sufficiently far downstream of SS, then by the time the Ub moiety emerges from the ribosome, the latter is already docked, and the nascent Ub moiety enters the ER before it can fold and/or be targeted by DUBs. Thus, the cleavage at the Ub moiety of an SS-bearing Ub fusion in the cytosol can serve as an *in vivo* kinetic marker and a tool for analyzing targeting in protein translocation (Johnsson and Varshavsky, 1994b). The temporal sensitivity of the UTA technique stems from rapid folding of the nascent Ub moiety that precludes its translocation and makes it a substrate of DUBs in the cytosol shortly after the emergence of the fusion's Ub moiety from the ribosome.

Split-Ubiquitin Technique for Detection of Protein–Protein Interactions *In Vivo*

Another Ub-based method, termed the split-protein sensor (SPS), makes it possible to detect and monitor a protein-protein interaction as a function of time, at the natural sites of this interaction in a living cell (Johnsson and Varshavsky, 1994a). These capabilities of the split-Ub technique distinguish it from the two-hybrid assay (Phizicky and Fields, 1995). The key idea of the split-Ub technique was applied by other groups to design a variety of split-protein assays, termed PCA (protein complementation assays), that use "split" versions of other proteins, such as dihydrofolate reductase (DHFR), β-lactamase, and green fluorescent protein (GFP) (Cabantous et al., 2005; Galarneau et al., 2002; Pelletier et al., 1998; Remy and Michnick, 1999; Zhang et al., 2004).

The split-Ub technique is based on the following observations: when a C-terminal fragment of the 76-residue Ub (C_{ub}) was expressed as a fusion to a reporter protein, the fusion was cleaved by DUBs only if an N-terminal fragment of Ub (N_{ub}) was also expressed in the same cell. This reconstitution of native Ub from its fragments, detectable by the *in vivo* cleavage assay, was not observed with a mutationally altered N_{ub}. However, if C_{ub} and the altered N_{ub} were each linked to polypeptides that interact *in vivo*, the cleavage of the fusion containing C_{ub} was restored, yielding a generally applicable assay for kinetic and equilibrium aspects of *in vivo* protein interactions (Johnsson and Varshavsky, 1994a).

The enhancement of Ub reconstitution by interacting polypeptides linked to fragments of Ub is caused by a local increase in concentration of one Ub fragment in the vicinity of its "complementing" counterpart. This, in turn, increases the probability that the two Ub fragments coalesce to form a quasinative Ub moiety, whose (at least) transient formation results in irreversible cleavage of the fusion by DUBs. This cleavage can be detected readily and can be followed as a function of time or at steady state (Dunnwald et al., 1999; Johnsson and Varshavsky, 1994a). Unlike the two-hybrid method, which is based on the apposition of two structurally independent protein domains whose folding and functions do not require direct interactions between the domains, the split-Ub assay and its descendants that use split versions of other proteins involve reconstituting the conformation of a relatively small, single-domain protein. One application of the split-Ub sensor has shown that this assay is capable of detecting transient *in vivo* interactions such as the binding of a signal sequence of a translocated protein to SEC62, a component of the ER channel (Dunnwald et al., 1999). Different reporter readouts and selection-based screens have been devised for the split-Ub assay, making it possible to use this method for discovering *in vivo* ligands of a protein of interest, similarly to the main application of the two-hybrid assay (Stagljar et al., 1998; Wittke et al., 1999). This and other uses of the split-Ub technique are described elsewhere (Dues et al., 2001; Eckert and Johnsson, 2003; Johnsson, 2002; Laser et al., 2000; Raquet et al., 2001; Wittke et al., 2000, 2002). As mentioned previously, the invention of the split-Ub method (Johnsson and Varshavsky, 1994a) led to the development of many other split-protein sensors, including those that use split DHFR and split GFP (Cabantous et al., 2005; Galarneau et al., 2002; Pelletier et al., 1998; Remy and Michnick, 1999; Tafelmeyer et al., 2004; Zhang et al., 2004).

Ubiquitin–Protein-Reference (UPR) Technique

Direct measurements of the *in vivo* degradation of intracellular proteins require a pulse-chase assay. It involves the labeling of nascent proteins for a short time with a radioactive precursor ("pulse"), the termination of labeling through the removal of radiolabel and/or the addition of a translation inhibitor, and the analysis of a labeled protein of interest at various times afterwards ("chase"), using immunoprecipation and SDS-PAGE, or analogous techniques. Its advantage of being direct notwithstanding, a conventional pulse-chase assay is fraught with sources of error. For example, immunoprecipitation yields may vary from sample to sample, and the volumes of samples loaded on a gel may vary as well. If the labeling for specific chase times is done with separate batches of cells (as is the case, for example, with anchorage-dependent mammalian cell cultures), the efficiency of labeling is yet another unstable parameter of the assay. As a result, pulse-chase data tend to be semiquantitative at best, lacking the means to correct for these errors.

A way to address these problems through an "internal-reference" strategy, termed the ubiquitin-protein-reference technique (UPR), was described in 1996 (Lévy *et al.*, 1996). This method, an extension of the Ub fusion technique, was applied to pulse-chase assays with mammalian cells and *S. cerevisiae* (Kwon *et al.*, 2002; Lévy *et al.*, 1999; Rao *et al.*, 2001; Suzuki and Varshavsky, 1999; Turner *et al.*, 2000), and with *Xenopus* oocytes as well (Sheng *et al.*, 2002). The UPR technique can compensate for several sources of data scatter (Fig. 4). This method uses a linear fusion in which Ub is located between a protein of interest and a reference protein moiety (Fig. 4A). The fusion is cotranslationally cleaved by DUBs after the last residue of Ub, producing equimolar amounts of the protein of interest and the reference protein bearing the C-terminal Ub moiety. If both the reference protein and the protein of interest are immunoprecipitated in a pulse-chase assay, the relative amounts of the protein of interest can be normalized against the reference protein in the same sample (Suzuki and Varshavsky, 1999; Turner *et al.*, 2000). The UPR technique can thus compensate for the scatter of immunoprecipitation yields, sample volumes, and other sources of sample-to-sample variation. The increased accuracy afforded by UPR underscores insufficiency of the "half-life" terminology, because the *in vivo* degradation of many proteins strongly deviates from a first-order kinetics. For a discussion of this issue and specific terminology for describing nonexponential decay, see Lévy *et al.* (1996) and Suzuki and Varshavsky (1999).

A more recent study (Qian *et al.*, 2002) with UPR-type constructs and transiently transfected mammalian cells has shown that a fusion containing a C-terminal Ub moiety (analogous to the DHFR-Ub^{R48} "reference" module in Fig. 4B) can be conjugated to other intracellular proteins, *a la* Ub itself. We also observed such a conjugation in transiently transfected mammalian cells that strongly overexpressed DHFR-Ub^{R48} (derived from DHFR-Ub^{R48}-X-e^{K}-βgal) (Fig. 4B) (Z. Xu and A.V., unpublished data). The extent of this conjugation seemed to be negligible at significantly lower levels of transient expression in mammalian cells (unpublished data) and was not observed at all, thus far, in similar UPR assays with *S. cerevisiae* (Suzuki and Varshavsky, 1999; Turner *et al.*, 2000, and data not shown). Thus, some Ub ligases in mammalian cells, but not in yeast, can use as a conjugation substrate (apparently at low efficiency) even a Ub moiety that bears both a protein-size N-terminal extension and the Lys → Arg alteration at position 48. Although the relative efficiency of this "undesirable" conjugation in mammalian cells seems to be low, it is a complication to be addressed. A way to do so would be to identify mutations of C-terminal Ub moiety (Fig. 4B) that render it completely inactive as a substrate for conjugation by Ub ligases without perturbing significantly its recognition by DUBs in the context of a protein fusion. It remains to be seen whether such a "deconvolution," through either targeted or random mutagenesis of Ub, is actually feasible.

Ubiquitin Sandwich Technique

Nascent polypeptides emerging from the ribosome may, in the process of folding, present degradation signals similar to those recognized by the Ub system in misfolded or otherwise damaged proteins. It has been a longstanding question whether a significant fraction of nascent polypeptides is cotranslationally degraded. Determining whether nascent polypeptides are actually degraded *in vivo* has been difficult, because at any given time the nascent chains of a particular protein species are of different sizes and, therefore, would not form a band on electrophoresis in a conventional pulse-chase assay. The Ub sandwich technique, published in 2000, makes it possible to detect cotranslational protein degradation by measuring the steady-state ratio of two reporter proteins whose relative abundance is established cotranslationally (Turner and Varshavsky, 2000).

Operationally, the Ub sandwich technique (Turner and Varshavsky, 2000) is a three-protein version of the UPR technique described previously. A polypeptide to be examined for cotranslational degradation, termed **B**, is sandwiched between two stable (lacking posttranslational degrons) reporter domains **A** and **C** in a linear fusion protein. The three polypeptides are connected through Ub moieties to yield a fusion protein of the form **A**Ub-**B**Ub-**C**Ub. The independent polypeptides **A**Ub, **B**ub, and **C**Ub that result from cotranslational cleavages of **AUb-BUb-CUb** by DUBs are called modules below. The DUB-mediated cleavages establish a kinetic competition between two mutually exclusive events during the synthesis of the **AUb-BUb-CUb** fusion: cotranslational cleavage at the **BUb-CUb** junction to release the long-lived **CUb** module or, alternately, cotranslational degradation of the entire **BUb-CUb** nascent chain by the 26S proteasome. In the latter case, the processivity of proteasome-mediated degradation results in the destruction of Ub moiety between **B** and **C** *before* it can be recognized by DUBs. The resulting drop in the level of **CUb** module relative to levels of **AUb**, referred to as the C/A ratio, reflects the cotranslational degradation of domain **B**. This measurement provides a *minimal* estimate of the total amount of cotranslational degradation, because nonprocessive cotranslational degradation events that do not extend into the **C** domain are not detected. The Ub sandwich method was recently used to demonstrate that more than 50% of nascent protein molecules bearing an N-degron can be degraded cotranslationally in *S. cerevisiae*, never reaching their mature size before their destruction by processive proteolysis (Turner and Varshavsky, 2000). Similar conclusions, through the use of other approaches, were also reached by other groups (Adachi *et al.*, 2004; Reits *et al.*, 2000; Schubert *et al.*, 2000).

If cotranslational protein degradation by the Ub system is found to be extensive for at least some wild-type proteins, it could be accounted for as an evolutionary tradeoff between the necessity of identifying and destroying degron-bearing mature proteins and the mechanistic difficulty of distinguishing between posttranslationally and cotranslationally presented degrons. Cotranslational protein degradation may also represent a previously unrecognized form of protein quality control, which destroys nascent chains that fail to fold correctly. These and other questions about physiological aspects of the cotranslational protein degradation can now be addressed directly in living cells through the Ub sandwich technique.

Concluding Remarks

The Ub fusion technique is made possible by the ability of DUBs to cleave a Ub fusion *in vivo* or *in vitro* after the last residue of Ub irrespective of sequence context downstream from the cleaved peptide bond. Since its development two decades ago, the Ub fusion technique gave rise to a number of applications whose common feature is use of the (largely) cotranslational and highly specific cleavage of a Ub-containing fusion by DUBs. Among these applications are the UPR technique, which increases the accuracy of pulse-chase assays, and the Ub sandwich technique, which makes it possible to determine the extent of cotranslational protein degradation *in vivo* for any protein of interest. One useful feature of the Ub moiety is its ability, as a part of linear fusions, to increase the yields and solubility of overexpressed proteins or short peptides in either eukaryotic or bacterial hosts. In yet another class of Ub-based applications, the demonstrated coalescence of peptide-size Ub fragments into a quasinative Ub fold yielded the split-Ub assay for detecting protein interactions *in vivo*, an advance that led to the development of several other split-protein assays. Other recent applications include a ubiquitin-based assay for detecting the uptake, by intact cells, of proteins containing "transduction" domains that enable the crossing of plasma membrane (Loison *et al.*, 2005). Ub fusions continue to be useful in a remarkable variety of ways.

Acknowledgments

I am most grateful to the former and current members of my laboratory, whose work made possible some of the advances described in this review. I also thank Rohan Baker (Australian National University) for Fig. 3 and his permission to publish it here. Our studies are supported by grants from the National Institutes of Health (GM31530 and DK39520) and the Ellison Medical Foundation.

References

Adachi, K., Lakka, V., Zhao, Y., and Surrey, S. (2004). Ubiquitylation of nascent globin chains in a cell-free system. *J. Biol. Chem.* **279,** 41767–41774.

Amerik, A. Y., and Hochstrasser, M. (2004). Mechanism and function of deubiquitinating enzymes. *Biochim. Biophys. Acta* **1695,** 189–207.

Amon, A. (1997). Regulation of B-type cyclin proteolysis by Cdc28-associated kinases in budding yeast. *EMBO J.* **16,** 2693–2702.

Aparicio, O. M. (2003). Tackling an essential problem in functional proteomics of *Saccharomyces cerevisiae*. *Genome Biol.* **4,** 230.

Bachmair, A., Finley, D., and Varshavsky, A. (1986). *In vivo* half-life of a protein is a function of its amino-terminal residue. *Science* **234,** 179–186.

Bachmair, A., and Varshavsky, A. (1989). The degradation signal in a short-lived protein. *Cell* **56,** 1019–1032.

Baker, R. T. (1996). Protein expression using ubiquitin fusion and cleavage. *Curr. Op. Biotechnol.* **7,** 541–546.

Baker, R. T., Catanzariti, A.-M., Karunasekara, Y., Soboleva, T. A., Sharwood, R., Whitney, S., and Board, P. G. (2005). Using deubiquitylating enzymes as research tools. *In* "Methods in Enzymology" (R. J. Deshaies, ed.). Academic Press, New York.

Baker, R. T., Smith, S. A., Marano, R., McKee, J., and Board, P. G. (1994). Protein expression using cotranslational fusion and cleavage of ubiquitin. Mutagenesis of the glutathione-binding site of human pi class glutathione S-transferase. *J. Biol. Chem.* **269,** 25381–25386.

Baumeister, W., Walz, J., Zühl, F., and Seemüller, E. (1998). The proteasome: Paradigm of a self-compartmentalizing protease. *Cell* **92,** 367–380.

Bradshaw, R. A., Brickey, W. W., and Walker, K. W. (1998). N-terminal processing: The methionine aminopeptidase and N alpha-acetyl transferase families. *Trends Biochem. Sci.* **23,** 263–267.

Butt, T. R., Jonnalagadda, S., Monia, B. P., Sternberg, E. J., Marsh, J. A., Stadel, J. M., Ecker, D. J., and Crooke, S. T. (1989). Ubiquitin fusion augments the yield of cloned gene products in *Escherichia coli. Proc. Natl. Acad. Sci. USA* **86,** 2540–2544.

Cabantous, S., Terwilliger, T. C., and Waldo, G. S. (2005). Protein tagging and detection with engineered self-assembling fragments of green fluorescent protein. *Nature Biotechnol.* **23,** 102–107.

Caponigro, G., and Parker, R. (1995). Multiple functions for the polyA-binding protein in mRNA decapping and deadenylation in yeast. *Genes Dev.* **9,** 2421–2432.

Catanzariti, A.-M., Soboleva, T. A., Jans, D. A., Board, P. G., and Baker, R. T. (2004). An efficient system for high-level expression and easy purification of authentic recombinant proteins. *Protein Sci.* **13,** 1331–1339.

Chau, V., Tobias, J. W., Bachmair, A., Marriott, D., Ecker, D. J., Gonda, D. K., and Varshavsky, A. (1989). A multiubiquitin chain is confined to specific lysine in a targeted short-lived protein. *Science* **243,** 1576–1583.

Coggan, M., Baker, R., Miloszewski, K., Woodfield, G., and Board, P. (1995). Mutations causing coagulation factor XIII subunit A deficiency: Characterization of the mutant proteins after expression in yeast. *Blood* **9,** 2455–2460.

Dantuma, N. P., Lindsten, K., Glas, R., Jellne, M., and Masucci, M. G. (2000). Short-lived green fluorescent proteins for quantifying ubiquitin/proteasome-dependent proteolysis in living cells. *Nature Biotechnol.* **18,** 494–496.

Deichsel, H., Friedel, S., Detterbeck, A., Coyne, C., Hamker, U., and MacWilliams, H. K. (1999). Green fluorescent proteins with short half-lives as reporters in Dictyostelium discoideum. *Dev. Genes Evol.* **209,** 63–68.

Dohmen, J., and Varshavsky, A. (2005). Heat-inducible degron and the making of conditional mutants. *In* "Methods in Enzymology" (R. J. Deshaies, ed.). Academic Press, New York.

Dohmen, R. J., Wu, P., and Varshavsky, A. (1994). Heat-inducible degron: A method for constructing temperature-sensitive mutants. *Science* **263,** 1273–1276.

Du, F., Navarro-Garcia, F., Xia, Z., Tasaki, T., and Varshavsky, A. (2002). Pairs of dipeptides synergistically activate the binding of substrate by ubiquitin ligase through dissociation of its autoinhibitory domain. *Proc. Natl. Acad. Sci. USA* **99,** 14110–14115.

Dues, G., Muller, S., and Johnsson, N. (2001). Detection of a conformational change in G-gamma upon binding G-beta in living cells. *FEBS Lett.* **505,** 75–80.

Dunnwald, M., Varshavsky, A., and Johnsson, N. (1999). Detection of transient *in vivo* interactions between substrate and transporter during protein translocation into the endoplasmic reticulum. *Mol. Biol. Cell* **10,** 329–344.

Ecker, D. J., Stadel, J. M., Butt, T. R., Marsh, J. A., Monia, B. P., Powers, D. A., Gorman, J. A., Clark, P. E., Warren, F., Shatzman, A., *et al.* (1989). Increasing gene expression in yeast by fusion to ubiquitin. *J. Biol. Chem.* **264,** 7715–7719.

Eckert, J. H., and Johnsson, N. (2003). Pex10p links the ubiquitin-conjugating enzyme Pex4p to the protein import machinery of the peroxisome. *J. Cell Sci.* **116,** 3623–3634.

Falnes, P. O., and Olsnes, S. (1998). Modulation of the intracellular stability and toxicity of diphtheria toxin through degradation by the N-end rule pathway. *EMBO J.* **17,** 615–625.

Falnes, P. O., Welker, R., Krausslich, H. G., and Olsnes, S. (1999). Toxins are activated by HIV-type-1 protease through removal of signal for degradation by the N-end rule pathway. *Biochemical J.* **343,** 199–207.

Fang, S., and Weissman, A. M. (2004). A field guide to ubiquitylation. *Cell Mol. Life. Sci.* **61,** 1546–1561.

Finley, D., Bartel, B., and Varshavsky, A. (1989). The tails of ubiquitin precursors are ribosomal proteins whose fusion to ubiquitin facilitates ribosome biogenesis. *Nature* **338**, 394–401.

Finley, D., Özkaynak, E., and Varshavsky, A. (1987). The yeast polyubiquitin gene is essential for resistance to high temperatures, starvation, and other stresses. *Cell* **48**, 1035–1046.

Galarneau, A., Primeau, M., Trudeau, L. E., and Michnick, S. W. (2002). Beta-lactamase protein fragment complementation assays as *in vivo* and *in vitro* sensors of protein-protein interactions. *Nature Biotechnol.* **20**, 619–622.

Gardner, R. G., and Hampton, R. Y. (1999). A 'distributed degron' allows regulated entry into the ER degradation pathway. *EMBO J.* **18**, 5994–6004.

Gehring, M. R., Condon, B., Margosiak, S. A., and Kan, C. C. (1995). Characterization of the Phe-81 and Val-82 human fibroblast collagenase catalytic domain purified from *Escherichia coli. J. Biol. Chem.* **270**, 22507–22513.

Ghislain, M., Dohmen, R. J., Levy, F., and Varshavsky, A. (1996). Cdc48p interacts with Ufd3p, a WD repeat protein required for ubiquitin-mediated proteolysis in *Saccharomyces cerevisiae. EMBO J.* **15**, 4884–4899.

Gilchrist, C. A., Gray, D. A., and Baker, R. T. (1997). A ubiquitin-specific protease that efficiently cleaves the ubiquitin-proline bond. *J. Biol. Chem.* **272**, 32280–32285.

Gonda, D. K., Bachmair, A., Wünning, I., Tobias, J. W., Lane, W. S., and Varshavsky, A. (1989). Universality and structure of the N-end rule. *J. Biol. Chem.* **264**, 16700–16712.

Han, K., Hong, J., Lim, H. C., Kim, C. H., Park, Y., and Cho, J. M. (1994). Tyrosinase production in recombinant *E. coli* containing trp promoter and ubiquitin sequence. *Ann. NY. Acad. Sci.* **721**, 30–42.

Hardy, C. F. (1996). Characterization of an essential Orc2p-associated factor that plays a role in DNA replication. *Mol. Cell. Biol.* **16**, 1832–1841.

Hemelaar, J., Galardy, P. J., Borodovsky, A., Kessler, B. M., and Ovaa, H. (2004). Chemistry-based functional proteomics: Mechanism-based activity-profiling tools for ubiquitin and ubiquitin-like specific proteases. *J. Proteome Res.* **3**, 268–276.

Hershko, A., Ciechanover, A., and Varshavsky, A. (2000). The ubiquitin system. *Nature Med.* **10**, 1073–1081.

Hicke, L., and Dunn, R. (2003). Regulation of membrane protein transport by ubiquitin and ubiquitin-binding proteins. *Annu. Rev. Cell Dev. Biol.* **19**, 141–172.

Hill, C. P., Johnston, N. L., and Cohen, R. E. (1993). Crystal structure of a ubiquitin-dependent degradation substrate: A three-disulfide form of lysozyme. *Proc. Natl. Acad. Sci. USA* **90**, 4136–4140.

Hondred, D., Walker, J. M., Mathews, D. E., and Vierstra, R. D. (1999). Use of ubiquitin fusions to augment protein expression in transgenic plants. *Plant Physiol.* **119**, 713–724.

Johnson, E. S., Bartel, B.,W., and Varshavsky, A. (1992). Ubiquitin as a degradation signal. *EMBO J.* **11**, 497–505.

Johnson, E. S., Ma, P. C., Ota, I. M., and Varshavsky, A. (1995). A proteolytic pathway that recognizes ubiquitin as a degradation signal. *J. Biol. Chem.* **270**, 17442–17456.

Johnsson, N. (2002). A split-ubiquitin-based assay detects the influence of mutations on the conformational stability of the p53 DNA-binding domain *in vivo. FEBS Lett.* **531**, 259–264.

Johnsson, N., and Varshavsky, A. (1994a). Split ubiquitin as a sensor of protein interactions *in vivo. Proc. Natl. Acad. Sci. USA* **91**, 10340–10344.

Johnsson, N., and Varshavsky, A. (1994b). Ubiquitin-assisted dissection of protein transport across cell membranes. *EMBO J.* **13**, 2686–2698.

Kanemaki, M., Sanchez-Diaz, A., Gambus, A., and Labib, K. (2003). Functional proteomic identification of DNA replication proteins by induced proteolysis *in vivo. Nature* **423**, 720–724.

Kapust, R. B., and Waugh, D. S. (1999). *Escherichia coli* maltose-binding protein is uncommonly effective at promoting solubility of polypeptides to which it is fused. *Protein Sci.* **8,** 1668–1674.

Kesti, T., McDonald, W. H., Yates, J. R., Jr., and Wittenberg, C. (2004). Cell cycle-dependent phosphorylation of the DNA polymerase epsilon subunit, Dpb2, by the Cdc28 cyclin-dependent protein kinase. *J. Biol. Chem.* **279,** 14245–14255.

Koegl, M., Hoppe, T., Schlenker, S., Ulrich, H. D., Mayer, T. U., and Jentsch, S. (1999). A novel ubiquitination factor, E4, is involved in multiubiquitin chain assembly. *Cell* **96,** 635–644.

Koken, M. H., Odijk, H. H., Van Duin, M., Fornerod, M., and Hoeijmakers, J. H. (1993). Augmentation of protein production by a combination of the T7 RNA polymerase system and ubiquitin fusion: Overproduction of the human DNA repair protein, ERCC1, as a ubiquitin fusion in *Escherichia coli*. *Biochem. Biophys. Res. Commun.* **195,** 643–653.

Kwon, Y. T., Kashina, A. S., Davydov, I. V., Hu, R.-G., An, J. Y., Seo, J. W., Du, F., and Varshavsky, A. (2002). An essential role of N-terminal arginylation in cardiovascular development. *Science* **297,** 96–99.

Kwon, Y. T., Xia, Z. X., An, J. Y., Davydov, I. V., Seo, J. W., Xie, Y., and Varshavsky, A. (2003). Female lethality and apoptosis of spermatocytes in mice lacking the UBR2 ubiquitin ligase of the N-end rule pathway. *Mol. Cell. Biol.* **23,** 8255–8271.

Labib, K., Tercero, J. A., and Diffley, J. F. (2000). Uninterrupted MCM2-7 function required for DNA replication fork progression. *Science* **288,** 1643–1647 (erratum in: *Science* 1289, 2052 [2000]).

Laser, H., Bongards, C., Schuller, J., Heck, S., Johnsson, N., and Lehming, N. (2000). A new screen for protein interactions reveals that the *Saccharomyces cerevisiae* high mobility group proteins Nhp6A/B are involved in the regulation of the GAL1 promoter. *Proc. Natl. Acad. Sci. USA* **97,** 13732–13737.

Lévy, F., Johnsson, N., Rümenapf, T., and Varshavsky, A. (1996). Using ubiquitin to follow the metabolic fate of a protein. *Proc. Natl. Acad. Sci. USA* **93,** 4907–4912.

Lévy, F., Johnston, J. A., and Varshavsky, A. (1999). Analysis of a conditional degradation signal in yeast and mammalian cells. *Eur. J. Biochem.* **259,** 244–252.

Loison, F., Nizard, P., Sourisseau, T., Le Goff, P., Deburre, L., Le Drean, Y., and Michel, D. (2005). A ubiquitin-based assay for the cytosolic uptake of protein transduction domains. *Mol. Ther.* **11,** 205–214.

Mak, P., McDonnell, D. P., Weigel, N. L., Schrader, W. T., and O'Malley, B. W. (1989). Expression of functional chicken oviduct progesterone receptors in yeast *Saccharomyces cerevisiae*. *J. Biol. Chem.* **264,** 21613–21618.

Moqtaderi, Z., Bai, Y., Poon, D., Weil, P. A., and Struhl, K. (1996). TBP-associated factors are not generally required for transcriptional activation in yeast. *Nature* **383,** 188–191.

Park, E. C., Finley, D., and Szostak, J. W. (1992). A strategy for the generation of conditional mutations by protein destabilization. *Proc. Natl. Acad. Sci. USA* **89,** 1249–1252.

Paz, I., Meunier, J.-R., and Choder, M. (1999). Monitoring dynamics of gene expression in yeast during stationary phase. *Gene* **236,** 33–42.

Pelletier, J. N., Campbell-Valois, F. X., and Michnick, S. W. (1998). Oligomerization domain-directed reassembly of active dihydrofolate reductase from rationally designed fragments. *Proc. Natl. Acad. Sci. USA* **95,** 12141–12146.

Petroski, M. D., and Deshaies, R. J. (2005). Function and regulation of cullin-RING ubiquitin ligases. *Nature Rev. Mol. Cell Biol.* **6,** 9–20.

Phizicky, E. M., and Fields, S. (1995). Protein-protein interactions: Methods for detection and analysis. *Microbiol. Rev.* **59,** 94–123.

Pickart, C. (2004). Back to the future with ubiquitin. *Cell* **116,** 181–190.

Pickart, C., and Cohen, R. E. (2004). Proteasomes and their kin: Proteases in the machine age. *Nature Rev. Mol. Cell Biol.* **5,** 177–187.

Pickart, C., and Fushman, D. (2004). Polyubiquitin chains: Polymeric protein signals. *Curr. Op. Chem. Biol.* **8,** 610–616.

Pilon, A., Yost, P., Chase, T. E., Lohnas, G., Burkett, T., Roberts, S., and Bentley, W. E. (1997). Ubiquitin fusion technology: Bioprocessing of peptides. *Biotechnol. Prog.* **13,** 374–379.

Qian, S.-B., Ott, D. E., Schubert, U., Bennink, J. R., and Yewdell, J. W. (2002). Fusion proteins with COOH-terminal ubiquitin are stable and maintain functionality *in vivo*. *J. Biol. Chem.* **277,** 38818–38826.

Rao, H., Uhlmann, F., Nasmyth, K., and Varshavsky, A. (2001). Degradation of a cohesin subunit by the N-end rule pathway is essential for chromosome stability. *Nature* **410,** 955–960.

Raquet, X., Eckert, J. H., Muller, S., and Johnsson, N. (2001). Detection of altered protein conformations in living cells. *J. Mol. Biol.* **305,** 927–938.

Rechsteiner, M., and Hill, C. P. (2005). Mobilizing the proteolytic machine: Cell biological roles of proteasome activators and inhibitors. *Trends Cell. Biol.* **15,** 27–33.

Redman, K. L., and Rechsteiner, M. (1989). Identification of the long ubiquitin extension as ribosomal protein S27a. *Nature* **338,** 438–440.

Reits, E. A. J., Vos, J. C., Grommé, M., and Neefjes, J. (2000). The major substrates for TAP *in vivo* are derived from newly synthesized proteins. *Nature* **404,** 774–778.

Remy, I., and Michnick, S. W. (1999). Clonal selection and *in vivo* quantitation of protein interactions with protein-fragment complementation assays. *Proc. Natl. Acad. Sci. USA* **96,** 5394–5399.

Rian, E., Jemtland, R., Olstad, O. K., Gordeladze, J. O., and Gautvik, K. M. (1993). Synthesis of human parathyroid hormone-related protein (10141) in *Saccharomyces cerevisiae*. A correct amino-terminal processing is obtained by the ubiquitin fusion approach. *Eur. J. Biochem.* **213,** 641–648.

Sabin, E. A., Lee-Ng, C. T., Shuster, J. R., and Barr, P. J. (1989). High-level expression and *in vivo* processing of chimeric ubiquitin fusion proteins in *Saccharomyces cerevisiae*. *BioTechnology* **7,** 705–709.

Schubert, U., Antón, L. C., Gibbs, J., Norbury, C. C., Yewdell, J. W., and Bennink, J. R. (2000). Rapid degradation of a large fraction of newly synthesized proteins by proteasomes. *Nature* **404,** 770–774.

Sheng, J., Kumagai, A., Dunphy, W. G., and Varshavsky, A. (2002). Dissection of c-MOS degron. *EMBO J.* **21,** 6061–6071.

Stagljar, I., Korostensky, C., Johnsson, N., and te Heesen, S. (1998). A genetic system based on split-ubiquitin for the analysis of interactions between membrane proteins *in vivo*. *Proc. Natl. Acad. Sci. USA* **95,** 5187–5192.

Suzuki, T., and Varshavsky, A. (1999). Degradation signals in the lysine-asparagine sequence space. *EMBO J.* **18,** 6017–6026.

Tafelmeyer, P., Johnsson, N., and Johnsson, K. (2004). Transforming a beta/alpha-barrel enzyme into a split-protein sensor through directed evolution. *Chem. Biol.* **11,** 589–591.

Tcherniuk, S. O., Chroboszek, J., and Balakirev, M. Y. (2004). Construction of tumor-specific toxins using ubiquitin fusion technique. *Mol. Ther.* **11,** 196–204.

Tobery, T., and Siliciano, R. F. (1999). Induction of enhanced CTL-dependent protective immunity *in vivo* by N-end rule targeting of a model tumor antigen. *J. Immunol.* **162,** 639–642.

Tobias, J. W., Shrader, T. E., Rocap, G., and Varshavsky, A. (1991). The N-end rule in bacteria. *Science* **254,** 1374–1377.

Townsend, A., Bastin, J., Gould, K., Brownlee, G., Andrew, M., Coupar, B., Boyle, D., Chan, S., and Smith, G. (1988). Defective presentation to class I-restricted cytotoxic T lymphocytes in vaccinia-infected cells is overcome by enhanced degradation of antigen. *J. Exp. Med.* **168,** 1211–1224.

Turner, G. C., Du, F., and Varshavsky, A. (2000). Peptides accelerate their uptake by activating a ubiquitin-dependent proteolytic pathway. *Nature* **405,** 579–583.

Turner, G. C., and Varshavsky, A. (2000). Detecting and measuring cotranslational protein degradation *in vivo. Science* **289,** 2117–2120.

Valasek, L., Mathew, A. A., Shin, B. S., Nielsen, K. H., Szamecz, B., and Hinnebusch, A. G. (2003). The yeast eIF3 subunits TIF32/a, NIP1/c, and eIF5 make critical connections with the 40S ribosome *in vivo. Genes Dev.* **17,** 786–799.

Varshavsky, A. (1991). Naming a targeting signal. *Cell* **64,** 13–15.

Varshavsky, A. (1995). Codominance and toxins: A path to drugs of nearly unlimited selectivity. *Proc. Natl. Acad. Sci. USA* **92,** 3663–3667.

Varshavsky, A. (1996a). The N-end rule. *Cold Spring Harb. Symp. Quant. Biol.* **60,** 461–478.

Varshavsky, A. (1996b). The N-end rule: Functions, mysteries, uses. *Proc. Natl. Acad. Sci. USA* **93,** 12142–12149.

Varshavsky, A. (1997). The ubiquitin system. *Trends Biochem. Sci.* **22,** 383–387.

Varshavsky, A. (1998). Codominant interference, antieffectors, and multitarget drugs. *Proc. Natl. Acad. Sci. USA* **95,** 2094–2099.

Varshavsky, A. (2000). Ubiquitin fusion technique and its descendants. *Meth. Enzymol.* **327,** 578–593.

Varshavsky, A. (2003). The N-end rule and regulation of apoptosis. *Nature Cell Biol.* **5,** 373–376.

Verma, R., L., A., Oania, R., McDonald, W. H., Yates, J. R. I., Koonin, E. V., and Deshaies, R. J. (2002). Role of Rpn11 metalloprotease in deubiquitination and degradation by the 26S proteasome. *Science* **298,** 611–615.

Walter, P., Gilmore, R., and Blobel, G. (1984). Protein translocation across the endoplasmic reticulum. *Cell* **38,** 5–8.

Wang, X., Ira, G., Tercero, J. A., Holmes, A. M., Diffley, J. F., and Haber, J. E. (2004). Role of DNA replication proteins in double-strand break-induced recombination in *Saccharomyces cerevisiae. Mol. Cell. Biol.* **24,** 6891–6899.

Webb, E. C. (ed.) (1992). "Enzyme Nomenclature, 1992." Academic Press, New York.

Wilkinson, K., and Hochstrasser, M. (1998). The deubiquitinating enzymes. *In* "Ubiquitin and the Biology of the Cell" (J.-M. Peters, J. R. Harris, and D. Finley, eds.), pp. 99–146. Plenum Press, New York.

Wilkinson, K. D. (2000). Ubiquitination and deubiquitination: Targeting of proteins for degradation by the proteasome. *Semin. Cell Dev. Biol.* **11,** 141–148.

Wittke, S., Dunnwald, M., Albertsen, M., and Johnsson, N. (2002). Recognition of a subset of signal sequences by Ssh1p, a Sec61p-related protein in the membrane of endoplasmic reticulum of yeast *Saccharomyces cerevisiae. Mol. Biol. Cell* **13,** 2223–2232.

Wittke, S., Dunnwald, M., and Johnsson, N. (2000). Sec62p, a component of the endoplasmic reticulum protein translocation machinery, contains multiple binding sites for the Sec-complex. *Mol. Biol. Cell* **11,** 3859–3871.

Wittke, S., Lewke, N., Müller, S., and Johnsson, N. (1999). Probing the molecular environment of membrane proteins *in vivo. Mol. Biol. Cell* **10,** 2519–2530.

Wolf, J., Nicks, M., Deitz, S., van Tuinen, E., and Franzusoff, A. (1998). An N-end rule destabilization mutant reveals pre-Golgi requirements for Sec7p in yeast membrane traffic. *Biochem. Biophys. Res. Commun.* **243,** 191–198.

Worley, C. K., Ling, R., and Callis, J. (1998). Engineering *in vivo* instability of firefly luciferase and *Escherichia coli* beta-glucuronidase in higher plants using recognition elements from the ubiquitin pathway. *Plant. Mol. Biol.* **37,** 337–347.

Yin, J., Kwon, Y. T., Varshavsky, A., and Wang, W. (2004). RECQL4, mutated in the Rothmund-Thomson and RAPADILINO syndromes, interacts with ubiquitin ligases UBR1 and UBR2 of the N-end rule pathway. *Hum. Mol. Genet.* **13,** 2421–2430.

Yoo, Y., Rote, K., and Rechsteiner, M. (1989). Synthesis of peptides as cloned ubiquitin extensions. *J. Biol. Chem.* **264,** 17078–17083.

Zhang, S., Ma, C., and Chalfie, M. (2004). Combinatorial marking of cells and organelles with reconstituted fluorescent proteins. *Cell* **119,** 137–144.

[52] Heat-Inducible Degron and the Making of Conditional Mutants

By R. Jürgen Dohmen and Alexander Varshavsky

Abstract

Conditional mutants retain the function of a specific gene under one set of conditions, called permissive, and lack that function under a different set of conditions, called nonpermissive; the latter must be still permissive for the wild-type allele of a gene. Such mutants make possible the analysis of physiological changes that follow controlled inactivation of a gene or gene product and can be used to address the function of any gene. Temperature-sensitive (*ts*) mutants, first used in functional studies more than half a century ago, remain a mainstay of genetic analyses. One limitation of the classical *ts* approach is the uncertainty as to whether a given gene can be mutated to yield a *ts* product. Another problem with conventional *ts* mutations is that they are often too leaky to be useful. In 1994, we described a new method, based on a heat-activated degradation signal (degron) that is targeted by the N-end–rule pathway in the yeast *Saccharomyces cerevisiae*. The corresponding mutants were termed *td* (temperature-activated degron) to distinguish them from conventional *ts* mutants. The *td* method requires neither a missense mutation in a gene of interest nor an alteration in its expression patterns. Arg-DHFR^{ts}, a *ts* variant of dihydrofolate reductase–bearing N-terminal Arg residue (a destabilizing residue in the N-end rule) was shown to function as a portable, heat-activated degron, in that Arg-DHFR^{ts} was long-lived at 23° but became short-lived at 37°, owing to activation of its previously cryptic degron. Linking, in a linear fusion, this portable *ts*-degron to a protein of interest results in destruction of the latter at 37°, thereby yielding a *ts* (*td*) mutant of a corresponding gene. Since the introduction of the *td* method in 1994, numerous studies have successfully used *td* alleles of specific genes in functional analyses.

Introduction

Genomic and proteomic investigations have yielded, and continue to produce, a large amount of information about genes and their protein products. In contrast, the evidence bearing on physiological roles of specific proteins is much more scarce. To address the functional part of biological inquiry, one would like to perturb, at will and selectively, the function of any protein of interest *in vivo* and to analyze the resulting phenotypic effects, thereby probing the protein's role in a cell. Ideally, a method for doing so should be applicable both to individual gene products and to a large collection of them. Conditional mutants provide one such technique and are particularly useful for analysis of genes whose functions are essential for the organism's viability. A conditional mutant retains the function of a gene under one set of conditions, called permissive, and lacks that function under a different set of conditions, called nonpermissive; the latter must be still permissive for the wild-type allele of a gene. Conditional mutants make possible the analysis of physiological changes that follow controlled inactivation of a gene or gene product and can be used to address the function of any gene. Conventional temperature-sensitive (*ts*) conditional mutants, produced through mutagenesis of a gene of interest, are cumbersome to make, are often too "leaky" (retain too much activity at nonpermissive temperature), and also tend to display a "phenotypic lag" between the imposition of nonpermissive conditions and the emergence of a relevant null phenotype. In 1994, we described a different method for making *ts* mutants, through the tagging of a protein of interest with a portable heat-activated degradation signal (degron) (Dohmen *et al.*, 1994). Degron-based strategies for producing *ts* mutants have several advantages over the standard approach, including the absence of necessity to search for a *ts* allele by random mutagenesis.

Conditional Mutants

Since the introduction of temperature-sensitive (*ts*) mutants 55 years ago (Horowitz, 1950), a variety of conditional mutants and methods for producing them have been developed. Thus, *ts*, *cs* (cold-sensitive) (Moir *et al.*, 1982; Pringle, 1975), *ds* (D_2O-sensitive) (Bartel and Varshavsky, 1988), osmolality-sensitive (Aiba *et al.*, 1998; Latterich and Watson, 1991), pH-sensitive (Colb and Shapiro, 1977), formamide-sensitive (Aguilera, 1994), and analogous mutants in a given gene can be produced by mutagenizing the gene and screening for its mutant alleles that are functional under permissive but not under nonpermissive conditions. Temperature-sensitive mutants can also be produced by using a cell carrying a *ts* suppressor tRNA that can suppress a nonsense mutation in a gene of interest and does so at permissive but not at nonpermissive temperature, (Marschalek *et al.*, 1990; Steege and Horabin, 1983). A cold-sensitive intron that fails to splice out from a pre-mRNA at nonpermissive temperature, or a temperature-sensitive intein (protein-splicing element) have also been used to construct conditional mutants (Yoshimatsu and Nagawa, 1989; Zeidler *et al.*, 2004). Yet another strategy, conditional expression of genes from inducible promoters, has been widely used to characterize gene functions in prokaryotes and eukaryotes (Hill and Bloom, 1987; Hu and Davidson, 1987; Mnaimneh *et al.*, 2004; Zhang and Moss, 1991). Advantages of this method include preservation of a gene's initial open reading frame (ORF) and applicability to any organism and any gene whose expression can be brought under control of an inducible (or repressible) promoter.

A frequent problem with conditional phenotypes produced by either the *ts/cs* or regulated-promoter methods is leakiness (i.e., an insufficiently low activity of either a *ts* protein at nonpermissive temperature or a gene of interest in the "off" state of a relevant promoter). In a study that used a tetracycline-repressible promoter to generate *S. cerevisiae* strains bearing promoter-shutoff versions of essential genes, it was found that ∼45% of more than 600 strains examined did not exhibit strong growth-impairment phenotypes in the presence of tetracycline (in the "off" state of promoter) (Mnaimneh *et al.*, 2004). In addition, ∼20% of strains with a detectable promoter-off phenotype were also growth impaired in the absence of tetracycline (in the "on" state of promoter) (Mnaimneh *et al.*, 2004), suggesting misregulation of these genes in the presence of nonnative promoter. Yet another frequent problem stems from "phenotypic lag," a significant time interval between the imposition of nonpermissive conditions and the emergence of a relevant null phenotype (Hann and Walter, 1991). The extent of delay is determined by the rate of disappearance of a gene product; this depends on metabolic stabilities of the corresponding mRNA and protein. A long phenotypic lag is more likely with proteins that act catalytically rather than stoichiometrically.

To address these problems, methods for producing conditional mutants that allow a regulated metabolic destabilization of a protein of interest have been developed. One useful feature of such methods is fast disappearance of mutant proteins under nonpermissive conditions, resulting in rapid onsets of null phenotypes. In combination with earlier strategies, these methods can also be used to reduce or eliminate the leakiness or phenotypic lag of *ts* and promoter-dependent conditional mutants capable of being regulated (Moqtaderi *et al.*, 1996; Park *et al.*, 1992). Proteolysis-based methods use protein fusions in which a protein of interest is linked to a portable domain that functions as a degradation signal or "degron" (Varshavsky, 1991). The earliest and still most useful mutants of this class use degrons that target proteins for degradation by the N-end–rule pathway. Another shorthand in use, largely for denoting degrons in cyclins and functionally related proteins, is "destruction box." Although "degron" and "destruction box" are essentially synonyms, "degron" is a shorter, one-word term and readily yields compact acronyms for specific degradation signals through the use of prefixes and other auxiliary notations (Varshavsky, 1991). The terminology of degradation signals would be simplified and clarified if degron-based terms, with appropriate prefixes, subscripts, or superscripts, are systematically used for denoting all degrons, including those named differently at present.

The N-End–Rule Pathway

The N-end rule relates the *in vivo* half-life of a protein to the identity of its N-terminal residue (Bachmair *et al.*, 1986; Varshavsky, 1996, 2003). Physiological functions of this universally present ubiquitin (Ub)-dependent proteolytic pathway (Fig. 1) continue to emerge and are described elsewhere (Davydov and Varshavsky, 2000; Ditzel *et al.*, 2003; Du *et al.*, 2002; Kwon *et al.*, 2000, 2001, 2002, 2003; Rao *et al.*, 2001; Turner *et al.*, 2000; Varshavsky, 2003; Yin *et al.*, 2004). One class of degrons recognized by the N-end–rule pathway is called N-degrons. An N-degron consists of a protein's destabilizing N-terminal residue (recognized by a specific E3 Ub ligase) and an internal Lys residue, the site of formation of a protein-linked poly-Ub chain (Bachmair and Varshavsky, 1989; Dohmen *et al.*, 1991a; Suzuki and Varshavsky, 1999; Varshavsky, 1996). A ubiquitylated protein is processively degraded by the 26S proteasome (Crews, 2003; Förster and Hill, 2003; Hochstrasser, 1996; Pickart, 2004; Rechsteiner, 1998; Varshavsky, 1997; Verma *et al.*, 2003; Zwickl *et al.*, 2000). In the yeast *Saccharomyces cerevisiae*, two substrate-binding sites of UBR1, a 225-kDa, RING-domain E3 Ub ligase, recognize (bind to) primary destabilizing N-terminal residues of two types, basic (type 1, Arg, Lys, and His) and bulky hydrophobic (type 2, Phe, Leu, Tyr, Trp, and Ile) (Du *et al.*, 2002; Kwon *et al.*, 2003). Several other N-terminal residues function as tertiary (Asn, Gln) and secondary (Asp, Glu) destabilizing residues, in that they are recognized

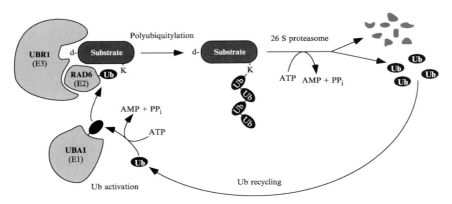

FIG. 1. The N-end–rule pathway. This simplified diagram of the pathway illustrates its dependence on ubiquitin (Ub) conjugation and the 26S proteasome, as well as its property of recognizing N-degrons. The diagram omits other pathway's features, such as, for example, the hierarchic structure of the N-end rule, with primary, secondary, and tertiary destabilizing residues (Varshavsky, 2003) and the fact that the UBR1 Ub ligase has at least three distinct substrate-binding sites, one of which recognizes internal (non-N-terminal) degrons (Du *et al.*, 2002). The conjugation of Ub to other proteins involves a preliminary ATP-dependent step, in which the last (Gly-76) residue of Ub is joined, through a thioester bond, to a Cys residue in the Ub-activating (E1) enzyme encoded by *UBA1*. The activated Ub is transferred to a Cys residue of RAD6, a Ub-conjugating (E2) enzyme that forms, together with the E3 component UBR1, the Ub ligase of the N-end–rule pathway (Varshavsky, 1996). The recognition of a substrate's destabilizing N-terminal residue ("**d**") by UBR1 leads to formation, by the UBR1/ RAD6 Ub ligase, of a substrate-linked poly-Ub chain. The resulting ubiquitylated protein is processively degraded to short peptides by the ATP-dependent 26S proteasome, with Ub recovered (through the activity of deubiquitylating enzymes (DUBs)) and reused.

by UBR1 after their enzymatic conjugation to Arg, a primary destabilizing residue (Davydov and Varshavsky, 2000; Kwon *et al.*, 2002; Varshavsky, 1996). In the case of N-terminal Asn and Gln (as well as Cys in metazoans), the conjugation of Arg is preceded by other enzymatic modifications of the acceptor residue (Baker and Varshavsky, 1995; Kwon *et al.*, 2000, 2002). UBR1, which operates in a complex with the Ub-conjugating (E2) enzyme RAD6 (Dohmen *et al.*, 1991a; Varshavsky, 1996; Xie and Varshavsky, 1999), also contains a third substrate-binding site that targets proteins such as CUP9, a transcriptional repressor, through its internal (non-N-terminal) degron (Byrd *et al.*, 1998; Du *et al.*, 2002; Turner *et al.*, 2000). In contrast to yeast, several functionally overlapping E3 Ub ligases mediate the N-end–rule pathway in mammals (Kwon *et al.*, 2001, 2003). None of the multiple functions of the *S. cerevisiae* N-end–rule pathway (Du *et al.*, 2002;

Rao *et al.*, 2001; Turner *et al.*, 2000) is essential for quasi-normal growth of cells in standard media (Bartel *et al.*, 1990; Rao *et al.*, 2001).

Using Ubiquitin Fusions to Make N-Degrons *In Vivo*

The ubiquitin (Ub) fusion technique was invented in 1986 through experiments in which a segment of DNA encoding the 76-residue Ub was joined, in-frame, to DNA encoding *E. coli* β-galactosidase (βgal) (Bachmair *et al.*, 1986; Varshavsky, 2000). When the resulting protein fusion was expressed in *S. cerevisiae* and detected by radiolabeling-immunoprecipitation, only the fusion's βgal moiety was observed, even if the labeling time was short enough to be roughly equal to the time it takes, *in vivo*, to translate the ORF of Ub-βgal (1–2 min). In yeast, and in all eukaryotes examined afterwards, the Ub moiety of a Ub fusion is cleaved off, cotranslationally or nearly so, after the last residue of Ub (Fig. 2). The proteases involved are called deubiquitylating enzymes (DUBs) (Hochstrasser, 2000; Wilkinson, 2000). A eukaryotic cell contains many distinct DUBs, all of which are specific for the Ub moiety. *In vivo* cleavage at the Ub-polypeptide junction of a Ub fusion has been shown to be largely cotranslational (Turner and Varshavsky, 2000). One physiological function of the cleavage reaction (Fig. 2) is to mediate the excision of Ub from its natural DNA-encoded precursors, either linear poly-Ub or fusions between Ub and specific ribosomal proteins (Finley *et al.*, 1987, 1989).

Another finding about the DUB-mediated cleavage reaction (Fig. 2) led to discovery of the N-end–rule pathway. It was shown that the cleavage of a Ub-X-polypeptide fusion after the last residue of Ub takes place regardless of the identity of a residue X at the C-terminal side of the cleavage site, the Pro residue being a single exception (Bachmair *et al.*, 1986). By allowing a bypass of "normal" N-terminal processing of a newly formed protein, this result yielded an *in vivo* method for placing different residues at the N-termini of otherwise identical proteins. It was then discovered that the *in vivo* half-lives of resulting test proteins were determined by the identities of their N-terminal residues, a relation referred to as the N-end rule (Bachmair *et al.*, 1986; Varshavsky, 1996).

The Ub fusion technique (Fig. 2) remains the method of choice for producing, *in vivo*, a desired N-terminal residue in a protein of interest. Owing to constraints of the genetic code, nascent proteins bear N-terminal Met (formyl-Met in prokaryotes). The known methionine aminopeptidases (MetAPs) that remove N-terminal Met do so if, and only if, the residue to be made N-terminal has a small enough side chain (Bradshaw *et al.*, 1998). Such residues are stabilizing ones in the N-end rule; the exceptions (only in metazoan cells, not in fungi) are Cys, Ser, Ala, and Thr (Varshavsky,

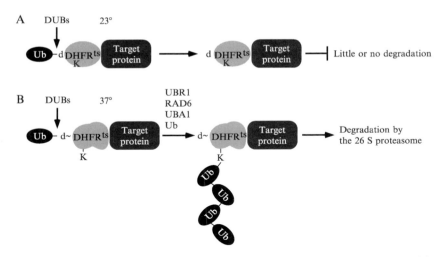

FIG. 2. Temperature-activated degron (*td*) and its use for making temperature-sensitive (*ts*) mutants. (A) The target protein is expressed as a fusion to a "*td* module," which consists of ubiquitin (Ub) followed by a residue ("**d**") such as Arg (which is destabilizing in the N-end rule), then by a particular *ts* mutant (Pro → Leu at position 66) of the mouse dihydrofolate reductase (DHFR) (Arg-DHFR^{ts}), and finally by the moiety of a protein of interest ("target protein"). On expression of this fusion, the Ub moiety is cleaved off, cotranslationally or nearly so, by deubiquitylating enzymes (DUBs), yielding a destabilizing residue ("**d**") at the N-terminus of resulting protein. The latter is not targeted for degradation by the N-end rule pathway at permissive temperature (23°), despite the presence of N-terminal "**d**", either because the N-terminal residue is not sufficiently exposed for recognition by UBR1 and/or because there is no sterically suitable Lys residue to serve as the second determinant of N-degron (see Fig. 1 and the main text). (B) The upshift to nonpermissive temperature (37°) causes a conformational perturbation of the Arg-DHFR^{ts} module (indicated in B by a changed shape of the module) that activates the previously cryptic N-degron. The resulting Ub-dependent, processive degradation of the Arg-DHFR^{ts} moiety by the N-end rule pathway also destroys the linked downstream moiety of a protein of interest ("target protein").

1996, 2003). In other words, MetAPs do not cleave off N-terminal Met if it is followed by any of 12 destabilizing residues in the yeast-type N-end rule. Ub-specific DUBs are free of this constraint, except when the residue X of a Ub-X-polypeptide is Pro, in which case the cleavage still takes place but at a much lower rate (Bachmair and Varshavsky, 1989; Bachmair *et al.*, 1986; Johnson *et al.*, 1992). A DUB was subsequently identified that can efficiently cleave at the Ub-Pro junction (Gilchrist *et al.*, 1997). Note that N-degrons can also be produced in proteins through *in vivo* cleavages by other proteases, such as separases, caspases, and calpains (Ditzel *et al.*, 2003; Rao *et al.*, 2001; Varshavsky, 2003).

Using Regulated Proteolysis to Produce Conditional Mutants

Controlled Expression of a Protein Bearing a Constitutive N-Degron

The first method using the N-end–rule pathway to produce conditional mutants in yeast employed a conditional (galactose-inducible) transcriptional promoter to express a Ub fusion to a protein of interest that contained a destabilizing residue at the Ub-protein junction (Park *et al.*, 1992). To ensure, in this approach, that a complete, two-determinant N-degron (see above) is present at the protein's N-terminus after the (cotranslational) removal of the Ub moiety, one can also place, at the Ub-protein junction, a portable carrier of the second determinant (a targetable Lys residue[s]). The carrier usually used is a ~45-residue N-terminal extension termed e^K (extension bearing lysine [K]) derived from previously engineered N-end–rule substrates (Bachmair and Varshavsky, 1989; Suzuki and Varshavsky, 1999; Varshavsky, 1996). In the original application of the "constitutive N-degron" technique (Park *et al.*, 1992), an N-degron, in the context of a Ub fusion, was tagged to the N-terminus of ARD1, a subunit of N-terminal acetyltransferase. The resulting (modified) *ARD1* ORF was expressed in an *ard1*Δ strain from the galactose-inducible, glucose-repressible P_{GAL1} promoter, yielding ARD1$^+$ cells on galactose media, despite metabolic instability of N-degron-bearing ARD1. Shutting off the expression of modified *ARD1* by removal of galactose and addition of glucose resulted in rapid disappearance of the N-degron-bearing ARD1 protein, thus reducing phenotypic lag. In addition, the presence of N-degron in ARD1 decreased the leakiness of conditional phenotype under nonpermissive (galactose-off) conditions (Park *et al.*, 1992).

Conditional Activity of the N-End Rule Pathway

A different way to implement conditional phenotypes using N-degrons is to regulate the activity of the N-end–rule pathway, instead of regulating the expression of a pathway's substrate. In one implementation of this approach, we constructed a *S. cerevisiae* strain in which UBR1, the E3 Ub ligase of the N-end–rule pathway, was expressed from P_{GAL1}. The resulting strain, JD54 (Ghislain *et al.*, 1996), was used to produce a conditional (galactose-mediated) mutant of SRP54, a subunit of the signal recognition particle (SRP). SRP54 was modified to contain a constitutive N-degron, in the context of a Ub-Arg-e^K-SRP54 fusion, with Arg (a destabilizing residue) being the junctional residue. (R. J. D., K. Madura, B. Bartel, and A. V., unpublished data). We termed such modified proteins (and the corresponding mutants) "*ns*" (N-end rule-sensitive). A DNA fragment containing the ORF of Ub-Arg-e^K-SRP54 downstream from the P_{CUP1} promoter was used to replace wild-type *SRP54*, using a gene-replacement strategy similar to the one illustrated in Fig. 3A. On glucose media (galactose-off, UBR1 off), a JD54 strain containing the *ns* allele of *SRP54* exhibited wild-type phenotype, whereas on galactose media it grew very slowly, owing to low steady-state levels of the (now short-lived) Arg-e^K-SRP54. In a more elaborate design, a tightly regulated version of P_{CUP1} was used to simultaneously induce *UBR1* and a repressor that specifically down-regulated the promoter of a gene of interest (Moqtaderi *et al.*, 1996). (The corresponding protein of interest was tagged with a constitutive N-degron.)

One difficulty with strategies that use regulated transcriptional promoters is that they require a change in media composition. Unless a promoter-inducing (or repressing) molecule is highly specific for the promoter in question, it may affect biological processes other than those impacted by a gene of interest. In addition, the expression of a gene from nonnative promoter may result in abnormal phenotypes under permissive conditions (Mnaimneh *et al.*, 2004).

Construction and Uses of Temperature-Sensitive N-Degron

To bypass the preceding and related drawbacks and to enable construction of *ts* mutants without a search for *ts* mutations in a gene of interest, we developed a portable, heat-inducible N-degron. The corresponding mutants, in which a protein becomes short-lived at nonpermissive temperature, were termed *td* (temperature-activated degron) to distinguish them from conventional *ts* mutants (Dohmen *et al.*, 1994). The *td* method requires neither a change in media composition nor an alteration in expression patterns of a gene of interest. The idea was to identify a thermolabile variant of a small, monomeric protein that becomes a short-lived substrate of the N-end–rule pathway only at nonpermissive temperature through a partial unfolding that activates the protein's cryptic N-degron. This can happen, at nonpermissive temperature, either through an increased mobility of a protein's relevant Lys residue(s) (i.e., the second determinant of N-degron) or through an increased steric exposure of a protein's "built-in" destabilizing N-terminal residue. Because proteolysis by the N-end–rule pathway is highly processive, one would be able to make, then, any protein conditionally short-lived by expressing it as a fusion to the thus engineered thermolabile protein, with the latter serving as a portable, heat-activated N-degron (Fig. 2).

Arg-DHFR, a variant of the 21-kDa mouse dihydrofolate reductase in which the wild-type N-terminal Val was replaced by Arg (using the Ub

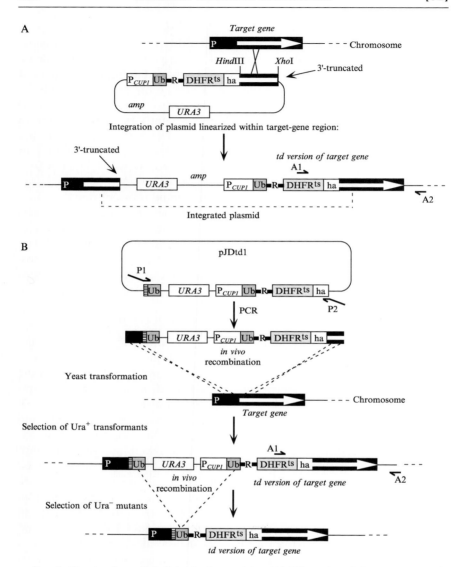

FIG. 3. Construction of *td* alleles in *S. cerevisiae*. (A) *Td* tagging of the target gene via plasmid integration. An integrative plasmid is produced, in which a 5'-region of the target gene (generated by PCR) is inserted, in-frame, downstream of the sequence coding for Ub-Arg-DHFR^{ts}. Expression of the resulting fusion is mediated by a promoter such as the copper-regulated P_{CUP1}. The plasmid is linearized by a single cut within the target gene's region for integration by means of homologous recombination at its genomic locus in *S. cerevisiae*. As a result, a 3'-truncated, nonfunctional derivative of the target gene forms upstream of its P_{CUP1}-

fusion technique), was a long-lived protein in *S. cerevisiae* ($t_{1/2} > 6$ h at 30°), even though Arg (unlike Val) is a destabilizing residue in the N-end rule (Bachmair and Varshavsky, 1989; Dohmen *et al.*, 1994). We searched for a *ts* allele of Arg-DHFR whose cryptic N-degron would be activated at 37° but not at 23°. A single missense mutation, Pro → Leu at position 66, proved sufficient for the purpose (Dohmen *et al.*, 1994). Later studies suggested that it was largely a change in steric exposure of N-terminal Arg in the mutant Arg-DHFR^{ts} (enabling the recognition of Arg by the UBR1 Ub ligase), rather than "mobilization" of an internal Lys residue(s), that caused the activation of cryptic N-degron in Arg-DHFR^{ts} at nonpermissive temperature (Lévy *et al.*, 1999) (J. Johnston and A. V., unpublished data). Fusions of Arg-DHFR^{ts} to reporters such as URA3 or CDC28 demonstrated that the Arg-DHFR^{ts} domain of these fusions could render the entire fusions short-lived at nonpermissive temperature, resulting in null phenotypes of *URA3* and *CDC28*, respectively (Dohmen *et al.*, 1994).

Since the introduction of this method (Fig. 2) (Dohmen *et al.*, 1994), numerous studies have successfully used *td* alleles of specific genes in functional analyses (Amon, 1997; Caponigro and Parker, 1995; Hardy, 1996; Kesti *et al.*, 2004; Labib *et al.*, 2000; Valasek *et al.*, 2003; Wang *et al.*, 2004; Wolf *et al.*, 1998). Moreover, a proteomic project has recently produced, using a PCR-based procedure, *td* alleles of more than 50% of essential genes in *S. cerevisiae* (Aparicio, 2003; Kanemaki *et al.*, 2003). The *td* technique has also been used with the fission yeast *Schizosaccharomyces pombe* (Rajagopalan *et al.*, 2004). This method should be applicable, in

---

controlled, functional *td* counterpart. Correct transformants can be identified by analytical PCR, using primers A1 and A2. A1 is specific for the DHFR-coding sequence. A2 hybridizes to the target gene downstream of its sequence that was present in the integrated plasmid. (B) PCR-based strategy of *td* tagging. The plasmid pJDtd1 is used as a template to amplify a *URA3*-marked *td* cassette using primers P1 and P2. As a result, short sequences (~45 bp) that are identical to sequences of the target gene are linked to the fragment encoding the *td* module. This PCR-produced linear DNA fragment is used to transform *S. cerevisiae*. Homologous recombination between the short flanking sequences of incoming fragment and the target gene yields an insertion of the *URA3*-linked *td* module between the target gene's promoter and ORF. The correct Ura$^+$ transformants can be identified, at this step, using analytical PCR as in A. The resulting locus expresses a *td* version of the target gene from the P$_{CUP1}$ promoter, which is preceded by the *URA3* marker flanked by two tandem repeats of Ub-coding sequence. Upon counter-selection for Ura$^-$ cells with FOA (see the main text), one can isolate cell clones in which *URA3* had been excised by homologous recombination between the two repeated Ub sequences. The resulting strains would bear a genetically stable *td* version of the target gene, at its natural chromosomal location and expressed from its native promoter (P). Note that the 5'-region of the first Ub-coding sequence is modified (the hatched rectangle) through multiple translationally silent mutations, to make possible specific annealing of the PCR primer P1.

principle, to any organism that allows the use of conventional *ts* mutants. Because the genetic screen that yielded Arg-DHFR^{ts} variant was far from exhaustive, one may be able to identify, through a similar screen, a *ts* allele (s) of Arg-DHFR that is even better (as a conditional N-degron) than the current Arg-DHFR^{ts}, or is activated, for example, at a lower temperature. The latter variant may be helpful for applications to poikilothermic organisms such as, for instance, *Drosophila*, where nonpermissive temperature must be lower than 37°.

One limitation of a *td* technique based on N-degrons is that it cannot be applied to proteins whose function is incompatible with N-terminal extensions. An alternative approach, in this case, would be to use a conditional C-terminal degron (Tongaonkar *et al.*, 1999). Other recent strategies involve the targeting of the Ub-proteasome system to conditionally destroy a protein of interest by linking a domain that recognizes the target protein to a system's specific component, for example, a Ub-conjugating (E2) enzyme or an F-box subunit of an SCF-type E3 Ub ligase (Gosink and Vierstra, 1995; Sakamoto *et al.*, 2003; Zhou *et al.*, 2000). The *td* method (Dohmen *et al.*, 1994) also would not work with proteins in compartments outside the cytosol and the nucleus. However, because the Ub system can target proteins that are retrotranslocated from compartments such as ER (Hampton, 2002; Kostova and Wolf, 2003), the principles of *td* strategy, with appropriate modifications, should be applicable to compartmentalized proteins as well.

Experimental Procedures

Construction of td Alleles by Targeted Plasmid Integration

A one-step strategy, in *S. cerevisiae*, for replacing a gene of interest with an integrated *td* allele of this gene involves site-specific integration, by means of homologous recombination, of a plasmid that contains an allele's 3′-truncated derivative (Fig. 3A) (Dohmen *et al.*, 1994). Integration of such a plasmid would result in a genomic locus that expresses the intended *td* allele from a non-native promoter such as, for example, P$_{CUP1}$, and also contains a truncated, nonfunctional version of the original gene expressed from its native promoter (Fig. 3A). In this case, the truncated 5′-fragment of the gene that adjoins the *td* allele of the same gene should be as short as possible, to minimize potential interference by the 3′-truncated gene product that would be expressed alongside the *td* allele.

A two-step strategy improves on this design. Specifically, a fragment containing the native promoter of a gene of interest is inserted into a plasmid instead of P$_{CUP1}$, resulting, on genomic integration of the plasmid, in expression of the *td* allele from the gene's native promoter. In the second step, a recombination event between two nearby identical (and identically oriented) regions of DNA containing native promoter would evict both the *URA3* marker and the gene's truncated fragment. The desired second-step recombinants can be selected for on plates containing fluoro-orotic acid (FOA), which inhibits the growth of cells expressing URA3 (Orr-Weaver *et al.*, 1981). (FOA is converted, by $URA3^+$ cells, to a toxic compound fluorodeoxyuridylate, an inhibitor of thymidylate synthase.) The final outcome would be a recombinationally stable locus expressing a *td* allele of interest from the gene's native promoter. Fig. 3B illustrates a PCR-based two-step strategy that yields the same result.

Construction of Integrative Plasmids

The plasmid pPW66R, which was constructed for making $cdc28^{td}$ mutants (Dohmen *et al.*, 1994), can be used as a starting point. From ∼0.5 to ∼1 kbp at the 5′ end of the target gene's ORF are amplified by PCR, using high-fidelity DNA polymerase and two primers. These oligonucleotides should be designed to contain sequences of ∼25 nucleotides (nts) that are identical to those of the target sequences, and in addition carry, at the oligos' 5′ ends, appropriate restriction sites for subsequent cloning. In our specific construct, the 5′-oligonucleotide introduces a *Hin*dIII site and two Gly codons upstream of the ATG start codon of a gene of interest (CGCCAA GCT TCC GGG GGG ATG...). The Gly residues serve as a spacer between the *td* domain (Arg-$DHFR^{ts}$) and the protein of interest. The 3′-oligonucleotide, which determines the site of 3′-truncation of the gene's ORF, introduces an *Xho*I site. By use of these primers, a 5′-segment of the target gene's ORF is amplified by PCR, with *S. cerevisiae* genomic DNA (or appropriate plasmid containing the target gene) as a template. The PCR-produced fragment and the vector pPW66R (Dohmen *et al.*, 1994) are cleaved with *Hin*dIII and *Xho*I, followed by ligation. The inserted fragment should contain a restriction site that would be unique in the resulting plasmid. This site is subsequently used to linearize the plasmid for its targeted integration into the yeast genome. After transformation of *E. coli* with ligation products, ampicillin-resistant clones are selected. A correct integrative plasmid is identified by restriction analysis. A PCR-generated insert should be, in addition, verified by DNA sequencing.

Genomic td Tagging by Plasmid Integration

A plasmid constructed as described previously is linearized by digesting it with a restriction enzyme that cleaves once within the target sequence, yielding a DNA fragment with recombinogenic ends that can integrate at the targeted genomic locus by means of homologous recombination (Orr-Weaver et al., 1981). If no unique site is available, a restriction enzyme that cuts in the ORF fragment as well as at another place in the plasmid should suffice, with a partial (instead of complete) digestion of the plasmid. Because a fraction of the cleaved plasmid would be linearized at the desired site, this procedure should yield enough yeast transformants with correctly integrated plasmid. S. cerevisiae transformants in which the linearized plasmid has integrated correctly are identified by phenotypic tests and analytical PCR as outlined in the following.

Construction of td Alleles Using a PCR-Based Genomic Transplacement Strategy

In another strategy (Fig. 3B), a genomic td allele is produced in a single step through a double homologous recombination between a gene of interest and a PCR product carrying the td tag (encoding Arg-DHFR^{ts}) downstream of the P_{CUP1} promoter and URA3 (selectable marker). As a result, the target gene's ORF is extended, at its 5'-end, by the td tag, and the resulting (integrated) fusion is expressed from P_{CUP1}. Similar approaches for generating td alleles were described that used related plasmids (Labib et al., 2000; Petracek and Longtine, 2002). A new aspect of the construction described previously (Fig. 3B) is that the "intermediate" integration product contains direct repeats of Ub-coding sequences that flank the URA3 marker and P_{CUP1}. The second-step excision, through homologous recombination between Ub-coding repeats, results in eviction of the (now unnecessary) URA3 and P_{CUP1}. Cells that lost URA3 can be selected on media containing FOA (Boeke et al., 1987). In the resulting S. cerevisiae strain, the td-tagged gene of interest is expressed from its native promoter.

This strategy (Fig. 3B) makes possible, using the same initial plasmid, to examine, at first, the expression of a td-tagged allele from the copper-inducible P_{CUP1} promoter. Then, if desired, one can produce a strain in which the same td-tagged gene is expressed from its native promoter. The latter configuration would be usually preferred, because it would bypass potential problems of misregulation of a td-tagged gene. However, with some genes, a feedback regulation may result in a strong up-regulation of the gene's native promoter on induction of degradation of the td-tagged gene product at nonpermissive temperature. The option, in such experimental settings, of using the unrelated P_{CUP1} promoter may provide a welcome alternative. This promoter has the additional advantage of allowing adjustments of its activity by varying the concentration of Cu^{2+} in the medium.

Vectors and Strategies to Generate PCR Fragments for td Tagging

The preceding strategy involves PCR to produce DNA fragments, which are then used to create *td* alleles by genomic transplacement. PCR joins desired DNA fragments to the *td* cassette that allow its targeting to a genomic location of a gene of interest. Several plasmids that can serve as PCR templates to produce *td* modules have been described. The basic strategy for amplifying *td* modules is the same for all of them, but the design of primers varies (Labib *et al.*, 2000; Petracek and Longtine, 2002). In the plasmid pKL187, the *td* tag is controlled by P_{CUP1} and is preceded by the kanamycin resistance marker (Labib *et al.*, 2000). In the plasmids pFA6a-kanMX6-tsDegron-3Ha, pFA6a-TRP1-tsDegron-3Ha, or pFA6a-His3MX6-tsDegron-3Ha, the P_{GAL1} promoter-linked fragment encoding Arg-DHFR^{ts} is preceded by a kanamycin resistance marker (conferring resistance to G418), a *TRP1* marker, or a *HIS3* marker, respectively (Petracek and Longtine, 2002). We produced the plasmid pJDtd1 (Fig. 3B), with *URA3* as a selectable and counter-selectable marker, and with P_{CUP1} upstream of the *td* module.

Two PCR strategies can be distinguished. One of them involves the attachment, on amplification, of "short flanking sequences," whereas the other attaches "long flanking sequences." In the former approach, ~45-nt sequences identical to those in targeted regions of a gene of interest are linked to both ends of *td* module as parts of PCR primers. Such 45-nt "flanking" sequences are usually sufficient for mediating specific targeting at high enough frequency on transformation of *S. cerevisiae* (Baudin *et al.*, 1993; Lorenz *et al.*, 1995; Manivasakam *et al.*, 1995; Wach *et al.*, 1994). If longer flanking sequences are required, they can be attached using a different PCR protocol (Labib *et al.*, 2000; Petracek and Longtine, 2002).

Design of Primers

In the "short flanking sequence" strategy, the forward primer P1 consists of ~45 nts whose sequence is identical to that at the 3' end of a target gene's promoter, followed by the 5' end of the Ub-coding sequence (5'-... ATGCAAATCTTTGTTAAAACCC-3'), into which silent mutations had been introduced to distinguish it from the second Ub-coding sequence in pJDtd1. Despite these sequence differences (7 of 22 nucleotide positions altered) between the Ub-coding repeats, PCR generally produces, even at relatively high temperatures, significant amounts of a smaller DNA fragment, derived from annealing of this primer to the second Ub-coding sequence. However, because this smaller fragment lacks the *URA3* marker, it would not interfere with recovery of correct integrants on yeast transformation. The reverse primer P2 consists of ~45 nts of the opposite strand at the 5′ end of target gene's ORF, followed by a sequence identical to that at the 3′ end of *td* module (5′-... CCCTCCTAAAAATGCAGCGTA-3′).

Occasionally, the "short flanking sequence" strategy fails to yield correct recombinants. In such cases, an assembly PCR protocol can be used that allows the attachment of longer gene-specific sequences (Petracek and Longtine, 2002; Wach, 1996). At first, two "long flanking sequences" DNA fragments are produced by PCR. One of them is a ~0.5 kbp copy of a sequence at the target gene's promoter. The other fragment contains a sequence of similar size that is identical to a sequence at the 5′ end of the target gene's ORF. The two fragments are produced using target gene–specific primers and either genomic DNA or a plasmid carrying the target gene as a template. The reversed primer for producing a promoter-specific PCR product comprises a junction between the target gene's promoter and the *td* module (5′-GGTTTTAACAAAGATTTGCAT, followed by 20–25 nts of the opposite strand at the promoter's 3′ end). The forward primer for PCR yielding the ORF fragment is used to produce a junction between the ORF and the 3′ end of the *td* module (5′-TACGCTGCATTTTTAG-GAGGG, followed by 20–25 nts of the 5′ end of the target gene's ORF). In a third, "assembly-PCR" reaction, the previous two fragments are purified and combined with either pJDtd1 DNA or a PCR fragment that contains just the *td* module, plus the two "distal" primers used in the first two PCRs. The resulting product would be similar to the one obtained using the "short flanking sequence" strategy, but with longer sequence stretches identical to those of the target gene. This should increase, on yeast transformation, the frequency of correctly targeted recombinants.

Materials

Yeast Media

Yeast synthetic minimal medium with dextrose (SD) is used to select transformants. The medium contains 6.7-g yeast nitrogen base without amino acids and 2% glucose (dextrose) per liter, as well as amino acids or other nutrients required by the yeast strain used.

Yeast Strains

A congenic strain set such as JD47-13C, JD51, JD55, and JD54 carrying suitable auxotrophic mutations such as *ura3*, *leu2*, *trp1*, or *his3* is recommended (Ghislain *et al.*, 1996). JD51 is an isogenic diploid derived from the haploid *MATa* strain JD47–13C on transformation with a plasmid carrying *HO*, which encodes endonuclease-mediating mating-type conversion (Herskowitz and Jensen, 1991). JD47-13C is a Gal$^+$ segregant from a cross of YPH500 (Sikorski and Hieter, 1989) and BBY45 (Bartel *et al.*, 1990). JD54 is a haploid strain expressing *UBR1* from galactose-inducible P$_{GAL1}$ (Ghislain *et al.*, 1996). JD55 is a congenic *ubr1*Δ control.

Genomic targeting of *td* modules, either by plasmid integration or PCR-based strategies, should be carried out in parallel with strains of the set described previously. If no correctly targeted transformants are obtained with a wild-type haploid strain (owing, for example, to functional inactivation of an essential gene because of incompatibility of its product with N-terminal extensions), the corresponding diploid strain may still yield such transformants. One can determine, then, by tetrad analysis, whether or not the *td* module is lethal in haploids. The *ubr1*Δ background should be used in such tests as a negative control (no N-end–rule pathway).

An alternative to comparing phenotypes of *UBR1*$^+$ and *ubr1*Δ strains that express a *td*-tagged protein is to use a strain with a tightly regulated *UBR1* gene. In the strain JD54, *UBR1* expression is controlled by P$_{GAL1}$. On glucose media, the expression is off, whereas on galactose media the expression of UBR1 is induced much above its wild-type levels (which are rate-limiting for proteolysis by the N-end–rule pathway (Bartel *et al.*, 1990; Petracek and Longtine, 2002), resulting in enhanced degradation of N-end–rule substrates and, consequently, in a tighter *td* phenotype, at least with some *td* module-linked proteins (Ghislain *et al.*, 1996; Labib *et al.*, 2000). *UBR1* can also be overexpressed constitutively, from high-copy plasmids expressing it either from native promoter (Petracek and Longtine, 2002) or from P$_{ADH1}$. These plasmids are available on request. Constructs for production of either *LEU2*- or *HIS3*-marked *ubr1*Δ alleles in other strain backgrounds are also available.

Plasmids and Template DNA

For integrative tagging strategy, vectors with selectable markers such as *URA3, LEU2, TRP1*, or *HIS3* (Gietz and Sugino, 1988; Sikorski and Hieter, 1989) are used that contain the cassette encoding Arg-DHFR^{ts} expressed from a promoter such as P$_{CUP1}$. In the strategy illustrated in Fig. 3A, pPW66R, a *URA3*-based integrative plasmid derived from pRS316 (Sikorski and Hieter, 1989), is used as a starting construct for producing *td* alleles (Dohmen *et al.*, 1994). For PCR-based tagging strategy, the plasmid pJDtd1 (Fig. 3B) or related plasmids (Labib *et al.*, 2000; Petracek and Longtine, 2002) can be used as templates. Sequence information on pPW66R and pJDtd1 is available at http://www.genetik.uni-koeln.de/groups/Dohmen/.

Experimental Protocols

Generation of td Cassettes by PCR

Standard PCR conditions can be used to produce *td* cassettes containing target gene–specific flanking sequences (Petracek and Longtine, 2002). Commercially available high-fidelity thermostable polymerase kits should be used to avoid incorporation of mismatched nucleotides.

Transformation of Frozen Competent Yeast Cells

For making *td* alleles, an easy and rapid transformation protocol is recommended that has the advantage of allowing the use of frozen competent cells (Dohmen *et al.*, 1991b). Transformation efficiency of this protocol, although lower than those with other procedures (Gietz and Woods, 2002), is usually sufficient.

1. Grow yeast strains in YPD (10 ml per transformation) to an OD_{600} of 0.6–1.0.
2. Spin down cells at 1200g for 5 min and resuspend the pellet in 0.5 volumes (5 ml per transformation) of solution A (1 M sorbitol, 3% ethylene glycol, 5% dimethylsulfoxide (DMSO), 10 mM bicine-NaOH, pH 8.4).
3. Spin down cells at 1200g for 5 min and resuspend the pellet in 0.02 volumes (200 μl per transformation) of solution A.
4. Freeze 200-μl samples of competent cells in Microfuge tubes by placing them directly in a $-80°$ freezer. (Freezing in liquid nitrogen reduces transformation efficiency.)
5. For transformation, a linearized plasmid or a PCR product described previously, plus 5 μl (10 mg/l) heat-denatured calf thymus DNA are added to a 0.2-ml sample of frozen competent cells, followed by thawing in water at either 37° or room temperature, with vigorous agitation for 5 min.
6. Add 1.2 ml of solution B (40% polyethylene glycol (PEG-1000) 1000, 0.2 M bicine-NaOH, pH 8.4), and mix gently, by inverting the tube several times. The source of PEG-1000 can strongly affect transformation efficiency. We use PEG-1000 from Roth, Karlsruhe, Germany.)
7. Incubate for 60 min at 30°, without shaking.

8. Spin down cells at 1200g for 5 min, and wash them with 1.2 ml of solution C (0.15 M NaCl, bicine-NaOH, pH 8.4).
9. After spinning down the cells again, decant the supernatant, resuspend cells in the remaining solution, and plate them onto selective media.
10. Select for transformants by incubation at 25° or 30° for 4–5 days.

Verification of Correct Targeting Using Analytical PCR

Td tagging of yeast cells by transformation either with the integrative plasmids or the PCR-generated cassettes should result in clones bearing modified versions of the target genes at their authentic genomic location. Analytical PCR can be used to identify transformants that carry a correctly targeted *td* module (Fig. 3) and to distinguish them from other transformants. One primer (A1) should be specific for the *td* tag, and the other (A2) should be specific for a sequence of the target gene outside of the region present in DNA that was used for integrative transformation (Fig. 3). The primer A1 can be a ~25-nt oligonucleotide whose sequence is identical to that at the 3' end of the DHFR ORF.

1. Transfer a small amount of cells (just covering the pipette tip) from a transformant's colony into a PCR tube using a sterile 200-μl pipette tip. Use untransformed yeast cells as a negative control.
2. Heat the sample in a microwave for 1 min.
3. Set up and run analytical PCR (up to 35 cycles), using primers A1 and A2, heat-treated cells, and Taq polymerase.
4. Analyze PCR products by agarose gel electrophoresis.

Transformants that yield an expected PCR product would be expected to carry the properly integrated *td* cassette. This PCR test, however, does not preclude the possibility that cells contain an additional, wild-type, copy of the target gene. Strains carrying both a *td*-tagged and a wild-type copy of the gene occasionally form on transformation of haploid cells as a result of cell fusion, which can be induced by PEG treatment. The resulting transformants, although mating as haploids, would be diploid and heterozygous in regard to *td* mutation. The loss of wild-type copy of the target gene can be verified either by a second analytical PCR or by Southern hybridization analysis. Such additional tests are unnecessary if, in addition to yielding a correct first-PCR product, the cells also exhibit the relevant ts phenotype. Complementation of this phenotype with a wild-type copy of the gene would then suffice to verify the correctness of strain construction.

Analysis of td Mutants

The first step in analyzing *td*-tagged strains is characterization of their growth phenotype at permissive and nonpermissive temperatures. If *td* tagging involved an essential gene, and if the Arg-DHFR^{ts} module confers a sufficiently short half-life on the corresponding protein at nonpermissive temperature, the mutant should be unable to grow at 37°. If no tight *ts* phenotype is observed in the wild-type (*UBR1*$^+$) background, one can try producing correct transformants with a strain overexpressing *UBR1*, such as JD54 (see earlier) (Labib *et al.*, 2000). An important control in these phenotypic analyses are otherwise identical *td* module-bearing transformants produced in a *ubr1*Δ strain, such as JD55, because such transformants should not exhibit a *ts* phenotype at nonpermissive temperature (Labib *et al.*, 2000). Heat-induced degradation of a *td*-tagged protein can be detected either by immunoblotting of proteins extracted from cells after various times after temperature upshift, or by pulse-chase analysis (Dohmen *et al.*, 1994; Labib *et al.*, 2000). Epitope tags in the *td* module (Fig. 3) allow detection of *td*-tagged proteins with commercially available antibodies.

Acknowledgments

We thank Kerstin Göttsche for excellent technical assistance in the construction of pJDtd1. We are grateful to Pei-Pei Wu for her contribution to the development of *td* method, and to Kiran Madura and Bonnie Bartel for their contribution to the development of *ns* method. Our studies are supported by grants from the DFG (Do 649/2) to R. J. D. and from the National Institutes of Health (GM31530 and DK39520) and the Ellison Medical Foundation to A. V.

References

Aguilera, A. (1994). Formamide sensitivity: A novel conditional phenotype in yeast. *Genetics* **136,** 87–91.

Aiba, H., Kawaura, R., Yamamoto, E., Yamada, H., Takegawa, K., and Mizuno, T. (1998). Isolation and characterization of high-osmolarity-sensitive mutants of fission yeast. *J. Bact.* **180,** 5038–5043.

Amon, A. (1997). Regulation of B-type cyclin proteolysis by Cdc28-associated kinases in budding yeast. *EMBO J.* **16,** 2693–2702.

Aparicio, O. M. (2003). Tackling an essential problem in functional proteomics of *Saccharomyces cerevisiae*. *Genome Biol.* **4,** 230.

Bachmair, A., Finley, D., and Varshavsky, A. (1986). *In vivo* half-life of a protein is a function of its amino-terminal residue. *Science* **234,** 179–186.

Bachmair, A., and Varshavsky, A. (1989). The degradation signal in a short-lived protein. *Cell* **56,** 1019–1032.

Baker, R. T., and Varshavsky, A. (1995). Yeast N-terminal amidase. A new enzyme and component of the N-end rule pathway. *J. Biol. Chem.* **270,** 12065–12074.

Bartel, B., and Varshavsky, A. (1988). Hypersensitivity to heavy water: A new conditional phenotype. *Cell* **52,** 935–941.

Bartel, B., Wünning, I., and Varshavsky, A. (1990). The recognition component of the N-end rule pathway. *EMBO J.* **9,** 3179–3189.

Baudin, A., Ozier-Kalogeropolous, O., Denoel, A., Lacroute, A., and Cullin, C. (1993). A simple and efficient method for direct gene deletion in *Saccharomyces cerevisiae*. *Nucl. Ac. Res.* **21,** 3329–3330.

Boeke, J. D., Trueheart, J., Natsoulis, G., and Fink, G. R. (1987). 5-Fluoroorotic acid as a selective agent in yeast molecular genetics. *Meth. Enzymol.* **154,** 164–175.

Bradshaw, R. A., Brickey, W. W., and Walker, K. W. (1998). N-terminal processing: The methionine aminopeptidase and N alpha-acetyl transferase families. *Trends Biochem. Sci.* **23,** 263–267.

Byrd, C., Turner, G. C., and Varshavsky, A. (1998). The N-end rule pathway controls the import of peptides through degradation of a transcriptional repressor. *EMBO J.* **17,** 269–277.

Caponigro, G., and Parker, R. (1995). Multiple functions for the polyA-binding protein in mRNA decapping and deadenylation in yeast. *Genes Dev.* **9,** 2421–2432.

Colb, M., and Shapiro, L. (1977). A pH-conditional mutant of *Escherichia coli*. *Proc. Natl. Acad. Sci. USA* **74,** 5637–5641.

Crews, C. M. (2003). Feeding the machine: Mechanisms of proteasome-catalyzed degradation of ubiquitinated proteins. *Curr. Op. Chem. Biol.* **7,** 534–539.

Davydov, I. V., and Varshavsky, A. (2000). RGS4 is arginylated and degraded by the N-end rule pathway *in vitro*. *J. Biol. Chem.* **275,** 22931–22941.

Ditzel, M., Wilson, R., Tenev, T., Zachariou, A., Paul, A., Deas, E., and Meier, P. (2003). Degradation of DIAP1 by the N-end rule pathway is essential for regulating apoptosis. *Nature Cell Biol.* **5,** 467–473.

Dohmen, R. J., Madura, K., Bartel, B., and Varshavsky, A. (1991a). The N-end rule is mediated by the UBC2 (RAD6) ubiquitin-conjugating enzyme. *Proc. Natl. Acad. Sci. USA* **88,** 7351–7355.

Dohmen, R. J., Strasser, A. W. M., Höner, C. B., and Hollenberg, C. P. (1991b). An efficient transformation procedure enabling long-term storage of competent cells of various yeast genera. *Yeast* **7,** 691–692.

Dohmen, R. J., Wu, P., and Varshavsky, A. (1994). Heat-inducible degron: A method for constructing temperature-sensitive mutants. *Science* **263,** 1273–1276.

Du, F., Navarro-Garcia, F., Xia, Z., Tasaki, T., and Varshavsky, A. (2002). Pairs of dipeptides synergistically activate the binding of substrate by ubiquitin ligase through dissociation of its autoinhibitory domain. *Proc. Natl. Acad. Sci. USA* **99,** 14110–14115.

Finley, D., Bartel, B., and Varshavsky, A. (1989). The tails of ubiquitin precursors are ribosomal proteins whose fusion to ubiquitin facilitates ribosome biogenesis. *Nature* **338,** 394–401.

Finley, D., Özkaynak, E., and Varshavsky, A. (1987). The yeast polyubiquitin gene is essential for resistance to high temperatures, starvation, and other stresses. *Cell* **48,** 1035–1046.

Förster, A., and Hill, C. P. (2003). Proteasome degradation: Enter the substrate. *Trends Cell. Biol.* **13,** 550–553.

Ghislain, M., Dohmen, R. J., Levy, F., and Varshavsky, A. (1996). Cdc48p interacts with Ufd3p, a WD repeat protein required for ubiquitin-mediated proteolysis in *Saccharomyces cerevisiae*. *EMBO J.* **15,** 4884–4899.

Gietz, R. D., and Sugino, A. (1988). New yeast-*Escherichia coli* shuttle vectors constructed with *in vitro* mutagenized yeast genes lacking six-base pair restriction sites. *Gene* **74**, 527–534.

Gietz, R. D., and Woods, R. A. (2002). Transformation of yeast by lithium acetate/single-stranded carrier DNA/polyethylene glycol method. *Meth. Enzymol.* **350**, 87–96.

Gilchrist, C. A., Gray, D. A., and Baker, R. T. (1997). A ubiquitin-specific protease that efficiently cleaves the ubiquitin-proline bond. *J. Biol. Chem.* **272**, 32280–32285.

Gosink, M. M., and Vierstra, R. D. (1995). Redirecting the specificity of ubiquitination by modifying ubiquitin-conjugating enzymes. *Proc. Natl. Acad. Sci. USA* **92**, 9117–9121.

Hampton, R. Y. (2002). ER-associated degradation in protein quality control and cellular regulation. *Curr. Op. Cell Biol.* **14**, 476–482.

Hann, B. C., and Walter, P. (1991). The signal recognition particle in *S. cerevisiae*. *Cell* **67**, 131–144.

Hardy, C. F. (1996). Characterization of an essential Orc2p-associated factor that plays a role in DNA replication. *Mol. Cell. Biol.* **16**, 1832–1841.

Herskowitz, I., and Jensen, R. E. (1991). Putting the *HO* gene to work: Practical uses for mating-type switching. *Meth. Enzymol.* **194**, 132–146.

Hill, A., and Bloom, K. (1987). Genetic manipulation of centromere function. *Mol. Cell. Biol.* **7**, 2397–2405.

Hochstrasser, M. (1996). Ubiquitin-dependent protein degradation. *Annu. Rev. Genet.* **30**, 405–439.

Hochstrasser, M. (2000). Evolution and function of ubiquitin-like protein-conjugation systems. *Nature Cell Biol.* **2**, 153–157.

Horowitz, N. H. (1950). Biochemical genetics of *Neurospora*. *Adv. Genet.* **3**, 33–71.

Hu, M. C., and Davidson, N. (1987). The inducible lac operator-repressor system is functional in mammalian cells. *Cell* **48**, 555–566.

Johnson, E. S., Bartel, B., W., and Varshavsky, A. (1992). Ubiquitin as a degradation signal. *EMBO J.* **11**, 497–505.

Kanemaki, M., Sanchez-Diaz, A., Gambus, A., and Labib, K. (2003). Functional proteomic identification of DNA replication proteins by induced proteolysis *in vivo*. *Nature* **423**, 720–724.

Kesti, T., McDonald, W. H., Yates, J. R. r., and Wittenberg, C. (2004). Cell cycle-dependent phosphorylation of the DNA polymerase epsilon subunit, Dpb2, by the Cdc28 cyclin-dependent protein kinase. *J. Biol. Chem.* **279**, 14245–14255.

Kostova, Z., and Wolf, D. H. (2003). For whom the bell tolls: Protein quality control of the endoplasmic reticulum and the ubiquitin-proteasome connection. *EMBO J.* **22**, 2309–2317.

Kwon, Y. T., Balogh, S. A., Davydov, I. V., Kashina, A. S., Yoon, J. K., Xie, Y., Gaur, A., Hyde, L., Denenberg, V. H., and Varshavsky, A. (2000). Altered activity, social behavior, and spatial memory in mice lacking the NTAN1 amidase and the asparagine branch of the N-end rule pathway. *Mol. Cell. Biol.* **20**, 4135–4148.

Kwon, Y. T., Kashina, A. S., Davydov, I. V., Hu, R.-G., An, J. Y., Seo, J. W., Du, F., and Varshavsky, A. (2002). An essential role of N-terminal arginylation in cardiovascular development. *Science* **297**, 96–99.

Kwon, Y. T., Xia, Z., Davydov, I. V., Lecker, S. H., and Varshavsky, A. (2001). Construction and analysis of mouse strains lacking the ubiquitin ligase UBR1 (E3α) of the N-end rule pathway. *Mol. Cell. Biol.* **21**, 8007–8021.

Kwon, Y. T., Xia, Z. X., An, J. Y., Davydov, I. V., Seo, J. W., Xie, Y., and Varshavsky, A. (2003). Female lethality and apoptosis of spermatocytes in mice lacking the UBR2 ubiquitin ligase of the N-end rule pathway. *Mol. Cell. Biol.* **23**, 8255–8271.

Labib, K., Tercero, J. A., and Diffley, J. F. (2000). Uninterrupted MCM2-7 function required for DNA replication fork progression. *Science* **288**, 1643–1647 (erratum in: Science 1289, 2052 (2000)).

Latterich, M., and Watson, M. D. (1991). Isolation and characterization of osmosensitive vacuolar mutants of *Saccharomyces cerevisiae*. *Mol. Microbiol.* **5**, 2417–2426.

Lévy, F., Johnston, J. A., and Varshavsky, A. (1999). Analysis of a conditional degradation signal in yeast and mammalian cells. *Eur. J. Biochem.* **259**, 244–252.

Lorenz, M. C., Muir, R. S., Lim, E., McElver, J., Weber, S. C., and Heitman, J. (1995). Gene disruption with PCR products in *Saccharomyces cerevisiae*. *Gene* **158**, 113–117.

Manivasakam, P., Weber, S. C., McElver, J., and Schiestl, R. H. (1995). Micro-homology mediated PCR targeting in *Saccharomyces cerevisiae*. *Nucl. Acids Res.* **23**, 2799–2800.

Marschalek, R., Kalpaxis, D., and Dingermann, T. (1990). Temperature-sensitive synthesis of transfer RNAs *in vivo* in *Saccharomyces cerevisiae*. *EMBO J.* **9**, 1253–1258.

Mnaimneh, S., Davierwala, A. P., Haynes, J., Moffat, J., Peng, W. T., Zhang, W., Yang, X., Pootoolal, J., Chua, G., Lopez, A., *et al.* (2004). Exploration of essential gene functions via titratable promoter alleles. *Cell* **118**, 31–44.

Moir, D., Stewart, S. E., Osmond, B. C., and Botstein, D. (1982). Cold-sensitive cell division-cycle mutants of yeast: Isolation, properties, and pseudoreversion studies. *Genetics* **100**, 547–563.

Moqtaderi, Z., Bai, Y., Poon, D., Weil, P. A., and Struhl, K. (1996). TBP-associated factors are not generally required for transcriptional activation in yeast. *Nature* **383**, 188–191.

Orr-Weaver, T. L., Szostak, J. W., and Rothstein, R. J. (1981). Yeast transformation: A model system for the study of recombination. *Proc. Natl. Acad. Sci. USA* **78**, 6354–6358.

Park, E. C., Finley, D., and Szostak, J. W. (1992). A strategy for the generation of conditional mutations by protein destabilization. *Proc. Natl. Acad. Sci. USA* **89**, 1249–1252.

Petracek, M. E., and Longtine, M. S. (2002). PCR-based engineering of yeast genome. *Meth. Enzymol.* **350**, 445–469.

Pickart, C. (2004). Back to the future with ubiquitin. *Cell* **116**, 181–190.

Pringle, J. R. (1975). Induction, selection, and experimental uses of temperature-sensitive and other conditional mutants in yeast. *Methods Cell Biol.* **12**, 233–272.

Rajagopalan, S., Liling, Z., Liu, J., and Balasubramanian, M. (2004). The N-degron approach to create temperature-sensitive mutants in *Schizosaccharomyces pombe*. *Methods Cell Biol.* **33**, 206–212.

Rao, H., Uhlmann, F., Nasmyth, K., and Varshavsky, A. (2001). Degradation of a cohesin subunit by the N-end rule pathway is essential for chromosome stability. *Nature* **410**, 955–960.

Rechsteiner, M. (1998). The 26S proteasome. *In* "Ubiquitin and the Biology of the Cell" (J. M. Peters, J. R. Harris, and D. Finley, eds.), pp. 147–189. Plenum Press, New York.

Sakamoto, K. M., Kim, K. B., Verma, R., Ransick, A., Stein, B., Crews, C. M., and Deshaies, R. J. (2003). Development of protacs to target cancer-promoting proteins for ubiquitination and degradation. *Mol. Cell. Proteomics* **2**, 1350–1358.

Sikorski, R. S., and Hieter, P. (1989). A system of shuttle vectors and yeast host strains designed for efficient manipulation of DNA in *S. cerevisiae*. *Genetics* **122**, 19–27.

Steege, D. A., and Horabin, J. I. (1983). Temperature-inducible amber suppressor: Construction of plasmids containing the *Escherichia coli* serU- (supD-) gene under control of the bacteriophage lambda pL promoter. *J. Bact.* **155**, 1417–1425.

Suzuki, T., and Varshavsky, A. (1999). Degradation signals in the lysine-asparagine sequence space. *EMBO J.* **18**, 6017–6026.

Tongaonkar, P., Beck, K., Shinde, U. P., and Madura, K. (1999). Characterization of a temperature-sensitive mutant of a ubiquitin-conjugating enzyme and its use as a heat-inducible degradation signal. *Anal. Biochem.* **272,** 263–269.

Turner, G. C., Du, F., and Varshavsky, A. (2000). Peptides accelerate their uptake by activating a ubiquitin-dependent proteolytic pathway. *Nature* **405,** 579–583.

Turner, G. C., and Varshavsky, A. (2000). Detecting and measuring cotranslational protein degradation *in vivo. Science* **289,** 2117–2120.

Valasek, L., Mathew, A. A., Shin, B. S., Nielsen, K. H., Szamecz, B., and Hinnebusch, A. G. (2003). The yeast eIF3 subunits TIF32/a, NIP1/c, and eIF5 make critical connections with the 40S ribosome *in vivo. Genes Dev.* **17,** 786–799.

Varshavsky, A. (1991). Naming a targeting signal. *Cell* **64,** 13–15.

Varshavsky, A. (1996). The N-end rule: Functions, mysteries, uses. *Proc. Natl. Acad. Sci. USA* **93,** 12142–12149.

Varshavsky, A. (1997). The ubiquitin system. *Trends Biochem. Sci.* **22,** 383–387.

Varshavsky, A. (2000). Ubiquitin fusion technique and its descendants. *Meth. Enzymol.* **327,** 578–593.

Varshavsky, A. (2003). The N-end rule and regulation of apoptosis. *Nature Cell Biol.* **5,** 373–376.

Verma, R., Oania, R., Graumann, J., and Deshaies, R. J. (2003). Multiubiquitin chain receptors define a layer of substrate selectivity in the ubiquitin-proteasome system. *Cell* **118,** 99–110.

Wach, A., Brachat, A., Pohlmann, R., and Philippsen, P. (1994). New heterologous modules for classical or PCR-based gene disruptions in *Saccharomyces cerevisiae. Yeast* **10,** 1793–1808.

Wach, A. Y. (1996). PCR-synthesis of marker cassettes with long flanking homology regions for gene disruptions in *S. cerevisiae. Yeast* **12,** 259–265.

Wang, X., Ira, G., Tercero, J. A., Holmes, A. M., Diffley, J. F., and Haber, J. E. (2004). Role of DNA replication proteins in double-strand break-induced recombination in *Saccharomyces cerevisiae. Mol. Cell. Biol.* **24,** 6891–6899.

Wilkinson, K. D. (2000). Ubiquitination and deubiquitination: Targeting of proteins for degradation by the proteasome. *Semin. Cell Dev. Biol.* **11,** 141–148.

Wolf, J., Nicks, M., Deitz, S., van Tuinen, E., and Franzusoff, A. (1998). An N-end rule destabilization mutant reveals pre-Golgi requirements for Sec7p in yeast membrane traffic. *Biochem. Biophys. Res. Commun.* **243,** 191–198.

Xie, Y., and Varshavsky, A. (1999). The E2-E3 interaction in the N-end rule pathway: The RING-H2 finger of E3 is required for the synthesis of multiubiquitin chain. *EMBO J.* **18,** 6832–6844.

Yin, J., Kwon, Y. T., Varshavsky, A., and Wang, W. (2004). RECQL4, mutated in the Rothmund-Thomson and RAPADILINO syndromes, interacts with ubiquitin ligases UBR1 and UBR2 of the N-end rule pathway. *Human Mol. Genet.* **13,** 2421–2430.

Yoshimatsu, T., and Nagawa, F. (1989). Control of gene expression by artificial introns in *Saccharomyces cerevisiae. Science* **244,** 1346–1348.

Zeidler, M. P., Tan, C., Bellaiche, Y., Cherry, S., Hader, S., Gayko, U., and Perrimon, N. (2004). Temperature-sensitive control of protein activity by conditionally splicing inteins. *Nature Biotechnol.* **22,** 871–876.

Zhang, Y. F., and Moss, B. (1991). Inducer-dependent conditional-lethal mutant animal viruses. *Proc. Natl. Acad. Sci. USA* **88,** 1511–1515.

Zhou, P., Bogacki, R., McReynolds, L., and Howley, P. M. (2000). Harnessing the ubiquitination machinery to target the degradation of specific cellular proteins. *Mol. Cell* **6,** 751–756.

Zwickl, P., Baumeister, W., and Steven, A. (2000). Dis-assembly lines: The proteasome and related ATP-assisted proteases. *Curr. Opin. Struct. Biol.* **10,** 242–250.

[53] Ectopic Targeting of Substrates to the Ubiquitin Pathway

By JIANXUAN ZHANG and PENGBO ZHOU

Abstract

Explanation of the physiological function of a cellular protein often requires targeted removal of that protein to reveal the associated biochemical and phenotypic alterations. A variety of technologies such as gene targeting and RNAi have been developed to abrogate the biosynthesis of the protein of interest. Recently, targeted protein degradation by harnessing the cellular ubiquitin–proteolytic machinery has emerged as a novel reverse genetic tool for loss-of-function studies. Targeted proteolysis operates at the posttranslational level to directly accelerate the turnover rate of the target protein and opens up new avenues for the dissection of complicated protein functions associated with posttranslational events, which are unattainable by a simple blocking of the biosynthesis of the target protein.

Introduction

The steady-state level of a protein is defined as an equilibrium state when its biosynthesis and degradation rates are equal. Understanding the function of a given cellular protein often requires reduction of its normal level so as to reveal effected alterations of biological activities and/or phenotype. A variety of such reverse genetic tools have been developed over the years that operate at the biosynthesis step, either at the DNA level by gene targeting or at the RNA level by means of RNAi, antisense oligodeoxynucleotides, ribozymes, or synthetic zinc finger transcriptional inhibitors. Here we describe a method, designated protein knockout, which operates at the degradation step to allow depletion of a cellular protein through an acceleration of its destruction by the ubiquitin-proteasome pathway.

The ubiquitin-proteasome pathway is a major piece of the cellular protein destruction machinery by which eukaryotic cells remove unwanted or harmful proteins. Protein ubiquitination requires the coordinated action of at least three classes of enzymes: the E1 ubiquitin-activating enzyme, the E2 ubiquitin-conjugating enzyme, and the E3 ubiquitin-protein ligase (reviewed in Hershko and Ciechanover [1998]). Among these enzymes, E3 is exclusively responsible for defining substrate specificity through direct binding to the target. Many of the E3s characterized thus far are multimeric protein complexes. One such E3 is the SCF ubiquitin ligase (Skp1, cullin 1, F-box–containing substrate receptor, and Hrt1/Rbx1/Roc1), which is constitutively active and targets a variety of cell cycle regulators, signaling molecules, and transcription factors for ubiquitination (reviewed in Deshaies [1999]). The substrate diversity of SCF is conferred by multiple F-box–containing substrate receptors, each of which recognizes several different substrates bearing defined recognition signals. The remaining subunits assemble into a core complex that coordinates the action of multiple F-box proteins in a dynamic manner.

The substrate receptors of the SCF complexes contain two modular domains for protein–protein interactions: the F-box that binds to Skp1 for entry into the core SCF, and a second domain (WD40, leucine-rich repeats [LRR], or Zn-finger) that recruits substrates (Fig. 1). Like the transcription factors with functionally separable DNA-binding and transcriptional activation domains, F-box proteins are evolved such that an independent protein–protein interaction module, when linked to an F-box, can confer

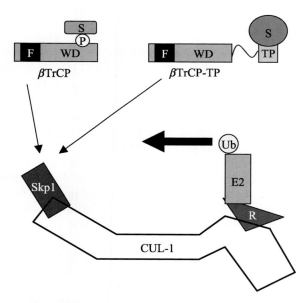

Fig. 1. Engineering the F-box-containing βTrCP substrate receptor for targeted degradation of specific cellular proteins. F, F-box domain that interacts with Skp1; WD, WD40 repeats for binding to phosphorylated (P) endogenous substrates (S) (top left). TP, targeting peptide; S (top right), target protein of interest; E2, Cdc34 ubiquitin-conjugating enzyme; Ub, ubiquitin; R, RING domain protein Hrt1/Rbx1/Roc1. (See color insert.)

new substrate specificity to the SCF machinery (Ptashne and Gann, 2002). These structural and operational characteristics offer the versatility of engineering F-box proteins to be able to target cellular proteins that are otherwise not SCF substrates (Zhou et al., 2000). The engineered F-box chimeras have been used successfully to target the degradation of a variety of cellular proteins both in cultured eukaryotic cells and in animal models (Chen et al., 2004; Cohen et al., 2004; Cong et al., 2003; Liu et al., 2004; Su et al., 2003; Zhang et al., 2003; Zhou et al., 2000). This chapter summarizes the procedures and general considerations of what should be taken into account in engineering the mammalian F-box-containing βTrCP substrate receptor to achieve targeted degradation of the desired cellular proteins.

Methods

Overview

The SCF ubiquitination machinery is present ubiquitously and is constitutively active throughout the cell cycle (Willems et al., 1996; Zhou and Howley, 1998). The multimeric SCF complex can be separated into two functional entities: the F-box-containing substrate receptor, which is exclusively responsible for substrate recruitment and the remaining subunits, Cullin 1, Skp1, and Hrt1/Rbx1/Roc1, which form a core complex to mediate ubiquitin transfer to the substrate (Fig. 1). The protein knockout system harnesses the SCF ubiquitin-proteolytic machinery to direct the degradation of cellular proteins of interest (Zhou et al., 2000) (Fig. 1). The strategy of protein knockout is to engineer a chimeric ubiquitin ligase with the binding peptide of the intended target fused to a specific F-box protein, and thereby redirect the constitutive SCF proteolytic machinery toward the degradation of cellular proteins that are otherwise not SCF targets (Zhou et al., 2000) (Fig. 1). The basic procedure involves cloning of the cDNA encoding the targeting peptide (TP) to the 3' end of βTrCP. The resulting βTrCP-TP chimera is expressed *in vitro* in the rabbit reticulocyte lysate or in cultured cells and tested for binding to the target protein by coimmunoprecipitation. The steady-state levels of the transfected epitope-tagged target protein are measured in the absence or presence of increasing doses of cotransfected βTrCP-TP to assess the efficacy of the targeted degradation. For the degradation of endogenous protein targets, the engineered βTrCP-TP can be delivered either by transient transfection and enrichment of transfected populations or by recombinant adenoviral, retroviral, or lentiviral vectors that are able to achieve highly efficient infection. Downregulation of the target protein is determined by Western blotting, and the alterations of biological activities or phenotype are subsequently assessed.

Selection of Targeting Peptides (TP)

Selection of a TP that links covalently to the F-box protein (βTrCP) is key to the success of targeted proteolysis. Because many proteins are either identified through protein–protein interactions (i.e., yeast two-hybrid assay or biochemical copurification) or characterized on the basis of their interaction with other cellular proteins, the available binding partners for the protein of interest are often described in the literature. This information is able to provide a convenient source for TPs. cDNAs encoding TPs can be either synthesized directly or obtained by polymerase chain reaction and then cloned into the pCDNA-F-TrCP-MCS plasmid to construct the βTrCP-TP ubiquitin ligase (Fig. 2). In the event that such information is not available, TPs can be obtained by screening phage display libraries or random peptide aptamer-based yeast two-hybrid libraries.

Two factors should be considered in the design of chimeric βTrCP-TP ubiquitin ligases: one is that the TP should bind to the substrate with relatively high affinity. Targeted degradation depends on the efficiency of substrate recruitment, which often correlates with the affinity between the TP and the intended substrate. We have found that an efficient co-immunoprecipitation (co-IP) between βTrCP-TP and the substrate is a prerequisite for targeted degradation. It should be noted that direct co-IP after transient transfection in target cells is not always reliably accurate in estimating the binding affinity, because such interactions may lead to a degradation of the substrate. We routinely conduct co-IP in 300 μl of NP-40 lysis buffer using proteins synthesized *in vitro* in rabbit reticulocyte lysate (Harlow and Lane, 1988; Zhou *et al.*, 2000). The second factor is the targeting specificity. TPs that contain binding sites for other cellular targets will likely recruit these proteins for destruction. Thus, the minimal binding domain, if available, is strongly preferred over large proteins or peptides. Such information can be obtained from x-ray crystallographic structures of protein–protein or protein–peptide complexes or from comprehensive deletion analysis for the purpose of mapping protein–protein interaction domains. Critical amino acid residues are often identified from these studies, which, when mutated, effectively abrogate the interactions. Such TPs with specific amino acid substitutions can serve as ideal negative controls for explaining the specificity of targeted proteolysis. It should be noted that selection of the minimal binding domains that can serve as

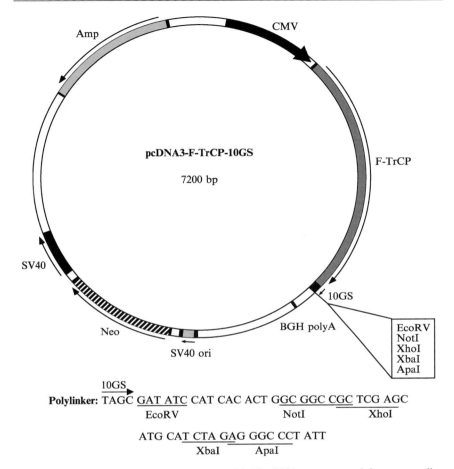

FIG. 2. Plasmid map of pCDNA-F-TrCP-10GS. The DNA sequence and the open reading frame within the multiple cloning sites for cloning cDNAs encoding TPs are indicated.

TPs will sometimes compromise the binding affinity, because the adjacent amino acids, although not involved in direct protein–protein contact, may nevertheless be required for optimal binding with the substrate. One solution is to increase the expression levels of βTrCP-TP to compensate for the loss of affinity. Another possibility to enhance substrate recruitment, which remains to be tested, would be to fuse multiple tandem copies of TPs to βTrCP.

Efficacy Testing and Optimization

To evaluate the efficacy of the engineered βTrCP-TP, the epitope-tagged substrate is transfected in HeLa cells to assess degradation in the presence of co-transfected βTrCP-TP. The amount of βTrCP-TP introduced relative to that of the epitope-tagged substrate is critical for targeted degradation and needs to be determined by titration. HeLa cells are transfected with a constant dose of the tagged substrate and increasing amounts of βTrCP-TP. At 48–60 h after transfection, cells are harvested, lysed in NP-40 or RIPA lysis buffer (Harlow and Lane, 1988), and subjected to SDS-PAGE and Western blotting to measure the levels of epitope-tagged substrate relative to the expression levels of βTrCP-TP. The ratio of the transfected plasmid DNAs encoding βTrCP-TP and the epitope-tagged substrate at which the greatest reduction of substrate is seen is selected to further evaluate targeted substrate degradation. This is carried out by a determination of the half-life by pulse-chase analysis and the ubiquitination status in the absence or presence of transfected βTrCP-TP. The detailed procedures for half-life determination and *in vivo* ubiquitination assays are described elsewhere (Chen *et al.*, 2001; Zhou, 2004).

Like the case for native SCF$^{\beta\text{TrCP}}$, efficient ubiquitination by the assembled SCF$^{\beta\text{TrCP-TP}}$ machinery requires that the recruited substrate should be properly positioned toward the Cdc34 ubiquitin-conjugating enzyme to accept ubiquitin (Wu *et al.*, 2003). However, it is often extremely difficult to predict the tertiary structure of the engineered βTrCP-TP. One solution is to insert flexible peptide linkers between βTrCP and TP to enhance the opportunity of correctly orienting the recruited substrate toward Cdc34. Such flexible linkers also serve to alleviate the potential steric hindrance between the F-box protein and the TP, thereby increasing the efficiency of substrate binding and recruitment. A linker consisting of 10 copies of Gly-Ser repeats has been shown to be necessary for targeted c-myc recruitment and degradation by the engineered βTrCP-Max(HLH/L) (Cohen *et al.*, 2004; Zhang and Zhou, unpublished data). It should be noted that the flexible linker serves to increase the chances of correct substrate orientation by the chimeric βTrCP-TP but may not be the optimal configuration for ubiquitin transfer. The native SCF$^{\beta\text{TrCP}}$ ubiquitin ligase is believed to assume a rigid structure that presents the substrate toward Cdc34 (Wu *et al.*, 2003). In contrast, conformational flexibility is essential for the catalytic activity of the WWP1 HECT domain E3 ligase (Verdecia *et al.*, 2003). Whether to choose a rigid or a flexible linker must ultimately be determined on an empirical, case-by-case basis. Based on the crystal structure of βTrCP, the TP within the βTrCP-TP fusion is predicted to protrude out from the top surface of the βTrCP WD40 propellor toward Cdc34.

Targeting Endogenous Substrates

Targeted degradation of endogenous substrates requires highly efficient delivery of the engineered βTrCP-TP into every target cell. Transient transfection provides a simple and usefully applicable method in many cell lines. In the event that low transfection efficiency is encountered, the following methods can be used to enrich the target cells containing the transfected βTrCP-TP: (1) cotransfection of a plasmid carrying a drug-resistant gene (such as puromycin, hygromycin, or neomycin), followed by drug selection to eliminate the nontransfected cells. Furthermore, stable βTrCP-TP–expressing cell lines can be generated by this method; (2) cotransfection of a plasmid expressing specific cell surface receptor (such as CD19), followed by cell selection using immunomagnetic beads conjugated to the anti-CD19 monoclonal antibody (Zhang *et al.*, 2003; Zhou *et al.*, 2000); (3) cotransfection of a plasmid expressing a fluorescent protein (GFP, YFP, or DS-Red), followed by flow cytometry–activated cell sorting. Cells enriched for transfected βTrCP-TP are lysed in NP-40 or RIPA buffer and subjected to SDS-PAGE and Western blotting to determine the steady-state levels of the endogenous target (Harlow and Lane, 1988).

Recombinant adenoviruses and retroviruses offer more efficient vehicles for the delivery of βTrCP-TP in both cell lines and primary cells that are refractory to transient transfection. The AdEasy system developed by He and Vogelstein has been used successfully for delivery of the engineered βTrCP-TP ubiquitin ligases and promotes efficient degradation of the retinoblastoma family proteins and c-myc (Cohen *et al.*, 2004; Zhang *et al.*, 2003; Zhou *et al.*, 2000). The detailed procedure for adenoviral production has been described elsewhere (Falck-Pederson, 1998; He *et al.*, 1998). βTrCP-TP is subcloned into the pAdTrackCMV vector and transfected into 293 cells together with AdEasy-1 for packaging into adenoviruses, which are then amplified and purified by CsCl gradient centrifugation to obtain high titer virus stock. We routinely produce viruses at an MOI (multiplicity of infection) greater than 10^{12}. Expression of GFP from the adenoviral vector allows convenient monitoring of the infection efficiency (He *et al.*, 1998). Adenovirus is capable of infecting a variety of mammalian cells and tissues at high efficiency and promotes high-level expression of βTrCP-TP. These characteristics of the adenoviral system provide tremendous versatility for improving targeted proteolysis. For instance, inefficient degradation because of low affinity of TP to substrate can be compensated by high MOI infection. Moreover, the capacity of introducing a wide range of virus into recipient cells can be exploited to achieve not only complete degradation but also fine-tuning of the steady-state levels of the substrate (Zhang et al., 2003) (see "Unique Properties and Applications"). It should be noted that high-level overexpression of βTrCP-TP can squelch the core SCF, resulting in the inhibition of targeted proteolysis (Zhang and Zhou, unpublished result). Therefore, adenoviruses should be carefully titrated to define the MOI for the desired knockdown levels of the intended substrates.

Certain primary or hematopoietic cells are refractory to adenoviral infection. Retroviral vectors provide an alternative vehicle for delivering βTrCP-TP in dividing cells. The pBMN-GFP retroviral vector (Orbigen Inc.) has demonstrated efficient transduction of βTrCP-TP into the HL-60 promyelocytic leukemia cells to degrade proteins of the Rb family (Zhang et al., 2003). The protocols for generation of the retrovirus and infection of the target cells are described in detail elsewhere (Persons et al., 1998). Because the βTrCP-TP transgene is transduced and integrated into the host genome, stable cell lines can be generated that maintain sustained degradation of the substrate. However, the expression level of βTrCP-TP is generally lower than that of adenoviral-mediated expression, resulting in less potent degradation of the intended substrate. The Epstein-Barr virus nuclear antigen (EBNA-1) in the pBMN-GFP vector allows the establishment of stable episomes (5–20 copies per cell) and thus increases the expression of βTrCP-TP. Retroviral-mediated gene transfer requires that the target cells should be mitotically active at the time of infection. For delivery of βTrCP-TP in non-dividing cells, a lentiviral-based delivery system can be used.

Unique Properties and Applications

Because βTrCP-TP is the limiting factor for targeted degradation and its expression level can be controlled through infection at different MOI levels of recombinant adenovirus, the protein knockout approach can be used not only to eliminate the substrate but also to fine-tune the intracellular protein levels (Zhang et al., 2003). Furthermore, βTrCP-TP can be expressed under the control of an inducible promoter to achieve "inducible knockout or knockdown" as desired (Cong et al., 2003).

Targeted proteolysis operates at the posttranslational level, whereas other reverse genetic approaches such as gene targeting, siRNA, antisense oligodeoxynucleotides, or ribozymes function to block the biosynthesis of cellular proteins. The operational mechanism of targeted proteolysis can be conveniently exploited to dissect the functional complexity of target proteins that are associated with specific posttranslational events:

(1) TPs that selectively bind only to posttranslationally modified forms of the substrate can be used to probe the function of such modifications by knocking down that subpopulation instead of inducing an overall degradation of the entire population of the substrate. The βTrCP-E7N ubiquitin-ligase has been shown to selectively eliminate the hypophosphorylated Rb while leaving the hyperphosphorylated form intact (Zhang et al., 2003). (2) βTrCP-TP can be directed to a specific subcellular compartment to achieve "compartmentalized " degradation. For instance, β-catenin is localized in two subcellular compartments with different functions: the cytoplasm/nuclear population serves as a component of the wnt-signaling pathway, and the cytoplasmic membrane subpopulation functions to connect E-cadherin and α-catenin in actin cytoskeleton organization. Knockout or knockdown approaches at the biosynthesis level would necessarily target both populations of β-catenin. The βTrCP-Ecad ubiquitin ligase was designed to target only the cytoplasm/nuclear β-catenin and spare the membrane subpopulations (Cong et al., 2003); (3) A single βTrCP-TP that recognizes the conserved domain(s) shared by a protein family can simultaneously knockout or knockdown the entire protein family, thereby ablating the underlying redundant functions that cannot be revealed by eliminating one member in isolation (Zhang et al., 2003). (4) βTrCP-TP can mediate the degradation of target proteins that are conserved across many eukaryotic species. In contrast, knockdown technologies operating at the biosynthesis level such as RNAi require precise sequence identity and may not tolerate variation of even a single nucleotide of the target genes between eukaryotic species.

Summary

This chapter describes a simple and efficient method to direct the robust SCF ubiquitination machinery toward degradation of specific cellular proteins and generate loss-of-function mutations in mammalian cells. The general procedures and specific considerations for each step outlined are derived from several years of intensive effort. The F-box-containing βTrCP has been extensively modified and a vector constructed that allows a single cloning step to assemble an engineered βTrCP-TP ubiquitin ligase. Plasmid- or viral-based gene delivery systems are in place for transducing the engineered βTrCP-TP for targeted degradation of the endogenous substrates in a variety of vertebrate cells and in animals. It should be noted that efficient degradation requires proper assembly of the SCF complex to orient the recruited substrate in a position optimal for accepting ubiquitin from Cdc34. Individual TPs and substrates are divergent in structural configurations, especially in the context of the engineered βTrCP and the other SCF components. Therefore, additional modifications that entail further protein engineering in practice might prove to be necessary to achieve the ideally desired levels of knockdown. Future studies are being taken up toward the construction of a panel of engineered βTrCP vectors that are able to accommodate the structural diversity of TPs and substrates. Specific TPs will be designed and grafted onto optimized F-box-containing substrate receptors to exquisitely probe the functional complexity of cellular proteins in vertebrate cells and in animals.

Acknowledgments

We thank Peter Howley for the generous support of this work in its initial stages and for helpful suggestions and Kevin Boru and Jennifer Lee for editing. P. Z. was a recipient of the Kimmel Scholar Award from the Sidney Kimmel Foundation for Cancer Research. This work was supported by National Institute of Health grant (5 R33 CA92792), the Speaker's Fund from the New York Academy of Medicine, the Dorothy Rodbell Cohen Foundation for Sarcoma Research, Susan G. Komen Foundation for Breast Cancer Research, the Academic Medicine Development Company Foundation, and the Mary Kay Ash Charitable Foundation.

References

Chen, W., Lee, J., Cho, S. Y., and Fine, H. A. (2004). Proteasome-mediated destruction of the cyclin a/cyclin-dependent kinase 2 complex suppresses tumor cell growth *in vitro* and *in vivo*. *Cancer Res.* **64,** 3949–3957.

Chen, X., Zhang, Y., Douglas, L., and Zhou, P. (2001). UV-damaged DNA-binding proteins are targets of CUL-4A-mediated ubiquitination and degradation. *J. Biol. Chem.* **276,** 48175–48182.

Cohen, J. C., Scott, D., Miller, J., Zhang, J., Zhou, P., and Larson, J. E. (2004). Transient in utero knockout (TIUKO) of C-MYC affects late lung and intestinal development in the mouse. *BMC Dev. Biol.* **4,** 4.

Cong, F., Zhang, J., Pao, W., Zhou, P., and Varmus, H. (2003). A protein knockdown strategy to study the function of beta-catenin in tumorigenesis. *BMC Mol. Biol.* **4,** 10.

Deshaies, R. J. (1999). SCF and Cullin/Ring H2-based ubiquitin ligases. *Annu. Rev. Cell Dev. Biol.* **15,** 435–467.

Falck-Pederson, E. (1998). Use and application of adenovirus expression vectors. *In* "Cells, a Laboratory Manual" (D. L. Spector, R. D. Goldman, and L. A. Leinwand, eds.), Vol. 2, pp. 90.1–90.28. Cold Spring Harbor Laboratory Press, Cold Spring Harbor.

Harlow, E., and Lane, D. (1988). "Antibodies, a Laboratory Manual." Cold Spring Harbor Laboratories, Cold Spring Harbor.

He, T. C., Zhou, S., da Costa, L. T., Yu, J., Kinzler, K. W., and Vogelstein, B. (1998). A simplified system for generating recombinant adenoviruses. *Proc. Natl. Acad. Sci. USA* **95,** 2509–2514.

Hershko, A., and Ciechanover, A. (1998). The ubiquitin system. *Annu. Rev. Biochem.* **67,** 425–479.

Liu, J., Stevens, J., Matsunami, N., and White, R. L. (2004). Targeted degradation of beta-catenin by chimeric F-box fusion proteins. *Biochem. Biophys. Res. Commun.* **313,** 1023–1029.

Persons, D. A., Mehaffey, M. G., Kaleko, M., Nienhuis, A. W., and Vanin, E. F. (1998). An improved method for generating retroviral producer clones for vectors lacking a selectable marker gene. *Blood Cells Mol. Dis.* **24,** 167–182.

Ptashne, M., and Gann, A. (2002). "Genes & Signals." Cold Spring Harbor Laboratory Press, Cold Spring Harbor.

Su, Y., Ishikawa, S., Kojima, M., and Liu, B. (2003). Eradication of pathogenic beta-catenin by Skp1/Cullin/F box ubiquitination machinery. *Proc. Natl. Acad. Sci. USA* **100,** 12729–12734.

Verdecia, M. A., Joazeiro, C. A., Wells, N. J., Ferrer, J. L., Bowman, M. E., Hunter, T., and Noel, J. P. (2003). Conformational flexibility underlies ubiquitin ligation mediated by the WWP1 HECT domain E3 ligase. *Mol. Cell* **11,** 249–259.

Willems, A. R., Lanker, S., Patton, E. E., Craig, K. L., Nason, T. F., Mathias, N., Kobayashi, R., Wittenberg, C., and Tyers, M. (1996). Cdc53 targets phosphorylated G1 cyclins for degradation by the ubiquitin proteolytic pathway. *Cell* **86,** 453–463.

Wu, G., Xu, G., Schulman, B. A., Jeffrey, P. D., Harper, J. W., and Pavletich, N. P. (2003). Structure of a beta-TrCP1-Skp1-beta-catenin complex: Destruction motif binding and lysine specificity of the SCF(beta-TrCP1) ubiquitin ligase. *Mol. Cell* **11,** 1445–1456.

Zhang, J., Zheng, N., and Zhou, P. (2003). Exploring the functional complexity of cellular proteins by protein knockout. *Proc. Natl. Acad. Sci. USA* **100,** 14127–14132.

Zhou, P. (2004). Determining protein half-lives. *In* "Signal Transduction Protocols" (R. C. Dickson and M. D. Mendenhall, eds.), Vol. 284, pp. 67–77. Humana Press, Totowa.

Zhou, P., Bogacki, R., McReynolds, L., and Howley, P. M. (2000). Harnessing the ubiquitination machinery to target the degradation of specific cellular proteins [In Process Citation]. *Mol. Cell* **6,** 751–756.

Zhou, P., and Howley, P. M. (1998). Ubiquitination and degradation of the substrate recognition subunits of SCF ubiquitin-protein ligases. *Mol. Cell* **2,** 571–580.

[54] Chimeric Molecules to Target Proteins for Ubiquitination and Degradation

By KATHLEEN M. SAKAMOTO

Abstract

Protein degradation is one of the tactics used by the cell for irreversibly inactivating proteins. In eukaryotes, ATP-dependent protein degradation in the cytoplasm and nucleus is carried out by the 26S proteasome. Most proteins are targeted to the 26S proteasome by covalent attachment of a multiubiquitin chain. A key component of the enzyme cascade that results in attachment of the multiubiquitin chain to the target or labile protein is the ubiquitin ligase that controls the specificity of the ubiquitination reaction. Defects in ubiquitin-dependent proteolysis have been shown to result in a variety of human diseases, including cancer, neurodegenerative diseases, and metabolic disorders.

The SCF (Skp1-Cullin-F-box-Hrt1) complex is a heteromeric ubiquitin ligase that multiubiquitinates proteins important for signal transduction and cell cycle progression. A technology was developed known as Protac (*Pro*teolysis *Ta*rgeting *C*himeric Molecule) that acts as a bridge, bringing together the SCF ubiquitin ligase with a protein target, resulting in its ubiquitination and degradation. The Protac contains an SCF-binding peptide moiety at one end that is recognized by SCF that is chemically linked to the binding partner or ligand of the target protein. The first demonstration of the efficacy of Protac technology was the successful recruitment, ubiquitination, and degradation of the protein methionine aminopeptidase-2 (MetAP-2) through a covalent interaction between MetAP-2 and Protac. Subsequently, we demonstrated that Protacs could effectively ubiquitinate and degrade cancer-promoting proteins (estrogen and androgen receptors) through noncovalent interactions *in vitro* and in cells. Finally, cell-permeable Protacs can also promote the degradation of proteins in cells. This chapter includes experiments to test the ability of Protacs to target proteins *in vitro* and in cells.

Introduction

Ubiquitin-dependent proteolysis is a major pathway that regulates intracellular protein levels. Posttranslational modification of proteins by E3 ubiquitin ligases results in multiubiquitin chain formation and subsequent degradation by the 26S proteasome (Ciechanover et al., 2000; Deshaies, 1999; Sakamoto, 2002). One potential approach to treating human disease is to recruit a disease-related protein to an E3 ligase for ubiquitination and subsequent degradation. To this end, a technology known as Protacs (*Pro*teolysis *Ta*rgeting *C*himeric Molecules) was developed. The goal of Protac therapy is to create a "bridging molecule" that could link together a disease-related protein to an E3 ligase. Protacs consist of one moiety (e.g., a peptide), which is recognized by the E3 ligase. This moiety or peptide is then chemically linked to a binding partner of the target. The idea is that Protacs would bring the target to the E3 ligase in close enough proximity for multiubiquitin attachment, which would then be recognized by the 26S proteasome (Fig. 1). The advantage of this approach is that it is catalytic and theoretically can be used to recruit any protein, even those that exist in a multisubunit complex.

Several applications for Protac therapy are possible. In cancer, the predominant approach to treating patients is chemotherapy and radiation. Both of these forms of therapy result in complications because of effects on normal cells. Therefore, development of therapeutic approaches to specifically target cancer-causing proteins without affecting normal cells is desirable.

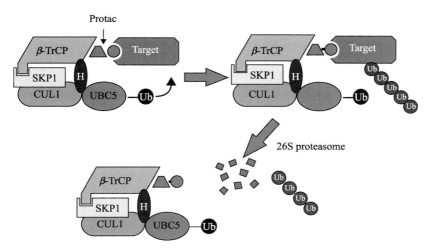

FIG. 1. Protac-1 targets MetAP-2 to SCF. Protac-1 is a chimeric molecule that consists of a phosphopeptide moiety and a small molecule moiety that interacts with the protein target (Sakamoto et al., 2001) (See color insert.)

To test the efficacy of Protacs *in vitro* and *in vivo*, several components are essential. First, a functional E3 ligase is necessary, either in purified form or isolated from cell extracts. Additional components of ubiquitination reaction, including ATP, E1, E2, and ubiquitin, are also required. Second, a small peptide or molecule recognized by the E3 ligase must be identified. Finally, a target with a well-characterized binding partner must be selected that will be chemically linked to the peptide. Finally, successful application of Protacs technology depends on the ability of the Protac to enter cells to target the protein for ubiquitination and degradation. For clinical application, therapeutic drug concentrations are usually considered to be in the nanomolar range.

In addition to the use of Protacs for the treatment of human disease, these molecules provide a chemical genetic approach to "knocking down" proteins to study their function (Schneekloth et al., 2004). The advantages of Protacs are that they are specific and do not require transfections or transduction. Protacs can be directly applied to cells or injected into animals without the use of vectors. Given the increased number of E3 ligases identified by the Human Genome Project, the possibilities for different combinations of Protacs that link specific targets to different ligases are unlimited. This chapter describes general strategies of testing the efficacy of Protacs using two E3 ligases as an example: SCF$^{\beta\text{-TRCP}}$ and Von Hippel

Lindau (VHL) complexes (Ivan et al., 2001; Kaelin, 2002). Three different targets will be described: methionine aminopeptidase-2 (MetAP-2), estrogen receptor (ER), and androgen receptor (AR). We will provide an overview of binding assays, transfections, immunoprecipitations, and ubiquitination and degradation assays of the proteins targeted to ubiquitin ligases by Protacs.

Strategies to Assess the Efficacy of Protacs In Vitro

As proof of concept, we generated a Protac molecule that targets the protein MetAP-2 for ubiquitination and degradation. MetAP-2 cleaves the N-terminal methionine from nascent polypeptides and is one of the targets of angiogenesis inhibitors fumagillin and ovalicin (Griffith et al., 1997; Sin et al., 1997). Ovalicin covalently binds to MetAP-2 at the His-231 active site. Inhibition of MetAP-2 is thought to block endothelial cell proliferation by causing G1 arrest (Yeh et al., 2000). MetAP-2 is a stable protein that has not been demonstrated to be ubiquitinated or an endogenous substrate of $SCF^{\beta\text{-TRCP}}$. For these reasons, Met-AP2 was chosen to be the initial target to test Protacs.

The heteromeric ubiquitin ligase, $SCF^{\beta\text{-TRCP}}$ (Skp1-Cullin-Fbox-Hrt1), was selected because the F-box protein β-TRCP/E3RS was previously shown to bind to IκBα (inhibitor of NFκBα) through a minimal phosphopeptide sequence, DRHDS*GLDS*M (phosphoserines indicated by asterisks) (Ben-Neriah, 2002; Karin and Ben-Neriah, 2000). This 10-amino acid phosphopeptide was linked to ovalicin to form the Protac (Protac-1) as previously described (Sakamoto et al., 2001).

MetAP-2-Protac Coupling Assay

MetAP-2 (9 μM) was incubated with increasing concentrations of Protac-1 for 45 min at room temperature (Fig. 2). Reactions were supplemented with SDS loading dye, fractionated on an SDS/10% polyacrylamide gel, transferred onto a nitrocellulose membrane, and immunoblotted with rabbit polyclonal anti-MetAP-2 antisera (Zymed, Inc.). Detection was performed using enhanced chemiluminescence (Amersham, Inc.).

Tissue Culture

293T cells were cultured in DMEM with 10% (vol/vol) FBS (Gibco, Inc.), penicillin (100 units/ml), streptomycin (100 mg/ml), and L-glutamine (2 mM). Cells were split 1:5 before the day of transfection and transiently transfected with 40 μg of plasmid. Cells were 60% confluent in 100-mm

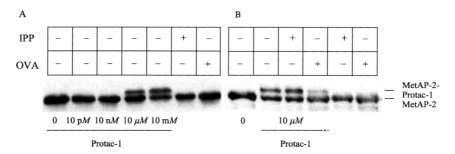

FIG. 2. MetAP-2 binds Protac specifically and in a concentration-dependent manner. (A) MetAP-2 (9 μM) was incubated with increasing concentrations of Protac-1 at room temperature for 45 min. The last two lanes depict MetAP-2 that was incubated with either free IκBα phosphopeptide (IPP; 50 μM) or free ovalicin (OVA; 50 μM), as indicated. After incubation, samples were supplemented with SDS-PAGE loading dye, fractionated by SDS-PAGE, and immunoblotted with MetAP-2 antiserum. (B) Same as (A), except MetAP-2 (9 μM) plus Protac-1 (10 μM) were supplemented with either IκBα phosphopeptide (50 μM) or ovalicin (50 μM) as indicated. Protac binding to MetAP-2 was inhibited by addition of ovalicin, but not phosphopeptide (B) (Sakamoto et al., 2001).

dishes on the day of transfection. DNA (20 μg of pFLAG-CUL1 and 20 μg of pFLAG-β-TRCP) was added. Cells were transfected using calcium phosphate precipitation as previously described (Lyapina et al., 1998). Cells were harvested 30 h after transfection. Five micrograms of pGL-1, a plasmid containing the cytomegalovirus (CMV) promoter linked to the green fluorescent protein (GFP) cDNA, was cotransfected into cells to determine transfection efficiency. In all experiments, greater than 80% of the cells were GFP-positive at the time of harvest, indicating high transfection efficiency.

Immunoprecipitations and Ubiquitination Assays

293T cells were lysed with 200 μl of lysis buffer (25 mM Tris-Cl, pH 7.5/150 mM NaCl/0.1% Triton X-100/5 mM NaF/0.05 mM EGTA/1 mM PMSF). Pellets were lysed by vortexing for 10 sec in a 4° cold room, then placed on ice for 15 min. After centrifugation at 13,000 rpm in a Microfuge for 5 min at 4°, the supernatant was added to 20 μl of FLAG M2 affinity beads (Sigma) and incubated for 2 h rotating at 4°. Beads were spun down at 13,000 rpm and washed with buffer A (25 mM Hepes buffer, pH 7.4/0.01% Triton X-100/150 mM NaCl) and one wash with buffer B (the same as buffer A but without Triton X-100). Four microliters of MetAP-2 (18 μM) stock, 4 μl of Protac-1 (100 μM), 0.5 μl of 0.1 μg/μl purified mouse E1 (Boston Biochem), 1 μl of 0.5 μg/μl human Cdc34 E2 (Boston Biochem), and 1 μl of

25 mM ATP were added to 20 μl (packed volume) of FLAG beads immunoprecipitated with SCF. Reactions were incubated for 1 h at 30° in a Thermomixer (Eppendorf) with constant mixing. SDS-PAGE loading buffer was added to terminate reactions, which were then evaluated by Western blot analysis as previously described (Sakamoto et al., 2001) (Fig. 3). Our results demonstrated that MetAP-2 bound to Protac could be ubiquitinated *in vitro* in the presence of SCF. These methods can be generalized to other ubiquitin ligases provided that a small molecule or peptide ligand exists to enable the synthesis of a suitable Protac and expression vectors that contain tagged versions of the protein or subunits are available. Alternative tags (e.g., myc or HA) have been used, and the resin can be cross-linked with an antibody, which can then be used to immunoprecipitate the E3 ligase from mammalian cells. Both the ER and AR are members of the steroid hormone receptor superfamily whose interactions with ligand (estradiol and testosterone, respectively) have been well characterized (Fig. 4). The ER has been implicated in the progression of breast cancer (Howell et al., 2003). Similarly, hormone-dependent prostate cancer cells grow in response to androgens (Debes et al., 2002). Therefore, both ER and AR are logical targets for cancer therapy. To target ER for ubiquitination and degradation, a Protac

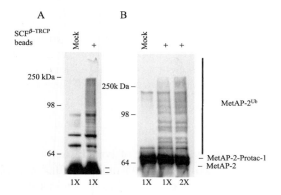

Fig. 3. Protac mediates MetAP-2 ubiquitination by SCF. (A) Ubiquitination of the 46-kDa fragment of MetAP-2. MetAP-2–Protac-1 mixture was added to either control (mock) or SCF$^{\beta\text{-TRCP}}$ beads (+) supplemented with ATP plus purified E1, E2 (Cdc34), and ubiquitin. UbcH5c (500 ng) was also tested as E2 in the reaction, which resulted in the same degree of ubiquitination as observed with Cdc34 (data not shown). Reactions were incubated for 1 h at 30°, and were evaluated by SDS-PAGE followed by Western blot analysis with anti-MetAP-2 antiserum. (B) Ubiquitination of full-length (67 kDa) MetAP-2. Same as (A), except that the 67-kDa preparation of MetAP-2 was used, and E1, E2, plus ubiquitin were either added at normal (1×) or twofold higher (2×) levels, as indicated (Sakamoto et al., 2001).

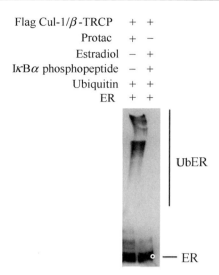

FIG. 4. Protac-2 activates ubiquitination of ER *in vitro*. Purified ER was incubated with recombinant E1, E2, ATP, ubiquitin, and immobilized SCF$^{\beta\text{-TRCP}}$ isolated from animal cells by virtue of Flag tags on cotransfected Cul1 and β-TRCP. Reactions were supplemented with the indicated concentration of Protac-2, incubated for 60 min at 30°, and monitored by SDS-PAGE followed by immunoblotting with an anti-ER antibody (Sakamoto *et al.*, 2003).

(Protac-2) was synthesized, containing the IκBα phosphopeptide linked to estradiol (the ligand for ER) (Sakamoto *et al.*, 2003).

Determination of Protein Degradation of Ubiquitinated Proteins In Vitro

The success of Protacs depends not only on efficient ubiquitination of the proposed target but also degradation of that target in cells. Several approaches can be used both *in vitro* and *in vivo* to demonstrate that the target is being destroyed. First, demonstration of degradation *in vitro* can be performed with purified 26S proteasome. For these experiments, we used purified yeast proteasomes as previously described (Verma *et al.*, 2000, 2002).

Ubiquitination assays were first performed with the immunoprecipitated E3 ligase, purified target, E1, E2, ATP, and ubiquitin with Protac. Purified yeast 26S proteasomes (40 μl of 0.5 mg/ml) were added to ubiquitinated protein (e.g., ER) on beads. The reaction was supplemented with 6 μl of 1 mM ATP, 2 μl of 0.2 M magnesium acetate, and ubiquitin aldehyde (5 μM final concentration). The reaction was incubated for 10 min at 30° with the occasional mixing in the Thermomixer (Eppendorf). To verify that degradation is due to proteasomes and not other proteases, purified 26S proteasomes were preincubated for 45 min at 30° with 1 mm of 1, 10 phenanthroline (Sigma) (a metal chelator and inhibitor of the RPN11 deubiquitinating enzyme in the 26S proteasome) (Fig. 5).

Strategies to Assess the Efficacy of Protacs In Vivo

Clinical application of Protacs is dependent on successful ubiquitination and degradation of the protein target by endogenous ubiquitin ligases and proteasomes within cells. There are several approaches to test the efficacy of Protacs using cell extracts or application directly to cells. Depending on the polarity of the Protac, efficiency of internalization in cells is variable. If Protacs are hydrophilic, such as the case with the Protac-1 that contains the IκBα phosphopeptide, extracts or microinjections are possible approaches. For cell-permeable Protacs, it is be possible to directly bath apply Protacs to cells.

Degradation Experiments with Xenopus Extracts

Extracts from unfertilized *Xenopus laevis* eggs were prepared on the day of the experiment as previously described (Murray, 1991). MetAP-2 (4 μl of 9 μM) was incubated with Protac-1 (50 μM) at room temperature for 45

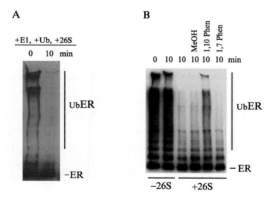

FIG. 5. Ubiquitinated ER is degraded by the 26S proteasome. (A) Ubiquitination reactions performed as described in the legend to Fig. 5A were supplemented with purified yeast 26S proteasomes. Within 10 min, complete degradation of ER was observed. (B) Purified 26S proteasome preparations were preincubated in 1,10 phenanthroline (1 mM) or 1,7 phenanthroline (1 mM) before addition. The metal chelator 1,10 phenanthroline inhibits the Rpn11-associated deubiquitinating activity that is required for substrate degradation by the proteasome. Degradation of ER was partially inhibited by addition of 1,10 phenanthroline, but not the inactive derivative 1,7 phenanthroline (Sakamoto *et al.*, 2003).

min. The MetAP-2–Protac-1 mixture was added to 10 μl of extract in addition to excess ovalicin (10 μM final concentration). The excess of ovalicin was added to saturate any free MetAP-2 in the reaction. Additional components in the reaction included constitutively active IKK (IKK-EE; 0.4 μg) and okadaic acid (10 μM final concentration) to maintain phosphorylation of the IκBα peptide moiety of Protac. To test for specificity of proteasomal degradation, various proteasome inhibitors were used, including N-acetyl-leu-leu-norleucinal (LLnL, 50 μM final) or epoxomicin (10 μM final). Protease inhibitors chymotrypsin, pepstatin, and leupeptin cocktail (15 μg/ml final concentration) were also added to the extracts. Reactions were incubated for time points up to 30 min at room temperature and terminated by adding 50 μl of SDS loading buffer. Samples were then evaluated by Western blot analysis using MetAP-2 antiserum (Fig. 6).

Microinjection as a Method to Study Effects of Protacs on Ubiquitination and Degradation of Target Proteins

Protacs that contain a phosphopeptide do not enter cells efficiently. Various protein transduction domains, lipid-based transfection reagents, and electroporation or other transient transfection methods can be tested. However, to facilitate Protacs entering into cells, microinjections were performed. For these experiments, Protac-3 (IκBα) phosphopeptide-testosterone was synthesized to target the AR (Sakamoto et al., 2003). As a readout of protein degradation, 293 cells stably expressing AR-GFP were selected using G418 (600 μg/ml). Before microinjections, cells were approximately 60% confluent in 6-cm dishes.

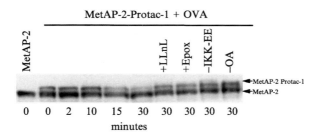

FIG. 6. MetAP-2-Protac but not free MetAP-2 is degraded in *Xenopus* extracts. The MetAP-2–Protac-1 mixture or MetAP-2 alone was added to *Xenopus* egg extract fortified with ovalicin (OVA; 100 μM), IKK-EE (0.4 μg), and okadaic acid (10 μM). Where indicated, reactions were either deprived of IKK-EE or okadaic acid (OA) or were further supplemented with 50 μM LLnL or 10 μM epoxomicin (Epox). Reactions were incubated for the indicated time points at room temperature, terminated by adding SDS-PAGE loading dye, and evaluated by SDS-PAGE followed by Western blotting with anti-MetAP-2 antiserum (Sakamoto et al., 2001).

Protac-3 diluted in a KCl solution (10 μM final) with rhodamine dextran (molecular mass 10,000 Da; 50 μg/ml) was injected into cells through a microcapillary needle using a pressurized injection system (Picospritzer II; General Valve Corporation). Coinjection with rhodamine dextran is critical to ensure that decrease in AR-GFP is not due to leakage of protein from cells after microinjection. The injected volume was 0.2 pl, representing 5–10% of the cell volume. GFP and rhodamine fluorescence can be visualized with a fluorescent microscope (Zeiss) and photographs taken with an attached camera (Nikon). Within 1 h after microinjection, disappearance of AR-GFP is visible (Fig. 7). Cells should remain rhodamine

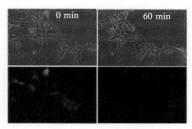

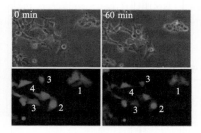

Degree of AR-GFP Disappearance	Percent (out of >200 cells)
1. NONE	4
2. MINIMAL	16
3. PARTIAL	29
4. COMPLETE	51

FIG. 7. Microinjection of Protac leads to AR-GFP degradation in cells. Protac-3 (10 μM in the microinjection needle) was introduced using a Picospritzer II pressurized microinjector into 293^{AR-GFP} cells in a solution containing KCl (200 μM) and rhodamine dextran (50 μg/ml). Approximately 10% of total cell volume was injected. (A) Protac-3 induces AR-GFP disappearance within 60 min. The top panels show cell morphology under light microscopy overlaid with images of cells injected with Protac as indicated by rhodamine fluorescence (pink color). The bottom panels show images of GFP fluorescence. By 1 h, GFP signal disappeared in almost all microinjected cells. To quantify these results, we injected more than 200 cells and classified the degree of GFP disappearance as being either none (1), minimal (2), partial (3), or complete (4). Examples from each category and the tabulated results are shown in (B). These results were reproducible in three independent experiments performed on separate days with 30–50 cells injected per day (Sakamoto et al., 2003). (See color insert.)

positive provided that injection has not caused lysis of cells or leakage of AR-GFP from cells. Greater than 200 cells per experiment (in three separate experiments) provide data demonstrating that Protacs induces degradation of the target. AR-GFP disappearance can then be quantitated by categorizing the intensity of GRP signal as indicative of complete disappearance, partial disappearance, minimal disappearance, or no disappearance. To verify that the disappearance of AR-GFP from cells is proteasome dependent, cells were pretreated with proteasome inhibitor epoxomicin (10 μM final) for 5 h before microinjections or were coinjected with epoxomicin (10 μM).

Methods to Test a Cell-Permeable Protac

Reagents capable of redirecting the substrate specificity of the ubiquitin-proteasome pathway in protein degradation would be useful experimental tools for modulating cellular phenotype and potentially acting as drugs to eliminate disease-promoting proteins. To use Protacs to remove a gene product at the posttranslational level, a cell-permeable reagent would be necessary. A HIF1α-DHT Protac was developed for this purpose. Given the lack of small molecule E3 ligase ligands, the seven amino acid sequence ALAPYIP from hypoxia-inducible factor 1α (HIF1α) was chosen for the E3 recognition domain of Protac-4 (Schneekloth et al., 2004). This sequence has been demonstrated to be the minimum recognition domain for the von Hippel–Lindau tumor suppressor protein (VHL) (Hon et al.,

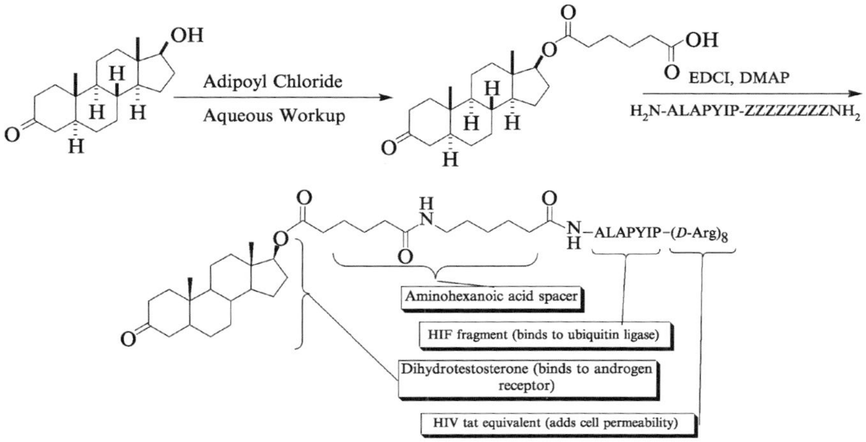

FIG. 8. Chemical structure of HIF-DHT Protac (Schneekloth et al., 2004).

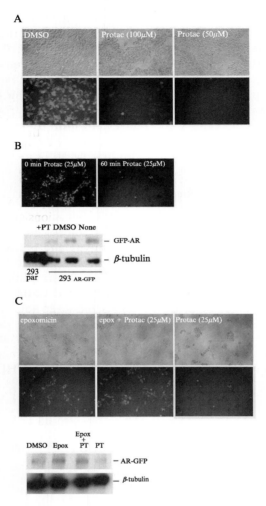

Fig. 9. HIF-DHT Protac mediates AR-GFP degradation in a proteasome-dependent manner; $293^{AR\text{-}GFP}$ cells (0.5×10^6 cells/ml) were plated at 50% confluence in a volume of 200 µl of media in a 96-well dish. (A and B) Protac induces AR-GFP disappearance within 60 min. Protac in a 100-, 50-, or 25-μM concentration or DMSO control in a volume of 0.6 µl was added. Cells were visualized under light (top) or fluorescent (bottom) microscopy 1 h after treatment. Photographs were taken with a SC35 type 12, 35-mm camera attached to an Olympus fluorescent inverted microscope. (B) AR-GFP protein is decreased in cells treated with Protac. Lysates were prepared from parental cells (293 par) or AR-GFP expressing cells treated with Protac (+PT), DMSO, or no treatment (None) for 60 min. Western blot analysis was performed with rabbit polyclonal anti-AR antisera (1:1000; UBI) or β-tubulin (1:200;

2002; Kaelin, 2002). VHL is part of the VBC-Cul2 E3 ubiquitin ligase complex. Under normoxic conditions, a proline hydroxylase catalyzes the hydroxylation of HIF1α at the (Epstein et al., 2001) central proline in the ALAPYIP sequence. This modification results in recognition and polyubiquitination by VHL. HIF1α is constitutively ubiquitinated and degraded under normoxic conditions (Kaelin, 2002). In addition, a poly-D-arginine tag derived from HIV tat was added to the carboxy terminus of the peptide sequence to confer cell permeability and prevent nonspecific proteolysis (Kirschberg et al., 2003; Wender et al., 2000) (Fig. 8). This Protac should then enter the cell, be recognized and hydroxylated by a prolyl hydroxylase, and subsequently be bound by both the VHL E3 ligase and the target, AR.

The 293 cells stably expressing AR-GFP were used to study the effects of HIF1α-DHT Protac on AR degradation. For these experiments, greater than 95% of cells expressed AR-GFP. On the day before experiments, cells were plated in 96-well plates with 200 μl of media at 60% confluence. Protac was dissolved in DMSO and was added to cells at concentrations ranging between 10 μM–100 μM. The presence or absence of GFP expression after Protac treatment was determined by fluorescent microscopy. A time course was performed, but for HIF1α-DHT Protac, the effects were observed within 2 h. To assess proteasome-dependent degradation, cells were pretreated with epoxomicin (10 μM final concentration) for 4 h before adding Protac. Western blot analysis was performed to determine levels of AR-GFP (Fig. 9).

To measure the protein levels of AR-GFP after Protac treatment, the cells were harvested, washed with PBS once, then pelleted at 1500 rpm. Cells were lysed with boiling SDS loading buffer (30 μl), then boiled for 5 min. Lysates were subjected to 8% polyacrylamide gel electrophoresis, and the proteins were transferred to nitrocellulose membrane. Western blot analysis was performed with antiandrogen receptor (1:1000) and anti-beta tubulin (1:200) antisera. Detection was determined using chemiluminescence.

Santa Cruz). (C) Epoxomicin inhibits Protac-induced degradation of AR-GFP. Cells were plated at a density of 0.3×10^6 cells/ml and treated with 10 μM epoxomicin (Calbiochem) or DMSO for 4 h before adding Protac (25 μM) for 60 min. (D) Western blot analysis was performed with cells in 96-well dishes treated with Protac (25 μM), DMSO (left), epoxomicin (10 μM), epoxomicin (10 μM) + Protac (50 or 25 μM), or Protac alone (50 or 25 μM) (Schneekloth et al., 2004). (See color insert.)

Acknowledgments

I would like to thank Ray Deshaies for his guidance and mentorship and Rati Verma for helpful discussions. This work was supported by the National Institutes of Health (R21CA108545), UCLA SPORE for Prostate Cancer (P50 CA92131), CaPCURE, Department of Defense (DAMD17-03-1-0220 and BC032613), UC BioSTAR Project (01-10232), Stein-Oppenheimer Award, and the Susan G. Komen Breast Cancer Foundation.

References

Ben-Neriah, Y. (2002). Regulatory functions of ubiquitination in the immune system. *Nat. Immunol.* **3,** 20–26.
Ciechanover, A., Orian, A., and Schwartz, A. L. (2000). Ubiquitin-mediated proteolysis: Biological regulation via destruction. *Bioessays* **22,** 442–451.
Debes, J. D., Schmidt, L. J., Huang, H., and Tindall, D. J. (2002). p300 mediates androgen-independent transactivation of the androgen receptor by interleukin 6. *Cancer Res.* **62,** 5632–5636.
Deshaies, R. J. (1999). SCF and Cullin/Ring H2-based ubiquitin ligases. *Annu. Rev. Cell Dev. Biol.* **15,** 435–467.
Epstein, A. C., Gleadle, J. M., McNeill, L. A., Hewitson, K. S., O'Rourke, J., Mole, D. R., Mukherji, M., Metzen, E., Wilson, M. I., Dhanda, A., Tian, Y. M., Masson, N., Hamilton, D. L., Jaakkola, P., Barstead, R., Hodgkin, J., Maxwell, P. H., Pugh, C. W., Schofield, C. J., and Ratcliffe, P. J. (2001). *C. elegans* EGL-9 and mammalian homologs define a family of dioxygenases that regulate HIF by prolyl hydroxylation. *Cell* **107,** 43–54.
Griffith, E. C., Su, Z., Turk, B. E., Chen, S., Chang, Y. H., Wu, Z., Biemann, K., and Liu, J. O. (1997). Methionine aminopeptidase (type 2) is the common target for angiogenesis inhibitors AGM-1470 and ovalicin. *Chem. Biol.* **4,** 461–471.
Hon, W. C., Wilson, M. I., Harlos, K., Claridge, T. D., Schofield, C. J., Pugh, C. W., Maxwell, P. H., Ratcliffe, P. J., Stuart, D. I., and Jones, E. Y. (2002). Structural basis for the recognition of hydroxyproline in HIF-1 alpha by pVHL. *Nature* **417,** 975–978.
Howell, A., Howell, S. J., and Evans, D. G. (2003). New approaches to the endocrine prevention and treatment of breast cancer. *Cancer Chemother. Pharmacol.* **52**(Suppl. 1), S39–S44.
Ivan, M., Kondo, K., Yang, H., Kim, W., Valiando, J., Ohh, M., Salic, A., Asara, J. M., Lane, W. S., and Kaelin, W. G., Jr. (2001). HIFalpha targeted for VHL-mediated destruction by proline hydroxylation: Implications for O2 sensing. *Science* **292,** 464–468.
Kaelin, W. G., Jr. (2002). Molecular basis of the VHL hereditary cancer syndrome. *Nat. Rev. Cancer* **2,** 673–682.
Karin, M., and Ben-Neriah, Y. (2000). Phosphorylation meets ubiquitination: The control of NF-[kappa]B activity. *Annu. Rev. Immunol.* **18,** 621–663.
Kirschberg, T. A., VanDeusen, C. L., Rothbard, J. B., Yang, M., and Wender, P. A. (2003). Arginine-based molecular transporters: The synthesis and chemical evaluation of releasable taxol-transporter conjugates. *Org. Lett.* **5,** 3459–3462.
Lyapina, S. A., Correll, C. C., Kipreos, E. T., and Deshaies, R. J. (1998). Human CUL1 forms an evolutionarily conserved ubiquitin ligase complex (SCF) with SKP1 and an F-box protein. *Proc. Natl. Acad. Sci. USA* **95,** 7451–7456.
Murray, A. W. (1991). Cell cycle extracts. *Methods Cell Biol.* **36,** 581–605.
Sakamoto, K. M. (2002). Ubiquitin-dependent proteolysis: Its role in human diseases and the design of therapeutic strategies. *Mol. Genet. Metab.* **77,** 44–56.

Sakamoto, K. M., Kim, K. B., Kumagai, A., Mercurio, F., Crews, C. M., and Deshaies, R. J. (2001). Protacs: Chimeric molecules that target proteins to the Skp1-Cullin-F box complex for ubiquitination and degradation. *Proc. Natl. Acad. Sci. USA* **98,** 8554–8559.

Sakamoto, K. M., Kim, K. B., Verma, R., Ransick, A., Stein, B., Crews, C. M., and Deshaies, R. J. (2003). Development of Protacs to target cancer-promoting proteins for ubiquitination and degradation. *Mol. Cell Proteomics.* **2,** 1350–1358.

Schneekloth, J. S., Jr., Fonseca, F. N., Koldobskiy, M., Mandal, A., Deshaies, R., Sakamoto, K., and Crews, C. M. (2004). Chemical genetic control of protein levels: Selective *in vivo* targeted degradation. *J. Am. Chem. Soc.* **126,** 3748–3754.

Sin, N., Meng, L., Wang, M. Q., Wen, J. J., Bornmann, W. G., and Crews, C. M. (1997). The anti-angiogenic agent fumagillin covalently binds and inhibits the methionine aminopeptidase, MetAP-2. *Proc. Natl. Acad. Sci. USA* **94,** 6099–6103.

Verma, R., Aravind, L., Oania, R., McDonald, W. H., Yates, J. R., 3rd, Koonin, E. V., and Deshaies, R. J. (2002). Role of Rpn11 metalloprotease in deubiquitination and degradation by the 26S proteasome. *Science* **298,** 611–615.

Verma, R., Chen, S., Feldman, R., Schieltz, D., Yates, J., Dohmen, J., and Deshaies, R. J. (2000). Proteasomal proteomics: Identification of nucleotide-sensitive proteasome-interacting proteins by mass spectrometric analysis of affinity-purified proteasomes. *Mol. Biol. Cell* **11,** 3425–3439.

Wender, P. A., Mitchell, D. J., Pattabiraman, K., Pelkey, E. T., Steinman, L., and Rothbard, J. B. (2000). The design, synthesis, and evaluation of molecules that enable or enhance cellular uptake: Peptoid molecular transporters. *Proc. Natl. Acad. Sci. USA* **97,** 13003–13008.

Yeh, J. R., Mohan, R., and Crews, C. M. (2000). The antiangiogenic agent TNP-470 requires p53 and p21CIP/WAF for endothelial cell growth arrest. *Proc. Natl. Acad. Sci. USA* **97,** 12782–12787.

Author Index

A

Aarsland, D., 99
Aas, T., 350
Abagyan, R. A., 635, 637, 638, 641, 642, 643, 646, 647, 648
Abbott, B. J., 532
Abdul-Karim, F. W., 107, 109
Abe, Y., 79
Abelson, J., 315
Aberle, H., 512
Abola, E. E., 641
Abraham, S., 513
Aburaki, S., 590, 599, 601
Adams, J., 491, 493, 495, 567, 586, 610, 683
Adams, P. D., 125
Adams, S. L., 290, 439, 446, 453
Adler, A. S., 219
Admon, A., 261, 262, 264, 278, 569
Aebersold, R., 13, 21, 369, 376
Affar, E. B., 708
Aguilar, R. C., 177, 392
Ahrens, P. B., 88
Ahringer, J., 314
Aida, N., 306
Aimi, Y., 96
Ajioka, J., 54
Akagi, T., 278
Akiguchi, I., 107
Akioka, H., 84
Akiyama, H., 94
Akiyama, S., 121
Akopian, T. N., 492
Alam, S. L., 136, 147
Alban, A., 92
Alberghina, L., 438
Albericio, F., 589, 595
Albert, I., 436, 570
Albert, T. K., 356, 358, 359, 361, 363, 365
Albrecht, S., 107
Alcasabas, A., 399
Aldridge, F., 101, 107
Alexander, J. M., 42

Alfarano, C., 290
Alien, I. V., 98
Alizadeh, A., 343, 345, 347
Alkalay, I., 570, 581
Al-Khedhairy, A., 93
Alkushi, A., 336, 350
Allaman, M. M., 136, 138, 147, 194
Allen, H., 436
Alley, M. C., 532
Allshire, R., 559
Alnemri, E. S., 634
Alper, P. B., 610, 613, 620
Al-Sarraj, S., 104
Altun, M., 496
Al-Tuwaijri, M., 514
Alvarez, R. B., 107
Amador, V., 15, 250, 252, 260, 261, 262, 278
Amati, B., 718
Amit, S., 250, 289, 290, 436, 439, 570, 581
An, J., 417, 635, 637, 647
Ancuta, P., 15
Andersen, J. R., 290
Andersen, J. S., 250, 289, 290, 379, 401, 436, 439, 570, 581
Anderson, D. C., 145, 416
Anderson, G. A., 285
Anderson, S., 306
Anderton, B. H., 96, 98, 101
Andrews, B., 314, 454
Angus, B., 345, 350
Ankel, H., 88
Annan, R. S., 66, 437, 439
Anselmo, A., 401
Anton, L. C., 79, 460, 560
Anton, R. C., 345
Appel, J. R., 613
Arabi, A., 560
Arai, H., 436
Arai, N., 107
Aravind, L., 17, 226, 471
Ardley, H. C., 80, 567
Arenzana-Seisdedos, F., 278
Argos, P., 641

Arima, K., 9, 23
Aristarkhov, A., 569
Arkin, M. R., 624
Armstrong, R. A., 104
Arnason, T., 5, 13, 52, 278
Arnold, J., 93
Aronica, E., 99
Arrio-Dupont, M., 552
Aruleba, S., 314
Arvai, A. S., 720
Ashcroft, M., 623
Asher, C., 165
Ashwell, J. D., 689
Askanas, V., 107
Assfalg, M., 23, 147, 148, 179, 181, 182, 183, 186, 190
Attanasio, A., 94, 101
Au, S. W., 303
Auburger, G., 101
Audhya, A., 198
Auger, K. R., 8, 623, 669, 702, 703, 704, 705, 706, 709, 712, 713, 714
Auld, K., 285
Auth, H., 587, 588, 590, 593, 597, 598
Autilio-Gambetti, L., 95, 101, 107, 109
Avalos, J. L., 451
Aviel, S., 261, 262
Avila, J., 92, 491, 495, 503
Axelrod, D., 555
Ayad, N. G., 250, 409, 411

B

Babbitt, R. W., 676
Baboshina, O. V., 22, 144
Babst, M., 136, 141, 194, 198
Babu, J. R., 143, 144
Bacci, B., 95
Bachant, J., 399
Bache, K., 194
Bachmair, A., 4, 6, 22, 23, 52, 54, 55, 56, 67, 266, 272, 368, 417
Bacik, I., 560
Backes, B. J., 610, 613, 620
Bader, G. D., 290, 314, 439, 446, 453, 454
Bae, S. H., 147, 187
Baeg, G. H., 570, 582
Baek, K. H., 664
Baek, S. H., 37, 495
Baer, R., 9, 23, 708

Baer, S., 471
Bai, C., 288, 434, 444
Bailly, E., 718
Bain, J. L. W., 5
Baisch, H., 351
Bajdik, C., 350
Bajorath, J., 637
Bak, T. H., 104
Baker, R. T., 17, 38, 54, 57, 121, 471
Baldi, A., 718
Baldi, F., 718
Baldi, L., 278
Baleux, F., 278
Balija, V., 314
Ballard, D. W., 266
Balzi, E., 466
Banaszak, L. J., 635, 648
Bancher, C., 96
Banerjee, A., 272
Bang, O., 38
Banik, N., 586
Baranov, E., 514
Barbey, R., 290, 449
Bardag-Gorce, F., 109
Barfoot, R., 120
Barford, D., 38, 303, 324
Barker, N., 335, 336
Barlev, N., 121
Barlund, M., 338
Barnett, J., 326, 686
Barnickel, G., 635, 648
Baron, A., 120
Barrio, J., 514
Barroga, C. F., 522
Bartel, B., 59, 60, 495, 496, 497, 515
Bartel, P., 138, 139
Bartholomew, D. U., 165
Bartolini, F., 15, 252, 261, 262, 278
Bartunik, H. D., 228
Baruch, A., 620
Bashir, T., 250, 260, 261
Basilico, C., 259, 260
Basilion, J., 514
Batalov, S., 137, 646
Bauer, A., 512
Baum, B., 314
Baumeister, W., 228, 229, 491, 497, 550
Bax, A., 146
Bays, N. W., 463, 497, 498
Beachy, P. A., 634

Beal, R., 11, 24, 33, 137, 141, 143, 179
Beaudenon, S., 356
Beaujeux, T. P., 101
Bebok, Z., 560
Bechmair, A., 54
Becker, F., 54, 55, 56
Becker, L. E., 107
Beddington, R., 505
Beemer, F. A., 120
Beer, D. G., 337
Beer-Romero, P., 256, 260, 261, 289, 290, 299, 512, 570, 661, 705
Beers, E., 52, 243
Beguinot, L., 512
Behnke, M., 586
Behrens, A., 436
Behrens, J., 570
Beinke, S., 436
Belgareh, N., 121
Belichenko, I., 394
Bell, J. E., 101
Bellen, H., 136
Belunis, C., 718, 721
Benaron, D., 514
Benarous, R., 436, 570
Bence, N. F., 15, 482, 484, 494, 496
Benckhuijsen, W., 494
Benfield, P. A., 8, 623, 669, 702, 703, 705, 706, 709, 712, 713, 714
Bengal, E., 261, 262, 264, 278
Benham, A. M., 551
Benichou, S., 436
Benmaamar, R., 416
Bennacer, B., 590, 601
Benne, R., 88, 96
Benner, J., 38, 41
Ben-Neriah, Y., 250, 289, 290, 436, 439, 521, 570, 581
Bennett, E. J., 484
Bennett, K., 290, 439, 446, 453
Bennett, M. K., 416, 567, 661, 664, 683
Bennetzen, J. L., 376
Bennink, J. R., 79, 460, 559, 560
Ben-Saadon, R., 13, 21, 261
Ben-Shushan, E., 570, 581
Benson, K., 513
Benveniste, H., 514
Berdo, I., 661, 705
Berg, T., 624, 634
Berger, S. L., 121

Berglund, L. E., 54
Berk, A., 514
Berliner, L. J., 186
Berlioz-Torrent, C., 436
Bernal, T., 336, 405, 408
Bernards, R., 350
Bernier-Villamor, V., 396
Berset, C., 436
Berti, G., 95
Bertolaet, B. L., 136, 158, 159
Bertwistle, D., 261
Besancon, F., 88
Best, J., 524
Betz, U. A., 524
Bevdekar, A. R., 109
Beyers, M., 16
Bharadwaj, R., 673
Bhaumik, S., 514
Bhorade, R., 514
Bickford, L. C., 568
Bieder, T., 462
Biegel, T. A., 326, 686
Biggs, P. J., 120
Bignell, G. R., 120
Bigio, E. H., 104
Bilak, M., 107
Bilbao, J. M., 107
Billett, M., 101, 107, 109
Bilodeau, P. S., 136, 138, 147, 194
Bingol, B., 15
Bird, T. D., 104
Birman, Y., 570, 581
Birmington, C. L., 560
Biserni, A., 503
Bishop, N., 194
Blackman, R. K., 244, 272
Blagosklonny, M. W., 512
Blair, E., 120
Blaiseau, P. L., 290
Blanchard, A., 101, 107, 109
Blasberg, R. G., 514, 531
Blat, Y., 399
Blethrow, J. D., 451, 452
Bliznyuk, A. A., 635, 647
Blobel, G., 269, 399
Bloecher, A., 13, 21, 304, 439
Blomhoff, R., 514
Blondel, M., 451
Blondelle, S. E., 613
Bloom, J., 15, 252, 261, 262, 278, 512

Blumbergs, P. C., 107
Board, P., 54
Bochtler, M., 228
Boehncke, W. H., 512
Boeke, J. D., 451
Boelens, R., 147, 185, 356, 358, 359, 360, 361, 363, 365
Boger, D. L., 620, 634
Boghosian-Sell, L., 121
Bogyo, M., 65, 493, 494, 498, 499, 586, 610, 611, 613, 615, 616, 620
Bohmann, D., 17, 52, 243, 244, 253, 661
Boisclair, M. D., 657, 658, 669, 671, 678, 685, 712
Boman, A. L., 136, 138, 147, 194
Bond, U., 54
Bonetti, B., 95
Bonin, M., 95
Bonneil, E., 261, 262
Bonvin, A. M., 147, 183, 185, 356, 358, 360
Boone, C., 314, 454
Boone, D. L., 17, 471
Borchers, C., 306
Bordallo, J., 462, 463
Bornstein, G., 256, 302, 439, 718
Borodovsky, A., 38, 47, 121, 122, 124, 470, 471, 494, 498, 499
Borresen-Dale, A. L., 350
Bos, P. R., 619
Bossi, G., 136, 143, 145
Bossy-Wetzel, E., 92, 575
Botstein, D., 312, 347, 350, 619
Bottomley, S., 657, 658, 669, 671, 678, 685, 712
Boucher, P., 4, 5, 14, 16, 22, 59, 60, 272
Bouldin, Y., 101
Bour, S. P., 436
Bourdon, J. C., 469, 702, 705
Bourne, Y., 720
Boutilier, K., 290, 439, 446, 453
Bouvet, M., 514
Bowen, Z., 638, 720
Bowerman, B., 306
Braak, E., 94
Braak, H., 94, 101
Brachmann, C. B., 451
Bradley, K. E., 162
Bradley, S., 513
Brady, G. P., Jr., 635, 647
Braendgaard, H., 95, 104

Braithwaite, K. L., 345
Brancolini, C., 575
Brand, A. H., 492
Brand, L., 671
Braun, A., 4, 6, 23, 279, 368
Braun, T., 313, 461
Brech, A., 194
Breitschopf, K., 261, 262, 264, 278
Brent, R., 363
Brett, T. J., 42
Bridge, J. A., 121
Bridges, L. R., 98
Bright, P. M., 13, 88
Brignone, C., 162, 708
Brion, J. P., 96
Brockman, J. A., 266
Brodsky, J. L., 177, 466
Broe, M., 104
Brooks, C. L., 685, 708
Bross, P. F., 513
Browder, Z., 54
Brower, M. E., 513
Brown, C., 120
Brown, D. R., 101
Brown, J., 101, 107
Brown, K., 278
Brown, M. S., 312, 321
Brown, P. O., 347, 350, 619
Brown, R. W., 345
Bruce, M., 101
Brumell, J. H., 560
Brun, J., 16
Brundin, P., 92, 491, 495
Brunger, A. T., 125
Brunner, C., 96
Bruseo, C., 634
Brzovic, P. S., 360
Bubendorf, L., 338
Buchanan, J. A., 466
Buchberger, A., 5, 24, 335
Bucher, P., 137
Buchsbaum, D., 514
Buchwald, M., 466
Buck, G., 54
Budka, H., 96, 107
Bujard, H., 598
Bulawa, C. E., 244, 272
Burch, A. D., 80, 560
Burke, D., 315, 363
Burke, J. R., 325, 332

Burlingame, A., 568, 620
Burn, D., 99, 104
Burton, J. L., 336
Buschhorn, B. A., 462, 465
Busino, L., 289
Bussey, H., 314, 454
Butt, T. R., 55, 60
Byers, B., 497
Bylebyl, G. R., 394

C

Cadrin, M., 109
Cagle, P. T., 345
Cagney, G., 159, 161, 163, 439
Cahill, S., 183, 184
Cai, S. Y., 676
Cairns, J., 350
Cairns, N. J., 104
Caligiuri, M., 623, 669, 703, 709, 712, 713, 714
Callis, J., 8, 17, 52, 54, 55, 57, 60, 61,
  205, 243, 244
Calogero, A., 514
Calvert, V., 337, 339
Calvo, D., 708
Camalier, R. F., 532
Cammarata, S., 101
Camosseto, V., 79
Camp, R. L., 344, 346
Campbell, A. K., 531
Campbell, R. E., 531, 550
Campbell, S. J., 647
Cannizzaro, L. A., 121
Cantley, L. C., 437
Cantor, C. R., 671
Cao, X., 436
Cao, Y., 512, 521, 522
Cao, Z., 657, 669, 678, 685, 712
Capati, C., 13, 15, 244, 266, 268, 272, 274
Cappello, F., 79
Capua, M. R., 136, 143, 145
Caputi, M, 718
Cardone, M., 647
Cardozo, C., 550, 604
Cardozo, T., 38, 289, 297, 638, 642, 643, 646
Carl, U. D., 443
Carlos Machado, J., 99
Carlsen, H., 514
Carney, D. S., 136, 144, 194
Carpino, L. A., 595, 596

Carr, S. A., 66, 437, 439
Carrano, A. C., 256, 258, 531, 638, 718, 720
Carroll, M., 121
Carson, F. L., 344, 345
Carson, M., 127
Carter, S., 16
Cartwright, H., 95
Carvajal, D., 567, 622, 623, 630, 631, 635
Casagrande, R., 47, 121, 122, 470
Cascio, T. C., 638
Case, A., 567
Case, C. P., 98
Casey, B., 104
Casp, C. B., 598
Castro, A., 406
Catanzariti, A., 54
Catarin, B., 439, 454
Catzavelos, C., 336
Cavallaro, T., 95
Cavasotto, C. N., 647
Celic, I., 451
Cerundolo, V., 550
Chakraborty, T., 443, 444
Chakravarti, A., 466
Chalfie, M., 491
Chan, H. M., 708
Chandler, J., 57
Chandler, J. S., 55
Chang, G. W., 512
Chang, H. T., 104
Chang, K., 314
Chang, Y. T., 568
Chantome, A., 141
Chao, C. C., 379, 401
Chatterjee, S., 586
Chau, V., 4, 5, 6, 8, 14, 16, 22, 23, 52,
  59, 60, 96, 243, 256, 260, 261, 266, 272,
  306, 368, 512
Chaudhuri, T., 514
Cheang, M. C., 350
Chelvanayagam, G., 641
Chen, D., 470, 708
Chen, D. J., 147
Chen, G., 337
Chen, G. Y., 619
Chen, H., 121, 136, 143, 145
Chen, I. S., 136
Chen, J., 625, 634
Chen, L., 158, 159, 161, 247
Chen, P., 311, 315, 317

Chen, Q., 289, 436, 437, 438, 439, 441, 445, 451
Chen, S., 221, 389, 586
Chen, X., 13, 21
Chen, Y., 95, 146, 147, 306
Chen, Y. N., 634
Chen, Z., 24, 46, 136, 144, 194, 266
Chen, Z. J., 4, 5, 6, 22, 23, 24, 136, 279, 289, 368, 664
Cheng, D., 369, 377, 378, 391, 416, 454
Cheng, D. M., 3, 8, 12, 13, 21, 22, 23, 24, 52, 53, 54, 60, 61, 137, 144, 145, 266, 268, 269, 279
Cherest, H., 290
Cherry, S., 514
Chi, Y., 13, 21, 66, 287, 434, 439, 451, 454
Chiaur, D. S., 289
Chiba, T., 38, 306, 640
Chicocca, E., 514
Chien, C. T., 138
Chiesa, M., 289
Childerstone, M., 665
Ching, K. A., 137
Chinowsky, T. M., 165
Chio, A., 101
Chishima, T., 514
Chiu, H., 99
Chiu, R. K., 14, 15, 16, 17
Chiu, Y. H., 136
Chock, P. B., 379, 401
Chodosh, L. A., 121
Choi, B. S., 147, 187
Choi, S., 65, 493, 567, 587
Chomsky, O., 482, 483, 496
Chong, S., 38, 41
Chou, S., 436
Choy, W. Y., 186
Christian, J. L., 570
Chu, C., 287, 289, 290, 296, 299, 302, 303, 305, 435, 436, 570, 638
Chu, T., 198
Chu, W., 718, 721
Chung, C. H., 37, 38, 84, 232, 495
Chung, G. G., 344, 346
Chung, J., 145
Chung, S. S., 38
Chung, T. D. Y., 726
Ciana, P., 503
Ciechanover, A., 8, 13, 21, 75, 88, 92, 120, 165, 177, 215, 228, 243, 249, 261, 262, 264, 278, 324, 356, 416, 417, 434, 469, 491, 495, 501, 655, 664, 702, 704
Cjeka, Z., 229
Clackson, T., 624
Clague, M. J., 16
Claridge, T. D., 640
Clark, K., 586
Clarke, D. J., 136, 158, 159
Claypool, J., 287, 434
Cleary, M., 314
Clegg, R. M., 670
Clemens, K. R., 145
Cleveland, M. G., 120
Clevers, H., 335, 336
Clifford, S. C., 512
Clore, G. M., 125, 146, 185
Clurman, B. E., 13, 21, 289, 299, 302, 304, 305, 436, 439, 458
Coccetti, P., 438
Cochran, A. G., 634
Cochran, E., 95
Cockman, M. E., 512
Coffey, D. M., 345
Coffino, P., 73, 568, 572, 577, 579, 581, 582
Cohen, M. L., 107, 121
Cohen, R. E., 7, 11, 24, 26, 31, 33, 158, 179, 215, 226, 243, 268, 470, 471
Cohen, S. B., 634
Cohen-Mansfield, J., 99
Colbert, D. L., 665
Cole, J. C., 635, 648
Cole, N. B., 557
Collart, M. A., 356, 358, 359, 361, 363, 365
Collins, T., 524
Colman, A., 38, 42
Colon, E., 8, 17, 52, 54, 60, 61, 205, 244
Comb, D. G., 38, 41
Conaway, J. W., 287, 434, 638
Conaway, R. C., 287, 434, 638
Concordet, J. P., 436, 570
Conklin, D. S., 314
Connell-Crowley, L., 305
Conover, D., 439
Conrad, M. N., 287, 434
Contag, C., 514, 531
Contag, P., 514
Cook, D. R. J., 122, 124
Cook, J., 47, 122, 470
Cook, L. A., 417
Cook, W. J., 185

Cooper, H. J., 454
Cope, G. A., 250
Copeland, N. G., 121
Copeland, R. A., 8, 623, 669, 702, 703, 704, 705, 706, 708, 709, 712, 713, 714
Corbin, E., 107
Cordon-Cardo, C., 340, 344, 345
Corey, E. J., 65, 493, 587, 601
Corless, C., 121
Corman, J. I., 669, 703, 704, 705
Correll, C. C., 287, 302, 434, 446
Corsi, D., 52
Cortez, S., 104
Costantini, F., 505
Cote, M., 165
Cottingham, I. R., 38, 42
Cottrell, J. S., 369
Couck, A. M., 96
Coulombe, P., 261, 262
Coulson, A., 314
Coux, O., 228, 229, 496, 610
Cowan, S. W., 125
Cowburn, D., 147, 183, 184
Cowley, S. M., 136
Cox, A. B., 94, 101
Cox, M. J., 54
Cox, T. K., 466
Craig, E. A., 138, 159, 161
Craig, K. L., 244, 287, 302, 304, 434, 439, 446, 448, 450
Craik, C. S., 610
Cravatt, B. F., 620
Crawford, A., 570
Crawford, J., 290
Creasy, D. M., 369
Crews, C. M., 493, 494, 498, 587, 588, 590, 593, 597, 598, 599, 603, 604, 605, 619
Crews, P., 634
Cribier, S., 552
Crinelli, R., 52
Crispino, J. D., 360, 365
Croce, C. M., 121, 634
Cronin, S., 311, 313, 318, 320, 321, 461, 463
Crooke, S. T., 4, 5, 14, 16, 22, 55, 59, 60, 272
Crooks, G. M., 514
Crosas, B., 121
Cruickshank, A. A., 586
Cruz-Sanchez, F. F., 104
Cryns, V. L., 405, 412, 416
Cuervo, J. H., 613

Culvenor, J. G., 107
Cummings, C. J., 99
Cummings, K. B., 198
Cummings, M., 5, 13
Cuny, G. D., 567

D

Da Costa, C., 436
Dahlheimer, J., 514
Dahmus, M. E., 8, 17, 52, 54, 60, 61, 205, 244
Dai, H., 350
Dai, M., 8
Dalakas, M. C., 107
d'Albis, A., 552
Dal Cin, P., 121
Dale, G. E., 98
Dalrymple, S. A., 326, 686
Damjanovich, S., 670
D'Andrea, A. D., 57, 121
Danenberg, K. D., 337
Danenberg, P. V., 337
Dang, L. C., 46, 669, 670, 677
Daniels, C. M., 136, 188, 390
Dantuma, N. P., 92, 481, 482, 491, 492, 493, 494, 495, 496, 497, 498, 499, 501, 502, 503, 508, 515, 588, 598, 604, 605
Dargemont, C., 121
Darji, A., 443, 444
Darula, Z., 620
Dasso, M., 569
Davie, T. R., 107
Davies, B. A., 136, 144, 194
Davies, K. J., 514
Davies, L., 95
Davies, P., 95
Davies, R. R., 104
Davis, D. R., 136, 147
Davis, L. J., 671
Davis, M., 250, 289, 290, 436, 439
Davydov, I. V., 416, 417, 420, 421, 658, 669, 686
Dawson, D., 363
Dawson, T. M., 92, 315
Day, P. M., 560
De, J. O., 99
Deak, P. M., 461, 462, 463, 466
De Alba, E., 148
Dearborn, D. G., 11
Deas, E., 417

Deato, M. E., 708
Debad, J. D., 686
DeBose-Boyd, R. A., 312, 321
De Camilli, P. V., 121, 136, 143, 145
De Crescenzo, G., 165
Dees, E. C., 512, 513
De Ferrari, G. V., 569, 570
de Girolami, U., 95
de Graaf, B., 104
deGroot, R. J., 417
Degterev, A., 647
de Hoon, M. J. L., 347
Deisenhofer, J., 673
Dejean, A., 252
Delaere, P., 96
DeLano, W. L., 125
del los Santos, R., 436, 570
Delpech, B., 120
Del Sal, G., 256, 260, 261, 512
De Luca, A., 718
Deluca, M., 571
DeMartino, G., 15, 252, 261, 262, 278
DeMartino, G. N., 83, 84
De Martino, G. N., 93
DeMartino, G. N., 560
Demo, S., 416, 567, 661, 664, 683
den Besten, W., 261
Deng, L., 4, 5, 6, 23, 24, 136, 144, 194, 279, 368
Deng, X. W., 638
Denk, H., 109
Dennery, P., 514
Dennis, S., 635, 647, 648
DeRisi, J., 620
de Ruwe, M. J., 358, 359, 361, 363, 365
Desai, R., 514
Deshaies, R. J., 8, 66, 73, 158, 215, 221, 225, 226, 231, 250, 266, 287, 288, 302, 306, 335, 389, 434, 437, 439, 446, 451, 454, 471, 483, 568, 572, 577, 579, 581, 582
Desharnais, J., 634
Desterro, J. M., 394
Devaux, P. F., 552
Deveraux, Q., 137, 141, 143
De Vos, R. A., 88
de Vries, E., 514
de Vrij, F. M., 88, 498
Dewar, D., 290
Diamond, M. A., 8, 623, 669, 702, 703, 705, 706, 709, 712, 713, 714
Diara, Y., 96

Dias, D. C., 250
Diaz-Hernandez, M., 92, 491, 495, 503
Dick, E. J., 94
Dick, E. J., Jr., 101
Dick, L. R., 586
Dickson, D., 99
Dickson, D. W., 94, 95, 99, 101, 104
Dieckmann, T., 136
Diffley, J. F., 290, 436
Di Fiore, P. P., 84, 121, 135, 136, 143, 145, 368
di Fonzo, A., 438
Dijkstra, K., 146
Dikic, I., 84, 368
Di Napoli, M., 512
Dinh, M., 326, 686
DiPersio, J. F., 514
Dittmar, G., 17, 279, 368, 370
Dittmar, G. A., 70
Ditzel, L., 228
Ditzel, M., 417
Divita, G., 136, 158, 159
Dixit, V. M., 17, 471
Djaballah, H., 676
Do, V. M., 570, 582
Dobeli, H., 57
Dobler, M. R., 590, 601
Dodson, A., 101
Doherty, F., 93, 101, 107
Doherty, F. J., 101, 107, 109
Dohlman, H. G., 369
Dohmen, J., 221, 389
Dohmen, R. J., 299, 311, 395, 396
Domann, E., 443
Dominguez, C., 147, 185, 356, 358, 359, 360, 361, 365
Donaldson, I., 290
Donaldson, K. M., 137, 390
Donaldson, L. W., 186
Dong, Y., 314
D'Onofrio, M., 73, 568, 572, 577, 579, 581, 582
Donohoe, M., 708
Donzelli, M., 289, 290, 291, 294
Dooley, C. T., 613
Doree, M., 406
Dorrello, N. V., 250, 260, 261, 289
Dou, Q. P., 512, 587
Dougherty, C. S., 669, 703, 704, 705
Dowling, R. L., 669, 703, 704, 705
Doze, P., 514

Drac, H., 107
Draetta, G. F., 121, 256, 258, 260, 261, 289, 290, 291, 294, 512, 531, 661, 705
Dreveny, I., 147, 148
Drijfhout, J. W., 494
Drury, L. S., 290, 436
Du, F., 417
Duan, H., 661
Duan, X., 136
Dube, M., 16
Dubois, B., 99
Duda, D., 37, 666, 667, 675
Duda, J. E., 99
Dulic, V., 718
Dunn, J., 326, 686
Dunn, R., 5, 21, 24, 193, 219, 368
Dunn, S. E., 350
Duong, T., 482, 493, 494
Dupuis, B., 336, 350
Durand, H., 436, 570
Durbin, R., 137, 314
Durocher, D., 290
Durr, E., 271
Duyckaerts, C., 96, 104

E

Ea, C.-K., 4, 5, 24, 136
Eames, B., 514
Earnshaw, D. L., 676
Ebel, F., 443
Eby, M., 17, 471
Echeverri, C., 314
Ecker, D. J., 4, 5, 6, 14, 16, 22, 23, 52, 55, 59, 60, 266, 272, 368
Eckstein, J., 661, 705
Eddy, S., 137
Edelsbrunner, H., 635, 648
Edidin, M., 557
Edling, C., 661
Ehlers, M. D., 92
Eigenbrot, C., 641
Einhaus, S. L., 94
Eisen, H., 54
Eisen, M. B., 347, 350, 619
Eisenman, R. N., 136, 289, 299, 302, 436, 439
Eissa, N. T., 15
Ekiel, I., 165
El-Deiry, W. S., 512, 531
Eldridge, A. G., 335

Elenbaas, B., 625, 631
Eleuteri, A. M., 550
Elez, R., 336
Elias, D., 165
Elias, J. E., 3, 8, 12, 13, 21, 22, 23, 24, 52, 53, 54, 60, 61, 137, 144, 145, 266, 268, 269, 279, 369, 375, 377, 378, 391, 416, 454
Elias, S., 8, 434
Elkind, J. L., 165
Elledge, S. J., 287, 288, 289, 290, 291, 292, 295, 296, 297, 299, 302, 303, 304, 305, 306, 323, 399, 434, 435, 436, 444, 446, 448, 570, 638, 642, 720
Ellemunter, H., 109
Ellgaard, L., 460
Elliott, P. J., 512
Elliott, T., 550
Ellis, S., 5
Ellis, W. G., 98
Ellison, K. S., 244, 272
Ellison, M. J., 5, 13, 17, 52, 54, 60, 243, 244, 272, 278
Ellman, J. A., 610
Elofsson, M., 493, 587, 588, 593, 597, 598, 599, 603, 605, 619
Elsasser, S., 70
Elson, E., 555
Emerson, S. D., 627, 628, 629, 630
Emr, S. C., 136, 139, 141, 143, 145, 193, 194, 198
Emr, S. D., 136, 141, 390
Emslie, E., 38, 42
Endo, Y., 345
Enenkel, C., 244, 272, 559
Eng, J., 271, 275, 369, 375, 376
Eng, L. F., 98
Engel, J., 620
Engel, W. K., 107
Errington, D. R., 95, 101, 107, 109
Esaki, Y., 98
Escalante-Semerena, J. C., 451
Espina, V., 337, 339
Esposito, V., 718
Estelle, M., 37
Estojak, J., 363
Esumi, T., 590, 601
Euskirchen, G., 491
Evangelista, M., 314, 454
Evans, D. G., 120
Evans, P. C., 471

Evans, T. C., Jr., 38, 41
Evdokimov, E., 379, 401
Everett, C., 99
Eystein Lonning, P., 350
Eytan, E., 256, 258, 531, 569, 718

F

Fabrizi, G. M., 95
Fabunmi, R. P., 560
Fairbrother, W. J., 146
Fairclough, R. H., 671
Fajerman, I., 261
Falbel, T. G., 57
Fallon, L., 120
Falquet, L., 136, 137
Fan, C., 482, 494
Fang, G., 409
Fang, S., 38, 52, 658, 664, 684, 685, 689, 701, 702
Fang, Y., 482, 493, 494
Fantus, D., 165
Fardeau, M., 107
Faretta, M., 136, 143, 145
Farooque, M., 94
Farquhar, C., 101, 107
Farr, A. L., 704
Farrell, A. T., 513
Fashena, S. J., 138
Fauck, J., 635, 648
Feany, M. B., 95
Feigon, J., 136, 147, 148, 187, 188
Feldman, H., 99, 104
Feldman, R., 221, 287, 302, 389, 434, 446
Feldmann, H., 497
Feng, M. T., 70, 689
Fenteany, G., 65, 493, 587
Ferdous, A., 559
Fergusson, J., 96, 107
Fernandez-Recio, J., 646
Ferry, D. K., 313, 318, 461
Ferry, K. V., 8, 702, 703, 705, 706, 709
Fidzianska, A., 107
Fiebig, K. M., 186
Fiebiger, E., 496
Fields, G. B., 589, 593
Fields, S., 138, 139, 159, 161, 163, 439
Figeys, D., 290
Figueiredo-Pereira, M. E., 586
Filipovic, Z., 567, 622, 623, 630, 631, 635

Finger, A., 313, 461
Fink, G. R., 60, 67, 245, 312, 315, 436, 439
Finley, D., 3, 4, 5, 6, 8, 12, 13, 14, 16, 17, 21, 22, 23, 24, 52, 53, 54, 59, 60, 61, 67, 70, 121, 137, 144, 145, 229, 266, 268, 269, 272, 278, 279, 368, 369, 370, 377, 378, 391, 416, 454, 496, 497, 664
Finnin, M. S., 638, 720
Fiore, F., 256, 258, 531
Fischbeck, K. H., 93
Fischer, D. F., 88, 96, 498
Fischer, G. Z., 436
Fischer, J. A., 38
Fisher, R. D., 136, 147
Fisher, R. J., 165
Fisk, H. A., 279
Flament Durand, J., 96
Fleener, C. A., 671
Fleming, J. A., 244, 272
Fleming, K. G., 5, 24, 32, 387
Fletcher, J. A., 121
Flick, K., 13, 15, 243, 244, 266, 268, 272, 274, 436
Flierman, D., 8
Flury, I., 321
Forman-Kay, J. D., 186
Foster, M. P., 145
Fotouhi, N., 567, 622, 623, 630, 631, 635
Foucault, G., 552
Fox, D. III, 360
Fox, N. C., 95, 104
Fradkov, A. F., 492
France, D. S., 634
Francis, S. A., 136, 137, 138, 141, 142, 143, 145, 188, 390
Frank, R., 440, 441, 443
Franz, K. J., 186
Franzoso, G., 278
Fraser, A. G., 314
Freed, E., 335
Freedman, D. A., 623, 702
Freedom, R. M., 107
Freeman, R., 146
Freemont, P., 147, 148
Freiesleben, W., 95
French, B. A., 109
French, S. W., 93, 109
Fried, V. A., 229, 497
Friedberg, A. S., 136, 144, 194
Friedländer, R., 460, 461

Friedman, R., 96
Friend, S. H., 350
Fromm, S., 514
Frommel, C., 635, 648
Frontali, L., 497
Fruh, K., 598
Fry, D. C., 627, 628, 629, 630
Fryer, C. J., 436
Fu, H., 496
Fuchsbichler, A., 109
Fujigasaki, H., 101
Fujii, Y., 121
Fujimoto, T., 64
Fujimura, T., 345
Fujimuro, M., 76, 79, 80, 81, 83, 84, 230, 279, 669
Fujino, J., 79
Fujino, M., 438
Fujiwara, T., 84
Fukada, D., 336
Fukai, I., 121
Fukuda, M., 9, 23, 360
Funabiki, H., 408
Funata, N., 101
Furstenthal, L., 335
Furukawa, M., 306
Fushida, S., 345
Fushman, D., 4, 5, 21, 23, 24, 73, 147, 148, 178, 179, 180, 181, 182, 183, 184, 185, 186, 189, 190, 568, 572, 577, 579, 581, 582
Futcher, B., 376, 450

G

Gaal, K., 109
Gabuzda, D., 15
Gaczynska, M., 65, 493, 550, 586, 611, 613
Gai, W. P., 107
Galan, J. M., 9, 14, 225, 279, 451
Galardy, P. J., 121
Gali, R. R., 17, 70, 279, 368, 370
Gallo, J. M., 101
Galloway, C. A., 40
Galluzzi, L., 52
Galova, M., 287, 434
Galser, S., 306
Galvin, J. E., 99
Gambetti, P., 95, 101, 107, 109
Gambhir, S., 514
Gammon, S. T., 514

Gan, Q.-F., 326, 686
Gan Erdene, T., 122, 124
Gan-Erdene, T., 38, 47, 48, 121, 122, 470, 471
Ganguly, M., 95, 104
Ganoth, D., 256, 289, 290, 291, 294, 302, 439, 718
Gansbacher, B., 514
Gao, X., 657, 669, 678, 685, 712
Garbutt, K. C., 638
Garcia-Mata, R., 560
Gardner, H. P., 121
Gardner, R. G., 311, 313, 318, 463
Garner-Hamrick, P., 570
Garrett, N., 71
Garrido, C., 141
Garrus, J. E., 165
Garty, H., 165
Gatti, E., 79
Gauthier, S., 99
Ge, S., 514
Geetha, T., 143, 144
Gehring, K., 165
Geiger, B., 512
Geisler, S., 350
Gekakis, N., 137, 390
Gelman, M. S., 482
Gentleman, S., 99
Gentz, R., 356
Gentzsch, M., 466
George, D. L., 623
Gerdes, J., 351
Gerez, L., 96
Gerritsen, M. E., 524
Gertler, F. B., 289, 436, 437, 438, 439, 441, 443, 445, 451
Geyer, R., 306
Gharib, T. G., 337
Ghirlando, R., 136
Ghosh, S., 521
Giardina, C., 523
Gibbs, J., 79, 460, 560
Gietz, R. D., 463
Gilbert, D. J., 121
Gilchrist, C. A., 16
Gilks, C. B., 336, 343, 345, 347, 350
Gill, G., 47, 122, 124, 162, 470
Gillet, V., 635, 648
Gilon, T., 482, 483, 496
Giordana, M. T., 94, 101
Giordano, A., 718

Giordano, G. G., 718
Giordano, T. J., 337
Giot, L., 439
Giralt, E., 589
Glas, R., 482, 493, 494, 495, 498, 499, 502, 515
Glass, S., 657, 658, 661, 669, 671, 678, 685, 705, 712
Gleeson, F., 290
Glezer, E. N., 686
Glick, M., 635, 647
Glickman, M. H., 75, 88, 120, 165, 229, 469, 496, 702, 704
Gloss, L. M., 11
Glotzer, M., 335, 404, 569, 572
Gluud, C., 109
Gnann, A., 466
Gnoj, L., 314
Gobburu, J. V., 513
Godbolt, A. K., 104
Godwin, B., 159, 161, 439
Goebel, H. H., 107
Goebl, M., 288, 434, 444
Goedert, M., 99
Goffeau, A., 466
Goh, A. M., 147, 148, 187
Goheer, A., 513
Gold, N. D., 647
Goldberg, A. L., 64, 65, 228, 466, 492, 493, 497, 517, 550, 567, 586, 587, 598, 610, 611, 613
Goldberg, J., 634
Goldenberg, S. J., 638
Goldman, J. E., 107
Goldstein, J. L., 312, 321
Goldstone, K., 570
Golemis, E. A., 138, 363
Gonatas, N. K., 16
Gonda, D. K., 4, 6, 22, 23, 52, 266, 272, 368, 417
Gondo, M. M., 345
Gong, L., 37
Gonzalez, F., 559
Goodrich, D. W., 305
Goradia, A., 497, 498
Gordon, C., 159, 161, 162, 165, 174, 387, 559
Gordon, E. J., 619
Gorina, S., 625, 631
Gossen, M., 598
Gotta, M., 314
Goudeau, B., 107

Goudreault, M., 290
Gould, S. J., 571
Gramm, C., 586
Graner, E., 120
Grant, G. H., 635, 647
Graumann, J., 158, 225, 226
Graves, B., 567, 622, 623, 630, 631, 635
Gray, D. A., 14, 15, 16, 17, 121
Gray, N. S., 568
Gray, T., 99, 101, 107, 109
Gready, J. E., 635, 647
Green, D. R., 575
Green, H., 120
Green, M. R., 661
Greenbaum, D., 620
Greenwald, I., 289
Greenwood, C. J., 676
Gregor, A., 107
Gregori, L., 272
Grenier, L., 586
Gresser, O., 79
Greten, F. R., 512, 521, 522
Grever, M. R., 532
Griac, P., 436
Gribbon, P., 552
Grierson, D., 590, 601
Griesinger, C., 186
Griffin, T. J., 376
Grim, J., 289, 299, 302, 304, 436, 439
Grimme, S., 186
Griot, L., 159, 161
Groger, A. M., 718
Groll, M., 228, 550, 588
Gronenborn, A. M., 146
Gros, P., 125
Grosse-Kunstleve, R. W., 125
Grossman, S. R., 162, 708
Groudine, M., 305
Groussin, L., 436
Grubmeyer, C., 451
Gruhler, A., 244, 272, 290, 439, 446, 453
Grundke-Iqbal, I., 96
Grune, T., 514
Gruneberg, S., 647
Grunwald, J., 186
Grzesiek, S., 146
Gsell, B., 147
Gstaiger, M., 336
Gu, W., 470, 471, 685, 708
Guardavaccaro, D., 250, 260, 261, 289

Guazzi, G. C., 95
Guenot, J., 641
Guidi, M., 136, 143, 145
Gullick, W. J., 350
Gullotta, F., 94, 101
Gumbiner, B., 570
Gupta, K., 121
Gurbuxani, S., 141
Gurdon, J. B., 71
Gurien-West, M., 304, 439
Gururaja, T., 145, 416
Guthrie, C., 60, 245, 315
Gwozd, C., 5, 13
Gygi, S. P., 3, 8, 12, 13, 21, 22, 23, 24, 52, 53, 54, 60, 61, 137, 144, 145, 266, 268, 269, 279, 285, 369, 375, 376, 377, 378, 391, 409, 416, 454
Gyuris, J., 623

H

Ha, S. I., 121
Haas, A. L., 5, 6, 13, 14, 23, 52, 53, 88, 144, 147, 266, 272, 278, 324, 370, 705
Haber, J., 497
Haber, M., 635
Haggarty, S. J., 569
Haglund, K., 84, 368
Haguenauer-Tsapis, R., 9, 14, 121, 225, 279
Hai, T., 436
Haigh, N. G., 460
Hakala, K., 673
Haldeman, M. T., 24, 29
Hall, B. D., 376
Hall, P. A., 702
Halladay, J., 138, 159, 161
Hallberg, E., 560
Halley, D., 120
Halliday, G. M., 95, 99, 104
Hamada, M., 587
Hamashin, V. T., 616
Hamilton, M. H., 417
Hamm, S., 471
Hamon, M., 471
Hampton, R. Y., 311, 313, 317, 318, 320, 321, 461, 463, 497, 498
Han, C., 570, 582
Han, X., 634
Hanada, M., 587
Hanash, S. M., 337
Hanazawa, H., 358, 359, 361, 363, 365
Hanes, R. N., 145
Hankovszky, H. O., 186
Hanna, J., 121
Hannon, G. J., 323
Hansen, J., 120
Hansen, L. H., 290
Hanzawa, H., 363
Hao, M. H., 642
Hao, Y., 684
Harbers, K., 358
Hardeland, U., 379, 401, 402
Hardingham, T. E., 552
Hardouin, C., 620
Haririnia, A., 23, 147, 148, 179, 182, 183, 186, 190
Harlos, K., 640
Harlow, E., 344
Harper, J. W., 287, 288, 289, 290, 291, 292, 295, 296, 297, 299, 302, 303, 304, 305, 306, 336, 434, 435, 436, 439, 444, 446, 448, 469, 570, 638, 642, 647, 648, 720
Harrell, J. M., 638
Harris, A. L., 350
Harris, J. L., 610, 613, 620
Harrison, S. C., 522
Hart, A. A., 350
Hart, M., 436, 570
Hartmann, E., 482
Hartmann-Petersen, R., 159, 161, 162, 165, 174, 387
Harumiya, Y., 531
Hasegawa, M., 98
Hasegawa, S., 514
Hashikawa, T., 278
Hashimoto, C., 684
Hashimoto, T., 345
Haspeslagh, D., 523
Hastie, T., 350
Hatakeyama, S., 4, 23, 38, 92, 278, 279, 289, 436, 531, 570, 590, 601, 684, 685, 689
Hatori, M., 587
Hattori, K., 570
Hattori, N., 147
Hatzubai, A., 250, 289, 290, 436, 439, 570, 581
Haug, J. S., 514
Haugland, R. P., 682
Hauglund, M. J., 136, 138, 147, 194
Haupt, Y., 623
Hauw, J. J., 96

Hay, R. T., 278, 379, 394, 401, 436, 454, 469
Hayashi, S. I., 231, 232
Hayes, M., 350
Hayrapetian, L., 620
Hays, L. G., 275
Hayward, P. A., 101
He, L., 635
He, X., 136, 570, 582
He, Y., 96
He, Y. J., 306
Heald, R., 568
Hearst, J. E., 670
Heasman, J., 570
Heath, J. K., 454
Heessen, S., 493, 497, 501
Heid, H., 109
Heilbut, A., 290, 439, 446, 453
Heilman, D. W., 661
Heim, J., 193
Heinrich, R., 570
Heinz-Erian, P., 109
Hekking, B. G., 496, 619
Heldin, C. H., 261
Helenius, A., 460
Heller, H., 8, 12, 326, 434
Hellman, U., 261
Hellmuth, K., 313, 461
Hemelaar, J., 47, 122, 124, 470
Hemmila, I., 657, 658, 669, 671, 678, 685, 712, 718
Henchoz, S., 439, 454
Hendil, K. B., 84
Hendlich, M., 635, 648
Hendrickson, R. C., 290
Hengst, L., 718
Hennecke, S., 718
Hense, M., 443, 444
Henze, M., 136, 158, 159
Herberts, C., 494
Hernandez, F., 92, 491, 495, 503
Herr, W., 365
Herschman, H., 514
Hershko, A., 6, 8, 12, 75, 177, 215, 228, 243, 249, 256, 258, 278, 289, 290, 291, 294, 302, 324, 326, 356, 417, 434, 439, 469, 491, 531, 569, 664, 702, 718
Herskowitz, I., 439, 454
Herz, F., 95
Herzig, M. C., 99
Heutink, P., 104

Hibi, K., 121
Hibi, N., 76, 81, 669
Hicke, L., 3, 5, 21, 24, 75, 120, 135, 136, 137, 138, 139, 141, 142, 143, 145, 147, 158, 188, 193, 219, 368, 390
Hideg, K., 186
Hieter, P., 581
Higginson, D. S., 136, 147
Hildebrandt, J. D., 417
Hill, C. P., 31, 136, 147, 178, 356, 664
Hill, M., 121
Hiller, M. M., 313, 461
Hillova, J., 121
Hilt, W., 244, 272
Hirai, A., 531
Hirai, S., 99, 101
Hirakawa, T., 76, 79, 81
Hirano, A., 95, 104, 107
Hirao, T., 640
Hirsch, C., 460
Hirvonen, C. A., 38, 41
Hisamatsu, H., 83
Hitchcock, A. L., 285
Hite, S., 95
Hitt, R., 462, 463, 466
Ho, Y., 290, 439, 446, 453
Hobo, B., 88
Hochstrasser, M., 16, 17, 37, 52, 54, 60, 121, 148, 243, 268, 311, 315, 317, 394, 395, 396, 397, 493, 497, 664, 669
Hochuli, E., 57
Hodges, J. R., 104
Hodgins, R., 5, 13, 244, 272
Hodgkiss-Harlow, K., 497, 498
Hoege, C., 4, 14, 22, 266, 368, 393, 394, 395, 396, 397, 399, 402
Hoffman, L., 4, 5, 11, 23, 24, 32, 33, 34, 71, 225, 368, 481
Hoffman, R., 514
Hoffmann, L., 179
Hofmann, B., 120
Hofmann, K., 135, 136, 137, 288, 434, 444
Hofmann, R. M., 6, 11, 17, 23, 24, 29, 32, 33
Hogan, B., 505
Hogue, C. W., 290, 314, 454
Hoi, E. M., 88, 96
Hol, E. M., 498
Holder, A. A., 620
Hollingshead, M. G., 532
Holloway, B. R., 685

Holton, J., 104
Holton, J. L., 95, 104
Holtzman, T., 67, 436, 439
Hon, W. C., 640
Honda, R., 702
Hong, C. A., 657, 669, 678, 685, 712
Hong, M., 4, 5, 24, 136
Hong, X., 494, 498, 499
Honig, B., 631
Hoog, C., 569
Hook, S. S., 136
Hoos, A., 340, 344, 345
Hope, J., 101, 107
Hoppe, T., 278, 335, 496
Horazdovsky, B. F., 136, 144, 194
Horiuchi, K. Y., 669, 703, 704, 705
Horman, A., 194
Horn, E. J., 577
Horne, C. H., 350
Horwich, A. L., 493
Horwitz, S. B., 635
Hoshi, H., 590, 599, 601
Hoskins, J. R., 493
Hospers, G., 514
Houghten, R. A., 613, 616
Howald-Stevenson, I., 194
Howell, K. E., 275
Howley, P. M., 52, 53, 147, 148, 162, 187, 356
Hsi, B. L., 121
Hsieh, H. M., 289, 336, 436
Hsu, F. D., 336, 350
Hsu, J. Y., 289, 335, 336, 352, 436
Hu, G. G., 417
Hu, J. S., 146
Hu, M., 471
Huang, C. C., 337, 482, 494
Huang, D. T., 37, 666, 667, 675
Huang, J., 416, 514, 567, 661, 664, 683, 712
Huang, K. S., 718, 721
Huang, L., 436
Huang, P. S., 671
Huang, S. G., 657, 669, 678, 685, 712
Huang, Z., 634
Hubbard, A. K., 523
Hubbard, E. J., 289
Hubbard, G. B., 94, 101
Huber, M., 326, 686
Huber, R., 228, 550, 588
Huddleston, M. J., 66, 437, 439
Hue, D., 96

Huebner, K., 121
Huesken, D., 120
Huibregtse, J. M., 52, 53, 356
Hunter, T., 215, 436
Huntsman, D., 336, 350
Hurley, J. H., 136, 137, 138, 141, 142, 143, 145
Hurt, E., 379, 401, 402
Huyer, G., 198
Hwang, D., 586

I

Iannettoni, M. D., 337
Ichihara, A., 76, 230, 231, 232, 235
Ichimura, Y., 469
Igarashi, K., 231
Ihara, Y., 98
Ihara, Y. Y., 101
Ii, K., 235
Iinuma, H., 587
Iizuka, T., 99
Ikawa, M., 504
Ikeda, A., 261
Ikeda, M., 261
Ikegami, T., 186
Imai, Y., 278
Imaki, H., 289, 436
Imamura, T., 261
Imoto, S., 347
Imperiali, B., 186
Inada, Y., 261
Inazawa, J., 121
Ince, P., 99
Inomata, M., 64
Inoue, M., 345
Ipsen, S., 99
Iqbal, K., 96
Iqbal, M., 586
Ironside, J. W., 101
Ishida, K., 101
Ishida, N., 279, 289, 436, 531, 570
Ishida, T., 345
Ishihara, N., 469
Ishikawa, K., 101
Ishioka, T., 684
Ishov, A. M., 121
Isono, E., 65, 69, 72, 84
Issakani, S. D., 567, 661, 664, 712
Ito, 68
Ito, K., 121

Ito, Y., 147
Iwabuchi, K., 101
Iwabuchi, Y., 590, 601
Iwai, K., 640
Iwaki, A., 107
Iwaki, T., 107
Iwamoto-Sugai, M., 147
Iwata, M., 101
Iwatsubo, T., 98
Iwaya, K., 345
Izeki, E., 95
Izumi, M., 345

J

Jackson, M., 99, 104
Jackson, P. K., 289, 335, 336, 352, 436
Jackson, R. M., 647
Jacob, R., 568
Jacobs, M. D., 522
Jacobson, S., 95
Jacquemot, C., 54
Jaffray, E., 379, 401, 454
Jahnke, W., 145
Jain, A. N., 635, 648
Jain, N. U., 185
Jaksch, M., 107
James, P., 138, 159, 161
Jans, D., 54
Janssen, L., 494
Janzen, J., 436
Janzer, R. C., 95
Jaros, E., 104
Jarosch, E., 460, 461
Jarosinski, M. A., 136
Jatkoe, T., 661
Javerzat, J. P., 559
Jebanathirajah, J., 409
Jefferson, D., 101, 107, 109
Jeffrey, L. C., 185
Jeffrey, P. D., 287, 289, 299, 302, 356, 435, 638, 647, 648, 720
Jeffrey, S. S., 350
Jellinger, K., 96
Jellne, M., 482, 493, 494, 495, 497, 499, 501, 502, 515
Jemmerson, R., 575
Jenkins, N. A., 121
Jenkins, T. M., 531

Jenner, P., 93
Jensen, D. E., 121
Jensen, J. P., 684, 685, 689, 701, 702
Jensen, K., 109
Jensen, O. N., 369
Jensen, P. H., 107
Jensen, T., 342
Jensen, T. J., 466, 493
Jentoft, N., 11
Jentsch, S., 4, 14, 22, 37, 120, 174, 266, 278, 311, 315, 317, 335, 368, 393, 394, 395, 396, 397, 399, 402, 469, 496, 497
Jesmok, G., 524
Jespersen, H., 290
Jiang, J. S., 125, 289, 641
Jiang, P., 514
Jiang, X., 482, 493, 494, 560
Jin, B., 436
Jin, C., 147
Jin, J., 289, 290, 291, 292, 295, 296, 297, 299, 302, 303, 436, 439, 642
Joachimiak, A., 136
Joazeiro, C. A., 137, 356, 390
Johansen, L. E., 290
Johnsen, H., 350
Johnson, A. E., 460
Johnson, E., 515
Johnson, E. S., 5, 22, 269, 379, 393, 394, 395, 396, 399, 401, 402, 495, 496, 497
Johnson, J. L., 306
Johnson, P., 311, 315, 317
Johnston, J. A., 93, 560
Johnston, M., 159, 161, 439
Johnston, S., 497
Johnston, S. A., 559
Johnston, S. C., 31
Jones, A. G., 104
Jones, C., 120
Jones, E. A., 136
Jones, E. Y., 640
Jones, K. A., 436
Jones, M. R., 314
Jones, S., 314
Jones, S. N., 623
Jones, T. A., 125, 641
Jordan, R., 336
Joseph, N., 121
Josephs, K. A., 95, 104
Josiah, S., 657, 658, 669, 671, 678, 685, 712

Juarez, L., 568
Judson, R. S., 159, 161, 439
Jung, D., 623
Jurewicz, A. J., 676

K

Kaelin, W. G., 335, 336
Kaelin, W. G., Jr., 250, 306, 411, 514, 531, 532, 538, 640
Kagawa, S., 84
Kahn, J. E., 107
Kahvejian, A., 165
Kaihara, A., 514
Kain, S. R., 482, 493, 494
Kairies, N., 588
Kaiser, B. K., 335
Kaiser, C. S., 446, 449
Kaiser, P., 13, 15, 243, 244, 266, 268, 272, 274, 289, 290, 302, 303, 439
Kaiser, R., 165
Kaji, M., 121
Kalashnikova, T. I., 436
Kalbfleisch, T., 159, 161, 439
Kallen, K. J., 524
Kallioniemi, A., 338
Kam, Z., 512
Kamath, R. S., 314
Kamei, H., 587
Kameji, T., 231
Kamerkar, S., 657, 658, 669, 671, 678, 685, 712
Kaminska, A. M., 107
Kamionka, M., 147, 148, 187, 188
Kamitani, T., 37
Kammlott, U., 567, 622, 623, 630, 631, 635
Kamphorst, W., 88, 104
Kamura, T., 4, 23, 287, 289, 434, 436
Kanapin, A., 314
Kanayama, A., 4, 5, 24, 136
Kandrac, J., 325, 332
Kane, R., 513
Kane, S. A., 671
Kaneta, K., 587
Kang, R. S., 136, 147, 188, 390
Kang, S. C., 101
Kaplan, K. B., 287, 302, 434, 446
Kapoor, T. M., 569
Kapusta, L., 718
Kardia, S. L., 337

Karin, M., 17, 279, 368, 370, 512, 521, 522
Karlsson, R., 165
Kasai, Y., 121
Kaseda, M., 101
Kashina, A. S., 417
Kasperek, E. M., 24, 29, 185
Katayama, R., 684
Katayama, S., 306
Kato, K., 147, 640
Kato, M., 95, 136
Kato, S., 84, 95, 107, 306
Katoh, Y., 138, 143, 144, 145
Katzmann, D. J., 136, 141, 144, 193, 194, 198, 390
Katzmann, K. J., 136, 139, 141, 143, 145
Kauer, J. C., 586
Kauffman, M., 610
Kaufman, J., 146
Kaur, K., 401
Kavallaris, M., 635
Kawabata, A., 684
Kawahara, H., 65, 72, 109, 147
Kawahata, K., 136
Kawamata, T., 96
Kawamura, M., 66, 84
Kawarai, T., 104
Kawashima, S., 64
Kawaski, H., 64
Kawasumi, M., 438
Kay, L. E., 186
Kaykas, A., 569, 570
Kazaz, A., 623
Kearney, W. R., 136, 138, 147, 194
Kedersha, N. L., 162
Keeffe, J. R., 360
Keetch, C., 147, 148
Kelly, A., 550
Kelly, P., 350
Kemler, R., 512
Kenna, M. A., 451
Kenten, J., 658
Kentsis, A., 250
Kenward, N., 101, 107
Kenworthy, A., 555
Kerem, B., 466
Kerkhoven, R. M., 350
Kertesz, A., 104
Kessler, B. M., 38, 47, 121, 122, 124, 470, 471, 494, 496, 498, 499, 619
Kettner, M., 101

Keyomarsi, K., 289, 290, 296, 299, 302, 303, 305, 436
Khan, M. I., 55, 60
Khan, N., 104
Khochbin, S., 141
Khoury, A., 313, 318, 461
Kiger, A. A., 314
Kikuchi, J., 147, 570
Kikuchi, Y., 396
Kikuta, H., 109
Kilshaw, P. J., 471
Kim, A. S., 568
Kim, B. K., 147, 187
Kim, C., 463
Kim, C. N., 575
Kim, J. H., 38
Kim, K. A., 147, 187
Kim, K. B., 494, 498, 587, 588, 593, 597, 598, 599, 604, 605
Kim, K. I., 37
Kim, W. Y., 306
Kim, Y. B., 8, 623, 669, 702, 703, 705, 706, 709, 712, 713, 714
Kimura, M., 336
King, R. W., 73, 405, 412, 416, 568, 569, 572, 577, 579, 581, 582
King Engel, W., 107
Kinnucan, E. R., 638, 720
Kinoshita, N., 587
Kintner, C., 570
Kinzler, K. W., 335, 336, 623
Kirisako, T., 469
Kiriyama, M., 121
Kirsch, J. F., 11
Kirschner, M. W., 250, 335, 336, 404, 405, 408, 409, 411, 412, 413, 416, 568, 569, 570, 571, 572, 581
Kiser, G. L., 466
Kispert, A., 512
Kisselev, A. F., 162, 492, 496, 517, 567, 587, 610
Kitagawa, M., 531, 570
Kitajewski, J., 289
Kitamura, N., 136
Kjeldgaard, 125
Klapisz, E., 144
Klaus, W., 147
Klebe, G., 647, 648
Kleckner, N., 399
Kleihues, P., 95

Kleijnen, M. F., 162
Klein, C., 524, 567, 622, 623, 630, 631, 635
Klein, P., 438
Klevit, R. E., 9, 23, 360
Kleywegt, G. J., 641
Kloetzel, P. M., 550, 559, 598
Klos, D. A., 219
Kloser, A. K., 466
Kluck, R. M., 575
Klunder, J. M., 586
Knecht, R., 336
Knight, J. R., 159, 161, 439
Knop, M., 313, 461
Ko, T. K., 256, 302, 439, 718
Kobayashi, H., 138, 143, 144, 145, 244
Kobayashi, K., 109
Kobayashi, R., 439, 638
Kobayashi, S., 587
Kobayashi, Y., 121
Kodadek, T., 559
Kodama, E., 79
Koegl, M., 139, 278, 335, 496
Koepp, D. M., 287, 289, 290, 296, 299, 302, 303, 305, 434, 435, 436, 638
Koguchi, Y., 587
Kogure, T., 550
Kohanski, R. A., 550
Kohler, G., 77
Kohn, A. D., 569, 570
Kohn, D. B., 514
Kohno, J., 587
Kohno, T., 147
Koike, M., 101
Kolb, J. M., 676
Kolli, N., 38, 47, 121, 122, 124, 470, 471
Kolling, R., 194
Kolodziej, P., 51, 52
Kolodziejski, P. J., 15
Komada, M., 136
Komatsubara, S., 587
Komi, N., 232
Kominami, E., 469
Kominami, K., 66, 83, 84, 306, 504
Kon, N., 685
Kondo, H., 147, 148
Kondo, J., 96
Kong, N., 567, 622, 623, 630, 631, 635
Koning, A., 311, 318
Konishi, M., 587, 590, 599, 601
Kononen, J., 338

Koo, J.-S., 15
Koonin, E. V., 17, 226, 471
Kopito, R. R., 15, 16, 93, 466, 482, 484, 494, 496, 560
Kopp, F., 194
Koppel, D. E., 555
Koptio, R. R., 15, 17
Korhonen, L., 92
Korinek, V., 335, 336
Kornitzer, D., 67, 436, 439
Kortvelyesi, T., 635, 647, 648
Kosen, P. A., 185
Kossiakoff, A. A., 641
Kostova, Z., 92, 460, 462, 465, 466
Kotani, S., 407
Koyano, S., 101
Kozlov, G., 165
Kranz, A., 194
Kraulis, P. J., 631
Krek, T. W., 439, 718
Krek, W., 336
Kress, Y., 94, 99, 101
Kril, J. J., 104
Krishna, N. R., 143, 144
Krogh, A., 137
Kroon, G. J., 146
Kruger, R., 570
Ksiezak Reding, H., 94, 101
Ku, J., 99
Kuang, J., 405
Kubbutat, M. H., 623
Kucera, R. B., 38, 41
Kuckelkorn, U., 514
Kudoh, T., 656, 660, 669, 685, 712
Kuhn, R. J., 417
Kulka, R. G., 482, 483, 496
Kumar, S., 162
Kumatori, A., 235
Kung, A. L., 708
Kuo, M. L., 261
Kuras, L., 287, 290, 449, 450
Kurimoto, E., 147
Kuriyan, J., 641
Kuroda, M., 345
Kurz, T., 306
Kushnirov, V. V., 68, 363
Kussie, P. H., 625, 631
Kuster, B., 379, 401, 402
Kuszewski, J., 125
Kuznetsov, D. A., 641

Kwok, B. H., 493, 587, 588, 593, 597, 598, 599
Kwon, S. W., 401
Kwon, Y. T., 417

L

Labas, Y. A., 492
Labbe, J. C., 406
Lach, B., 107
Lacy, E., 505
Lafourcade, C., 451
Lagrazon, K., 9, 23
Lai, Z., 8, 623, 669, 702, 703, 704, 705, 706, 709, 712, 713, 714
Lain, S., 661
Lakhani, S. R., 120
Lakowicz, J. R., 671
Laleli-Sahin, E., 147
Lam, T. T., 454
Lam, Y. A., 34
Lambert, K., 121
Lambertson, D., 247
Lamond, A. I., 379, 401
Landon, M., 37, 88, 92, 93, 95, 96, 98, 101, 107, 109
Landry, D., 38, 41
Lane, D., 344, 622, 702
Lane, D. P., 469, 623, 625, 661, 705
Lane, W. S., 65, 417, 493, 587
Laney, J. D., 243, 493, 669
Lang, V., 436
Langdon, W. Y., 512
Langner, C., 109
Lanker, S., 244, 436, 439, 450
Lansbury, P. T., Jr., 120, 567
Lantos, P. L., 104, 107
Larsen, B., 256, 302, 439, 718
Larsen, C. N., 70, 124, 147
Larsson, A., 165
La Rue, J., 436
Lashuel, H. A., 120, 567
Laskowski, R. A., 635, 648
Lassmann, H., 96
Lassot, I., 436, 570
Lassota, P., 514, 531, 532, 538
Laszlo, L., 92, 93, 101, 107
Latres, E., 289
Laurenzi, V. D., 702
Laursen, H., 95, 104
Lavon, I., 250, 289, 290, 436, 439

Lawlor, B., 99
Lawrence, C. W., 461, 462
Lawson, T. G., 34
Layfield, R., 37, 92, 93
Lazarevic, D., 575
Le Bot, N., 314
Lech, P. J., 147, 148
Lechpammer, M., 120
Leclerc, A., 107
Lee, D. H., 64, 497
Lee, E., 570, 571
Lee, H., 54
Lee, I., 326, 686
Lee, K. J., 147, 187
Lee, M. G., 560
Lee, S., 326, 686
Lee, S. J., 436, 531, 640, 718
Lee, S. L., 513
Lee, T., 449
Lee, V. M., 99
Leffler, H., 185
Leggett, D. S., 70, 121
Legtenberg, Y. I. A., 358, 359, 361, 363, 365
Lehmann, A., 559
Lehner, C., 406
Leigh, I., 120
Leigh, P. N., 98, 101
Leighton, J., 513
Leighton, S., 338
Leisti, J., 120
Leland, J. K., 686
Lelouard, H., 79
Lelyveld, V. S., 122
Lemaire, M., 356
Lemaire, P., 71
Lemeer, S., 144
Lemke, H., 351
Lenk, U., 482
Lennox, G., 92, 93, 95, 98, 99, 101, 107, 109
Leonchiks, A., 501
Leonetti, F., 610
Lepourcelet, M., 634
Leskovac, V., 325, 332
Leth, P., 109
Leung, D., 620
Levine, A. J., 622, 623, 625, 631, 702
Levitskaya, J., 501
Levitt, D. G., 635, 648
Levkowitz, G., 512
Ley, S. C., 436

Li, B., 512, 673
Li, G. L., 94
Li, J., 93, 109, 610, 613, 620
Li, L., 514
Li, M., 314, 470, 471, 685, 708
Li, M. Z., 296
Li, P., 471
Li, Q., 521
Li, R., 101
Li, S. J., 394, 395, 396, 397
Li, T., 379, 401
Li, W., 137, 145, 416, 514
Li, X., 482, 493, 494
Li, X.-M., 514
Li, Y., 159, 161, 439, 570, 582
Li, Z. W., 512, 521, 522
Liang, C. Y., 513
Liang, J., 635, 648
Liao, X., 147
Licklider, L. J., 375
Lidsten, K., 498
Lightcap, E. S., 244, 272
Lim, M., 336
Lim, N. S., 165
Lima, C. D., 47, 122, 124, 396, 470
Lin, D., 641
Lin, X., 570, 582
Lin, Z., 290
Lindholm, D., 92
Lindsten, K., 92, 482, 491, 493, 494, 495, 496, 499, 501, 502, 503, 508, 515, 588, 598, 604, 605
Ling, R. L., 8, 17, 52, 54, 60, 61, 205, 244
Linsley, P. S., 350
Linsten, K., 494, 498
Liotta, L. A., 337, 339
Lippa, C., 95, 99
Lippincott-Schwartz, J., 555, 557, 560
Lippman, M. E., 637
Lipton, A. M., 104
Lipton, S. A., 92
Lis, J. T., 54
Liu, C., 570, 582
Liu, C. F., 136
Liu, C. L., 336, 343, 345, 347, 350
Liu, C. M., 627, 628, 629
Liu, D., 296, 524, 634
Liu, E. A., 567, 622, 623, 630, 631, 635
Liu, F., 524
Liu, H., 436

Liu, J., 638
Liu, Q., 147, 296
Liu, W. K., 99
Liu, X., 405, 411, 575
Liu, Y., 120, 567
Liu, Z., 120
Livingston, D. M., 708
Llena, J. F., 95
Lloyd, T. E., 136
Lloyd-Williams, P., 589
Locher, M., 315, 317
Lockhart, D. J., 460, 461
Lockshon, D., 159, 161, 439
Loda, M., 120, 718
Lodish, M. B., 360, 365
Loeb, K., 88
Loeb, K. R., 304, 439
Logan, C. Y., 569, 570
Lokey, R. S., 568
Loktev, A., 289, 336, 436
Longaretti, C., 451
Longnecker, R., 261
Loo, M. A., 466, 493
Lopez, O. L., 99
Lorca, T., 406
Lorick, K. L., 684, 685, 689, 702
Losko, S., 194
Lostritto, R. T., 513
Love, S., 98
Lovering, R. C., 289, 297, 642
Lowe, J., 88, 92, 93, 95, 96, 98, 99, 101, 104, 107, 109, 228
Lowe, L., 101, 107
Lowry, O. H., 704
Lu, C., 436
Lucas, J. J., 92, 491, 495, 503
Lucignani, G., 503
Ludwig, R. L., 701, 702
Lue, Y. H., 109
Lugovskoy, A., 647
Lukacs, C., 567, 622, 623, 630, 631, 635
Lukas, J., 352
Luke, M. P., 708
Luker, G., 514, 515, 518, 519
Luker, G. D., 503, 514, 531
Luker, K. E., 503, 514, 515, 518, 519, 531
Lukyanov, S. A., 492
Lundgren, J., 498, 499
Lunec, J., 345
Lungo, W., 109

Luo, J., 470
Lustig, K. D., 405, 412, 416
Luthert, P., 98
Lyapina, S., 287, 434
Lykke-Andersen, K., 638

M

Ma, A., 17, 471
Ma, J., 8, 669, 702, 703, 704, 705, 706, 709
Ma, L., 288, 434, 444
Ma, P. C., 5, 22
Ma, Y. T., 586
MacCoss, M. J., 271
MacKay, J. P., 360, 365
Mackenzie, I. R., 104
Maclaren, D., 514
Mac Lennan, K., 93
Madore, S., 661
Madura, K., 158, 159, 161, 247, 497
Maeda, Y., 101
Maggi, A., 503
Magnani, M., 52
Magni, F., 438
Maher, E. R., 512
Mahesh, U., 619
Mahmood, U., 514
Mahrus, S., 610
Mains, P. E., 306
Maiorov, V., 642
Maiti, T., 634
Mak, A. Y., 326, 686
Mak, H. Y., 360, 365
Makifiichi, T., 101
Makretsov, N. A., 350
Malandrini, A., 95
Mallamo, J. P., 586
Mallari, R., 657, 669, 678, 685, 712
Malspeis, L., 532
Manetto, V., 107, 109
Maniatis, T., 266
Manly, S. P., 676
Mann, C., 497
Mann, M., 140, 250, 289, 290, 369, 376, 379, 401, 436, 439, 570, 581
Manning, A. M., 250, 289, 290, 436, 439
Manning, N. O., 641
Manns, M. P., 524
Mansfield, T. A., 159, 161, 439
Mao, M., 350

Mao, Y., 136
Marceau, N., 109
Marechal, V., 625, 631
Margottin, F., 436, 570
Margottin-Goguet, F., 336, 436
Marguet, F., 289
Mariage-Samson, R., 121
Marino, C. R., 560
Markcker, K. A., 54
Markelov, M. L., 492
Markiewicz, D., 466
Markovic-Plese, S., 93
Marotti, L. A., Jr., 369
Marquis, S. T., 121
Marriott, D., 4, 6, 22, 23, 52, 266, 272, 368
Marsh, J., 55, 60
Marshall, A. G., 454
Marshesi, V. T., 676
Marsischky, G., 3, 8, 12, 13, 21, 22, 23, 24, 52, 53, 54, 60, 61, 137, 144, 145, 266, 268, 269, 279, 369, 377, 378, 391, 416, 454
Martinez, A. M., 406
Martinez-Noel, G., 358
Martinovic, S., 285
Marton, M. J., 350
Maslyar, D. J., 634
Masson, P., 498, 499
Massoud, T., 514
Massucci, M., 261, 262
Masterson, R. V., 54, 55, 56
Mastrandrea, L. D., 11
Masucci, M. G., 92, 121, 482, 493, 494, 495, 497, 498, 499, 501, 502, 508, 515
Matese, J. C., 350
Mathias, N., 244, 439
Mathis, G., 718
Matoskova, B., 121
Matsufuji, S., 231, 232
Matsumoto, M., 4, 23, 279, 570
Matsunaga, K., 336
Matsushita, H., 101
Matsuzaki, K., 64
Mattecucci, C., 121
Matthews, S., 147, 148
Matthiesen, J., 290
Mattiace, L. A., 95
Matunis, M. J., 396
Matyus, L., 670
Matz, M. V., 492
Maul, G. G., 121

Maurizi, M. R., 493
Maxwell, P. H., 512, 640
Maya, R., 623
Mayer, J., 93
Mayer, L., 570, 571
Mayer, R. J., 37, 88, 92, 93, 95, 96, 98, 101, 107, 109
Mayer, T. U., 278, 335, 496, 568, 569
Mayo, J. G., 532
Maytal, V., 469
McArdle, B., 55, 57
McBride, H. J., 67, 436, 439
McBride, P., 101
McBride, T., 101
McCardle, L., 101
McClure, C., 657, 658, 669, 671, 678, 685, 712
McCombie, W. R., 323
McConnell, I., 101
McConnell, R., 98
McCormack, A., 271
McCormack, A. L., 369, 375
McCormack, E. A., 303
McCormack, W. T., 598
McCracken, A. A., 177, 466
McCrea, P., 570
McCusker, J., 497
McDermott, H., 88, 101, 107
McDevitt, H. O., 551
McDonald, H., 215, 231, 271
McDonald, W. H., 13, 15, 226, 244, 266, 268, 271, 272, 274, 471
McDonnell, J. M., 165
McGarry, T. J., 336, 405, 408, 412, 416
McGeer, E. G., 96
McGeer, P. L., 94, 96
McGuinn, W. D., 513
McIntosh, J. R., 559
McKee, C., 38, 42
McKeith, I. G., 99, 104
McKeith, L., 99
McKeown, C., 147, 148
McMahon, A. P., 570
McMaster, J. S., 65, 493, 586, 611, 613, 616
McNamara, G., 514
McNaught, K. S., 93
McPhaul, L. W., 109
McPike, M., 15
McQuaid, S., 98
McQuire, D., 101, 107, 109
McRackan, T. R., 417

McRitchie, D. A., 95
Mead, P. E., 405, 412, 416
Medaglia, M. V., 165
Medicherla, B., 462, 465
Medzihradszky, K., 620
Mehle, A., 15
Mehlum, A., 194
Mehta, A. I., 337, 339
Meier, P., 417
Meimoun, A., 67, 436, 439
Meissner, L. C., 98
Melandri, F. D., 46, 669, 670, 677
Melchior, F., 393
Melino, G., 702
Mellon, J. K., 345
Meloche, S., 261, 262
Melter, M., 109
Melton, D. A., 570
Meltzer, P. S., 623
Memet, S., 88
Mendenhall, M. D., 289, 436, 437, 438, 439, 441, 445, 450, 451
Menendez-Benito, V., 493, 508
Meng, L., 493, 587, 588, 590, 593, 597, 598, 599
Mercurio, F., 250, 289, 290, 436, 439
Merkley, N., 147
Mersha, F. B., 38, 41
Messina, P. A., 586
Messing, J., 56
Mestan, J., 336
Metuzals, J., 109
Meyer, H. H., 136, 143, 147
Meyers, C. A., 325, 332
Michael, D., 623
Michaelis, S., 198, 446, 449, 466
Michalickova, K., 290
Michalik, A., 501
Michaud, C., 604
Midgley, C A., 623
Migheli, A., 94, 101
Migita, T., 120
Mihatsch, M. J., 338
Millar, A., 38, 42, 290, 439, 446, 453
Miller, J. J., 287, 435, 638
Miller, K., 89
Miller, K. R., 325, 332
Miller, P., 661
Miller, S. L., 145
Milstein, C., 77

Minna, J., 121
Minshull, J., 569
Mintzer, J., 99
Miranker, A. D., 493
Misek, D. E., 337
Misra, S., 136
Misteli, T., 557
Mistl, C., 99
Mitchell, A., 446, 449
Mitchison, G., 137
Mitchison, T., 647
Mitchison, T. J., 568, 569
Mitsui, A., 52
Miura, T., 147
Miwa, K., 345
Miyamato, C., 656, 660, 669, 685, 712
Miyamoto, C., 147, 531
Miyamoto, K., 360
Miyano, S., 347
Miyashita, N., 64
Miyashita, T., 345
Miyawaki, A., 550
Mizoi, J., 396
Mizuno, E., 136
Mizuno, Y., 95, 147
Mizusawa, H., 101, 107
Mizushima, N., 37, 469
Mizushima, T., 228, 640
Mizutani, T., 95
Mocellini, C., 101
Moghe, A., 569
Mohan, R., 493, 587, 588, 597, 598, 599, 603, 605, 619
Mohney, R. P., 145
Moldovan, G. L., 4, 14, 22, 266, 368, 394, 395, 396, 397, 399, 402
Molineaux, S., 567, 661, 664
Moll, U. M., 512
Momand, J., 623
Momose, I., 587
Monaco, J. J., 551
Monaco, S., 95
Monia, B. P., 4, 5, 14, 16, 22, 59, 60, 272
Montagnoli, A., 256, 258, 531
Montgomery, K., 343, 345, 347
Montgomery, R. O., 109
Monura, N., 121
Moomaw, C. R., 83
Moon, R. T., 569, 570
Moore, A., 514

Moore, K. J., 676
Moore, L., 290, 439, 446, 453
Moore, S. A., 524
Moossa, A., 514
Moran, M. F., 290
Moreau, J., 625, 631
Moren, A., 261
Moreno, S., 314
Moretto, G., 95
Morgan, D. O., 451, 452, 568, 717
Morham, S. G., 165
Mori, H., 95, 96, 101, 107, 336
Moriguchi, R., 64
Morimoto, Y., 228
Morin, N., 406
Morin, P. J., 335, 336
Morita, M., 104
Moriyama, S., 121
Morphew, M., 559
Morrell, K., 98, 99, 101, 107, 109
Morris, J. G., 95
Morris, J. R., 4, 23
Morrison, J. H., 99
Morse, D. E., 513
Mosimann, U. P., 99
Moskaug, J., 514
Moslehi, J. J., 306
Mountz, J., 514
Moustakas, A., 261
Muallem, S., 560
Mueller, T. D., 147, 148, 187, 188
Muhammad, S., 451
Muhandiram, D. R., 186
Mukai, K., 345
Mulder, N., 514
Mulkins, M. A., 326, 686
Muller, B., 70
Muller, S., 393
Muller, T., 109
Muller, U., 358
Muller, W., 109, 524
Muller-Hocker, J., 109
Mulley, B., 647
Munoz, D. G., 104
Murakami, M., 409
Murakami, Y., 231, 232
Muraoka, Y., 587
Murata, S., 38, 229
Muratani, M., 177
Murayama, S., 101

Muro, T., 101
Murphy, E. G., 107
Murphy, G. M., Jr., 98
Murray, A. W., 335, 404, 407, 408, 569, 572, 575
Murray, K. J., 676
Musial, A., 15
Muskat, B., 290
Musselman, A. L., 671
Myszka, D. G., 136, 147, 165
Myung, J., 494, 498, 587, 588, 598, 599, 603, 604, 605, 619

N

Nagamalleswari, K., 47, 48
Naganawa, H., 587
Nagasaka, H., 109
Nagasaka, T., 121
Nagase, T., 121
Nahm, S., 592
Naito, M., 684
Nakagawa, T., 587
Nakamichi, I., 570
Nakamura, A., 101
Nakamura, H., 684
Nakamura, T., 121
Nakanishi, T., 504
Nakano, S., 107
Nakao, A., 121
Nakashima, Y., 121
Nakatani, Y., 708
Nakayama, H., 121
Nakayama, K., 138, 143, 144, 145, 278, 289, 436, 531, 570
Nakayama, K. I., 4, 23, 38, 92, 279, 289, 436
Nalepa, G., 289, 290, 291, 292, 295, 296, 297, 299, 302, 303
Nam, S., 587
Nan, L., 109
Narayan, V., 159, 161, 439
Nash, P., 289, 436, 437, 438, 439, 441, 445, 451
Nasmyth, K., 287, 417, 434
Nason, T. F., 244, 439
Nasu, H., 76, 81, 669
Nateri, A. S., 436
Natkunam, Y., 343, 345, 347
Naviglio, S., 121

Nazif, T., 610, 611, 613, 615, 620
Neal, D. E., 345
Nedelkov, D., 165
Neefjes, J., 481, 492, 493, 494, 496, 550, 551, 555
Neijssen, J., 494
Nelson, M. D., 514
Nelson, R. W., 165
Neuberg, D. S., 514, 531, 532, 538
Newitt, R., 369
Newmeyer, D. D., 575
Newport, J. W., 569
Nguyen, V., 109
Ni, J., 567
Nicholls, A., 631
Nicholson, S., 350
Nickitenko, A., 136
Niebuhr, K., 443, 444
Nielsen, E., 290
Nielsen, P. A., 290
Nielsen, T. O., 336, 350
Niermeijer, M. F., 104
Nihira, S., 656, 660, 669, 685, 712
Nijbroek, G., 466
Nikolaev, A. Y., 470
Niland, J., 623
Niles, E. G., 11, 24
Nilges, M., 125
Ninomiya, I., 345
Nishihara, Y., 436
Nishikawa, H., 9, 23, 360
Nishikiori, T., 587
Nishimoto, T., 259, 260
NIshimune, Y., 504
Nishimura, G., 345
Nishio, M., 587
Nishiyama, M., 289, 436
Nishiyama, Y., 587
Nishizawa, M., 438
Nitz, M., 186
Noble, W. S., 145, 416
Noda, C., 84
Noda, T., 469, 684
Nolan, J., 347
Norbury, C. C., 79, 460
Noro, C. Y., 570
Norris, M. D., 635
North, A., 657, 669, 678, 685, 712
Nunzi, M. G., 95
Nusse, R., 569, 570

O

Oania, R., 158, 225, 226, 471
Oberoi, P., 658, 669
O'Brien, J., 99
O'Bryan, J. P., 145
Ochesky, J., 514
O'Conner, L. B., 54
O'Connor-McCourt, M., 165
Oda, M., 95
Odorizzi, G., 193, 194
Oechler, M., 336
Ogawa, H., 345
Ogg, S. C., 379, 401
Oguchi, T., 65, 72
Ogura, T., 230
Ohama, E., 95
Ohata, H., 684
Ohba, M., 84
Ohh, M., 306
Ohkawa, K., 76, 79, 81, 669
Ohnuki, T., 587
Ohnuma, T., 590, 599, 601
Ohsumi, M., 469
Ohsumi, Y., 37, 469
Ohta, M., 109
Ohta, T., 9, 23, 345, 360
Okabe, M., 504
Okamoto, J., 9, 23
Okamoto, K., 99, 104
Okan, Y., 336
Okawa, Y., 76, 79, 81
Okeda, R., 101
Oki, T., 587, 590, 599, 601
Okuda, T., 587
Okumura, F., 4, 23, 289, 436
Okura, N., 84
Olanow, C. W., 93
Oldenburg, K. R., 726
Oldham, C. E., 145
Oliner, J. D., 623
Oliveira, A. M., 121
Omar, R., 94
Omura, S., 15, 17, 64, 466, 512
O'Neil, B., 512, 513
Ong, A. M., 684, 685, 689
Onno, M., 121
Oooka, S., 9, 23
Oprea, T. I., 625
Oren, M., 261, 262, 623

Orian, A., 136, 289, 299, 302, 436, 439
Orkin, S. H., 360, 365
Orlicky, S., 289, 435, 436, 437, 438, 439, 441, 443, 445, 451, 647
Orlov, I., 5, 24, 32, 387
Orlowski, M., 550, 586, 604
Orlowski, R. Z., 512, 513
O'Rourke, K. M., 17, 471
Orringer, M. B., 337
Orry, A. J., 647
Osada, H., 436
Osaka, F., 306
Osaka, H., 95
O'Shaughnessy, A., 314
Oshiro, M., 514
Osley, M. A., 177
Ostresh, J. M., 616
Ota, I. M., 5, 22
Otoguro, K., 64
Ott, D. E., 559
Otto, R. T., 461
Ottobrini, L., 503
Ouerfelli, O., 568
Ouni, I., 13, 15, 244, 266, 268, 272, 274
Ovaa, H., 38, 47, 121, 122, 124, 470, 471
Oved, S., 512
Overkleeft, H. S., 47, 121, 122, 470, 494, 496, 498, 499, 619
Owicki, J. C., 665
Oyanagi, K., 104
Ozawa, T., 95, 104, 514
Ozkan, E., 673
Ozkaynak, E., 13, 59, 60

P

Paddison, P. J., 314
Pagano, M., 15, 38, 250, 252, 256, 258, 260, 261, 262, 278, 287, 289, 290, 291, 294, 297, 302, 416, 435, 439, 512, 531, 638, 642, 661, 705, 718, 720
Page, N., 314, 454
Palme, S., 627, 628, 629
Palmer, I., 146
Pamphlett, R., 95
Pan, T., 101
Pan, Z. Q., 47, 48, 250
Pance, A., 141
Pandit, A., 109
Pannu, N. S., 125

Panse, V. G., 379, 401, 402
Papa, F. R., 121
Papp, M. I., 107
Pappin, D. J., 369
Pappolla, M. A., 94
Paraz, M., 451, 452
Parcellier, A., 141
Parisi, J. E., 95, 104
Park, K. C., 38
Park, M. R., 94
Park, S. H., 462, 466
Parker, A. E., 661
Parker, M. G., 360, 365
Parlati, F., 416, 567, 661, 664, 683
Parry, G., 37
Parsons, A. B., 314, 454
Pasa-Tolic, L., 285
Passmore, L. A., 38, 303, 324
Patel, S. D., 551
Patil, C. K., 460, 461
Patrick, G. N., 15
Patterson, G. H., 560
Patton, A., 95
Patton, E. E., 244, 287, 434, 439, 450
Paul, A., 303, 417
Paulmurugan, R., 514
Paulus, H., 38, 41
Pavletich, N. P., 287, 289, 299, 302, 356, 435, 625, 631, 638, 640, 647, 648, 720
Pawson, T., 289, 290, 436, 437, 438, 439, 441, 445, 451
Payan, D. G., 145, 416, 567, 661, 664, 683
Payne, M., 136, 147
Pazdur, R., 513
Peacock, M., 350
Pearl, G. S., 95, 104
Pedersen, M. S., 54
Pelham, H. R., 194
Pelletier, J. J., 38, 41
Pellman, D., 57
Peng, J., 266, 268, 269, 279, 369, 375, 377, 378, 391, 416, 454
Peng, J. M., 3, 8, 12, 13, 21, 22, 23, 24, 52, 53, 54, 60, 61, 137, 144, 145
Penman, S., 514
Perez-Atayde, A. R., 121
Pericin, D., 325, 332
Perkins, D. N., 369
Perkins, G., 290, 436
Perl, D. P., 93

Perler, F. B., 38, 41
Perou, C. M., 350
Perret, C., 436, 570
Perrimon, N., 314
Perrin, A. J., 560
Perry, G., 96, 98, 101, 107, 109
Perry, P., 559
Perry, R. H., 104
Pescatore, B., 186
Peter, M., 306, 439, 451, 454
Peters, J. M., 569, 581
Peters, K. P., 635, 648
Peters, N. R., 73, 405, 568, 572, 577, 579, 581, 582
Peterse, H. L., 350
Petersen, F., 634
Petersen, R. C., 95, 104
Peterson, D., 514
Peterson, J. R., 568
Peterson, P. A., 598
Petrenko, O., 512
Petricoin, E. F. III, 337, 339
Petroski, M. D., 8, 266, 483
Petrovic, J., 494, 498, 499
Petrucelli, L., 92
Pezzulo, T., 94, 101
Pfander, B., 4, 14, 22, 266, 368, 394, 395, 396, 397, 399, 402
Pfeifer, G., 229
Pfleger, C. M., 404, 409, 572
Phair, R. D., 557
Phelps, M., 514
Phillips, C. L., 178
Piao, Y. S., 104
Pica, C. M., 503, 514, 515, 518, 519, 531
Picard, D., 312, 387
Pickart, C. M., 3, 4, 5, 6, 7, 9, 11, 12, 17, 21, 23, 24, 26, 29, 32, 33, 34, 71, 75, 92, 137, 141, 143, 147, 148, 158, 165, 177, 178, 179, 180, 182, 183, 185, 186, 189, 190, 215, 225, 226, 228, 243, 249, 268, 279, 356, 368, 387, 393, 434, 481, 493, 664, 676, 684, 704
Picksley, S. M., 623, 625
Pierce, J. W., 524
Pierre, P., 79
Pietenpol, J. A., 623
Piette, F., 96
Pijnenburg, M., 144
Pike, I., 95, 101, 107, 109
Pind, S., 466, 493
Pinilla, C., 613
Pintard, L., 306
Piotrowski, J., 11, 24, 33, 179
Piper, R. C., 136, 138, 147, 194
Piwnica-Worms, D., 503, 514, 515, 518, 519, 531
Piwnica-Worms, H., 514
Place, R. F., 523
Plamondon, L., 586
Playfer, J., 99
Plemper, R. K., 461, 462, 463
Plenchette, S., 141
Ploegh, H. L., 38, 47, 65, 121, 122, 124, 470, 471, 493, 494, 496, 498, 499, 586, 611, 613, 616, 619
Plougastel, B., 551
Pochart, P., 159, 161, 439
Podlaski, F., 567, 622, 623, 627, 628, 629, 630, 631, 635, 718, 721
Podtelejnikov, A., 290
Polakis, P., 436, 569, 570
Pollack, H., 514
Pollok, B., 515
Polo, S., 84, 121, 135, 136, 143, 145
Poorkaj, P., 104
Pope, A. J., 676
Pornillos, O., 136, 147
Pornillos, O. W., 165
Porro, D., 438
Poths, S., 95
Potts, W., 247
Poulin, G., 314
Poulsen, V., 290
Power, D., 96
Power, D. M., 101
Power, J. H., 107
Pradhan, A. M., 109
Prag, G., 136, 137, 138, 141, 142, 143, 145
Prapong, W., 343, 345, 347
Prasher, D. C., 491
Pray, T., 416, 664, 683
Pray, T. R., 567, 661, 664
Prayson, R. A., 107
Prendergast, G. C., 121
Prestegard, J. H., 185
Price, J. S., 124
Prilusky, J., 641
Princiotta, M. F., 560
Prior, J., 514
Prior, J. L., 514

Prives, C., 702
Probst, A., 99, 104
Proctor, M., 121
Proper, E. A., 88
Pugh, C. W., 512, 640
Pyrowolakis, G., 4, 14, 22, 37, 120, 174, 266, 368, 393, 394, 395, 396, 397, 399, 402

Q

Qian, S. B., 559
Qin, J., 289, 290, 291, 292, 295, 296, 297, 299, 302, 303, 470
Qu, F., 683
Qu, K., 416, 664
Quackenbush, J., 339, 348, 350
Quang, P. N., 451, 452
Quinn, J. G., 165
Quinn, N. P., 95, 104
Quiocho, F. A., 136
Quist, H., 350
Qureshi-Emili, A., 159, 161, 439

R

Raasi, S., 5, 23, 24, 29, 32, 33, 34, 147, 148, 158, 165, 179, 182, 183, 186, 190, 387
Radhakrishnan, I., 136, 145, 147, 188, 390
Raghibizadeh, S., 314, 454
Rahman, A., 513
Raiborg, C., 93
Raine, C. S., 95
Rajadas, J., 484
Rajesh, S., 147
Ramesha, C. S., 326, 686
Rana, T. M., 670
Randal, M., 641
Randall, K., 99
Randall, R. J., 704
Rankin, J., 99
Rankin, S., 409
Rao, H., 158, 159, 161, 417
Rao-Naik, C., 55
Rapley, E., 120
Rapoport, T. A., 8
Rasmussen, K. J., 290
Rasmussen, S., 120
Ratcliffe, P. J., 512, 640
Rauscher, F. J. III, 121

Ravid, R., 104
Ravid, T., 16
Raymond, C. K., 194
Read, M. A., 306
Read, R. J., 125
Realini, C. A., 498, 499, 560
Reboul, J., 306
Rechsteiner, M., 4, 5, 23, 32, 34, 71, 137, 141, 143, 225, 368, 481, 560, 610, 613, 620
Reed, S. I., 136, 158, 159, 243, 289, 290, 302, 303, 305, 379, 401, 402, 439, 497, 718, 720, 721
Reggiori, F., 194
Reiboldt, A., 586
Reid, B. G., 493
Reid, W. D., 99
Reimann, J. D., 289, 335, 336, 352, 436
Reinhard, M., 443
Reinheckel, T., 514
Reinstein, E., 261, 262
Reits, E. A., 494, 550, 551, 555
Reizer, J., 146
Resch, H., 107
Resnitzky, D., 718
Rettig, M. P., 514
Reverter, D., 47, 122, 470
Revesz, T., 95, 104
Reymond, M., 145
Reynard, G., 437
Reynolds, L., 101, 107, 109
Rice, L. M., 125
Rich, R. L., 136, 147, 165
Richard, V., 436
Richards, W. G., 635, 647
Riddle, S. M., 31
Riera-Sans, L., 436
Riess, O., 95
Riley, D. A., 5
Riley, N. E., 93, 109
Rimm, D. L., 344, 346
Rine, J., 311, 313, 318
Rinner, W., 107
Riordan, J. R., 466, 493
Rippmann, F., 635, 648
Rippstein, P., 107
Rist, B., 376
Ritchey, J. K., 514
Ritter, O., 641
Rivalle, C., 590, 601
Rizzuto, N., 95

Roberts, C., 350
Roberts, G. W., 98
Roberts, J. M., 13, 21, 304, 305, 439, 717
Roberts, R. F., 686
Roberts, W. L., 686
Robertson, A. D., 136, 138, 147, 194
Robertson, J. T., 94
Robillard, G. T., 146
Robinson, C., 147, 148
Robinson, C. R., 522
Robinson, D. D., 635, 647
Robinson, H., 136, 147
Robinson, M., 314, 454
Robinson, P. A., 80, 567, 664
Robitschek, J., 93
Rock, K. L., 550, 586, 598
Rodems, S., 515
Rodier, G., 261, 262
Rodriguez, J., 575
Rodriguez, M. S., 278, 379, 394, 401
Roelofs, J., 3, 8, 12, 13, 21, 22, 23, 24, 52, 53, 54, 60, 61, 137, 144, 145, 266, 268, 269, 279, 369, 377, 378, 391, 416, 454
Rogaeva, E., 104
Rogers, B., 514
Rolen, U., 121
Rolfe, M., 252, 256, 260, 261, 512, 581, 623, 661, 669, 703, 705, 709, 712, 713, 714
Rommens, J. M., 466
Rosa, P., 95, 104
Rosario, L. A., 513
Rose, I. A., 6, 705
Rose, S. A., 80
Rosebough, N. J., 704
Rosen, J., 321
Rosen, M. K., 568
Rosenberg, A. E., 121
Rosol, M., 514
Rossi, R. L., 438
Rossi, S., 120
Rosso, S. M., 104
Rossor, M. N., 95, 104
Rothberg, J. M., 159, 161, 439
Rothnagel, J. A., 93, 109
Rothstein, L., 586
Rouillon, A., 449
Roussel, M. F., 261
Roy, G., 641
Rub, U., 101
Rubin, D. M., 229, 496, 497

Rubinfeld, B., 436, 570
Rubinsztein, D. C., 99
Ruderman, J. V., 569
Rudin, M., 531
Rämenapf, T., 417
Ruppert, J., 635, 648
Russell, N. S., 141, 144
Russo, G. L., 438
Rusten, T. E., 93
Rustum, C., 560
Rutishauser, J., 460, 466
Ryan, A., 439, 718
Ryan, C., 514
Ryan, J. J., 379, 399, 401, 402
Ryu, K. S., 147, 187
Ryzhikov, S., 638

S

Sa, D., 287, 450
Sachidanandam, R., 323
Sadis, S., 4, 5, 6, 14, 16, 22, 52, 53, 59, 60, 244, 266, 272, 278, 370, 496
Sadygov, R. G., 271
Saeki, M., 84, 306
Saeki, Y., 65, 69, 72, 74, 84, 219, 221
Safiran, Y. J., 658, 669
Safran, M., 514, 531, 532, 538
Saier, M. H., Jr., 146
Sainsbury, J. R., 350
Saito, Y., 64
Saitoh, A., 84
Sakai, N., 79
Sakamoto, K. M., 73, 568, 572, 577, 579, 581, 582
Sakamoto, T., 147
Sakata, E., 147
Sakhaii, P., 186
Sala-Newby, G., 531
Salerno, W. J., 136, 188, 390
Salic, A., 570, 571
Salmeron, A., 436
Salonga, D., 337
Salvadori, C., 95
Salvestroni, R., 95
Samejima, I., 162, 174
Sampat, R. M., 15, 482, 484, 494, 496
Sampson, D. A., 396
Sandmeyer, S., 287, 434
Sandoval, C. M., 93

Sangfelt, O., 289, 290, 302, 303, 439
Sanwai, B., 54
Saraf, A., 271
Saran, B., 94
Saraste, M., 635
Sarkar, B., 186
Sarkar, S., 198
Sasagawa, T., 345
Sasaki, H., 121
Sasaki, T., 345
Sasaki, Y., 64
Sassi, H., 290
Sastry, A., 158, 159, 161
Sato, K., 9, 23
Sato, M., 514
Satomi, Y., 469
Satymurthy, N., 514
Sauer, R. T., 522
Sauter, G., 338
Saville, M. K., 469, 661, 705
Savitsky, A. P., 492
Sawada, H., 76, 79, 80, 81, 229, 230, 279, 669
Sawano, A., 550
Saxena, K., 186
Sayre, L. M., 95
Scarafia, L. E., 326, 686
Schacher, A., 57
Schaefer, A., 462, 465
Schaefer, L. K., 289, 290, 296, 299, 302, 303, 305, 436
Schandorff, S., 290
Schauber, C., 247
Scheek, R. M., 146
Scheffner, M., 52, 53, 261, 262, 356, 492
Schell, J., 54, 55, 56
Schelper, R. L., 104
Schena, M., 312, 387
Scheraga, H. A., 642
Scherer, D. C., 266
Schey, K. L., 417
Schieltz, D., 221, 389
Schiffer, D., 94, 101
Schlabach, M., 314
Schlenck, B., 109
Schlenker, S., 278, 335, 496
Schlesinger, M. J., 54
Schlessinger, J., 555
Schmidt, M., 121
Schmidt, U., 587, 590, 592, 599, 601
Schmitt, E., 141

Schneider, B. L., 450
Schneider, C., 575
Schneider, G., 625, 626
Schneider, S., 337
Schnell, J. D., 135, 136, 139, 141, 143, 145, 158, 390
Schnieke, A. E., 38, 42
Schoenleber, R., 524
Schofield, C. J., 640
Schraml, P., 338
Schrander-Stumpel, C., 120
Schreiber, G. J., 350
Schreiber, S. L., 65, 493, 569, 587
Schubert, U., 79, 460, 559, 560
Schulman, B. A., 37, 287, 289, 299, 302, 435, 638, 647, 648, 666, 667, 675, 720
Schultz, C., 94, 101
Schultz, D. C., 121
Schultz, P. G., 568
Schuman, E. M., 15
Schutz, G., 524
Schwab, M., 434, 437, 453
Schwab, U., 351
Schwalbe, H., 186
Schwartz, A. L., 261, 262
Schwartz, D., 3, 8, 12, 13, 21, 22, 23, 24, 37, 52, 53, 54, 60, 61, 137, 144, 145, 148, 266, 268, 269, 279, 369, 377, 378, 391, 416, 454
Schwarz, E. M., 522
Schwarzenbacher, R., 92
Schweiger, M., 313, 461
Schweitzer, J. B., 94
Schwickart, M., 70
Schwienhorst, I., 395, 396
Schwieters, C. D., 185
Sciaky, N., 557
Scobie, K., 314
Scott, G. B., 80
Scott, M. E., 38, 41
Scott, P. M., 136, 138, 147, 194
Seal, S., 120
Seeger, M., 159, 161, 162, 165, 174, 387
Seeler, J. S., 252
Seelig, L., 463
Seemuller, E., 228, 491
Segel, I. H., 324, 332, 709, 713
Segeral, E., 436
Seibenhener, M. L., 143, 144
Seidensticker, M. J., 570
Seigneurin-Berny, D., 141

Seipmann, T. J., 324
Seitelberger, F., 96
Sekido, Y., 121
Sekine, K., 684
Sekizawa, R., 587
Self, T., 101
Selig, L., 436
Selkoe, D. J., 107
Selvin, P. R., 664
Semenov, M., 570, 582
Semino-Mora, C., 107
Semple, C., 159, 161, 165, 174, 385, 387, 416
Sen, P., 288, 434, 444
Senn, H., 147
Seo, J. W., 417
Seol, J. H., 287, 434
Sepp-Lorenzino, L., 604
Serdaroglu, P., 107
Serebriiskii, I. G., 138
Seshagiri, S., 17, 471
Seth, R. B., 4, 5, 24, 136
Seufert, W., 495, 496, 497, 515
Sever, N., 312, 321
Sfeir, A. J., 136, 144, 194
Shah, K., 451, 452
Shaito, A., 4, 5, 24, 136
Shan, S., 634
Shapir, R., 54
Shapiro, G., 514, 531, 532, 538
Sharfstein, S., 514
Sharipo, A., 501
Sharma, V., 514
Sharp, K. A., 631
Sharp, P. A., 52
Shashidharan, P., 93
Shaw, G., 96
Shaw, G. S., 147
Shaw, J. D., 198
Shaw, P. J., 101
She, J. X., 598
Sheaff, R. J., 305
Shearer, A., 321
Shekhtman, A., 147
Shen, R. F., 379, 401
Shen, Z., 147
Sheng, J., 417
Sherman, F., 17, 279, 368, 370, 460, 462
Sherr, C. J., 261, 717
Sheung, J., 712
Shevchenko, A., 287, 376, 434

Shewnarane, J., 290
Shi, X., 436
Shi, Y., 471, 708
Shiba, T., 138, 143, 144, 145
Shiba, Y., 138, 143, 144, 145
Shibata, T., 147
Shih, A. H., 162
Shih, G. H., 147
Shih, S. C., 3, 136, 137, 138, 139, 141, 142, 143, 145, 188, 390
Shiloh, A., 470
Shim, E. H., 638
Shimada, H., 514
Shimada, M., 65, 72
Shimbara, N., 83, 84
Shimizu, K., 345
Shimizu, Y., 83
Shimonishi, Y., 469
Shimuzu, Y., 84
Shin, H. W., 138, 143, 144, 145
Shin, S., 586, 611, 613, 616
Shin, T. H., 306
Shirane, M., 531, 570
Shirogane, T., 289, 290, 291, 292, 295, 296, 297, 299, 302, 303
Shivdasani, T. A., 634
Shoji, S., 107
Shokat, K. M., 451, 452
Shumway, S. D., 638
Sibbald, P. R., 635
Sicheri, F., 289, 435, 436, 437, 438, 439, 441, 443, 445, 451, 647
Siddiqui, N., 165
Siebenlist, U., 278
Siebert, R., 120
Sielecki, T. M., 623, 669, 703, 709, 712, 713, 714
Sigal, G. B., 686
Sigismund, S., 84, 136, 143, 145
Signoretti, S., 120
Sigrist, S., 406
Sikorska, M., 107
Silberstein, M., 635, 648
Silva, J. M., 314
Silver, P., 497
Silver, P. A., 285
Siman, R., 586
Simon, J. A., 54
Simon, M., 315
Simonson, T., 125

Simpson, P., 147, 148
Sin, N., 493, 587, 588, 590, 593, 597, 598, 599
Singer, J., 304, 439
Singer, T., 244, 272
Singh, S. K., 493
Sinha, I., 165
Sitte, N., 514
Skorodumov, C., 646
Skowyra, D., 287, 302, 304, 434, 446, 448
Skrynnikov, N. R., 186
Skullerud, K., 104
Slaughter, C., 4, 6, 23, 279, 368
Slaughter, C. A., 83, 84
Slingerland, J., 336, 718
Sloper-Mould, K. E., 3, 141, 143
Sluijs, J. A., 88
Smith, A. P., 439, 718
Smith, C. L., 557
Smith, D. M., 587
Smith, H. C., 40
Smith, J. S., 451
Smith, M. C., 514
Smith, R. D., 285
Smith, T. S., 471
Smith, T. W., 95
Snapp, E., 555
So, N. S., 560
So, O.-Y., 326, 686
Soboleva, T., 38, 54
Soccio, R. E., 162
Soda, M., 278
Sohrmann, M., 314
Sokol, S., 570
Solary, E., 141
Solbach, C., 336
Solomon, E., 4, 23
Solomon, M. J., 336
Somerville, R. A., 101
Sommer, T., 311, 315, 317, 460, 461, 462, 482, 497
Sonderegger, C., 634
Sone, T., 65, 72, 74, 84, 221
Soneji, Y., 436
Sonenberg, N., 165
Song, B. L., 312, 321
Song, J., 503, 515, 518, 519, 531
Song, L., 287, 435, 638
Song, O., 138
Song, Y., 524
Songyang, Z., 437

Sonnemans, M. A., 88
Sorensen, B. D., 290
Sorensen, C. S., 352
Sorensen, E., 514, 531, 532, 538
Sørlie, T., 350
Sorokina, I., 144
Sorscher, E. J., 560
Sotriffer, C., 648
Sowden, M. P., 40
Soylemezoglu, F., 95
Sparaco, M., 95
Sparks, A., 625, 705
Sparks, A. B., 335, 336
Spellman, P. T., 347, 619
Spence, J., 6, 14, 17, 22, 52, 53, 266, 272, 278, 279, 368, 370
Spencer, E., 4, 6, 23, 279, 289, 368
Spendlove, I., 95, 101, 107, 109
Spielewoy, N., 436
Spies, T., 550
Spiess, M., 460, 466
Spillantini, M. G., 99, 104
Spilman, S., 514
Splittgerber, U., 588, 603, 605, 619
Spormann, D. O., 193
Spruck, C., 439, 718
Spruck, C. H., 289, 290, 302, 303, 439
Squadrito, M., 290, 291, 294
Srayko, M., 306
Srinivasan, M., 159, 161, 439
Srinivasula, S. M., 634
St. George-Hyslop, P., 104
Stack, J., 515
Stadtman, E. R., 379, 401
Stahl, M., 625, 626
Stahl, S. J., 146
Staines, W., 107
Stamenova, S. D., 136, 147
Standaert, R. F., 65, 493, 587
Stappert, J., 512
Starai, V. J., 451
Staszewski, L. M., 17, 52, 243, 244, 253, 661
Staub, O., 37
Stearns, T., 315, 363
Stebbins, C. E., 640
Steen, H., 140, 369
Stein, H., 351
Stein, R. L., 46, 567, 586, 669, 670, 677

Stemmler, T. L., 136, 147
Stenmark, H., 93, 194
Stephenson, M. A., 345
Sternglanz, R., 138
Sternieri, F., 438
Sternlieb, I., 109
Steur, E. N., 88
Stevens, T. H., 194
Stevenson, D., 514
Stevenson, D. K., 531
Stevenson, J. K., 522
Stevenson, L. F., 705
Stezowski, J. J., 42
Stieber, A., 16
Stillman, D. J., 67, 436, 439
Stock, D., 228
Stockhammer, G., 514
Stommel, J. M., 685
Stone, M., 159, 161, 165, 174, 387
Stopa, E. G., 104
Stouten, P. F., 635, 647
Strack, B., 15
Strack, P., 289, 290, 299, 570, 623, 669, 703, 709, 712, 713, 714
Strand, K., 104
Strauss, J. H., 417
Stray, K. M., 165
Strebel, K., 436
Strebhardt, K., 336
Streetz, K. L., 524
Strobel, T., 107
Strohmaier, H., 289, 290, 302, 303, 439, 718
Stuart, D. I., 640
Stubbs, M. T., 647
Stukenberg, P. T., 405, 412, 416
Stumptner, C., 109
Stutz, F., 121
Subramani, S., 571
Sudol, M., 215
Suenaga, T., 107
Sugawara, K., 587
Sugihara, T., 590, 601
Sui, G., 708
Sullivan, M. L., 466
Sun, F.-X., 514
Sun, H., 436, 531, 569, 638, 718
Sun, L., 4, 5, 22, 24, 136, 559, 664
Sun, Y., 416, 661
Sundquist, W. I., 136, 147, 165

Supek, F., 137, 390
Surdin-Kerjan, Y., 290
Sussman, J. L., 641
Sutanto, M., 136, 137, 138, 139, 141, 142, 143, 145, 390
Suzuki, K., 101
Suzuki, S., 587
Suzuki, T., 306
Swagell, C., 93, 109
Swanson, K. A., 136, 147
Swanson, R., 315, 317
Swarbrick, G. M., 463
Swaroop, M., 661
Swash, M., 101
Swearingen, E., 657, 669, 678, 685, 712
Swindle, J., 54
Swinney, D. C., 326, 686
Sy, M. S., 101
Szollosi, J., 670
Sztul, E. S., 560
Szymkiewicz, I., 84

T

Tabaton, M., 101, 107, 109
Tabb, D. L., 271
Tabias, J. W., 368
Tachihara, K., 514
Tagami, H., 708
Taggart, J., 290
Taguchi, H., 76, 79, 81
Tainer, J. A., 720
Taipale, J., 634
Tait, A. S., 635
Takada, K., 76, 79, 81, 669
Takahashi, E., 84
Takahashi, H., 104
Takahashi, J., 101
Takahashi, K., 587
Takahashi, M., 120
Takahashi, R., 278
Takahashi, T., 121, 259, 260
Takahashi, Y., 79, 396
Takai, S., 232
Takao, T., 469
Takase, T., 121
Takatsu, H., 138, 143, 144, 145
Takeuchi, J., 66, 84
Takeuchi, T., 587
Takizawa, S., 79

Tall, G. G., 136, 144, 194
Tam, S. W., 256, 260, 261, 512
Tamaoka, A., 107
Tamura, T., 76, 230, 231, 232
Tan, C. F., 104
Tan, N. G. S., 80
Tan, Y., 570, 582
Tanabe, T., 101
Tanahashi, N., 76, 84, 230, 231, 232
Tanaka, E., 79
Tanaka, F., 93
Tanaka, H., 64, 407, 702
Tanaka, K., 38, 76, 84, 147, 228, 229, 230, 231, 232, 235, 306, 497, 610, 640
Tanaka, M., 104
Tanaka, T., 336
Tanashi, N., 83
Tang, D., 120
Tang, J., 136
Tang, M., 16
Tang, X., 289, 435, 436, 437, 438, 439, 441, 443, 445, 451, 647
Tang, Y., 436
Tang, Z., 673
Tani, T., 345
Tanida, I., 469
Taniguchi, M., 436
Tanner, M. S., 109
Tansey, W. P., 177, 365
Taroni, C., 625, 626
Tasaki, T., 417
Tasuoka, N., 228
Tateishi, J., 107
Tatsuta, K., 587
Taxis, C., 462, 463, 466
Taye, A., 93
Taylor, J. M., 337
Taylor, J. P., 93
Taylor, M. S., 162, 174
Taylor, P., 290, 439, 446, 453
Taylor, R. S., 275
Tekle, E., 379, 401
Tempel, R., 436
Tenev, T., 417
Tennant, L., 145
Teodoro, J. G., 661
Terasaki, M., 557
Terzyan, S., 136
Tessarz, P., 493, 497, 501

Tezel, E., 121
Theodoras, A., 256, 260, 261, 512, 661, 705
Theorell, H., 714
Thiagalingam, S., 623
Thibault, P., 261, 262
Thoma, S., 57
Thomas, D., 278, 287, 290, 436, 449, 450
Thomas, P. J., 560
Thompson, J., 278
Thoreen, C. C., 3, 8, 12, 13, 21, 22, 23, 24, 52, 53, 54, 60, 61, 137, 144, 145, 266, 268, 269, 279, 369, 375, 377, 378, 391, 416, 454
Thoroddsen, V., 244, 272
Thorsen, T., 350
Thorsness, P., 497
Thress, K., 305
Thrower, J. S., 4, 5, 23, 32, 34, 71, 178, 225, 368, 481
Thurig, S., 16
Tibshirani, R., 350
Timmers, H. T. M., 356, 358, 359, 360, 361, 363, 365
Ting, A. Y., 531, 550
Tinklenberg, J. R., 98
Tipler, C., 93
Tjandra, N., 148
Tjuvajev, J. G., 514
Tobias, J. W., 4, 6, 22, 23, 52, 57, 266, 272, 417
Tochtrop, G. P., 73, 568, 572, 577, 579, 581, 582
Toda, S., 587
Toda, T., 306
Todokoro, K., 407
Toh, E. A., 396
Toh-e, A., 65, 66, 69, 72, 74, 84, 221, 306, 438, 497
Tokosawa, H., 76, 81
Tolnay, M., 99, 104
Tome, F. M., 107
Tomisugi, Y., 228
Tomita, K., 587
Tommerup, N., 121
Tomokane, N., 107
Tomonaga, M., 101
Tong, A. H., 314, 454
Tongaonkar, P., 247
Tonks, N. K., 569

Toogood, P. L., 624
Topp, J. D., 136, 144, 194
Torhorst, J., 338
Tortorella, D., 65, 493, 496, 586, 611, 613
Toru, S., 101
Totrov, M. M., 635, 637, 638, 641, 643, 646, 647, 648
Totykuni, T., 514
Townsend, A., 550
Toyoshima, Y., 104
Traub, L. M., 193
Trautwein, C., 524
Travers, K. J., 460, 461
Treier, M., 17, 52, 243, 244, 253, 661
Trempe, J. F., 165
Trivic, S., 325, 332
Trojanowski, J. Q., 99
Tronche, F., 524
Trottier, Y., 101
Trowsdale, J., 550, 551
Tsai, C. C., 137
Tsien, R. Y., 492, 493, 531, 550
Tsihlias, J., 718
Tsirigotis, M., 14, 15, 16, 17
Tsubuki, S., 64
Tsuchiya, K., 587
Tsui, C., 29
Tsui, L. C., 466
Tsukada, Y., 76, 81, 669
Tsukihara, T., 228, 640
Tsunematsu, R., 289, 436
Tsurumi, C., 84
Tsuruo, T., 684
Tsvetkov, L. M., 436, 531, 718
Tu, Y., 491
Tubing, F., 70
Tuckwell, D. S., 107
Tugendreich, S., 581
Turner, D. L., 405
Turner, G., 120
Tut, V. M., 345
Tyers, M., 244, 256, 287, 289, 290, 302, 304, 306, 314, 434, 435, 436, 437, 438, 439, 441, 443, 445, 446, 448, 449, 450, 451, 453, 454, 647, 718
Tytler, E. M., 436

U

Ubersax, J. A., 451, 452
Uchida, K., 337
Uchihara, T., 101
Uchiyama, K., 147, 148
Ueda, R., 121
Uehara, H., 514
Uetz, P., 159, 161, 163, 439
Ugai, S., 232
Uhlmann, F., 417
Ullrich, O., 514
Ulrich, H. D., 278, 335, 496
Umahara, T., 95
Umemoto, K., 185
Umezawa, Y., 514
Unno, M., 228
Urban, J., 460, 461
Urbanowski, J. L., 194
Urbauer, J., 147
Urbe, S., 16

V

Vaalburg, W., 514
Vaglio, P., 306
Vajda, S., 635, 647, 648
Vajsar, J., 107
van Bergen en Henegouwen, P. M., 144
van Berkum, N. L., 356
Van Broeckhoven, C., 501
Van Demark, A. P., 29, 356
van den Heuvel, F. A. J., 358, 359, 361, 363, 365
Van Denmark, A. P., 664
van Den Ouweland, A., 120
Vanderhyden, B. C., 16
van de Rijn, M., 336, 343, 345, 347, 350
van der Kooy, K., 350
van der Vliet, P. C., 363
van de Vijver, M. J., 350
van Dijk, A. D. J., 183, 185
Van Dijk, R., 88
van Leeuwen, F. W., 88, 93, 96, 109, 498
van Nocker, S., 496
van Nuland, N. A., 146
Vanoni, M., 438
van Roessel, P., 492
van Schaik, F. M., 356, 358, 360
van Swieten, J. C., 104

van't Veer, L. J., 350
van Waarde, A., 514
Varadan, R., 23, 73, 147, 148, 179, 180, 181, 182, 183, 186, 189, 190, 568, 572, 577, 579, 581, 582
Varshavsky, A., 4, 5, 6, 13, 22, 23, 52, 54, 57, 59, 60, 67, 243, 266, 272, 299, 311, 315, 368, 416, 417, 420, 421, 453, 495, 496, 497, 515, 559, 664, 700
Vassilev, L. T., 567, 622, 623, 630, 631, 635, 718, 721
Vater, C. A., 194
Vaux, E. C., 512
Veenstra, T. D., 285
Vega, I., 247
Velayutham, M., 34
Vence, L. M., 38, 41
Venot, A., 185
Verbois, S. L., 513
Verdier, L., 186
Verdonk, M. L., 635, 648
Verhage, M. C., 88
Verhoef, L. G., 498, 501
Verkleij, A. J., 144
Verma, I. M., 521, 522
Verma, R., 73, 158, 215, 221, 225, 226, 231, 389, 437, 471, 568, 572, 577, 579, 581, 582
Vertegaal, A. C., 379, 401
Vicart, P., 107
Vickers, S. M., 436
Vidal, M., 306
Vieira, J., 56
Vierstra, R. D., 52, 53, 54, 57, 496, 497
Vigliani, M. C., 94, 101
Vigneron, S., 406
Vijayadamodar, G., 159, 161, 439
Vinitsky, A., 604
Virelizier, J. L., 278
Vissing, H., 121
Vlach, J., 718
Vo, M., 290
Vogelstein, B., 335, 336, 622, 623, 702
Voges, D., 550
Vogt, P. K., 634
Vojtesek, B., 625
Volkel, P., 326, 686
Volkwein, C., 460, 461, 462
von Ilberg, C., 336

Vonsattel, J. P., 104
von Schwedler, U. K., 165
Vorhees, P. M., 512, 513
Vorm, O., 376
Vousden, K. H., 623, 658, 669, 701, 702
Vu, B. T., 567, 622, 623, 627, 628, 629, 630, 631, 635
Vuust, J., 54
Vyberg, M., 109

W

Wacker, H. H., 351
Wada, K., 95
Waddell, I. D., 685
Wagner, G., 647
Wahl, G. M., 685
Waldemar, G., 95, 104
Walden, H., 37, 666, 667, 675
Walker, J., 57
Walker, J. R., 436
Walker, O., 23, 147, 148, 179, 180, 181, 182, 183, 186, 189
Wall, N. R., 708
Wallace, M., 159, 161, 162, 165, 174, 387, 559
Walling, L. A., 577
Wallis, R., 147, 148
Walsh, B. J., 635
Walsh, S. T., 147
Walter, J., 482
Walter, P., 460, 461
Walter, U., 443
Walters, K. J., 147, 148, 187
Walz, J., 228, 491
Walz, T., 121
Wan, M., 436
Wan, Y., 250, 405, 411
Wang, A. J., 147
Wang, B., 136, 147
Wang, C., 4, 6, 23, 279, 368
Wang, E. W., 610
Wang, G., 619
Wang, H., 634
Wang, H. E., 165
Wang, J. L., 634
Wang, K., 598
Wang, P., 287, 356, 435, 638
Wang, Q., 147, 148, 187

Wang, S., 637
Wang, T., 379, 401
Wang, W., 531
Wang, X., 514, 575
Wang, Y., 661
Wang, Y. C., 513
Wang, Y. L., 95, 136, 143, 369
Wanington, E. K., 104
Wanschit, J., 107
Ward, C. L., 15, 17, 93, 466, 560
Ward, L., 96, 99
Ward, W. W., 491
Wardrop, J., 705
Warlo, I., 107
Warren, G. L., 125, 136, 143
Warren, W., 120
Washburn, M. P., 375
Wasowicz, M., 96
Watanabe, K., 514, 656, 660
Watanabe, M., 99, 436
Watanabe, S., 669, 685, 712
Waterman, H., 512
Watkins, S. C., 466
Watson, M. H., 136, 158, 159, 439, 718
Watson, P., 635, 648
Wearsch, P. A., 560
Webb, W. W., 555
Weber, D., 718, 721
Weber-Ban, E. U., 493
Wee, K. E., 8, 669, 702, 703, 704, 705, 706, 709
Wee, S., 306
Wefes, I., 229, 496
Wehland, J., 443, 444
Wei, N., 638
Wei, W., 250, 411, 514, 531, 532, 538
Wei, Y., 306
Weinreb, S. M., 592
Weinstein, L. J., 120
Weintraub, H., 405
Weis, W. I., 641
Weiss, S., 443, 444
Weissleder, R., 514, 531
Weissman, A. M., 38, 52, 158, 176, 356, 367, 658, 683, 684, 685, 689, 701, 702
Weissman, J. S., 460, 461
Weissman, Z., 67, 436, 439
Weissmann, A. M., 664
Welch, W., 635, 648
Welchman, D. P., 314
Welcker, M., 289, 299, 302, 304, 436, 439, 458

Weld, H. A., 15
Weller, S. K., 80, 560
Wells, J. A., 624
Wendland, B., 177, 193, 198, 392
Weremowicz, S., 121
Werner, T., 379, 401, 402
Wertkin, A., 94, 101
Wertz, I. E., 17, 471
Westbrook, T., 314
Westhead, D. R., 647
Wettstein, D. A., 165
White, C. L. III, 104
White, J. B., 436
White, M., 139
White, M. A., 401
Whiteman, M., 101
Whitney, M., 515
Wiborg, O., 54
Wickner, S., 493
Widiger, G. N., 601
Wiesener, M. S., 512
Wiesmann, C., 17, 471
Wigley, W. C., 560
Wignall, S. M., 568
Wiitenberg, C., 244
Wilczynski, S., 623
Wilhovsky, S. K., 463, 497, 498
Wilk, S., 586
Wilkinson, C. R., 162, 165, 174, 387, 559
Wilkinson, C. R. M., 159, 161
Wilkinson, K. D., 11, 24, 33, 38, 47, 48, 54, 121, 122, 124, 141, 144, 147, 179, 243, 268, 324, 367, 470, 471, 664, 676
Willems, A. R., 244, 287, 289, 290, 306, 434, 435, 437, 439, 443, 450, 453, 647
Willemsen, R., 104
Willett, P., 635, 648
Williams, D., 720
Williams, D. B., 466, 493
Williams, G., 513
Willis, J. H., 306
Willison, K. R., 303
Willwohl, D., 101
Wilm, M., 376
Wilson, M. I., 640
Wilson, R., 417
Wimmer, G., 107
Winberg, G., 261, 262

Wind, A., 54
Wing, S. S., 470
Wingfield, P. T., 146
Winistorfer, S. C., 136, 138, 147, 194
Winkle, J. H., 616
Winkler, G. S., 356, 358, 359, 360, 361, 365
Winston, J. T., 289, 290, 299, 570
Winters, M. E., 337, 339
Wisniewski, H. M., 96
Wisniewski, T., 101
Withers-Ward, E. S., 136
Wittenberg, C., 243, 436, 439, 450
Witteveen, A. T., 350
Wittinghofer, A., 635
Wodicka, L., 460, 461
Wohlschlegel, J. A., 13, 15, 244, 266, 268, 272, 274, 379, 401, 402
Wohlstadter, J., 686
Wohnert, J., 186
Wojcik, C., 512
Wolberger, C., 29, 451
Wolf, D. A., 306, 497
Wolf, D. H., 92, 193, 244, 272, 313, 460, 461, 462, 463, 465, 466, 497
Wolff, M., 136, 158, 159
Wolters, D., 375
Wolters, G. K., 146
Wolting, C., 290
Won, K. A., 289, 290, 302, 303, 305, 439
Wong, A., 121
Wong, B. R., 416, 567, 661, 664, 683
Wong, B. S., 101
Wong, H. C., 143, 144
Wong, K. T., 98
Woo, K. M., 492
Wood, A. W., 634
Wood, J. D., 101
Wood, M. K., 325, 332
Wood, N. W., 99
Woodbury, E. L., 451, 452
Woodman, P., 194
Woods, D., 658, 669
Woods, R. A., 463
Woods, Y. L., 705
Woodward, C., 635, 648
Wooten, M. W., 143, 144
Worrall, S., 93, 109
Woulfe, J., 16, 104
Wouters, B. G., 14, 15, 16, 17

Wright, A. P., 560
Wright, C., 350
Wright, J., 278
Wright, P. E., 145, 146
Wright, R., 311, 318
Wu, C. C., 275
Wu, E., 95
Wu, G., 289, 299, 302, 435, 638, 647, 648
Wu, G. S., 512
Wu, J. W., 471
Wu, K., 47, 48, 250
Wu, L., 514, 623, 702
Wu, M. N., 136
Wu, P., 17, 299, 311, 471, 671
Wu-Baer, F., 9, 23, 708
Wulfkuhle, J., 337, 339
Wänning, I., 417, 496, 497
Wustefeld, T., 524
Wuttke, D. S., 145
Wykoff, C. C., 512
Wylie, C., 570
Wynn, R., 669, 703, 704, 705
Wyttenbach, A., 99

X

Xia, G., 24, 29, 141, 143
Xia, Z., 417
Xie, Y., 417
Xing, X., 567
Xiong, Y., 306, 638
Xirodimas, D. P., 469, 661, 705
Xu, G., 289, 299, 302, 435, 638, 647, 648
Xu, H., 314, 454
Xu, K., 718, 721
Xu, L., 289, 290, 291, 292, 295, 296, 297, 299, 302, 303, 306
Xu, M. Q., 38, 41
Xu, Y., 272
Xu, Y.-Z., 326, 686
Xuereb, J. H., 104

Y

Yabe, D., 312, 321
Yabuki, N., 656, 660, 669, 685, 712
Yada, M., 279, 289, 436
Yaffe, M. P., 279
Yagi, M., 345

Yagishita, S., 101
Yamada, S., 147
Yamada, T., 94
Yamaguchi, Y., 147, 640
Yamakami, M., 84, 138, 143
Yamakawa, Y., 121
Yamamoto, H., 587
Yamamoto, K. R., 312, 387
Yamanaka, G., 676
Yamasaki, M., 84
Yamazaki, T., 121
Yamoah, K., 250
Yan, N., 57
Yanagisawa, T., 99
Yang, D. C., 379, 401, 637
Yang, J., 575
Yang, L., 4, 6, 23, 279, 290, 368, 436
Yang, M. J., 159, 161, 439, 514
Yang, T., 8, 623, 669, 702, 703, 705, 706, 709, 712, 713, 714
Yang, X., 638
Yang, Y., 598, 658, 669, 689
Yankner, B. A., 570, 582
Yano, Y., 101
Yao, S. Q., 619
Yao, T., 470, 471
Yarden, Y., 512
Yaron, A., 250, 289, 290, 436, 439
Yashiroda, H., 229
Yasuda, H., 407, 702
Yasuhara, O., 96
Yates, J. R., 13, 15, 221, 244, 266, 268, 271, 272, 274, 306, 389
Yates, J. R. III, 215, 226, 231, 271, 275, 369, 375, 379, 401, 402, 471
Ye, X., 289, 290, 296, 299, 302, 303, 305, 436
Ye, Y., 8
Yeh, E. T., 37
Yeh, K. H., 436, 531, 718
Yeh, L. A., 567
Yen, S. H., 94, 95, 99, 101, 107
Yew, P. R., 256, 260, 261, 512
Yewdell, J. W., 79, 460, 550, 559, 560
Yin, H., 137, 390
Yin, L., 47, 48
Yin, P., 708
Yochim, J. M., 337
Yokoji, M., 101

Yokosawa, H., 65, 69, 72, 74, 76, 79, 80, 81, 83, 84, 138, 143, 147, 221, 229, 230, 279, 669
Yokosawa, N., 84
Yokota, K., 84
Yokoyama, S., 147
Yonemura, Y., 345
Yonetani, T., 714
Yorida, E., 350
Yoshida, Y., 640
Yoshimori, T., 84, 138, 143
Yoshino, K., 138, 143, 144, 145
You, J., 4, 6, 7, 9, 11, 12, 17, 23, 26, 279, 368
Young, K. E., 634
Young, P., 137, 498, 499
Young, R. A., 51, 52, 66, 439
Young, S. W., 657, 669, 678, 685, 712
Yu, H., 15, 409, 482, 569, 673
Yuan, J., 405, 412, 416, 647
Yuan, Q. X., 109
Yuan, W., 9, 23
Yuan, X., 147, 148
Yukiue, H., 121

Z

Zachariae, W., 287, 434
Zachariou, A., 417
Zamir, E., 512
Zaraisky, A. G., 492
Zatloukal, K., 109
Zavitz, K. H., 165
Zhai, P., 136
Zhang, C., 15
Zhang, G. J., 250, 411, 514, 531, 532, 538
Zhang, H., 436, 531, 638, 718
Zhang, J., 531, 550
Zhang, J.-H., 726
Zhang, M., 14, 15, 16, 17, 73, 568, 572, 577, 579, 581, 582
Zhang, R., 136
Zhang, X., 66, 147, 148, 439, 684
Zhang, X. C., 136
Zhang, Y., 466
Zhang, Y. S., 671
Zhang, Z., 570, 582
Zhang, Z. J., 634
Zhao, X., 482, 493, 494
Zhao, Y., 401

Zhdankina, O., 136, 138, 147, 194
Zhelev, N., 514, 531, 532, 538
Zheng, J., 638
Zheng, N., 287, 356, 435, 638
Zhou, H., 17, 379, 399, 401, 402, 471
Zhou, P., 162
Zhou, W., 379, 399, 401, 402
Zhu, G., 136
Zhu, H., 673
Zhu, Y., 121
Zhu, Z., 413
Zimmermann, J., 120
Zinn, K., 514
Zipperlen, P., 314
Ziv, T., 261, 262, 264, 278
Zoghbi, H. Y., 99
Zollner, T. M., 512
Zon, L. I., 405, 412, 416
Zou, H., 336, 405, 408
Zou, J. Y., 125
Zouambia, M., 88
Zuhl, F., 228, 491
Zuiderweg, E. R., 189, 360
Zukin, R. S., 99
Zweier, J. L., 34
Zwickl, P., 550

Subject Index

A

Alexander's disease, ubiquitin immunohistochemistry, 107
Alzheimer's disease, ubiquitin immunohistochemistry, 96, 98–99
Amyotrophic lateral sclerosis, ubiquitin immunohistochemistry, 101, 104
Anaphase-promoting complex
  autoubiquitination, 742
  functional overview, 741
  small molecule inhibitor screening in *Xenopus* egg extracts, *see* Proteasome inhibitors
  structure, 741
  therapeutic targeting rationale, 742
  tumor dysregulation, 336
  ubiquitination assay for high-throughput screening of inhibitors
    inhibitor identification, 751–752
    overview, 742–743
    plate-based assay, 748, 750
    principles, 746, 748
    protein expression and purification
      expression systems, 743
      FLAG-tagged protein, 745–746
      glutathione *S*-transferase fusion protein, 745
      histidine-tagged proteins, 744–745
    screening conditions, 750–751
  in vitro expression cloning for substrate identification
    degradation conditions, 408
    G1 phase substrate identification, 409
    modification for other E3 enzymes, 413–414
    overview, 405–407
    polyacrylamide gel electrophoresis, 412–413
    somatic cell extract studies, 410–412
    substrate pool preparation, 407–408
    Xenopus egg extract preparation, 407
APC, *see* Anaphase-promoting complex

B

Biacore, *see* Surface plasmon resonance
Bioluminescence imaging, *see* Luciferase
Bi-substrate kinetic analysis, ubiquitin ligases, 330–333
Bortezomib, structure and properties, 586–587

C

β-Catenin
  proteolysis
    inhibitor high-throughput screening, 578
    pathway, 569–570
  tumor dysregulation, 335–336
CavSearch, MDM2 inhibitor analysis, 625–627
Cdc4, *see* Skp1-Cull-F-box complexes
CDK, *see* Cyclin-dependent kinase
Cks1
  homogeneous time-resolved fluorescence high-throughput screening of Cks1–Skp2 interaction inhibitors
    binding conditions, 721, 725–726
    FLAG-tagged Cks1 preparation, 720–721
    flow chart, 725
    glutathione *S*-transferase fusion protein preparation, 720
    instrumentation, 721, 724–725
    materials, 719
    plate reader, 726–727
    principles, 718–719
    p27 ubiquitination upregulation, 718
COD1, *see* HRD pathway
Conditional mutant, *see* N-Degron
Confocal laser scanning microscopy
  LMP2-green fluorescent protein tagging for proteasome imaging
    activity effects of incorporation, 553

889

Confocal laser scanning microscopy *(cont.)*
  living cell imaging using fluorescence recovery after photobleaching, 555–557, 559
    localization patterns, 560–561
    overview, 550–551
    potential fusion targets, 559–560
    proteasome incorporation assays, 551–553
    rationale, 551
Cyclin-dependent kinase
  multisite cyclin-dependent kinase substrates as potential SCFCdc4 substrates, 451–453
  p27 inhibitor, *see* p27

D

Degradation of alpha2 pathway, fusion reporters for gene identification
  deg1-lacZ screen, 315–316
  deg1-Ura3p screen, 31
  experimental design principles
    degradation phenotype considerations, 312–313
    directed genetic screens, 314
    mode and stability of expression, 311–312
    mutant elimination in screens, 313–314
    pathway fidelity and degrons, 310–311
    yeast mutant sources, 314
  overview of pathway, 315
Degron, *see also* N-degron; Phosphodegron
  definition, 435, 777
  fusion reporter design for degradation studies, 310–311
  green fluorescent protein GFP^u reporters for ubiquitin-proteasome system function studies
    advantages, 489
    mammalian cells
      comparative analysis with second protein marker, 486–487
      flow cytometry and proteasome inhibitor studies, 484, 486
      fluorescence microscopy imaging, 487–489
      transfection, 484
    overview, 481–482
    reporter features, 482–484

DELFIA, *see* Dissociation-enhanced lanthanide fluoroimmunoassay
Deubiquitinating enzymes
  abundance of types, 120
  active site-directed probe synthesis
    expressed protein ligation
      cysteine-reactive group synthesis and purification for chemical ligation, 474–475
      HAUb$_{75}$–MESNa synthesis, 474
      overview, 472
      plasmid construction, 473–474
      materials, 473
      Michael acceptor synthesis, 472–473
      structural considerations, 471–472
    thiol-reactive probes
      liquid chromatography–mass spectrometry analysis, 476–477
      purification, 476
      radioiodination, 477
      synthesis, 475–476
  classification, 470–471
  functional overview, 470–471
  NEDD8 C-terminus derivative studies
    assays
      optimization, 46–47
      reagent verification, 45–46
      steady-state assays, 46
    site-directed labeling
      affinity purification of labeled proteins, 48–49
      optimization, 47, 49
      principles, 47
  regulation mechanisms, 121
  tissue microarray analysis of tumors, 352
  ubiquitin fusion technique, *see* Ubiquitin fusion technique
  ubiquitin/ubiquitin-like protein probes for activity profiling and identification
    EL-4 cell extract preparation and labeling with active site-directed probes, 127
    immunoprecipitation for tandem mass spectrometry analysis, 127
    labeling of conjugating enzymes, 122–124
    labeling reactions in cell extracts and detection
      epitope-tagged probes, 129
      radiolabeled probes, 128–129

materials, 124
overview, 120–122
protein expression and purification, 124–125
X-ray crystallography, 125, 127
UCH-L3, see UCH-L3
Dissociation-enhanced lanthanide fluoroimmunoassay, high-throughput screening of E3 inhibitors
advantages and limitations, 658
applications, 658
principles, 657–658
DOA pathway, see Degradation of alpha2 pathway
Dsk2, yeast two-hybrid system for binding partner identification, 161–163
DUBs, see Deubiquitinating enzymes

E

E2
E3 enzyme specificity, see E3
gene abundance, 683–684
inhibitor high-throughput screening, see High-throughput screening
substrate specificity, 23–24
UBC domain, 356
E3, see also Anaphase-promoting complex; MDM2; RING finger proteins; Rsp5; Skp1-Cull-F-box complexes
E2 enzyme specificity and engineering of pairs
binding assays
glutathione S-transferase pull-down assay, 362–363
yeast two-hybrid system, 363–364
E2 enzyme
expression and purification, 361
nitrogen-15 labeling, 361–362
homology-based approach
non-Ubc4/5–E3 interactions, 359–360
Ubc4/5–E3 interactions, 358–359
nuclear magnetic resonance chemical shift perturbation of protein–protein interactions, 360
overview, 356, 358
site directed mutagenesis, 362
ubiquitination assay of functional interaction, 364–365
functional overview, 654–655, 683–684
gene abundance, 683–684
high-throughput screening of inhibitors
complexity of assays, 661–662
dissociation-enhanced lanthanide fluoroimmunoassay
advantages and limitations, 658
applications, 658
principles, 657–658
electrochemiluminescence assay
advantages and limitations, 659
applications, 659
principles, 658–659
fluorescence resonance energy transfer
advantages and limitations, 657
p53 ubiquitination, 656
principles, 655–656
TRAF6 ubiquitination, 657, 685
ubiquitin transfer from E2 to E3, 656–657
laboratory-based ubiquitination assays, 661
scintillation proximity assay
principles, 660
applications, 660
advantages and limitations, 661
kinetic mechanism analysis
bi-substrate kinetic analysis, 330–333
IκB mutant
description, 324
ubiquitination assay and optimization, 326–330
materials, 326
overview, 324–326
PY motif recognition, 216
substrate identification, see Anaphase-promoting complex
substrate specificity, 215–216
therapeutic targeting rationale, 685
ECL, see Electrochemiluminescence
Electrochemiluminescence
high-throughput screening for E2 and E3 inhibitors
applications, 699–700
E2 assay
direct detection, 697–698
enzyme preparation, 697
indirect detection, 698–699
materials, 695, 697

Electrochemiluminescence *(cont.)*
  E3 assay
    glutathione S-transferase fusion protein expression, 689–690
    plate assay, 690–695
    materials, 689
    principles, 686–689
    rationale, 685–686
  high-throughput screening for E3 inhibitors
    advantages and limitations, 659
    applications, 659
    principles, 658–659
  N-end rule-dependent ubiquitination substrate screening *in vitro* with plate assay
    complementary DNA library, 421–422
    expression system, 416–417, 422
    high-throughput automation, 422
    inhibitor studies, 423, 427
    plates, 422
    primary screening for ubiquitination, 424–425, 427
    principles, 415–421
    reader, 422
Endoplasmic reticulum-associated degradation, *see* Endoplasmic reticulum quality control and associated degradation
Endoplasmic reticulum quality control and associated degradation
  conservation of pathway, 460–461
  functions, 460
  machinery, 460
  yeast mutant studies
    gene deletion library, 463
    gene identification in endoplasmic reticulum-associated degradation, 465–466
    library transformation and genomic screening, 463, 465
    mutagenesis, 461
    substrate choice and generation, 461
Endosomal sorting, ubiquitin-dependent
  binding assays for membrane recruitment
    crosslinking of membrane proteins, 208–209
    membrane preparation, 205–206
    recombinant protein binding assays
      materials, 207
      principles, 206–207
      purified recombinant protein binding, 207–208
      in vitro transcription/translation product binding, 208
  green fluorescent protein fusion protein studies, 194
  multivesicular body pathway, 193–194
  prospects for study, 210
  protein turnover rate measurement
    cycloheximide chase, 194, 200–202
    pulse–chase immunoprecipitation, 195
  ubiquitin binding by candidate sorting machinery, 202
  ubiquitinated cargo co-purification with ubiquitin-binding machinery, protein A chimera-facilitated affinity purification
    isolation of binding complexes with immunoglobulin G Sepharose, 204–205
    lysis and clearing, 203–204
    materials, 202–203
    spheroplasting, 203
Epifluorescence microscopy
  green fluorescent protein GFP^u reporter degradation imaging, 487–489
  LMP2-green fluorescent protein tagging for proteasome imaging
    activity effects of incorporation, 553
    epifluorescence microscopy, 554–555
    localization patterns, 560–561
    overview, 550–551
    potential fusion targets, 559–560
    proteasome incorporation assays, 551–553
    rationale, 551
Eponemycin
  epoxomicin/dihydroeponemycin chimera synthesis, 599, 601, 603
  structure, 587
  structure-activity relationship studies, 598–599
Epoxomicin
  epoxomicin/dihydroeponemycin chimera synthesis, 599, 601, 603
  proteasome inhibition
    assay, 587–588, 605–606
    mechanism, 588
  structure, 587

SUBJECT INDEX

structure-activity relationship studies, 598–599
synthesis
 chemical characterization, 597
 coupling of fragments, 595–596
 high-performance liquid chromatography, 597
 left-hand fragment, 593, 595
 right-hand fragment, 590, 592–593
 strategy, 589
ERQD, *see* Endoplasmic reticulum quality control and associated degradation
Expressed protein ligation, deubiquitinating enzyme active site-directed probe synthesis
 cysteine-reactive group synthesis and purification for chemical ligation, 474–475
 HAUb$_{75}$–MESNa synthesis, 474
 overview, 472
 plasmid construction, 473–474

F

F-box proteins, *see* Skp1-Cul1-F-box complexes
Flow cytometry
 green fluorescent protein GFP^u reporter and proteasome inhibitor studies, 484, 486
 real-time monitoring of ubiquitin-dependent degradation using green fluorescent protein reporters
  stable transfection and analysis of mammalian cell lines, 498–501
  yeast, 497–498
Fluorescence microscopy, *see* Confocal laser scanning microscopy; Epifluorescence microscopy; Immunofluorescence microscopy
Fluorescence recovery after photobleaching, *see* Confocal laser scanning microscopy
Fluorescence resonance energy transfer
 high-throughput screening of E3 inhibitors
  advantages and limitations, 657
  p53 ubiquitination, 656
  principles, 655–656
  TRAF6 ubiquitination, 657, 685
  ubiquitin transfer from E2 to E3, 656–657
 ubiquitin chain dynamics analysis for high-throughput screening
  comparison with other ubiquitin transfer assays, 669–670, 680
  cysteine-containing ubiquitin preparation, 665–666, 672
  depolymerization assay, 668, 676–678
  design of assay, 670–671
  deubiquitinating enzyme preparation, 667
  fluorophore tagging of ubiquitin, 667, 673
  overview, 664–665
  pitfalls, 678, 680
  polymerization assay, 667–668, 673–676
  ubiquitin pathway enzyme preparation, 666–667, 672
FRET, *see* Fluorescence resonance energy transfer

G

Genetic screens, *see* Degradation of alpha2 pathway; HRD pathway
GFP, *see* Green fluorescent protein
Green fluorescent protein
 advantages as reporter substrate, 492–493
 androgen receptor constructs for Protac testing, 841–843, 845
 degron GFP^u reporters for ubiquitin-proteasome system function studies
  advantages, 489
  mammalian cells
   comparative analysis with second protein marker, 486–487
   flow cytometry and proteasome inhibitor studies, 484, 486
   fluorescence microscopy imaging, 487–489
   transfection, 484
  overview, 481–482
  reporter features, 482–484
 LMP2-green fluorescent protein tagging for proteasome imaging
  activity effects of incorporation, 553
  confocal laser scanning microscopy of living cells using fluorescence recovery after photobleaching, 555–557, 559

Green fluorescent protein (cont.)
  epifluorescence microscopy, 554–555
  localization patterns, 560–561
  overview, 550–551
  potential fusion targets, 559–560
  proteasome incorporation assays, 551–553
  rationale, 551
photobleaching, 550–551
real-time monitoring of ubiquitin-dependent degradation
  caveats, 494–495
  degradation signal incorporation into reporter, 493, 495
  design of reporters, 495–496
  flow cytometry analysis of yeast, 497–498
  fusion with proteasome substrates, 493
  overview, 492–493
  proteasome inhibitor utilization, 493–494
  stabilization signal identification
    mammalian cells, 502–503
    overview, 501–502
    yeast, 502
  stable transfection and analysis of mammalian cell lines, 498–501
  ubiquitin fusion, 495–496
ubiquitin^{676V}–green fluorescence protein transgenic mice
  applications, 503
  breeding, 504
  generation, 504
  genotyping, 504–505
  immunofluorescence microscopy, 506–508
  immunohistochemistry, 506–507
  primary cell culture studies, 508
  proteasome inhibitor administration, 505–506

H

High-throughput screening
  anaphase-promoting complex ubiquitination assay for inhibitor screening
    inhibitor identification, 751–752
    overview, 742–743
    plate-based assay, 748, 750
    principles, 746, 748
    protein expression and purification
      expression systems, 743
      FLAG-tagged protein, 745–746
      glutathione S-transferase fusion protein, 745
      histidine-tagged proteins, 744–745
    screening conditions, 750–751
  Cks1–Skp2 interaction inhibitor screening using homogeneous time-resolved fluorescence
    binding conditions, 721, 725–726
    FLAG-tagged Cks1 preparation, 720–721
    flow chart, 725
    glutathione S-transferase fusion protein preparation, 720
    instrumentation, 721, 724–725
    materials, 719
    plate reader, 726–727
    principles, 718–719
  E3 inhibitors
    complexity of assays, 661–662
    dissociation-enhanced lanthanide fluoroimmunoassay
      advantages and limitations, 658
      applications, 658
      principles, 657–658
    electrochemiluminescence assay
      advantages and limitations, 659
      applications, 659
      principles, 658–659
    fluorescence resonance energy transfer
      advantages and limitations, 657
      p53 ubiquitination, 656
      principles, 655–656
      TRAF6 ubiquitination, 657, 685
      ubiquitin transfer from E2 to E3, 656–657
    laboratory-based ubiquitination assays, 661
    scintillation proximity assay
      advantages and limitations, 661
      applications, 660
      principles, 660
  electrochemiluminescence assay for E2 and E3 inhibitor screening
    applications, 699–700
    E2 assay
      direct detection, 697–698

enzyme preparation, 697
indirect detection, 698–699
materials, 695, 697
E3 assay
glutathione S-transferase fusion
protein expression, 689–690
plate assay, 690–695
materials, 689
principles, 686–689
rationale, 685–686
fluorescence resonance energy transfer of ubiquitin chain dynamics
comparison with other ubiquitin transfer assays, 669–670, 680
cysteine-containing ubiquitin preparation, 665–666, 672
depolymerization assay, 668, 676–678
design of assay, 670–671
deubiquitinating enzyme preparation, 667
fluorophore tagging of ubiquitin, 667, 673
overview, 664–665
pitfalls, 678, 680
polymerization assay, 667–668, 673–676
ubiquitin pathway enzyme preparation, 666–667, 672
IkBα ubiquitination inhibitors
incubation conditions, 731–732, 735–736
materials, 730
optimization, 737–739
phosphorylated substrate preparation, 731
plate assay, 732
principles, 732–733
protein preparation, 731
Western blot, 732
MDM2 inhibitors, 709, 712
proteasome small molecule inhibitor screening in *Xenopus* egg extracts
β-catenin proteolysis inhibitor high-throughput screening, 578
cyclin proteolysis inhibitor high-throughput screening, 576–578
Skp1-Cul1-F-box complexes substrate interface inhibitors, 637–638
HMG-CoA reductase, *see* HRD pathway
HRD pathway, fusion reporters for gene identification

COD1 regulator identification
two-gene screen, 319–320
two-protein screening, 318–319
experimental design principles
degradation phenotype considerations, 312–313
directed genetic screens, 314
mode and stability of expression, 311–312
mutant elimination in screens, 313–314
pathway fidelity and degrons, 310–311
yeast mutant sources, 314
HMG-CoA reductase degradation, 317–318
optical screen for HMG-CoA reductase degradation blockers, 320–321
HTS, *see* High-throughput screening
Huntington's disease, ubiquitin immunohistochemistry, 99, 101, 104
Hypoxia-inducible factor-α (HIF-α), tumor dysregulation, 336

I

IkBα
luciferase fusion reporter for degradation studies
bioluminescence imaging
cells, 522–523
hepatocyte xenografts *in vivo*, 524, 526
construction, 522
controls for ligand-induced degradation studies, 523
proteasome inhibitor studies, 526
stable expression in HeLa cells, 522
tumor necrosis factor-α response, 523–524
validation of degradation, 523–524
ubiquitination
assay for high-throughput screening of inhibitors
incubation conditions, 731–732, 735–736
materials, 730
optimization, 737–739
phosphorylated substrate preparation, 731
plate assay, 732
principles, 732–733

IkBα (cont.)
    protein preparation, 731
        Western blot, 732
    nuclear factor-kB activation, 730
Immunoaffinity chromatography
    polyubiquitins with monoclonal antibodies, 81, 83–84
    ubiquitinated proteins for modification site identification
        antiubiquitin column preparation, 280–282
        cell culture with proteasome inhibitor, 282
        denaturing affinity chromatography, 283
Immunofluorescence microscopy
    polyubiquitin characterization with monoclonal antibodies, 80
    ubiquitin^{676V}–green fluorescence protein in transgenic mice, 506–508
Immunohistochemistry, ubiquitin
    aggresomal response, 93
    Alexander's disease, 107
    Alzheimer's disease, 96, 98–99
    amyotrophic lateral sclerosis, 101, 104
    antibodies, 88
    antigen retrieval, 91
    axonal damage and spheroids, 94–95
    cytoskeletal system, 92–93
    endosome-lysosome system, 93–94
    fixation
        overview, 88–89
        paraffin wax sections, 89
    inclusion bodies in disease, 95–97, 107
    labeled avidin/avidin-biotin technique, 90–91
    Mallory bodies, 108–110
    Marinesco bodies, 101
    multiple system atrophy, 104, 107
    myopathies, 107
    neurofilament inclusion disease, 104
    neuronal swelling, 95
    Parkinson's disease, 99
    principles, 89
    prion disease, 101
    rationale and applications, 87–88, 110
    solutions, 91–92
    tissue microarrays, 343–344
    trinucleotide repeat disorders, 99, 101, 104
    ubiquitin^{676V}–green fluorescence protein in transgenic mice, 506–507

Immunoprecipitation
    polyubiquitins with monoclonal antibodies, 81
    pulse–chase immunoprecipitation for protein turnover rate measurement, 195
    ubiquitin hydrolases, 127
Inclusion body myositis, ubiquitin immunohistochemistry, 107
Intein, *see* NEDD8
IVEC, *see* In Vitro expression cloning

K

Kennedy's syndrome, ubiquitin immunohistochemistry, 99, 101, 104

L

Lactacystin, proteasome inhibition, 64–65
LMP2, *see* 26S Proteasome
Luciferase
    advantages in molecular imaging studies, 515
    β-catenin reporter preparation for small molecule proteasome inhibitor screening in *Xenopus* egg extracts, 570–571
    IkBα–luciferase reporter for ligand-induced degradation studies
        bioluminescence imaging
            cells, 522–523
            hepatocyte xenografts *in vivo*, 524, 526
        controls, 523
        proteasome inhibitor studies, 526
        reporter construction, 522
        stable expression in HeLa cells, 522
        tumor necrosis factor-α response, 523–524
        validation, 523–524
    p27–luciferase reporter for measuring Cdk2 activity
        expression plasmid construction, 532–533
        hollow fiber assay
            bioluminescence imaging *in vitro*, 543–544
            bioluminescence imaging *in vivo*, 546–548
            encapsulation of cells, 539, 541–543

hollow fiber preparation, 539
implantation of hollow fibers into
  nude mice, 544–546
overview, 531–532
stable transfection of HeLa cells, 533,
  535–536
ubiquitination of p27, 531
validation using small interfering RNA
  against cyclin A or Cdk2, 537–538
Western blot of expression, 536
tetraubiquitinated reporter for proteasome
  activity studies
  bioluminescence imaging
    cells, 517
    xenografts *in vivo*, 517–519, 521
  plasmids, 515
  proteasome inhibitor studies, 517–518
  stable expression in HeLa cells, 515–516

M

Mallory bodies, ubiquitin
  immunohistochemistry, 108–110
Marinesco bodies, ubiquitin
  immunohistochemistry, 101
Mass spectrometry
  polyubiquitin chain structure confirmation,
    12–13
  ubiquitin chain architecture analysis
    cell lysate preparation, 274
    gel electrophoresis and peptide
      digestion, 275
    high-performance liquid
      chromatography of peptide
      digests, 275
    mass spectrometry, 275–276
    nickel affinity chromatography of
      recombinant protein, 274
    overview, 272–274
  ubiquitinated protein identification
    gel electrophoresis–liquid
      chromatography–tandem mass
      spectrometry, 375–376
    liquid chromatography–tandem mass
      spectrometry, 374–375
    nickel affinity chromatography,
      370, 372, 374
    overview, 369
    ubiquitin-like protein applications, 379
    validation and controls

gel electrophoresis, 377
histidine-rich protein interference,
  376–377
ubiquitination site identification,
  377–379
Western blot, 379
yeast growth and harvest, 370
ubiquitination site determination
  affinity-purified proteins
    advantages, 284–285
    antiubiquitin column preparation,
      280–282
    caveats, 284
    cell culture with proteasome
      inhibitor, 282
    denaturing affinity
      chromatography, 283
    mass spectrometry and database
      searching, 284
    materials, 280
    overview, 279–280
    peptide digestion, 283–284
  histidine-tagged proteins
    cell lysate preparation, 268–269
    high-performance liquid
      chromatography of peptide
      digests, 270–271
    nickel affinity chromatography of
      recombinant protein, 269–270
    overview, 267–268, 369
    peptide digestion, 270
    tandem mass spectrometry,
      271–272
    principles, 369–370
  ubiquitin hydrolase tandem mass
    spectrometry analysis, 127
MDM2
  E3 activity assay for inhibitor development
    advantages of assay, 714–715
    distinguishing polyubiquitination from
      multiple monoubiquitination, 708
    E2 conjugation with labeled ubiquitin,
      705–706
    high-throughput screening, 709, 712
    incubation conditions and fluorescence
      detection, 706–708
    inhibitor identification and
      characterization, 712–714
    mechanism studies, 709
    principles, 703

MDM2 (cont.)
  ubiquitin labeling with Oregon Green, 703–704
  p53 interactions
    affinity, 624
    nutlin inhibition of p53 interaction, 623
    regulation of p53, 623
    small molecule inhibitor development
      cavSearch analysis, 625–627
      hit assessment by nuclear magnetic resonance, 627–630
      structural considerations, 624
      X-ray crystallography, 630–632
    specificity, 624
    structural aspects, 624–627
    therapeutic targeting rationale, 702–703
Methionine aminopeptidase-2, Protac efficacy testing, see Protac
MG115, structure and properties, 586–587
MG132
  proteasome inhibition, 64
  structure and properties, 586–587
Molecular imaging
  bioluminescence imaging, see Luciferase
  overview, 513–515, 530–531
MS, see Mass spectrometry
Multiple system atrophy, ubiquitin immunohistochemistry, 104, 107
Multivesicular body pathway, see Endosomal sorting, ubiquitin-dependent

N

N-degron
  conditional mutant generation
    advantages, 800–801
    conditional activity of N-end rule pathway, 807
    controlled expression of protein with constitutive N-degron, 806
    temperature-activated N-degron
      construction and uses
        materials, 814–815
        mutant analysis, 818
        overview, 807, 809–810
        polymerase chain reaction-based genomic transplacement strategy, 812–814, 816
        targeted plasmid integration, 810–812
        transformation of competent yeast, 816–817
        verification of correct targeting, 817
      ubiquitin fusion technique, 784–785, 804–806
    conditional toxins, 785–786
    definition, 778
    N-end rule pathway, 802–804
    structure, 781–782, 802
NEDD8, thiol ester production and C-terminus derivatization using inteins
  aldehyde derivative synthesis, 43
  amidomethyl coumarin derivative synthesis, 43
  application to other ubiquitin-like proteins, 42
  deubiquitinating enzyme assays
    optimization, 46–47
    reagent verification, 45–46
    steady-state assays, 46
  deubiquitinating enzyme site-directed labeling
    affinity purification of labeled proteins, 48–49
    optimization, 47, 49
    principles, 47
  FLAG fusion protein generation, 40–41
  N-hydroxysuccinimide as catalyst, 42–43
  monitoring with high-performance liquid chromatography, 43–44
  overview, 38, 40
  purification, 41
  purification of derivatives
    hydrophobic interaction chromatography, 45
    ion-exchange chromatography, 44–45
  reagent solubility, 42
  side reaction minimization during coupling, 43
  splicing reaction, 41
  vinylsulfone derivative synthesis, 43
N-end rule-dependent ubiquitination
  green fluorescent protein reporter design, 495
  N-degron, see N-degron
  overview, 802–804
  substrate characteristics, 417
  substrate screening in vitro
    advantages, 430–431
    complementary DNA library, 421–422

denaturing gel electrophoresis, 424, 429–430
electrochemiluminescence plate assay
 high-throughput automation, 422
 inhibitor studies, 423, 427
 plates, 422
 primary screening for ubiquitination, 424–425, 427
 reader, 422
expression system, 416–417, 422
immunoprecipitation and Western blot, 423–424
principles, 415–421
RGS4 studies, 420, 424–425
sulfo-TAG-labeled antibodies, 420
sulfo-TAG-labeled ubiquitin detection, 420
NF-kB, see Nuclear factor-kB
NMR, see Nuclear magnetic resonance
Nuclear factor-kB
 activation, 729–730
 defects in disease, 521
 functional overview, 729
 inhibitor, see IkBα
Nuclear magnetic resonance
 chemical shift perturbation studies
  E2–E3 interactions, 360
  monoubiquitin–protein interactions
   binding parameter quantification, 149–152
   challenges of structure determination in fast exchange, 152–153
   principles, 145–147
   residual dipolar couplings, 149
   validation, 147, 149
 MDM2 inhibitor analysis, 627–630
 polyubiquitin conformation studies
  chemical shift perturbation mapping of interdomain interface in polyubiquitin, 180–181
  interdomain orientation of diubiquitin, 181–182, 184–185
  overview, 177–178
  segmental isotope labeling, 178–179
  site-specific spin labeling for long-distance information, 185–186
 ubiquitin-binding protein interaction studies
  modeling challenges, 187
  overview, 186–187

stoichiometry of binding using spin relaxation, 188–189
UBA2/Lys63-Ub2 binding affinity studies, 189–190
UBA2/ubiquitin complex structure, 187–188

P

p27
 Cks1 upregulation of ubiquitination, 718
 luciferase fusion reporter for measuring Cdk2 activity
  expression plasmid construction, 532–533
  hollow fiber assay
   bioluminescence imaging *in vitro*, 543–544
   bioluminescence imaging *in vitro*, 546–548
   encapsulation of cells, 539, 541–543
   hollow fiber preparation, 539
   implantation of hollow fibers into nude mice, 544–546
   overview, 531–532
   stable transfection of HeLa cells, 533, 535–536
   ubiquinination of p27, 531
   validation using small interfering RNA against cyclin A or Cdk2, 537–538
   Western blot of expression, 536
 prognostic value in cancer, 717–718
 Skp1-Cul1-F-box complex ubiquitination, 531, 718
 structure, 229–230
 subunits, 228–229
p53
 fluorescence resonance energy transfer assay of ubiquitination, 656
 MDM2 binding
  affinity, 624
  nutlin inhibition, 623
  overview, 623
  small molecule inhibitor development
   cavSearch analysis, 625–627
   hit assessment by nuclear magnetic resonance, 627–630
   structural considerations, 624
   X-ray crystallography, 630–632
  specificity, 624

p53 (cont.)
    structural aspects, 624–627
    mutation in cancer, 702
    stress response, 622–623
Parkinson's disease, ubiquitin immunohistochemistry, 99
Phosphodegron
    definition, 435
    F-box protein identification for particular degradation targets
        binding assay, 301
        overview, 299–300
        phosphopeptide immobilization on agarose beads, 300
        reticulocyte lysate translation of F-box proteins, 301
    genome-wide scan of motifs using Spot peptide arrays
        amino acid stock solutions, 441
        Cdc4-Skp1 complex preparation and probing of arrays, 443–444
        membrane regeneration, 444
        overview, 440–441, 453
        phosphodegron peptide characteristics, 444–445
        side-chain deprotection, 442–443
        synthesis, 441–442
    sequences and F-box protein binding, 436, 438–439
Polyubiquitin
    affinity purification of proteins from yeast
        affinity chromatography, 389–390
        cell culture and lysis, 387–389
        identification of proteins, 391
        multiubiquitin chain-binding protein immobilization on Sepharose, 387
        overview, 385–386
        recovery, 390
        yeast strain design, 386–387
    conjugating enzyme specificity, 23–24
    domains in interacting protein recognition, 24, 144
    fluorescence resonance energy transfer assay of chain dynamics, see Fluorescence resonance energy transfer
    functions, 3–5, 21–23, 75, 177–178
    linkage types, 368
    mass spectrometry, see Mass spectrometry
    metabolism by deubiquitinating enzymes, 5–6
    monoclonal antibodies
        applications, 76, 84
        cross-reactivity assays, 79
        immunofluorescence microscopy of polyubiquitins, 80
        polyubiquitin isolation
            immunoaffinity chromatography, 81, 83–84
            immunoprecipitation, 81
        production
            immunization, 77
            negative screening of hybridoma cells, 77, 79
            polyubiquitin-conjugated protein preparation, 76–77
            positive screening of hybridoma cells, 77
        Western blot characterization of polyubiquitins, 79–80
    N-end rule pathway, see N-end rule-dependent ubiquitination
    nuclear magnetic resonance, see Nuclear magnetic resonance
    structure modification
        linear chains, 13
        epitope tagging, 17
        lysine-to-arginine point mutants
            dominant negative mutant studies in mammalian cells, 14–16
            enzyme interactions, 8–9
            expression, 7–8, 14
            heteropolymeric linkage, 8
            monoubiquitin conjugation, 8
            Ub-K48R mutants, 6
            Ub-K63R mutants, 22
            wild-type ubiquitin interference in crude extracts, 8
        mass spectrometry for chain structure confirmation, 12–13
        single-lysine ubiquitin mutants
            applications, 9–10
            chain structure probing in living cells, 16–17
            chemical modification for generation, 10–12
            purification, 10

synthesis
  conjugating enzyme expression and purification, 29
  K48-linked chains
    deblocking reactions, 31
    K48-Ub$_2$ synthesis and purification, 29–30
    K48-Ub$_4$ synthesis and purification, 31–32
    long-chain synthesis, 32
  K63-linked chains, 32
  overview, 24–25
  specialized chain applications, 33–34
  ubiquitin
    expression in *Escherichia coli*, 26
    purification, 26–29
    yield and recovery, 33
  tandem ubiquitin, *see* Tandem ubiquitin
  variations, 3, 21, 23
Prion disease, ubiquitin immunohistochemistry, 101
Protac
  applications, 834–835
  efficacy testing
    assessment *in vivo*
      cell-permeable Protac testing, 843, 845
      microinjection of androgen receptor constructs, 841–843
      Xenopus extract assays, 840–841
    methionine aminopeptidase-2-Protac coupling assay *in vitro*
      degradation demonstration, 839–840
      immunoprecipitation, 837–838
      tissue culture, 836–837
      ubiquitination assay, 838–839
  overview, 835–836
20S Proteasome
  mechanism of action, 229–230
  purification from bovine reticulocyte lysates for proteasome inhibition assay, 605–606
  purification from rat liver
    large-scale purification, 233–234
    small-scale purification, 238
26S Proteasome
  architecture, 228
  bioluminescence imaging of degradation, *see* Luciferase
  components, 228, 550
  fluorogenic peptide substrate assay using succinyl-LLVY-4-methyl-coumaryl-7-amide, 230, 232
  green fluorescent protein reporters, *see* Green fluorescent protein
  inhibitors, *see* Proteasome inhibitors
  LMP2-green fluorescent protein tagging for imaging
    activity effects of incorporation, 553
    confocal laser scanning microscopy of living cells using fluorescence recovery after photobleaching, 555–557, 559
    epifluorescence microscopy, 554–555
    localization patterns, 560–561
    overview, 550–551
    potential fusion targets, 559–560
    proteasome incorporation assays, 551–553
    rationale, 551
  lysozyme degradation assay, 230–231
  ornithine decarboxylase degradation assay, 231–232
  purification from rat liver
    large-scale purification, 234–235
    small-scale purification
      conventional chromatography, 236–237
      immunoaffinity chromatography, 237–238
      nuclear extract preparation, 235–236
      storage, 239
  Sic1^{PY} degradation assay
    gel electrophoresis, 224
    overview, 231
    26S proteasome isolation, 221–223
    reaction conditions, 223–224
    ubiquitination, 219–221
    ubiquitin chain specificity studies, 225–226
  tandem ubiquitin inhibition, *see* Tandem ubiquitin
Proteasome inhibitors, *see also specific inhibitors*
  clinical application, 567
  development of inhibitors with novel activities
    caspase-like activity inhibitors, 604–605

Proteasome inhibitors *(cont.)*
   chymotrypsin-like activity inhibitors, 603–604
   epoxomicin/dihydroeponemycin chimera synthesis, 599, 601, 603
   overview, 598
   structure-activity relationship studies, 598–599
   green fluorescent protein GFP^u reporter degradation studies, 484, 486
   IkBα–luciferase reporter degradation studies, 526
   N-end rule-dependent ubiquitination substrate screening *in vitro*, 423, 427
   peptide-based covalent inhibitor screening using competition-based screening
      advantages and limitations, 620–621
      competition assay design using selective active site label, 612–613
      inhibitor library expansion, 619–620
      positional scanning libraries
         design, 613–615
         Fmoc-Asp(OtBu)-dimethyl hydroxyl amide synthesis, 615
         Fmoc-Asp(OtBu)-H synthesis, 615
         Fmoc-Asp(OtBu)-vinyl sulfone synthesis, 616
         Fmoc-Asp-vinyl sulfone coupling to resin, 616
         Fmoc-Asp-vinyl sulfone synthesis, 616
         screening conditions, 617, 619
         synthesis, 616–617
      principles, 610–612
      subunit selective inhibitor identification, 619
   real-time monitoring of ubiquitin-dependent degradation, 493–494
   small molecule inhibitor screening in *Xenopus* egg extracts
      active compound mechanism characterization, 579–582
      β-catenin proteolysis inhibitor high-throughput screening, 578
      cell cycle progression and cyclin B proteolysis, 568–569
      cyclin B activator preparation, 572
      cyclin proteolysis inhibitor high-throughput screening, 576–578
      extract preparation
         egg collection and extraction, 573–575
         ovulation induction, 573
         quality assessment, 575–576
         solutions, 572–573
      forward chemical genetics, 568–570
      luciferase–β-catenin reporter preparation, 570–571
      Wnt pathway and β-catenin proteolysis, 569–570
   tetraubiquitinated luciferase reporter for proteasome activity studies, 517–518
Proteolysis targeting chimeric molecule, *see* Protac
Proteomics, *see* Mass spectrometry
PY motif, *see* Sic1^{PY}

R

RGS4, N-end rule-dependent ubiquitination, 420, 424–425
RING finger proteins
   abundance in humans, 356
   domain structure, 356
   E2 enzyme specificity, 356, 358
RNA interference
   F-box protein identification in target degradation, 297–299
   p27–luciferase reporter for measuring Cdk2 assay validation using small interfering RNA against cyclin A or Cdk2, 537–538
Rsp5, polyubiquitination of Sic1^{PY}
   enzyme preparation, 219–220
   materials, 220
   reaction conditions, 220–221

S

SCF, *see* Skp1-Cul1-F-box complexes
Scintillation proximity assay
   construction and preparation from *Escherichia coli*, 216–218
   degradation assay
      gel electrophoresis, 224
      overview, 231
      26S proteasome isolation, 221–223
      reaction conditions, 223–224
      ubiquitin chain specificity studies, 225–226
   E3 recognition, 216

SUBJECT INDEX 903

high-throughput screening of E3 inhibitors
  advantages and limitations, 661
  applications, 660
  principles, 660
polyubiquitination
  materials, 220
  reaction conditions, 220–221
  Rsp5 ubiquitin ligase preparation, 219–220
PY domain insertion in other proteins, 226
Sic1^{PY}, tandem ubiquitin and proteasome inhibition studies, 70–71
Skp1-Cul1-F-box complexes
  components and functions, 287–288, 434
  druggability of substrate interfaces
    drug discovery approaches
      high-throughput screening, 637–638
      structure-based drug design, 637–638, 647
      virtual library screening, 637–638, 647–648
    overview, 634–637
    structural data for druggability assessment
      crystal-induced local conformational variability, 645
      local quality of structural data, 641–642, 646
      local structural reliability assessment, 642–645
      surface pocket identification, 647–649
      three-dimensional structures, 638, 641
  ectopic targeting of substrates to ubiquitin-proteasome pathway by engineering F-box substrate receptor
    applications, 830–832
    efficacy testing and optimization, 828
    endogenous substrate targeting, 829–830
    principles, 823–826
    targeting peptide selection, 826–827
  F-box proteins
    abundance in eukaryotes, 289
    dominant negative Cul1 mutant for target identification
      cycloheximide time course, 294
      principles, 291
      pulse–chase, 292–294
      steady-state protein accumulation, 292
      stimulus-induced turnover, 294–295
      transfection, 291–292
    identification for particular degradation targets
      Cul1 involvement, 290
      overview of approaches, 289–290
      RNA interference for validation, 297–299
    phosphodegrons for F-box protein identification
      binding assay, 301
      overview, 299–300
      phosphopeptide immobilization on agarose beads, 300
      reticulocyte lysate translation of F-box proteins, 301
    prospects for study, 306
    screening for substrate interactions *in vivo*
      F-box protein cloning and expression, 295–296
      glutathione S-transferase pull-down assay, 296–297
      overview, 295
      transfection, 296
  Skp2–Cks1 interactions, *see* Cks1
  IkBα as substrate, *see* IkBα
  p27 as substrate, *see* p27
  phosphorylation-dependent substrate identification
    genome-wide scan of phosphodegron motifs using Spot peptide arrays
      amino acid stock solutions, 441
      Cdc4-Skp1 complex preparation and probing of arrays, 443–444
      membrane regeneration, 444
      overview, 440–441, 453
      phosphodegron peptide characteristics, 444–445
      side-chain deprotection, 442–443
      synthesis, 441–442
    multisite cyclin-dependent kinase substrates as potential SCFCdc4 substrates, 451–453
    phosphodegrons and F-box protein binding, 435–436, 438–439
    principles, 439
    secondary tests of candidate Cdc4 substrates
      capture of substrate *in vitro*, 447

Skp1-Cul1-F-box complexes *(cont.)*
  Cdc4-dependent protein instability in yeast strains, 449–450
  overview, 451
  recombinant substrate preparation, 446–447
  ubiquitination of substrate *in vitro*, 447–449
 reconstitution for F-box protein analysis
  overview, 301–302
  SCF$^{\beta\text{-TRCP}}$ from reticulocye lysate
   Cdc25A phosphorylation, 303
   radiolabeled Cdc25A expression, 302
   ubiquitination reactions, 303
  SCF^{Fbw7} from baculovirus–insect cell system
   cyclin E ubiquitination, 305–306
   purification of complex, 304–305
 structure, 434–435
 tumor dysregulation, 335–336
 yeast substrates, 437–438
Skp2
 homogeneous time-resolved fluorescence high-throughput screening of Cks1-Skp2 interaction inhibitors
  binding conditions, 721, 725–726
  FLAG-tagged Cks1 preparation, 720–721
  flow chart, 725
  glutathione *S*-transferase fusion protein preparation, 720
  instrumentation, 721, 724–725
  materials, 719
  plate reader, 726–727
  principles, 718–719
 tissue microarray analysis of tumors, 350–351
SPA, *see* Scintillation proximity assay
Spinocerebellar ataxia, ubiquitin immunohistochemistry, 99, 101, 104
Split-ubiquitin technique
 competition assay for verification, 771–772
 5-fluoroorotic acid resistance testing, 768–770
 genome-wide interaction screen, 765–766
 library transformation in yeast, 765–768
 N-terminal peptide mutant selection and reporter level optimization, 763–765

positive plasmid isolation and sequencing, 770–771
principles, 757–759, 789–790
reporters
 molecular weight reporter, 759–760
 N-end rule substrate arginine-Ura3p, 761–762
 transcription-based readouts, 762
 Ura3p, 760
reverse split-ubiquitin technique for conformational probing, 772–775
Spot peptide array, *see* Phosphodegron
SPR, *see* Surface plasmon resonance
Structure-based drug design, protein–protein interaction inhibitor development, 637–638, 647
SUMOylation
 conjugating enzymes, 123
 conjugation reaction, 394
 consensus sequence, 395–396
 deconjugation, 394–396
 functions, 393–394
 isoforms and processing of SUMO, 393–394
 protein conjugate purification from yeast
  cell growth and lysis, 398
  hexahistidine-FLAG-tagged protein utilization, 402
  identification of proteins, 399–401
  immunoprecipitation, 399
  liquid chromatography–mass spectrometry analysis, 402
  nickel affinity chromatography, 399
  overview, 396–397
  pull-down assay and Western blotting, 401
  yeast strain, 397–398
 regulation, 402–403
 targets, 393, 395
Surface plasmon resonance
 kinetic constants, 166
 principles, 166–168
 thermodynamic parameters, 166
 ubiquitin-proteasome system studies
  activity determination, 172–173
  advantages, 165–166
  analyte preparation, 169–170
  kinetic analysis and data interpretation, 173–175
  ligand

immobilization on sensor chip,
170–172
preparation, 169
limitations, 166
regeneration conditions, 173
troubleshooting, 172, 175

T

Tandem ubiquitin
  gene construction, 65–67
  proteasome inhibition studies
    in vitro
      histidine-tagged tandem ubiquitin studies, 71–72
      materials, 70
      overview, 65, 70
      Sic1^{PY} as substrate, 70–71
    in vivo
      β-galactosidase reporter degradation inhibition, 69–70
      Gcn4 degradation inhibition, 69
      materials, 68–69
      overview, 65–68
      Xenopus embryos, 72–74
Tissue microarray
  advantages in tumor studies, 339, 352–353
  antibodies for immunohistochemistry, 343–344
  availability, 339–341
  caveats, 344–345
  construction
    core size, 343
    planning, 342–343
    punching, 342
  data analysis
    archiving, 346–347
    clustering, 347–348, 350
    scoring, 345–356
    software, 347–348
  principles, 338–339
  ubiquitin proteasome system component studies
    comparison with other techniques, 337–338, 352–353
    tumor analysis, 350–352
TMA, see Tissue microarray
TRAF6, see Tumor necrosis factor receptor-associated factor 6

Tumor necrosis factor receptor-associated factor 6, fluorescence resonance energy transfer assay of ubiquitination, 657, 685
Tumors, see Tissue microarray
Two-hybrid system, see Yeast two-hybrid system

U

UbcH5b, CNOT4 interactions, 356, 358
Ubiquitin
  abundance in cells, 87
  binding domains, see Ubiquitin-binding domains
  functional overview of conjugation, 228, 278–279, 367–368, 469, 702
  immunohistochemistry, see Immunohistochemistry, ubiquitin
  nomenclature, 777
  pathways for ubiquitination, 495–496
  polymers, see N-end rule-dependent ubiquitination; Polyubiquitin; Tandem ubiquitin
  protein ubiquitination status determination
    hexahistidine-ubiquitin expression
      cell lysate preparation, 245–246
      expression plasmids, 244
      nickel affinity chromatography purification of ubiquitinated proteins, 246–247
      principles, 243–244
    indirect assays
      E1 temperature sensitive cell line assays, 260
      materials, 259
      overview, 259
      substrate half-life analysis in presence of Ub(K0), 260–261
    in vitro analysis
      materials, 257
      overview, 255–257
      purified recombinant substrate ubiquitination, 258–259
      sulfur-35-labeled substrate ubiquitination, 257–258
    in vivo analysis
      histidine-tagged ubiquitin utilization, 252–254

Ubiquitin (cont.)
  immunoprecipitation and Western
      blot of affinity-tagged
      ubiquitinated proteins, 254–255
    materials, 251
    overview, 250–251
    Western blot, 251–252
  site determination
    mass spectrometry, see Mass
        spectrometry
    N-terminal methionine versus lysine
      materials, 262
      N-terminal methionine
          modification with myc tag
          or carbamylation, 262
      overview, 261
      Western blot, 263–264
  tagging
    epitope tagging, 17
    histidine tagging
      nickel affinity chromatography,
          57–58, 60–62
      rationale, 53
      vector construction, 55–57
      yeast expression, 59–60
    overview, 51–52, 54–55
    vector design, 52–53
  targeting of substrates, see Protac;
      Skp1-Cul1-F-box complexes
Ubiquitin-binding domains
  classification, 135–136, 158
  monoubiquitin-binding protein
      identification
    affinity chromatography, 139–141
    bioinformatics, 135, 137
    yeast two-hybrid system, 137–139
  monoubiquitination role studies, 144–145
  polyubiquitin chain specificity studies, 144
  protein interaction studies, see Nuclear
      magnetic resonance; Surface plasmon
      resonance; Yeast two-hybrid system
  pull-down assay principles, 141–143
  ubiquitin mutant studies of binding
      residues, 143–144
Ubiquitin deconjugation, see
    Deubiquitinating enzymes
Ubiquitin-dependent endosomal sorting, see
    Endosomal sorting, ubiquitin-dependent
Ubiquitin fusion technique
  N-degron

    conditional mutants based on
        N-degrons, 784–785
    conditional toxins, 785–786
    definition, 778
    structure, 781–782
  principles, 777–779, 781
  reporters, 782–783
  ubiquitin fusion overexpression,
      786–788
Ubiquitin ligase, see E3
Ubiquitin–protein reference technique,
    principles, 790–791
Ubiquitin sandwich technique, principles,
    791–792
Ubiquitin translocation assay,
    principles, 788
UCH-L3, UbVME derivative
    adduct formation
  materials, 124
  overview, 121
  synthesis and purification, 124–125
  X-ray crystallography, 125, 127
UPR, see Ubiquitin–protein reference
    technique
UTA, see Ubiquitin translocation assay

V

Velcade, see Bortezomib
Virtual library screening, protein–protein
    interaction inhibitor development,
    637–638, 647–648
In Vitro expression cloning
  anaphase-promoting complex substrate
      identification
    degradation conditions, 408
    G1 phase substrate identification, 409
    modification for other E3 enzymes,
        413–414
    overview, 405–407
    polyacrylamide gel electrophoresis,
        412–413
    somatic cell extract studies,
        410–412
    substrate pool preparation,
        407–408
    Xenopus egg extract
        preparation, 407
  principles, 405
VLS, see Virtual library screening

W

Western blot
  IkBα ubiquitination, 732
  N-end rule-dependent ubiquitination substrates, 423–424
  p27–luciferase reporter, 536
  polyubiquitin characterization with monoclonal antibodies, 79–80
  ubiquitinated protein identification, 379
  ubiquitination status determination
    direct Western blot, 251–252
    immunoprecipitation and Western blot of affinity-tagged ubiquitinated proteins, 254–255
    site determination with N-terminal methionine blocking, 263–264

X

X-ray crystallography
  MDM2 inhibitor analysis, 630–632
  Skp1-Cul1-F-box complexes, druggability of substrate interfaces
    structural data for druggability assessment
      crystal-induced local conformational variability, 645
      local quality of structural data, 641–642, 646
      local structural reliability assessment, 642–645
      surface pocket identification, 647–649
      three-dimensional structures, 638, 641
  UCH-L3–UbVME adduct, 125, 127

Y

Yeast two-hybrid system
  E2–E3 interactions, 363–364
  monoubiquitin-binding protein identification, 137–139
  UBA domain–ubiquitin interaction studies
    Dsk2 binding partner identification, 161–163
    materials, 159
    overview, 158
    ubiquitin variant studies, 159–161
YU101
  proteasome inhibition assay, 605–606
  synthesis and characterization, 603–604

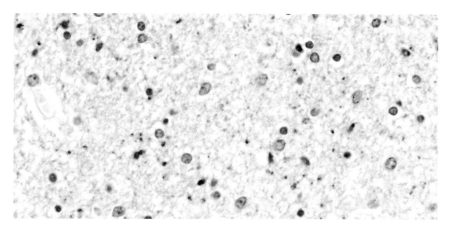

LOWE *ET AL.*, CHAPTER 7, FIG. 1. Dot-like staining in the white matter of the brain. This is an age-related abnormality seen in most mammalian species. The intensity of this type of staining increases in many neurodegenerative diseases.

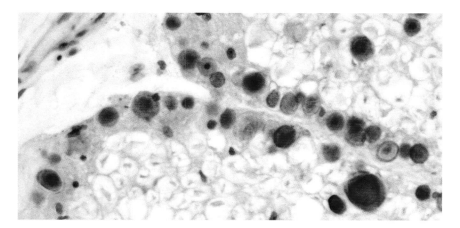

LOWE *ET AL.*, CHAPTER 7, FIG. 2. Corpora amylacea are mainly composed of polyglucosan material with a small protein component. They are spherical bodies with a lamellar structure. It is common for these to be stained with antiubiquitin. It remains uncertain whether this is nonspecific antibody binding to polyglucosan or represents a biologically significant accumulation. The main importance of recognizing this pattern of staining is to avoid misinterpretation with other inclusions.

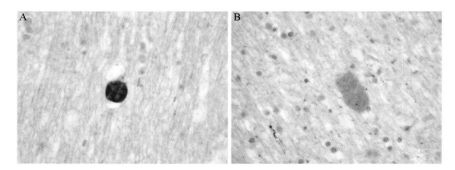

LOWE *ET AL.*, CHAPTER 7, FIG. 3. (A) An axonal spheroid seen as a spherical swelling in the white matter. This example shows relatively dense immunostaining for antiubiquitin. (B) Many spheroids show less intense ubiquitin immunoreactivity and have a granular pattern of staining, possibly related to localization in lysosome-related vesicles.

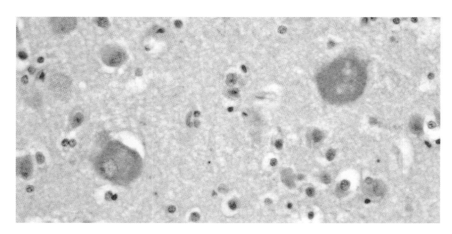

LOWE *ET AL.*, CHAPTER 7, FIG. 4. Swollen neurons in the cerebral cortex generally show weak ubiquitin immunoreactivity. In some instances, swollen neurons do not show enhanced ubiquitin immunoreactivity.

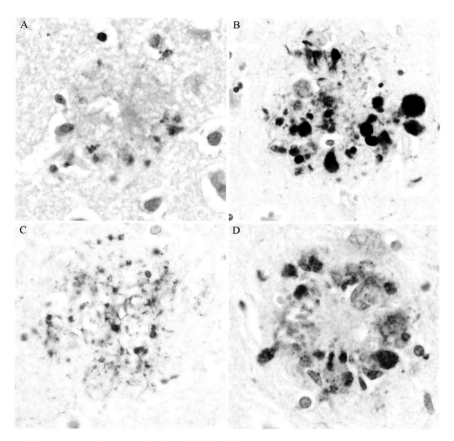

LOWE ET AL., CHAPTER 7, FIG. 5. Ubiquitin immunoreactivity in relation to amyloid plaques in Alzheimer's disease. The amyloid component is not specifically stained in these preparations. Ubiquitin immunoreactivity is either in lysosome-related structures or with accumulated T protein in nerve cell processes (neurites). (A) In early and loose deposits of amyloid, dot-like staining is seen together with fine linear neurites. (B) Large bulbous, dense-staining neurites are seen in relation to some amyloid plaques. (C) This pattern of thin, wispy neurite staining in relation to plaques is generally associated with T immunoreactivity. (D) A classical cored neuritic plaque showing a mixture of bulbous and linear ubiquitin-immunoreactive neurites.

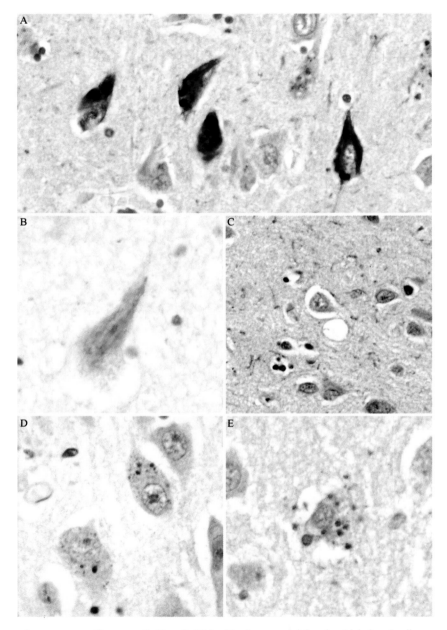

LOWE ET AL., CHAPTER 7, FIG. 6. Ubiquitin immunostaining in Alzheimer's disease. (A) Neurofibrillary tangles can stain strongly with antiubiquitin. Most are not stained. (B) In many cases of Alzheimer's disease, tangles that stain for ubiquitin do so weakly. (C) Neuropil threads are nerve cell processes containing abnormally accumulated *T* protein. Ubiquitin immunostaining can detect a proportion of such neurites. (D) Granulovacuolar degeneration seen in pyramidal cells of the hippocampus can be stained for ubiquitin, but this is inconsistently present. (E) A pyramidal cell in the hippocampus surrounded by ubiquitin-immunoreactive perisomatic granules.

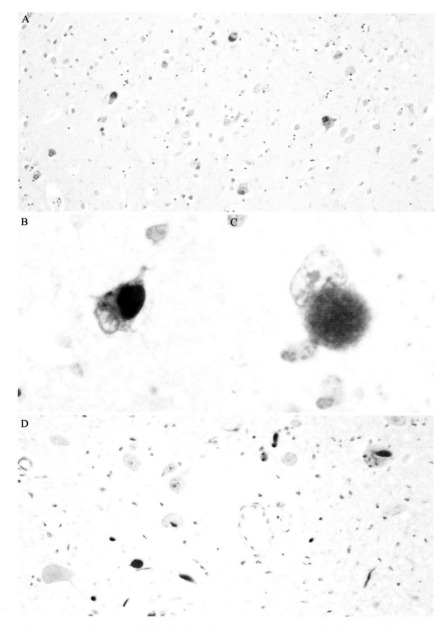

LOWE ET AL., CHAPTER 7, FIG. 7. Lewy body pathology. Lewy bodies are related to the pathological accumulation of α-synuclein. (A) Cortical Lewy bodies are readily detected with antiubiquitin. (B) At high magnification, some cortical Lewy bodies are densely stained for ubiquitin. (C) Many Lewy bodies show less intense staining with a somewhat granular pattern. (D) Ubiquitin staining of the dorsal vagal nucleus showing Lewy neurites as linear, ovoid, and beaded structures. Such neurites are seen in regions affected by Lewy bodies and are also based on the accumulation of α-synuclein.

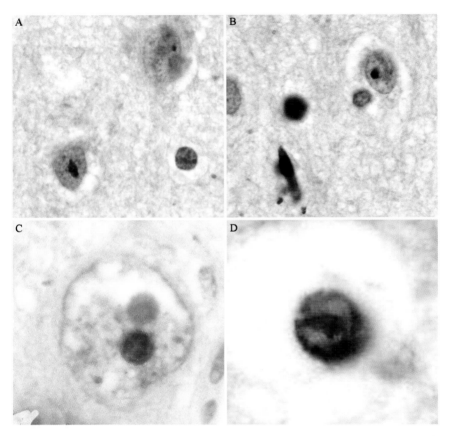

LOWE ET AL., CHAPTER 7, FIG. 8. (A) Huntington's disease neuronal nuclei contain ubiquitin-immunoreactive inclusions. These would also immunostain with antibodies to huntingtin or polyglutamine. (B) Huntington's disease showing a nuclear inclusion and accumulation of ubiquitinated protein in nerve cell processes as Huntington neurites. (C) A neuronal nucleus showing a Marinesco body immunoreactive for ubiquitin. The blue-stained round profile next to it is the nucleolus. (D) Neuronal inclusions immunoreactive for ubiquitin in a case of frontotemporal lobar degeneration associated with amyotrophic lateral sclerosis.

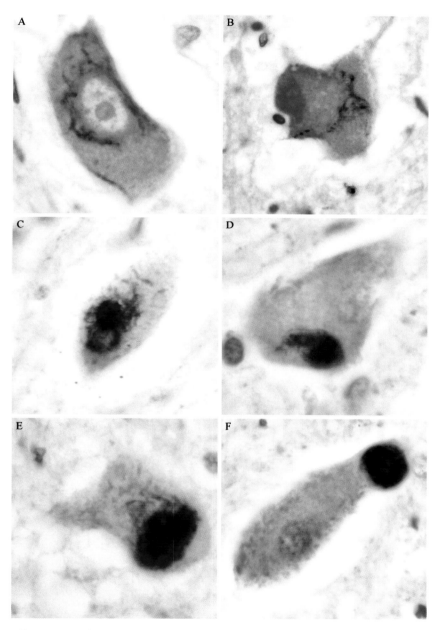

Lowe ET AL., CHAPTER 7, FIG. 9. Inclusions in amyotrophic lateral sclerosis. (A and B) Filamentous inclusions termed *skeins* are seen in affected motor neurons. (C–E) Many inclusions appear as solid masses of material with frayed filamentous margins. (F) A minority of inclusions have a dense-stained solid spherical appearance.

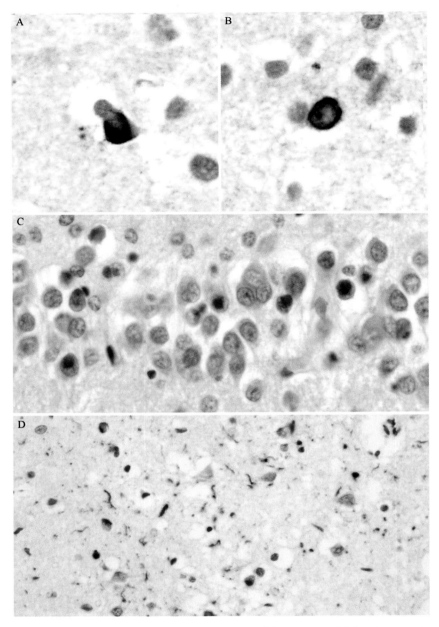

Lowe *et al.*, Chapter 7, Fig. 10. In some cases of frontotemporal lobar degeneration, patients have pathological changes detected with antiubiquitin in nonmotor areas of the brain that are similar to those seen in amyotrophic lateral sclerosis. (A and B) Ubiquitin-immunoreactive inclusions seen in layer 2 cortical neurons. Some are paranuclear, whereas others form rings around the nucleus. © Many small ubiquitin-immunoreactive inclusions in the neurons of the hippocampal dentate granule cells. (D) Ubiquitin-immunoreactive neurites seen in the outer cortical layers of frontal and temporal lobe from patients with frontotemporal lobar degeneration.

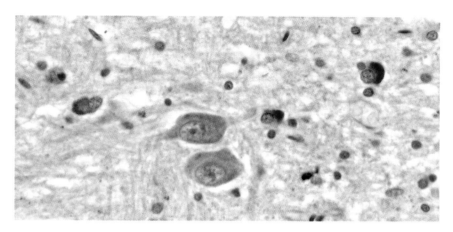

LOWE ET AL., CHAPTER 7, FIG. 11. Ubiquitin staining showing inclusions in oligodendroglial cells in multiple system atrophy.

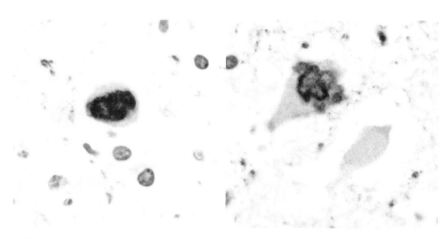

LOWE ET AL., CHAPTER 7, FIG. 12. Ubiquitin immunostaining showing inclusions in inferior olivary neurons. These are age-related and have no disease-specific association. The nature of these inclusions remains uncertain.

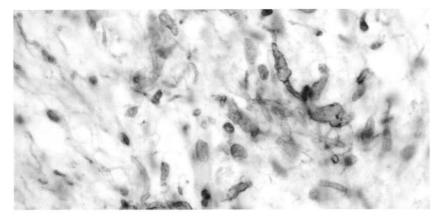

Lowe ET AL., CHAPTER 7, FIG. 13. Rosenthal fibers showing staining of the periphery with antiubiquitin. These inclusions are seen in astrocytic cells and are based on accumulation of αB-crystallin and glial fibrillary acidic protein.

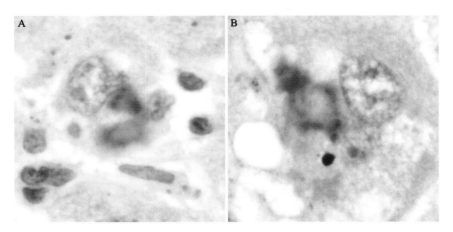

Lowe ET AL., CHAPTER 7, FIG. 14. Mallory's hyaline (A and B) seen in mouse liver after induction by griseofulvin treatment. Ubiquitin immunoreactivity is located at the periphery of inclusions.

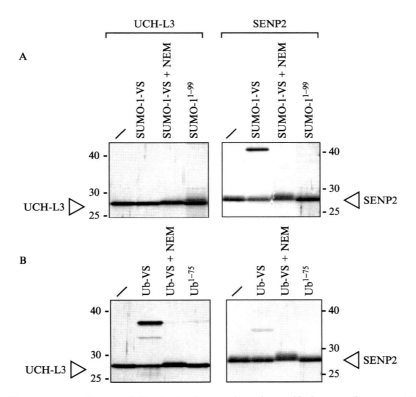

GALARDY ET AL., CHAPTER 8, FIG. 1. *In vitro* reaction using purified enzymatic components. The SUMO specific protease SENP2 reacts with a SUMO-based but not with an Ub-based site-directed probe. All labeling is inhibited by *N*-ethylmaleimide (NEM) and, therefore, cysteine dependent. Reversely, the USP UCH-L3 reacts with an Ub probe but not with a SUMO-based one.

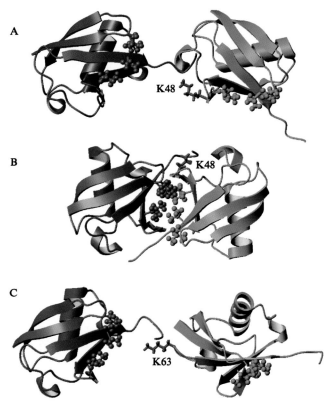

VARADAN *ET AL.*, CHAPTER 12, FIG. 3. Solution conformations of Lys48-linked Ub$_2$ at (A) pH 4.5, (B) pH 6.8, and (C) pH 6.8 of Lys63-linked Ub$_2$. The ribbons are colored blue and green for the distal and proximal domains, respectively; the side chains of the hydrophobic patch residues Leu8-Ile44-Val70 are shown in ball-and-stick representation. Also shown in red is the side chain of the linkage lysine. Shown in (B) is the NMR structure of Lys48-linked Ub$_2$ (PDB entry 2BGF) obtained by domain docking on the basis of a combination of RDCs and CSP mapping data (van Dijk *et al.*, 2005). (A and C) represent interdomain orientations (the positioning of the domains is arbitrary) obtained by aligning the rotational diffusion tensors reported by both domains (A) or the alignment tensors derived from RDCs (C). The ^{15}N relaxation measurements used the standard protocols [*e.g.*, Fushman *et al.* (1997)]; the RDCs were measured in the liquid-crystalline phase of n-alkyl-poly(ethyleneglycol)/n-hexanol mixtures as detailed in (Varadan *et al.*, 2002, 2004). The diffusion and alignment tensors were derived from the experimental data using in-house computer programs ROTDIF (Walker *et al.*, 2004) and ALTENS (Varadan *et al.*, 2002).

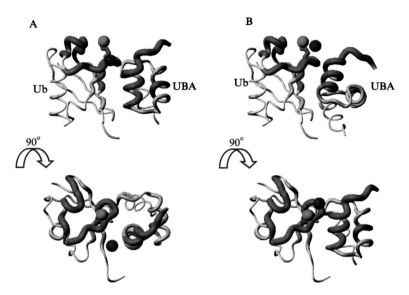

VARADAN ET AL., CHAPTER 12, FIG. 4. Experimental verification of the existing models for the monoUb/UBA2 complex using site-directed spin labeling. (A) represents the Ub/UBA2 structure modeled by the Ub/CUE complex (Kang et al., 2003), whereas the docked structure (Mueller et al., 2004) is shown in (B). The bottom row shows views of the same structures from the top. The ribbons represent the backbone of Ub and UBA2; the atom coordinates are from Mueller et al. (2004). The ribbon width increases proportionally to the observed paramagnetic line broadening (hence closer distance to the spin label) and is color-coded by this distance as red (closest, <17 Å), orange (17–26 Å), and yellow (>26 Å). The spheres represent the reconstructed positions of the spin label as "seen" by Ub (green) and by UBA2 (blue). Also shown is the side chain of Cys48 (Ub), where the spin label was covalently attached. The coordinates of the spin label were obtained using a three-dimensional search algorithm aimed at satisfying all available amide–MTSL distance constraints.

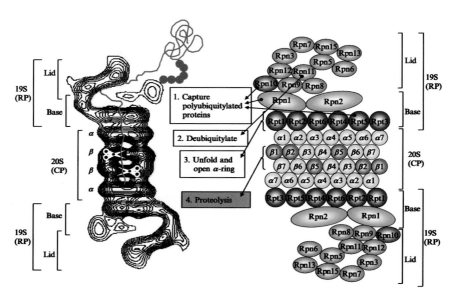

HIRANO ET AL., CHAPTER 15, FIG. 1. Molecular organization of 26S proteasomes. (Left panel) Averaged image of the 26S proteasome complex of rat based on electron micrographs. The α and β rings of the 20S proteasome are indicated. Photograph kindly provided by W. Baumeister. (Right panel) Schematic drawing of the subunit structure. CP, core particle (alias 20S proteasome); RP, 19S regulatory particle (alias PA700) consisting of the base and lid subcomplexes; Rpn, RP non-ATPase; Rpt, RP triple–ATPase. Note that relative positions of 19S subunits have not been established.

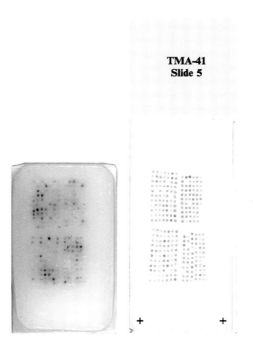

LEHMAN ET AL., CHAPTER 23, FIG. 1. TMA Construction. A finished TMA paraffin block and corresponding H&E–stained TMA slide.

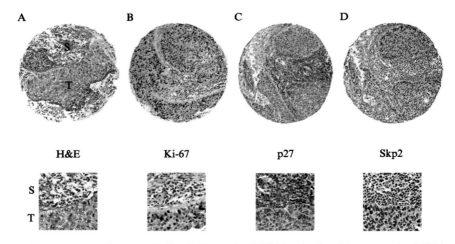

LEHMAN ET AL., CHAPTER 23, FIG. 2. Example of H&E–stained and immunostained TMA cores. The first micrograph (A) is an H&E–stained core section of a squamous cell carcinoma demonstrating areas of tumor (T) and surrounding stroma (S). The latter consists mostly of chronic inflammatory cells in this case. The micrograph below each core is an enlargement of the stomal/tumor interface. Ki-67 is a general marker of cellular proliferation. The second core (B) shows that the tumor cells are immunoreactive for Ki-67, whereas the benign inflammatory cells within the stroma are immunonegative. p27 is a cyclin-dependent kinase inhibitor most often significantly expressed in benign nonproliferating cells. The third core (C) shows p27 immunopositivity of the stromal inflammatory cells. The tumor cells are p27 immunonegative. The last core (D) is immunostained for Skp2, the SCF ubiquitin ligase adapter subunit responsible for binding p27 and initiating the ubiquitin-dependent degradation of p27. Skp2 has been shown to be overexpressed in cancers correlating with low p27 levels and cell cycle deregulation. Note that the immunostaining pattern is the reverse of that of p27; the benign inflammatory cells are Skp2 immunonegative, whereas the tumor cells are Skp2 immunopositive. The immunostains are counterstained with a blue stain (Mayer's hematoxylin) so that individual cells can be visualized. The arrangement of tumor and stromal cells appears somewhat different in each core, because the corresponding TMA sections are from different levels (depths) of the TMA block. Original magnifications of the cores are 100×.

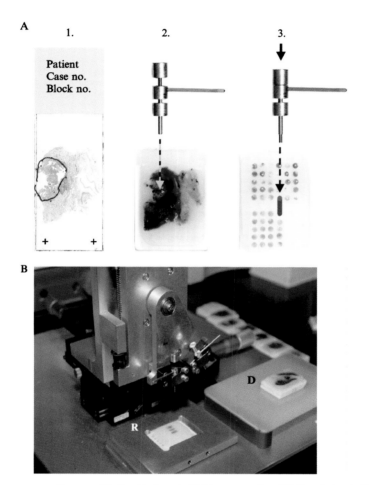

LEHMAN ET AL., CHAPTER 23, FIG. 3. Steps in TMA construction. (A) Step 1, A suitable area of tumor is identified on an H&E–stained microscope slide and circled. Step 2, A core is then punched from the corresponding area of the paraffin block used to make the slide (donor block). Step 3, Next the core is extruded into a prepunched hole in a new block (recipient block). Specimen cores are placed in orderly rows and columns in this manner. The block is then used to cut thin sections, which are adhered onto new microscope slides as depicted in Fig. 1. The punches shown are from the Beecher manual arrayer. (B) To facilitate block construction, we use the Beecher Manual Arrayer. Here, a core is taken from the donor block (D) and repositioned in the recipient block (R).

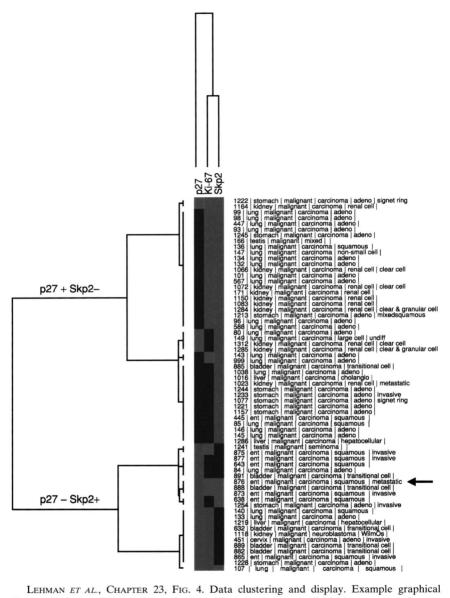

LEHMAN ET AL., CHAPTER 23, FIG. 4. Data clustering and display. Example graphical display of clustered TMA data using TreeView™ software. Individual tumor samples are shown on the vertical axis, and protein markers (p27, Ki-67 and Skp2) are shown on the horizontal axis above. Green indicates no expression; dark red, low to intermediate expression (3 to <30% of tumor cells immunopositive); bright red, high expression (>30% of tumor cells immunopositive). The tumors are separated into p27-positive/Skp2-negative and p27-negative/Skp2-positive groups and then further subdivided based on Ki67 expression. Ki67 and Skp2 cluster more closely together than with p27 as indicated by the shorter branches linking the two proteins. The arrow indicates the tumor (#876) shown in Fig. 2.

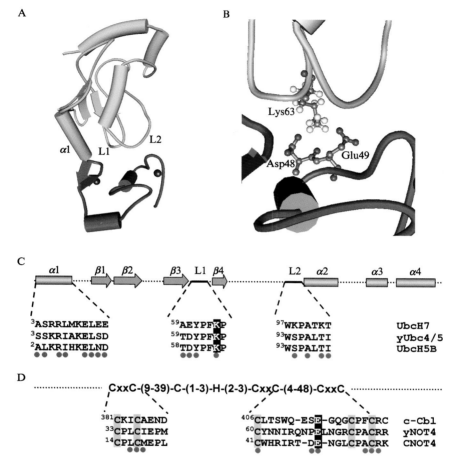

WINKLER AND TIMMERS, CHAPTER 24, FIG. 1. (A) Overview of the UbcH5b–NOT4 RING structure (PDB accession number 1UR6). UbcH5b is shown in yellow, the CNOT4 RING domain in red, and the Zn^{2+} ions in gray. Indicated are the three UbcH5b regions making contact with the CNOT4 RING. (B) Detail of the UbcH5b–CNOT4 RING structure. The UbcH5b residue Lys63 interacts with acidic residues Asp48 and Glu49 of the CNOT4 RING domain. (A) and (B) were generated using WebLab Viewer Lite (Molecular Simulations Inc). Note that not all structures in the ensemble of five best solutions contain β-sheets in the CNOT4 RING domain. (C) Schematic diagram of the UBC domain. Regions important for the interaction with E3 enzymes are expanded. Shown are amino acid sequences of UbcH5b, yeast (y) Ubc4 and Ubc5, and UbcH7. Circles indicate UbcH5b residues involved in the interaction with the CNOT4 RING domain as identified by chemical-shift perturbation experiments (combined chemical shift differences >0.1 ppm) (Dominguez *et al.*, 2004). The conserved loop L1 residue corresponding to UbcH5b Lys63 is highlighted. (D) Schematic diagram of the RING domain. Expanded are amino acid sequences of CNOT4, yeast Not4, and c-Cbl. Circles indicate CNOT4 residues involved in the interaction with UbcH5b as identified by chemical-shift perturbation experiments (combined chemical shift differences >0.05 ppm) (Albert *et al.*, 2002). Indicated in gray are cysteine residues involved in Zn^{2+}-coordination. The conserved acidic residue corresponding to CNOT4 Glu49 is highlighted.

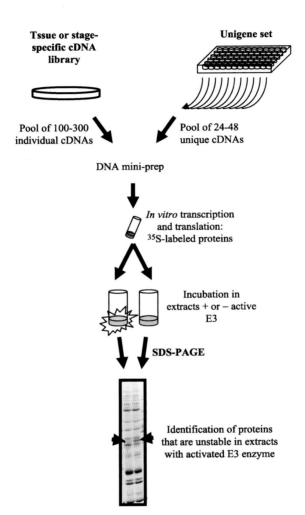

AYAD ET AL., CHAPTER 28, FIG. 1. Schematic illustration of *in vitro* expression cloning (IVEC). Small pools of *in vitro* transcribed and translated radiolabeled proteins are incubated in extracts that have been treated to activate the E3 enzyme of interest. The proteins that are specifically ubiquitinated by the activated E3 are degraded in the extract and can be identified when the pools are run on SDS-PAGE gels adjacent to the same pool incubated in control extracts. See text for details.

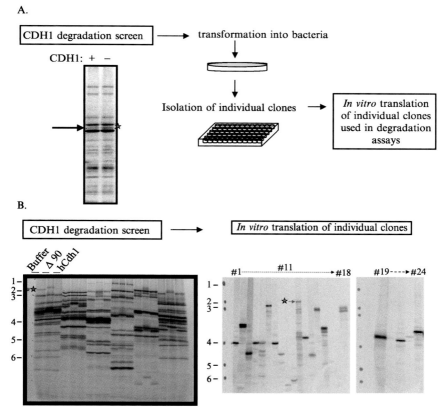

AYAD ET AL., CHAPTER 28, FIG. 2. Sib selection versus isolating positive clones from unigene set. (A) SDS-PAGE of pools of *in vitro* translated proteins incubated in *Xenopus* egg extracts supplemented with CDH1 or buffer (+ or −, respectively). Once the desired pool is identified, a sib selection procedure is performed to isolate the single clone of interest labeled with star. (B) Alternately, if a unigene set of cDNAs is used, sib selection is not necessary, because the single clone can be identified simply by matching the molecular weight of *in vitro* translation products by SDS-PAGE. An example of this technique is shown for a protein degraded in *Xenopus* egg extracts supplemented with Cdh1 and stable in extracts supplemented with buffer or a nondegradable version of cyclin B (Δ90). The band of interest is labeled with a star and has a slightly lower electrophoretic mobility than the #2 molecular weight standard.

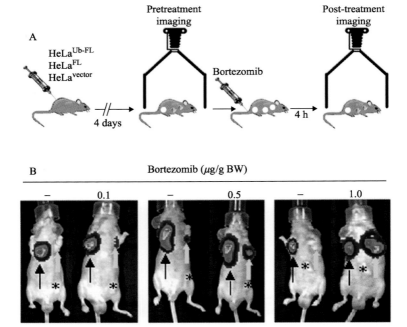

GROSS AND PIWNICA-WORMS, CHAPTER 35, FIG. 2. *In vivo* bioluminescence imaging of Ub-FL monitors proteasome function and inhibition in living mice. (A) Schematic representation of the experimental timeline. (B) Mice bearing size-matched tumors were imaged one day before (−) and 4 h after tail vein injection of the indicated doses of bortezomib. FL, Ub-FL, and vector control tumors are denoted by black arrows, yellow arrows, and asterisks, respectively (Modified from Luker *et al.*, 2003a).

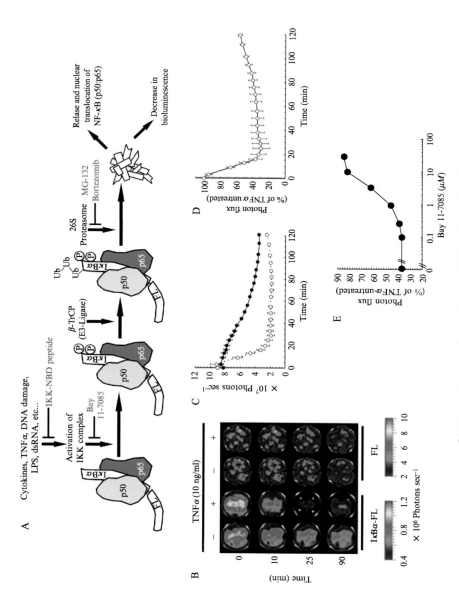

GROSS AND PIWNICA-WORMS, CHAPTER 35, FIG. 3. *(continued)*

Gross and Piwnica-Worms, Chapter 35, Fig. 3. Validation of IκBα-FL functionality *in vitro*. (A) Schematic representation of stimulus-induced degradation of IκBα-FL. In resting cells, IκBα-FL binds NF-κB (p50:p65 dimer). On stimulation (e.g., TNFα binding to its cognate receptor), the upstream kinase complex (IKK) is activated and in turn phosphorylates IκBα-FL. This double phosphorylation renders IκBα-FL a substrate for the specific E3-ligase (β-TrCP) that polyubiquitinates IκBα-FL. Polyubiquitinated IκBα-FL is then selectively degraded by the 26S proteasome, producing a decrease in bioluminescence. NF-κB then freely translocates to the nucleus to promote κB-dependent gene transcription. Molecular targets of various NF-κB modulators are shown in red: IKK-NBD peptide specifically interferes with activation of the IKK complex. Bay 11-7085 inhibits the kinase activity of IKK, and MG-132 and bortezomib inhibit the 26S proteasome. (B) Bioluminescence imaging of HeLa$^{IB\alpha\text{-}FL}$ (left) and HeLa^{FL} (right) cells before and at the indicated time points after addition of TNFα (10 ng/ml) or vehicle (PBS). Images show color-coded maps of photon flux superimposed on black-and-white photographs of the assay plates. (C) Changes in raw bioluminescence plotted as a function of time after addition of TNFα (○) or vehicle (●) to HeLa$^{IB\alpha\text{-}FL}$ cells. (D) TNFα-induced net degradation and resynthesis of IκBα-FL over time calculated from the photon flux ratio of treated HeLa$^{IB\alpha\text{-}FL}$ cells over unstimulated control values. (E) Concentration-dependent inhibition of TNFα-induced degradation of IκBα-FL by the IKK inhibitor Bay 11-7085.

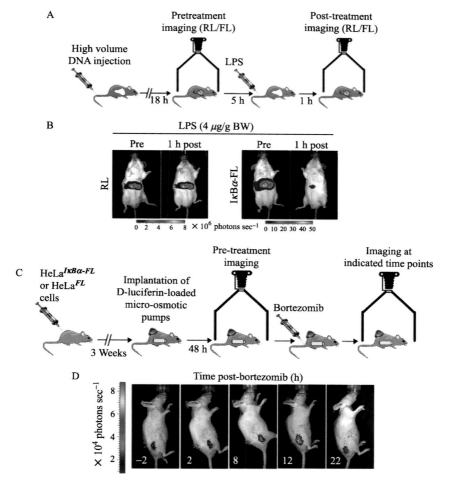

GROSS AND PIWNICA-WORMS, CHAPTER 35, FIG. 4. Imaging proteasomal-dependent IκBα degradation in living mice. (A, B) Imaging pharmacological modulation of LPS-induced IκBα degradation. (A) Schematic representation of the experimental timeline. (B) Representative bioluminescence images of RL (left two panels) and IκBα-FL (right two panels) taken before or 1 h after LPS stimulation. All images correspond to an individual mouse. (C, D) Real-time imaging of IκBα accumulation in tumors of bortezomib-treated mice. (C) Schematic representation of the experimental timeline. (D) Representative bioluminescence images of a HeLa$^{I B \alpha\text{-}F L}$ tumor-bearing mouse, taken at the indicated time points (h) before and after i.p. administration of bortezomib (1 μg/g BW). Note that D-luciferin was continuously delivered by a subcutaneously implanted microosmotic pump.

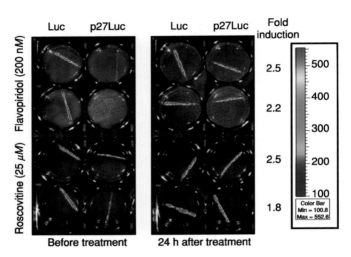

ZHANG AND KAELIN, CHAPTER 36, FIG. 2. Bioluminescent imaging of hollow fibers *in vitro*. Hollow fibers were filled with polyclonal U2OS cells producing Luc or p27Luc and cultivated in 6-well plates. Bioluminescent images were acquired before and after treatment with 200 nM flavopiridol or 25 μM roscovitine using Xenogen IVIS™ imaging system. Fold induction was calculated as p27Luc/Luc$_{posttreatment}$ ÷ p27Luc/Luc$_{pretreatment}$.

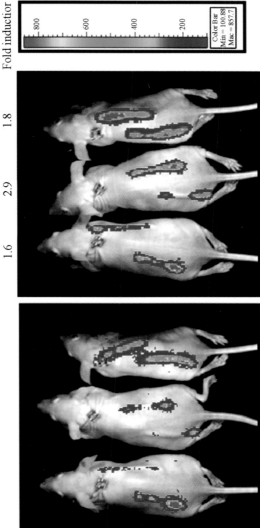

ZHANG AND KAELIN, CHAPTER 36, FIG. 3. Bioluminescent imaging of hollow fibers *in vivo*. Hollow fibers filled with polyclonal U2OS cells producing Luc (left flank) or p27Luc (right flank) were implanted subcutaneously into nude mice. Seven days later, baseline bioluminescent images were acquired using Xenogen IVIS™ imaging system (left). After two doses of flavopiridol (5 mg/kg, once a day by i.p.), repeat images were obtained. Fold induction was calculated as p27Luc/Luc$_{posttreatment}$ ÷ p27Luc/Luc$_{pretreatment}$.

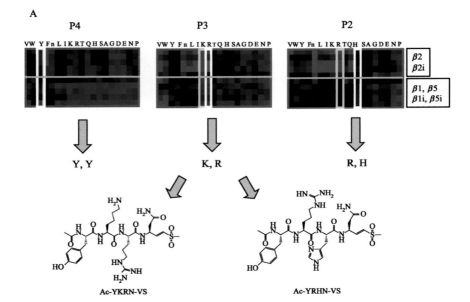

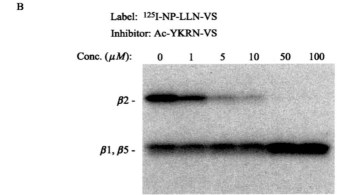

BOGYO, CHAPTER 40, FIG. 4. Designing a subunit-selective inhibitor based on library inhibition data. (A) Data from P2, P3, and P4 library competition screens (see Fig. 2) are compiled and visualized using expression array analysis software (see text). The resulting graph groups binding data for the $\beta 2$ and $\beta 2i$ subunits at the top and binding data for the $\beta 1$, $\beta 1i$, $\beta 5$, $\beta 5i$ at the bottom. Red blocks indicate potent binders, whereas green blocks represent poor binders. The constant residues used in the inhibitors are listed across the top. Residues at each position that show selectivity for the $\beta 2$ subunit are highlighted with white and yellow boxes. The resulting two inhibitors designed on the basis of these data are shown below. (B) A typical competition assay in which the $\beta 2$-specific inhibitor Ac-YKRN-VS from (A) is used to treat crude NIH-3T3 lysates at a range of concentrations. Addition of the general label ^{125}I-NP-LLN-VS shows the selective binding of the inhibitor to the $\beta 2$ subunit at concentrations as high as 100 μM.

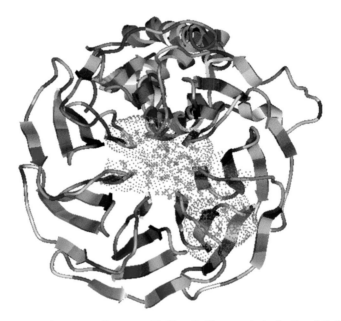

CARDOZO AND ABAGYAN, CHAPTER 42, FIG. 2. Energy strain in the β-TrCP protein backbone. Face-on view of the WD-40 domain of β-TrCP. The protein backbone is displayed as a ribbon depiction with each amino acid position shaded according to energy strain score in the backbone: red = high energy; blue = low energy. The substrate-binding peptide in this complex is displayed as a green stippled cloud to identify the location of the substrate-binding interface of β-TrCP.

CARDOZO AND ABAGYAN, CHAPTER 42, FIG. 3. B-factor mapping of β-TrCP structure. Face-on view of the WD-40 domain of β-TrCP. Left panel, All atoms are shown and colored according to B-factor (temperature factor): red = high B-factor, blue = low B-factor. Right panel, The protein backbone is displayed as a ribbon depiction with each amino acid position shaded according to B-factor for the entire residue: red = high B-factor, blue = low B-factor. The substrate-binding peptide in this complex is displayed as a green stippled cloud in both panels to identify the location of the substrate-binding interface of β-TrCP. The entire structure is different shades of blue, indicating high experimental local reliability throughout.

CARDOZO AND ABAGYAN, CHAPTER 42, FIG. 4. Crystal packing analysis of β-TrCP substrate-binding interface. Depiction of the unit cell of the crystal used to solve the structure of β-TrCP. The index monomer is colored black, and the substrate peptide in the complex is displayed as green space-filling spheres to identify the substrate-binding surface. Symmetry-related molecules in the crystal packing around the index structure are shown in yellow ribbon depiction. Left panel, View of the entire assembly. Right panel, Close-up of the substrate-binding domain showing that only the tail of the substrate peptide (arrow) contacts a crystal neighbor, whereas the substrate-binding interface is free in solution (dashed circled).

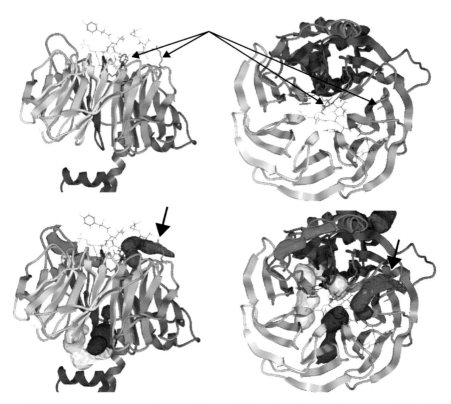

CARDOZO AND ABAGYAN, CHAPTER 42, FIG. 5. Ligand pockets on the surface of β-TrCP. Side (left column) and face-on (right column) view of β-TrCP. The protein backbone is shown in ribbon depiction, the substrate peptide is displayed in thin wire depiction, and four residues known to abrogate ubiquitylation by of β-TrCP are displayed as stick depictions and indicated by the thin arrows. Ligand-binding pockets above the threshold size established by the plots shown in Fig. 1 are depicted as geometric objects. One ligand-binding pocket (red: thick arrows) is directly at the substrate-binding interface and in contact with the key functional residues. The method described in the text to perform this pocket analysis was also used for the analysis shown in Fig. 1.

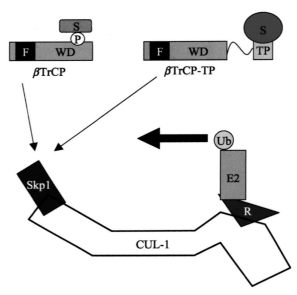

ZHANG AND ZHOU, CHAPTER 53, FIG. 1. Engineering the F-box-containing β TrCP substrate receptor for targeted degradation of specific cellular proteins. F, F-box domain that interacts with Skp1; WD, WD40 repeats for binding to phosphorylated (P) endogenous substrates (S) (in purple). TP, targeting peptide; S (in red), target protein of interest; E2, Cdc34 ubiquitin-conjugating enzyme; Ub, ubiquitin; R, RING domain protein Hrt1/Rbx1/Roc1.

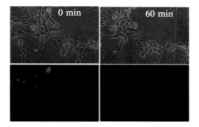

Degree of AR-GFP Disappearance	Percent (out of >200 cells)
1. NONE	4
2. MINIMAL	16
3. PARTIAL	29
4. COMPLETE	51

SAKAMOTO, CHAPTER 54, FIG. 7. Microinjection of Protac leads to AR-GFP degradation in cells. Protac-3 (10 μM in the microinjection needle) was introduced using a Picospritzer II pressurized microinjector into 293$^{AR\text{-}GFP}$ cells in a solution containing KCl (200 μM) and rhodamine dextran (50 $\mu g/ml$). Approximately 10% of total cell volume was injected. (A) Protac-3 induces AR-GFP disappearance within 60 min. The top panels show cell morphology under light microscopy overlaid with images of cells injected with Protac as indicated by rhodamine fluorescence (pink color). The bottom panels show images of GFP fluorescence. By 1 h, GFP signal disappeared in almost all microinjected cells. To quantify these results, we injected more than 200 cells and classified the degree of GFP disappearance as being either none (1), minimal (2), partial (3), or complete (4). Examples from each category and the tabulated results are shown in (B). These results were reproducible in three independent experiments performed on separate days with 30–50 cells injected per day (Sakamoto et al., 2003).

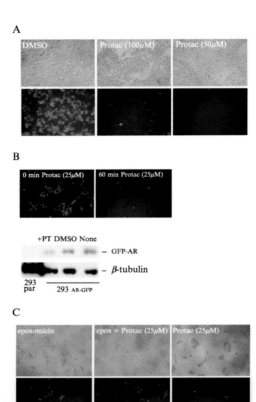

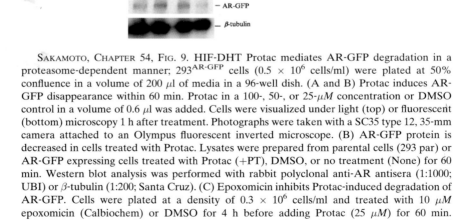

SAKAMOTO, CHAPTER 54, FIG. 9. HIF-DHT Protac mediates AR-GFP degradation in a proteasome-dependent manner; 293^{AR-GFP} cells (0.5×10^6 cells/ml) were plated at 50% confluence in a volume of 200 μl of media in a 96-well dish. (A and B) Protac induces AR-GFP disappearance within 60 min. Protac in a 100-, 50-, or 25-μM concentration or DMSO control in a volume of 0.6 μl was added. Cells were visualized under light (top) or fluorescent (bottom) microscopy 1 h after treatment. Photographs were taken with a SC35 type 12, 35-mm camera attached to an Olympus fluorescent inverted microscope. (B) AR-GFP protein is decreased in cells treated with Protac. Lysates were prepared from parental cells (293 par) or AR-GFP expressing cells treated with Protac (+PT), DMSO, or no treatment (None) for 60 min. Western blot analysis was performed with rabbit polyclonal anti-AR antisera (1:1000; UBI) or β-tubulin (1:200; Santa Cruz). (C) Epoxomicin inhibits Protac-induced degradation of AR-GFP. Cells were plated at a density of 0.3×10^6 cells/ml and treated with 10 μM epoxomicin (Calbiochem) or DMSO for 4 h before adding Protac (25 μM) for 60 min. (D) Western blot analysis was performed with cells in 96-well dishes treated with Protac (25 μM), DMSO (left), epoxomicin (10 μM), epoxomicin (10 μM) + Protac (50 or 25 μM), or Protac alone (50 or 25 μM) (Schneekloth et al., 2004).